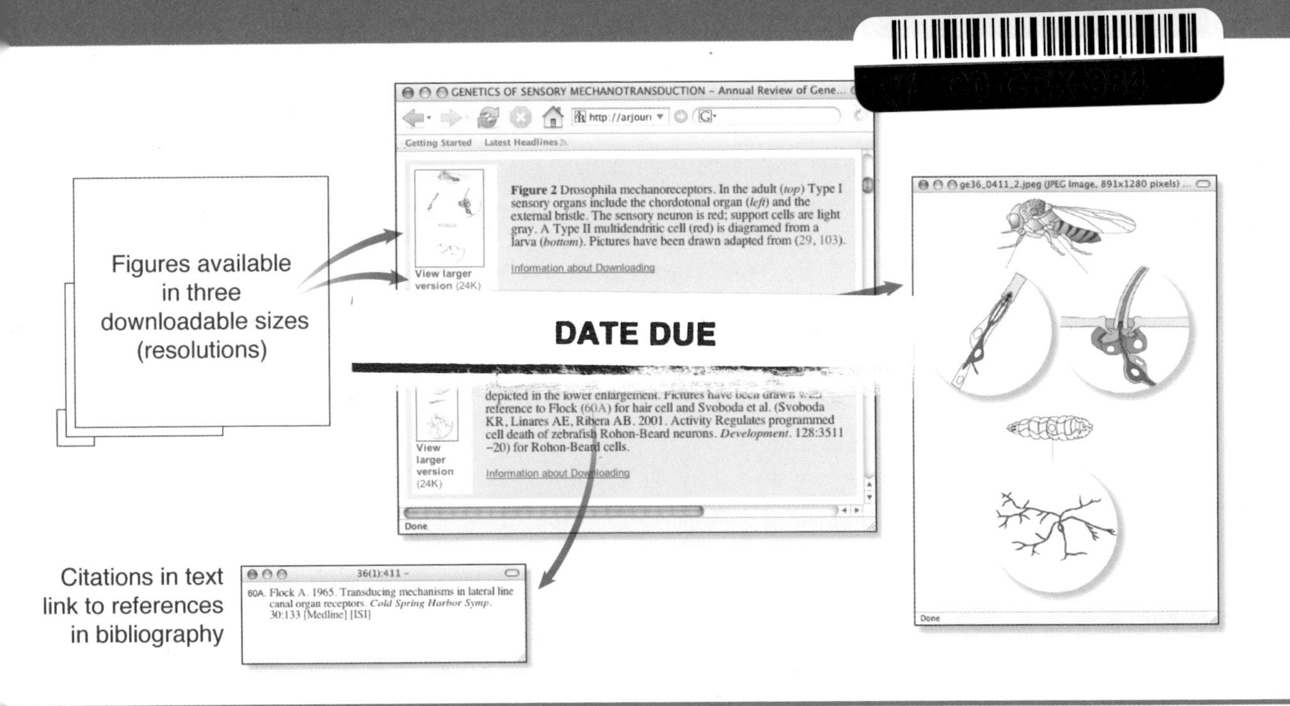

Figures available in three downloadable sizes (resolutions)
Citations in text link to references in bibliography
GENETICS OF SENSORY MECHANOTRANSDUCTION – Annual Review of Gene...
Figure 2 Drosophila mechanoreceptors. In the adult (top) Type I sensory organs include the chordotonal organ (left) and the external bristle. The sensory neuron is red; support cells are light gray. A Type II multidendritic cell (red) is diagramed from a larva (bottom). Pictures have been drawn adapted from (29, 103).
Information about Downloading
60A. Flock A. 1965. Transducing mechanisms in lateral line canal organ receptors. Cold Spring Harbor Symp. 30:133 [Medline] [ISI]

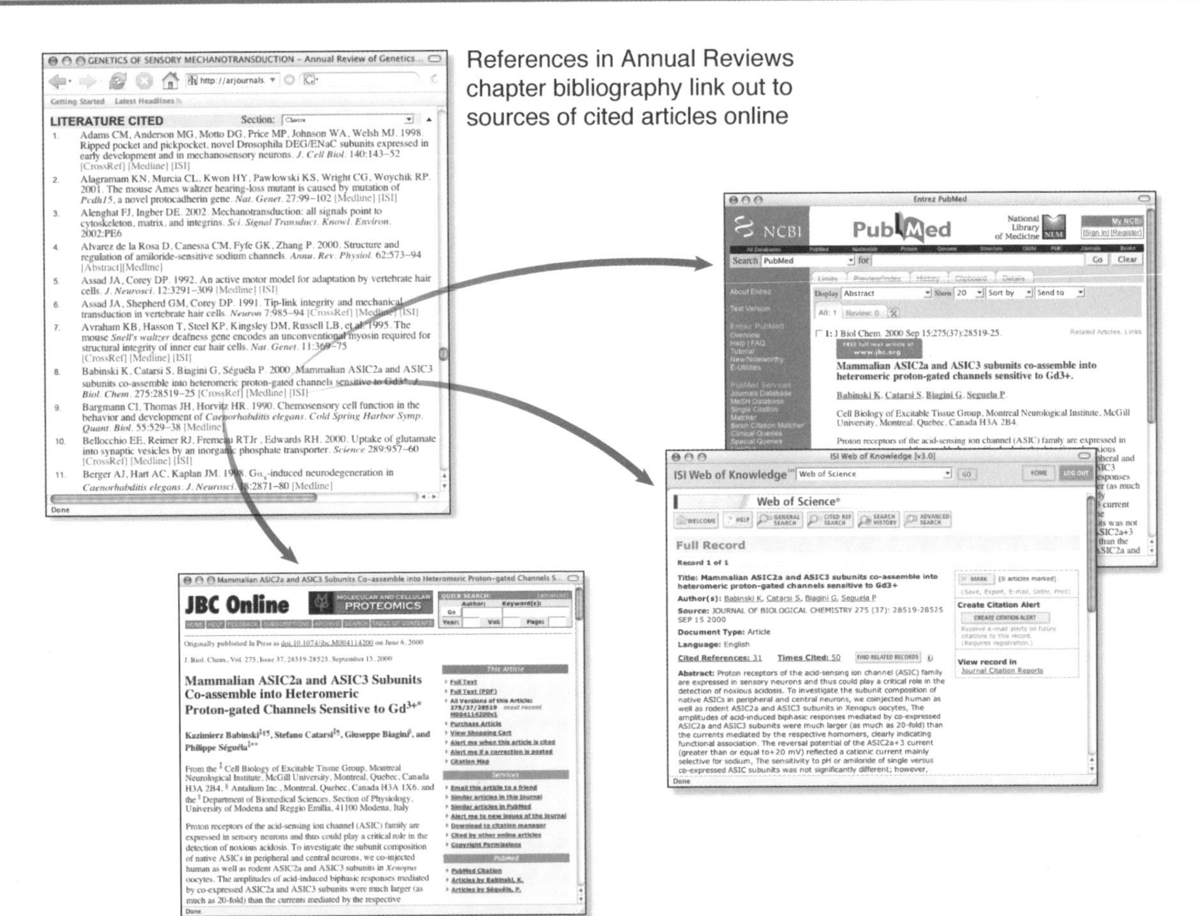

References in Annual Reviews chapter bibliography link out to sources of cited articles online
LITERATURE CITED
JBC Online
PubMed
ISI Web of Knowledge
Web of Science
Full Record

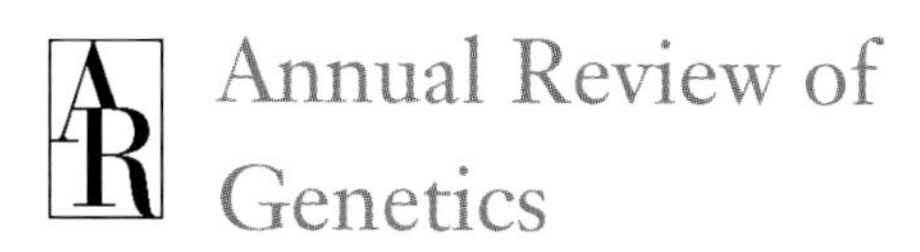
Annual Review of
Genetics

Production Editor: Hilda Gould
Bibliographic Quality Control: Mary A. Glass
Electronic Content Coordinator: Suzanne K. Moses
Subject Indexer: Kyra Kitts

Annual Review of Genetics

Volume 39, 2005

Allan Campbell, *Editor*
Stanford University, Stanford

Wyatt W. Anderson, *Associate Editor*
University of Georgia, Athens

Elizabeth W. Jones, *Associate Editor*
Carnegie Mellon University, Pittsburgh

www.annualreviews.org • science@annualreviews.org • 650-493-4400

Annual Reviews
4139 El Camino Way • P.O. Box 10139 • Palo Alto, California 94303-0139

Annual Reviews
Palo Alto, California, USA

International Standard Serial Number: 0066-4197
International Standard Book Number: 0-8243-1239-2
Library of Congress Catalog Card Number: 63-8847

∞ The paper used in this publication meets the minimum requirements of American National Standards for Information Sciences—Permanence of Paper for Printed Library Materials, ANSI Z39.48-1992.

TYPESET BY TECHBOOKS, FAIRFAX, VA
PRINTED AND BOUND BY QUEBECOR WORLD, KINGSPORT, TN

Annual Review of Genetics

Volume 39, 2005

Contents

INDEXES

ERRATA

An online log of corrections to *Annual Review of Genetics* chapters may be found at http://genet.annualreviews.org/errata.shtml

Related Articles

From the ***Annual Review of Biochemistry***, Volume 74 (2005)

From the ***Annual Review of Cell and Developmental Biology,*** Volume 21 (2005)

Mechanisms of Apoptosis through Structural Biology
Nieng Yan and Yigong Shi

Assembly of Variant Histones into Chromatin
Steven Henikoff and Kami Ahmad

Molecular Mechanisms of Steroid Hormone Signaling in Plants
Grégory Vert, Jennifer L. Nemhauser, Niko Geldner, Fangxin Hong, and Joanne Chory

RNA Transport and Local Control of Translation
Stefan Kindler, Huidong Wang, Dietmar Richter, and Henri Tiedge

RNA Silencing Systems and Their Relevance to Plant Development
Frederick Meins, Jr., Azeddine Si-Ammour, and Todd Blevins

Quorum Sensing: Cell-to-Cell Communication in Bacteria
Christopher M. Waters and Bonnie L. Bassler

Pushing the Envelope: Structure, Function and Dynamics of the Nuclear Periphery
Martin Hetzer, Tobias C. Walther, and Iain W. Mattaj

Phagocytosis: At the Crossroads of Innate and Adaptive Immunity
Isabelle Jutras and Michel Desjardins

Retinotectal Mapping: New Insights from Molecular Genetics
Greg Lemke and Michaël Reber

Stem Cell Niche: Structure and Function
Linheng Li and Ting Xie

Specificity and Versatility in TGF-β Signaling Through Smads
Xin-Hua Feng and Rik Derynck

The Great Escape: When Cancer Cells Hijack the Genes for Chemotaxis and Motility
John Condeelis, Robert Singer, and Jeffrey E. Segall

From the ***Annual Review of Genomics and Human Genetics,*** Volume 6 (2005)

A Personal Sixty-Year Tour of Genetics and Medicine
Alfred G. Knudson

Complex Genetics of Glaucoma Susceptibility
Richard T. Libby, Douglas B. Gould, Michael G. Anderson, and Simon W.M. John

Noonan Syndrome and Related Disorders: Genetics and Pathogenesis
Marco Tartaglia and Bruce D. Gelb

Silencing of the Mammalian X Chromosome
Jennifer C. Chow, Ziny Yen, Sonia M. Ziesche, and Carolyn J. Brown

From the ***Annual Review of Microbiology,*** Volume 59 (2005)

Genome Trees and the Nature of Genome Evolution
Berend Snel, Martijn A. Huynen, and Bas E. Dutilh

Cellular Functions, Mechanism of Action, and Regulation of FtsH Protease
Koreaki Ito and Yoshinori Akiyama

The Genetics of the Persistent Infection and Demyelinating Disease Caused by Theiler's Virus
Michel Brahic, Jean-François Bureau, and Thomas Michiels

Genome-Wide Responses to DNA-Damaging Agents
Rebecca C. Fry, Thomas J. Begley, and Leona D. Samson

Regulation of Bacterial Gene Expression by Riboswitches
Wade C. Winkler and Ronald R. Breaker

Opportunities for Genetic Investigation Afforded by Acinetobacter baylyi, A Nutritionally Versatile Bacterial Species that Is Highly Competent for Natural Transformation
David M. Young, Donna Parke, and L. Nicholas Ornston

The Origins of New Pandemic Viruses: The Acquisition of New Host Ranges by Canine Parvovirus and Influenza A Viruses
Colin R. Parrish and Yoshihiro Kawaoka

From the ***Annual Review of Phytopathology,*** Volume 43 (2005)

Replication of Alfamo- and Ilarviruses: Role of the Coat Protein
John F. Bol

Viroids and Viroid-Host Interactions
Ricardo Flores, Carmen Hernández, A. Emilio Martínez de Alba, José-Antonio Daròs, and Francesco Di Serio

Plant Pathology and RNAi: A Brief History
John A. Lindbo and William G. Doughtery

Mechanisms of Fungal Speciation
Linda M. Kohn

Exploiting Chinks in the Plant's Armor: Evolution and Emergence of Geminiviruses
Maria R. Rojas, Charles Hagen, William J. Lucas, and Robert L. Gilbertson

Regulation of Secondary Metabolism in Filamentous Fungi
Jae-Hyuk Yu and Nancy Keller

RNA Silencing in Productive Virus Infections
Robin MacDiarmid

Genetics of Plant Virus Resistance
Byoung-Cheorl Kang, Inhwa Yeam, and Molly M. Jahn

Annual Reviews is a nonprofit scientific publisher established to promote the advancement of the sciences. Beginning in 1932 with the *Annual Review of Biochemistry*, the Company has pursued as its principal function the publication of high-quality, reasonably priced *Annual Review* volumes. The volumes are organized by Editors and Editorial Committees who invite qualified authors to contribute critical articles reviewing significant developments within each major discipline. The Editor-in-Chief invites those interested in serving as future Editorial Committee members to communicate directly with him. Annual Reviews is administered by a Board of Directors, whose members serve without compensation.

Annual Reviews of

Anthropology
Astronomy and Astrophysics
Biochemistry
Biomedical Engineering
Biophysics and Biomolecular Structure
Cell and Developmental Biology
Clinical Psychology
Earth and Planetary Sciences
Ecology, Evolution, and Systematics
Entomology
Environment and Resources
Fluid Mechanics
Genetics
Genomics and Human Genetics
Immunology
Law and Social Science
Materials Research
Medicine
Microbiology
Neuroscience
Nuclear and Particle Science
Nutrition
Pathology: Mechanisms of Disease
Pharmacology and Toxicology
Physical Chemistry
Physiology
Phytopathology
Plant Biology
Political Science
Psychology
Public Health
Sociology

SPECIAL PUBLICATIONS
Excitement and Fascination of Science, Vols. 1, 2, 3, and 4

John Maynard Smith

Richard E. Michod

Department of Ecology and Evolutionary Biology, University of Arizona, Tucson, Arizona 85721; email: michod@u.arizona.edu

Annu. Rev. Genet. 2005. 39:1–8

First published online as a Review in Advance on August 30, 2005

The *Annual Review of Genetics* is online at http://genet.annualreviews.org

doi: 10.1146/annurev.genet.39.040505.114723

0066-4197/05/1215-0001$20.00

Key Words

fitness, game theory, evolutionary transitions, sex, recombination

Abstract

John Maynard Smith was one of the most original thinkers in evolutionary biology of the post neo-Darwinian synthesis age. He was able to define new problems with clarity and by doing so open up new research directions. He did this in a number of areas including game theory and evolution, the evolution of sex, animal behavior, evolutionary transitions and molecular evolution. Although he is best known for his research and his ideas, he was a great expositor and wrote many books, including introductory texts in the areas of evolution and genetics, ecology and mathematical modeling, as well as advanced expositions of research problems.

John Maynard Smith was one of the most original thinkers in evolutionary biology of the post neo-Darwinian synthesis age. He was able to define new problems with clarity and by doing so open up new research directions. He was able to solve problems and then move on to new problems. He was a free spirit in the world of ideas, and he pursued his ideas wherever they led even if they conflicted with his personal views (as was sometimes the case with sociobiology). He believed in an objective world and science's capacity to approach it. His life was a celebration of what it means to be curious about this world and to understand it on its own terms. He was highly creative and made rational discovery an art form. Simple mathematical models and concepts were his tools (some concepts, such as the evolutionarily stable strategy or ESS, he invented). An ESS is a strategy or population state in which rare mutant phenotypes are at a fitness disadvantage relative to the common phenotype (an important point is that the fitnesses of both rare and common phenotypes are evaluated using the ecological properties determined by the common phenotype). His subjects were organisms and their phenotypes, which under his analysis revealed general characteristics of the evolutionary process. He generated a large number of deeply penetrating ideas on a wide range of topics.

Although he is best known for his research and his ideas, he was a great expositor and wrote many books, including introductory texts in the areas of evolution and genetics (20, 26, 40) and ecology and mathematical modeling (24, 27), as well as advanced expositions of research problems, such as the evolution of sex (31), game theory and evolution (33), and evolutionary transitions (48). Three collections of papers already exist in his honor (1, 2, 8) and at least two are currently in preparation (*Journals of Theoretical Biology* and *Biology and Philosophy*). He was active and productive in research until he died; his last book on the evolution of animal signals appeared in 2003 (44). He lived a long and fulfilling life and was a happy man.

He did not go in for the usual accolades of academic distinction; for example, he did not cultivate a following of dedicated students nor did he seek large sums of funding for his work; he was approached for consideration for knighthood, but he would not allow his name to go forward. Those who admire him do so out of a deep respect for his ideas and his science, not out of a shared intellectual pedigree or training. Still, he was a very social man; if he was in attendance at a conference it was sure to be a success! He had an active group at Sussex (where he spent most of his career) attracting scholars and students from all around the world.

John Maynard Smith was born in London in 1920 and died quietly at his home last year in Lewes, England. His early education was at Eton College where he read works of his future mentor, J.B.S. Haldane (one of the founding fathers of the neo-Darwinian theory of evolution, with S. Wright and R.A. Fisher). In 1947, John began studying with Haldane at University College, London, after studying engineering at Trinity College, Cambridge and working in aircraft design during the Second World War. At Cambridge he met his wife and life-long companion and some-time collaborator Sheila. He practiced Haldane's approach to science, which John characterized as a "combination of the abstract and the particular" (34), and like Haldane, Maynard Smith used simple mathematical models to clarify concrete biological problems. While studying with Haldane, John began an academic position at University College and never actually took his Ph.D. During the years 1939–1946 he was active in the British Communist Party; he left the party in 1956 following a period of progressive disenchantment; in particular he was troubled with the Communist Party's behavior during the Lysenko affair and with the Russian invasion of Hungary (34). In 1965 he became the founding Dean of the School of Biological Sciences at the University of Sussex, where he remained for the rest of his life.

Maynard Smith was awarded many prizes including the Crafoord Prize in 1999 (along with Ernst Mayr and George C. Williams) for his "fundamental contributions to the conceptual development of evolutionary biology," and in 2001 the Kyoto Prize in Basic Sciences for his application of game theory to biology and his idea of the evolutionarily stable strategy. The awarding foundation pointed out that the ESS idea has not only revolutionized evolutionary biology but also such diverse fields as economics, business sciences, and politics. He was a Fellow of the Royal Society in Great Britain and a Foreign Associate of the National Academy of Sciences in the United States.

Maynard Smith made his way in the world of science by the force of his ideas and the clarity of his thought. There was no misty profundity in his work or writings. Throughout his life, he sought to understand fitness; what it means in concrete terms for organisms, how it can be used to understand their phenotypes, and how new levels of fitness are created during evolutionary transitions in individuality (transitions that create new kinds of evolutionary units, for example, genes, cells, multicellular organisms, etc.). He said (34), "my pleasure comes from seeing the same mathematical structures emerging from such diverse problems as the evolution of behavior and the origin of life." His early work during the 1950s and 1960s involved experimental genetics on inbreeding and aging in *Drosophila subobscura* (e.g., 5, 18, 19, 21). Among his results during that period was the demonstration of a trade-off between the basic fitness components, longevity and fecundity (21), results that anticipated much of the recent work on aging and life history evolution.

Maynard Smith's work then became progressively more theoretical. He explained this shift as an outcome of his taking on administrative roles at Sussex in 1965 and also to the fact that, when his mentor Haldane departed for India, John became more confident in his own theoretical work (34). Although his mathematical techniques were basic, he had an uncanny ability to ask important biological questions and to use simple mathematical models to clarify the issues involved. He was also a good naturalist. He stuck with a problem until it yielded to his analysis, and then he moved on to a new problem. He was not big on multitasking (unless by "multitasking" we include talking science while drinking!) and he was not big on technology (I couldn't imagine him with a cell phone or a Palm Pilot), although he did move from slide rules to computers when they became available. At about that time, I remember him saying that he had stopped writing computer code or doing "sums" (as he called mathematical modeling) in the afternoon, as he would usually have a few pints at lunch and this often led to faulty programs and analyses. As a result, he usually wrote papers or worked on administrative tasks after lunch.

In his theoretical work, Maynard Smith was especially interested in the levels at which natural selection acted; for example, for the benefit of what unit, the individual or the group, may a trait be best explained (22, 28). And although he argued decisively for individual selection in most cases, say for the evolution of sex (23, 29–35) or for ritualized animal conflict (33, 45), two problems that were previously explained by group selection (as being good for the species), he was equally interested in how groups become individuals during evolutionary transitions. He viewed the major events in evolution as a series of transformations in the way in which information is coded and transferred—genes, chromosomes, networks of genes, genes in cells, cells-in-cells (eukaryotic cells), cells in groups (multicellular organisms), kin groups, societies, and language (36, 39, 48, 52, 53).

Maynard Smith developed game theory as a tool in evolutionary biology and with G. Price, the concept of an evolutionarily stable strategy (45). An ESS is characterized by the situation in which rare mutant phenotypes are at a fitness disadvantage relative to the common phenotype (an important point is that the fitnesses of both rare and common phenotypes are evaluated using the ecological properties determined by the common phenotype). The ESS tool is in wide use today to study evolution in situations in which an organism's fitness is a frequency-dependent function of other organisms in the population. The ESS approach is a kind of short cut for predicting the outcome of natural selection in situations involving frequency-dependent selection and avoids writing down explicit gene frequency equations for the underlying traits, many of which, like animal behaviors, have unknown genetics anyway. Although approaches similar to the ESS had been used on occasion by R.A Fisher and W.D. Hamilton, Maynard Smith generalized the ESS approach and studied systematically how natural selection could replace rational human decision making in predicting phenotypes in conflictual situations (33).

Maynard Smith was the grand master of the evolution of sex problem; he clarified the many costs of sex and defined the problem of sex in an especially clear and compelling way so that other biologists could both appreciate its fundamental importance and contribute to its solution. Sex is everywhere, yet what is most obvious about the trait is the high cost it imposes on the parents. What are the benefits of sex that offset these costs? The evolution of sex interested John because sexual reproduction was often interpreted as a trait that evolved because it benefited the species, by increasing the species' rate of adaptation to its environment. He showed how incomplete this good-for-the-species explanation was, by showing how hard it was to make this argument formal (indeed, under some conditions models show the argument is false) and by demonstrating in quantitative form the advantages of asexual reproduction. He also classified and studied the various kinds of benefits that could possibly overcome the intrinsic advantage of asexual reproduction to increase in sexual populations (this intrinsic advantage comes from their avoiding the so-called twofold cost of sex). His book on the evolution of sex (31) is a classic and still one of the best treatments of the problem even though an immense amount of work has been done since it was published 27 years ago (indeed, much of this work was done because of the book). In reading his book today, we see an especially clear example of how John was able to identify the essential elements of a problem area in a rather complete way and by so doing, invite others into the area to help solve the problem.

Maynard Smith did pioneering work on sex and recombination in bacterial populations and showed that recombination has been more significant in prokaryotes than many had suspected (38, 41, 42, 46, 47, 51). This was his major research interest late in life. He also anticipated and contributed to developments in the theory of molecular evolution. For example, he pioneered the notion of natural selection in protein sequence space (25) and discovered genetic hitchhiking and its effect on linked variation (9, 43), something that is fundamental to our understanding of molecular variation today. In addition, he created the only clear mathematical model of epigenetic gene regulation that I am aware of (37).

Maynard Smith is known for his recognition and development of evolutionary transitions as a problem in evolutionary biology. Transforming our understanding of life is the realization that evolution occurs not only through mutational change in populations but also during evolutionary transitions in individuality (ETIs)—when groups become so integrated that they evolve into a new higher-level individual. The major landmarks in the diversification of life and the hierarchical organization of the living world are consequences of a series of ETIs: from nonlife to life, from networks of cooperating genes to the first prokaryotic-like cell, from prokaryotic to eukaryotic cells, from unicellular to multicellular organisms, from asexual to sexual populations, and from solitary to social organisms. It is a major challenge to understand why (environmental selective pressures) and how (underlying genetics, physiology, and development) the basic features of an evolutionary individual, such as fitness heritability, indivisibility, and evolvability, shift their reference from the old to the new level. This is the ETI problem that Maynard Smith more than anyone else helped to define.

What is special about the ETI problem for evolutionary biology is that cooperation plays a central role. Maynard Smith was always interested in the evolution of altruism and cooperation, because at first glance such behaviors seem counter to what organisms should do to maximize their fitness. Throughout most of the development of ecology and evolution, the study of cooperation received much less attention than other forms of ecological interaction, such as competition and predation. Scholars generally viewed cooperation to be of limited interest, of special relevance to certain groups of organisms to be sure, as in the social insects, birds, our own species, and our primate relatives, but not of general significance to life on earth. All that has changed with the study of ETIs. What began as the study of animal social behavior some 40 years ago has now embraced the study of interactions at all biological levels. Instead of being seen as a special characteristic clustered in certain groups of social animals, cooperation is now seen as the primary creative force behind ever-greater levels of complexity and organization in all of biology. Cooperation plays this central role in ETIs because it exports fitness from the lower level (its costs) to the new higher level (its benefits).

Recognizing the importance of cooperation in the history of life on earth has taken some time, especially for neo-Darwinians and population biologists. Darwin (6), Wilson (56), and Hamilton (10, 11, 13) all understood the importance of cooperation for social organisms. There was pioneering work done as early as 1902 on the importance of cooperation in the struggle for existence (14), and there was the now widely accepted theory of Margulis (15, 16) and others on the endosymbiotic origins of mitochondria and chloroplasts in the eukaryotic cell. However, cooperation was also viewed as a destabilizing force in ecological communities and likely of limited significance because of the positive feedback loops it creates (17). Sociobiology had defined altruism as its core problem (56), but the altruism problem was not viewed as general to life on Earth until others began applying cooperation thinking to the evolution of interactions at other levels in the hierarchy of life in addition to social organisms, such as to the level of genes within gene groups (e.g., 7) and to the level of cells within cell groups (e.g., 3). Concomitant with the generalization of the cooperation problem was the development of multilevel selection theory (e.g., 12, 22, 49, 50, 54, 55). The ETI problem grew out of these two developments that, in effect, extended the sociobiology revolution to all kinds of replicating units in the hierarchy of life.

What Maynard Smith did was to synthesize a diverse body of work into a comprehensive framework for ETIs; in addition, he mapped out the problem area with a clarity that only he could produce. He did this initially in two papers (36, 39) and later in a much more systematic and complete way in his book with E. Szathmáry (48). While not single-handedly defining the ETI problem, nor solving it, Maynard Smith and Szathmáry have mapped out the important issues, and this has stimulated an increasing number of biologists to enter this area. This is yet another example of how he was able to define a problem area with clarity so that others could contribute to it.

Maynard Smith was a lot of fun to be around. He treated all people with equal respect, but he was quick to expose sloppy thinking and unreceptive to pomposity. As a result, he was a feared debater of creationists. He had a child-like wonder about him—this was my first impression of him, as a second-year graduate student visiting his group at Sussex in 1975. I have many memories of JMS (as he was often called)—unkempt appearance (the hair!), that sparkle in his eyes, the pub crawls over the South Downs, and the many long discussions over tea and beer. As a young student I wanted to be just like him, and in many ways, I still do. Now years later as I sit and reflect on this wonderful human being and what his life and work means for the rest of us, I still think mainly of how much fun he was to do science with. I miss him very much.

There are many resources available for those who want to know more about this extraordinary man. JMS was asked to write a short autobiographical sketch emphasizing his work in animal behavior (34). There is also a perspective in *Genetics* (4). A good place to begin browsing for material is the JMS web site maintained by the Center for the Study of Evolution at the University of Sussex (**http://www.lifesci.sussex.ac.uk/CSE/members/jms/jms.htm**). I have used this resource extensively in preparing this biography and have also benefited from feedback on the manuscript and insights into the man from Paul Harvey, Brian Charlesworth and Joel Peck. Matt Herron and David Harper also provided helpful comments on the manuscript.

LITERATURE CITED

1. *Selection* 1[1–3]. 2000. ed. E Szathmary. Oxford, UK: Blackwell Sci.
2. *Philos. Trans. R. Soc. London Ser. B* 355.2000:1551–684
3. Buss LW. 1987. *The Evolution of Individuality*. Princeton, NJ: Princeton Univ. Press. 201 pp.
4. Charlesworth B. 2004. Anecdotal, historical and critical commentaries on genetics. John Maynard Smith: January 6, 1920–April 19, 2004. *Genetics* 168:1105–9
5. Clarke JM, Maynard Smith J. 1966. Increase in the rate of protein synthesis with age in *Drosophila subobscura*. *Nature* 209:627–29
6. Darwin C. 1859. *The Origin of Species by Means of Natural Selection, or Preservation of Favoured Races in the Struggle for Life*. London: John Murray
7. Eigen M, Schuster P. 1979. *The Hypercycle, a Principle of Natural Self-Organization*. Berlin: Springer-Verlag
8. Greenwood PJ, Harvey PH, Slatkin M. 1985. *Evolution: Essays in Honor of John Maynard Smith*. London: Cambridge Univ. Press. 328 pp.
9. Haigh J, Maynard Smith J. 1976. The hitch-hiking effect—a reply. *Genet. Res.* 27:85–87
10. Hamilton WD. 1963. The evolution of altruistic behavior. *Am. Nat.* 97:354–56
11. Hamilton WD. 1964. The genetical evolution of social behaviour. I. *J. Theor. Biol.* 7:1–16
12. Hamilton WD. 1975. Innate social aptitudes of man: an approach from evolutionary genetics. In *Biosocial Anthropology*, ed. R Fox, pp. 133–55. New York: Wiley
13. Hamilton WD. 1964. The genetical evolution of social behaviour. II. *J. Theor. Biol.* 7:17–52
14. Kropotkin P. 1902. *Mutual Aid: A Factor in Evolution*. London: Allen Lane
15. Margulis L. 1970. *Origin of Eukaryotic Cells*. New Haven: Yale Univ. Press
16. Margulis L. 1981. *Symbiosis in Cell Evolution*. San Francisco: Freeman. 419 pp.
17. May RM. 1973. *Stability and Complexity in Model Ecosystems*. Princeton: Princeton Univ. Press
18. Maynard Smith J. 1956. Fertility, mating behaviour and sexual selection in *Drosophila subobscura*. *J. Genet.* 54:261–79
19. Maynard Smith J. 1958. Prolongation of the life of *Drosophila subobscura* by a brief exposure of adults to a high temperature. *Nature* 181:496–97
20. Maynard Smith J. 1958. *Theory of Evolution*. Harmondsworth, UK: Penguin. 320 pp.
21. Maynard Smith J. 1959. Sex-limited inheritance of longevity in *Drosophila subobscura*. *J. Genet.* 56:1–9

22. Maynard Smith J. 1964. Group selection and kin selection. *Nature* 201:145–47
23. Maynard Smith J. 1968. Evolution in sexual and asexual populations. *Am. Nat.* 102:469–73
24. Maynard Smith J. 1968. *Mathematical Ideas in Biology*. Cambridge: Cambridge Univ. Press. 152 pp.
25. Maynard Smith J. 1970. Natural selection and the concept of protein space. *Nature* 225:563–64
26. Maynard Smith J. 1972. *On Evolution*. Edinburgh: Edinburgh Univ. Press. 125 pp.
27. Maynard Smith J. 1974. *Models in Ecology*. London: Cambridge Univ. Press. 146 pp.
28. Maynard Smith J. 1976. Group selection. *Q. Rev. Biol.* 51:277–83
29. Maynard Smith J. 1976. A short term advantage for sex and recombination through sib-competition. *J. Theor. Biol.* 63:245–58
30. Maynard Smith J. 1977. The sex habit in plants and animals. In *Measuring Selection in Natural Populations*, ed. FB Christiansen, TM Fenchel, pp. 265–73. Berlin: Springer-Verlag
31. Maynard Smith J. 1978. *The Evolution of Sex*. London: Cambridge Univ. Press
32. Maynard Smith J. 1981. Will a sexual population evolve to an ESS? *Am. Nat.* 117:1015–18
33. Maynard Smith J. 1982. *Evolution and the Theory of Games*. London: Cambridge Univ. Press
34. Maynard Smith J. 1985. In Haldane's footsteps. In *Leaders in the Study of Animal Behavior: Autobiographical Perspectives*, ed. DA Dewsbury, pp. 347–54. Lewisburg, PA: Bucknell Univ. Press
35. Maynard Smith J. 1988. The evolution of recombination. In *The Evolution of Sex, An Examination of Current Ideas*, ed. RE Michod, BR Levin, pp. 106–25. Sunderland, MA: Sinauer
36. Maynard Smith J. 1988. Evolutionary progress and levels of selection. In *Evolutionary Progress*, ed. MH Nitecki, pp. 219–30. Chicago: Univ. Chicago Press
37. Maynard Smith J. 1990. Models of a dual inheritance system. *J. Theor. Biol.* 143:41–53
38. Maynard Smith J. 1990. The evolution of prokaryotes: Does sex matter? *Annu. Rev. Ecol. Syst.* 21:1–12
39. Maynard Smith J. 1991. A Darwinian view of symbiosis. In *Symbiosis as a Source of Evolutionary Innovation*, ed. L Margulis, R Fester, pp. 26–39. Cambridge: MIT Press
40. Maynard Smith J. 1998. *Evolutionary Genetics*. Oxford: Oxford Univ. Press. 325 pp.
41. Maynard Smith J, Dowson CG, Spratt BG. 1991. Localized sex in bacteria. *Nature* 249:29–31
42. Maynard Smith J, Feil EJ, Smith NH. 2000. Population structure and evolutionary dynamics of pathogenic bacteria. *BioEssays* 22:1115–22
43. Maynard Smith J, Haigh J. 1974. The hitch-hiking effect of a favourable gene. *Genet. Res.* 23:23–35
44. Maynard Smith J, Harper D. 2003. *Animal Signals*. Oxford: Oxford Univ. Press
45. Maynard Smith J, Price GR. 1973. The logic of animal conflict. *Nature* 246:15–18
46. Maynard Smith J, Smith NH. 1996. Site-specific codon bias in bacteria. *Genetics* 142:1037–43
47. Maynard Smith J, Smith NH, O'Rourke M, Spratt BG. 1993. How clonal are bacteria? *Proc. Natl. Acad. Sci. USA* 90:4384–88
48. Maynard Smith J, Szathmáry E. 1995. *The Major Transitions in Evolution*. San Francisco: Freeman. 346 pp.
49. Price GR. 1972. Extension of covariance selection mathematics. *Ann. Hum. Genet.* 35:485–90
50. Price GR. 1970. Selection and covariance. *Nature* 227:529–31

51. Smith NH, Dale J, Inwald J, Palmer S, Gordon SV, et al. 2003. The population structure of *Mycobacterium bovis* in Great Britain: clonal expansion. *Proc. Natl. Acad. Sci. USA* 100:15271–75
52. Szathmáry E, Maynard Smith J. 1995. The major evolutionary transitions. *Nature* 374:227–32
53. Szathmáry E, Maynard Smith J. 1997. From replicators to reproducers: the first major transitions leading to life. *J. Theor. Biol.* 187:555–71
54. Wade MJ. 1978. A critical review of models of group selection. *Q. Rev. Biol.* 53:101–14
55. Wilson DS. 1980. *The Natural Selection of Populations and Communities*. Menlo Park, CA: Benjamin/Cummings
56. Wilson EO. 1975. *Sociobiology: The New Synthesis*. Cambridge, MA: Belknap

The Genetics of Hearing and Balance in Zebrafish

Teresa Nicolson

Oregon Hearing Research Center and Vollum Institute, Oregon Health and Science University, Portland, Oregon 97239; email: nicolson@ohsu.edu

Annu. Rev. Genet. 2005. 39:9–22

The *Annual Review of Genetics* is online at http://genet.annualreviews.org

doi: 10.1146/annurev.genet.39.073003.105049

0066-4197/05/1215-0009$20.00

Key Words

deafness, development, hair cells, inner ear, lateral line organ, mechanotransduction

Abstract

The zebrafish is an excellent model system for studying the molecular basis of inner ear development and function. The eggs develop *ex utero* and the ear is transparent for the first few weeks of life. Forward genetic screens and antisense technology have helped to elucidate the signaling pathways and molecules required for inner ear development and function. This review addresses the most recent advances in our understanding of how the ear forms and discusses the molecules in hair cells that are essential for sensing sound and movement in the zebrafish.

Contents

THE INNER EAR RECEPTOR: THE SENSORY HAIR CELL

The inner ear detects motion and sound. The sensory patches dedicated to this task are termed maculae, cristae, or the organ of Corti. Maculae contain receptors coupled to dense crystal structures known as otoconia or otoliths. Cristae are situated in the semicircular canals. The organ of Corti is a specialization for hearing found only in higher vertebrates. The sensory receptors of the inner ear, the hair cells, are exquisitely sensitive to the mechanical stimuli of movement and sound (20). Their highly specialized apical surfaces consist of unusual processes, termed stereocilia. Each stereocilium is densely packed with actin filaments in a paracrystalline array. The hair-like appearance of the bundle of apical stereocilia gave rise to the name "hair cells." Within each hair bundle, stereocilia are arranged with remarkable precision in rows, with increasing height toward one end of the bundle. Next to the tallest stereocilia in vestibular or lower vertebrate hair cells is a true cilium known as the kinocilium. Its role appears to couple the bundle of stereocilia to the overlying extracellular matrix or the otolithic membrane. In mammalian auditory hair cells, kinocilia appear during development but are reabsorbed in mature hair cells. In the cochlea, the tallest stereocilia of the outer hair cells are instead attached to the tectorial membrane, a gelatinous membrane very similar to the otolithic or cupula membranes present in maculae or cristae. All hair cells are bathed in a special fluid, the endolymph, a high potassium fluid unique to the ear. When hair cells are mechanically stimulated, stereocilia pivot about their bases and move together as a unit because of numerous extracellular linkages. Movement allows cations to flow into the hair bundles, resulting in depolarization.

Crista: sensory epithelium of hair cells and supporting cells inside the semicircular canals

Macula: sensory epithelium associated with otoliths

Otolith: calcium carbonate stone situated above the macula

It has been estimated that a movement of three angstroms is sufficient to excite hair cells. Unfortunately, this sensitivity may underlie the vulnerability to mutations. Deafness is one of the most common inherited diseases (31). In addition, age-related hearing loss occurs at a high frequency (11). There are many things that can go wrong in the ear (see Hereditary Hearing Loss Homepage at **http://webhost.ua.ac.be/hhh/**). The endolymph can be thrown out of balance by mutations affecting ion transport and homeostasis. Hair bundles can be malformed or degenerate over time. The calcium carbonate biominerals, the otoconia or otoliths, may be absent or dislodged from the sensory epithelia. Using genetics as a tool, we are beginning to grasp the molecular basis of how the ear develops and how these pathologies may arise.

ANATOMY AND FUNCTION OF THE ZEBRAFISH EAR

Orientation with respect to gravity and the environment is a key concern for both aquatic and terrestrial animals. In vertebrates, the anatomical features of the vestibular labyrinth are highly conserved in terms of structure and function. These features include three semicircular canals and two or more macular organs (**Figure 1**). At the base of each canal is a rounded mound of sensory epithelium called the crista. Within each crista, the long kinocilia of the hair cells are embedded in a gelatinous membrane called the cupula that spans the canal. Movements of the head cause the fluid inside the canals to impinge upon the cupula or gelatinous membrane, deflecting the embedded kinocilia and hair bundles. Signals are subsequently sent from the excited hair cells to the brain, providing information about the position of the head and angular acceleration. All vertebrates possess two macular organs, the saccule and utricle.

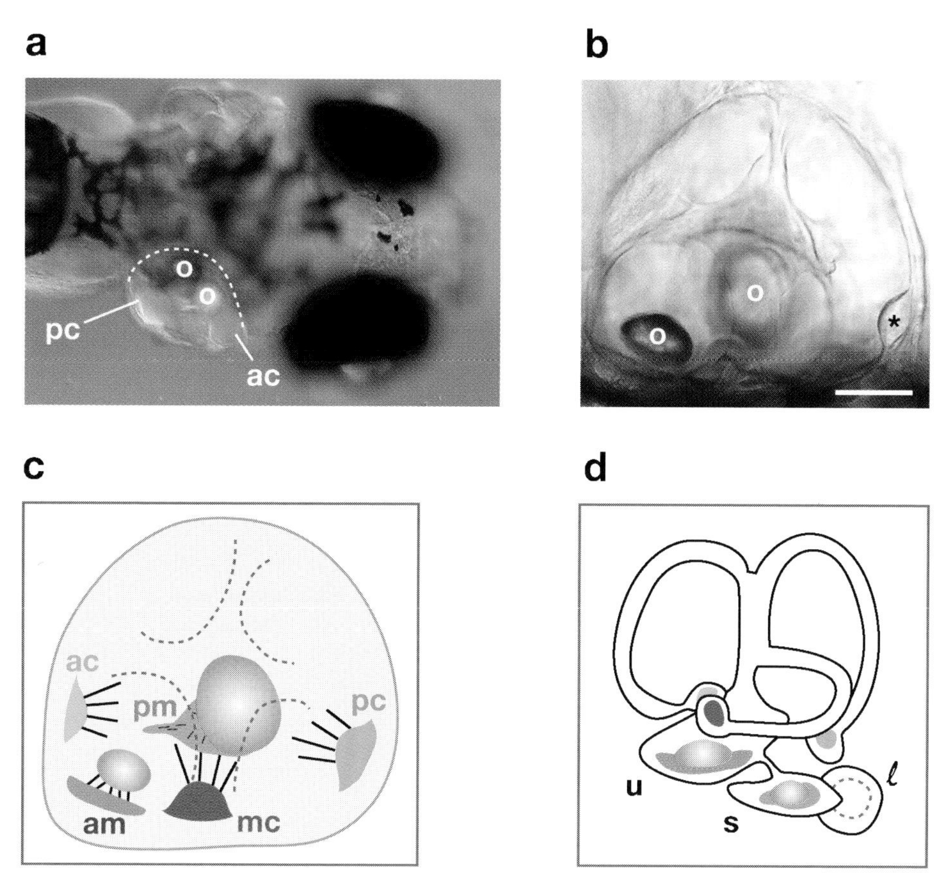

Figure 1

Larval and adult inner ear of the zebrafish. (*A*) Dorsal view of a live, 5 day-old zebrafish larvae. The plane of focus includes the anterior and posterior canals (*solid lines*). The dotted line indicates the medial wall of the developing otocyst. Both maculae are located relatively deep within the otocyst (out of focus otoliths). (*B*) Lateral view of the larval inner ear. The focal plane includes the anterior otolith and the neuroepithelium within the posterior canal (*asterisk*). The epithelial columns that join in the middle are also visible. (*C*) Diagrammatic representation of the larval ear. All five sensory patches are indicated in color. Dotted lines indicate the epithelial structures around which the semicircular canals form. (*D*) Diagram of the adult zebrafish ear. A third macular organ, the lagena (dotted area), forms later during the juvenile stage. Colors correspond to those in panel *C*. Scale bar indicates 100 μm in *A* and 60 μm in *B*. ac, anterior crista; am, anterior macula; l, lagena; mc, medial crista; o, otolith; pc, posterior crista; pm, posterior macula; s, saccule; u, utricle.

These pouch-like organs contain a bed of hair cells coupled to calcium carbonate crystals. In fish, the crystals coalesce to form a single large polycrystalline mass called an otolith (**Figure 1**). Forces impinging upon the otoliths cause them to move, which in turn deflects the underlying hair bundles. Macular organs are important for sensing linear acceleration and gravity. In lower vertebrates such as fish and frogs, they also are used in hearing (33). Fish use their saccular organs to detect frequencies between 10–4000 Hz. Some fish are classified as hearing specialists based on the presence of a series of bones known as Weberian ossicles that connect the swim bladder to the saccule. Sound can set the air-filled swim bladder into motion and this motion is transmitted to the sensory epithelium via the ossicles, in effect amplifying the sound. The zebrafish, along with many related species within its clade, are hearing specialists. Additional endorgans include the macular organ called the lagena and the macula neglecta that develop after the larval stage. Their function is primarily auditory (33, 50).

Lateral line organ: system of superficial and canal hair cells that detects water movements

Neuromast: group of hair cells at the surface of the skin

In addition to the ear, fish (and frogs) possess another organ that employs sensory hair cells, the lateral line organ. This organ allows the detection of low-frequency stimuli such as water movements and is important in schooling, prey detection, and other behaviors. It is called the lateral line because the groupings of hair cells known as neuromasts (**Figure 2**) are located in a series extending along the trunk. Although an attractive hypothesis because of the anatomical simplicity of lateral line organ, still under debate is whether the lateral line organ was the first to evolve and then later give rise to the ear (34).

The organization and morphology of the inner ear neuroepithelium in fish resembles

a

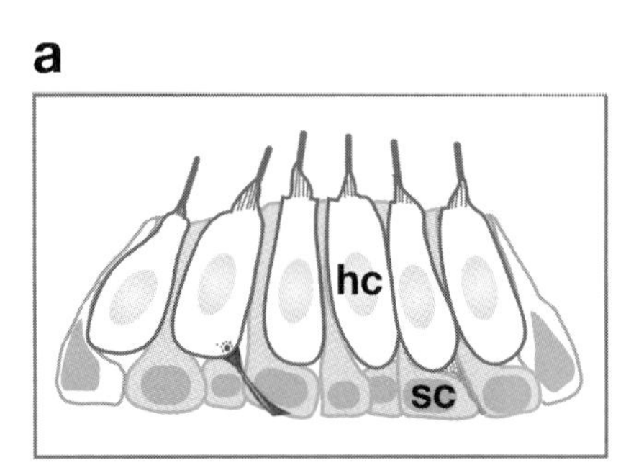

b

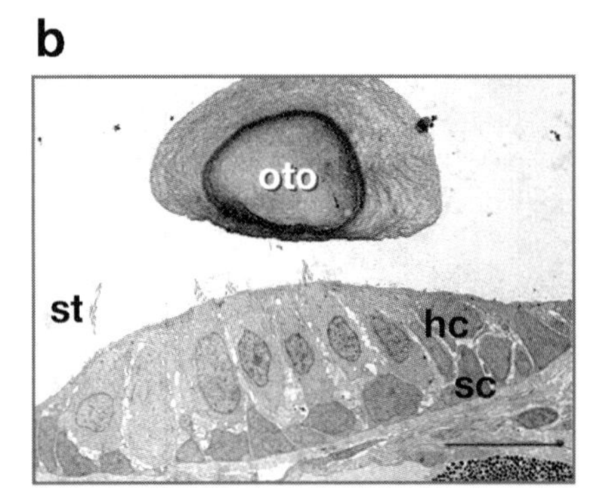

c

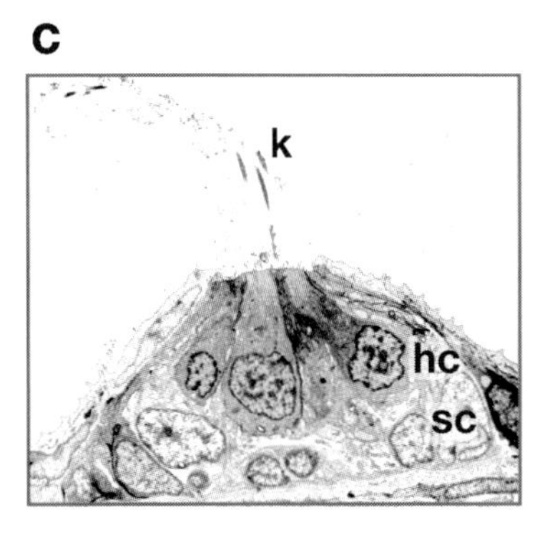

Figure 2

Organization of the sensory epithelium. (*A*) Diagram of cross section of inner ear neuroepithelium (anterior macula). The oval-shaped hair cells (*yellow*) are present in the upper two thirds of the epithelium. The lower supporting cells (*blue*) are situated between hair cells. The epithelium is bathed in a high potassium solution, the endolymph. The outer cells (*white*) are presumably dark cells that regulate the potential of the endolymph. Also depicted are an example of an afferent synapse (*red*) and an efferent synapse (*orange*). Transmission electron micrographs of an anterior macula (*B*) and a neuromast (*C*) at 120 hpf. The cylindrical oval-shaped sensory hair cells lie above the dark nuclei in the soma of the supporting cells. Scale bar in *B* indicates 7 μm in *B* and 3.5 μm in *C*. hc, hair cells; k, kinocilia; oto, otolith; sc, supporting cells; st, stereocilia.

that found in higher vertebrates (**Figure 2**). At the surface of the neuroepithelium, hair bundles project into the inner ear lumen or in the lateral line, directly into the aqueous environment. Each hair cell is surrounded by 5 to 6 supporting cells whose apical surfaces are covered with microvilli. The oval-shaped cell bodies of the hair cells are situated above the more cuboidal-like nuclei of the supporting cells. The eighth cranial nerve fibers create multiple synapses along the basolateral membrane of the hair cells. The specialized calyx surrounding Type I vestibular hair cells in higher vertebrates is not present in fish, though long protrusions or extensions of the basal surface into individual nerve fibers have been detected (5). At afferent synapses, one or more synaptic ribbons or dense bodies are located next to the basolateral membrane of the hair cell (**Figure 3**). These electron-dense structures are surrounded by synaptic vesicles and are thought to play a role in synaptic transmission. The same organization characterizes the mounded clusters of hair cells or neuromasts that comprise the superficial lateral line organ (**Figure 2**). Similar to hair cells in the semicircular canals, the kinocilia of lateral line hair cells are long and embedded in a gelatinous cupula.

DEVELOPMENT OF THE ZEBRAFISH EAR

The development of the zebrafish ear is similar to that observed in other vertebrates (35, 50). The one striking exception is the mechanism by which the lumen of the otocyst forms. In zebrafish, the otic placode forms a hollow space within the cluster of precursor cells. In chicks and mice, the otic placode invaginates, creating a cup that eventually closes to form a lumen. Other differences include the way in which the semicircular canals form and the lack of extensive cell death in remodeling during morphogenesis. Despite these differences, studies of development of the zebrafish ear have increased our knowledge about the general mechanisms employed. Overall, the development of the ear proceeds quite rapidly in zebrafish. Within 16 hours post fertilization (hpf), a thickening of the ectoderm, the otic placode, becomes visible (13). Soon after cavitation (18 hpf), the first hair cells appear at opposite poles of the ventral portion of the lumen. These two sensory patches will give rise to the anterior and posterior maculae. When the first hair cells appear, otoliths attach to the kinocilia around 20 hpf (35). During this period neuroblasts delaminate from the otic vesicle to form the first-order neurons of the auditory/vestibular (VIIIth) nerve (13). At approximately 45 hpf, epithelial columns sprout from the otocyst wall and grow into the lumen. About 10 h later, these projections meet and fuse. Eventually they will form the central parts or hubs around which the semicircular canals develop (**Figure 1**). The sensory patches or cristae within the semicircular canals form about a day later than the anterior and posterior maculae. By 96 hpf, the larvae are free swimming and have fully functional ears (**Figure 1**).

Otic vesicle: developing inner ear fluid-filled, sac-like structure lined with inner ear progenitor cells

hpf: hours post fertilization

The following sections focus on otic induction, semicircular canal development, and hair-cell specification. For more information on related developmental topics such as lateral line or otolith development, please see related reviews by Ghysen & Dambly-Chaudiere (2004) and Söllner & Nicolson (2004).

Genes Implicated in Otic Induction

An in-depth review on this topic can be found in an excellent review by Whitfield et al. (see related reviews). The purpose of this section is to summarize recent progress in understanding otic induction. The accessibility of the zebrafish inner ear coupled with forward genetics and antisense technology have allowed a big step forward in defining signaling events that give rise to placode development. Both early and late events have been described

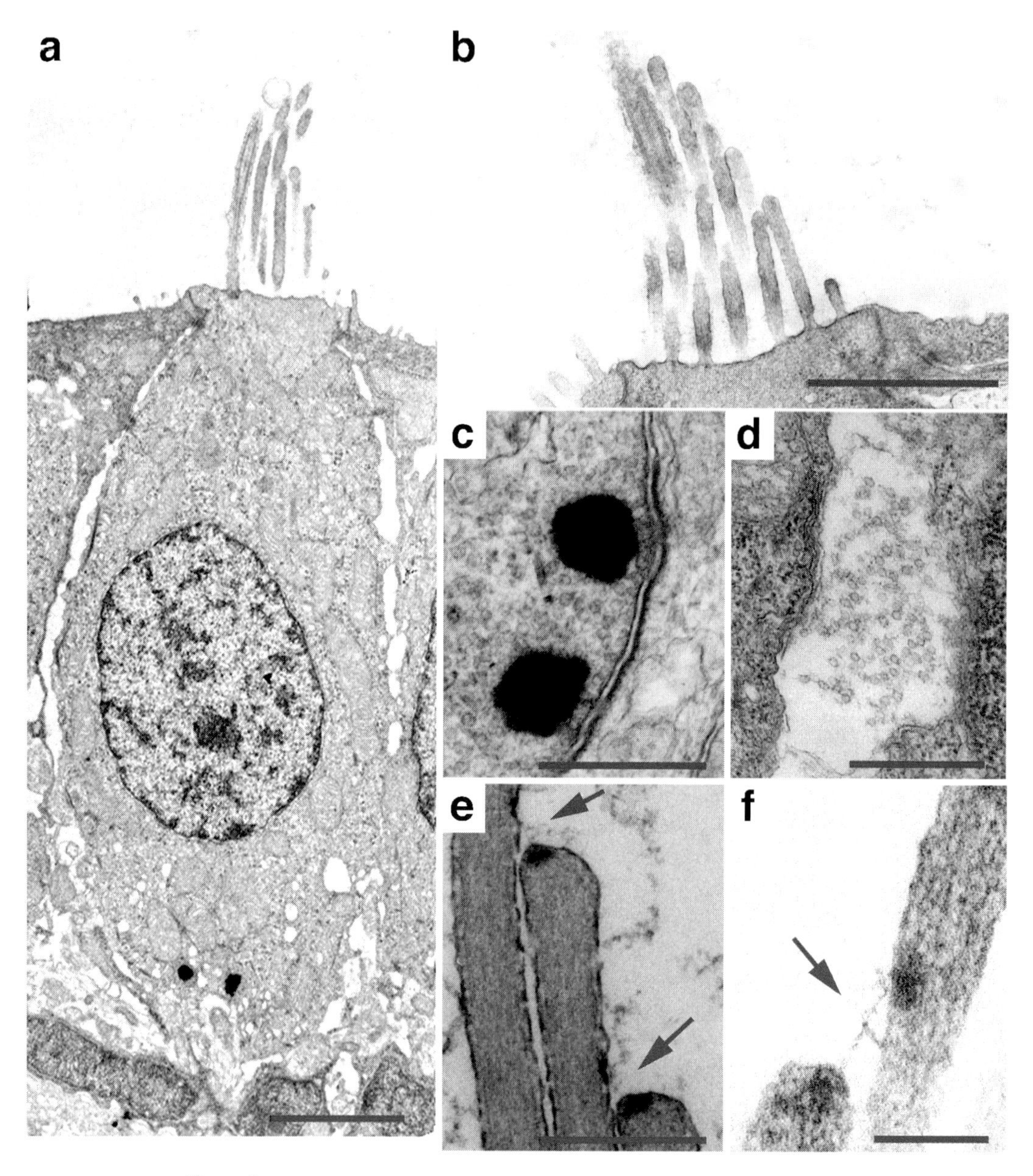

Figure 3

Fine structure of zebrafish inner ear hair cells (120 hpf). (*A*) A sensory hair cell of the anterior macula. Scale bar, 2 μm. (*B*) Transverse section of a hair cell bundle. The increase in height of the closely arranged stereocilia can be seen (only a third of the stereocilia are present in this section). Scale bar, 1 μm. (*C*) Two synaptic bodies surrounded by secretory vesicles at the basolateral end of the cell. (*D*) An efferent synapse filled with vesicles. (*E*) Very fine extracellular linkages at the tips of two stereocilia. The darkened areas of insertion plaques at the upper insertion site and the caps of the stereocilia are readily visible. (*F*) A single tip link interconnecting neighboring stereocilia. Scale bars, 0.5 μm in *C-E*, and 0.125 μm in *F*.

at the molecular level, leading the field of development in terms of our knowledge of otic induction.

Very early transplantation studies have suggested that the source of inductive cues for placode development was hindbrain tissue (47). Fibroblast growth factors (Fgfs) have been candidates for signaling as they are expressed in the region adjacent to the otic placode. More definitive evidence, however, was lacking until recent studies by several laboratories showing that *fgf3* and *fgf8* are required for otic induction (25, 28, 32). Recessive loss-of-function mutations in the *acerebellar/fgf8* gene cause the development of small ears, as does the removal of *fgf8* activity. If *fgf3* is knocked down using antisense oligonucleotides or morpholinos in the mutant *acerebellar* background, then the otic placode is completely absent.

Targets of the Fgf3 and Fgf8 pathway have also been identified by examining embryos harboring the deficiency Df^{b380} that lacks *sox9a*, *dlx3b*, and *dlx4b* (26). The absence of otic placodes in Df^{b380} mutants can be rescued by injection of *sox9a*, *dlx3b*, and *dlx4b* mRNA, and is mimicked by combined injection of morpholinos against all three genes. Only expression of *sox9a* is dependent on Fgf3 and Fgf8 signaling. The paired box transcription factors, Pax2 and Pax8, are also downstream targets of Fgf signaling (16). Onset of *pax8* expression in the preotic region occurs before the duplicates *pax2a* and *pax2b* are expressed. Loss of *pax2a* or *pax8* alone results in abnormally small ears. As with concomitant loss of *fgf3* and *fgf8* activity, loss of both *pax2a* and *pax8* causes otic induction to fail. The *hearsay* gene encodes a forkhead transcription factor, Foxi1, and mutations affect otic placode development (45). Analysis of both *hearsay* and Df^{b380} mutants reveals that Foxi1 influences the levels of *pax8* expression (16, 45). In addition, Dlx3b was shown to regulate expression of the later acting *pax2a* gene (16). Removal of both *foxi1* and *dlx3b* blocks the development of the otic placode.

Genes Implicated in Morphogenesis of the Otic Vesicle

Fgf: fibroblast growth factor

Ugdh: Uridine 5′-diphosphate glucose dehydrogenase

HA: hyaluronic acid

The structure of the embryonic otocyst is very simple in comparison to the mature adult ear. To detect movement of the head in various dimensions, the semicircular canals are positioned in three different planes. How this remarkable arrangement is achieved at the molecular level is only now beginning to be understood. We do, however, have some hints at what is required to establish the initial outgrowth of the epithelial columns that are destined to shape and position the semicircular canals. Mutagenesis screens have uncovered a class of mutants with jaw defects that fail to fuse the inner ear epithelial columns near the center of the lumen (29). The canal rudiments initially form, but then fail to grow and project away from the wall of the otocyst. *jekyll* is one such mutant. The gene was cloned and found to encode an enzyme, Uridine 5′-diphosphate glucose dehydrogenase (Ugdh), that produces one of the subunit building blocks of hyaluronic acid (HA) (49). Another study found that the zebrafish orthologue of a human deafness gene, *dfna5*, was also required for proper canal and jaw development (4). If *dfna5* activity was knocked down, then expression of *ugdh* was reduced in the ear and the pharyngeal arches of the jaw. It is not clear what the function of Dfna5 is because it shares no homology to other proteins, nor has it been localized; however, some evidence suggests that it may act at the transcriptional level (12), presumably upstream of *ugdh*. In both cases, removal of *ugdh* or *dfna5* results in loss of HA production. What does HA synthesis have to do with outgrowth of epithelial columns in the ear? It is a mystery at this point, though an interesting idea has been put forth proposing that HA acts as a propellant while it is secreted from the cells that extend forward into the lumen (14).

Disorganized epithelial columns are also a prominent feature of *dog-eared/eya1* mutants (24). Most of the otic vesicles in these mutants are abnormally small and the cristae fail to develop due to apoptosis of sensory neural precursors. In addition, cell death occurs within the migrating lateral line primordia, leading to loss of neuromasts at the end of the larval trunk. Larvae carrying *chameleon* ^{tf18b} and *slow muscle omitted* ^{b641}/*smoothened* mutations also have disorganized canals and abnormal ear morphology (15). The abnormalities are due to loss of inductive signals to the posterior region of the otic vesicle, resulting in a duplication of anterior structures. Overexpression of *patched1* mRNA phenocopies the mutants, demonstrating a role for Hedgehog signaling from underlying midline structures in patterning of the otic vesicle.

Genes Implicated in Specification of Hair Cells

Studies mainly from the mouse reveal that lateral inhibition plays a major role in determining the fate of the neuroepithelial cells (see related reviews, Bryant et al., 2002). A recent advance in zebrafish suggests that ubiquitination by the Mind bomb/ubiquitin ligase controls endocytosis of the Delta ligand in presumptive hair cells (21). Surprisingly, endocytosis of Delta leads to enhanced signaling by the Delta-Notch pathway.

Other recent studies have implicated glial cells in preventing premature development of lateral line hair cells (10, 27). Sox10 is required for specification of glial cells, and in the *colourless/sox10* mutant, hair cells develop precociously. Although the signaling events from glial cells to the lateral line primordial cells have yet to be described, this discovery reveals a novel role for glial cells in development. It will be interesting to determine whether such a mechanism is conserved in the development of other organs or in other vertebrates.

HAIR-CELL FUNCTION IN ZEBRAFISH

The overall morphology of the inner ear and lateral line hair cells closely resembles that seen in vestibular hair cells in other vertebrates (**Figures 2** and **3**). To date, a number of genes required for hair-cell function in the zebrafish have been associated with auditory defects in mice and humans, thus revealing the conservation of function of these genes (**Table 1**). The phenotype of zebrafish mutants with balance and hearing defects is characterized by a swimming behavior in circular motions, hence the name "circler" mutants (30). In some cases, a complete lack or partial response to acoustic stimuli suggests that the animals are deaf. This can be easily tested by examining the acoustic startle reflex, which is very robust and consistently evoked at the free-swimming larval stage (>80 hpf).

Genes Implicated in Mechanotransduction

Mechanotransduction is the conversion of mechanical energy into electrical signals. At the hair bundle, the mechanical energy of sound or movement is converted to electrical impulses that propagate along the basolateral membrane of the hair cell body. These signals are then transmitted via first-order neurons to the brain. Transmission electron microscopic data of zebrafish hair bundles reveals that they possess what is thought to be a transduction apparatus at the tips of stereocilia (**Figure 3**) (44). The morphological correlates include an obliquely oriented tip link or fine extracellular filament of approximately 150 nm in length, and electron densities at either end of the tip link where mechanosensitive channels are presumed to be present. In addition, sensitivity of mechanotransduction to aminoglycosidic antibiotics or the calcium chelator EGTA has been demonstrated at the physiological level (30, 39). Over the past two decades, data from electrophysiological and biophysical studies of bullfrog hair cells suggest that the tip link pulls open channels when hair bundles are deflected in the excitatory direction (19). The search for the components required for mechanotransduction has been an ongoing effort. In recent studies using

Table 1 Genes required for inner ear development and function in zebrafish

Mutant/Gene	Expression pattern	Phenotype	References
	Developmental		
fgf3	Hindbrain	Small malformed ear	(25, 26, 28, 32)
acerebellar/fgf8	Hindbrain	Small malformed ear	(25, 26, 28, 32)
chameleon^{tf18b}	Unknown	Anteriorized ear	(15)
colourless/sox10	Neural crest	Small malformed ear	(16)
dfna5	Inner ear	Defective canals	(4)
Df^{b380} (*sox9a, dlx3b-4b*)	Preotic region	Ear absent	(16)
dog-eared/eya1	Otic anlage	Fewer hair cells	(24)
hearsay/foxi1	Otic anlage	Small malformed ear	(45)
jekyll/ugdh1	Jaw, inner ear	Defective canals	(49)
mind bomb/ubeE3-like	n.d.	Extra hair cells	(21)
no isthmus/pax2a	Preotic region	Extra hair cells	(16)
pax8	Preotic region	Small malformed ear	(16)
slow muscle omitted^{b641}*/smo*	Otic vesicle, Hindbrain	Anteriorized ear	(15)
	Hair-cell function/survival		
gemini/cav1.3	Hair cells, brain, retina	Hair-cell defect synaptic transmission affected?	(41)
mariner/myoVIIa	Hair cells	Hair-bundle defect	(7)
orbiter/pcdh15a	Hair cells, brain	Hair-bundle defect	(38)
ru848/choroideremia	Ubiquitous	Hair-cell death	(46)
satellite/myo6b	Hair cells	Hair-bundle defect abnormal membrane fusion	(22, 39)
sputnik/cdh23	Hair cells, brain, retina	Hair-bundle defect mechanotransduction affected	(44)
trpn1	Hair cells, neural crest, retina, brain, gills	Mechanotransduction affected	(42)

n.d. not done.

either forward genetics or antisense technology, two key players of the transduction apparatus have been identified in zebrafish, the tip link and transduction channel.

A candidate for the tip link was discovered by analysis of *sputnik/cadherin23* mutants (44). In homozygous larvae, the tip link is absent. In addition, an antibody against Cadherin 23 labeled the tips of inner ear hair bundles and produced some punctate labeling of the bundle. EM gold labeling was not performed in zebrafish; however, in murine and bullfrog hair bundles, labeling appears to be restricted to the tip links and the kinocilial links, a presumably related structural connection (43). The collective data argue for a role for Cadherin 23 in transduction, suggesting that its unusually long extracellular domain (27 cadherin repeats) is the fine tip link filament spanning between neighboring stereocilia.

A candidate gene approach to identify the transduction channel arose from studies of mechanosensory bristle mutants in *Drosophila* (23). One particular mutant, *nompC*, had an interesting defect in transduction currents, and the gene encodes a novel transient receptor potential (TRP) channel (48). The zebrafish orthologue, *trpn1*, is expressed in inner ear hair cells and knock-down of its activity results

TRP: transient receptor potential

in deafness and loss of receptor potentials in lateral line hair cells (42).

Although Trpn1 is a promising candidate for the transduction channel in zebrafish inner ear hair cells, the subcellular localization of the channel is not known. In addition, a genetic mutation in *trpn1* has not been reported. The reason may be that saturation screens for circler mutants have not been performed, or mutations in *trpn1* may result in early embryonic lethality. Transient expression of *trpn1* in presumptive migrating neural crest cells lends some support to this idea.

Genes Affecting Integrity of Hair Bundles

Besides the above-mentioned tip links and kinocilial links, the hair bundle is interconnected by horizontal links and ankle links. Stabilization of the bundle is also provided by the dense actin meshwork known as the cuticular plate, located just below the apical surface. The actin filaments of the stereocilia converge at the base of each stereocilium and then insert their long rootlets into the cuticular plate. The structural integrity of the links and cuticular plate is crucial for mechanotransduction by the hair bundle.

Recent studies of *satellite/myosin VI* or *ru920/myosin VI* mutants suggest that Myosin VI is required for the integrity of the cuticular plate and prevents fusion of stereocilia (22, 37). A similar phenotype was detected in Snell's waltzer/myosin VI mice mutants (3, 40), suggesting that the role of myosin VI in bundle integrity is highly conserved.

Along with *cadherin23* and *myosin VIIA* (7, 44), more severe mutations in *protocadherin 15* cause splaying of bundles (38). Although Protocadherin 15 has been localized to the hair bundles in other vertebrates (1), it is not known where it is located within the zebrafish stereocilia. Based on its long extracellular domain, one might also expect it to be one of the linkages interconnecting stereocilia.

Genes Affecting Hair-Cell Synaptic Transmission

As sensory receptors, hair cells have the remarkable capability of producing prolonged, tonic synaptic transmission (8). How this is accomplished at the molecular level is not well understood. An unusual structure, the synaptic ribbon, tethers synaptic vesicles next to the synaptic cleft. Many hundreds of vesicles are located near the ribbon body at the basolateral membrane. Given its putative role in promoting vesicle fusion, it was not surprising to find that mutations in the L-type calcium channel, Cav1.3, caused deafness in mutant *gemini* larvae (41). Indeed, the Cav1.3 channel was found to localize in ring-like structures near the basolateral membrane of zebrafish lateral line hair cells. What is lacking in the zebrafish is the means to directly measure synaptic transmission or to perform patch clamp analysis of hair cell currents. However, transgenic lines carrying genetic calcium indicators such as chameleon (2, 9, 18) will be an important tool for studying transmission or perhaps even transduction in zebrafish hair cells. Such experiments can be done in live, undissected larvae, providing real advantages over studies that rely on explants or isolated cells. As the collection of zebrafish auditory/vestibular mutants includes genes that appear to be acting downstream of mechanotransduction (T. Nicolson, unpublished observations), it is likely that more proteins involved in synaptic transmission will be found.

Genes Affecting Hair-Cell Survival

Many forms of human hearing loss are due to hair-cell degeneration, especially among the elderly (11). Genes that are required for hair-cell survival are therefore of great interest; however, there are few that have been reported thus far. Mutations in the *dog-eared/eya1* gene result in the reduction of hair cells, but this appears to be mainly a failure in differentiation or death of progenitors. In *dog-eared* mutants, cell death occurs within the developing otocyst and lateral line primordia, but does not appear to be restricted to hair cells (24). Mutations in the *choroidermia* gene also cause hair cell death during an early developmental stage such that otic vesicles and neuromasts in *ru848/choroidermia* fish have very few hair cells remaining at 120 hpf (46). Additional effects include abnormal pigmentation and disorganization of the retinal pigment epithelium and photoreceptors in the eye. The *choroidermia* gene is named after an X-linked disorder that causes photoreceptor degeneration in humans and encodes a Rab GTPase escort protein (6, 36). The effect on hair cell survival in *choroidermia* mutants remains to be explored. Efforts to understand the molecular basis of hair cell survival and regeneration are under way (17) and may lead to the discovery of important players in this process.

SUMMARY POINTS

1. FGF signaling from the hindbrain to the overlying ectoderm plays an important role in otic induction.
2. After induction, modeling of the ear is accomplished by diverse mechanisms such as Hedgehog signaling and the production of hyaluronic acid that leads to outgrowth of epithelial columns that form the semicircular canals.
3. Hair cells are specified by the Delta-Notch signaling pathway that is enhanced through endocytosis of the Delta ligand by the ubiquitin ligase, Mind bomb.
4. Mechanotransduction in zebrafish hair cells requires Cadherin 23 and Trpn1, two candidates for components of the transduction apparatus: the tip link and transduction channel, respectively.
5. The transducing organelle of the hair cell, the hair bundle, is dependent upon unconventional myosins and novel cadherins for integrity and function as seen in higher vertebrates such as mice and humans.

ACKNOWLEDGMENT

I thank John Brigande for his help and comments on the information on ear development.

LITERATURE CITED

1. Ahmed ZM, Riazuddin S, Ahmad J, Bernstein SL, Guo Y, et al. 2003. PCDH15 is expressed in the neurosensory epithelium of the eye and ear and mutant alleles are responsible for both USH1F and DFNB23. *Hum. Mol. Genet.* 12:3215–23
2. Ashworth R. 2004. Approaches to measuring calcium in zebrafish: focus on neuronal development. *Cell Calcium* 35:393–402
3. Avraham KB, Hasson T, Steel KP, Kingsley DM, Russell LB, et al. 1995. The mouse Snell's waltzer deafness gene encodes an unconventional myosin required for structural integrity of inner ear hair cells. *Nat. Genet.* 11:369–75
4. Busch-Nentwich E, Sollner C, Roehl H, Nicolson T. 2004. The deafness gene *dfna5* is crucial for *ugdh* expression and HA production in the developing ear in zebrafish. *Development* 131:943–51

5. Chang JS, Popper AN, Saidel WM. 1992. Heterogeneity of sensory hair cells in a fish ear. *J. Comp. Neurol.* 324:621–40
6. Cremers FP, van de Pol DJ, van Kerkhoff LP, Wieringa B, Ropers HH. 1990. Cloning of a gene that is rearranged in patients with choroideraemia. *Nature* 347:674–7
7. Ernest S, Rauch GJ, Haffter P, Geisler R, Petit C, Nicolson T. 2000. Mariner is defective in myosin VIIA: a zebrafish model for human hereditary deafness. *Hum. Mol. Genet.* 9:2189–96
8. Fuchs PA, Glowatzki E, Moser T. 2003. The afferent synapse of cochlear hair cells. *Curr. Opin. Neurobiol.* 13:452–58
9. Gahtan E, Baier H. 2004. Of lasers, mutants, and see-through brains: functional neuroanatomy in zebrafish. *J. Neurobiol.* 59:147–61
10. **Grant KA, Raible DW, Piotrowski T. 2005. Regulation of latent sensory hair cell precursors by glia in the zebrafish lateral line. *Neuron* 45:69–80**

Establishes a role for glial cells in repressing hair-cell differentiation.

11. Gratton MA, Vazquez AE. 2003. Age-related hearing loss: current research. *Curr. Opin. Otolaryngol. Head Neck Surg.* 11:367–71
12. Gregan J, Lindner K, Brimage L, Franklin R, Namdar M, et al. 2003. Fission yeast Cdc23/Mcm10 functions after pre-replicative complex formation to promote Cdc45 chromatin binding. *Mol. Biol. Cell* 14:3876–87
13. Haddon C, Lewis J. 1996. Early ear development in the embryo of the zebrafish, *Danio rerio*. *J. Comp. Neurol.* 365:113–28
14. Haddon CM, Lewis JH. 1991. Hyaluronan as a propellant for epithelial movement: the development of semicircular canals in the inner ear of Xenopus. *Development* 112:541–50
15. **Hammond KL, Loynes HE, Folarin AA, Smith J, Whitfield TT. 2003. Hedgehog signalling is required for correct anteroposterior patterning of the zebrafish otic vesicle. *Development* 130:1403–17**

Example of an anteriorized otic vesicle phenotype in a zebrafish mutant.

16. Hans S, Liu D, Westerfield M. 2004. Pax8 and Pax2a function synergistically in otic specification, downstream of the Foxi1 and Dlx3b transcription factors. *Development* 131:5091–102
17. Harris JA, Cheng AG, Cunningham LL, MacDonald G, Raible DW, Rubel EW. 2003. Neomycin-induced hair cell death and rapid regeneration in the lateral line of zebrafish (*Danio rerio*). *J. Assoc. Res. Otolaryngol.* 4:219–34
18. Higashijima S, Masino MA, Mandel G, Fetcho JR. 2003. Imaging neuronal activity during zebrafish behavior with a genetically encoded calcium indicator. *J. Neurophysiol.* 90:3986–97
19. Hudspeth AJ. 1989. How the ear's works work. *Nature* 341:397–404
20. Hudspeth AJ. 1992. Hair-bundle mechanics and a model for mechanoelectrical transduction by hair cells. *Soc. Gen. Physiol. Ser.* 47:357–70
21. Itoh M, Kim CH, Palardy G, Oda T, Jiang YJ, et al. 2003. Mind bomb is a ubiquitin ligase that is essential for efficient activation of Notch signaling by Delta. *Dev. Cell* 4:67–82
22. Kappler JA, Starr CJ, Chan DK, Kollmar R, Hudspeth AJ. 2004. A nonsense mutation in the gene encoding a zebrafish myosin VI isoform causes defects in hair-cell mechanotransduction. *Proc. Natl. Acad. Sci. USA* 101:13056–61
23. Kernan M, Zuker C. 1995. Genetic approaches to mechanosensory transduction. *Curr. Opin. Neurobiol.* 5:443–48
24. **Kozlowski DJ, Whitfield TT, Hukriede NA, Lam WK, Weinberg ES. 2005. The zebrafish dog-eared mutation disrupts *eya1*, a gene required for cell survival and differentiation in the inner ear and lateral line. *Dev. Biol.* 277:27–41**

Animal model for human *eya1*-associated vestibular and hearing defects.

25. Leger S, Brand M. 2002. Fgf8 and Fgf3 are required for zebrafish ear placode induction, maintenance and inner ear patterning. *Mech. Dev.* 119:91–108

26. Liu D, Chu H, Maves L, Yan YL, Morcos PA, et al. 2003. Fgf3 and Fgf8 dependent and independent transcription factors are required for otic placode specification. *Development* 130:2213–24

27. Lopez-Schier H, Hudspeth AJ. 2005. Supernumerary neuromasts in the posterior lateral line of zebrafish lacking peripheral glia. *Proc. Natl. Acad. Sci. USA* 102:1496–501

28. Maroon H, Walshe J, Mahmood R, Kiefer P, Dickson C, Mason I. 2002. Fgf3 and Fgf8 are required together for formation of the otic placode and vesicle. *Development* 129:2099–108

29. Neuhauss SC, Solnica-Krezel L, Schier AF, Zwartkruis F, Stemple DL, et al. 1996. Mutations affecting craniofacial development in zebrafish. *Development* 123:357–67

30. Nicolson T, Rusch A, Friedrich RW, Granato M, Ruppersberg JP, Nusslein-Volhard C. 1998. Genetic analysis of vertebrate sensory hair cell mechanosensation: the zebrafish circler mutants. *Neuron* 20:271–83

31. Petit C. 1996. Genes responsible for human hereditary deafness: symphony of a thousand. *Nat. Genet.* 14:385–91

32. Phillips BT, Bolding K, Riley BB. 2001. Zebrafish fgf3 and fgf8 encode redundant functions required for otic placode induction. *Dev. Biol.* 235:351–65

33. Popper AN, Fay RR. 1993. Sound detection and processing by fish: critical review and major research questions. *Brain Behav. Evol.* 41:14–38

34. Popper AN, Fay RR. 1997. Evolution of the ear and hearing: issues and questions. *Brain Behav. Evol.* 50:213–21

35. Riley BB, Phillips BT. 2003. Ringing in the new ear: resolution of cell interactions in otic development. *Dev. Biol.* 261:289–312

36. Seabra MC, Brown MS, Goldstein JL. 1993. Retinal degeneration in choroideremia: deficiency of rab geranylgeranyl transferase. *Science* 259:377–81

37. Seiler C, Ben-David O, Sidi S, Hendrich O, Rusch A, et al. 2004. Myosin VI is required for structural integrity of the apical surface of sensory hair cells in zebrafish. *Dev. Biol.* 272:328–38

38. Seiler C, Finger-Baier KC, Rinner O, Makhankov YV, Schwarz H, et al. 2005. Duplicated genes with split functions: independent roles of protocadherin15 orthologues in zebrafish hearing and vision. *Development* 132:615–23

39. Seiler C, Nicolson T. 1999. Defective calmodulin-dependent rapid apical endocytosis in zebrafish sensory hair cell mutants. *J. Neurobiol.* 41:424–34

40. Self T, Sobe T, Copeland NG, Jenkins NA, Avraham KB, Steel KP. 1999. Role of myosin VI in the differentiation of cochlear hair cells. *Dev. Biol.* 214:331–41

41. Sidi S, Busch-Nentwich E, Friedrich R, Schoenberger U, Nicolson T. 2004. *gemini* encodes a zebrafish L-type calcium channel that localizes at sensory hair cell ribbon synapses. *J. Neurosci.* 24:4213–23

42. Sidi S, Friedrich RW, Nicolson T. 2003. NompC TRP channel required for vertebrate sensory hair cell mechanotransduction. *Science* 301:96–99

43. Siemens J, Lillo C, Dumont RA, Reynolds A, Williams DS, et al. 2004. Cadherin 23 is a component of the tip link in hair-cell stereocilia. *Nature* 428:950–55

44. Sollner C, Rauch GJ, Siemens J, Geisler R, Schuster SC, et al. 2004. Mutations in cadherin 23 affect tip links in zebrafish sensory hair cells. *Nature* 428:955–59

45. Solomon KS, Kudoh T, Dawid IB, Fritz A. 2003. Zebrafish foxi1 mediates otic placode formation and jaw development. *Development* 130:929–40

Establishes an Fgf-independent mechanism for otic induction.

First example of a TRP channel required for mechanotransduction in a vertebrate hair cell.

Key reports regarding a candidate molecule for the tip link in hair cells.

46. Starr CJ, Kappler JA, Chan DK, Kollmar R, Hudspeth AJ. 2004. Mutation of the zebrafish choroideremia gene encoding Rab escort protein 1 devastates hair cells. *Proc. Natl. Acad. Sci. USA* 101:2572–77
47. Van de Water TR, Represa J. 1991. Tissue interactions and growth factors that control development of the inner ear. Neural tube-otic anlage interaction. *Ann. NY Acad. Sci.* 630:116–28
48. Walker RG, Willingham AT, Zuker CS. 2000. A Drosophila mechanosensory transduction channel. *Science* 287:2229–34
49. Walsh EC, Stainier DY. 2001. UDP-glucose dehydrogenase required for cardiac valve formation in zebrafish. *Science* 293:1670–73
50. Whitfield TT, Riley BB, Chiang MY, Phillips B. 2002. Development of the zebrafish inner ear. *Dev. Dyn.* 223:427–58

RELATED REVIEWS

Development of the zebrafish inner ear. T Whitfield, B Riley, M Chiang, B Phillips. 2002. *Dev. Dyn.* 223:427–58

Development of the zebrafish lateral line. 2004. A Ghysen, C Dambly-Chaudiere *Curr. Opin. Neurobiol.* 14:67–73

Sensory organ development in the inner ear: molecular and cellular mechanisms. J Bryant, R Goodyear, G Richardson. 2002. *Br. Med. Bull.* 63:39–57

Sound detection and processing by fish: critical review and major research questions. A Popper, R Fay. 1993. *Brain Behav. Ecol.* 41:14–38

The zebrafish as a genetic model to study otolith formation. C Söllner, T Nicolson. In *Biomineralization, Progress in Biology, Molecular Biology, and Application*, ed. E Bauerlein, 2:229–41. Weinhein: Wiley

---

LINKS

http://webhost.ua.ac.be/hhh/
http://zfin.org

Immunoglobulin Gene Diversification

Nancy Maizels

Departments of Immunology and Biochemistry, University of Washington Medical School, Seattle, Washington 98195-7650, email: maizels@u.washington.edu

Annu. Rev. Genet.
2005. 39:23–46

First published online as a Review in Advance on June 21, 2005

The *Annual Review of Genetics* is online at http://genet.annualreviews.org

doi: 10.1146/annurev.genet.39.073003.110544

0066-4197/05/1215-0023$20.00

Key Words

antibody, deamination, mutation, recombination, repair

Abstract

Three processes alter genomic sequence and structure at the immunoglobulin genes of B lymphocytes: gene conversion, somatic hypermutation, and class switch recombination. Though the molecular signatures of these processes differ, they occur by a shared pathway which is induced by targeted DNA deamination by a B cell–specific factor, activation induced cytidine deaminase (AID). Ubiquitous factors critical for DNA repair carry out all downstream steps, creating mutations and deletions in genomic DNA. This review focuses on the genetic and biochemical mechanisms of diversification of immunoglobulin genes.

Contents

DIVERSIFICATION CREATES A ROBUST AND DYNAMIC IMMUNE RESPONSE

The immune system must recognize an immense variety of different pathogens and respond dynamically as microorganisms evolve during the course of infection. To accomplish this, lymphocytes carry out targeted and regulated alterations of genomic structure and sequence. Early in B cell and T cell development, antigen receptor variable (V) regions are constructed by V(D)J recombination, a site-specific recombination process initiated by the lymphocyte-specific proteins RAG1 and RAG2 (8, 37), as shown in **Figure 1**. Following successful V(D)J recombination, a B cell expresses an immunoglobulin (Ig) molecule composed of two distinct chains, a heavy chain and light chain, encoded at separate alleles. The variable regions of the heavy and light chains together form a domain that recognizes antigen, and the heavy chain constant (C) region determines how antigen is removed from the body. Subsequently, B cells increase the diversity of the repertoire and tailor Ig molecules to optimize the response to a specific pathogen, by modifying genomic sequence and structure of the rearranged and expressed Ig genes in characteristic and striking ways.

These processes occur only in B cells, and they are regulated by developmental and environmental stimuli. They are essential, and immunodeficiency results if they are impaired. Although the molecular signatures are distinct, all these processes are initiated by a single B cell–specific factor, activation-induced deaminase (AID). AID is a cytidine deaminase that carries out targeted deamination of C → U at expressed Ig genes. Uracil in DNA is normally repaired by conserved, ubiquitous, and high-fidelity pathways, but at the Ig genes of B lymphocytes, repair is somehow diverted to ensure a mutagenic outcome. AID is the only B cell–specific factor involved, and downstream steps depend on ubiquitous repair factors that, paradoxically, promote mutagenesis of B cell Ig genes. This review discusses our current mechanistic understanding of the pathways of Ig gene diversification.

Gene conversion (templated mutation): produces sequence changes with matches in germline donor genes or pseudogenes

Somatic hypermutation (nontemplated mutation): produces point mutations with no matches in germline DNA

Class switch recombination: joins a new constant region to the expressed heavy chain variable region, promoting deletion of many kb of chromosomal DNA

MOLECULAR SIGNATURES OF Ig GENE DIVERSIFICATION

The signatures of Ig gene diversification are different at different loci, in different species, and in response to distinct developmental and environmental regulation. As background, three of the best-characterized examples are described below.

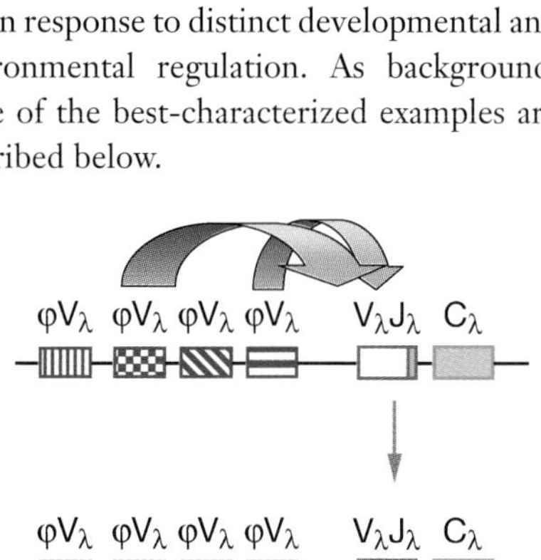

Figure 1

Combinatorial and mutagenic diversification of V regions. *Left*, combinatorial diversification at the murine κ light chain locus. In each B cell, one of many Vκ segment recombines with one of several J_κ segments to produce a functional κ variable region. *Right*, templated diversification at the chicken λ light chain locus. The $V_\lambda J_\lambda$ variable region created by recombination of the single functional V_λ and J_λ segments undergoes gene conversion templated by upstream V_λ pseudogenes (φV_λ).

Gene Conversion (Templated Mutation) of Chicken Ig V Regions

Chickens have only a single functional light chain V segment, V_λ, which recombines with a single J_λ segment very early in B cell development. The rearranged and expressed $V_\lambda J_\lambda$ then undergoes gene conversion, or templated mutation. An array of 25 nonfunctional, "pseudo-V_λ" regions, located just upstream of the functional V_λ, provide donors for sequence transfer. Diversified $V\lambda$ regions contain tracts of sequence changes with clear matches in germline DNA (110, 127). **Figure 1** compares combinatorial diversification of murine V regions and templated diversification of chicken V regions.

Somatic Hypermutation (Nontemplated Mutation) in Murine and Human B Cells

In humans and mice, a multitude of V, D, and J segments recombine to create the functional Ig genes of the pre-immune repertoire (8, 37). Following antigen activation, the expressed heavy and light chain V regions are altered by somatic hypermutation, which obliterates existing sequence information and inserts nontemplated (point) mutations (61, 64, 83). Hypermutation is targeted to the transcribed V regions; the C regions and the rest of the genome remain untouched. Hypermutation occurs following activation with antigen, and is coupled with selection for B cell clones that express high-affinity antibodies. Together, hypermutation and clonal selection not only increase antibody affinity, but also provide a dynamic response to pathogens that are simultaneously undergoing continuous mutation and selection.

MAMMALIAN Ig ISOTYPES

Mammals produce four isotypes (or classes) of Ig: IgM, IgG, IgE, and IgA, encoded by the μ, γ, ε, and α constant regions, respectively. Related IgG subclasses are encoded by distinct $C\gamma$ regions. Each Ig isotype is specialized for particular modes of antigen removal. IgM, the first isotype synthesized by a B cell, activates complement. IgG, the most abundant isotype in serum, binds receptors on phagocytic cells. IgG antibodies cross the placenta to provide maternal protection to the fetus. IgA antibodies are abundant in secretions, such as tears and saliva; they coat invading pathogens to prevent proliferation. IgE antibodies can provide protection against parasitic nematodes, but in developed countries they are the bad guys: They bind basophils and mast cells, activating histamine release and allergy.

Class Switch Recombination in Mammalian B Cells

Class switch recombination detaches an expressed heavy chain variable (VDJ) region from one constant (C) region and joins it to a downstream C region, deleting the DNA between (19, 61, 66) (**Figure 2**). The C region determines how antigen is removed from the body, and each class of C region is specialized for clearance of specific kinds of pathogens, via specific pathways (SIDEBAR). Thus, switch recombination modifies how an Ig molecule removes antigen without affecting specificity of antigen-recognition by literally *switching* C regions. Recombination junctions form within switch (S) regions, repetitive and degenerate guanine-rich regions 2–10 kb in length. Recombination is region-specific, not sequence-specific, and produces junctions at heterogeneous sites and sequences within the S regions (29). Switch junctions are within introns, so junctional heterogeneity does not affect the sequence or structure of the heavy chain polypeptide. A dedicated enhancer and promoter upstream of each S region activate S region transcription in response to extracellular signals delivered by T cells and cytokines, which thereby regulate production of specific classes of Ig. The critical signal for recombination is transcription in *cis* of the S regions targeted for recombination.

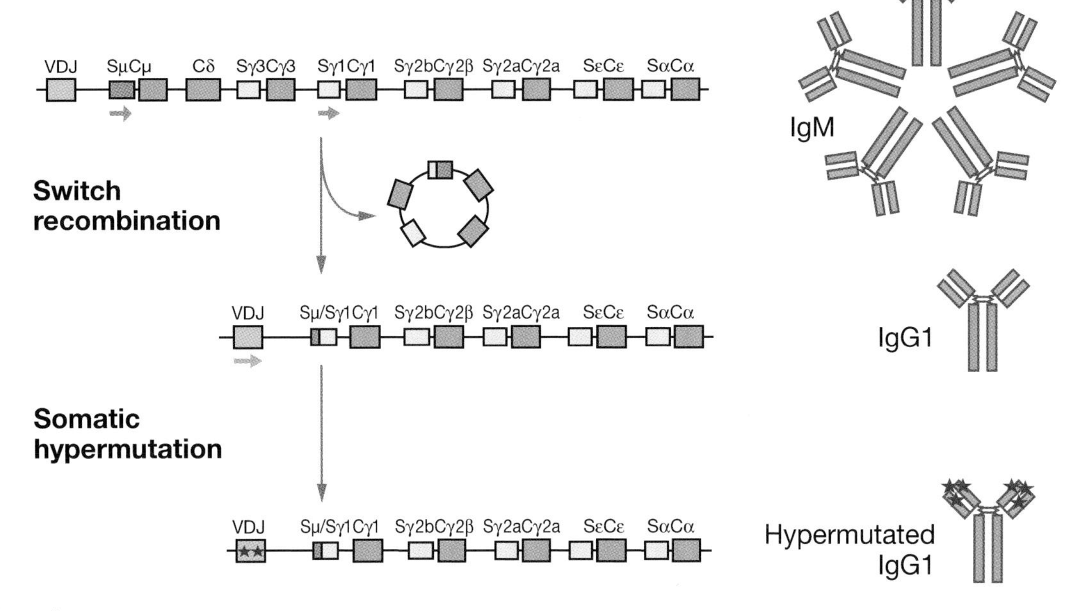

Figure 2

Class switch recombination and somatic hypermutation at the murine heavy chain locus. Switch recombination joins a rearranged and expressed heavy chain variable region (VDJ) to a new downstream constant region (C), forming junctions within switch (S) regions. Somatic hypermutation alters V region sequence (*stars*). V and S regions are activated for diversification by transcription (*colored arrows*). The figure illustrates recombination from $S\mu$ to $S\gamma 1$, which results in a switch from IgM to IgG1 antibody expression; changes in Ig molecules are shown at the right.

A Shared Pathway Produces the Distinct Signature Outcomes of Ig Gene Diversification

That the pathways of Ig gene diversification-are mechanistically related was first suggested by absence of a clear evolutionary pattern to the use of templated or nontemplated mutation. Nontemplated mutation alters Ig V regions in humans and mice (61, 64, 83), but templated mutation alters V regions in closely related mammals, including common farm animals such as sheep, cattle, and pigs (16, 57, 72), as well as chicken and other fowl (70); and both templated and nontemplated mutation can occur in a single species, most notably rabbit (57). This led to the proposal that targeted mutagenesis in all organisms is initiated by a DNA lesion leading to a break, which could then be processed by an error-prone repair pathway, to produce nontemplated mutation; or by recombinational repair to produce templated mutation (63). This model has been borne out by evidence that, in chicken cells, impaired homologous recombination due to deficiencies in RAD51 paralogs (114) or BRCA2 (43) causes a shift from templated to nontemplated mutation. Gene conversion, somatic hypermutation, and switch recombination were united within a single diversification pathway by experiments showing that all three processes depend upon a single B cell–specific enzyme, AID (2, 42, 76, 109).

AID DEAMINATES DNA TO INITIATE Ig GENE DIVERSIFICATION

AID was first identified by Honjo and collaborators in a screen for genes induced specifically upon activation of mammalian B cells

(77). Targeted deletion of the murine *Aid* gene was then shown to essentially abolish both somatic hypermutation and class switch recombination in activated murine B cells (76). In humans, AID is the product of *HIGM2*, a well-characterized locus associated with severe immunodeficiency characterized by lymph node hyperplasia, the absence of somatic hypermutation, and impaired switch recombination evident as low levels of serum antibodies of IgG, IgA, and IgE classes, and elevated levels of IgM antibodies (109). Elevated IgM levels often are evident as one physiological consequence of impaired switch recombination, reflecting an attempt by the immune system to compensate for the deficiencies in other Ig isotypes.

AID is Expressed in Germinal Center B Cells

AID expression is restricted to brief windows in B cell development during which Ig gene diversification occurs. In chicken, sheep, and rabbit, diversification of the limited combinatorial repertoire is especially active in young animals. In newly hatched chicks, diversification occurs in a specialized lymphoid organ, the bursa, which has no homolog in mammals (the term "B" lymphocyte originally distinguished "bursa-derived" lymphocytes from "thymus-derived" or "T lymphocytes" in the chicken); and AID expression is coordinated with bursal diversification (136). In humans and mice, AID is induced in antigen-activated B cells (77, 90). These cells populate germinal centers, histologically distinctive microenvironments within spleen, lymph nodes, tonsils, appendix, and intestinal Peyer's patches, where B cells switch, hypermutate, and undergo selection for high-affinity clones. BCL6 is required for germinal center formation, and in germinal center B cells, BCL6 downregulates p53, thereby creating a permissive environment for DNA diversification (96).

A chicken bursal lymphoma line, DT40, carries out constitutive Ig gene conversion; and hypermutation is ongoing in a considerable subset of human B cell lines derived from germinal center tumors. These lines have proven to be valuable models for studies of Ig gene diversification.

AID Overrides Uracil DNA Repair to Mutate C → U in *Escherichia coli*

AID is related to APOBEC1 (77), a cytidine deaminase that edits a specific cytidine in the apolipoprotein B transcript to produce a nonsense codon that results in expression of a truncated polypeptide. This initially suggested that AID might target a specific mRNA encoding a master regulator in the form of a critical nuclease or transcription factor, rather than participate directly in the diversification mechanism. Uracil in DNA is a frequent and spontaneous lesion, which can originate by deamination of C → U at C/G base pairs, or by incorporation of dUTP opposite A during replication (6, 52). It is repaired by an efficient base excision repair pathway, in which uracil nucleoside glycosylases remove the uracil base, abasic endonucleases nick the phosphodiester backbone at the abasic site, DNA polymerase β synthesizes new DNA, and ligases seal the nick, as shown in **Figure 3** (6, 52).

The hypothesis that AID directly deaminates DNA in living cells was tested by asking if expression of AID in *E. coli* stimulates mutation of reporter genes (94). Cytidine deamination at a C/G base pair will produce a U•G mismatch, and upon replication a C/G → T/A transition mutation will be fixed in genomic DNA. Consistent with AID-initiated C → U deamination as the source of mutations, reporter genes carried a high proportion of transition mutations at C/G pairs, and mutations levels increased in a *ung*- strain, deficient in the major uracil nucleoside glycosylase activity (94). Thus, AID expression can overwhelm the normally efficient uracil DNA repair pathway to cause mutagenesis in *E. coli*.

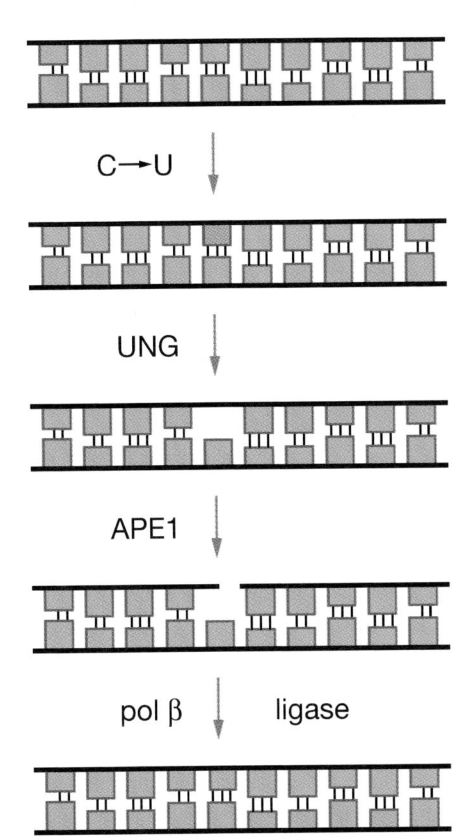

Figure 3

Faithful repair of uracil in DNA. Uracil is repaired by a conserved and ubiquitous pathway: uracil nucleoside glycosylase (UNG) removes the uracil base (*orange*), AP endonuclease 1 (APE1) cleaves the phosphodiester backbone at the abasic site, and DNA polymerase β and ligases repair the gap.

AID Deaminates Single-Stranded DNA: Transcriptional Targeting of Diversification

Biochemical analysis provided a second line of evidence that AID participates in the mutagenic mechanism. Recombinant AID had been shown to deaminate cytidine in vitro (77). Nonetheless, the demonstration that AID deaminates C → U in single-stranded DNA, within an artificial transcription bubble, or within a transcribed region, but not in duplex DNA (13, 21, 28, 103, 123) provided a breakthrough in understanding, immediately explaining why transcription is critical for Ig gene diversification: Transient denaturation accompanying transcription produces the single-stranded DNA substrate for AID.

Recombinant AID displays a 7- to 15-fold preference for deamination of Cs in the nontemplate DNA strand in vitro (12, 21, 95, 103). This was initially puzzling. There is no clear strand bias in products of hypermutation (73), so AID almost certainly deaminates both DNA strands. Two tactics may provide access to the template strand. Interaction with the eukaryotic single-strand binding factor, RPA, facilitiates deamination of the nontemplate strand in transcribed substrates in vitro (20). AID will also deaminate both strands within regions of negative DNA supercoiling (120), which could contribute to targeting AID to both DNA strands in the region ahead of an advancing RNA polymerase.

Hypermutation Hotspots are Deamination Hotspots

AID preferentially deaminates C's within the sequence context WRC (W = A/T; R = purine) both in vitro and in vivo (12, 95). Analysis of sequences of vast numbers of hypermutated mammalian V regions had shown that a significant fraction of mutations occurred within motifs conforming to the consensus WRCY (complement, RGYW; Y = pyrimidine) (111), which includes the preferential motif for AID deamination. Thus, hypermutation hotspots are deamination hotspots.

Hypermutation hotspots were initially interpreted to reflect selection for high-affinity antibodies because they appeared to cluster within the complementarity-determining regions (CDRs), which make direct contact with antigen. The discovery that they reflect substrate preference of AID explains why the same hotspot appears in antigen-selected V regions, unselected V regions, and in non-Ig reporter genes (47). Moreover, V regions may have evolved to take advantage of AID-induced hypermutation, as motifs for AID deamination concentrate within the CDRs, where mutation can fine-tune antigen recognition, and are depleted within framework regions, where mutation can destroy antibody function. Further suggestive of coevolution of V regions and AID, each of the two serine codons, AGY and TCN, is favored in different subdomains of the V regions: AGY, which more closely approximates the deamination consensus, is favored in CDRs; and TCN in framework regions (47). This localization of AGY serine codons in the CDRs extends back to pufferfish, which encode an AID homolog with a preferred deamination motif the same as that of mammalian AID (9, 22).

PARADOXICAL MUTAGENIC FUNCTIONS FOR UBIQUITOUS REPAIR FACTORS

The evidence that AID deaminates DNA directly provided a mechanism for the first step in mutagenesis, but raised a new question: How is repair of uracil in DNA diverted from faithful to mutagenic pathways at B cell Ig genes? Ectopic expression of AID is sufficient to induce switch recombination (85) or somatic hypermutation (139) in engineered minigenes in mammalian fibroblasts, so downstream steps in the AID-initiated mutagenic pathway must therefore be carried out by ubiquitous factors. Indeed, the factors shown to participate in this pathway are highly conserved repair factors that are normally essential for maintenance of genomic stability. Paradoxically, they create irreversible changes in genomic sequence and structure in the process of repairing AID-initiated DNA damage at B cell Ig genes.

UNG Promotes Ig Gene Diversification

Neuberger and collaborators reasoned that if UNG (uracil nucleoside glycosylase) attacks U produced by AID deamination, then diminished levels of UNG would affect Ig gene diversification. This hypothesis was tested in an *XRCC2−/−* derivative of the chicken B cell line, DT40, which accumulates both nontemplated and templated mutations at the Ig loci (114). Consistent with function of UNG in the mutagenic pathway, decreased levels of UNG activity in DT40 *XRCC2−/−* mutants resulted in an increase in transition mutations at C/G pairs (25). Moreover, in *Ung−/−* mice, switch recombination is impaired, and there is an increase in transition mutations at C/G pairs (101). Transitions at C/G pairs can occur when a U•G mismatch generated upon deamination is fixed upon replication (**Figure 4**). In contrast to these results, one recent report claimed that class switch recombination does not depend upon UNG deglycosylase activity (10). However, the deglycosylase activity of the mutants tested in those experiments appears not to have been completely impaired (124), which calls into question the conclusions of that other report.

Analysis of human genetic disease provided additional support for function of UNG in the AID-initiated pathway of targeted diversification (44). A severe human immunodeficiency, HIGM ("Hyper IgM"), characterized by elevated IgM and low or absent serum IgG and IgA, results from mutations in the *UNG* gene (44). As in *Ung-/-* mice (101), in this human genetic disease deficient switch recombination is accompanied by a shift in the hypermutation spectrum, increasing the fraction of transition mutations at C/G base pairs. Notably, the absence of UNG activity in affected individuals had not been detected as a DNA repair deficiency before it was revealed as an immunodeficiency (44).

What is the fate of the abasic site created by UNG? Cells contain numerous enzymes that can cleave DNA containing abasic sites, including abasic endonucleases like APE1, which cleave the phosphodiester backbone; and lyases, which attack the deoxyribose 1′ C (6, 52). The cleaved DNA could then participate directly in gene conversion or class switch recombination (**Figure 4**). Alternatively, low-fidelity repair at the cleaved end or opposite the abasic site ("translesion" synthesis) could result in point mutations.

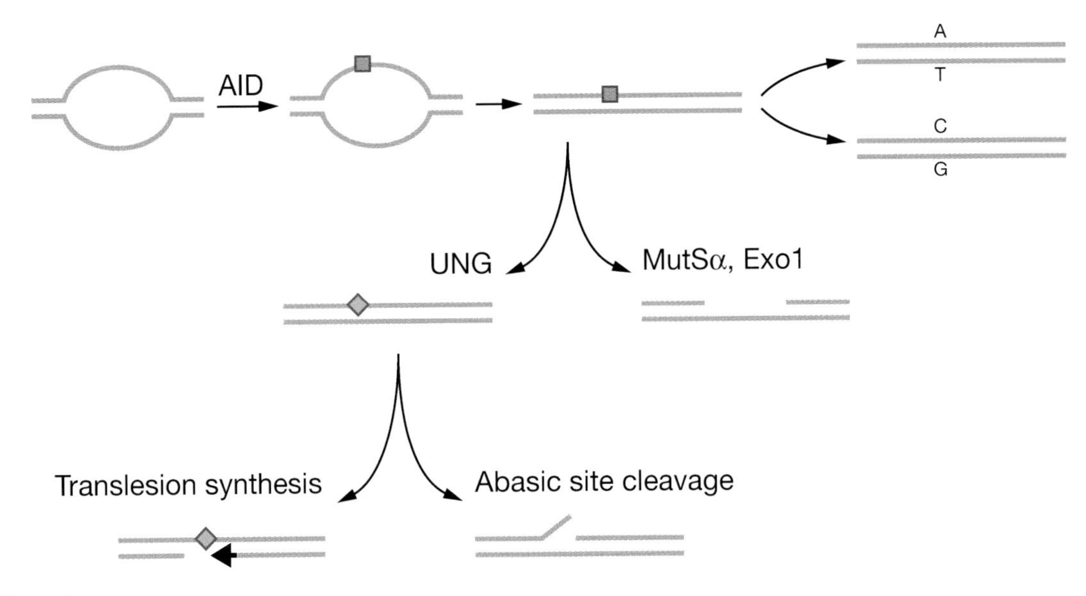

Figure 4

Mutagenic outcomes of AID-initiated deamination. AID deaminates C → U (*orange square*) in transcribed DNA (represented as a bubble to emphasize the transcriptional requirement for deamination). The duplex containing the U•G mismatch has three possible fates: replication will fix C/G → T/A transition mutation (*top right*); resection by MutSα-Exo1 will reveal a patch or gap (*center right*); and deglycosylation by UNG will create an abasic site (*blue diamond*) which undergoes further mutagenic repair (*center*). Mutations may be introduced opposite the abasic site by low-fidelity DNA synthesis (translesion synthesis, *bottom left*); or an abasic endonuclease or lyase may cleave and generate a DNA break (abasic site cleavage, *bottom right*).

NBS1 and the MRE11/RAD50/NBS1 Complex in Ig Gene Diversification

The conserved MRE11/RAD50/NBS1 (MRN) complex is involved in the response to DNA damage and cell-cycle checkpoint control. NBS1 is the regulatory subunit, which responds to and regulates ATM kinase and governs nuclear localization and activities of the MRE11/RAD50 nuclease. NBS1 displays AID-dependent localization to S regions in activated B cells (93), and conditional knockout of *Nbs1* in murine B cells results in impaired switch recombination (51, 108). As a member of the MRN complex, NBS1 promotes both somatic hypermutation in human cells and gene conversion in chicken cells (138). MRN is also necessary for recombination-associated mutagenesis within Sμ and at switch junctions, as the spectra of junction mutations are altered in B cells from individuals with Nijmegen break syndrome, due to deficient NBS1, and AT-like disease, resulting from deficient MRE11 (56); and junctional mutagenesis is absent in ataxia telangiectasia (AT), due to deficient ATM (86, 87).

MutSα (MSH2/MSH6) and Exo1 Promote Switching and Hypermutation

In single-base mismatch repair, MutSα (the MSH2/MSH6 heterodimer) recognizes the mismatch, recruits MutLα (MLH1/PMS2), which recruits and promotes DNA excision by ExoI, to generate a gap, which undergoes high-fidelity repair and religation (41, 53, 115, 125). One might therefore have predicted that mismatch repair factors would correct AID-induced mutations and prevent mutagenesis.

Counterintuitively, hypermutation and switch recombination decrease upon targeted disruption of *Msh2*, *Msh6*, and *Exo1* in mice AID (5, 32, 60, 68, 100). In wild-type mice, hypermutated sites include both C/G pairs and also T/A pairs, but in the absence of MutSα or Exo1, mutations focus at C/G pairs (45, 97, 133) and at sequences conforming to the AID deamination motif, WRCY (100). Switch junctions similarly focus at the WRCY motif (32, 117, 131), a deamination consensus. Mutation of the *Msh2* ATPase, which abolishes function in mismatch repair, recapitulates changes in hypermutation spectrum evident in *Msh2−/−* mice, but has little effect on switch recombination (67). Deficiencies in *Pms2* and *Mlh1* (MutLα) alter levels or spectra of hypermutation (17, 48, 49, 135), and affect levels of switching only modestly but cause an increased dependence on microhomology-mediated end joining (33, 119).

Taken together, these results identify key functions in hypermutation and switch recombination for MutSα and Exo1, and suggest more subtle roles for other mismatch repair factors. Purified recombinant hMutSα binds specifically to duplex DNA carrying U•G mismatches (58, 134), suggesting that mismatch repair factors may collaborate to resect the region containing U (**Figure 4**). Resection will reveal a patch of DNA that could be the template for low-fidelity DNA synthesis, and generate ends that could participate in gene conversion or switch recombination.

Ung and *Msh2* Define Two Paths to Mutagenesis

The differences between mutation spectra in wild-type and MutSα-deficient mice suggested that there might be two distinct phases of hypermutation, the first MutSα-independent and focused at hotspots; and the second MutSα-dependent, and more evenly distributed (100). This proposal was validated by studying Ig gene diversification in *Msh2−/−Ung−/−* mice (99). In the double mutant there is no switch recombination and no hypermutation at T/A pairs, and all somatic hypermutation consists of transition mutations at C/G pairs, apparently produced upon replication of U•G mismatches created by AID. Thus, *Ung* and *Msh2* are defining elements in two pathways for mutagenic repair of AID-initiated damage. The ability of MutSα to recognize U•G mismatches in vitro (58, 134) suggests that MutSα and UNG may compete for repair of uracil at Ig genes (**Figure 4**).

S REGIONS ARE ACTIVE PARTNERS IN SWITCH RECOMBINATION

We tend to think of proteins acting on DNA substrates. However, the transcriptional requirement for Ig gene diversification suggests a more active role, and structures formed within transcriptionally activated S regions prove to be specific targets for recombination factors.

S Regions: A Mix of G-Rich Repeats and Deamination Hotspots

The active role of the S regions in switching is shown by the observation that frequencies of switch recombination correlate with S region target length (145). S regions carry two characteristic sequence motifs, G-rich degenerate repeats, and hotspot motifs for AID deamination. These are distributed differently in different S regions. An extreme case is a Xenopus switch region, which is AT-rich and contains reiterations of the tetramer AGCT motif (78). This sequence conforms to the AID deamination consensus, WRC, and can substitute effectively for a murine S region by providing a dense array of sites for AID (144). Similarly, the murine Sμ region contains an extended GAGTC repeat. Deletion of the GAGTC repeat to leave mainly G-rich repeats reduces the efficiency of switching somewhat, and makes recombination dependent on *Msh2* (74).

G-Loops in Transcribed S Regions are Deaminated by AID and Synapsed by MutSα

S regions are G-rich on the nontemplate (top) strand, and must be transcribed in *cis* and in the correct physiological orientation to support recombination (122). Switch region transcription causes formation of characteristic structures that have been identified in electron microscopic images as extended G-loops (30). G-loops contain a cotranscriptional RNA:DNA hybrid on the template strand, as anticipated by gel electrophoresis (23, 104, 105, 128, 140) and atomic force microscopy (75); and regions of G4 DNA interspersed with single-stranded DNA on the G-rich strand (30). Two nucleases active in transcription-coupled repair, ERCC1/XPF and XPG, can cleave the upstream and downstream junctions between the RNA:DNA hybrid and duplex DNA in transcribed S regions in vitro (128). However, analyses of murine models revealed no effect on switch recombination resulting from mutation of *Xpf* (130) or *Xpg* (129), and at most a modest effect of deletion of *Ercc1* (118).

Two factors shown by genetic analysis to be key to switch recombination do recognize G-loops in transcribed switch regions. AID interacts specifically with and deaminates the G-rich strand of G-loops (31, 141), consistent with its preference for single-stranded DNA. MutSα binds to G4 DNA with affinity considerably higher than for heteroduplex (T•G or U•G) mismatches; and MutSα bound to G-loops promotes DNA synapsis between transcriptionally activated S regions (58). The RNA:DNA hybrid within a G-loop may prolong DNA denaturation, thereby increasing accessibility of the DNA to AID or MutSα.

Many Factors Collaborate in Switch Recombination

Switch recombination depends upon four factors involved in the DNA damage response: the cell cycle regulator ATM (62, 106); the phosphorylated variant histone γ-H2AX (107); NBS1, the regulatory component of the MRE11/RAD50/NBS1 (MRN) complex (51, 108); and the p53-interacting protein, 53BP1 (65, 132). ATM and γ-H2AX are required only for inter-S region recombination and not for intra-S region recombination, leading to the suggestion that they facilitate synapsis by promoting higher-order chromatin remodeling (106, 107). Switch recombination rejoins the chromosome and generates switch circles carrying the excised fragment (19, 61, 66), as shown in **Figure 2**. It is therefore not surprising that switch recombination also depends upon factors involved in nonhomologous end-joining, including Ku and DNA-PKcs (113). Interactions with these or other factors may explain the observation that switch recombination depends upon a 10 amino acid region at the very C terminus of AID, a region unnecessary for cytidine deamination or for gene conversion or somatic hypermutation (7, 121, 126).

DNA BREAKS AND DIVERSIFYING Ig GENES

Figure 5 presents a model for how DNA breaks produced UNG-dependent abasic site cleavage or MutSα-Exo1-dependent resection can be processed to produce the three signature outcomes of Ig gene diversification, by pathways that combine elements of error-prone DNA synthesis, homologous recombination, and non-homologous end-joining.

Error-Prone Polymerases and Point Mutations

Error-prone polymerases are a likely source of nontemplated mutations that accompany gene conversion, hypermutation, and switch recombination. Among the error-prone polymerases characterized in recent years (36, 38, 54), the strongest candidate for function in hypermutation is pol η, a Y-family

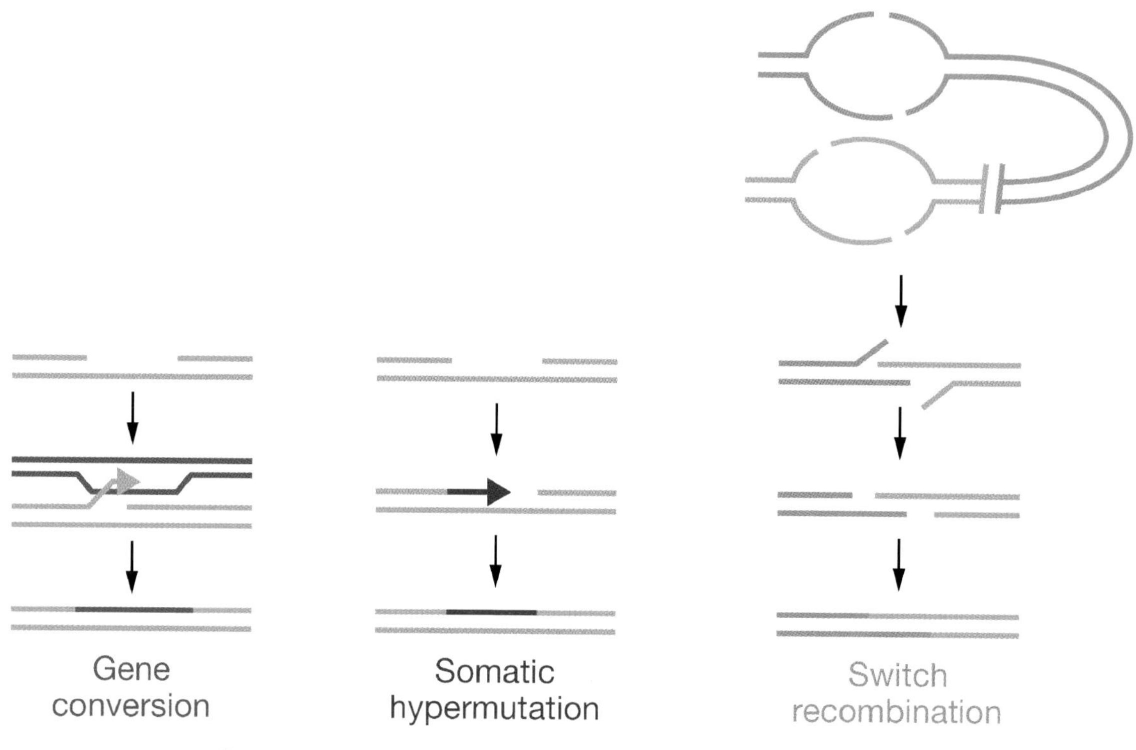

Figure 5

DNA breaks give rise to the signature products of Ig gene diversification. *Left*, a single break in the nontemplate DNA strand of a V region initiates gene conversion (*blue*) with the appropriate 5′ to 3′ directionality. *Center*, error-prone repair using a DNA end as primer creates nontemplated mutations at a V region, leading to hypermutation (*pink*). For other paths to hypermutation, see text and **Figure 4**. *Right*, two cleaved and synapsing switch regions (*gray and orange*) undergo nonhomologous end-joining.

polymerase encoded by the *XPV* gene. Mutations in *XPV* result in xeroderma pigmentosum variant disease, characterized by faulty repair of UV damage. In individuals deficient in pol η, hypermutation at T/A pairs is diminished without apparent immunodeficiency (35, 146, 147). The pol η mutagenesis spectrum in vitro supports possible function as an T/A mutator in somatic hypermutation (92, 112). Moreover, MutSα, which is necessary for hypermutation at T/A pairs (82, 99), stimulates pol η activity in vitro (134). Another Y-family polymerase, pol ι, appeared to be a strong contender as a G/C mutator (24, 34, 98), but enthusiasm dampened with the discovery that strain 129 mice naturally lack pol ι but hypermutation is not altered (71). Pol ζ, a B-family polymerase, can extend from a mismatched primer and may work in collaboration with other mutagenic polymerases (27, 46, 81, 142). Error-prone polymerases interact with one another and the replication apparatus, which may facilitate redundancy and make it difficult to establish critical function of any single polymerase.

Nontemplated base changes in hypermutation have been assumed to originate by misincorporation by low-fidelity polymerases, using the four canonical deoxynucleoside triphosphates. An intriguing alternative hypothesis is that mutation results from incorporation of nonstandard bases, such as dUTP, and subsequent repair by either high- or low-fidelity polymerases (82).

A Break is Critical for Gene Conversion

A simple model for gene conversion is that AID and UNG together create an abasic site, which is cleaved; and the resulting 3′ end undergoes homologous repair, promoted by RAD51 paralogs and BRCA2 (43, 114), copying new sequence information from a donor pseudogene (**Figure 5**, left). The documented 5′ to 3′ directionality of gene conversion (69) suggests that a single break in the top strand creates the primer for gene conversion. There has as yet been no genetic test of function of MutSα or Exo1 in Ig gene conversion, although these factors could in principle provide a 3′ end for strand invasion.

Gene conversion at chicken Ig genes produces not only templated but also nontemplated mutations, like those generated by somatic hypermutation (1). Moreover, the balance shifts to nontemplated point mutations as a result of inhibition of UNG (26), or interference with homologous recombination by mutation of *RAD51* paralogs or *BRCA2* (43, 114) or deletion of pseudogene sequence donors (3). Thus, the DNA end that is the primer for gene conversion (**Figure 5**, left) can alternatively undergo repair by hypermutation (**Figure 5**, center).

Breaks May Produce Somatic Hypermutation

The U•G mismatches introduced by AID can be processed in three ways leading to nontemplated mutations. Repair by an error-prone polymerase at a break (**Figure 5**, center) would produce nontemplated muation, as is found in hypermutation. Two other routes to mutagenesis do not involve a DNA break: replication to generate C/G → T/A transition mutations—the only mutations observed in *Msh2-/-Ung-/-* double knockouts (99); and mutagenic synthesis opposite an abasic site produced by UNG (**Figure 4**). Breaks almost certainly accompany hypermutation in vivo, as unselected hypermutated V regions contain short DNA deletions, like those found in S regions (39).

Switch Recombination Requires Two DNA Ends

Switch recombination joins two S regions (**Figure 5**, right), deleting a large stretch of chromosomal DNA. Switching requires breaks in both DNA strands, which could arise from UNG or Mutsα-Exo1-dependent cleavage of both strands, or by replication at a single-strand break to create a blunt or staggered end. Switching, like V(D)J recombination, depends on nonhomologous end-joining factors (113) to process DNA ends. Following switch recombination, the excised S regions can be recovered as nonreplicating circles (**Figure 2**). In essentially all processes of programmed recombination that involve free DNA ends, the ends are sequestered until recombination is completed. This suggests that switch circles may be products of reciprocal recombination, a question that could almost certainly be addressed using single-cell PCR to compare junctions in switch circles and chromosomal DNA.

Once a B cell has completed switching, switch regions resemble products of hypermutation: Sμ switch regions accumulate AID-dependent point mutations (79), and switch junctions are flanked by recombination-associated point mutations (29, 116). Whether the mechanism of S region mutagenesis is entirely or only partially shared with that of hypermutation is not yet resolved.

Single-Strand Breaks, Not Double-Strand Breaks, in AID-Initiated Diversification

Experiments seeking to define DNA intermediates of Ig gene diversification have identified double-strand breaks in mammalian B cells

(14, 18, 88, 143). However, these breaks appear not to depend upon ongoing hypermutation (14) or AID expression (15, 89, 143), and may not be relevant to AID-initiated diversification. Single-strand DNA breaks (4, 50) and staggered ends (143) have been identified, and are consistent with current understanding of the cleavage mechanism. Staggered ends could result from independent cleavage of both DNA strands, or replication of a region containing a nick or gap on one strand, which generates one double-strand break and one staggered end.

TARGETING AND MISTARGETING OF GENE DIVERSIFICATION

How is Diversification Targeted to Ig Genes?

AID is targeted (almost) exclusively to the expressed Ig genes. The requirement for a single-stranded substrate produced upon transcriptional activation provides one key to AID targeting. Targeting does not depend upon specific *cis*-elements or chromosomal position, but may depend on modifications in chromatin or specific interactions of AID with RNA polymerase II (59, 80, 137). Key open questions are how actively transcribing non-Ig genes are protected from deamination, and how deamination is limited to a few kb downstream of the promoters, protecting the C region from mutagenesis.

Mistargeting: Hypermutation and Translocation of Proto-Oncogenes

The c-*MYC* proto-oncogene undergoes both aberrant hypermutation and translocation in B cell tumors (91). This genetic instability reflects targeting of AID to c-*MYC*. c-*MYC* is transcribed in activated B cells, making it a potential substrate for deamination by AID. Moreover, AID is required for c-*Myc* translocation leading to tumorigenesis in a murine model for Burkitt's lymphoma (102). Electron micrographic imaging reveals that G-loops form within the transcribed c-*MYC* gene, with structures similar to those that form in transcribed S regions (30, 58), and that these G-loops are targets for AID (31). G-loops map to a G-rich region within the first exon and intron of the transcribed c-*MYC* gene, the same zone that undergoes aberrant hypermutation and translocations. In addition to c-*MYC*, a small subset of other non-Ig genes undergo aberrant hypermutation or translocation in B cells, including the *BCL6*, *CD95/FAS*, *RHO/TTF*, *PAX-5*, and *PIM1* proto-oncogenes (55, 91); and the B29 and MB1 B cell receptor genes (40). Further analysis of mistargeting of AID to these genes may provide insights into how AID is targeted with considerable accuracy to the Ig loci.

AID is But One Member of the APOBEC Gene Family

AID has some interesting relatives in the APOBEC family of cytidine deaminases, which may function as effectors of innate immunity and as sources of somatic mutagenesis in non-B cells. Two of these, APOBEC3F and APOBEC3G, prevent retroviral replication and deaminate retroviral cDNAs during second-strand synthesis (11), although the antiviral effects may not depend entirely upon deamination (84). Each APOBEC has a distinctive, context-dependent mutation pattern (9), raising the interesting possiblity that viral targets for specific APOBECs may have coevolved, like V regions and AID. Some APOBECs have a wider expression profile than AID, which is restricted to activated B cells (9, 22), suggestive of very general functions. An expansion of APOBEC3-related genes has occurred during human evolution (9, 22), creating a handful of related deaminases—orphan enzymes, the targets of which are yet to be discovered.

SUMMARY POINTS

1. Ig gene diversification is essential to the immune response and produces three signature outcomes: gene conversion (templated mutation), somatic hypermutation (non-templated mutation) and class switch recombination (region-specific DNA deletion).
2. AID initiates Ig gene diversification by deaminating cytidines in transcribed Ig genes at consensus motifs, identical to hypermutation hotspots. AID deaminates single-stranded DNA but not double-stranded DNA, explaining how deamination is targeted to transcribed genes, but not how transcribed non-Ig genes escape deamination.
3. Conserved factors UNG, MRN, MutSα, and Exo1 play paradoxical roles in Ig gene diversification, promoting—rather than preventing—genetic instability.
4. Targeting of diversification depends upon specificity of AID for single-stranded DNA, which restricts deamination to transcribed genes; and formation of characteristic structures, G-loops, within transcribed S regions. How many transcribed regions escape attack by AID is not understood. Proto-oncogenes like c-*MYC* are mistargeted, which can lead to aberrant hypermutation, translocation, and tumorigenesis in B cells.

PERSPECTIVE AND FUTURE DIRECTIONS

In the past few years Ig gene diversification has been pried out of the metaphorical black box and defined in genetic and biochemical terms. We have learned that a single, simple DNA modification, cytidine deamination, can drive three processes that had once seemed unrelated: hypermutation, gene conversion, and class switch recombination. We have seen how these specialized processes relate to ubiquitous and conserved pathways of DNA repair and recombination that maintain genomic stability in all organisms, and we have been forced to appreciate that factors which normally promote genomic integrity can be active participants in its opposite. Some future directions are clear. We need to learn more about targeting of deamination and functions of deaminases in human biology. We need to learn more about how Ig gene diversification occurs, not only at the molecular level but also within the larger context of chromatin structure and nuclear organization. These details of immune diversification will have implications for our broader understanding of biology. Most important, if recent progress has shown us anything, it is to expect surprises. As we pose questions for the future, we can look forward to answers that go beyond our ability to hypothesize and speculate.

ACKNOWLEDGMENTS

I thank members of my laboratory and my colleagues for valuable discussions, and the NIH for supporting our research on Ig gene diversification (GM39799, GM41712).

LITERATURE CITED

1. Arakawa H, Buerstedde JM. 2004. Immunoglobulin gene conversion: insights from bursal B cells and the DT40 cell line. *Dev. Dyn.* 229:458–64

2. Arakawa H, Hauschild J, Buerstedde JM. 2002. Requirement of the Activation-Induced Deaminase (AID) gene for immunoglobulin gene conversion. *Science* 295:1301–6
3. Arakawa H, Saribasak H, Buerstedde JM. 2004. Activation-induced cytidine deaminase initiates immunoglobulin gene conversion and hypermutation by a common intermediate. *PLoS Biol.* 2:E179
4. Arudchandran A, Bernstein RM, Max EE. 2004. Single-stranded DNA breaks adjacent to cytosines occur during Ig gene class switch recombination. *J. Immunol.* 173:3223–29
5. Bardwell PD, Woo CJ, Wei K, Li Z, Martin A, et al. 2004. Altered somatic hypermutation and reduced class-switch recombination in exonuclease 1-mutant mice. *Nat. Immunol.* 5:224–29
6. Barnes DE, Lindahl T. 2004. Repair and genetic consequences of endogenous DNA base damage in mammalian cells. *Annu. Rev. Genet.* 38:445–76
7. Barreto V, Reina-San-Martin B, Ramiro AR, McBride KM, Nussenzweig MC. 2003. C-terminal deletion of AID uncouples class switch recombination from somatic hypermutation and gene conversion. *Mol. Cell* 12:501–8
8. Bassing CH, Swat W, Alt FW. 2002. The mechanism and regulation of chromosomal V(D)J recombination. *Cell* 109(Suppl.):S45–55
9. Beale RC, Petersen-Mahrt SK, Watt IN, Harris RS, Rada C, Neuberger MS. 2004. Comparison of the differential context-dependence of DNA deamination by APOBEC enzymes: correlation with mutation spectra in vivo. *J. Mol. Biol.* 337:585–96
10. Begum NA, Kinoshita K, Kakazu N, Muramatsu M, Nagaoka H, et al. 2004. Uracil DNA glycosylase activity is dispensable for immunoglobulin class switch. *Science* 305:1160–63
11. Bishop KN, Holmes RK, Sheehy AM, Davidson NO, Cho SJ, Malim MH. 2004. Cytidine deamination of retroviral DNA by diverse APOBEC proteins. *Curr. Biol.* 14:1392–96
12. Bransteitter R, Pham P, Calabrese P, Goodman MF. 2004. Biochemical analysis of hypermutational targeting by wild type and mutant AID. *J. Biol. Chem.* 279:51612–21
13. Bransteitter R, Pham P, Scharff MD, Goodman MF. 2003. Activation-induced cytidine deaminase deaminates deoxycytidine on single-stranded DNA but requires the action of RNase. *Proc. Natl. Acad. Sci. USA* 100:4102–7
14. Bross L, Fukita Y, McBlane F, Demolliere C, Rajewsky K, Jacobs H. 2000. DNA double-strand breaks in immunoglobulin genes undergoing somatic hypermutation. *Immunity* 13:589–97
15. Bross L, Muramatsu M, Kinoshita K, Honjo T, Jacobs H. 2002. DNA double-strand breaks: prior to but not sufficient in targeting hypermutation. *J. Exp. Med.* 195:1187–92
16. Butler JE, Sun J, Kacskovics I, Brown WR, Navarro P. 1996. The V_H and C_H immunoglobulin genes of swine: implications for repertoire development. *Vet. Immunol. Immunopathol.* 54:7–17
17. Cascalho M, Wong J, Steinberg C, Wabl M. 1998. Mismatch repair co-opted by hypermutation. *Science* 279:1207–10
18. Catalan N, Selz F, Imai K, Revy P, Fischer A, Durandy A. 2003. The block in immunoglobulin class switch recombination caused by activation-induced cytidine deaminase deficiency occurs prior to the generation of DNA double strand breaks in switch mu region. *J. Immunol.* 171:2504–9
19. Chaudhuri J, Alt FW. 2004. Class-switch recombination: interplay of transcription, DNA deamination and DNA repair. *Nat. Rev. Immunol.* 4:541–52
20. Chaudhuri J, Khuong C, Alt FW. 2004. Replication protein A interacts with AID to promote deamination of somatic hypermutation targets. *Nature* 430:992–98

21. Chaudhuri J, Tian M, Khuong C, Chua K, Pinaud E, Alt FW. 2003. Transcription-targeted DNA deamination by the AID antibody diversification enzyme. *Nature* 422:726–30
22. Conticello SG, Thomas CJ, Petersen-Mahrt SK, Neuberger MS. 2005. Evolution of the AID/APOBEC family of polynucleotide (deoxy)cytidine deaminases. *Mol. Biol. Evol.* 22:367–77
23. Daniels GA, Lieber MR. 1995. RNA:DNA complex formation upon transcription of immunoglobulin switch regions: implications for the mechanism and regulation of class switch recombination. *Nucleic Acids Res.* 23:5006–11
24. Denepoux S, Razanajaona D, Blanchard D, Meffre G, Capra JD, et al. 1997. Induction of somatic mutation in a human B cell line in vitro. *Immunity* 6:35–46
25. Di Noia J, Neuberger MS. 2002. Altering the pathway of immunoglobulin hypermutation by inhibiting uracil-DNA glycosylase. *Nature* 419:43–48
26. Di Noia JM, Neuberger MS. 2004. Immunoglobulin gene conversion in chicken DT40 cells largely proceeds through an abasic site intermediate generated by excision of the uracil produced by AID-mediated deoxycytidine deamination. *Eur. J. Immunol.* 34:504–8
27. Diaz M, Verkoczy LK, Flajnik MF, Klinman NR. 2001. Decreased frequency of somatic hypermutation and impaired affinity maturation but intact germinal center formation in mice expressing antisense RNA to DNA polymerase zeta. *J. Immunol.* 167:327–35
28. Dickerson SK, Market E, Besmer E, Papavasiliou FN. 2003. AID mediates hypermutation by deaminating single stranded DNA. *J. Exp. Med.* 197:1291–96
29. Dunnick W, Hertz GZ, Scappino L, Gritzmacher C. 1993. DNA sequences at immunoglobulin switch region recombination sites. *Nucleic Acids Res.* 21:365–72
30. Duquette ML, Handa P, Vincent JA, Taylor AF, Maizels N. 2004. Intracellular transcription of G-rich DNAs induces formation of G-loops, novel structures containing G4 DNA. *Genes Dev.* 18:1618–29
31. Duquette ML, Pham P, Goodman MF, Maizels N. 2005. AID binds to transcription-induced structures in c-*MYC* that map to regions associated with translocation and hypermutation. *Oncogene*. In press
32. Ehrenstein MR, Neuberger MS. 1999. Deficiency in Msh2 affects the efficiency and local sequence specificity of immunoglobulin class-switch recombination: parallels with somatic hypermutation. *EMBO J.* 18:3484–90
33. Ehrenstein MR, Rada C, Jones AM, Milstein C, Neuberger MS. 2001. Switch junction sequences in PMS2-deficient mice reveal a microhomology-mediated mechanism of Ig class switch recombination. *Proc. Natl. Acad. Sci. USA* 98:14553–58
34. Faili A, Aoufouchi S, Flatter E, Gueranger Q, Reynaud CA, Weill JC. 2002. Induction of somatic hypermutation in immunoglobulin genes is dependent on DNA polymerase iota. *Nature* 419:944–47
35. Faili A, Aoufouchi S, Weller S, Vuillier F, Stary A, et al. 2004. DNA polymerase eta is involved in hypermutation occurring during immunoglobulin class switch recombination. *J. Exp. Med.* 199:265–70
36. Fischhaber PL, Friedberg EC. 2005. How are specialized (low-fidelity) eukaryotic polymerases selected and switched with high-fidelity polymerases during translesion DNA synthesis? *DNA Repair* 4:279–83
37. Gellert M. 2002. V(D)J recombination: RAG proteins, repair factors, and regulation. *Annu. Rev. Biochem.* 71:101–32
38. Goodman MF. 2002. Error-prone repair DNA polymerases in prokaryotes and eukaryotes. *Annu. Rev. Biochem.* 71:17–50

39. Goossens T, Klein U, Kuppers R. 1998. Frequent occurrence of deletions and duplications during somatic hypermutation: implications for oncogene translocations and heavy chain disease. *Proc. Natl. Acad. Sci. USA* 95:2463–68
40. Gordon MS, Kanegai CM, Doerr JR, Wall R. 2003. Somatic hypermutation of the B cell receptor genes B29 (Igbeta, CD79b) and mb1 (Igalpha, CD79a). *Proc. Natl. Acad. Sci. USA* 100:4126–31
41. Harfe BD, Jinks-Robertson S. 2000. DNA mismatch repair and genetic instability. *Annu. Rev. Genet.* 34:359–99
42. Harris RS, Sale JE, Petersen-Mahrt SK, Neuberger MS. 2002. AID is essential for immunoglobulin V gene conversion in a cultured B cell line. *Curr. Biol.* 12:435–38
43. Hatanaka A, Yamazoe M, Sale JE, Takata M, Yamamoto K, et al. 2005. Similar effects of Brca2 truncation and Rad51 paralog deficiency on immunoglobulin V gene diversification in DT40 cells support an early role for Rad51 paralogs in homologous recombination. *Mol. Cell. Biol.* 25:1124–34
44. Imai K, Slupphaug G, Lee WI, Revy P, Nonoyama S, et al. 2003. Human uracil-DNA glycosylase deficiency associated with profoundly impaired immunoglobulin class-switch recombination. *Nat. Immunol.* 4:1023–28
45. Jacobs H, Fukita Y, van der Horst GT, de Boer J, Weeda G, et al. 1998. Hypermutation of immunoglobulin genes in memory B cells of DNA repair-deficient mice. *J. Exp. Med.* 187:1735–43
46. Johnson RE, Washington MT, Haracska L, Prakash S, Prakash L. 2000. Eukaryotic polymerases iota and zeta act sequentially to bypass DNA lesions. *Nature* 406:1015–19
47. Jolly CJ, Wagner SD, Rada C, Klix N, Milstein C, Neuberger MS. 1996. The targeting of somatic hypermutation. *Semin. Immunol.* 8:159–68
48. Kim N, Bozek G, Lo JC, Storb U. 1999. Different mismatch repair deficiencies all have the same effects on somatic hypermutation: intact primary mechanism accompanied by secondary modifications. *J. Exp. Med.* 190:21–30
49. Kong Q, Maizels N. 1999. PMS2-deficiency diminishes hypermutation of a $\lambda 1$ transgene in young but not older mice. *Mol. Immunol.* 36:83–91
50. Kong Q, Maizels N. 2001. DNA breaks in hypermutating immunoglobulin genes: evidence for a break and repair pathway of somatic hypermutation. *Genetics* 158:369–78
51. Kracker S, Bergmann Y, Demuth I, Frappart PO, Hildebrand G, et al. 2005. Nibrin functions in Ig class-switch recombination. *Proc. Natl. Acad. Sci. USA* 102:1584–89
52. Krokan HE, Drablos F, Slupphaug G. 2002. Uracil in DNA—occurrence, consequences and repair. *Oncogene* 21:8935–48
53. Kunkel TA, Erie DA. 2005. DNA mismatch repair. *Annu. Rev. Biochem.* 74:681–710
54. Kunkel TA, Pavlov YI, Bebenek K. 2003. Functions of human DNA polymerases eta, kappa and iota suggested by their properties, including fidelity with undamaged DNA templates. *DNA Repair* 2:135–49
55. Kuppers R, Dalla-Favera R. 2001. Mechanisms of chromosomal translocations in B cell lymphomas. *Oncogene* 20:5580–94
56. Lahdesmaki A, Taylor AM, Chrzanowska KH, Pan-Hammarstrom Q. 2004. Delineation of the role of the Mre11 complex in class switch recombination. *J. Biol. Chem.* 279:16479–87
57. Lanning D, Zhu X, Zhai SK, Knight KL. 2000. Development of the antibody repertoire in rabbit: gut-associated lymphoid tissue, microbes, and selection. *Immunol. Rev.* 175:214–28

58. Larson ED, Duquette ML, Cummings WJ, Streiff RJ, Maizels N. 2005. MutSα binds to and promotes synapsis of transcriptionally activated immunoglobulin switch regions. *Curr. Biol.* 15:470–74
59. Li Z, Luo Z, Scharff MD. 2004. Differential regulation of histone acetylation and generation of mutations in switch regions is associated with Ig class switching. *Proc. Natl. Acad. Sci. USA* 101:15428–33
60. Li Z, Scherer SJ, Ronai D, Iglesias-Ussel MD, Peled JU, et al. 2004. Examination of Msh6- and Msh3-deficient mice in class switching reveals overlapping and distinct roles of MutS homologues in antibody diversification. *J. Exp. Med.* 200:47–59
61. Li Z, Woo CJ, Iglesias-Ussel MD, Ronai D, Scharff MD. 2004. The generation of antibody diversity through somatic hypermutation and class switch recombination. *Genes Dev.* 18:1–11
62. Lumsden JM, McCarty T, Petiniot LK, Shen R, Barlow C, et al. 2004. Immunoglobulin class switch recombination is impaired in Atm-deficient mice. *J. Exp. Med.* 200:1111–21
63. Maizels N. 1995. Somatic hypermutation: How many mechanisms diversify V region sequences? *Cell* 83:9–12
64. Maizels N, Scharff MD. 2004. Molecular mechanisms of hypermutation. In *Molecular Biology of B Cells*, ed. T Honjo, M Neuberger, FW Alt, pp. 327–38. Amsterdam: Elsevier
65. Manis JP, Morales JC, Xia Z, Kutok JL, Alt FW, Carpenter PB. 2004. 53BP1 links DNA damage-response pathways to immunoglobulin heavy chain class-switch recombination. *Nat. Immunol.* 5:481–87
66. Manis JP, Tian M, Alt FW. 2002. Mechanism and control of class-switch recombination. *Trends Immunol.* 23:31–39
67. Martin A, Li Z, Lin DP, Bardwell PD, Iglesias-Ussel MD, et al. 2003. Msh2 ATPase activity is essential for somatic hypermutation at A-T basepairs and for efficient class switch recombination. *J. Exp. Med.* 198:1171–78
68. Martomo SA, Yang WW, Gearhart PJ. 2004. A role for Msh6 but not Msh3 in somatic hypermutation and class switch recombination. *J. Exp. Med.* 200:61–68
69. McCormack WT, Thompson CB. 1990. Chicken IgL variable region gene conversions display pseudogene donor preference and 5′ to 3′ polarity. *Genes Dev.* 4:548–58
70. McCormack WT, Tjoelker LW, Thompson CB. 1991. Avian B-cell development: generation of an immunoglobulin repertoire by gene conversion. *Annu. Rev. Immunol.* 9:219–41
71. McDonald JP, Frank EG, Plosky BS, Rogozin IB, Masutani C, et al. 2003. 129-derived strains of mice are deficient in DNA polymerase iota and have normal immunoglobulin hypermutation. *J. Exp. Med.* 198:635–43
72. Meyer A, Parng CL, Hansal SA, Osborne BA, Goldsby RA. 1997. Immunoglobulin gene diversification in cattle. *Int. Rev. Immunol.* 15:165–83
73. Milstein C, Neuberger MS, Staden R. 1998. Both DNA strands of antibody genes are hypermutation targets. *Proc. Natl. Acad. Sci. USA* 95:8791–94
74. Min IM, Schrader CE, Vardo J, Luby TM, D'Avirro N, et al. 2003. The Smu tandem repeat region is critical for Ig isotype switching in the absence of Msh2. *Immunity* 19:515–24
75. Mizuta R, Iwai K, Shigeno M, Mizuta M, Ushiki T, Kitamura D. 2002. Molecular visualization of immunoglobulin switch region RNA/DNA complex by atomic force microscope. *J. Biol. Chem.* 278:4431–34

76. Muramatsu M, Kinoshita K, Fagarasan S, Yamada S, Shinkai Y, Honjo T. 2000. Class switch recombination and somatic hypermutation require Activation-Induced Deaminase (AID), a member of RNA editing cytidine deaminase family. *Cell* 553–63
77. Muramatsu M, Sankaranand VS, Anant S, Sugai M, Kinoshita K, et al. 1999. Specific expression of activation-induced cytidine deaminase (AID), a novel member of the RNA-editing deaminase family in germinal center B cells. *J. Biol. Chem.* 274:18470–76
78. Mussman R, Courtet M, Schwaer J, Du Pasquier L. 1997. Microsites for immunoglobulin switch recombination breakpoints from *Xenopus* to mammals. *Eur. J. Immunol.* 27:2610–19
79. Nagaoka H, Muramatsu M, Yamamura N, Kinoshita K, Honjo T. 2002. Activation-induced deaminase (AID)-directed hypermutation in the immunoglobulin Smu region: implication of AID involvement in a common step of class switch recombination and somatic hypermutation. *J. Exp. Med.* 195:529–34
80. Nambu Y, Sugai M, Gonda H, Lee CG, Katakai T, et al. 2003. Transcription-coupled events associating with immunoglobulin switch region chromatin. *Science* 302:2137–40
81. Nelson JR, Lawrence CW, Hinkle DC. 1996. Thymine-thymine dimer bypass by yeast DNA polymerase zeta. *Science* 272:1646–49
82. Neuberger MS, Di Noia JM, Beale RC, Williams GT, Yang Z, Rada C. 2005. Somatic hypermutation at A.T pairs: polymerase error versus dUTP incorporation. *Nat. Rev. Immunol.* 5:171–78
83. Neuberger MS, Harris RS, Di Noia J, Petersen-Mahrt SK. 2003. Immunity through DNA deamination. *Trends Biochem. Sci.* 28:305–12
84. Newman EN, Holmes RK, Craig HM, Klein KC, Lingappa JR, et al. 2005. Antiviral function of APOBEC3G can be dissociated from cytidine deaminase activity. *Curr. Biol.* 15:166–70
85. Okazaki IM, Kinoshita K, Muramatsu M, Yoshikawa K, Honjo T. 2002. The AID enzyme induces class switch recombination in fibroblasts. *Nature* 416:340–45
86. Pan Q, Petit-Frere C, Lahdesmaki A, Gregorek H, Chrzanowska KH, Hammarstrom L. 2002. Alternative end joining during switch recombination in patients with ataxia-telangiectasia. *Eur. J. Immunol.* 32:1300–8
87. Pan-Hammarstrom Q, Dai S, Zhao Y, van Dijk-Hard IF, Gatti RA, et al. 2003. ATM is not required in somatic hypermutation of V_H, but is involved in the introduction of mutations in the switch mu region. *J. Immunol.* 170:3707–16
88. Papavasiliou FN, Schatz DG. 2000. Cell-cycle-regulated DNA double-stranded breaks in somatic hypermutation of immunoglobulin genes. *Nature* 408:216–21
89. Papavasiliou FN, Schatz DG. 2002. The activation-induced deaminase functions in a postcleavage step of the somatic hypermutation process. *J. Exp. Med.* 195:1193–98
90. Pasqualucci L, Guglielmino R, Houldsworth J, Mohr J, Aoufouchi S, Ploakiewicz R. 2004. Expression of the AID protein in normal and neoplastic B cells. *Blood* 104:3318–25
91. Pasqualucci L, Neumeister P, Goossens T, Nanjangud G, Chaganti RS, et al. 2001. Hypermutation of multiple proto-oncogenes in B-cell diffuse large-cell lymphomas. *Nature* 412:341–46
92. Pavlov YI, Rogozin IB, Galkin AP, Aksenova AY, Hanaoka F, et al. 2002. Correlation of somatic hypermutation specificity and A-T base pair substitution errors by DNA polymerase eta during copying of a mouse immunoglobulin kappa light chain transgene. *Proc. Natl. Acad. Sci. USA* 99:9954–59

93. Petersen S, Casellas R, Reina-San-Martin B, Chen HT, Difilippantonio MJ, et al. 2001. AID is required to initiate Nbs1/gamma-H2AX focus formation and mutations at sites of class switching. *Nature* 414:660–65
94. Petersen-Mahrt SK, Harris RS, Neuberger MS. 2002. AID mutates *E. coli* suggesting a DNA deamination mechanism for antibody diversification. *Nature* 418:99–104
95. Pham P, Bransteitter R, Petruska J, Goodman MF. 2003. Processive AID-catalysed cytosine deamination on single-stranded DNA simulates somatic hypermutation. *Nature* 424:103–7
96. Phan RT, Dalla-Favera R. 2004. The BCL6 proto-oncogene suppresses p53 expression in germinal-centre B cells. *Nature* 432:635–39
97. Phung QH, Winter DB, Cranston A, Tarone RE, Bohr VA, et al. 1998. Increased hypermutation at G and C nucleotides in immunoglobulin variable genes from mice deficient in the MSH2 mismatch repair protein. *J. Exp. Med.* 187:1745–51
98. Poltoratsky V, Woo CJ, Tippin B, Martin A, Goodman MF, Scharff MD. 2001. Expression of error-prone polymerases in BL2 cells activated for Ig somatic hypermutation. *Proc. Natl. Acad. Sci. USA* 98:7976–81
99. Rada C, Di Noia JM, Neuberger MS. 2004. Mismatch recognition and uracil excision provide complementary paths to both Ig switching and the a/t-focused phase of somatic mutation. *Mol. Cell* 16:163–71
100. Rada C, Ehrenstein MR, Neuberger MS, Milstein C. 1998. Hot spot focusing of somatic hypermutation in MSH2-deficient mice suggests two stages of mutational targeting. *Immunity* 9:135–41
101. Rada C, Williams GT, Nilsen H, Barnes DE, Lindahl T, Neuberger MS. 2002. Immunoglobulin isotype switching is inhibited and somatic hypermutation perturbed in UNG-deficient mice. *Curr. Biol.* 12:1748–55
102. Ramiro AR, Jankovic M, Eisenreich T, Difilippantonio S, Chen-Kiang S, et al. 2004. AID is required for c-Myc/IgH chromosome translocations in vivo. *Cell* 118:431–38
103. Ramiro AR, Stavropoulos P, Jankovic M, Nussenzweig MC. 2003. Transcription enhances AID-mediated cytidine deamination by exposing single-stranded DNA on the nontemplate strand. *Nat. Immunol.* 4:452–56
104. Reaban ME, Griffin JA. 1990. Induction of RNA-stabilized DNA conformers by transcription of an immunoglobulin switch region. *Nature* 348:342–44
105. Reaban ME, Lebowitz J, Griffin JA. 1994. Transcription induces the formation of a stable RNA-DNA hybrid in the immunoglobulin alpha switch region. *J. Biol. Chem.* 269:21850–57
106. Reina-San-Martin B, Chen HT, Nussenzweig A, Nussenzweig MC. 2004. ATM is required for efficient recombination between immunoglobulin switch regions. *J. Exp. Med.* 200:1103–10
107. Reina-San-Martin B, Difilippantonio S, Hanitsch L, Masilamani RF, Nussenzweig A, Nussenzweig MC. 2003. H2AX is required for recombination between immunoglobulin switch regions but not for intra-switch region recombination or somatic hypermutation. *J. Exp. Med.* 197:1767–78
108. Reina-San-Martin B, Nussenzweig MC, Nussenzweig A, Difilippantonio S. 2005. Genomic instability, endoreduplication, and diminished Ig class-switch recombination in B cells lacking Nbs1. *Proc. Natl. Acad. Sci. USA* 102:1590–95
109. Revy P, Muto T, Levy Y, Geissmann F, Plebani A, et al. 2000. Activation-induced cytidine deaminase (AID) deficiency causes the autosomal recessive form of the Hyper-IgM syndrome (HIGM2). *Cell* 102:565–75

110. Reynaud CA, Anquez V, Grimal H, Weill JC. 1987. A hyperconversion mechanism generates the chicken light chain preimmune repertoire. *Cell* 48:379–88
111. Rogozin IB, Kolchanov NA. 1992. Somatic hypermutagenesis in immunoglobulin genes. II. Influence of neighbouring base sequences on mutagenesis. *Biochim. Biophys. Acta* 1171:11–18
112. Rogozin IB, Pavlov YI, Bebenek K, Matsuda T, Kunkel TA. 2001. Somatic mutation hotspots correlate with DNA polymerase eta error spectrum. *Nat. Immunol.* 2:530–36
113. Rooney S, Chaudhuri J, Alt FW. 2004. The role of the non-homologous end-joining pathway in lymphocyte development. *Immunol. Rev.* 200:115–31
114. Sale JE, Calandrini DM, Takata M, Takeda S, Neuberger MS. 2001. Ablation of XRCC2/3 transforms immunoglobulin V gene conversion into somatic hypermutation. *Nature* 412:921–26
115. Schofield MJ, Hsieh P. 2003. DNA mismatch repair: molecular mechanisms and biological function. *Annu. Rev. Microbiol.* 57:579–608
116. Schrader CE, Bradley SP, Vardo J, Mochegova SN, Flanagan E, Stavnezer J. 2003. Mutations occur in the Ig Smu region but rarely in Sgamma regions prior to class switch recombination. *EMBO J.* 22:5893–903
117. Schrader CE, Edelmann W, Kucherlapati R, Stavnezer J. 1999. Reduced isotype switching in splenic B cells from mice deficient in mismatch repair enzymes. *J. Exp. Med.* 190:323–30
118. Schrader CE, Vardo J, Linehan E, Twarog MZ, Niedernhofer LJ, et al. 2004. Deletion of the nucleotide excision repair gene Ercc1 reduces immunoglobulin class switching and alters mutations near switch recombination junctions. *J. Exp. Med.* 200:321–30
119. Schrader CE, Vardo J, Stavnezer J. 2002. Role for mismatch repair proteins Msh2, Mlh1, and Pms2 in immunoglobulin class switching shown by sequence analysis of recombination junctions. *J. Exp. Med.* 195:367–73
120. Shen HM, Storb U. 2004. Activation-induced cytidine deaminase (AID) can target both DNA strands when the DNA is supercoiled. *Proc. Natl. Acad. Sci. USA* 101:12997–3002
121. Shinkura R, Ito S, Begum NA, Nagaoka H, Muramatsu M, et al. 2004. Separate domains of AID are required for somatic hypermutation and class-switch recombination. *Nat. Immunol.* 5:707–12
122. Shinkura R, Tian M, Smith M, Chua K, Fujiwara Y, Alt FW. 2003. The influence of transcriptional orientation on endogenous switch region function. *Nat. Immunol.* 4:435–41
123. Sohail A, Klapacz J, Samaranayake M, Ullah A, Bhagwat AS. 2003. Human activation-induced cytidine deaminase causes transcription-dependent, strand-biased C to U deaminations. *Nucleic Acids Res.* 31:2990–94
124. Stivers JT. 2004. Comment on "Uracil DNA glycosylase activity is dispensable for immunoglobulin class switch." *Science* 306:2042; author reply 42
125. Stojic L, Brun R, Jiricny J. 2004. Mismatch repair and DNA damage signalling. *DNA Repair* 3:1091–101
126. Ta VT, Nagaoka H, Catalan N, Durandy A, Fischer A, et al. 2003. AID mutant analyses indicate requirement for class-switch-specific cofactors. *Nat. Immunol.* 4:843–48
127. Thompson CB, Neiman PE. 1987. Somatic diversification of the chicken immunoglobulin light chain gene is limited to the rearranged variable gene segment. *Cell* 48:369–78
128. Tian M, Alt FW. 2000. Transcription-induced cleavage of immunogloublin switch regions by nucleotide excision repair nucleases in vitro. *J. Biol. Chem.* 275:24163–72

129. Tian M, Jones DA, Smith M, Shinkura R, Alt FW. 2004. Deficiency in the nuclease activity of xeroderma pigmentosum G in mice leads to hypersensitivity to UV irradiation. *Mol. Cell. Biol.* 24:2237–42
130. Tian M, Shinkura R, Shinkura N, Alt FW. 2004. Growth retardation, early death, and DNA repair defects in mice deficient for the nucleotide excision repair enzyme XPF. *Mol. Cell. Biol.* 24:1200–5
131. Vora KA, Tumas-Brundage KM, Lentz VM, Cranston A, Fishel R, Manser T. 1999. Severe attenuation of the B cell immune response in Msh2-deficient mice. *J. Exp. Med.* 189:471–81
132. Ward IM, Reina-San-Martin B, Olaru A, Minn K, Tamada K, et al. 2004. 53BP1 is required for class switch recombination. *J. Cell Biol.* 165:459–64
133. Wiesendanger M, Kneitz B, Edelmann W, Scharff MD. 2000. Somatic hypermutation in MutS homologue (MSH)3-, MSH6-, and MSH3/MSH6-deficient mice reveals a role for the MSH2-MSH6 heterodimer in modulating the base substitution pattern. *J. Exp. Med.* 191:579–84
134. Wilson TM, Vaisman A, Martomo SA, Sullivan P, Lan L, et al. 2005. MSH2-MSH6 stimulates DNA polymerase eta, suggesting a role for A:T mutations in antibody genes. *J. Exp. Med.* 201:637–45
135. Winter DB, Phung QH, Umar A, Baker SM, Tarone RE, et al. 1998. Altered spectra of hypermutation in antibodies from mice deficient for the DNA mismatch repair protein PMS2. *Proc. Natl. Acad. Sci. USA* 95:6953–58
136. Withers DR, Davison TF, Young JR. 2005. Developmentally programmed expression of AID in chicken B cells. *Dev. Comp. Immunol.* 29:651–62
137. Woo CJ, Martin A, Scharff MD. 2003. Induction of somatic hypermutation is associated with modifications in immunoglobulin variable region chromatin. *Immunity* 19:479–89
138. Yabuki M, Fujii M, Maizels N. 2005. The MRE11/RAD50/NBS1 complex accelerates somatic hypermutation and gene conversion of immunoglobulin variable regions. *Nat. Immunol.* 6:730–36
139. Yoshikawa K, Okazaki IM, Eto T, Kinoshita K, Muramatsu M, et al. 2002. AID enzyme-induced hypermutation in an actively transcribed gene in fibroblasts. *Science* 296:2033–36
140. Yu K, Chedin F, Hsieh CL, Wilson TE, Lieber MR. 2003. R-loops at immunoglobulin class switch regions in the chromosomes of stimulated B cells. *Nat. Immunol.* 4:442–51
141. Yu K, Roy D, Bayramyan M, Haworth IS, Lieber MR. 2005. Fine-structure analysis of activation-induced deaminase accessibility to class switch region R-loops. *Mol. Cell. Biol.* 25:1730–36
142. Zan H, Komori A, Li Z, Cerutti A, Schaffer A, et al. 2001. The translesion DNA polymerase zeta plays a major role in Ig and bcl-6 somatic hypermutation. *Immunity* 14:643–53
143. Zan H, Wu X, Komori A, Holloman WK, Casali P. 2003. AID-dependent generation of resected double-strand DNA breaks and recruitment of Rad52/Rad51 in somatic hypermutation. *Immunity* 18:727–38
144. Zarrin AA, Alt FW, Chaudhuri J, Stokes N, Kaushal D, et al. 2004. An evolutionarily conserved target motif for immunoglobulin class-switch recombination. *Nat. Immunol.* 5:1275–81
145. Zarrin AA, Tian M, Wang J, Borjeson T, Alt FW. 2005. Influence of switch region length on immunoglobulin class switch recombination. *Proc. Natl. Acad. Sci. USA* 102:2466–70

146. Zeng X, Negrete GA, Kasmer C, Yang WW, Gearhart PJ. 2004. Absence of DNA polymerase eta reveals targeting of C mutations on the nontranscribed strand in immunoglobulin switch regions. *J. Exp. Med.* 199:917–24
147. Zeng X, Winter DB, Kasmer C, Kraemer KH, Lehmann AR, Gearhart PJ. 2001. DNA polymerase eta is an A-T mutator in somatic hypermutation of immunoglobulin variable genes. *Nat. Immunol.* 2:537–41

Complexity in Regulation of Tryptophan Biosynthesis in *Bacillus subtilis*

Paul Gollnick,[1] Paul Babitzke,[2] Alfred Antson,[3] and Charles Yanofsky[4]

[1]Department of Biological Sciences, State University of New York, Buffalo, New York 14260; email: gollnick@acsu.buffalo.edu

[2]Department of Biochemistry and Molecular Biology, The Pennsylvania State University, University Park, Pennsylvania 16802; email: pxb28@psu.edu

[3]York Structural Biology Laboratory, Department of Chemistry, University of York, Y0101 5YW, York, United Kingdom; email: fred@ysbl.york.ac.uk

[4]Department of Biological Sciences, Stanford University, Stanford, California 94305; email: yanofsky@cmgm.stanford.edu

Annu. Rev. Genet.
2005. 39:47–68

First published online as a Review in Advance on June 21, 2005

The *Annual Review of Genetics* is online at http://genet.annualreviews.org

doi: 10.1146/annurev.genet.39.073003.093745

0066-4197/05/1215-0047$20.00

Key Words

trp operon regulation, attenuation, tryptophan sensing, $tRNA^{Trp}$ sensing, TRAP regulatory protein, AT regulatory protein

Abstract

Bacillus subtilis uses novel regulatory mechanisms in controlling expression of its genes of tryptophan synthesis and transport. These mechanisms respond to changes in the intracellular concentrations of free tryptophan and uncharged $tRNA^{Trp}$. The major *B. subtilis* protein that regulates tryptophan biosynthesis is the tryptophan-activated RNA-binding attenuation protein, TRAP. TRAP is a ring-shaped molecule composed of 11 identical subunits. Active TRAP binds to unique RNA segments containing multiple trinucleotide (NAG) repeats. Binding regulates both transcription termination and translation in the *trp* operon, and translation of other coding regions relevant to tryptophan metabolism. When there is a deficiency of charged $tRNA^{Trp}$, *B. subtilis* forms an anti-TRAP protein, AT. AT antagonizes TRAP function, thereby increasing expression of all the genes regulated by TRAP. Thus *B. subtilis* and *Escherichia coli* respond to identical regulatory signals, tryptophan and uncharged $tRNA^{Trp}$, yet they employ different mechanisms in regulating *trp* gene expression.

Contents

INTRODUCTION

Bacillus subtilis, like many micoorganisms, can form all of the enzymes required for biosynthesis of the amino acid tryptophan. Organisms with this capacity employ essentially the same sequence of reactions, catalyzed by the same seven enzyme domains. It is likely, therefore, that the genes specifying the *trp* biosynthetic enzymes evolved just once. Despite this genetic and functional conservation, the regulatory mechanisms controlling expression of the genes of tryptophan synthesis vary greatly. This variation reflects genomic differences in operon organization, as well as the utilization of intermediates or the product of the pathway for different purposes. In addition, the initial precursor of the tryptophan biosynthetic pathway, chorismic acid, serves as precursor of the other two aromatic amino acids, phenylalanine and tyrosine, as well as several less abundant essential aromatic compounds, e.g., folic acid. Therefore, tryptophan biosynthesis necessarily must be related to synthesis of these other metabolites. Furthermore, three other compounds, glutamine, phosphoribosylpyrophosphate (PRPP), and serine, provide an amino group or carbon atoms during tryptophan synthesis.

The primary role of tryptophan within living organisms is as a novel residue within many proteins. However, tryptophan serves other roles as well; for example, in eukaryotes it is often the principal precursor of niacin and other metabolites produced along the niacin pathway. In addition, in many bacterial species tryptophan is degraded to indole, and the indole serves other purposes. Several organisms have enzymes that degrade tryptophan to anthranilate, and can reutilize the anthranilate for tryptophan synthesis. These additional reactions and functions undoubtedly contributed to each organism's choices during the course of its evolution, and led to the development of appropriate regulatory strategies for tryptophan biosynthesis. Ancestry undoubtedly also influenced the regulatory mechanisms that were adopted.

For tryptophan to be incorporated into protein, it must be charged onto its cognate tRNA by tryptophanyl-tRNA synthetase. Since the cellular concentrations of tryptophan, $tRNA^{Trp}$, and tryptophanyl-tRNA synthetase, each must vary, it is logical that organisms, such as *B. subtilis*, sense the availability of charged $tRNA^{Trp}$ and tryptophan in regulating tryptophan biosynthesis. As we describe in this article, *B. subtilis* exploits features of the leader segment of the *trp* operon transcript—the segment immediately preceding the *trpE* coding region—in sensing both tryptophan and $tRNA^{Trp}$ as regulatory signals. Tryptophan also serves as a feedback inhibitor of anthranilate synthase, the enzyme catalyzing the first reaction in the tryptophan biosynthetic pathway. As we describe, these as well as several additional events influence *trp* operon expression in *B. subtilis*.

In the following sections of this article we initially describe our present understanding of the organization and functions of the *trp* genes of *B. subtilis*. We then focus on the regulatory mechanisms and cell components that this organism uses in controlling tryptophan formation.

FEATURES OF THE *trp* OPERON AND ITS REGULATION

Tryptophan Pathway

The sequence of seven enzymatic reactions leading to the formation of tryptophan from chorismate is presented in **Figure 1*A***. Six of the seven genes encoding the proteins that provide these functions are organized in the *trpEDCFBA* (*trp*) operon, which is a suboperon within the aromatic supraoperon (**Figure 1*B***). This 12-gene supraoperon also contains genes involved in general aromatic amino acid and histidine biosynthesis including *aroF-aroB-aroH* and *hisH-tyrR-aroE* (22, 25). The seventh *trp* gene, *trpG* (*pabA*), specifying a glutamine amidotransferase, is located within the folic acid biosynthetic operon (49). This amidotransferase participates in the biosynthesis of both tryptophan and folic acid. This arrangement of the *trp* genes differs from that in *Escherichia coli* and many other bacterial species (67), but this order is conserved in several other bacilli including *B. stearothermophilus* (13), *B. pumilus* (31), *B. halodurans* (54), and *B. licheniformis* (43).

The anthranilate synthase enzyme complex that catalyzes the initial reaction in the tryptophan pathway (**Figure 1*A***) is a complex consisting of two polypeptides, TrpE and TrpG. The second, third, and fourth reactions of the pathway are catalyzed by the individual polypeptide products of the *trpD*, *trpF*, and *trpC* genes, respectively (**Figure 1*A***). The last two-step reaction in the pathway is catalyzed by tryptophan synthase, a complex composed of the TrpB and TrpA polypeptides (**Figure 1*A***).

trp operon: the operon containing six of the seven genes required for tryptophan synthesis

Transcription attenuation: a common mechanism used by bacteria exploiting features of an RNA transcript to regulate transcription of the gene or genes downstream in the same operon

TRAP: the tryptophan activatable RNA-binding regulatory protein

Regulation of the *trp* Operon by Transcription Attenuation

Two promoters are used to transcribe the *trp* operon segment of the *aro* supraoperon (**Figure 1*B***). One promoter is located at the beginning of the supraoperon, preceding *aroF*, and the second, *trp* promoter, is located approximately 200 bp upstream of *trpE*. Transcription initiation at the upstream *aroF* promoter is regulated in response to aromatic amino acids by an as yet uncharacterized mechanism (R. Khurana & C. Yanofsky, unpublished). Transcription initiation at the *trp* promoter is not known to be regulated in response to changes in the cellular concentration of tryptophan (30). Rather, transcription of the structural gene region of the *trp* operon, whether initiated at either promoter, is regulated by a transcription attenuation mechanism based on formation of either of two alternative RNA secondary structures in the transcript segment immediately preceding *trpE* (**Figure 2*A***). These structures include an intrinsic transcription terminator and an upstream antiterminator structure. Formation of these structures is mutually exclusive because they overlap by four nucleotides (5, 30, 48). An RNA-binding regulatory protein called TRAP (*trp* RNA-binding Attenuation Protein) controls which structure forms in response to changes in the intracellular tryptophan concentration. In cells containing excess

a Tryptophan biosynthetic pathway

chorismate —(Gln; TrpE + TrpG)→ anthranilate —(PRPP; TrpD)→ phosphoribosyl-anthranilate —(TrpF)→ carboxyphenylamino-deoxyribulose-5-phosphate —(TrpC)→ indole-3-glycerol phosphate ↔ indole —(L-Ser; TrpA + TrpB)→ L-tryptophan

b *trp* operon - *aro* supraoperon

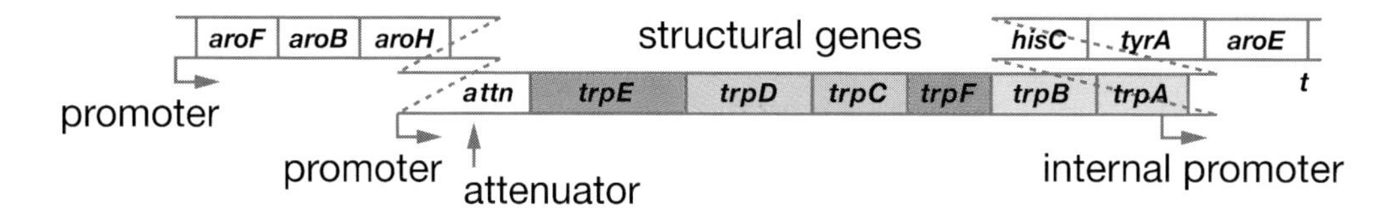

folate operon

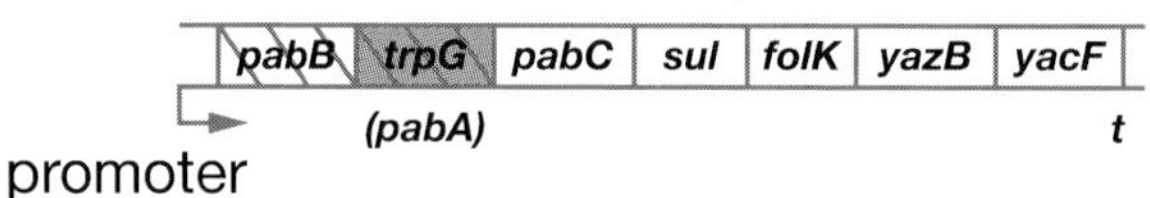

Figure 1

(*A*) Tryptophan biosynthetic pathway showing the polypeptides involved in catalysis of each reaction. (*B*) Organization of the tryptophan biosynthetic genes within the aromatic and folate operons. Positions of the promoters, intrinsic transcription terminators (*t*) and the attenuator are shown. The genes in (*B*) are color coded to match the polypeptides in (*A*). TrpG forms complexes with both TrpE (Trp biosynthesis) and PabB (folate biosynthesis, indicated by hatched bars through *pabB* and *trp*G). Tryptophan synthase is composed of TrpA and TrpB.

Terminator hairpin structure: a base-paired segment of a leader RNA followed by a sequence of U residues that is recognized by a transcribing RNA polymerase molecule as an intrinsic terminator, a transcription stop signal

tryptophan, TRAP is activated, and it binds to a segment of the nascent *trp* leader transcript that consists of 11 GAG and UAG repeats (4). This binding site overlaps the 5′ segment of the antiterminator structure. TRAP binding therefore inhibits formation of the antiterminator, hence promoting formation of the terminator hairpin, which terminates transcription in the *trp* leader region, prior to the *trpE* structural gene. When tryptophan is growth-limiting, TRAP is not activated and does not bind to *trp* mRNA. Under these conditions, the antiterminator structure forms, preventing formation of the terminator, and allows transcription to continue into the *trp* operon structural gene region.

In addition to the antiterminator and terminator structures, another RNA hairpin can form, at the 5′ end of the *trp* leader transcript (5′ stem-loop) (**Figure 2*A***) (17, 52). This structure also participates in the transcription attenuation mechanism. Disruption of this structure increases *trp* operon expression in vivo and transcriptional readthrough in vitro. The mechanism by which this stem-loop structure functions is not clear; however, existing evidence suggests that TRAP interaction with this 5′ stem-loop reduces the number of (G/U)AG repeats required for stable association of TRAP with leader mRNA (17, 52). The 5′ stem-loop structure is conserved in *B. pumilus*, *B. stearothermophilus*, *B. caldotenax*, *B. licheniformis*, and *Geobacillus kaustophilus*. Thus, the features of *trp* operon attenuation in *B. subtilis* are conserved in many *Bacillus* species.

The 75 amino acid TRAP polypeptide is encoded by *mtrB*, which is the second gene

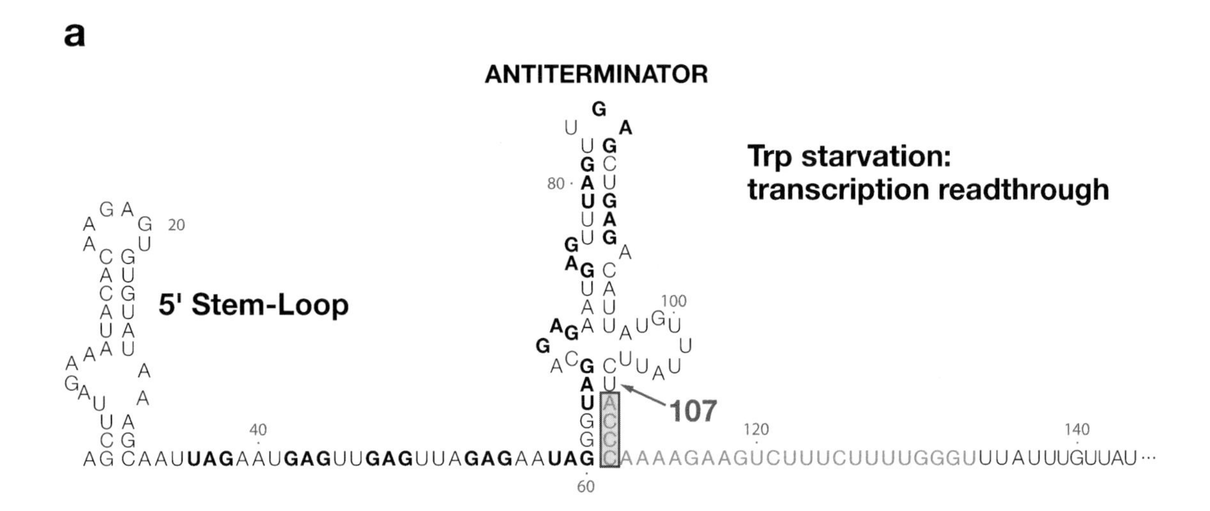

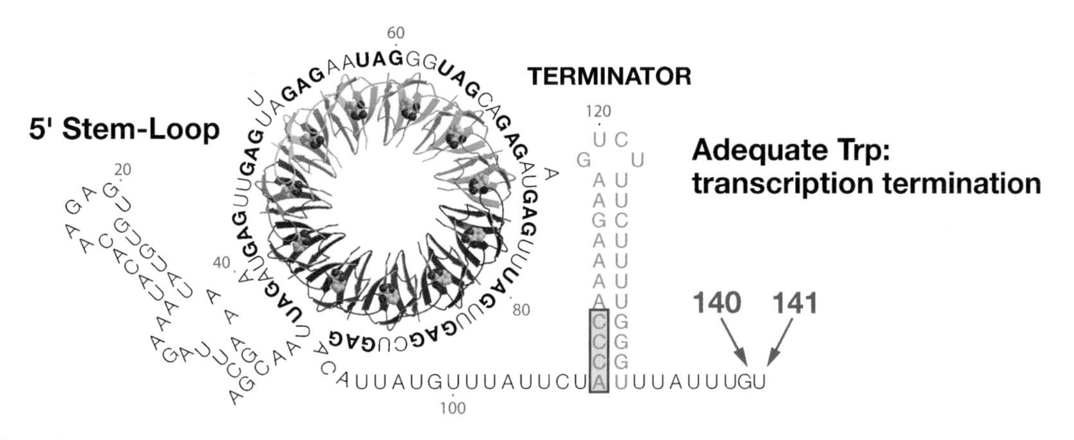

Figure 2

Models of the transcription attenuation and *trpE* translation control mechanisms. (*A*) During transcription, RNA polymerase pauses following addition of U107. Under tryptophan-limiting conditions, formation of the antiterminator promotes transcription readthrough into the *trp* operon structural genes. In the presence of tryptophan, TRAP binds to the (G/U)AG repeats, thereby releasing the paused RNA polymerase and preventing antiterminator formation. As transcription proceeds, formation of the overlapping terminator causes transcription termination following synthesis of G140 or U141. (*B*) During transcription, RNA polymerase pauses upon synthesis of U144. Under tryptophan-limiting conditions, the RNA adopts a structure such that the *trpE* SD sequence is single stranded and available for ribosome binding. In the presence of excess tryptophan, TRAP binding prevents formation of the large secondary structure. As transcription proceeds, the *trpE* SD becomes sequestered in a hairpin, which inhibits ribosome binding and translation initiation.

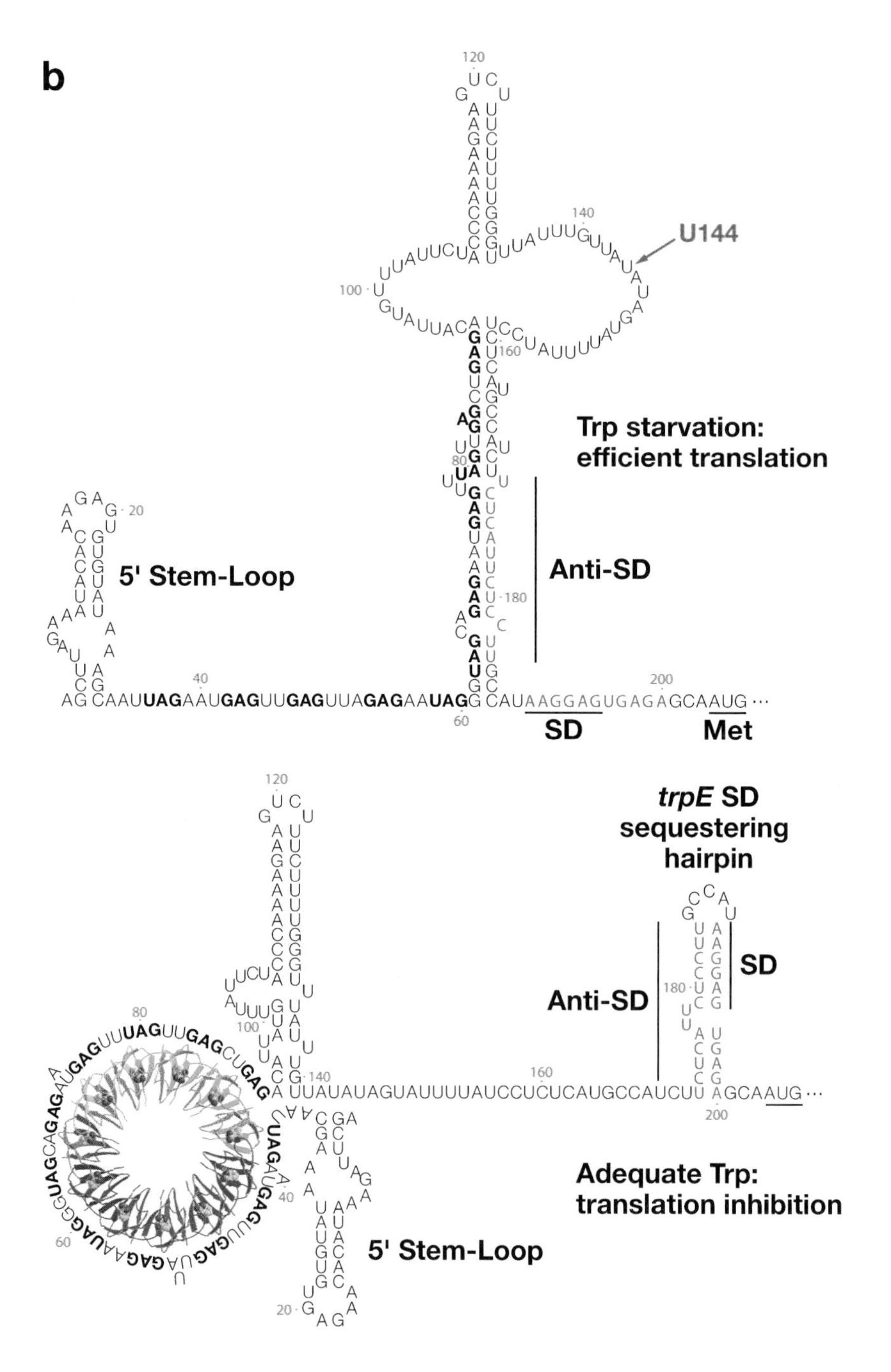

Figure 2
(*Continued*)

in a two-gene operon, *mtrAB*. This operon is located ≈8 kbp upstream of the *trp* operon (23). *mtrA* encodes GTP cyclohydrolase I, an enzyme involved in folic acid biosynthesis (3). The *mtrAB* operon is not regulated in response to tryptophan nor is TRAP involved in controlling its own expression; cellular TRAP levels are relatively constant, and independent of tryptophan starvation or the growth phase (33, 38, 65). Since *mtrA* encodes GTP cyclohydrolase I, an enzyme involved in folic acid biosynthesis, it is conceivable that folic acid or an intermediate in its biosynthesis is involved in regulating expression of the *mtr* operon.

The amino acid sequence of TRAP does not have significant homology with other characterized RNA-binding protein motifs. To date, the only other protein identified with significant sequence homology to TRAP is SplA, a regulator of the spore photoproduct lyase gene in *B. subtilis* (20). Addition of purified TRAP (plus tryptophan) to an in vitro transcription system using *B. subtilis* RNA polymerase (RNAP) enhanced termination in the leader region of the *trp* operon (5). Similar results were obtained using T7, SP6, or *E. coli* RNAP (42), consistent with a model in which TRAP functions by interacting with RNA rather than with RNAP. Mobility shift assays have been used to demonstrate directly that TRAP specifically binds *trp* leader RNA and that its binding is tryptophan dependent (42).

RNA Polymerase Pausing

Synchronization of TRAP binding, and/or RNA folding, with the position of the RNA polymerase molecule transcribing the *trp* leader region, is essential for the attenuation process. Pausing of RNA polymerase in the leader region is crucial, as it is in other attenuation mechanisms (32). Two NusA-stimulated pause sites were identified in the *B. subtilis trp* leader region, at positions U107 and U144 (59). U107 immediately precedes the critical 4-nt overlap between the antiterminator and terminator structures (**Figure 2*A***). Hence RNA polymerase pausing at this site could provide a delay that would allow TRAP to bind to the nascent *trp* transcript before formation of the antiterminator structure is completed. Pausing at U144 likely plays a role in translational control of *trpE* expression (see below).

TRAP-RNA Recognition

Both in vitro (4–7, 10) and in vivo studies (2) have demonstrated that TRAP recognizes RNAs containing multiple NAG trinucleotide repeats, with principal binding to AG and a preference for N of $G \approx U > A > C$. These triplet repeats are separated from each other in a transcript by nonconserved "spacer" nucleotides. Two nucleotide spacers are optimal (7, 10), and pyrimidines are favored over purines (10, 58). TRAP binds preferentially to single-stranded RNA; the presence of secondary structure in the RNA inhibits its binding (7, 58). The TRAP binding site in the *B. subtilis trp* leader RNA consists of 11 triplet repeats (7 GAGs and 4 UAGs), each separated by 2 or 3 nucleotide spacers (**Figure 2*A***). Five binding triplets are located in the single-stranded region preceding the antiterminator structure (**Figure 2*A***). The *trp* leader RNAs of several other bacilli contain similar TRAP binding sites composed of multiple (G/U)AG repeats (13, 26, 54). TRAP binding sites involved in translational control of *B. subtilis* genes (see below), *trpG* (16, 64), *trpP (yhaG)* (44, 45, 61), and *ycbK* (45) each contain nine triplet repeats, including several AAG sequences. In some of these binding sites there are spacers containing up to eight nucleotides separating adjacent triplets.

Antiterminator hairpin structure: a base-paired segment of leader RNA that includes several nucleotides that must be unpaired for the terminator to form. Formation of the antiterminator prevents formation of the terminator

Triplet repeats: an RNA segment containing a series of nucleotide triplets with the following sequence: NAG. The RNA segment with these triplet repeats is recognized by the TRAP protein as a binding site. TRAP binds and wraps the RNA segment with these repeats around its periphery, preventing formation of the antiterminator

TRAP-Dependent Regulation of *trpE* and *trpD* Translation

In addition to regulating transcription of the *trp* operon, TRAP regulates translation of *trpE* approximately 15-fold. When TRAP binds to a *trp* operon transcript, but this transcript escapes termination, the *trpE* Shine-Dalgarno (SD) sequence becomes sequestered in an RNA secondary structure (*trpE* SD sequestering hairpin, **Figure 2*B***), resulting in inhibition of ribosome binding (15, 30, 40). In the absence of bound TRAP, the *trp* leader transcript forms a large secondary structure in which the *trpE* SD sequence is single-stranded and available for ribosome binding (**Figure 2*B***) (15, 40, 46). Formation of the *trpE* SD sequestering hairpin also regulates expression of *trpD*, the second gene in the operon, about sevenfold. The coding sequences for *trpE* and *trpD* overlap by 29 nucleotides, resulting in translational coupling of these adjacent coding regions (62). In addition, the inhibition of *trpE* and *trpD* translation caused by the *trpE* SD sequestering hairpin allows Rho access to the nascent transcript, causing transcriptional polarity (62).

NusA stimulates RNA polymerase pausing in the *trp* leader region, following synthesis of U144 (59). Since this pause position is just downstream from sites of transcription termination (at G140 and U141), pausing at U144 does not participate in the transcription attenuation mechanism. Instead, it is likely that pausing at this position plays a role in the *trpE* translation control mechanism. Perhaps pausing at U144 provides additional time for tryptophan-activated TRAP to bind to the nascent transcript and promote formation of the *trpE* SD sequestering hairpin. Thus pausing by RNA polymerase, followed by TRAP binding, may increase the likelihood that translation is inhibited when a sufficient concentration of tryptophan is present in the cell.

In addition to the RNA secondary structures described thus far, a Mg^{2+}-dependent RNA tertiary structure can form in the *trp* leader transcript. Existing evidence suggests that this structure participates in the *trpE* translation control mechanism by interfering with TRAP-mediated formation of the *trpE* SD sequestering hairpin (46). Under limiting tryptophan conditions, the majority of TRAP molecules would not be activated, resulting in increased transcription readthrough. The leader RNA would then be capable of forming this tertiary structure, which would inhibit subsequent TRAP binding.

TRAP Structure

The structures of TRAP from *B. subtilis* (2) and *B. stearothermophilus* (13) in complex with tryptophan have been determined by X-ray crystallography. The amino acid sequences of the two proteins are 77% identical and their structures are also very similar, consisting of 11 identical subunits arranged in a ring (**Figure 3**). Neither polypeptide contains tryptophan. The TRAP oligomer is composed of 11 seven-stranded antiparallel β-sheets, each containing 4 β-strands from one subunit and 3 β-strands from the adjacent subunit. This structural arrangement generates extensive interfaces between adjacent subunits with multiple main chain–main chain hydrogen bonding interactions that stabilize the oligomer structure; TRAP has been shown to be very thermostable (2, 10, 13). TRAP is activated for RNA recognition by the binding of up to 11 molecules of L-tryptophan (2, 6). The crystal structure reveals that each tryptophan molecule is bound between adjacent subunits with its indole ring buried in a hydrophobic pocket between adjacent β-sheets. Nine hydrogen bonds are formed between the amino group, the carboxyl group, and the indole nitrogen of each bound tryptophan, with amino acids residing in two loops on adjacent subunits. The hydrogen bonds formed between Thr30, Gln47, and Thr49 and tryptophan are essential for TRAP function (60). Conversely, the hydrogen bond between Ser53 and the carboxyl group of tryptophan is dispensable for TRAP activity (60).

The precise mechanism by which tryptophan binding activates TRAP for RNA binding is not known because the detailed structure of tryptophan-free TRAP has not been determined. However, tryptophan binding is not required for oligomerization of the protein; several studies have established that TRAP remains an 11-mer in the absence of bound tryptophan (35, 50). The observation that bound tryptophan residues are nearly completely buried within the protein suggests that the binding pockets must exist in a different, more open conformation, prior to tryptophan binding (1, 2, 13). Moreover, the RNA-binding site is distinct from—and distant to—the tryptophan binding sites on the protein (1). Together, these observations suggest that tryptophan binding induces a conformational change in TRAP that affects its RNA-binding sites; a recent analysis supports this hypothesis (34). NMR studies to

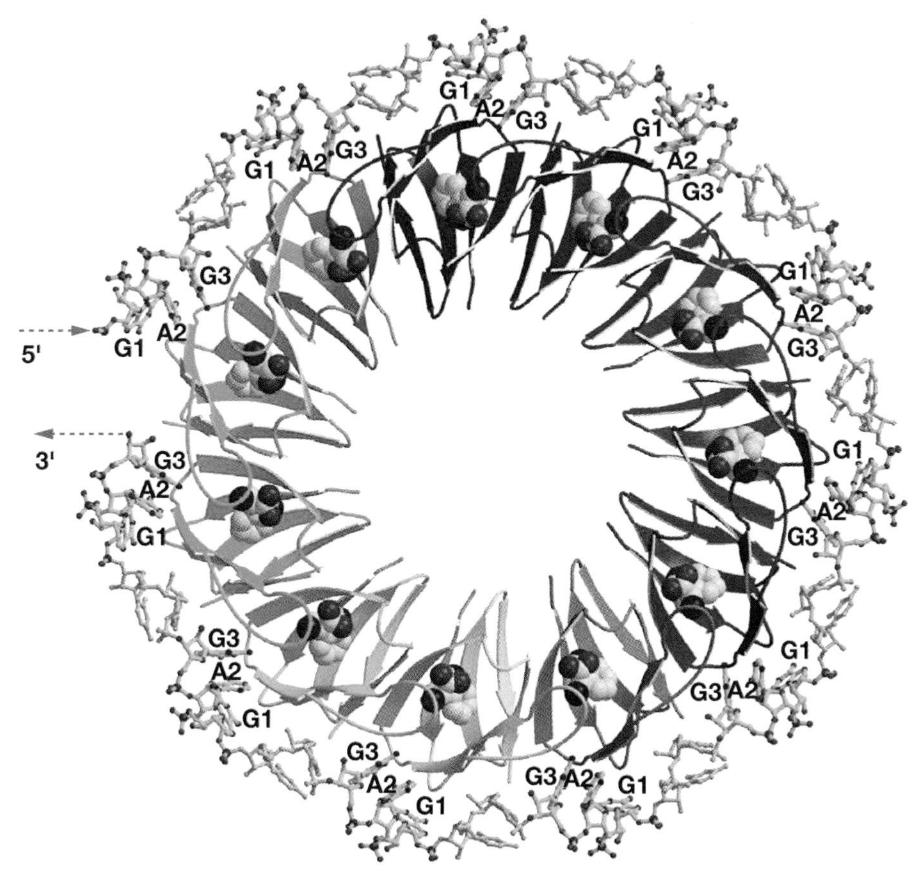

Figure 3

Structure of the TRAP 11-mer bound to RNA. Each protein subunit is shown in a different color and the bound tryptophan molecules are shown in van der Waals spheres. The nucleotides of the GAG repeats in the RNA are shown in light green ball-and-stick models and are labeled for each repeat. The spacer nucleotides, which can be of variable sequence and conformation, are shown in gray ball-and-stick models. In the crystal structures the beginning and ends of the bound RNA are not discernible, and hence we have arbitrarily chosen one site to designate the 5′ and 3′ ends of the RNA.

investigate the dynamic properties of TRAP in both the nonliganded (apo state) and activated states showed that apo-TRAP is more flexible than the tryptophan-bound protein (39), implying that tryptophan-modulated protein flexibility (dynamics) plays a crucial role in TRAP function by altering TRAP's RNA-binding affinity.

A recent study using TRAP hetero-11-mers, composed of both wild-type subunits and mutant subunits defective in binding tryptophan, indicated that each bound tryptophan activates only the RNA-binding site associated with the two subunits that contain this tryptophan binding site (34). It was also observed that binding of only a few tryptophan molecules is sufficient to activate significantly TRAP's RNA-binding activity.

Structure of the TRAP-RNA Complex

Mutagenesis studies identified three amino acid residues of each subunit (Lys37, Lys56 and Arg58), which, when replaced by alanine, specifically interfered with RNA binding

(63). These three residues are aligned on the perimeter of the TRAP ring, suggesting that the bound RNA wraps around the outside of the protein ring with the (G/U)AG repeats interacting with the 11 Lys Lys Arg patches on the protein. Nucleoside analog studies identified several important functional groups on the second A and third G of each triplet repeat that are necessary for interacting with TRAP (19). These studies also suggested that the bases in the first position (G or U) as well as the spacer bases are not crucial for TRAP recognition and binding.

Several crystal structures have been solved for *B. stearothermophilus* TRAP complexed with RNAs consisting of 11 GAG or UAG repeats separated by various spacers (**Figure 3**) (1, 27, 28). As predicted, the RNA wraps around the TRAP protein ring with the bases pointing toward the protein and the phosphodiester backbone exposed to the solvent. As a result, most of the direct hydrogen-bonding interactions are with the RNA bases and there are no contacts to phosphates. The only direct hydrogen bond to the RNA backbone is to the 2′ OH of the ribose of the third G in each repeat. This contact, which was predicted to be critical for binding by deoxyribonucleoside substitution studies (19), explains how TRAP distinguishes RNA from DNA. As predicted, Lys37, Lys56, and Arg58, each specifically interacts with the RNA. Lys37 hydrogen bonds to the second A of each triplet repeat, while both Lys56 and Arg58 hydrogen bond with the third G of each repeat. In addition, Glu36 forms two hydrogen bonds with the base of the third G of each repeat, which also forms stacking interactions with the base of the adjacent upstream A residue and with the side chain of Phe32. The spacer bases do not directly contact the protein except for a few nonspecific van der Waals interactions. Structural data indicate that three-nucleotide spacers are more flexible than two-nucleotide spacers. Depending on the sequence, the second or both nucleotides of a two-nucleotide spacer region have a flexible conformation. This is consistent with the lack of sequence conservation in spacers of natural TRAP binding sites (13) and with findings indicating that their composition is generally not critical for TRAP binding (7, 9, 58). Observed destabilization of an ordered structure within the spacer region (27, 28), including both unstacking of bases and induction of conformational flexibility, could contribute to the increase in entropy upon RNA binding (9) and promote degradation of RNA and recycling of TRAP (see section below). The simple mechanism of wrapping the RNA around the circular TRAP complex explains both the specificity of the protein for its RNA targets as well as how TRAP binding alters RNA structure, functioning in regulating both transcription and translation.

Mechanism of TRAP Binding to RNA

Results from several studies suggest that the mechanism of TRAP binding to RNA involves an initial interaction between the protein and a small subset of triplet repeats, followed by wrapping of the remainder of the RNA-binding site around the protein (18, 34, 35). Furthermore, footprinting studies indicate that TRAP binds to its RNA target by a 5′ to 3′ mechanism in which the initial complex is formed with the triplet repeats in the 5′ segment of the binding site (8). This binding mechanism may be important for TRAP's function in attenuation because TRAP must bind and prevent formation of the antiterminator RNA structure before RNA polymerase proceeds beyond the terminator region. Since only 20 nt separate the TRAP binding site from the 3′ end of the antiterminator, TRAP has only a short time in which to control the regulatory decision; this may also be affected by pausing of RNA polymerase (see above). The first five triplet repeats of the TRAP binding site in the *trp* leader region are unpaired and do not participate in the formation of any of the secondary structures involved in attenuation (**Figure 2*A***), nor is their presence required for high-affinity binding to TRAP (19). Nevertheless, altering as few as two of these repeats nearly eliminated attenuation control of the *trp* operon in vivo (M. Milescu & P. Gollnick, submitted). One explanation for this observation is that these repeats are essential to allow proper timing of TRAP binding to the nascent *trp* leader RNA, relative to the position of the transcribing RNA polymerase. In vitro transcription studies are consistent with this proposal (M. Milescu & P. Gollnick, submitted).

Recycling of RNA-Bound TRAP

Proper regulation of *trp* operon expression by transcription attenuation depends upon polynucleotide phosphorylase (PNPase) degradation of terminated *trp* leader RNA transcripts, freeing TRAP for binding to other transcripts (14). In a *B. subtilis* strain lacking PNPase, for example, the *trp* operon is overexpressed even in the presence of excess tryptophan. An RNA fragment from the *trp* leader region accumulates, and has a long half-life, in a *pnp* mutant but not in a wild-type strain. In vitro analysis demonstrated that PNPase is required to degrade *trp* leader RNA bound to TRAP. Hence loss of *trp* operon attenuation control in the PNPase-deficient strain is due to the inability of ribonucleases other than PNPase to efficiently degrade TRAP-bound RNA, resulting in sequestration of the limiting amount of TRAP in the cell. The absence of ordered structure within several three-nucleotide spacer regions of *trp* leader RNA (27) suggests that these regions could facilitate RNA degradation by presenting these unstructured spacer regions to PNPase. This hypothesis is supported by footprinting studies showing that the 5′- and 3′-most triplets were least protected by bound TRAP (4).

Four operons of *B. subtilis* yield transcripts with TRAP binding sites, and there usually are only 200–400 molecules of TRAP per cell (38). Thus, in *B. subtilis*, specific ribonuclease degradation of the RNA in TRAP-RNA complexes is required for recycling of bound TRAP. Such a mechanism may be relevant to other systems in which the concentration of an RNA-binding protein is limiting, making it necessary that this protein be freed, to keep pace with ongoing transcription (14).

TRAP REGULATION OF *trpG* (*pabA*), *trpP*, AND *ycbK* TRANSLATION

trpG (*pabA*) is the second gene in the folate operon (**Figure 1*B***) (49). The TrpG polypeptide functions as a glutamine amidotransferase in the biosynthesis of both tryptophan, as TrpG-TrpE (**Figure 1*A***), and folic acid, as TrpG-PabB (29). Translation of *trpG* is regulated by TRAP in response to tryptophan, whereas translation of *pabB* is not (64). TRAP binds to nine triplet repeats (seven GAG, one UAG, and one AAG) that surround and overlap the *trpG* SD sequence and inhibits TrpG synthesis about 15-fold (16, 61). Because the TRAP and ribosome binding sites overlap, these findings establish that TRAP inhibits translation of *trpG* by directly competing with ribosome binding. This translation control mechanism is in stark contrast to the mechanisms responsible for regulating translation of *trpE* and *trpD*.

TRAP also regulates translation of *trpP*, which encodes a putative tryptophan transport protein (44). In this case, TRAP interaction with nine triplet repeats (five GAG, three UAG, and one AAG) that overlap the *trpP* SD sequence and translation initiation region inhibits TrpP synthesis 900-fold. As is the case for *trpG* (*pabA*), TRAP inhibits translation of *trpP* by directly competing with ribosome binding (61).

Another putative TRAP binding site consisting of 11 triplet repeats was identified that overlaps the SD sequence and translation initiation region of *ycbK*, a gene of unknown function in the *at* operon (45). Surprisingly, TRAP only protected the central 7 triplets (5 GAG, 1 AAG, and 1 CAG) from ribonuclease cleavage (H. Yakhnin & P. Babitzke, unpublished results). Thus, while it appears that TRAP regulates YcbK synthesis by a translation control mechanism similar to that of *trpG* and *trpP*, the footprinting results suggest that TRAP does not physically interact with the *ycbK* SD sequence. Instead, TRAP apparently binds to a segment of RNA that lies entirely within the *ycbK* coding sequence and inhibits its translation.

TRAP-dependent regulation is far more extensive for *trpP* (900-fold) than for *trpG* (15-fold) despite TRAP having a five- to tenfold higher affinity for the *trpG* transcript (61) (H. Yakhnin & P. Babitzke, unpublished results). What could account for this apparent discrepancy? The last two triplet repeats in the *trpP* transcript are within the coding region, whereas all nine repeats in *trpG* are upstream of the start codon. Thus, the relative positions of the triplet repeats in these two transcripts with respect to their cognate SD sequences and translation initiation regions may be responsible for TRAP having tighter control over *trpP* expression. Perhaps extending the TRAP binding site into the *trpP* coding region results in more effective inhibition of ribosome binding. The relative gene arrangement of the *trpG* and *trpP* operons may also contribute to the difference in TRAP-dependent control. While *trpP* is contained within a single-gene operon, *trpG* is the second gene in the folate operon (**Figure 1*B***). Interestingly, the *pabB* translation stop codon lies within the *trpG* SD sequence. Since 7 of 9 triplet repeats in the *trpG* TRAP binding site lie upstream of the *pabB* stop codon, translation of *pabB* may lead to displacement of bound TRAP, thereby allowing translation initiation of *trpG*. Since *trpP* is a single-gene operon, ribosome-mediated displacement of bound TRAP would not be a factor (61).

REGULATORY MECHANISMS ALTERING ENZYME ACTION

The TrpE subunit of anthranilate synthase contains a tryptophan binding site (41), although the affinity for this interaction is not known. Tryptophan binding to the TrpE-TrpG complex provides feedback inhibition of anthranilate synthase activity, thereby regulating entry of chorismate into the tryptophan biosynthetic pathway (25, 41). As a consequence, chorismate can be more efficiently utilized in the biosynthesis of phenylalanine, tyrosine, and folate.

REGULATORY CROSS-TALK

The TrpG polypeptide functions as the common glutamine amidotransferase subunit of anthranilate synthase (TrpE-TrpG) and para-aminobenzoate synthase (PabB-TrpG) in tryptophan and folate biosynthesis, respectively (29). While translation of *trpG* is inhibited by tryptophan-activated TRAP, the translation initiation region for *trpG* is designed to allow only 15-fold regulation in response to tryptophan. Apparently, this modest level of regulation is necessary to provide sufficient TrpG synthesis in the presence of excess tryptophan to maintain folate biosynthesis (16, 61). The organization of the *mtrAB* operon provides another interesting link between folate and tryptophan biosynthesis. *mtrA* encodes GTP cyclohydrolase I, an enzyme involved in folate biosynthesis, while *mtrB* encodes TRAP, the regulator of the *trp* operon and of *trpG*, *trpP*, and *ycbK* (3). It is tempting to speculate that the relationship between folate and tryptophan biosynthesis stems from the fact that chorismate is a common precursor for both compounds.

In addition to chorismate, PRPP, glutamine, and serine serve as substrates in the tryptophan biosynthetic pathway. PRPP is used in the biosynthesis of both tryptophan and histidine, whereas glutamine and serine are both required for general protein synthesis and numerous other reactions in intermediary metabolism. Thus, the availability of each of these compounds is likely to influence tryptophan synthesis (25). Histidine availability affects expression of genes of the aromatic amino acid biosynthetic pathway. Histidine can repress the synthesis of both

DAHP synthase and prephenate dehydrogenase; however, the biochemical basis for this cross-pathway regulation is not known (25). In addition, unpublished studies suggest that tyrosine may be the principal compound that influences transcription initiation at the *aroF* promoter (R. Khurana & C. Yanofsky, unpublished). However, regulation of transcription at the *aroF* promoter, the relative contribution of *aroF*-initiated transcripts to *trp* operon transcription, and *aroF* expression under different physiological conditions have not yet been thoroughly analyzed.

REGULATORY SENSING OF UNCHARGED tRNA^{Trp}

Although it was evident in the early 1970s that uncharged tRNA^{Trp} probably plays a role in regulation of *trp* operon expression in *B. subtilis* (51), the existence of TRAP, and its mechanism of action, were as yet unknown. Therefore it was not possible to ask whether tRNA^{Trp} was operating through the TRAP system or by an independent mechanism. Studies being conducted at that time with the *trp* operon of *E. coli* also suggested that uncharged tRNA^{Trp} as well as tryptophan was recognized as a signal in regulation of *trp* operon expression. Subsequent investigations with *E. coli* defined the role of tRNA^{Trp} and alternative antiterminator/terminator leader RNA structures in the transcription attenuation process used by this organism (32, 66). Transcription initiation in the *trp* operon of *E. coli* is also regulated by sensing the cellular tryptophan concentration; this is achieved by tryptophan activation of the *trp* aporepressor (68).

Uncharged tRNA^{Trp} Accumulation Activates *at* Operon Expression

The role of tryptophan-activated TRAP in tryptophan-mediated regulation of transcription termination in the *trp* operon of *B. subtilis* was established in the early 1990s (21, 22). It was also shown that this bacterium uses a common "T Box" transcription attenuation mechanism to sense uncharged tRNA in regulating transcription termination in the leader regions of many of its tRNA synthetase operons, including the *trpS* (tryptophanyl-tRNAsynthetase) operon (24). It seemed likely, therefore, that if uncharged tRNA^{Trp} is recognized as a signal in regulation of *trp* operon transcription in *B. subtilis*, the T box mechanism would be employed. Accordingly, a genome-wide search was conducted for an operon of *B. subtilis*, other than the *trpS* operon, that was regulated by a leader region designed to sense uncharged tRNA^{Trp} by the T box mechanism. One operon was identified, *yczA-ycbK*, an operon of previously unknown function (45). Further studies established that this operon, subsequently named the *at* operon, was responsible for this organism's response to uncharged tRNA^{Trp} in regulating transcription of the *trp* operon (55, 56). The role of the first gene of this operon, *yczA*, was determined, and this gene was renamed *rtpA* (for regulator of TRAP), since its product regulated TRAP protein function. The *rtpA* polypeptide contains 53 residues, and lacks tryptophan.

T box mechanism: a mechanism of transcription attenuation in which an uncharged tRNA can pair with leader RNA, favoring the formation of an RNA antiterminator structure rather than a terminator structure

at operon: the operon specifying the anti-TRAP protein, AT. This operon is regulated both transcriptionally and translationally by uncharged tRNA^{Trp}. When uncharged tRNA^{Trp} accumulates it leads to the synthesis of the anti-TRAP protein, AT

Features of the *at* Operon

The organization and features of the *at* operon are summarized in **Figure 4** (12). This operon contains a T box leader region that specifies a transcript segment that can fold to form alternative RNA antiterminator and terminator structures (45). This leader transcript is designed to sense uncharged tRNA^{Trp}, which, when accumulated, would pair with a specific segment of the leader transcript, stabilizing formation of the antiterminator structure. The antiterminator would prevent formation of the terminator, hence transcription would continue into the structural genes of the operon. The *at* operon leader region was observed to contain an additional important regulatory feature. Immediately downstream of the T box terminator, there is a 10-codon reading frame, *rtpLP*, containing

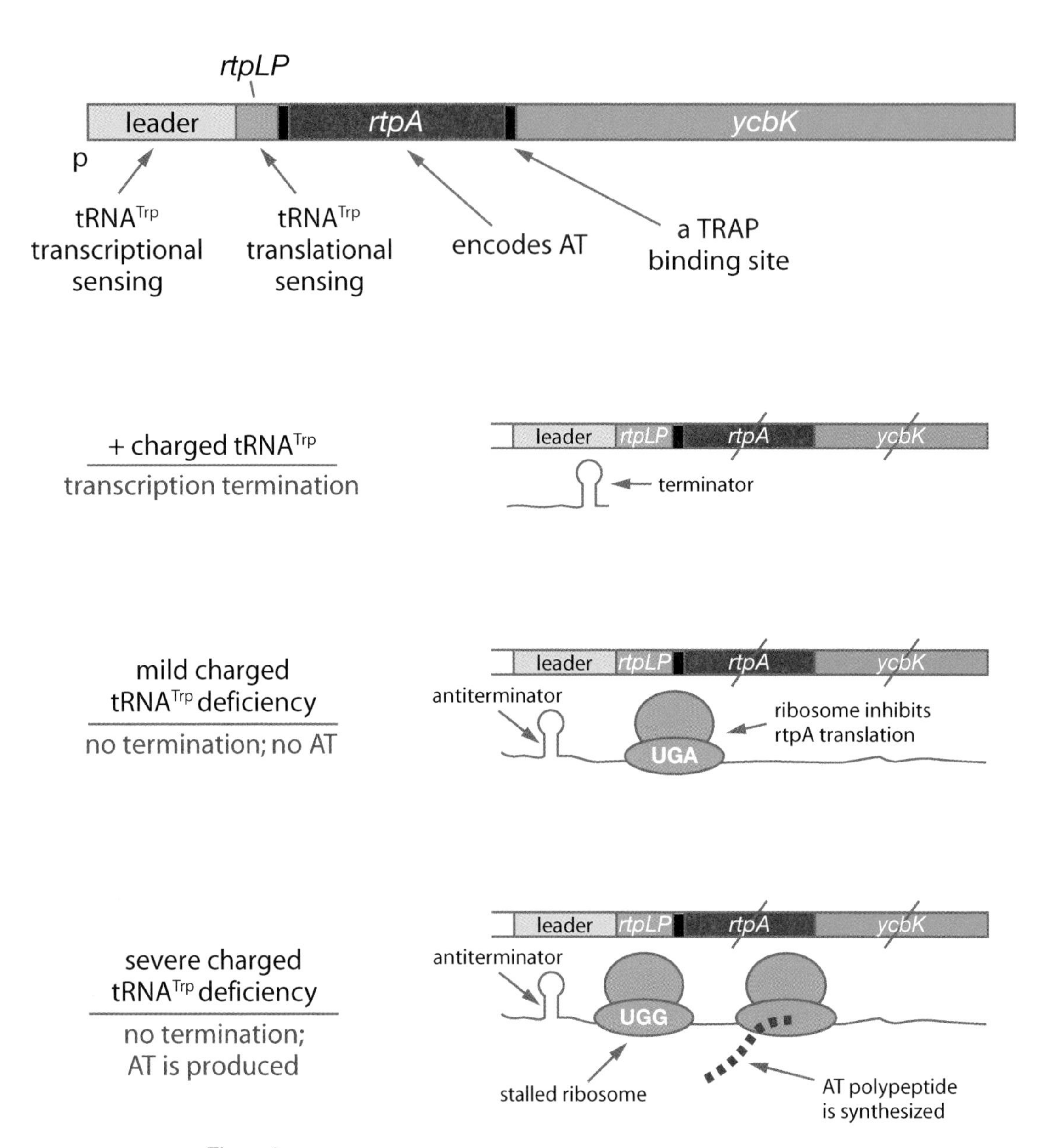

Figure 4

Organization, regulation, and functions of the *at* operon. Transcription of the *at* operon is regulated by the T box mechanism of transcription attenuation, in response to the accumulation of uncharged $tRNA^{Trp}$. When there is a mild charged-$tRNA^{Trp}$ deficiency, transcription of the *at* operon proceeds and the leader peptide coding region, *rtpLP*, is translated. The translating ribosome reaches the *rtpLP* stop codon where it inhibits translation initiation at the *rtpA* start codon. When there is a severe charged-$tRNA^{Trp}$ deficiency the ribosome translating the *rtpLP* coding region stalls at one of its three Trp codons, allowing initiation of translation of *rtpA*, and synthesis of the AT polypeptide.

three consecutive Trp codons (12). The SD sequence for the *rtpA* coding region is only 6 nucleotides downstream from the *rtpLP* stop codon. The position of the *rtpLP* leader peptide coding region—and the location of its three Trp codons—have been shown to play a role in increasing synthesis of the *rtpA* product, protein AT (12). Whenever there is insufficient charged $tRNA^{Trp}$ to allow translation of the three *rtpLP* Trp codons, the translating ribosome is believed to stall at one of these Trp codons, exposing the *rtpA* SD sequence for efficient ribosome binding and initiation of translation of *rtpA* (12). However, whenever there is sufficient charged $tRNA^{Trp}$ to allow completion of translation of *rtpLP*, the ribosome reaching the *rtpLP* stop codon presumably blocks the *rtpA* SD sequence and inhibits translation initiation at the *rtpA* start codon. In vivo studies have confirmed that a translating ribosome stalled at one of the Trp codons of *rtpLP* does activate AT synthesis, whereas a ribosome reaching the *rtpLP* stop codon inhibits *rtpA* translation (12).

AT, the *rtpA* Product, is an Anti-TRAP Protein

Studies in vitro have shown that the purified AT protein can bind to tryptophan-activated TRAP, and, when bound, can prevent TRAP from binding to its target RNAs (56). Analysis of the features of TRAP required for AT binding revealed that AT binds to TRAP only when TRAP is tryptophan activated. Consistent with this finding, replacing residues in TRAP necessary for tryptophan binding prevents AT from binding to TRAP (55). In addition, replacing TRAP residues required for target RNA recognition also prevents AT binding (55, 65). It is likely, therefore, that the surface features of TRAP generated by tryptophan binding—features that are essential for RNA binding—also are required for AT binding. AT appears to exist in equilibrium between trimers and dodecamers in solution (47, 50). It has not yet been established which is the predominant form in vivo, or in which of these conformations it binds to TRAP. Estimates of the number of TRAP molecules per cell have given values around 200 molecules/cell (38). Estimates of the number of AT trimers formed per cell when expression of the *at* operon may be maximal have given values of approximately two AT trimers per molecule of TRAP (65). Under these conditions, *trp* operon expression was 70% of the level observed in a TRAP-deficient strain (65).

AT: the anti-TRAP protein.

AT, Its Structure and Possible Mechanism of Action

AT is a small 53 amino acid polypeptide. The central portion of the AT chain contains two cysteine-/glycine-rich motifs, C-X-X-C-X-G-X-G, with sequence homology to DnaJ (56). In DnaJ two such motifs form a Zn-binding domain (37), and AT, like DnaJ, was shown to contain bound Zn (57). Substitution of any of several cysteine residues with alanine results in conversion of the AT into inactive monomers in vitro and degradation of AT in vivo (57), indicating that zinc is required for maintaining the functionally active quaternary structure of AT.

The X-ray structure of AT (M.B. Shevtso, Y. Chen, P. Gollnick & F. Antson, manuscript in preparation) shows that each Zn atom is coordinated by four cysteine side chains. Three subunits of AT form tight trimers, and four trimers further associate into a 12-subunit complex with individual trimers positioned in the corners of a tetrahedron (**Figure 5**).

In solution, AT exists in equilibrium between trimer and 12-subunit forms with equal weight fractions of AT_3 and AT_{12} observed at 20 μM [AT_3] (50). Ultracentrifugation studies performed in the range of 12–48 μM [AT_3] and various molar ratios of AT:TRAP suggest that in the AT:TRAP complex, four trimers of AT are bound to TRAP (50). Even though a significant proportion of AT exists as trimers at such concentrations, no complexes corresponding to single trimers of AT bound to TRAP were detected. This suggests

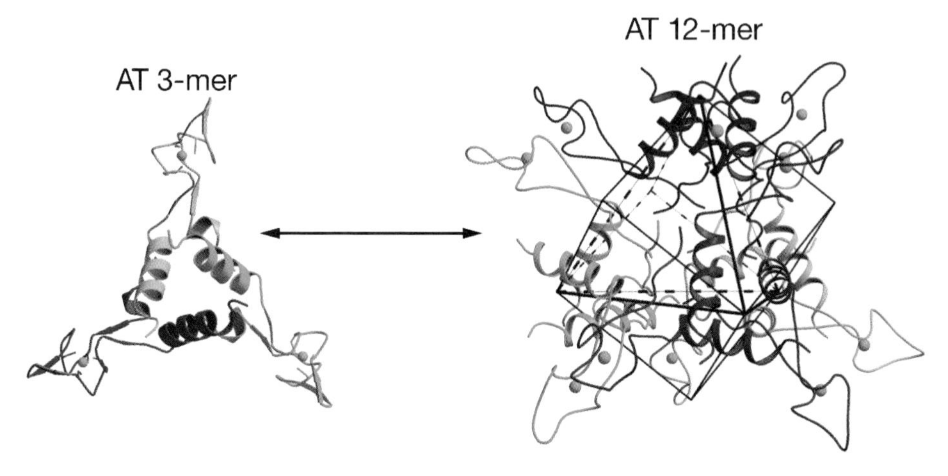

Figure 5

Structure of the AT protein. Ribbon diagrams are displayed for the AT 3-mer and 12-mer. In the trimer each polypeptide is in a separate color, whereas in the 12-mer each trimer is shown in a different color. Zinc atoms in both trimer and 12-mer are indicated as blue spheres. The four trimers of the 12-mer are arranged in the corners of a tetrahedron, which is displayed together with an associated cube.

that TRAP interacts either with a preassembled AT 12-mer or its interaction with trimers of AT is cooperative, promoting formation of the AT 12-mer assembly. The data also suggest that at physiological concentrations of AT (65), which are about two orders of magnitude lower than those required for formation of the 12-mer (50), the functional complex with TRAP may contain a 12-mer of AT.

The AT 12-mer observed in the X-ray structure may represent the functionally active state of AT, although further studies, preferably at the same concentrations that exist in vivo, are needed to demonstrate this. In this conformation the 12 Zn-binding domains are exposed at the surface of both the trimer and the 12-subunit structures (M.B. Shevtsov, Y. Chen, P. Gollnick & F. Antson, manuscript in preparation). In DnaJ, one Zn-binding domain binds to denatured protein substrates (53) and the second was shown to interact with the chaperone protein DnaK (36). It is tempting to speculate that the Zn-binding domains of AT interact with TRAP, although at this stage there is no direct experimental evidence for this assumption. It also remains to be seen how the 11-subunit TRAP protein interacts with either the 3-subunit or 12-subunit AT molecule and whether the symmetry of either protein is broken or adjusted during complex formation.

CONCLUSIONS

The organization of the *trp* operon of *B. subtilis*, and the mechanisms used in regulating this operon's expression, appear to reflect the intent of this organism to interrelate many of the enzymatic functions needed for synthesis of all the aromatic amino acids as well as related aromatic compounds (11). This objective is perhaps logical, since the common aromatic pathway provides the branch-point intermediate, chorismate, which serves as precursor of each of these aromatic compounds. In providing tryptophan for protein synthesis, the cellular components whose concentrations would be most crucial would be tryptophan itself and tryptophan-charged $tRNA^{Trp}$. As described in this article, changes in the intracellular concentration of either of these two molecules serves as a key regulatory signal. Their presence determines whether the *trp* operon will be transcribed, whether the *trp* transcript will be translated, and whether chorismate will enter the tryptophan biosynthetic pathway. Organization of the *trp* operon within the aromatic supraoperon, and the location of the regulatory elements influencing *trp* operon transcription, reflect a desire to establish a regulatory relationship between synthesis of enzymes of the common aromatic pathway and synthesis of the enzymes providing phenylalanine, tyrosine, and tryptophan (11). Ancestry, evolution, and operon design established that one of the *trp* gene polypeptides, TrpG, functions as a component of two enzyme complexes, one concerned with o-aminobenzoic acid (anthranilate) synthesis, involved in tryptophan formation, and the second participating in synthesis of p-aminobenzoic acid, a precursor of folic acid. Fine control is achieved by placing the structural gene for tryptophanyl-tRNA synthetase under uncharged $tRNA^{Trp}$ control (33).

As described in this article, tryptophan is principally sensed as a metabolic signal by the TRAP regulatory protein. Tryptophan activation of TRAP causes transcription termination in the *trp* operon leader region and inhibition of *trpE* translation. Active TRAP also inhibits translation of three additional relevant coding regions. Uncharged $tRNA^{Trp}$ is also recognized as a regulatory signal during transcription of the leader region of the *at* operon, and during translation of a peptide coding region in this operon's leader transcript. Accumulation of uncharged $tRNA^{Trp}$ promotes transcription antitermination in the *at* operon leader region, as well as increasing translation of the *rtpA* coding region. When AT is overproduced it inactivates TRAP, resulting in elevated expression of all operons subject to TRAP regulation. Uncharged $tRNA^{Trp}$ also increases transcription of the *trpS* operon, leading to the production of additional trytophanyl-tRNA synthetase, thereby increasing the likelihood that charged $tRNA^{Trp}$ will be provided for protein synthesis. An additional organizational feature relating folate synthesis to tryptophan synthesis is the location of *mtrB*, the structural gene for TRAP. *mtrB* is the second gene in a two-gene operon with *mtrA*, the structural gene for the enzyme that catalyzes the first reaction in pterin formation during folate biosynthesis. In conclusion, genetic, biochemical, and physiological studies with the *trp* operon and related operons of *B. subtilis* have elucidated the many interactions and interrelations crucial to regulation of tryptophan biosynthesis in this organism.

SUMMARY POINTS

1. The *trp* operon of *B. subtilis* resides within an aromatic supraoperon that contains other genes concerned with aromatic amino acid biosynthesis.
2. Transcription of the *trp* operon is regulated by transcription attenuation, in response to the availability of tryptophan, and charged $tRNA^{Trp}$.
3. Tryptophan activates an RNA-binding protein, TRAP, which regulates transcription termination in the leader region of the *trp* operon.
4. The unique trinucleotide repeat sequence that serves as a TRAP binding site was identified.
5. TRAP binding also regulates translation of the first coding region of the operon, *trpE*.
6. TRAP also regulates initiation of translation of *trpG*, *trpP*, and *ycbK*.
7. Transcription pausing in the *trp* operon leader region is essential to TRAP action.

8. Transcription of the *trp* operon is also regulated upon accumulation of uncharged tRNA^{Trp}.
9. Uncharged tRNA^{Trp} activates transcription and translation of the *at* operon, leading to the synthesis of the anti-TRAP protein, AT.
10. AT binds to TRAP and prevents TRAP from functioning.
11. The structures of the TRAP-RNA complex and of the AT protein reveal how they may act.

LITERATURE CITED

1. **Antson AA, Dodson EJ, Dodson GG, Greaves RB, Chen X-P, Gollnick P. 1999. Structure of the *trp* RNA-binding attenuation protein, TRAP, bound to RNA. *Nature* 401:235–42**
2. Antson AA, Otridge JB, Brzozowski AM, Dodson EJ, Dodson GG, et al. 1995. The three dimensional structure of *trp* RNA-binding attenuation protein. *Nature* 374:693–700
3. Babitzke P, Gollnick P, Yanofsky C. 1992. The *mtrAB* operon of *Bacillus subtilis* encodes GTP cyclohydrolase I (MtrA), an enzyme involved in folic acid biosynthesis, and MtrB, a regulator of tryptophan biosynthesis. *J. Bacteriol.* 174:2059–64
4. Babitzke P, Stults JT, Shire SJ, Yanofsky C. 1994. TRAP, the *trp* RNA-binding attenuation protein of *Bacillus subtilis*, is a multisubunit complex that appears to recognize G/UAG repeats in the *trpEDCFBA* and *trpG* transcripts. *J. Biol. Chem.* 269:16597–604
5. **Babitzke P, Yanofsky C. 1993. Reconstitution of *Bacillus subtilis trp* attenuation in vitro with TRAP, the *trp* RNA-binding attenuation protein. *Proc. Natl. Acad. Sci. USA* 90:133–37**
6. Babitzke P, Yanofsky C. 1995. Structural features of L-tryptophan required for activation of TRAP, the *trp* RNA-binding attenuation protein of *Bacillus subtilis*. *J. Biol. Chem.* 270:12452–56
7. Babitzke P, Yealy J, Campanelli D. 1996. Interaction of the *trp* RNA-binding attenuation protein (TRAP) of *Bacillus subtilis* with RNA: effects of the number of GAG repeats, the nucleotides separating adjacent repeats and RNA structure. *J. Bacteriol.* 178:5159–63
8. Barbolina MV, Li X, Gollnick P. 2005. *Bacillus subtilis* TRAP binds to its RNA target by a 5′ to 3′ directional mechanism. *J. Mol. Biol.* 345:667–79
9. Baumann C, Otridge J, Gollnick P. 1996. Kinetic and thermodynamic analysis of the interaction between TRAP (*trp* RNA-binding attenuation protein) and *trp* leader RNA from *Bacillus subtilis*. *J. Biol. Chem.* 271:12269–74
10. Baumann C, Xirasagar S, Gollnick P. 1997. The *trp* RNA-binding attenuation protein (TRAP) from *B. subtilis* binds to unstacked *trp* leader RNA. *J. Biol. Chem.* 272:19863–69
11. Berka RM, Cui X, Yanofsky C. 2003. Genomewide transcriptional changes associated with genetic alterations and nutritional supplementation affecting tryptophan metabolism in *Bacillus subtilis*. *Proc. Natl. Acad. Sci. USA* 100:5682–87
12. **Chen G, Yanofsky C. 2003. Tandem transcription and translation regulatory sensing of uncharged tryptophan tRNA. *Science* 301:211–13**
13. Chen X-P, Antson AA, Yang M, Baumann C, Li P, et al. 1999. Regulatory features of the *trp* operon and crystal structure of the *trp* RNA-binding attenuation protein from *Bacillus stearothermophilus*. *J. Mol. Biol.* 289:1003–16

The authors describe the high-resolution crystal structure of TRAP complexed with an RNA composed of 11 GAG repeats separated by AU spacers.

The *mtrB* gene product, TRAP, when in the presence of tryptophan, was shown to bind to the antiterminator segment of *trp* leader RNA, resulting in transcription termination.

Production of the AT protein was shown to be regulated translationally as well as transcriptionally in response to the accumulation of uncharged tryptophanyl-tRNA.

14. Deikus G, Babitzke P, Bechhofer DH. 2004. Recycling of a regulatory protein by degradation of the RNA to which it binds. *Proc. Natl. Acad. Sci. USA* 101:2747–51
15. Du H, Babitzke P. 1998. *trp* RNA-binding attenuation protein-mediated long distance RNA refolding regulates translation of *trpE* in *Bacillus subtilis*. *J. Biol. Chem.* 273:20494–503
16. Du H, Tarpey R, Babitzke P. 1997. The *trp* RNA-binding attenuation protein regulates TrpG synthesis by binding to the *trpG* ribosome binding site of *Bacillus subtilis*. *J. Bacteriol.* 179:2582–86
17. Du H, Yakhnin AV, Subramanian D, Babitzke P. 2000. *trp* RNA-binding attenuation protein–5′ stem-loop RNA interaction is required for proper transcription attenuation control of the *Bacillus subtilis trpEDCFBA* operon. *J. Bacteriol.* 182:1819–27
18. Elliott M, Gottlieb P, Gollnick P. 2001. The mechanism of RNA binding to TRAP: initiation and cooperative interactions. *RNA* 7:85–93
19. Elliott MB, Gottlieb PA, Gollnick P. 1999. Probing the TRAP-RNA interaction with nucleoside analogs. *RNA* 5:1277–89
20. Fajardo-Cavazos P, Nicholson WL. 2000. The TRAP-like SplA protein is a trans-acting negative regulator of spore photoproduct lyase synthesis during *Bacillus subtilis* sporulation. *J. Bacteriol.* 182:555–60
21. Gollnick P, Babitzke P. 2002. Transcription attenuation. *Biochim. Biophys. Acta* 1577:240–50
22. Gollnick P, Babitzke P, Merino E, Yanofsky C. 2002. Aromatic amino acid metabolism in *Bacillus subtilis*. In *Bacillus subtilis and its Closest Relatives*, ed. AL Sonenshein, JA Hoch, R Losick, pp. 233–44. Washington, DC: ASM press
23. **Gollnick P, Ishino S, Kuroda MI, Henner D, Yanofsky C. 1990. The *mtr* locus is a two gene operon required for transcription attenuation in the *trp* operon of *Bacillus subtilis*. *Proc. Natl. Acad. Sci. USA* 87:8726–30**
24. Henkin TM, Yanofsky C. 2002. Regulation by transcription attenuation in bacteria: How RNA provides instructions for transcription termination/antitermination decisions. *BioEssays* 24:700–7
25. Henner D, Yanofsky C. 1993. Biosynthesis of aromatic amino acids. In Bacillus subtilis *and Other Gram-Positive Bacteria: Biochemistry, Physiology, and Molecular Genetics*, ed. AL Sonenschein, JA Hoch, R Losick, pp. 269–80. Washington, DC: Am. Soc. Microbiol.
26. Hoffman R, Gollnick P. 1995. The *mtrB* gene of *Bacillus pumilus* encodes a protein with sequence and functional homology to the *trp* RNA-binding attenuation protein (TRAP) of *Bacillus subtilis*. *J. Bacteriol.* 177:839–42
27. Hopcroft NH, Manfredo A, Wendt AL, Brzozowski AM, Gollnick P, Antson AA. 2004. The interaction of RNA with TRAP: the role of triplet repeats and separating spacer nucleotides. *J. Mol. Biol.* 338:43–53
28. Hopcroft NH, Wendt AL, Gollnick P, Antson AA. 2002. Specificity of TRAP-RNA interactions: crystal structures of two complexes with different RNA sequences. *Acta Crystallogr. D Biol. Crystallogr.* 58:615–21
29. Kane JF. 1977. Regulation of a common amidotransferase subunit. *J. Bacteriol.* 132:419–25
30. **Kuroda MI, Henner D, Yanofsky C. 1988. *cis*-Acting sites in the transcript of the *Bacillus subtilis trp* operon regulate expression of the operon. *J. Bacteriol.* 170:3080–88**

A key report describing cloning and sequencing the *mtr* operon, which encodes TRAP and GTP cyclohydrolase I.

This work characterized the terminator and antiterminator RNA structures involved in attenuation control of the *B. subtilis trp* operon as well as provided the initial indication that an RNA binding protein (the product of the *mtr* locus) is involved in controlling which structure forms.

TRAP binding to trp operon leader mRNA was shown not only to promote transcription termination but cause structural changes in the fraction of un-terminated leader RNAs, blocking initiation of translation of trpE, the first structural gene of the operon.

These studies provided the first direct demonstration that TRAP binds to *trp* leader RNA and that binding is tryptophan dependent.

Computer predictions identified the operon responsible for the increase in *trp* operon expression associated with a deficiency of charged tryptophanyl-tRNA.

This paper describes the first findings indicating that the *B. subtilis trp* operon is regulated by a different attenuation mechanism than the well-characterized mechanism that controls the *E. coli trp* operon.

31. Kuroda MI, Shimotsu H, Henner DJ, Yanofsky C. 1986. Regulatory elements common to the *Bacillus pumilus* and *Bacillus subtilis trp* operons. *J. Bacteriol.* 167:792–98
32. Landick R, Turnbough CL, Yanofsky C. 1996. Transcription attenuation. In *Escherichia coli and Salmonella Cellular and Molecular Biology*, ed. FC Neidhardt, pp. 1263–86. Washington, DC: ASM Press
33. Lee AI, Sarsero JP, Yanofsky C. 1996. A temperature-sensitive *trpS* mutation interferes with *trp* RNA-binding attenuation protein (TRAP) regulation of *trp* gene expression in *Bacillus subtilis*. *J. Bacteriol.* 178:6518–24
34. Li PTX, Gollnick P. 2002. Using heter-11-mers composed of wild type and mutant subunits to study tryptophan binding to TRAP and its role in activating RNA binding. *J. Biol. Chem.* 277:35567–73
35. Li PTX, Scott DJ, Gollnick P. 2002. Creating hetero-11-mers composed of wild-type and mutant subunits to study RNA binding to TRAP. *J. Biol. Chem.* 277:11838–44
36. Linke K, Wolfram T, Bussemer J, Jakob U. 2003. The roles of the two zinc binding sites in DnaJ. *J. Biol. Chem.* 278:44457–66
37. Martinez-Yamout M, Legge GB, Zhang O, Wright PE, Dyson HJ. 2000. Solution structure of the cysteine-rich domain of the *Escherichia coli* chaperone protein DnaJ. *J. Mol. Biol.* 300:805–18
38. McCabe BC, Gollnick P. 2004. Cellular levels of *trp* RNA-binding attenuation protein in *Bacillus subtilis*. *J. Bacteriol.* 186:5157–59
39. McElroy C, Manfredo A, Wendt A, Gollnick P, Foster M. 2002. TROSY-NMR studies of the 91 kDa TRAP protein reveal allosteric control of a gene regulatory protein by ligand-altered flexibility. *J. Mol. Biol.* 323:463–73
40. **Merino E, Babitzke P, Yanofsky C. 1995. *trp* RNA-binding attenuation protein (TRAP)-*trp* leader RNA interactions mediate translational as well as transcriptional regulation of the *Bacillus subtilis trp* operon. *J. Bacteriol.* 177:6362–70**
41. Nester EW, Jensen RA. 1966. Control of aromatic acid biosynthesis in *Bacillus subtilis*: sequenial feedback inhibition. *J. Bacteriol.* 91:1594–98
42. **Otridge J, Gollnick P. 1993. MtrB from *Bacillus subtilis* binds specifically to *trp* leader RNA in a tryptophan dependent manner. *Proc. Natl. Acad. Sci. USA* 90:128–32**
43. Rey MW, Ramaiya P, Nelson BA, Brody-Karpin SD, Zaretsky EJ, et al. 2004. Complete genome sequence of the industrial bacterium *Bacillus licheniformis* and comparisons with closely related Bacillus species. *Genome Biol.* 5:R77
44. Sarsero JP, Merino E, Yanofsky C. 2000. A *Bacillus subtilis* gene of previously unknown function, *yhaG*, is translationally regulated by tryptophan-activated TRAP, and appears to be concerned with tryptophan transport. *J. Bacteriol.* 182:2329–31
45. **Sarsero JP, Merino E, Yanofsky C. 2000. A *Bacillus subtilis* operon containing genes of unknown function senses $tRNA^{trp}$ charging and regulates expression of the genes of tryptophan biosynthesis. *Proc. Natl. Acad. Sci.* 97:2656–61**
46. Schaak JE, Yakhnin H, Bevilacqua PC, Babitzke P. 2003. A Mg^{2+}-dependent RNA tertiary structure forms in the *Bacillus subtilis trp* operon leader transcript and appears to interfere with *trpE* translation control by inhibiting TRAP binding. *J. Mol. Biol.* 332:555–74
47. Shevtsov MB, Chen Y, Gollnick P, Antson AA. 2004. Anti-TRAP protein from *Bacillus subtilis*: crystallization and internal symmetry. *Acta Crystallogr. D Biol. Crystallogr.* 60:1311–14
48. **Shimotsu H, Kuroda MI, Yanofsky C, Henner DJ. 1986. Novel form of transcription attenuation regulates expression of the *Bacillus subtilis* tryptophan operon. *J. Bacteriol.* 166:461–71**

49. Slock J, Stahly DP, Han C-Y, Six EW, Crawford IP. 1990. An apparent *Bacillus subtilis* folic acid biosynthetic operon containing *pab*, an amphibolic *trpG* gene, a third gene required of synthesis of *para*-aminobenzoic acid, and the dihydropteroate synthase gene. *J. Bacteriol.* 172:7211–26
50. Snyder D, Lary J, Chen Y, Gollnick P, Cole JL. 2004. Interaction of the *trp* RNA-binding attenuation protein (TRAP) with anti-TRAP. *J. Mol. Biol.* 338:669–82
51. **Steinberg W. 1974. Temperature-induced derepression of tryptophan biosynthesis in a tryptophanyl-transfer ribonucleic acid synthetase mutant of *Bacillus subtilis*. *J. Bacteriol.* 117:1023–34**
52. Sudershana S, Du H, Mahalanabis M, Babitzke P. 1999. A 5′ RNA stem-loop participates in the transcription attenuation mechanism that controls expression of the *Bacillus subtilis trpEDCFBA* operon. *J. Bacteriol.* 181:5742–49
53. Szabo A, Korszun R, Hartl FU, Flanagan J. 1996. A zinc finger-like domain of the molecular chaperone DnaJ is involved in binding to denatured protein substrates. *EMBO J.* 15:408–17
54. Szigeti R, Milescu M, Gollnick P. 2004. Regulation of the tryptophan biosynthetic genes in *Bacillus halodurans*: common elements but different strategies than those used by *Bacillus subtilis*. *J. Bacteriol.* 186:818–28
55. Valbuzzi A, Gollnick P, Babitzke P, Yanofsky C. 2002. The anti-*trp* RNA-binding attenuation protein (anti-TRAP), AT, recognizes the tryptophan-activated RNA binding domain of the TRAP regulatory protein. *J. Biol. Chem.* 277:10608–13
56. **Valbuzzi A, Yanofsky C. 2001. Inhibition of the *B. subtilis* regulatory protein TRAP by the TRAP-inhibitory protein, AT. *Science* 293:2057–59**
57. Valbuzzi A, Yanofsky C. 2002. Zinc is required for assembly and function of the anti-trp RNA-binding attenuation protein, AT. *J. Biol. Chem.* 277:48574–78
58. Xirasagar S, Elliott MB, Bartolini W, Gollnick P, Gottlieb P. 1998. RNA structure inhibits the TRAP (*trp* RNA-binding attenuation protein)-RNA interaction. *J. Biol. Chem.* 273:27146–53
59. Yakhnin AV, Babitzke P. 2002. NusA-stimulated RNA polymerase pausing and termination participates in the *Bacillus subtilis trp* operon attenuation mechanism in vitro. *Proc. Natl. Acad. Sci. USA* 99:11067–72
60. Yakhnin AV, Trimble JJ, Chiaro CR, Babitzke P. 2000. Effects of mutations in the L-tryptophan binding pocket of the *trp* RNA-binding attenuation protein of *Bacillus subtilis*. *J. Biol. Chem.* 275:4519–24
61. Yakhnin H, Zhang H, Yakhnin AV, Babitzke P. 2004. The *trp* RNA-binding attenuation protein of *Bacillus subtilis* regulates translation of the tryptophan transport gene *trpP* (yhaG) by blocking ribosome binding. *J. Bacteriol.* 186:278–86
62. Yakhnin H, Babiarz JE, Yakhnin AV, Babitzke P. 2001. Expression of the *Bacillus subtilis trpEDCFBA* operon is influenced by translational coupling and Rho termination factor. *J. Bacteriol.* 183:5918–26
63. Yang M, Chen X-P, Millitello K, Hoffman R, Fernandez B, et al. 1997. Alanine-scanning mutagenesis of *Bacillus subtilis trp* RNA-binding attenuation protein (TRAP) reveals residues involved in tryptophan binding and RNA binding. *J. Mol. Biol.* 270:696–710
64. Yang M, de Saizieu A, van Loon APGM, Gollnick P. 1995. Translation of *trpG* in *Bacillus subtilis* is regulated by the *trp* RNA-binding attenuation protein (TRAP). *J. Bacteriol.* 177:4272–78

A mutant producing a temperature-sensitive tryptophanyl-tRNA synthetase exhibited elevated expression of the *trp* operon despite growth in the presence of tryptophan, implicating uncharged tryptophanyl-tRNA as a regulatory signal.

The AT protein was shown to bind directly to tryptophan-activate TRAP, preventing TRAP from binding to *trp* leader mRNA and promoting transcription termination.

65. Yang W-J, Yanofsky C. 2005. Effects of tryptophan starvation on levels of the trp RNA-binding attenuation protein (TRAP) and Anti-TRAP (AT) regulatory protein and their influence on *trp* operon expression in *Bacillus subtilis*. *J. Bacteriol.* 187:1884–91
66. Yanofsky C. 2004. The different roles of tryptophan transfer RNA in regulating *trp* operon expression in *E. coli* versus *B. subtilis*. *Trends Genet.* 20:367–74
67. Yanofsky C, Crawford IP. 1987. The tryptophan operon. In *Escherichia coli and Salmonella typhimurium Cellular and Molecular Biology*, ed. F Neidhardt, pp. 1453–72. Washington, DC: ASM Press
68. Yanofsky C, Miles EW, Kirschner K, Bauerle R. 1999. *TRP* OPERON. *Encycl. Mol. Biol.* 4:2676–89

ACKNOWLEDGMENTS

The authors wish to thank past and current members of their labs who have contributed to the work described in this chapter as well as the National Institutes of Health, National Science Foundation, American Cancer Society, American Heart Association, and the Welcome Trust for providing financial support to the authors.

Cell-Cycle Control of Gene Expression in Budding and Fission Yeast

Jürg Bähler

Wellcome Trust Sanger Institute, Hinxton, Cambridge CB10 1SA, United Kingdom; email: jurg@sanger.ac.uk

Annu. Rev. Genet. 2005. 39:69–94

First published online as a Review in Advance on June 21, 2005

The *Annual Review of Genetics* is online at http://genet.annualreviews.org

doi: 10.1146/annurev.genet.39.110304.095808

0066-4197/05/1215-0069$20.00

Key Words

Saccharomyces cerevisiae, *Schizosaccharomyces pombe*, transcription, forkhead, cell division, microarray

Abstract

Cell-cycle control of transcription seems to be a universal feature of proliferating cells, although relatively little is known about its biological significance and conservation between organisms. The two distantly related yeasts *Saccharomyces cerevisiae* and *Schizosaccharomyces pombe* have provided valuable complementary insight into the regulation of periodic transcription as a function of the cell cycle. More recently, genome-wide studies of proliferating cells have identified hundreds of periodically expressed genes and underlying mechanisms of transcriptional control. This review discusses the regulation of three major transcriptional waves, which roughly coincide with three main cell-cycle transitions (initiation of DNA replication, entry into mitosis, and exit from mitosis). I also compare and contrast the transcriptional regulatory networks between the two yeasts and discuss the evolutionary conservation and possible roles for cell cycle-regulated transcription.

Contents

INTRODUCTION

Transcription factor (TF): a protein that regulates the initiation of transcription for a specific set of genes

Proliferation of all cells is mediated through the cell-division cycle, which consists of four main phases: genome duplication (S-phase) and nuclear division (mitosis or M-phase), separated by two gap phases (G1 and G2). Transcription of a number of genes peaks at specific cell-cycle phases. This first became evident with the discovery that histone mRNA levels oscillate as a function of the cell cycle and peak during DNA replication in both budding and fission yeast (67, 106). Since this finding, hundreds of genes that are periodically expressed during the cell cycle have been identified in yeast and other organisms. Many of these genes have specific roles in the cell cycle, and their expression peaks coincide with the phase during which they function. Much has also been learned about transcription factors (TFs) that are involved in the regulated expression of periodic genes, and about mechanisms to integrate TF activity with overall cell-cycle control. Phase-specific control of gene expression appears to be a universal feature of the cell-division cycle. Genome-wide studies of transcription and its regulation in proliferating cells are now starting to provide a global perspective on this aspect of cell-cycle control (29, 63, 159, 168a, 171).

This review summarizes the current knowledge of cell cycle–regulated transcription in both budding yeast (*Saccharomyces cerevisiae*) and fission yeast (*Schizosaccharomyces pombe*), revealing similarities as well as intriguing differences between these two distantly related model organisms. Cell cycle–regulated transcription and its integration with overall cell-cycle control has been worked out in greater detail for budding yeast, but fission yeast provides a valuable complementary system that can give unique insight and help to define universal principles, as it does in other fields (59). A main focus is on recent data obtained through global genomic approaches; for more comprehensive surveys of the older data, the reader is referred to previous reviews (26, 99, 157). In both yeasts, transcriptional control during the G1/S transition (initiation of DNA replication) has been most widely studied, but recently much progress has also been made in dissecting the regulation of transcription during other cell-cycle phases such as the G2/M transition (entry into mitosis) and the M/G1 transition (exit from mitosis).

REGULATION OF GENE EXPRESSION PEAKING DURING G1/S

MBF and SBF Transcription Factor Complexes

At the end of G1-phase, cells decide whether to commit to cell division in a process called Start in yeast or restriction point in mammalian cells. The transcript levels of several genes increase during late G1, promoting the initiation of DNA replication and other events of the G1/S transition. In budding yeast, two related TF complexes activate the expression of these genes: MBF (MCB-binding factor; also known as DSC1, for DNA synthesis control complex) and SBF (SCB-binding factor). The MBF complex consists of two protein components, Mbp1p and Swi6p, and binds to regulatory promoter sites named MCB (*MluI* cell-cycle box), because they are related to the recognition site of the *MluI* restriction enzyme (**Table 1**) (26, 81). These MCB elements are both necessary and sufficient for cell-cycle–regulated transcription (94, 113). The SBF complex consists of Swi4p and Swi6p and binds to promoter sites named SCB (Swi4/6-dependent cell-cycle box) (**Table 1**), which are sufficient to promote G1/S-specific transcription in heterologous genes (8, 9, 27, 124). The highly similar Mbp1p and Swi4p proteins are the DNA-binding components of MBF and SBF, respectively, while Swi6p plays a regulatory role in both TF complexes (52, 93, 124, 133). All three proteins contain ankyrin repeats within a central domain, indicating a common ancestry (23, 27).

In fission yeast, a related MBF/DSC1 TF complex also activates gene expression during the G1/S transition by binding to MCB promoter sites, which are conserved between the two yeasts (**Table 1**) (10, 32, 95, 102, 116, 155, 174). The fission yeast MBF complex contains Cdc10p and at least two additional, related ankyrin-repeat proteins, Res1p and Res2p, which bind to Cdc10p at their C termini, whereas their N termini contain DNA-binding domains and are involved in heterodimer formation with each other (11, 12, 165, 175). The fission yeast MBF complex also plays an important role during premeiotic DNA replication, after minor modifications in its composition and with altered target gene specificity (11, 44, 50, 105, 110). During budding yeast meiosis, a subset of DNA replication genes are regulated by MBF, while the meiotic regulation of the cyclin gene *CLB5* is independent of MBF yet still requires MCB promoter elements (134).

MCB: *MluI* cell-cycle box

MBF: MCB-binding factor

SCB: Swi4/6-dependent cell-cycle box

SBF: SCB-binding factor

MBF/SBF: yeast TF complexes that regulate the initiation of DNA replication; functional equivalent of human E2F-DP complexes

Table 1 Selected binding sites for cell-cycle transcription factors (TFs)

		Budding yeast		Fission yeast	
Name	**Phase**	**Binding site[a]**	**TF**	**Binding site[a]**	**TF**
MCB	G1/S	ACGCGT	MBF	ACGCGT	MBF
SCB	G1/S	CRCGAAA[b]	SBF	Not described	Not described
SFF/Forkhead	G2/M	Mcm1[c] + GTAAACAA[b]	Mcm1p + Fkh2p	GTAAACAA[b]	Sep1p? Fkh2p?
PCB	G2/M	Not described	Not described	GNAACR	PBF, Mbx1?
ECB	M/G1	YAATTA + Mcm1[c]	Yox1/Yhp1p + Mcm1p	Not described	Not described
Swi5/Ace2	M/G1	ACCAGCR[b]	Swi5p, Ace2p	ACCAGCCNT[b]	Ace2p

[a]Different versions of binding sites have been described, and typical sequence patterns were selected for this table to represent each binding site.

[b]These sites are also present in reverse orientation.

[c]A typical Mcm1 site is: TTWCCYAAWNNGGWAA.

N is any base; R is A or G; Y is C or T; W is A or T. See main text for references on the various binding sites.

CDK: cyclin-dependent kinase

Binding site: short regulatory DNA motif located upstream of a protein coding region (promoter) that is recognized by a specific TF to regulate gene activity

Cell-Cycle Regulation of MBF and SBF Activity

Knowledge of specific roles for various subunits of the MBF and SBF complexes as well as the cell-cycle regulation of their activities is still limited. Regulation is complex and coordinated at different levels (**Figure 1**). Both *S. cerevisiae* Swi6p and *S. pombe* Cdc10p appear to have negative as well as positive roles in MCB-mediated transcription (26, 111). In budding yeast, the Cln3p-Cdk1p/Cdc28p CDK (cyclin-dependent kinase) is a key regulator for transcriptional activation of MBF- and SBF-dependent genes (51, 149, 160), and genetic data point to Swi6p as a critical, but indirect, target of Cln3p (168). SBF and MBF bind to their target genes in early G1, but the RNA polymerase II machinery is only recruited to these promoters upon activation of Cln3p-Cdk1p in late G1-phase (40). In a recent breakthrough, Whi5p has been identified as the long-sought intermediate factor between Cln3p and SBF/MBF: It inhibits G1/S transcription through association with SBF until CDK-dependent phosphorylation during late G1 drives Whi5p out of the nucleus and relieves its inhibition (42, 46). This regulatory pathway is satisfyingly analogous to the cyclin D-Rb-E2F circuit that controls G1/S transcription in mammals and is deregulated in most cancer cells (140). Other activators such as Bck2p (167) and Stb1p (43, 68) regulate G1/S transcription in parallel to Cln3p, but their roles are less clear. The onset of SBF- and MBF-mediated transcription is dependent on a minimal cell-size threshold. Relatively poorly understood is how the cell measures and signals cell size requirements to SBF and MBF, although recent research gives fascinating insight into the coordination of cell division with cell growth (reviewed in 77).

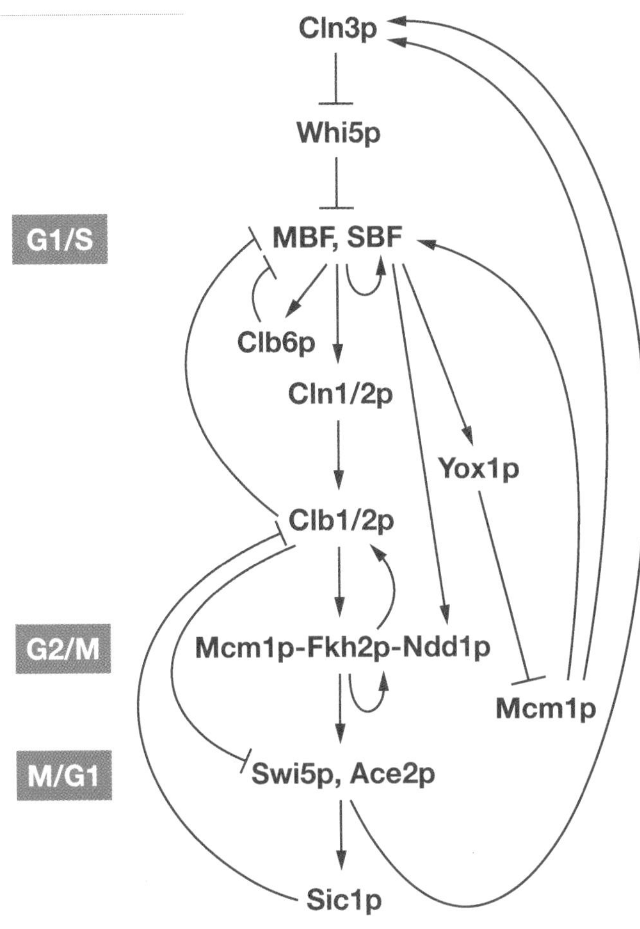

Figure 1

Scheme of main regulatory circuits that drive the gene expression program during the budding yeast cell cycle. TFs are shown in red. Red and blue lines indicate transcriptional and posttranslational regulation, respectively. Arrows: positive regulation; bars: negative regulation. The cell-cycle transitions where the main TFs act are indicated on the left side. See main text for details and references.

Swi4p is also regulated at the transcriptional level, both by ECB binding sites in its promoter acting during M/G1 progression (see below) and by a positive feedback involving SBF and MBF (60, 108, 112, 145). In G2-phase, MBF- and SBF-mediated transcription is switched off through the Clb1/2p-Cdk1p CDK complex (6, 82, 148), which is itself activated by a combination of events following transcriptional activation of *CLN1* and *CLN2* by SBF and MBF (**Figure 1**) (reviewed in 63). Moreover, phosphorylation of Swi6p by the Clb6p-Cdk1p CDK, which is also transcriptionally activated by SBF and MBF, leads to nuclear export of Swi6p during S-phase, whereas dephosphorylation of Swi6p by the mitotic exit phosphatase Cdc14p leads to nuclear import of Swi6p during G1-phase

(64, 143). The Paf1-RNA polymerase II complex is specifically required for full expression of many SBF- and MBF-regulated genes but not for their periodic regulation (131).

In fission yeast, MBF is constitutively bound throughout the cell cycle to the MCB site in the promoter of the *cdc18* target gene (169), although it activates transcription specifically in late G1-phase. The role of CDK in MBF regulation is less clear than for budding yeast. It has been suggested that activation of MBF-dependent gene expression is triggered by Cdk1p/Cdc2p CDK during G1-phase (135), and Cdk1p activity has been reported to be required for Cdc10p/Res1p complex formation (38). However, other data indicate that MBF activity itself is unlikely to be controlled by CDK (18). Instead, the Pcl-like cyclin Pas1p, when associated with its kinase partner Pef1p, may activate MBF during the G1/S transition (154). Additional proteins such as Rep2p are crucial for regulating the activity of the MBF complex (18, 122, 151, 164). Transcript levels of Rep2p, Cdc10p, and Res2p slightly increase coincident with MBF target genes (126, 138). It has also been hypothesized that an activator of MBF-regulated transcription is degraded by the SCF ubiquitin ligase during S-, G2-, and M-phases, when MBF is inactive (172). Moreover, the cyclin Cig2p, whose periodic transcription is dependent on MBF, inhibits MBF activity in G2-phase by negative feedback through binding to Res2p and phosphorylation of Res1p by Cdk1p-Cig2p CDK (13). *cig2* expression is slightly delayed relative to the expression of other MBF target genes, suggesting that *cig2* may be subject to somewhat different cell-cycle controls (17).

Genes Regulated by SBF and MBF

A number of SBF and MBF target genes have been identified by traditional approaches in budding yeast (reviewed in 26). More recently, the genomic binding sites of MBF and SBF have been globally mapped using a combination of chomatin immunoprecipitation and microarrays (75, 145). These studies have identified about 200 new putative targets, although genome-wide datasets can be noisy when used on their own. MBF and SBF have partially redundant functions as reflected by the substantial overlap between their target genes. However, both factors also play distinct roles: Genes bound by SBF predominantly function in budding as well as membrane and cell-wall biosynthesis, while MBF binds many genes involved in DNA replication and repair (75, 145). The functional specialization of MBF and SBF may allow for independent regulation of distinct molecular processes that normally occur at the same time during the budding yeast cell cycle.

Microarray: technology for genome-wide measurements of transcript levels or TF binding regions by hybridization to DNA spots arrayed at high density on slides.

Although SBF and MBF bind to the promoters of more than 200 genes, they are associated with only a minority of the genes whose transcript levels peak at the G1/S transition in budding yeast. Additional regulators, such as Skn7p and Stb1p, appear to contribute directly to G1/S transcription (15, 25, 43, 90, 91). Some functions of Mbp1p such as bud-emergence require its interaction with the response regulator Skn7p instead of Swi6p (25, 118). Moreover, SBF activates at least nine additional TFs that in turn induce the expression of further genes during the G1/S transition (71). Although these factors functionally overlap to some degree with SBF and MBF, they each also have specialized functions. Most of these TFs bind to the promoters of yet other TFs involved in cell-cycle progression and/or differentiation. Thus, SBF launches a complex transcriptional cascade, and a large network of TFs is involved in coordinating processes at the beginning of the budding yeast cell cycle (71).

In fission yeast, more than ten MBF target genes have been reported (102, and references cited therein). Global gene expression profiling recently identified dozens of additional potential MBF target genes, although the number of genes regulated by fission yeast MBF seems to be substantially lower than in

MADS box TF: TF with conserved MADS box domain, including *S. cerevisiae* Mcm1p and *S. pombe* Mbx1, that often function in combination with various accessory factors

SFF: Swi five factor

Forkhead TF: member of a TF family that contains a conserved DNA-binding domain of winged helix structure (including *S. cerevisiae* Fkh1p and Fkh2p and *S. pombe* Sep1p and Fkh2p), named after the *Drosophila* forkhead factor; forkhead TFs function during development, cell differentiation, and the cell cycle

budding yeast (127, 138). Genes whose expression is regulated by MBF mainly function in DNA replication or cell-cycle control. Unlike in budding yeast, genes for membrane and cell wall formation do not appear to be regulated, possibly reflecting differences in the growth cycles and the absence of an SBF complex in fission yeast.

REGULATION OF GENE EXPRESSION PEAKING DURING G2/M

Although the factors regulating G1/S transcription have been known for many years, the identification of factors that regulate the G2/M transcriptional program promoting entry into mitosis was more difficult, because genetic analyses proved elusive. Studies of the *SWI5* promoter in budding yeast revealed that its periodic expression requires Mcm1p, which binds to promoters as a dimer together with a ternary complex factor initially termed SFF (Swi five factor) (4, 97, 100). Mcm1p is a ubiquitous MADS-box TF similar to human serum response factor; MADS box proteins are combinatorial TFs that often derive their regulatory specificity from accessory factors (115, 142, 158).

Forkhead Transcription Factor Complexes

At the beginning of the new millenium, a series of elegant studies unmasked the identity and control of SFF in budding yeast, making it the best understood mechanism of cell cycle–regulated transcription. The two proteins Fkh1p and Fkh2p, members of the conserved forkhead family of TFs (33), are required to regulate transcription of genes during the G2/M transition (69, 83, 85, 129, 148, 173; reviewed in 28, 62, 76). These TFs recognize Forkhead/SFF binding sites (4, 79, 129, 173), which are frequently located next to Mcm1 binding sites in the promoters of *CLB2* cluster genes that are expressed during G2/M (**Table 1**) (2, 24, 62, 129, 148, 170). Both Forkhead and Mcm1 binding sites are conserved in Metazoa (79, 129, 169). Fkh1p and Fkh2p appear to have partially redundant yet distinct functions in the control of G2/M-regulated genes. Fkh2p is the main factor that forms a ternary complex with Mcm1p on gene promoters, while Fkh1p can bind promoters in the absence of Mcm1p and also plays additional roles in cell-type determination (69, 70, 150). DNA bending induced by Mcm1p and a region upstream of the FKH DNA-binding domain of Fkh2p are instrumental in recruiting Fkh2p into the ternary transcription complex (24, 92). Surprisingly, both Fkh1p and Fkh2p also associate with coding regions of active genes and regulate, in opposing ways, transcriptional elongation and termination (120), probably reflecting a function independent of their roles in cell-cycle transcription.

Cell Cycle-Regulation of Forkhead Activity

Fkh2p and Mcm1p remain bound at promoters of G2/M-regulated genes throughout the cell cycle (4, 83). The periodic activity of this transcription complex is regulated by an additional factor, the coactivator Ndd1p (83, 96, 136). Ndd1p is essential for the G2/M transition and for activating gene expression of G2/M-regulated genes, and it associates with promoters of these genes in a Fkh2p-Mcm1p-dependent manner. Ndd1p is regulated both transcriptionally and via protein stability, and protein levels peak during S- and G2-phases. Loss of Fkh2p function suppresses the requirement for Ndd1p, indicating that Fkh2p also plays a negative regulatory role in transcription (83).

Amon et al. (6) have proposed that the Clb1/2p-Cdk1p kinase functions in a positive feedback loop to induce transcription of the cyclin genes *CLB1* and *CLB2* (**Figure 1**). Recent data give insight into the mechanism of this regulatory circuit. Fkh2p is phosphorylated by Clb5p-Cdk1p CDK, thus triggering recruitment of Ndd1p to promoters bound by the Fkh2p-Mcm1p complex (128). The Ndd1p-Fkh2p interaction is further stabilized upon phosphorylation of Ndd1p by Clb2p-Cdk1p CDK, leading to a specific association with the forkhead-associated (FHA) domain of Fkh2p (45, 136), which is a conserved phosphopeptide recognition site (56, 57). Transcription of *CLB1* and *CLB2* during the G2/M transition is directly activated through this association of Ndd1p with Fkh2p-Mcm1p (28, 62, 76). In addition, Fkh2p itself becomes phosphorylated before mitosis (129), but whether this contributes to the regulation of G2/M transcription is not known. Morris et al. (121) showed that Cks1p, a conserved CDK-interacting protein, activates G2/M transcription of the *CLB2* cluster gene Cdc20p by recruiting the proteasome to its promoter. The proteasome has been implicated in transcriptional control in addition to its well-known proteolytic role (65), but whether Cks1p also contributes to the regulation of other *CLB2* cluster genes remains to be determined.

Forkhead-Dependent Transcription in Fission Yeast

Knowledge of the transcriptional regulation during the G2/M transition is more limited in fission yeast, although a partially similar picture is emerging. The fission yeast genome contains four TFs of the forkhead family: Mei4p appears to function specifically in gene transcription during the meiotic nuclear divisions (1, 72, 105). SPAC1142.08 is highly similar to budding yeast Fhl1p that controls ribosomal protein gene expression (104; and references therein), and it does not appear to affect cell-cycle transcription (30; G. Rustici & J. Bähler, unpublished data). Evidence is accumulating that the two other forkhead proteins, Sep1p (137, 146) and SPBC16G5.15c (Fkh2p), are involved in transcriptional activation during M-phase. Sep1p is required for periodic expression of *cdc15* whose transcript levels peak during early mitosis (58, 176). Microarray analyses revealed that Sep1p is involved in activating transcription of dozens of additional genes during mitosis (**Figure 2**) (138). These genes are also significantly enriched for TF binding sites similar to the conserved binding sites of forkhead proteins (**Table 1**) (72, 79, 129, 130, 138, 173). Fkh2p, which is the closest homolog to budding yeast Fkh2p, also appears to function in transcriptional regulation during the cell cycle, because several mitotically induced genes show constitutive expression in its absence (30, 31). Intriguingly, basal transcription levels of mitotic genes are increased in the absence of Fkh2p, whereas they are decreased in the absence of Sep1p (31, 138). This indicates that Fkh2p, unlike Sep1p, has roles in repressing transcription at some stages of the cell cycle. Overproduction of Fkh2p is lethal in wild-type cells but not in *sep1* mutants, suggesting that Sep1p is required for Fkh2p function (30). Further work is needed to establish whether and how Sep1p and Fkh2p function together to regulate periodic cell-cycle transcription during mitosis in fission yeast.

As in budding yeast, there is evidence that a MADS-box TF (Mbx1p) functions together with Fkh2p in transcriptional regulation (30). Unlike in budding yeast, however, Forkhead binding sites do not seem to be accompanied by Mcm1 sites (127, 138). Mbx1p is part of an uncharacterized TF complex named PBF (PCB binding factor) that binds to a short promoter site called PCB (pombe cell-cycle box; **Table 1**) (7, 30). The genes with a PCB site also contain Forkhead binding sites (3, 31, 138), and it is possible that the combination of these two motifs provides the platform required for combinatorial control by forkhead and MADS-box TFs. The polo kinase Plo1p, whose transcript is induced during mitosis, regulates PBF activity and may therefore provide a positive feedback mechanism to activate mitotic gene transcription (7).

Fission yeast contains no clear orthologs of budding yeast Ndd1p, and it is not known how transcription during mitosis is activated.

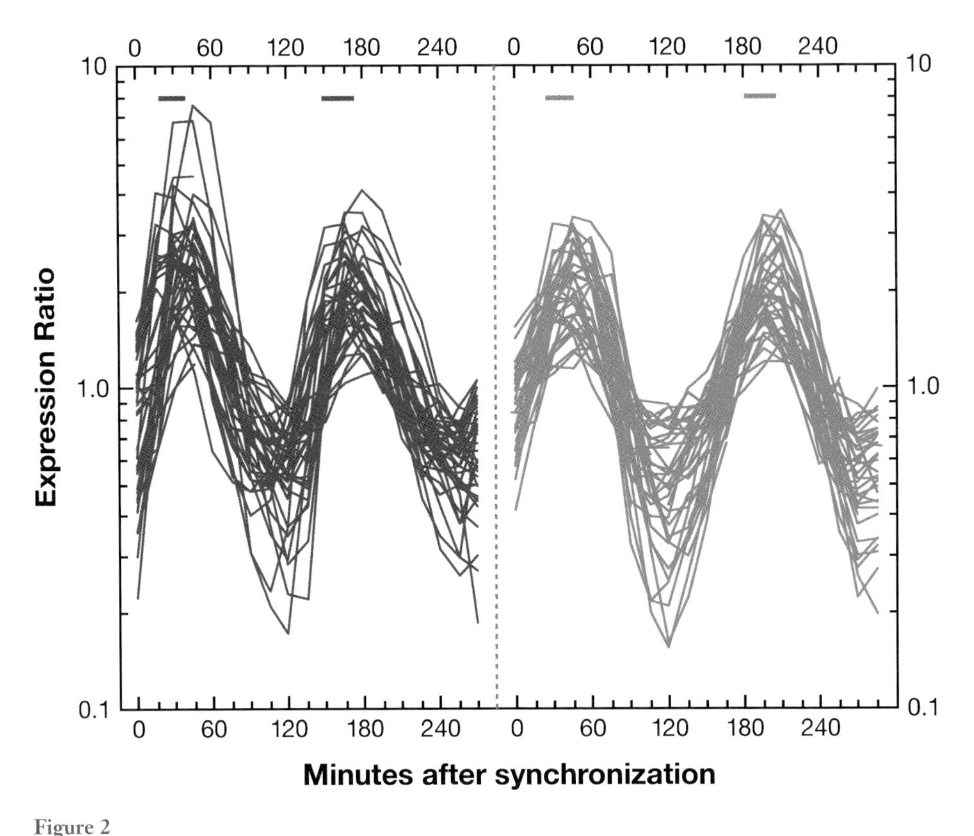

Figure 2

Expression profiles of 39 Sep1p-dependent genes with peak transcript levels around mitosis in fission yeast. Some of these genes are direct targets of Sep1p, whereas others are indirectly activated by Sep1p via its target gene Ace2p. Blue profiles: cell-cycle synchronization using *cdc25* block-release (whole culture synchronization); green profiles: cell-cycle synchronization using centrifugal elutriation (selective synchronization). Two cell cycles were followed in each experiment, and transcript levels relative to unsynchronized cells were measured in 15-min intervals using DNA microarrays (138). The approximate timing and duration of M-phases are indicated by bars above the graphs.

Intriguingly, only Fkh2p contains a potential FHA domain while Sep1p does not (31). Transcription of *fkh2* but not of *sep1* is regulated during the cell cycle, although expression levels peak relatively late, coinciding with those of the target genes (31, 138, 176). Accordingly, the *fkh2* promoter contains a Forkhead binding site (138), raising the possibility that Sep1p and/or Fkh2p activate *fkh2* transcription, either as a positive feedback or, given the negative role of Fkh2p in transcription, to switch off mitotic transcription in a negative feedback. Similarly, there is also evidence in budding yeast that Fkh1p and Fkh2p themselves activate *FKH2* transcription (70). As in budding yeast, Fkh2p is phosphorylated during mitosis coincident with increased transcription of target genes (31). It has also been proposed that the Cdk1p CDK controls the activity of Sep1p based on genetic interactions between *sep1* mutants and mutants in *cdk1* or in genes for Cdk1p regulators (*cdc25* and *wee1*) (66). The TFIIH-associated CDK-activating kinase (Mcs6-complex) seems to be particularly important for Sep1p-regulated gene expression in fission yeast, a function that is probably independent of CDK activation (89).

REGULATION OF GENE EXPRESSION PEAKING DURING M/G1

The Mcm1p Transcription Factor

In budding yeast, the MADS-box protein Mcm1p also plays a role in the transcription of genes whose expression levels peak during mitotic exit. A promoter element named ECB (early cell-cycle box) contains a Mcm1 binding site and functions in the periodic transcription of genes during the M/G1 transition (112, 148). Genes for regulatory proteins such as the SBF component *SWI4* and the cyclin *CLN3*, which are transcriptionally activated in late mitosis through ECB elements, are rate limiting for G1 progression (see above) (98, 101, 112). Mcm1p binds constitutively to ECB sites (101), and M/G1 specificity may be conferred by timed transcriptional repression: Two homeodomain proteins, Yox1p and Yhp1p, bind to typical homeodomain binding sites upstream of the Mcm1 binding sites in ECB (**Table 1**), and they act as repressors restricting gene transcription from ECB promoters to the M/G1-phase of the cell cycle (132). Yox1p itself is transcriptionally activated by the SBF complex (71, 75, 145); Mcm1p is thus setting up a negative feedback loop that determines the duration of ECB activity (**Figure 1**). Yhp1p is transcribed later in the cell cycle and helps to maintain ECB repression until late M-phase (132).

The Ace2p and Swi5p Transcription Factors

The two related TFs Swi5p and Ace2p also activate transcription of genes during the M/G1 transition in budding yeast (53). Like the MBF/SBF and Fkh1p/Fkh2p TF pairs described above, Swi5p and Ace2p have overlapping but also specialized functions in transcriptional control (54, 55, 108, 145). Swi5p and Ace2p target genes are associated with a Swi5 binding site (**Table 1**) (54, 80, 148). During the M/G1 transition, cytokinesis separates mother and daughter cells, and cellular identity or fate can change. Swi5p and Ace2p activate several genes with roles in cell separation (20, 53, 55). Moreover, these two TFs also function in specifying the identities of mother and daughter cells, together with one of their target genes, the TF Ash1p. Both Ace2p and Ash1p accumulate exclusively in daughter cells and control daughter cell–specific gene expression (19, 22, 36, 144, 153, 162). Ace2p also functions in the G1 delay specific to daughter cells (87). G1-specific transcription of the *HO* endonuclease gene that initiates mating-type switching requires both Swi5p and SBF; the ordered stepwise recruitment of chromatin remodeling complexes, TFs, and RNA polymerase II complexes to the *HO* promoter during the cell cycle has been studied in some detail (41, 84).

Ace2p/Swi5p: yeast TF paralogs that regulate mitotic exit, cell division, and daughter cell identity

ECB: early cell-cycle box

Cell Cycle-Regulation of Ace2p and Swi5p Activity

ACE2 and *SWI5* are themselves transcriptionally activated during the G2/M transition by the Mcm1p-Fkh2p complex described above (**Figure 1**) (97, 145, 173). In addition, the nuclear localization of Swi5p is negatively regulated by Clb2p-Cdk1p CDK: Proteolysis of Clb2p during anaphase leads to dephosphorylation of Swi5p, which then accumulates in the nucleus to activate transcription of target genes (117, 123, 161). Ace2p, on the other hand, is positively regulated by phosphorylation: A signaling network named RAM (regulation of Ace2p activity and cellular morphogenesis) culminating in activation of the Mob1p-Cbk1p kinase complex activates both Ace2p and polarized growth (20, 125, 162). The functional architecture of RAM signaling is similar to the mitotic exit network (MEN) and septation initiation network (SIN) in budding and fission yeast, respectively (16, 109). RAM acts together with MEN to promote daughter cell–specific localization of Ace2p, and it coordinates Ace2p-dependent transcription with MEN activation at the end of mitosis (162).

Transcriptional regulatory network: map of complex connections between TFs, proteins that regulate TFs, and TF target genes that together control a biological process

Ace2p in Fission Yeast

A fission yeast homolog of Ace2p has recently been characterized and given the same name (103). As in budding yeast, Ace2p is required for cell separation, and it activates the transcription of *eng1* that functions in degradation of the cell division septum. Ace2p controls the expression of several additional genes whose transcript levels peak in M/G1-phase (**Figure 2**), and these genes are enriched for a binding site similar to the Swi5 site in budding yeast (**Table 1**) (3, 138). Moreover, *ace2* expression is itself regulated during mitosis by the forkhead TFs Sep1p and Fkh2p (3, 30, 138). This transcriptional cascade is reminiscent of the one in budding yeast (see above), and it explains that *sep1* and *ace2* mutants show similar defects in cell separation (103, 146). It is possible that the Sep1p-Ace2p pathway contributed to the evolution of single-celled fungi (14).

TRANSCRIPTIONAL REGULATORY NETWORKS

The three main transcriptional waves described above do not act in isolation but are integrated with the cell-cycle machinery, with each other, and with additional clusters of periodically transcribed genes at other phases of the cell cycle. In recent years, much has been learned about transcriptional regulatory networks of the cell cycle through elegant genome-wide approaches pioneered with budding yeast (reviewed in 29, 63, 159, 168a, 171). This work revealed a serial regulation of TFs, whereby transcriptional activators functioning during one stage of the cell cycle regulate activators functioning during the next stage, thus resulting in a continuous cycle of interdependent waves of transcription (**Figure 3**). In late G1, the MBF and SBF complexes activate transcription of *NDD1*, whose gene product in turn activates

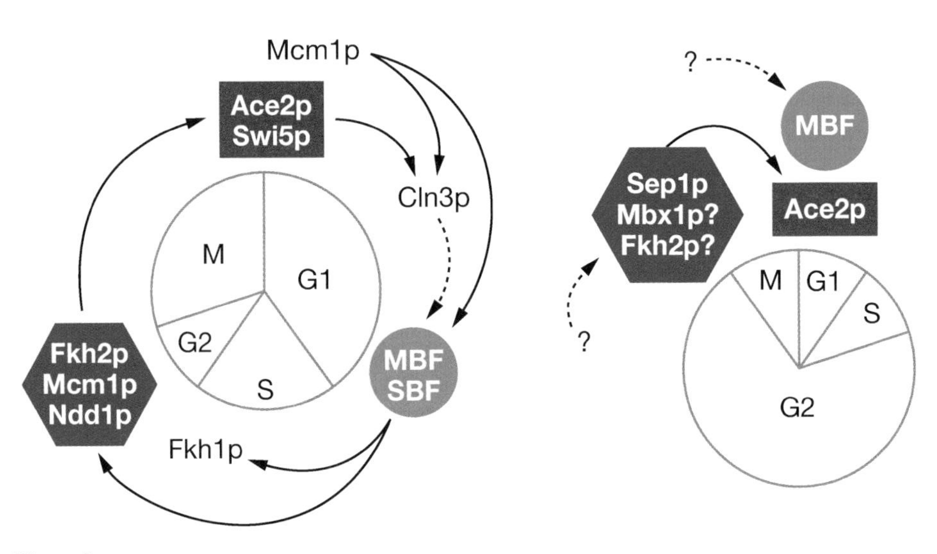

Figure 3

Transcriptional regulatory networks during the cell cycles in budding and fission yeast. Orthologous TFs or complexes are indicated by corresponding frames and colors, and approximate cell-cycle phases where TFs function are given within the circles. Continuous and hatched lines indicate transcriptional and posttranslational regulation, respectively. See main text for details and references.

the Fkh2p-Mcm1p complex leading to transcription of *SWI5* and *ACE2* during G2/M. Mcm1p, Swi5p, and Ace2p, which all act in M/G1, then close the circle either directly by inducing transcription of the SBF gene *SWI4*, or indirectly by inducing transcription of the cyclin gene *CLN3*, whose product then activates both SBF and MBF. In addition, the transcriptional network is driven by complex overlying layers of posttranslational regulation such as phosphorylation, protein degradation, and protein localization, some details of which were described in the previous sections (**Figure 1**). As illustrated by the regulation of the SBF/MBF and forkhead TF complexes described above, promoter prebinding of inhibited TFs that wait for a cell-cycle signal to activate transcription seems to be a general regulatory principle (78, 168a, 171).

Recent microarray experiments with fission yeast identified periodically transcribed genes clustered into four major waves of expression according to the time of their peak expression levels (127, 138). The forkhead protein Sep1p regulates mitotic genes in the first cluster including Ace2p, which then activates transcription in the second cluster during the M/G1 transition. Other genes in the second cluster required for G1/S progression are regulated by MBF independently of Sep1p and Ace2p (**Figure 3**). The third cluster coincides with S-phase, while a fourth cluster contains genes weakly regulated during G2-phase.

The three main cell-cycle transcriptional regulators or complexes are conserved between fission and budding yeast, and at least some of these TFs and promoter binding sites are also conserved in Metazoa. For example, forkhead proteins regulate the expression of mitotic genes such as cyclin B and polo kinase also in mammalian cells (5, 88), and many parallels are evident between the yeast MBF complex and the human E2F-DP heterodimer with regard to structure, function, target genes, and binding sites (29, 86, 156). In budding yeast, pairs of transcriptional activators exhibit partially redundant functions (Ace2p and Swi5p, MBF and SBF), while fission yeast appears to rely on only one complex at each stage (**Figure 3**). Despite the similarities between the main transcriptional activators, the regulatory networks show major differences in circuitry reflecting differences in cell-cycle phasing between the two yeasts (**Figure 3**). In fission yeast, M- and G1-phases take place within ~20% of total cell-cycle duration (99). Accordingly, peak transcription levels of ~80% of the highly regulated periodic genes are concentrated in a short window, during which all known cell-cycle TFs function (138). In budding yeast, the functions of the main transcriptional regulators are spread more evenly over the cell cycle due to longer G1- and M-phases but short G2-phase (**Figure 3**). The transcriptional cascade from forkhead to Ace2p regulators acting during M- and G1-phases is conserved in fission yeast (3, 30, 138), whereas other aspects of the transcriptional network are clearly different. The MBF complex, which regulates G1/S progression in both yeasts, acts downstream of Ace2p and Swi5p and upstream of the forkhead complex in budding yeast (145, and references cited therein), while it functions in parallel to, and independently of, Ace2p and Sep1p in fission yeast (138). The absence of a distinct G1-phase in fission yeast means that mitotic exit (controlled by Ace2p) and initiation of S-phase (controlled by MBF) coincide, which is reflected in the regulatory circuit (**Figure 3**). Unlike in budding yeast, transcriptional activators in fission yeast probably do not form a fully connected cyclic network. The long G2-phase (~70% of cell cycle), during which mainly weakly regulated genes and no obvious transcriptional activators are expressed, is unlikely to be bridged by transcriptional control (138). Moreover, Ace2p is the only known cell-cycle TF in fission yeast whose transcription is strongly regulated and also peaking ahead of its target genes. It is therefore likely that in fission yeast posttranslational mechanisms are relatively more important to regulate periodic transcription during the cell cycle.

There are still large gaps in our knowledge, and the real picture of the transcriptional cell-cycle network is certainly more complex than conveyed in **Figures 1** and **3**. For many periodically expressed genes, the mechanisms responsible for their cell-cycle regulation are not known. For example, the TFs activating histone gene expression are not known in either yeast, even though these were the first genes that had been identified as periodically expressed, and potential TF binding sites have been described (10, 61, 67, 106, 107). Unlike the binding sites for the main transcriptional regulators described above (**Table 1**), the histone promoter sites are not conserved between budding and fission yeast. It is possible that SBF contributes to histone transcription in budding yeast as it binds to several histone gene promoters (145). Two proteins, Hir1p and Hir2p, act as transcriptional repressors that function in periodic transcription of histone genes in budding yeast (141, 147). Fission yeast Hip1p, which is closely related to Hir1p and Hir2p, is similarly required for repression of histone gene expression outside of S-phase, but histone mRNA levels also appear to be positively regulated by an unknown factor (21). Little is known about the regulation of genes whose transcript levels peak during G2-phase in fission yeast. It is possible that the transcript levels of some genes are regulated at the level of mRNA stability.

Furthermore, many genes are regulated by different combinations of TFs, and several regulators such as Mcm1p can act at various cell-cycle stages, adding additional gene expression modules to the network (15, 71, 73, 90, 145). For example, some genes appear to be regulated by a combination of SBF and forkhead, and some genes are regulated by forkhead independently of Mcm1p. *CDC20* is an example for genes with a promoter that contains binding sites for the Mcm1p-Fkh2p complex together with an ECB element that binds Yox1p/Yhp1p and Mcm1p. These genes are expressed at intermediate times between the two transcriptional waves promoted by either individual regulator complex, and they continue to be periodically transcribed if either regulatory pathway is disrupted (132, 173). Kato et al. (78) recently showed that regulation is frequently relayed from a TF acting earlier to a TF acting later via combinatorial, joint-phase control of the two TFs to regulate gene expression at intermediate times; this means that two cooperating TFs can generate at least three peaks of gene expression, which could explain the continuum of peaks observed over the cell cycle. Thus, combinatorial control of cell cycle–regulated transcription greatly contributes to the complexity and sophistication of periodic gene expression.

CONSERVATION AND IMPORTANCE OF CELL CYCLE–REGULATED TRANSCRIPTION

Genome-wide studies of the gene expression program during the budding yeast cell cycle identified about 400 to 800 periodically transcribed genes (35, 148). A corresponding study in fission yeast revealed about 400 genes, but only 136 of these genes exhibited high-amplitude cell cycle–dependent changes comparable to those in budding yeast (138). Another microarray analysis reported a less conservative list of >700 cell cycle–regulated *S. pombe* genes (127), although no data on reproducibility are available and many of these genes show marginal periodicity in expression profiles. Overall, more genes appear to be strongly regulated during the cell cycle in budding yeast compared with fission yeast. For example, transcriptional regulation of nine different cyclin genes helps to drive both the transcriptional network (**Figure 1**) and the cell cycle in budding yeast (29, 63, 119, 145, 148). In fission yeast, only one cyclin gene (*cig2*) is strongly regulated and two others (*cdc13*, *cig1*) show weak regulation (37, 126, 138). The weak transcriptional regulation of *cdc13* and *cig1* might be an evolutionary remnant, or this regulation could become more important, and more pronounced, when the cell cycle slows down, e.g., during fluctuations in available nutrients. These observations indicate that fission yeast relies less on transcriptional control for cell-cycle regulation than does budding yeast, and therefore posttranslational mechanisms may play relatively more important roles.

Periodic gene expression that has been conserved through evolution is likely of more general biological significance for proper cell-cycle function. It is striking that genes for glycosylphosphatidylinositol-modified (GPI) cell surface proteins (47) are significantly enriched among periodically transcribed genes in both yeasts (14 of 33 *S. pombe* genes and 27 of 66 *S. cerevisiae* genes encoding GPI proteins; $P < 10^{-4}$; J. Mata & J. Bähler, unpublished data). De Lichtenberg et al. (49) reported that periodically expressed budding yeast genes encode proteins that tend to share combinations of certain protein features such as phosphorylation, glycosylation, and instability that help to distinguish these proteins with high confidence. On the other hand, periodic gene expression is not necessarily conserved between budding and fission yeast (163). To obtain a conservative estimate of orthologous genes that are periodically expressed in both yeasts, Rustici et al. (138) determined the overlap between the genes reported as periodic in both studies of budding yeast (35, 147) and the genes periodically expressed with high amplitude in fission yeast (**Figure 4**). Although the resulting overlap of about 40 genes is highly significant, it is also surprisingly small. Less conservative comparisons reveal ~80–140 genes that are cell cycle–regulated in both yeasts, but a substantial number of these would be expected by chance given the sizes of gene lists involved (**Figure 4**) (127, 138). This suggests that cell cycle–regulated transcription of the majority

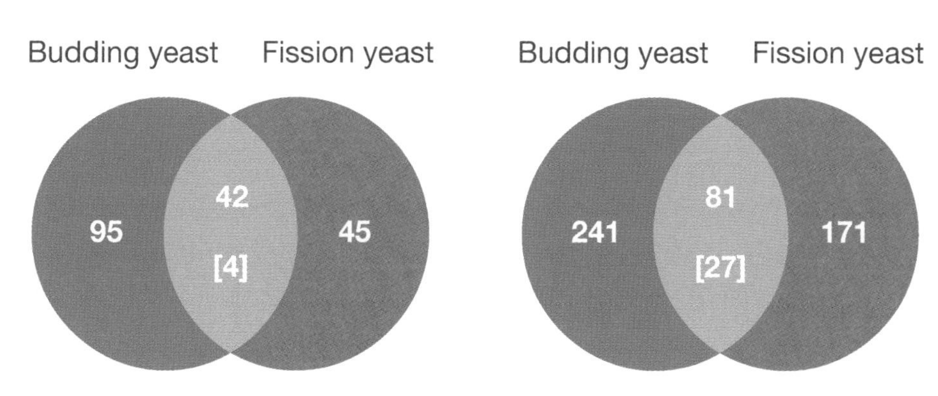

Figure 4

Conservation of cell cycle–regulated transcription between budding and fission yeast. Left Venn diagram, conservative estimate: The 301 genes that were periodically expressed in both budding yeast microarray studies (35, 148) were compared with the 136 highly regulated genes that were periodically expressed in a fission yeast microarray study (138). Among the fission yeast genes that have budding yeast orthologs, 42 genes were also periodic in the budding yeast studies. Right Venn diagram, less conservative estimate: The 800 genes that were periodically expressed in the study of Spellman et al. (147) were compared with all 407 periodic genes identified by Rustici et al. (138). Among the fission yeast genes that have budding yeast orthologs, 81 genes were also periodic in the budding yeast studies. In brackets: numbers of genes expected to overlap by chance, given the sizes of the gene sets compared and the total number of 2981 genes with orthologs. The overlaps are highly significant ($P \sim$ 2e-35 and 2e-22 for left and right diagrams, respectively). Only numbers of genes with orthologs in the other yeast are shown in the comparisons.

Table 2 Selected core genes periodically expressed in both budding and fission yeast

Budding yeast ortholog	Fission yeast ortholog	Function
DNA replication		
POL1 and *POL2*	*pol1* and *cdc20*	DNA polymerases α and ε
RFA1	*ssb1*	Single-stranded DNA-binding protein
CDC6	*cdc18*	Regulator of DNA replication initiation
MRC1	*mrc1*	DNA replication checkpoint protein
RNR1	*cdc22*	Ribonucleotide reductase
SMC3 and *MCD1*	*psm3* and *rad21*	Cohesins
HTZ1	*pht1*	Histone variant
8 histone genes	9 histone genes	Histones H2A, H2B, H3, and H4
Mitosis and cell division		
CDC5, *IPL1*, and *KIN3*	*plo1*, *ark1*, and *fin1*	Polo, Aurora, and NimA kinases
CDC20	*slp1*	Activator of anaphase promoting complex
SPO12	*wis3*	Putative cell-cycle regulator
KAR3, *KIP1*	*klp5*, *klp6*, *klp8*	Kinesin microtubule motor
MOB1 and *DBF2*	*mob1* and *sid2*	Proteins involved in mitotic exit/septation
MYO1	*myo3*	Myosin II heavy chain
BUD4	*mid2*	Protein involved in cytokinesis
ACE2	*ace2*	Transcription factor
HOF1	*imp2*	Protein involved in cell division
DSE4	*eng1*	Glucanase for cell separation
CHS2	*chs2*	Protein involved in septum formation
TOS7	*mac1*	Putative role in cell separation
Others		
SWE1	*mik1*	Kinase inhibiting cyclin-dependent kinase
CLB1-CLB6	*cig2*	B-type cyclins
MSH6	*msh6*	Mismatch-repair protein
RAD51	*rhp51*	DNA repair protein

Selected genes that have orthologs and are periodically expressed in both yeasts. For details of data analysis and a full list of core cell cycle-regulated genes, see Rustici et al. (138).

of genes has been poorly conserved through evolution. Genes listed in **Table 2** are part of a core set of cell cycle–regulated transcripts. Most of these genes have well-characterized and conserved regulatory functions in basic processes such as DNA replication, mitosis, and/or cell division. The majority of the genes in **Table 2** are conserved in Metazoa and are also cell cycle–regulated in human cells (166, and references cited therein).

Why has the transcriptional regulation of this relatively small core set of genes been conserved through evolution? It is possible that periodic transcription of these genes is critically important for driving progression through the cell cycle. Perhaps they are genes that must be highly regulated to ensure orderly progression, or they are required in relatively large amounts and with peak demands such as the histones. Some genes (e.g., those encoding histones) could also be regulated to provide a fresh pool of unmodified proteins to override previous posttranslational modifications, thus allowing the cells to start with a blank sheet. Consistent with this, transcriptionally regulated proteins tend to be regulated also at posttranscriptional levels (48). Uncoupling of histone transcription from the cell cycle is not necessarily lethal for cells (141), but production of histones in correct stochiometric amounts is important as increased expression of one class of histone relative to another causes increased chromosome loss (114). In addition, restriction of histone gene transcription to S-phase could be necessary to prevent regular histones from competing with histone variants for incorporation into chromosomes of nonreplicating cells. Consistent with this idea, the gene for the fission yeast centromere-specific histone variant CENP-A (*cnp1*) is periodically expressed earlier in the cell cycle than the regular histones (152).

What is the biological significance of cell cycle–regulated transcription, especially for those genes whose periodic expression is not conserved? The standard experiment to check the importance of transcription, constitutive expression of genes that are normally cell cycle–regulated, has produced only a few examples of its significance (26). In a *fkh1 fkh2* double mutant, cell-cycle regulation of many G2/M-phase genes becomes constitutive, but the cells remain viable and are quite healthy (173). It is therefore possible that regulation of most periodic genes is not absolutely required, but together they may enhance the fidelity and efficiency of the cell-division cycle. It is reasonable to assume that timely supply of products reflecting the demand of the cells will help the cell cycle to run smoothly and efficiently, which is as important for a reiterated essential process as execution of the process itself.

Clearly, many cell-cycle proteins are regulated at multiple and partly redundant levels, and it is the overall regulation that is likely to be important. Recent data show that protein complexes functioning during the cell cycle consist of both periodically and constitutively expressed subunits in budding yeast; the former are frequently also regulated by additional mechanisms such as CDK phosphorylation and targeted degradation and may thus play crucial and dynamic roles in controlling protein complex activity during the cell cycle (48). This system ensures the supply of new, unmodified CDK targets with special regulatory roles in each cell cycle. Transcriptional control should therefore not be considered in isolation. Although the absence of transcriptional control in rapidly proliferating yeast cells could be buffered by faster acting posttranscriptional and posttranslational regulation, transcription might become more important in driving the cell cycle when nutrient limitations slow down growth and proliferation, and/or it might help to kick start the cell cycle when nutrients increase.

It has been argued that the large number of reported cell cycle–regulated genes should be treated with scepticism, because the observed periodicity of many genes might reflect a stress response induced by cell-cycle synchronization (39). In fission yeast, this could be the case for the mainly weakly regulated genes whose transcripts peak during G2-phase, and which overlap significantly with genes modulated during environmental stress (34, 138). However, this would imply that cells are more responsive or more sensitive to stress during G2-phase and periodically repeat the stress response in subsequent cell cycles.

It has been suggested that parsimony may explain much of cell cycle–regulated transcription (148, 166), whereby genes are expressed when there is a special need for their products at a particular phase in the cell cycle, and these needs can differ between organisms. For example, the periodic induction of several genes functioning in metabolism and growth occurs during early G2-phase in fission yeast, at a time when these cells increase their overall growth rate after a switch from monopolar to bipolar growth mode (138). On the other hand, genes for bud emergence and growth in budding yeast are expressed during G1-phase, reflecting the particular biology of these cells (148).

Genes for spatial landmark proteins in budding yeast illustrate that periodic transcription can regulate organism-specific roles such as cellular morphogenesis (138). Bud8p and Bud9p are homologous membrane proteins that mark the distal and proximal cell poles, respectively, and their transcripts peak at different times in the cell cycle. The importance of transcriptional timing was shown by elegant promoter-swap experiments between *BUD8* and *BUD9*, which results in mislocalization of both gene products to the opposite poles of the cell (139).

An intriguing possibility is that CDKs and cyclins originally functioned in transcriptional control before evolving into the key regulators capable of driving the cell cycle through a free-running oscillation in activity. This transition could have led to a cell-cycle machinery that relies less on transcriptional regulation to drive cell proliferation. In accordance with this hypothesis, some members of the CDK family can control the initiation and elongation of transcription, and cyclins are structurally similar to the basal transcription factor TFIIB and the RB transcriptional regulator (reviewed in 74, 119). Even the main cell-cycle CDK in *S. cerevisiae*, Cdk1p, has additional roles in transcriptional regulation (121). In *S. pombe*, the Mcs6 CDK complex not only activates the Cdk1p CDK but is also critical for cell cycle–regulated transcription (89).

SUMMARY POINTS

1. Major cell-cycle TFs (MBF, forkhead, and Ace2p) and their binding sites are conserved in budding and fission yeast and to varying degree also in Metazoa, and they function at corresponding cell-cycle phases (initiation of DNA replication, entry and exit from mitosis).
2. Despite functional conservation of three major cell-cycle TFs, there are clear differences in the circuitry of the regulatory networks between budding and fission yeast, probably reflecting a rewiring during evolution to adjust for differences in cell-cycle phasing between the two yeasts.
3. More genes appear to be strongly regulated during the cell cycle in budding yeast compared to fission yeast, raising the possibility that the latter relies less on transcriptional control for cell-cycle regulation, and posttranscriptional mechanisms may play relatively more important roles.
4. Surprisingly, periodic transcription of most genes is not conserved between the two yeasts, except for a relatively small core set of genes that mainly function in basic cell-cycle processes such as mitosis, cytokinesis, and DNA replication. Most genes of this core set appear to be universally regulated during the eukaryotic cell cycle, and they may play key roles in controlling cell-cycle progression.
5. DNA microarrays and other genome-wide methods are now providing global perspectives on the system underlying cell cycle–regulated gene expression and its integration with other aspects of the cell-cycle machinery.
6. As in other research areas, comparative studies between the two distantly related yeasts are proving fruitful for the elucidation of basic features and critical targets of transcriptional networks during eukaryotic cell cycles. Such studies also aid our understanding of the evolutionary plasticity of transcriptional regulation, and provide insight into the overall biological significance of cell cycle–regulated transcription.

ACKNOWLEDGMENTS

I thank Christopher McInerny, Brian Morgan, and Mike Tyers for communicating data prior to publication, and José Ayté, Christopher McInerny, Brian Morgan, and Brian Wilhelm for comments on the manuscript. Research in the Bähler laboratory is funded by Cancer Research UK [CUK], Grant No. C9546/A5262 and by DIAMONDS, a EU FP6 Lifescihealth STREP.

LITERATURE CITED

1. Abe H, Shimoda C. 2000. Autoregulated expression of *Schizosaccharomyces pombe* meiosis-specific transcription factor Mei4 and a genome-wide search for its target genes. *Genetics* 154:1497–508
2. Acton TB, Zhong H, Vershon AK. 1997. DNA-binding specificity of Mcm1: operator mutations that alter DNA-bending and transcriptional activities by a MADS box protein. *Mol. Cell Biol.* 17:1881–89
3. Alonso-Nunez ML, An H, Martin-Cuadrado AB, Mehta S, Petit C, et al. 2005. Ace2p controls the expression of genes required for cell separation in *Schizosaccharomyces pombe. Mol. Biol. Cell.* 16:2003–17
4. Althoefer H, Schleiffer A, Wassmann K, Nordheim A, Ammerer G. 1995. Mcm1 is required to coordinate G2-specific transcription in *Saccharomyces cerevisiae. Mol. Cell Biol.* 15:5917–28
5. Alvarez B, Martínez AC, Burgering BM, Carrera AC. 2001. Forkhead transcription factors contribute to execution of the mitotic programme in mammals. *Nature* 413:744–47
6. Amon A, Tyers M, Futcher B, Nasmyth K. 1993. Mechanisms that help the yeast cell cycle clock tick: G2 cyclins transcriptionally activate G2 cyclins and repress G1 cyclins. *Cell* 74:993–1007
7. Anderson M, Ng SS, Marchesi V, MacIver FH, Stevens FE, et al. 2002. *plo1+* regulates gene transcription at the M-G(1) interval during the fission yeast mitotic cell cycle. *EMBO J.* 21:5745–55
8. Andrews BJ, Herskowitz I. 1989. Identification of a DNA binding factor involved in cell-cycle control of the yeast HO gene. *Cell* 57:21–29
9. Andrews BJ, Herskowitz I. 1989. The yeast *SWI4* protein contains a motif present in developmental regulators and is part of a complex involved in cell-cycle-dependent transcription. *Nature* 342:830–33
10. Aves SJ, Durkacz BW, Carr A, Nurse P. 1985. Cloning, sequencing and transcriptional control of the *Schizosaccharomyces pombe* cdc10 'start' gene. *EMBO J.* 4:457–63
11. Ayté J, Leis JF, DeCaprio JA. 1997. The fission yeast protein p73res2 is an essential component of the mitotic MBF complex and a master regulator of meiosis. *Mol. Cell Biol.* 17:6246–54
12. Ayté J, Leis JF, Herrera A, Tang E, Yang H, DeCaprio JA. 1995. The *Schizosaccharomyces pombe* MBF complex requires heterodimerization for entry into S phase. *Mol. Cell Biol.* 15:2589–99
13. Ayté J, Schweitzer C, Zarzov P, Nurse P, DeCaprio JA. 2001. Feedback regulation of the MBF transcription factor by cyclin Cig2. *Nat. Cell Biol.* 3:1043–50
14. Bähler J. 2005. A transcriptional pathway for cell separation in fission yeast. *Cell Cycle* 4:39–41
15. Banerjee N, Zhang MQ. 2003. Identifying cooperativity among transcription factors controlling the cell cycle in yeast. *Nucleic Acids Res.* 31:7024–31

16. Bardin AJ, Amon A. 2001. Men and sin: What's the difference? *Nat. Rev. Mol. Cell Biol.* 2:815–26
17. Baum B, Nishitani H, Yanow S, Nurse P. 1998. Cdc18 transcription and proteolysis couple S phase to passage through mitosis. *EMBO J.* 17:5689–98
18. Baum B, Wuarin J, Nurse P. 1997. Control of S-phase periodic transcription in the fission yeast mitotic cycle. *EMBO J.* 16:4676–88
19. Bertrand E, Chartrand P, Schaefer M, Shenoy SM, Singer RH, Long RM. 1998. Localization of ASH1 mRNA particles in living yeast. *Mol. Cell* 2:437–45
20. Bidlingmaier S, Weiss EL, Seidel C, Drubin DG, Snyder M. 2001. The Cbk1p pathway is important for polarized cell growth and cell separation in *Saccharomyces cerevisiae*. *Mol. Cell Biol.* 21:2449–62
21. Blackwell C, Martin KA, Greenall A, Pidoux A, Allshire RC, Whitehall SK. 2004. The *Schizosaccharomyces pombe* HIRA-like protein Hip1 is required for the periodic expression of histone genes and contributes to the function of complex centromeres. *Mol. Cell Biol.* 24:4309–20
22. Bobola N, Jansen RP, Shin TH, Nasmyth K. 1996. Asymmetric accumulation of Ash1p in postanaphase nuclei depends on a myosin and restricts yeast mating-type switching to mother cells. *Cell* 84:699–709
23. Bork P. 1993. Hundreds of ankyrin-like repeats in functionally diverse proteins: mobile modules that cross phyla horizontally? *Proteins* 17:363–74
24. Boros J, Lim FL, Darieva Z, Pic-Taylor A, Harman R, et al. 2003. Molecular determinants of the cell-cycle regulated Mcm1p-Fkh2p transcription factor complex. *Nucleic Acids Res.* 31:2279–88
25. Bouquin N, Johnson AL, Morgan BA, Johnston LH. 1999. Association of the cell cycle transcription factor Mbp1 with the Skn7 response regulator in budding yeast. *Mol. Biol. Cell* 10:3389–400
26. Breeden L. 1996. Start-specific transcription in yeast. *Curr. Top. Microbiol. Immunol.* 208:95–127
27. Breeden L, Nasmyth K. 1987. Cell cycle control of the yeast HO gene: *cis*- and *trans*-acting regulators. *Cell* 48:389–97
28. Breeden LL. 2000. Cyclin transcription: Timing is everything. *Curr. Biol.* 10:R586–88
29. **Breeden LL. 2003. Periodic transcription: a cycle within a cycle. *Curr. Biol.* 13:R31–38**

A good survey of recent data on cell cycle-regulated transcription that also addresses possible biological functions.

30. Buck V, Ng SS, Ruiz-Garcia AB, Papadopoulou K, Bhatti S, et al. 2004. Fkh2p and Sep1p regulate mitotic gene transcription in fission yeast. *J. Cell Sci.* 117:5623–32
31. Bulmer R, Pic-Taylor A, Whitehall SK, Martin KA, Millar JB, et al. 2004. The forkhead transcription factor Fkh2 regulates the cell division cycle of *Schizosaccharomyces pombe*. *Eukaryot. Cell* 3:944–54
32. Caligiuri M, Beach D. 1993. Sct1 functions in partnership with Cdc10 in a transcription complex that activates cell cycle START and inhibits differentiation. *Cell* 72:607–19
33. Carlsson P, Mahlapuu M. 2002. Forkhead transcription factors: key players in development and metabolism. *Dev. Biol.* 250:1–23
34. Chen D, Toone WM, Mata J, Lyne R, Burns G, et al. 2003. Global transcriptional responses of fission yeast to environmental stress. *Mol. Biol. Cell* 14:214–29
35. Cho RJ, Campbell MJ, Winzeler EA, Steinmetz L, Conway A, et al. 1998. A genome-wide transcriptional analysis of the mitotic cell cycle. *Mol. Cell.* 2:65–73
36. Colman-Lerner A, Chin TE, Brent R. 2001. Yeast Cbk1 and Mob2 activate daughter-specific genetic programs to induce asymmetric cell fates. *Cell* 107:739–50

37. Connolly T, Beach D. 1994. Interaction between the Cig1 and Cig2 B-type cyclins in the fission yeast cell cycle. *Mol. Cell Biol.* 14:768–76
38. Connolly T, Caligiuri M, Beach D. 1997. The Cdc2 protein kinase controls Cdc10/Sct1 complex formation. *Mol. Biol. Cell* 8:1105–15
39. Cooper S, Shedden K. 2003. Microarray analysis of gene expression during the cell cycle. *Cell Chrom.* 2:1
40. Cosma MP, Panizza S, Nasmyth K. 2001. Cdk1 triggers association of RNA polymerase to cell cycle promoters only after recruitment of the mediator by SBF. *Mol. Cell.* 7:1213–20
41. Cosma MP, Tanaka T, Nasmyth K. 1999. Ordered recruitment of transcription and chromatin remodeling factors to a cell cycle– and developmentally regulated promoter. *Cell* 97:299–311
42. Costanzo M, Nishikawa JL, Tang X, Millman JS, Schub O, et al. 2004. CDK activity antagonizes Whi5, an inhibitor of G1/S transcription in yeast. *Cell* 117:899–913
43. Costanzo M, Schub O, Andrews B. 2003. G1 transcription factors are differentially regulated in *Saccharomyces cerevisiae* by the Swi6-binding protein Stb1. *Mol. Cell Biol.* 23:5064–77
44. Cunliffe L, White S, McInerny CJ. 2004. DSC1-MCB regulation of meiotic transcription in *Schizosaccharomyces pombe*. *Mol. Genet. Genom.* 271:60–71
45. Darieva Z, Pic-Taylor A, Boros J, Spanos A, Geymonat M, et al. 2003. Cell cycle-regulated transcription through the FHA domain of Fkh2p and the coactivator Ndd1p. *Curr. Biol.* 13:1740–45
46. de Bruin RA, McDonald WH, Kalashnikova TI, Yates J 3rd, Wittenberg C. 2004. Cln3 activates G1-specific transcription via phosphorylation of the SBF bound repressor Whi5. *Cell* 117:887–98
47. De Groot PW, Hellingwerf KJ, Klis FM. 2003. Genome-wide identification of fungal GPI proteins. *Yeast* 20:781–96
48. **de Lichtenberg U, Jensen LJ, Brunak S, Bork P. 2005. Dynamic complex formation during the yeast cell cycle. *Science* 307:724–27**

An elegant analysis of the dynamics and regulation of protein complexes during the cell cycle, integrating expression profiling and physical interaction data.

49. de Lichtenberg U, Jensen TS, Jensen LJ, Brunak S. 2003. Protein feature based identification of cell cycle regulated proteins in yeast. *J. Mol. Biol.* 329:663–74
50. Ding R, Smith GR. 1998. Global control of meiotic recombination genes by *Schizosaccharomyces pombe* rec16 (rep1). *Mol. Gen. Genet.* 258:663–70
51. Dirick L, Bohm T, Nasmyth K. 1995. Roles and regulation of Cln-Cdc28 kinases at the start of the cell cycle of *Saccharomyces cerevisiae*. *EMBO J.* 14:4803–13
52. Dirick L, Moll T, Auer H, Nasmyth K. 1992. A central role for SWI6 in modulating cell cycle Start-specific transcription in yeast. *Nature* 357:508–13
53. Dohrmann PR, Butler G, Tamai K, Dorland S, Greene JR, et al. 1992. Parallel pathways of gene regulation: homologous regulators SWI5 and ACE2 differentially control transcription of HO and chitinase. *Genes Dev.* 6:93–104
54. Dohrmann PR, Voth WP, Stillman DJ. 1996. Role of negative regulation in promoter specificity of the homologous transcriptional activators Ace2p and Swi5p. *Mol. Cell Biol.* 16:1746–58
55. Doolin MT, Johnson AL, Johnston LH, Butler G. 2001. Overlapping and distinct roles of the duplicated yeast transcription factors Ace2p and Swi5p. *Mol. Microbiol.* 40:422–32
56. Durocher D, Henckel J, Fersht AR, Jackson SP. 1999. The FHA domain is a modular phosphopeptide recognition motif. *Mol. Cell* 4:387–94

57. Durocher D, Jackson SP. 2002. The FHA domain. *FEBS Lett.* 513:58–66
58. Fankhauser C, Reymond A, Cerutti L, Utzig S, Hofmann K, Simanis V. 1995. The *S. pombe* cdc15 gene is a key element in the reorganization of F-actin at mitosis. *Cell* 82:435–44
59. Forsburg SL. 1999. The best yeast? *Trends Genet.* 15:340–44
60. Foster R, Mikesell GE, Breeden L. 1993. Multiple SWI6-dependent *cis*-acting elements control SWI4 transcription through the cell cycle. *Mol. Cell Biol.* 13:3792–801
61. Freeman KB, Karns LR, Lutz KA, Smith MM. 1992. Histone H3 transcription in *Saccharomyces cerevisiae* is controlled by multiple cell cycle activation sites and a constitutive negative regulatory element. *Mol. Cell Biol.* 12:5455–63
62. Futcher B. 2000. Microarrays and cell cycle transcription in yeast. *Curr. Opin. Cell Biol.* 12:710–15
63. **Futcher B. 2002. Transcriptional regulatory networks and the yeast cell cycle. *Curr. Opin. Cell Biol.* 14:676–83**
64. Geymonat M, Spanos A, Wells GP, Smerdon SJ, Sedgwick SG. 2004. Clb6/Cdc28 and cdc14 regulate phosphorylation status and cellular localization of swi6. *Mol. Cell Biol.* 24:2277–85
65. Gonzalez F, Delahodde A, Kodadek T, Johnston SA. 2002. Recruitment of a 19S proteasome subcomplex to an activated promoter. *Science* 296:548–50
66. Grallert A, Grallert B, Ribar B, Sipiczki M. 1998. Coordination of initiation of nuclear division and initiation of cell division in *Schizosaccharomyces pombe*: genetic interactions of mutations. *J. Bacteriol.* 180:892–900
67. Hereford LM, Osley MA, Ludwig TR 2nd, McLaughlin CS. 1981. Cell-cycle regulation of yeast histone mRNA. *Cell* 24:367–75
68. Ho Y, Costanzo M, Moore L, Kobayashi R, Andrews BJ. 1999. Regulation of transcription at the *Saccharomyces cerevisiae* Start transition by Stb1, a Swi6-binding protein. *Mol. Cell Biol.* 19:5267–78
69. Hollenhorst PC, Bose ME, Mielke MR, Muller U, Fox CA. 2000. Forkhead genes in transcriptional silencing, cell morphology and the cell cycle. Overlapping and distinct functions for FKH1 and FKH2 in *Saccharomyces cerevisiae*. *Genetics* 154:1533–48
70. Hollenhorst PC, Pietz G, Fox CA. 2001. Mechanisms controlling differential promoter-occupancy by the yeast forkhead proteins Fkh1p and Fkh2p: implications for regulating the cell cycle and differentiation. *Genes Dev.* 15:2445–56
71. **Horak CE, Luscombe NM, Qian J, Bertone P, Piccirrillo S, et al. 2002. Complex transcriptional circuitry at the G1/S transition in *Saccharomyces cerevisiae*. *Genes Dev.* 16:3017–33**
72. Horie S, Watanabe Y, Tanaka K, Nishiwaki S, Fujioka H, et al. 1998. The *Schizosaccharomyces pombe* mei4+ gene encodes a meiosis-specific transcription factor containing a forkhead DNA-binding domain. *Mol. Cell Biol.* 18:2118–29
73. Ihmels J, Friedlander G, Bergmann S, Sarig O, Ziv Y, Barkai N. 2002. Revealing modular organization in the yeast transcriptional network. *Nat. Genet.* 31:370–77
74. Ingolia NT, Murray AW. 2004. The ups and downs of modeling the cell cycle. *Curr. Biol.* 14:R771–77
75. **Iyer VR, Horak CE, Scafe CS, Botstein D, Snyder M, Brown PO. 2001. Genomic binding sites of the yeast cell-cycle transcription factors SBF and MBF. *Nature* 409:533–38**

A useful overview that points out important issues pertinent to the dissection of transcriptional regulatory networks.

A comprehensive study of the complex transcriptional network that regulates the G1/S transition in budding yeast, involving numerous TFs downstream of MBF and SBF.

The first genome-wide analysis of genomic binding sites of cell-cycle TFs, revealing numerous putative target genes and functional specialization of SBF and MBF.

76. Jorgensen P, Tyers M. 2000. The fork'ed path to mitosis. *Genome Biol.* 1: reviews1022
77. Jorgensen P, Tyers M. 2004. How cells coordinate growth and division. *Curr. Biol.* 14:R1014–27
78. **Kato M, Hata N, Banerjee N, Futcher B, Zhang MQ. 2004. Identifying combinatorial regulation of transcription factors and binding motifs. *Genome Biol.* 5:R56**
79. Kaufmann E, Knöchel W. 1996. Five years on the wings of forkhead. *Mech. Dev.* 57:3–20
80. Knapp D, Bhoite L, Stillman DJ, Nasmyth K. 1996. The transcription factor Swi5 regulates expression of the cyclin kinase inhibitor p40SIC1. *Mol. Cell Biol.* 16:5701–7
81. Koch C, Moll T, Neuberg M, Ahorn H, Nasmyth K. 1993. A role for the transcription factors Mbp1 and Swi4 in progression from G1 to S phase. *Science* 261:1551–57
82. Koch C, Schleiffer A, Ammerer G, Nasmyth K. 1996. Switching transcription on and off during the yeast cell cycle: Cln/Cdc28 kinases activate bound transcription factor SBF (Swi4/Swi6) at start, whereas Clb/Cdc28 kinases displace it from the promoter in G2. *Genes Dev.* 10:129–41
83. Koranda M, Schleiffer A, Endler L, Ammerer G. 2000. Forkhead-like transcription factors recruit Ndd1 to the chromatin of G2/M-specific promoters. *Nature* 406:94–98
84. Krebs JE, Kuo MH, Allis CD, Peterson CL. 1999. Cell cycle-regulated histone acetylation required for expression of the yeast HO gene. *Genes Dev.* 13:1412–21
85. Kumar R, Reynolds DM, Shevchenko A, Goldstone SD, Dalton S. 2000. Forkhead transcription factors, Fkh1p and Fkh2p, collaborate with Mcm1p to control transcription required for M-phase. *Curr. Biol.* 10:896–906
86. La Thangue NB, Taylor WR. 1993. A structural similarity between mammalian and yeast transcription factors for cell-cycle-regulated genes. *Trends Cell Biol.* 3:75–76
87. Laabs TL, Markwardt DD, Slattery MG, Newcomb LL, Stillman DJ, Heideman W. 2003. ACE2 is required for daughter cell–specific G1 delay in *Saccharomyces cerevisiae*. *Proc. Natl. Acad. Sci. USA* 100:10275–80
88. Laoukili J, Kooistra MR, Bras A, Kauw J, Kerkhoven RM, et al. 2005. FoxM1 is required for execution of the mitotic programme and chromosome stability. *Nat. Cell Biol.* 7:126–36
89. Lee KM, Miklos I, Du H, Watt S, Szilagyi Z, et al. 2005. Impairment of the TFIIH-associated CDK-activating kinase selectively affects cell. *Mol. Biol. Cell.* 16:2734–45
90. Lee TI, Rinaldi NJ, Robert F, Odom DT, Bar-Joseph Z, et al. 2002. Transcriptional regulatory networks in *Saccharomyces cerevisiae*. *Science* 298:799–804
91. Liao JC, Boscolo R, Yang YL, Tran LM, Sabatti C, Roychowdhury VP. 2003. Network component analysis: reconstruction of regulatory signals in biological systems. *Proc. Natl. Acad. Sci. USA* 100:15522–27
92. Lim FL, Hayes A, West AG, Pic-Taylor A, Darieva Z, et al. 2003. Mcm1p-induced DNA bending regulates the formation of ternary transcription factor complexes. *Mol. Cell Biol.* 23:450–61
93. Lowndes NF, Johnson AL, Breeden L, Johnston LH. 1992. SWI6 protein is required for transcription of the periodically expressed DNA synthesis genes in budding yeast. *Nature* 357:505–8
94. Lowndes NF, Johnson AL, Johnston LH. 1991. Coordination of expression of DNA synthesis genes in budding yeast by a cell-cycle regulated *trans* factor. *Nature* 350:247–50
95. Lowndes NF, McInerny CJ, Johnson AL, Fantes PA, Johnston LH. 1992. Control of DNA synthesis genes in fission yeast by the cell-cycle gene cdc10+. *Nature* 355:449–53
96. Loy CJ, Lydall D, Surana U. 1999. NDD1, a high-dosage suppressor of cdc28-1N,

A good example for the insight that can be obtained by mining and integrating different global data sets, uncovering mechanisms of TF interactions and combinatorial control.

is essential for expression of a subset of late-S-phase-specific genes in *Saccharomyces cerevisiae*. *Mol. Cell Biol.* 19:3312–27

97. Lydall D, Ammerer G, Nasmyth K. 1991. A new role for MCM1 in yeast: cell cycle regulation of SWI5 transcription. *Genes Dev.* 5:2405–19
98. MacKay VL, Mai B, Waters L, Breeden LL. 2001. Early cell cycle box-mediated transcription of CLN3 and SWI4 contributes to the proper timing of the G1-to-S transition in budding yeast. *Mol. Cell Biol.* 21:4140–48
99. MacNeill SA, Nurse P. 1997. Cell cycle control in fission yeast. In *The Molecular and Cellular Biology of the Yeast Saccharomyces: Life Cycle and Cell Biology*, ed. JR Pringle, J Broach, EW Jones, pp. 697–763. Cold Spring Harbor, NY: Cold Spring Harbor Lab. Press
100. Maher M, Cong F, Kindelberger D, Nasmyth K, Dalton S. 1995. Cell cycle-regulated transcription of the CLB2 gene is dependent on Mcm1 and a ternary complex factor. *Mol. Cell Biol.* 15:3129–37
101. Mai B, Miles S, Breeden LL. 2002. Characterization of the ECB binding complex responsible for the M/G1-specific transcription of CLN3 and SWI4. *Mol. Cell Biol.* 22:430–41
102. Maqbool Z, Kersey PJ, Fantes PA, McInerny CJ. 2003. MCB-mediated regulation of cell cycle-specific cdc22+ transcription in fission yeast. *Mol. Genet. Genom.* 269:765–75
103. Martín-Cuadrado AB, Dueñas E, Sipiczki M, Vázquez de Aldana CR, Del Rey F. 2003. The endo-beta-1,3-glucanase eng1p is required for dissolution of the primary septum during cell separation in *Schizosaccharomyces pombe*. *J. Cell Sci.* 116:1689–98
104. Martin DE, Soulard A, Hall MN. 2004. TOR regulates ribosomal protein gene expression via PKA and the Forkhead transcription factor FHL1. *Cell* 119:969–79
105. Mata J, Lyne R, Burns G, Bähler J. 2002. The transcriptional program of meiosis and sporulation in fission yeast. *Nat. Genet.* 32:143–47
106. Matsumoto S, Yanagida M. 1985. Histone gene organization of fission yeast: a common upstream sequence. *EMBO J.* 4:3531–38
107. Matsumoto S, Yanagida M, Nurse P. 1987. Histone transcription in cell cycle mutants of fission yeast. *EMBO J.* 6:1093–97
108. McBride HJ, Yu Y, Stillman DJ. 1999. Distinct regions of the Swi5 and Ace2 transcription factors are required for specific gene activation. *J. Biol. Chem.* 274:21029–36
109. McCollum D, Gould KL. 2001. Timing is everything: regulation of mitotic exit and cytokinesis by the MEN and SIN. *Trends Cell Biol.* 11:89–95
110. McInerny CJ. 2004. Cell cycle-regulated transcription in fission yeast. *Biochem. Soc. Trans.* 32:967–72
111. McInerny CJ, Kersey PJ, Creanor J, Fantes PA. 1995. Positive and negative roles for cdc10 in cell cycle gene expression. *Nucleic Acids Res.* 23:4761–68
112. McInerny CJ, Partridge JF, Mikesell GE, Creemer DP, Breeden LL. 1997. A novel Mcm1-dependent element in the SWI4, CLN3, CDC6, and CDC47 promoters activates M/G1-specific transcription. *Genes Dev.* 11:1277–88
113. McIntosh EM, Atkinson T, Storms RK, Smith M. 1991. Characterization of a short, cis-acting DNA sequence which conveys cell cycle stage-dependent transcription in *Saccharomyces cerevisiae*. *Mol. Cell Biol.* 11:329–37
114. Meeks-Wagner D, Hartwell LH. 1986. Normal stoichiometry of histone dimer sets is necessary for high fidelity of mitotic chromosome transmission. *Cell* 44:43–52
115. Messenguy F, Dubois E. 2003. Role of MADS box proteins and their cofactors in combinatorial control of gene expression and cell development. *Gene* 316:1–21

116. Miyamoto M, Tanaka K, Okayama H. 1994. res^{2+}, a new member of the $cdc10^{+}$/SWI4 family, controls the 'start' of mitotic and meiotic cycles in fission yeast. *EMBO J.* 13:1873–80
117. Moll T, Tebb G, Surana U, Robitsch H, Nasmyth K. 1991. The role of phosphorylation and the CDC28 protein kinase in cell cycle-regulated nuclear import of the S. cerevisiae transcription factor SWI5. *Cell* 66:743–58
118. Morgan BA, Bouquin N, Merrill GF, Johnston LH. 1995. A yeast transcription factor bypassing the requirement for SBF and DSC1/MBF in budding yeast has homology to bacterial signal transduction proteins. *EMBO J.* 14:5679–89
119. Morgan DO. 1997. Cyclin-dependent kinases: engines, clocks, and microprocessors. *Annu. Rev. Cell Dev. Biol.* 13:261–91
120. Morillon A, O'Sullivan J, Azad A, Proudfoot N, Mellor J. 2003. Regulation of elongating RNA polymerase II by forkhead transcription factors in yeast. *Science* 300:492–95
121. Morris MC, Kaiser P, Rudyak S, Baskerville C, Watson MH, Reed SI. 2003. Cks1-dependent proteasome recruitment and activation of CDC20 transcription in budding yeast. *Nature* 424:1009–13
122. Nakashima N, Tanaka K, Sturm S, Okayama H. 1995. Fission yeast Rep2 is a putative transcriptional activator subunit for the cell cycle 'start' function of Res2-Cdc10. *EMBO J.* 14:4794–802
123. Nasmyth K, Adolf G, Lydall D, Seddon A. 1990. The identification of a second cell cycle control on the HO promoter in yeast: cell cycle regulation of SW15 nuclear entry. *Cell* 62:631–47
124. Nasmyth K, Dirick L. 1991. The role of SWI4 and SWI6 in the activity of G1 cyclins in yeast. *Cell* 66:995–1013
125. Nelson B, Kurischko C, Horecka J, Mody M, Nair P, et al. 2003. RAM: a conserved signaling network that regulates Ace2p transcriptional activity and polarized morphogenesis. *Mol. Biol. Cell* 14:3782–803
126. Obara-Ishihara T, Okayama H. 1994. A B-type cyclin negatively regulates conjugation via interacting with cell cycle 'start' genes in fission yeast. *EMBO J.* 13:1863–72
127. Peng X, Murthy Karuturi RK, Miller LD, Lin K, Jia Y, et al. 2005. Identification of cell cycle-regulated genes in fission yeast. *Mol. Biol. Cell* 16:1026–42
128. Pic-Taylor A, Darieva Z, Morgan BA, Sharrocks AD. 2004. Regulation of cell cycle-specific gene expression through cyclin-dependent kinase-mediated phosphorylation of the forkhead transcription factor Fkh2p. *Mol. Cell Biol.* 24:10036–46
129. Pic A, Lim FL, Ross SJ, Veal EA, Johnson AL, et al. 2000. The forkhead protein Fkh2 is a component of the yeast cell cycle transcription factor SFF. *EMBO J.* 19:3750–61
130. Pierrou S, Hellqvist M, Samuelsson L, Enerback S, Carlsson P. 1994. Cloning and characterization of seven human forkhead proteins: binding site specificity and DNA bending. *EMBO J.* 13:5002–12
131. Porter SE, Washburn TM, Chang M, Jaehning JA. 2002. The yeast pafl-RNA polymerase II complex is required for full expression of a subset of cell cycle-regulated genes. *Eukaryot. Cell* 1:830–42
132. Pramila T, Miles S, GuhaThakurta D, Jemiolo D, Breeden LL. 2002. Conserved homeodomain proteins interact with MADS box protein Mcm1 to restrict ECB-dependent transcription to the M/G1 phase of the cell cycle. *Genes Dev.* 16:3034–45
133. Primig M, Sockanathan S, Auer H, Nasmyth K. 1992. Anatomy of a transcription factor important for the start of the cell cycle in *Saccharomyces cerevisiae*. *Nature* 358:593–97

134. Raithatha SA, Stuart DT. 2005. Meiosis-specific regulation of the *S.cerevisiae* S-phase cyclin CLB5 is dependent upon MCB elements in its promoter but is independent of MBF activity. *Genetics* 169:1329–42
135. Reymond A, Marks J, Simanis V. 1993. The activity of *S. pombe* DSC-1-like factor is cell cycle regulated and dependent on the activity of p34cdc2. *EMBO J.* 12:4325–34
136. Reynolds D, Shi BJ, McLean C, Katsis F, Kemp B, Dalton S. 2003. Recruitment of Thr 319-phosphorylated Ndd1p to the FHA domain of Fkh2p requires Clb kinase activity: a mechanism for CLB cluster gene activation. *Genes Dev.* 17:1789–802
137. Ribar B, Banrevi A, Sipiczki M. 1997. sep1+ encodes a transcription-factor homologue of the HNF-3/forkhead DNA-binding-domain family in *Schizosaccharomyces pombe*. *Gene* 202:1–5
138. Rustici G, Mata J, Kivinen K, Lio P, Penkett CJ, et al. 2004. Periodic gene expression program of the fission yeast cell cycle. *Nat. Genet.* 36:809–17

A comprehensive microarray study on periodic transcription and its control in *S. pombe*, comparing cell-cycle transcriptomes and regulatory networks to *S. cerevisiae*.

139. Schenkman LR, Caruso C, Page N, Pringle JR. 2002. The role of cell cycle-regulated expression in the localization of spatial landmark proteins in yeast. *J. Cell Biol.* 156:829–41
140. Sherr CJ, McCormick F. 2002. The RB and p53 pathways in cancer. *Cancer Cell* 2:103–12
141. Sherwood PW, Tsang SV, Osley MA. 1993. Characterization of HIR1 and HIR2, two genes required for regulation of histone gene transcription in *Saccharomyces cerevisiae*. *Mol. Cell Biol.* 13:28–38
142. Shore P, Sharrocks AD. 1995. The MADS-box family of transcription factors. *Eur. J. Biochem.* 229:1–13
143. Sidorova JM, Mikesell GE, Breeden LL. 1995. Cell cycle-regulated phosphorylation of Swi6 controls its nuclear localization. *Mol. Biol. Cell* 6:1641–58
144. Sil A, Herskowitz I. 1996. Identification of asymmetrically localized determinant, Ash1p, required for lineage-specific transcription of the yeast HO gene. *Cell* 84:711–22
145. Simon I, Barnett J, Hannett N, Harbison CT, Rinaldi NJ, et al. 2001. Serial regulation of transcriptional regulators in the yeast cell cycle. *Cell* 106:697–708

This study determines global binding sites of several TFs and provides a valuable framework to understand the transcriptional regulatory network of the cell cycle.

146. Sipiczki M, Grallert B, Miklos I. 1993. Mycelial and syncytial growth in *Schizosaccharomyces pombe* induced by novel septation mutations. *J. Cell Sci.* 104:485–93
147. Spector MS, Raff A, DeSilva H, Lee K, Osley MA. 1997. Hir1p and Hir2p function as transcriptional corepressors to regulate histone gene transcription in the *Saccharomyces cerevisiae* cell cycle. *Mol. Cell Biol.* 17:545–52
148. Spellman PT, Sherlock G, Zhang MQ, Iyer VR, Anders K, et al. 1998. Comprehensive identification of cell cycle-regulated genes of the yeast *Saccharomyces cerevisiae* by microarray hybridization. *Mol. Biol. Cell* 9:3273–97

A pioneering microarray paper on periodic gene expression that provided the fruitful basis for numerous follow-up studies.

149. Stuart D, Wittenberg C. 1995. CLN3, not positive feedback, determines the timing of CLN2 transcription in cycling cells. *Genes Dev.* 9:2780–94
150. Sun K, Coic E, Zhou Z, Durrens P, Haber JE. 2002. Saccharomyces forkhead protein Fkh1 regulates donor preference during mating-type switching through the recombination enhancer. *Genes Dev.* 16:2085–96
151. Tahara S, Tanaka K, Yuasa Y, Okayama H. 1998. Functional domains of rep2, a transcriptional activator subunit for Res2-Cdc10, controlling the cell cycle "start". *Mol. Biol. Cell* 9:1577–88
152. Takahashi K, Chen ES, Yanagida M. 2000. Requirement of Mis6 centromere connector for localizing a CENP-A-like protein in fission yeast. *Science* 288:2215–19

153. Takizawa PA, Sil A, Swedlow JR, Herskowitz I, Vale RD. 1997. Actin-dependent localization of an RNA encoding a cell-fate determinant in yeast. *Nature* 389:90–93
154. Tanaka K, Okayama H. 2000. A pcl-like cyclin activates the Res2p-Cdc10p cell cycle "start" transcriptional factor complex in fission yeast. *Mol. Biol. Cell* 11:2845–62
155. Tanaka K, Okazaki K, Okazaki N, Ueda T, Sugiyama A, et al. 1992. A new cdc gene required for S phase entry of *Schizosaccharomyces pombe* encodes a protein similar to the cdc 10+ and SWI4 gene products. *EMBO J.* 11:4923–32
156. Taylor IA, Treiber MK, Olivi L, Smerdon SJ. 1997. The X-ray structure of the DNA-binding domain from the *Saccharomyces cerevisiae* cell-cycle transcription factor Mbp1 at 2.1 Å resolution. *J. Mol. Biol.* 272:1–8
157. Toone WM, Aerne BL, Morgan BA, Johnston LH. 1997. Getting started: regulating the initiation of DNA replication in yeast. *Annu. Rev. Microbiol.* 51:125–49
158. Treisman R. 1994. Ternary complex factors: growth factor regulated transcriptional activators. *Curr. Opin. Genet. Dev.* 4:96–101
159. Tyers M. 2004. Cell cycle goes global. *Curr. Opin. Cell Biol.* 16:602–13

An excellent overview and perspective on the power of genome-wide approaches to understand cell-cycle control.

160. Tyers M, Tokiwa G, Futcher B. 1993. Comparison of the *Saccharomyces cerevisiae* G1 cyclins: Cln3 may be an upstream activator of Cln1, Cln2 and other cyclins. *EMBO J.* 12:1955–68
161. Visintin R, Craig K, Hwang ES, Prinz S, Tyers M, Amon A. 1998. The phosphatase Cdc14 triggers mitotic exit by reversal of Cdk-dependent phosphorylation. *Mol. Cell* 2:709–18
162. Weiss EL, Kurischko C, Zhang C, Shokat K, Drubin DG, Luca FC. 2002. The *Saccharomyces cerevisiae* Mob2p-Cbk1p kinase complex promotes polarized growth and acts with the mitotic exit network to facilitate daughter cell–specific localization of Ace2p transcription factor. *J. Cell Biol.* 158:885–900
163. White JH, Barker DG, Nurse P, Johnston LH. 1986. Periodic transcription as a means of regulating gene expression during the cell cycle: contrasting modes of expression of DNA ligase genes in budding and fission yeast. *EMBO J.* 5:1705–9
164. White S, Khaliq F, Sotiriou S, McInerny CJ. 2001. The role of DSC1 components cdc10+, rep1+ and rep2+ in MCB gene transcription at the mitotic G1-S boundary in fission yeast. *Curr. Genet.* 40:251–59
165. Whitehall S, Stacey P, Dawson K, Jones N. 1999. Cell cycle-regulated transcription in fission yeast: Cdc10-Res protein interactions during the cell cycle and domains required for regulated transcription. *Mol. Biol. Cell* 10:3705–15
166. Whitfield ML, Sherlock G, Saldanha AJ, Murray JI, Ball CA, et al. 2002. Identification of genes periodically expressed in the human cell cycle and their expression in tumors. *Mol. Biol. Cell* 13:1977–2000
167. Wijnen H, Futcher B. 1999. Genetic analysis of the shared role of CLN3 and BCK2 at the G1-S transition in *Saccharomyces cerevisiae*. *Genetics* 153:1131–43
168. Wijnen H, Landman A, Futcher B. 2002. The G1 cyclin Cln3 promotes cell cycle entry via the transcription factor Swi6. *Mol. Cell Biol.* 22:4402–18
168a. Wittenberg C, Reed SI. 2005. Cell cycle-dependent transcription in yeast: promoters, transcription factors, and transcriptomes. *Oncogene* 24:2746–55
169. Wuarin J, Buck V, Nurse P, Millar JB. 2002. Stable association of mitotic cyclin B/Cdc2 to replication origins prevents endoreduplication. *Cell* 111:419–31
170. Wynne J, Treisman R. 1992. SRF and MCM1 have related but distinct DNA binding specificities. *Nucleic Acids Res.* 20:3297–303

171. Wyrick JJ, Young RA. 2002. Deciphering gene expression regulatory networks. *Curr. Opin. Genet. Dev.* 12:130–36
172. Yamano H, Kitamura K, Kominami K, Lehmann A, Katayama S, et al. 2000. The spike of S phase cyclin Cig2 expression at the G1-S border in fission yeast requires both APC and SCF ubiquitin ligases. *Mol. Cell* 6:1377–87
173. Zhu G, Spellman PT, Volpe T, Brown PO, Botstein D, et al. 2000. Two yeast forkhead genes regulate the cell cycle and pseudohyphal growth. *Nature* 406:90–94
174. Zhu Y, Takeda T, Nasmyth K, Jones N. 1994. pct1+, which encodes a new DNA-binding partner of p85cdc10, is required for meiosis in the fission yeast *Schizosaccharomyces pombe*. *Genes Dev.* 8:885–98
175. Zhu Y, Takeda T, Whitehall S, Peat N, Jones N. 1997. Functional characterization of the fission yeast Start-specific transcription factor Res2. *EMBO J.* 16:1023–34
176. Zilahi E, Salimova E, Simanis V, Sipiczki M. 2000. The *S. pombe* sep1 gene encodes a nuclear protein that is required for periodic expression of the cdc15 gene. *FEBS Lett.* 481:105–8

RELATED RESOURCES

The Regulation of Histone Synthesis in the Cell Cycle
MA Osley
Annual Review of Biochemistry. Volume 60, Page 827–861, July 1991

Yeast Transcriptional Regulatory Mechanisms
K Struhl
Annual Review of Genetics. Volume 29, Page 651–674, December 1995

Getting Started: Regulating the Initiation of DNA Replication in Yeast
WM Toone, BL Aerne, BA Morgan, LH Johnston
Annual Review of Microbiology. Volume 51, Page 125–149, October 1997

Cyclin-Dependent Kinases: Engines, Clocks, and Microprocessors
DO Morgan
Annual Review of Cell and Developmental Biology. Volume 13, Page 261–291, November 1997

Transcription of Eukaryotic Protein-Coding Genes
TI Lee, RA Young
Annual Review of Genetics. Volume 34, Page 77–137, December 2000

The Dynamics of Chromosome Organization and Gene Regulation
DL Spector
Annual Review of Biochemistry. Volume 72, Page 573–608, July 2003

Comparative Developmental Genetics and the Evolution of Arthropod Body Plans

David R. Angelini[1] and Thomas C. Kaufman[2]

[1]Department of Ecology and Evolutionary Biology, University of Connecticut, Storrs, Connecticut 06269-3043; email: david.angeliniguconn.edu

[2]Department of Biology, Indiana University, Bloomington, Indiana 47405-3700; email: kaufman@bio.indiana.edu

Annu. Rev. Genet.
2005. 39:95–119

First published online as a Review in Advance on June 21, 2005

The *Annual Review of Genetics* is online at http://genet.annualreviews.org

doi: 10.1146/annurev.genet.39.073003.112310

0066-4197/05/1215-0095$20.00

Key Words

comparative developmental genetics, Arthropoda, body plan, Hox genes, insect wing modifications

Abstract

The arthropods display a wide range of morphological diversity, varying tagmosis, as well as other aspects of the body plan, such as appendage and cuticular morphology. Here we review the roles of developmental regulatory genes in the evolution of arthropod morphology, with an emphasis on what is known from morphologically diverse species. Examination of tagmatic evolution reveals that these changes have been accompanied by changes in the expression patterns of Hox genes. In contrast, review of the modifications to wing morphology seen in insects shows that these body plan changes have generally favored alterations in downstream target genes. These and other examples are used to discuss the evolutionary implications of comparative developmental genetic data.

Contents

INTRODUCTION

The diversity of animal life is one of our world's most mysterious and intriguing qualities. Individual species have multiplied to fill ecological niches through a dizzying array of physiological and morphological modifications. Modern evolutionary biology has begun to understand the mechanisms involved in speciation (28, 98). However, one of the principal and persistent mysteries remains the origins and evolution of the novel morphologies and body plans that arise among diverse species (71). In recent decades, investigators of comparative developmental genetics have applied tools and ideas from molecular and developmental biology to some of the questions of morphological evolution with heartening success. This new field has also been known as phylontogenetics, or more commonly "evo-devo"—the truncated catchall named for two of its most influential parent disciplines, evolutionary and developmental biology.

One of the animal groups that has been a major beneficiary of comparative developmental genetics is the Arthropoda. [The others are vertebrates, basal chordates, and deuterostomes (e.g., 43, 66, 79).] Since their appearance in the Cambrian, approximately 530 mya, arthropods have dominated the animal world. They have evolved myriad variations on an anatomical theme. In practical, experimental terms, arthropod evo-devo has flourished as it has drawn on the well-established fields of entomology, carcinology (the study of crustaceans), and genetics. The first two have described a wealth of diverse morphology and body plans, while the latter has provided tools and new developmental hypotheses for their investigation.

Comparative Developmental Genetics and the Hox Genes

The homeotic complex (Hox) genes have emerged from genetic studies of the fruitfly, *Drosophila melanogaster*, as crucial early regulators of segment identity, and thus body plan organization, in arthropods. Hox genes are homeodomain transcription factors, remarkably well conserved in sequence and expression across the arthropods and other animals. Therefore, they provide a reliable and accessible experimental inroad to the study of diverse body plans. In general, Hox genes are expressed alone or in overlapping domains of adjacent body segments. They exhibit the intiguing feature of "colinearity," appearing in the gene complex in the order in which they are expressed along the anterior-posterior (AP) body axis. Ten Hox genes are expressed along the body of most arthropods, where they are usually named for their orthologues in *Drosophila*. From anterior to posterior, these are *labial* (*lab*), *proboscipedia* (*pb*), *Hox3/zen*, *Deformed* (*Dfd*), *Sex combs reduced* (*Scr*), *Hox6/ftz*, *Antennapedia* (*Antp*), *Ultrabithorax* (*Ubx*), *abdominal-A* (*abd-A*), and *Abdominal-B* (*Abd-B*). In the insects, *Hox3/zen* and *Hox6/ftz* have been modified and do not function as typical Hox genes in this group (48).

Hox genes specify the identity of body segments and structures where they are expressed, and mutations result in a homeotic transformation to some other fate, often to a more posterior identity (reviewed in 47). In 1978, characterization of the homeotic *bithorax* mutations of *Drosophila* led Ed Lewis to presage the growth of comparative Hox work that would come decades later:

> Flies almost certainly evolved from insects with four wings instead of two and insects are believed to have come from arthropod forms with many legs instead of six. During the evolution of the fly, two major groups of genes must have evolved: "leg-suppressing" genes which removed legs from abdominal segments of millipede-like ancestors followed by "haltere-promoting" genes which suppressed the second pair of wings of four-winged ancestors. If evolution indeed proceeded in this way, then mutations in the latter group of genes should produce four-winged flies and mutations in the former group, flies with extra legs. (57)

This evolutionary scenario described by Lewis has not been borne out quite as he envisioned it—rather than the evolution of new genes, the evolution of regulatory interactions appears to have been key to body plan changes. As we discuss below, the details of Hox expression domains and timing, as well as the target genes controlled by specific Hox genes, have been associated with greater and lesser aspects of body plan evolution in a range of arthropod groups.

Several excellent reviews covering different aspects of Hox genes and their connections to arthropod evolution have been published in recent years (8, 47, 69). These articles have emphasized the commonalities and themes seen across the arthropods and other animals. Here, we have attempted to organize our discussion in terms of several of the major novelties in arthropod body plan evolution. In the course of this, we revisit some of the same topics and update their consideration with recent data. Principally, we hope to illustrate the diversity of arthropod morphology and review developmental genetic data relevant to its evolutionary plasticity. With this in mind, we do not limit ourselves to discussions of tagmosis or to the Hox genes. These have been fruitful lines of research, but they are necessarily just the beginning.

What is the Meaning of the "Body Plan" Concept?

A body plan is a basic pattern of anatomical organization shared by a group of animals (71). However, there is sometimes disagreement over what constitutes a body plan. Part of this confusion may be historical, but much

AP: anterior-posterior, as in anterior-posterior body axis

Hox: homeotic complex. A cluster of homeodomain transcription factors required to specify the identity of body segments along the AP body axis of arthropods and other animals

Homeosis: the transformation of the normal identity of an anatomical structure or body segment to another's identity. Homeosis is usually considered in the context of mutations in developmental regulatory genes. Homeotic, regulatory mutations in these genes have also been proposed as a factor in some instances of body plan evolution

Tagmosis: the specialization and/or anatomical unification of adjacent body segments into "tagma" to facilitate certain behaviors, often through similar modification of appendages. Such behaviors may include gathering sensory information, feeding, locomotion of some sort, gas exchange, or brooding

of it doubtlessly stems from ambiguity inherent in the term. The first such conceptual grouping based on anatomy was the archetype defined by Richard Owen, who also established the more enduring idea of homology. An archetype was envisioned as all the possible variations upon an anatomical theme (64). However, in rejecting Darwin's theory of evolution by natural selection, Owen's archetype never addressed the relationships between distinct morphologies. In 1945, Joseph Henry Woodger first proposed the *bauplan* (literally a "structural" or "building" plan) as the collection of homologous anatomical features seen across the natural history of a group (70). This recast Owen's idea in an evolutionary context, and became translated as "body plan." The term is still sometimes erroneously used to denote an anatomical grade of organization. However, body plans remain a useful concept because they summarize a collection of ancestral and synapomorphic characters within a group, while accepting their various derivations, and asserting an implied hypothesis that these similarities appear due to the monophyly of the group.

Given this definition, where is it appropriate to apply the concept? Does it only apply at greater levels of classification, such as phyla? Can we speak of the arthropod or chordate body plans, but not apply the term to the anatomy of insects or tetrapods? We suggest that a useful concept should not be artificially limited, and see no problem in speaking of the body plan of any presumptively monophyletic group sharing a characteristic anatomy. It should be possible to consider "greater" or "lesser" levels of body plans. Indeed, this seems appropriate, given that significant morphological innovation has appeared within many phyla since their appearance in the Cambrian, and these may be no less significant to their natural history. Fitch & Sudhaus have made a similar argument based on changes in the body axes of nematode groups (35). Other examples of such innovations include the evolution of jointed limbs in sarcopterigid vertebrates, the appearance of wings in pterygote insects, and the loss of segmentation in higher mollusks. Therefore, body plans may be related by degrees to synapomorphies seen within any clade.

What Defines the Arthropod Body Plan?

The arthropods have traditionally been defined as segmented, appendage-bearing protostomes, protected by a cuticle that is periodically shed with growth (22). They are further distinguished from related groups, such as the onychophorans and tardigrades, by the fact that the appendages consist of podomeres with separate musculature and innervation (82). The specialization and/or anatomical unification of adjacent body segments, or tagmosis, helps to facilitate certain behaviors and varies greatly among arthropod groups. Tagmosis may also have followed convergent patterns along separate arthropod lineages, and several recent studies, based on molecular sequences and cladistic treatments of morphology, have questioned traditional arthropod groups, such as the Uniramia (54), Mandibulata (49), and Hexapoda (62). It has also fueled much debate over the phylogenetic relationships and monophyly of the extant arthropod classes (20, 21, 49, 73, 103). We frame our discussions in what we favor as the least controversial and most consensual of these phylogenies (20, 37, 72), summarized in **Figure 1.** Our favored phylogenetic hypothesis of arthropod relationships groups the insects, crustaceans, and myriapods into the Mandibulata. This group is unified by mouthpart homologies, and excludes the chelicerates.

DEVELOPMENTAL GENETICS OF ARTHROPOD TAGMOSIS

Evolution of the Arthropod Head

The union of anterior body segments into a well-developed head characterizes many of the extant arthropod groups. This presumably provides the advantages of gathering the sensory and feeding appendages at the forward end of these active animals. Cephalization seems to have been associated with the recruitment of posterior segments into the head, coupled with a reduction of anterior segments. A recent morphological study by Budd (23) examined the head anatomy of several fossil arthropods in an attempt to address the homology of the large frontal appendages that characterize fossil crustacean-like species, such as *Yohoia* and *Fortiforceps*, as well as lobopods, such as *Aysheaia*. This phylogenetic study included specimens that could be confidently assigned as basal members of the Chelicerata and Mandibulata, thereby including representatives of the extant crown groups without the interference of too many modern synapomorphies. In Budd's analysis, the crown groups (Mandibulata and Chelicerata) and trilobites formed a well-supported clade that includes fossil species lacking the frontal appendages, such as *Emeraldella* and *Cambropachycope*. Based on this and other anatomical data, he suggests that the labrum seen in Mandibulata and Chelicerata may be evolutionarily derived from the frontal appendage. This is consistent with the expression of appendage patterning genes in the labrum (41) and with their functional requirement for proper labral development in insects and chelicerates (9, 80).

TAGMOSIS IN MAJOR ARTHROPOD GROUPS

Insects are the most tagmatically consistent arthropod class. They bear a head of fused segments, a thorax of three segments, followed by an abdomen of 10 or 11 segments.

The myriapods possess a well-organized head, similar to that of insects, followed by a varying number of homonymous trunk segments. Chilopoda (centipedes) bear one pair of legs on each trunk segment, while Diplopoda (millipedes) bear two pairs of legs on most segments. Pauropoda represent a curious intermediate state, where a segment as seen from the dorsal side spans what are apparently two segments ventrally.

Presumably, basal crustaceans consist of head and trunk tagmata. Among Malacostraca, the body plan consists of three tagmata: cephalon, pereon, and pleon. The appendages of the pereon and pleon are usually divided functionally to tasks such as walking, swimming, respiration, or brooding eggs, but these tasks do not always fall to the same tagma in different groups.

The basal chelicerates possess a body plan organized into three tagmata: prosoma, mesosoma, and metasoma. The prosoma bears the mouthparts and legs, whereas the mesosoma bears respiratory appendages. Arachnids have apparently eliminated the metasoma (16), and consist of two tagmata: prosoma and opisthosoma.

Hox Genes and Tagmosis

From genetic studies in *Drosophila*, Hox genes have been well characterized as high-level regulatory transcription factors, which act to impart specific identities upon the body segments and other structures in which they are expressed (reviewed in 47). In certain instances, the overlap of two or more Hox genes can produce fates distinct from those specified by either gene alone. For example, in the *Drosophila* labial imaginal disc, *pb* and *Scr* interact to direct proboscis development, where alone these genes specify only maxillary palp or leg (2). Therefore, the 10 ancestral arthropod Hox genes are theoretically capable of specifying at least 20 unique body regions. (Assuming colinearity, if each Hox gene has an area where it is uniquely expressed and another in overlap with its neighbors, 2n-1 regions can be demarcated. An anterior Hox-free region adds one additional possible identity.) It is possible that such extensive differentiation exists within the central nervous system, but this has not been carefully examined. However, in the embryonic ectoderm of most arthropods, several Hox genes typically overlap in a given segment, such that far fewer than the theoretical maximum number of body regions is initially distinguished.

What is usually observed is a correlation between the tagmatic boundaries of an arthropod's body plan and the overlap of Hox genes in that region (**Figure 1**). For example, in arachnids (7, 31, 86), *lab*, *pb*, *Hox3*, *Dfd*,

Scr, and *Hox6* are expressed in a nested pattern within the prosoma, where they overlap broadly. Similarly, in the opisthosoma expression of *Antp*, *Ubx*, *abd-A*, and *Abd-B* orthologues also overlap in a nested manner. In the Mandibulata, a distinct head usually gathers several pairs of appendages, which perform separate gnathal functions. This reaches an extreme in the decapod crustaceans, which may have as many as seven pairs of appendages that function in feeding. In these arthropods, the Hox genes overlap very little in the head (5, 46, 47). It seems likely that this facilitates the specification of a greater number of distinct identities, although this has only been tested functionally in a handful of insect species.

Nevertheless, it is relatively uncommon for Hox genes to cross a tagmatic boundary. This is most clearly seen in the arachnids,

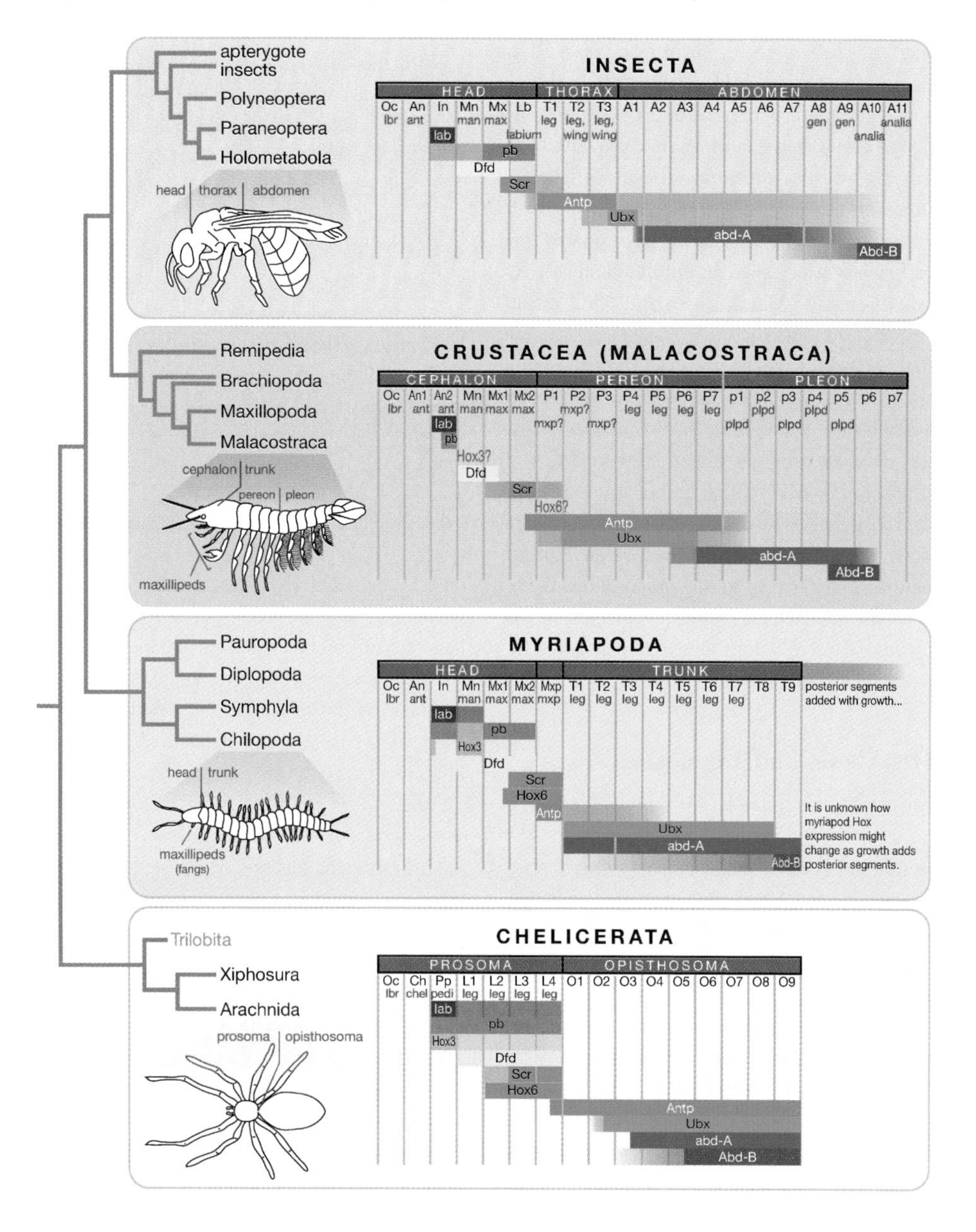

where segments of the prosoma and opisthosoma express separate sets of Hox genes. Only the *Antp* orthologue shows a small degree of crossover into the L4 prosomal segment, while it is predominantly expressed in the opisthosoma. In myriapods, the trunk is characterized by the expression of *Ubx* and *abd-A* (46). Again, *Antp* is the only Hox gene to cross this tagmatic boundary, although unlike its pattern in chelicerates, in the centipede it is predominantly expressed in the maxillipedial segment and tapers off posteriorly into the trunk. In the insects, *Antp* and *abd-A* chiefly specify the thorax and abdomen, respectively.

Among different crustacean body plans, the Hox genes of the trunk correlate in expression with tagmosis. Brachiopod crustaceans have a homonymous trunk, and in the brachiopod *Artemia franciscana*, *Antp*, *Ubx*, and *abd-A* overlap extensively in their expression (11). However, in the decapod *Procambarus clarkii* (5) and the isopod *Porcellio scaber* (4), expression of *Antp* and *Ubx* is restricted to the pereon, while *abd-A* expression appears in the pleon. Therefore, the boundaries of these tagmata are respected by the Hox genes.

So, does the overlap of Hox genes provide a molecular definition of tagmata? Not reliably, it appears. There are many instances where Hox expression crosses tagmatic boundaries, often at later stages of development to modify the fate of individual segments within a tagma. For example, in later stages of *Drosophila* embryogenesis, *Scr* and *Ubx* expression expand from neighboring tagmata into T1 and T3, respectively, to modify their identities. As is discussed below, these domains of expression likely evolved in connection with the placement and specialization of wings on the insect body plan.

T1: first thoracic segment or prothorax

The Recruitment of Maxillipeds in Crustacea

Hox genes are known to specify segment identity in *Drosophila*, alone or in combination. However, viewed from an evolutionary perspective, do Hox genes act passively to specify the identity of segments, or can they play an instructive role? In other words, how easily can modules of effector genes come under the regulation of different Hox genes or combinations of Hox genes? If Hox genes were to

Figure 1

The four most commonly recognized arthropod classes differ in body plan and Hox gene expression. A consensus tree of major arthropod groups is shown here. This tree is not meant to be exhaustive, and numerous taxa have been omitted. For each class, a representative of the most well-examined body plan is shown. The consensus patterns of Hox gene expression are also shown for these representative groups. Body segments and the appendages they typically bear are abbreviated. For segments: Oc, ocular; Ch, cheliceral; Pd; pedipalpal; L1, etc., first leg-bearing, etc.; O1, etc., first opisthosomal, etc.; An1, etc., first antennal, etc.; In, intercalary; Mn, mandibular; Mx1, etc., first maxillary, etc.; Mxp, maxillipedial; T1, etc., first thoracic, etc., P1, etc., first pereonic, etc.; p1, etc., first pleonic, etc.; A1, etc., first abdominal, etc. Appendage abbreviations: lbr, labrum; chel, chelicerae; pedi, pedipalps; ant, antennae; man, mandibles; max, maxillae; mxp, maxillipeds; plpd, pleopods; gen, genitalia. The Holometabola include insect orders with true metamorphosis. Hemimetabolous pterygote insects comprise the Paraneoptera, which includes Hemiptera and allied orders, and the Polyneoptera, which includes Orthoptera, Phasmatodea, and others. Apterygote insects are a paraphyletic assemblage that includes the firebrat *Thermobia* (Zygentoma). Malacostracan crustaceans include the isopods, decapods, and the "true shrimp." Maxillopoda and Brachiopoda are diverse and possibly paraphyletic groups. Remipedia includes the barnacles and related crustaceans. Myriapoda includes Chilopoda (true centipedes), Symphyla (garden centipedes), Diplopoda (millipedes), and Pauropoda. The chelicerates are considered to be basal among the arthropods. Arachnida includes most extant chelicerates, including the Araneae (spiders), Acari (mites), as well as scorpions and others. The Xiphosura include extinct chelicerates as well as extant horseshoe crabs of the genus *Limulus*. The extinct trilobites are likely a basal lineage within Chelicerata.

passively identify segments, then evolutionary changes in tagmatic boundaries should be possible without shifts in Hox expression along the AP axis. In this case, network association between Hox genes and their target effectors must be flexible. However, if these network associations are not plastic, and individual Hox genes maintain stable regulatory connections to target genes, then tagmatic shifts would require changes in Hox expression. Apparently, body plan evolution has proceeded by both routes, as we will show. The following example suggests that evolutionary changes in Hox expression may be instructive.

Among the crustaceans, numerous lineages, including the Maxillipoda, Isopoda, and Decapoda, have modified thoracic appendages to roles in feeding. The body segments bearing these maxillipeds may also be incorporated into the cephalic carapace, as in eucarid decapods. In effect, this is a modification of body plan tagmosis at the level of these crustacean groups. Therefore, investigators have examined how the Hox genes are expressed in crustaceans with and without maxillipeds. Segments of the crustacean trunk express *Ubx* and *abd-A* (4, 5, 11). An antibody to the Ubx and Abd-A proteins with broad phylogenetic cross-reactivity allowed Averof & Patel to survey seven diverse crustacean species (13). This study included species with maxillipeds independently derived among the Maxillipoda and Malacostraca, and illustrated that the anterior boundary of Ubx/Abd-A correlates with the most anterior segment bearing walking legs. That is, segments with appendages recruited to function as maxillipeds during their evolutionary history no longer appear to express *Ubx*. Instead, maxilliped-bearing segments express the more anterior Hox genes, *Scr* and/or *Antp* (5). The shift in tagmatic boundary has been accompanied by a shift in Hox expression. Furthermore, different lineages have evidently employed similar modifications in the evolution of these convergent structures.

A possible reason for this may be that Hox genes anterior of *Ubx*, such as *Scr* or *Hox6*, may have already been in control of target genes important to the function of gnathal appendages. If so, mutations changing Hox expression might have been more likely than those bringing these targets under the regulation of a new, more posterior Hox gene. In this way, the evolution of maxillipeds may be canalized by network architecture.

A related story provides further evidence for the instructive activity of Hox genes. The isopod crustacean *Porcellio scaber* has evolved a single pair of maxillipeds along an independent lineage from the decapods. In this species, *Ubx* expression also makes its anterior boundary at the maxilliped-bearing segment. However, Abzhanov & Kaufman have shown that during early embryogenesis, the appendage appears to develop as a leg and only later transforms into a distinct maxilliped (3). *Scr* is expressed in the maxilliped segment, but the protein is not detectible by antibody in the appendages until stages after the morphological transformation. Presumably, this delay in maxilliped identity results from the suppression of *Scr* mRNA translation. Although the purpose of the delay remains unknown, it provides a developmental example of how Hox genes may specify segment identity in a non-model organism.

Reduction of Tagmata in Some Lineages

While most arthropods have successfully exploited an elaborate tagmosis, many others have evolved a secondarily simplified body plan. These include many parasites, species with reduced mobility, as well as those modified for microscopic habitats. Rhizocephala (Crustacea, Remipedia) are parasites of other crustaceans, and lack most typical crustacean structures, including most body segmentation (22). The "trunk" or abdomen of the body is extremely reduced in larval stages, and absent in adults. In the rhizocephalan *Sacculina carcini*, the larval abdomen expresses *Abd-B* before it degenerates at the end of larval development (19). Strangely, *abd-A* is apparently not expressed. This is intriguing, since if selection favored a reduced abdomen in the lineage leading to *Sacculina*, it could have been possible for Hox genes in this tagma to activate apoptotic pathways to eliminate these segments. Apoptotic targets are regulated by *Dfd* and *Abd-B* in the *Drosophila* embryo to produce intersegmental furrows (59). It is possible that in an ancestor of *Sacculina*, *Abd-B* had similar regulatory connections to the cell death pathway, which were co-opted to the elimination of the abdomen.

Similarly, acarid mites of the suborder Parasitiformes lack obvious segmentation, and the opisthosoma is fused and dramatically reduced (22). Such extreme reductions are also seen in Hexapoda. Collembolans are a basal hexapod lineage, and the presumed sister group of the insects. The family Sminthuridae are noted for an abdomen that is reduced and fused to the thorax, giving the animals a globular appearance (29). Reduction of this kind even appears among the pterygote insects. The Coccoidea (scale bugs) are sexually dimorphic, and females are often sessile. The female abdomen is reduced and lacks obvious segmentation (29). Numerous other examples exist, and these groups have certainly reduced their body plans independently. Therefore, it is tempting to speculate whether these reductions of the abdomen have also been accompanied by modifications in the roles of *abd-A* and *Abd-B* reminiscent of the Rhizocephala. Does *Sacculina* represent an extreme case, or might the elimination of Hox expression and/or the activation of apoptotic pathways provide a common evolutionary route to tagmatic reduction?

EVOLUTION OF LESSER BODY PLAN FEATURES IN INSECTS

Insect tagmosis is remarkably consistent. However, these arthropods have evolved amazing variations from this pattern. The developmental genetic mechanisms responsible for these varied morphologies are just now being explored. Much of this work has focused on modifications to wing morphology, and a guide to some of the taxa discussed below is given in **Figure 2**.

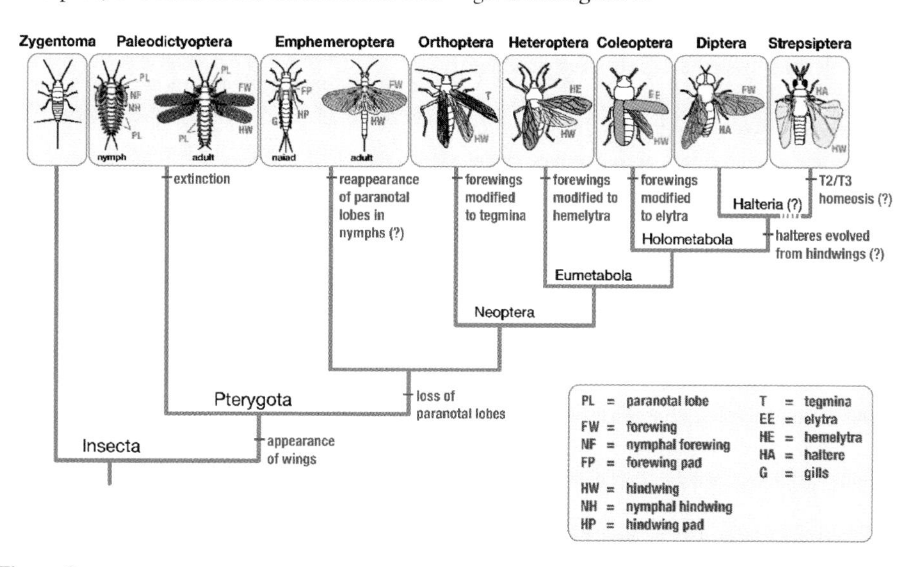

Figure 2

Some of the wing modifications seen among insects. The paranotal lobes seen in primitive insects are shown in blue. Forewings and their derivatives are shown in orange, while hindwings and derived structures are shown in yellow. One evolutionary scenario for the origins of these wing modifications is mapped onto a tree of these groups, based on the phylogeny of Wheeler and coworkers (97).

The Origin of Wings

One of the most distinctive features of the insects is their ability to fly. Debate over the evolutionary origins of insect wings has been a subject of much consideration in the past century. Two main hypotheses have been considered: The first asserts that wings evolved from outgrowths of the dorsolateral cuticle (36), first as an adaptation to parachuting or gliding (81), before articulation and muscle connections allowed powered flight. The second idea postulates that wings were modified from ancestral exites (dorsal projections) on the proximal legs and abdominal appendages of early insects (53). In turn, these exites are thought to derive from gill-like epipods of primitive crustacean-grade ancestors (12), and possibly homologous to the gills seen in aquatic larvae of some extant Ephemeroptera (mayflies).

Genetic studies of *Drosophila* have cataloged a number of ways in which wing development resembles development of the other appendages (reviewed in 61). In this species, the wing and leg imaginal primordia are derived from a shared pool of precursor cells. Wings utilize some of the same signaling mechanisms to establish axial polarity. These data seem to support the hypothesis that wings evolved from an appendage derivative.

However, it is possible that in evolution from a body wall outgrowth, appendicular developmental networks were co-opted for use in protowings. Thus, Averof & Cohen reasoned that it would be very unlikely to find genes principally involved in insect wing development in the epipods of crustaceans. To test this hypothesis, they determined the protein accumulation of two "wing gene" products in two crustaceans: the brachiopod *Artemia franciscana* and the malacostracan crayfish *Pacifastacus leniusculus* (12). The genes chosen were *pdm*, which is required early to specify the fate of the *Drosophila* wing disc but has only limited expression in the legs, and *apterous* (*ap*), which functions in determination of dorsal-ventral (DV) polarity in the wing. In *Artemia*, Pdm and Ap proteins were found throughout the more distal of the two epipods, whereas in *Pacifastacus*, Pdm was detected in the single epipod of appendages in this species. These data provide molecular evidence for an evolutionary connection between epipods and wings, and support anatomical claims of their homology (53).

These researchers later expanded their investigation and found that *pdm* and *ap* are expressed in chelicerate structures, which have also been suggested as homologues of crustacean epipods (30). The Xiphosura are unique among modern chelicerates in that the opisthosomal segments bear appendages. In species such as the horseshoe crab *Limulus polyphemus*, the opisthosomal appendages are modified into gas exchange structures called book gills. These structures have been internalized as book lungs in the evolution of arachnids, such as the spider *Cupiennius salei*. Both book gills and book lungs express *pdm* and *ap* during embryonic development. In *Cupiennius*, these genes are expressed in the spinnerets of the opisthosoma (30), which are also thought to be an appendicular derivative (6, 22, 80). These studies suggest that derived structures such as insect wings, crustacean epipods, xiphosuran book gills, and the arachnid book lungs and spinnerets may all share a common ancestry. If so, this provides a remarkable example of the evolutionary flexibility of the arthropods.

These studies are useful, but should be interpreted cautiously. They represent only the expression data of two genes in four divergent species. One paradox of the wings-from-epipods hypothesis is that it requires each of the four independent lineages of primitively wingless hexapods to have lost wings or their precursors in parallel. Jockusch & Ober (50) have recently presented a critical review of the hypothesis of appendicular wing origin. They also used molecular markers and histology to show that, unlike *Drosophila*, the wing and leg primordia are not descended from a common pool of cells in *Tribolium* and the grasshopper, *Schistocerca*. Instead these authors propose that the appearance of *pdm* and *ap* in wings and epipods may represent convergent co-option.

Suppression of Prothoracic and Abdominal Wings in Modern Insects

Early wings are thought to have been present on all thoracic and abdominal segments. The oldest fossil species identifiable as pterygote insects possess primitive wings on T2 and T3, and lateral winglets or "paranotal lobes" on T1 and abdominal segments A1 to A9 (53) (depicted in **Figure 2**). Unlike most modern insects, the wings and paranotal lobes of Paleozoic species were also present in the juvenile instars (101). Wing venation was very extensive in fossil specimens (55). Therefore, in several ways, modern insects have reduced the prominence of wings in their body plan (29). Venation has been reduced, while wing number has been restricted to two and limited to adults in most extant species.

In *Drosophila*, input from thoracic Hox genes, such as *Antp*, is not required for the activation of wing development (25). Instead, it appears that several Hox genes act to repress wing development elsewhere along the body. In the prothorax (T1) of the *Drosophila* embryo, *Scr* is expressed in the dorsal ectoderm where it is required to suppress the development of wing primordia (25, 67). Dorsal T1 expression domains of *Scr* have been noted in other insects (77, 93). Similar expression also appears in the primitively wingless insect *Thermobia* (77) and the isopod crustacean *Porcellio scaber* (D.R.A. & T.C.K., unpublished data). Expression of *Scr* in these primitively wingless groups has led to the idea that the function of *Scr* in T1 wing suppression is an exaptation of pterygote insects. However, without functional data, we can only speculate about the potential function of this dorsal *Scr* expression domain in primitively wingless species.

In the insect abdomen, *Ubx* and *abd-A* apparently suppress wing development. Genes such as *snail*, which mark early wing discs in *Drosophila*, become ectopically expressed in the abdomen of embryos lacking *abd-A* (25). In the milkweed bug *Oncopeltus*, depletion of *abd-A* causes ectopic abdominal leg development (10). Abdominal segments of *abd-A*-depleted individuals also acquire dorsal pigmentation similar to T3. This pigmentation marks the location of the juvenile hindwing pads in *Oncopeltus*, suggesting that *abd-A* depletion in this species may also relieve suppression of abdominal wing development.

A1, etc.: first abdominal segment, etc.

T2: second thoracic segment or mesothorax

T3: third thoracic segment or metathorax

Differentiation of Forewings and Hindwings

Most winged insects bear two pairs of wings: forewings on the mesothorax (T2) and hindwings on the metathorax (T3). However, the forewings and hindwings have distinct morphologies, which may allow the wings to cooperate aerodynamically to enhance flight proficiency (29). Alternatively, one pair of wings may be modified for the purposes of protection or display, as with the coleopteran elytra, or for balance, as with the halteres of Diptera and Strepsiptera. These more derived examples are discussed below, but first let us consider the basic issue of more subtle distinctions between forewing and hindwing.

For several decades, it has been known that the Hox gene *Ubx* was required to distinguish the identity of the T3 body segment from T2 in *Drosophila* (57). In *Drosophila* and other Diptera, the hindwings are modified into small balancing organs called halteres, and *Ubx* acts to repress wings on T3. So what is the situation of other insects in which wings are present on T3? Would *Ubx* expression be absent from the developing hindwings? Warren and colleagues (94) investigated this possibility in the butterfly *Precis coenia*, where they found that Ubx protein accumulated at high levels in the imaginal hindwing (T3) discs, but was absent from forewing (T2) discs. Weatherbee et al. showed that several wing patterning genes suppressed by Ubx in the *Drosophila* haltere disc are expressed in patterns similar to the *Drosophila* wing disc in both the forewing and hindwing discs of

Precis (96). Furthermore, a spontaneous mutation was identified in *Precis* in which *Ubx* fails to be expressed in patches of the hindwing (96). In the absence of Ubx protein, these patches adopt pigmentation patterns characteristic of the forewing. Conversely, hindwing pigmentation could be induced in the forewings of *Precis* by constitutive expression of *Ubx* from a viral vector (56). Therefore, rather than acting to repress wing development, it appears that *Ubx* acts to distinguish hindwing from forewing.

The Evolution of Dipteran Halteres from Hindwings

The differentiation of hindwings from forewings is taken to an extreme in the Diptera. In this group the hindwings are modified into a pair of balancing organs called halteres. In *Drosophila* the Hox gene *Ubx* is required to specify the halteres, and certain *Ubx* mutant alleles produce flies without halteres, but rather with two pairs of wings (57). *Ubx* intervenes at several levels to direct haltere development, by inhibiting genes involved in dorsal-ventral specification of the disc (60), organ size and shape, and bristle formation (95). *Ubx* is expressed in regions of the T3 body segment in a wide range of insects with varied modifications of fore- and hindwings (10, 17, 25, 102), and this may be an ancestral trait shared by pterygote insects. Therefore, it is likely that the evolution of halteres proceeded by a slow process as *Ubx* regulation was acquired within the *cis*-regulatory elements of genes functioning at each of these levels.

The Evolution of Elytra from Forewings in Coleoptera

The Coleoptera are perhaps the most successful lineage of animals, with over 350,000 described species (29). They are characterized by the modification of the forewings into elytra, a protective covering over the abdomen. Therefore, in Coleoptera, such as the red flour beetle, *Tribolium castaneum*, flight is dependent on the hindwings (T3), which are similar to the single pair of wings (T2) in Diptera.

What is the role, if any, of *Ubx* in these modified forewings, the elytra? It is possible that *Ubx* acts to modify wing development to produce elytra from the T2 segment in *Tribolium*, just as it does in the T3 disc of *Drosophila* to produce halteres. Alternatively, *Ubx* may simply identify T3 structures and regulate a separate set of target genes to allow hindwing development in the beetle. Recently, these hypotheses were tested by Tomoyasu and coworkers, who used RNA interference to suppress *Ubx* function in *Tribolium* larvae (87). They found that with the reduction of *Ubx* activity the T2 and T3 dorsal imaginal discs develop as elytra. Similarly, mutations in the *Tribolium Scr* orthologue, which is expressed in the T1 segment, produce ectopic elytra-like structures on T1, rather than wing-like structures (14, 15). The *Antp* orthologue in *Tribolium* appears to have no influence on the identity of elytra or hindwings, since its suppression causes no defects in these structures (87). Therefore, in the absence of input from Hox genes, such as *Ubx*, the dorsal imaginal discs develop as elytra. This contrasts the situation in *Drosophila*, where in the lack of *Ubx* the imaginal discs develop as two pairs of wings (57). Furthermore, Tomoyasu and coworkers have shown that in *Tribolium*, *Ubx* does not repress target genes as it does in the haltere disc of *Drosophila*. Instead, these genes are expressed in unique patterns in the elytra that are independent of *Ubx* regulation. It seems likely that the ancestral role of *Ubx* has been to identify T3. During evolution, as T2 and T3 structures were modified differently in separate lineages, different target genes came under *Ubx* regulation in order to produce distinct morphologies unique to T3.

In this example, the unregulated (or at least Hox-free) pathway has diverged to produce the elytra developmental program. However, for the T2 discs to develop as elytra and to retain a functional pair of hindwings, *Ubx* would have to acquire regulatory control over wing development in order to preserve this developmental mechanism. This is very different from the evolutionary scenario imagined for the Diptera, in which the evolution of a novel structure (halteres) was accompanied by the appearance of novel regulatory relationships to *produce* that structure.

Forewing modifications appear in other insect groups as well. The forewings of Orthoptera are modified into a leathery protective form called tegmina. Many species of Heteroptera feed on plants, from which they acquire toxic substances used to deter predators. The forewings or hemelytra are often thickened proximally and brightly colored to advertise the insects' toxicity. It is interesting to speculate whether modifications in wing patterning seen in the evolution of coleopteran elytra might also have been paralleled in the evolution of these other modified forewings, and whether the development of the more typical hindwings is dependent on suppression of the modified developmental mechanisms by *Ubx*.

Phylogenetic Homeosis Among the Strepsiptera?

A more complex instance of wing modification exists in a little-known group of insects: the Strepsiptera. Adult males of these endoparasitic insects have halteres similar to those of Diptera. Strangely, strepsipteran halteres are found on T2, whereas T3 bears functional wings. The phylogenetic position of the Strepsiptera among insect orders has been controversial, but molecular and morphological data exist supporting a sister-group relationship between Strepsiptera and Diptera, termed the Halteria (97, 100). If the Halteria are indeed a monophyletic group, then it is interesting to consider the evolution of halteres among this proposed clade. Did halteres evolve separately in each lineage? From hindwings in Diptera and forewings in Strepsiptera? And if so, were these insects somehow predisposed, genetically or otherwise, for such anatomical modifications? Alternatively, Whiting & Wheeler have suggested that the haltere developmental mechanism was somehow co-opted from one thoracic segment to the other after the divergence of these lineages (100). They point out that mutations in *Ubx* in the early strepsipteran lineage, similar to the *bithorax* and *postbithorax* (57) regulatory alleles of *Drosophila Ubx*, could cause *Ubx* expression in T1 rather than in T3, leading to a phylogenetic homeosis of wings and halteres. Bennett and colleagues have examined *Ubx* expression in embryos of one strepsipteran species (18; R. Bennett, personal communication). Unexpectedly, *Ubx* expression was limited to the first abdominal segment, and did not appear in either T2 or T3. Although it is possible that later expression could appear in imaginal discs, these data suggest that an ancestral homeotic-type mutation in the Strepsiptera appears unlikely.

However, it remains possible that Strepsiptera diverged from Diptera after the evolution of halteres along a dipteran-style body plan. In such a scenario, the loss of *Ubx* expression in the T3 segment may have permitted the development of T3 wings. The existing haltere developmental mechanism may then have become activated by some other factor expressed in T2. *Antp* is one candidate for this role. More studies in these experimentally challenging but evolutionarily intriguing insects will be necessary to resolve these issues.

The Appearance of Winglessness Among Insects

Reduction and loss of structures is one of the more common modifications seen in various taxa. From the nearly ubiquitous paranotal lobes of Paleozoic insects to modern forms, insects have exploited the advantages of reduction in wings. Winglessness has also offered other niches, which have been invaded by members of the Pterygota.

In the social Hymenoptera, nonreproductive castes of ants are wingless, facilitating a subterranean lifestyle (44). Sufficient larval nutrition produces an increase in juvenile

JH: juvenile hormone. An insect hormone used to trigger developmental events, such as the survival of wing discs in juvenile ants

hormone (JH) levels, which is required for development of queens. Part of this developmental program is the maturation of T2 and T3 wings. Individuals receiving less food (and hence producing less JH) develop as sterile, wingless workers (63). The mechanisms of wing specification and patterning are considered to be well conserved, based on the similar expression of orthologues in *Drosophila* (Diptera) and the lepidopteran *Precis* (24, 96). Therefore, working with four species of ants, Abouheif & Wray investigated the expression of genes known for their roles in the wing development of other insects (1). Their study confirmed that these orthologues were conserved in their expression patterns in the wings of reproductive castes. Surprisingly, their data from nonproductive castes also revealed that wing development was disrupted at different points in these different species.

Ants of *Pheidole morrisi* appear in two nonreproductive wingless castes: soldiers and minor workers. The fate of the T2 wing discs in nonreproductive castes of *Pheidole* is determined by a second JH requirement. JH exposure at the second checkpoint specifies soldiers, in which the T2 wing discs evert at the prepupal stage, but proceed to die apoptotically, whereas the wing discs of workers disintegrate before the end of larval development (78). Abouheif & Wray also found that the wing development pathway is disrupted much earlier in *Pheidole* workers than in soldiers (1). It was unexpected to find that wing development could be interrupted in so many different ways in species for which wingless castes are a shared ancestral trait. This may have been due to the neutral drift of developmental networks (see below).

At least two possible routes for the evolution of these interruptions of wing development are possible. We can presume that since all extant and fossil ant species possess wingless nonreproductive and winged reproductive castes (44), selection has acted to maintain this social arrangement. Therefore, winglessness in workers must not come at the cost of winglessness in all castes. This would imply a genomic conservation of the wing developmental network. It is possible that this network receives input from the reception of JH and wing patterning is actively blocked at that point in its absence. If so, the point of interface between the JH and wing development pathways must be fairly flexible because it apparently varies between species and castes. Alternatively, when JH is not supplied at a critical stage, pathways related to JH reception (about which very little is known) could directly activate cell death in the wing discs. In this scenario, the default developmental pathway would be to produce wings in the style of the reproductive castes. Variation in the timing of JH checkpoints, between species and castes, would also vary the timing of cell death relative to the stage of wing patterning. This second hypothesis seems more likely, since castes that are determined by later JH checkpoints, like *Pheidole* soldiers, generally have a later point at which wing development is interrupted.

Winglessness has also evolved in solitary insects. Unlike polyphenic social insects, which apparently maintain the wing developmental pathways, wingless solitary insects could in theory lose the wing developmental network due to accumulations of neutral mutations. The phasmids, or stick insects (Phasmatodea), exhibit elaborate camouflage resembling vegetation. Winglessness also appears in the majority of species, while others possess a range of reduction in wing morphology. Partial wing loss or reduction is sometimes sexually dimorphic, with males retaining wings and flight ability. A recent molecular phylogeny by Whiting (99) found that, whereas winged forms are basal among the Phasmatodea, the majority of wingless species are derived from a single lineage. However, remarkably, wings also appear to have reappeared within this largely wingless group.

Flight involves the coordination of nerves and muscles, as well as the aerodynamics of wing shape. Genetic studies of *Drosophila* wing development have shown that while key genes involved in wing patterning may function in multiple developmental processes, the genetic network they comprise to facilitate wing development is not used elsewhere (26). Without selection to maintain these regulatory connections, it is expected that mutations eliminating them should be neutral and accumulate with relative speed. Therefore, once lost, the reappearance of a complex trait, such as winged flight, has been considered very unlikely. Nevertheless, Whiting concludes based on parsimony that the reappearance of wings may have occurred on as many as four separate occasions during phasmid evolution (99). Acceptance of these findings implies that our understanding of genetic network evolution may be somewhat incomplete.

The Evolution of Foreleg Combs

Insect diversity encompasses far more than wing modifications. For other traits, few genetic data may be available. However, one area of some research is the occurrence of foreleg combs. These are most familiar as the sex combs of *D. melanogaster*. Sex combs are a row of large anterior-ventral bristles at the distal end of the first tarsal segment (the basitarus) of *Drosophila* males. Males use these structures in courtship, where they stroke the female's abdomen to stimulate ovulation (27). This fairly minor anatomical feature is nonetheless specified by a Hox gene and lends itself to the name *Sex combs reduced* (*Scr*). *Scr* is expressed in the T1 leg disc, including a domain corresponding to the location of the adult sex comb in males. Loss of *Scr* activity in the T1 leg discs of *Drosophila* results in legs lacking sex combs, while ectopic expression can produce sex combs on the T2 and T3 legs (67).

A similar structure appears in the milkweed bug *Oncopeltus fasciatus*. In this species, combs appear on the distal foretibia of both sexes and are used for grooming their long rostrum. *Oncopeltus Scr* is also expressed in an anterior-ventral patch of the T1 legs, which appears to correspond with the location of the foretibial comb (77). Depletion of *Scr* activity by RNAi in *Oncopeltus* eliminates the foretibial combs (45).

RNAi: RNA interference. A method for depleting gene activity by the introduction of a double-stranded length of RNA transcript.

The similarity of combs in these species is somewhat surprising. Foreleg combs appear relatively rarely and are found among the Heteroptera and Diptera, as well as some Coleoptera and Hymenoptera. Therefore, *Oncopeltus* and *Drosophila* have apparently evolved foreleg combs independently. *Scr* is expressed in a patch of the distal embryonic T1 legs of the cricket *Acheta domestica*, despite the lack of any comb-like structures in this species (77). Since this domain of *Scr* expression may be widely conserved, it has been suggested to act as an exaptation, facilitating the evolution of combs (77). However, an obvious question becomes, why combs?

A broader survey of insects reveals that the distal tibia or proximal tarsi of the T1 legs often bear unique specialized structures. Although some Coleoptera have foreleg combs, similar to *Oncopeltus*, others bear large spine-like cuticular outgrowths on the T1 basitarsi. The foretibia of Embiidina (web spinners) bear glands used to produce silk threads. Furthermore, some Orthoptera, including *Acheta*, possess tympanal hearing organs on the distal tibia of the forelegs. It seems likely that, rather than being uniquely associated with combs, *Scr* expression in the T1 legs may specify many if not all of these structures. However, confirmation of such a regulatory relationship must await functional analyses in these other species.

The expression of *Scr* in the T1 legs may be an evolutionary innovation of the pterygote insects. *Scr* expression is absent from the T1 limbs of the apterygote *Thermobia* (77), and this species lacks any distinctive morphology on the distal forelegs. However, as noted, all pterygotes in which *Scr* expression has been examined show a correlation of the T1 leg domain with specialized structures, such as combs or tympanal organs. Since *Scr* is a regulator of transcription, it is plausible that this conserved domain has been predisposed to acquiring regulation over some aspects of morphology in the context of the T1 leg. Many of these, such as the large bristles of foreleg combs, are likely to come at a fairly mild fitness cost to individuals. If they appear with relative frequency over evolutionary time, selection may favor them occasionally, especially when sexual pressures apply (65), as with *Drosophila*.

DSD: developmental systems drift. The process by which neutral mutations may cause random changes in regulatory networks over evolutionary time

DISCUSSION

Implications of Developmental Systems Drift

All genetic material is subject to random mutation. Often mutations fail to alter DNA in any deleterious way. Kimura has described how these neutral mutations may accumulate, unaffected by selection since they do not influence phenotypic fitness (51). In time, as two lineages diverge from an ancestor, the similarity of orthologous sequences will drift apart from one another. True & Haag have proposed a similar form of drift acting at the level of ontogenetic networks (88), and arising from two factors. The first is Kimura's theory of neutral evolution in genes. The second comes from the insights of developmental genetics that most phenotypes are the product of complex networks of interacting regulatory genes (32). As a simple example, imagine that two lineages diverge from an ancestor in whom a phenotype is the product of a developmental network in which one regulatory gene product, A, activates an intermediate regulatory gene, B, that then activates an effector gene, C. If in one lineage, the *cis*-regulatory elements of gene C, the effector, change such that they may bind protein A, then gene B is functionally redundant. A subsequent mutation may then eliminate gene B or its binding site in the regulatory region of gene C without an effect on the phenotype. After the fact, examination of the network in each lineage would reveal that despite identical phenotypes, the network architecture in each is different. Although this is a simplistic example, such divergence in developmental networks may have occurred in the suppression of wing development in worker ants, as reviewed above. True & Haag have catalogued a number of other instances acting at various phylogenetic levels (88), and they have dubbed this phenomenon developmental systems drift (DSD).

The implications of DSD to comparative developmental genetics are quite important. DSD suggests that species may possess vastly more diverse genetic networks than can be predicted from their morphologies. One example from our own work involves the activation of the gene *spalt* in the antennae of insects. In *Drosophila* and the milkweed bug *Oncopeltus*, *spalt* is expressed in similar domains where it is required for formation of antennal joints (34, 92). *spalt* is activated cooperatively by *Distal-less* and *homothorax* in the antennae of *Drosophila* (34); however, these genes appear not to act in this way to specify *spalt* expression in *Oncopeltus* (9a). We suspect that in *Oncopeltus*, *spalt* is specified by a different unidentified factor. Since *spalt* performs similar developmental functions in each species, it seems unlikely that selection would directly alter the network architecture upstream of *spalt*. It is possible these network changes are related to other adaptive morphological changes, but random drift in the network architecture seems the simpler scenario.

Another important caveat of the DSD concept is that the divergence of networks can confound the candidate gene approach. As an example, let us reconsider our hypothetical 3-gene regulatory network from above. If the species that retains gene B as an intermediate were a model organism, then an investigator might choose to explore its orthologue's function in the other lineage. However, this study would be fruitless since gene B is no longer a component of the pathway in the second lineage. While this information is useful in understanding the regulatory evolution of these species, these sorts of negative data are rarely reported by themselves. Furthermore, if gene A is unknown in either species, then the investigator has no more candidates for which to examine the regulation of gene C.

Fortunately, as genomic technology becomes more universally applicable, DSD may present fascinating opportunities for research. Theoretical modelers of developmental networks have been puzzled over the seeming robustness that networks exhibit in models, and have met with a frustrating lack of experimental data that describe the range of possible values for parameters such as gene expression rates (89). One problem may be that species are likely to exhibit a much narrower range of values than the total number of workable sets possible. However, DSD suggests that a diverse collection of species will provide a wide range of developmental parameter sets. Such experiments have been done with at best a few species and have been labor intensive (42, 74). However, a high-throughput method is needed for such analyses of large species numbers. If a large enough number of species could be assayed for the transcription levels of genes involved in one developmental process, such data would describe the range of transcription levels that are functionally possible to facilitate that process. Importantly, relationships between genes would also become obvious from such data. For example, perhaps genes A and B are consistently expressed strongly while gene C is at low levels, or vice versa, but never at high levels for all three. It is likely that such a relationship would be meaningful in the context of the network's architecture. Such large-scale genomic experiments would help explain, from principle, rather than post hoc explanations, how genetic networks specify morphology.

Integrating Comparative Developmental Conclusions into Evolutionary Biology

Most comparative developmental genetic data have been collected from a wide phylogenetic sampling. In contrast, few studies have examined morphological and developmental genetic changes that might be relevant to speciation or at least resulting from relatively short-term isolation. This leads to an important question: Can comparative developmental genetics address the origins of novel morphologies?

This is a frequent criticism of macroevolutionary evo-devo. It is true that at such a phylogenetic scale, the origins of morphology may be only a matter of informed speculation. However, it would be wrong to then dismiss macroscale developmental studies as irrelevant to evolutionary biology. Like physiology and anatomy, the development of an organism is the end product of its evolution to date. Such studies describe the possible mechanisms by which organisms may be patterned, allowing us to begin to understand the limitations and language that genetic networks use in patterning organisms.

However, investigators of comparative developmental genetics have increasingly turned to smaller phylogenetic scales. Here, evo-devo can provide useful data on the origins of morphological novelty. Aside from exploring specific instances of morphological radiation, these studies also bear on the two major hypotheses of morphological evolution. The first is the traditional infinitesimal evolutionary model of Fisher, by which morphology changes by small quantitative degrees over generations due to selection and adaptive fitness advantages of some segregating alleles (34a). The second hypothesis arises from the observation that evolution of new morphology has occurred rapidly in the fossil record. This has led to speculation that spontaneous homeotic mutations or "hopeful monsters" might account for such changes (39, 40). This issue is still open, as it now seems that evidence exists to support morphological evolution by infinitesimal changes as well as mutations in developmental regulatory genes. However, these mechanisms appear in different contexts, which may mark an important distinction between them.

The concept of the hopeful monster was first suggested in 1940 by Goldschmidt (39), and arose from his work on mutations in the *Drosophila* homeotic genes. He proposed that variably penetrant homeotic mutations might introduce the changes in morphology seen between species and higher taxa. The idea was derided, partly because Goldschmidt used it in support of more eccentric genetic theories (33). However, recent comparative studies of drosophilids have emerged that seem to suggest sudden mutation in developmental regulatory genes or networks can produce morphological differences between closely related species.

Evidence for Hopeful Monsters

Abdominal pigmentation varies widely within the *D. melanogaster* species group. It is sexually dimorphic in some species, in which it may play a role in mate choice. Kopp and colleagues have shown that the transcription factor *bric-a-brac* (*bab*) is required for specification of the melanic pattern in the abdomen (52). *bab* is an important developmental regulator, which also functions in the development of the limbs. In *D. melanogaster*, where black pigmentation is found over much of the male abdomen, *bab* is also activated by *Abd-B* and *doublesex* (*dsx*), the primary identifier of somatic sex. However, in other species, the expression of *bab* correlates with pigmentation rather than with sex (in sexually monomorphic species) or *Abd-B* expression (in species where pigmentation is restricted to more posterior segments). This suggests that since the radiation of these species, estimated to be within the last 5–10 million years (58), mutations in the regulatory regions of *bab* have altered inputs from *dsx* and the Hox genes. Therefore, mutations in this developmental regulatory gene (*bab*) have resulted in significant morphological divergence among these species.

Another example comes from the larval cuticle morphology of drosophilids. The larvae of *Drosophila sechellia* have a naked cuticle, where *D. melanogaster* possess trichomes. Sucena & Stern have shown by quantitative genetic analysis that remarkably the *shaven-baby* (*svb*) locus is the exclusive source of this morphological difference. The *D. sechellia* naked cuticle "phenotype" is not complemented in crosses to *D. melanogaster svb* mutants, which also have a naked cuticle phenotype (85), suggesting that *D. sechellia svb* bears a spontaneous mutation similar to the experimentally generated *D. melanogaster svb* mutant. Naked cuticle morphology has appeared independently several times among drosophilids, and using similar methods, the same group has shown that similar noncomplementing mutations at the *svb* locus have accompanied the morphology each time (84). Furthermore, among two populations of *D. borealis* the *svb* locus possess wild-type (that is, *D. melanogaster*-like) and mutant (*D. sechellia*-like) alleles in different interfertile populations (84). The *svb* locus encodes a zinc-finger transcription factor, which is also required for female fertility. Mutations in the coding sequence produce fertility defects. These *ovo* alleles are genetically separable and complement *svb* alleles, which are due to mutations in the regulatory regions of the gene. This suggests that evolution of naked cuticle may be channeled to proceed via *svb*-type regulatory mutations.

Genetic Evidence for Infinitesimal Morphological Evolution

It is easy to imagine evolutionary scenarios for pigmentation and cuticle morphology that include hopeful monsters. Presumably, mutations affecting these traits come at some initial fitness cost, but it is likely not so great that they could not segregate under some conditions, be tested by selection, and come to fixation in some lineages. Therefore, evolutionary experiments in lesser body plan features could be fairly common. However, can the same occur for greater body plan features? Because of the high mortality associated with homeotic transformation of body plan, mutations of the type studied by geneticists seem unlikely. It may also be an unnecessary possibility.

Polymorphisms exist among wild populations for at least one Hox gene, *Ubx*. Often these polymorphisms are only manifest phenotypically in extraordinary environmental circumstances. For example, Gibson & Hogness have shown that naturally occurring *Ubx* polymorphisms in *D. melanogaster* produce varying haltere defects when larvae develop in the presence of ether (38a). Stern has also shown that *Ubx* polymorphisms can lead to variation in the bristle patterns of the T3 leg in hybrids with *D. simulans* (83). Nor is this phenomenon unique to *Ubx* or *Drosophila*, as revealed by recent studies of polymorphism in human populations (76). Variation in human interleukin-4 appears to have phenotypic trade-offs for immune fitness, and changes at this locus relative to those of the great apes appear to indicate positive selection (75). Therefore, it appears that variation normally segregates in "genes of large effect."

These examples deal with traits that are manifest only under extreme environmental circumstances. (We may consider hybridization an extreme cellular environment for a haploid genome. As for humans—modern life has often been called extremely taxing, but this may be especially true considering the historical influence of disease on survival.) However, it is possible for environmentally induced traits to become phenotypically fixed under strong selection. This concept of "genetic assimilation" was proposed by Waddington (91), based on experiments he conducted in *Drosophila*. Waddington heat-shocked a line of flies and observed that wing crossveins were absent in a moderate percentage of individuals. However, after 12 generations of heat shock followed by selection for a crossveinless phenotype, flies developed without crossveins even in the absence of the heat stimulus (90). Other examples of genetic assimilation have been reviewed by Pigliucci & Murren (68). It should be pointed out that genetic assimilation is not usually associated with infinitesimal evolution. From the standpoint of phenotype, it is easy to understand why—since morphology changes rapidly under strong selection. However, it operates on the same infinitesimal molecular changes and the interactions of fitness and selection.

To summarize, polymorphisms exist in high-level regulatory genes, and even if they are not of immediately apparent fitness value, it may be possible for the environment to select and fix novel morphologies arising from this variation. Alternatively, for traits with a potentially low fitness cost, mutations affecting developmental regulatory genes may produce novel morphologies in the style of hopeful monsters. In general, mutations in *cis*-regulatory elements appear more commonly than those in coding sequences, and it follows then that evolution may tend to favor alterations in the regulatory networks of developmental genes. Taken together, these processes may contribute to the adaptive success of populations that ultimately result in isolation and speciation (28, 68).

Genetic assimilation: the phenomenon in which environmentally induced traits become expressed in the absence of the original environmental stimulus. It is thought to require persistent strong selection for the trait.

Canalization: the theory formulated by Waddington (91) that developmental genetics may constrain evolution, such that certain phenotypes are more likely than others. He also suggested that during the course of parallel morphological evolution in separate lineages, similar genetic changes might be required or more likely in the evolution from one state to another.

Future Directions

The remaining task of comparative developmental genetics is to explain why differences in developmental networks produce differing morphologies. The on-going comparisons of network architecture in morphologically diverse and phylogenetically distant species will help to describe the range of possible mechanisms for conserved and derived structures. This will provide the first insights into how diverse structures may differ developmentally. However, this still will not explain why species differ morphologically. To pose a similar question as an example, does our current understanding of metathorax specification or wing patterning really tell us why the *D. melanogaster* wing adopts its particular adult morphology, as opposed to that of *D. virilis*? Not really. Although we hope to have demonstrated here that inroads are being made into this next level, illustrating how regulatory genes can influence the detailed aspects of morphology and change with their evolution. It will be necessary to identify downstream effector genes, characterize their functions in diverse species, and understand their connections to the regulators that ultimately direct them in the production of cells comprising diverse morphological structures.

SUMMARY POINTS

1. Extensive morphological diversity exists among arthropods, in their tagmosis, appendage modifications, as well as lesser features, such as the placement of sensory structures, endocrine organs, and cuticular structures. All of these morphological traits may be considered part of the body plan at various taxonomic levels. Very little of this lesser morphological variation has been studied with developmental genetic methods.
2. Arthropod tagmosis correlates with the expression of Hox genes, and evolutionary changes in tagmatic boundaries have correlated with shifts in expression of some Hox genes.
3. In insects, the Hox genes *Scr*, *Ubx*, and *abd-A* have been central to the evolution of wing placement on the body. Modifications to wing morphology have involved changing downstream targets for *Ubx*.
4. Morphological evolution may proceed by two proposed mechanisms: infinitesimal mutation or homeotic (hopeful monster-type) mutation.
5. Available comparative genetic data suggest that the former mechanism may be prominent in the evolution of tagmatic body plan changes, while spontaneous homeotic mutations may contribute to the evolution of lesser body plan features.
6. Evolutionary changes in developmental genes seem preferential to regulatory sequences, which conserve the structure of the encoded protein, but may change the architecture of developmental networks.

FUTURE DIRECTIONS

The extensive exploration of expression patterns in diverse species must be followed up by functional analyses. This will help test earlier conclusions and provide insights into the mechanisms producing biological diversity. Unlike expression studies, functional experiments can also allow analysis of epistatic relationships.

Genetic studies have made *Drosophila* by far the most well understood arthropod. However, to investigate structures and processes that may be absent or highly modified in *Drosophila*, it will be necessary to take forward genetic approaches with other species. Such screens are currently under way with species such as *Tribolium*.

Genetic studies have identified high-level regulatory genes important in development, but these genes represent a small fraction of the genome. To understand how they truly specify morphology, it will be necessary to determine their connections to downstream effector genes.

Development of high-throughput methods will be needed for quantitative analysis of expression levels. This will provide rapid comparisons of genetic networks in separate genotypes or species.

ACKNOWLEDGMENTS

D.R.A. would like to thank Cynthia Hughes for a number of conversations on arthropod diversity and evolution that influenced and encouraged this work.

LITERATURE CITED

1. Abouheif E, Wray GA. 2002. Evolution of the gene network underlying wing polyphenism in ants. *Science* 297:249–52
2. Abzhanov A, Holtzman S, Kaufman TC. 2001. The *Drosophila* proboscis is specified by two Hox genes, *proboscipedia* and *Sex combs reduced*, via repression of leg and antennal appendage genes. *Development* 128:2803–14
3. Abzhanov A, Kaufman TC. 1999. Novel regulation of the homeotic gene *Scr* associated with a crustacean leg-to-maxilliped appendage transformation. *Development* 126:1121–28
4. Abzhanov A, Kaufman TC. 2000. Crustacean (malacostracan) Hox genes and the evolution of the arthropod trunk. *Development* 127:2239–49
5. Abzhanov A, Kaufman TC. 2000. Embryonic expression patterns of the Hox genes of the crayfish *Procambarus clarkii* (Crustacea, Decapoda). *Evol. Dev.* 2:271–83
6. Abzhanov A, Kaufman TC. 2000. Homologs of *Drosophila* appendage genes in the patterning of arthropod limbs. *Dev. Biol.* 227:673–89
7. Abzhanov A, Popadic A, Kaufman TC. 1999. Chelicerate Hox genes and the homology of arthropod segments. *Evol. Dev.* 1:77–89
8. Akam M. 1998. Hox genes, homeosis and the evolution of segment identity: no need for hopeless monsters. *Int. J. Dev. Biol.* 42:445–51
9. Angelini DR, Kaufman TC. 2004. Functional analyses in the hemipteran *Oncopeltus fasciatus* reveal conserved and derived aspects of appendage patterning in insects. *Dev. Biol.* 271:306–21

9a. Angelini DR, Kaufman TC. 2006. Depletion of *spalt* in the milkweed bug *Oncopeltus fasciatus* (Hemiptera) causes antennal defects unrelated to Johnston's organ and does not require wildtype *Distal-less* activity. *Evol. Dev.* In review

10. Angelini DR, Liu PZ, Hughes CL, Kaufman TC. submitted. Hox gene functions and interactions in milkweed bug *Oncopeltus fasciatus* (Heteroptera). *Development*
11. Averof M, Akam M. 1995. Hox genes and the diversification of insect and crustacean body plans. *Nature* 376:420–23
12. Averof M, Cohen SM. 1997. Evolutionary origin of insect wings from ancestral gills. *Nature* 385:627–30
13. Averof M, Patel NH. 1997. Crustacean appendage evolution associated with changes in Hox gene expression. *Nature* 388:682–86
14. Beeman RW. 1987. A homeotic gene cluster in the red flour beetle. *Nature* 327:247–49
15. Beeman RW, Stuart JJ, Haas MS, Denell RE. 1989. Genetic analysis of the homeotic gene complex (HOM-C) in the beetle Tribolium castaneum. *Dev. Biol.* 133:196–209

16. Beklemishev WN. 1964. *Principles of the Comparative Anatomy of Invertebrates*. Chicago, IL: Univ. Chicago Press
17. Bennett RL, Brown SJ, Denell RE. 1999. Molecular and genetic analysis of the *Tribolium Ultrabithorax* ortholog, *Ultrathorax*. *Dev. Genes Evol.* 209:608–19
18. Bennett RL, Cheatham J, Vaughan M, Whiting MF, Brown SJ, Denell RE. 1998. *Molecular and genetic analysis of the Ubx homologs in beetles and Strepsiptera*. Presented at Soc. Dev. Biol. 57th Annu. Meet., San Francisco
19. Blin M, Rabet N, Deutsch JS, Mouchel-Vielh E. 2003. Possible implication of Hox genes Abdominal-B and abdominal-A in the specification of genital and abdominal segments in cirripedes. *Dev. Genes Evol.* 213:90–96
20. Boore JL, Collins TM, Stanton D, Daehler LL, Brown WM. 1995. Deducing the pattern of arthropod phylogeny from mitochondrial DNA rearrangements. *Nature* 376:163–65
21. Briggs DEG, Fortey RA. 1989. The early radiation and relationships of the major arthropod groups. *Science* 246:241–43
22. **Brusca RC, Brusca GJ. 1990. *The Invertebrates*. Sunderland, MA: Sinauer**

An excellent and extensive introductory text to invertebrate zoology.

23. Budd GE. 2002. A palaeontological solution to the arthropod head problem. *Nature* 417:271–75
24. Carroll SB, Gates J, Keys DN, Paddock SW, Panganiban GE, et al. 1994. Pattern formation and eyespot determination in butterfly wings. *Science* 265:109–14
25. Carroll SB, Weatherbee SD, Langeland JA. 1995. Homeotic genes and the regulation and evolution of insect wing number. *Nature* 375:58–61
26. Cohen SM. 1993. Imaginal disc development. In *The Development of Drosophila*, ed. M Bate, A Martinez-Arías, pp. 747–843. Cold Spring Harbor, NY: Cold Spring Harbor Lab. Press
27. Cook RM. 1977. Behavioral role of the sexcombs in *Drosophila melanogaster* and *Drosophila simulans*. *Behav. Genet.* 7:349–57
28. Coyne JA, Orr HA. 1998. The evolutionary genetics of speciation. *Philos. Trans. R. Soc. London Ser. B* 353:287–305
29. Daly HV, Doyen JT, Purcell AHI. 1998. *Introduction to Insect Biology and Diversity*. Oxford: Oxford Univ. Press
30. Damen W, Saridaki T, Averof M. 2002. Diverse adaptations of an ancestral gill. A common evolutionary origin for wings, breathing organs, and spinnerets. *Curr. Biol.* 12:1711
31. Damen WGM, Hausdorf M, Seyfarth EA, Tautz D. 1998. A conserved mode of head segmentation in arthropods revealed by the expression pattern of Hox genes in a spider. *Proc. Natl. Acad. Sci. USA* 95:10665–70
32. Davidson EH, McClay DR, Hood L. 2003. Regulatory gene networks and the properties of the developmental process. *Proc. Natl. Acad. Sci. USA* 100:1475–80
33. Dietrich MR. 2003. Richard Goldschmidt: hopeful monsters and other 'heresies'. *Nat. Rev. Genet.* 4:68–74
34. Dong PD, Todi SV, Eberl DF, Boekhoff-Falk G. 2003. Drosophila spalt/spalt-related mutants exhibit Townes-Brocks' syndrome phenotypes. *Proc. Natl. Acad. Sci. USA* 100:10293–98

34a. Fisher RA. 1930. *The Genetical Theory of Natural Selection.* Oxford: Clarendon Press

35. Fitch DHA, Sudhaus W. 2002. One small step for worms, one giant leap for "Bauplan?" *Evol. Dev.* 4:243–46
36. Flower JW. 1964. On the origin of flight in insects. *J. Insect Physiol.* 10:81–88
37. Friedrich M, Tautz D. 1995. Ribosomal DNA phylogeny of the major extant arthropod classes and the evolution of myriapods. *Nature* 376:165–67

38. Gerhart J. 2000. Inversion of the chordate body axis: Are there alternatives? *Proc. Natl. Acad. Sci. USA* 97:4445–48

38a. Gibson G, Hogness DS. 1996. Effect of polymorphism in the Drosophila regulatory gene Ultrabithorax on homeotic stability. *Science* 271:200–3

39. Goldschmidt RB. 1940. *The Material Basis of Evolution*. New Haven, CT: Yale Univ. Press

40. Gould SJ. 1977. The return of hopeful monsters. *Nat. Hist.* 86:22–30

41. Haas MS, Brown SJ, Beeman RW. 2001. Pondering the procephalon: the segmental origin of the labrum. *Dev. Genes Evol.* 211:89–95

42. Hinman VF, Nguyen AT, Cameron RA, Davidson EH. 2003. Developmental gene regulatory network architecture across 500 million years of echinoderm evolution. *Proc. Natl. Acad. Sci. USA* 100:13356–61

43. Holland LZ, Laudet V, Schubert M. 2004. The chordate amphioxus: an emerging model organism for developmental biology. *Cell Mol. Life Sci.* 61:2290–308

44. Hölldobler B, Wilson EO. 1990. *The Ants*. Cambridge, MA: Harvard Univ. Press

45. Hughes CL, Kaufman TC. 2000. RNAi analysis of *Deformed*, *proboscipedia* and *Sex combs reduced* in the milkweed bug *Oncopeltus fasciatus*: novel roles for Hox genes in the Hemipteran head. *Development* 127:3683–94

An excellent source of methods of RNA interference in a non-model insect.

46. Hughes CL, Kaufman TC. 2002. Exploring the myriapod body plan: expression patterns of the ten Hox genes in a centipede. *Development* 129:1225–38

47. Hughes CL, Kaufman TC. 2002. Hox genes and the evolution of the arthropod body plan. *Evol. Dev.* 4:459–99

This recent review summarizes Hox expression data from a range of arthropods, and provides a useful reference to primary genetic data from *Drosophila* and other genetic model insects.

48. Hughes CL, Liu PZ, Kaufman TC. 2004. Expression patterns of the rogue Hox genes Hox3/zen and fushi tarazu in the apterygote insect *Thermobia domestica*. *Evol. Dev.* 6:393–401

49. Hwang UW, Friedrich M, Tautz D, Park CJ, Kim W. 2001. Mitochondrial protein phylogeny joins myriapods with chelicerates. *Nature* 413:154–57

50. Jockusch EL, Ober KA. 2004. Hypothesis testing in evolutionary developmental biology: a case study from insect wings. *J. Hered.* 95:382–96

51. Kimura M. 1968. Evolutionary rate at the molecular level. *Nature* 217:624–26

52. Kopp A, Duncan I, Carroll SB. 2000. Genetic control and evolution of sexually dimorphic characters in *Drosophila*. *Nature* 408:553–59

53. Kukalová-Peck J. 1983. Origin of the insect wing and wing articulation from the arthropodan leg. *Can. J. Zool.* 61:1618–69

54. Kukalová-Peck J. 1992. The "Uniramia" do not exist: the ground plan of Pterygota as revealed by Permian Diaphanoptera from Russia (Insecta: Paleodictyopteroidea). *Can. J. Zool.* 70:236–55

55. Kukalová-Peck J, Richardson ES Jr. 1983. New Homoiopteridae (Insecta: Paleodictyoptera) with wing articulation from Upper Carboniferous strata of Mazon Creek, Illinois. *Can. J. Zool.* 61:1670–87

56. Lewis DL, DeCamillis M, Bennett RL. 2000. Distinct roles of the homeotic genes *Ubx* and *abd-A* in beetle embryonic abdominal appendage development. *Proc. Natl. Acad. Sci. USA* 97:4504–9

57. Lewis EB. 1978. A gene controlling segmentation in *Drosophila*. *Nature* 276:565–70

58. Li Y-J, Satta Y, Takahata N. 1999. Paleo-demography of the *Drosophila melanogaster* subgroup: application of the maximum likelihood method. *Genes Genet. Syst.* 74:117–27

59. Lohmann I, McGinnis N, Bodmer M, McGinnis W. 2002. The Drosophila Hox gene *deformed* sculpts head morphology via direct regulation of the apoptosis activator reaper. *Cell* 110:457

60. Mohit P, Bajpai R, Shashidhara LS. 2003. Regulation of *Wingless* and *Vestigial* expression in wing and haltere discs of *Drosophila*. *Development* 130:1537–47
61. Morata G. 2001. How Drosophila appendages develop. *Nat. Rev. Mol. Cell Biol.* 2:89–97
62. Nardi F, Spinsanti G, Boore JL, Carapelli A, Dallai R, Frati F. 2003. Hexapod origins: monophyletic or paraphyletic? *Science* 299:1887–89
63. Ono S. 1982. Effect on juvenile hormone on the caste determination in the ant *Pheidole fervida* (Hymenoptera, Formicidae). *Appl. Entomol. Zool.* 17:1–7
64. Owen R. 1843. *Lectures on the Comparative Anatomy and Physiology of the Invertebrate Animals, delivered at the Royal College of Surgeons in 1843*. London: Longman, Brown, Green & Longmans
65. Panhuis TM, Butlin R, Zuk M, Tregenza T. 2001. Sexual selection and speciation. *Trends Ecol. Evol.* 16:364–71
66. Parichy DM. 2003. Pigment patterns: fish in stripes and spots. *Curr. Biol.* 13:R947–50
67. Pattatucci AM, Kaufman TC. 1991. The homeotic gene Sex combs reduced of *Drosophila melanogaster* is differentially regulated in the embryonic and imaginal stages of development. *Genetics* 129:443–61
68. **Pigliucci M, Murren CJ. 2003. Perspective: genetic assimilation and a possible evolutionary paradox: Can macroevolution sometimes be so fast as to pass us by? *Evolution* 57:1455–64**

This article reviews much of the present data and theory on genetic assimilation.

69. Popadic A, Abzhanov A, Rusch D, Kaufman TC. 1998. Understanding the genetic basis of morphological evolution: the role of homeotic genes in the diversification of the arthropod bauplan. *Int. J. Dev. Biol.* 42:453–61
70. Puelles L, Medina L. 2002. Field homology as a way to reconcile genetic and developmental variability with adult homology. *Brain Res. Bull.* 57:243–55
71. **Raff RA. 1996. *The Shape of Life*. Chicago: Univ. Chicago Press. 520 pp.**

Raff's text frames the essential questions of evo-devo and presents an extensive collection of the classical issues in the field.

72. Regier JC, Shultz JW. 1997. Molecular phylogeny of the major arthropod groups indicates polyphyly of crustaceans and a new hypothesis for the origin of hexapods. *Mol. Biol. Evol.* 14:902–13
73. Regier JC, Shultz JW. 2001. Elongation factor-2: a useful gene for arthropod phylogenetics. *Mol. Phylogenet. Evol.* 20:136–48
74. Revilla-i-Domingo R, Davidson EH. 2003. Developmental gene network analysis. *Int. J. Dev. Biol.* 47:695–703
75. Rockman MV, Hahn MW, Soranzo N, Goldstein DB, Wray GA. 2003. Positive selection on a human-specific transcription factor binding site regulating IL4 expression. *Curr. Biol.* 13:2118–23
76. Rockman MV, Wray GA. 2002. Abundant raw material for cis-regulatory evolution in humans. *Mol. Biol. Evol.* 19:1991–2004
77. Rogers BT, Peterson MD, Kaufman TC. 1997. Evolution of the insect body plan as revealed by the Sex combs reduced expression pattern. *Development* 124:149–57
78. Sameshima SY, Miura T, Matsumoto T. 2004. Wing disc development during caste differentiation in the ant *Pheidole megacephala* (Hymenoptera: Formicidae). *Evol. Dev.* 6:336–41
79. Satoh N, Satou Y, Davidson B, Levine M. 2003. *Ciona intestinalis*: an emerging model for whole-genome analyses. *Trends Genet.* 19:376–81
80. Schoppmeier M, Damen WG. 2001. Double-stranded RNA interference in the spider *Cupiennius salei*: the role of *Distal-less* is evolutionarily conserved in arthropod appendage formation. *Dev. Genes Evol.* 211:76–82
81. Smart J. 1971. Paleoecological factors affecting the origin of winged insects. *Proc. Int. Congr. Entomol., 13th, Moscow, 1968* 1:304–6

82. **Snodgrass RE. 1935. *Principles of Insect Morphology*. New York: McGraw-Hill**
83. Stern DL. 1998. A role of *Ultrabithorax* in morphological differences between *Drosophila* species. *Nature* 396:463–66
84. Sucena E, Delon I, Jones I, Payre F, Stern DL. 2003. Regulatory evolution of *shaven-baby/ovo* underlies multiple cases of morphological parallelism. *Nature* 424:935–38
85. Sucena E, Stern DL. 2000. Divergence of larval morphology between *Drosophila sechellia* and its sibling species caused by *cis*-regulatory evolution of *ovo/shaven-baby*. *Proc. Natl. Acad. Sci. USA* 97:4530–34
86. Telford MJ, Thomas RH. 1998. Expression of homeobox genes shows chelicerate arthropods retain their deutocerebral segment. *Proc. Natl. Acad. Sci. USA* 95:10671–75
87. Tomoyasu Y, Wheeler SR, Denell RE. 2005. Ultrabithorax is required for membranous wing identity in the beetle *Tribolium castaneum*. *Nature*. 433:643–47
88. **True JR, Haag ES. 2001. Developmental system drift and flexibility in evolutionary trajectories. *Evol. Dev.* 3:109–19**
89. Von Dassow G, Odell GM. 2002. Design and constraints of the *Drosophila* segment polarity module: robust spatial patterning emerges from intertwined cell state switches. *J. Exp. Zool.* 294:179–215
90. Waddington CH. 1942. Canalization of development and the inheritance of acquired characters. *Nature* 150:563–65
91. **Waddington CH. 1953. Epigenetics and evolution. In *Evolution (SEB Symposium VIII)*, ed. R Brown, JF Danielli. Cambridge, UK: Cambridge Univ. Press**
92. Wagner-Bernholz JT, Wilson C, Gibson G, Schuh R, Gehring WJ. 1991. Identification of target genes of the homeotic gene Antennapedia by enhancer detection. *Genes Dev.* 5:2467–80
93. Walldorf U, Binner P, Fleig R. 2000. Hox genes in the honey bee *Apis mellifera*. *Dev. Genes Evol.* 210:483–92
94. Warren RW, Nagy L, Selegue J, Gates J, Carroll S. 1994. Evolution of homeotic gene regulation and function in flies and butterflies. *Nature* 372:458–61
95. Weatherbee SD, Halder G, Kim J, Hudson A, Carroll S. 1998. Ultrabithorax regulates genes at several levels of the wing-patterning hierarchy to shape the development of the Drosophila haltere. *Genes Dev.* 12:1474–82
96. Weatherbee SD, Nijhout HF, Grunert LW, Halder G, Galant R, et al. 1999. Ultrabithorax function in butterfly wings and the evolution of insect wing patterns. *Curr. Biol.* 9:109–15
97. Wheeler WC, Whiting M, Wheeler QD, Carpenter JM. 2001. The phylogeny of the extant hexapod orders. *Cladistics* 17:113–69
98. White MJD. 1968. Models of speciation. *Science* 159:1065–70
99. Whiting MF, Bradler S, Maxwell T. 2003. Loss and recovery of wings in stick insects. *Nature* 421:264–67
100. Whiting MF, Wheeler WC. 1994. Insect homeotic transformation. *Nature* 368:696
101. Wootton RJ. 1972. Nymphs of Palaeodictyoptera (Insecta) from the Westphalian of England. *Paleontology* 15:662–75
102. Zheng Z, Khoo A, Fambrough D Jr, Garza L, Booker R. 1999. Homeotic gene expression in the wild-type and a homeotic mutant of the moth *Manduca sexta*. *Dev. Genes Evol.* 209:460–72
103. Zrzavy J, Hypsa V, Vlaskova M. 1997. Arthropod phylogeny: taxonomic congruence, total evidence and conditional combination approaches to morphological and molecular data sets. In *Arthropod Relationships*, ed. RA Fortey, RH Thomas, pp. 97–108. London: Chapman & Hall

Snodgrass' 1935 text remains an authority on insect anatomy.

True & Haag introduce the concept of developmental systems drift, and discuss examples from the literature.

Together with *The Strategy of The Genes* (1957, London, Allen & Unwin), this text outlined many of Waddington's ideas, such as canalization and developmental constraint, which have been revisited again in recent decades.

Concerted and Birth-and-Death Evolution of Multigene Families*

Masatoshi Nei[1] and Alejandro P. Rooney[2]

[1]Institute of Molecular Evolutionary Genetics and1 Department of Biology, Pennsylvania State University, University Park, Pennsylvania 16802; email: nxm2@psu.edu

[2]U.S. Department of Agriculture, Agricultural Research Service, National Center for Agricultural Utilization Research, Peoria, Illinois 61604; email: rooney@ncaur.usda.gov

Annu. Rev. Genet.
2005. 39:121–52

First published online as a Review in Advance on June 22, 2005

The *Annual Review of Genetics* is online at http://genet.annualreviews.org

doi: 10.1146/annurev.genet.39.073003.112240

0066-4197/05/1215-0121$20.00

Key Words

birth-and-death evolution, concerted evolution, origins of new genetic systems, gene conversion, MHC genes, ribosomal RNA

Abstract

Until around 1990, most multigene families were thought to be subject to concerted evolution, in which all member genes of a family evolve as a unit in concert. However, phylogenetic analysis of MHC and other immune system genes showed a quite different evolutionary pattern, and a new model called birth-and-death evolution was proposed. In this model, new genes are created by gene duplication and some duplicate genes stay in the genome for a long time, whereas others are inactivated or deleted from the genome. Later investigations have shown that most non-rRNA genes including highly conserved histone or ubiquitin genes are subject to this type of evolution. However, the controversy over the two models is still continuing because the distinction between the two models becomes difficult when sequence differences are small. Unlike concerted evolution, the model of birth-and-death evolution can give some insights into the origins of new genetic systems or new phenotypic characters.

Contents

Multigene family: a group of genes that have descended from a common ancestral gene and therefore have similar functions and similar DNA sequences

INTRODUCTION

A multigene family is a group of genes that have descended from a common ancestral gene and therefore have similar functions and similar DNA sequences. A group of related multigene families is sometimes called a supergene family. A well-known example of a supergene family is the globin superfamily that is composed of the α-like, β-like, and some other gene families (134a). A gene family may also be subdivided into subfamilies whenever convenient.

The evolution of multigene families has been the subject of controversy for many years. The paradigm of evolution of multigene families before 1970 was that of hemoglobin α, β, γ, and δ chains and myoglobin (54). The genes encoding these polypeptides or proteins are phylogenetically related and have diverged gradually as the duplicate genes acquired new gene functions. This mode of evolution may be called "divergent evolution" (**Figure 1*a***). Around 1970, however, a number of researchers showed that ribosomal RNAs (rRNAs) in *Xenopus* are encoded by a large number of tandemly repeated genes and that the nucleotide sequences of the intergenic regions of the genes are more similar within a species than between two related species (11).

These observations were difficult to explain by the model of divergent evolution, and a new model called "concerted evolution" was proposed (**Figure 1*b***). In this model all the members of a gene family are assumed to evolve in a concerted manner rather than independently, and a mutation occurring in a repeat spreads through the entire member genes by repeated occurrence of unequal crossover or gene conversion. This model is capable of explaining previously puzzling observations about the evolution of rRNA genes.

This apparent success led many authors to believe that most multigene families evolve following the model of concerted evolution, and a number of authors investigated the evolutionary modes of various multigene families (48, 110, 178). Later, however, the applicability of concerted evolution to some gene families was questioned as both DNA and amino acid sequence data became available, and another model called birth-and-death evolution (98) was proposed. In this model new genes are created by gene duplication, and some duplicated genes are maintained in the genome for a long time, whereas others are deleted or become nonfunctional

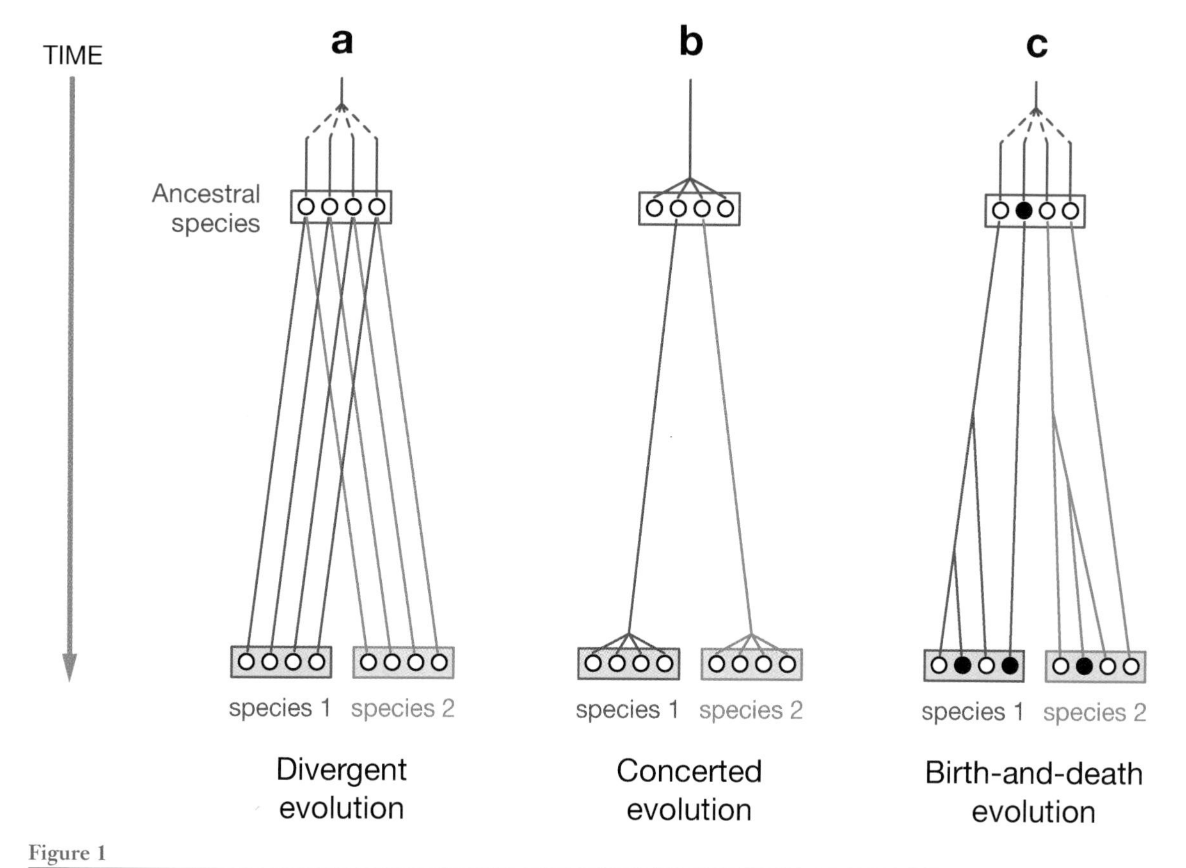

Figure 1

Three different models of evolution of multigene families. Open circles stand for functional genes and closed circles for pseudogenes.

through deleterious mutations (**Figure 1*c***). This model applies to most multigene families concerned with immune systems such as immunoglobulins and major histocompatibility complex (MHC) (53, 116) and disease-resistance genes (173). In recent years, even a highly conserved gene family such as the ubiquitin family was shown to be subject to birth-and-death evolution. Yet the controversy over the evolution of multigene families continues, partly because there are so many different types of gene families and partly because the general mechanism of gene conversion is still unclear.

In this paper we review recent studies of evolution of multigene families with some historical backgrounds.

CONCERTED EVOLUTION

Ribosomal RNA and Other RNA Genes

One of the best-known examples of concerted evolution comes from the study of rRNA genes. The rRNA gene family of the African toads *Xenopus laevis* and *X. mulleri* consists of about 450 repeats or members. Each member consists of the 18S, 5.8S, and 28S RNA genes, external transcribed spacers (ETS1 and ETS2), internal transcribed spacers (ITS1 and ITS2), and an intergenic spacer (IGS) (**Figure 2**) (15). Using DNA or RNA hybridization techniques, Brown et al. (11) showed that the nucleotide sequences of IGS are very similar among member genes of the

Concerted evolution: a form of multigene family evolution in which all the member genes are assumed to evolve as a unit in concert and a mutation occurring in a repeat spreads through the entire member genes by repeated occurrence of unequal crossover or gene conversion

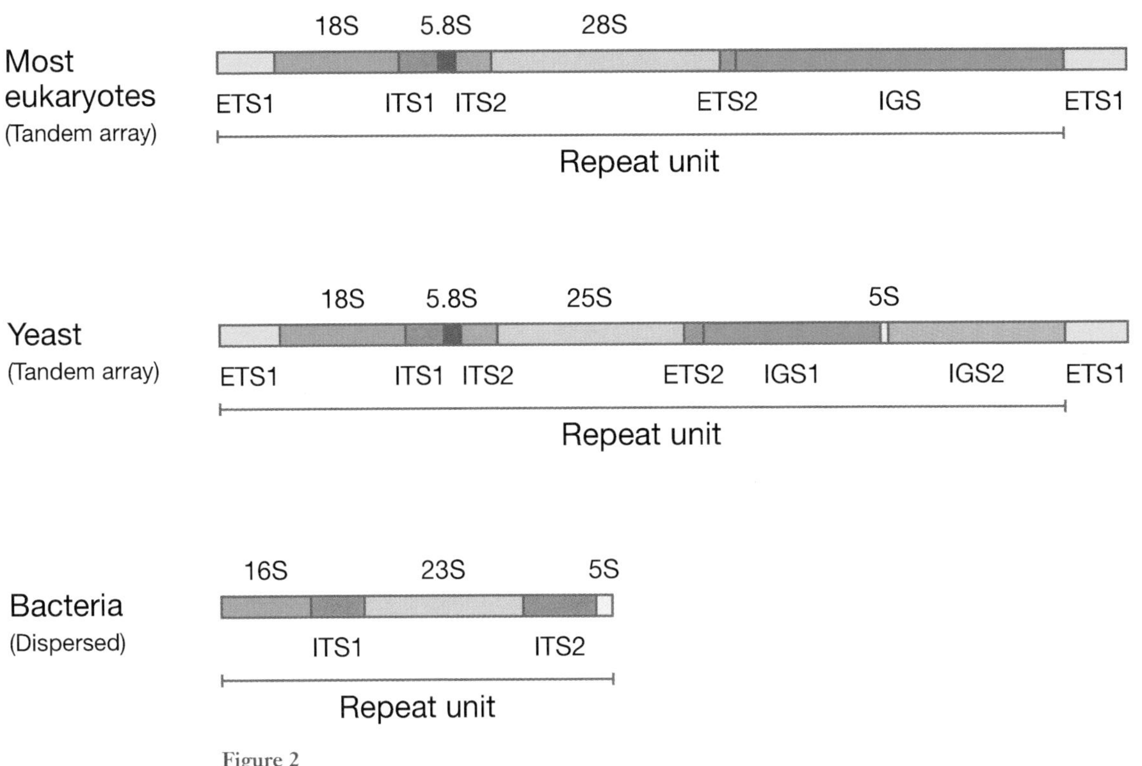

Figure 2

Molecular structures of rRNA gene repeats in different organisms.

Gene conversion: a form of nonreciprocal recombination in which a DNA segment of a recipient gene is copied from a donor gene

same species but differ by about 10% between *X. laevis* and *X. mulleri*. This observation could not be explained by the then popular model of divergent evolution. According to this model, the differences in nucleotide sequence between different repeats of the same species are expected to be as large as those between repeats of different species. The explanation becomes more difficult to accept if we note that the nucleotide sequences of the 18S and 28S coding regions are virtually identical between *X. laevis* and *X. mulleri*. Actually, the 18S and 28S coding regions are very similar even among distantly related organisms such as animals and plants.

This puzzling observation can be explained by the model of horizontal or concerted evolution originally proposed by Brown et al. (11). According to this model, unequal crossover occurs randomly among members of a gene family, and repeated occurrence of unequal crossover has an effect of homogenizing the member genes, as mentioned above. In this case, the number of member genes may increase or decrease by chance, but a certain range of the number of genes is maintained because of the functional requirement. In the absence of mutation, this process will eventually lead to the homogeneity of all member genes of a family. In reality, of course, mutation always occurs, so that a gene family is expected to have some variant genes. It should now be clear that when a species diverges into two species and the gene cluster in each descendant species evolves independently, the cluster within each species tends to have similar gene copies because of unequal crossover, whereas the genes belonging to the cluster of the two species gradually diverge by mutation (**Figure 1*b***). This is exactly what we observe with the IGS regions of rRNA genes in *Xenopus*. Later, Smith (144, 145) conducted computer simulations to show that concerted evolution indeed can explain the observation about the evolutionary change of the IGS region. As mentioned above, the 18S and 28S coding regions are virtually identical between *X. laevis* and *X. mulleri* as well as between different copies of the same species. This identity has apparently been maintained by strong purifying selection that operates for the coding regions. Thus, we can explain the entire set of observations about the rRNA gene family of *Xenopus* in terms of unequal crossover, mutation, and purifying selection.

In addition to these factors, gene conversion (55, 143) was proposed as a mechanism of homogenizing the member genes of multigene families. The role of gene conversion is similar to that of unequal crossover. The only difference is that the latter may increase or decrease the number of genes, whereas the former does not. The idea of gene conversion later became popular among theoretical population geneticists, partly because it is easy to develop mathematical models of concerted evolution (8, 90–92, 111–113, 161). In reality, the molecular mechanism of gene conversion in multigene families is not well understood, particularly when sequence identity is patchy, though gene conversion in yeast can be explained by a DNA breakage followed by invasive DNA replication (120). Furthermore, in most mathematical formulations the effect of purifying selection operating in the coding regions of 18S and 28S genes has been neglected. Therefore, caution should be exercised in the application of the mathematical formulas to real data. The gene conversion theory has also become popular among researchers of MHC polymorphism, as is discussed below.

Note that the relative contributions of unequal crossover (or gene conversion) and purifying selection to the homogenization of the rRNA genes have rarely been discussed. For this reason, the homogeneity of the rRNA-coding regions (18S and 28S) was often attributed to unequal crossover rather than to purifying selection. Actually, even the IGS regions appear to be subject to purifying selection in *Xenopus* because this region contains elements of promoters and enhancers (15, 132). Furthermore, the ITS regions have a level of variability as low as that of the rRNA-coding regions in the fungus *Fusarium graminearum* (K. O'Donnell, personal communication). This has probably occurred because they are closely linked to the highly conserved rRNA-coding regions or because they have some important functions. It is therefore necessary to keep in mind that concerted evolution applies primarily to the IGS region, and even in this region a substantial proportion of mutations may be eliminated by purifying selection.

In recent years, the so-called complete genome sequence has been published for many different organisms. Unfortunately, the rRNA gene regions are usually excluded from the sequence, mainly because of the difficulty of sequencing a large number of similar genes. In humans, chimpanzees, gorillas, and orangutans, however, some sequence data are available for a small portion of the rRNA gene regions (35). The hominoid genome has about 500 rRNA gene repeats, and the molecular structure of each repeat unit is similar to that of *Xenopus* (15). However, hominoid rRNA genes are clustered in the telomeric regions of five different chromosomes. The pattern of sequence similarity among the IGS, 18S, and 28S regions of each chromosome is very similar to that of *Xenopus* genes. The 18S and 28S gene regions are virtually identical within and between hominoid species, and the IGS regions from the same chromosomes are also similar and show only about 0.5% nucleotide differences.

The rRNA gene clusters closely located at the chromosomal ends are also very similar in each hominoid species, but the IGS regions distal to the telomere are somewhat differentiated (35). This observation suggests that genes are exchanged between different chromosomes through unequal crossover or gene conversion that occurs primarily in the chromosomal region proximal to the telomere.

Birth-and-death evolution: a form of multigene family evolution in which new genes are created by gene duplication and some duplicate genes are maintained in the genome for a long time, whereas others are deleted or inactivated through deleterious mutations

In fact, the distal IGS regions often showed a substantial amount of sequence difference (3 ~ 7% per site).

The 5S RNA genes form separate gene clusters and are located on a different genomic region. This gene family includes 9000~24,000 member genes in *Xenopus* (10) and about 500 members in humans (35). These 5S rRNA genes are also known to undergo concerted evolution (10). Furthermore, the gene families of small nuclear RNAs (snRNA) involved in intron-splicing and other important cellular metabolisms apparently undergo concerted evolution. An extensive study of concerted evolution of U2 snRNA genes in primates has been conducted by Weiner and his group (70, 122, 123), and these authors have shown that the coding regions of U2 snRNA gene members are very similar to one another but the intergenic regions are heterogeneous within each species. These results again demonstrate the importance of purifying selection in the coding regions.

In yeast, the basic unit of rRNA gene repeats includes an additional RNA gene (5S gene) in the middle of the IGS region, but IGS is again subject to concerted evolution (125, 146, 151, 170). In bacterial genomes there are only a few rRNA gene repeats, and they are generally dispersed in the genome (69). For example, *Escherichia coli* has 7 copies of the rRNA operon, which is composed of 16S, 23S, and 5S rRNA genes (**Figure 2**). The spacer (ITS1) between the 16S and 23S genes usually contains one or two tRNA genes, and these tRNA genes are not necessarily the same among different copies of the operon.

The rRNA-coding regions are again very similar among different operons. The sequence difference of the 16S gene among repeats is 0.0055 per site in *E. coli* and 0 in *Haemophilus influenzae*. However, the ITS1 is quite heterogeneous. This region often contains unique nucleotide sequences shared by only a few operons. These sequences are patchy and could represent traces of gene conversion events that occurred in the past. On the basis of these observations, Liao (69) concluded that the rRNA genes in bacteria are generally homogenized by gene conversion. However, these observations can also be explained by strong purifying selection and occasional unequal crossover. If unequal crossover occurs in the ITS1 regions as well as outside the rRNA operons, a unique nucleotide sequence in an ITS can be transferred to other ITSs. In this case, we would expect that the sequence length of ITS varies from operon to operon. In fact, this is the case in *E. coli*, and the length of the ITS1 between the 16S and 23S genes varies from 500 bp to 800 bp.

The above literature survey indicates that most RNA genes in both prokaryotes and eukaryotes are subject to concerted evolution. However, there are exceptions to this rule. The most conspicious is the rRNA genes in species of the malaria parasite plasmodia. In these species the number of rRNA gene is low (a few to a dozen copies), and they are dispersed in different chromosomes (68a, 86a). These genes are grouped into a few different classes in terms of function and structure. These different classes of rRNA genes are used in different stages of the life cycle of plasmodia, which infest both insects and vertebrates (68a). The gene sequences are similar within each class of rRNA genes but are different between different classes. The inter-class differences are often substantial and amount to about 10 percent. Conducting a phylogenetic analysis of the sequences from several closely related species, Rooney (135) concluded that the gene family in plasmodia actually evolve following the birth-and-death model. The existence of different classes of rRNA genes in the same genome has been reported in several other organism such as flatworms (14a) and oak tree (89a).

The 5S RNA gene family is also heterogeneous in some species. The best known example is that of filamentus fungi. The number of copies of 5S RNA genes in these species is 50 ~ 100, and they are dispersed in the genome rather than organized as a tandem array. They can also be grouped into a few different classes by means of sequence similarity. Studying the evolutionary pattern of the sequences from four species of this group of fungi, Rooney & Ward (137) concluded that this 5S RNA gene family is subject to birth-and-death evolution. They found that 18% to 83% of genes are pseudogenes. There are several other species in which heterogeneous 5S genes exists. Examples are *Xenopus laevis* (162a) and wheat (58a).

Tandemly Arrayed Histone Genes

Not long after the evolution of rRNA gene families was explained by the model of concerted evolution, many researchers began to assume that this model applies to various other multigene families (17, 110, 115, 144, 145). The general view then was that a gene family that produces a large amount of gene products is subject to concerted evolution to homogenize the genes. One such family was the histone gene family of sea urchins (20, 21, 44, 47, 57). This is a large multigene family with several hundred members that are divided into four classes on the basis of developmental and tissue-specific expression patterns (**Figure 3**): (*a*) "early histone genes" that are active during late oogenesis through the blastula stage of embryogenesis, (*b*) "cleavage stage histone genes" that encode the first histones expressed after fertilization, (*c*) "late histone genes" that are active from the late blastula stage onwards, and (*d*) "sperm histone genes" that are expressed only during spermatogenesis (19, 78, 83, 101, 102).

The early histone genes are present in about 300–500 repeat units in most sea urchin species. In the sea urchin *Lytechinus pictus*, they are arranged in 3 tandem arrays that consist of virtually identical repeating units of the 5 histone genes (H1, H2A, H2B, H3, and H4), each of which is separated by noncoding IGS regions (21, 47) (**Figure 3**). In this species, only the early genes are arranged in tandem array, whereas the other three classes of genes appear to be dispersed throughout the genome and present in significantly fewer copy numbers (16, 78). Using DNA heteroduplex and restriction mapping analyses, it was demonstrated that the IGS regions of the early gene tandem arrays in *L. pictus* show a considerable amount of variation, whereas the protein-encoding regions are highly conserved (20, 21, 47). This was taken as evidence for concerted evolution in the early histone genes of this species. Not long afterwards, researchers studying the sequence divergence of late histone genes of *L. pictus* claimed that these genes also undergo concerted evolution (130), as did another study on the late genes of the sea urchin *Strongylocentrotus purpuratus* (83). However, it was later shown that histone genes generally evolve following a birth-and-death process with strong purifying selection (127, 136), as discussed below.

Pseudogenes: nonfunctional genes generated by nonsense mutation, frameshift mutation, or partial nucleotide deletion

Like the sea urchin genes, the 5 histone genes in *Drosophila melanogaster* are arranged in a repeating unit (71). This unit is repeated about 110 times in a tandem array found on chromosome 2L (121). Two types of repeat units that differ with respect to length (5.0 kb and 4.8 kb) have apparently arisen owing to differences in the noncoding region (71). In addition to the tandem array, so-called variant histone genes are located in other parts of the genome (119). On the basis of restriction fragment patterns, Coen et al. (18) argued that the histone genes of *D. melanogaster* and its close relatives underwent concerted evolution. According to these authors, the absence of different banding patterns within a species was evidence for concerted evolution, because a restriction site must have spread to all other repeat units after it had arisen subsequent to the divergence of *Drosophila* species. However, this line of inference based on negative data should not be used as support for concerted evolution.

Matsuo & Yamazaki (82) later obtained nucleotide sequences of several different histone H3 genes and their flanking regions from *Drosophila*. They obtained data from 10 clones of a single chromosome (a single individual), 10 clones from different chromosomes (a

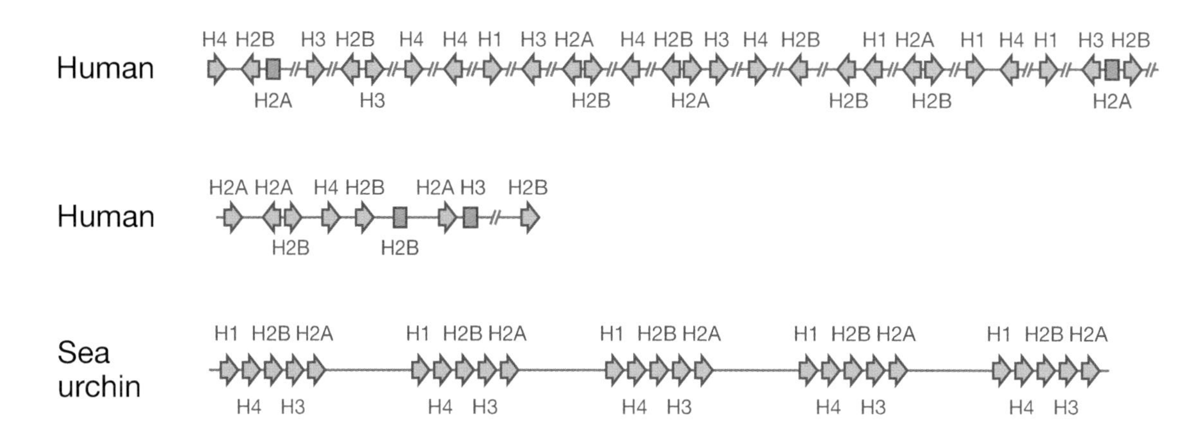

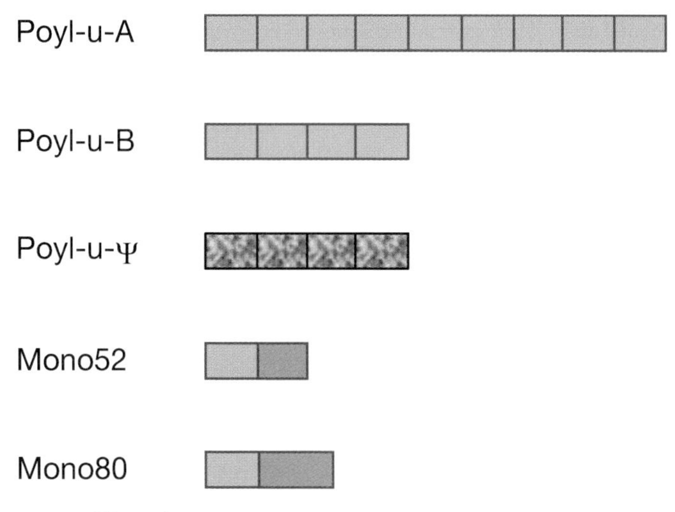

Figure 3

Genomic structures of histone gene (*a*) and ubiquitin gene repeats (*b*). The red square box in human histone gene repeats represents a pseudogene. Expressed histone genes are shown in blue, and an arrow indicates the direction of transcription when known. Each box in the poly-u gene complex represents one copy of ubiquitin gene. Poly-uψ shows a poly-u pseudogene. A monomeric ubiquitin locus is composed of a ubiquitin gene and a ribosomal protein gene with 52 codons (Mono52) or 80 codons (Mono80).

population sample) of *D. melanogaster*, and a single clone from a single chromosome of a sibling species, *D. simulans*. They found that variability within an individual was 1.7 times less than that of a population sample and 14 times less than the interspecific variation. Using complex mathematical models, they argued that these observations can be explained by concerted evolution. However, their sequence data indicate that the number of polymorphic sites is quite small and is not substantially different between the protein coding and flanking regions.

By the end of the 1980s, most researchers in the field had concluded that virtually all multigene families evolve in a concerted fashion. Therefore, the studies with sea urchin and *Drosophila* histone genes mentioned above were received as confirmation of the general view. No one considered the possibility that purifying selection is an alternative explanation for the similarity of histone sequences. This happened because the above studies, which were often performed using restriction fragment analysis, did not have sufficient power to distinguish between intralocus and interlocus variation of histone genes. Recently, this view has changed, as discussed below.

Other Genes

There are many other genes for which concerted evolution has been reported. A well-known example is the heatshock protein gene (*hsp70* genes) in *Drosophila*. *D. melanogaster* has a relatively large family of *hsp* genes, of which two, *hsp70Aa* and *hsp70Ab*, are a pair of inverted tandem repeats, and the nucleotide sequences of the two genes are virtually identical. *D. simulans*, a sibling species of *D. melanogaster*, also has a similar tandem pair of nearly identical *hsp70* genes. Finding these pairs of genes, Leigh Brown & Ish-Horowicz (68) and Bettencourt & Feder (7) proposed that the within-species identity of the two genes is caused by gene conversion. This is certainly a plausible explanation, but if we consider the evolution of the entire members of this gene family, the evolutionary pattern does not necessarily conform to concerted evolution (see below).

There are many reports about possible gene conversion in small multigene families. One example is that between the two loci of Rh blood group genes in hominoids (14b, 60). Conducting a detailed statistical analysis, Kitano & Saitou (60) reported that several gene conversions occurred in each of humans, chimpanzees, and gorillas. However, it is very difficult to distinguish between gene conversion and independent nucleotide substitution in their case (65a). Their results are also dependent on the phylogenetic tree of the genes used. Therefore, their conclusion is questionable. Furthermore, the sequence identity of the two genes was rather low compared with the average identity of genes from humans and chimpanzees. A more detailed study is necessary about this gene family.

Many authors have claimed that the genetic variability of MHC loci is caused by gene conversion, and this was thought to be a source of genetic variability within loci rather than a homogenizing factor (36, 73, 86, 112, 164). Similar hypotheses were presented to explain the high degree of antibody diversity (110). These views remained popular until a more realistic model of maintenance of polymorphism was proposed.

BIRTH-AND-DEATH EVOLUTION

MHC Genes

The birth-and-death model of evolution of multigene families was first proposed to explain the unusual pattern of evolution of MHC genes in mammals (51, 65, 96, 98). The function of MHC genes is to bind self or foreign peptides and present them to T lymphocytes, thereby triggering an immune response (63). MHC genes can be divided into class I and class II genes on the basis of molecular structure and function of the polypeptide encoded. Class I genes can be further divided into classical and nonclassical genes. The classical class I (Ia) genes are highly polymorphic, and the number of alleles per locus sometimes exceeds 100. This high degree of polymorphism is important for protecting the host from attack by various types of parasites (viruses, bacteria, fungi, and others), which are always changing with time. By contrast, the nonclassical class I (Ib) genes are less polymorphic and their functions may be quite different from those of Ia genes.

In the 1970s and 1980s when most investigators believed that multigene families were generally subject to concerted evolution, the MHC gene family was no exception. Therefore, some authors attempted to explain the polymorphism of MHC genes by means of unequal crossover or gene conversion (74, 111, 112, 164). In particular, Ohta (112) and Weiss et al. (164) proposed that the high degree of polymorphism at Ia loci could be explained by gene conversion. This view is based on the idea that if some parts of a sequence at a monomorphic locus are converted by another nucleotide sequence from another locus, polymorphism is generated at the first locus. Enhancement of polymorphism would occur even if both loci are polymorphic to some extent as long as the nucleotide sequences between the two loci are sufficiently different and gene conversion occurs in both ways between the two loci. The problem with this idea is that it does not explain why gene conversion starts to occur between two previously differentiated loci suddenly at an evolutionary time. The coexistence of Ia and Ib genes in the same DNA region is also difficult to explain. If gene conversion occurs continuously between the two loci, the extent of polymorphism should be essentially the same for the two loci. Furthermore, if phylogenetic analysis is conducted for the alleles from different loci, there would be no monophyletic clades formed for each locus. In reality, this is not the case. Klein & Figueroa (62) and Kriener et al. (65a) critically examined data that seemingly supported the idea of concerted evolution and concluded that the evidence is weak. They argued that some of the data showing the identical gene segments between paralogous pairs of genes can be explained by co-ancestry of the segments or even clustered mutations (65a, 169).

The idea of gene conversion was weakened considerably when Hughes & Nei (50, 52) showed that MHC polymorphism is primarily caused by overdominant selection that operates at the peptide-binding region of MHC molecules. This finding made it unnecessary to invoke gene conversion as an explanation for MHC polymorphism. It also provided a theoretical basis for the concept of *trans*-specific (long-term) polymorphism previously discovered (5, 28, 62, 66, 85). In a phylogenetic analysis of MHC class I and class II genes, Hughes & Nei (52, 53, 97) showed that the evolutionary pattern of these genes was very different from what would be expected under concerted evolution. The tree for class I genes from a number of vertebrate species is presented in **Figure 4*a***. It indicates that different orders or families of mammals often have different genes or genetic loci. For example, the classical loci A, B, and C are shared only by hominoid species (e.g., human, gorilla, and orangutan), but the New World monkeys (e.g., tamarin) and nonprimate mammals do not have the genes. Similarly, cats and mice have different Ia loci. In other words, different families or orders of mammals do not have truly orthologous genes. This evolutionary pattern indicates that some genes were generated by gene duplication and some duplicate genes were lost after the divergence of mammalian orders (51, 97). Actually, the genomic regions of human and mouse MHC genes contain a large number of pseudogenes (58), exactly as would be expected under the birth-and-death model (**Figure 1*c***). Also, this conclusion makes sense biologically because the genetic variability at MHC loci is generated to defend the host from many new types of parasites. Gene conversion and unequal crossover are not essential for this purpose, though they may occur.

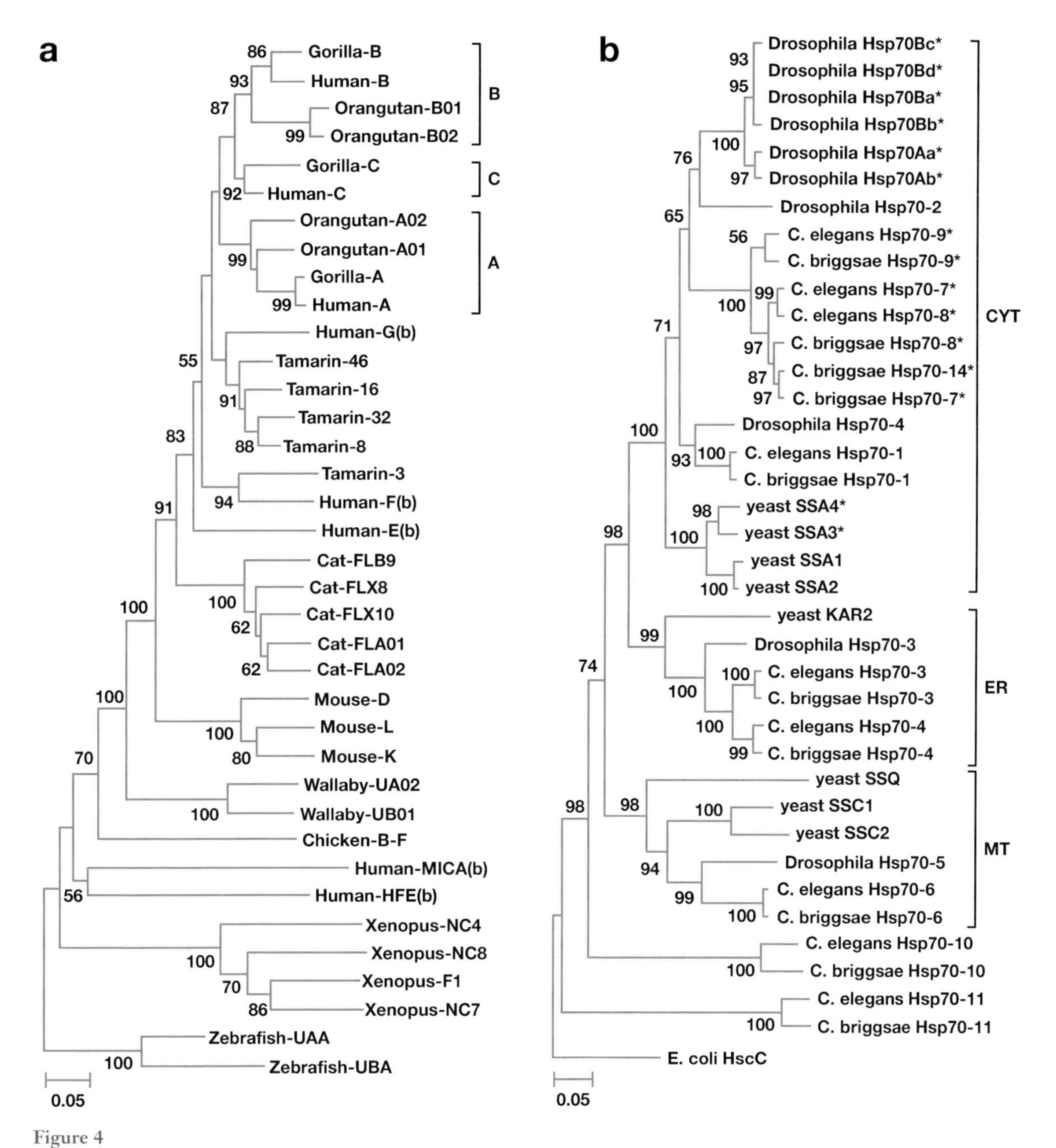

Figure 4

(*a*) Phylogenetic tree of MHC class I genes from vertebrates. (*b*) Phylogenetic tree of heat shock protein genes (*hsp70*) from two species of nematodes, *Drosophila melanogaster*, and yeast. CYT, ER, and MT stand for the genes expressed in cytoplasm, endoplasmic reticulum, and mitochondria, respectively. The ∗ sign indicates a highly heat-inducible gene. Modified from Nei et al. (96) and Nikolaidis & Nei (109).

Figure 4*a* also shows that a few nonclassical loci as indicated by (*b*) diverged from Ia loci a long time ago and now have different functions, as discussed below. This type of acquisition of new functions by duplicate genes is also an important feature of birth-and-death evolution that is not shared by concerted evolution. Phylogenetic analysis of class II region genes showed essentially the same evolutionary pattern as that of class I genes (53). However, the rate of gene birth and gene death is much lower in the class II gene family than in the class I gene family (97, 126, 152).

Despite these findings, a substantial number of authors maintain that gene conversion or unequal crossover is an important factor for generating polymorphism (12, 114, 124). Some investigators invoked gene conversion to explain the identity of a single or a few nucleotides between different alleles at the same locus or different paralogous sequences (24, 162, 167). Actually, there is evidence that intralocus gene conversion or recombination occurs occasionally (6, 46, 79, 84, 138, 162, 172). Exon exchange between different loci has also been documented (40, 51). Furthermore, there is some molecular evidence that interlocus gene conversion occurs at MHC loci (2, 33, 41, 45, 46, 56, 168). Hogstrand & Bohme (45) reported that the frequency of occurence of gene conversion between the *Ab* and *Eb* loci in mice is about 2×10^{-6} per sperm per generation. If this estimate is right, it is much higher than the rate of nucleotide substitution (of the order of 10^{-9} per generation). Therefore, the nucleotide sequences of the two loci, *Ab* and *Eb*, would become nearly identical in the long run. In reality, however, a phylogenetic analysis of allelic sequences from loci *Ab* and *Eb* has shown that alleles from each locus form a monophyletic clade and the alleles from different loci do not intermingle (40). This is also generally true with the genes from different organisms (40), indicating that the effect of gene conversion on MHC polymophism is quite minor. These results suggest that either the current estimate of gene conversion frequency is too high or many gene conversions observed in the mouse experiment are not fixed in the natural population because of selective disadvantage compared with wild-type alleles (40). Martinsohn et al. (80) conducted a critical examination of papers claiming gene conversions and concluded that gene conversions or interlocus recombinations do occur but those that enhance polymorphism are not proven. This conclusion agrees with that of Gu & Nei (40). Therefore, MHC polymorphism appears to be primarily generated by nucleotide substitution and positive selection.

Immunoglobulin and Related Gene Families

Immunoglobulins are composed of heavy (H) chains and light chains, and the latter chains can be further divided into λ and κ chains. Each of the three groups of chains consists of constant and variable regions (63), and the polypeptides of variable regions in each category are encoded by a genomic cluster, called a variable region gene family. There are about $50 \sim 100$ genes in each of the H, λ, and κ variable region gene families in human. About 50% of these variable region genes are pseudogenes (81, 118). All of these multigene families are subject to birth-and-death evolution (116–118, 141, 142, 148). The molecular structure and the genomic organization of T-cell receptors are similar to those of immunoglobulins, and the variable region gene families for different classes of T-cell receptor genes are also known to be subject to birth-and-death evolutions (147, 149, 150). Most of these gene families include many pseudogenes. Therefore, the death rate of these genes is quite high. In both immunoglobulin and T-cell receptor families, individual functional genes apparently maintain their own identity without much effect from interlocus recombination or gene conversion, because the branch lengths of individual genes are usually very long and sometimes correspond to tens of millions of years. In the case of immunoglobulin κ variable region genes, the human genome contains about two times as many genes as does the chimpanzee because of a DNA block duplication that occurred in the human lineage about 2 mya. In this case, the orthologous genes between the original DNA segment and the new duplicated segment can easily be identified (141). This supports the idea that intergenic recombination and gene conversion have little effect on sequence evolution. Note also that the extent of polymorphism in these genes is much smaller than that of MHC genes.

In addition to the above multigene families, the gene families concerned with innate immunity have also been shown to undergo birth-and-death evolution (42, 43, 107, 108). The evolutionary pattern of these genes is more complex than those of the adaptive immune system. For example, the natural killer (NK) cell receptors of humans [KIR: killer cell immunoglobulin-like receptor (KIR genes)] are composed of immunoglobulin-like domains, but those of rodent receptors (Ly49) genes are of lectin-type, and the molecular structures of these two groups of genes are very different (63). It is unclear how these different types of NK cell receptors originated in two different orders of mammals. Both KIR and Ly49 gene families are known to be subject to birth-and-death evolution (42, 43). KIR genes are also subject to domain shuffling as well as to mutational changes (9, 42, 129, 156, 166). Furthermore, the number of member genes of these gene families has expanded very rapidly by gene duplication during the past 20–30 million years (42, 43, 59). Yet, about a half of these genes are apparently nonfunctional. For example, the rat genome contains 33 Ly49 genes, but 16 of them are pseudogenes (42).

Many other immune systems gene families undergo birth-and-death evolution (**Table 1**). In fact, almost all immune systems genes except solitary or small-sized gene families appear to be subject to this model of evolution, and the rate of gene turnover is generally quite high, as expected from their function.

Table 1 Some examples of gene families that undergo birth-and-death evolution

Multigene family	Organism	Reference
(A) Immune system		
MHC	Vertebrates	(51, 96, 98)
Immunoglobulins	Vertebrates	(96, 116)
T-cell receptors	Vertebrates	(149)
Natural killer cell receptors	Mammals	(42, 43)
Eosinophilic RNases	Rodents	(173)
Disease resistance (R) loci	Plants	(87)
Cecropins	*Drosophila*	(128)
α-Defensins	Mammals	(75)
β-Defensins	Mammals	(89)
(B) Sensory system		
Chemoreceptors	Nematodes	(131)
Taste receptors	Mammals	(22)
Sex pheromone desaturases	Insects	(72, 133)
Olfactory receptors	Mammals	(103, 165)
(C) Development		
Homeobox genes	Animals	(160)
MADS-box	Plants	(94)
WAK-like kinase	*Arabidopsis*	(158)
(D) Highly conserved		
Histones	Eukaryotes	(26, 127, 136)
Amylases	*Drosophila*	(177)
Peroxidases	All kingdoms	(171)
Ubiquitins	Eukaryotes	(99)
Nuclear ribosomal RNA	Protists, Fungi	(135, 137)
(E) Miscellaneous		
DUP240 genes	Yeast	(67)
Polygalacturonases	Fungi	(37)
3-Finger venom toxins	Snakes	(30)
Replication proteins	Nanoviruses	(49)
ABC transporters	Eukaryotes	(3, 4, 23)

Olfactory Receptor Genes

The MHC and immunoglobulin variable gene families are quite large, but the largest gene family in mammals is the olfactory receptor (OR) gene family. Olfactory (odor molecules) receptors are G-protein–coupled receptors that contain 7 α-helical transmembrane regions. OR genes are about 310 codons long and have no introns in the coding regions. These genes are expressed in sensory neurons of olfactory epithelia in nasal cavities. The human and mouse genomes contain about 800 and 1400 genes, respectively; over 60% of human OR genes are pseudogenes whereas only about 25% of mouse OR genes are pseudogenes (**Table 2**) (32, 34, 103, 106, 168, 175, 176, 179). These genes are scattered over many different locations of almost all chromosomes, and they are generally arranged as a tandem array in each genomic location. It is relatively easy to identify the orthologous gene pairs between humans and mouse by conducting phylogenetic analysis. This suggests that gene conversion or unequal crossover has not occurred frequently and that the number of OR genes apparently has increased by tandem gene duplication and chromosomal rearrangement (**Figure 5**). There are not many traces of transposition of genes mediated by transposons.

As is shown in **Table 2**, the human and mouse genomes have about 390 and 1040 functional genes, respectively. To examine whether this is due to the loss of OR genes in the human lineage or the gain of genes in the mouse lineage, Niimura & Nei (105) estimated the number of functional OR genes that existed in the most recent common ancestor (MRCA) of humans and mice. The number obtained was approximately 754 genes. This indicates that the human lineage lost many functional OR genes, whereas the mouse lineage gained a substantial number of genes (**Figure 6*a***). Some authors have suggested that mice require a larger number of OR genes because they are nocturnal and heavily dependent on olfaction, whereas humans do not need so many genes because they often use the visual sense for finding mates and food.

In reality, the ability to smell is controlled not only by the number of OR genes but also by the brain function for odor recognition, and the human brain possibly has a higher power of distinguishing between subtle differences of odor molecules than the mouse brain (140). At present, however, the mechanism of odor recognition in the brain is virtually unknown, and therefore we do not consider this factor in this paper.

As indicated in **Table 2**, mammalian OR genes can be divided into class I and class II genes. Previously, class I genes were thought to be for aquatic odorants, and class II genes for airborne odorants (29). In mammals and chickens, most genes belong to class II genes, whereas class I genes make up only 2.5% ~ 13%. It was later found that the zebrafish has one class II OR genes, although the search for OR genes in fish is still incomplete. Conducting phylogenetic analysis of OR genes from fish, *X. tropicalis*, chickens, and humans,

Table 2 Numbers of 9 group ($\alpha, \beta, \ldots, \kappa$) OR genes in 6 different species of vertebrates

	Zebrafish		Pufferfish		*X. tropicalis*		Chicken		Mouse		Human	
	Func	All	Func	All	Func	All	Func	All	Func	All	Func	All
α (I)	0	0	0	0	2	6	9	14	115	163	57	102
β (I)	1	2	1	1	5	19	0	0	0	0	0	0
γ (II)	1	1	0	0	370	802	72	543	922	1228	331	700
δ (I)	44	55	28	61	22	36	0	0	0	0	0	0
ε (I)	11	14	2	2	6	17	0	0	0	0	0	0
ζ (I)	27	40	6	8	0	0	0	0	0	0	0	0
η	16	23	5	24	3	6	0	0	0	0	0	0
θ	1	1	1	1	1	1	1	1	0	0	0	0
κ	1	1	1	1	1	1	0	0	0	0	0	0
Total	102	137	44	98	410	888	82	558	1037	1391	388	802

I; class I genes. II; class II genes. Func; functional genes. All; functional genes plus pseudogenes. From Niimura & Nei (106).

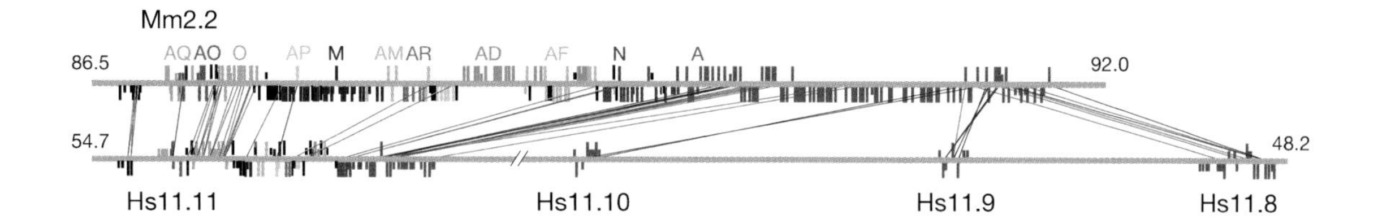

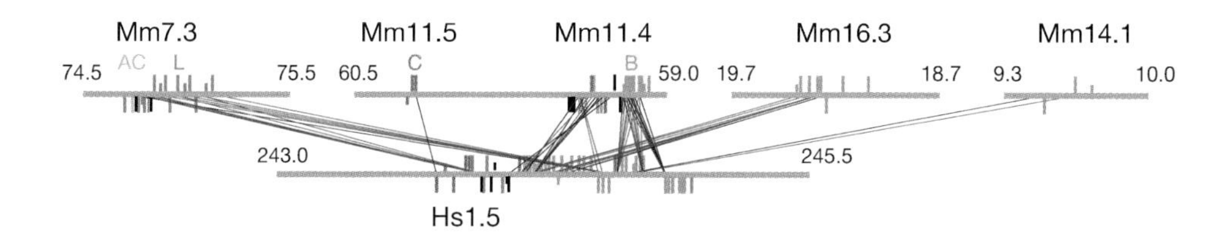

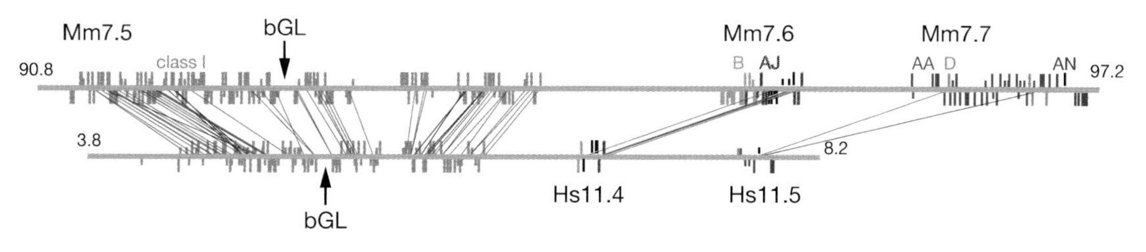

Figure 5

Orthologous relationships of class II OR genes between mouse (Mm) and human (Hs) genomic clusters. OR gene genomic clusters are indicated by chromosome number and Giemsa-stained band number in each species. Only a few chromosomes are presented. Long and short vertical bars show the locations of functional and nonfunctional OR genes, respectively. A vertical bar above a horizontal line indicates the transcriptional direction opposite to that below a horizontal line. Different colors represent different phylogenetic clades. Red and blue lines connecting mouse and human OR genes represent orthologous gene pairs. A red line indicates that transcriptional directions of orthologous genes are conserved between mice and humans, whereas a blue line indicates that they are inverted. Alphabetical letters such as A, J, G, and AQ represent different phylogenetic clades. Arrows show the location of the β-globin gene cluster (βGL). From Niimura & Nei (103).

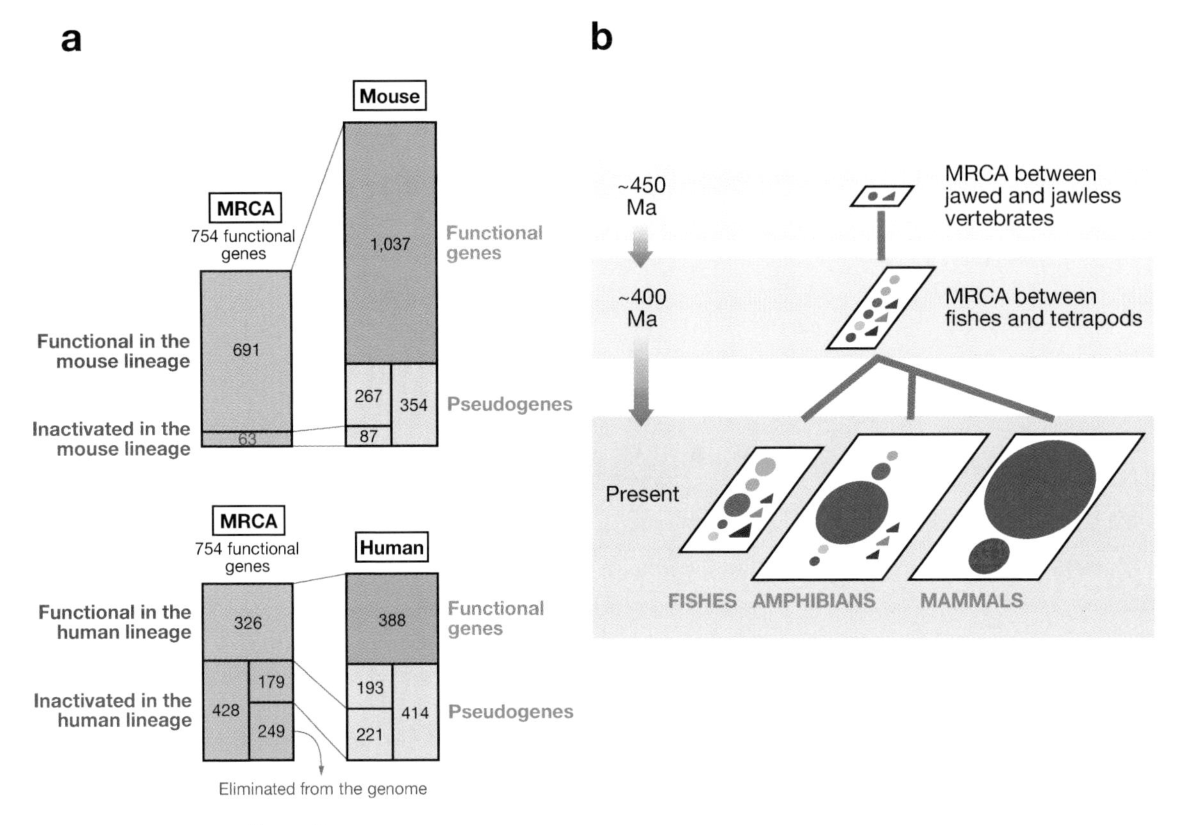

Figure 6

Evolutionary changes of functional OR genes. (*a*) The most recent common ancestor (MRCA) between humans and mice had approximately 754 functional OR genes. The number of functional OR genes increased to 1037 in the mouse lineage, whereas it declined to 388 in the human lineage. (*b*) The MRCA between jawed and jawless fishes had at least two functional genes, whereas the MRCA between fishes and tetrapods had at least 9 functional OR genes. Class II genes (*blue circle*) expanded enormously in amphibians and mammals.

Niimura & Nei (106) estimated that the MRCA of these species had at least 9 genes and that one of them generated mammalian class I genes and another generated class II genes (**Figure 6*b***). Amphibians also have additional class I and class II genes, as well as five other groups of genes. Fish appear to have eight different groups of genes, but none of them appears to be very common. These results indicate that the currently dominant class I and class II genes in mammals are relatively recent products of multiple gene duplication events.

Pheromone Receptor and Synthesis Genes

Pheromones are water-soluble chemicals emitted and are sensed by individuals of the same species to elicit reproductive behaviors or physiological character changes. In terrestrial vertebrates they are perceived by the vomeronasal organ (VNO), which is located at the base of the nasal cavity and is separated from the main olfactory epithelium. One supergene family that controls the VNO receptors is called the V1R (vomeronasal receptor 1) gene family, a member of which consists of about 330 codons without introns (25). The pheromone receptors are G-protein–coupled receptors as in the case of olfactory receptors, but there is little sequence similarity between the two proteins.

The mouse genome has about 350 genes, but the number of functional genes is 187. The rat genome has 102 functional genes and about 50 pseudogenes (39). Similarly, opossum and cow have a substantial number of functional genes. By contrast, the human has only 4 functional genes and nearly 200 pseudogenes. The number of functional genes in dogs is also quite small. The difference between the mouse and human genomes apparently occurred by massive pseudogenization of V1R genes in the human. In fact, some primate species including humans do not have functional VNOs and therefore are thought to have no perception of vomeronasal pheromone. The small number of functional V1R genes in humans seem to be relics of V1R gene pseudogenization. This pseudogenization in humans apparently occurred because humans use visual and auditory senses for sexual and physiological behavior. This example indicates that the number of copies of a multigene family can vary enormously even among different orders of mammals.

In insects a more advanced pheromone recognition system has developed. Mate finding in most moth species is achieved by female-emitted sex pheromones that are dispersed in a wide area. The key enzymes responsible for producing the pheromones are acetyl COA desaturases, which are encoded by a medium-sized gene family (133, 134). This gene family is also known to be subject to birth-and-death evolution.

BIRTH-AND-DEATH EVOLUTION WITH STRONG PURIFYING SELECTION

Histone Genes

As noted above, when genetic variation of multigene families was studied by restriction enzyme analysis, many gene families that are required to produce a large quantity of gene products were assumed to be subject to concerted evolution. One example is the histone gene family. In this gene family, even authors who studied nucleotide sequences maintained that histone gene families are subject to concerted evolution (82). This view arose from their preconception about gene conversion, as well as the fact that the number of sequences studied was small.

By the 1990s, however, a substantial number of sequences of histone genes had been accumulated from various species of animals, plants, fungi, and protists. Rooney et al. (136) and Piontkivska et al. (127) conducted an extensive statistical analysis of these data to examine whether the histone gene families are subject to concerted evolution or birth-and-death evolution. They reasoned that if concerted evolution is the main factor, both the number of synonymous differences per synonymous site (p_S) and the number of nonsynonymous differences per nonsynonymous site (p_N) must be virtually 0 for any pair of genes because gene conversion affects both synonymous and nonsynonymous sites in the same way. By contrast, if protein similarity is caused by purifying selection but every member gene evolves independently, p_S is expected to be greater than p_N because in this case synonymous substitutions accumulate continuously whereas nonsynonymous substitutions do not.

When this approach was applied to histone H3 and H4 genes from diverse groups of organisms, p_S was clearly higher than p_N in almost all cases (**Table 3**) (59a, 127, 136). Similar results were also obtained from an extensive study of the histone H1 gene family (26). These results therefore clearly show that the histone gene families are subject to strong purifying selection but all member genes evolve according to a birth-and-death process. Of course, occasionally there were cases in which p_S and p_N were both 0 or close to 0. For such a case, there is a possibility that gene conversion has occurred. It can also be explained by

Table 3 The p_S values (below diagonal) and p_N values (above diagonal) ($\times$ 100) observed between representative histones H3 sequences from human and mouse

	Human 1	Human 4	Human 16	Mouse 2	Mouse 8	Mouse 10
Human 1	—	4.0	0.0	0.3	0.7	0.7
Human 4	53.6	—	4.0	0.3	0.7	1.4
Human 16	54.3	49.1	—	0.3	0.7	1.0
Mouse 2	50.0	40.1	46.2	—	4.0	7.0
Mouse 8	53.6	39.2	47.0	10.4	—	0.0
Mouse 10	57.0	40.9	48.7	19.2	16.7	—

[a] p_S and p_N were estimated by the modified Nei-Gojobori method (174). The pairwise deletion option was used with the transition/transversion ratio (R) = 0.98. From Rooney et al. (136).

recent gene duplication. Since the number of cases in which $p_S > p_N$ were overwhelmingly large, Rooney et al. and Piontkivska et al. concluded that histone genes are generally subject to birth-and-death evolution with strong purifying selection.

Ubiquitin Genes

Ubiquitin is a small protein consisting of 76 amino acids that plays a major role in both cellular processes and protein degradation in eukaryotes. It is one of the most conserved proteins, and 72 of the 76 amino acids appear to be invariant in animals, plants, and fungi (38). Ubiquitins are encoded by a small- to medium-size gene family, which comprises monomeric and polymeric genes. Monomeric genes consist of 228 nucleotides (76 codons) with an additional sequence that encodes a ribosomal protein (**Figure 3*b***). By contrast, polymeric genes known as polyubiquitins (poly-u) are composed of tandem repeats of a 76-codon gene with no spacer sequence between them. The number of ubiquitin units in a poly-u locus, the number of poly-u loci, and the number of monomeric genes per genome vary considerably among eukaryotic species. Yet all ubiquitin genes encode the same amino acid sequence in each species.

These properties of the ubiquitin genes clearly indicate the importance of purifying selection. However, this multigene family was thought to be subject to concerted evolution until recently (88, 100, 139, 153, 159). This view rested partly on the preconception by many authors that homogenization of member genes of a family is caused by concerted evolution and partly on the fact that they generally worked with the genes from a few closely related species.

This view changed after 2000 when Nei et al. (99) conducted an extensive statistical analysis of all available data. The results of this study clearly indicated that it is purifying selection rather than concerted evolution that homogenizes protein sequences. In most species, nonsynonymous nucleotide differences among the member genes was 0, whereas the synonymous differences were virtually saturated (see **Table 4**). Of course,

Table 4 The p_S values ($\times$ 100) for intraspecific and interspecific comparisons of ubiquitin poly-u B loci from humans (H1, H2, and H3) and mice (M1, M2, M3, and M4). All the p_N values were 0. From Nei et al. (99)

Gene	H1	H2	H3	M1	M2	M3
H1						
H2	12					
H3	19	7				
M1	28	24	25			
M2	30	24	25	7		
M3	31	25	26	4	9	
M4	27	22	24	15	7	16

some pairs of member genes showed small nucleotide differences apparently because of recent gene duplication.

MIXED PROCESS OF CONCERTED AND BIRTH-AND-DEATH EVOLUTION

As noted above, the pair of inverted repeat genes *hsp70Aa* and *hsp70Ab* for heat shock proteins in *Drosophila* has virtually identical nucleotide sequences and therefore it probably represents one of the best cases of gene conversion. However, this does not mean that all *hsp70* genes for this highly conserved protein gene family are subject to concerted evolution. **Figure 4*b*** shows a phylogenetic tree for *hsp70* genes from *D. melanogaster*, two species of nematodes (*Caenorhabditis elegans* and *C. briggsae*), and yeast. The sequence identity of *hsp70Aa* and *hsp70Ab* is clear from this tree, but other *Drosophila hsp70* genes do not necessarily show a pattern of concerted evolution, though heat-inducible genes (marked *) are generally closely related. For example, *hsp70-2*, *hsp70-3*, and *hsp70-4* are quite different from *hsp70Aa* or *hsp70Ab*.

In *C. elegans*, the protein sequences of *hsp70-7* and *hsp70-8*, a pair of inverted repeats, are identical, but other *hsp70* genes are not necessarily closely related. Similarly, in *C. briggsae*, genes *hsp70-7* and *hsp70-14* are closely related to each other but other genes are not necessarily so. Actually, there are many *C. briggsae* genes that pair with *C. elegans* genes. These pairs of genes diverged before the separation of the two species, and therefore they have evolved independently for a long time (50 ~ 100 million years). Furthermore, yeast genes show no indication of concerted evolution. We can therefore conclude that the heat shock gene families have been subject to both concerted and birth-and-death evolution.

Essentially the same situation was observed with another highly conserved family of amylase genes in *Drosophila* (177). Some species (e.g., *D. orena*, *D. lini*, and *D. kikkawai*) of *Drosophila* have a pair of identical genes, but most other species do not. A phylogenetic analysis of 49 active genes by Zhang et al. (177) showed that some gene lineages were lost from the genome while some other genes were duplicated. These observations suggest that this gene family is subject to a mixture of concerted and birth-and-death evolution.

Historically, a well-known gene family that is apparently subject to a mixed process of concerted and birth-and-death evolution is the α-like globin gene family. The human genome contains 4 functional genes (α, α_2, θ_1, and ζ) and 3 pseudogenes (ψa_1, $\psi\alpha_2$, and $\psi\zeta$). The α_1 and α_2 genes are virtually identical within hominoids (178), but other genes are considerably differentiated. In practice, the evolution of the globin families is much more complicated than previously thought (A.P. Rooney, unpublished).

ORIGINS OF NEW GENETIC SYSTEMS GENERATED BY BIRTH-AND-DEATH EVOLUTION

Although single-locus genes can be the major determinants of some phenotypic characters such as color vision and the oxygen-transporting function of hemoglobins, most genetic systems or phenotypic characters are controlled by the interaction of multigene families. Here genetic systems mean any functional units of biological organization such as the olfaction (odor recognition) and adaptive immune system in vertebrates, flower development in plants, meiosis, and mitosis. Evolution of these genetic systems is obviously very complicated, but it appears that the only way to understand it is to know the evolution of each component multigene family and their interaction with other gene families using the simplest possible organisms. Note that if multigene families evolve following the model of birth-and-death processes, groups of new functional genes often evolve, whereas concerted evolution does not allow the functional differentiation of genes because all member genes are supposed to evolve as a unit in concert.

Previously, we noted that some MHC class Ib genes have functions different from those of Ia genes. For example, gene HFE in humans (**Figure 4*a***) evolved independently of other class I genes and has acquired a new function. It now has the ability to form complexes with the receptor for iron-binding transferrins and thus regulate the uptake of dietary iron by cells of the intestine (27). A mutation of this gene is known to cause the genetic disease hemochromatosis. This is one example that was generated by a birth-and-death process. **Figure 4*b*** shows that heat shock protein genes are expressed in three different locations of a cell. These are also products of gene duplication and acquisition of new functions. Similarly, we have seen that olfactory genes diverged extensively from fish to mammals. In the following section, a few examples are considered in some details.

Adaptive Immune System

One of the best-studied cases of acquisition of new gene functions is the evolution of adaptive immune system in jawed vertebrates. In the adaptive immune system, lifelong immunity is maintained for certain groups of parasites (viruses, bacteria, fungi, and others) once the host is attacked by them (e.g., the immunity to smallpox viruses). However, the jawless vertebrates and other nonvertebrate animals do not have this system, though most animals have the so-called innate immune system, which defends the host from parasites but does not retain memory of past attacks (63).

How did the adaptive immune system evolve? This is still a mystery and is currently under active investigation (14, 64, 157). However, it is well known that this system works through the interaction of many different multigene families such as the MHC, immunoglobulin, and T-cell receptor gene families. Most of these multigene families are evolutionarily related and are apparently products of long-term birth-and-death evolution. Therefore, it seems that continuous operation of birth-and-death evolution has generated a new genetic system. Klein & Nikolaidis (64) argued that the adaptive immune system evolved by assembling elements that have evolved primarily to serve other functions and incorporated existing molecular cascades, resulting in the appearance of new organs and new types of cells.

Animal Development

Homeobox genes are member genes of an important supergene family that control animal and plant development. They encode transcription factors and can be divided into two groups; Typical and Atypical groups. Typical homeobox genes contain a homeobox of 60 codons, whereas Atypical group genes have a homeobox with either a few more or less codons (13). The Typical group includes several dozen gene families. A well-known example is the HOX gene family that plays a key role in animal body patterning. This family has also undergone birth-and-death evolution (1). Another example is the PAX6 gene family that controls the eye development (31). The Atypical group includes about seven gene families, five of which are called the TALE group. The TALE group genes are characterized by three extra codons between the helix 1 and helix 2 regions. These gene families were all concerned with some aspects of development in eukaryotes and were derived from a common ancestor that existed before the separation of animals, plants, and fungi. In other words, these diverse gene families were generated by successive gene duplication and differentiation over a long period of time. Interestingly, occasional loss of paralogous genes also appears to contribute to the differentiation of phenotypic characters (J. Nam & M. Nei, unpublished).

Olfactory System in Vertebrates

Another example of the origin of new genetic system by birth-and-death evolution is the evolution of the olfactory system in vertebrates. We have seen that fish and mammals have quite different OR gene families, which have apparently adapted to receive different odorants that are available. Furthermore, class II OR genes are a relatively new invention in vertebrate evolution, and this gene family contained an enormously large number of genes. Yet, many of the genes are pseudogenes. It is now known that the class II OR genes can be divided into many subfamilies, each specialized to receive a given group of orders such as good smell or bad smell (77, 140). Because this system falls within the realm of neuroscience (175), it is outside the scope of this review.

Flower Development in Plants

Flowers of angiosperms (flowering plants) are composed of sepals, petals, stamens, pistils, etc., and differ from poorly developed flower-like organs in gynosperms (seed plants). The development of flower organs are controlled by transcription factor genes called MADS-box genes. There are several classes of MADS-box genes that are essential for flower development (76, 155, 163). In a phylogenetic analysis of MADS-box genes, Nam et al. (93) predicted that MADS-box genes controlling flower development (floral MADS-box genes) originated about 650 mya. Tanabe et al. (154) identified floral MADS-box like genes in three species of green algae, which are believed to have originated about 700 mya. If we note that the oldest fossil records of angiosperms and gymnosperms are about 150 and 300 million years old, respectively, it appears that the ancestral genes of floral MADS-box existed a long time ago before the flowering system evolved. Tanabe et al. speculate that this group of genes originally controlled the development of haploid and diploid stages of green algae. MADS-box genes are ancient genes and are known to exist in plants, animals, and fungi. In animals they control muscle development. In the process of evolution of gymnosperms and angiosperms, however, different MADS-box genes appear to have evolved to form flowers (93).

CONCLUSIONS

In this paper we have discussed the controversy over the models of concerted and birth-and-death evolution. The model of concerted evolution was originally thought to apply to gene families that are responsible for producing a large quantity of the same gene products, as in the case of rRNA genes. However, the production of a large quantity of the same gene products can also be achieved by strong purifying selection without concerted evolution. In fact, the histone and ubiquitin genes use this strategy, and the underlying DNA evolution occurs as a birth-and-death process. It is therefore important to distinguish between purifying selection and concerted evolution in producing homogeneous gene products. In the past, a number of authors have assumed that gene products are homogenized only by concerted evolution.

There are some exceptions to the above statement. Relatively small gene families with strong purifying selection such as the heat shock protein and amylase gene families undergo a mixed process of concerted and birth-and-death evolution. In these families, a pair of inverted gene sequences appears to be particularly susceptible to gene conversion or gene conversion-like events that homogenize the pair of genes. It is possible that gene conversion occurs more easily for inverted gene pairs. However, other member genes are apparently subject to birth-and-death evolution.

The gene families that produce a variety of gene products are usually subject to birth-and-death evolution. This is quite reasonable, because this model of evolution promotes genetic variation. There are reports indicating that the extent of genetic variability is enhanced by gene conversion or unequal crossover. However, these processes are primarily for homogenization of member genes, and the interpretation of data supporting this idea should be reexamined. This does not mean that gene conversion or unequal crossover never occurs in these genes. Actually, there is some evidence for the occurrence of gene conversion or domain shuffling in MHC genes. However, the contribution of these events to the diversification of multigene families in long-term evolution seems to be minor. The controversy over the concerted and birth-and-death evolution has occurred partly because of misunderstandings and misconceptions. Since each author usually works with one gene family from a limited number of species, the results cannot be blindly extended to other genes or other organisms. It is important to examine each gene family carefully and derive objective conclusions.

We have indicated that the model of birth-and-death evolution would give a reasonable explanation of generation of new gene families but the model of concerted evolution cannot. However, how different subgroups of a gene family acquire new functions is not yet well understood. Since the generation of new gene clusters occurs largely by chance, the initial stage of evolution of multigene families could be fortuitous. However, the newly generated duplicate genes or gene families may evolve to interact with other existing gene families and promote the adaptation of organisms to new environments. In the future it will be important to examine the relative contribution of positive selection and random genetic drift in the evolution of multigene families.

We have also indicated that many genetic systems or important phenotypic characters are controlled by the interaction of a number of multigene families and therefore we may be able to understand the evolution of new genetic systems or new phenotypic characters by studying the evolution of component multigene families. We may then be able to study the interaction between different gene families. This will be one of the most important problems in evolutionary biology in the future.

In the 1960s little was known about the multigene families. Studying the rates of gene duplication and formation of pseudogenes, Nei (95) stated "there may be a great deal of duplicate genes and also nonsense DNA in today's vertebrates." He also stated that "higher organisms including man have ample scope to evolve into various directions." The first prediction is now confirmed, but the validity of the second prediction remains to be seen.

SUMMARY POINTS

1. The controversy over the models of concerted and birth-and-death evolution of multigene families was reviewed with some historical background.
2. The model of concerted evolution is capable of explaining the evolutionary pattern of highly conserved ribosomal or other RNA genes that are essential for cell metabolism. However, the high conservation of the coding regions of these genes is primarily caused by purifying selection rather than by concerted evolution.
3. Until recently, the high conservation of some proteins such as histones was considered to be due to concerted evolution. Recent studies indicate that the conservation of these proteins is primarily caused by purifying selection, and their DNA sequences evolve following the model of birth-and-death evolution.
4. Some genes, such as heat shock protein and *Drosophila* amylase genes, appear to be subject to a mixed process of concerted and birth-and-death evolution.

5. Most gene families, particularly those producing variable gene products, are subject to birth-and-death evolution.
6. Although some investigators maintain that MHC polymorphism is enhanced by gene conversion, phylogenetic analysis of polymorphic alleles shows that the effect of this factor is minor in comparison to mutation and overdominant selection.
7. The model of birth-and-death evolution gives a reasonable explanation of the generation of new gene families, whereas the model of concerted evolution does not.
8. Most genetic systems such as the adaptive immune system in vertebrates are controlled by the interaction of many multigene families each of which undergoes a birth-and-death process. Therefore, the evolution of new genetic systems or phenotype characters is also caused by a large-scale birth-and-death evolution.

FUTURE DIRECTIONS

1. The controversy over concerted and birth-and-death evolution is partly due to the fact that the molecular mechanism of gene conversion is not well understood except in fungi and bacteria and it is not clear how gene conversion occurs between distantly located member genes as in the case of MHC loci. It is therefore important to clarify the molecular mechanism of gene conversion in vertebrates.
2. Evolutionary relationships and functional differentiation of multigene families should be studied at the statistical and biochemical levels.
3. Evolution of new genetic systems or phenotypic characters should be studied by consideration of the interaction of multigene families.

ACKNOWLEDGMENT

We would like to thank Dan Graur, Li Hao, Jan Klein, Eddie Holmes, Jongmin Nam, Yoshihito Niimura, Naruya Saitou, Yoshiyuki Suzuki, and Jianzhi Zhang for their comments on an earlier version of the manuscript. This was supported by NIH grant GM020293-32 to M.N.

LITERATURE CITED

1. Amores A, Suzuki T, Yan Y-L, Pomeroy J, Singer A, et al. 2004. Developmental roles of pufferfish Hox clusters and genome evolution in ray-fin fish. *Genome Res.* 14:1–10
2. Andersson L, Gustafsson K, Jonsson AK, Rask L. 1991. Concerted evolution in a segment of the first domain exon of polymorphic MHC class II beta loci. *Immunogenetics* 33:235–42
3. Anjard C, Loomis WF. 2002. Evolutionary analyses of ABC transporters of *Dictyostelium discoideum*. *Eukaryot. Cell* 1:643–52
4. Annilo T, Dean M. 2004. Degeneration of an ATP-binding cassette transporter gene, ABCC13, in different mammalian lineages. *Genomics* 84:34–46

5. Arden B, Klein J. 1982. Biochemical comparison of major histocompatibility complex molecules from different subspecies of *Mus musculus*: evidence for *trans*-specific evolution of alleles. *Proc. Natl. Acad. Sci. USA* 79:2342–46
6. Belich MP, Madrigal JA, Hildebrand WH, Zemmour J, Williams RC, et al. 1992. Unusual HLA-B alleles in two tribes of Brazilian Indians. *Nature* 357:326–29
7. Bettencourt BR, Feder ME. 2002. Rapid concerted evolution via gene conversion at the Drosophila hsp70 genes. *J. Mol. Evol.* 54:569–86
8. Birky CWJ, Skavaril RV. 1976. Maintenance of genetic homogeneity in systems with multiple genomes. *Genet. Res.* 27:249–65
9. Brennan J, Lemieux S, Freeman JD, Mager DL, Takei F. 1996. Heterogeneity among Ly-49C natural killer (NK) cells: characterization of highly related receptors with differing functions and expression patterns. *J. Exp. Med.* 184:2085–90
10. Brown DD, Sugimoto K. 1974. The structure and evolution of ribosomal and 5S DNAs in *Xenopus laevis* and *Xenopus mulleri* 505. *Cold Spring Harbor Symp. Quant. Biol.* 38:501–8
11. Brown DD, Wensink PC, Jordan E. 1972. *Xenopus laevis* and *Xenopus mulleri*: the evolution of tandem genes. *J. Mol. Biol.* 63:57–73
12. Brunsberg U, Edfors-Lilja I, Andersson L, Gustafsson K. 1996. Structure and organization of pig MHC class II DRB genes: evidence for genetic exchange between loci. *Immunogenetics* 44:1–8
13. Burglin TR. 1997. Analysis of TALE superclass homeobox genes (MEIS, PBC, KNOX, Iroquois, TGIF) reveals a novel domain conserved between plants and animals. *Nucleic Acids Res.* 25:4173–80
14. Cannon JP, Haire RH, Litman GW. 2002. Identification of diversified genes that contain immunoglobulin-like variable regions in a protochordate. *Nat. Immunol.* 3:1200–7

14a. Carranza S, Giribet G, Ribera C, Baguna J, Riutort M. 1996. Evidence that two types of 18S rRNA coexist in the genome of *Dugesia* (*Schmidtea*) *mediterranea* (Platyhelminthes, Turbellaria, Tricladida). *Mol. Biol. Evol.* 13:824–32

14b. Carritt B, Kemp TJ, Poulter M. 1997. Evolution of the human RH (rhesus) blood group genes: a 50 year old prediction (partially) fulfilled. *Hum. Mol. Genet.* 6:843–50

15. Caudy AA, Pikaard CS. 2002. Xenopus ribosomal RNA gene intergenic spacer elements conferring transcriptional enhancement and nucleolar dominance-like competition in oocytes. *J. Biol. Chem.* 277:31577–84
16. Childs G, Nocente-McGrath C, Lieber T, Holt C, Knowles JA. 1982. Sea urchin (*Lytechinus pictus*) late-stage histone H3 and H4 genes: characterization and mapping of a clustered but nontandemly linked multigene family. *Cell* 31:383–93
17. Coen E, Strachan T, Dover G. 1982. Dynamics of concerted evolution of ribosomal DNA and histone gene families in the melanogaster species subgroup of Drosophila. *J. Mol. Biol.* 158:17–35
18. Coen ES, Thoday JM, Dover G. 1982. Rate of turnover of structural variants in the rDNA gene family of *Drosophila melanogaster*. *Nature* 18:564–68
19. Cohen LH, Newrock KM, Zweidler A. 1975. Stage-specific switches in histone synthesis during embryogenesis of the sea urchin. *Science* 190:994–97
20. Cohn RH, Kedes LH. 1979. Nonallelic histone gene clusters of individual sea urchins (*Lytechinus pictus*): mapping of homologies in coding and spacer DNA. *Cell* 18:855–64
21. Cohn RH, Kedes LH. 1979. Nonallelic histone gene clusters of individual sea urchins (*Lytechinus pictus*): polarity and gene organization. *Cell* 18:843–53

22. Conte C, Ebeling M, Marcuz A, Nef P, Andres-Barquin J. 2003. Evolutionary relationships of the Tas2r receptor gene families in mouse and human. *Physiol. Genomics* 14:73–82
23. Dean M, Rhetzky A, Allikmets R. 2001. The human ATP-Binding Cassette (ABC) transporter superfamily. *Genome Res.* 11:1156–66
24. Dormoy A, Reviron DV, Froelich N, Weiller PJ, Mercier PJ, Tongio MM. 1997. Birth of a new allele in a sibling: *cis* or *trans* gene conversion during meiosis? *Immunogenetics* 46:520–23
25. Dulac C, Axel R. 1995. A novel family of genes encoding putative pheromone receptors in mammals. *Cell* 83:195–206
26. Eirin-Lopez JM, Gonzalez-Tizon AM, Martinez A, Mendez J. 2004. Birth-and-death evolution with strong purifying selection in the histone H1 multigene family and the origin of orphon H1 genes. *Mol. Biol. Evol.* 21:1992–2003
27. Feder JN, Penny DM, Irrinki A, Lee VK, Lebron JA, et al. 1998. The hemochromatosis gene product complexes with the transferrin receptor and lowers its affinity for ligand binding. *Proc. Natl. Acad. Sci. USA* 95:1472–77
28. Figueroa F, Gunther E, Klein J. 1988. MHC polymorphism pre-dating speciation. *Nature* 335:265–67
29. Freitag J, Ludwig G, Andreini I, Rossler P, Breer H. 1998. Olfactory receptors in aquatic and terrestrial vertebrates. *J. Comp. Physiol. A* 183:635–50
30. Fry BG, Wuster W, Kini RM, Brusic V, Khan S, et al. 2003. Molecular evolution and phylogeny of elapid snake venom three-finger toxins. *J. Mol. Evol.* 57:110–29
31. Gehring W. 1998. *Master Control Genes in Development and Evolution: The Homeobox Story*. New Haven/London: Yale Univ. Press
32. Glusman G, Yanai I, Rubin I, Lancet D. 2001. The complete human olfactory subgenome. *Genome Res.* 11:685–702
33. Go Y, Satta Y, Kawamoto Y, Rakotoarisoa G, Randrianjafy A, et al. 2003. Frequent segmental sequence exchanges and rapid gene duplication characterize the MHC class I genes in lemurs. *Immunogenetics* 55:450–61
34. Godfrey PA, Malnic B, Buck LB. 2004. The mouse olfactory receptor gene family. *Proc. Natl. Acad. Sci. USA* 101:2156–61
35. Gonzalez IL, Sylvester JE. 2001. Human rDNA: evolutionary patterns within the genes and tandem arrays derived from multiple chromosomes. *Genomics* 736:255–63
36. Gorski J, Mach B. 1986. Polymorphism of human Ia antigens: gene conversion between two DR beta loci results in a new HLA-D/DR specificity. *Nature* 322:67–70
37. Gotesson A, Marshall JS, Jones DA, Hardham AR. 2002. Characterization and evolutionary analysis of a large polygalacturonase gene family in the oomycete plant pathogen *Phytophthora cinnamomi*. *Mol. Plant-Microbe Interact.* 15:907–21
38. Graham RW, Jones D, Candido EP. 1989. UbiA, the major polyubiquitin locus in *Caenorhabditis elegans*, has unusual structural features and is constitutively expressed. *Mol. Cell Biol.* 9:268–77
39. Grus WE, Shi P, Zhang Y-P, Zhang J. 2005. Dramatic variation of the vomeronasal pheromone receptor gene repertoire among five orders of placental and marsupial mammals. *Proc. Natl. Acad. Sci. USA* 102:5767–72
40. Gu X, Nei M. 1999. Locus specificity of polymorphic alleles and evolution by a birth-and-death process in mammalian MHC genes. *Mol. Biol. Evol.* 16:147–56
41. Hanneman WH, Schimenti KJ, Schimenti JC. 1997. Molecular analysis of gene conversion in spermatids from transgenic mice. *Gene* 200:185–92

42. Hao L, Nei M. 2004. Genomic organization and evolutionary analysis of Ly49 genes encoding the rodent natural killer cell receptors: rapid evolution by repeated gene duplication. *Immunogenetics* 56:343–54
43. Hao L, Nei M. 2005. Rapid expansion of killer cell immunoglobulin-like receptor genes in primates and their coevolution with MHC Class I genes. *Gene* 347:149–59
44. Hentschel CC, Birnstiel ML. 1981. The organization and expression of histone gene families. *Cell* 25:301–13
45. Hogstrand K, Bohme J. 1994. A determination of the frequency of gene conversion in unmanipulated mouse sperm. *Proc. Natl. Acad. Sci. USA* 91:9921–25
46. Hogstrand K, Bohme J. 1999. Gene conversion of major histocompatibility complex genes is associated with CpG-rich regions. *Immunogenetics* 49:446–55
47. Holt CA, Childs G. 1984. A new family of tandem repetitive early histone genes in the sea urchin *Lytechinus pictus*: evidence for concerted evolution within tandem arrays. *Nucleic Acids Res.* 12:6455–71
48. Hood L, Campbell JH, Elgin SCR. 1975. The organization, expression, and evolution of antibody genes and other multigene families. *Annu. Rev. Genet.* 9:305–53
49. Hughes AL. 2004. Birth-and-death evolution of protein-coding regions and concerted evolution of non-coding regions in the multi-component genomes of nanoviruses. *Mol. Phylogenet. Evol.* 30:287–94
50. Hughes AL, Nei M. 1988. Pattern of nucleotide substitution at major histocompatibility complex class I loci reveals overdominant selection. *Nature* 335:167–70
51. Hughes AL, Nei M. 1989. Evolution of the major histocompatibility complex: independent origin of nonclassical class I genes in different groups of mammals. *Mol. Biol. Evol.* 6:559–79
52. Hughes AL, Nei M. 1989. Nucleotide substitution at major histocompatibility complex II loci: evidence for overdominant selection. *Proc. Natl. Acad. Sci. USA* 86:958–62
53. Hughes AL, Nei M. 1990. Evolutionary relationships of class II major-histocompatibility-complex genes in mammals. *Mol. Biol. Evol.* 7:491–514
54. Ingram VM. 1961. Gene evolution and the haemoglobins. *Nature* 189:704–8
55. Jeffreys A. 1979. DNA sequence variants in the $G\gamma$-, $A\gamma$-, δ- and β-globin genes of man. *Cell* 18:1–10
56. Jeffreys AJ, May CA. 2004. Intense and highly localized gene conversion activity in human meiotic crossover hot spots. *Nat. Genet.* 36:151–56
57. Kedes LH. 1979. Histone genes and histone messengers. *Annu. Rev. Biochem.* 48:837–70
58. Kelley J, Walter L, Trowsdale J. 2005. Comparative genomics of major histocompatibility complexes. *Immunogenetics* 56:683–95

58a. Kellogg EA, Appels R. 1995. Intraspecific and interspecific variation in 5S RNA genes are decoupled in diploid wheat relatives. *Genetics* 140:325–43

59. Khakoo SI, Rajalingam R, Shum BP, Weidenbach K, Flodin L, et al. 2000. Rapid evolution of NK cell receptor systems demonstrated by comparison of chimpanzees and humans. *Immunity* 12:687–98

59a. Kim HN, Yamazaki T. 2004. Nonconcerted evolution of histone 3 genes in a liverwort, *Conocephalum conicum*. *Genes Genet. Syst.* 79:331–44

60. Kitano T, Saitou N. 1999. Evolution of Rh blood group genes have experienced gene conversions and positive selection. *J. Mol. Evol.* 49:615–26
61. Deleted in proof
62. Klein J, Figueroa F. 1986. Evolution of the major histocompatibility complex. *Crit. Rev. Immunol.* 6:295–386

63. Klein J, Horejsi V. 1997. *Immunology*. London: Blackwell Sci. 2nd ed.
64. Klein J, Nikolaidis N. 2005. The descent of the antibody-based immune system by gradual evolution. *Proc. Natl. Acad. Sci. USA* 102:169–74
65. Klein J, Ono H, Klein D, O'hUigin C. 1993. The accordion model of MHC evolution. *Prog. Immunol.* 8:137–43
65a. Kriener K, O'hUigin C, Klein J. 2000. Conversion or convergence? Introns of primate DRB genes tell the true story. In *Major Histocompatibility Complex; Evolution, Structure, and Function*, ed. M Kasahara, pp. 354–76. Tokyo: Springer-Verlag
66. Lawlor DA, Ward FE, Ennis PD, Jackson AP, Parham P. 1988. *HLA-A* and *B* polymorphisms predate the divergence of humans and chimpanzees. *Nature* 335:268–71
67. Leh-Louis V, Wirth B, Despons L, Wain-Hobson S, Potier S, Souciet JL. 2004. Differential evolution of the *Saccharomyces cerevisiae* DUP240 paralogs and implication of recombination in phylogeny. *Nucleic Acids Res.* 32:2069–78
68. Leigh Brown AJ, Ish-Horowicz D. 1981. Evolution of the 87A and 87C heat-shock loci in Drosophila. *Nature* 290:677–82
68a. Li J, Gutell RR, Damberger SH, Wirtz RA, Kissinger JC, et al. 1997. Regulation and trafficking of three distinct 18 S ribosomal RNAs during development of the malaria parasite. *J. Mol. Biol.* 269:203–13
69. Liao D. 2000. Gene conversion drives within genic sequences: converted evolution of ribosomal RNA genes in bacteria and archaea. *J. Mol. Evol.* 51:305–17
70. Liao D, Pavelitz T, Weiner AM. 1998. Characterization of a novel class of interspersed LTR elements in primate genomes: structure, genomic distribution, and evolution. *J. Mol. Evol.* 46:649–60
71. Lifton RP, Goldberg ML, Karp RW, Hogness DS. 1977. The organization of the histone genes in *Drosophila melanogaster*; functional and evolutionary implications. *Cold Spring Harbor Symp. Quant. Biol.* 2:1047–51
72. Liu W, Rooney AP, Xue B, Roelofs WL. 2004. Desaturases from the spotted fireworm moth (*Choristoneura parallela*) shed light on the evolutionary origins of novel moth sex pheromone desaturases. *Gene* 342:303–11
73. Loh DY, Baltimore D. 1984. Sexual preference of apparent gene conversion events in MHC genes of mice. *Nature* 309:639–40
74. Lopez de Castro JA, Strominger JL, Strong DM, Orr HT. 1982. Structure of cross-reactive human histocompatibility antigens HLA-A28 and HLA-A2: possible implications for the generation of HLA polymorphism. *Proc. Natl. Acad. Sci. USA* 79:3813–17
75. Lynn DJ, Lloyd AT, Fares MA, O'Farrelly C. 2004. Evidence of positively selected sites in mammalian alpha-defensins. *Mol. Biol. Evol.* 21:819–27
76. Ma H, de Pamphilis CW. 2000. The ABCs of floral evolution. *Cell* 101:5–8
77. Malnic B, Godfrey PA, Buck LB. 2004. The human olfactory receptor gene family. *Proc. Natl. Acad. Sci. USA* 101:2584–89
78. Mandl B, Brandt WF, Superti-Furga G, Graninger PG, Birnstiel ML, Busslinger M. 1997. The five cleavage-stage (CS) histones of the sea urchin are encoded by a maternally expressed family of replacement histone genes: functional equivalence of the CS H1 and frog H1M(B4) proteins. *Mol. Cell Biol.* 17:1189–200
79. Marcos CY, Fernandez-Vina MA, Lazaro AM, Nulf CJ, Raimondi EH, Stastny P. 1997. Novel HLA-B35 subtypes: putative gene conversion events with donor sequences from alleles common in native Americans (HLA-B*4002 or B*4801). *Hum. Immunol.* 53:148–55

80. Martinsohn JT, Sousa AB, Guethlein LA, Howard JC. 1999. The gene conversion hypothesis of MHC evolution: a review. *Immunogenetics* 50:168–200
81. Matsuda F, Ishii K, Bouvagnet P, Kuma K, Hyashida H, et al. 1998. The complete nucleotide sequence of the human immunoglobulin heavy chain variable region locus. *J. Exp. Med.* 188:2151–62
82. Matsuo Y, Yamazaki T. 1989. tRNA derived insertion element in histone gene repeating unit of *Drosophila melanogaster*. *Nucleic Acids Res.* 17:225–38
83. Maxson R, Cohn RH, Kedes L, Mohun T. 1983. Expression and organization of histone genes. *Annu. Rev. Genet.* 17:239–77
84. McAdam SN, Boyson JE, Liu X, Garber TL, Hughes AL, et al. 1994. A uniquely high level of recombination at the HLA-B locus. *Proc. Natl. Acad. Sci. USA* 91:5893–97
85. McConnell TJ, Talbot WS, McIndoe RA, Wakeland EK. 1988. The origin of MHC class II gene polymorphism within the genus *Mus*. *Nature* 332:651–54
86. Mellor AL, Weiss EH, Ramachandran K, Flavell RA. 1983. A potential donor gene for the bm1 gene conversion event in the C57BL mouse. *Nature* 306:792–95
86a. Mercereau-Puijalon O, Barale JC, Bischoff E. 2002. Three multigene families in plasmodium parasites: facts and questions. *Int. J. Parasitol.* 32:1323–44
87. Michelmore RW, Meyers BC. 1998. Clusters of resistance genes in plants evolve by divergent selection and a birth-and-death process. *Genome Res.* 8:1113–30
88. Mita K, Ichimura S, Nenoi M. 1991. Essential factors determining codon usage in ubiquitin genes. *J. Mol. Evol.* 33:216–25
89. Morrison C, Vagnarelli P, Sonoda E, Takeda S, Earnshaw WC. 2003. Sister chromatid cohesion and genome stability in vertebrate cells. *Biochem. Soc. Trans.* 31:263–65
89a. Muir G, Fleming CC, Schlotterer C. 2001. Three divergent rDNA clusters predates species divergence in *Quercus petraea* (Matt.) Liebl. and *Quercus robur* L. *Mol. Biol. Evol.* 18:112–19
90. Nagylaki T. 1984. Evolution of multigene families under interchromosomal gene conversion. *Proc. Natl. Acad. Sci. USA* 81:3796–800
91. Nagylaki T. 1984. The evolution of multigene families under intrachromosomal gene conversion. *Genetics* 106:529–48
92. Nagylaki T, Petes TD. 1982. Intrachromosomal gene conversion and the maintenance of sequence homogeneity among repeated genes. *Genetics* 100:315–37
93. Nam J, de Pamphilis CW, Ma H, Nei M. 2003. Antiquity and evolution of the MADS-box gene family controlling flower development in plants. *Mol. Biol. Evol.* 20:1435–47
94. Nam J, Kim J, Lee S, An G, Ma H, Nei M. 2004. Type I MADS-box genes have experienced faster birth-and-death evolution than type II MADS-box genes in angiosperms. *Proc. Natl. Acad. Sci. USA* 101:1910–15
95. Nei M. 1969. Gene duplication and nucleotide substitution in evolution. *Nature* 221:40–42
96. Nei M, Gu X, Sitnikova T. 1997. Evolution by the birth-and-death process in multigene families of the vertebrate immune system. *Proc. Natl. Acad. Sci. USA* 94:7799–806
97. Nei M, Hughes AL. 1991. Polymorphism and evolution of the major histocompatibility complex loci in mammals. In *Evolution at the Molecular Level*, ed. R Selander, A Clark, T Whittam, pp. 222–47. Sunderland, MA: Sinauer
98. Nei M, Hughes AL. 1992. Balanced polymorphism and evolution by the birth-and-death process in the MHC loci. In *11th Histocompatibility Workshop and Conference*, ed. K Tsuji, M Aizawa, T Sasazuki, pp. 27–38. Oxford, UK: Oxford Univ. Press

99. Nei M, Rogozin IB, Piontkivska H. 2000. Purifying selection and birth-and-death evolution in the ubiquitin gene family. *Proc. Natl. Acad. Sci. USA* 97:10866–71
100. Nenoi M, Mita K, Ichimura S, Kawano A. 1998. Higher frequency of concerted evolutionary events in rodents than in man at the polyubiquitin gene VNTR locus. *Genetics* 148:867–76
101. Newrock KM, Alfageme CR, Nardi RV, Cohen LH. 1978. Histone changes during chromatin remodeling in embryogenesis. *Cold Spring Harbor Symp. Quant. Biol.* 1:421–31
102. Newrock KM, Cohen LH, Hendricks MB, Donnelly RJ, Weinberg ES. 1978. Stage-specific mRNAs coding for subtypes of H2A and H2B histones in the sea urchin embryo. *Cell* 14:327–36
103. Niimura Y, Nei M. 2003. Evolution of olfactory receptor genes in the human genome. *Proc. Natl. Acad. Sci. USA* 100:12235–40
104. Niimura Y, Nei M. 2005. Comparative evolutionary analysis of olfactory receptor gene clusters between humans and mice. *Gene* 346:13–21
105. Niimura Y, Nei M. 2005. Evolutionary changes of the number of olfactory receptor genes in the human and mouse lineages. *Gene* 346:23–28
106. Niimura Y, Nei M. 2005. Evolutionary dynamics of olfactory receptor genes in fishes and tetrapods. *Proc. Natl. Acad. Sci. USA*. In press
107. Nikolaidis N, Klein J, Nei M. 2005. Origin and evolution of the Ig-like domains present in mammalian leukocyte receptors: insights from chicken, frog, and fish homologues. *Immunogenetics* 57:151–57
108. Nikolaidis N, Makalowska I, Chalkia D, Makalowski W, Klein J, Nei M. 2005. Origin and evolution of the chicken leukocyte receptor complex. *Proc. Natl. Acad. Sci. USA* 102:4057–62
109. Nikolaidis N, Nei M. 2004. Concerted and nonconcerted evolution of the Hsp70 gene superfamily in two sibling species of nematodes. *Mol. Biol. Evol.* 21:498–505
110. Ohta T. 1980. *Evolution and Variation of Multigene Families*. Berlin: Springer-Verlag
111. Ohta T. 1982. Allelic and nonallelic homology of a supergene family. *Proc. Natl. Acad. Sci. USA* 79:3251–54
112. Ohta T. 1983. On the evolution of multigene families. *Theor. Popul. Biol.* 23:216–40
113. Ohta T. 1985. A model of duplicative transposition and gene conversion for repetitive DNA families. *Genetics* 110:513–24
114. Ohta T. 1991. Role of diversifying selection and gene conversion in evolution of major histocompatibility complex loci. *Proc. Natl. Acad. Sci. USA* 88:6716–20
115. Ohta T, Dover GA. 1984. The cohesive population genetics of molecular drive. *Genetics* 108:501–21
116. Ota T, Nei M. 1994. Divergent evolution and evolution by the birth-and-death process in the immunoglobulin V_H gene family. *Mol. Biol. Evol.* 11:469–82
117. Ota T, Nei M. 1995. Evolution of immunoglobulin VH pseudogenes in chickens. *Mol. Biol. Evol.* 12:94–102
118. Ota T, Sitnikova T, Nei M. 2000. Evolution of vertebrate immunoglobulin variable gene segments. *Curr. Top. Microbiol. Immunol.* 248:221–45
119. Palmer D, Snyder LA, Blumenfeld M. 1980. Drosophila nucleosomes contain an unusual histone-like protein. *Proc. Natl. Acad. Sci. USA* 77:2671–75
120. Paques F, Haber JE. 1999. Multiple pathways of recombination induced by double-strand breaks in *Saccharomyces cerevisiae*. *Microbiol. Mol. Biol. Rev.* 63:349–404

121. Pardue ML, Kedes LH, Weinberg ES, Birnstiel ML. 1977. Localization of sequences coding for histone messenger RNA in the chromosomes of *Drosophila melanogaster*. *Chromosoma* 63:135–51
122. Pavelitz T, Liao D, Weiner AM. 1999. Concerted evolution of the tandem array encoding primate U2 snRNA (the RNU2 locus) is accompanied by dramatic remodeling of the junctions with flanking chromosomal sequences. *EMBO J.* 18:3783–92
123. Pavelitz T, Rusche L, Matera AG, Scharf JM, Weiner AM. 1995. Concerted evolution of the tandem array encoding primate U2 snRNA occurs in situ, without changing the cytological context of the RNU2 locus. *EMBO J.* 14:169–77
124. Pease LR, Horton RM, Pullen JK, Yun TJ. 1993. Unusual mutation clusters provide insight into class I gene conversion mechanisms. *Mol. Cell Biol.* 13:4374–81
125. Petes TD. 1980. Unequal meiotic recombination within tandem arrays of yeast ribosomal DNA genes. *Cell* 19:765–74
126. Piontkivska H, Nei M. 2003. Birth-and-death evolution in primate MHC class I genes: divergence time estimates. *Mol. Biol. Evol.* 20:601–9
127. Piontkivska H, Rooney AP, Nei M. 2002. Purifying selection and birth-and-death evolution in the histone H4 gene family. *Mol. Biol. Evol.* 19:689–97
128. Quesada H, Ramos-Onsins SE, Aguade M. 2005. Birth-and-death evolution of the Cecropin multigene family in Drosophila. *J. Mol. Evol.* 60:1–11
129. Rajalingam R, Parham P, Abi-Rached L. 2004. Domain shuffling has been the main mechanism forming new hominoid killer cell Ig-like receptors. *J. Immunol.* 172:356–69
130. Roberts SB, Weisser KE, Childs G. 1984. Sequence comparisons of non-allelic late histone genes and their early stage counterparts. Evidence for gene conversion within the sea urchin late stage gene family. *J. Mol. Biol.* 174:647–62
131. Robertson HM. 1998. Two large families of chemoreceptor genes in the nematodes *Caenorhabditis elegans* and *Caenorhabditis briggsae* reveal extensive gene duplication, diversification, movement, and intron loss. *Genome Res.* 8:449–63
132. Robinett CC, O'Connor A, Dunaway M. 1997. The repeat organizer, a specialized insulator element within the intergenic spacer of the Xenopus rRNA genes. *Mol. Cell Biol.* 17:2866–75
133. Roelofs WL, Liu W, Hao G, Jiao H, Rooney AP, Linn CEJ. 2002. Evolution of moth sex pheromones via ancestral genes. *Proc. Natl. Acad. Sci. USA* 99:13621–26
134. Roelofs WL, Rooney AP. 2003. Molecular genetics and evolution of pheromone biosynthesis in Lepidoptera. *Proc. Natl. Acad. Sci. USA* 100:9179–84
134a. Roesner A, Fuchs C, Hankeln T, Burmester T. 2005. A globin gene of ancient evolutionary origin in lower vertebrates: evidence for two distinct globin families in animals. *Mol. Biol. Evol.* 22:12–20
135. Rooney AP. 2004. Mechanisms underlying the evolution and maintenance of functionally heterogeneous 18S rRNA genes in apicomplexans. *Mol. Biol. Evol.* 21:1704–11
136. Rooney AP, Piontkivska H, Nei M. 2002. Molecular evolution of the nontandemly repeated genes of the histone 3 multigene family. *Mol. Biol. Evol.* 19:68–75
137. Rooney AP, Ward TJ. 2005. Evolution of a large ribosomal RNA multigene family in filamentous fungi: birth and death of a concerted evolution paradigm. *Proc. Natl. Acad. Sci. USA* 102:5084–89
138. Schaschl H, Suchentrunk F, Hammer S, Goodman SJ. 2005. Recombination and the origin of sequence diversity in the DRB MHC class II locus in chamois (*Rupicapra* spp.). *Immunogenetics* 57:108–15

139. Sharp PM, Li W-H. 1987. The codon adaptation index a measure of directional synonymous codon usage bias, and its potential applications. *Nucleic Acids Res.* 15:1281–95
140. Shepherd GM. 2004. The human sense of smell: Are we better than we think? *PLOS Biol.* 2:572–75
141. Sitnikova T, Nei M. 1998. Evolution of immunoglobulin kappa chain variable region genes in vertebrates. *Mol. Biol. Evol.* 15:50–60
142. Sitnikova T, Su C. 1998. Coevolution of immunoglobulin heavy- and light-chain variable-region gene families. *Mol. Biol. Evol.* 15:617–25
143. Slightom JL, Blechl AE, Smithies O. 1980. Human fetal $G_{\gamma-}$ and $A_{\gamma-}$ globin genes: complete nucleotide sequences suggest that DNA can be exchanged between these duplicated genes. *Cell* 21:627–38
144. Smith GP. 1974. Unequal crossover and the evolution of multigene families. *Cold Spring Harbor Symp. Quant. Biol.* 38:507–13
145. Smith GP. 1976. Evolution of repeated DNA sequences by unequal crossovers. *Science* 191:528–34
146. Srivastava AK, Schlessinger D. 1991. Structure and organization of ribosomal DNA. *Biochimie* 73:631–38
147. Su C, Jakobsen IB, Gu X, Nei M. 1999. Diversity and evolution of T-cell receptor variable region genes in mammals and birds. *Immunogenetics* 50:301–8
148. Su C, Nei M. 1999. Fifty-million-year old polymorphism at an immunoglobulin variable region gene locus in the rabbit evolutionary lineage. *Proc. Natl. Acad. Sci. USA* 96:9710–15
149. Su C, Nei M. 2001. Evolutionary dynamics of the T-cell receptor VB gene family as inferred from the human and mouse genomic sequences. *Mol. Biol. Evol.* 18:503–13
150. Su C, Nguyen VK, Nei M. 2002. Adaptive evolution of variable region genes encoding an unusual type of immunoglobulin in camelids. *Mol. Biol. Evol.* 19:205–15
151. Szostak JW, Wu R. 1980. Unequal crossing over in the ribosomal DNA of *Saccharomyces cerevisiae*. *Nature* 284:426–30
152. Takahashi K, Rooney AP, Nei M. 2000. Origins and divergence times of mammalian class II MHC gene clusters. *J. Hered.* 19:198–204
153. Tan Y, Bishoff ST, Riley MA. 1993. Ubiquitins revisited: further examples of within- and between-locus concerted evolution. *Mol. Phylogenet. Evol.* 2:351–60
154. Tanabe Y, Hasebe M, Sekimoto H, Nishiyama T, Kitani M, et al. 2005. Characterization of MADS-box genes in charophycean green algae and its implication for the evolution of MADS-box genes. *Proc. Natl. Acad. Sci. USA* 102:2436–41
155. Theissen G. 2001. Development of floral organ identity: stories from the MADS house. *Curr. Opin. Plant Biol.* 4:75–85
156. Trowsdale J, Barten R, Haude A, Stewart CA, Beck S, Wilson MJ. 2001. The genomic context of natural killer receptor extended gene families. *Immunol. Rev.* 181:20–38
157. van den Berg TK, Yoder JA, Litman GW. 2004. On the origins of adaptive immunity: innate immune receptors join the tale. *Trends Immunol.* 25:11–16
158. Verica JA, Chae L, Tong H, Ingmire P, He ZH. 2003. Tissue-specific and developmentally regulated expression of a cluster of tandemly arrayed cell wall-associated kinase-like kinase genes in Arabidopsis. *Plant Physiol.* 133:1732–46
159. Vrana PB, Wheeler WC. 1996. Molecular evolution and phylogenetic utility of the polyubiquitin locus in mammals and higher vertebrates. *Mol. Phylogenet. Evol.* 6:259–69
160. Wagner GP, Amemiya C, Ruddle F. 2003. Hox cluster duplications and the opportunity for evolutionary novelties. *Proc. Natl. Acad. Sci. USA* 100:14603–6

161. Walsh JB. 1987. Sequence-dependent gene conversion—can duplicated genes diverge fast enough to escape conversion. *Genetics* 117:543–57
162. Watkins DI, McAdam SN, Liu XM, Strang CR, Milford EL, et al. 1992. New recombinant HLA-B alleles in a tribe of South-American Amerindians indicate rapid evolution of MHC class-1 loci. *Nature* 357:329–33
162a. Wegnez M, Monier R, Denis H. 1972. Sequence heterogeneity of 5S rRNA in *Xenopus laevis*. *FEBS Lett.* 25:13–20
163. Weigel D, Meyerowitz EM. 1994. The ABCs of floral homeotic genes. *Cell* 78:203–9
164. Weiss EH, Mellor AL, Golden L, Fahrner K, Simpson E, et al. 1983. The structure of a mutant H-2 gene suggests that the generation of polymorphism in H-2 genes may occur by gene conversion-like events. *Nature* 301:671–74
165. Whinnett A, Mundy NI. 2003. Isolation of novel olfactory receptor genes in marmosets (*Callithrix*): insights into pseudogene formation and evidence for functional degeneracy in non-human primates. *Gene* 304:87–96
166. Wilhelm BT, Gagnier L, Mager DL. 2002. Sequence analysis of the Ly49 cluster in C57BL/6 mice: a rapidly evolving multigene family in the immune system. *Genomics* 80:646–61
167. Wu S, Saunders TL, Bach FH. 1986. Polymorphism of human Ia antigens generated by reciprocal intergenic exchange between two DR loci. *Nature* 324:676–79
168. Young JM, Friedman C, Williams EM, Ross JA, Tonnes-Priddy L, Trask BJ. 2002. Different evolutionary processes shaped the mouse and human olfactory receptor gene families. *Hum. Mol. Genet.* 11:535–46
169. Yun TJ, Melvold RW, Pease LR. 1997. A complex major histocompatibility complex D locus variant generated by an unusual recombination mechanism in mice. *Proc. Natl. Acad. Sci. USA* 94:1384–89
170. Zamb TJ, Petes TD. 1982. Analysis of the junction between ribosomal RNA genes and single-copy chromosomal sequences in the yeast *Saccharomyces cerevisiae*. *Cell* 28:355–64
171. Zamocky M. 2004. Phylogenetic relationships in class I of the superfamily of bacterial, fungal, and plant peroxidases. *Eur. J. Biochem.* 271:3297–309
172. Zangenberg G, Huang MM, Arnheim N, Erlich H. 1995. New HLA-DPB1 alleles generated by interallelic gene conversion detected by analysis of sperm. *Nat. Genet.* 10:407–14
173. Zhang J, Dyer KD, Rosenberg HF. 2000. Evolution of the rodent eosinophil-associated RNase gene family by rapid gene sorting and positive selection. *Proc. Natl. Acad. Sci. USA* 97:4701–6
174. Zhang J, Rosenberg HF, Nei M. 1998. Positive Darwinian selection after gene duplication in primate ribonuclease genes. *Proc. Natl. Acad. Sci. USA* 95:3708–13
175. Zhang X, Firestein S. 2002. The olfactory receptor gene superfamily of the mouse. *Nat. Neurosci.* 5:124–33
176. Zhang Z, Carriero N, Gerstein M. 2004. Comparative analysis of processed pseudogenes in the mouse and human genomes. *Trends Genet.* 20:62–67
177. Zhang Z, Inomata N, Yamazaki T, Kishino H. 2003. Evolutionary history and mode of the amylase multigene family in *Drosophila*. *J. Mol. Evol.* 57:702–9
178. Zimmer EA, Martin SL, Beverley SM, Kan YW, Wilson AC. 1980. Rapid duplication and loss of genes coding for the chains of hemoglobin. *Proc. Natl. Acad. Sci. USA* 77:2158–62
179. Zozulya S, Echeverri F, Nguyen T. 2001. The human olfactory receptor repertoire. *Genome Biol.* 2:RESEARCH0018

Drosophila as a Model for Human Neurodegenerative Disease

Julide Bilen and Nancy M. Bonini[1]

Department of Biology, University of Pennsylvania, [1]Howard Hughes Medical Institute, Philadelphia, Pennsylvania 19104; email: nbonini@sas.upenn.edu

Annu. Rev. Genet.
2005. 39:153–71

First published online as a Review in Advance on June 28, 2005

The *Annual Review of Genetics* is online at http://genet.annualreviews.org

doi: 10.1146/annurev.genet.39.110304.095804

0066-4197/05/1215-0153$20.00

Key Words

neurodegenerative diseases, *Drosophila*, polyglutamine, Parkinson disease, Alzheimer disease, chaperone

Abstract

Among many achievements in the neurodegeneration field in the past decade, two require special attention due to the huge impact on our understanding of molecular and cellular pathogenesis of human neurodegenerative diseases. First is defining specific mutations in familial neurodegenerative diseases and second is modeling these diseases in easily manipulable model organisms including the fruit fly, nematode, and yeast. The power of these genetic systems has revealed many genetic factors involved in the various pathways affected, as well as provided potential drug targets for therapeutics. This review focuses on fruit fly models of human neurodegenerative diseases, with emphasis on how fly models have provided new insights into various aspects of human diseases.

Contents

INTRODUCTION

Human neurodegenerative diseases cause devastating clinical symptoms including ataxia, tremor and movement disorders, or cognitive and memory loss due to dysfunction and loss of specific neurons in the brain. Few cures and treatments are available. These diseases include PD, AD, FTDP, trinucleotide expansion diseases of type I (polyQ) diseases or type II noncoding trinucleotide expansion diseases such as Fragile X syndrome (4, 14, 16, 30, 48, 49, 61, 67, 70, 104). These diseases are typically sporadic, but select cases are inherited familially in a recessive or dominant manner depending upon the disease. One common pathological feature of most of these diseases is abnormal protein accumulation into inclusions, suggesting a structural and potentially functional alteration of the disease protein.

PD: Parkinson disease

AD: Alzheimer disease

FTDP: frontotemporal dementia with parkinsonism

polyQ: polyglutamine

The approach to model human neurodegenerative diseases in the simple fruit fly *Drosophila* offers many advantages for studying molecular and cellular pathology of human disease. These benefits include a faster time frame due to the shorter life span of the fly, large number of progeny, availability of many techniques and tools to manipulate gene expression, and relatively well-known anatomy and phenotypes (56, 91). Comparisons between the fly and human genomes indicate a high degree of conservation in fundamental biological pathways (74). With the availability of the fly genome sequence and vast array of molecular genetic assays, flies have great value not only in forward genetic screens, but also in reverse genetic approaches for functional analysis of human disease genes. Moreover, large-scale pharmalogical screens are also possible since flies have a complex nervous system and brain, but minimal barrier to prevent access to central nervous system tissue. Here, we focus on fly models of select human neurodegenerative diseases including polyQ diseases, noncoding triplet repeat diseases, PD, AD, and FTDP, and approaches and insight from modifier genes and screens.

FLY MODELS OF HUMAN POLYGLUTAMINE DISEASES

The molecular basis of at least nine human neurodegenerative diseases known as polyQ diseases has been defined as a CAG repeat expansion within the open reading frame of the respective genes (28, 103). These diseases include SBMA (also known as Kennedy disease), HD, SCA1, 2, 3, 6, 7, and 17, and DRPLA. The CAG expansion encodes a polyQ domain that confers a dominant toxicity to the disease protein, leading to dysfunction and late-onset neurodegeneration in select neurons. Transgenic models in the mouse of HD, SCA1, and a CAG repeat expansion in the coding region of an unrelated protein lead to neurodegeneration, suggesting that the polyQ domain per se has the toxicity (7, 36, 53, 65).

Modeling Human Polyglutamine Diseases in *Drosophila melanogaster*

To model the SCA type 3 (also known as Machado Joseph disease) (SCA3/MJD) in the fly, the C-terminally truncated domain of the pathogenic protein (SCA3tr-Q78) and the control counterpart (SCA3tr-Q27) were expressed in the fly (94). Expression can be directed to specific tissues using the GAL4/UAS system, which activates the transgene of interest via an upstream activating sequence (UAS) in a GAL4 transcription activity-dependent manner (5). Strikingly, expression of SCA3tr-Q78 in the eye causes loss of pigmentation and collapse of internal retinal structure (**Figure 1*A*,*B***) (94), whereas SCA3tr-Q27 has no effect. The severity of SCA3-induced neurodegeneration in the fly depends on the expression level and the polyQ repeat length of the pathogenic protein (**Figure 1*B*,*C*,*D***). In addition, the pathogenic ataxin-3 protein forms accumulations known as nuclear inclusions, while the control protein remains diffuse within the cytoplasm. Subsequent studies have shown that an expanded polyQ domain in the full-length ataxin-3 protein context also causes late-onset progressive neurodegeneration (**Figure 1*E***) (93). The pathogenicity of the truncated ataxin-3 protein is more severe than that of the full-length protein, due at least in part to the protective nature of functional domains of the normal protein (**Figure 1*F*,*G*,*H***) (93).

Subsequent studies have shown that directed expression of the N-terminal portion of the HD gene also causes late-onset progressive neurodegeneration in the fly (37). Directed expression of exon1 of the HD gene bearing 75 or 120 polyQ repeats indicates that pathogenicity of the disease protein is

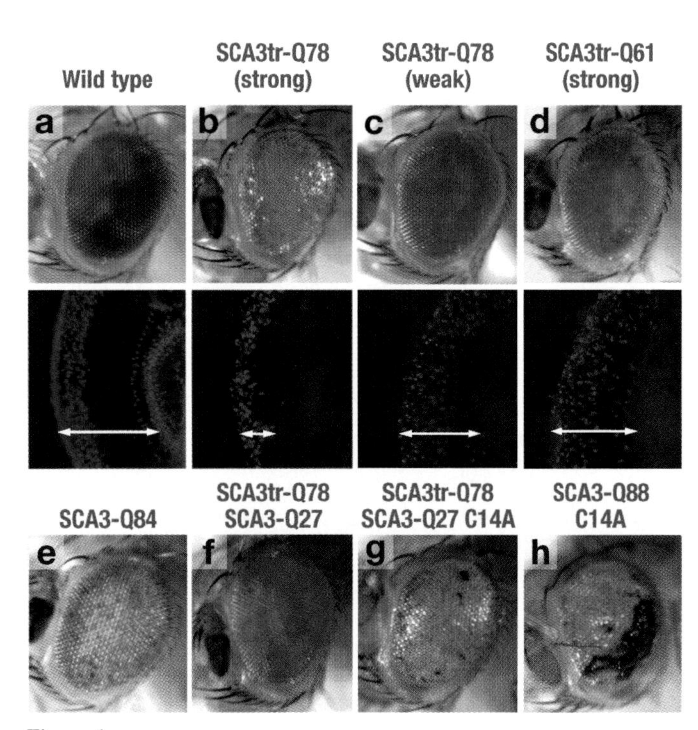

Figure 1

Fly models of SCA3 show that neurodegeneration is dependent upon protein expression level, polyQ length and non-polyQ content of the toxic protein. (*A–D*) External eye and internal retinal sections of 1 day-old flies. (*A*) Control of external eye and normal retinal thickness indicated with white arrow. Genotype is Oregon-R. (*B–C*) Directed expression of SCAtr-Q78 causes expression level-dependent external and internal eye degeneration associated with pigment loss and reduced retinal thickness. Strong expression (*B*) shows more severe degeneration than the weaker expression (*C*) (94). Genotypes: *w; UAS-SCA3tr-Q78(S) gmr-gal4/+* (*B*) and *w; gmr-gal4/+; UAS-SCA3tr-Q78(W)/+* (*C*). (*D*) Directed expression of shorter polyQ length protein SCA3tr-Q61 causes less severe neurodegeneration compared with (*B*) (9). Genotype: *w; gmr-gal4 UAS-SCA3tr-Q61(S)/+*. This contraction occurred in the transgene shown in (*B*). (*E–H*) External eye pictures of 1 day-old flies. (*E*) Directed expression of full length pathogenic ataxin-3 disease protein causes mild external eye degeneration (93). Genotype: *w; UAS-SCA3-Q84 gmr-gal4/+*. (*F*) Co-expression of full-length nonpathogenic form of ataxin-3 suppresses degeneration (compare *B* and *F*). Genotype: *w; UAS-SCA3tr-Q78 gmr-gal4* in *trans* to *UAS-SCA3-Q27*. (*G*) Mutation of the ubiquitin protease domain of the ataxin-3 protein abolishes its protective effect. Genotype: *w; UAS-SCA3tr-Q78 gmr-gal4* in *trans* to *UAS-SCA3-Q27 C14A*. (*H*) Expression of the pathogenic protein with a point mutation in the ubiquitin protease domain induces strikingly severe degeneration, similar to that of the truncated disease protein lacking the ubiquitin protease domain entirely (compare *E* with *H* and *B* with *H*) Genotype: *w; gmr-gal4 UAS-SCA3-Q88 C14A/+*.

polyQ repeat length dependent, as is typical of human disease. Unlike the SCA3 model, pathogenic HD protein accumulates in the nucleus but does not form obvious inclusions. Fernandez-Funez et al. (19) similarly found that full-length SCA1 protein bearing a pathogenic polyQ repeat (SCA1Q82) shows progressive neurodegeneration. Interestingly, the control SCA1 protein with a normal-length polyQ repeat (SCA1Q30) also causes a rough eye phenotype, suggesting that pathogenicity of SCA1 disease protein may depend on both the polyQ repeat length and expression level of the protein. An SBMA fly model using a pathogenic form of the androgen receptor shows ligand-dependent neurodegeneration (86). Strikingly, feeding an agonist or antagonist ligand to flies induces neurodegeneration, indicating that nuclear localization is key. A pure polyQ repeat alone is highly toxic in flies, inducing severe degeneration, underscoring the fundamental toxic nature of a polyQ domain (41, 54).

Forward genetics: an unbiased approach for identification of genes based on phenotype and interaction with the process of interest

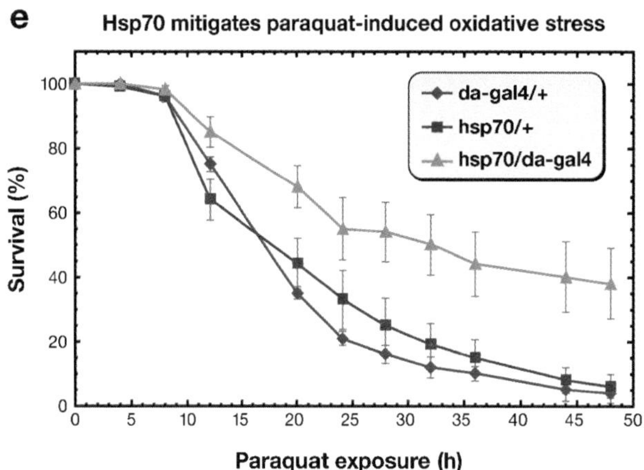

Figure 2

The chaperone Hsp70 suppresses polyQ-induced neurodegeneration, general protein misfolding, and paraquat-induced oxidative stress. (*A–D*) External eye pictures of 1 day-old flies. (*A–B*) Co-expression of human Hsp70 suppresses (*B*) SCA3tr-Q78 induced neurodegeneration (92). Genotype: *w; UAS-SCA3tr-Q78(S) gmr-gal4* in *trans* to + (*A*) and *UAS-hsp70* (*B*). (*C–D*) Degenerate-like eye phenotype induced upon expression of dominant negative Hsp70 (Hsp70.K71E) (3). (*C*) This phenotype is mitigated by added Hsp70 activity (*D*). Genotype: *w; UAS-Hsp70.K71E/+* (*C*) and *w; UAS-Hsp70.K71E* in *trans* to *UAS-Hsp70* (*D*). (*E*) Feeding the oxidative stress inducer paraquat to flies causes toxicity. Ubiquitous expression of hsp70 mitigates toxicity, indicating a protective role of Hsp70 in situations of oxidative stress. Flies (1–3 days old) were fed sucrose-based food with 20 mM paraquat. Genotypes: *w; da-gal4/+* (driver line control), *w; UAS-Hsp70* (transgene control) and *w; da-gal4* in *trans* to *UAS-Hsp70*.

Both overexpression and loss of function of the *Drosophila* homolog of SCA2 (dSCA2) disease gene causes tissue degeneration (76). This is hypothesized to be due to actin filament disorganization, suggesting that the normal cellular function of dSCA2 may be necessary for cell integrity. RNAi-mediated loss of function of the *Drosophila* homolog of HD causes axonal transport defects and degeneration in the eye, yielding a phenotype remarkably similar to overexpression of pathogenic forms of the human HD gene (27).

These findings indicate that, in addition to dominant toxicity of the polyQ expansion, the non-polyQ content of each disease protein may contribute to pathogenesis in a unique way, leading to disease-specific characteristics. These data also suggest that there are cases where pathogenicity may interfere or reduce the normal function of the disease gene, which contributes to the phenotype.

Modification Pathways of Polyglutamine Degeneration

Chaperones as suppressors of polyQ pathogenesis. A role of chaperones in polyQ toxicity was suggested with the finding that ubiquitinated nuclear inclusions sequester the chaperone heat shock protein 40 (Hsp40) in postmortem patient tissue sections and cell culture models (12). Strikingly, directed expression of the major stress-induced chaperone *hsp70* suppresses both SCA3 and HD induced neurodegeneration in the fly (**Figure 2*A,B***) (9, 92). Chaperones function to help fold proteins and to prevent protein aggregation, refold, and/or promote degradation of abnormal proteins via protein turnover pathways (6, 55, 66). In cell assays, heat shock proteins decrease polyQ accumulations into inclusions (39). Western immunoblot analysis indicates an increase in the solubility of polyQ aggregates in the presence of Hsp70 (9). Moreover, directed expression of a dominant negative form of *hsp70* (DN-Hsp70) causes a degenerate-like eye phenotype, indicating that general misfolding defects due to compromised chaperoning activity may lead to loss of cell integrity (**Figure 2*C***) (3). The DN-Hsp70-induced degenerate-like eye phenotype is mitigated by added Hsp70 activity (**Figure 2*D***). These findings suggest that chaperones may be essential for the maintenance and survival of cells, such that inhibition or limiting of chaperone activity by the pathogenic disease protein may be detrimental. Further studies may reveal additional insight into the role of chaperones in neurodegeneration.

Besides Hsp70, two cochaperones, Hsp40 and the *Drosophila* homologue of tetratricopeptide repeat protein 2, were found as suppressors of polyQ-induced degeneration in a genome-wide screen for the modifiers of polyQ toxicity (41). Hsp40 and heat shock response element hsr-ω were identified in a genome-wide screen of SCA1 toxicity (19). Although these screens were not saturated, only select chaperones may play roles in suppression of polyQ toxicity. One *Drosophila* homologue of hsp40, DNAJ1, suppresses neurodegeneration synergistically with Hsp70, but DNAJ2 does not, indicating possible substrate selectivity of chaperone activities (9).

Protein degradation and ubiquitin pathways in polyQ toxicity. Fly models of polyQ diseases show an expression-level-dependent neurodegeneration phenotype. Stronger expression lines have more severe degeneration and earlier onset of protein accumulation (19, 94, 103). Several modifiers appear associated with reduced protein accumulation (10, 73, 75, 87), suggesting that suppression may be linked to cellular degradation pathways, either ubiquitin proteasome or lysosomal degradation pathways. Studies done with fly models suggest that both protein degradation pathways are involved in the mechanisms of neuronal degeneration. Indeed, several components of the ubiquitin proteasome pathway have been identified in genome-wide modifier screens as enhancers or suppressors of polyQ neurodegeneration (19). Loss of function of ubiquitin enhances SCA1 toxicity, indicating that ubiquitination may normally help mitigate the toxicity of the SCA1 pathogenic protein (19). Consistent with a neuroprotective role of ubiquitin, loss of function of two different ubiquitin conjugases, dUbcD1 and dUbc-E2H, were also found as enhancers in the same screen. Recent findings indicate that the disease protein itself may be a target of ubiquitin (82).

Strikingly, full-length ataxin-3 protein has been found to suppress polyQ neurodegeneration (**Figure 1*F***), suggesting that the normal function of ataxin-3 may influence its own pathogenesis (93). In addition to the polyQ domain, ataxin-3 has a ubiquitin protease domain and ubiquitin-interacting motifs. The normal ataxin-3 protein colocalizes to nuclear inclusions when coexpressed with a toxic polyQ protein, and decreases the accumulation of the pathogenic protein in flies. Mutation of the ubiquitin protease domain abolishes the ability of the protein to suppress, and causes the pathogenic protein to become strikingly more toxic (**Figure 1*G*,*H***). Suppression also depends on the proteasome, linking protein quality control pathways with the protective effect of ataxin-3. Ataxin-3 is also effective against the toxicity of other polyQ proteins, including HD.

Other ubiquitin-like peptides (SUMO) may also be involved in polyQ pathogenesis. Steffan et al. (82) showed that ubiquitin and SUMO compete for the same lysine

Reverse genetics: a candidate gene, chosen for its predicted molecular functions, is tested for potential role in a process of interest

SBMA: spinobulbar muscular atropy

HD: Huntington disease

SCA: spinocerebellar ataxia

DRPLA: dentatorubral-pallidoluysian atrophy

GAL4/UAS system: a commonly used, two-component expression system in the fly where the tissue-specific expression of the yeast GAL4 transcription factor activates the expression of the transgene of interest via an upstream activator sequence (UAS)

Dominant: one mutant copy of gene (or transgene) exhibits the phenotype

Dominant negative mutant: a mutant protein that appears to reduce activity of the endogeneous pathway

Modifier screens: an initial phenotype is chosen and is screened for modulation, either suppression or enhancement, by changes in other genes, or treatment with agents such as drugs

residues on the HD exon 1 fragment in cell culture assays. A 50% downregulation of SUMO activity in flies or mutations in lysine residues that are targets for SUMOylation mitigates HD-induced neurodegeneration, suggesting that SUMOylation normally promotes pathogenicity in flies. However, reducing the SUMO-1 activity via a reduced amount of SUMO-1 activating enzyme enhances androgen receptor-induced neurodegeneration (8), suggesting that SUMOylation may be protective or detrimental depending on the disease protein and relative balance of different pathways.

The autophagy-inducing drug rapamycin has been shown to decrease both polyQ- and PD-associated α-synuclein aggregates in cell culture models (72, 95). Recently, Ravikumar et al. (73) reported that the rapamycin target, mTOR, is recruited to protein aggregates in HD, SCA2, SCA3, SCA7, and DRPLA patient postmortem tissue and in HD transgenic mouse models. Strikingly, feeding HD fly or mouse models with rapamycin mitigates neurodegeneration and behavioral abnormalities, respectively. Rapamycin suppression is associated with reduced accumulation of the pathogenic HD protein, thus is coupling lysosomal degradation of the protein to reduced polyQ pathogenesis.

These findings emphasize that players of protein quality control pathways in chaperones and protein degradation machineries play a significant role in molecular and cellular mechanisms of polyQ pathogenesis, although many components and details remain to be defined.

Transcriptional activity in polyQ diseases. Reduced acetyltransferase activity of transcription factors including cAMP response element binding protein (CBP) and p300/CBP-associated factor has been found in the polyQ disease background (83). This may be due to direct interactions between the polyQ domain and acetyltransferase domains of these proteins, which compromises the acetylation level of the chromatin. Strikingly, inhibiting the activity of a counteracting group of enzymes (histone deacetylases) genetically or pharmaceutically mitigates HD-induced neurodegeneration in flies and mice (83), suggesting that transcription misregulation may contribute to polyQ pathogenesis.

A modifier screen for SCA1-induced neurodegeneration revealed several additional transcription regulators including Sin3A, Rpd3, dCtBP, dSir2, Pap/Trap, and Tara as enhancers of SCA1 toxicity (19). Taken together these data indicate that transcription may be a target in polyQ disease, although the precise events that occur may depend on the particular disease protein involved. However, the ability to mitigate the phenotype with a general drug such as an HDAC inhibitor provides great promise for treatment.

Signaling pathways. Phosphorylation has been identified as a posttranslational modification that regulates the pathogenesis of ataxin-1. Akt kinase phoshorylates ataxin-1 at serine 776, which promotes interaction with the intracellular signaling protein 14-3-3, that was found to selectively bind to the phosphoserine residue 776 of ataxin-1 (10, 17). Directed expression of both 14-3-3 and Akt kinase gene enhances SCA1-induced neurodegeneration in the fly. It will be interesting to define upstream components of this pathway to understand how this signaling pathway can be regulated to modulate disease course.

Studies of Ravikumar et al. (73) implicated mTOR in various aspects of polyQ pathogenesis. They suggest that loss of function due to recruitment of mTOR to the aggregates releases the inhibition of mTOR on autophagy and may deregulate the mTOR-dependent signaling pathways including S6K and 4E-BP1. Their data show reduced mTOR kinase activity in the disease background, but downstream targets of mTOR linking to neurodegeneration and autophagic pathways to cause an eventual decrease in accumulation of the pathogenic protein remain to be defined.

Axonal transport defects in polyQ toxicity. Several groups have tested whether polyQ degeneration leads to axonal transport defects that contribute to neuronal dysfunction and loss (27, 50, 85). RNAi-induced loss of function of the *Drosophila* homologue of the HD gene causes organelle accumulations in larval neurons. PolyQ aggregates accumulate in axons and dentrites blocking the axonal transport in larval motor neurons, depleting motor proteins, and causing morphological changes in the axons. Directed expression of pathogenic polyQ proteins in neurons causes irregular movements in larvae and is associated with TUNEL staining indicative of programmed cell death (27). Axonal transport defects only occur if polyQ protein is localized to the cytoplasm, but not the nucleus, indicating that aspects of cytoplasmic and nuclear toxicity of polyQ protein are distinct. Such disruption of axonal transport can also lead to profound functional electrophysiological defects in the neurons (50). Since mutations in motor proteins or proteins associated with motors can cause degeneration in vertebrate models and humans (60), these data may indicate a critical role of axonal transport in disease.

Cell death and survival pathways. The cell death mechanisms of polyQ disease in neurodegeneration are not well elucidated. Morphological analysis of polyQ mouse models and human patient tissue reveals condensed cells with dark swollen mitochondria, a condition that is distinct from programmed cell death and necrosis (90). A general caspase inhibitor baculoviral protein p35 has limited effect on SCA3 toxicity, and no effect on HD-induced neurodegeneration, indicating that caspase activity may not be involved (37, 94). Recently, Jackson and colleagues reported loss of function of an upstream component of programmed cell death pathway, *Drosophila* CED4 homologue, dApaf1, strikingly suppresses polyQ neurodegeneration (75). Loss of *dApaf1* appears associated with decreased inclusions of polyQ protein as well as rescue of the cells from death, suggesting that Apaf1 regulated death pathways, as well as a potentially novel role of Apaf1 in protein quality control, may be involved.

Higashiyama et al. (33) implicated the *ter94* gene, which encodes the fly homologue of VCP/CDC48, a member of AAA+ class of ATPase proteins, in SCA3-induced neurodegeneration. Loss of function of *ter94* suppresses SCA3 toxicity whereas overexpression of ter94 causes a phenotype similar to polyQ-induced neurodegeneration. Ter94 has no obvious effect on protein aggregates. Further studies are required to address how ter94 activity modulates polyQ toxicity. Another protein shown to suppress polyQ toxicity without changing the accumulation of polyQ protein is human myeloid leukemia factor 1 (dMLF) (42). dMLF colocalizes to polyQ inclusions. Current data suggest that both ter94 and dMLF may act downstream of protein aggregation to prevent neurodegeneration. Potentially, they are regulators of cell loss mechanisms of polyQ toxicity.

UTR: untranslated region

FLY MODELS OF NONCODING TRINUCLEOTIDE REPEAT DISEASES

Noncoding Trinucleotide Repeat Diseases

Expansion of trinucleotide repeats including CGG, CTG, CAG, GCC, and GAA within the 5′ or 3′ UTR or an intron of the respective genes define the type II trinuclotide diseases (13). These diseases include Fragile X syndrome, Fragile XE syndrome, Friedreich's ataxia, myotonic dystrophytype I, SCA8, and SCA12. In type II diseases, repeat expansion within the noncoding region is thought to lead to loss of function of the disease gene, or gain of function of the disease-associated mRNA or both, causing neuronal dysfunction and degeneration. Fly models of some such diseases have been generated.

Protein accumulations: a common feature of many neurodegenerative diseases. The disease protein, along with other proteins, forms insoluble accumulations typically rich in β-pleated sheet structure. Typically referred to as inclusion bodies in polyQ diseases, Lewy bodies in Parkinson disease, and neurofibrillary tangles and amyloid plaques in Alzheimer disease

Modeling Fragile X in the Fly

Fragile X syndrome is the most common form of inherited mental retardation, affecting one in 6000 people (22). The symptoms of Fragile X syndrome, learning disabilities, mental retardation and progressive decline of cognitive abilities, are linked to a CGG repeat expansion in the 5′ UTR of the fragile X mental retardation 1 (*fmr1*) gene. Hypermethylation of the *fmr1* promotor due to a CGG repeat expansion and subsequent silencing of *fmr1* gene transcription is thought to be causative. Fly models of Fragile X syndrome may reveal new insight into both loss of function of *fmr1* and possible roles of the expanded CGG mRNA in neurons.

Characterization of behavioral and morphological defects in null alleles of the single *Drosophila* homologue of human FMR1 gene (*dfmr1*) revealed several phenotypes including eclosion defects, circadian rhythm abnormalities, locomotor defects, reduced courtship behavior, memory defects, abnormalities in neurite extention, axon guidance, and branching (15, 57, 59, 102). Abnormalities of the neuronal processes are thought to account for various behavioral phenotypes in *dfmr1* null flies. A mutation in *futsch*, encoding the *Drosophila* homologue of human microtubule-associated protein MAP1B, phenocopies synaptic growth defects of *dfmr1* loss of function (102). Strikingly, double mutants of *dfmr1* and *futsch* have no problem in synaptic growth and function, suggesting that Futsch is a target of the *dfmr1* gene, which regulates synaptic activity. Recently, McBride et al. (57) reported that feeding *dfmr1* null flies with metabotropic glutamate receptor (mGluR) antagonists or lithium rescues synaptic plasticity, courtship behavior, and memory defects, implicating misregulation of mGluR signaling in the disease state and a potential drug for therapeutics.

A novel progressive neurodegenerative ataxia associated with neuronal and glial nuclear protein accumulations was identified with premutation (60–200 CGG) repeat expansions within the 5′ UTR of the *fmr-1* gene (29). To address the potential effect of such *fmr1* premutations in *Drosophila*, Jin et al. (40) made transgenic lines bearing a 60 or 90 CGG repeat upstream of translation start codon of the reporter protein eGFP. Remarkably, such flies expressing 60 or 90 CGG repeats show a progressive neurodegenerative phenotype, similar to flies expressing pathogenic polyQ protein. The CGG-associated RNA-induced neurodegeneration results in ubiquitin and Hsp70-positive nuclear inclusions, also noted in human patients. Furthermore, overexpression of *hsp70* rescues RNA-induced progressive neurodegeneration. These striking similarities between the phenotype in flies of polyQ–mediated and RNA-mediated neurodegeneration suggest a common pathway between these situations. Identification of genetic factors associated with RNA-induced neurodegeneration may reveal insight into how protein- and RNA-induced toxicities are linked.

A Model for SCA8

SCA8 results from the expansion of a CTG repeat within the 3′ UTR of a noncoding gene (63). Because the gene overlaps a second gene, the human homologue of *Drosophila kelch*, SCA8 pathogenesis may be due to a toxic effect of the CUG repeat expansion in the SCA8 RNA or an antisense-dependent loss of function of human *kelch*. SCA8 disease has been modeled in *Drosophila* using a human SCA8 cDNA bearing 112CTG and 9CTG repeats (62). Directed expression of both SCA8-112CTG and the control SCA8-9CTG causes mild external eye roughness and progressive internal eye degeneration, suggesting that the expression level of the gene is key to toxicity in the fly. Several mRNA binding proteins including Staufen, Muscleblind and Split ends have been identified in a loss of function screen for modifiers of the SCA8-CTG112-mediated rough eye phenotype. Staufen is shown to be recruited to the expanded CUG mRNA, but how Staufen or other mRNA binding proteins function in RNA-induced toxicity remain to be answered. RNA binding proteins have been found to modulate SCA1-Q82 toxicity as well (19). More recently, a CTG expansion has been modeled in the fly, revealing accumulations of the RNA, although without a pathogenic phenotype (34). These data indicate that, whereas this is a promising direction to yield novel information regarding these diseases, fundamental aspects of mechanisms of these diseases remain to be revealed.

FLY MODELS OF PARKINSON DISEASE

Parkinson Disease

PD is a common movement disorder, affecting about 1% of the population older than 65 (47). Pathologically, the disease is characterized by loss of dopaminergic neurons in the substantia nigra, and protein accumulations known as Lewy bodies (LB). A number of genes found to be associated with familial PD have provided insight into both familial and sporadic forms of disease. Several pieces of evidence implicate the ubiquitin proteasome system (UPS) and chaperones in PD pathology. LBs are immunopositive for ubiquitin as well as chaperones (2, 58). Mutations in *parkin*, encoding an E3-specific ubiquitin ligase, have been linked to autosomal recessive juvenile onset parkinsonism (AR-JP) (43, 51). E3 ubiquitin ligases are required to transfer ubiquitin to specific target proteins for their degradation. Loss of *parkin*, and thus potentially the buildup of select target proteins, might cause disease. Other hypotheses for PD include dysfunction of mitochondrial complex I and subsequent oxidative stress, and compromised energy level of DA neurons. Supporting this hypothesis, compounds inhibiting mitochondrial complex I activity and inducing oxidative stress, including 1-methyl-4-phenyl-1,2,3,6 tetrahydropyridine, paraquat, and rotenone, cause selective dopaminergic neuronal loss (71, 79, 88). Exposing rodents or primates to these compounds also results in protein accumulations immunopositive for α-synuclein, a protein in which mutations, including A30P and A53T, have been found linked to dominantly inherited PD (46, 69).

Fly Models for Parkinson Disease

Feany & Bender (18) modeled PD in *Drosophila* using normal human α-synuclein and two familial mutant forms (A30P and A53T). Directed expression of α-synuclein in the brain does not cause gross morphological changes, but rather leads to adult onset loss of tyrosine hydroxylase (TH) immunostaining in select DA neurons. α-synuclein forms LB-like accumulations in the fly, similar to human PD, and causes climbing defects indicating motor dysfunction.

Loss of function of the *parkin* gene in flies causes sensitivity to oxidative stress and an abnormal wing phenotype (24, 68). Muscle morphology indeed shows dramatic degeneration of structure over time, with swollen, sick-looking mitochondria. Although DA neurons are normal, these phenotypes may prove helpful for understanding the role of parkin.

More recently, rotenone has been shown to cause effects in flies, including some loss of DA neurons and abnormal locomoter function, indicating that mechanisms of environmental as well as genetic influences on PD may be approachable in flies (11).

Modification of Pathways of Parkinson Disease

Auluck et al. (2) examined the role of chaperones in PD to show that coexpression of hsp70 in the DA neurons rescues α-synuclein-induced neurotoxicity whereas compromised levels of Hsp70 enhance toxic events in flies. Hsp70 and the cochaperone Hsp40 localize to LBs and LNs in human patient tissue from PD and other synucleinopathies (2). These findings emphasize the significance of chaperones in PD pathology. Polymorphisms in

APP: amyloid precursor protein

BACE: beta site APP cleaving enzyme

NFT: neurofibrillary tangles

Hsp70 have now been found to be a risk factor for PD (98), and Hsp70 has been shown to mitigate PD-like toxicity in mouse models (44). Upregulation of the heat shock response via the drug geldanamycin (GA) suppresses degenerative events of α-synuclein in vivo in flies (1). These studies indicate that such drug approaches may prove promising to protect against deleterious events of PD.

Impairment of oxidative stress is a main hypothesis in the PD field. To investigate the possibility that Hsp70 may modulate this aspect of PD, we examined whether Hsp70 could protect against paraquat-induced oxidative stress. Adult-onset exposure of flies to 20mM paraquat severely shortens the lifespan of the animals such that 50% of flies die by 17–19 h of exposure. However, ubiquitous expression of hsp70 confers protection such that 50% loss of flies does not occur until 38 ± 11.5 h. These data support a protective role of Hsp70 in oxidative stress, in addition to modulation of protein misfolding and ubiquitin-dependent proteolytic pathways (**Figure 2*E***).

Microarray approaches have been performed on flies expressing α-synuclein (77) to reveal that select genes are affected. These include genes involved in catecholamine and DA synthesis, as well as mitochondrial proteins. These genes were strikingly different from those altered in flies expressing tau, supportive of distinct pathogenesis in different disease states. Functional analysis of these genes may provide futher insights into the pathogenesis of α-synuclein-induced neuronal dysfunction. Comparison of microarray data of α-synuclein- and tau-induced toxicity revealed few overlaps, suggesting that these diseases may modulate distinct pathways.

Additional studies have suggested interactions between parkin and α-synuclein, such that overexpression of parkin protects against deleterious consequences of α-synuclein to DA neurons (31, 100). A more recent microarray approach has revealed upregulation of oxidative stress components and innate immune response genes upon loss of function of *Drosophila parkin*, but no significant change in endoplasmic reticulum induced stress or cell cycle components (23).

FLY MODELS OF ALZHEIMER AND RELATED DISEASES

Alzheimer and Related Diseases

AD is associated with progressive memory loss, and is the most common neurodegenerative disease. Mutations in the genes that encode APP, Presenilin 1 and 2 (PS1 and PS2), are associated with familial AD (78). Pathological features include extracellular filamentous proteinaceous aggregates called senile plaques that contain proteolytically processed forms of APP (Aβ peptides including Aβ40 and mainly Aβ42). These forms are generated through processing of APP by γ-secretase, of which PS1 and PS2 are subunits, and BACE. AD is also characterized by NFT of tau within neurons (16, 30, 96).

NFT containing hyperphosphorylated tau are also associated with other neurodegenerative diseases, including frontotemperal dementia and parkinsonism linked to chromosome 17 (FTDP-17). Tau is expressed within neurons and functions in microtubule stabilization. Flies have a homologue of tau expressed in the nervous system (32).

Modeling Alzheimer Disease with Aβ and APP in *Drosophila*

Modeling AD in flies is more challenging than other neurodegenerative diseases because not all components of APP proteolytic processing machinery appear conserved. The *Drosophila* homologue of APP (dAPPL) does not have the Aβ domain and the fly has no BACE activity, although it has γ-secretase components, including Presenilin (Psn) and Nicastrin (4, 16, 45). However, fly models still provide insight into various aspects of AD pathology. Overexpression of wild-type Psn and AD-linked mutant forms causes programmed cell death in flies, providing evidence of cell loss mechanisms that may be associated with AD neurodegeneration (101).

One phenotype observed upon the directed expression of wild-type human APP (hAPP), pathogenic APP mutation (hAPP-Swedish), and dAPPL is a blistered wing (21), suggesting that cell-cell adhesion functions are conserved among the family members. Directed expression of hAPP similarly colocalizes to the synaptic terminals in the nervous system, like endogenous dAPPL localization (99). Overexpression of either hAPP or dAPPL does not result in degeneration. However, interestingly dAPPL is recruited to amyloid-like protein aggregates in another fly neurodegeneration mutant, *blue cheese* (20).

Although Aβ cannot be produced normally in flies, directed expression of the transgenes encoding peptides Aβ42 and Aβ40 in the nervous system causes various phenotypes. Both Aβ42 and Aβ40 cause progressive loss of learning ability but only Aβ42 reduces life span, causes degeneration in Kenyon cells of the brain, and forms diffuse accumulations similar to amyloid plaques (35).

Coexpression of human BACE and hAPP in flies produces Aβ40 and AB42 fragments and leads to progressive degeneration in the retina (25). Overexpression of Psn (part of the component of γ-secretase activity) enhances neurodegeneration. These data indicate that the fly can be used to study aspects of AD, including drug effects—for example, BACE inhibitors can mitigate aspects of the phenotype in flies.

Gunawardana & Goldstein (26) examined vesicle trafficking and found that deletion of *dAPPL* and directed expression of hAPP, two pathogenic forms (hAPP-LONDON and hAPP-Swedish), or dAPPL in larval motor neurons causes accumulation of organelles along axonal tracts. This is a specific phenotype, typically observed only upon loss of function of motor proteins. The organelle jam phenotype is modulated by the anterograde motor protein kinesin and retrograde transport protein dynein, supporting the idea that APP family members may function as vesicle receptors through interactions with kinesin-1. Evidence for problems in axonal transport has now been seen in transgenic mouse models as well (84).

Tau Models in *Drosophila*

Directed expression of human tau in flies leads to degeneration, axonal tranport defects, and axonal accumulations (38, 64, 89, 97). Wittmann et al. (97) created a fly model of tau with wild-type and FTDP-linked mutant forms of tau expressed in the nervous system and in cholinergic neurons. Both normal and pathogenic tau cause adult-onset progressive neurodegeneration, although the mutant forms are more severe. The degeneration shows vacuolization, but no obvious NFT suggesting that formation of neurofibrillary tangles may not be necessary for degeneration. Tau expression directed to the fly eye gives a striking phenotype of a reduced and rough eye, making tau particularly amenable to genetic screen approaches.

Modification Pathways of Alzheimer Disease and Taupathies in the Fly

The serine-threonine kinase GSK-3β is thought to play a role in the hyperphosphorylation of tau in neurofibrillary tangles in AD and other tauopathies (49, 52, 81). Jackson et al. (38) examined the effect of the *Drosophila* homologue of GSK-3, *shaggy*, a component of Wingless signaling pathway, on tau toxicity in the fly. Loss of function of *shaggy* suppresses tau degeneration, whereas upregulation enhances in association with neurofibrillary tangle-like protein accumulations. These data suggest that tangle formation promotes tau toxicity. In the tau fly model, directed expression of anti-apoptotic genes including p35, DIAP1, and DIAP2 mitigates tau toxicity (38, 80), suggesting that programmed cell death may play a key role in tau-induced degeneration.

An overexpression modifier screen (80) against the tau-induced rough eye phenotype revealed several suppressors and enhancers, including a number of kinases and phosphatases, emphasizing the significance of phosphorylation on tau toxicity. Other categories of modifiers include apoptosis genes, cytoskeleton components, and cation transporters. Interestingly, overexpression of the *Drosophila* homologue of SCA2 and *dfmr1* enhances tau degeneration, revealing a potential common mechanism between tau pathology and Fragile X syndrome and polyQ diseases. Overall, this genomic screen approach revealed little overlap between modifiers of SCA1 pathology and tau pathology, suggesting distinct mechanisms.

Further studies on tau phosphorylation have shown that mutation of PAR-1 kinase sites in tau to prevent phosphorylation mitigates the toxicity of tau in flies (64). Par-1, independently identified as an enhancer of tau by Shulman & Feany (80), is a serine-threonine kinase that modulates tau at the select sites S262/S356. Whereas upregulation of par-1 promotes tau toxicity, prevention of phosphorylation mitigates tau toxicity. These data indicate that modulation of tau by phosphorylation may be a critical aspect of tau-associated degeneration. *Drosophila* is particularly amenable to detailed studies of the significance and role of phosphorylation of specific sites, due to the ease of transgenic approaches with modified protein forms.

SUMMARY POINTS

1. The success of modeling of fundamental aspects of several distinct human neurodegenerative diseases in fly has been an enormous advance toward discovering new molecular and cellular pathogenic events associated with disease.
2. Features of neurodegenerative diseases modeled in the fly include accumulation of disease proteins in abnormal aggregates, toxicity of the proteins to induce neuronal dysfunction and loss, among other features reflective of the human diseases.
3. Genome-wide forward genetic analysis, candidate gene approaches, and microarray analysis using these models have revealed modifiers of neurodegenerative phenotypes, including chaperones, components of ubiquitin pathways, transcription factors, signal transduction components, components of axonal trafficking, and oxidative stress pathways.
4. Striking mitigation of neurodegeneration by drugs and compounds including the histone deacetylase inhibitor SAHA, stress upregulator geldanamycin, and rapamycin, which induces autophagy, indicates that fly models are providing a powerful approach for therapeutic agents that have application to human disease.

FUTURE ISSUES TO BE RESOLVED

1. With the availability of many fly models, it is now possible to compare similarities and differences of neurodegenerative disease phenotypes, including details of mechanism and selectivity of a large number of modifiers.
2. Modeling in the fly is being extended to other human diseases, to apply the power of the vast array of molecular genetic approaches available in *Drosophila* to broader questions of human disease and their potential treatment.

3. Genome-wide screens for modifiers of neurodegenerative phenotypes have only begun, such that future screens promise to reveal additional novel aspects of neurodegeneration.
4. The future holds promise to provide additional information on how these modifiers and mechanisms function to modulate neurodegeneration; these studies must be performed hand in hand with models in higher organisms to reach the ultimate goal of potential new treatments for the human situation.

LITERATURE CITED

1. Auluck PK, Bonini NM. 2002. Pharmacological prevention of Parkinson disease in *Drosophila*. *Nat. Med.* 8:1185–86
2. **Auluck PK, Chan HY, Trojanowski JQ, Lee VM, Bonini NM. 2002. Chaperone suppression of alpha-synuclein toxicity in a *Drosophila* model for Parkinson's disease. *Science* 295:865–68**

This manuscript provides initial in vivo evidence indicating a potential role of chaperones in Parkinson disease.

3. Bonini NM. 2002. Chaperoning brain degeneration. *Proc. Natl. Acad. Sci. USA* 99(Suppl.) 4:16407–11
4. Bonini NM, Fortini ME. 2003. Human neurodegenerative disease modeling using *Drosophila*. *Annu. Rev. Neurosci.* 26:627–56
5. Brand AH, Perrimon N. 1993. Targeted gene expression as a means of altering cell fates and generating dominant phenotypes. *Development* 118:401–15
6. Bukau B, Horwich AL. 1998. The Hsp70 and Hsp60 chaperone machines. *Cell* 92:351–66
7. Burright EN, Clark HB, Servadio A, Matilla T, Feddersen RM, et al. 1995. SCA1 transgenic mice: a model for neurodegeneration caused by an expanded CAG trinucleotide repeat. *Cell* 82:937–48
8. Chan HY, Warrick JM, Andriola I, Merry D, Bonini NM. 2002. Genetic modulation of polyglutamine toxicity by protein conjugation pathways in *Drosophila*. *Hum. Mol. Genet.* 11:2895–904
9. Chan HY, Warrick JM, Gray-Board GL, Paulson HL, Bonini NM. 2000. Mechanisms of chaperone suppression of polyglutamine disease: Selectivity, synergy and modulation of protein solubility in *Drosophila*. *Hum. Mol. Genet.* 9:2811–20
10. Chen HK, Fernandez-Funez P, Acevedo SF, Lam YC, Kaytor MD, et al. 2003. Interaction of Akt-phosphorylated ataxin-1 with 14-3-3 mediates neurodegeneration in spinocerebellar ataxia type 1. *Cell* 113:457–68
11. Coulom H, Birman S. 2004. Chronic exposure to rotenone models sporadic Parkinson's disease in *Drosophila melanogaster*. *J. Neurosci.* 24:10993–98
12. Cummings CJ, Mancini MA, Antalffy B, DeFranco DB, Orr HT, Zoghbi HY. 1998. Chaperone suppression of aggregation and altered subcellular proteasome localization imply protein misfolding in SCA1. *Nat. Genet.* 19:148–54
13. Cummings CJ, Zoghbi HY. 2000. Fourteen and counting: unraveling trinucleotide repeat diseases. *Hum. Mol. Genet.* 9:909–16
14. Dawson TM, Dawson VL. 2003. Molecular pathways of neurodegeneration in Parkinson's disease. *Science* 302:819–22
15. Dockendorff TC, Su HS, McBride SM, Yang Z, Choi CH, et al. 2002. *Drosophila* lacking dfmr1 activity show defects in circadian output and fail to maintain courtship interest. *Neuron* 34:973–84

A fly model for Parkinson disease, showing tissue specificity of α-synuclein toxicity in dopaminergic neurons, accumulation of α-synuclein in Lewy body-like accumulations, and loss of climbing ability.

This study is a genome-wide screen for modifiers of pathogenic SCA1 neurodegeneration in the fly, suggesting key roles for components of ubiquitin and chaperone pathways, among others.

Loss of function and overexpression analysis of APP revealed that APP has a role in axonal transport in *Drosophila*; these findings have been extended to polyQ diseases, suggesting that axonal transport defects may be a common feature of several neurodegenerative situations.

16. Driscoll M, Gerstbrein B. 2003. Dying for a cause: Invertebrate genetics takes on human neurodegeneration. *Nat. Rev. Genet.* 4:181–94
17. Emamian ES, Kaytor MD, Duvick LA, Zu T, Tousey SK, et al. 2003. Serine 776 of ataxin-1 is critical for polyglutamine-induced disease in SCA1 transgenic mice. *Neuron* 38:375–87
18. **Feany MB, Bender WW. 2000. A *Drosophila* model of Parkinson's disease. *Nature* 404:394–98**
19. **Fernandez-Funez P, Nino-Rosales ML, de Gouyon B, She WC, Luchak JM, et al. 2000. Identification of genes that modify ataxin-1-induced neurodegeneration. *Nature* 408:101–6**
20. Finley KD, Edeen PT, Cumming RC, Mardahl-Dumesnil MD, Taylor BJ, et al. 2003. blue cheese mutations define a novel, conserved gene involved in progressive neural degeneration. *J. Neurosci.* 23:1254–64
21. Fossgreen A, Bruckner B, Czech C, Masters CL, Beyreuther K, Paro R. 1998. Transgenic *Drosophila* expressing human amyloid precursor protein show gamma-secretase activity and a blistered-wing phenotype. *Proc. Natl. Acad. Sci. USA* 95:13703–8
22. Gao FB. 2002. Understanding fragile X syndrome: insights from retarded flies. *Neuron* 34:859–62
23. Greene JC, Whitworth AJ, Andrews LA, Parker TJ, Pallanck LJ. 2005. Genetic and genomic studies of Drosophila parkin mutants implicate oxidative stress and innate immune responses in pathogenesis. *Hum. Mol. Genet.* 14:799–811
24. Greene JC, Whitworth AJ, Kuo I, Andrews LA, Feany MB, Pallanck LJ. 2003. Mitochondrial pathology and apoptotic muscle degeneration in *Drosophila* parkin mutants. *Proc. Natl. Acad. Sci. USA* 100:4078–83
25. Greeve I, Kretzschmar D, Tschape JA, Beyn A, Brellinger C, et al. 2004. Age-dependent neurodegeneration and Alzheimer-amyloid plaque formation in transgenic *Drosophila*. *J. Neurosci.* 24:3899–906
26. **Gunawardena S, Goldstein LS. 2001. Disruption of axonal transport and neuronal viability by amyloid precursor protein mutations in *Drosophila*. *Neuron* 32:389–401**
27. Gunawardena S, Her LS, Brusch RG, Laymon RA, Niesman IR, et al. 2003. Disruption of axonal transport by loss of huntingtin or expression of pathogenic polyQ proteins in *Drosophila*. *Neuron* 40:25–40
28. Gusella JF, MacDonald ME. 2000. Molecular genetics: unmasking polyglutamine triggers in neurodegenerative disease. *Nat. Rev. Neurosci.* 1:109–15
29. Hagerman RJ, Leehey M, Heinrichs W, Tassone F, Wilson R, et al. 2001. Intention tremor, parkinsonism, and generalized brain atrophy in male carriers of fragile X. *Neurology* 57:127–30
30. Hardy J, Selkoe DJ. 2002. The amyloid hypothesis of Alzheimer's disease: progress and problems on the road to therapeutics. *Science* 297:353–56
31. Haywood AF, Staveley BE. 2004. parkin counteracts symptoms in a *Drosophila* model of Parkinson's disease. *BMC Neurosci.* 5:14
32. Heidary G, Fortini ME. 2001. Identification and characterization of the *Drosophila* tau homolog. *Mech. Dev.* 108:171–78
33. Higashiyama H, Hirose F, Yamaguchi M, Inoue YH, Fujikake N, et al. 2002. Identification of ter94, *Drosophila* VCP, as a modulator of polyglutamine-induced neurodegeneration. *Cell Death Differ.* 9:264–73

34. Houseley JM, Wang Z, Brock GJ, Soloway J, Artero R, et al. 2005. Myotonic dystrophy associated expanded CUG repeat muscleblind positive ribonuclear foci are not toxic to Drosophila. *Hum. Mol. Genet.* 14:873–83
35. Iijima K, Liu HP, Chiang AS, Hearn SA, Konsolaki M, Zhong Y. 2004. Dissecting the pathological effects of human Abeta40 and Abeta42 in *Drosophila*: a potential model for Alzheimer's disease. *Proc. Natl. Acad. Sci. USA* 101:6623–28
36. Ikeda H, Yamaguchi M, Sugai S, Aze Y, Narumiya S, Kakizuka A. 1996. Expanded polyglutamine in the Machado-Joseph disease protein induces cell death in vitro and in vivo. *Nat. Genet.* 13:196–202
37. **Jackson GR, Salecker I, Dong X, Yao X, Arnheim N, et al. 1998. Polyglutamine-expanded human huntingtin transgenes induce degeneration of *Drosophila* photoreceptor neurons. *Neuron* 21:633–42**

An initial study modeling Huntington disease in the fly.

38. Jackson GR, Wiedau-Pazos M, Sang TK, Wagle N, Brown CA, et al. 2002. Human wild-type tau interacts with wingless pathway components and produces neurofibrillary pathology in *Drosophila*. *Neuron* 34:509–19
39. Jana NR, Nukina N. 2003. Recent advances in understanding the pathogenesis of polyglutamine diseases: involvement of molecular chaperones and ubiquitin-proteasome pathway. *J. Chem. Neuroanat.* 26:95–101
40. **Jin P, Zarnescu DC, Zhang F, Pearson CE, Lucchesi JC, et al. 2003. RNA-mediated neurodegeneration caused by the fragile X premutation rCGG repeats in *Drosophila*. *Neuron* 39:739–47**

This study reports evidence for RNA-induced toxicity in vivo; the toxicity is associated with ubiquitin- and Hsp70-positive accumulations and is modulated by Hsp70.

41. **Kazemi-Esfarjani P, Benzer S. 2000. Genetic suppression of polyglutamine toxicity in *Drosophila*. *Science* 287:1837–40**

This study is a forward genetic modifier screen for suppressors of polyQ toxicity in fly, revealing J-domain containing proteins as suppressors of polyQ degeneration.

42. Kazemi-Esfarjani P, Benzer S. 2002. Suppression of polyglutamine toxicity by a *Drosophila* homolog of myeloid leukemia factor 1. *Hum. Mol. Genet.* 11:2657–72
43. Kitada T, Asakawa S, Hattori N, Matsumine H, Yamamura Y, et al. 1998. Mutations in the parkin gene cause autosomal recessive juvenile parkinsonism. *Nature* 392:605–8
44. Klucken J, Shin Y, Masliah E, Hyman BT, McLean PJ. 2004. Hsp70 reduces alpha-synuclein aggregation and toxicity. *J. Biol. Chem.* 279:25497–502
45. Kopan R, Goate A. 2002. Aph-2/nicastrin: an essential component of gamma-secretase and regulator of Notch signaling and Presenilin localization. *Neuron* 33:321–24
46. Kruger R, Kuhn W, Muller T, Woitalla D, Graeber M, et al. 1998. Ala30Pro mutation in the gene encoding alpha-synuclein in Parkinson's disease. *Nat. Genet.* 18:106–8
47. Lang AE, Lozano AM. 1998. Parkinson's disease. (Part 1). *N. Engl. J. Med.* 339:1044–53
48. Lang AE, Lozano AM. 1998. Parkinson's disease. (Part 2). *N. Engl. J. Med.* 339:1130–43
49. Lee VM, Goedert M, Trojanowski JQ. 2001. Neurodegenerative tauopathies. *Annu. Rev. Neurosci.* 24:1121–59
50. Lee WC, Yoshihara M, Littleton JT. 2004. Cytoplasmic aggregates trap polyglutamine-containing proteins and block axonal transport in a *Drosophila* model of Huntington's disease. *Proc. Natl. Acad. Sci. USA* 101:3224–29
51. Leroy E, Boyer R, Auburger G, Leube B, Ulm G, et al. 1998. The ubiquitin pathway in Parkinson's disease. *Nature* 395:451–52
52. Mandelkow E. 1999. Alzheimer's disease: the tangled tale of tau. *Nature* 402:588–89
53. Mangiarini L, Sathasivam K, Seller M, Cozens B, Harper A, et al. 1996. Exon 1 of the HD gene with an expanded CAG repeat is sufficient to cause a progressive neurological phenotype in transgenic mice. *Cell* 87:493–506

54. Marsh JL, Walker H, Theisen H, Zhu YZ, Fielder T, et al. 2000. Expanded polyglutamine peptides alone are intrinsically cytotoxic and cause neurodegeneration in *Drosophila*. *Hum. Mol. Genet.* 9:13–25
55. Mathew A, Morimoto RI. 1998. Role of the heat-shock response in the life and death of proteins. *Ann. NY Acad. Sci.* 851:99–111
56. Matthews KA, Kaufman TC, Gelbart WM. 2005. Research resources for Drosophila: the expanding universe. *Nat. Rev. Genet.* 6:179–93
57. McBride SM, Choi CH, Wang Y, Liebelt D, Braunstein E, et al. 2005. Pharmacological rescue of synaptic plasticity, courtship behavior, and mushroom body defects in a Drosophila model of fragile x syndrome. *Neuron* 45:753–64
58. McLean PJ, Kawamata H, Shariff S, Hewett J, Sharma N, et al. 2002. TorsinA and heat shock proteins act as molecular chaperones: suppression of alpha-synuclein aggregation. *J. Neurochem.* 83:846–54
59. Morales J, Hiesinger PR, Schroeder AJ, Kume K, Verstreken P, et al. 2002. *Drosophila* fragile X protein, DFXR, regulates neuronal morphology and function in the brain. *Neuron* 34:961–72
60. Deleted in proof
61. Muqit MM, Feany MB. 2002. Modelling neurodegenerative diseases in Drosophila: a fruitful approach? *Nat. Rev. Neurosci.* 3:237–43
62. Mutsuddi M, Marshall CM, Benzow KA, Koob MD, Rebay I. 2004. The spinocerebellar ataxia 8 noncoding RNA causes neurodegeneration and associates with staufen in *Drosophila*. *Curr. Biol.* 14:302–8
63. Nemes JP, Benzow KA, Moseley ML, Ranum LP, Koob MD. 2000. The SCA8 transcript is an antisense RNA to a brain-specific transcript encoding a novel actin-binding protein (KLHL1). *Hum. Mol. Genet.* 9:1543–51
64. **Nishimura I, Yang Y, Lu B. 2004. PAR-1 kinase plays an initiator role in a temporally ordered phosphorylation process that confers tau toxicity in *Drosophila*. *Cell* 116:671–82**

This study indicates that phosphorylation of tau plays a fundamental role in its toxicity.

65. Ordway JM, Tallaksen-Greene S, Gutekunst CA, Bernstein EM, Cearley JA, et al. 1997. Ectopically expressed CAG repeats cause intranuclear inclusions and a progressive late onset neurological phenotype in the mouse. *Cell* 91:753–63
66. Parsell DA, Taulien J, Lindquist S. 1993. The role of heat-shock proteins in thermotolerance. *Philos. Trans. R. Soc. London Ser. B* 339:279–85
67. Perutz MF. 1999. Glutamine repeats and neurodegenerative diseases: molecular aspects. *Trends Biochem. Sci.* 24:58–63
68. Pesah Y, Pham T, Burgess H, Middlebrooks B, Verstreken P, et al. 2004. *Drosophila* parkin mutants have decreased mass and cell size and increased sensitivity to oxygen radical stress. *Development* 131:2183–94
69. Polymeropoulos MH, Lavedan C, Leroy E, Ide SE, Dehejia A, et al. 1997. Mutation in the alpha-synuclein gene identified in families with Parkinson's disease. *Science* 276:2045–47
70. Price DL, Tanzi RE, Borchelt DR, Sisodia SS. 1998. Alzheimer's disease: genetic studies and transgenic models. *Annu. Rev. Genet.* 32:461–93
71. Przedborski S, Jackson-Lewis V, Djaldetti R, Liberatore G, Vila M, et al. 2000. The parkinsonian toxin MPTP: action and mechanism. *Restor. Neurol. Neurosci.* 16:135–42
71a. Puls I, Jonnakuty C, LaMonte BH, Holzbaur EL, Tokito M, et al. 2003. Mutant dynactin in motor neuron disease. *Nat. Genet.* 33:455–56

72. Ravikumar B, Duden R, Rubinsztein DC. 2002. Aggregate-prone proteins with polyglutamine and polyalanine expansions are degraded by autophagy. *Hum. Mol. Genet.* 11:1107–17
73. Ravikumar B, Vacher C, Berger Z, Davies JE, Luo S, et al. 2004. Inhibition of mTOR induces autophagy and reduces toxicity of polyglutamine expansions in fly and mouse models of Huntington disease. *Nat. Genet.* 36:585–95
74. Rubin GM, Yandell MD, Wortman JR, Gabor Miklos GL, Nelson CR, et al. 2000. Comparative genomics of the eukaryotes. *Science* 287:2204–15
75. Sang TK, Li C, Liu W, Rodriguez A, Abrams JM, et al. 2005. Inactivation of Drosophila Apaf-1 related killer suppresses formation of polyglutamine aggregates and blocks polyglutamine pathogenesis. *Hum. Mol. Genet.* 14:357–72
76. Satterfield TF, Jackson SM, Pallanck LJ. 2002. A *Drosophila* homolog of the polyglutamine disease gene SCA2 is a dosage-sensitive regulator of actin filament formation. *Genetics* 162:1687–702
77. Scherzer CR, Jensen RV, Gullans SR, Feany MB. 2003. Gene expression changes presage neurodegeneration in a *Drosophila* model of Parkinson's disease. *Hum. Mol. Genet.* 12:2457–66
78. Selkoe DJ. 2004. Cell biology of protein misfolding: the examples of Alzheimer's and Parkinson's diseases. *Nat. Cell. Biol.* 6:1054–61
79. Sherer TB, Kim JH, Betarbet R, Greenamyre JT. 2003. Subcutaneous rotenone exposure causes highly selective dopaminergic degeneration and alpha-synuclein aggregation. *Exp. Neurol.* 179:9–16
80. Shulman JM, Feany MB. 2003. Genetic modifiers of tauopathy in *Drosophila*. *Genetics* 165:1233–42
81. Sperber BR, Leight S, Goedert M, Lee VM. 1995. Glycogen synthase kinase-3 beta phosphorylates tau protein at multiple sites in intact cells. *Neurosci. Lett.* 197:149–53
82. Steffan JS, Agrawal N, Pallos J, Rockabrand E, Trotman LC, et al. 2004. SUMO modification of Huntingtin and Huntington's disease pathology. *Science* 304:100–4
83. **Steffan JS, Bodai L, Pallos J, Poelman M, McCampbell A, et al. 2001. Histone deacetylase inhibitors arrest polyglutamine-dependent neurodegeneration in *Drosophila*. *Nature* 413:739–43**
84. Stokin GB, Lillo C, Falzone TL, Brusch RG, Rockenstein E, et al. 2005. Axonopathy and transport deficits early in the pathogenesis of Alzheimer's disease. *Science* 307:1282–88
85. Szebenyi G, Morfini GA, Babcock A, Gould M, Selkoe K, et al. 2003. Neuropathogenic forms of huntingtin and androgen receptor inhibit fast axonal transport. *Neuron* 40:41–52
86. Takeyama K, Ito S, Yamamoto A, Tanimoto H, Furutani T, et al. 2002. Androgen-dependent neurodegeneration by polyglutamine-expanded human androgen receptor in *Drosophila*. *Neuron* 35:855–64
87. Taylor JP, Taye AA, Campbell C, Kazemi-Esfarjani P, Fischbeck KH, Min KT. 2003. Aberrant histone acetylation, altered transcription, and retinal degeneration in a Drosophila model of polyglutamine disease are rescued by CREB-binding protein. *Genes Dev.* 17:1463–68
88. Thiruchelvam M, Brockel BJ, Richfield EK, Baggs RB, Cory-Slechta DA. 2000. Potentiated and preferential effects of combined paraquat and maneb on nigrostriatal dopamine systems: environmental risk factors for Parkinson's disease? *Brain Res.* 873:225–34
89. Torroja L, Chu H, Kotovsky I, White K. 1999. Neuronal overexpression of APPL, the *Drosophila* homologue of the amyloid precursor protein (APP), disrupts axonal transport. *Curr. Biol.* 9:489–92

This study shows that pathogenic Huntington disease protein compromises histone acetylation, suggesting transcriptional dysregulation occurs in disease, and uses a pharmacological approach in a fly model to reveal that histone deacetylase inhibitors like SAHA are potent suppressors of neurodegeneration.

90. Turmaine M, Raza A, Mahal A, Mangiarini L, Bates GP, Davies SW. 2000. Nonapoptotic neurodegeneration in a transgenic mouse model of Huntington's disease. *Proc. Natl. Acad. Sci. USA* 97:8093–97
91. Venken KJ, Bellen HJ. 2005. Emerging technologies for gene manipulation in *Drosophila melanogaster*. *Nat. Rev. Genet.* 6:167–78
92. Warrick JM, Chan HY, Gray-Board GL, Chai Y, Paulson HL, Bonini NM. 1999. Suppression of polyglutamine-mediated neurodegeneration in *Drosophila* by the molecular chaperone HSP70. *Nat. Genet.* 23:425–28
93. Warrick JM, Morabito LM, Bilen J, Gordesky-Gold B, Faust LZ, et al. 2005. Ataxin-3 suppresses polyglutamine neurodegeneration in *Drosophila* by a ubiquitin-associated mechanism. *Mol. Cell* 18:37–48
94. **Warrick JM, Paulson HL, Gray-Board GL, Bui QT, Fischbeck KH, et al. 1998. Expanded polyglutamine protein forms nuclear inclusions and causes neural degeneration in *Drosophila*. *Cell* 93:939–49**

The initial study modeling a human neurodegenerative disease in flies describes late-onset progressive neurodegeneration induced by a human mutant version of the ataxin-3 protein.

95. Webb JL, Ravikumar B, Atkins J, Skepper JN, Rubinsztein DC. 2003. Alpha-Synuclein is degraded by both autophagy and the proteasome. *J. Biol. Chem.* 278:25009–13
96. Wilson CA, Doms RW, Lee VM. 2003. Distinct presenilin-dependent and presenilin-independent gamma-secretases are responsible for total cellular Abeta production. *J. Neurosci. Res.* 74:361–69
97. **Wittmann CW, Wszolek MF, Shulman JM, Salvaterra PM, Lewis J, et al. 2001. Tauopathy in *Drosophila*: neurodegeneration without neurofibrillary tangles. *Science* 293:711–14**

A fly model for tauopathies, showing that neurodegeneration occurs in the absence of visible accumulations in tangles.

98. Wu YR, Wang CK, Chen CM, Hsu Y, Lin SJ, et al. 2004. Analysis of heat-shock protein 70 gene polymorphisms and the risk of Parkinson's disease. *Hum. Genet.* 114:236–41
99. Yagi Y, Tomita S, Nakamura M, Suzuki T. 2000. Overexpression of human amyloid precursor protein in *Drosophila*. *Mol. Cell Biol. Res. Comm.* 4:43–49
100. Yang Y, Nishimura I, Imai Y, Takahashi R, Lu B. 2003. Parkin suppresses dopaminergic neuron-selective neurotoxicity induced by Pael-R in *Drosophila*. *Neuron* 37:911–24
101. Ye Y, Fortini ME. 1999. Apoptotic activities of wild-type and Alzheimer's disease-related mutant presenilins in *Drosophila melanogaster*. *J. Cell Biol.* 146:1351–64
102. Zhang YQ, Bailey AM, Matthies HJ, Renden RB, Smith MA, et al. 2001. *Drosophila* fragile X-related gene regulates the MAP1B homolog Futsch to control synaptic structure and function. *Cell* 107:591–603
103. Zoghbi HY, Botas J. 2002. Mouse and fly models of neurodegeneration. *Trends Genet.* 18:463–71
104. Zoghbi HY, Orr HT. 2000. Glutamine repeats and neurodegeneration. *Annu. Rev. Neurosci.* 23:217–47

RELATED RESOURCES

1. Zoghbi HY, Orr HT. 2000. Glutamine repeats and neurodegeneration. *Annu. Rev. Neurosci.* 23:217–47
2. Lee VG, Goedert M, Trojanowski JQ. 2001. Neurodegenerative tauopathies. *Annu. Rev. Neurosci.* 24:1121–59

3. Selkoe DJ, Podlisny MB. 2002. Deciphering the genetic basis of Alzheimer's disease. *Annu. Rev. Genomics Hum. Genet.* 3:67–99
4. Caughey B, Lansbury PT. 2003. Protofibrils, pores, fibrils, and neurodegeneration: separating the responsible protein aggregates from the innocent bystanders. *Annu. Rev. Neurosci.* 26:667–98

Molecular Mechanisms of Germline Stem Cell Regulation

Marco D. Wong, Zhigang Jin, and Ting Xie

Stowers Institute for Medical Research, Kansas City, Missouri 64110;
email: tgx@stowers-institute.org

Annu. Rev. Genet.
2005. 39:173–95

First published online as a Review in Advance on August 30, 2005

The *Annual Review of Genetics* is online at http://genet.annualreviews.org

doi: 10.1146/annurev.genet.39.073003.105855

0066-4197/05/1215-0173$20.00

Key Words

germline stem cells, self-renewal, differentiation, niche, asymmetric cell division

Abstract

Germline stem cells (GSCs), which can self-renew and generate differentiated progeny, are unique stem cells in that they are solely dedicated to reproduction and transmit genetic information from generation to generation. Through the use of genetic techniques in *Drosophila*, *Caenorhabditis elegans*, and mouse, exciting progress has been made in understanding molecular mechanisms underlying interactions between stem cells and niches. The knowledge gained from studying GSCs has provided an intellectual framework for defining niches and molecular regulatory mechanisms for other adult stem cells. In this review, we summarize recent progress and discuss conserved mechanisms underlying GSC self-renewal and differentiation by comparing three GSC systems. Because GSCs and other adult stem cells share "stemness," we hope this review will help define fundamental principles of stem cell regulation and provide further guidance for future studies of other adult stem cells.

Contents

INTRODUCTION

With their remarkable ability to self-renew and undergo differentiation, stem cells are crucial to development and tissue homeostasis (68, 88). Interest in stem cell research has burgeoned since the successful culture of human embryonic stem cells (hESCs), which are able to generate various differentiated cell types (81, 92). In addition to ESCs, stem cells in a variety of adult tissues are also able to generate one or several differentiated cell types throughout an individual's lifetime. Germline stem cells (GSC) are dedicated to producing gametes for transmission of genetic information from generation to generation and, therefore, are true "immortal stem cells." The sterility resulting from GSC loss can be easily recognized, and it facilitates the identification of extrinsic signals and intrinsic factors in genetic model systems such as *Drosophila*, *C. elegans*, and mouse. Furthermore, GSCs are easily identified in the anatomically simple *Drosophila* ovary and testis and have enabled the first elucidation of relationships between stem cells and their microenvironment or "niche" (48, 95, 105). The investigation of the GSC niche and the regulatory mechanisms of stem cell self-renewal in *Drosophila* has provided guiding principles for study of adult stem cells in other systems, because relationships between stem cells and their niches are conserved. Cultured mouse GSCs and stem cell transplantation make it feasible to elucidate molecular regulatory mechanisms of mammalian GSCs (44, 53). This review summarizes our current understanding of GSC regulation, highlights conserved molecular mechanisms, and predicts future challenges.

REGULATORY MECHANISMS OF GERMLINE STEM CELL SELF-RENEWAL AND DIFFERENTIATION IN THE *DROSOPHILA* OVARY

General Features of the *Drosophila* Ovarian GSCs and their Niche

The identification of stem cells poses unique challenges particularly in mammalian systems because stem cells are rare and indistinguishable from early differentiated progeny (68, 88). The *Drosophila* ovarian GSC system circumvents this problem by virtue of its simple anatomy, unique molecular markers, and a linear arrangement between stem cells and their differentiated progeny (30, 60, 102). At the tip of the germarium, the anterior end of the ovarioles, 2 to 3 GSCs can be identified by their anteriorly located spectrosome (spherical fusome) and anchorage to cap cells through DE-cadherin-mediated cell adhesion (**Figure 1**; **Table 1**) (87). The fusome is a germ cell-specific organelle rich in membrane skeletal proteins such as Spectrins and Hu-li tai-shao (Hts) (21) (62). The cap cells form a niche that regulates the behavior of GSCs, perhaps with some contributions from terminal filament (TF) and inner germarial sheath (IGS) cells close to GSCs (105). These GSCs undergo asymmetric division: The daughter that remains in the niche retains GSC identity, while the other daughter cell moves away from the niche to differentiate into a cystoblast. The cystoblast also has a spectrosome and undergoes four rounds of synchronous division with incomplete cytokinesis to form a 16-cell cyst containing a branched fusome and ring canals connecting individual cystocytes (21).

The well-defined morphology and the linear fashion in which GSCs and their progeny progress throughout the germarium, along with available genetic tools in *Drosophila*, have facilitated the investigation of GSC maintenance and differentiation. For example, by using a heat-inducible FLP (flippase, a DNA recombinase)-FRT (FLP recognition target)-mediated recombination technique to produce a marked mutant GSC clone that can be compared with the control GSC clone side-by-side in the same germarium, we can determine the role of a particular gene in stem cell self-renewal, differentiation, and division (104). With a well-characterized niche and readily identifiable GSCs, the *Drosophila* ovary represents an excellent model system to investigate stem cell biology *in vivo* at the molecular and cellular level (30, 60, 102, 103).

BMP, Piwi, and Yb Function in the Niche to Regulate the Maintenance and Division of GSCs

The GSC asymmetric cell division can be partially attributed to the extrinsic factors emanating from the niche. Regulatory molecules,

Stem cell: a unique undifferentiated cell that has the ability to self-renew and generate differentiated cell types

GSC: germline stem cell

Intrinsic factors: factors that act within the cell to control its behavior

Niche: the molecular milieu or microenvironment formed by support cells that express regulatory molecules that promote stem cell self-renewal and block differentiation

Self-renewal: the ability of a stem cell to regenerate itself

IGS: inner germarial sheath

Asymmetric cell division: cell division generating two daughter cells that have different cell fates

Extrinsic factors: factors that function outside the cell for controlling its behavior

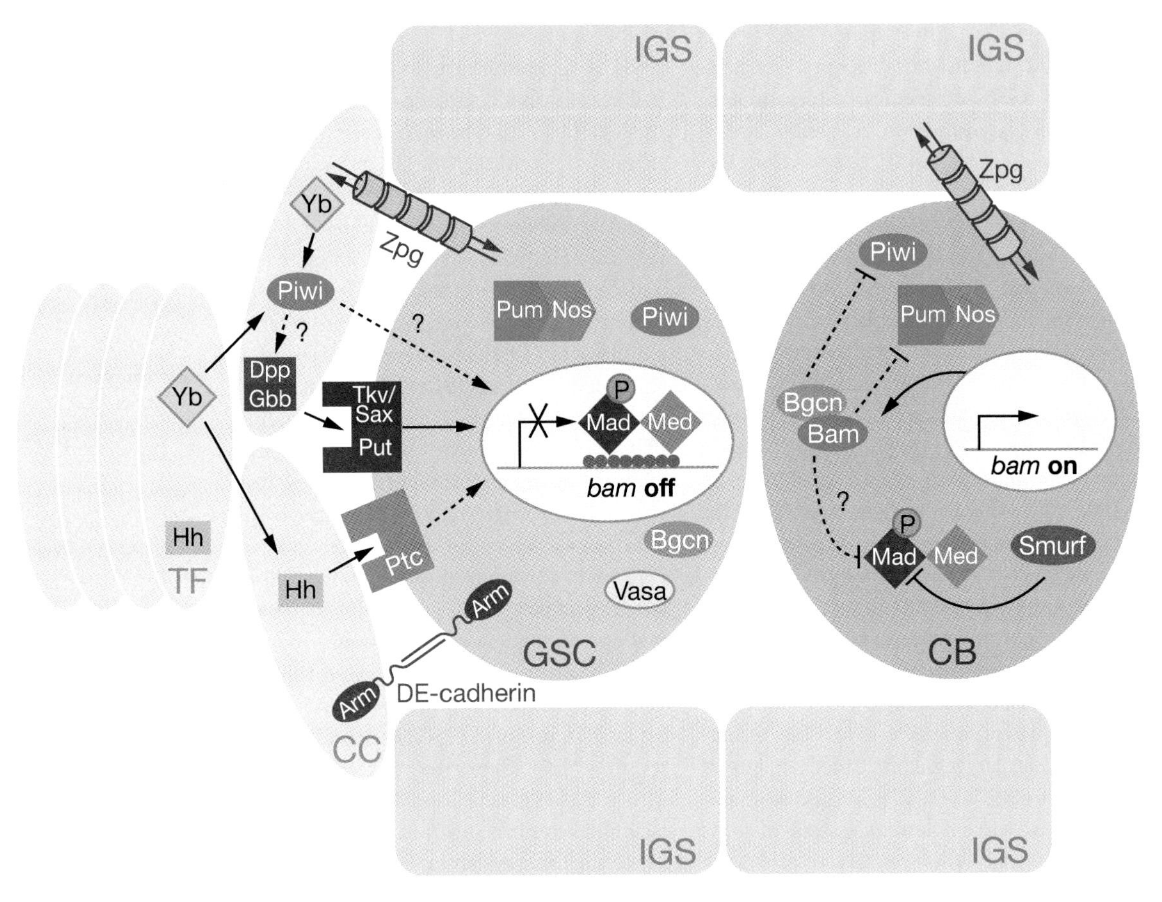

Figure 1

Major signaling pathways and intrinsic factors for GSC self-renewal and differentiation in the *Drosophila* ovary. The BMPs (Dpp and Gbb), Yb, Piwi, and Hh are expressed in the TF/cap cells and control GSC self-renewal. In the GSC, Pum/Nos and Vasa are required for controlling self-renewal, and Piwi regulates GSC division. GSCs are anchored to their niche by DE-cadherin-mediated cell adhesions. Gap junctions formed by Zpg are important for GSC maintenance and CB differentiation. Bam is required for cystoblast differentiation but is repressed in GSCs by Mad/Med complexes. However, this repression of Bam is overcome in CB, partially through Smurf.

BMP: bone morphogenetic protein

bone morphogenetic proteins (BMPs) and Piwi, are produced from cap cells to modulate ovarian GSC maintenance and division via intracellular signaling pathways (**Figure 1**; **Table 1**). Two BMPs, Dpp and Gbb, are expressed primarily in cap cells (86, 104) and serve as short-range signals to activate BMP signaling in GSCs through type I (Tkv and Sax) and type II (Punt) BMP receptors to nuclear complexes, Mad [a founding member of SMA and MAD (SMAD) family proteins; a BMP-specific SMAD], and Medea (SMAD4, also known as Co-SMAD) to control their self-renewal and division (42, 104). Mutations in *dpp*, *gbb*, or downstream components lead to GSC loss by premature differentiation and slower division rates (86, 104, 105). Overexpression of *dpp*, but not *gbb*, completely blocks cystoblast differentiation, resulting in proliferation of GSC-like tumors throughout the germarium (86, 104). These studies indicate that BMP signaling is necessary and

Table 1 Extrinsic signals and intrinsic factors that are required for regulating GSC function

Species	Pathways and genes	Functions: Ovary	Functions: Testis
D. melanogaster			
	BMP: *dpp*, *gbb* (niche) *tkv*, *put*, *mad*, *Med* (GSC)	GSC self-renewal/division (43, 86, 104)	GSC self-renewal (45, 85)
	smurf (CB)	CB differentiation (10, 11, 13)	Unknown
	JAK/STAT: *upd* (niche) *hop*, *stat92*^{E} (GSC)	Unknown	GSC self-renewal (5, 48, 95)
	Piwi/Yb: *piwi* (niche)	GSC self-renewal (16, 17)	GSC self-renewal (16)
	Yb (niche)	GSC self-renewal (51, 52)	Unknown
	EGF: *stet* (germline cells)	CB differentiation (80)	Gonialblast differentiation (80)
	Egfr, *raf* (somatic cyst cells)	Unknown	Gonialblast/spermatogonia differentiation (49, 93)
	Translational regulation:		
	pum/nos (GSC)	GSC self-renewal (2, 12, 25, 31, 61, 98)	Unknown
	vasa (germline cells)	GSC maintenance/survival (89)	Unknown
	bam/bgcn (CB)	CB differentiation (28, 59, 64, 65, 69)	Gonialblast differentiation (45, 79) Spermatogonial proliferation (33, 65)
	sxl, *orb* (CB)	CB differentiation (4, 14, 58)	Unknown
	Cell adhesion: *E-cadherin/arm* (niche and GSC)	Niche anchorage of GSC (87)	Potential niche anchorage of GSC GSC spindle orientation (107)
	Gap junction: *zpg* (germline cells)	GSC maintenance/survival and differentiation (29, 91)	GSC survival and differentiation (29, 91)
	Cell cycle regulators: *APC1*, *APC2*, *cnn* (GSC)	Unknown	GSC spindle orientation (107)
	cycB (PGC and GSC)	PGC/GSC proliferation and GSC maintenance (99)	GSC maintenance (99)
C. elegans	**Notch**: *lag-2* (niche) *glp-1*, *lag-1*, lag-3 (GSC)	GSC and germline cell proliferation (1, 3, 20, 23, 37, 56, 75)	
	Translational regulation: *fbf-1/fbf-2* (GSC)	GSC self-renewal and proliferation (19, 57, 100, 109)	
	gld-1, *gld-2*, *gld-3*, *nos-3* (differentiated germline cells)	Promoting meiosis (24, 35, 38, 40, 41, 55, 63, 97)	
M. musculus	**GDNF**: *GDNF* (Sertoli cells) *Ret*, *GFRα1* (GSC)		GSC self-renewal (44, 53, 66)
	BMP: BMP4, BMP7, BMP8a, BMP8b (germline cells)		Germline cell viability (76, 110)
	SCF/c-Kit: *SCF* (Sertoli cells) *c-Kit* (differentiated spermatogonia)	Unknown	Spermatogonia differentiation (22, 71, 108)
	Transcriptional regulation: Plzf (spermatogonia)		GSC maintenance (9, 15)
	Translational regulation: nos-2 (germline cells)		GSC maintenance (94)

GSC, germline stem cell; PGC, primordial germ cell; CB, cystoblast.

sufficient for GSC self-renewal. Since Gbb and Dpp likely use common signal transducers, it remains to be determined why *dpp*, but not *gbb*, is sufficient to block germline cell differentiation when overexpressed (34, 46).

Although GSCs normally undergo asymmetric cell division, they are also capable of dividing symmetrically to generate two GSCs when both of the daughters remain in the niche (105). Surprisingly, partially differentiated cells such as 2-, 4-, and 8-cell cysts revert back to GSCs when the niche signal Dpp is provided (43). It remains to be determined how Dpp signaling can completely turn off the active differentiation program in cystocytes. Understanding this phenomenon would provide more insight into how GSC self-renewal and differentiation is regulated, and it may allow the regeneration of stem cells from differentiated cells in future regenerative medicine, if such cell fate reversal also exists in mammalian stem cells.

Like BMPs, Piwi and Yb are expressed in niche cells and are also involved in GSC maintenance (**Figure 1**; **Table 1**). Loss-of-function mutations in *piwi* and *Yb* cause rapid loss of GSCs, while *Yb* and *piwi* overexpression increases GSC-like or cystoblast-like germ cells (16, 17, 51, 52, 61). Yb is a novel intracellular protein that regulates *piwi* and *hh* (*hedgehog*) expression (51, 52), whereas Piwi is the founding member of the *piwi* family genes containing conserved PAZ and Piwi domains that bind to RNAs (16). Hh signaling appears to play a redundant role with Piwi to control GSC maintenance (52). Although the requirement of Piwi in the niche for GSC maintenance is well established, how this is accomplished remains unclear.

As a cystoblast moves away from the cap cells, it becomes surrounded by cellular processes of IGS cells, raising a possibility that IGS cells regulate cystoblast differentiation. Indeed, a study on *stet* function has revealed the link between IGS cells and germ cell differentiation (80). In *stet* mutant ovaries, the cellular processes of IGS cells are severely reduced, and spectrosome-containing single germ cells accumulate, suggesting that *stet* is required for cystoblast differentiation. *stet*, expressed in germ cells, encodes a membrane protease similar to Rhomboid that can cleave and release EGF ligands (96). Supporting a role of EGF signaling in IGS cells, the activated MAP kinase accumulates in wild-type IGS cells but is reduced in *stet* mutant IGS cells. However, it remains to be determined which EGF ligand is activated by Stet in germ cells and how IGS cells control germ cell differentiation.

Intrinsic Factors Play Essential Roles in GSC Maintenance and Differentiation

Two classes of intrinsic factors govern GSC self-renewal or differentiation: self-renewing factors and differentiation-promoting factors (**Figure 1**; **Table 1**). Pumilio (Pum) and Nanos (Nos) are defined as intrinsic self-renewing factors because mutations in these genes cause premature GSC loss (25, 31, 61, 98). They are RNA binding proteins that form protein complexes that repress translation of mRNAs in *Drosophila* embryos (2). *vasa* (*vas*), encoding a *Drosophila* homologue of eukaryotic initiation factor 4A, is also likely required for ovarian GSC self-renewal, because *vas* mutant germaria contain few degenerate or growth-arrested germ cells (89). These studies indicate that translational regulation plays a critical role in GSC self-renewal.

Furthermore, E-cadherin-mediated cell adhesion and Cyclin B also participate in controlling GSC self-renewal (**Figure 1**; **Table 1**). E-cadherin and its interacting partner Armadillo (Arm, β-catenin), expressed in GSCs and cap cells, form adherens junctions, which anchor GSCs to cap cells during niche formation and help recruit GSCs to the niche (87). The main function of adherens junctions is to keep GSCs in close proximity to niche cells to receive maximal BMP signaling for self-renewal. Cadherin-mediated cell adhesion represents a conserved mechanism for anchoring stem cells in the niche in a variety of systems (26). Cyclin B is specifically required in germ cells for promoting division of PGCs and GSCs and possibly for controlling GSC maintenance (99). Presumably, Cyclin B promotes PGC and GSC division through regulating cell-cycle progression by activating CDK1. However, it remains unclear whether it controls GSC maintenance indirectly by regulating cell-cycle progression or directly by interacting with the GSC maintenance machinery. Taken together, different classes of intrinsic factors play distinct roles in controlling GSC self-renewal and proliferation.

Several differentiation-promoting factors are involved in cystoblast differentiation in the ovary; these include Bam (Bag of marbles), Bgcn (Benign gonial cell neoplasm), Orb, and Sxl (Sex lethal) (4, 14, 58, 59, 64, 69). Among these, Bam, a novel protein, and Bgcn, related to DExH-box RNA binding proteins, are essential differentiation factors: Mutations in *bam* and *bgcn* germline cells abolish cystoblast differentiation, leading to a cystoblast-like germ cell tumor phenotype (28, 65, 69); overexpresssion of *bam* triggers GSC differentiation and consequently germ cell depletion in the ovary (70). Genetic interactions between *bam* and *bgcn* suggest that their gene products may form protein complexes that regulate the mRNA stability or translation. Mutations in *orb* and *Sxl*, encoding proteins involved in the regulation of mRNA polyadenylation and translation, respectively, cause an accumulation of GSC-like or cystoblast-like cells mixed with early differentiated cysts, implying that these genes play unessential roles in cystoblast differentiation (4, 14, 58). In addition, a gap junction connexin, Zpg (Zero population growth), is present in cytoplasmic membranes of GSCs and their differentiated progeny in the ovary and testis and is required to maintain GSCs and to promote germ cell differentiation (29, 91). Loss of *zpg* function results in partial GSC loss due to cell death and accumulation of a few undifferentiated single cells. Because Zpg is a gap junction component, it likely helps transport small molecules from supporting somatic cells to control the survival of GSCs and the differentiation of their progeny. *zpg* may function in parallel with *bam* to regulate germ cell differentiation (29). Pum and Nos control GSC self-renewal through repressing a Bam-independent differentiation pathway (12, 31, 90). *bam* transcription is not upregulated in *pum* and *nos* mutant GSCs, although *bam* mutant germ cells that are also mutant for *pum* can differentiate. In summary, different classes of differentiation factors work synergistically to drive cystoblast differentiation by negatively regulating functions of self-renewing factors.

Extrinsic Signals Regulate the Function of Intrinsic Factors to Control GSC Self-Renewal and Differentiation

Recent studies on *Drosophila* ovarian GSCs have revealed that extrinsic signals impinge on intrinsic factors to control their functions. In GSCs, *bam* is actively repressed through a transcriptional silencer in the *bam* promoter (13). BMP signals, Dpp and Gbb, from niche cells activate BMP signaling, leading to formation of SMAD complexes, which directly bind to the *bam* silencer to repress *bam* expression in GSCs (11, 86). Since BMPs function as short-range signals, cystoblasts with insufficient BMP signaling fail to repress *bam* expression and consequently upregulate *bam* to promote differentiation. Similarly, via an unidentified mechanism, the niche Piwi is required to repress *bam* in GSCs for their self-renewal (12, 90). As both the BMP- and Piwi-mediated signaling pathways are essential for controlling GSC self-renewal, Piwi may converge with BMP signaling by regulating the stability, production, processing, and/or activation of BMP molecules in the niche or BMP signal transduction in GSCs to repress *bam* expression (12, 90). The observation that a mutation in *smurf*, encoding a ubiquitin E3 ligase that regulates degradation of phosphorylated Mad, can rescue a *piwi* GSC loss phenotype indicates that Piwi-mediated signaling and BMP signaling must intersect at the level of, or above, *smurf* (12). Therefore, extrinsic and intrinsic factors work in a coordinated manner to promote GSC fate over cystoblast fate, when instructive signals are received from the niche.

In cystoblasts, Bam initiates a differentiation program and promotes cyst formation. It is proposed that Bam interacts with Bgcn and represses Pum/Nos complexes to promote differentiation (**Figure 1**) (12, 90). Although Dpp and Gbb only activate BMP signaling in GSCs due to their short-range action, there is also a backup system for actively repressing BMP signaling in the cystoblast. Namely, upregulated Bam in the cystoblast can serve as a negative regulator of BMP signaling, and Smurf, which functions in CBs and descendents, negatively regulates BMP signaling activities by targeting phosphorylated Mad for degradation (10). Therefore, precise control of BMP signaling in GSCs and cystoblasts is crucial for these cells to achieve a critical balance between self-renewal and differentiation.

REGULATORY MECHANISMS OF GERMLINE STEM CELL SELF-RENEWAL AND DIFFERENTIATION IN THE *DROSOPHILA* TESTIS

General Features of the *Drosophila* Testicular GSCs and their Niche

As in the *Drosophila* ovary, GSCs and their niche are also well defined in the *Drosophila* testis by virtue of its simple anatomy and availability of molecular markers (48, 95, 106, 107). Seven to nine spectrosome-containing GSCs are located at the apical tip of the testis and directly contact the hub cells that function as a GSC niche (106). Similar to ovarian GSCs, they divide asymmetrically and give rise to one GSC that remains in the niche and one differentiated gonialblast that moves away from the niche. The gonialblast, a counterpart of the cystoblast, also undergoes four rounds of synchronous division with incomplete cytokinesis to form a 16-cell cluster with a branched fusome (27, 47). Each gonialblast or its descendent is surrounded by two somatic cyst cells (SCCs) that control its continuous differentiation (32, 36). These cysts cells are functionally similar to IGS cells in the ovary and are important for germ cell differentiation (49, 80, 93).

Niche Signals, BMP and Unpaired, Control Testicular GSC Self-Renewal

In the *Drosophila* testis, Janus Kinase-Signal Transduction and Activator of Transcription (JAK-STAT) and BMP signaling are indispensable for GSC self-renewal (45, 48, 85, 95) (**Figure 2; Table 1**). Upd produced by the hub acts as a short-range signal to activate JAK (Hopscotch, Hop) and STAT (STAT92E) downstream components in GSCs. Removal of either *hop* or *stat92E* causes GSC differentiation by disrupting self-renewal, whereas ectopic expression of Upd increases the number of GSC-like or gonialblast-like cells in the testis, indicating that JAK-STAT signaling is both sufficient and necessary for GSC renewal. Therefore, JAK-STAT signaling plays an instructive role in GSC self-renewal in the testis similar to that of the Dpp/BMP signaling pathway in the ovary. Remarkably, differentiated germ cells revert back into GSCs when *stat* function is restored in a temperature-sensitive *stat* mutant (5). As in the ovary (43), this study suggests that niche signals maintain GSCs not only by preventing differentiation but also by reprogramming differentiated cells back into stem cells when all GSCs are lost from the niche. It is important to identify downstream target genes of JAK-STAT signaling in GSCs to understand how it controls GSC self-renewal.

gbb appears to have a stronger effect on GSC maintenance than *dpp* in the testis,

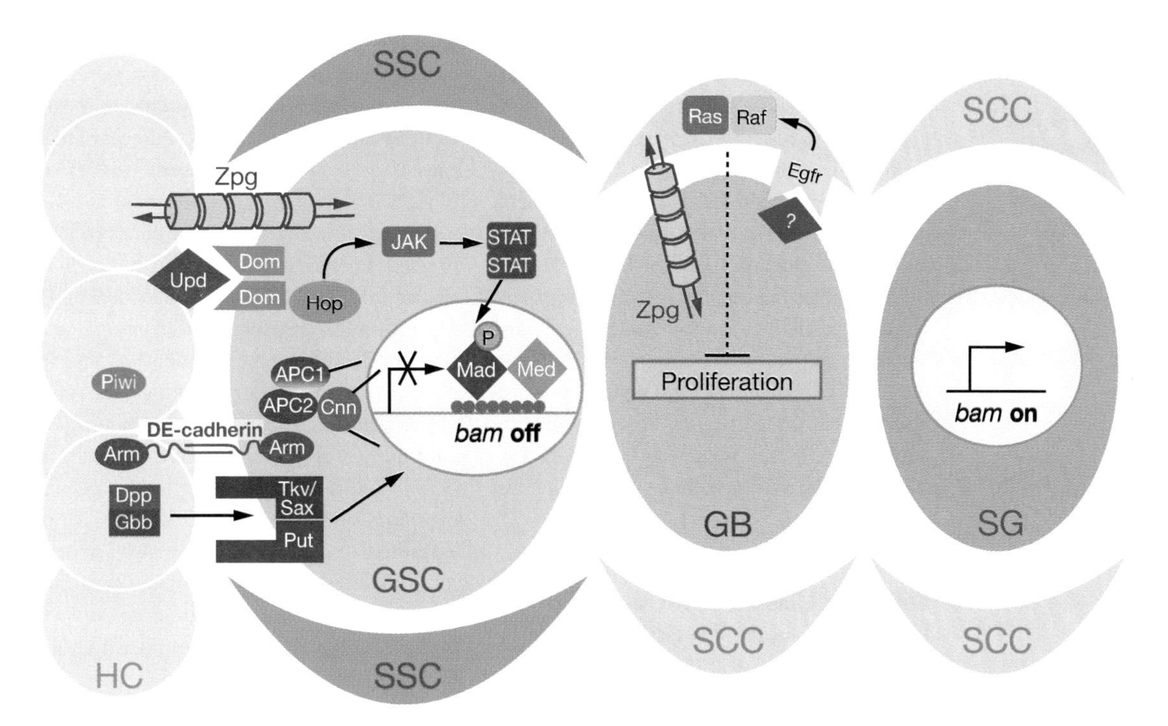

Figure 2

Major signaling pathways and intrinsic factors for testicular GSC self-renewal and differentiation in the *Drosophila* testis. The GSCs are likely anchored to the niche via DE-cadherin-mediated adhesions and are enveloped by SSCs. Gbb/Dpp and Upd, expressed in hub cells (HC), activate BMP and JAK-STAT signaling for GSC self-renewal, respectively. Bam is repressed in GSCs and gonialblasts by Gbb/Dpp signaling and is expressed in spermatogonial cells. In the GSC, APC1, APC2, and Cnn are required for correct spindle orientation. Gap junctions formed by Zpg are important for GSC maintenance. EGFR signaling in somatic cyst cells (SCC) is activated by an unknown ligand and is important for gonialblast differentiation and spermatogonial cell (SG) proliferation.

which is consistent with their relative expression levels in the hub and SCCs (45, 85). Removal of BMP downstream components (*punt*, *tkv*, *mad*, and *Med*) causes severe GSC loss similar to the disruption of JAK-STAT signaling in the testis, indicating that both BMP and JAK-STAT signaling play essential roles in controlling male GSCs either by regulating each other or working in parallel. On the other hand, *dpp* overexpression only partially suppresses differentiation of gonialblasts, whereas *gbb* overexpression has no obvious effect. Therefore, BMP signaling plays a permissive role for GSC self-renewal in males.

Although Piwi-mediated signaling is also required for GSC maintenance in the testis, it remains unclear which cells require its function for GSC self-renewal and whether it interacts genetically with JAK-STAT and BMP pathways. *piwi* is expressed in early germ cells, hub cells, and somatic cysts of the adult testis, and mutations in *piwi* lead to premature loss of GSCs (61). As discussed earlier, Piwi-mediated signaling may interact with the BMP pathway potentially through Smurf and thereby work cooperatively to sustain testicular GSCs (12).

The IGS equivalent cells, SCCs, are also involved in regulating gonialblast differentiation (49, 80, 93). An unidentified SCC signal received by the gonialblast requires the EGFR and Raf-mediated MAP kinase signaling cascade in SCCs (49, 93) (**Figure 2**; **Table 1**). Genetic mosaic

SCC: somatic cyst cell

JAK-STAT: Janus Kinase-Signal Transduction and Activator of Transcription

Symmetric cell division: cell division generating two identical daughter cells

analysis reveals the requirement of *Egfr* and *raf* functions in SCCs for regulating gonialblast differentiation/proliferation. The GSC population in the *Egfr* or *raf* mutant testis remains active longer than wild type. Finally, *stet* also functions in the germ cells to control gonialblast differentiation since *stet* mutant testes have more gonialblast-like cells than wild type (80). Because EGFR signaling is involved in regulating functions of IGS cells and SCCs, it will be interesting to see whether the same EGF ligand in the germ cells is responsible for activating EGFR signaling in both IGS cells and SCCs and what IGS/SSC signal(s) controls cystoblast or gonialblast differentiation.

Different Classes of Intracellular Factors Regulate Testicular GSCs

As for the ovarian GSCs, E-cadherin and β-catenin also accumulate between GSCs and hub cells (107), and are likely involved in anchoring GSCs (**Figure 2**). The accumulated E-cadherin at the GSC and hub interface may serve as a platform for binding APC2, a *Drosophila* Adenomatous Polyposis Coli (APC) homolog, to the GSC cortex to orient the mitotic spindles perpendicular to the niche (107). APC2 and two integral centrosome components, centrosomin (cnn) and APC1, are required for the orientation of the stem cell spindle. Loss-of-function mutations in *apc1*, *apc2*, and *cnn* result in mispositioned centrosomes and misoriented spindles, giving rise to symmetric cell division and the accumulation of more GSCs around the hub (107). This study demonstrates that intrinsic control of spindle orientation is crucial for maintaining a stable number of stem cells in the niche in addition to extrinsic niche signals. It remains to be seen whether this intrinsic mechanism of orienting the stem cell spindle is a universal mechanism for ensuring that only one of the two stem cell daughters maintains stem cell identity in other stem cell systems.

Although *bam* and *bgcn* are essential for cystoblast differentiation, they are dispensable for gonialblast differentiation because *bam* or *bgcn* mutant gonialblasts form germ cell cysts (33). However, *bam* alone is sufficient to cause male GSCs to differentiate because *bam* overexpression leads to GSC depletion (45, 79). In order to maintain GSC self-renewal in the testis, *bam* needs to be repressed. Indeed, it is actively repressed in testicular GSCs by BMP signaling initiated by Dpp and Gbb similar to the *Drosophila* ovary (**Figure 2**; **Table 1**). However, *bam* and *bgcn* play essential roles in restricting four rounds of cyst division as *bam* and *bgcn* mutant cysts continue to divide after the fourth division (33). In addition, communication between gonialblasts and their surrounding SCCs via Zpg-mediated gap junctions is also important for gonialblast differentiation and early germ cell survival (91). Although Pum and Nos in the *Drosophila* ovary, Pumilio homologs in *C. elegans*, and Nos in mice have been demonstrated to be required for controlling GSC self-renewal (19, 31, 94, 98), its role in GSC self-renewal in the *Drosophila* testis awaits determination. Furthermore, little is known about how niche signals control GSC self-renewal except for the involvement of BMP signaling in repressing *bam* expression in GSCs. The genetic and molecular relationships between extrinsic signals and intrinsic factors in the testis will require thoroughgoing investigation in the near future.

REGULATORY MECHANISMS OF GERM CELL FATE SPECIFICATION IN *C. ELEGANS*

General Features of the *C. elegans* GSCs and their Niche

The gonad of the *C. elegans* hermaphrodite consists of two symmetrical, U-shaped tubes that are connected to a common uterus. GSCs are located in the mitotic region (MR) that directly contacts the distal tip cell (DTC), although specific individual GSCs have not yet been precisely defined. GSCs and their early progeny continue to proliferate in the MR, and the germ cells moving away from the DTC terminate their mitotic activities, enter meiosis, and eventually develop into mature eggs or sperm (20). The fact that only the germ cells that contact the DTC maintain GSC identity suggests that the DTC functions as a GSC niche. Consistent with this idea, the somatic DTC is shown to be required for maintaining germline mitosis by laser ablation experiments (50). Strikingly, the *C. elegans* gonad displays an arrangement of stem cells and differentiated cell types resembling those in the *Drosophila* ovary and testis. Powerful genetics and the simple gonadal anatomy have made *C. elegans* GSCs a productive system to study the regulation of stem cell self-renewal, proliferation, and differentiation.

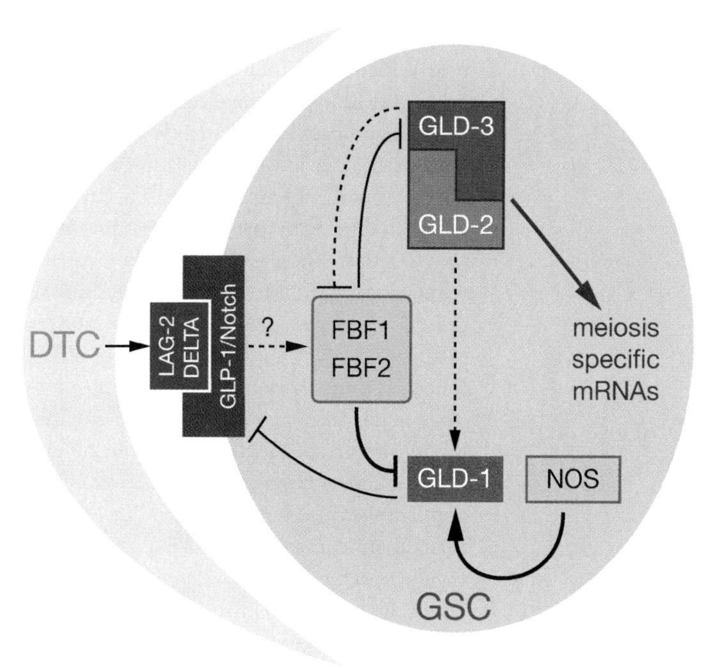

Figure 3

Signaling pathways regulating the mitosis and meiosis switch in *C. elegans* germline. The LAG-2/Delta ligand expressed from the DTC binds to the GLP-1/Notch receptor on the germ cell to activate GLP-1/Notch signaling, which may act on FBF-1 and -2 to ensure mitosis through inhibiting two regulatory branches that promote meiosis; GLD-3/GLD-2 and GLD-1/NOS. GLD-3/GLD-2 may activate meiosis-promoting mRNAs like *gld*-1 mRNA and GLD-1/NOS promotes meiosis by repressing mRNAs critical for mitosis. GLD-1 facilitates meiosis by blocking translation of *glp*-1 mRNA.

A Notch-Like Signal from the DTC is Both Necessary and Sufficient for Controlling GSC Self-Renewal and Proliferation

Genetic studies have identified a Notch signaling cascade essential for maintaining GSCs in *C. elegans* (**Figure 3**; **Table 1**). The DTC expresses a Delta-like Notch ligand, LAG-2, whereas the mitotic germ cells express a Notch-type receptor, GLP-1 (20, 37). LAG-2 binding triggers proteolytic cleavage of GLP-1 to generate a truncated intracellular domain for transport to the nucleus, where it forms protein complexes with other transcription factors, LAG-1 and LAG-3, to control target gene expression. LAG-1 is a CSL [CBF-1, Su(H), Lag-1]-type transcriptional regulator, and LAG-3 is a glutamine-rich protein that possibly tethers LAG-1 and the cleaved GLP-1 intracellular domain (23, 75). Loss of GLP-1 and LAG-2 function causes GSC loss and consequently premature entry into meiosis (1, 37, 56). In contrast, constitutive GLP-1 activity prevents entry into meiosis and causes germ cell overproliferation (3). Together with the DTC ablation experiments, these studies show that the DTC functions as a GSC niche and GLP-1/Notch signaling activated by LAG-2 from the niche directly controls GSC self-renewal and proliferation. The direct targets of the GLP-1/Notch signaling pathway remain to be determined. Furthermore, it is unclear whether the *C. elegans* niche maintains GSCs through a population mechanism or a stereotypic asymmetric division mechanism as in *Drosophila*, although current data favor the former.

DTC: distal tip cell

Intrinsic Factors Regulating RNA Stability and Translation Play Essential Roles in GSC Maintenance and Differentiation

As in *Drosophila*, two classes of intrinsic factors control GSC maintenance or differentiation (**Figure 3**; **Table 1**): self-renewing factors and differentiation-promoting factors. Two nearly identical Pumilio-like RNA binding proteins, FBF-1 and FBF-2, control GSC self-renewal and proliferation. In *fbf-1 fbf-2* double mutants, germline proliferation is initially normal, but GSCs are prematurely depleted owing to differentiation and entry into meiosis (19, 109). Pumilio-like proteins usually regulate protein translation by binding to the 3′ untranslated region (UTR) of a target mRNA to inhibit its translation (101). FBF1 functions redundantly with FBF2 to promote mitosis but has an opposite effect on fine-tuning the size of the mitotic region, reflecting a regulatory circuit for maintaining a GSC population (57). The size of the mitotic region is precisely controlled as FBF-1 and FBF-2 regulate each other's expression, and this reciprocal repression is probably direct through FBF binding sites located in *fbf*-1 and *fbf*-2 3′ UTRs.

Several differentiation genes have been identified for controlling the entry into meiosis: *gld-1*, *gld-2*, *gld-3*, and *nos-3*. *gld-1* functions redundantly with *gld-2* to promote meiotic entry and/or inhibit germ cell proliferation (41) (**Figure 3**; **Table 1**). Unlike other *nos* genes in *Drosophila* and mice, *nos-3*, one of three *C. elegans nos* genes, promotes differentiation by enhancing GLD-1 accumulation (35). *gld-2 gld-1* and *gld-2 nos-3* double mutants have a tumorous germline phenotype due to a defect in meiotic entry, whereas *gld-1*, *gld-2*, and *nos-3* single mutants exhibit normal meiotic entry (35, 41). The GLD-1 and GLD-2 pathways are both thought to regulate expression of target genes at a posttranscriptional level on the basis of their molecular identities. GLD-1 is a STAR/KH domain translational repressor (38, 40, 78), while GLD-2 is a catalytic subunit of a poly(A) RNA polymerase (55, 97). GLD3, a Bicaudal-C homolog, genetically interacts with GLD-2 to promote meiosis by activating mRNAs critical for meiosis (24). Moreover, GLD-3 and GLD-2 form a heterodimeric enzyme that polyadenylates and may activate meiosis-promoting mRNAs such as *gld*-1 mRNA. As expected, two classes of intrinsic factors regulate each other. For example, FBF2 regulates the size of the mitotic region by means of repressing the translation of meiosis-promoting *gld*-1 and *gld-3* mRNAs (57). Similar to *Drosophila*, intrinsic factors that regulate mRNA stability and/or translation play key roles in controlling GSC maintenance and germ cell differentiation in *C. elegans*.

Interplay Between Notch-Like Signaling and Intrinsic Factors Is Critical for Controlling GSC Self-Renewal and Differentiation

In *Drosophila*, niche signals repress expression of differentiation-promoting genes and thereby maintain stem cell self-renewal. In *C. elegans*, GLP-1/Notch signaling also inhibits the activities of meiosis-promoting genes, *gld-1*, *gld-2*, and *nos-3*, although the mechanism currently is unclear (35, 41). Notch signaling may activate FBF translational repressors that repress the functions of differentiation-promoting genes (**Figure 3**) (24, 100, 109). In order for germ cells to differentiate, GLP-1/Notch signaling has to be shut off by differentiation genes. Indeed, GLD-1 facilitates meiosis through down-regulating GLP-1/Notch signaling in the distal region by binding to the 3′ UTR of *glp*-1 mRNA to block translation (63). Therefore, GLP-1/Notch signaling and intrinsic factors must precisely balance positive and negative regulatory actions to determine whether germ cells remain in the mitotic cycle or enter the meiotic cycle.

REGULATORY MECHANISMS OF GERMLINE STEM CELL SELF-RENEWAL AND DIFFERENTIATION IN MAMMALS

General Features of the Mouse GSCs and their Putative Niche

Stem cell transplantation, simple anatomy, and genetics make the mouse testis a powerful model to study complex regulatory networks of mammalian GSCs and their niche. The spermatogonial stem cells (referred to as GSCs here for consistency) are a subset of single cells (A_{single} or A_s), which are located along the basement membrane of seminiferous tubules in the mouse testis (7). GSCs self-renew or produce a differentiated A_s daughter that divides to form a pair of interconnected spermatogonial cells called A_{pair}. The A_{pair} spermatogonial cells can divide synchronously to form a chain of interconnected spermatogonial cells that subsequently become differentiating spermatogonia, spermatocytes, spermatids, and sperm cells, which is reminiscent of cyst formation in the *Drosophila* ovary and testis. GSCs and early differentiated spermatogonia are morphologically alike and thus indistinguishable. GSCs are very rare (1 in 5000) cells in the adult mouse testis based on transplantation studies (7). Fluorescence-activated cell sorting (FACS) in conjunction with the transplantation assay for GSCs has identified the antigenic profile of GSCs as α_V-integrin$^{-/dim}$ α_6-integrin$^+$ Thy-1$^{lo/+}$ C-Kit$^-$ (54).

Sertoli cells, the somatic cells in the seminiferous tubules that physically interact with GSCs, likely constitute a functional GSC niche by providing growth factors that promote GSC self-renewal and/or proliferation. Several studies support the idea that these cells regulate the maintenance of the stem cell pool. First, transplantation of GSCs into infertile male mice has shown that Sertoli cells can indeed support GSC maintenance and spermatogenesis (8). Second, Sertoli cells produce GDNF (glial cell line-derived neutrophic factor) that controls GSC maintenance in a dosage-dependent manner (66). Furthermore, transplantation of Sertoli cells into the mouse testes that are defective in Sertoli cell function demonstrates that Sertoli cells are indeed essential for maintaining spermatogenesis (84), but it is still not clear whether and how Sertoli cells alone can constitute the GSC niche. A series of transplantation experiments in mice and rats suggest that the GSC niche in the mammalian testis is very dynamic during development (72, 73, 77, 82, 83). The most urgent issue in studying GSCs in the mouse testis is to define the physical structure of the niche and its associated signals.

Although most mammalian females were believed to lose the capacity for germ cell self-renewal during fetal life, a recent study argues that juvenile and adult mouse ovaries possess mitotically active germ cells that continuously replenish the follicle pool (39). Despite compelling experimental evidence, it remains to be seen whether these observations can be duplicated in other mammals. If those GSCs exist in the peripheral epithelial layer of the ovary, a new avenue will be opened to study GSCs in mammals and further investigate how GSCs are regulated differently in both sexes.

GDNF: glial cell line-derived neurotrophic factor

Signals from Sertoli Cells Control the Maintenance and Differentiation of GSCs

One extrinsic factor involved in GSC self-renewal and proliferation is GDNF, which is released from Sertoli cells (66) (**Figure 4**; **Table 1**). GDNF binds two heterologous receptors, Ret and GFR-α1 which are expressed in spermatogonial cells. *gdnf* $^{+/-}$ mice lose their GSCs prematurely in testes, indicating that GDNF is essential for GSC self-renewal. Overexpression of GDNF causes accumulation of GSC-like cells. Two recent studies show that male mouse GSCs can be cultured and expanded *in vitro* in the presence of GDNF for more than 6 months and can reconstitute long-term spermatogenesis and restore fertility when transplanted to sterile recipient mice (44, 53). These findings support the hypothesis that Sertoli cells must contribute to the function of the GSC niche. In addition to GDNF signaling, BMP signaling also has a role in GSC maintenance (76, 110). Multiple BMPs, BMP4, BMP7, and BMP8 are expressed in male germ cells, while BMP4 is also expressed in Sertoli cells (76, 110), which is in

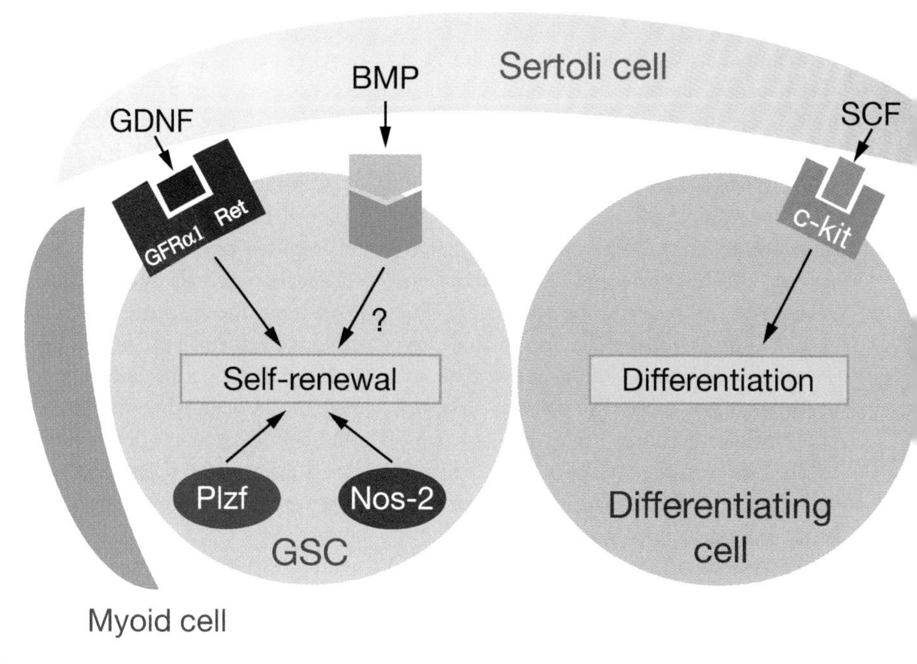

Figure 4

Extrinsic and intrinsic factors for GSC self-renewal/differentiation in the mouse testis. The putative niche cell, Sertoli cells, which exhibit functional polarity, express GDNF, BMPs, and SCF to promote self-renewal and differentiation by binding to their receptors GDFα1/Ret, BMPs and C-Kit, respectively. The myoid cell may assist the niche function. Intrinsic factors for self-renewal include Plzf and Nos-2.

contrast to restricted expression of BMPs only in somatic niche cells in both the *Drosophila* ovary and testis. Intriguingly, targeted disruption of these genes has revealed that they all play important but redundant roles in maintaining the viability of germ cells, including GSCs (76, 110). It remains unclear whether they are required for GSC self-renewal as well. Therefore, from *C. elegans*, *Drosophila* to mice, extrinsic signals from their niches play instructive roles in controlling GSC self-renewal.

Differentiation of GSC progeny in mice also depends on extrinsic signals. The stem cell factor (SCF) produced by Sertoli cells activates c-Kit, a tyrosine kinase receptor for SCF, to promote the differentiation of GSC progeny (22, 71, 108) (**Figure 4**; **Table 1**). Loss-of-function mutations in *c-kit* cause an arrest in an early step of spermatogonia differentiation, suggesting that the SCF/c-Kit pathway is required for germ cell differentiation and survival (22, 71, 108). The identification of immediate target genes controlled by SCF/c-Kit signaling will be crucial for understanding how germ cell differentiation is regulated in mice.

Different Classes of Intrinsic Factors Control GSC Maintenance and Differentiation in the Mouse Testis

Two different intrinsic self-renewing factors have been identified in mice, *nanos2* and Plzf (promyelocytic leukemia zinc-finger) (9, 15, 94). *nanos2* is expressed predominantly in male germ cells, and a *nanos2* mutation results in complete GSC loss (94). In contrast, *nanos3* is expressed in primordial germ cells (PGCs) of both sexes, and a *nanos3* mutation leads to complete loss of germ cells in both sexes. As discussed earlier, one *nanos* gene maintains PGCs as well as GSCs in the *Drosophila* ovary, which is in contrast to two genes sharing these roles in mice. It would be interesting to see whether *pumilio*-like genes are also required for maintaining PGCs and GSCs in mice since its *Drosophila* counterpart functions in the same protein complex with Nanos to repress translation and maintain GSCs. *plzf*-null mice have lost their ability to maintain GSCs in the testis, and its function is required only in GSCs (9, 15). Plzf, a member of the POK (POZ and Krüppel) family of transcriptional repressors, is expressed in early undifferentiated spermatogonial cells. It can potentially recruit members of the mammalian Polycomb family, such as BMI1, to link epigenetic modifications to transcriptional control. Since BMI1 has recently been shown to be required for self-renewal and proliferation of hematopoietic and neural stem cells (67, 74), it would be worthwhile to see whether BMI1 is also required to maintain GSCs in the testis. It is equally important to know how Plzf and Nanos2 interact with the known GDNF signaling pathway to control GSC self-renewal in the testis and to understand how the transition from a GSC to a differentiating A_s is regulated intrinsically.

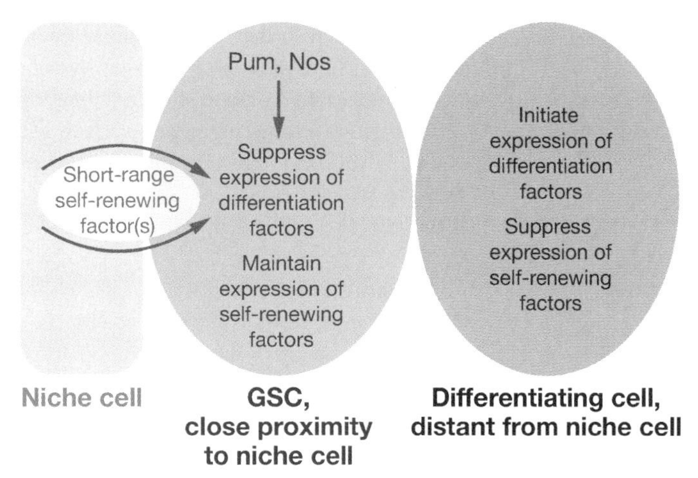

Figure 5

A conceptual relationship between GSCs and their niche. The niche cell produces short-range self-renewing signals, and self-renewing intrinsic factors such as Pum/Nos promote self-renewal by repressing expression of differentiation-promoting factors. In differentiating cells, however, this process is reversed to initiate differentiation.

SUMMARY AND FUTURE DIRECTIONS

A comparative analysis of GSC systems in *Drosophila*, *C. elegans*, and mice has elucidated some fundamental principles and strategies for GSC self-renewal and differentiation (**Figure 5**). First, GSCs are situated in specialized regulatory niches to ensure self-renewal. Cap cells function as a GSC niche in the *Drosophila* ovary, hub cells form a GSC niche in the *Drosophila* testis, and the DTC is a GSC niche in *C. elegans*, as demonstrated by cell biological and genetic studies (18, 106). In the mouse testis, GSC transplantation experiments have revealed the existence of a GSC niche but its structure has not yet been defined (6). Second, GSC niches in *Drosophila* and *C. elegans* exhibit structural asymmetry to ensure that the GSC daughters remaining inside the niche self-renew, and the others outside the niche generate differentiated cells. Third, short-range niche signals prevent GSCs from differentiation and thereby maintain stem cell self-renewal. Niche signals, BMPs in the *Drosophila* testis and ovary, Upd in the *Drosophila* testis, and LAG-2 in *C. elegans* gonad all function over a short distance to maintain GSCs. However, it is unknown whether GDNF is also a short-range signal that acts specifically on GSCs in the mouse testis. Fourth, intrinsic factors regulating mRNA stability and/or translation are conserved for their ability to regulate GSC self-renewal and differentiation. In *Drosophila* and *C. elegans*, the majority of intrinsic factors are related to regulation of mRNA stability and/or translation. Notably, Pum and Nos are conserved translational repressors that are involved in GSC self-renewal from *Drosophila*, *C. elegans* to mice. Finally, the interplay between extrinsic signals and intrinsic factors is critical for GSC self-renewal. In the *Drosophila* ovary and testis, BMP signaling directly represses expression of a differentiation-promoting factor, *bam*, in order to maintain GSCs, whereas, in *C. elegans*, GLP-1/Notch signaling is implicated in repressing functions of differentiation-promoting genes, such as *gld* genes. It will be interesting to see whether these conserved GSC regulatory mechanisms are also utilized by other adult stem cells.

Despite the commonalities of GSCs in different systems, obvious differences exist between them. First, a different combination of extrinsic signals is needed for different GSCs. For example, only one GLP-1/Notch signaling pathway is required for GSC self-renewal in *C. elegans*, Piwi and BMP signaling are needed for controlling GSC self-renewal in the *Drosophila* ovary, while Piwi, BMP and JAK-STAT signaling contribute to GSC maintenance in the *Drosophila* testis, and GDNF and, likely, BMPs work together to control GSC maintenance in the mouse testis. Because these different combinations of signaling pathways are sufficient to prevent GSC differentiation, these differences may reflect distinct developmental histories of different GSCs. Second, the same extrinsic signals exhibit some differences in their ability to regulate GSCs. One example is how Dpp plays slightly different roles in the *Drosophila* ovary and testis. Dpp plays an instructive role in controlling GSC self-renewal in the *Drosophila* ovary, but it plays a minor role in GSCs of the *Drosophila* testis. Third, some intrinsic factors are needed in only one GSC system but not in others to direct GSC self-renewal and differentiation. GSCs in *Drosophila*, *C. elegans*, and mice form differently during early development; their different developmental histories may confer GSCs distinct properties, which thus require the use of different intrinsic factors to control their self-renewal and differentiation. For example, *bam* has not been identified and shown to be required for germ cell differentiation in *C. elegans*, while many intrinsic factors identified in *C. elegans*, such as *gld* genes, have not been shown to be required for GSC differentiation in *Drosophila*. Together, these differences in the mechanisms regulating GSC behavior have revealed that different combinations of extrinsic signals and intrinsic factors can achieve the same purposes of controlling self-renewal and differentiation.

The knowledge gained from studies on GSCs and their niches in *Drosophila* and *C. elegans* has provided an intellectual framework for defining stem cells and their niches in mammalian systems. Moreover, the signaling pathways identified for controlling GSC self-renewal have also been shown to regulate different adult stem cell types in mammals. Although we have learned so much from studying GSCs in different systems, many questions remain. What constitutes a GSC niche in the mouse testis? How is GSC niche formation regulated in different species? How does signaling initiated by niche cells interact with intrinsic factors to control GSC self-renewal and differentiation? Whether and how do GSCs regulate their niche function? Is GSC aging due to intrinsic aging or niche aging? The answers to these questions will greatly advance our understanding of GSC regulation and will also provide insight into how adult stem cells are regulated in general.

SUMMARY POINTS

1. The regulatory microenvironment or "niche" directly controls GSC asymmetric division and self-renewal.
2. GSC niches exhibit structural asymmetry to ensure that only one of the two GSC daughters remains in the niche for self-renewal.

3. Niche signals function at short range to act directly on GSCs to prevent them from differentiation and thereby control self-renewal.
4. Different combinations of niche signals are needed for GSC self-renewal in different systems.
5. The same niche signal may function differently in different GSC systems.
6. Differentiation of GSC daughters is controlled by extrinsic signals from their surrounding somatic cells.
7. Several classes of intrinsic factors are involved in controlling GSC self-renewal and differentiation. Pumilio and Nanos families of proteins are conserved intrinsic factors for GSC self-renewal.
8. Intimate interplay between extrinsic factors and intrinsic factors is critical for GSC self-renewal.

ACKNOWLEDGMENTS

We would like to thank Daniel Kirilly for making all the figures and the other Xie laboratory members for stimulating scientific discussions. Owing to space constraints, we apologize to colleagues whose work was not mentioned or discussed in detail. The authors are supported by Stowers Institute for Medical Research and N.I.H. (1R01 GM64428-01).

LITERATURE CITED

1. **Austin J, Kimble J. 1987. *glp-1* is required in the germ line for regulation of the decision between mitosis and meiosis in *C. elegans*. *Cell* 51:589–99**
2. Barker DD, Wang C, Moore J, Dickinson LK, Lehmann R. 1992. Pumilio is essential for function but not for distribution of the *Drosophila* abdominal determinant Nanos. *Genes Dev.* 6:2312–26
3. Berry LW, Westlund B, Schedl T. 1997. Germ-line tumor formation caused by activation of *glp-1*, a *Caenorhabditis elegans* member of the Notch family of receptors. *Development* 124:925–36
4. Bopp D, Horabin JI, Lersch RA, Cline TW, Schedl P. 1993. Expression of the Sex-lethal gene is controlled at multiple levels during *Drosophila* oogenesis. *Development* 118:797–812
5. **Brawley C, Matunis E. 2004. Regeneration of male germline stem cells by spermatogonial dedifferentiation in vivo. *Science* 304:1331–34**
6. Brinster R, Avarbock M. 1994. Germline transmission of donor haplotype following spermatogonial transplantation. *Proc. Natl. Acad. Sci. USA* 91:11303–7
7. Brinster RL. 2002. Germline stem cell transplantation and transgenesis. *Science* 296:2174–76
8. **Brinster RL, Zimmermann JW. 1994. Spermatogenesis following male germ-cell transplantation. *Proc. Natl. Acad. Sci. USA* 91:11298–302**
9. **Buaas F, Kirsh A, Sharma M, McLean D, Morris J, et al. 2004. Plzf is required in adult male germ cells for stem cell self-renewal. *Nat. Genet.* 36:647–52**

The first report demonstrating that the GLP-1/Notch receptor is required for maintaining GSCs and repressing meiosis.

This paper shows that differentiated germ cell cysts can revert back into GSCs and reclaim the niche in the *Drosophila* testis. This reversion depends on JAK-STAT signaling.

The first paper reporting the

development of germ cell transplantation for studying germ cell development. The transplantation has been proven to be a powerful tool for studying GSC regulation in the mouse testis.

This paper reports that a nuclear factor, Plzf, is essential for controlling GSC self-renewal.

This paper shows that niche signals maintain GSCs by directly repressing expression of differentiation-promoting genes.

This paper also reports that Plzf is essential for controlling GSC self-renewal and proliferation.

This paper shows that *piwi* functions in somatic supporting cells to control GSC maintenance and proliferation in the *Drosophila* ovary.

This paper reports the identification of two closely related Pumilio-like genes that are essential for controlling GSC maintenance in *C. elegans*.

10. Casanueva MO, Ferguson EL. 2004. Germline stem cell number in the *Drosophila* ovary is regulated by redundant mechanisms that control Dpp signaling. *Development* 131:1881–90
11. **Chen D, McKearin D. 2003. Dpp signaling silences bam transcription directly to establish asymmetric divisions of germline stem cells. *Curr. Biol.* 13:1786–91**
12. Chen D, McKearin D. 2005. Gene circuitry controlling a stem cell niche. *Curr. Biol.* 15:179–84
13. Chen D, McKearin DM. 2003. A discrete transcriptional silencer in the bam gene determines asymmetric division of the *Drosophila* germline stem cell. *Development* 130:1159–70
14. Christerson LB, McKearin DM. 1994. *orb* is required for anteroposterior and dorsoventral patterning during *Drosophila* oogenesis. *Genes Dev.* 8:614–28
15. **Costoya J, Hobbs R, Barna M, Cattoretti G, Manova K, et al. 2004. Essential role of Plzf in maintenance of spermatogonial stem cells. *Nat. Genet.* 36:653–59**
16. **Cox DN, Chao A, Baker J, Chang L, Qiao D, Lin H. 1998. A novel class of evolutionarily conserved genes defined by piwi are essential for stem cell self-renewal. *Genes Dev.* 12:3715–27**
17. Cox DN, Chao A, Lin H. 2000. *piwi* encodes a nucleoplasmic factor whose activity modulates the number and division rate of germline stem cells. *Development* 127:503–14
18. Crittenden S, Eckmann CR, Wang L, Bernstein DS, Wickens MJK. 2003. Regulation of the mitosis/meiosis decision in the *Caenorhabditis elegans* germline. *Philos. Trans. R. Soc. London Ser. B* 358: 1359–62
19. **Crittenden SL, Bernstein DS, Bachorik JL, Thompson BE, Gallegos M, et al. 2002. A conserved RNA-binding protein controls germline stem cells in *Caenorhabditis elegans*. *Nature* 417:660–63**
20. Crittenden SL, Troemel ER, Evans TC, Kimble J. 1994. GLP-1 is localized to the mitotic region of the *C. elegans* germ line. *Development* 120:2901–11
21. de Cuevas M, Lilly MA, Spradling AC. 1997. Germline cyst formation in *Drosophila*. *Annu. Rev. Genet.* 31:405–28
22. deRooij D, Okabe M, Nishimune Y. 1999. Arrest of spermatogonial differentiation in jsd/jsd, Sl17H/Sl17H, and cryptorchid mice. *Biol. Reprod.* 61:842–47
23. Doyle TG, Wen C, Greenwald I. 2000. SEL-8, a nuclear protein required for LIN-12 and GLP-1 signaling in *Caenorhabditis elegans*. *Proc. Natl. Acad. Sci. USA* 97:7877–81
24. Eckmann C, Crittenden SL, Suh N, Kimble J. 2004. GLD-3 and control of the mitosis/meiosis decision in the germline of *Caenorhabditis elegans*. *Genetics* 168:147–60
25. Forbes A, Lehmann R. 1998. Nanos and Pumilio have critical roles in the development and function of *Drosophila* germline stem cells. *Development* 125:679–90
26. Fuchs E, Tumbar T, Guasch G. 2004. Socializing with the neighbors: stem cells and their niche. *Cell* 116:769–78
27. Fuller MT. 1993. Spermatogenesis. In *The Development of Drosophila*, ed. M Bate, A Martinez Arias, pp. 71–147. Cold Spring Harbor: Cold Spring Harbor Lab. Press
28. Gateff E, Kurzik-Dumke U, Wismar J, Loffler T, Habtemichael N, et al. 1996. *Drosophila* differentiation genes instrumental in tumor suppression. *Int. J. Dev. Biol.* 40:149–56
29. Gilboa L, Forbes A, Tazuke SI, Fuller MT, Lehmann R. 2003. Germ line stem cell differentiation in *Drosophila* requires gap junctions and proceeds via an intermediate state. *Development* 130:6625–34
30. Gilboa L, Lehmann R. 2004. How different is Venus from Mars? The genetics of germline stem cells in *Drosophila* females and males. *Development* 131:4895–905

31. Gilboa L, Lehmann R. 2004. Repression of primordial germ cell differentiation parallels germ line stem cell maintenance. *Curr. Biol.* 14:981–86

32. Gonczy P, DiNardo S. 1996. The germ line regulates somatic cyst cell proliferation and fate during Drosophila spermatogenesis. *Development* 122:2437–47

33. Gonczy P, Matunis E, DiNardo S. 1997. *bag-of-marbles* and *benign gonial cell neoplasm* act in the germline to restrict proliferation during *Drosophila* spermatogenesis. *Development* 124:4361–71

34. Haerry TE, Khalsa O, O'Connor MB, Wharton KA. 1998. Synergistic signaling by two BMP ligands through the SAX and TKV receptors controls wing growth and patterning in Drosophila. *Development* 125:3977–87

35. Hansen D, Hubbard EJ, Schedl T. 2004. Multi-pathway control of the proliferation versus meiotic development decision in the *Caenorhabditis elegans* germline. *Dev. Biol.* 268:342–57

36. Hardy RW, Tokuyasu KT, Lindsley DL, Garavito M. 1979. The germinal proliferation center in the testis of *Drosophila melanogaster*. *J. Ultrastruct. Res.* 69:180–90

37. Henderson ST, Gao D, Lambie EJ, Kimble J. 1994. *lag-2* may encode a signaling ligand for the GLP-1 and LIN-12 receptors of C. elegans. *Development* 120:2913–24

38. Jan E, Motzny CK, Graves LE, Goodwin EB. 1999. The STAR protein, GLD-1, is a translational regulator of sexual identity in *Caenorhabditis elegans*. *EMBO J.* 18:258–69

39. Johnson J, Canning J, Kaneko T, Pru J, Tilly J. 2004. Germline stem cells and follicular renewal in the postnatal mammalian ovary. *Nature* 428:145–50

40. Jones AR, Francis R, Schedl T. 1996. GLD-1, a cytoplasmic protein essential for oocyte differentiation, shows stage- and sex-specific expression during *Caenorhabditis elegans* germline development. *Dev. Biol.* 180:165–83

41. Kadyk LC, Kimble J. 1998. Genetic regulation of entry into meiosis in *Caenorhabditis elegans*. *Development* 125:1803–13

42. Kai T, Spradling A. 2003. An empty *Drosophila* stem cell niche reactivates the proliferation of ectopic cells. *Proc. Natl. Acad. Sci. USA* 100:4633–38

43. Kai T, Spradling A. 2004. Differentiating germ cells can revert into functional stem cells in *Drosophila melanogaster* ovaries. *Nature* 428:564–69

44. Kanatsu-Shinohara M, Miki H, Inoue K, Ogonuki N, Toyokuni S, et al. 2005. Long-term culture of mouse male germline stem cells under serum-or feeder-free conditions. *Biol. Reprod.* 72:985–91

45. Kawase E, Wong MD, Ding BC, Xie T. 2004. Gbb/Bmp signaling is essential for maintaining germline stem cells and for repressing bam transcription in the Drosophila testis. *Development* 131:1365–75

46. Khalsa O, Yoon JW, Torres-Schumann S, Wharton KA. 1998. TGF-beta/BMP superfamily members, Gbb-60A and Dpp, cooperate to provide pattern information and establish cell identity in the Drosophila wing. *Development* 125:2723–34

47. Kiger AA, Fuller MT. 2001. Male germ-line stem cells. In *Stem Cell Biology*, ed. DR Marshak, RL Gardner, D Gottlieb, pp. 149–88. Cold Spring Harbor, NY: Cold Spring Harbor Lab. Press

48. Kiger AA, Jones DL, Schulz C, Rogers MB, Fuller MT. 2001. Stem cell self-renewal specified by JAK-STAT activation in response to a support cell cue. *Science* 294:2542–45

49. Kiger AA, White-Cooper H, Fuller MT. 2000. Somatic support cells restrict germline stem cell self-renewal and promote differentiation. *Nature* 407:750–54

This paper reports that Nos and Pum may function together to repress PGC and GSC differentiation in a Dpp-independent manner.

This paper provides experimental evidence challenging the dogma that GSCs do not persist in postnatal mammalian ovary.

This study shows that differentiated germ cell cysts can revert back into GSCs in early female gonads and in the adult ovary. The GSCs regenerated from differentiated germ cells function normally.

This study shows that the hub cells function as a GSC niche in the *Drosophila* testis by demonstrating that they produce Upd that activates JAK-STAT signaling to control GSC self-renewal.

This study uses laser ablation to show that the DTC is required for maintaining germ cells in *C. elegans*.

The first paper showing that GSCs in the mouse testis can be cultured and expanded *in vitro*. These cultured GSCs provide a powerful system for studying their regulation *in vitro*.

The first paper showing that Pumilio is required for maintaining GSCs in the *Drosophila* ovary.

This paper identified the first signal from Sertoli cells, GDNF, that is necessary and sufficient for controlling GSC self-renewal and proliferation.

50. **Kimble JE, White JG. 1981. On the control of germ cell development in *Caenorhabditis elegans*. *Dev. Biol.* 81:208–19**
51. King FJ, Lin H. 1999. Somatic signaling mediated by *fs(1)Yb* is essential for germline stem cell maintenance during *Drosophila* oogenesis. *Development* 126:1833–44
52. King FJ, Szakmary A, Cox DN, Lin H. 2001. *Yb* modulates the divisions of both germline and somatic stem cells through *piwi*- and *hh*-mediated mechanisms in the Drosophila ovary. *Mol. Cell.* 7:497–508
53. **Kubota H, Avarbock M, Brinster R. 2004. Growth factors essential for self-renewal and expansion of mouse spermatogonial stem cells. *Proc. Natl. Acad. Sci.* 101:16489–94**
54. Kubota H, Avarbock MR, Brinster RL. 2003. Spermatogonial stem cells share some, but not all, phenotypic and functional characteristics with other stem cells. *Proc. Natl. Acad. Sci. USA* 100:6487–92
55. Kwak JE, Wang L, Ballantyne S, Kimble J, Wickens M. 2004. Mammalian GLD-2 homologs are poly(A) polymerases. *Proc. Natl. Acad. Sci. USA* 101:4407–12
56. Lambie EJ, Kimble J. 1991. Two homologous regulatory genes, lin-12 and glp-1, have overlapping functions. *Development* 112:231–40
57. Lamont LCS, Bernstein D, Wickens M, Kimble J. 2004. FBF-1 and FBF-2 regulate the size of the mitotic region in the *C. elegans* germline. *Dev Cell.* 7:697–707
58. Lantz V, Chang JS, Horabin JI, Bopp D, Schedl P. 1994. The *Drosophila orb* RNA-binding protein is required for the formation of the egg chamber and establishment of polarity. *Genes Dev.* 8:598–613
59. Lavoie CA, Ohlstein B, McKearin DM. 1999. Localization and function of Bam protein require the *benign gonial cell neoplasm* gene product. *Dev. Biol.* 212:405–13
60. Lin H. 2002. The stem-cell niche theory: lessons from flies. *Nat. Rev. Genet.* 3:931–40
61. **Lin H, Spradling AC. 1997. A novel group of *pumilio* mutations affects the asymmetric division of germline stem cells in the *Drosophila* ovary. *Development* 124:2463–76**
62. Lin H, Yue L, Spradling AC. 1994. The *Drosophila* fusome, a germline-specific organelle, contains membrane skeletal proteins and functions in cyst formation. *Development* 120:947–56
63. Marin V, Evans TC. 2003. Translational repression of a *C. elegans* Notch mRNA by the STAR/KH domain protein GLD-1. *Development* 130:2623–32
64. McKearin D, Ohlstein B. 1995. A role for the *Drosophila bag-of-marbles* protein in the differentiation of cystoblasts from germline stem cells. *Development* 121:2937–47
65. McKearin DM, Spradling AC. 1990. *bag-of-marbles*: a *Drosophila* gene required to initiate both male and female gametogenesis. *Genes Dev.* 4:2242–51
66. **Meng X, Lindahl M, Hyvonen ME, Parvinen M, deRooij DG, et al. 2000. Regulation of cell fate decision of undifferentiated spermatogonia by GDNF. *Science* 287:1489–93**
67. Molofsky AV, Pardal R, Iwashita T, Park IK, Clarke MF, Morrison SJ. 2003. Bmi-1 dependence distinguishes neural stem cell self-renewal from progenitor proliferation. *Nature* 425:962–67
68. Morrison SJ, Shah NM, Anderson DJ. 1997. Regulatory mechanisms in stem cell biology. *Cell* 88:287–98
69. Ohlstein B, Lavoie CA, Vef O, Gateff E, McKearin DM. 2000. The *Drosophila* cystoblast differentiation factor, benign gonial cell neoplasm, is related to DExH-box proteins and interacts genetically with bag-of-marbles. *Genetics* 155:1809–19

70. Ohlstein B, McKearin D. 1997. Ectopic expression of the *Drosophila* Bam protein eliminates oogenic germline stem cells. *Development* 124:3651–62
71. Ohta H, Yomogida K, Dohmae K, Nishimune Y. 2000. Regulation of proliferation and differentiation in spermatogonial stem cells: the role of c-kit and its ligand SCF. *Development* 127:2125–31
72. Orwig KE, Ryu BY, Avarbock MR, Brinster RL. 2002. Male germ-line stem cell potential is predicted by morphology of cells in neonatal rat testes. *Proc. Natl. Acad. Sci. USA* 99:11706–11
73. Orwig KE, Shinohara T, Avarbock MR, Brinster RL. 2002. Functional analysis of stem cells in the adult rat testis. *Biol. Reprod.* 66:944–49
74. Park IK, Qian D, Kiel M, Becker MW, Pihalja M, et al. 2003. Bmi-1 is required for maintenance of adult self-renewing haematopoietic stem cells. *Nature* 423:302–5
75. Petcherski AG, Kimble J. 2000. LAG-3 is a putative transcriptional activator in the *C. elegans* Notch pathway. *Nature* 405:364–68
76. Puglisi R, Montanari M, Chiarella P, Stefanini M, Boitani C. 2004. Regulatory role of BMP2 and BMP7 in spermatogonia and Sertoli cell proliferation in the immature mouse. *Eur. J. Endocrinol.* 151:511–20
77. Ryu BY, Orwig KE, Avarbock MR, Brinster RL. 2003. Stem cell and niche development in the postnatal rat testis. *Dev. Biol.* 263:253–63
78. Saccomanno L, Loushin C, Jan E, Punkay E, Artzt K, Goodwin EB. 1999. The STAR protein QKI-6 is a translational repressor. *Proc. Natl. Acad. Sci. USA* 96:12605–10
79. Schulz C, Kiger AA, Tazuke SI, Yamashita YM, Pantalena-Filho LC, et al. 2004. A misexpression screen reveals effects of bag-of-marbles and TGF beta class signaling on the Drosophila male germ-line stem cell lineage. *Genetics* 167:707–23
80. Schulz C, Wood CG, Jones DL, Tazuke SI, Fuller MT. 2002. Signaling from germ cells mediated by the rhomboid homolog stet organizes encapsulation by somatic support cells. *Development* 129:4523–34
81. Shamblott MJ, Axelman J, Wang S, Bugg EM, Littlefield JW, et al. 1998. Derivation of pluripotent stem cells from cultured human primordial germ cells. *Proc. Natl. Acad. Sci. USA* 95:13726–31
82. Shinohara T, Orwig KE, Avarbock MR, Brinster RL. 2001. Remodeling of the postnatal mouse testis is accompanied by dramatic changes in stem cell number and niche accessibility. *Proc. Natl. Acad. Sci. USA* 98:6186–91
83. Shinohara T, Orwig KE, Avarbock MR, Brinster RL. 2002. Germ line stem cell competition in postnatal mouse testes. *Biol. Reprod.* 66:1491–97
84. Shinohara T, Orwig KE, Avarbock MR, Brinster RL. 2003. Restoration of spermatogenesis in infertile mice by Sertoli cell transplantation. *Biol. Reprod.* 68:1064–71
85. Shivdasani AA, Ingham PW. 2003. Regulation of stem cell maintenance and transit amplifying cell proliferation by *tgf-beta* signaling in *Drosophila* spermatogenesis. *Curr. Biol.* 13:2065–72
86. **Song X, Wong MD, Kawase E, Xi R, Ding BC, et al. 2004. Bmp signals from niche cells directly repress transcription of a differentiation-promoting gene, bag of marbles, in germline stem cells in the *Drosophila* ovary. *Development* 131:1353–64**
87. **Song X, Zhu CH, Doan C, Xie T. 2002. Germline stem cells anchored by adherens junctions in the *Drosophila* ovary niches. *Science* 296:1855–57**
88. Spradling A, Drummond-Barbosa D, Kai T. 2001. Stem cells find their niche. *Nature* 414:98–104

This paper shows that niche signals maintain GSCs by directly repressing expression of differentiation-promoting genes. It also shows that *gbb*, a *Drosophila* homolog of BMP5-8, is also essential for maintaining GSCs and repressing *bam* expression.

The first paper showing that E-cadherin-mediated cell adhesion is essential for anchoring and recruiting GSCs to their niche. Cadherins are also involved in anchoring adult stem cells in mammalian systems.

89. Styhler S, Nakamura A, Swan A, Suter B, Lasko P. 1998. vasa is required for GURKEN accumulation in the oocyte, and is involved in oocyte differentiation and germline cyst development. *Development* 125:1569–78

90. Szakmary A, Cox DN, Wang Z, Lin H. 2005. Regulatory Relationship among piwi, pumilio, and bag-of-marbles in *Drosophila* germline stem cell self-renewal and differentiation. *Curr. Biol.* 15:171–78

91. Tazuke SI, Schulz C, Gilboa L, Fogarty M, Mahowald AP, et al. 2002. A germline-specific gap junction protein required for survival of differentiating early germ cells. *Development* 129:2529–39

92. Thomson JA, Itskovitz-Eldor J, Shapiro SS, Waknitz MA, Swiergiel JJ, et al. 1998. Embryonic stem cell lines derived from human blastocysts. *Science* 282:1145–47

93. Tran J, Brenner TJ, DiNardo S. 2000. Somatic control over the germline stem cell lineage during *Drosophila* spermatogenesis. *Nature* 407:754–57

94. Tsuda M, Sasaoka Y, Kiso M, Abe K, Haraguchi S, et al. 2003. Conserved role of nanos proteins in germ cell development. *Science* 301:1239–41

95. Tulina N, Matunis E. 2001. Control of stem cell self-renewal in Drosophila spermatogenesis by JAK- STAT signaling. *Science* 294:2546–49

96. Urban S, Lee JR, Freeman M. 2001. *Drosophila rhomboid-1* defines a family of putative intramembrane serine proteases. *Cell* 107:173–82

97. Wang L, Eckmann CR, Kadyk LC, Wickens M, Kimble J. 2002. A regulatory cytoplasmic poly(A) polymerase in *Caenorhabditis elegans*. *Nature* 419:312–16

98. Wang Z, Lin H. 2004. Nanos maintains germline stem cell self-renewal by preventing differentiation. *Science* 303:2016–19

99. Wang Z, Lin H. 2005. The division of *Drosophila* germline stem cells and their precursors requires a specific cyclin. *Curr. Biol.* 15:328–33

100. Wickens M, Bernstein D, Crittenden S, Luitjens C, Kimble J. 2001. PUF proteins and 3'UTR regulation in the *Caenorhabditis elegans* germ line. *Cold Spring Harb Symp. Quant. Biol.* 66:337–43

101. Wickens M, Bernstein DS, Kimble J, Parker R. 2002. A PUF family portrait: 3'UTR regulation as a way of life. *Trends Genet.* 18:150–57

102. Xie T, Kawase E, Kirilly D, Wong MD. 2005. Intimate relationships with their neighbors: Tales of stem cells in *Drosophila* reproductive systems. *Dev. Dyn.* 232:775–90

103. Xie T, Spradling A. 2001. The Drosophila ovary: an in vivo stem cell system. In *Stem Cell Biology*, ed. DR Marshak, RL Gardner, D Gottlieb, pp. 129–48. Cold Spring Harbor, NY: Cold Spring Harbor Lab. Press

104. Xie T, Spradling AC. 1998. *decapentaplegic* is essential for the maintenance and division of germline stem cells in the *Drosophila* ovary. *Cell* 94:251–60

105. Xie T, Spradling AC. 2000. A niche maintaining germ line stem cells in the *Drosophila* ovary. *Science* 290:328–30

106. Yamashita Y, Fuller M, Jones DL. 2005. Signaling in stem cell niches: lessons from the *Drosophila* germline. *J. Cell Sci.* 15:665–72

107. Yamashita YM, Jones DL, Fuller MT. 2003. Orientation of asymmetric stem cell division by the APC tumor suppressor and centrosome. *Science* 301:1547–50

108. Yoshinaga K, Nishikawa S, Ogawa M, Hayashi S, Kunisada T, et al. 1991. Role of c-kit in mouse spermatogenesis: identification of spermatogonia as a specific site of c-kit expression and function. *Development* 113:689–99

94. This paper shows that two mouse *nanos* genes are involved in regulation of PGC and GSC functions in mice. *nano2* is required for GSC maintenance in the testis, while *nanos3* is required for PGC maintenance in both sexes.

95. This study shows that the hub cells function as a GSC niche in the *Drosophila* testis by providing Upd that activates JAK-STAT signaling in GSCs for self-renewal.

98. This study shows that *nanos* is required for preventing the differentiation of PGCs and GSCs in the *Drosophila* female gonads.

104. This study demonstrates that Dpp/BMP2-4 is essential for controlling GSC self-renewal and division. The first study using mutant clonal analysis to study gene function in the control of GSC self-renewal.

109. Zhang B, Gallegos M, Puoti A, Durkin E, Fields S, et al. 1997. A conserved RNA-binding protein that regulates sexual fates in the *C. elegans* hermaphrodite germ line. *Nature* 390:477–84
110. Zhao G, Chen YX, Liu XM, Xu Z, Qi X. 2001. Mutation in *Bmp7* exacerbates the phenotype of *Bmp8a* mutants in spermatogenesis and epididymis. *Dev. Biol.* 240:212–22

105. The first study directly demonstrating that GSCs are situated in the niche in the *Drosophila* ovary. A GSC can undergo symmetric cell division to produce two GSCs if both daughters remain in the niche.

107. This study demonstrates that regulation of GSC spindle orientation is critical for asymmetric GSC division in the *Drosophila* testis. The authors also show that APC-1, APC-2 and Centrosomin play crucial roles in orienting the GSC spindle.

Molecular Signatures of Natural Selection

Rasmus Nielsen

Center for Bioinformatics and Department of Evolutionary Biology, University of Copenhagen, 2100 Copenhagen Ø, Denmark; email: rasmus@binf.ku.dk

Annu. Rev. Genet.
2005. 39:197–218

First published online as a Review in Advance on August 31, 2005

The *Annual Review of Genetics* is online at http://genet.annualreviews.org

doi: 10.1146/annurev.genet.39.073003.112420

0066-4197/05/1215-0197$20.00

Key Words

Darwinian selection, neutrality tests, genome scans, positive selection, phylogenetic footprinting

Abstract

There is an increasing interest in detecting genes, or genomic regions, that have been targeted by natural selection. The interest stems from a basic desire to learn more about evolutionary processes in humans and other organisms, and from the realization that inferences regarding selection may provide important functional information. This review provides a nonmathematical description of the issues involved in detecting selection from DNA sequences and SNP data and is intended for readers who are not familiar with population genetic theory. Particular attention is placed on issues relating to the analysis of large-scale genomic data sets.

Contents

INTRODUCTION

Population geneticists have for decades been occupied with the problem of quantifying the relative contribution of natural selection in shaping the genetic variation observed among living organisms. In one school of thought, known as the neutral theory, most of the variation within and between species is selectively neutral, i.e., it does not affect the fitness of the organisms (58, 59). New mutations that arise may increase in frequency in the population due to random factors, even though they do not provide a fitness advantage to the organisms carrying them. The process by which allele frequencies change in populations due to random factors is known as genetic drift.

SNP: single nucleotide polymorphism

A second school of thought maintains that a large proportion of the variation observed does affect the fitness of the organisms and is subject to Darwinian selection (39). These issues have not been settled with the availability of large-scale genomic data, but the debate has shifted from a focus on general laws or patterns of molecular evolution to the description of particular instances where natural selection has shaped the pattern of variation. This type of analysis is increasingly being done because it has become apparent that inferences regarding the patterns and distribution of selection in genes and genomes may provide important functional information. For example, in the human genome, the areas where disease genes are segregating should be under selection (assuming that the disease phenotype leads to a reduction in fitness). Even very small fitness effects may, on an evolutionary time scale, leave a very strong pattern. Therefore, in theory it may be possible to identify putative genetic disease factors by identifying regions of the human genome that currently are under selection (7). In general, positions in the genome that are under selection must be of functional importance. Inferences regarding selection have therefore been used extensively to identify functional regions or protein residues (12, 91). The purpose of this paper is to review the current knowledge regarding the effect of selection on a genome and to discuss methods for detecting selection using molecular data, especially genomic DNA sequence and single nucleotide polymorphism (SNP) data.

The Nomenclature of Selection Models

Much confusion exists in the literature regarding how various types of selection are defined, in particular because some of the terminology is used slightly differently within different scientific communities. At the risk of contributing further to this confusion, I propose here some simple definitions for some of the common terms used in the discussion of selection models before moving on to the main topics of this review.

The basic population genetic terms are well-defined. The classical population genetic models that students of biology will first encounter are models with two alleles, typically denoted A and a. Selection then occurs if the fitnesses of the three possible genotypes (w_{AA}, w_{Aa}, and w_{aa}) are not all equal. There is directional selection if the fitnesses of the three genotypes are not all equal and if $w_{AA} > w_{Aa} > w_{aa}$ or $w_{AA} < w_{Aa} < w_{aa}$. Directional selection tends to eliminate variation within populations and either increase or decrease variation between species depending on whether A or a is the new mutant. Overdominance occurs if the heterozygote has the highest fitness if $w_{AA} < w_{Aa} > w_{aa}$. Overdominance is a case of balancing selection where variability is maintained in the population due to selection. In haploid organisms, selection occurs if $w_A \neq w_a$ and overdominance is not possible. The difference in fitness between alleles is the selection coefficient, i.e., for the haploid model the selection coefficient could be defined as $s_A = w_A - w_a$.

In the molecular evolution literature, it has been common to use the terminology of positive selection, negative selection, purifying selection, and diversifying selection. Here we define negative selection as any type of selection where new mutations are selected against. Likewise, we define positive selection as any type of selection where new mutations are advantageous (have positive selection coefficients). In the context of the simple two-allele models, both directional selection and overdominance can be cases of positive selection. Purifying selection is identical to negative selection in that it describes selection against new variants. Diversifying selection has in the population genetics literature been synonymous with disruptive selection, a type of selection where two or more extreme phenotypic values are favoured simultaneously. This type of selection will often increase variability, and diversifying selection has, therefore, in the molecular evolution literature recently been used more generically to describe any type of selection that increases variability. However, as disruptive selection may reduce genetic variability when one of the extreme types becomes fixed in the population, and since there are many other forms of selection that can increase levels of genetic variability, the more generic use of the term "diversifying selection" should probably be avoided.

When a new mutant does not affect the fitness of the individual in which it arises (i.e., $w_{AA} = w_{Aa} = w_a$), it is said to be neutral. In general, neutrality describes the condition where the loci under consideration are not affected by selection. A statistical method aimed at rejecting a model of neutral evolution is called a neutrality test.

Balancing selection: selection that increases variability within a population

Positive selection: selection acting upon new advantageous mutations

Negative selection: selection acting upon new deleterious mutation

Neutrality test: a statistical test of a model which assumes all mutations are either neutral or strongly deleterious

Neutral mutation: a mutation that does not affect the fitness of individuals who carry it in either heterozygous or homozygous condition

POPULATION GENETIC PREDICTIONS

One of the main interests in molecular population genetics is to distinguish molecular variation that is neutral (only affected by random genetic drift) from variation that is subject to selection, particularly positive selection. An important point is that neutral models usually allow for the presence of strongly deleterious mutations that have such strong negative fitness consequences that they are immediately eliminated from the population (58). If selection only involves such mutations of very strong effect, the only mutations that will actually segregate in the population are the neutral mutations. Therefore, neutral models include the possible existence of pervasive strong negative selection. Although negative or purifying selection may be of great interest because it may help detect regions or residues of functional importance, much interest in the evolutionary literature focuses on positive selection because it is associated with adaptation and the evolution of new form or function. One of the main points of contention in population genetics has been the degree to which positive selection is important in explaining the pattern of variability within and between species (39, 59).

Selective sweep: the process by which a new advantageous mutation eliminates or reduces variation in linked neutral sites as it increases in frequency in the population

Much of the theoretical literature in population genetics over the past 50 years has focused on developing and analyzing models that generalize the previously mentioned basic di-allelic models to models where more than two alleles may be segregating, where multiple mutations may arise and interact—possibly in the presence of recombination, where the environment may be changing through time, and where random genetic drift may be acting in populations subject to various demographic forces (25, 39). From theory alone we have gained many valuable insights, including the fact that the efficacy of selection depends not only on the selection coefficient, but primarily on the product of the selection coefficient and the effective population size. An increased effect of selection may be due to either an increased population size or a larger selection coefficient. Among other important findings is that balancing selection may occur for many reasons other than overdominance, (e.g., fluctuating environmental conditions) and could therefore, potentially, be quite common (38, 39). However, the efficacy of selection will tend to be reduced when multiple selected alleles are segregating simultaneously in the genome. The mutations will tend to interfere with each other and reduce the local effective population size (8, 29, 40, 57). Many population geneticists used to believe that the number of selective deaths required to maintain large amounts of selection would have to be so large that selection would probably play a very small role in shaping genetic variation (43, 60, 61). These types of arguments, known as genetic load arguments, were instrumental in the development of the neutral theory. However, the amount of selection that a genome can permit depends on the way mutations interact in their effect on organismal fitness and on several other critical model assumptions (25, 62, 71, 107). Population genetic theory does not exclude the possibility that selection is very pervasive and cannot alone determine the relative importance and modality of selection in the absence of data from real living organisms (25, 39).

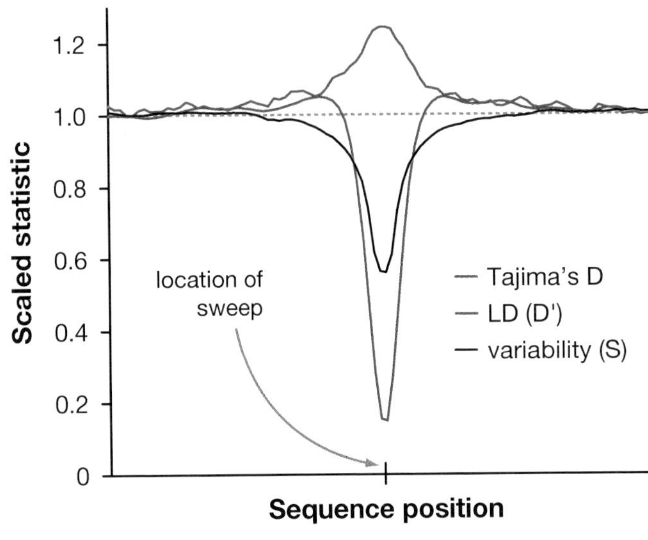

Figure 1

The effect of a selective sweep on genetic variation. The figure is based on averaging over 100 simulations of a strong selective sweep. It illustrates how the number of variable sites (variability) is reduced, LD is increased, and the frequency spectrum, as measured by Tajima's D, is skewed, in the region around the selective sweep. All statistics are calculated in a sliding window along the sequence right after the advantageous allele has reached frequency 1 in the population. All statistics are also scaled so that the expected value under neutrality equals one.

Much excitement currently exists in the population genetics communities over the fact that many predictions generated from the theory may now be tested in the context of the large genomic data sets. In particular, we should be able to detect the molecular signatures of new, strongly selected advantageous mutations that have recently become fixed (reached a frequency of one in the population). As these mutations increase in frequency, they tend to reduce variation in the neighboring region where neutral variants are segregating (13, 51, 52, 68). This process, by which a selected mutation reduces variability in linked sites as it goes to fixation, is known as a selective sweep (**Figure 1**). The hope is that by analysis of large comparative genomic data sets and large SNP data sets we will be able to determine how and where both positive and negative selection has affected variation in humans and other organisms.

POPULATION GENETIC SIGNATURES OF SELECTION

One of the main effects of selection is to modify the levels of variability within and between species (**Table 1**). A selective sweep tends to drastically reduce variation within a population, but will not lead to a reduction in species-specific differences. Conversely, negative selection acting on multiple loci will tend to reduce variability between species more drastically then variability within species. **Table 1** summarizes how various types of selection affect variability. Note that changes in the mutation rate alone will have the same effect on interspecific (between-species) and intraspecific (within-species) variability. However, selection affects intraspecific and interspecific variability differently. Many of the common population genetic methods for detecting selection are therefore based on comparing variation with and between species, most famously the HKA test (48). In this test, the rate of polymorphisms to divergence is compared for multiple genes. If the ratio varies more among genes than expected on a neutral model, neutrality is rejected.

Population Differentiation

Selection may in many cases increase the degree of differentiation among populations. In particular, recent theory shows that a selective sweep can have a dramatic impact on the level of population subdivision, particularly when the sweep has not yet spread to all populations within a species (20, 65, 97). When a locus shows extraordinary levels of genetic population differentiation, compared with other loci, this may then be interpreted as evidence for positive selection.

One of the first neutrality tests proposed, the Lewontin-Krakauer (63) test, takes advantage of this fact. This test rejects the neutral model for a locus if the level of genetic differentiation among populations is larger than predicted by a specific neutral model. It has recently been resurrected in various forms (1, 9, 10, 53, 92, 114), primarily driven by the availability of large-scale genomic data. For example, Akey et al. (1) looked at variation in F_{ST} (the most common

Table 1 The effect of selection and mutation on variability within and between species

Evolutionary factor	Intraspecific variability[a]	Interspecific variability	Ratio of interspecific to intraspecific variability	Frequency spectrum
Increased mutation rate	Increases	Increases	No effect	No effect
Negative directional selection	Reduced	Reduced	Reduced if selection is not too strong	Increases the proportion of low frequency variants
Positive directional selection	May increase or decrease	Increased	Increased	Increases the proportion of high frequency variants
Balancing selection	Increases	May increase or decrease	Reduced	Increases the proportion of intermediate frequency variants
Selective sweep (linked neutral sites)	Decreased	No effect on mean rate of substitution, but the variance increases	Increased	Mostly increases the proportion of low frequency variants

[a]Note that selection also affects other features of the data not mentioned here, such as levels of LD, haplotype structure, and levels of population subdivision.

measure of population differentiation) among human populations genome-wide. Beaumont & Balding (9) developed a sophisticated statistical method for identifying loci that may be outliers in terms of levels of population subdivision.

Frequency spectrum: the allelic sample distribution in independent nucleotide sites

LD: linkage disequilibrium

The Frequency Spectrum

Selection also affects the distribution of alleles within populations. For DNA sequence or SNP data, some of the most commonly applied tests are based on summarizing information regarding the so-called frequency spectrum. The frequency spectrum is a count of the number of mutations that exist in a frequency of $x_i = i/n$ for $i = 1, 2, \ldots, n-1$, in a sample of size n. In other words, it represents a summary of the allele frequencies of the various mutations in the sample. In a standard neutral model (i.e., a model with random mating, constant population size, no population subdivision, etc), the expected value of x_i is proportional to $1/i$. Selection against deleterious mutations will increase the fraction of mutations segregating at low frequencies in the sample. A selective sweep has roughly the same effect on the frequency spectrum (13). Conversely, positive selection will tend to increase the frequency in a sample of mutations segregating at high frequencies. The effect of selection on the frequency spectrum is summarized in **Figure 2**.

Many of the classic neutrality tests, therefore, focus on capturing information regarding the frequency spectrum. The most famous example is the Tajima's D test (112). In this test, the average number of nucleotide differences between pairs of sequences is compared with the total number of segregating sites (SNPs). If the difference between these two measures of variability is larger than what is expected on the standard neutral model, this model is rejected. The effect of a selective sweep on Tajima's D is shown in **Figure 1**. Fu & Li (34) extended this test to take information regarding the polarity of the information into account by the use of an evolutionary outgroup (e.g., a chimpanzee in the analysis of human genetic variation), and more refinements were introduced by Fu (32, 33). Fay & Wu (28) suggested a test that weights information from high-frequency derived mutations higher. These tests are probably the most commonly applied neutrality tests to date.

Models of Selective Sweeps

The pattern of variability left by a selective sweep is a rather complicated spatial pattern (**Figure 1**). By taking information regarding this pattern into account, the power of the neutrality tests can be improved, and it may even be possible to pinpoint the location of a selective sweep. Kim & Stephan (56) developed a method based on an explicit population genetic model of a selective sweep. Using this model, they could calculate the expected frequency spectrum in a site as a function of its distance to an advantageous mutation. By fitting the data to this model, they could estimate the location of the selective sweep and the strength of the selective sweep, and perform hypothesis tests regarding the presence of a sweep. This method is particularly useful in that it takes advantage of the spatial pattern left by the sweep along the sequence.

LD and Haplotype Structure

Levels of linkage disequilibrium (LD), the correlation among alleles from different loci, will increase in selected regions. Regions containing a polymorphism under balancing selection will tend to reduce LD if the polymorphism is old, but may increase LD in a transient phase. Selective sweeps also increase levels of LD in a transient phase (**Figure 1**), although this phase may be relatively short (82). Recently, there has been increased awareness that an incomplete sweep (when the adaptive mutation has not yet been fixed in the population) leaves a distinct pattern in the haplotype structure (87). This has led to the development of many statistical methods for detecting selection based on LD. Hudson et al. (47)

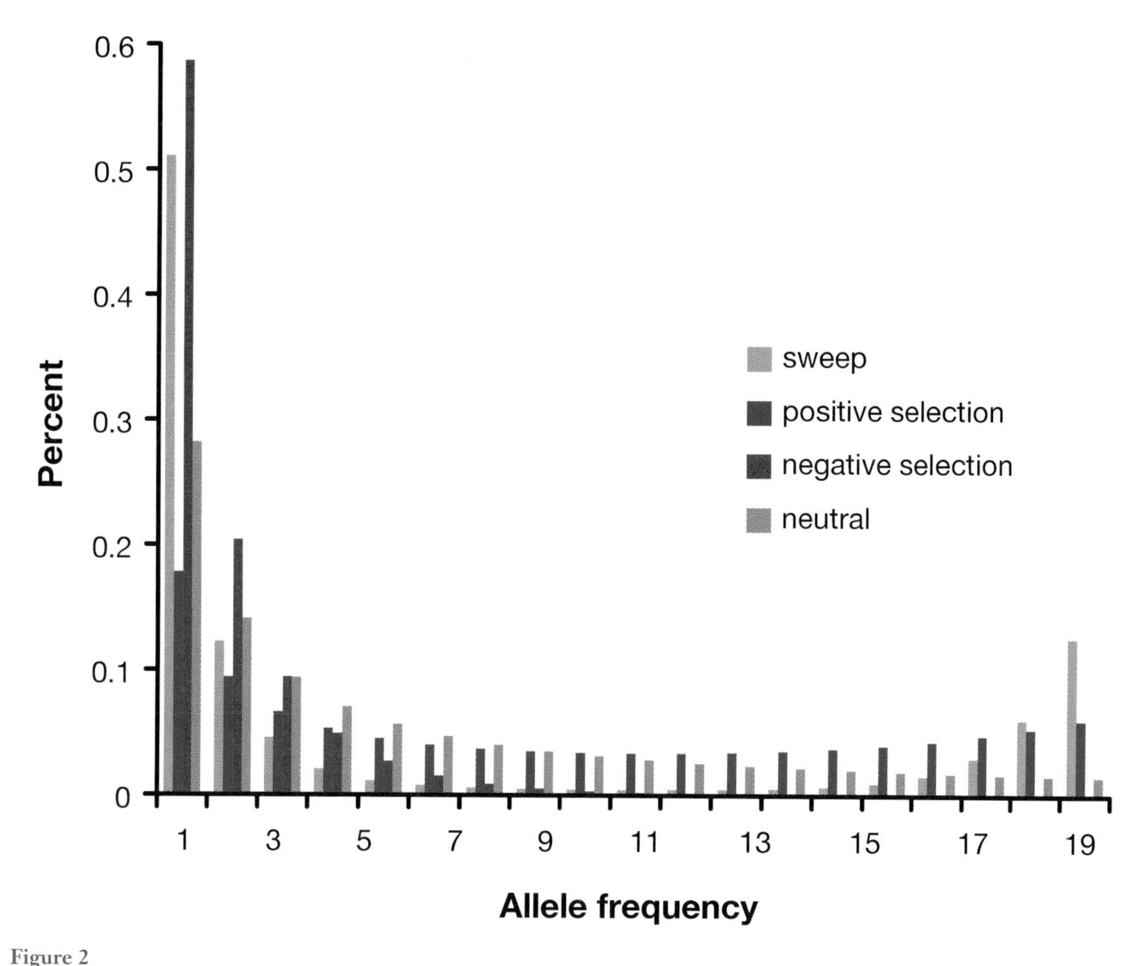

Figure 2

The frequency spectrum under a selective sweep, negative selection, neutrality, and positive selection. The frequency spectra under negative and positive selection are calculated using the PRF model by Sawyer & Hartl (88) for mutations with $2Ns = -5$ and 5, respectively, where N is the population size and s is the selection coefficient. For the selective sweep, the frequency spectrum is calculated in a window around the location of the adaptive mutation immediately after it has reached fixation in the population. In all cases, a demographic model of a population of constant size with no population subdivision is assumed.

developed a test based on the number of alleles occurring in a sample. Andolfatto et al. (4) developed a related test to determine whether any subset of consecutive variable sites contains fewer haplotypes than expected under a neutral model. A similar test was also proposed by Depaulis & Veuille (23). A variation on this theme was proposed by Sabeti et al. (87) who considered the increase in the number of distinct haplotypes away from the location of a putative selective sweep. Kelly (54) considered the level of association between pairs of loci. Kim & Nielsen (55) extended the method of Kim & Stephan (56) to include pairs of sites to incorporate information regarding linkage disequilibrium.

MacDonald-Kreitman Tests

Finally, the MacDonald-Kreitman test (69) explores the fact that mutations in coding regions come in two different flavors:

nonsynonymous mutations and synonymous mutations. It summarizes the data in what has become known as a MacDonald-Kreitman table, which contains counts of the number of nonsynonymous and synonymous mutations within and between species. If selection only affects the nonsynonymous mutations, negative selection will reduce the number of nonsynonymous mutations and positive selection will increase the number of nonsynonymous mutations, relative to the number of synonymous mutations. However, the effect will be stronger in divergence data than in polymorphism data. A test similar to the HKA test can therefore be constructed comparing the ratios of nonsynonymous to synonymous mutations within and between species. If these ratios differ significantly, this provides evidence for selection.

STATISTICAL CONCERNS

The neutrality tests are all tests of complicated population genetic models that make specific assumptions about the demography of the populations, in particular a constant population size and no population structure. In addition, in some of the tests there may be other implicit assumptions regarding distributions of recombination rates and mutation rates. Many of these tests have long been known to be highly sensitive to the demographic assumptions. For example, Simonsen et al. (96) showed that Tajima's D test (112) would reject a neutral model very frequently in the presence of population growth. The molecular signature of population growth is in many ways similar to the local effect of a selective sweep, and neutrality tests are often used as a method to detect population growth (85). Nielsen (73), Przeworski (82), and Ingvarsson (50) also argued that simple models of population subdivision can lead the commonly used neutrality tests to reject the neutral model with high probability, even in the absence of selection. In addition, even if the presence of selection can be established, in many cases it can be difficult to distinguish between the pattern left by selective sweeps and selection on slightly deleterious mutations (so-called background selection) (18, 19).

Tests based on patterns of LD may be particularly sensitive to the underlying model assumptions, because they (in addition to assumptions regarding demography) contain strong assumptions regarding the underlying recombination rates. Recent studies suggest that recombination rates are highly variable among regions (70) and among closely related species (83, 117). If that is true, it may not be advisable to focus attention toward patterns of LD when attempting to detect selection. Nonetheless, haplotype structure can be highly informative, particularly in detecting incomplete selective sweeps (87). Further research into how haplotype patterns can be used robustly to infer selection may be warranted.

Because of the effect of demographic assumptions on the population genetic neutrality tests, the results of these tests have often been contentious and often have not led to firm conclusions regarding the action of selection. One exception is the MacDonald-Kreitman (69) test. This test has increased robustness because the sites in which synonymous and nonsynonymous mutations occur are interspersed among each other and therefore similarly affected by demography and genetic drift. In fact, the MacDonald-Kreitman (69) test is robust to any demographic assumption (73). Unfortunately, it may not be very suitable for detecting recent selective sweeps because both nonsynonymous and synonymous mutations, linked to the beneficial mutation, will be similarly affected by the selective sweep. Also, the MacDonald-Kreitman (69) test cannot distinguish between past and present selection. Reducing the information in the data simply to the number of nonsynonymous mutations and synonymous mutations leads to a significant loss of information.

One possible way to circumvent the problem of demographic confounding effects is to compare multiple loci. For example, Galtier et al. (35) have implemented a statistical method, applicable to microsatellite loci, to test whether the signature of population growth is constant among loci or varies among loci. If the effect varies significantly among loci, beyond what can be explained by the demographic model, this may be interpreted as evidence for a selective sweep. In general, one can assume that if strong departures from the neutral model are seen only on one or a few outlier loci, this may be interpreted as evidence for selection on these loci. However, certain demographic factors, such as population subdivision, may increase the variance among loci (73). Certain demographic models may be more likely than others to produce outlier loci even in the absence of selection.

The application of population genetic tests other than the MacDonald-Kreitman test requires careful consideration of the possible range of demographic factors that may affect the results (2, 73). It is not very meaningful in itself to reject the standard neutral model using these methods without paying careful attention to the underlying demographics. Even the interpretation of significant results of the MacDonald-Kreitman test requires attention to demography if the directionality (positive versus negative) of selection is to be inferred (26). Fortunately, many recent studies go to great lengths in trying to exclude the possibility that rejections of a neutral model may be caused by demographic effects (3, 116).

SIGNATURES OF SELECTION IN COMPARATIVE DATA

While population genetic approaches aim at detecting ongoing selection in a population, comparative approaches, involving data from multiple different species, are suitable for detecting past selection. The major tool used to detect selection from comparative data is to compare the ratio of nonsynonymous mutations per nonsynonymous site to the number of synonymous mutations per nonsynonymous site (d_N/d_S). If there is no selection, not even strongly deleterious mutations, synonymous and nonsynonymous substitutions should occur at the same rate and we would expect $d_N/d_S = 1$. If there is negative selection, $d_N/d_S < 1$ and if there is positive selection, $d_N/d_S > 1$. The d_N/d_S ratio is therefore a proxy for the effect of selection that helps to identify not only selection, but also the directionality of selection. It is therefore a very commonly used tool for detection of positive selection and has been used in a variety of cases, for example, to demonstrate the presence of positive selection on HIV sequences (78) and on the human major histocompatibility locus (MHC) (49). However, as negative selection will tend to dominate in evolution, comparing the average rate of synonymous and nonsynonymous substitution in aligned sequences is a very conservative tool. If the gene is functional so that many or most mutations will disrupt function, the amount of positive selection needed to elevate the d_N/d_S above one is enormous. To overcome this problem, methods have been devised for detecting positive selection that takes variation in the d_N/d_S ratio into account (78, 127). The basic idea is to allow the d_N/d_S ratio to follow a statistical distribution among sites. If a distribution that allows values of $d_N/d_S > 1$ fits the data significantly better than a model that does not allow for such values, this is interpreted as evidence for positive selection. The methodology has been widely used and has led to a sharp increase in the number of loci where researchers have detected the presence of positive selection (31, 100, 125). This has also led to some skepticism toward this methodology (105, 106), although it has been found to perform well in simulation studies and is based on well-established statistical principles (5, 120, 124).

Several different statistical methods allow site-specific inferences regarding positive selection (30, 78, 104). The objective of these methods is to determine if specific sites have been targeted by positive (or negative) selection. In several cases, these methods have been used to make functional prediction regarding particular protein residues (91).

d_N: number of nonsynonymous mutations per nonsynonymous site

d_s: number of synonymous mutations per synonymous site

d_N/d_S ratio: the rate ratio of nonsynonymous to synonymous substitutions

The same type of methodology used to model variation in the d_N/d_S ratio among sites has also been used to model estimates of d_N/d_S along particular lineages of a phylogeny (123, 126, 128). This allows the testing of hypotheses regarding selective pressures on particular evolutionary lineages. Models have also been developed that allow site-specific inferences on a particular group of lineages on a phylogeny (128). Several excellent recent reviews describe the statistical methods used to detect selection from comparative data in more detail (124, 125). A summary of the different tests of neutrality is given in **Table 2**.

Targets of Positive Selection

Using analyses of comparative data, a clear picture emerges of the systems that most often are involved in positive selection of the kind that leads to increases in the d_N/d_S ratio (75). Typically, it involves an interaction between two organisms, or two different genetic components within the same organism, that compete or interact in such a way that an equilibrium is never reached. The best known examples are host-pathogen interactions that lead to positive selection of genes in pathogens (27, 30, 45, 78, 100) or in host immune and defense systems (49, 75, 90, 100). Other examples include genes involved in gametogenesis or expressed on the surface of gametes (75, 109, 110, 122). The forces creating positive selection in these genes may include sperm competition (122) and genetic conflicts between sperm and egg-cell (108). Positive selection also seems to be common in cases where selfish genes have the opportunity to create segregation distortion, potentially

Table 2 A very incomplete list of methods for detecting selection from DNA sequence and SNP data

Test	Data	Pattern	Requires multiple loci	Robust to demographic factors?	References
Tajima's D and related	Population genetic data	Frequency spectrum	No	No	(28, 32–34, 112)
Modeling of selective sweep—spatial pattern	Population genetic data	Frequency spectrum/spatial pattern	No	No	(55, 56)
Tests based on LD	Population genetic data	LD and/or haplotype structure	No	No	(4, 23, 47, 54, 87)
F_{ST} based and related tests	Population genetic data	Amount of population subdivision	Yes	No[a]	(1, 9, 10, 53, 92, 114)
HKA test	Population genetic and comparative data	Number of polymorphisms/substitutions	Yes	No	(48)
Macdonald-Kreitman-type tests	Population genetic and comparative data	Number of nonsynonymous and synonymous polymorphisms	No	Yes	(16, 69)
d_N/d_S ratio tests	Comparative data or population genetic data without recombination (6)	Nonsynonymous and synonymous substitutions	No	Yes	(49, 78, 104, 123, 128, 129)

[a]The degree to which these tests are robust to the underlying demographic assumptions is controversial and has not been fully explored.

reducing the fitness of the organism (46, 75). This type of genomic conflict may, for example, occur in loci associated with centromeres (46, 66, 67) or involved in apoptosis during spermatogenesis (75). Positive selection in terms of elevated d_N/d_S ratios tend to detect selection situations where repeated selective fixations have occurred in the same gene or in the same site, due to a continued dynamic interaction. In contrast, population genetic methods have the ability to detect selection on a single adaptive mutation that recently has swept through the population.

So far, very little research has been done to detect positive selection in noncoding regions based on comparative data. Although methods similar to those used to detect elevated d_N/d_S ratios can be devised for noncoding regions (119), sites in noncoding regions cannot easily be divided into possible selected sites and nonselected sites, similarly to nonsynonymous and synonymous sites in coding regions. Nonetheless, the presence of highly variable sites in noncoding regions may be signs of positive selection, and methods to identify such sites may find good use in the analysis of comparative genomic data. A serious practical problem that may arise in the application of such methods is the possibility of confounding misalignments with hypervariable regions.

Most of the literature on statistical methods for detecting selection from comparative data (e.g., from d_N/d_S ratios) and from population genetic data has been poorly connected. Although the comparative approaches have provided the most unambiguous evidence for positive selection, results have rarely been interpreted in terms of population genetic theory. One probable reason is that multiple population genetic models could generate the same pattern of observed d_N/d_S ratios, and that any detailed inferences of population genetic processes using comparative data would be based on a very strong assumptions regarding the way fitnesses are assigned to mutations (79). Comparative data in themselves are, therefore, unlikely to provide more detailed information regarding population genetic processes but relatively vague assertions of positive and negative selection and their distribution in the genome. Inferences regarding the type of negative or positive selection operating (e.g., balancing versus positive directional selection) must involve population genetic data. Moreover, comparative approaches cannot alone determine if selection is currently acting in a population. For such inferences population genetic data are also needed.

GENOMIC APPROACHES

The availability of large-scale genomic data has created new challenges and opportunities, especially in allowing for more nonparametric outlier analyses. Genes with increased levels of LD, reduced or enhanced levels of variability, increased levels of population differentiation, or skewed allele frequency spectra may be good candidates for selected loci. Recently, there has been heightened interest particularly in using increased population subdivision among populations as a method for detecting selection (1, 9, 44, 53, 64, 92, 93, 101, 102, 114). For example, Akey et al. (1) used variation in F_{ST} (a common measure of population subdivision) in the human genome to identify regions of increased population subdivision.

However, the availability of genomic data does not solve the fundamental problem that population-level demographic processes and selection are confounded. Many demographic processes, such as certain types of population subdivision, may increase the variance in the statistics used to detect selection. Certain demographic models are, therefore, more likely than other models to produce outliers. The outlier approach in population genetics does not solve the problem that a postulated signature of selection, inferred from population genetic data, may instead be the product of complicated demographics. Nonetheless, certain approaches based on detecting extreme levels of population subdivision seem to have some robustness to the model assumptions (9, 114).

PRF Models

PRF: Poisson random field

The simultaneous analysis of multiple genomic loci allows the estimation of parameters that are common among loci, potentially leading to increased power and robustness. For example, Bustamante et al. (16) analyzed MacDonald-Kreitman tables from *Arabidopsis* and *Drosophila* in a statistical framework that allows the divergence time between species to be a shared parameter among all loci, leading to increased statistical power. Similar approaches can be used to increase the robustness of the statistical methods by explicitly estimating demographic parameters, thereby taking the uncertainty introduced by the unknown demographic processes into account. This is particularly convenient in the framework of Poisson random field (PRF) models introduced by Sawyer & Hartl (88). These models assume that all loci (individual SNP sites) are independent, i.e., effectively unlinked. This implies that they may provide a good approximation in the analysis of SNP data from multiple locations throughout the genome, but less so in the analysis of DNA sequence data from a single or a few loci. In these models, the expected frequency spectrum (or the entries of a MacDonald-Kreitman table) can be calculated directly using mathematical models. This means that selection coefficients for particular classes of mutations can be estimated directly, and various hypotheses regarding selection can be tested in a rigorous statistical framework (15–17, 89). For example, it is possible to estimate which types of amino acid-changing mutations have the largest effect on fitness (15, 118). Such methods may eventually be very useful when designing statistical methods for predicting which mutations are most likely to cause disease. However, inferences based on PRF models differ fundamentally from most other methods for identifying selection, because the effect of selection on linked neutral sites is not incorporated into the models. Whereas most methods for detecting positive selection in terms of selective sweeps consider the effect of a positively selected mutation on the nearby neutral variation, PRF models provide predictions regarding the selected mutation itself. In most applications, estimates based on PRF models will, therefore, be biased (17). Nonetheless, the PRF models provide a convenient computationally tractable statistical framework for examining the effect of selection on different classes of mutations.

Williamson et al. (118) used PRF models to estimate the average selection coefficient acting on different classes of mutations in the human genome. The novelty of their approach (118) was that a demographic model was fitted to the data from synonymous mutations, while selection coefficients were estimated for the same demographic model applied to nonsynonymous mutations. The resulting test was shown to be robust to many different assumptions regarding demographic processes. By explicitly incorporating demography into the model, a high degree of robustness was achieved. Unfortunately, there are no similar approaches for detecting selection from individual loci containing multiple linked mutations. The current methods for taking demographic processes into account when analyzing data from loci with linked mutations involve extensive simulations of data under various demographic models (3, 75, 116).

SNP Data

With the availability of large-scale SNP data sets, it should, in principle, be possible to provide detailed selection maps in humans and other organisms. Standard methods for detecting selection from population genetics can, in principle, be applied to provide a detailed picture of the regions of the genome that may have been targeted by selection. However, most SNP data have been obtained through a complicated SNP discovery process that minimally involves the discovery (or ascertainment) of SNPs in a small sample followed by genotyping in a larger sample. The process by which the SNPs have been selected affects levels of LD observed in the data (77), the frequency spectrum (77), and levels of population subdivision (74, 115). It also affects the variance in these statistics, complicating genomic methods based on outlier detection. The solution to this problem is to explicitly take the ascertainment process into account. Most statistical methods can be corrected relatively easily (76, 77), leading to new valid methods for detecting selection that take the SNP ascertainment process into account. Unfortunately, most current SNP databases and large-scale SNP genotyping efforts (37) are not associated with sufficiently detailed information regarding the ascertainment process necessary for appropriate ascertainment bias corrections. At present, it is difficult or impossible to make valid inferences regarding selection from most large-scale SNP data sources. It is to be hoped that this will change in the future as researchers become more aware of the importance in maintaining detailed records regarding SNP ascertainment processes.

Comparative Genomic Data

As more and more genomes are sequenced, comparative approaches for detecting positive selection at a genome-wide scale are becoming increasingly common (22, 75). The standard methods for detecting positive (or negative) selection using d_N/d_S ratios can be applied directly in studies on a genomic scale. However, current methods can be improved by establishing models that take advantage of the fact that (ignoring within-species variability) all genes in a phylogeny share the same evolutionary tree.

FUNCTIONAL INFERENCES

In the field of bioinformatics there has been a long tradition of using conserved sites in comparative data to infer function. The implicit assumption is that high levels of conservation are caused by negative selection against new deleterious selection, i.e., functional constraints. In the absence of site-specific suppression of the biological mutation rate, highly reduced levels of variability must be caused by negative selection.

Phylogenetic Footprinting

Although there exist many methods for quantifying how conserved a site, or a set of sites, is, the most statistically solid methods for identifying conserved sites are known as phylogenetic footprinting. In these methods, the rate of substitution in a particular site (or collection of sites) is estimated by considering the pattern of mutation along the underlying phylogeny. This is typically done by mapping mutations onto the phylogenetic tree using parsimony (12) and is complicated by the fact that the alignment may be ambiguous in noncoding regions for divergent species. These methods have been used for a variety of purposes and have been particularly successful in identifying regulatory elements in noncoding DNA (24, 111). The advantage of these methods is that they explicitly take the underlying evolutionary correlations (the phylogeny) into account, leading to increased statistical power and accuracy over methods that do not consider the phylogeny.

One of the most exciting recent discoveries in the field of genomics is the presence of extremely conserved regions, with no known function, in mammalian genomes (11). Such regions may be regulatory regions, containing conserved structural features or unannotated protein-coding genes or RNA genes. To determine if these regions are truly under selection, neutrality tests comparing intraspecific and interspecific variability could be used. There is even the possibility of positive selection in noncoding regions. More research is needed to develop appropriate statistical methods for identifying selection outside coding regions from genomic scale comparative and population genetic data.

Disease Genetics

In disease genetics, there is an increased awareness that regions of the human genome that have been targeted by positive selection may be disease associated (7). Disease-causing mutations should affect organismal fitness, except if the age of onset of the disease is very late. There is, therefore, an intimate relationship between disease and selection that potentially can be exploited in identifying candidate disease loci and candidate SNPs.

A very promising application is in the identification of putative disease-causing SNPs. Evolutionary inferences from comparative and population genetic data, in combination with functional and structural information, can be used to predict which mutations most likely have negative fitness consequences. The mutations with the most severe fitness consequences are obviously the mutations that are most likely to be disease causing. Several different methods have already been described that allow predicting of potential disease-causing mutation (72, 84). These methods may potentially be improved by using explicit population genetic models. This seems to be a particularly promising application of PRF models as these models can describe explicitly the selection coefficients acting on particular classes of mutations (15).

Positive Selection

While there has long been a focus on the use of conservation (negative selection) to find functional elements, increased attention has recently been directed toward the possibility of using inferences regarding positive selection to elucidate functional relationships. In human genetics, several cases are known where recessive disease-causing mutations were thought to be carried to high frequencies in the populations, because they confer a fitness advantage in the heterozygote condition. Diseases that have been hypothesized to have been targeted by this type of overdominant selection include sickle-cell anemia (42), glucose-6-phosphate dehydrogenase deficiency (86), Tay-Sachs disease (99), cystic fibrosis (94), and Phenylketnonuria (121). Not known is how many of the common disease factors have been influenced by overdominant selection, but these observations do suggest that regions of the human genome that have been targeted by balancing selection may contain disease-causing variants worth exploring.

In virology, site-specific inferences regarding positive selection have been used in several cases to identify functionally important sites. In the HIV virus, site-specific inferences of d_N/d_S ratios have been used to identify positions that may be involved in drug resistance (21). In HIV and other viruses, sites that may interact with the host immune system have been identified by detecting site-specific selective pressures, and it has been proposed that such methods may assist in the development of vaccines (36, 95). It has also been proposed that site-specific inferences of d_N/d_S ratios may help predict the evolution of virulent strains of influenza (14). Recently, site-specific inferences of d_N/d_S ratios from different primate species were used to identify a new species-specific retroviral restriction domain (91).

EVIDENCE FOR SELECTION

There is an increasing amount of evidence that selection is important in shaping variation within and between species. In human SNP data, there is a clear difference in the frequency spectrum between nonsynonymous and synonymous mutations (103, 118). This observation in itself shows that a large proportion of the mutations that are segregating in humans (and presumably in other species as well) are affected by selection. In addition, there is a rapidly growing list of specific genes that show evidence for positive selection in both humans and other organisms (7, 31, 98, 113, 125). This explosion of results showing a presence of positive selection may in fact suggest that positive selection is much more common than previously believed. Positively selected mutations may just have remained hidden among all the negatively selected mutations. In addition, ambiguity in the interpretation of classical population genetic neutrality tests, due to the presence of confounding demographic factors, may have precluded the establishment of firm conclusions regarding the pervasiveness of selection. As more large-scale data have accumulated, and methods that are robust to demographic assumptions have been applied, a clearer picture of the pervasiveness of positive selection has been established. Modern versions of the neutral theory (80, 81) allow for a substantial amount of negative selection, and even some positive selection. As the evidence for selection accumulates, the debate regarding the causes of molecular evolution should focus on whether selection is so dominating that effective population sizes and standing levels of variation are best described by the models of repeated selective sweeps favored by Gillespie (40, 41), or whether classical models of genetic drift are most appropriate. In the models that Gillespie has proposed, known as genetic draft models, mutations causing species differences are not neutral mutations increasing in frequency due to genetic drift, but primarily neutral mutations increasing in frequency due to linkage with adaptive mutations sweeping through the population. Even though only few mutations are adaptive, the population genetic dynamics is determined by the selective forces acting on the adaptive mutations, not by genetic drift. There is no mathematical or empirical evidence to suggest that this model is unrealistic, and as the evidence in favor of positive selection accumulates, the question arises whether models of draft should replace models of drift.

With the new availability of very large population genetic and comparative genomic data sets, we should soon be able to determine how many genes, and how big a proportion of mutations, have been affected by positive and negative selection. This will also lead to more evolutionary explorations into the molecular nature of adaptation, help predict which SNPs in humans may be disease associated, and lead to improved functional annotations of genomic data. Methods that combine comparative and population genetic data, and methods that have a high degree of robustness to the underlying demographic factors may be particularly useful in this endeavor.

SUMMARY POINTS

1. Both positive and negative selection leave distinctive signatures at the molecular level that can be detected using statistical tests.
2. In population genetic data, selection may affect levels of variability, linkage disequilibrium, haplotype structure and allelic distribution in each nucleotide site (frequency spectrum). In comparative data, selection has a strong effect on the d_N/d_S ratio.
3. Statistical methods for detecting selection differ in the assumptions they make and how powerful they are. Most methods applicable to population genetic data rely on strong assumptions regarding the demography of the populations, while comparative methods are free of such assumptions.
4. An increasing amount of evidence suggests that positive selection is much more pervasive than previously thought.
5. Inferences regarding selection provide a powerful tool in functional studies, for example for the prediction of possible disease-related genomic regions.

UNRESOLVED ISSUES

1. Can robust statistical population genetic tests be developed that can help identify genomic regions targeted by positive selection?
2. Will inferences regarding selection help identify disease loci in humans and other organisms?
3. Should we focus on genetic draft instead of genetic drift?

LITERATURE CITED

1. Akey JM, Zhang G, Zhang K, Jin L, Shriver MD. 2002. Interrogating a high-density SNP map for signatures of natural selection. *Genome Res.* 12:1805–14
2. **Andolfatto P. 2001. Adaptive hitchhiking effects on genome variability. *Curr. Opin. Genet. Dev.* 11:635–41**
3. Andolfatto P, Przeworski M. 2000. A genome-wide departure from the standard neutral model in natural populations of Drosophila. *Genetics* 156:257–68
4. Andolfatto P, Wall JD, Kreitman M. 1999. Unusual haplotype structure at the proximal breakpoint of In(2L)t in a natural population of *Drosophila melanogaster*. *Genetics* 153:1297–311
5. Anisimova M, Bielawski JP, Yang ZH. 2001. Accuracy and power of the likelihood ratio test in detecting adaptive molecular evolution. *Mol. Biol. Evol.* 18:1585–92
6. Anisimova M, Nielsen R, Yang ZH. 2003. Effect of recombination on the accuracy of the likelihood method for detecting positive selection at amino acid sites. *Genetics* 164:1229–36
7. Bamshad M, Wooding SP. 2003. Signature of natural selection in the human genome. *Nat. Rev. Genet.* 4:99
8. Barton NH. 1995. Linkage and the limits to natural selection. *Genetics* 140:821–41
9. Beaumont MA, Balding DJ. 2004. Identifying adaptive genetic divergence among populations from genome scans. *Mol. Ecol.* 13:969–80
10. Beaumont MA, Nichols RA. 1996. Evaluating loci for use in the genetic analysis of population structure. *Proc. R. Soc. London Ser. B* 263:1619–26
11. Bejerano G, Pheasant M, Makunin I, Stephen S, Kent WJ, et al. 2004. Ultraconserved elements in the human genome. *Science* 304:1321–25
12. Blanchette M, Tompa M. 2002. Discovery of regulatory elements by a computational method for phylogenetic footprinting. *Genome Res.* 12:739–48
13. Braverman JM, Hudson RR, Kaplan NL, Langley CH, Stephan W. 1995. The hitchhiking effect on the site frequency-spectrum of DNA polymorphisms. *Genetics* 140:783–96
14. Bush RM, Bender CA, Subbarao K, Cox NJ, Fitch WM. 1999. Predicting the evolution of human influenza A. *Science* 286:1921–25
15. Bustamante CD, Nielsen R, Hartl DL. 2003. Maximum likelihood and Bayesian methods for estimating the distribution of selective effects among classes of mutations using DNA polymorphism data. *Theor. Popul. Biol.* 63:91–103
16. Bustamante CD, Nielsen R, Sawyer SA, Olsen KM, Purugganan MD, Hartl DL. 2002. The cost of inbreeding in Arabidopsis. *Nature* 416:531–34
17. Bustamante CD, Wakeley J, Sawyer S, Hartl DL. 2001. Directional selection and the site-frequency spectrum. *Genetics* 159:1779–88

A related review focusing on the problem of distinguishing background selection from selective sweeps, with particular focus on *Drosophila* populations.

18. Charlesworth B. 1994. The effect of background selection against deleterious mutations on weakly selected, linked variants. *Genet. Res.* 63:213–27
19. Charlesworth B, Morgan MT, Charlesworth D. 1993. The effect of deleterious mutations on neutral molecular variation. *Genetics* 134:1289–303
20. Charlesworth B, Nordborg M, Charlesworth D. 1997. The effects of local selection, balanced polymorphism and background selection on equilibrium patterns of genetic diversity in subdivided populations. *Genet. Res.* 70:155–74
21. Chen L, Perlina A, Lee CJ. 2004. Positive selection detection in 40,000 human immunodeficiency virus (HIV) type 1 sequences automatically identifies drug resistance and positive fitness mutations in HIV protease and reverse transcriptase. *J. Virol.* 78:3722–32
22. Clark AG, Glanowski S, Nielsen R, Thomas PD, Kejariwal A, et al. 2003. Inferring nonneutral evolution from human-chimp-mouse orthologous gene trios. *Science* 302:1960–63
23. Depaulis F, Veuille M. 1998. Neutrality tests based on the distribution of haplotypes under an infinite-site model. *Mol. Biol. Evol.* 15:1788–90
24. Duret L, Bucher P. 1997. Searching for regulatory elements in human noncoding sequences. *Curr. Opin. Struct. Biol.* 7:399–406
25. **Ewens WJ. 2004. *Mathematical Population Genetics. I. Theoretical Introduction.* Berlin/Heidelberg/New York: Springer**

A great introduction to population genetic theory for the mathematically literate.

26. Eyre-Walker A. 2002. Changing effective population size and the McDonald-Kreitman test. *Genetics* 162:2017–24
27. Fares MA, Moya A, Escarmis C, Baranowski E, Domingo E, Barrio E. 2001. Evidence for positive selection in the capsid protein-coding region of the foot-and-mouth disease virus (FMDV) subjected to experimental passage regimens. *Mol. Biol. Evol.* 18:10
28. Fay JC, Wu CI. 2000. Hitchhiking under positive Darwinian selection. *Genetics* 155:1405–13
29. Felsenstein J. 1974. Evolutionary advantage of recombination. *Genetics* 78:737–56
30. Fitch WM, Bush RM, Bender CA, Cox NJ. 1997. Long term trends in the evolution of H(3) HA1 human influenza type A. *Proc. Natl. Acad. Sci. USA* 94:7712–18
31. Ford MJ. 2002. Applications of selective neutrality tests to molecular ecology. *Mol. Ecol.* 11:1245–62
32. Fu YX. 1996. New statistical tests of neutrality for DNA samples from a population. *Genetics* 143:557–70
33. Fu YX. 1997. Statistical tests of neutrality of mutations against population growth, hitchhiking and background selection. *Genetics* 147:915–25
34. Fu YX, Li WH. 1993. Statistical tests of neutrality of mutations. *Genetics* 133:693–709
35. Galtier N, Depaulis F, Barton NH. 2000. Detecting bottlenecks and selective sweeps from DNA sequence polymorphism. *Genetics* 155:981–87
36. Gaschen B, Taylor J, Yusim K, Foley B, Gao F, et al. 2002. Diversity considerations in HIV-1 vaccine selection. *Science* 299:1515–18
37. Gibbs RA, Belmont JW, Hardenbol P, Willis TD, Yu FL, et al. 2003. The International HapMap Project. *Nature* 426:789–96
38. Gillespi JH, Langley CH. 1974. General model to account for enzyme variation in natural populations. *Genetics* 76:837–84
39. Gillespie JH. 1991. *The Causes of Molecular Evolution*. New York: Oxford Univ. Press. 336 pp.
40. Gillespie JH. 2000. Genetic drift in an infinite population: the pseudohitchhiking model. *Genetics* 155:909–19

41. Gillespie JH. 2001. Is the population size of a species relevant to its evolution? *Evolution* 55:2161–69
42. Haldane JBS. 1949. Disease and evolution. *Ricerca Sci.* 19:3–10
43. Haldane JBS. 1957. The cost of natural selection. *Genetics* 55:511–24
44. Harr B, Kauer M, Schlotterer C. 2002. Hitchhiking mapping: a population-based fine-mapping strategy for adaptive mutations in *Drosophila melanogaster*. *Proc. Natl. Acad. Sci. USA* 99:12949–54
45. Haydon DT, Bastos AD, Knowles NJ, Samuel AR. 2001. Evidence for positive selection in foot-and-mouth disease virus capsid genes from field isolates. *Genetics* 157:7
46. Henikoff S, Malik HS. 2002. Selfish drivers. *Nature* 417:227
47. Hudson RR, Bailey K, Skarecky D, Kwiatowski J, Ayala FJ. 1994. Evidence for positive selection in the superoxide-dismutase (Sod) region of *Drosophila-melanogaster*. *Genetics* 136:1329–40
48. Hudson RR, Kreitman M, Aguade M. 1987. A test of neutral molecular evolution based on nucleotide data. *Genetics* 116:153–59
49. Hughes AL, Nei M. 1988. Pattern of nucleotide substitution at major histocompatibility complex class-I loci reveals overdominant selection. *Nature* 335:167–70
50. Ingvarsson PK. 2004. Population subdivision and the Hudson-Kreitman-Aguade test: testing for deviations from the neutral model in organelle genomes. *Genet. Res.* 83:31–39
51. Kaplan NL, Darden T, Hudson RR. 1988. The coalescent process in models with selection. *Genetics* 120:819–29
52. Kaplan NL, Hudson RR, Langley CH. 1989. The hitchhiking effect revisited. *Genetics* 123:887–99
53. Kayser M, Brauer S, Stoneking M. 2003. A genome scan to detect candidate regions influenced by local natural selection in human populations. *Mol. Biol. Evol.* 20:893–900
54. Kelly JK. 1997. A test of neutrality based on interlocus associations. *Genetics* 146:1197–206
55. Kim Y, Nielsen R. 2004. Linkage disequilibrium as a signature of selective sweeps. *Genetics* 167:1513–24
56. Kim Y, Stephan W. 2002. Detecting a local signature of genetic hitchhiking along a recombining chromosome. *Genetics* 160:765–77
57. Kim Y, Stephan W. 2003. Selective sweeps in the presence of interference among partially linked loci. *Genetics* 164:389–98
58. Kimura M. 1968. Evolutionary rate at the molecular level. *Nature* 217:624
59. Kimura M. 1983. *The Neutral Theory of Molecular Evolution*. New York: Cambridge Univ. Press. 367 pp.
60. Kimura M. 1995. Limitations of Darwinian selection in a finite population. *Proc. Natl. Acad. Sci. USA* 92:2343–44
61. Kimura M, Crow J. 1964. The number of alleles that can be maintained in a finite population. *Genetics* 40:725–38
62. Kondrashov AS. 1982. Selection against harmful mutations in large sexual and asexual populations. *Genet. Res.* 40:325–32
63. Lewontin RC, Krakauer J. 1973. Distribution of gene frequency as a test of theory of selective neutrality of polymorphisms. *Genetics* 74:175–95
64. Luikart G, England PR, Tallmon D, Jordan S, Taberlet P. 2003. The power and promise of population genomics: from genotyping to genome typing. *Nat. Rev. Genet.* 4:981–94

65. Majewski J, Cohan FM. 1999. Adapt globally, act locally: the effect of selective sweeps on bacterial sequence diversity. *Genetics* 152:1459–74
66. Malik HS, Henikoff S. 2001. Adaptive evolution of cid, a centromere-specific histone in Drosophila. *Genetics* 157:1293–98
67. Malik HS, Henikoff S. 2002. Conflict begets complexity: the evolution of centromeres. *Curr. Opin. Genet. Dev.* 12:711–18
68. Maynard Smith J, Haigh J. The hitch-hiking effect of a favourable gene. *Genet. Res.* 23:23–35
69. McDonald JH, Kreitman M. 1991. Adaptive protein evolution at the Adh locus in Drosophila. *Nature* 351:652–54
70. McVean GAT, Myers SR, Hunt S, Deloukas P, Bentley DR, Donnelly P. 2004. The fine-scale structure of recombination rate variation in the human genome. *Science* 304:581–84
71. Milkman RD. 1967. Heterosis as a major cause of heterozygosity in nature. *Genetics* 55:493–95
72. Ng PC, Henikoff S. 2001. Predicting deleterious amino acid substitutions. *Genome Res.* 11:863–74
73. Nielsen R. 2001. Statistical tests of selective neutrality in the age of genomics. *Heredity* 86:641–47
74. Nielsen R. 2004. Population genetic analysis of ascertained SNP data. *Human Genomics* 3:218–24
75. Nielsen R, Bustamante CD, Clark AG, Glanowski S, Sackton TB, et al. 2005. A scan for positively selected genes in the genomes of humans and chimpanzees. *PLoS Biol.* In press
76. Nielsen R, Hubisz MJ, Clark AG. 2004. Reconstituting the frequency spectrum of ascertained single-nucleotide polymorphism data. *Genetics* 168:2373–82
77. Nielsen R, Signorovitch J. 2003. Correcting for ascertainment biases when analyzing SNP data: applications to the estimation of linkage disequilibrium. *Theor. Popul. Biol.* 63:245–55
78. Nielsen R, Yang ZH. 1998. Likelihood models for detecting positively selected amino acid sites and applications to the HIV-1 envelope gene. *Genetics* 148:929–36
79. Nielsen R, Yang ZH. 2003. Estimating the distribution of selection coefficients from phylogenetic data with applications to mitochondrial and viral DNA. *Mol. Biol. Evol.* 20:1231–39
80. Ohta T. 1992. The nearly neutral theory of molecular evolution. *Annu. Rev. Ecol. Syst.* 23:263–86
81. Ohta T. 2002. Near-neutrality in evolution of genes and gene regulation. *Proc. Natl. Acad. Sci. USA* 99:16134–37
82. Przeworski M. 2002. The signature of positive selection at randomly chosen loci. *Genetics* 160:1179–89
83. Ptak SE, Roeder AD, Stephens M, Gilad Y, Paabo S, Przeworski M. 2004. Absence of the TAP2 human recombination hotspot in chimpanzees. *PLoS Biol.* 2:849–55
84. Ramensky V, Bork P, Sunyaev S. 2002. Human non-synonymous SNPs: server and survey. *Nucleic Acids Res.* 30:3894–900
85. Ramos-Onsins SE, Rozas J. 2002. Statistical properties of new neutrality tests against population growth. *Mol. Biol. Evol.* 19:2092–100
86. Ruwende C, Khoo SC, Snow AW, Yates SNR, Kwiatkowski D, et al. 1995. Natural-selection of hemizygotes and heterozygotes for G6pd deficiency in Africa by resistance to severe malaria. *Nature* 376:246–49

The classic paper introducing the idea of a selective sweep.

87. Sabeti PC, Reich DE, Higgins JM, Levine HZP, Richter DJ, et al. 2002. Detecting recent positive selection in the human genome from haplotype structure. *Nature* 419:832–37

88. Sawyer SA, Hartl DL. 1992. Population genetics of polymorphism and divergence. *Genetics* 132:1161–76

89. Sawyer SA, Kulathinal RJ, Bustamante CD, Hartl DL. 2003. Bayesian analysis suggests that most amino acid replacements in Drosophila are driven by positive selection. *J. Mol. Evol.* 57(Suppl.) 1:S154–64

90. Sawyer SL, Emerman M, Malik HS. 2004. Ancient adaptive evolution of the primate antiviral DNA-editing enzyme APOBEC3G. *PLoS Biol.* 2:1278–85

91. Sawyer SL, Wu LI, Emerman M, Malik HS. 2005. Positive selection of primate TRIM5 alpha identifies a critical species-specific retroviral restriction domain. *Proc. Natl. Acad. Sci. USA* 102:2832–37

92. Schlotterer C. 2002. A microsatellite-based multilocus screen for the identification of local selective sweeps. *Genetics* 160:753–63

93. Schlotterer C. 2003. Hitchhiking mapping—functional genomics from the population genetics perspective. *Trends Genet.* 19:32–38

94. Schroeder SA, Gaughan DM, Swift M. 1995. Protection against bronchial-asthma by Cftr Delta-F508 mutation—a heterozygote advantage in cystic-fibrosis. *Nat. Med.* 1:703–5

95. Sheridan I, Pybus OG, Holmes EC, Klenerman P. 2004. High-resolution phylogenetic analysis of Hepatitis C virus adaptation and its relationship to disease progression. *J. Virol.* 78:3447–54

96. Simonsen KL, Churchill GA, Aquadro CF. 1995. Properties of statistical tests of neutrality for DNA polymorphism data. *Genetics* 141:413–29

97. Slatkin M, Wiehe T. 1998. Genetic hitch-hiking in a subdivided population. *Genet. Res.* 71:155–60

98. Smith NGC, Eyre-Walker A. 2002. Adaptive protein evolution in Drosophila. *Nature* 415:1022–24

99. Spyropoulos B, Moens PB, Davidson J, Lowden JA. 1981. Heterozygote advantage in Tay-Sachs carriers. *Am. J. Hum. Genet.* 33:375–80

100. Stahl EA, Bishop JG. 2000. Plant-pathogen arms races at the molecular level. *Curr. Opin. Plant Biol.* 3:299

101. Storz JF. 2005. Using genome scans of DNA polymorphism to infer adaptive population divergence. *Mol. Ecol.* 14:671–88

102. Storz JF, Payseur BA, Nachman MW. 2004. Genome scans of DNA variability in humans reveal evidence for selective sweeps outside of Africa. *Mol. Biol. Evol.* 21:1800–11

103. Sunyaev SR III WCL, Ramensky VE, Bork P. 2000. SNP frequencies in human genes an excess of rare alleles and differing modes of selection. *Trends Genet.* 16:335–37

104. Suzuki Y, Gojobori T. 1999. A method for detecting positive selection at single amino acid sites. *Mol. Biol. Evol.* 16:1315–28

105. Suzuki Y, Nei M. 2002. Simulation study of the reliability and robustness of the statistical methods for detecting positive selection at single amino acid sites. *Mol. Biol. Evol.* 19:1865–69

106. Suzuki Y, Nei M. 2004. False-positive selection identified by ML-based methods: examples from the Sig1 gene of the diatom *Thalassiosira weissflogii* and the tax gene of a human T-cell lymphotropic virus. *Mol. Biol. Evol.* 21:914–21

107. Sved JA, Reed TE, Bodmer WF. 1967. The number of balanced polymorphisms that can be maintained in a natural population. *Genetics* 55:469–81

The first paper introducing PRF models as a statistical framework for population genetic inferences.

An elegant paper showing how inferences regarding selection can be used to make functional predictions that can be tested in the lab.

A related review of methods for detecting selection in genomic scans, with particular focus on statistics based on population differentiation.

108. Swanson WJ, Aquadro CF, Vacquier VD. 2001. Polymorphism in abalone fertilization proteins is consistent with the neutral evolution of the egg's receptor for lysin (VERL) and positive Darwinian selection of sperm lysin. *Mol. Biol. Evol.* 18:376–83
109. Swanson WJ, Nielsen R, Yang QF. 2003. Pervasive adaptive evolution in mammalian fertilization proteins. *Mol. Biol. Evol.* 20:18–20
110. Swanson WJ, Zhang ZH, Wolfner MF, Aquadro CF. 2001. Positive Darwinian selection drives the evolution of several female reproductive proteins in mammals. *Proc. Natl. Acad. Sci. USA* 98:2509–14
111. Tagle D, Koop B, Goodman M, Slightom J, Hess D, Jones R. 1988. Embryonic ε and γ globin genes of a prosimian primate (*Galago crassicaudatus*); nucleotide and amino acid sequences, developmental regulation and phylogenetic footprints. *J. Mol. Biol.* 203:439–55
112. Tajima F. 1989. Statistical method for testing the neutral mutation hypothesis by DNA polymorphism. *Genetics* 123:585–95
113. Vallender EJ, Lahn BT. 2004. Positive selection on the human genome. *Hum. Mol. Genet.* 13:R245–54
114. Vitalis R, Dawson K, Boursot P. 2001. Interpretation of variation across marker loci as evidence of selection. *Genetics* 158:1811–23
115. Wakeley J, Nielsen R, Liu-Cordero SN, Ardlie K. 2001. The discovery of single-nucleotide polymorphisms—and inferences about human demographic history. *Am. J. Hum. Genet.* 69:1332–47
116. Wall JD, Andolfatto P, Przeworski M. 2002. Testing models of selection and demography in *Drosophila simulans*. *Genetics* 162:203–16
117. Wall JD, Frisse LA, Hudson RR, Di Rienzo A. 2003. Comparative linkage-disequilibrium analysis of the beta-globin hotspot in primates. *Am. J. Hum. Genet.* 73:1330–40
118. Williamson SH, Hernadez R, Fledel-Alon A, Zhu L, Nielsen R, Bustamante CD. 2005. Simultaneous inference of selection and population growth from patterns of variation in the human genome. *Proc. Natl. Acad. Sci. USA.* 102:7882–87
119. Wong WSW, Nielsen R. 2004. Detecting selection in noncoding regions of nucleotide sequences. *Genetics* 167:949–58
120. Wong WSW, Yang ZH, Goldman N, Nielsen R. 2004. Accuracy and power of statistical methods for detecting adaptive evolution in protein coding sequences and for identifying positively selected sites. *Genetics* 168:1041–51
121. Woolf LI, McBean MS, Woolf FM, Cahalane SF. 1975. Phenylketonuria as a balanced polymorphism—nature of heterozygote advantage. *Ann. Hum. Genet.* 38:461–69
122. Wyckoff GJ, Wang W, Wu CI. 2000. Rapid evolution of male reproductive genes in the descent of man. *Nature* 403:304
123. Yang ZH. 1998. Likelihood ratio tests for detecting positive selection and application to primate lysozyme evolution. *Mol. Biol. Evol.* 15:568–73
124. Yang ZH. 2002. Inference of selection from multiple species alignments. *Curr. Opin. Genet. Dev.* 12:688–94
125. Yang ZH, Bielawski JP. 2000. Statistical methods for detecting molecular adaptation. *Trends. Ecol. Evol.* 15:496–503
126. Yang ZH, Nielsen R. 1998. Synonymous and nonsynonymous rate variation in nuclear genes of mammals. *J. Mol. Evol.* 46:409–18
127. Yang ZH, Nielsen R. 2000. Estimating synonymous and nonsynonymous substitution rates under realistic evolutionary models. *Mol. Biol. Evol.* 17:32–43

Tajima's classic paper introducing his well-known neutrality test.

A nice study showing how both selection and demography must be taken into account when interpreting genetic data.

A review of the statistical methodology used to detect positive selection using d_N/d_S ratios.

128. Yang ZH, Nielsen R. 2002. Codon-substitution models for detecting molecular adaptation at individual sites along specific lineages. *Mol. Biol. Evol.* 19:908–17
129. Yang ZH, Nielsen R, Goldman N, Pedersen AMK. 2000. Codon-substitution models for heterogeneous selection pressure at amino acid sites. *Genetics* 155:431–49

RELATED RESOURCES

Fay JC, Wu C-I. 2003. Sequence divergence, functional constraint, and selection in protein evolution. *Annu. Rev. Genomics Hum. Genet.* 4:213–35
Lewontin RC. 2002. Directions in evolutionary biology. *Annu Rev. Genet.* 36:1–18
Kreitman M. 2000. Methods to detect selection in populations with applications to the human. *Annu. Rev. Genomics Hum. Genet.* 1:539–59

T-Box Genes in Vertebrate Development

L.A. Naiche,[1] Zachary Harrelson,[1]
Robert G. Kelly,[1,2] and Virginia E. Papaioannou[1]

[1]Department of Genetics and Development, College of Physicians and Surgeons of Columbia University, New York, New York 10032; email: nla2@columbia.edu, zh42@columbia.edu, vep1@columbia.edu

[2]Developmental Biology Institute of Marseilles, Campus de Luminy Case 907, 13288 Marseille Cedex 9, France; email: kelly@ibdm.univ-mrs.fr

Annu. Rev. Genet.
2005. 39:219–39

First published online as a Review in Advance on July 5, 2005

The *Annual Review of Genetics* is online at http://genet.annualreviews.org

doi: 10.1146/annurev.genet.39.073003.105925

0066-4197/05/1215-0219$20.00

Key Words

Tbx, transcription factor, *Brachyury*, organogenesis, heart, limbs

Abstract

The myriad developmental roles served by the T-box family of transcription factor genes defy easy categorization. Present in all metazoans, the T-box genes are involved in early embryonic cell fate decisions, regulation of the development of extraembryonic structures, embryonic patterning, and many aspects of organogenesis. They are unusual in displaying dosage sensitivity in most instances. In humans, mutations in T-box genes are responsible for developmental dysmorphic syndromes, and several T-box genes have been implicated in neoplastic processes. T-box transcription factors function in many different signaling pathways, notably bone morphogenetic protein and fibroblast growth factor pathways. The few downstream target genes that have been identified indicate a wide range of downstream effectors.

Contents

INTRODUCTION

The T-box family, defined by a common DNA binding domain known as the T-box, is evolutionarily ancient, probably arising in the common ancestor of metazoan organisms (2) (**Figure 1**). T-box genes first came to the attention of geneticists in 1927 with the discovery of a mutation, *Brachyury* (or *T*, for short-tail), which caused truncated tails in mice (30). In recent years, both spontaneous and induced mutations in T-box genes have demonstrated that these genes are important regulators of a wide range of tissues and organs during development, as well as major contributors to several human syndromes (**Table 1**). As this family was discovered quite recently, comparatively little is known about transcriptional regulatory capabilities and signaling interactions of its members. Nonetheless, its importance in an array of developing tissues has led to the rapid expansion of the field. In this review, we explore recent literature on T-box gene function, concentrating on mammalian development.

TBE: T-box binding element

T-BOX TRANSCRIPTION FACTORS

T-Box Proteins and DNA Binding

The T-box DNA binding sequence, the T-site or T-box binding element (TBE), was first defined as the sequence with the highest affinity for Brachyury (57). Brachyury binds this palindromic sequence as a dimer (77), with each monomer of Brachyury binding half of the sequence, or T-half site (5′-AGGTGTGAAATT-3′). Extensive studies have demonstrated that all T-box proteins tested are capable of binding the T-half site as monomers (15, 49, 61, 77, 78, 97, 98), although some have different optimal target sequences (36, 65). Comparisons between T-box proteins have shown preference for different synthetic combinations of T-half sites in varying orientations, numbers, and spacing (27, 97), which may help create promoter specificity for target genes.

The crystal structures of both Brachyury and TBX3 T-domain homodimers bound to the canonical T-site have been elucidated (26, 71). Both T-box proteins make the same DNA contacts with the same amino acids, indicating strong conservation of the underlying DNA binding functions between T-box subfamilies. However, whereas the Brachyury dimer is stabilized by a hydrophobic patch and a salt bridge, TBX3 dimers are oriented differently on the DNA and are weakly connected. These differences in ternary structure probably underlie the differences in half site preference among different T-box family members.

Transcriptional Regulation

As well as binding DNA, T-box genes have been shown to regulate transcription. Activation domains have been mapped to the

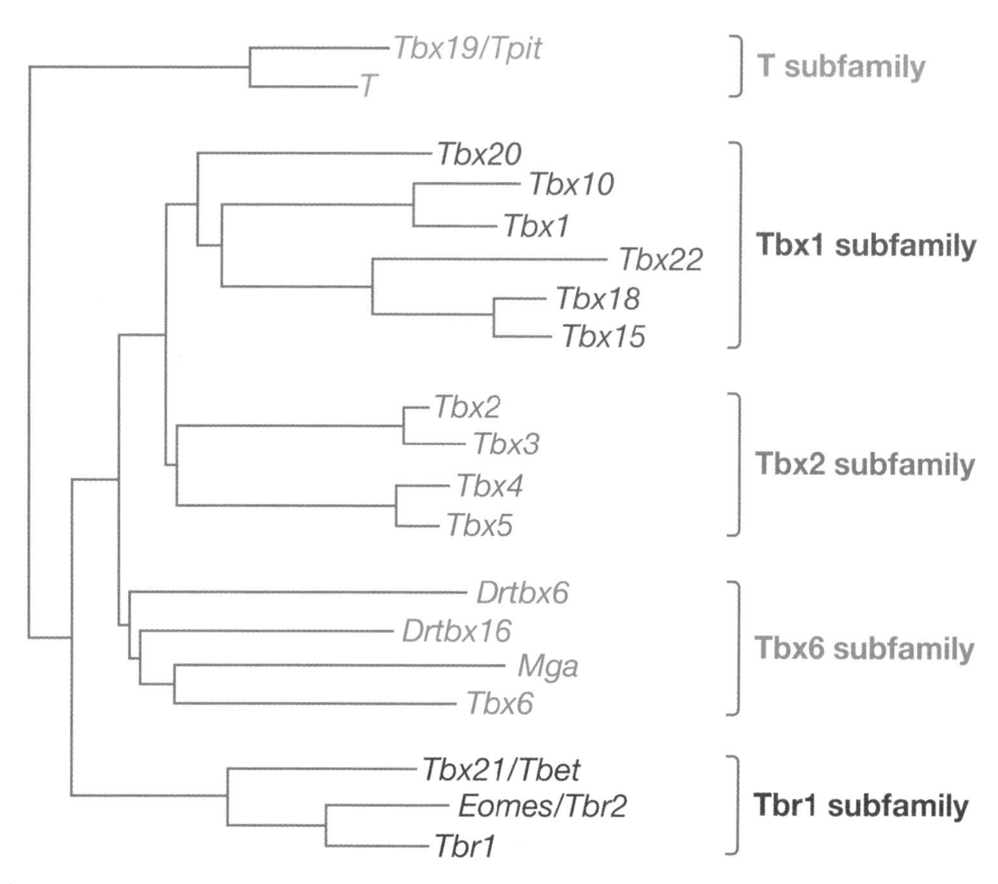

Figure 1

Schematic phylogenetic tree of the T-box gene family of vertebrates, based on the phylogenetic analysis in Reference (76) showing the relationship of genes in the five subfamilies indicated by brackets on the right. All of these genes are present in mammals with the exception of the zebrafish genes *Drtbx6* and *Drtbx16*, which do not have orthologs in mammals.

C-terminal domains of several T-box proteins (56, 98, 120). In some cases, the mechanism of activation is known: Tbx19 activates transcription by recruiting SRC/p160 coactivators to the promoter (68), while Tbr1 forms a complex with nucleosome assembly proteins (112). T-box proteins can also repress transcription, as has been shown for Tbx2 and Tbx3 (21, 41, 65). Some T-box genes contain both activation and repression domains in their C-terminal domains (56, 98) and Tbx2 has been reported to act in either fashion, depending on promoter context (78).

Interactions with Other Transcription Factors

Athough some T-box gene targets appear to be regulated by T-box proteins alone (60), there is a growing body of work demonstrating that target genes are controlled in combination with other transcription factors. Cooperative binding of promoters and synergistic upregulation of target gene expression is seen with T-box factors and homeodomain (13, 61, 98), GATA zinc finger (31, 35, 98), and LIM domain proteins (59). Frequently, these interactions enhance target gene activation, and

Table 1 Comparison of the effects of mutations in all known human and mouse T-box genes illustrating the prevalence of dosage sensitivity of the phenotypes[a]

Mouse gene; human gene	Human syndrome	Mouse heterozygous phenotype	Mouse homozygous phenotype
T; *T*	Not known	Viable, short/no tail (30)	Embryonic lethal, failure of posterior mesoderm (45)
Tbx19; *TBX19* (*TPIT*)	Recessive isolated ACTH deficiency (84)	Normal (84)	ACTH deficiency, pigment defects (84)
Tbx1; *TBX1*	DiGeorge, craniofacial, glandular, vascular, and heart abnormalities (4)	Viable, thymus and vascular abnormalities (52, 64)	Neonatal lethal; craniofacial, glandular, vascular, and heart abnormalities (52, 64)
Tbx10; *TBX10*	Not known	Susceptibility to cleft lip and palate (*Dancer*: ectopic gain-of-function) (17)	Cleft lip and palate (*Dancer*: ectopic gain-of-function) (17)
Tbx15; *TBX15*	Not known	Normal (19)	Craniofacial viable, malformations and pigment pattern alterations (*droopy ear*) (19)
Tbx18; *TBX18*	Not known	Normal (18)	Postnatal lethal, vertebral malformations (18)
Tbx20; *TBX20*	Not known	Heart contractile function defects (99)	Embryonic lethal, heart abnormalities (99)
Tbx22; *TBX22*	X-linked cleft palate with ankyloglossia (11)	Not known	Not known
Tbx2; *TBX2*	Not known	Normal (43)	Embryonic lethal, heart and limb abnormalities (43)
Tbx3; *TBX3*	Ulnar-mammary: hypoplastic mammary glands, abnormal external genitalia, limb abnormalities (5)	Hypoplastic mammary glands, abnormal external genitalia (28, 53)	Embryonic lethal, yolk sac, limb and mammary gland defects (28)
Tbx4; *TBX4*	Small patella (10)	Reduced allantois growth rate (72)	Embryonic lethal, allantois and hindlimb defects (72)
Tbx5; *TBX5*	Holt-Oram, heart and hand abnormalities (7)	Heart abnormalities, reduced viability (14)	Embryonic lethal, severe heart malformations (14)
Tbx6; *TBX6*	Not known	Normal (24)	Embryonic lethal, somite abnormalities (24)
Tbr1; *TBR1*	Not known	Normal (16, 46)	Olfactory bulb and cortical defects (16, 46)
Eomes; *EOMES*	Not known	Normal (89)	Embryonic lethal, trophoblast and mesoderm failure (89)
Tbx21; *TBX21* (*TBET*)	Not known	Airway hyperresponsiveness, intermediate INF-γ levels in Th1 cells (32, 104)	Airway hyperresponsiveness, no Th1 cells (32, 104)

[a]Text colors indicate genes in different subfamilies as indicated in **Figure 1** (see text for additional references).

probably contribute to promoter specificity. Some of these interactions can be generalized to transcription factor subfamilies—both T subfamily (but not Tbx1 subfamily) proteins directly interact with all members of the Pitx family (but not the closely related Otx family) of homeodomain proteins (61). Some of these interactions are exquisitely specific—Tbx20 interacts with GATA5 but not the related transcription factor GATA4 (98). In an even more extreme case of specificity, LMP4 binds both of the most closely related vertebrate T-box proteins, Tbx4 and Tbx5, but interacts with each via a different LIM domain repeat (59).

The observed in vitro interactions have biological relevance. Mutations in *TBX5* cause Holt-Oram syndrome (HOS), which results in multiple heart defects. While some patients have truncation mutations in *TBX5* that result in loss of DNA binding or activation, others have only point mutations. Analyses of such mutant proteins have shown that loss of interaction with cardiac transcription factor NKX2-5 is sufficient to cause disease, even when the mutant TBX5 is otherwise intact (31). Likewise, point mutations in NKX2–5, which ablate TBX5 binding, have been shown to cause heart disease in humans (35), further demonstrating the biological requirements for cooperative binding between T-box proteins and other factors.

Predicting T-Box Gene Function

Reliable prediction of T-box gene function is complicated by considerable functional lability. Depending on context, T-box proteins may homodimerize, heterodimerize, or cooperatively bind other transcription factors. Individual T-box proteins can exhibit either activation or repression activities, and single genes may do both depending on promoter context. In some cases, one T-box protein is capable of competing off another at a particular promoter (40, 41), making it difficult to predict which is relevant to a particular target gene in cells where the T-box genes are coexpressed.

Several labs have attempted to inhibit individual T-box gene function with putative dominant negative proteins, e.g., a truncated version containing only the T-box domain (29, 87, 90, 101, 106a). The action of these engineered proteins may be complex and may influence other T-box genes or targets, so caution must be used when interpreting such experiments. Although dominant negative proteins may be specific, in some cases a presumed dominant negative protein produces phenotypic consequences more severe (93) or different (90) from what is observed in the genetic null, indicating interference with other protein(s).

HOS: Holt-Oram syndrome

TE: trophectoderm

dpc: days post coitus

T-BOX GENES DURING EARLY DEVELOPMENT

T-box gene expression is widespread during embryonic development and has been noted in all stages, from the oocyte (12, 115) to the adult (102). In the early embryo, T-box genes are required for both evolutionarily ancient processes, such as gastrulation [recently reviewed by (94)] and comparatively recent developments such as uterine implantation and umbilicus formation.

Roles in Extraembryonic Tissues

The earliest demonstrated role of T-box genes in mammalian embryos is that of *Eomesodermin*. In the mouse embryo, the first lineage decision occurs when the outer cells of the morula differentiate into the trophectoderm (TE), which will form placental structures. TE dramatically upregulates *Eomesodermin*. In the absence of *Eomesodermin*, TE cells are defective and neither proliferate in vitro nor participate in uterine implantation in vivo (89), resulting in embryonic death shortly after implantation.

Brachyury is expressed in the caudal primitive streak and is responsible for posterior mesodermal development. *Brachyury* expression extends into the base of the allantois, an extraembryonic mesodermal outgrowth that will eventually become the umbilical cord. Deletion of *Brachyury* leads to stunted growth of the allantois; however, this is likely secondary to a more general defect in mesoderm production (8). In contrast, *Tbx4* is expressed in the allantois and has allantois-specific effects when mutated. Specifically, *Tbx4* is expressed at the site of origin of the allantois and continues to be expressed in the allantois and umbilical cord through at least 13.5 days post coitus (dpc). Embryos homozygous for a null mutation in *Tbx4*

BMP: bone morphogenetic protein

FGF: fibroblast growth factor

PSM: presomitic mesoderm

have multiple abnormalities in the allantois, including upregulated apoptosis, failure to undergo characteristic morphological changes, and failure to express known markers of allantois differentiation such as *Tbx2* and *VCAM1*. Loss of *Tbx4* also disrupts allantoic vasculogenesis after endothelial cells differentiate from the allantoic mesoderm, but before they coalesce into patent blood vessels (72). Little is known about the growth or patterning of the allantois, so it is difficult to firmly tie the diverse defects into known signaling pathways. Mutation of the bone morphogenetic protein (BMP) *Bmp4* in the epiblast causes similar allantois defects (33), but *Bmp4* is expressed normally in *Tbx4* mutants, suggesting that *Tbx4* may be a downstream effector of BMP signaling in this tissue.

Tbx3 is expressed in the yolk sac in both endoderm and mesoderm layers. Disruption of *Tbx3* results in an aberrant yolk sac with variably diminished vascular development and a highly apoptotic endoderm layer (28). It is unclear whether the primary defect is in one or both layers and it is possible that these defects are secondary to heart defects in the embryo (Z.H., R.G.K., V.E.P., unpublished).

T-Box Gene Function in Mesoderm

Many T-box genes serve important functions during mesoderm formation and patterning in the vertebrate gastrula: *T* and *Eomesodermin* in the mouse; *Xbra*, *Xeomesodermin*, and *XvegT* in *Xenopus*; *no tail*, *spadetail*, *eomesodermin*, and *tbx6* in zebrafish (75). These genes are critical for mesoderm formation at various axial levels and regulate such key factors as fibroblast growth factor (FGF) signaling and cell migration. However, T-box gene functions during gastrulation are recursive and overlapping, creating a complex web of developmental roles and signaling interactions. This field has been recently reviewed (94).

The role of T-box genes in patterning the embryonic mesoderm, however, is not restricted to gastrulation. Several T-box genes are involved in the patterning of nascent mesoderm following its ingression through the primitive streak. *Tbx6* is expressed in the paraxial, presomitic mesoderm (PSM) of the mouse embryo after its exit from the primitive streak (22a). Embryos homozygous for a null mutation in *Tbx6* die mid-gestation. Rostral somites are present but morphologically abnormal, indicating that *Tbx6* is not absolutely required for somite formation (24). However, the posterior embryo is progressively more affected: Caudal somites are replaced with ectopic paraxial neural tubes, and the tail bud, the ongoing source of new mesoderm, is aberrantly expanded. The *Tbx6* homozygous mutant phenotype lends itself to two nonexclusive interpretations. First, the presence of ectopic neural tubes in place of caudal somites suggests that *Tbx6* is required to promote a mesodermal fate in posterior paraxial tissue, and that in its absence a default neural program predominates. In vivo teratoma analysis revealed a conspicuous absence of skeletal muscle in tumors derived from *Tbx6* null tail bud cells, implying a potential requirement for myogenic specification or differentiation. In vitro differentiation assays, however, show the lack of an absolute requirement for *Tbx6* in myogenesis (22). A second potential explanation for the *Tbx6* mutant phenotype concerns the enlarged tail bud. Levels of cellular proliferation and programmed cell death are unaltered in *Tbx6* null tail buds (22), suggesting that *Tbx6* mutant mesoderm may fail to migrate out of the tail bud. Although this hypothesis is not incompatible with a role for *Tbx6* in mesodermal specification, the formation of ectopic neural tubes might result from an insufficient supply of mesoderm to the paraxial regions as a secondary defect of deficient migration from the tail bud.

Additional information on the role of *Tbx6* comes from transgenic rescue experiments and a naturally occurring *Tbx6* mutation called *rib-vertebrae* (*rv*) (113, 114). Homozygous *rv* mutant embryos, as well as null *Tbx6* mutant embryos rescued with a transgene that partially restores *Tbx6* expression, live past mid-gestation only to display later defects in rib and vertebrae development. These abnormalities are preceded by reduced expression of markers of the anterior compartment of the somite, with reciprocally expanded expression of markers normally restricted to the posterior somite (114), showing that *Tbx6* is required to maintain the anterior compartment. *Tbx6* is therefore required not only to specify somite formation in the paraxial mesoderm, but also for somite patterning. Severity of the vertebral defects increases as levels of *Tbx6* are progressively reduced with *rv* and *Tbx6* null alleles, implicating dosage as an important aspect of *Tbx6* function in later somite development (113).

Similar to *Tbx6*, *Tbx18* is required to maintain the fate of the anterior compartment of somites. Homozygous *Tbx18* null mutant mice die perinatally and display a range of rib and vertebrae defects. Molecular analysis of these mutant embryos shows that, in the absence of *Tbx18*, anterior and posterior somite compartment specification occurs correctly in the PSM but fails to be maintained during somite maturation (18). Studies in zebrafish (9), chick (42a, 107), and mouse (58) show that *Tbx18* is expressed in the anterior region of somites. Although the mechanism through which this expression is achieved varies between species, all studies agree that *Tbx18* transcription is initiated in the PSM and progressively restricted to the anterior compartment as each somite matures. PSM injected with a *Tbx18* expression vector can induce the formation of somite boundaries when grafted into PSM caudal to where *Tbx18* is normally expressed, suggesting a role for the gene in somite segmentation (107).

tbx24 in zebrafish is a third T-box gene directly involved in the regulation of vertebrate somite development. *tbx24* does not cluster with any of the T-box gene subfamilies and there is no mammalian ortholog. Expression is first detected in paraxial mesoderm of the zebrafish gastrula and is later confined to the anterior and intermediate PSM. Morpholino-mediated *tbx24* knockdown experiments yield embryos with morphological and molecular evidence of disrupted somite segmentation. An early role in mesoderm specification is unlikely as mesodermal and neural markers are expressed normally in *tbx24* morphants. Rescue experiments show that the naturally occurring *fused somites* phenotype is caused by mutations in *tbx24* (74).

IFT: inflow tract

OFT: outflow tract

T-BOX GENES DURING ORGANOGENESIS

Complex Functions During Cardiogenesis

Many T-box genes are expressed in specific chambers or regions of the developing vertebrate heart, including *Tbx1*, *Tbx2*, *Tbx3*, *Tbx5*, *Tbx18*, and *Tbx20* (75). Despite overlapping expression patterns (**Figure 2**), experimental studies reveal that each gene has unique developmental functions. The identification of cardiac transcriptional binding partners with differential affinities for individual T-box factors has contributed to understanding how the combined regulatory influences of multiple, related and coexpressed genes generate unique downstream target gene expression patterns during organogenesis.

Targeted mutagenesis in mouse has revealed essential roles for *Tbx5*, *Tbx1*, *Tbx2*, and *Tbx20* in cardiac development. Mouse embryos homozygous for a null mutation in *Tbx5* exhibit abnormal development of posterior heart structures, including hypoplastic left ventricle, atria, and inflow tract (IFT) (15). These defects are accompanied by the reduced expression of critical cardiac genes, including *GATA4*, *MLC2v*, and *Irx4* (15). Ectopic expression of *Tbx5* in both chick and mouse also implicates *Tbx5* in the development and positioning of the interventricular septum (106). Homozygous loss of *Tbx1* produces abnormalities of anterior cardiac development, including shortening of the outflow tract (OFT), the absence of OFT septation, ventricular septal defects, and abnormal remodeling of the aortic arch arteries

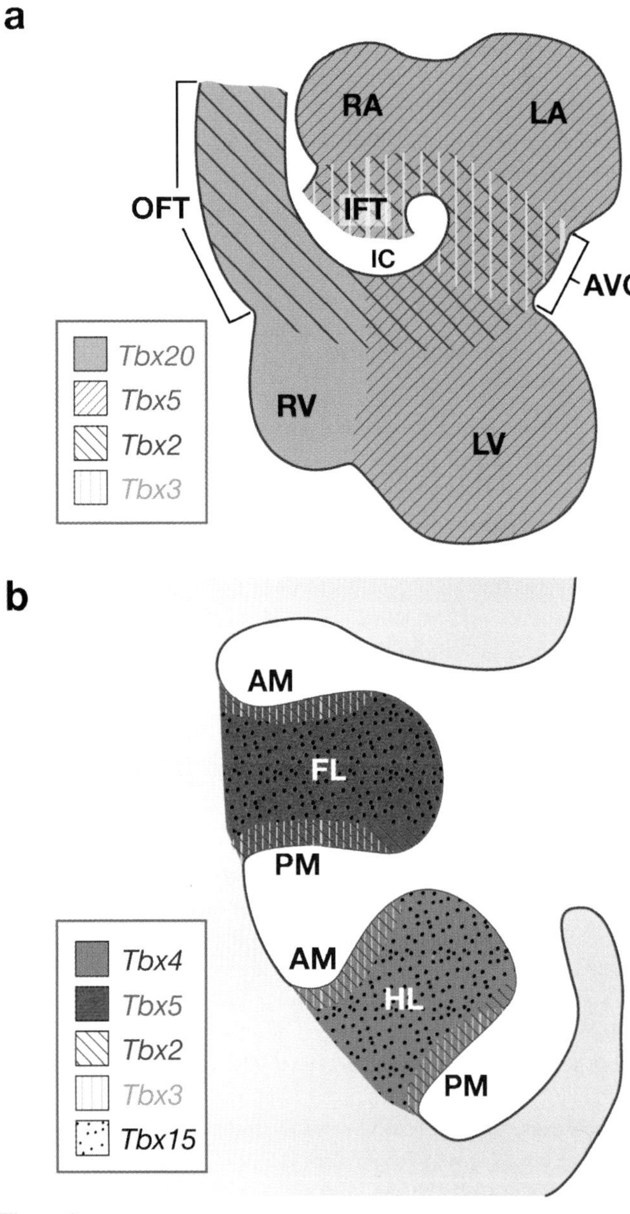

Figure 2

Diagrams of overlapping T-box gene expression in selected organs. (*A*) Cardiac T-box gene expression in 9.5 dpc–10.5 dpc embryos. RA, right atrium; LA, left atrium; RV, right ventricle; LV, left ventricle; AVC, atrioventricular canal; OFT, outflow tract; IFT, inflow tract; IC, inner curvature. (*B*) Limb T-box gene expression in 10.5–12.5 dpc mouse embryos. FL, forelimb; HL, hindlimb; AM, anterior margin; PM, posterior margin.

(52, 64, 109). Conversely, ectopic expression of *Tbx1* in the heart tube can lead to an elongated OFT (50). *Tbx1* mutants display reduced proliferation in the anterior heart field (AHF), which contributes to the OFT (116), demonstrating a role in proliferation of cardiac progenitor cells. Embryos homozygous for a null mutation in *Tbx2* reveal a role for the gene in repressing chamber differentiation in the atrioventricular canal (AVC) during functional specialization of the ventricular and atrial compartments. Several chamber-specific markers are ectopically expressed in the AVC of *Tbx2* homozygous mutants, including *Nppa* and *Cx40* (43). In a complementary experiment, the expression of these genes is undetectable when *Tbx2* is ectopically expressed throughout the heart tube (25). Many *Tbx2* homozygous null mutants also exhibit defects in OFT septation and remodeling of the aortic arch arteries (43). Homozygous null *Tbx20* mutants die midgestation due to defective hearts, which fail to loop and which display many morphological and molecular abnormalities, including widespread upregulation of *Tbx2*. The presence of two poorly developed chamber-like structures and the reduction of *Nppa* expression suggests that *Tbx20* plays a role during cardiac chamber formation in the early heart tube (18a, 96a, 99). Together, T-box genes therefore control development of most regions of the heart and regulate multiple aspects of cardiogenesis including chamber differentiation, suppression of differentiation in nonchamber regions, and cell proliferation.

As in the case of *Tbx4* in the allantois, T-box genes in the heart appear to be regulated by members of the BMP family. Bead implantation experiments in chick show that BMP2 can promote *Tbx2* and *Tbx3* expression (117). Chick explant cultures confirm that *Tbx2* and *Tbx3* expression can be enhanced by the presence of BMP2, whereas *Tbx2* message is reduced in the presence of the BMP inhibitor, noggin. Furthermore, cardiac *Tbx2* expression is greatly reduced in homozygous *Bmp2* mouse mutants (117). In a second pathway common to several T-box genes [see (94) and next section], *Tbx1* regulates cardiac FGF signaling. *Fgf8* and *Fgf10* expression is reduced in the AHF of embryos with reduced or absent *Tbx1* expression (50, 116). A genetic interaction between *Tbx1* and *Fgf8* is also indicated by an increase in the frequency and severity of aortic arch artery remodeling defects in doubly heterozygous mutants compared to single heterozygotes (110). Recent work suggests that T-box genes also have the capacity to regulate their own expression. T-half sites in the *Tbx5* promoter were protected in fingerprinting assays after incubation with nuclear extracts, suggesting that T-box factors bind these endogenous sites. Furthermore, transfection of *Tbx5* activates a *Tbx5* expression reporter in cultured cell lines (100), supporting a potential auto-regulatory or T-box gene cross-regulatory loop.

The regulation of *Nppa* transcription during cardiac chamber formation provides an interesting example of how multiple inputs from T-box factors become integrated to generate a complex expression profile. The *Nppa* promoter contains Nkx binding elements (NKE) for the homeodomain factor and cardiac lineage marker Nkx2-5, in addition to several T-half sites (15, 41, 47). Biochemical studies show that Tbx5 and Nkx2-5 not only interact with each other (47), but also specifically and cooperatively bind the *Nppa* promoter to synergistically activate *Nppa*-reporter expression (15, 47). The success of this regulatory influence depends on intact NKEs and TBEs (15, 41). Tbx2, a demonstrated transcriptional repressor, is also capable of cooperatively binding the *Nppa* promoter with Nkx2-5. Tbx5-Nkx2-5–mediated activation of the *Nppa*-reporter is incrementally repressed by increasing amounts of Tbx2 (41). In the 9.5 dpc mouse embryo, *Tbx2* is expressed in the AVC (25, 41, 43) within a subdomain of the cardiac expression domain of *Tbx5* (23) (**Figure 2**). Considered together, this information has led to the hypothesis that *Nppa* expression is regulated by the competing interactions of Tbx5 and Tbx2 with Nkx2-5 and the T-half sites in the *Nppa* promoter (41). The progressive reduction of *Nppa* transcription in *Tbx5* heterozygous and homozygous null mutants (15) and the ectopic *Nppa* expression in the AVC of *Tbx2* homozygous mutants (43) support this hypothesis. In vitro reporter assays show that *Tbx3*, which is also expressed in the AVC of 9.5 dpc mouse embryos, can repress Tbx5-Nkx2-5–mediated activation of *Nppa* (48). Tbx20 can similarly interact with Nkx2–5 and synergistically activate reporter expression driven from a *Cx40* promoter fragment or a synthetic promoter containing only NKEs (98). Thus, Tbx5, Tbx2, Tbx3, and Tbx20 all potentially participate in the global regulation of *Nppa* expression, and their interactions are likely to control other aspects of cardiac development.

AHF: anterior heart field

AVC: atrioventricular canal

NKE: Nkx binding element

Cardiac development, like somitogenesis, is also sensitive to the dosage of T-box gene activity. Elimination of *Tbx5* leads to embryonic lethality at 9.5 dpc preceded by severe hypoplasia of posterior cardiac structures (15). Diminished *Tbx5* activity in heterozygous embryos leads to less severe cardiac abnormalities, including conduction and septal defects that contribute to compromised survival varying with genetic background. *Tbx5* heterozygous mutants exhibit a reduction in *Nppa* expression that is intermediate between wild-type and homozygous mutant embryos (15), demonstrating a correlation between gene dosage and target gene expression levels. Similar effects are observed in *Tbx1* mutant phenotypes. *Tbx1* heterozygous null mutants lack fourth arch arteries at 10.5 dpc, a defect less severe than that found in homozygous mutants (52, 64). Furthermore, combinations of null and hypomorphic mutations comprising an allelic series demonstrate differential sensitivity to *Tbx1* dosage. Small reductions in *Tbx1* dosage are sufficient to produce incompletely penetrant aortic arch artery defects. Progressively greater reductions in dosage increase the severity and frequency of arch artery remodeling defects and OFT septation anomalies (50). The range of developmental responses to varying

GENE DOSAGE

The first T-box gene mutation was discovered by virtue of a phenotype affecting the tail of heterozygous newborns concomitant with the failure of homozygous mice to survive to birth, demonstrating a semidominant mode of inheritance (30). All of the known autosomal human T-box gene syndromes, with the exception of the recessive isolated ACTH deficiency, occur in heterozygous individuals, whereas homozygotes have not been identified, presumably due to embryonic lethality. In mice, heterozygous effects have been identified for half of the genes mutated, but because these defects are sometimes quite subtle, further investigations may yet identify additional examples of dosage sensitivity. Mouse studies indicate that the full spectrum of tissues affected in homozygous mutants is not necessarily affected by haploinsufficiency. In comparing human and mouse phenotypes, there is a high degree of similarity in tissues affected and the nature of the mutant defects. However, the dose sensitivity may vary considerably—leading to examples of genes that have much more severe or extensive heterozygous phenotypes in humans than that observed in mice (**Table 1**).

T-box protein levels supports the idea that dosage-sensitivity is a common theme within the T-box gene family (see sidebar).

Limb Outgrowth and Patterning

Limb development has long been a major interest of T-box gene researchers. As early as 1996 it was noted that all four members of the Tbx2 subfamily are expressed during limb development in provocative expression domains (37) (**Figure 2**). *Tbx4* and *Tbx5* are expressed in similar patterns but in different limbs—the hindlimbs and forelimbs, respectively. Their expression initiates in the limb fields before morphological limb buds are apparent, and is maintained in their respective limbs until late in development. One of the only other genes known to be expressed specifically in one set of limbs from such an early stage is *Pitx1*, which is expressed only in the hindlimbs (62, 92), and which has been shown to cooperatively interact with another T-box gene, *Tbx19*, in the pituitary (see below). Thus, it was widely hypothesized that *Tbx4* and *Tbx5* conferred hindlimb and forelimb identities, possibly in cooperation with *Pitx1*. Ectopic expression studies in chick reinforced this hypothesis. Electroporation of *Tbx4* into the forelimb or *Tbx5* into the hindlimb produced a variety of malformations, some of which suggested transformation to opposing limb fates (88, 105). However, these experiments were complicated by the presence of the endogenous T-box gene activity. Electroporation of *Tbx4* or *Tbx5* into the limb where each is normally expressed also created severe abnormalities, further confounding analysis and again demonstrating the dosage sensitivity characteristic of this gene family.

Mutations in both *Tbx4* and *Tbx5* are associated with dominant limb defect syndromes in humans: Heterozygous mutations in *TBX4* cause small patella syndrome, characterized by minor hip, knee, and foot defects (10); heterozygous mutations in *TBX5* cause HOS, characterized by moderate-to-severe arm defects in addition to heart abnormalities (7). Targeted mutations in mouse have shown that *Tbx4* and *Tbx5* play similar, but not identical, roles in limb outgrowth. In *Tbx5* homozygous mutants, the forelimb field is defined and appropriate anterior/posterior and dorsoventral patterning is established (1). However, *Fgf10*, a gene critical for bud formation and outgrowth, is not expressed in the forelimb bud mesenchyme and a morphological bud is never observed. In *Tbx4* homozygous mutants, limb bud formation progresses somewhat further (72). The hindlimb bud of *Tbx4* mutant embryos is morphologically evident and mesenchymal and ectodermal FGF signaling is initiated. *Fgf10* is rapidly lost in the mesenchyme, however, and the limb does not progress beyond the bud stage. The mechanism behind the differences between *Tbx4* and *Tbx5* is not known.

The failure of limb outgrowth prevented assessment of forelimb/hindlimb identity in the *Tbx4* and *Tbx5* null embryos. A recent study using a conditional *Tbx5* allele in combination with limb-specific recombinase transgenes has definitively negated the role of these genes in determining limb identity (70). *Tbx4* misexpressed in the forelimb field is capable of rescuing *Tbx5* homozygous mutant forelimb outgrowth, demonstrating functional overlap between these two genes. However, the *Tbx4*-rescued limbs are forelimb-like and show no morphological or molecular hindlimb characteristics. Conversely, misexpression of *Pitx1* with either *Tbx4* or *Tbx5* in the forelimb induces hindlimb-like morphology, indicating that *Pitx1* is indeed a regulator of fore- versus hindlimb identity, whereas *Tbx4* and *Tbx5* are not.

Tbx2 subfamily members *Tbx2* and *Tbx3* are also expressed in the developing limb in largely overlapping domains along the anterior and posterior margins of the limb mesenchyme (37, 38). In chick, *Tbx2* is restricted slightly more posteriorly than *Tbx3* in the posterior limb margin, and is also expressed in the apical ectodermal ridge (AER) at the distal limb margin. Overexpression and dominant negative studies in chick have shown that both *Tbx2* and *Tbx3* induce posterior digit identities, with *Tbx2* having the stronger posteriorizing activity (101). In the mouse and human, both the expression patterns and roles of *Tbx2* and *Tbx3* appear to be reversed relative to chick. In mouse, *Tbx3* is more restricted in the posterior margin and is expressed in the AER, while *Tbx2* is not in the AER (37). In contrast to the chick studies, genetic nulls of each gene in mouse indicate that *Tbx3* is required for development of the posterior margin of the limb (28), while *Tbx2* has only minor effects (43). In humans, *TBX3* is associated with ulna-mammary syndrome (5), which causes defects in posterior limb formation, but no mutations have been identified for *TBX2*.

In addition to the Tbx2 subfamily genes, *Tbx15*, a member of the Tbx1 subfamily, is also expressed in limb mesenchyme in a pattern complementary to the expression domains of *Tbx2* and *Tbx3* (3, 96). Mice mutant for *Tbx15* are viable but have widespread minor defects in the skeletal elements of the limbs, characterized by abnormally short and thin bones with poor articulation. These defects may be caused by diminished proliferation in the early limb bud resulting in smaller precartilaginous mesenchymal condensations. Proliferation is reduced only in the core limb mesenchyme, where *Tbx15* is expressed, but is normal in the limb margins, suggesting that *Tbx2* and/or *Tbx3* may maintain proliferation levels in the latter tissues (see sidebar).

AER: apical ectodermal ridge

Multiple Craniofacial Effects

TBX1 has been identified as a major gene underlying DiGeorge or del22q11.2 syndrome in humans, a syndrome characterized by craniofacial and cardiovascular defects and heterozygous multigene deletions on chromosome 22 [reviewed in (4)]. In addition to cardiovascular defects, mice lacking *Tbx1* exhibit multiple craniofacial defects similar to DiGeorge patients, including defects in ear and mandible development, cleft palate, and defects in branchiomeric muscles (52, 55, 109). Mutations in the zebrafish homologue reveal that the role of *Tbx1* in craniofacial and pharyngeal development (82) is evolutionarily conserved.

Craniofacial development involves complex interactions between ectoderm, anterior mesoderm, endoderm, and neural crest-derived mesenchyme. *Tbx1* is expressed in both pharyngeal endoderm and anterior mesoderm (23) and has been shown to regulate growth factor expression in the pharyngeal region, providing an explanation for the hypoplasia of pharyngeal mesodermal derivatives in mutants (50, 110, 116). Potential targets of *Tbx1* in the pharyngeal region include *Fgf8*, *Fgf10*, and the forkhead transcription factor *Foxa2*, which in combination with *Shh* also acts upstream of *Tbx1*, mediating an autoregulatory loop (50, 110, 116, 118). Because *Tbx1* is expressed in anterior mesoderm rather than neural crest-derived mesenchyme, craniofacial defects affecting bone development observed in *Tbx1* homozygous mutant mice,

T-BOX GENES AND CANCER

Mounting evidence indicates a functional link between T-box genes, cellular proliferation/survival, and tumorigenesis, particularly breast tumors. A direct causal link between T-box genes and cancer, however, has yet to be unequivocally proven. *TBX2* is located within a human chromosomal segment, 17q22–24, which is frequently amplified in breast and pancreatic cancers (6, 51, 67). Several oncogenic candidates have been identified within this segment, including *TBX2*, which are frequently overexpressed in a subset of primary tumors and cancer cell lines (6, 51, 67). Unlike many of the other 17q22–24 candidates, *TBX2* is also amplified and overexpressed in a large fraction of *BRCA1*- and *BRCA2*-related breast tumors (95). *TBX2* and *TBX3* have been molecularly linked to the cell cycle machinery via the ARF-MDM2-p53 pathway. *TBX2* and *TBX3* can rescue a premature senescence phenotype in murine embryonic fibroblasts (20, 51). Recent work in melanoma cell lines has shown that a dominant negative Tbx2 is able to induce senescence (108). Both genes can also bind a putative TBE in the human $p14^{ARF}$ promoter and *TBX2* can repress a $p14^{ARF}$ expression reporter in vitro (65). *TBX2* is capable of repressing other transcripts from the *Cdkn* gene family, including $p15^{INK4b}$, $p16^{INK4a}$ (51), and p21 (83). There is no evidence, however, of $p15^{INK4b}$, $p16^{INK4a}$, ARF, or p21 overexpression in *Tbx2* homozygous null mutant embryos, and no evidence of a genetic interaction between *Tbx2* (or *Tbx3*) and p53 (43, 53). Nonetheless, the impaired proliferation of the mammary gland ductal tree in *Tbx3* heterozygous mice and the exacerbation of this phenotype in double heterozygous *Tbx2*; *Tbx3* mice (53) demonstrate the important contribution of these genes to normal mammary development and suggest that further work is required to determine how misregulation of *TBX2* or *TBX3* can contribute to mammary tumorigenesis.

such as micrognathia (small mandible), may be secondary to failure of mesodermal specification and/or mesodermal or endodermal-derived signaling to surrounding cell types.

Overexpression of human *TBX1* and three adjacent genes in transgenic mice leads to a spectrum of craniofacial and pharyngeal defects that overlaps with the *Tbx1* null phenotype (34). Dosage compensation using a *Tbx1* null allele reveals that the majority of defects, including cleft palate, thymic hypoplasia, and cardiovascular defects, are due specifically to overexpression of *Tbx1*, suggesting that closely regulated levels of Tbx1 are required for normal development and that either too much or too little can incur similar phenotypic defects (63) (see sidebar). However, inner ear defects and hearing loss observed in mice overexpressing *Tbx1* are not complemented by *Tbx1* null alleles, suggesting a different mechanism of *Tbx1* action in the otic epithelium. *Tbx1* specifies regional identity and fate boundary formation in the otic epithelium at early stages of inner ear development and acts cell autonomously to expand a population of cells that normally give rise to vestibular and auditory organs (86, 111). Overexpression of *Tbx1* leads to ectopic inner ear sensory organs, whereas loss leads to failure of sensory organ development and expanded neurogenesis. *Tbx1* therefore operates as a selector gene in the otocyst, controlling sensory versus neural cell fate (86).

Tbx1 also plays a central role in craniofacial muscle development. Branchiomeric skeletal muscles are derived from the mesodermal core of the pharyngeal arches. The most anterior, or mandibular, pharyngeal arch forms normally in the absence of *Tbx1* (52, 55, 109); however, the myogenic determination genes *Myf5* and *MyoD* fail to be activated, resulting in defects in jaw muscles (55). *Tbx1* regulates, but is not absolutely required for, myogenic specification in the mandibular arch since a low level of myogenic determination gene activation occurs sporadically in the absence of *Tbx1*, producing severely hypoplastic or unilateral jaw muscles. *Tbx1* therefore confers robustness on myogenic specification in the mandibular arch.

Another T-box gene, *TBX22*, is also required for palate development and mutations cause X-linked cleft palate and ankyloglossia (11, 69). The expression pattern of murine *Tbx22* is consistent with a role in the nasal septum during palatal shelf fusion, and in the tissue bridge between the tongue and floor of the mouth (44). It is not known whether *Tbx22* interacts epistatically with *Tbx1* during palate development.

Cell Fate in the Pituitary

Tbx19 (119), also known as *Tpit*, is expressed in the two pro-opiomelanocortin (POMC)-expressing lineages of the pituitary gland, the adrenocorticotrophic hormone (ACTH) producing corticotrophs of the anterior lobe and the melanotrophs of the intermediate lobe. Loss-of-function mutations in human *TBX19* cause early onset ACTH deficiency. These mutations appear to be completely recessive as heterozygous humans and mice have no ACTH deficiency and heterozygous mice have normal numbers of POMC-expressing cells in their pituitaries (84, 85). *Tbx19* expression initiates prior to POMC expression during embryonic development and is also present in the adult pituitary. Although not required for lineage commitment of POMC-expressing cells, *Tbx19* is required for the terminal differentiation of corticotrophs and melanotrophs and for successful upregulation of POMC. *Tbx19* also represses gonadotroph differentiation, leading to the idea that this factor is responsible for alternative cell fates during pituitary development (61, 66, 85). *Tbx19* activates POMC transcription in cooperation with *Pitx1* (61, 66) through the recruitment of SRC/p160 coactivators to its cognate DNA target, the Tpit/Pitx regulatory element in the POMC promoter. This recruitment is mediated by direct binding between Tpit and SRC-2 and is upregulated by mitogen-activated protein kinase activity (68).

T-Box Genes in T Cells

Two T-box genes of the Tbr1 subfamily, *Eomesodermin* and *Tbx21* (also known as *T-bet* for T-box expressed in T cells), are important in the differentiation of both T and B cells of the immune system [reviewed by (39, 103)]. Adult mice with a null mutation in *Tbx21* show a complete lack of the T helper cell subset Th1, as well as effects on B cells and natural killer cells (39, 103, 104). Homozygous *Eomesodermin* mutants die in early gestation (see section on extraembryonic tissues), but heterozygous mutants display defects in $CD8^+$ T cells. There are indications that *Eomesodermin* and *Tbx21* work cooperatively in governing cellular immunity (79).

Tbx21 has a critical role in initiating Th1 development from naïve precursors both by inducing the Th1 terminal differentiation pathway and also by repressing the alternative Th2 differentiation pathway. Although it is not entirely clear how this effect is mediated, it very likely involves regulation of the *interferon-γ* locus, a definitive hallmark of the Th1 lineage, and may also involve chromatin remodeling (103). Because of this pivotal role in the differentiation of lymphoid cells, *Tbx21* is an important player in immune responses to infection and in autoimmune disease processes. For example, it has key roles in the pathogenesis of lupus (80), T-cell mediated colitis and Crohn's disease (73), and metastasis of primary prostate cancer (81). There is an association between type 1 diabetes and *Tbx21* polymorphisms in humans (54, 91). Mice with a null mutation in *Tbx21* show airway inflammatory features characteristic of asthma, whereas human asthma patients have reduced *TBX21* expression (32).

POMC: pro-opiomelanocortin

ACTH: adrenocorticotrophic hormone

SUMMARY POINTS

1. The T-box gene family, present in all metazoans, consists of 17 genes in mammals organized into 5 subfamilies.
2. T-box genes are expressed throughout development in dynamic patterns with both unique and overlapping areas of expression.

3. T-box proteins bind a common element, the T-site or TBE, but promoter specificity is provided by interactions with other transcription factors and different binding kinetics.
4. Mutations in T-box genes are responsible for developmental defects in many organisms, including several dominant or semidominant human developmental syndromes.
5. Dosage sensitivity to T-box proteins is frequently observed, with too much protein as disruptive as too little.
6. T-box factors function in a wide range of signaling pathways, but common themes put T-box genes downstream of BMP signaling and upstream of FGF signaling.

FUTURE DIRECTIONS

In vitro data have shown interactions between T-box proteins and other transcription factors in cooperative binding experiments. T-box proteins also have the potential to heterodimerize and are often expressed in overlapping patterns. A challenge for the future is to investigate the in vivo relevance of these interactions.

Little is known about how the complex expression profiles of T-box genes or the tightly regulated protein levels are achieved.

Because of the complex expression patterns of T-box genes, mutational analysis has not yet addressed all the possible developmental roles, and because early lethality often limits the exploration of later stages, new alleles, including conditional and hypomorphic alleles, will be needed.

While some developmental requirements for T-box genes have been demonstrated, relatively few downstream target genes effecting these functions have been identified.

ACKNOWLEDGMENTS

This work was supported by the National Institutes of Health grants GM06561 and HD033082 (V.E.P.). R.K. is an INSERM Research Fellow. We thank George Stratigopoulos for helpful discussions and Elinor Pisano for bibliographic assistance.

LITERATURE CITED

1. Agarwal P, Wylie JN, Galceran J, Arkhitko O, Li C, et al. 2003. *Tbx5* is essential for forelimb bud initiation following patterning of the limb field in the mouse embryo. *Development* 130:623–33
2. Agulnik SI, Garvey N, Hancock S, Ruvinsky I, Chapman DL, et al. 1996. Evolution of mouse *T-box* genes by tandem duplication and cluster dispersion. *Genetics* 144:249–54
3. Agulnik SI, Papaioannou VE, Silver LM. 1998. Cloning, mapping and expression analysis of *TBX15*, a new member of the T-box gene family. *Genomics* 51:68–75
4. Baldini A. 2003. DiGeorge's syndrome: a gene at last. *Lancet* 362:1342–43
5. Bamshad M, Lin RC, Law DJ, Watkins WS, Krakowiak PA, et al. 1997. Mutations in human *TBX3* alter limb, apocrine and genital development in ulnar-mammary syndrome. *Nat. Genet.* 16:311–15

6. Bärlund M, Monni O, Kononen J, Cornelison R, Torhorst J, et al. 2000. Multiple genes at 17q23 undergo amplification and overexpression in breast cancer. *Cancer Res.* 60:5340–44
7. Basson CT, Bachinsky DR, Lin RC, Levi T, Elkins JA, et al. 1997. Mutations in human cause limb and cardiac malformations in Holt-Oram syndrome. *Nat. Genet.* 15:30–35
8. Beddington RSP, Rashbass P, Wilson V. 1992. Brachyury—a gene affecting mouse gastrulation and early organogenesis. *Development* (*Suppl.*):157–65
9. Begemann G, Gilbert Y, Meyer A, Ingham PW. 2001. Cloning of zebrafish T-box genes *tbx15* and *tbx18* and their expression during embryonic development. *Mech. Dev.* 114:137–41
10. Bongers EMHF, Duijf PHG, van Beersum SEM, Schoots J, van Kampen A, et al. 2004. Mutations in the human *TBX4* gene cause small patella syndrome. *Am. J. Hum. Genet.* 74:1239–48
11. Braybrook C, Doudney K, Marcano ACB, Arnason A, Bjornsson A, et al. 2001. The T-box transcription factor gene *TBX22* is mutated in X-linked cleft palate and ankyloglossia. *Nat. Genet.* 29:179–83
12. Bruce AEE, Howley C, Zhou Y, Vickers SL, Silver LM, et al. 2003. The maternally expressed zebrafish T-box gene *eomesodermin* regulates organizer formation. *Development* 130:5503–17
13. Bruneau BG. 2002. Mouse models of cardiac chamber formation and congenital heart disease. *Trends Genet.* 18:S15–20
14. Bruneau BG, Logan M, Davis N, Levi T, Tabin CJ, et al. 1999. Chamber-specific cardiac expression of *Tbx5* and heart defects in Holt-Oram syndrome. *Dev. Biol.* 211:100–8
15. Bruneau BG, Nemer G, Schmitt JP, Charron F, Robitaille L, et al. 2001. A murine model of Holt-Oram syndrome defines roles of the T-box transcription factor Tbx5 in cardiogenesis and disease. *Cell* 106:709–21
16. Bulfone A, Wang F, Hevner R, Anderson S, Cutforth T, et al. 1998. An olfactory sensory map develops in the absence of normal projection neurons or GABAergic interneurons. *Neuron* 21:1273–82
17. Bush JO, Lan Y, Jiang R. 2004. The cleft lip and palate defects in *Dancer* mutant mice result from gain of function of the *Tbx10* gene. *Proc. Natl. Acad. Sci. USA* 101:7022–27
18. Bussen M, Petry M, Schuster-Gossler K, Leitges M, Gossler A, Kispert A. 2004. The T-box transcription factor Tbx18 maintains the separation of anterior and posterior somite compartments. *Genes Dev.* 18:1209–21

18a. Cai CL, Zhou W, Yang L, Bu L, Qyang Y, et al. 2005. T-box genes coordinate regional rates of proliferation and regional specification during cardiogenesis. *Development* 132:2475–87

19. Candille SI, Van Raamsdonk CD, Chen C, Kuijper S, Chen-Tsai Y, et al. 2004. Dorsoventral patterning of the mouse coat by *Tbx15*. *PLoS Biol.* 2:0030–42
20. Carlson H, Ota S, Campbell CE, Hurlin PJ. 2001. A dominant repression domain in Tbx3 mediates transcriptional repression and cell immortalization: relevance to mutations in Tbx3 that cause ulnar-mammary syndrome. *Hum. Mol. Genet.* 10:2403–13
21. Carreira S, Dexter TJ, Yavuzer U, Easty DJ, Goding CR. 1998. Brachyury-related transcription factor Tbx2 and repression of the melanocyte-specific TRP-1 promoter. *Mol. Cell. Biol.* 18:5099–108

22a. Chapman DL, Agulnik I, Hancock S, Silver LM, Papaioannou VE. 1996. *Tbx6*, a mouse T-box gene implicated in paraxial mesoderm formation at gastrulation. *Dev. Biol.* 180:534–42

22. Chapman DL, Cooper-Morgan A, Harrelson Z, Papaioannou VE. 2003. Critical role for *Tbx6* in mesoderm specification in the mouse embryo. *Mech. Dev.* 120:837–47
23. Chapman DL, Garvey N, Hancock S, Alexiou M, Agulnik S, et al. 1996. Expression of the T-box family genes, *Tbx1–Tbx5*, during early mouse development. *Dev. Dyn.* 206:379–90
24. Chapman DL, Papaioannou VE. 1998. Three neural tubes in mouse embryos with mutations in the T-box gene, *Tbx6*. *Nature* 391:695–97
25. Christoffels VM, Hoogaars WMH, Tessari A, Clout DEW, Moorman AFM, Campione M. 2004. Tbox transcription factor Tbx2 represses differentiation and formation of the cardiac chambers. *Dev. Dyn.* 229:763–70
26. Coll M, Seidman JG, Muller CW. 2002. Structure of the DNA-bound T-box domain of human TBX3, a transcription factor responsible for ulnar-mammary syndrome. *Structure* 10:343–56
27. Conlon FL, Fairclough L, Price BMJ, Casey ES, Smith JC. 2001. Determinants of T box protein specificity. *Development* 128:3749–58
28. Davenport TG, Jerome-Majewska LA, Papaioannou VE. 2003. Mammary gland, limb, and yolk sac defects in mice lacking *Tbx3*, the gene mutated in human ulnar mammary syndrome. *Development* 130:2263–73
29. Dheen T, Sleptsova-Friedrich I, Xu Y, Clark M, Lerach H, et al. 1999. Zebrafish *tbx-c* functions during formation of midline structures. *Development* 126:2703–13
30. Dobrovolskaia-Zavadskaia N. 1927. Sur la mortification spontanée de la queue chez la souris nouveau-née et sur l'existence d'un caractère (facteur) héréditaire "non viable". *C. R. Acad. Sci. Biol.* 97:114–16
31. Fan C, Liu M, Wang Q. 2003. Functional analysis of TBX5 missense mutations associated with Holt-Oram syndrome. *J. Biol. Chem.* 278:8780–85
32. Finnotto S, Neurath MF, Glickman JN, Qin S, Lehr HA, et al. 2002. Development of spontaneous airway changes consistent with human asthma in mice lacking T-bet. *Science* 295:336–38
33. Fujiwara T, Dunn NR, Hogan BLM. 2001. Bone morphogenetic protein 4 in the extraembryonic mesoderm is required for allantois development and the localization and survival of primordial germ cells in the mouse. *Proc. Natl. Acad. Sci. USA* 98:13739–44
34. Funke B, Epstein JA, Kochilas LK, Lu MM, Pandita RK, et al. 2001. Mice overexpressing genes from the 22q11 region deleted in velo-cardio-facial syndrome/DiGeorge syndrome have middle and inner ear defects. *Hum. Mol. Genet.* 10:2549–56
35. Garg V, Kathiriya IS, Barnes R, Schluterman MK, King IN, et al. 2003. GATA4 mutations cause human congenital heart defects and reveal an interaction with TBX5. *Nature* 424:443–47
36. Ghosh TK, Packham EA, Bonser AJ, Robinson TE, Cross SJ, Brook JD. 2001. Characterization of the TBX5 binding site and analysis of mutations that cause Holt-Oram syndrome. *Hum. Mol. Genet.* 10:1983–94
37. Gibson-Brown JJ, Agulnik SI, Chapman DL, Alexiou M, Garvey N, et al. 1996. Evidence of a role for T-box genes in the evolution of limb morphogenesis and the specification of forelimb/hindlimb identity. *Mech. Dev.* 56:93–101
38. Gibson-Brown JJ, Agulnik SI, Silver LM, Niswander L, Papaioannou VE. 1998. Involvement of T-box genes *Tbx2–Tbx5* in vertebrate limb specification and development. *Development* 125:2499–509

39. Glimcher LH, Townsend MJ, Sullivan BM, Lord GM. 2004. Recent developments in the transcriptional regulation of cytolytic effector cells. *Nat. Rev. Immunol.* 4:900–11
40. Goering LM, Hoshijima K, Hug B, Bisgrove B, Kispert A, Grunwald DJ. 2003. An interacting network of T-box genes directs gene expression and fate in the zebrafish mesoderm. *Proc. Natl. Acad. Sci. USA* 100:9410–15
41. Habets PEMH, Moorman AFM, Clout DEW, van Roon MA, Lingbeek M, et al. 2002. Cooperative action of Tbx2 and Nkx2.5 inhibits ANF expression in the atrioventricular canal: implications for cardiac chamber formation. *Genes Dev.* 16:1234–46
42a. Haenig B, Kispert A. 2004. Analysis of *TBX18* expression in chick embryos. *Dev. Genes Evol.* 214:407–11
42. Haenig B, Schmidt C, Kraus F, Pfordt M, Kispert A. 2002. Cloning and expression analysis of the chick ortholog of *TBX22*, the gene mutated in X-linked cleft palate and akyloglossia. *Mech. Dev.* 117:321–25
43. Harrelson Z, Kelly RG, Goldin SN, Gibson-Brown JJ, Bollag RJ, et al. 2004. *Tbx2* is essential for patterning the atrioventricular canal and for morphogenesis of the outflow tract during heart development. *Development* 131:5041–52
44. Herr A, Meunier D, Muller I, Rump A, Fundele R, et al. 2003. Expression of mouse Tbx22 supports its role in palatogenesis and glossogenesis. *Dev. Dyn.* 226:579–86
45. Herrmann BG. 1995. The mouse *Brachyury* (*T*) gene. *Semin. Dev. Biol.* 6:385–94
46. Hevner RF, Shi L, Justice N, Hsueh Y-P, Sheng M, et al. 2001. Tbr1 regulates differentiation of the preplate and layer 6. *Neuron* 29:353–66
47. Hiroi Y, Kudoh S, Monzen K, Ikeda Y, Yazaki Y, et al. 2001. Tbx5 associates with Nkx2–5 and synergistically promotes cardiomyocyte differentiation. *Nat. Genet.* 28:276–80
48. Hoogaars WM, Tessari A, Moorman AF, de Boer PA, Hagoort J, et al. 2004. The transcriptional repressor Tbx3 delineates the developing central conduction system of the heart. *Cardiovasc Res.* 62:489–99
49. Hsueh Y-P, Wang T-F, Yang F-C, Sheng M. 2000. Nuclear translocation and transcription regulation by the membrane-associated guanylate kinase CASK/LIN-2. *Nature* 404:298–302
50. Hu T, Yamagishi H, Maeda J, McAnally J, Yamagishi C, Srivastava D. 2004. Tbx1 regulates fibroblast growth factors in the anterior heart field through a reinforcing autoregulatory loop involving forkhead transcription factors. *Development* 131:5491–502
51. Jacobs JJL, Keblusek P, Robanus-Maandag E, Kristel P, Lingbeek M, et al. 2000. Senescence bypass screen identifies *TBX2*, which represses *Cdkn2a* (*p19*^{ARF}) and is amplified in a subset of human breast cancers. *Nat. Genet.* 26:291–99
52. Jerome LA, Papaioannou VE. 2001. DiGeorge syndrome phenotype in mice mutant for the T-box gene, *Tbx1*. *Nat. Genet.* 27:286–91
53. Jerome-Majewska LA, Jenkins GP, Ernstoff E, Zindy F, Sherr CJ, Papaioannou VE. 2005. *Tbx3*, the ulnar mammary syndrome gene, and *Tbx2* interact in mammary gland development through a p19Arf/p53-independent pathway. *Dev. Dyn.* In press
54. Juedes AE, Rodrigo E, Togher L, Glimcher LH, von Herrath MG. 2004. T-bet controls autoaggressive CD8 lymphocyte responses in type 1 diabetes. *J. Exp. Med.* 199:1153–62
55. Kelly RG, Jerome-Majewska LA, Papaioannou VE. 2004. The del 22q11.2 candidate gene *Tbx1* regulates branchiomeric myogenesis. *Hum. Mol. Genet.* 13:2829–40
56. Kispert A. 1995. The Brachyury protein: a T-domain transcription factor. *Semin. Dev. Biol.* 6:395–403
57. Kispert A, Herrmann BG. 1993. The *Brachyury* gene encodes a novel DNA binding protein. *EMBO J.* 12:3211–20

58. Kraus F, Haenig B, Kispert A. 2001. Cloning and expression analysis of the mouse T-box gene *Tbx18*. *Mech. Dev.* 100:83–86
59. Krause A, Zacharias W, Camarata T, Linkhart B, Law E, et al. 2004. Tbx5 and Tbx4 transcription factors interact with a new chicken PDZ-LIM protein in limb and heart development. *Dev. Biol.* 273:106–20
60. Kusch T, Storck T, Walldorf U, Reuter R. 2002. Brachyury proteins regulate target genes through modular binding sites in a cooperative fashion. *Genes Dev.* 16:518–29
61. Lamolet B, Pulichino A-M, Lamonerie T, Gauthier Y, Brue T, et al. 2001. A pituitary cell-restricted T box factor, Tpit, activates POMC transcription in cooperation with Pitx homeoproteins. *Cell* 104:849–59
62. Lanctot C, Lamolet B, Drouin J. 1997. The *bicoid*-related homeoprotein *Ptx1* defines the most anterior domain of the embryo and differentiates posterior from anterior lateral mesoderm. *Development* 124:2807–17
63. Liao J, Kochilas L, Nowotschin S, Arnold JS, Aggarwal VS, et al. 2004. Full spectrum of malformations in velo-cardio-facial syndrome/DiGeorge syndrome mouse models by altering Tbx1 dosage. *Hum. Mol. Genet.* 13:1577–85
64. Lindsay EA, Vitelli F, Su H, Morishima M, Huynh T, et al. 2001. *Tbx1* haploinsufficiency in the DiGeorge syndrome region causes aortic arch defects in mice. *Nature* 410:97–101
65. Lingbeek ME, Jacobs JJL, van Lohuizen M. 2002. The T-box repressors *TBX2* and *TBX3* specifically regulate the tumor-suppressor *p14^{ARF}* via a variant T-site in the initiator. *J. Biol. Chem.* 277:26120–27
66. Liu J, Lin C, Gleiberman A, Ohgi KA, Herman T, et al. 2001. *Tbx19*, a tissue-selective regulator of POMC gene expression. *Proc. Natl. Acad. Sci. USA* 98:8674–79
67. Mahlamäki EH, Bärlund M, Tanner M, Gorunova L, Höglund M, et al. 2002. Frequent amplification of 8q24, 11q, 17q, and 20q-specific genes in pancreatic cancer. *Genes Chromosomes Cancer* 35:353–58
68. Maira M, Couture C, Le Martelot G, Pulichino A-M, Bilodeau S, Drouin J. 2003. The T-box factor Tpit recruits SRC/p160 co-activators and mediates hormone action. *J. Biol. Chem.* 278:46523–32
69. Marçano AC, Doudney K, Braybrook C, Squires R, Patton MA, et al. 2004. *TBX22* mutations are a frequent cause of cleft palate. *J. Med. Genet.* 41:68–74
70. Minguillon C, Del Buono J, Logan MP. 2005. *Tbx5* and *Tbx4* are not sufficient to determine limb-specific morphologies but have common roles in initiating limb outgrowth. *Dev. Cell* 8:75–84
71. Muller CW, Herrmann BG. 1997. Crystallographic structure of the T domain-DNA complex of the *Brachyury* transcription factor. *Nature* 389:884–88
72. Naiche LA, Papaioannou VE. 2003. Loss of *Tbx4* blocks hindlimb development and affects vascularization and fusion of the allantois. *Development* 130:2681–93
73. Neurath MF, Weigmann B, Finotto S, Glickman JN, Nieuwenhuis E, et al. 2002. The transcription fator T-bet regulates mucosal T cell activation in experimental colitis and Crohn's disease. *J. Exp. Med.* 195:1129–43
74. Nikaido M, Kawakami A, Sawada A, Furutani-Seiki M, Takeda H, Araki K. 2002. *Tbx24*, encoding a T-box protein, is mutated in the zebrafish somite-segmentation mutant *fused somites*. *Nat. Genet.* 31:195–99
75. **Papaioannou VE. 2001. T-box genes in development: from hydra to humans. *Int. Rev. Cytol.* 207:1–70**

A comprehensive review of publications through 2000 of vertebrate and invertebrate T-box genes, their expression patterns, mutations and functions.

76. Papaioannou VE, Goldin SN. 2003. Introduction to the *T-box* genes and their roles in developmental signaling pathways. In *Inborn Errors of Development. The Molecular Basis of Clinical Disorders of Morphogenesis. Oxford Monogr. Med. Genet. No. 49*, ed. CJ Epstein, RP Erickson, A Wynshaw-Boris, pp. 686–98. Oxford: Oxford Univ. Press
77. Papapetrou C, Edwards YH, Sowden JC. 1997. The T transcription factor functions as a dimer and exhibits a common human polymorphism Gly-177-Asp in the conserved DNA-binding domain. *FEBS Lett.* 409:201–6
78. Paxton C, Zhao H, Chin Y, Langner K, Reecy J. 2002. Murine Tbx2 contains domains that activate and repress gene transcription. *Gene* 283:117–24
79. Pearce EL, Mullen AC, Martins GA, Krawczyk CM, Hutchins AS, et al. 2003. Control of effector $CD8^+$ T Cell function by the transcription factor *Eomesodermin*. *Science* 302:1041–43
80. Peng SL, Szabo SJ, Glimcher LH. 2002. T-bet regulates IgG class switching and pathogenic autoantibody production. *Proc. Natl. Acad. Sci. USA* 99:5545–50
81. Peng SL, Townsend MJ, Hecht JL, White IA, Glimcher LH. 2004. T-bet regulates metastasis rate in a murine model of primary prostate cancer. *Cancer Res.* 64:452–55
82. Piotrowski T, Ahn D-G, Schilling TF, Nair S, Ruvinsky I, et al. 2003. The zebrafish *van gogh* mutation disrupts *tbx1*, which is involved in the DiGeorge deletion syndrome in humans. *Development* 130:5043–52
83. Prince S, Carreira S, Vance KW, Abrahams A, Goding CR. 2004. Tbx2 directly represses the expression of the $p21^{WAF1}$ cyclin-dependent kinase inhibitor. *Cancer Res.* 64:1669–74
84. Pulichino AM, Vallette-Kasic S, Couture C, Gauthier Y, Brue T, et al. 2003. Human and mouse *TPIT* gene mutations cause early onset pituitary ACTH deficiency. *Genes Dev.* 17:711–16
85. Pulichino AM, Vallette-Kasic S, Tsai JP-Y, Couture C, Gauthier Y, Drouin J. 2003. Tpit determines alternate fates during pituitary cell differentiation. *Genes Dev.* 17:738–47
86. Raft S, Nowotschin S, Liao J, Morrow BE. 2004. Suppression of neural fate and control of inner ear morphogenesis by *Tbx1*. *Development* 131:1801–12
87. Rallis C, Bruneau BG, Del Buono J, Seidman CE, Seidman JG, et al. 2003. *Tbx5* is required for forelimb bud formation and continued outgrowth. *Development* 130:2741–51
88. Rodriguez-Esteban C, Tsukui T, Yonei S, Magallon J, Tamura K, Izpisua Belmonte JC. 1999. The T-box genes *Tbx4* and *Tbx5* regulate limb outgrowth and identity. *Nature* 398:814–18
89. Russ AP, Wattler S, Colledge WH, Aparicio SAJR, Carlton MBL, et al. 2000. *Eomesodermin* is required for mouse trophoblast development and mesoderm formation. *Nature* 404:95–98
90. Sakiyama J, Yamagishi A, Kuroiwa A. 2003. *Tbx4-Fgf10* system controls lung bud formation during chicken embryonic development. *Development* 130:1225–34
91. Sasaki Y, Ihara K, Matsuura N, Kohno H, Nagafuchi S, et al. 2004. Identification of a novel type 1 diabetes susceptibility gene, *T-bet*. *Hum. Genet.* 115:177–84
92. Shang J, Luo Y, Clayton DA. 1997. *Backfoot* is a novel homeobox gene expressed in the mesenchyme of developing hind limb. *Dev. Dyn.* 209:242–53
93. Shedlovsky A, King TR, Dove WF. 1988. Saturation germ line mutagenesis of the murine *t* region including a lethal allele at the quaking locus. *Proc. Natl. Acad. Sci. USA* 85:180–84
94. Showell C, Binder O, Conlon FL. 2004. T-box genes in early embryogenesis. *Dev. Dyn.* 229:201–18

95. Sinclair CS, Adem C, Naderi A, Soderberg CL, Johnson M, et al. 2002. *TBX2* is preferentially amplified in *BRCA1*- and *BRCA2*-related breast tumors. *Cancer Res.* 62:3587–91
96. Singh MK, Petry M, Haenig B, Lescher B, Leitges M, Kispert A. 2005. The T-box transcription factor Tbx15 is required for skeletal development. *Mech. Dev.* 122:131–44
96a. Singh MK, Christoffels VM, Dias JM, Trowe MO, Petry M, et al. 2005. *Tbx20* is essential for cardiac chamber differentiation and repression of *Tbx2*. *Development* 132:2697–707
97. Sinha S, Abraham S, Gronostajski RM, Campbell CE. 2000. Differential DNA binding and transcription modulation by three T-box proteins, T, TBX1 and TBX2. *Gene* 258:15–29
98. Stennard FA, Costa MW, Elliott DA, Rankin S, Haast SJ, et al. 2003. Cardiac T-box factor Tbx20 directly interacts with Nkx2–5, GATA4, and GATA5 in regulation of gene expression in the developing heart. *Dev. Biol.* 262:206–24
99. Stennard FA, Costa MW, Lai D, Biben C, Furtado MB, et al. 2005. Murine T-box transcription factor Tbx20 acts as a repressor during heart development,and is essential for adult heart integrity, function and adaption. *Development* 132:2451–62
100. Sun G, Lewis LE, Huang X, Nguyen Q, Prince C, Huang T. 2004. *TBX5*, a gene mutated in Holt-Oram syndrome, is regulated through a GC box and T-box binding elements (TBEs). *J. Cell. Biochem.* 92:189–99
101. Suzuki T, Takeuchi J, Koshiba-Takeuchi K, Ogura T. 2004. *Tbx* genes specify posterior digit identity through Shh and BMP signaling. *Dev. Cell* 6:43–53
102. Szabo SJ, Kim ST, Costa GL, Zhang X, Fathman CG, Glimcher LH. 2000. A novel transcription factor, T-bet, directs Th1 lineage commitment. *Cell* 100:655–69
103. Szabo SJ, Sullivan BM, Peng SL, Glimcher LH. 2003. Molecular mechanisms regulating TH1 immune responses. *Annu. Rev. Immunol.* 21:713–58
104. Szabo SJ, Sullivan BM, Stemmann C, Satoskar AR, Sleckman BP, Glimcher LH. 2002. Distinct effects of T-bet in T_H1 lineage commitment and IFN-gamma production in CD4 and CD8 T cells. *Science* 295:338–42
105. Takeuchi JK, Koshiba-Takeuchi K, Matsumoto K, Vogel-Höpker A, Naitoh-Matsuo M, et al. 1999. *Tbx5* and *Tbx4* genes determine the wing/leg identity of limb buds. *Nature* 398:810–14
106. Takeuchi JK, Ohgi M, Koshiba-Takeuchi K, Shiratori H, Sakaki I, et al. 2003. *Tbx5* specifies the left/right ventricles and ventricular septum position during cardiogenesis. *Development* 130:5953–64
106a. Takeuchi JK, Koshiba-Takeuchi K, Suzuki T, Kamimura M, Ogura K, Ogura T. 2003. Tbx5 and Tbx4 trigger limb initiation through activation of the Wnt/Fgf signaling cascade. *Development* 130:2729–39
107. Tanaka M, Tickle C. 2004. Tbx18 and boundary formation in chick somite and wing development. *Dev. Biol.* 268:470–80
108. Vance KW, Carreira S, Brosch G, Goding CR. 2005. Tbx2 is overexpressed and plays an important role in maintaining proliferation and suppression of senescence in melanomas. *Cancer Res.* 65: In press
109. Vitelli F, Morishima M, Taddei I, Lindsay EA, Baldini A. 2002. *Tbx1* mutation causes multiple cardiovascular defects and disrupts neural crest and cranial nerve migratory pathways. *Hum. Mol. Genet.* 11:915–22
110. Vitelli F, Taddei I, Morishima M, Meyers EN, Lindsay EA, Baldini A. 2002. A genetic link between *Tbx1* and fibroblast growth factor signaling. *Development* 129:4605–11

111. Vitelli F, Viola A, Morishima M, Pramparo T, Baldini A, Lindsay E. 2003. *TBX1* is required for inner ear morphogenesis. *Hum. Mol. Genet.* 12:2041–48
112. Wang G-S, Hong C-J, Yen T-Y, Huang H-Y, Ou Y, et al. 2004. Transcriptional modification by a CASK-interacting nucleosome assembly protein. *Neuron* 42:113–28
113. Watabe-Rudolph M, Schlautmann N, Papaioannou VE, Gossler A. 2002. The mouse rib-vertebrae mutation is a hypomorphic *Tbx6* allele. *Mech. Dev.* 119:251–56
114. White PH, Farkas DR, McFadden EE, Chapman DL. 2003. Defective somite patterning in mouse embryos with reduced levels of *Tbx6*. *Development* 130:1681–90
115. Xanthos JB, Kofron M, Wylie C, Heasman J. 2001. Maternal VegT is the initiator of a molecular nework specifying endoderm in *Xenopus laevis*. *Development* 128:167–80
116. Xu H, Morishima M, Wylie JN, Schwartz RJ, Bruneau BG, et al. 2004. *Tbx1* has a dual role in the morphogenesis of the cardiac outflow tract. *Development* 131:3217–27
117. Yamada M, Revelli J-P, Eichele G, Barron M, Schwartz RJ. 2000. Expression of chick *Tbx-2*, *Tbx-3*, and *Tbx-5* genes during early heart development: evidence for BMP2 induction of *Tbx2*. *Dev. Biol.* 228:95–105
118. Yamagishi H, Maeda J, Hu T, McAnally J, Conway SJ, et al. 2003. *Tbx1* is regulated by tissue-specific forkhead proteins through a common Sonic hedgehog-responsive enhancer. *Genes Dev.* 17:269–81
119. Yi C-H, Terrett JA, Li Q-Y, Ellington K, Packham EA, et al. 1999. Identification, mapping, and phylogenomic analysis of four new human members of the T-box gene family: *EOMES*, *TBX6*, *TBX18*, and *TBX19*. *Genomics* 55:10–20
120. Zaragoza MV, Lewis LE, Sun G, Wang E, Li L, et al. 2004. Identification of the TBX5 transactivating domain and the nuclear localization signal. *Gene* 330:9–18

RELATED RESOURCES

Szabo SJ, Sullivan BM, Peng SL, Glimcher LH. 2003. Molecular mechanisms regulating Th1 immune responses. *Annu. Rev. Immunol.* 21:713–58

Connecting Mammalian Genome with Phenome by ENU Mouse Mutagenesis: Gene Combinations Specifying the Immune System

Peter Papathanasiou and Christopher C. Goodnow

Australian Cancer Research Foundation Genetics Laboratory and Australian Phenomics Facility, John Curtin School of Medical Research, Australian National University, Canberra, ACT 2601 Australia; email: petepapa@stanford.edu, chris.goodnow@anu.edu.au

Annu. Rev. Genet. 2005. 39:241–62

First published online as a Review in Advance on July 29, 2005

The *Annual Review of Genetics* is online at http://genet.annualreviews.org

doi: 10.1146/annurev.genet.39.110304.095817

0066-4197/05/1215-0241$20.00

Key Words

mouse, genome, phenome, mutagenesis, allele, T lymphocyte

Abstract

The human and mouse genome sequences bring closer the goal of understanding how characteristics of adult mammalian physiology and pathology are encoded by DNA. Here we review the challenge of understanding how genes specify mammalian traits, with particular focus on the cells and behavior of the immune system. Summarized is the emerging experience, advantages, and limitations of using ethylnitrosourea (ENU) to modify the mouse genome and select informative variants by phenotypic screens, yielding two main conclusions. First, ENU-induced variation provides an eminently feasible route to understanding how the genome encodes important mammalian processes without any prior assumptions about genes, their chromosomal locations, or expression patterns. Second, ENU alleles match those arising by natural variation. By changing individual protein domains or splice products, these alleles reveal separate gene functions specified through protein combinations.

Contents

THE CHALLENGE OF BRIDGING THE GULF BETWEEN MAMMALIAN GENOME AND PHENOME

Phenome: the whole organism translation of the genome into cell, tissue and systemic phenotypes or characteristics

The completed draft sequence of two mammals, humans and mice, brings closer the goal of understanding how the specific characteristics of mammalian physiology and pathology are encoded in our DNA. The term phenome describes this whole organism translation of the genome into cell, tissue, and systemic phenotypes or characteristics. In short, the phenome is what the genome does. An understanding of all the DNA sequences necessary for healthy functions of cells and tissues would illuminate the molecular circuitry needed for rational preventive or curative medicine targeted to the root causes of disease.

This review discusses the challenges faced in bridging the gulf between genome and phenome focusing on the task of identifying how the genome encodes mammalian resistance to infection and immune system pathology such as allergy and autoimmune disease. There are two reasons to focus on this example. First, mammalian immunity is dramatically different and more complex than invertebrate immunity, so that little can be extrapolated from the latter. Second, the formation of blood cells, especially T lymphocytes, is perhaps the best understood mammalian trait in terms of contributing genes and proteins. Mice represent the Rosetta stone for deciphering the genome-phenome code because of the scope for combining genetic, biochemical, cellular, and whole-system analyses in this organism. Obtaining mice with informative, defined changes in their DNA on a sufficiently large scale is the immediate challenge, and we specifically discuss the advantages and limitations of chemical mutagenesis with ethylnitrosourea (ENU) relative to other approaches for studying the impact of genetic variation.

ENU: ethylnitrosourea

Signalosome: a multiprotein body within the cell responsible for intracellular transmission of extracellular signals

TCR: T cell receptor for antigen

Relationship Between the Number of Components Encoded in the Mammalian Genome and the Complexity of the Phenome

Mammals have a surprisingly small increase in gene number compared with invertebrates. The human genome contains 22,000–28,000 genes at a conservative estimate, and around 40,000 genes when computer-predicted sequences are factored in (36, 65). The mouse genome has the same number, and for the great majority it is now possible to align the syntenic regions and firmly identify the exact orthologues of their human counterparts (70). When compared against predicted gene numbers in the genomes of lower organisms, these tallies are approximately twice the number of genes needed to make a fruitfly [14,000 genes (1)] or a worm [19,000 genes (9)]. The mammalian count is around five times that of a single-celled eukaryote, the yeast *Saccharomyces cerevisiae* [6000 genes (17)].

Given that the gene count is only up by a factor of two, is the mammalian phenome more than twice as complex as that of an invertebrate? Mammals certainly have more than twice the diversity of cell types found in invertebrates, as exemplified by the diversity of cells in the mammalian immune, nervous, and endocrine systems. Fruitflies resist infection with a single blood cell type, a hemolymph cell, whereas mammalian blood cells differentiate into ten major lineages comprising the erythrocytes, platelets, macrophages, neutrophils, eosinophils, mast cells, dendritic cells, natural killer cells, T lymphocytes, and B lymphocytes. Remarkable diversity exists within each of the mammalian blood cell lineages, such as the stark differences between nucleated embryonic erythrocytes compared with anucleate fetal and adult red cells, each specialized to match distinct problems of oxygen-transport in a small embryo versus a fast-running adult. Sublineage diversity reaches its pinnacle in B and T lymphocytes, where an infinite number of unique cell types are produced, each specialized with a unique antigen receptor to react against a different infectious trigger.

How is this large increase in phenomic diversity encoded by a genome with only a modest increase in the gene count? This appears to be a paradox if one assumes the Beadle/Tatum/Garrod perspective of "one gene - one protein - one trait." From a biochemical perspective, however, specific traits are rarely encoded by a single protein but are usually produced by assemblies of many proteins working in concert. For example, the diversity of T lymphocytes depends upon an antigen receptor signalosome (71) comprising (*a*) two alternative pairs of T cell receptor chains, TCRα and β or TCRγ and δ, that are unique in each different T cell because of somatic recombination of their genes; (*b*) two different coreceptors, CD4 and CD8, that are alternatively expressed in different T cell subsets; and (*c*) at least 20 different signal-transduction proteins that are shared by most T cells. Equally complex protein combinations are critical for upstream, downstream, and parallel processes specifying T cell formation and response (**Figure 1**). Many of the proteins specifying T cells also specify unique characteristics of other blood cells or other cells in the body (**Table 1** and Supplementary **Table 1** (Follow the Supplemental Material link from the Annual Reviews home page at **http://www.annualreviews.org**)). Many of the signaling subunits of the TCR signalosome also assemble with other receptors in other blood cell types so that changing one or several subunits is sufficient to impart an entirely new sensory function. This type of combinatorial specification of cell diversity provides a simple explanation for mammalian complexity achieved with a modest doubling in protein-coding genes.

KEY

Viable, clearly essential
Viable, partial phenotype may be detectable
Viable, phenotype too subtle
Viable, redundant
Not viable
Poor viability

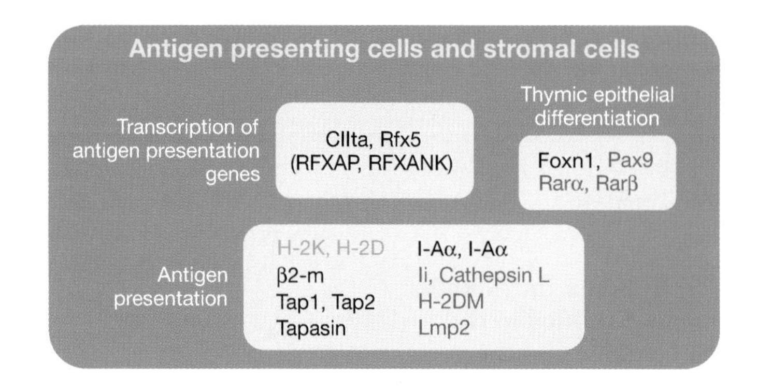

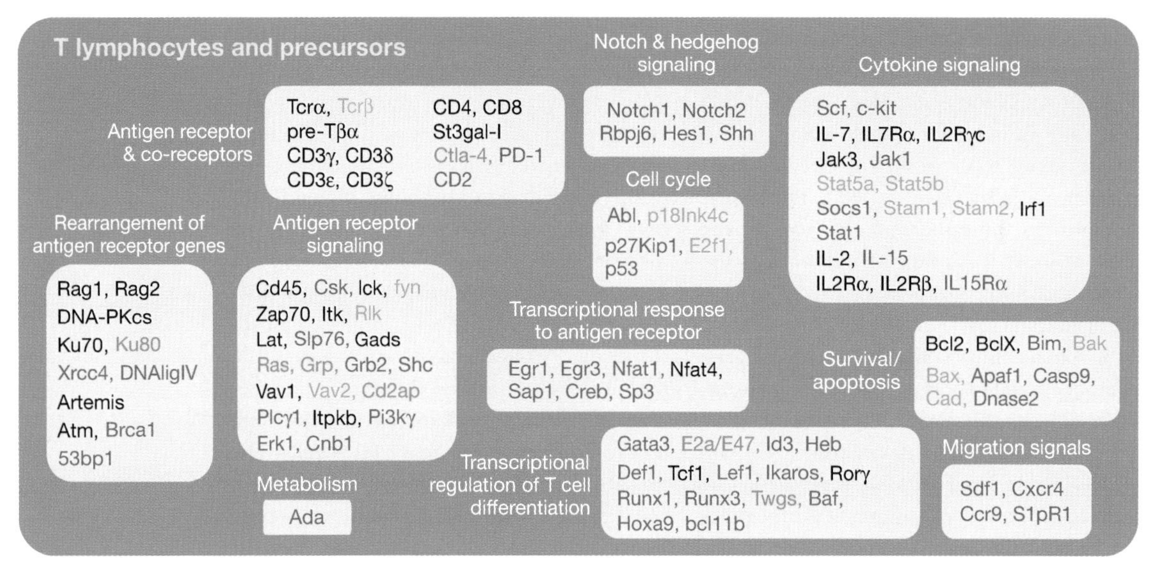

Figure 1

Gene and protein combinations known from mouse or human mutations to be required to establish normal numbers of circulating T lymphocytes grouped according to the biochemical function they serve. Genes are color-coded according to the phenotype of the null allele with respect to viability and impact on T cell numbers. More detailed information and references for each of these genes is given in Supplementary Table 1.

The magnitude of the potential increase in phenomic diversity created by combinatorial assembly and a doubling of the gene count can be estimated by making the following conservative assumptions: (*a*) the human or mouse genomes comprise 28,000 genes encoding 28,000 proteins (the issue of alternate splicing is discussed below); (*b*) the *Drosophila* genome comprises ~14,000 genes and proteins; and (*c*) the average multiprotein assembly specifying a particular biochemical trait comprises products of 10 different genes. The number of different potential combinations of 10-protein traits that can be produced can then be calculated:

$$\text{Human: } 28{,}000!/(10!\,27{,}990!) = 8.15 \times 10^{37}$$

$$\textit{Drosophila}\text{: } 14{,}000!/(10!\,13{,}990!) = 7.95 \times 10^{34}$$

In other words, a doubling of gene/protein numbers encoded in the genome yields a 1000-fold increase in possible 10-mer protein combinations available to specify the phenome.

Table 1 Examples of essential proteins acting combinatorially to specify T lymphocyte development as revealed by loss-of-function mutations, but also serving in other protein combinations to specify other cells and responses

Process	Protein	Other functions	Supplementary references[a]
Lineage commitment	Ikaros	B and NK cells, leukemia, hematopoietic stem cells, erythrocytes, embryonic development	(1–4)
	PU.1	Embryonic development, all hematopoietic lineages, osteopetrosis	(5–7)
	Notch1	Embryonic development, hematopoietic stem cells, leukemia, learning/memory	(8–13)
	Pax5	B cells, midbrain patterning	(14)
	GATA3	Embryonic development, nervous system, neural cells, ear morphogenesis, trophoblast giant cells, placental function, TH2 differentiation	(15–19)
	IL-15Rα	NK and NK T cells, TCR$\gamma\delta$ intraepithelial lymphocytes	(20, 21)
Growth signals	c-kit	Embryonic development, erythrocytes, hematopoietic stem/progenitor cells, mast cells, primordial germ cells, melanocytes, Cajal interstitial cells, gastrointestinal stromal tumors	(22–36)
	IL-7	B cells, $\gamma\delta$ T cells	(37–40)
	γ_c	B cells, $\gamma\delta$ T cells, NK cells	(41, 42)
	JAK3	B cells, NK cells	(43–49)
TCR rearrangement	RAG1&2	B cells	(50, 51)
	Ku70 Ku80 DNA-PKcs	DNA damage repair, genome integrity, B cells, leukemia	(52–55)
TCR signalosome	CD45	All leukocyte subsets, immune responses, lymphoproliferation, autoimmunity, mast cells, macrophages	(56–61)
	Zap70	T cell differentiation, activation, immunity and self-tolerance mechanisms, B cell chronic lymphocytic leukemia	(62)
	Slp76	Platelet function, vascular-lymphatic development	(63, 64)
	Lck	Retinal function	(65)
	Fyn	Pheromone perception, spatial learning, hippocampal development, epilepsy, CNS myelination, audiogenic seizures, fear-response, ethanol-sensitivity, mast cell degranulation, B cells, NK T cells	(66–80)
	Csk	Embryonic development, neural tube	(81)
	Vav1	B cells, NK cells, calcium mobilization in mast cells	(82–85)
	LAT	Lymphoproliferation, autoimmunity, mast cell activation, pre-B cells	(86–88)

[a]Reference numbers refer to the supplementary references listed online. Follow the Supplemental Material link from the Annual Reviews home page at **http://www.annualreviews.org.**

As opposed to subcellular biochemical traits such as TCR signalosomes specified by protein combinations, cellular traits such as T cell differentiation reflect "combinations of protein combinations" such as those required for T cell antigen receptor gene rearrangement, T cell gene expression, antigen presentation, and cytokine signaling summarized in **Figure 1**. If one assumes that 10 subcellular traits (each specified by a set of 10 proteins) combine to specify a cellular trait, then the doubling of the protein set in mammals compared with flies yields an increase in the number of potential cellular traits by a factor of 10^{30} (from 10^{333} to 10^{363}; **Table 2**). These rudimentary calculations emphasize that the doubling of gene count in mammals allows for more additional phenome complexity than is needed, and indeed the cardinal challenge for mammalian cells must be to channel these potential combinations into a manageable subset by extinguishing expression of the majority of genes in any one cell.

Aside from the increase in gene number between invertebrates and mammals, two changes in the quality of the genes—an increase in alternative mRNA splicing and an increase in the number of functional domains within encoded proteins—also potentially increase the number of multiprotein combinations encoded in our DNA. Based upon mapping of expressed-sequence tags (ESTs) to mRNA sequences and whole genome sequencing, 35%–60% of human and mouse genes show at least one alternative splice form (6, 24, 36, 45, 46, 65, 70). Is there more alternate mRNA splicing from mammalian genomes compared with invertebrates? Normalized bioinformatics protocols using identically sized random mRNA/EST (650/100,000) subsets from seven eukaryotic organisms—human, mouse, rat, cow, fly, nematode, and plant—showed a similar rate of alternative splicing with respect to both the number of genes affected and the number of variations per gene (6). Thus, whereas alternative splicing increases the complexity of the proteome, it may not account for increasing complexity between animal species. If one assumes that on average two different proteins are produced by alternative splicing from each *Drosophila* and mammalian gene, the effect on number of potential cellular traits specified can be seen by comparing the lines for 28,000 and 56,000 proteins in **Table 2**. The magnitude of difference (10^{30}) is unchanged relative to the numbers for one protein per gene.

Sequencing of the human genome revealed only modest innovation at the level of new vertebrate protein domains, whereas there was substantial innovation in the creation of new vertebrate protein architectures defined as distinct linear arrangements of domains within a polypeptide by shuffling, adding or deleting domains. Human proteins contain on average 1.8 times more domain architectures than worm or fly proteins, and 5.8 times more than yeast (36). As a specific

Table 2 Number of unique subcellular protein combinations that can potentially be formed from a genome encoding 7000 (e.g., yeast), 14,000 (e.g., fruitfly), 28,000 (e.g., human/mouse), 56,000, and 112,000 genes/proteins

Genes/proteins per genome	Number of unique subcellular protein combinations possible[a]			Number of cellular traits
	10-protein combinations	20-protein combinations	40-protein combinations	(combinations of 10-protein combinations)[b]
7000	7.73×10^{31}	3.19×10^{58}	6.98×10^{105}	$\sim 10^{303}$
14,000	7.95×10^{34}	3.39×10^{64}	8.11×10^{117}	$\sim 10^{333}$
28,000	8.15×10^{37}	3.58×10^{70}	9.17×10^{129}	$\sim 10^{363}$
56,000	8.35×10^{40}	3.77×10^{76}	1.02×10^{142}	$\sim 10^{393}$
112,000	8.56×10^{43}	3.96×10^{82}	1.13×10^{154}	$\sim 10^{423}$

[a] Calculations assume the average subcellular biochemical trait is specified by combinations of 10, 20, or 40 different proteins.

[b] Cellular traits are assumed to be specified by combinations of ten different subcellular traits, each specified by a 10-protein combination.

example, the CARMA1 protein, a critical scaffolding molecule participating in the TCR signalosome (discussed in more detail below), is a member of the membrane-associated guanylate kinase (MAGUK) family of proteins. The ancestral member of this family in invertebrates is represented by Disks Large in *Drosophila* containing three protein-interaction domains: a PZD domain, an SH3 domain, and a guanylate kinase (GUK) domain. The protein family is extensively radiated in mammals by addition of other protein-interaction domains so that the lymphocyte-specialized CARMA1 carries two added protein interaction domains at its N terminus: a caspase recruitment domain (CARD) and a coiled-coil (CC) domain. The CARD domain recruits a new protein complex—the Bcl10/Malt1/IκB kinase signalosome—enabling CARMA1 to scaffold a more elaborate multiprotein combination than its *Drosophila* ancestor. If one assumes that the near-doubling in potential protein docking sites on the average mammalian protein compared to flies increases from 10 to 20 the number of proteins that can assemble to specify a biochemical trait (**Table 2**), this translates into a 10^{32}-fold increase in potential biochemical traits independent of any increase in gene number.

The discussion above takes into account only the protein-coding parts of genes. Clearly, there are also genes encoding noncoding RNAs that are active and central for animals. These include the RNA components of protein translation (ribosomal and tRNAs), and the growing class of regulatory RNAs such as Xist, micro-RNAs, and antisense transcripts. Based on human-mouse genome comparisons, there are at least as many noncoding conservons in the mammalian genome as there are exons encoding protein, and these have comparable levels of sequence conservation (44). If the majority serve regulatory roles by specifying which theoretical combinations of proteins are actually made in the same cell, this would not increase the potential number of biochemical or cellular traits encoded in the genome. Instead, the likely role of these nonprotein elements is to manage the huge number of potential proteins encoded in the mammalian genome by controlling gene expression so that only specific combinations of proteins are made in a given cell.

Implications of Combinatorial Specification for Drug Development and Genetic Analysis

Current pharmacological efforts to use the genome sequence as a starting point for identifying antagonists of specific processes are dominated by the view that "one gene makes one protein makes one trait." If the diversity of the mammalian genome-phenome code is specified through protein combinations, then this is where drug development should best focus. Protein-protein interactions nevertheless remain the most difficult and least attractive targets for pharmaceutical intervention.

Similar practical implications follow for genetic analysis since systematic efforts to knock out or knock down all of the mammalian genes in mice starts with the assumption of "one gene - one protein - one trait." By the calculations above, however, most of the genome-phenome code will only be revealed by analyzing genetic variants that interfere with specific protein-protein interactions, notably point mutations arising by natural variation or induced experimentally by chemicals such as ENU.

ENU MUTAGENESIS AS A TOOL TO REVEAL GENOME SEQUENCES SPECIFYING KEY MAMMALIAN CELLS AND TRAITS

Nature and Frequency of Mouse Genome Sequence Variants Produced

ENU is emerging as a superb tool for dissecting mammalian genome-phenome links both for the rate and nature of genetic variants

Mb: megabase

Missense mutation: a mutation type that results in the substitution of an amino acid in the protein encoded by the mutated gene

G1, G2, G3: first-, second-, and third-generation offspring from ENU-treated male mice

it produces. ENU has been estimated to induce one new loss-of-function mutation per gene in every 700 gametes of treated male mice (19) and one nucleotide change in every one megabase (Mb) of genomic DNA (53). Other efficient germline chemical mutagens such as chlorambucil or radiation usually cause large deletions or inversions that often involve more than one gene because this acts on postmeiotic round spermatids (58). By contrast, ENU targets premeiotic spermatogonial stem cells inducing discrete lesions, specifically point mutations, through alkylation of nucleic acids. The most common mutations induced by ENU are A/T → T/A transversions and A/T → G/C transitions (28). More than 82% of the mutations sequenced show these types of mutations, most likely the result of mispairing after the alkylation at O4 and O2 thymine. When translated into a protein product, these changes result in 64% missense mutations, 10% nonsense, and 26% cause errors in mRNA splicing (28). Our own experience identifying immunological mutants matches this distribution. Genetic variation induced in mice by ENU is thus ideally suited to reveal discrete actions of proteins working combinatorially by selectively altering individual protein domains and splicing products in the same way as natural variation, while at the same time inducing a very low frequency of linked irrelevant sequence changes. In the past, single nucleotide changes were the most difficult to reveal, but this is no longer the case with the availability of the draft mouse genome sequence and efficient resequencing methods. The chief issue thus becomes one of selecting ENU-induced genetic variants of interest, and the alternative experimental strategies are outlined below.

ENU-Induced Allelic Series at Known Visible Loci

Large sets of informative allelic variants of specific genes have been isolated by breeding ENU-mutagenized wild-type mice directly with partners that are homozygous for recessive visible mutations, or by identifying dominant visible phenotypes and then analyzing a range of cell types in homozygous offspring. In both cases, new mutations are revealed in the first generation (G1) offspring of the ENU-treated males, allowing as many mutagenized gametes to be screened as G1 mice can be bred. For example, the different functions of the *microphthalamia*-associated transcription factor (*mitf*) in neural crest-derived melanocytes and in B lymphocytes were revealed by studying the consequences of different mutations in the *microphthalmia* locus (38, 63). An allelic series of *c-kit* (*W, White spotting*) and *kit ligand/stem cell factor* (*Steel*), either spontaneous or induced by ENU or other mutagens, has been critical for separating out distinct functions of this growth factor system in hematopoietic stem cells, erythrocytes, T cells, mast cells, germ cells, gastrointestinal motility, and melanocytes (37, 55). Similarly, analysis of a large set of ENU-induced *myosin VA* (*dilute*) alleles revealed distinct roles of N- and C-terminal domains in melanocytes and Purkinje cells (23). Equivalent studies have revealed more subtle alleles and phenotypes following ENU mutagenesis of the *Bmp5* (43) and *quaking* (11) loci. Another use of this strategy has been to isolate a second allele to confirm that a sequence change found in a spontaneous mutant is indeed responsible for a phenotype (10). The limitation of this approach, however, is that it focuses on a small subset of genes associated with visible traits so that it is not well suited for systematic analysis of internal systems such as immunity.

Chromosomal Region-Specific Screens for Gene Variants

Breeding ENU-mutagenized wild-type mice with partners bearing chromosomal deletions is an efficient strategy to identify mutations that are recessive and encompass genes essential for internal processes, development, and viability. The starting point has been mouse strains with heterozygous viable chromosomal deletions spanning visible marker loci such as the chromosome 7 albino (*Tyr*; *c*) and pink-eyed dilution (*p*) loci. By crossing heterozygous carriers of a deletion with first-generation (G1) offspring of ENU-mutagenized males, new ENU-induced mutations affecting individual genes within the chromosomal deletion were revealed because of hemizygosity in the second-generation (G2) offspring (57). Visible markers have been used in these studies to help identify offspring that are hemizygous for the mutagenized chromosome segment over the deletion.

Poor fertility or lethality of most monosomic deletions of appreciable size limits the general use of the strategy above. Extension of region-specific screening to any genome region has been made possible, however, by chromosome-engineering strategies employing gene targeting and Cre-lox recombination to generate chromosome inversions or balancer chromosomes (73). Meiotic recombination between wild-type and inverted chromosome segments is suppressed within and near to the inverted segment. ENU-induced point mutations in an initially wild-type chromosome segment can therefore be outcrossed with the balancer strain and then intercrossed to bring the point mutations to homozygosity in 25% of the third-generation (G3) offspring of the ENU-treated males. By using different visible markers linked to the mutagenized wild-type and inverted chromosome segment, homozygotes and carriers for new mutations generated by ENU can be identified and followed from generation to generation by simple visual typing, whereas absence of homozygotes indicates a lethal mutation in the balancer interval.

Kile et al. (32) elegantly demonstrated the use of this strategy to systematically discover genes in an inverted 24-cM/34-Mb portion of chromosome 11 representing 2% of the genome. They screened 735 pedigrees and isolated 88 mutant strains in the region marked by the balancer: 55 were embryonic or postnatal lethal; the others fell into various classes affecting hematopoiesis, the skin, nervous system, craniofacial development, and fertility. Complementation crosses showed that few of the lethal mutations were allelic, indicating that ~10% of the 700 potential protein-coding transcription units within the interval could be revealed as lethal mutations. A comparison with known spontaneous and targeted mutations in this region suggested that the ENU-induced phenotypes arise from genes not previously connected to a trait or represent functionally distinct variants of known genes.

One of the major advantages of this combination of ENU and visible genetics is that each new variant is tightly linked to a visible marker, enabling lethal recessive mutations to be propagated easily by breeding visibly distinct heterozygous carriers. Moreover, the fact that new mutations are premapped to a chromosomal region facilitates rapid triage of mutations in known genes, which can be quickly resequenced to identify the causative substitution. As the cost of sequencing falls, it should become economic to sequence all the exons in a 10–20-Mb interval, thus avoiding time-consuming fine-resolution meiotic mapping entirely. On the other hand, the limitation of regional mutagenesis is its focus on a small fraction of genes selected by genomic location rather than by function, and it can only be carried out when there is an appropriately marked balancer available on a suitable strain background.

Balancer chromosome: a chromosome with one or more inverted segments that suppress recombination allowing lethal mutations to be maintained without selection

cM: centiMorgan

Genome-Wide Screens for Induced Gene Variants

When the primary aim is to understand how the genome specifies key aspects of mammalian physiology such as immunity, the relevant genes are spread across all chromosomes and thus need to be pursued by a genome-wide approach.

Genome-wide dominant screens. Dominant ENU-induced mutations have been isolated on an unparalleled scale by testing over 40,000 first-generation (G1) offspring of mutagenized males for a range of visible,

Haploinsufficiency: a phenotype arising in diploid organisms owing to the loss of one functional copy of a gene

Antimorphic allele: a gene variant that interferes with wild-type allele function in a semidominant manner (also known as "dominant negative")

B6: C57BL/6 mouse strain

behavioral, metabolic, and immunological parameters (22, 50). Each new mutant stock must be propagated initially by tedious phenotypic testing of progeny until it can be mapped to a chromosomal region in an outcross. Once mapped, however, carriers can be identified by flanking DNA polymorphisms such as simple sequence length polymorphisms or single-nucleotide polymorphisms that are more readily tested, thus allowing new variants to be readily archived, distributed, and studied.

A key advantage of screening for dominant mutations genome-wide is that they are revealed in the G1 offspring from an ENU-treated male and wild-type females so that a large number of variant genome sequences can be screened. Another important advantage is the potential for haploinsufficient, antimorphic, or gain-of-function mutations to reveal gene functions in adult physiology separately from essential roles for the same genes during development. However, the limitation is that haploinsufficiency is subtle in most genes, making it difficult or impossible to recognize or map such variants, while strong dominant mutations are likely to be rare in most genes leading to recurrent isolation of variants in a few critical rate-limiting regulators (7, 31, 59). One invertebrate strategy to facilitate detection of haploinsufficient defects is to sensitize a process of interest with another mutation so that heterozygous loss-of-function variants act as enhancers and yield a striking phenotype (62). This has recently been used successfully to reveal ENU-induced modifiers of blood cell differentiation (7), epigenetic inheritance (52), and susceptibility to diabetes (66; G. Hoyne & C.C.G., unpublished data).

Genome-wide recessive screens. Genome-wide detection of recessive mutations requires two additional generations of inbreeding to bring any unique mutation carried by a G1 animal to homozygosity in its third-generation (G3) offspring. Two typical pedigree strategies are used to bring ENU-induced variants to homozygosity: either backcrossing G2 daughters with their G1 father, or intercrossing G2 brothers and sisters (47). This has two major consequences: (*a*) it greatly increases the logistical complexity of housing, breeding, and tracking the animals; and (*b*) a large burden of mutations across the genome are potentially brought to homozygosity in the G3 animals, potentially causing lethality and obscuring detection and propagation of informative traits with unrelated lethal mutations and complicating mutations. These issues were minimized in a genome-wide recessive screen of newborn G3 offspring for defects in phenylalanine metabolism, revealing mutations in phenylalanine hydroxylase in 1 of every 170 pedigrees (5, 60). This frequency of loss-of-function mutations is higher than predicted based on the specific locus test, perhaps because the encoded enzyme is intolerant to amino acid substitution or because there is no alternative pathway to process the screened metabolite. A spectacularly successful screen pioneered by Anderson et al. (30) tested for developmental patterning defects in G3 embryos at 9.5 days gestation, thus allowing each pedigree to be screened with only one or two boxes of mice (the G1 founder and several daughters). In 380 pedigrees, they isolated 42 recessive lethal variants with clear morphological abnormalities, and by genomic mapping and gene identification showed that the majority were in genes not previously connected to developmental patterning (14).

Feasibility and Efficiency of Genome-Wide Recessive Screens for Sequences Specifying the Production and Control of T Lymphocytes

In contrast to errors of metabolism and development, many important mammalian traits such as immunity, behavior, cancer, lactation, or fertility require an analysis of healthy adult animals. Our own studies focused initially on screening G3 offspring of ENU-treated C57BL/6 (B6) males for abnormalities in the formation of T lymphocyte subsets. This screen provided an ideal test of feasibility and efficacy because the list of known essential T cell genes (**Figure 1**) is almost certainly more comprehensive than any other mammalian cell type, thereby providing an unparalleled test to measure how efficiently a genome-wide recessive strategy identifies mutations in essential genes. Unlike visible traits, most T cell genes were identified initially by nongenetic means, typically protein biochemistry and molecular biology, and only subsequently shown to be essential through production of knockout mice or candidate gene sequencing in human immunodeficiency (Supplementary **Table 1**). They are thus unlikely to have been selected as particularly mutable loci. Because T cells are unique to higher vertebrates, and each of the mutations could be readily ascertained, the screen also tested the possibility that specific mammalian traits stem from the large class of conserved elements that distinguish the mammalian genome from invertebrates, such as noncoding conservons (44).

We used four different ENU dose regimes in B6 male mice: either 85 mg/kg or 100 mg/kg ENU, given either three or four times one week apart. This regime yielded 185 pedigrees founded by a single G1 male and successfully bred to G3 offspring. We identified 11 strains of mice with clear and reproducible T cell formation defects by flow cytometry analysis of blood and successfully propagated each as true-breeding Mendelian variants. The strains arose from all of the mutagenic regimes except the highest dose (4×100 mg/kg), which had poor recovery of fertility in treated males and poor fertility among the offspring, whereas the 3×85 mg/kg dose appeared optimal. All of the mutations were phenotypically recessive except one, *Plastic*, affecting the Ikaros protein, which caused a dominant phenotype of incompletely penetrant T cell leukemia (51). We have been successful in mapping and identifying the mutation in each of these strains so that general conclusions can be drawn about the efficacy and nature of mutations found without ascertainment bias.

QTL: quantitative trait loci

Three chief conclusions can be drawn. First, genome-wide screening for recessive variants in adult traits is highly feasible and effective. Ten mutations were identified in the 55 genes expected to have been detected by viable adult screens for clear T cell abnormalities (denoted by black or dark blue in **Figure 1**). This represents an 18% rate of loss-of-function mutations in 185 pedigrees. Taking into account that the daughter-father pedigree structure can only bring ~70% of mutations to homozygosity, this rate corresponds precisely with the measurement of 1 hit per locus per 700 gametes for 7 visible loci (19).

Second, all of the mutations identified are in protein-coding genes and affect the protein-coding region in one way or another. This argues against a major phenotype-specifying role for nonprotein-coding conservons. However, because the method of mutation detection depends upon having a large effect on the phenotype, by altering T cell numbers by at least two or three standard deviations, the nonprotein-encoding parts of the mammalian genome may have smaller effects on phenotype akin to quantitative trait loci (QTL). Given the potential for small effects to compound over successive infectious challenges and generations, these may nevertheless be important and require a different approach to reveal.

Third, the variants identified provide a unique tool to separate multiple functions of individual genes. Several illustrative examples are discussed below.

Alleles separating developmental from adult-specific gene roles. Forty-two essential T cell genes also serve other processes and were not expected in the screen because null alleles of these genes cause developmental or perinatal lethality (denoted red or green in **Figure 1**). One hit was nevertheless detected in this set, a hypomorphic allele of

Separation of function allele: a gene variant that affects some functions of a gene while preserving others

ES: embryonic stem

the T cell signaling adaptor *Slp76* which preserves its vascular morphogenesis functions so that the animals are born and grow normally, but partially cripples the roles of *Slp76* in T cell differentiation and completely eliminates its T cell inhibitory actions so that the animals develop paradoxical autoimmunity and allergy (L. Miosge and CCG, manuscript in preparation). A similar example is described by Cooke and colleagues (59) where a null allele of the transcription factor c-Myb causes embryonic lethality whereas a hypomorphic ENU-induced missense allele that interferes with binding to p300 reveals adult functions of c-Myb in differentiation of T and B lymphocytes and other blood cells.

Outside of the known T cell gene set, a specific role in T cell differentiation was revealed for an obscure gene contributing to a phylogenetically ancient protein complex with an essential role in chromosome maintenance (A. Fahrer and CCG, manuscript in preparation). A null allele of this gene would be expected to be early embryonic lethal, yet the missense variant apparently preserves these functions and separates out a discrete role in T cells that may have been added during evolution. These examples clearly illustrate the unique opportunity to tease out the combinatorial code through ENU-induced variants selected by adult phenotypes. It is important to note that this class of separation-of-function variants are isolated less efficiently, with 1 hit in 42 known lethal genes compared to 10 hits in 55 viable genes. It will thus be difficult to systematically identify all components of an adult trait through this approach, although the separation of function alleles obtained are revealing entry points.

Separating discrete functions of specific protein combinations: CARMA1. Like the *Slp76* variant, an even more severe and paradoxical allergy syndrome develops in the *unmodulated* mouse strain (27) which illustrates the complementarity between ENU-induced variants and gene-targeted alleles for dissecting out multiple gene functions. This recessive mutation causes subtle changes in circulating T and B cell subsets, but profoundly decreases their activation by particular stimuli and subsequent antibody responses. The *unmodulated* mutation was mapped to a ~1 Mb chromosomal region and resequencing of genes within the interval revealed a single T → A nucleotide substitution resulting in amino acid change L298Q in the CC domain of CARMA1/CARD11, a novel member of the MAGUK family of proteins. MAGUK proteins are 'synaptic' in neurons because they associate with the cell membrane and function as molecular scaffolds that tether and organize receptors to intracellular signaling machinery (13, 61). Like other MAGUK proteins, CARMA1 serves as a scaffold to assemble a signalplex of other proteins through its string of protein-protein interaction domains: CARD, CC, PDZ, SH3 and GUK (26). The *unmodulated* point mutation alters a conserved leucine residue in CARMA1's CC domain but preserves normal protein abundance in the cells. Because CC domains promote protein folding, dimerization or multimerization by forming bundles of interwound α helices, the substitution may prevent assembly into higher-order macromolecular signaling scaffold complexes or it may prevent binding to another partner protein. The biochemical consequence is a complete loss of phosphorylation of an essential adaptor linking to the NFκB signaling pathway, Bcl10, and diminished activation of the NFκB and JNK signaling pathways upon antigen receptor stimulation (26).

In parallel with our discovery of the CC domain, missense mutation in *unmodulated*, *Carma1^{null}* mutant mice were generated in parallel by gene-targeting in embryonic stem (ES) cells (12, 18, 48). Lymphocyte responses to some key activating receptors, CD28 and CD40, are equally disrupted by the missense and null mutations. By contrast, lymphocyte responses to other key receptors, TCR and TLR4 are crippled when Carma1 protein is absent but appear fully intact in cells with the missense mutation. By disrupting all lymphocyte activation pathways, the null mutation causes profound immunodeficiency, whereas the point mutant leaves IgG1 and IgE antibody responses intact and these ultimately overshoot to give rise to massive allergic antibody production and dermatitis (27). The *unmodulated* ENU allele thus reveals additional roles for CARMA1 as a scaffold for inhibitory signaling pathways which apparently are separated from its activating roles by a single nucleotide substitution.

A recessive niche-filling allele of ikaros. The *Plastic* strain was identified in the T cell screen through dominantly inherited T cell leukemia/lymphoma developing at 3–4 months of age (51). Homozygotes died embryonically at 15.5 days gestation with severe aplastic anemia due to a cell autonomous failure of erythroblasts to grow and terminally differentiate. The gene encoding the Ikaros transcription factor was a strong candidate because the *Plastic* mutation mapped to the same genomic interval, and mice heterozygous for a targeted internal deletion of two of Ikaros' four N-terminal DNA-binding zinc fingers (exons 3 and 4) develop T cell neoplasia at a similar age (72). Resequencing of the *Ikaros* cDNA in *Plastic* mice revealed a single A → G transition resulting in a missense amino acid change (H191R) in one of the histidine residues comprising the Krüppel Cys_2His_2 consensus motifs that coordinate a zinc ion to stabilize the third DNA-binding finger. Biochemically, the point mutant exhibits a complete loss of DNA-binding to target sequences but retains other domains for mutimerization with Ikaros and related proteins such as Aiolos (51).

The embryonically lethal erythrocyte differentiation defect in *Plastic* homozygotes was novel and surprising because analysis of null or other alleles created by homologous recombination in ES cells had ascribed Ikaros a selective role in lymphocyte formation. Mice homozygous for an Ikaros C-terminal mutation (exon 7) lack all cellular protein ($Ikaros^{null}$) and fail to form B lymphocytes while their T lymphocyte development is delayed (68). An N-terminal internal deletion (exons 3 and 4) forces overexpression of short dominant-negative isoforms ($Ikaros^{DN}$) that are normally rare and causes homozygotes to lack all lymphocytes (15). A third, hypomorphic Ikaros mutant ($Ikaros^{L}$) has also been generated by gene targeting: a deletion (exon 2) that retains the N-terminal DNA-binding and C-terminal dimerization domains yields a milder lymphoid phenotype (34). Broader hematopoietic roles for Ikaros had been considered, however, based upon the presence of mRNA at the earliest sites of hemopoiesis during ontogeny (16) and in all differentiated blood cell types, with variable abnormalities in erythroid/myeloid cell numbers in the targeted mouse strains (15, 40, 49, 68) and minor changes in embryonic/adult globin ratios and erythrocyte/macrophage gene expression (40). It was unclear whether these subtle abnormalities in nonlymphoid blood cells of adult $Ikaros^{null}$ and $Ikaros^{DN}$ mice were secondary to opportunistic infections or to deficiencies in hemopoietic stem cells as measured by transfer/reconstitution experiments (49). Thus, a discrepancy existed between the widespread hemopoietic expression of Ikaros and the narrower functional role as defined by protein-ablation experiments prior to analysis of the $Ikaros^{Plastic}$ allele.

Neomorphic allele: a gene variant that co-opts the gene into novel functions independent of wild-type function

Why does a point mutation in the Ikaros DNA-binding domain yield a more severe, widespread failure of blood cell formation than alleles that ablate all or some of the protein? This illustrates a key issue arising from mutant missense alleles: do they reveal loss of function beyond what can be seen with a null, or has the altered protein acquired new (neomorphic) or toxic/interfering (antimorphic) cellular effects that are informative with respect to normal functions? Genetically, the $Ikaros^{Plastic}$ point mutant protein appears to act as a loss-of-function rather than an antimorphic or neomorphic allele because the hematopoietic failure is recessive to wild-type Ikaros.

Antimorphic or neomorphic mutations are typically dominant, although there are examples of recessive antimorphic proteins that are suppressed by coexpressed wild-type protein because the mutant is less efficiently assembled into a complex/organelle. For example, missense collagens distort body shape in *Caenorhabditis elegans*, but exhibit this abnormal function only in homozygotes because wild-type products are preferentially incorporated into fibers in heterozygotes (35). By contrast, *Ikaros^{Plastic}* protein can be shown to retain normal capacity to dimerize and multimerize efficiently (51). Alternatively, recessive antimorphic or neomorphic effects beyond the null could potentially arise when the threshold amount of mutant protein needed to produce abnormal effects is achieved through inheritance of two copies of the gene (25, 56). However, a requirement for two doses of the mutant gene is ruled out by finding an equally severe hemopoietic defect in compound heterozygotes with a single missense *Ikaros^{Plastic}* allele and an *Ikaros^{null}* allele on the other chromosome (51). Collectively, these results indicate that the missense allele causes a more complete loss of normal *Ikaros* function than the null.

Like the other ENU mutants described above, the more complete loss of function from a missense than a null allele of *Ikaros* can again be explained by separation of function, in this case preserving Ikaros' role as a component of a higher-order complex while inactivating its role as a DNA-binding protein. Ikaros operates as a component of larger multimolecular complexes, such as the nucleosome remodelling and deacetylation complex (33), so that protein-interaction domains and scaffolding functions add biological specificity to the core activities of DNA binding, chromatin remodeling and histone deacetylation. When part or all of the Ikaros protein is eliminated by null/deletion mutations, this presumably creates a void in larger macromolecular complexes normally containing Ikaros that can potentially be filled by a family of related proteins. By contrast, this void is filled by the 'recessive niche-filling' point mutant allele which serves to inactivate Ikaros' normal functions without creating the opportunity for compensatory substitution by related family members.

Interestingly, this valuable type of genetic variant has not been recognized in invertebrate mutagenesis, with a single exception in the case of opportunistic compensation between yeast mitogen-activated protein (MAP) kinase homologues (41, 42). Two MAP kinases, Fus3 and Kss1, appear to be redundant as they substitute for one another if one protein is eliminated from the Ste5 signaling scaffold by a null mutation. The two kinases in fact perform discrete roles, but this can only be revealed if one is rendered enzymatically inactive but still able to assemble into its signaling scaffold by a point mutation. A similar mechanism may in fact account for many other examples of recessive alleles with greater severity than their homozygous null counterparts. Given the exponential increase in mammalian genetic complexity and possible phenotypic combinations, such examples may soon become the norm as opposed to the exception.

Entirely novel immune regulatory genes and mechanisms: Roquin. Based on the success of screening for genes controlling the well-studied process of T cell formation, we have now embarked upon screens of much less well understood traits such as self-tolerance and autoimmunity. One such screen tests for mature-onset formation of antibodies against self antigens in the cell nucleus, so-called antinuclear antibodies, as these are known to be a sensitive clinical indicator of either clinical or subclinical autoimmune diseases such as systemic lupus (67). This is a much more logistically difficult screen since G3 animals must be aged to 12–15 weeks old and sufficient animals per pedigree tested to identify two or more antinuclear antibody-positive individuals, in order to minimize false positives due to a low rate of antinuclear antibody formation in normal B6 mice. A series of such mutants was identified, and the first of these—*san roque*—has now been mapped and the gene variant identified (66). Both the gene and the mechanism it controls appear entirely novel. The gene, *Roquin*, encodes a member of the large RING-type E3 ubiquitin ligase protein family, but an unusual one in that it contains an RNA-binding type of zinc finger and localizes to cytoplasmic granules that control mRNA stability and translation. The gene is highly conserved through to invertebrates, yet nothing was known previously of its function. At the same time, the *san roque* variant reveals a new mechanism for inhibiting T cells with self-reactive receptors, such as the cells causing type 1 diabetes, by repressing the receptor ICOS which controls their differentiation into a potent subset of follicular helper T cells.

Genome-Wide Recessive Screens for Mammalian Innate Immune Responses

While T and B lymphocytes account for the extraordinarily adaptive immune responses mammals mount against any new chemical antigen an invading microbe might evolve, a complementary arm of immunity responds to phylogenetically conserved chemicals made by particular classes of microbes. Examples of these invariant chemicals include the lipopolysaccharide cell wall component and motile flagella of gram-negative bacteria, double-stranded RNA intermediates in viral replication, and unmethylated CpG nucleotides of prokaryotic DNA. These are sensed by a sophisticated system of receptors, signaling proteins, and immune cell responses that has been revealed very efficiently through spontaneous and ENU-induced mouse variants (4). By producing three generation pedigrees of ENU mutagenized B6 mice, Hoebe, Beutler and colleagues have systematically screened adult G3 mice from a large number of pedigrees for recessive variants interfering with the capacity of peritoneal macrophages to produce cytokines when exposed to one or more of these microbial stimuli. This screen has been extraordinarily successful, identifying novel signaling molecules that explain how individual receptors signal different responses (20), and revealing roles in sensing of specific chemicals for well-known genes that were nevertheless previously unconnected to the process of innate immunity (21). A large series of variants have been isolated already by this screen and over 50% identify novel genes or functions (B. Beutler, personal communication).

RELATIVE MERITS OF DIFFERENT APPROACHES TO DECIPHER HOW THE MAMMALIAN GENOME SPECIFIES THE PHENOME

The crucial material for bridging the gulf between genome and phenome is the opportunity to analyze the biochemical, cellular and systemic phenotype of individuals with defined variant DNA sequences at the scale demanded by a problem with $\sim 10^{363}$ possible phenotypic combinations (**Table 2**). Given the scale of the problem relative to the size of the research community, a key question is the relative merits of following a phenotype-driven approach, by identifying individuals with a different phenotype and then linking the phenotype to DNA sequence variants, or to follow a sequence-driven approach focusing first on specific DNA sequence variants and then identifying phenotypic consequences.

Phenotype-Driven Approaches

Mendelian variants. Humans with inherited Mendelian traits are the cardinal example of the phenotype-driven approach, drawing upon a huge population of people and naturally occurring DNA sequence variations that are individually phenotyped through primary, secondary, and tertiary levels of the medical system. One of the great attractions of

RNAi: RNA interference

Mendelian trait analysis in humans is that it makes no assumptions about what a "gene" is, whether it encodes protein or only RNA, or neither. Since natural variation often alters individual amino acids or splicing patterns, this approach is well suited to separating discrete functions of genes in humans. A number of the T cell specifying genes in **Figure 1** were discovered by analysis of Mendelian immunodeficiency syndromes, while many others have subsequently been linked to these syndromes by candidate gene sequencing. While these provide crucial anchor points for understanding human gene-phenotype links, it is nevertheless difficult to study how (or if) a specific DNA sequence variant results in a particular phenotype in humans because only isolated individuals exist and key tissues and cells cannot be analyzed. The ability to fill in the links between gene and systemic phenotype in mice with corresponding gene variants has proven essential, as illustrated by mutations in the AIRE gene which cause Autoimmune Polyendocrine Syndrome 1. The mechanism of action of AIRE, initiating a critical pathway for eliminating self-reactive T cells in the thymus, could only be deciphered through analysis of mice with corresponding mutations (2, 39).

The combination of ENU-induced variation in mice with phenotypic screening by clinically relevant tests can scan a similar scale of genome sequence diversity with four advantages: (*a*) large families are readily arranged; (*b*) there is a very low frequency of irrelevant DNA differences to confuse interpretation; (*c*) there is ready access to pure cell types for analyzing cellular and biochemical consequences of a given DNA variant; and (*d*) it is possible to reintroduce the wild-type copy of the gene by retroviral vectors or transgenesis. One of the added potential advantages of this approach in mice over humans is that it can in principle be tailored to discover disease-preventing sequence variants by mutagenizing a starting strain that is genetically or environmentally challenged to be susceptible to a disease like autoimmunity, cancer, or obesity. To date the only mutagen capable of producing the necessary scale and diversity of mouse gene variants is ENU.

Quantitative trait loci. As opposed to naturally occurring Mendelian traits where there is a clear and simple relationship between a genetic variant and phenotypic change, analysis of naturally occurring QTL represents a parallel phenotype-driven strategy. The advantage of QTL is that they also make no assumptions about the nature of a gene, but contribute to many common traits in humans and mice and encompass genes and genetic variants that have more subtle effects than Mendelian traits. As noted in **Figure 1**, 13 genes that are essential for T cell formation would not be expected in a genome-wide ENU screen because null alleles at these loci have modest effects on T cell numbers. Variants of genes in this class may nevertheless be revealed by analysis of QTL. The disadvantage of QTL is that the correlation between gene variant and phenotype is low, and ambiguity in linking variant sequences to genes arises because of the high background of irrelevant DNA sequence variations (~1 per 1000 bases) in the genomic intervals containing QTL (**Table 3**).

Sequence-Driven Approaches

Sequence-driven approaches to generate mouse mutants, such as gene targeting or gene-trapping in ES cells and more recently transgenic RNA interference (RNAi), are more limited in the number of sequence variants that can be explored for phenotypic consequences (**Table 3**), but have the advantage that they can focus on particular sequences where a knowledge framework is emerging from other sources. The advantages of a sequence-driven strategy and the missense variants generated by ENU are potentially combined by banking sperm and DNA from several thousand offspring of ENU-treated mice, and screening the DNA for nonsynonymous nucleotide substitutions in exons of

Table 3 Comparison of different approaches for deciphering genome sequences specifying the mammalian phenome

	Targeted null mutant	RNA interference	ENU point mutant	Quantitative trait loci
Prior knowledge about gene needed	Yes	No	No	No
Gene functions eliminated	All	Most	Selective	Selective
Genotype-phenotype correlation	High	Intermediate	High-intermediate	Low
Ambiguity due to differences in other genes	Low	Intermediate	Low (2 base variants/Mb)	High (1000 base variants/Mb)

interest (3, 53). A fraction of the missense mutations yield informative traits after retrieving frozen sperm and deriving homozygous offspring.

A key advantage of gene targeting is the production of conditional null alleles to separate out functions of individual genes in different tissues or stages of life. For example, the Notch receptor signaling pathway is required for specifying blood progenitor cell differentiation into T lymphocytes rather than B lymphocytes, and for later stages of B and T cell regulation, but is also required for embryonic developmental patterning of many tissues. A complete loss-of-function mutation in Notch1 causes embryonic lethality that obscures analysis of the other functions (8, 64). A simple hypomorphic allele generated by complete loss of one Notch allele to halve the protein expressed provided a partial solution because the reduced amount of protein was sufficient to support embryonic development, but had a subtle yet significant effect on T cell differentiation (69). A targeted allele of Notch1 that is flanked by lox sites and is selectively deleted by Cre recombinase only in lymphocytes, however, provided much clearer evidence that the gene is absolutely necessary for T cell differentiation (54).

The use of RNAi has blossomed in recent times, combining sequence- and phenotype-driven approaches with great effect in screens, for example, by *C. elegans* laboratories in seeking to assign function to every one of the 19,000 genes in the worm's genome (29). Whether such studies translate to the more complex mammalian combinatorial system will largely be determined by protocol optimization for nonvariegated expression or short-hairpin RNA in mice, confusion by off-target effects, and the inability of this approach to interfere with discrete protein domains and interactions. While the cost of producing such knock-down alleles may ultimately be lower, this may be dwarfed by the additional cost and time spent separating on-target from off-target phenotypic effects in whole animals.

CONCLUDING REMARKS

As summarized above, there is now sufficient experience with the use of ENU to induce changes in the mouse genome sequence and to select informative variants by phenotypic screens to conclude two key points. First, it provides an eminently feasible route to understanding how the genome encodes important processes in adult mammals, such as immunity, without prior assumptions about the nature of a gene, its chromosomal location, or its expression pattern. It is thus a powerful way to reveal new genes, and new functions for "known" genes. Second, the kinds of alleles produced by ENU mirror those arising by natural variation in humans and other mammals, and are a natural complement to null alleles produced by gene targeting. By changing the function of individual protein domains or splice products, these alleles reveal discrete functions of genes specified through specific protein-protein combinations. This ability to tease out selective and specific functions of proteins for a given mammalian process from the many combinatorial functions specified by each gene will be the most valuable product of ENU mutagenesis as it is applied more widely to the mammalian phenome.

SUMMARY POINTS

1. The doubling of protein-coding genes between invertebrates and mammals is more than sufficient to encode the expansion of cell diversity and complexity in systems such as mammalian immunity because this doubling leads to an estimated 1000 times increase in potential biochemical combinations of proteins, and an increase of 10^{30} in the cellular combinations of these biochemical conglomerates.
2. The corollary of combinatorial specification is that genetic dissection cannot assume "one gene - one protein - one trait" and that most of the mammalian genome-phenome code will be revealed by analyzing genetic variants that interfere with specific protein-protein interactions, notably point mutations arising by natural variation or induced experimentally by ENU. Examples from the immune system are provided.
3. ENU-induced variation provides an eminently feasible route to understanding how the genome encodes important processes in adult mammals without prior assumptions about the nature of a gene, its chromosomal location, or its expression pattern.
4. By changing individual protein domains or splice products, the kinds of alleles produced by ENU reveal separate functions of genes specified through protein-protein combinations.

LITERATURE CITED

1. Adams MD, Celniker SE, Holt RA, Evans CA, Gocayne JD, et al. 2000. The genome sequence of *Drosophila melanogaster*. *Science* 287:2185–95
2. Anderson MS, Venanzi ES, Klein L, Chen Z, Berzins S, et al. 2002. Projection of an immunological self-shadow within the thymus by the Aire protein. *Science* 298:1395–403
3. Beier DR. 2000. Sequence-based analysis of mutagenized mice. *Mamm. Genome* 11:594–97
4. Beutler B. 2004. Inferences, questions and possibilities in Toll-like receptor signalling. *Nature* 430:257–63
5. Bode VC, McDonald JD, Guenet JL, Simon D. 1988. hph-1: a mouse mutant with hereditary hyperphenylalaninemia induced by ethylnitrosourea mutagenesis. *Genetics* 118:299–305
6. Brett D, Pospisil H, Valcarcel J, Reich J, Bork P. 2002. Alternative splicing and genome complexity. *Nat. Genet.* 30:29–30
7. **Carpinelli MR, Hilton DJ, Metcalf D, Antonchuk JL, Hyland CD, et al. 2004. Suppressor screen in Mpl−/− mice: c-Myb mutation causes supraphysiological production of platelets in the absence of thrombopoietin signaling. *Proc. Natl. Acad. Sci. USA* 101:6553–58**
8. Conlon RA, Reaume AG, Rossant J. 1995. Notch1 is required for the coordinate segmentation of somites. *Development* 121:1533–45

The results of a "second generation" ENU modifier screen identifying mutations that ameliorate a mouse model of a human disease. See also Reference 62.

9. Consortium TCeS. 1998. Genome sequence of the nematode *C. elegans*: a platform for investigating biology. *Science* 282:2012–21
10. Cordes SP, Barsh GS. 1994. The mouse segmentation gene kr encodes a novel basic domain-leucine zipper transcription factor. *Cell* 79:1025–34
11. Cox RD, Hugill A, Shedlovsky A, Noveroske JK, Best S, et al. 1999. Contrasting effects of ENU induced embryonic lethal mutations of the quaking gene. *Genomics* 57:333–41
12. Egawa T, Albrecht B, Favier B, Sunshine MJ, Mirchandani K, et al. 2003. Requirement for CARMA1 in antigen receptor-induced NF-κB activation and lymphocyte proliferation. *Curr. Biol.* 13:1252–58
13. Fanning AS, Anderson JM. 1999. Protein modules as organizers of membrane structure. *Curr. Opin. Cell Biol.* 11:432–39
14. **Garcia-Garcia MJ, Eggenschwiler JT, Caspary T, Alcorn HL, Wyler MR, et al. 2005. Analysis of mouse embryonic patterning and morphogenesis by forward genetics. *Proc. Natl. Acad. Sci. USA* 102:5913–19**
15. Georgopoulos K, Bigby M, Wang JH, Molnar A, Wu P, et al. 1994. The Ikaros gene is required for the development of all lymphoid lineages. *Cell* 79:143–56
16. Georgopoulos K, Moore DD, Derfler B. 1992. Ikaros, an early lymphoid-specific transcription factor and a putative mediator for T cell commitment. *Science* 258:808–12
17. Goffeau A, Barrell BG, Bussey H, Davis RW, Dujon B, et al. 1996. Life with 6000 genes. *Science* 274:546, 63–67
18. Hara H, Wada T, Bakal C, Kozieradzki I, Suzuki S, et al. 2003. The MAGUK family protein CARD11 is essential for lymphocyte activation. *Immunity* 18:763–75
19. Hitotsumachi S, Carpenter DA, Russell WL. 1985. Dose-repetition increases the mutagenic effectiveness of N-ethyl-N-nitrosourea in mouse spermatogonia. *Proc. Natl. Acad. Sci. USA* 82:6619–21
20. **Hoebe K, Du X, Georgel P, Janssen E, Tabeta K, et al. 2003. Identification of Lps2 as a key transducer of MyD88-independent TIR signalling. *Nature* 424:743–48**
21. **Hoebe K, Georgel P, Rutschmann S, Du X, Mudd S, et al. 2005. CD36 is a sensor of diacylglycerides. *Nature* 433:523–27**
22. **Hrabé de Angelis MH, Flaswinkel H, Fuchs H, Rathkolb B, Soewarto D, et al. 2000. Genome-wide, large-scale production of mutant mice by ENU mutagenesis. *Nat. Genet.* 25:444–47**
23. Huang JD, Cope MJ, Mermall V, Strobel MC, Kendrick-Jones J, et al. 1998. Molecular genetic dissection of mouse unconventional myosin-VA: head region mutations. *Genetics* 148:1951–61
24. Huang YH, Chen YT, Lai JJ, Yang ST, Yang UC. 2002. PALS db: Putative Alternative Splicing database. *Nucleic Acids Res.* 30:186–90
25. Hughes M, Arundhati A, Lunness P, Shaw PJ, Doonan JH. 1996. A temperature-sensitive splicing mutation in the bimG gene of Aspergillus produces an N-terminal fragment which interferes with type 1 protein phosphatase function. *EMBO J.* 15:4574–83
26. Jun JE, Goodnow CC. 2003. Scaffolding of antigen receptors for immunogenic versus tolerogenic signaling. *Nat. Immunol.* 4:1057–64
27. **Jun JE, Wilson LE, Vinuesa CG, Lesage S, Blery M, et al. 2003. Identifying the MAGUK protein Carma-1 as a central regulator of humoral immune responses and atopy by genome-wide mouse mutagenesis. *Immunity* 18:751–62**
28. Justice MJ, Noveroske JK, Weber JS, Zheng B, Bradley A. 1999. Mouse ENU mutagenesis. *Hum. Mol. Genet.* 8:1955–63

Along with Reference 30 outlines a seminal ENU screen identifying recessive mutations producing easily visible disruptions in the morphology of the midgestation mouse embryo.

Along with Reference 21 describes two key ENU-induced recessive variants identified by macrophage hyporesponsiveness to microbial stimuli.

See annotation to Reference 20

Along with Reference 50 outlines the initial findings of two large European ENU projects screening over 40,000 G1 offspring for a range of dominant phenotypes.

Increased lymphocyte surface IgM expression identified an ENU-induced point mutation in Carma-1 and, when compared with the

null mutant, revealed a role for this protein as a scaffold for inhibitory signaling pathways separate from its function in lymphocyte activation.

See annotation to Reference 14.

Applied the balancer chromosome with great effect to efficiently isolate ENU-induced mutations within chromosome 11.

First described the inappropriate cross-talk that can occur through empty molecular niches and genetic redundancy, and highlighted the importance of point mutant proteins to more completely reveal gene function.

See annotation to Reference 22.

29. Kamath RS, Fraser AG, Dong Y, Poulin G, Durbin R, et al. 2003. Systematic functional analysis of the *Caenorhabditis elegans* genome using RNAi. *Nature* 421:231–37

30. Kasarskis A, Manova K, Anderson KV. 1998. A phenotype-based screen for embryonic lethal mutations in the mouse. *Proc. Natl. Acad. Sci. USA* 95:7485–90

31. Kiernan AE, Erven A, Voegeling S, Peters J, Nolan P, et al. 2002. ENU mutagenesis reveals a highly mutable locus on mouse Chromosome 4 that affects ear morphogenesis. *Mamm. Genome* 13:142–48

32. Kile BT, Hentges KE, Clark AT, Nakamura H, Salinger AP, et al. 2003. Functional genetic analysis of mouse chromosome 11. *Nature* 425:81–86

33. Kim J, Sif S, Jones B, Jackson A, Koipally J, et al. 1999. Ikaros DNA-binding proteins direct formation of chromatin remodeling complexes in lymphocytes. *Immunity* 10:345–55

34. Kirstetter P, Thomas M, Dierich A, Kastner P, Chan S. 2002. Ikaros is critical for B cell differentiation and function. *Eur. J. Immunol.* 32:720–30

35. Kramer JM, Johnson JJ. 1993. Analysis of mutations in the sqt-1 and rol-6 collagen genes of *Caenorhabditis elegans*. *Genetics* 135:1035–45

36. Lander ES, Linton LM, Birren B, Nusbaum C, Zody MC, et al. 2001. Initial sequencing and analysis of the human genome. *Nature* 409:860–921

37. Lev S, Blechman JM, Givol D, Yarden Y. 1994. Steel factor and c-kit protooncogene: genetic lessons in signal transduction. *Crit. Rev. Oncog.* 5:141–68

38. Lin L, Gerth AJ, Peng SL. 2004. Active inhibition of plasma cell development in resting B cells by microphthalmia-associated transcription factor. *J. Exp. Med.* 200:115–22

39. Liston A, Lesage S, Wilson J, Peltonen L, Goodnow CC. 2003. Aire regulates negative selection of organ-specific T cells. *Nat. Immunol.* 4:350–54

40. Lopez RA, Schoetz S, DeAngelis K, O'Neill D, Bank A. 2002. Multiple hematopoietic defects and delayed globin switching in Ikaros null mice. *Proc. Natl. Acad. Sci. USA* 99:602–7

41. Madhani HD, Fink GR. 1998. The riddle of MAP kinase signaling specificity. *Trends Genet* 14:151–55

42. Madhani HD, Styles CA, Fink GR. 1997. MAP kinases with distinct inhibitory functions impart signaling specificity during yeast differentiation. *Cell* 91:673–84

43. Marker PC, Seung K, Bland AE, Russell LB, Kingsley DM. 1997. Spectrum of Bmp5 mutations from germline mutagenesis experiments in mice. *Genetics* 145:435–43

44. Mattick JS. 2003. Challenging the dogma: the hidden layer of non-protein-coding RNAs in complex organisms. *BioEssays* 25:930–39

45. Mironov AA, Fickett JW, Gelfand MS. 1999. Frequent alternative splicing of human genes. *Genome Res.* 9:1288–93

46. Modrek B, Resch A, Grasso C, Lee C. 2001. Genome-wide detection of alternative splicing in expressed sequences of human genes. *Nucleic Acids Res.* 29:2850–59

47. Nelms KA, Goodnow CC. 2001. Genome-wide ENU mutagenesis to reveal immune regulators. *Immunity* 15:409–18

48. Newton K, Dixit VM. 2003. Mice lacking the CARD of CARMA1 exhibit defective B lymphocyte development and impaired proliferation of their B and T lymphocytes. *Curr. Biol.* 13:1247–51

49. Nichogiannopoulou A, Trevisan M, Neben S, Friedrich C, Georgopoulos K. 1999. Defects in hemopoietic stem cell activity in Ikaros mutant mice. *J. Exp. Med.* 190:1201–14

50. Nolan PM, Peters J, Strivens M, Rogers D, Hagan J, et al. 2000. A systematic, genome-wide, phenotype-driven mutagenesis programme for gene function studies in the mouse. *Nat. Genet.* 25:440–43

51. Papathanasiou P, Perkins AC, Cobb BS, Ferrini R, Sridharan R, et al. 2003. Widespread failure of hematolymphoid differentiation caused by a recessive niche-filling allele of the Ikaros transcription factor. *Immunity* 19:131–44
52. Preis JI, Downes M, Oates NA, Rasko JE, Whitelaw E. 2003. Sensitive flow cytometric analysis reveals a novel type of parent-of-origin effect in the mouse genome. *Curr. Biol.* 13:955–59
53. Quwailid MM, Hugill A, Dear N, Vizor L, Wells S, et al. 2004. A gene-driven ENU-based approach to generating an allelic series in any gene. *Mamm. Genome* 15:585–91
54. Radtke F, Wilson A, Stark G, Bauer M, van Meerwijk J, et al. 1999. Deficient T cell fate specification in mice with an induced inactivation of Notch1. *Immunity* 10:547–58
55. Rajaraman S, Davis WS, Mahakali-Zama A, Evans HK, Russell LB, Bedell MA. 2002. An allelic series of mutations in the Kit ligand gene of mice. II. Effects of ethylnitrosourea-induced Kitl point mutations on survival and peripheral blood cells of $Kitl^{Steel}$ mice. *Genetics* 162:341–53
56. Rasooly RS, New CM, Zhang P, Hawley RS, Baker BS. 1991. The lethal(1)TW-6^{cs} mutation of *Drosophila melanogaster* is a dominant antimorphic allele of nod and is associated with a single base change in the putative ATP-binding domain. *Genetics* 129:409–22
57. Rinchik EM, Carpenter DA, Selby PB. 1990. A strategy for fine-structure functional analysis of a 6- to 11-centimorgan region of mouse chromosome 7 by high-efficiency mutagenesis. *Proc. Natl. Acad. Sci. USA* 87:896–900
58. Russell LB, Hunsicker PR, Cacheiro NL, Bangham JW, Russell WL, Shelby MD. 1989. Chlorambucil effectively induces deletion mutations in mouse germ cells. *Proc. Natl. Acad. Sci. USA* 86:3704–8
59. Sandberg ML, Sutton SE, Pletcher MT, Wiltshire T, Tarantino LM, et al. 2005. c-Myb and p300 regulate hematopoietic stem cell proliferation and differentiation. *Dev. Cell* 8:153–66
60. Shedlovsky A, McDonald JD, Symula D, Dove WF. 1993. Mouse models of human phenylketonuria. *Genetics* 134:1205–10
61. Sheng M, Sala C. 2001. PDZ domains and the organization of supramolecular complexes. *Annu. Rev. Neurosci.* 24:1–29
62. Simon MA, Bowtell DD, Dodson GS, Laverty TR, Rubin GM. 1991. Ras1 and a putative guanine nucleotide exchange factor perform crucial steps in signaling by the sevenless protein tyrosine kinase. *Cell* 67:701–16
63. Steingrimsson E, Moore KJ, Lamoreux ML, Ferre-D'Amare AR, Burley SK, et al. 1994. Molecular basis of mouse microphthalmia (mi) mutations helps explain their developmental and phenotypic consequences. *Nat. Genet.* 8:256–63
64. Swiatek PJ, Lindsell CE, del Amo FF, Weinmaster G, Gridley T. 1994. Notch1 is essential for postimplantation development in mice. *Genes Dev.* 8:707–19
65. Venter JC, Adams MD, Myers EW, Li PW, Mural RJ, et al. 2001. The sequence of the human genome. *Science* 291:1304–51
66. Vinuesa CG, Cook MC, Angelucci C, Athanasopoulos V, Rui L, et al. 2005. A novel RING-type ubiquitin ligase family member essential to repress follicular helper T cells and autoimmunity. *Nature* 435:452–58
67. Vinuesa CG, Goodnow CC. 2004. Illuminating autoimmune regulators through controlled variation of the mouse genome sequence. *Immunity* 20:669–79
68. Wang JH, Nichogiannopoulou A, Wu L, Sun L, Sharpe AH, et al. 1996. Selective defects in the development of the fetal and adult lymphoid system in mice with an Ikaros null mutation. *Immunity* 5:537–49

Described a unique ENU-induced point mutant allele of Ikaros yielding a more severe hematopoietic defect than deletion mutants due to the ENU allele selectively inactivating Ikaros' normal functions without creating the opportunity for compensation by genetic redundancy. See also Reference 42.

By screening blood counts in ENU-induced recessive mutants, these authors identified a point mutation in c-Myb disrupting its interaction with p300 and revealed a remarkable increase in hematopoietic stem cell numbers and cycling.

An outstanding "second generation" modifier screen, here in *Drosophila* to identify mutations decreasing the effectiveness of signaling by a key protein tyrosine kinase.

A complex and sensitive autoimmunity ENU screen utilizing antinuclear

antibody levels identified a new ubiquitin ligase, denoted *roquin*, that is an essential negative regulator of follicular helper T cells and responses to self antigens.

A landmark technical advance through the application of established *Drosophila* genetics methodology to generate a mouse balancer chromosome.

69. Washburn T, Schweighoffer E, Gridley T, Chang D, Fowlkes BJ, et al. 1997. Notch activity influences the $\alpha\beta$ versus $\gamma\delta$ T cell lineage decision. *Cell* 88:833–43
70. Waterston RH, Lindblad-Toh K, Birney E, Rogers J, Abril JF, et al. 2002. Initial sequencing and comparative analysis of the mouse genome. *Nature* 420:520–62
71. Werlen G, Palmer E. 2002. The T-cell receptor signalosome: a dynamic structure with expanding complexity. *Curr. Opin. Immunol.* 14:299–305
72. Winandy S, Wu P, Georgopoulos K. 1995. A dominant mutation in the Ikaros gene leads to rapid development of leukemia and lymphoma. *Cell* 83:289–99
73. **Zheng B, Sage M, Cai WW, Thompson DM, Tavsanli BC, et al. 1999. Engineering a mouse balancer chromosome. *Nat. Genet.* 22:375–78**

Evolutionary Genetics of Reproductive Behavior in *Drosophila*: Connecting the Dots

Therese Ann Markow[1] and Patrick M. O'Grady[2]

[1]Department of Ecology and Evolutionary Biology, University of Arizona, Tucson, Arizona 85721; email: tmarkow@arl.arizona.edu

[2]Department of Biology, University of Vermont, Burlington, Vermont 05405; email: pogrady@uvm.edu

Annu. Rev. Genet. 2005. 39:263–91

First published online as a Review in Advance on August 30, 2005

The *Annual Review of Genetics* is online at http://genet.annualreviews.org

doi: 10.1146/annurev.genet.39.073003.112454

0066-4197/05/1215-0263$20.00

Key Words

evolution, genomics, sexual behavior, ecological genetics, sensory signals

Abstract

Species of the genus *Drosophila* exhibit enormous variation in all of their reproductive behaviors: resource use and specialization, courtship signaling, sperm utilization, and female remating. The genetic bases of this variability and its evolution are poorly understood. At the same time, *Drosophila* comparative genomics now has developed to a point at which approaches previously only possible with *D. melanogaster* can be exploited to address these questions. We have taken advantage of the known phylogenetic relationships of this group of flies not only to place these behaviors in an evolutionary framework, but to provide a roadmap for future genetic studies.

Contents

INTRODUCTION

Researchers using *Drosophila* as a model system to address evolutionary questions are standing on the threshold of a new era. One species, *D. melanogaster*, has emerged as a premier model organism for elucidating basic principles of eukaryotic genetics. Research utilizing *D. melanogaster* has been primarily laboratory based, relying increasingly on sophisticated molecular tools to understand the genetic bases of fundamental biological processes. A major milestone in the history of this model system was the sequencing of its entire genome in 2000 (1), the annotation of which is still being fine-tuned.

The wealth of interspecific diversity in this genus typically has long been attractive to evolutionary researchers hoping to understand the genetic basis of sexual selection and the process of speciation (18–20, 111, 137, 144; see also 35). Observations and experiments over nearly a century have revealed an astounding breadth of morphological, ecological, and behavioral diversity from hundreds of *Drosophila* species. Issues such as the role of reproductive behavior, including ecological and sexual isolation, relative to postzygotic incompatibilities or barriers, remain at the heart of many controversies about the process of speciation.

Evolutionary genetics now sits at the confluence of several biological disciplines, and advances in each will enable the next generation of researchers to ask and answer specific questions about *Drosophila* reproductive behavior. Given the whole genome sequence, the bulk of ecological and behavioral data, and the refinement in phylogenetic relationships of *Drosophila* species, the wealth of interspecific diversity in reproductive traits can now be placed in contexts that allow hypotheses to be generated not only about their evolution, but about the genetic mechanisms underlying them. Recently, these disciplines have begun to cross-fertilize and yield detailed hypotheses about the evolutionary genetics of morphological diversity, ecological adaptations, and reproductive isolation (61, 75, 76, 78, 157). As DNA sequencing technology has become more efficient, the potential for comparative genome sequencing from a number of closely related taxa has been realized. By the end of 2005, sequencing of the genomes of 12 *Drosophila* species will have been completed and the ability to employ molecular techniques previously available only for *D. melanogaster* will become increasingly accessible for these 12 species and their relatives. This means that we can now finally examine the interspecific diversity in such a way as to understand its origins and genetic bases.

Reproductive behavior actually represents a broad array of traits. For this review, reproductive behaviors are organized into two subgroups, premating and postmating. Premating reproductive behaviors include the full range of behaviors of both sexes, including mate location as well as courtship itself, which lead to successful copulation. Postmating reproductive behaviors refer to behaviors of inseminated females, primarily oviposition and receptivity to remating. Within these broad subdivisions of behavior, each category is itself a complex set of behaviors. This review has two parts. In the first section, we examine interspecific variability in reproductive behaviors in a broad and comprehensive evolutionary framework. All behaviors with which we are concerned involve the detection of signals, either from the environment or from another fly, and the responses of individuals to those signals. Thus in the second part of the review, we explore the potential sources and organization of genetic variability in the signals and the sensory systems that receive and process them in order to frame future experiments to elucidate their evolution.

REPRODUCTIVE BEHAVIORS IN *DROSOPHILA*

Extensive studies have demonstrated that the pre- and postmating reproductive behaviors referred to briefly above exhibit great interspecific variability in *Drosophila*. It is this variability that is treated here, accompanied by discussion of approaches for its study. Space limitations prevent the inclusion of all categories of variability. We therefore focus upon behavioral variants that appear to be more or less discrete phenotypes and for which a sufficient number of species have been characterized. In order to present the changes in reproductive behaviors in an evolutionary framework, we capitalize upon the known phylogenetic relationships among species. The genus *Drosophila* is divided into a number of species groups, radiations, and subgenera (156). Recent phylogenetic work (e.g., 119, 147) has provided a framework of evolutionary relationships within this genus that can be used to examine behaviors important to reproduction. The fine points of many relationships among groups are continually being refined, but the lack of resolution regarding those details should not detract from the goals of this review.

Premating Reproductive Behaviors

Locating breeding sites. For any given species, mating takes place at particular locations and at specific times of the year and/or day. Thus, an important part of premating behavior involves locating sites, via long-range signal detection-response systems, where prospective mates will be encountered. Signaling and response mechanisms underlying the location and utilization of such sites constitute an important reproductive behavioral process. *Drosophila* species vary widely in the resources they use, and thus the signaling mechanisms are expected to exhibit genetically based differences.

For most species of *Drosophila*, resources used for adult feeding are at or near the oviposition sites. Host or resource use thus can be considered to be both a premating and a postmating behavior and is treated as such here for the sake of economy. *Drosophila* species range from generalists (oligophagy, polyphagous) to specialists (monophagy). In layman's terms, *Drosophila* are referred to as fruit flies. However, many *Drosophila* species have become associated with decaying plant material including cacti, flowers, mushrooms or other fungi, tree sap or slime fluxes, and even with the excretory organs of land crabs (118). Most of these resources are associated with a unique microbe fauna that provides both larval and adult *Drosophila* with nutrition. This ecological diversity raises questions about the phylogenetic distribution of resource localization strategies and the genetic mechanisms that control their identification and utilization by flies.

Figure 1 is an overview of resource use by different species groups in the genus *Drosophila*. There is a considerable degree of conservation of resource type within *virilis-repleta* and *immigrans-tripunctata* radiations. Based on this phylogeny, the ancestor of the *virilis-repleta* radiation bred in sap or slime fluxes. There was then a switch to flowers and small, dry fruits in the lineage leading to the *repleta* radiation, with a subsequent switch to

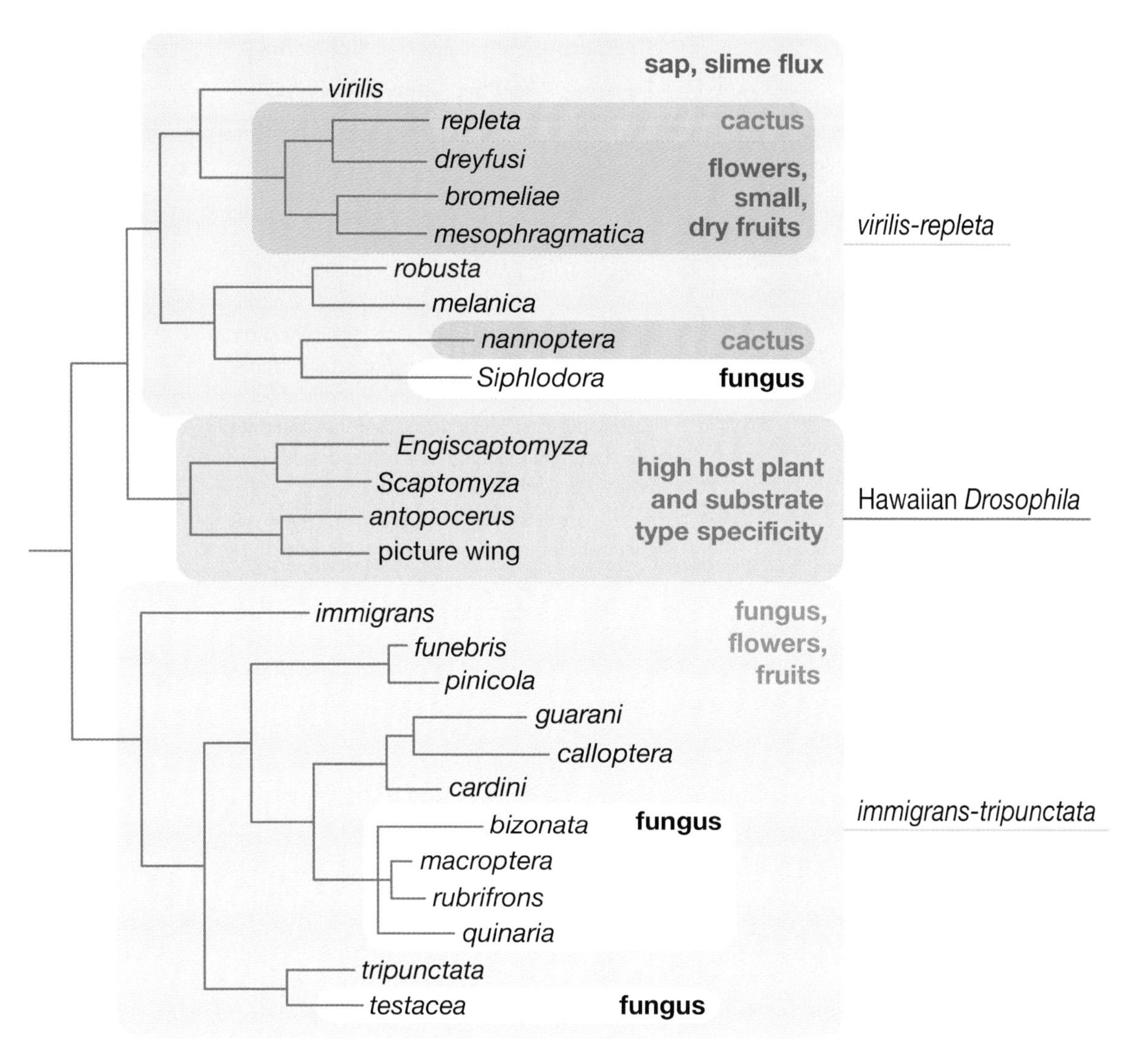

Figure 1

Phylogenetic distribution of host-resource mapped onto the phylogeny of *Drosophila*. The ancestor of the *virilis-repleta* radiation bred in sap and slime flux (*blue*). One lineage then evolved to use flowers and small, dry fruits (*orange*). Cactus use (*green*) evolved independently at least twice and fungal specialization (*white*) at least once. The ancestor of the *immigrans-tripunctata* radiation was a generalist on fungus, flowers and fruits (*yellow*). Specialization on fungi (*white*) evolved independently at least twice. (Modified after 119, 136, 156.)

cactus in the *repleta* species group. Based on this phylogeny, the *nannoptera* group represents an independent exploitation of cactus as a host substrate. Patterns of host switching within the *immigrans-tripunctata* radiation are less clear. Many of these species are fungus specialists, but several taxa also are regularly associated with flowers or small fruits, suggesting that this group is able to utilize smaller, more temporally restricted substrates. The Hawaiian Drosophilidae, in contrast to the remainder of the genus *Drosophila*, are highly host plant specific, with roughly 80% of picture wing species utilizing a single family as oviposition substrate (67, 77, 104).

A number of *Drosophila* species are known to specialize on a single host resource (**Figure 2*a*,*b***). In some cases, the genetics of this phenomenon is quite well understood. *Drosophila sechellia*, for example, has specialized on morinda fruit (*Morinda citrifolia*), a resource that is toxic to all other members of the *melanogaster* species group (**Figure 2*a***). *Drosophila sechellia* is resistant to these toxins and, in fact, requires them for full stimulation of oviposition behavior. Similarly, *D. pachea* of the *nannoptera* group has specialized on senita (*Lophocereus schottii*), which is highly toxic to other *Drosophila* (**Figure 2*b***). Thus evolution of specialization seems to have taken place on two separate scales. First, the radiation of an entire group of species onto a certain type of host has taken place. Within these clades, however, species are able to further specialize on particular types of resources within the host taxon.

Once they are at the breeding site, however, not all species orchestrate their courtship activities in the same way. Some species

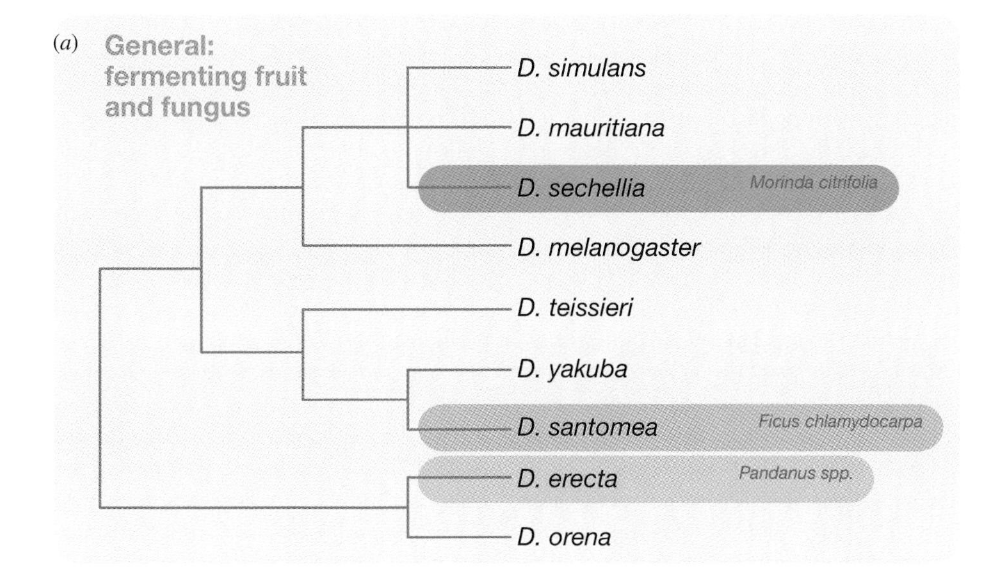

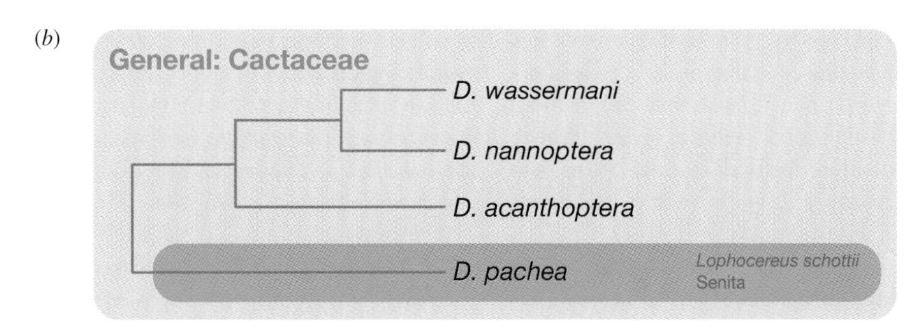

Figure 2

Examples of host specialization within the *melanogaster* and *nannoptera* species groups. (*a*) In the *melanogaster* subgroup the ancestral condition is the general use of rotting fruit and fungus (*yellow*) Subsequent specializations on *Morinda* (*red*), *Ficus* (*green*), and *Pandanus* (*violet*) evolved separately in three species. (*b*) The nannoptera species are all cactophilic and use a suite of host species. One taxon, *D. pachea*, has specialized on senita (*Lophocereus*). (Modified from 68, 122, 163.)

display at sites away from where they feed or oviposit (67, 104, 138). In others, such as *D. nigrospiracula* (93), males defend mating territories on parts of the plant that are away from the feeding locations. Species such as *D. melanogaster* and *D. simulans*, on the other hand, mate right on the rotting fruit where they feed. Mating location, relative to food resources, has not yet been sufficiently documented in enough species to permit any meaningful phylogenetic mapping.

Regardless of where courtship takes place, it will involve signaling between individuals that takes place at shorter range than those signals that attract flies to aggregation sites. Courtship has several components, starting with the identification, at the mating site, of conspecific members of the opposite sex, following which courtship and all of its various components can proceed. For species of *Drosophila* that have an exclusive association with a particular resource, once they arrive at the feeding site, the only other *Drosophila* they will encounter will be conspecific females and males. In such specialist taxa, courtship involves discriminating between conspecific males and females, and engaging in species-specific courtship processes in ways that will ensure their reproductive success. For *Drosophila* species that mate at resources utilized by congeners, additional systems must be present that allow them to discriminate members of their own from other species prior to investing energy and time in the courtship process. Reproductive behavior includes male-male interactions as well as those between the sexes, although the former are less well studied. Sexual signaling takes place in three sensory modalities: visual, auditory, and chemosensory.

Visual sexual signals. The role of visual signals during courtship can be inferred from several observational studies. In order for signals to have a visual component, they must be performed in the light and conducted within the visual field (i.e., in front) of the receiving individual. For many species, laboratory and field observations have documented the relative positions of males and females with respect to each other during courtship. For species in which courtship has not been observed directly, morphological or coloration patterns may be such that visual signaling can be inferred. For some *Drosophila* species, the sexes differ with respect to the potential for certain aspects of visual signaling. This difference is a function of the fact that, during the specific part of courtship involving male attempts to mount, or his licking of the female's genitalia, the male is behind the female and can receive, but not transmit visual information. Male visual information, then, will only be transmitted when males leave this position and move in front of females, or, as in some lekking Hawaiian species, perform ritualized displays to attract females to a mating site. For this reason, we examine the phylogenetic distribution of visual displays separately for males and females.

Figure 3 illustrates the phylogenetic distribution of whether or not males tend to position themselves in front of females during courtship as opposed to remaining behind them, out of view. There is a tendency for males of species in the *virilis-repleta* radiation to court behind the females, suggesting that male visual displays are not the primary form of sexual signaling in these taxa. This interpretation is consistent with the fact that these species show effectively no sexual dimorphism in coloration, wing pattern, or other visible morphological traits. It is also supported by the existence of exceptional taxa, such as *D. acanthoptera* in the *nannoptera* group, in which there is a sexual dimorphism for body color and in which males court in front of females.

Females are not merely recipients of visual signals during courtship. In a large number of species, females indicate their receptivity to males by a characteristic spreading of their wings (**Figure 4*a***). This behavior is typical of species in the *virilis* (162) and *repleta* groups (99), but has also has arisen in several other groups of *Drosophila* (137). The distribution of this behavior is variable,

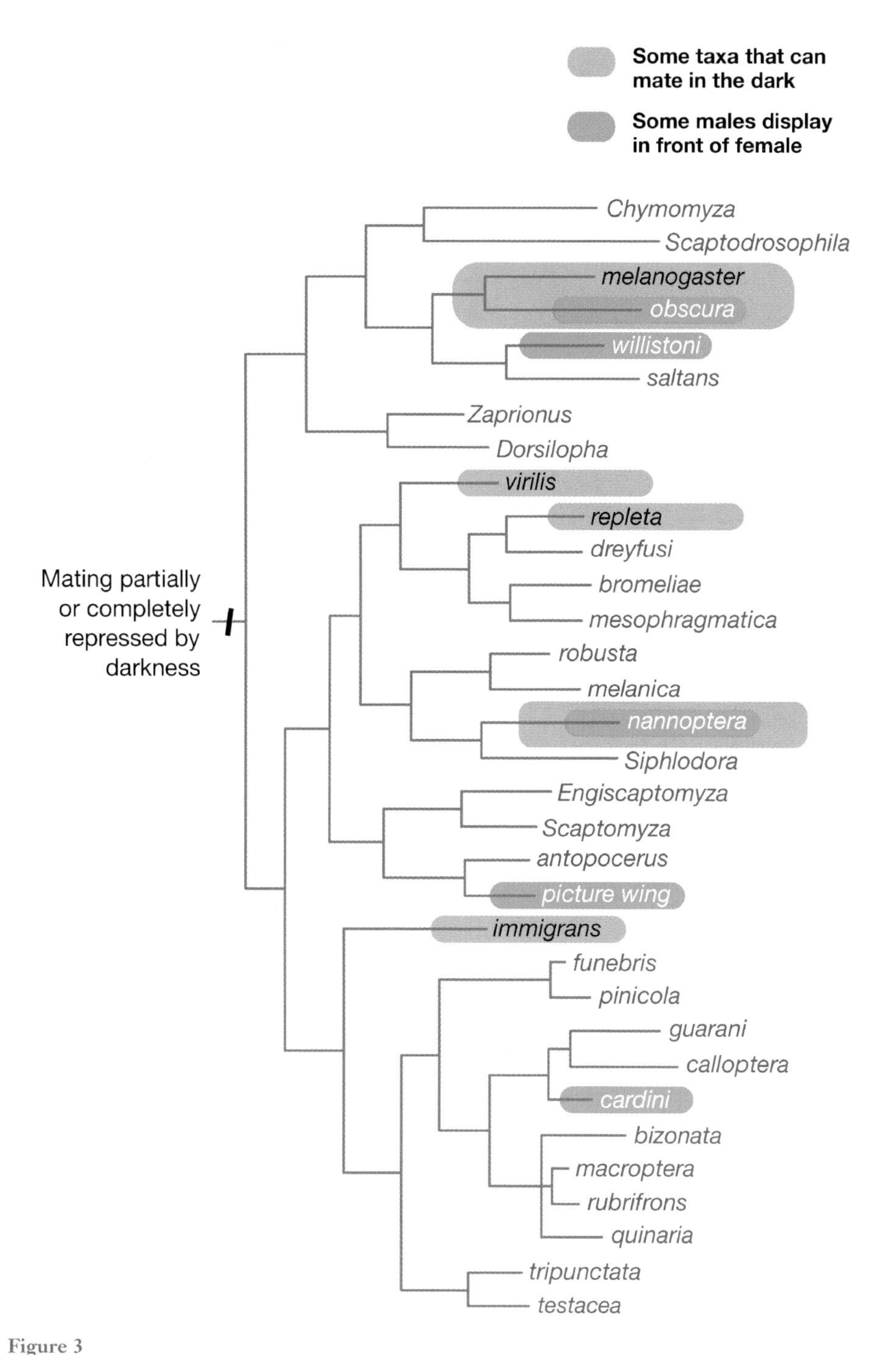

Figure 3

The evolution of male visual signals in courtship behavior in response to light availability. The ancestral condition is to have mating that is either partially or completely repressed by darkness. Those taxa that can mate in the dark are shown in tan. Taxa where males display in front of females are indicated in blue (Modified from 63, 65, 137).

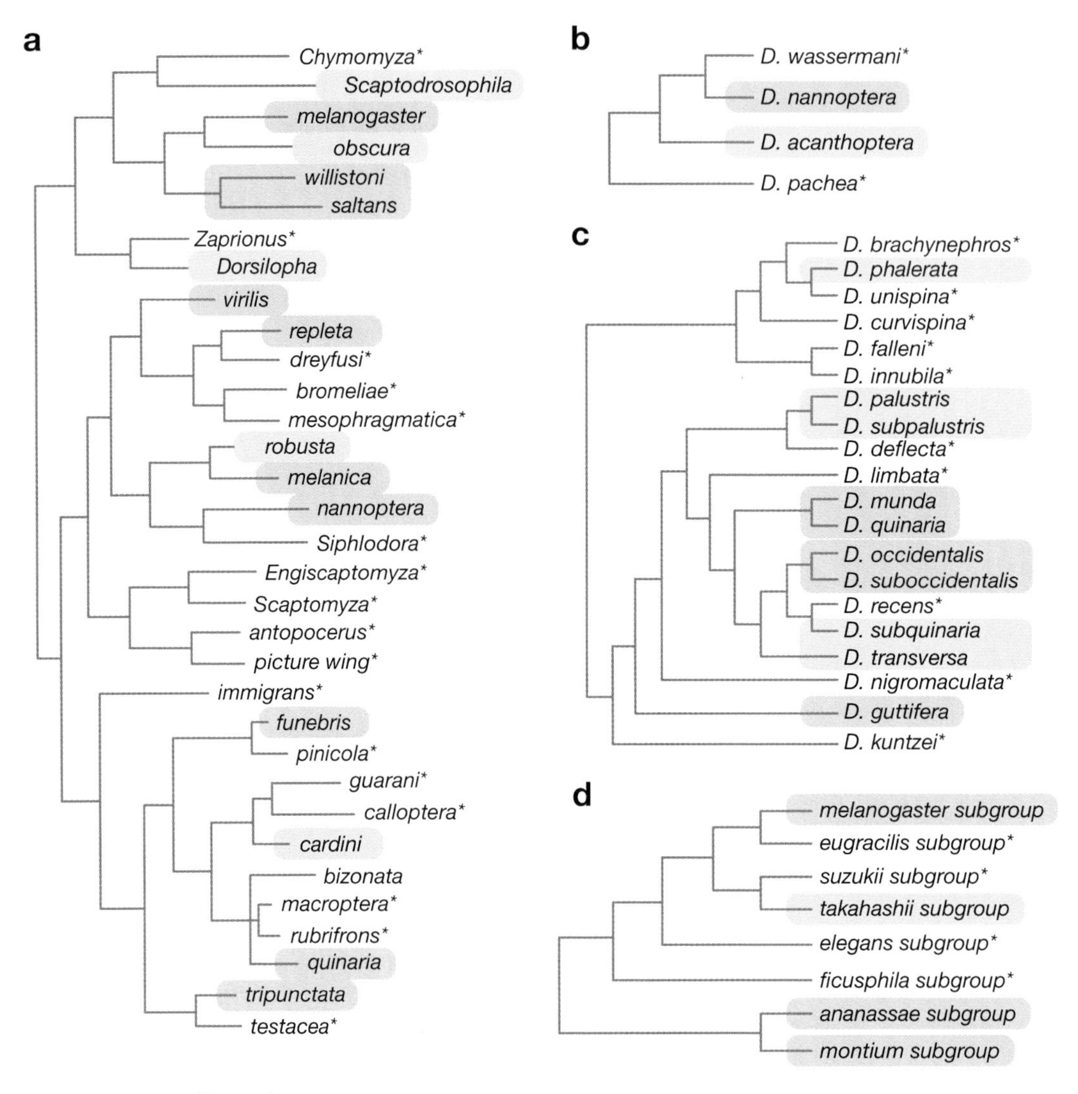

Figure 4

Phylogenetic distribution of female wing spreading display during courtship (modified from 137). No data are available for taxa with an asterisk. (*a*) Overview of the evolution of female wing spreading displays in the genus *Drosophila* and relatives. Clades where females of some species display (*green*) and do not display (*yellow*) are shown. (*b*) Data from only two taxa, *D. nannoptera* and *D. acanthoptera*, are available in the *nannoptera* group so the exact character transition is uncertain (163). (*c*) The wing spreading display has evolved at least three times in the *quinaria* species group (phylogeny modified after 128). (*d*) Data in the large *melanogaster* species group are highly variable, particularly within the *melanogaster* and *montium* species group, where female wing spreading may have been gained or lost several times.

particularly in the *nannoptera*, *quinaria*, and *melanogaster* groups (**Figure 4*b*–*d***). Likewise, males of these species typically will not attempt to mount a female until she has given this signal, whereas males in those species groups where female wing-spreading is absent will attempt intromission on a repeated and constant basis. Females of any species may perform other behaviors that indicate their receptivity, such as simply slowing down their locomotion (154), or spreading their vaginal plates (137), but these behaviors are not always strictly visual or even discrete, and thus less comparative information is available about them.

In addition to those visual behaviors that can be directly scored, the importance of unspecified visual cues can be inferred from studies in which insemination rates have been compared between pairs of the same species placed together in darkness and in light (63–65). Based upon the outcomes of this type of study, species can be considered to be either light independent, partially light dependent, or completely light dependent in their mating behavior (**Figure 3**).

Auditory signals. Auditory signals are utilized in the courtship of most *Drosophila* species. Courtship songs have been studied in over a hundred *Drosophila* species, and the majority of these have focused upon male songs and upon song variability at the inter- and intraspecific levels (70). Females of species in the *virilis-repleta* radiation and the *nannoptera* group also regularly produce songs while being courted (36, 48, 109, 124), and in a number of these species, an actual dialogue occurs between the sexes during courtship, referred to as "dueting" (10, 43). Males of many species also produce songs that are unlike courtship songs, but rather appear to be utilized to dissuade the amorous advances of other males (145).

The diversity observed among *Drosophila* species in male courtship song is so variable that it is difficult to describe in manageable terms. Most songs are composed of various pulses or bursts that have different structural and temporal features that distinguish them at the species level. Some species utilize one "type" of song, whereas others may perform four or five. Some song types are performed earlier in courtship than others. One caution should be entertained when examining songs that occur later in courtship. Recordings of courtship songs are conducted in small chambers in which female decamping is not one of the options available to unreceptive females. Courtships observed under these conditions are thus likely to last longer than those in nature where uninterested females can depart. Under confined conditions, therefore, males may become frustrated, and as courtships continue, exhibit behavioral components rarely observed in nature. A further complication is that different investigators have employed different terminologies to describe song parameters. There do not appear to be any homologies between lineages for particular song elements. In rare cases, a species will produce no sounds at all. Nonetheless, we have attempted to capture this variation in a meaningful way. Because there is no simple way to reduce all of the variation to character states that then can be placed in a phylogenetic context, we have scored members of a group or subgroup as to the number of the courtship song components typical of the group.

For each species group, we have summarized the number of different song types a species has been reported to produce (**Figures 5*a–e***). The *melanogaster* species group is a large clade that shows a highly variable array of auditory mating strategies. **Figure 5*a*** shows the distribution of song number within the *melanogaster* subgroup. Males of most species produce two song types, a sine song (158) and a pulse song (48), the latter of which exhibits important interspecific differences. Interestingly, males of one species, *D. yakuba*, have lost the sine song and only rely on the pulse song for identification of conspecific individuals. Males of the *montium* subgroup (not shown) produce one type of song, which varies among species, but is largely produced once copulation has begun (150, 151). This has reached its extreme proportion in two species of this subgroup, *D. birchii* and *D. serrata*, where males produce song only during copulation itself (71). In males of *D. ananassae* and others of the subgroup, males produce either one or two types of pulse songs (36, 168).

The *obscura* group is the sister clade of the *melanogaster* group yet produces completely

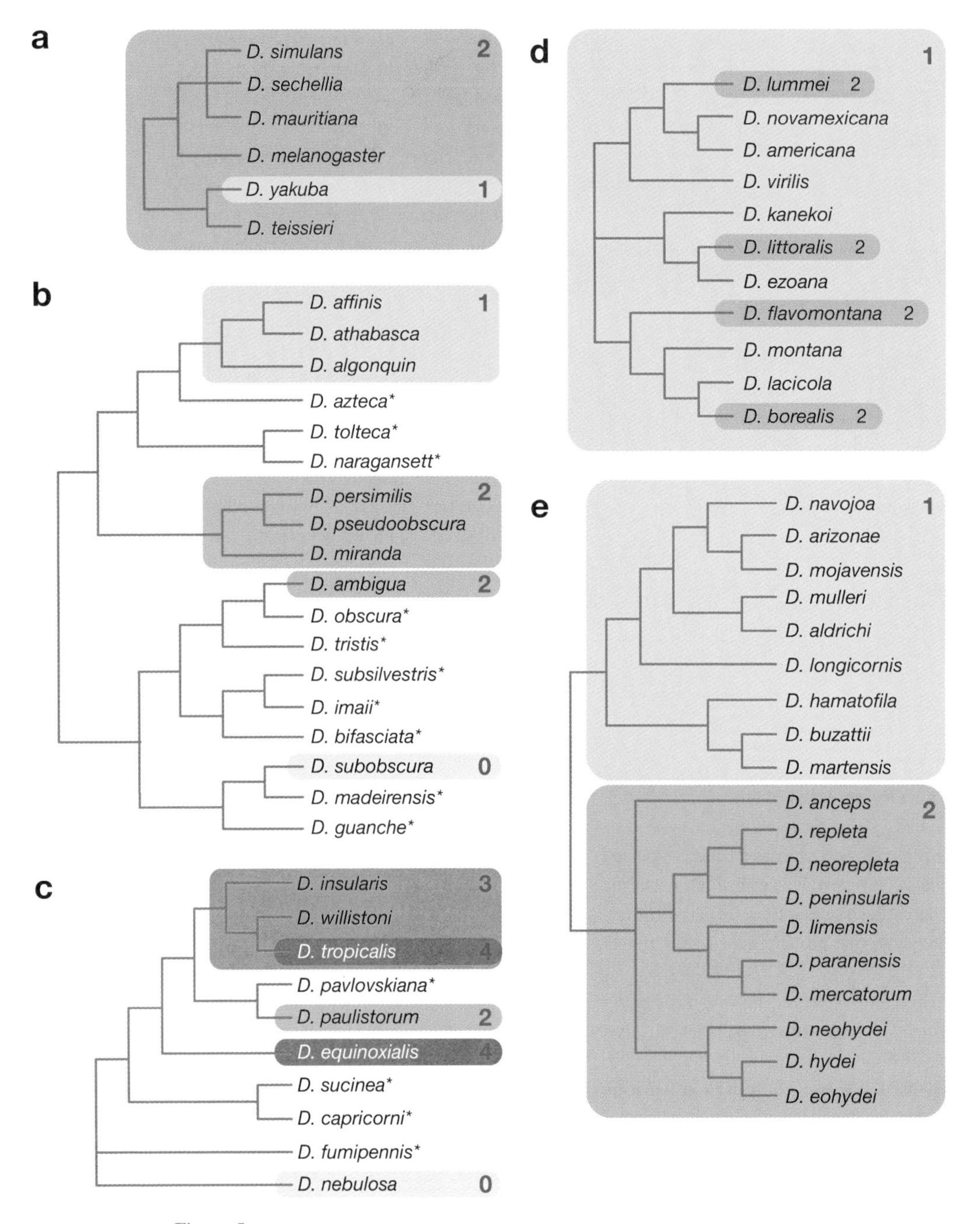

Figure 5

Evolution of the number of male courtship songs in the (*a*) the *melanogaster* species subgroup (modified from 48, 128); (*b*) the *obscura* species group (48, 108); (*c*) the *willistoni* species group (modified from 58); (*d*) the *virilis* species group (modified from 70, 135); and (*e*) the *repleta* species group (modified from 45, 49). Increasing color intensity from pale tan (0) to red (4) indicate number of male courtship songs observed. Information is lacking for species with an asterisk.

different types of song (**Figure 5*b***), referred to as high- and low-repetition songs (107). Neither of these is homologous to the pulse and sine songs observed in the *melanogaster* species. It is difficult to determine how these different signaling strategies may have evolved. Some species in the *affinis* subgroup (*D. affinis*, *D. athabasca*, *D. algonquin*) seem to use only a single type of song, whereas others in the *pseudoobscura* (*D. pseudoobscura*, *D. persimilis*, *D. miranda*) and *obscura* (*D. ambigua*) subgroup use two. Based on the current data, however, it is not clear whether the ancestral condition in the *obscura* group was one or two song types or whether the two-song strategy seen in the *pseudoobscura* and *obscura* subgroups was derived once or twice. Perhaps the most interesting song strategy observed in this group is found in *D. subobscura*, where males do not sing at all (48). It is likely that *D. subobscura* has shifted to an entirely visual mate recognition strategy since these species do not mate in the dark and males display in front of females (**Figure 3**).

Males of *willistoni* group species can produce up to four types of song (**Figure 5*c***), although the exact evolutionary history of song loss and acquisition is not completely clear (120). Although they are not sister taxa, both *D. tropicalis* and *D. equinoxialis* utilize four courtship songs, suggesting a complex series of gains and losses of song type and number for the intervening taxa. Interestingly, here again, there is a species, *D. nebulosa*, that does not sing at all. This species may rely on visual more than on auditory signaling. *Drosophila nebulosa* and *D. fumipennis*, the other basal member of this group, both have pigmented wings, a character not seen in the other *willistoni* taxa. Both of these species also display in front of the female, unlike all other *willistoni* taxa (**Figure 3**).

Males in the *virilis* group produce either one or two types of pulse songs. The evolution of this behavior, however, is quite complex. It appears that the use of a second pulse song has evolved at least four times in this group (**Figure 5*d***). In males of the *repleta* group (49), we also see a pattern in which there are either one or two types of songs produced (**Figure 5*e***).

Chemosensory signals. Chemical communication during courtship in *Drosophila* is thought to be mediated by the hydrocarbons (HCs) found in the adult epicuticle. Because they consist largely of long chain compounds that are not volatile, these HCs likely function at short range, through contact. Some HCs serve as aggregation pheromones (7–9, 66, 103, 127). Hydrocarbons can exhibit a remarkable degree of variability. They can differ in chain length, in the presence or absence of double bonds, and in the positions of the double bonds. Among *Drosophila* species, chain lengths range from between 20 and 40 carbons and for the most part are composed of various alkanes and alkenes (single and double bonds). Most *Drosophila* species produce a blend of HCs, and the characteristics of this blend can vary with age, sex, diet, and geographic origin within a species. Sexual dimorphisms in HCs can range from subtle differences in relative quantities of one or more molecules to the presence of completely different HCs between the sexes. Interspecific differences are also both quantitative and qualitative in nature. Considerable evidence exists that HCs play a role in sexual signaling within a species as well as for species recognition (reviewed in 52). Furthermore, because HCs are known to be important in water balance, these molecules and the genes controlling their production can be under both sexual and natural selection (99).

HC length is fairly well conserved in the genus *Drosophila* and can be roughly divided into three classes: short, intermediate, and long chains. The short chain morphology (23–29 carbons) is found in two groups, the subgenus *Sophohpora* (*melanogaster* and *willistoni* species) and the Hawaiian *Drosophila*. Intermediate length chains (22–31 carbons) are seen in the *virilis* group. The *repleta* group has the longest chains, from 28–40 carbons. Where these HC differ is in degree of sexual dimorphism and presence of unique molecules. The Hawaiian *Drosophila*, for example, have a unique, sex-specific HC profile. Other groups, such as the *repleta* species, show very small quantitative differences between the sexes.

HC variability in the *virilis* group roughly follows the phylogenetic relationships of the phylads (**Figure 6*a***). The *virilis* phylad, with the exception of *D. lummei*, shows some dimorphism in HCs (8). The *littoralis* and *montana* phylads, with the exception of *D. kanekoi*, are not dimorphic. The pattern of dimorphism in HC profiles is not as simple in the *melanogaster* subgroup (**Figure 6*b***). Although the ancestral reconstructions are equivocal, it is clear that dimorphism in HC characters is highly plastic in the *melanogaster* group and has shifted back and forth several times.

a

D. novamexicana
D. americana
D. texana
D. lummei
D. virilis ▶ **Unique 21C HC**
D. kanekoi ▶ **Amount of two pentacosadienes**
D. ezoana
D. littoralis
D. canadiana
D. flavomontana
D. montana
D. lacicola
D. borealis

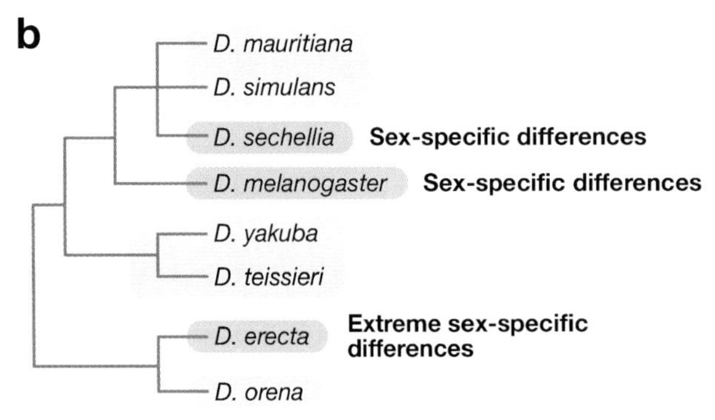

Figure 6

Evolution of hydrocarbon profiles in (*a*) the *virilis* species group (modified from 6, 135) and (*b*) the *melanogaster* species group (modified from 52, 74, 128). Species dimorphic for HC characters are shown in blue, those shown in yellow do not display sexual dimorphism in HC profile. Whether HC profiles are dimorphic or not seems to track phylogenetic relationships in the *virilis* group but are much more variable in the *melanogaster* subgroup.

Postmating Reproductive Behaviors

How do the reproductive behaviors of various species differ once mating has occurred? Processes occurring within the mated female can have significant effects on the reproductive success of both the female and male. For example, if females remate, it may create the opportunity for sperm competition. The ultimate fate of sperm inside the female reproductive tract can be the product of the continuing influence of a mate on the female's behavior, as can the propensity to remate and to oviposit, utilizing the sperm of that specific male. Finally, mated females must locate and utilize suitable oviposition sites.

Female remating. Species of *Drosophila* exhibit enormous variation in the frequency at which females remate (reviewed in 94, 96). In some species such as *D. subobscura*, *D. acanthoptera*, and *D. silvestris*, females effectively mate only once in their lifetimes, whereas in other species such as *D. hydei* or *D. nigrospiracula*, they have been observed to remate up to four times in a given morning. Frequencies at which females remate are presented in **Figure 7**.

The frequency at which females remate has important implications for their own reproductive fitness as well as that of their mates. In *D. melanogaster*, substances transferred to females during mating have been shown to reduce lifespan (21, 55), and it has been proposed, though not demonstrated, that

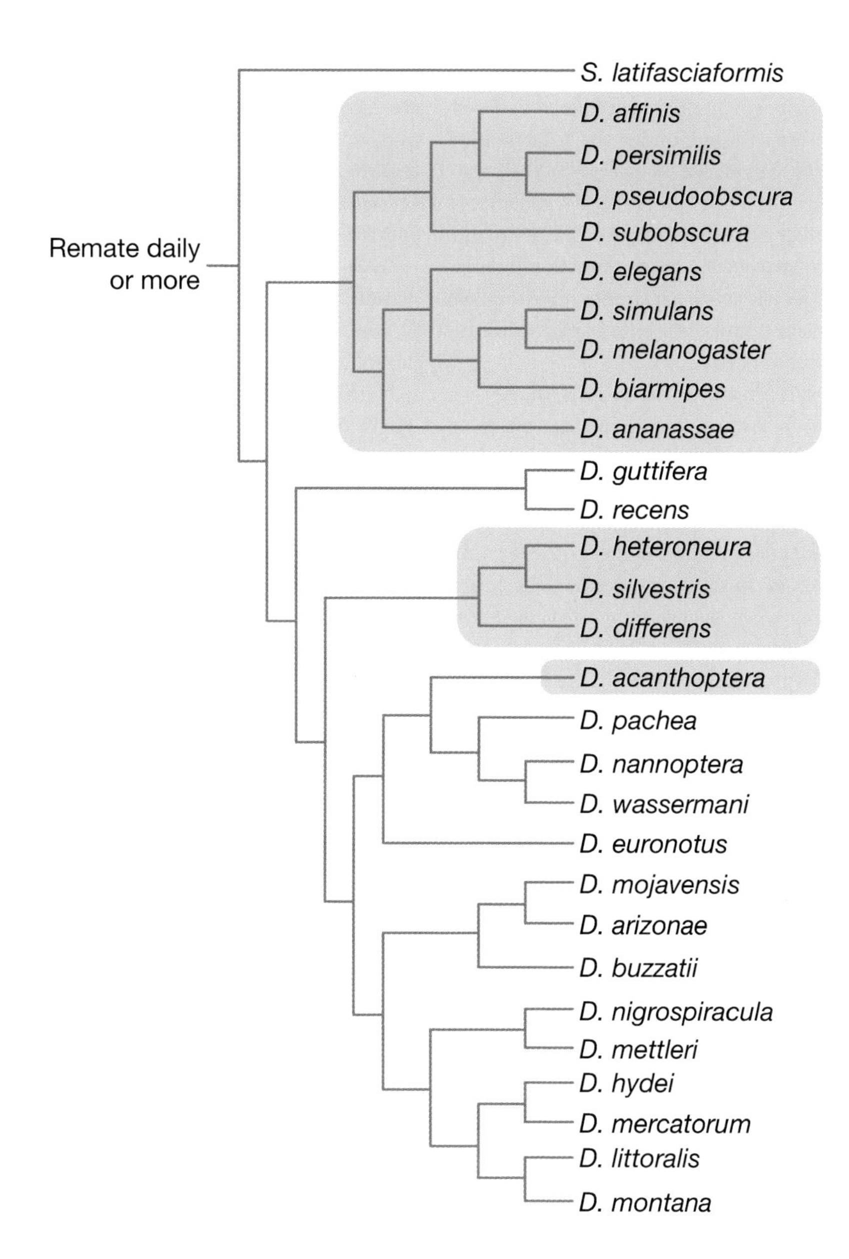

Figure 7

The evolution of female remating frequency. Although the ancestral condition in the group is to remate daily or more often several lineages have evolved taxa where remating occurs less than daily (*blue*). (Modified from 96, 132.)

copulating pairs are more vulnerable to predation and parasitism (117).

Remating and sperm utilization appear to be under the control of many factors: the characteristics of the ejaculate typically passed to females on a given mating and the interaction between the ejaculate components and the female's reproductive system. While these factors have been most widely studied in *D. melanogaster*, this species turns out not to be representative of the nature of these factors in the other species (94, 96). *Drosophila* species exhibit tremendous variation in their sperm-storage organs and in the use of these organs for the storage and retrieval of sperm (116). Species differ, as well, in the number of sperm males typically transfer during a single mating, from as few as 14, in *D. pachea*

(114), to 25,000, in *D. pseudoobscura* (133). For *D. melanogaster*, female remating is considered to be at least in part under the influence of the physical presence of sperm in the storage organs (89), but more recently, the effects of a number of different male accessory gland products have been demonstrated to influence not only female remating latency but the onset of oviposition and differential sperm utilization as well (26, 82, 167).

Sperm utilization patterns are such that a tendency toward sperm mixing exists in species in which females remate rapidly, but last male precedence exists in those species where females remate less often. This is likely to be a function of the numbers of sperm involved, as in species characterized by rapid female remating, where females receive few sperm on any given copulation, whereas relatively large numbers are delivered during a single copulation in species where females take longer to remate (94, 96, 113).

Seminal fluid components other than sperm appear to be pivotal to many of the processes that occur inside the female after she mates. A rapidly evolving group of approximately 80 proteins (25) are transferred to females in *D. melanogaster*. As these become characterized, it is clear that they play important roles in sperm storage and recovery, in female remating latency, and in the onset of oviposition (82, 167). In *D. melanogaster*, a species in which females remate after approximately 5 days, the receipt of certain seminal fluid components has been found to be responsible for shortening the life spans of females (21, 55). These studies suggest that seminal fluid components have evolved in ways that permit males to control female behavior long after the copulation is over. Mating systems of other *Drosophila*, however, suggest that females of some species have evolved different processes for dealing with seminal fluid compared with *D. melanogaster* (94, 96). Females of a number of *Drosophila* species actually extract large quantities of ejaculatory proteins (97, 98, 114, 115) and phosphorus (100) and use them to produce eggs. Because accessory gland proteins in *Drosophila* are rapidly evolving (25, 79), the identification and characterization of seminal fluid proteins in these other species has lagged behind *D. melanogaster*, and thus there are few comparative data at this time.

One obvious question, given the detrimental effect of mating on life span in *D. melanogaster* females, is whether females in rapidly remating species have even greater reductions in their life spans, given their frequent mating, or whether they have evolved some mechanism to escape this cost of mating. We addressed this question by examining the longevities of females from a set of other *Drosophila* species in which female mating frequencies are greater, as well as less, than those of *D. melanogaster*; *D. nigrospiracula*, in which the remating frequency is four times a day (91); *D. pachea*, *D. mettleri*, and *D. mojavensis*, in which females remate daily (91, 113); and *D. acanthoptera*, in which they mate once in their lifetime (114). We also included *D. melanogaster* as a control. The influence of mating was examined by placing females in food vials, which were changed daily, at which time any dead flies were counted. Females were placed in three treatment groups, with 25 individuals per treatment: single virgin females, females paired continually with males, and females paired with other females. In the female-female group, survival of focal females was scored. When either a male or a companion female died, it was replaced until the death of the focal fly was observed. The experiment was repeated twice. Mean age at death is presented in **Table 1**. The survival rates of mated versus virgin or control *D. melanogaster* females are consistent with earlier studies showing a significant effect of mating in reducing female longevity. Of the other species, for *D. acanthoptera*, in which females only mate once in their life times, survival was significantly reduced in mated females. In the remaining four species, however, females did not suffer any reduction in longevity, despite continual mating. These four species, *D. pachea*, *D. mettleri*, *D. nigrospiracula*, and *D. mojavensis*, are unrelated, and their mating systems have very different characteristics. For example, *D. mojavensis* females incorporate large amounts of seminal proteins from males into their somatic tissues and ovarian acolytes, whereas females of the other two do not. *Drosophila pachea* males pass very few sperm per mating and females are highly sperm limited. Therefore, in each lineage, different mechanisms have likely evolved to overcome or disarm the detrimental effects of mating, but the nature of those mechanisms remains a mystery for now.

Table 1 Mean age at death for six species of *Drosophila* with different female remating frequencies. Flies were housed either alone (SF), with another female (FF), or with a male (MF) and scored daily for survival. Significant differences are shown with an asterisk. Flies were changed every other day to fresh food vials until the focal female died

		Mean age at death		
Species	Rep	SF	FF	MF
D. melanogaster	1	70.9 ± 3.8	71.9 ± 3.6	53.3 ± 5.1*
	2	74.6 ± 3.3	67.5 ± 7.9	53.1 ± 5.0*
	3	63.5 ± 4.7	59.8 ± 5.8	43.6 ± 7.1*
D. nigrospiracula	1	83.0 ± 4.7	76.4 ± 7.7	81.0 ± 4.6
	2	72.0 ± 5.5	75.8 ± 2.8	68.8 ± 7.1
D. acanthoptera	1	47.3 ± 2.0	44.9 ± 1.31	36.2 ± 1.3*
	2	46.7 ± 2.9	43.2 ± 1.5	37.4 ± 2.2*
D. pachea	1	32.2 ± 3.0	43.5 ± 2.6	35.0 ± 3.0
	2	51.4 ± 3.7	52.6 ± 3.2	47.7 ± 4.1
	3	33.6 ± 2.8	42.2 ± 3.4	42.2 ± 3.2
D. mojavensis	1	50.4 ± 3.4	50.0 ± 3.0	57.9 ± 3.5
	2	44.7 ± 4.4	46.7 ± 2.8	41.1 ± 3.4
D. mettleri	1	57.0 ± 2.9	46.9 ± 6.8	57.4 ± 2.5
	2	36.7 ± 2.9	39.1 ± 3.2	41.4 ± 2.6

Oviposition behavior. The basis for oviposition site selection may overlap, in part, with identification of mating sites, but not necessarily completely. Long-range signals, largely volatile but possibly also visual, as well as microclimatic factors, attract the flies to potential sites. Females use contact signals, however, to make oviposition decisions, and genetic variation exists within a species (4, 73, 88) as well as among related species (50, 122) for oviposition preferences. It is largely assumed that at a gross level, female oviposition preferences, in terms of resource type, will be governed by long-range resource location, discussed above, which probably explains the paucity of data at the oviposition level. Other aspects of oviposition, such as periodicity, have not been studied in enough species to examine this reproductive character in a comparative or evolutionary context (50, 131).

EVOLUTIONARY GENETICS OF REPRODUCTIVE BEHAVIOR

From the foregoing descriptions, several conclusions can be drawn. One is that *Drosophila* species exhibit considerable variability in all aspects of their reproductive behaviors. Some of the divergence occurred at the time major lineages were forming, but some involves more recent differentiation. In all cases, the patterns necessarily involve shifts in the signals different species use to successfully reproduce. These shifts have occurred within particular sensory modalities, for example from attraction to one type of host versus another, or using one type of courtship song versus two, as well as with respect to the relative importance of multiple sensory modalities, such as vision versus olfaction during courtship. What this indicates is that, if signals themselves vary, the systems involved in their reception and response must also vary. What remains unclear, however, is how the differences among species and species groups have arisen and what their genetic, cellular, and physiological bases are. Historically, approaches to understanding the genetics of particular traits have used the tools of classical genetics: intra- and interspecific crosses, QTL mapping, and mutagenesis. These approaches have been useful in demonstrating genetic bases to intra- and interspecific differences in behaviors and in some cases, in identifying the genetic architecture or an area of the nervous system involved. In terms of identifying specific signals and signal detection systems, and having the ability to examine their coevolution, however, we have far to go. Given recent developments in genomics, including the sequencing of the genomes of multiple species of *Drosophila*, new approaches now can be exploited to address the evolutionary genetics of reproductive behaviors.

Host Use

Host use, in terms of both resource location and oviposition site preference, is mediated by the chemosensory system. We do not mean to imply that other factors, such as host microclimate, play no role in resource location, but these variables are beyond the scope of this review. Chemosensory information is classified as either olfactory or gustatory, depending upon whether the signals are volatile or contact. Although there may be some degree of functional overlap, the olfactory system is more likely to be involved in longer-range location of resources, whereas the gustatory system figures more prominently in close-range signaling between individuals during courtship and in oviposition decisions.

Olfactory information is received and processed by olfactory receptor neurons (ORNs), which are found in the two olfactory organs, the antennae and the maxillary palps. Antennae contain approximately 1200 ORNs, whereas the maxillary palps contain only about 120. The ORNs fall into 16 functional classes based upon the their odor response spectra (38), which are thought to depend, in turn, upon the expression of approximately 60 different odor receptor genes (160). Gustatory or taste receptor neurons (GRNs) most likely to be involved with oviposition and with sexual behavior are those on the abdomen, forelegs, and mouthparts, which are in contact with substrates and with flies of the opposite sex during the tapping and licking phases of courtship. With respect to the bristles on the forelegs, male *D. melanogaster* have nearly twice as many taste bristles on their forelegs as do females (102, 106, 143). Based upon the degrees of sequence similarity, the olfactory and gustatory receptor genes are likely to have a common evolutionary origin. Chemical information received by flies has two origins: the host resources and other flies. Candidate sensory processes for host location, therefore, are likely to be associated with the ORNs, whereas those mediating sexual behavior and oviposition are more likely associated with morphological structures in contact with other flies and food, the GRNs.

Each type of host resource provides a different chemical profile based not only upon the host's own chemistry, but upon the microbial community responsible for its breakdown, making it suitable as a *Drosophila* breeding site. Volatile profiles have been characterized for several *Drosophila* resources (51, 54, 87, 141). Stensmyr et al. (142) utilized the ecologically relevant volatiles to examine evolutionary conservation and divergence in the olfactory code among nine members of the *D. melanogaster* group of species. The group includes *D. sechellia*, which, in addition to being an island endemic, has specialized upon the fruit of *Morninda citrofolia*, which has a distinct chemical profile compared with the broader range of fruits utilized by other members of the group. The ab2 type (38) sensillum and its neurons were found to be missing in *D. sechellia*, apparently replaced by a higher number of the ab3 type, such that the overall number of sensilla of the large basiconic (LB) class was the same among the species in the group. There was also a shift in the key ligand for the ab3A-type neuron, from ethyl to methyl hexanoate. The bases for sensitivity shifts within given ORNs is unclear, but could be due to substitutions in their receptors. For example, ab3A ORNs in *D. melanogaster* express the receptor *Or22a*. *Drosophila simulans* has an orthologous counterpart, *DsOr22a*, whose sequence homology with that of *D. melanogaster* is 94% (40). Another group of proteins, the odorant binding proteins (161), which are thought to bind and present the odorants to the receptors, may also be found to contribute to observed species differences.

Long-distance location of appropriate hosts is only the first step in host utilization. Inseminated females then must make decisions about oviposition. When host shifts involve novel compounds, there is the potential for a mismatch between oviposition site and larval performance or fitness. The evolution of preference-performance correlations (31, 73) is a long-standing problem that can potentially be resolved using the *Drosophila* model system. An obvious approach is to identify the genes involved in oviposition decisions as well as in performance on a specific host and look for genetic and physiological relationships between the two processes. Oviposition preferences of a number of *Drosophila* species have been characterized (3, 88, 136). In most cases, there is a clear preference for certain types of hosts, and even for the yeasts associated with those hosts (4). Where interspecific crosses have been possible, as in the case of *D. sechellia*, the genetics of oviposition preference also has been analyzed (3). With respect to oviposition decisions, the mechanisms are likely to involve the gustatory reception system described under *Chemosensory signals and courtship*. Jones (75, 76) and Cariou et al. (17) have examined the genetic basis for resistance to *Morinda* toxins (octanoic acid and other compounds) by *D. sechellia* adults and larvae, revealing that resistance is due to a few semidominant alleles present only in this species. Two other taxa in the *melanogaster* subgroup, *D. santomea* and *D. erecta*, also appear to be restricted to single hosts (*Ficus* and *Pandanus*, respectively), in spite of more generalist sister species (**Figure 3*a***). The *melanogaster* subgroup offers an excellent opportunity to resolve genetic questions concerning preference-performance correlations and the evolution of host shifts (31).

Chemosensory Signals and Courtship

Contact chemoreceptors on other adult structures mediate the signaling involved in courtship and mating. A family of about 70 gustatory receptors, coded for by gustatory receptor genes (Gr), are found in sensilla on the proboscis, legs, and anterior wing margins (27, 44, 130). Some of these sensilla are male specific, occurring in twice the number on male forelegs as on those of females (106). One of the taste receptor genes, Gr68a, is expressed in 10 of the males-specific foreleg bristles of *D. melanogaster* (14). When expression of this gene is disrupted (14), male courtship is interrupted in a way that suggests the gene product is a putative receptor of female pheromones in this species.

Chemical signals received from other flies could come from several sources. The most likely, however, are the hydrocarbons associated with the epicuticle of flies of both sexes. As seen earlier, these can vary in a wide range of ways within and between species. Epicuticular hydrocarbons originate in the large polyploidy oenocytes located in the subepidermal layer (123). In most cases, once flies are sexually mature, their hydrocarbon profiles are fairly constant. In some species, it appears that females may emit pulses of pheromones by extruding their ovipositors, as this behavior can result in the inhibition of male courtship (137). In some species, such as *D. subobscura* (140) or *D. adiastola* (139), males present a liquid drop to females during courtship, the pheromonal properties of which remain unclear.

Many genetic studies of intra- and interspecific hydrocarbon variability strongly support not only the role of these compounds in sexual selection as well as in behavioral isolation, but also give an idea of the genetic architecture of the variability (reviewed in 52). For example, crosses among species of the *melanogaster* group indicate that female-specific pheromones are controlled by at least five different genes in chromosome 3, whereas male differences are interspersed across all three major chromosomes (32–34). Differences in the hydrocarbon profiles of *D. virilis* and *D. novamexicana* reside in two chromosomes, including one gene of major effect (41).

In *D. melanogaster*, mutants and other genetic manipulations have revealed that sex differences in the hydrocarbon profile are ultimately under the control of the sex determination hierarchy of genes (53, 125, 152, 153). With respect to species and sex differences in characteristics like HC chain length and double bond positions, the specific biosynthetic pathways and the genes controlling them are still largely unknown (reviewed in 52). Two desaturase genes, *desat1* and *desat2*, discovered in mutagenesis screens, are involved in hydrocarbon biosynthesis (37, 90, 146), and mutations in these genes disrupt courtship.

Visual Signals and Courtship

Visual signals and their reception and processing may prove more complicated to study. For chemical or auditory communication, the signals can be categorized as to features such as chemical composition or sound wavelength and nature and frequency of pulses. In addition, chemical and auditory communication can be manipulated at the levels of the signal and signal detection not only through genetic manipulation, but by nongenetic means as well. Although a role for visual signaling can be assumed with confidence for some species, the existence of variability in visual signaling within and between species is based largely upon inference. It may be that for many species the role of visual signals is less specific than the roles of chemosensation or audition. For example, when a given species will not mate in the dark (**Figure 3**), visual signals can be assumed to be critical to the process. Whether the critical signals are produced by the female, the male, or both, is not known. Also unknown is whether the relevant signals are the same as in a related species, but simply are less critical to the process or are perceived in a different way. Clues could be obtained through observations under infrared light, or by utilizing genetically blind flies of one sex or the other and determining the stage at which courtship fails.

Auditory Signals and Courtship Behavior

Courtship songs of most *Drosophila* species are generated by vibrations of the wings, through mechanisms related to the flight neuromuscular circuitry. The signals, produced by way of air displacement (46), give rise to "near-field sound," perceived only within a short distance of the source. The motor patterns producing the songs are different, however, from those for flight, and for species producing more than one type of courtship sound, there may be multiple motor patterns involved (149). Furthermore, it appears that the motor patterns involved in song production involve a feedback component (148).

Audition, the receipt and processing of auditory information, is thought to have evolved in species-specific ways in *Drosophila*. The auditory apparatus in *Drosophila* is the Johnson's organ, a type I mechanoreceptor (46) in the antenna, which functions though particle displacement generated by wing vibrations. What is known of the process for *Drosophila* is based upon studies with *D. melanogaster* (62). The antennal complex is comprised of three segments. The segment most proximal to the head is the scape, the middle segment is the pedicel, and the most distal segment, the funiculus gives rise to the arista, a long structure with multiple branches. The arista receives sound-induced vibrations and behaves like a stiff rod, which, because it is tightly connected to the funiculus, also stimulates this structure (62). The two in fact function as one mechanoreceptor unit. The auditory receptor itself, the Johnson's organ, lies within the pedicel. A process from the funiculus inserts into the pedicel, transferring the vibrations it receives via the arista to the auditory receptors (62).

The importance of these structures in audition has been verified by mutations in *D. melanogaster* that have modified or eliminated some aspect of their structure or function. Mutants such as *aristless (al)* and *thread (th)* alter the perception of courtship songs and courtship behavior (92) by modifying the external structures (15, 16), whereas others, such as *atonal (ato)*, *beethoven (btv)*, and *touch-insensitive-larvae-B (tilB)* affect neural structure and function as well as other developmental processes (47).

A considerable number of genes have been discovered to influence song in *D. melanogaster*, but none exclusively so. These genes and their actions are nicely reviewed by Gleason (57), who groups them by function into regulatory, ion channel, sex determination, and flight genes. Regulatory genes include *period (per)* (85) and *no-on-transient-A (nonA)*, of which *dissonance* is an allele (84). Ion channel genes include *cacophony (cac)* (83, 158, 159, 165) and *slowpoke* (112). Genes in the sex-determination hierarchy, *transformer (tra)*, *doublesex (dsx)*, *fruitless (fru)*, also have been found to influence song (reviewed in 11). Finally, some of the genes affecting flight, such as *croaker (cro)* (169) and *ariadne (ari-1)* (2) because of the role of the flight musculature in sound production, also influence songs, whereas some flightless mutants do not (5).

The relationship between the loci for which song aberrations have been found and those underlying naturally occurring variation in song production having evolutionary potential is unknown. Genetic variability for interpulse interval (IPI) in natural populations of *D. melanogaster* responds rapidly to directional selection and is thought to have an additive polygenic basis (121). Naturally occurring intraspecific variability in IPI has been localized to chromosome 3 of *D. melanogaster* (30), and QTL analysis of inbred lines revealed three significant QTLs (59). Similar studies have been undertaken in *D. virilis* and *D. littoralis* (72), *D. pseudoobscura* (134, 166) and *D. polios*, and its sibling *D. ananassae* (42). A major difference between song production in the Sophophoran subgenus and that observed for many flies in the subgenus *Drosophila* is that males of many species in the latter vibrate both wings simultaneous when singing (137; T.A. Markow, unpublished observations). Examining the effects of genes identified in *D. melanogaster* that modify wing position or cause the simultaneous use of both wings should provide attractive candidates for evolutionary studies.

Postmating Control of Reproductive Behavior

Copulation produces specific changes in *D. melanogaster* that have been well studied.

The two principal effects of mating in this species are the stimulation of oviposition and the delay of female remating. These are not identical in other species, as shown above. In a number of species, female remating is not delayed. Although there is less comparative information on the onset of oviposition, the few data that there are suggest that species differ in this character as well. For example, unlike *D. melanogaster*, where females begin to lay eggs within a few hours of mating, in some species a large mass, called the insemination reaction, forms in the uterus after mating, and oviposition does not commence until the mass subsides, which in some cases is the next day (164).

In *D. melanogaster*, postmating effects have been firmly connected with various male accessory gland proteins (Acps) passed to the female (reviewed in 82, 167). Some of these Acps also have been examined in the *D. simulans* complex of species (79). One protein in particular, the sex peptide, or Acp 70a, has been characterized in the greatest number of species (23, 24, 129), and its function and sites of action are becoming better known than those for other Acps. For example, by incubation of cryostat tissue sections of *D. melanogaster* females with a radioactive synthetic form of the sex peptide, binding was observed in parts of both the central and peripheral nervous systems as well as in the female genital tract (39, 82). Similar observations were also made for another seminal protein, DUP99B, which is made in the male's ejaculatory duct (82). In addition, the sex peptide has been found bound to the tails of sperm (82), which would explain the importance of sperm, as well as accessory gland proteins to the postmating behaviors of females (22). Male-derived proteins can be thought of in the same way as other signals that pass from one sex to the other. In this case, the signal is internal, but can be traced to sites of action in the female, including the CNS. Unlike in marine invertebrates where the female receptor is known (56) the mode of action of these Acps is not yet well understood. Given their number, in *D. melanogaster*, at least, there is likely to be either some redundancy in their action or specialization in function (22). Because of the number of processes that actually occur between copulation and oviposition, male-derived substances are likely to have a role in a variety of them. For example, sperm must find their way into storage, and they must be recovered, a process which, even within a species, is not random with respect to the male and female genotype (95). Females must recognize that they are inseminated, and then begin to release oocytes. Sperm must remain viable until an oviposition site is found, and they may also need to resist displacement or preferential use by females. The idea that the Acps are at least in part specialized in their functions is supported by the data: Some Acps have been localized to certain parts of the female reproductive tract involved with sperm storage, and there are indications that they are involved in the sperm storage or utilization process (167). The sex peptide seems to have multiple functions, as indicated above. Still another set of putative postmating behavior control genes has been identified from mutagenesis screens designed to disrupt oviposition. These genes are likely to act downstream of any male-induced signal, but until further study their functional role in the mating-oviposition cascade will not be known. The action of downstream genes, such as the oviposition mutants *dissatisfaction* (*dsf*) and Tyrosine Beta Hydroxylase (TBH), required for production of octopamine, necessary for oviposition (29, 86) and *cyclophorin-like (Cypl)* discovered in a screen (105), is expressed in the oviduct and has been proposed to be required for oviposition in the mated female.

Evidence for genetic divergence between the signals and receptors mediating postmating reproductive behaviors is inferred from the increased size and duration of the insemination reaction mass observed following matings between genetically differentiated populations of *D. mojavensis* (80) and, to an even greater degree, between related species, which is accompanied by increases in the time until females oviposit (110; T.A. Markow, unpublished observations). In extreme cases, those involving interspecific mating, the mass may remain forever, effectively preventing the female from ever remating or laying eggs (111).

CONNECTING THE DOTS

What are the nature and number of the genetic changes underlying the diversification of reproductive behavior in *Drosophila*? To what extent has the evolution of these species differences involved changes in regulatory rather than structural genes (60, 69)? Do the loci or chromosomal regions identified in QTL studies correspond to any of the candidate genes discovered through mutagenesis screens in *D. melanogaster* (81), and if so, do they retain the same functions in other species? If the identical function of a gene is retained across unrelated species, what is the level of sequence divergence observed in that gene? What are the levels of variation in natural populations of any of the species at the loci implicated in the interspecific differences?

One of the goals of mapping characters onto a phylogeny is to learn which states of the characters are ancestral and which are derived. With respect to many of the reproductive behaviors in *Drosophila* discussed above, data are available for large numbers of species. There are clear gaps, however, as not all species have been equally popular or easy to rear and study. Filling in some of these gaps will be important in ultimately understanding the evolution of the phenotypes and the genetic systems controlling them. Such studies can bridge the often disparate disciplines of systematics and genomics. When there is a change in a character state, for example, is it more commonly attributable to regulatory changes or to changes in the function or number of structural proteins? The switch-points seen on the phylogenetic maps of reproductive behavioral characters, where an entire lineage has undergone a major shift in something like host use or the use of female auditory signals or visual signals, would appear to be a juncture at which some major change in a signaling process has occurred. For example, the use of necrotic cacti might be associated with the appearance of a functionally different sent of olfactory receptors, odorant binding proteins, or arrangement of types of neurons in the sensillae. The same question could be asked of signalizing that occurs inside of the mated female. In several lineages, female remating frequency has shifted between monogamy and frequent remating. Remating of females, in *D. melanogaster* at least, is delayed by the action of one or more seminal fluid proteins. What is the nature or number of changes that have occurred in those lineages where seminal fluid clearly does not produce this effect? Additional and as yet unresolved questions include whether genes mediating intraspecific sexual selection are important to the evolution of sexual isolation between species (13) and the extent to which there is coevolution between characters such as male and female sexual signals (12), or host preference and larval performance during the evolution of host shifts (31). With the availability of genome sequences and the development of expression profiling systems and other tools for 12 *Drosophila* species (101), differing to various degrees in their evolutionary distances, we can begin to address these questions.

LITERATURE CITED

1. Adams MD, Celnicker SE, Holt RA, Evans CA, Gocayne JD, et al. 2000. The genome sequence of *Drosophila melanogaster*. *Science* 287:2185–96
2. Aguilera M, Oliveros M, Martinez-Padron M, Barbas JA, Ferrus A. 2000. *Ariadne-1*. A vital Drosophila gene is required in development and defines a new conserved family of ring-finger proteins. *Genetics* 155:1231–44
3. Amlou M, Moreteau B, David JR. 1998. Genetic analysis of *Drosophila sechellia* specialization: oviposition behaviour toward the major aliphatic acids of its host plant. *Behav. Genet.* 28:455–64
4. Barker JSF, Starmer WT, Fogleman JC. 1994. Genotype-specific habitat selection for oviposition sites in the cactophilic species *Drosophila buzzatii*. *Heredity* 72:384–95

4a Barker JSF, Starmer WT, MacIntyre RJ, eds. 1994. *Ecological and Evolutionary Genetics of Drosophila*. New York: Plenum

5. Barnes PT, Sullivan L, Villella A. 1998. Wing-beat frequency mutants and courtship behavior in *Drosophila melanogaster* males. *Behav. Genet.* 28:137–51
6. Bartelt RJ, Arnold MT, Schaner A, Jackson LL. 1986. Comparative analysis of cuticular hydrocarbons in the *Drosophila virilis* species group. *Comp. Biochem. Physiol.* 83:731–42
7. Bartelt RJ, Schaner A, Jackson LL. 1985. *cis*-vaccenyl acetate as an aggregation pheromone in *Drosophila melanogaster*. *J. Chem. Ecol.* 11:1747–56
8. Bartelt RJ, Schaner A, Jackson LL. 1988. Aggregation pheromones in *Drosophila borealis* and *Drosophila littoralis*. *J. Chem. Ecol.* 14:1319–28
9. Bartelt RJ, Schaner A, Jackson LL. 1989. Aggregation pheromone components in *Drosophila mulleri*. A chiral ester and an unsaturated ketone. *J. Chem. Ecol.* 15:399–412
10. Bennet-Clark, HC, Leroy Y, Tsacas L. 1980. Species and sex-specific song and courtship behavior in the genus *Zaprionus* (Diptera: Drosophilidae). *Anim. Behav.* 28:230–55

11. Billeter JC, Goodwin SF, O'Dell KMC. 2001. Genes mediating sex-specific behaviors in *Drosophila*. *Adv. Genet.* 47:87–116
12. Boake C. 1991. Coevolution of senders and receivers of sexual signals: genetic coupling and genetic correlations. *Trends Ecol. Evol.* 6:225–27
13. Boake C. 2002. Sexual signaling and speciation, a microevolutionary perspective. *Genetica* 116:205–14
14. Bray S, Amrein H. 2003. A putative *Drosophila* pheromone receptor expressed in male-specific taste neurons is required for efficient courtship. *Neuron* 39:1019–29
15. Burnet B, Connolly K, Dennis L. 1971. The function and processing of auditory information in the courtship behaviour of *Drosophila melanogaster*. *Anim. Behav.* 19:409–15
16. Burnet B, Eastwood L, Connolly K. 1977. The courtship song of male *Drosophila* lacking aristae. *Anim. Behav.* 25:460–64
17. Cariou ML, Silvain JF, Daubin V, Da Lage JL, Lachaise D. 2001. Genetic analysis by interspecific crosses of the tolerance of *Drosophila sechellia* to major aliphatic acids of its host plant. *Mol. Ecol.* 10:649–60
18. Carson HL. 1987. The genetic system, the deme, and the origin of species. *Annu. Rev. Genet.* 21:405–23
19. Carson HL, Kaneshiro KY. 1976. *Drosophila* of Hawaii: systematics and ecological genetics. *Annu. Rev. Ecol. Syst.* 7:311–45
20. Carson HL, Templeton AR. 1984. Genetic revolutions in relation to speciation phenomena: the founding of new populations. *Annu. Rev. Ecol. Syst.* 15:97–131
21. Chapman T, Liddle LF, Kalb JM, Wolfner MF, Partridge L. 1995. Cost of mating in *Drosophila melanogaster* females is mediated by male accessory gland products. *Nature* 373:241–44
22. Chapman T, Bangham J, Vinti G, Seifried B, Lung O, et al. 2003. The sex peptide of *Drosophila melanogaster*: female post-mating responses analyzed by using RNA interference. *Proc. Natl. Acad. Sci. USA* 100:9923–28
23. Chen PS, Balmer J. 1989. Secretory proteins and sex peptides of the male accessory gland in *Drosophila sechellia*. *J. Insect Physiol.* 35:759–64
24. Cirera S, Aguadé M. 1998. Molecular evolution of a duplication: the SP (Acp 70)A gene region of *D. subobscura* and *D. maderiensis*. *Mol. Biol. Evol.* 210:247–54
25. Civetta A, Singh R. 1995. High divergence of reproductive tract proteins and their association with postzygostic reproductive isolation in *Drosophila melanogaster* and *Drosophila virilis* group species. *J. Mol. Evol.* 41:1085–95
26. Clark AG, Aguade M, Prout T, Harshman LG, Langley CH. 1995. Variation in sperm displacement and its association with accessory gland protein loci in *Drosophila melanogaster*. *Genetics* 139:189–201
27. Clyne PJ, Warr CG, Freeman MR, Lessing D, Kim J, Carlson JR. 1999. A novel family of divergent seven-transmembrane proteins: candidate odorant receptors in *Drosophila*. *Neuron* 22:327–38
28. Cobb M, Jallon JM. 1990. Pheromones, mate recognition and courtship stimulation in the *Drosophila melanogaster* species sub-group. *Anim. Behav.* 39:1058–67
29. Cole SH, Carney GE, McClung CA, Willard SS, Taylor BJ, Hirsh J. 2005. Two functional but non-complementing *Drosophila* tyrosine decarboxylase genes: distinct roles for neural tyramine and octopamine in female fertility. *J. Biol. Chem.* 280:14948–55
30. Colegrave N, Hollocher H, Hinton K, Ritchie MG. 1999. The courtship song of African *Drosophila melanogaster*. *J. Evol. Biol.* 13:143–50

31. Courtney SP, Kibota T. 1990. Mother doesn't know best: host selection by ovipositing insects. In *Insect-Plant Relationships*, ed. EA Bernays, 2:161–88. Boca Raton, FL: CRC Press
32. Coyne JA. 1996. Genetics of differences in pheromonal hydrocarbons between *Drosophila melanogaster* and *D. simulans*. *Genetics* 143:353–64
33. Coyne JA. 1996. Genetics of a difference in male cuticular hydrocarbons between two sibling species, *Drosophila simulans* and *D. sechellia*. *Genetics* 143:1689–98
34. Coyne JA, Charlesworth B. 1997. Genetics of a pheromonal difference affecting sexual isolation between *Drosophila mauritiana* and *D. sechellia*. *Genetics* 145:1015–30
35. Coyne JA, Orr HA. 2004. *Speciation*. Sunderland, MA: Sinauer
36. Crossley SA. 1986. Courtship sounds and behaviour in the four species of the *Drosophila bipectinata* complex. *Anim. Behav.* 34:1146–59
37. Dallerac R, Labeur C, Jallon JM, Knipple DC, Roelofs WL, Wicker-Thomas C. 2000. A Δ9 desaturase gene with a different substrate specificity is responsible for the cuticular diene hydrocarbon polymorphism in *Drosophila melanogaster*. *Proc. Natl. Acad. Sci. USA* 97:9449–54
38. De Bruyne M, Foster K, Carlson J. 2001. Odor coding in the *Drosophila* antenna. *Neuron* 30:537–52
39. Ding Z, Haussmann I, Ottinger M, Kubli E. 2003. Sex peptides bind to two molecularly different targets in *Drosophila melanogaster* females. *J. Neurobiol.* 55:372–84
40. Dobritsa AA, van der Goes van Naters W, Warr CG, Steinbrecht RA, Carlson JR. 2001. Odor receptor expression and olfactory coding in *Drosophila*. *Curr. Commun. Mol. Biol.* 7
41. Doi M, Tomaru M, Matsubayashi H, Yamanoi K, Oguma Y. 1996. Genetic analysis of *Drosophila virilis* sex pheromone: genetic mapping of the locus producing Z-(11)-pentacosene. *Genet. Res.* 68:17–21
42. Doi M, Matsuda M, Tomaru M, Matsubayashi H, Oguma Y. 2001. A locus for female discrimination behavior causing sexual isolation in *Drosophila*. *Proc. Natl. Acad. Sci. USA* 98:6714–19
43. Donegan J, Ewing AW. 1980. Dueting in *Drosophila* and *Zaprionus* species. *Anim. Behav.* 20:1289
44. Dunipace L, Meister S, McNealy C, Amrein H. 2001. Spatially restricted expression of candidate taste receptors in the *Drosophila* gustatory system. *Curr. Biol.* 11:822–35
45. Durando CM, Baker RH, Etges WJ, Heed WB, Wasserman M, DeSalle R. 2000. Phylogenetic analysis of the *repleta* species group of the genus *Drosophila* using multiple sources of characters. *Mol. Phylogenet. Evol.* 16:296–307
46. Eberl DF. 1999. Feeling the vibes: chordotonal mechanisms in insect hearing. *Curr. Opin. Neurobiol.* 9:389–93
47. Eberl DF, Hardy RW, Kernan MJ. 2000. Genetically similar transduction mechanisms for touch and hearing in *Drosophila*. *J. Neurosci.* 20:5981–88
48. Ewing AW, Bennet-Clark HC. 1968. The courtship songs of *Drosophila*. *Behaviour* 31:288–301
49. Ewing AW, Miyan JA. 1986. Sexual selection, sexual isolation and the evolution of song in the *Drosophila repleta* group of species. *Anim. Behav.* 34:421–29
50. Fanara JJ, Hasson E. 2001. Oviposition acceptance and fecundity schedule in the cactophilic sibling species *Drosophila buzzatii* and *D. koepferi* on their natural hosts. *Evolution* 55:2615–19
51. Farine JP, Legal L, Moreteau B, Le Quere JL. 1996. Volatile components of ripe fruit of *Morinda citrifolia* and their effects on *Drosophila*. *Phytochemistry* 41:433–38

52. Ferveur JF. 2005. Cuticular hydrocarbons: their evolution and roles in *Drosophila* pheromonal communication. *Behav. Genet.* In press
53. Ferveur JF, Savarit F, O'Kane CJ, Sureau G, Greenspan RJ, Jallon JM. 1997. Genetic feminization of pheromones and its behavioral consequences in *Drosophila* males. *Science* 276:1555–58
54. Fogleman JC, Abril JR. 1990. Ecological and evolutionary importance of host plant chemistry. See Ref. 4a, pp. 121–43
55. Fowler K, Partidge L. 1989. A cost of mating in female fruitflies. *Nature* 338:760–61
56. Galindo BE, Vaquier VD, Swanson WD. 2003. Positive selection in the egg receptor for abalone sperm lysin. *Proc. Natl. Acad. Sci. USA* 100:4639–43
57. Gleason J. 2005. Mutations and natural variation in the courtship song of *Drosophila*. *Behav. Genet.* In press
58. Gleason J, Ritchie MA. 1998. Evolution of courtship song and reproductive isolation in the *Drosophila willistoni* species complex: Do sexual signals diverge the most quickly? *Evolution* 52:1493–500
59. Gleason JM, Nuzhdin SV, Ritchie MG. 2002. Quantitative trait loci affecting a courtship signal in *Drosophila melanogaster*. *Heredity* 89(1):1–6
60. Gompel N, Prud'homme B, Wittkopp PJ, Kassner VA, Carroll SB. 2005. Chance caught on the wing: *cis*-regulatory evolution and the origin of pigment patterns in *Drosophila*. *Nature* 433:481–87
61. Gompel N, Carroll SB. 2003. Genetic mechanisms and constraints governing the evolution of correlated traits in drosophilid flies. *Nature* 424:931–35
62. Göpfert MC, Robert D. 2002. The mechanical basis of *Drosophila* audition. *J. Exp. Biol.* 205:1199–208
63. Grossfield J. 1966. The influence of light on the mating behavior of *Drosophila*. *Univ. Tex. Publ. Stud. Genet.* 3(6615):147–76
64. Grossfield J. 1968. The relative importance of wing utilization in light dependent courtship in *Drosophila*. *Univ. Tex. Publ. Stud. Genet.* 4 (6818):147–56
65. Grossfield J. 1971. Geographic distribution and light dependent behavior in *Drosophila*. *Proc. Natl. Acad. Sci. USA* 68:2669–73
66. Hedlund K, Bartelt RJ, Dicke M, Vet LEM. 1996. Aggregation pheromones of *Drosophila immigrans*, *D. phalerata*, and *D. subobscura*. *J. Chem. Ecol.* 22:1835–44
67. Heed WB. 1968. Ecology of the Hawaiian Drosophilidae. *Univ. Tex. Publ. Stud. Genet.* 4(6818):387–419
68. Heed WB. 1982. The origin of *Drosophila* in the Sonoran Desert. In *Ecological Genetics and Evolution: The Cactus-Yeast-Drosophila Model*, ed. WT Starmer, JSF Barker, pp. 65–80. New York: Academic
69. Hersh BM, Carroll SB. 2005. Direct regulation of knot gene expression by Ultrabithorax and the evolution of *cis*-regulatory elements in *Drosophila*. *Development* 132:1567–77
70. Hoikkala A. 2005. Inheritance of male sound characteristics in *Drosophila species*. In *Insect Sounds and Communication; Physiology, Ecology and Evolution*, ed. S Drosopoulos, M Claridge. Boca Raton, FL:CRC. In Press
71. Hoikkala A, Crossley SA. 2000. Copulatory courtship in *Drosophila*: behaviour and songs of *D. birchii* and *D. serrata*. *J. Insect Behav.* 13:71–86
72. Hoikkala A, Paallysaho S, Aspi J, Lumme J. 2000. Localization of genes affecting species differences in male courtship song between *Drosophila virilis* and *D. littoralis*. *Genet. Res.* 75:37–45
73. Jaenike J. 1987. Genetics of ovisposition site preference in *Drosophila tripunctata*. *Heredity* 59:363–69

74. Jallon JM, David JR. 1987. Variations in cuticular hydrocarbons along the eight species of the *Drosophila melanogaster* subgroup. *Evolution* 41:487–502
75. Jones CD. 1998. The genetic basis of *Drosophila sechellia's* resistance to a host plant toxin. *Genetics* 149:1899–908
76. Jones CD. 2001. The genetic basis of larval resistance to a host plant toxin in *Drosophila sechellia*. *Genet. Res.* 78:225–33
77. Kambysellis MP, Ho KF, Craddock EM, Piano F, Parisi M, Cohen J. 1995. Pattern of ecological shifts in the diversification of Hawaiian *Drosophila* inferred from a molecular phylogeny. *Curr. Biol.* 5:1129–39
78. Kaufman TC, Severson DW, Robinson GE. 2002. The *Anopheles* genome and comparative insect genomics. *Science* 298:97–115
79. Kern A, Jones CD, Begun DJ. 2004. Molecular population genetics of male accessory gland proteins in the *Drosophila simulans* complex. *Genetics* 167:725–35
80. Knowles LL, Markow TA. 2001. Sexually antagonistic coevolution of a postmating-prezygotic reproductive character in desert *Drosophila*. *Proc. Natl. Acad. Sci. USA* 98:8692–96
81. Kopp A, Graze RM, Xu S, Carroll SB, Nuzhdin SV. 2003. Quantitative trait loci responsible for variation in sexually dimorphic traits in *Drosophila melanogaster*. *Genetics* 163:771–83
82. Kubli E. 2003. Sex-peptides: seminal peptides of the *Drosophila* male. *Cell. Mol. Life Sci.* 60:1689–704
83. Kulkarni SJ, Hall JC. 1987. Behavioral and cytogenetic analysis of the cacophony courtship song mutant and interacting genetic variants in *Drosophila melanogaster*. *Genetics* 116:461–75
84. Kulkarni SJ, Steinlauf AF, Hall JC. 1988. The *dissonance* mutant of *Drosophila melanogaster*: isolation, behavior and cytogenetics. *Genetics* 118:267–85
85. Kyriacou CP, Hall JC. 1989. Spectral analysis of *Drosophila* courtship song rhythms. *Anim. Behav.* 37:850–59
86. Lee HG, Seong CS, Kim YC, Davis RL, Han KA. 2003. Octopamine receptor OAMB is required for ovulation in *Drosophila melanogaster*. *Dev. Biol.* 264:179–90
87. Legal L, David JR, Jallon JM. 1992. Toxicity and attraction effects produced by *Morinda citrifolia* fruits on the *Drosophila melanogaster* complex of species. *Chemoecology* 3:125–29
88. Lofdahl KL. 1986. A genetic analysis of habitat selection in the cactophilic species, *Drosophila mojavensis*. In *The Evolutionary Genetics of Invertebrate Behavior*, ed. M Huettel, pp. 153–62. New York: Plenum
89. Manning A. 1962. A sperm factor controlling female receptivity in *Drosophila melanogaster*. *Nature* 194:253–54
90. Marcillac F, Bousquet F, Alabouvette J, Savarit F, Ferveur JF. 2005. Genetic and molecular characterization of a mutation that largely affects the production of sex pheromones in *Drosophila melanogaster*. *Genetics*. In press. doi: 10.1534/genetics
91. Markow TA. 1982. Mating systems of cactophilic *Drosophila*. In *Ecological Genetics and Evolution: The Cactus-Yeast-Drosophila Model*, ed. WT Starmer, JSF Barker, pp. 273–87. New York: Academic
92. Markow TA. 1987. Behavioral and sensory basis of courtship success in *Drosophila*. *Proc. Natl. Acad. Sci. USA* 84:6200–5
93. Markow TA. 1988. Reproductive behavior of *Drosophila* in the laboratory and in the field. *J. Comp. Psychol.* 102:169–74
94. Markow TA. 1996. Evolution of *Drosophila* mating systems. *Evol. Biol.* 29:73–106

95. Markow TA. 1997. Assortative fertilization in *Drosophila*. *Proc. Natl Acad. Sci. USA* 94:7756–60
96. Markow TA. 2002. Female remating, operational sex ratio, and the arena of sexual selection in *Drosophila*. *Evolution* 59:1725–34
97. Markow TA. Ankney PF. 1984. *Drosophila* males contribute to oogenesis in a multiple mating species. *Science* 224:302–3
98. Markow TA. Ankney PF. 1988. Insemination reaction in *Drosophila*: a copulatory plug in species showing male contribution to offspring. *Evolution* 42:1097–100
99. Markow TA, Toolson EC. 1990. Temperature effects on epicuticular hydrocarbons and sexual isolation in *Drosophila mojavensis*. See Ref. 4a, pp. 315–31
100. Markow TA, Coppola A, Watts TD. 2001. How *Drosophila* males make eggs: It is elemental. *Proc. R. Soc. Biol. Sci.* 268:1527–32
101. Matthews KA, Kaufman TC, Gelbart WM. 2005. Research resources for *Drosophila*: the expanding universe. *Nat. Rev.* 6:179–93
102. Meunier N, Ferveur J, Marion-Poll F. 2001. Sex-specific non-pheromonal taste receptors in Drosophila. *Curr. Biol.* 10:1583–86
103. Moats RA, Bartelt RJ, Jackson LL, Schaner A. 1987. Ester and ketone components of aggregation pheromone of *Drosophila hydei* (Diptera: Drosophilidae). *J. Chem. Ecol.* 13:451–62
104. Montgomery SL. 1975. Comparative breeding site ecology and the adaptive radiation of picture-winged *Drosophila* (Diptera: Drosophilidae) in Hawaii. *Proc. Hawaii. Entomol. Soc.* 22:65–103
105. Nakayama S, Aigaki T. 2001. A novel cyclophilin-like gene required for ovulation/oviposition in *Drosophila*. *Ann. Drosoph. Res. Conf.* 42:662B
106. Nayak SV, Singh RN. 1983. Sensilla on the tarsal segments and mouthparts of adult *Drosophila melanogaster* Meigen (Diptera: Drosophilidae). *Int. J. Insect Morphol. Embryol.* 12:273–91
107. Noor MAF, Aquadro CF. 1998. Courtship songs of *Drosophila pseudoobscura* and *D. persimilis*: analysis of variation. *Anim. Behav.* 56:115–25
108. O'Grady PM. Reevaluation of phylogeny in the *Drosophila obscura* species group. *Mol. Phylogenet. Evol.* 12(2):124–39
109. Paillette M, Ikeda H, Jallon JM. 1991. A new acoustic message in *Drosophila*: the rejection signal (R.S.) of *Drosophila melanogaster* and *Drosophila simulans*. *Bioacoustics* 3:247–52
110. Patterson JT. 1947. The insemination reaction and its bearing on the problem of speciation in the mulleri subgroup. *Univ. Tex. Publ. Stud. Genet.* (4720):41–77
111. Patterson JT. Stone WS. 1952. *Evolution in the Genus Drosophila*. New York: MacMillan. 610 pp.
112. Peixoto AA, Hall JC. 1998. Analysis of temperature-sensitive mutants reveals new genes involved in the courtship song of *Drosophila*. *Genetics* 148:827–38
113. Pitnick S. 1993. Operational sex ratios and sperm limitation in populations of *Drosophila pachea*. *Behav. Ecol. Sociobiol.* 33:383–91
114. Pitnick S, Markow TA. 1994. Male gametic strategies: sperm production and the allocation of ejaculate among successive mates by the sperm-limited fly, *Drosophila pachea* and its relatives. *Am. Nat.* 143:785–819
115. Pitnick S, Spicer G, Markow TA. 1997. A phylogenetic analysis of male ejaculatory donations in *Drosophila*. *Evolution* 51:833–45
116. Pitnick S, Markow TA, Spicer G. 1999. Evolution of sperm storage organs in *Drosophila*. *Evolution* 53:1804–22

117. Polak M, Markow TA. 1995. Effect of ectoparasitic mites, *Macrocheles subbadius* (Acarina: Macrochelidae), on sexual selection in a Sonoran Desert fruit fly, *Drosophila nigrospiracula* (Diptera: Drosophilidae). *Evolution* 49:660–69
118. Powell JR. 1997. *Progress and Prospects in Evolutionary Biology: The Drosophila Model.* New York: Oxford Univ. Press. 562 pp.
119. Remsen J, O'Grady PM. 2002. Phylogeny of Drosophilidae (Diptera), with comments on combined analysis and character support. *Mol. Phylogenet. Evol.* 24:248–63
120. Ritchie MG, Gleason JM. 1995. Rapid evolution of courtship song pattern in *Drosophila willistoni* sibling species. *J. Evol. Biol.* 8:463–79
121. Ritchie MG, Kyriacou CP. 1996. Artificial selection for a courtship signal in *Drosophila melanogaster*. *Anim. Behav.* 52:603–11
122. R'Kha S, Capy P, David JR. 1991. Host-plant specialization in the *Drosophila melanogaster* species complex: a physiological, behavioral, and genetical analysis. *Proc. Natl Acad. Sci. USA* 88:1835–39
123. Romer F. 1991. The oenocytes of insects: differentiation, changes during molting, and their possible involvement in the secretion of the molting hormone. In *Recent Advances in Comparative Arthropod Morphology, Physiology and Development*, ed. AP Gupta, pp. 542–66. New Brunswick, NJ: Rutgers Univ. Press
124. Satokangas P, Liimatainen JO, Hoikkala A. 1994. Songs produced by the females of the *Drosophila virilis* group of species. *Behav. Genet.* 24:263–72
125. Savarit F, Ferveur JF. 2002. Genetic study of the production of sexually dimorphic cuticular hydrocarbons in relation with the sex-determination gene transformer in *Drosophila melanogaster*. *Genet. Res.* 79:23–40
126. Savarit F, Sureau G, Cobb M, Ferveur JF. 1999. Genetic elimination of known pheromones reveals the fundamental chemical bases of mating and isolation in *Drosophila*. *Proc. Natl. Acad. Sci. USA* 96:9015–20
127. Schaner A, Bartelt RJ, Jackson LL. 1987. (Z)-11-Octadenyl acetate, an aggregation pheromone in *Drosophila simulans*. *J. Chem. Ecol.* 13:1777–86
128. Schawaroch V. 2002. Phylogeny of a paradigm lineage: the *Drosophila melanogaster* species group (Diptera: Drosophilidae). *Biol. J. Linn. Soc.* 76:21–37
129. Schmidt, T, Choffat Y, Schneider M, Hunziker P, Fuyama Y, Kubli E. 1993. *Drosophila suzukii* contains a peptide homologous to the *Drosophila melanogaster* sex-peptide and functional in both species. *Insect Biochem. Mol. Biol.* 23:571–79
130. Scott K, Brady R, Cravchik A, Morozov P, Rzhetsky A, et al. 2001. A chemosensory gene family encoding candidate gustatory and olfactory receptors in *Drosophila*. *Cell* 104:661–73
131. Sheeba V, Chandrashekaran MK, Joshi A, Sharma VK. 2001. Persistence of oviposition rhythm in individuals of *Drosophila melanogaster* reared in an aperiodic environment for several hundred generations. *J. Exp. Zool.* 290:541–49
132. Singh SR, Singh BN, Hoenigsberg HF. 2002. Female remating, sperm competition and sexual selection in *Drosophila*. *Genet. Mol. Res.* 1:178–215
133. Snook RR, Markow TA. 2001. Mating system evolution in sperm hetermorphic *Drosophila*. *J. Insect Physiol.* 4:957–64
134. Snook RR, Robertson, A, Crudgington HS, Ritchie MG. 2005. Experimental manipulation of sexual selection and the evolution of courtship song in *Drosophila pseudoobscura*. *Behav. Genet.* 35:245–55
135. Spicer GS. 1992. Reevaluation of the phylogeny of the *Drosophila virilis* species group (Diptera: Drosophilidae). *Ann. Entomol. Soc. Am.* 85(1):11–25

136. Spicer G, Jaenike J. 1996. Phylogenetic analysis of breeding site use and alpha amanatin tolerance within the *Drosophila quinaria* species group. *Evolution* 50:2328–37
137. Spieth HT. 1952. Mating behavior within the genus *Drosophila*. *Bull. Am. Mus. Nat. Hist.* 99:395–474
138. Spieth HT. 1966. Courtship behavior of endemic Hawaiian *Drosophila*. *Univ. Tex. Publ. Stud. Genet.* 3(6615):245–313
139. Spieth HT. 1978. Courtship patterns and evolution of the Adiastola and Planitibia species groups. *Evolution* 32:435–32
140. Steele RH. 1986. Courtship feeding in *D. subobscura* I. The nutritional significance of courtship feeding. *Anim. Behav.* 34:1087–98
141. Stnesmyr MC, Dekker M, Hansson BS. 2003. Evolution of the olfactory code in the *Drosophila melanogaster* subgroup. *Proc. R. Soc. B* 270:2333–40
142. Stensmyr MC, Giordano E, Balloi, A, Angioy AM, Hansson BS. 2003. Novel natural ligands for *Drosophila olfactory* receptor neurones. *J. Exp. Biol.* 206:715–24
143. Stocker RF. 1994. The organization of the chemosensory system in *Drosophila melanogaster*: a review. *Cell Tissue Res.* 275:3–26
144. Sturtevant AH, Dobzhansky TH. 1936. Inversions in the third chromosome of wild races of *Drosophila pseudoobscura* and their use in the study of the history of the species. *Proc. Natl. Acad. Sci. USA* 22:448–50
145. Suvanto L, Hoikkala A, Liimatainen J. 1994. Secondary courtships songs and inhibitory songs of *Drosophila virilis* group males. *Behav. Genet.* 24:85–94
146. Takahashi A, Tsaur SC, Coyne JA, Wu CI. 2001. The nucleotide changes governing cuticular hydrocarbon variation and their evolution in *Drosophila melanogaster*. *Proc. Natl. Acad. Sci. USA* 98:3920–25
147. Tatarenkov A, Ayala FJ. 2001. Phylogenetic relationships among species groups of the virilis-repleta radiation of *Drosophila*. *Mol. Phylogenet. Evol.* 21:327–31
148. Tauber E, Eberl DF. 2001. Song production in auditory mutants of *Drosophila:* the role of sensory feedback. *J. Comp. Physiol. A* 187:341–48
149. Tauber E, Eberl DF. 2003. Acoustic communication in *Drosophila*. *Behav. Process.* 64:197–210
150. Tomaru M, Oguma Y. 1994. Differences in courtship song in the species of the *Drosophila auraria* complex. *Anim. Behav.* 47:133–40
151. Tomaru M, Oguma O. 1994. Genetic basis and evolution of species-specific courtship song in the *Drosophila auraria* complex. *Genet. Res. Camb.* 63:11–17
152. Tompkins L, McRobert SP. 1985. The effect of transformer, doublesex and intersex mutations on the sexual behavior of *Drosophila melanogaster*. *Genetics* 111:89–96
153. Tompkins L, McRobert SP. 1989. Regulation of behavioral and pheromonal aspects of sex determination in *Drosophila melanogaster* by the Sex-lethal gene. *Genetics* 123:535–41
154. Tompkins L, Gross AC, Hall JC, Gailey DA, Siegel RW. 1982. The role of female movement in the sexual behavior of *Drosophila melanogaster*. *Behav. Genet.* 12:295–307
155. Tompkins L, McRobert SP, Kaneshiro KY. 1993. Chemical communication in Hawaiian *Drosophila*. *Evolution* 45:1407–19
156. Throckmorton LH. 1975. The phylogeny, ecology and geography of *Drosophila*. In *Handbook of Genetics 3: Invertebrates of Genetic Interest*, ed. RC King, pp. 421–69. New York: Plenum
157. True JR, Edwards KA, Yamamoto D, Carroll SB. 1999. *Drosophila* wing melanin patterns form by vein-dependent elaboration of enzymatic prepatterns. *Curr. Biol.* 9:1382–91
158. von Schilcher F. 1976. The function of pulse song and sine song in the courtship of *Drosophila melanogaster*. *Anim. Behav.* 24:622–25

159. von Schilcher F. 1977. A mutation which changes courtship song in *Drosophila melanogaster*. *Behav. Genet.* 7:251–59
160. Vosshall LB. 2001. The molecular logic of olfaction in *Drosophila*. *Chem. Senses* 26:207–15
161. Vosshall LB, Stensmyr MC. 2005. Wake up and smell the pheromones. *Neuron* 45:179–81
162. Viruostso M, Isoherranen E, Hoikkala A. 1996. Female wing spreading as an acceptance signal in the *Drosophila virilis* group of species. *J. Insect Behav.* 9:505–16
163. Ward BL, Heed WB. 1970. Chromosome phylogeny of *Drosophila pachea* and related species. *J. Hered.* 61:248–58
164. Wheeler MR. 1947. The insemination reaction in intraspecific pairings in *Drosophila*. *Univ. Tex. Publ. Stud. Genet.* 4720:78–115
165. Wheeler DA, Kulkarni SJ, Gailey DA, Hall JC. 1989. Spectral analysis of courtship songs in behavioral mutants of *Drosophila melanogaster*. *Behav. Genet.* 19:503–28
166. Williams MA, Blouin AG, Noor MAF. 2001. Courtship songs of *Drosophila pseudoobscura* and *D. persimilis*. *Heredity* 86:68–77
167. Wolfner MF. 1997. Tokens of love: functions and regulation of *Drosophila* male accessory gland products. *Insect Biochem. Mol. Biol.* 27:179–92
168. Yamada H, Sakai T, Tomaru M, Doi M, Matsuda M, Oguma Y. 2002. Search for species-specific mating signal in courtship songs of sympatric sibling species, *Drosophila ananassae* and *D. pallidosa*. *Genes Genet. Syst.* 77:97–106
169. Yokokura T, Ueda R, Yamamoto D. 1995. Phenotypic and molecular characterization of croaker, a new mating behavior mutant of *Drosophila melanogaster*. *Jpn. J. Genet.* 70:103–17

Sex Determination in the Teleost Medaka, *Oryzias latipes*

Masaru Matsuda

PRESTO, Japan Science and Technology Corporation (JST), Laboratory of Reproductive Biology, National Institute for Basic Biology, Okazaki 444-8585, Japan; email: matsuda@nibb.ac.jp

Annu. Rev. Genet.
2005. 39:293–307

The *Annual Review of Genetics* is online at http://genet.annualreviews.org

doi: 10.1146/annurev.genet.39.110304.095800

0066-4197/05/1215-0293$20.00

Key Words

sexual differentiation of gonads, sex-determining gene, medaka, *DMY DMRT1*, *Oryzias* species

Abstract

Although the sex of most animals is determined by genetic information, sex-determining genes had been identified only in mammals, several flies, and the worm *Caenorhabditis elegans* until the recent discovery of *DMY* (*DM*-domain gene on the *Y* chromosome) in the sex-determining region on the Y chromosome of the teleost fish medaka, *Oryzias latipes*. Functional and expression analyses of *DMY* have shown it to be the master gene for male sex determination in the medaka. The only sex-determining genes found so far in vertebrates are *Sry* and *DMY*. Therefore, the medaka is expected to become a good experimental animal for investigating the precise mechanisms involved in primary sex determination in nonmammalian vertebrates. This article reviews the origin of *DMY* and the sexual development of gonads in the medaka. The putative functions of *DMY* are also discussed.

Contents

INTRODUCTION

DMY: DM-domain gene on the Y chromosome

The sex of an individual is established by the sex of the gonad, and in most cases, whether a gonad becomes a testis or an ovary is determined by the genome of that individual. In most mammals, several flies, and the worm *Caenorhabditis elegans*, the gene at the top of the sex-determination cascade is known. In mammals, the sex-determining gene *SRY* is located on the Y chromosome: *SRY* was first identified from a deletion analysis of the human Y chromosome (63). Furthermore, *Sry*, the mouse homologue of *SRY*, was shown to be sufficient for male development in transgenic mice (28). Nonmammalian vertebrates also have a male heterogametic (XX-XY) sex-determination system, but no homolog of *Sry* could be found. Recently, we have identified the gene at the top of the sex-determination cascade in the teleost medaka, *Oryzias latipes*. This gene, named *DMY*, is the first sex-determining gene to be found among nonmammalian vertebrates.

In nonmammalian vertebrates, sex-determining systems are quite diverse. Sex is determined by heredity, environment, or both, and the pathway of sex determination can be manipulated by administering exogenous sex steroids during gonadal sex differentiation. This suggests that sex steroids play a critical role in recruiting the undifferentiated gonad to develop either as a testis or an ovary in nonmammalian vertebrates. In fish species, all the various sex-determining systems that exist in other vertebrate classes have been observed, including male heterogametic, female heterogametic, and temperature-dependent systems. Furthermore, fish constitute the only vertebrate class that shows natural hermaphroditism, either simultaneous or sequential, in a significant number of species. Although much is known about the process of sex differentiation in fish, the precise mechanisms involved in primary sex determination remain undefined (7). Sex-determining systems in fish appear to be at a primitive stage of evolution (65). Therefore, knowledge of the relationships between sex-determining genes and sex steroids should help us to understand animal sex determination and sex differentiation in general, and in particular to elucidate the conserved mechanisms that operate behind sex determination in vertebrates.

MEDAKA AS A MODEL ORGANISM

The characteristics of medaka as a model organism have been reviewed by Ishikawa (15) and Wittbrodt et al. (78). Briefly, this species, a small, egg-laying freshwater fish native to Asia, has been established as a model experimental animal in Japan. In addition to its small size and short generation time like zebrafish, medaka has three advantages: a large interstrain diversity among inbred strains; many related species, which can be mated with medaka to provide F_1 progeny; and a small genome size (800 Mb), half that of zebrafish and one third those of human and mouse. Furthermore, the sex of medaka is genetically determined, whereas the sex-determining systems in zebrafish and fugu, the other model fish, are unclear. Hence the medaka is a useful experimental fish for analyses of the mechanisms underlying sex determination and sex differentiation.

Wild populations of medaka have been divided into four genetically distinct groups (the Northern, Southern, East Korean, and China-West Korean populations), which have a large genomic diversity (32, 49, 71, 72). The nucleotide sequences of exons and introns of these populations differ by about 1% and 3%, respectively (43, 46). Furthermore, many inbred strains have been established from these populations; examples include the Hd-rR strain (**Figure 1*a***) from the Southern population and the HNI strain (**Figure 1*b***) from the Northern population (13, 14).

Many *Oryzias* species have been used to clone many genes. DNA sequences derived from 13 species of *Oryzias* have been submitted to public DNA databases and may be confirmed by the Taxonomy Browser of the NCBI (National Center for Biotechnology Information). Some species can be mated with each other to produce F_1 progeny. This feature has wide-ranging applications in genetic and evolutionary studies of *Oryzias* species (78). For example, sterile males can be obtained by interbreeding different *Oryzias* species (9). When *Oryzias latipes* males are mated with *O. curvinotus* (**Figure 1*c***) females, the male offspring are sterile but still do well in the mating dance. They can therefore induce the spawning of unfertilized eggs in *O. latipes* females. Using hybrid mortality differences between *O. latipes*–*O. luzonensis* diploids and triploids, Sato et al. (52) performed gene-centromere mapping of medaka sex chromosomes.

EST: expressed sequence tag

In addition to these characteristics, genome sequence and EST (expressed sequence tag) information have been expanded rapidly to take advantage of the compact genome size of medaka. Sequencing of the approximately 800-Mb medaka genome was an important target of the group grant project "Genome Science" (Grant-in-Aid for Scientific Research on Priority Areas supported by the Ministry of Education, Culture, Sports, Science and Technology of Japan) that was started at the Academia Sequencing Center of the National Institute of Genetics (NIG) in mid-2002. The Hd-rR strain chosen for this purpose is an inbred strain that belongs to the Southern population, and sequencing is being conducted by the whole-genome shotgun strategy. The initial plan was to assemble six- to eightfold coverage of 2-Kb

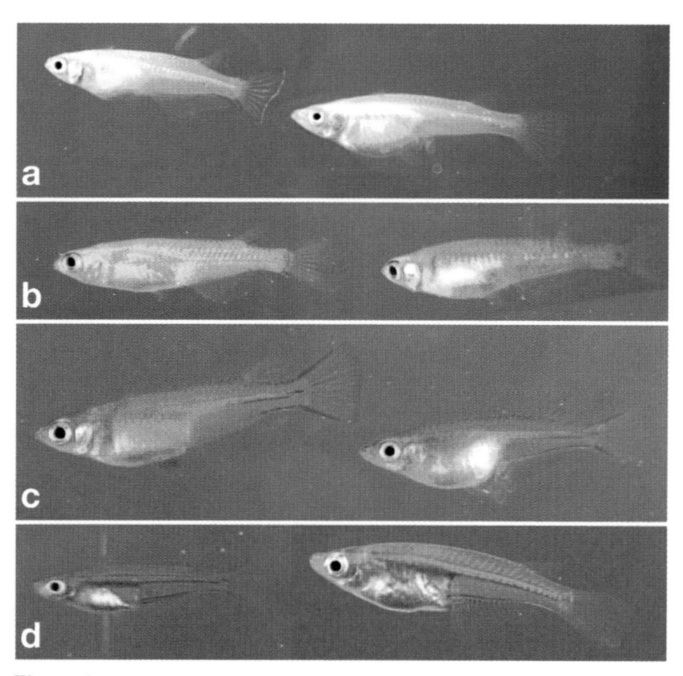

Figure 1

Photographs of medaka and related species. (*a*) The Hd-rR strain of *Oryzias latipes*, male (*left*) and female (*right*); (*b*) the HNI strain of *O. latipes*, male (*left*) and female (*right*); (*c*) *O. curvinotus*, male (*left*) and female (*right*); and (*d*) *O. hubbsi*, male (*left*) and female (*right*).

shotgun libraries together with longer insert libraries (10 Kb, 40 Kb, etc.); this undertaking is currently in progress to produce a set of high-quality scaffolds. To carry out this project, a brand-new genome assembler was developed, along with a new genome browser. Over 3000 SNP markers are nearing completion and will be used to arrange most of the scaffolds properly on the genome. It is anticipated that an extremely high-quality draft sequence of the medaka genome will result. Information on this procedure is available at **http://dolphin.lab.nig.ac.jp/medaka/**.

SEX-DETERMINATION AND SEX-DETERMINING GENE OF MEDAKA

Sex-Determining System of Medaka

The sex-determining system of the medaka is male heterogametic (the XX-XY system), as in mammals (1). In the d-rR strain, a cross between female (X^rX^r) and heterozygous orange-red male (X^rY^R) produces white females (X^rX^r) and orange-red males (X^rY^R) in equal numbers. The homozygous (*rr*) recessive condition for the orange-red body color locus gives rise to white females, and the heterozygous (*rR*) condition with the dominant allele for the same results in orange-red males since the dominant allele (*R*) is located on the Y chromosome. Hence, the *r* and *R* can be used as markers for identification of the genotypic sex, XX and XY, respectively. Using this strain, the first sex reversals were reported in 1953 by Yamamoto (80). Treatment with steroid sex hormones during the larval stage generated YY males, XY females, XX males, and even YY females (82). However, in medaka, as in many fish species, sex chromosomes are not morphologically differentiated (75). Although recombination occurs between the X and Y chromosomes near the sex-determining region on the sex chromosomes, the recombination rates in male meiosis are lower than those in female meiosis (25, 37, 81).

Identification of the Sex-Determining Gene of Medaka

Identification of the sex-determining gene in nonmammalian vertebrates remains an enigma despite the discovery of the sex-determining gene (*SRY/Sry*) in mammals in 1990 (63). The search for homologs of *Sry* and genome comparison between males and females has not succeeded in nonmammalian vertebrates, and positional cloning was thought to be an orthodox, stable option in identifying the sex-determining gene in these animals. This method, which was successfully used to identity the sex-determining gene in humans (63), requires that the experimental animal model have two characters: genetically determined sex and a genome that can be easily mapped. Among nonmammalian vertebrates, the teleost fish medaka, *Oryzias latipes*, satisfies both requirements.

A Y congenic strain, Hd-rR.Y^{HNI}, has been generated to highlight the genetic differences on the sex chromosomes between inbred strains (32). In the Hd-rR.Y^{HNI} strain, the sex-determining region is derived from the HNI strain on the Y chromosome, with the genetic background of the Hd-rR strain. Using this strain, we isolated sex-linked DNA markers (32, 33), made genetic and cytogenetic maps (33, 52), and constructed a bacterial artificial chromosome (BAC) genomic library (34).

By using recombinant breakpoint analysis of the congenic strain, we located a unique gene in the short sex-determining region on the Y chromosome. This gene consists of six exons and encodes a putative protein of 267 amino acids containing a DM domain. The DM domain was originally described as a DNA-binding motif shared between *doublesex* (*dsx*) in *Drosophila melanogaster* and *mab-3* in *C. elegans*. This Y-specific DM-domain gene was named *DMY* (34), or *Dmrt1b(Y)* (42, 77).

We assumed that because mutations occurring to the sex-determining gene might be non-lethal and cause a simple sex reversal, they must be maintained in wild populations.

This assumption was confirmed by screening wild populations (34, 45, 61). Shinomiya et al. (61) surveyed 40 localities of wild populations and 69 wild stocks, which were collected from wild populations and subsequently maintained. From this survey they found 26 XY females from 13 localities. Ohtake et al. (45) reported that 12 fish from 6 localities have a frameshift mutation in the 3rd exon of the protein-coding region of *DMY*, and that this resulted in the subsequent truncation of the DMY protein. All offspring that inherited this mutant allele of *DMY* were female. The other mutant exhibited a very low expression of *DMY*, with a high proportion of XY females in its progeny. These results further strengthen the idea that *DMY* is required for normal male development in medaka (34).

To demonstrate that *DMY* is sufficient for normal male development, gain-of-function (overexpression) studies are necessary. Therefore, we injected one-cell-stage embryos of medaka with a genomic DNA fragment carrying *DMY*, containing about 56 kb of the coding region, about 60 kb of the upstream noncoding region, and about 1.4 kb of the downstream noncoding region. In these transgenic progeny we found that 25% of XX fish develop into males (36). Furthermore, some males were found to be fertile. These data demonstrate that *DMY* is sufficient to induce male development in XX medaka. When combined with previous data, these findings indicate that *DMY* is the sex-determining gene of medaka. *DMY* is thus the first sex-determining gene to be found in nonmammalian vertebrates.

DMY mRNA and protein are expressed specifically in the somatic cells surrounding primordial germ cells (PGCs) in the early gonadal primordium, before morphological sex differences are discernible (23). When PGCs become localized in the coelomic epithelium under the nephric duct, specific signals for *DMY* were also found to be localized in the somatic cells surrounding the PGCs at stage 36, i.e., 2–3 days before hatching. However, somatic cells surrounding PGCs never express *DMY* during the early migratory period. Expression of *DMY* persists in Sertoli cell lineage cells, from PGC-supporting cells to Sertoli cells. Therefore, *DMY* expression may be found in adult testis (23). A slight expression of *DMY* was detected in many tissues by using reverse transcriptase (RT)-PCR (44).

PGC: primordial germ cell

DMRT1: DM-related transcription factor 1

EVOLUTION OF SEX-DETERMINING GENES IN *ORYZIAS* SPECIES

Origin of *DMY*

Where could *DMY*, the sex-determining gene of medaka, have originated? *DMY* has retained the conserved DM domain. DM-related genes have been identified from virtually all species examined, and in vertebrate species, *DMRT1* (*DM-r*elated *t*ranscription factor *1*), the DM-related gene most homologous to *DMY* (about 80% in amino-acid level), correlates with male development (3, 4, 6, 8, 11, 12, 22, 31, 39, 40, 47, 48, 55–57, 64, 67, 74, 76). Two lines of evidence suggest that *DMY* arose from a recent duplication event of the autosomal *DMRT1* genomic region. First, Y chromosome-linked *DMY* appears to have originated from a duplicate copy of autosomal *DMRT1* (30, 42) and then, in the Y chromosome, the duplicated *DMRT1* acquired a new function as a sex-determining gene, *DMY*. Second, *DMY* is also found in *Oryzias curvinotus*, which is most closely related to medaka (35), but is not found in other *Oryzias* species (*O. celebensis*, *O. mekongensis*) or in other fishes (guppy, tilapia, zebrafish, and fugu) (26). These results suggest that *DMY* is a recently evolved gene specific to some species of the genus *Oryzias*.

Phylogenetic trees have been constructed by analyzing *DMRT1* and *DMY* sequences of *Oryzias* species (27, 30, 35, 50, 83). A phylogenetic relationship of four species related to medaka and a summary of gene trees constructed from *DMY* and *DMRT1* are shown in **Figure 2*a*** and **Figure 2*b***, respectively. This gene topology demonstrates that *DMY* has

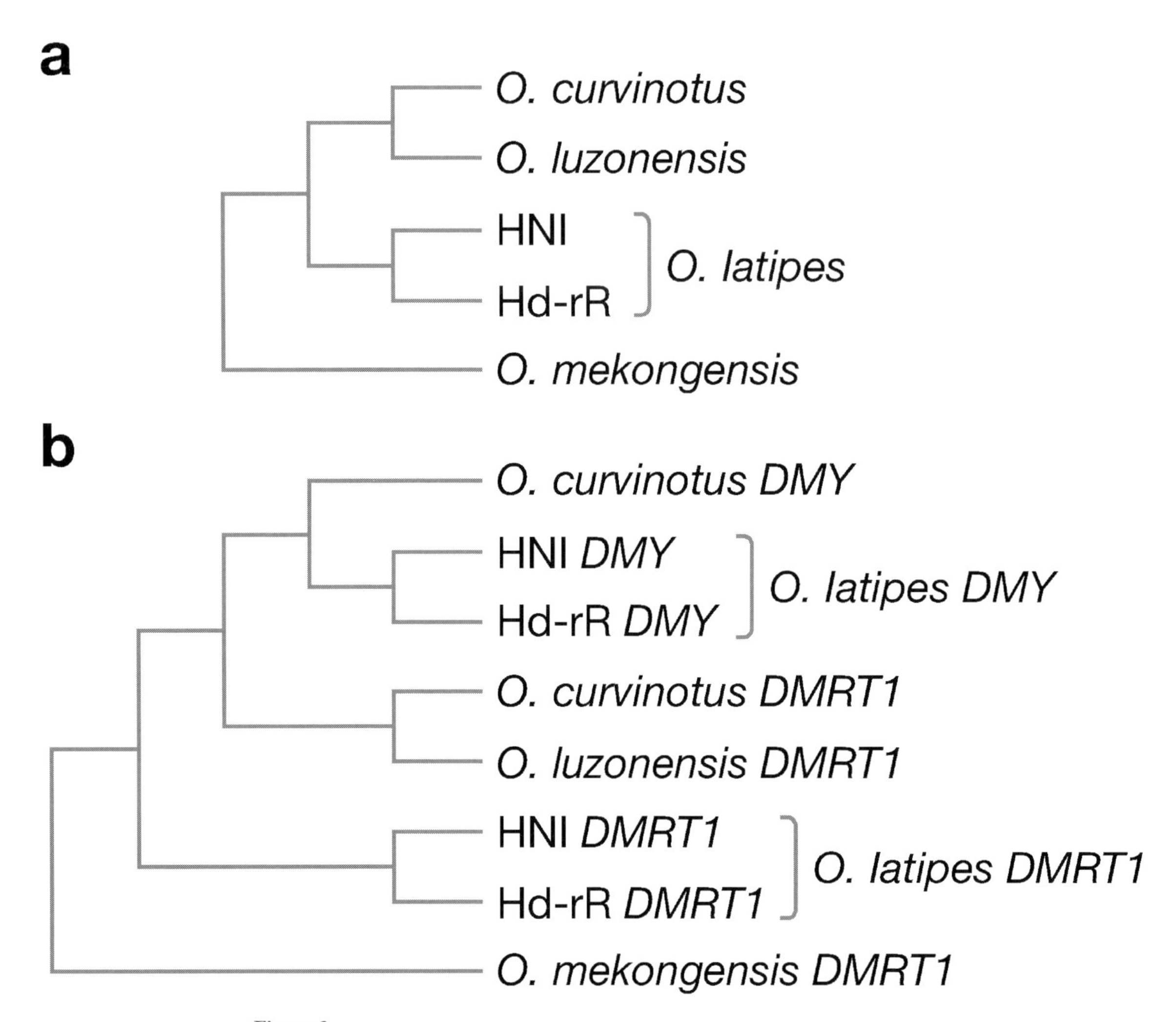

Figure 2

Phylogenetic trees. (*a*) The phylogenetic relationship between four *Oryzias* species. (*b*) A summary of gene trees constructed from *DMY* and *DMRT1* of four *Oryzias* species.

diverged from *DMRT1* of a common ancestor of *O. curvinotus* and *O. luzonensis* after separation from *O. latipes*. One attractive hypothesis is that *DMY* originated in the common ancestor of *O. curvinotus/O. luzonensis* by duplication of *DMRT1*, and was horizontally transferred from the common ancestor of *O. curvinotus/O. luzonensis* to an ancestor of *O. latipes* (50). After speciation of *O. curvinotus* and *O. luzonensis*, *O. luzonensis* might have lost *DMY* and utilized a different sex-determining gene (27). *Oryzias* fishes except *O. latipes* and *O. curvinotus* appear to have a different sex-determining gene from *DMY* (50). The possibility of horizontal transfer between *O. latipes* and *O. curvinotus* has been also suggested by analysis of the *Tol2* transposable element (24).

Sex-Determining Systems of *Oryzias* Species

As noted above, the sex-determining gene of *O. latipes* is *DMY*, located on the Y chromosome. In medaka linkage groups (LGs), the sex chromosome corresponds to LG 1. The closely related *O. curvinotus* also has *DMY*, and some genes located on medaka LG 1 also link to *DMY* in the *O. curvinotus* genome (27, 35), suggesting that the sex chromosome of *O. curvinotus* is coincident with that of *O. latipes*.

To elucidate the sex-determination system prevalent in the *Oryzias* species, mating experiments have used sex-reversed males and females (10, 21, 70, 73). Within the androgen-treated males of species that have the male heterogametic (XX-XY) sex-determining system, the sex-reversed males (XX males) will produce all female (all XX) progeny. On the other hand, within the estrogen-treated females of species that have the female heterogametic (ZZ-ZW) sex-determining system, the sex-reversed females (ZZ females) will produce all male (all ZZ) progeny. By this strategy, it has been suggested that the sex-determining system of *O. luzonensis* and *O. mekongensis* is the male heterogametic (XX-XY) system, and that of *O. hubbsi* (**Figure 1*d***) is the female heterogametic (ZZ-ZW) system (**Table 1**).

Another strategy for investigating the sex-determination system is to use the medaka EST data that are available. Since protein-coding regions are more conservative than the noncoding region, most of the primers of medaka EST would work well in other species. Searches for the EST markers, which are sex-linked, have been successful in three *Oryzias* species (21, 70, 73). These results show not only their sex-determination system, but also the syntenic linkage groups of medaka (**Table 1**). These results demonstrate that the sex chromosomes of each species are different from each other, suggesting that these species may have different sex-determining genes.

FUNCTIONS OF *DMY*

Morphological Development of Medaka Gonads

During the normal development of medaka, fertilized eggs develop to the hatching stage (stage 39) within 10 days at 26°C (16). The first morphological sex difference of gonads appears in the number of gonial-type germ cells at stage 38, one or two days before hatching (23) (**Figure 3**). From this stage, the activity of germ cell division in XX embryos becomes higher than that of XY embryos, and then male germ cells arrest in mitosis (23, 53). In males, the first male-specific structure of somatic cells is the acinous structure, which is the precursor of the testicular seminiferous tubules and can be distinguished at 10 days after hatching (19). In females, ovarian follicles are the first female-specific structure and become evident around the diplotene oocytes about 20 days after hatching (19). After these structures have developed, efferent ducts in the testes and ovarian cavities in ovaries become apparent.

Gene Expression During Normal Sexual Development of Medaka Gonads

In addition to these morphological data, some information on gene expression is also available in the medaka (**Figure 3**). Some female-specific genes in the germ cell lineage, such

Table 1 **Sex-determination systems in *Oryzias* species**

Scientific name	Sex-determination system	Linkage group[a]	Sex-determining gene
Oryzias latipes	XX-XY (1)	1 (34)	*DMY* (34)
Oryzias curvinotus	XX-XY (35)	1 (35)	*DMY* (35)
Oryzias luzonensis	XX-XY (10)	12 (73)	
Oryzias mekongensis	XX-XY (10)	2 (21)	
Oryzias hubbsi	ZZ-ZW (70)	5 (70)	

[a]Linkage group corresponds to the sex chromosomes of each species.

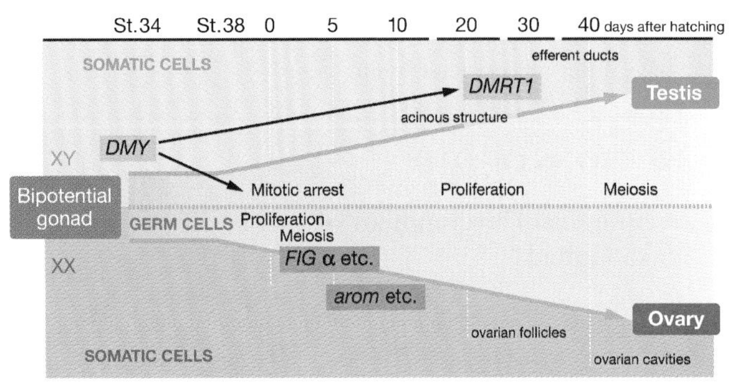

Figure 3

Conceptual illustration of normal sexual development of gonads in the medaka. Morphological events in germ cells are shown in the inner area of gray arrows, whereas those in somatic cells are shown in the outer area of gray arrows. Sex-specific genes expressed in the XY gonad and in the XX gonad are shown in green and pink boxes with black characters, respectively. Black arrows represent putative functions of DMY.

as *FIGα* (*f*actor *i*n the *g*erm line *α*) and the zona pellucida domain genes, are related to early oogenesis (18). Three sex-specific genes are expressed in somatic cells during sexual development of gonads in medaka: *aromatase* in females, and *DMY* and *DMRT1* in males (**Figure 3**). During normal development of XX females, *FIGα* shows sex-specific expression at one day after hatching (18), whereas in somatic cells of female gonads, *aromatase* mRNAs have been detected in XX fry at 4–10 days after hatching by in situ hybridization (68).

DMY appears to be closely related to *DMRT1* not only in nucleotide sequence (93% identity) but also in function. Because of the sequence similarity, expression of *DMY* and *DMRT1* cannot be distinguished by in situ hybridization, but can be distinguished by RT-PCR using specific primers. RT-PCR analyses have shown that expression of *DMY* in the testis is continuous from embryonic to adult stage (23, 42). At the early developmental stage (before onset of morphological development of the testis), *DMY* is expressed exclusively in XY embryos (because it exists only on the Y chromosome). By RT-PCR, mRNAs of autosomal *DMRT1* are detected as faint bands, and no difference in expression is observed between males and females. On the other hand, when testicular development begins to accelerate around 20 days after hatching, *DMRT1* expression increases only in XY males and reaches the same level as *DMY* expression (23).

Sexual Development in Sex-Reversed Medaka Gonads

In androgen-induced XX males, which have no *DMY*, *FIGα* is not expressed at the juvenile stage (54). In the adult testis of sex-reversed XX males, *DMRT1* signals are observed in the same cells as *DMY/DMRT1* signals of normal XY males by in situ hybridization (23). Therefore, androgens may up-regulate *DMRT1*, and *DMRT1* may subsequently promote testicular development. On the other hand, when estrogens are used for inducing XY sex-reversed females, *FIGα* expression is detected (42, 54). *DMY* expression is normally activated in XY bipotential gonads and is maintained in the XY ovary (42). Male development of XX medaka has also been found in closed colony breeding stocks and their highly inbred lines (41), and in the field survey (61). These results suggest that *DMY* is not necessary for gonadal differentiation, but *DMRT1* is required for testicular differentiation. Expressions of *FIGα* and *DMRT1* are correlated with the phenotypic differentiation of the gonads.

Genetic analyses of sex reversal have suggested the presence of autosomal modifiers for sexual differentiation of gonads (20, 41, 51, 59, 60). These modifiers led some XX individuals to male development. Fifteen *DMY* negative males from eight localities were found from the field survey (61). For example, one of these eight lines was genetically analyzed carefully and one locus was found to be related to sex reversal. The fish homozygous for a recessive mutant allele of this locus developed into males in the absence of *DMY* (59). In this line, the sex-determining system is female heterogametic (ZZ-ZW) and the locus might have a new sex-determining allele.

PUTATIVE FUNCTIONS OF *DMY*

We can attribute two functions to DMY. The first is its involvement in germ cell proliferation. *DMY* expression starts in the somatic cells surrounding the germ cells, which are found in the coelomic epithelium under the nephric duct at stage 34. This is before the occurrence of any sex differences in the germ cell number (after stage 38). When DMY was unable to work (*DMY* mutant), germ cells in XY embryos started to proliferate and then entered into meiosis just like XX embryos (34). The involvement of estrogen in the germ cell proliferation was reported by various studies on male germ cells of fish (2, 38, 66), chicken (79), and mammals (5, 29). In medaka, although aromatase expression is not found in the germ cell proliferation stage of the XX embryo (68), it has been shown that the fertilized egg contains estrogen (17). If this endogenous estrogen acts as a natural inducer of germ cell proliferation in the XX embryo, *DMY* might be inhibiting this cascade between estrogen (estrogen receptors) and germ cell proliferation.

The second function of DMY is to induce development of pre-Sertoli cells (somatic cells surrounding PGC) into Sertoli cells in XY gonads. In mammals, the transient nature of *Sry* expression in the gonad suggests that it acts as a switch that determines Sertoli cell fate, and that it is not involved in the maintenance of cell identity or cell function. Therefore, *Sry* must in some way activate other genes that are involved in defining and maintaining Sertoli cell identity (69). Three lines of evidence suggest that *DMY* triggers the development of Sertoli cells. First, in medaka, *DMY* expression is continuous in the testis from the embryonic to the adult stage in the Sertoli cell lineage. Second, because cell-autonomous differentiation of pre-Sertoli cells has been observed in XY gonads in which germ cells had been destroyed by busulfan treatment, signals from PGC do not appear to be necessary for the differentiation of pre-Sertoli cells (58). Third, some XX recipient- and XY donor-chimeras of medaka have developed into males that have only XX germ cells, suggesting that XY somatic cells induced sex reversal of the XX gonads, which contain XX germ cells and XX somatic cells (62).

CONCLUSIONS

Morphological development of the gonads in all vertebrate groups appears to have been conserved through evolution. Many genes that are important in gonadal sex differentiation in mammals, such as *DMRT1*, are conserved in other vertebrates and show gonad-specific expression during the period of sexual differentiation. However, the genetic mechanisms triggering sex determination appear to be diverse in nonmammalian vertebrates. Analyses of the sex-determining locus in *O. latipes* and related species suggest that *DMY* exists in only two species, and that other species may have another sex-determining gene. Such diversity of sex-determining genes may exist in other orders of fish species also. These facts suggest that in fish species, the sex-determining genes could be diverse, like the sex-determining systems. In other words, each species may have its own unique sex-determining gene. Identification of sex-determining genes in *Oryzias* species will be useful in understanding evolutionary trends in sex-determining systems.

DMY seems to have two different functions for germ cells and somatic cells separately. Whether the germ cell proliferation pattern is female type or not, the small number of somatic cells expressing *DMY* can induce a bipotential gonad to testis. Since *DMY* is expressed in the Sertoli cell lineage, it seems likely that the essence of testis development lies in the proper development of the Sertoli cells. DMY might act as a natural inducer for developing Sertoli cells. We hope that the picture of conserved mechanisms that underline sex determination in vertebrates might become clearer with analyses of the gene cascade, starting from *DMY* expression in the Sertoli cell lineage and from the autosomal modifier genes for sexual differentiation in medaka.

SUMMARY POINTS

1. *DMY*, the sex-determining gene of medaka, is the first sex-determining gene found among the nonmammalian vertebrates.
2. *DMY* is only found in the genomes of two species.
3. *DMY* appears to arise from a genome duplication event of *DMRT1* genomic region.
4. DMY seems to have two different functions—for germ cells and somatic cells separately.

ACKNOWLEDGMENTS

The author thanks Dr. Bindhu Paul-Prasanth for critical reading of the manuscript and valuable suggestions.

LITERATURE CITED

1. Aida T. 1921. On the inheritance of color in a freshwater fish, *Aplocheilus latipes* Temminck and Schlegel, with special reference to sex-linked inheritance. *Genetics* 6:554–73
2. Amer MA, Miura T, Miura C, Yamauchi K. 2001. Involvement of sex steroid hormones in the early stages of spermatogenesis in Japanese huchen (*Hucho perryi*). *Biol. Reprod.* 65:1057–66
3. Aoyama S, Shibata K, Tokunaga S, Takase M, Matsui K, Nakamura M. 2003. Expression of Dmrt1 protein in developing and in sex-reversed gonads of amphibians. *Cytogenet. Genome Res.* 101:295–301
4. Brunner B, Hornung U, Shan Z, Nanda I, Kondo M, et al. 2001. Genomic organization and expression of the doublesex-related gene cluster in vertebrates and detection of putative regulatory regions for *dmrt1*. *Genomics* 1:8–17
5. Carreau S, Lambard S, Delalande C, Denis-Galeraud I, Bilinska B, Bourguiba S. 2003. Aromatase expression and role of estrogens in male gonad: a review. *Reprod. Biol. Endocrinol.* 1:35
6. De Grandi A, Calvari V, Bertini V, Bulfone A, Peverali G, et al. 2000. The expression pattern of a mouse *doublesex*-related gene is consistent with a role in gonadal differentiation. *Mech. Dev.* 90:323–26
7. Devlin RH, Nagahama Y. 2002. Sex determination and sex differentiation in fish: an overview of genetic, physiological and environmental influences. *Aquaculture* 208:191–364
8. Guan G, Kobayashi T, Nagahama Y. 2000. Sexually dimorphic expression of two types of DM (Doublesex/Mab-3)-domain genes in a teleost fish, the Tilapia (*Oreochromis niloticus*). *Biochem. Biophys. Res. Commun.* 272:662–66
9. Hamaguchi S, Sakaizumi M. 1992. Sexually differentiated mechanisms of sterility in interspecific hybrids between *Oryzias latipes* and *O. curvinotus*. *J. Exp. Zool.* 263:323–29

10. Hamaguchi S, Toyazaki Y, Shinomiya A, Sakaizumi M. 2004. The XX-XY sex-determination system in *Oryzias luzonensis* and *O. mekongensis* revealed by the sex ratio of the progeny of sex-reversed fish. *Zool. Sci.* 21:1015–18
11. He CL, Du JL, Wu GC, Lee YH, Sun LT, Chang CF. 2003. Differential *Dmrt1* transcripts in gonads of the protandrous black porgy, *Acanthopagrus schlegeli*. *Cytogenet. Genome Res.* 101:309–13
12. Huang X, Cheng H, Guo Y, Liu L, Gui J, Zhou R. 2002. A conserved family of doublesex-related genes from fishes. *J. Exp. Zool.* 294:63–67
13. Hyodo-Taguchi Y, Egami N. 1985. Establishment of inbred strains of the medaka, *Oryzias latipes* and the usefulness of the strains for biomedical research. *Zool. Sci.* 2:305–16
14. Hyodo-Taguchi Y, Sakaizumi M. 1993. List of inbred strains of the medaka, *Oryzias latipes*, maintained in the Division of Biology, National Institute of Radiological Sciences. *Fish Biol. J. MEDAKA* 5:29–30
15. Ishikawa Y. 2000. Medakafish as a model system for vertebrate developmental genetics. *BioEssays* 22:487–95
16. Iwamatsu T. 1994. Stages of normal development in the medaka *Oryzias latipes*. *Zool. Sci.* 11:825–39
17. Iwamatsu T, Kobayashi H, Hamaguchi S, Sagegami R, Shuo T. 2005. Estradiol-17beta content in developing eggs and induced sex reversal of the medaka (*Oryzias latipes*). *J. Exp. Zool. A* 303:161–67
18. Kanamori A. 2000. Systematic identification of genes expressed during early oogenesis in medaka. *Mol. Reprod. Dev.* 55:31–36
19. Kanamori A, Nagahama Y, Egami N. 1985. Development of the tissue architecture in the gonads of the medaka, *Oryzias latipes*. *Zool. Sci.* 2:695–706
20. Kato M, Shinomiya A, Sakaizumi M, Hamaguchi S. 2003. The locus on medaka linkage group 17 controls XY sex reversal in interspecific hybrids between *Oryzias latipes* and *O. curvinotus*. *Genet. Zool. Sci.* 20:1612–15
21. Kawaguchi A, Shinomiya A, Sakaizumi M, Hamaguchi S. 2004. The sex determining gene of *Oryzias mekongensis* is located on a chromosome homologous to an autosomal region of *O. latipes*. *Zool. Sci.* 21:1344–47
22. Kettlewell JR, Raymond CS, Zarkower D. 2000. Temperature-dependent expression of turtle *Dmrt1* prior to sexual differentiation. *Genesis* 26:174–78
23. Kobayashi T, Matsuda M, Kajiura-Kobayashi H, Suzuki A, Saito N, et al. 2004. Two DM domain genes, *DMY* and *DMRT1*, involved in testicular differentiation and development in the medaka, *Oryzias latipes*. *Dev. Dyn.* 231:518–26
24. Koga A, Shimada A, Shima A, Sakaizumi M, Tachida H, Hori H. 2000. Evidence for recent invasion of the medaka fish genome by the *Tol2* transposable element. *Genetics* 155:273–81
25. Kondo M, Nagao E, Mitani H, Shima A. 2001. Differences in recombination frequencies during female and male meioses of the sex chromosomes of the medaka, *Oryzias latipes*. *Genet. Res.* 78:23–30
26. Kondo M, Nanda I, Hornung U, Asakawa S, Shimizu N, et al. 2003. Absence of the candidate male sex-determining gene *dmrt1b*(Y) of medaka from other fish species. *Curr. Biol.* 13:416–20
27. Kondo M, Nanda I, Hornung U, Schmid M, Schartl M. 2004. Evolutionary origin of the medaka Y chromosome. *Curr. Biol.* 14:1664–69
28. Koopman P, Gubbay J, Vivian N, Goodfellow P, Lovell-Badge R. 1991. Male development of chromosomally female mice transgenic for *Sry*. *Nature* 351:117–21

29. Li H, Papadopoulos V, Vidic B, Dym M, Culty M. 1997. Regulation of rat testis gonocyte proliferation by platelet-derived growth factor and estradiol: identification of signaling mechanisms involved. *Endocrinology* 138:1289–98
30. Lutfalla G, Roest Crollius H, Brunet FG, Laudet V, Robinson-Rechavi M. 2003. Inventing a sex-specific gene: a conserved role of *DMRT1* in teleost fishes plus a recent duplication in the medaka *Oryzias latipes* resulted in *DMY*. *J. Mol. Evol.* 57(Suppl.)1:S148–53
31. Marchand O, Govoroun M, D'Cotta H, McMeel O, Lareyre J, et al. 2000. *DMRT1* expression during gonadal differentiation and spermatogenesis in the rainbow trout, *Oncorhynchus mykiss*. *Biochim. Biophys. Acta* 1493:180–87
32. Matsuda M, Kusama T, Oshiro T, Kurihara Y, Hamaguchi S, Sakaizumi M. 1997. Isolation of a sex chromosome-specific DNA sequence in the medaka, *Oryzias latipes*. *Genes Genet. Syst.* 72:263–68
33. Matsuda M, Matsuda C, Hamaguchi S, Sakaizumi M. 1998. Identification of the sex chromosomes of the medaka, *Oryzias latipes*, by fluorescence *in situ* hybridization. *Cytogenet. Cell Genet.* 82:257–62
34. **Matsuda M, Nagahama Y, Shinomiya A, Sato T, Matsuda C, et al. 2002. *DMY* is a Y-specific DM-domain gene required for male development in the medaka fish. *Nature* 417:559–63**

Identification of *DMY* (sex-determining gene in medaka).

35. Matsuda M, Sato T, Toyazaki Y, Nagahama Y, Hamaguchi S, Sakaizumi M. 2003. *Oryzias curvinotus* has *DMY*, a gene that is required for male development in the medaka, *O. latipes*. *Zool. Sci.* 20:159–61
36. Matsuda M, Shinomiya A, Kinoshita M, Kobayashi T, Hamaguchi S, et al. 2003. *DMY* induce male development in XX transgenic medaka fish. *Zool. Sci.* 20:1612–15
37. Matsuda M, Sotoyama S, Hamaguchi S, Sakaizumi M. 1999. Male-specific restriction of recombination frequency in the sex chromosomes of the medaka, *Oryzias latipes*. *Genet. Res. Cambridge* 73:225–31
38. Miura T, Miura C, Ohta T, Nader MR, Todo T, Yamauchi K. 1999. Estradiol-17beta stimulates the renewal of spermatogonial stem cells in males. *Biochem. Biophys. Res. Commun.* 264:230–34
39. Moniot B, Berta P, Scherer G, Sudbeck P, Poulat F. 2000. Male-specific expression suggests role of *DMRT1* in human sex determination. *Mech. Dev.* 91:323–25
40. Murdock C, Wibbels T. 2003. Expression of *Dmrt1* in a turtle with temperature-dependent sex determination. *Cytogenet. Genome Res.* 101:302–8
41. Nanda I, Hornung U, Kondo M, Schmid M, Schartl M. 2003. Common spontaneous sex-reversed XX males of the medaka *Oryzias latipes*. *Genetics* 163:245–51
42. Nanda I, Kondo M, Hornung U, Asakawa S, Winkler C, et al. 2002. A duplicated copy of *DMRT1* in the sex-determining region of the Y chromosome of the medaka, *Oryzias latipes*. *Proc. Natl. Acad. Sci. USA* 99:11778–83
43. Naruse K, Fukamachi S, Mitani H, Kondo M, Matsuoka T, et al. 2000. A detailed linkage map of medaka, *Oryzias latipes*: comparative genomics and genome evolution. *Genetics* 154:1773–84
44. Ohmuro-Matsuyama Y, Matsuda M, Kobayashi T, Ikeuchi T, Nagahama Y. 2003. Expression of *DMY* and *DMRT1* in various tissues of the medaka (*Oryzias latipes*). *Zool. Sci.* 20:1395–98
45. Ohtake H, Togashi K, Ohtsuki H, Shinomiya A, Matsuda M, et al. 2003. Genetic analysis of XY female from wild populations of the medaka, *Oryzias latipes*. *Zool. Sci.* 20:1612–15

46. Ohtsuka M, Makino S, Yoda K, Wada H, Naruse K, et al. 1999. Construction of a linkage map of the medaka (*Oryzias latipes*) and mapping of the *Da* mutant locus defective in dorsoventral patterning. *Genome Res.* 9:1277–87
47. Raymond CS, Kettlewell JR, Hirsch B, Bardwell VJ, Zarkower D. 1999. Expression of *Dmrt1* in the genital ridge of mouse and chicken embryos suggests a role in vertebrate sexual development. *Dev. Biol.* 215:208–20
48. Raymond CS, Murphy MW, O'Sullivan MG, Bardwell VJ, Zarkower D. 2000. *Dmrt1*, a gene related to worm and fly sexual regulators, is required for mammalian testis differentiation. *Genes Dev.* 14:2587–95
49. Sakaizumi M, Moriwaki K, Egami N. 1983. Allozymic variation and regional differentiation in wild populations of the fish *Oryzias latipes*. *Copeia* 1983:311–18
50. Sato T, Matsuda M, Nagahama Y, Hamaguchi S, Sakaizumi M. 2003. Origin of *DMY*, a sex-determining gene of the medaka, *Oryzias latipes*. *Zool. Sci.* 20:1612–15
51. Sato T, Otsuki H, Hamaguchi S, Sakaizumi M. 2004. Genetic analysis of an XY female medaka derived from Shirone, Niigata, suggests an autosomal modifier. *Zool. Sci.* 21:1344–47
52. Sato T, Yokomizo S, Matsuda M, Hamaguchi S, Sakaizumi M. 2001. Gene-centromere mapping of medaka sex chromosomes using triploid hybrids between *Oryzias latipes* and *O. luzonensis*. *Genetica* 111:71–75
53. Satoh N, Egami N. 1972. Sex differentiation of germ cells in the teleost, *Oryzias latipes*, during normal embryonic development. *J. Embryol. Exp. Morphol.* 28:385–95
54. Scholz S, Rosler S, Schaffer M, Hornung U, Schartl M, Gutzeit HO. 2003. Hormonal Induction and stability of monosex populations in the medaka (*Oryzias latipes*): expression of sex-specific marker genes. *Biol. Reprod.* 69:673–78
55. Shan Z, Nanda I, Wang Y, Schmid M, Vortkamp A, Haaf T. 2000. Sex-specific expression of an evolutionarily conserved male regulatory gene, *DMRT1*, in birds. *Cytogenet. Cell Genet.* 89:252–57
56. Shetty S, Kirby P, Zarkower D, Graves JA. 2002. *DMRT1* in a ratite bird: evidence for a role in sex determination and discovery of a putative regulatory element. *Cytogenet. Genome Res.* 99:245–51
57. Shibata K, Takase M, Nakamura M. 2002. The *Dmrt1* expression in sex-reversed gonads of amphibians. *Gen. Comp. Endocrinol.* 127:232–41
58. Shinomiya A, Hamaguchi S, Shibata N. 2001. Sexual differentiation of germ cell deficient gonads in the medaka, *Oryzias latipes*. *J. Exp. Zool.* 290:402–10
59. Shinomiya A, Matsuda M, Nagahama Y, Hamaguchi S, Sakaizumi M. 2002. A strain having a ZZ/ZW sex determination system derived from a naturally occurring recessive mutation in the medaka, *Oryzias latipes*. *Zool. Sci.* 19:1496–500
60. Shinomiya A, Ohmori T, Hamaguchi S, Sakaizumi M. 2003. Mapping of a factor that causes male sex-reversal of XX medaka. *Zool. Sci.* 20:1612–15
61. Shinomiya A, Otake H, Togashi K, Hamaguchi S, Sakaizumi M. 2004. Field survey of sex-reversals in the medaka, *Oryzias latipes*: genotypic sexing of wild populations. *Zool. Sci.* 21:613–19
62. Shinomiya A, Shibata N, Sakaizumi M, Hamaguchi S. 2002. Sex reversal of genetic females (XX) induced by the transplantation of XY somatic cells in the medaka, *Oryzias latipes*. *Int. J. Dev. Biol.* 46:711–17

Identification of *Sry* (sex-determining gene in humans).

63. **Sinclair AH, Berta P, Palmer MS, Hawkins JR, Griffiths BL, et al. 1990. A gene from the human sex-determining region encodes a protein with homology to a conserved DNA-binding motif. *Nature* 346:240–44**
64. Smith CA, McClive PJ, Western PS, Reed KJ, Sinclair AH. 1999. Conservation of a sex-determining gene. *Nature* 402:601–2
65. Solari AJ. 1994. Sex chromosomes and sex determination in fishes. In *Sex Chromosomes and Sex Determination in Vertebrates*, ed. AJ Solari, pp. 233–47. Tokyo: CRC Press
66. Song M, Gutzeit HO. 2003. Effect of 17-alpha-ethynylestradiol on germ cell proliferation in organ and primary culture of medaka (*Oryzias latipes*) testis. *Dev. Growth Differ.* 45:327–37
67. Sreenivasulu K, Ganesh S, Raman R. 2002. Evolutionarily conserved, *DMRT1*, encodes alternatively spliced transcripts and shows dimorphic expression during gonadal differentiation in the lizard, *Calotes versicolor*. *Mech. Dev.* 119(Suppl. 1):S55–64
68. Suzuki A, Tanaka M, Shibata N, Nagahama Y. 2004. Expression of aromatase mRNA and effects of aromatase inhibitor during ovarian development in the medaka, *Oryzias latipes*. *J. Exp. Zool A.* 301:266–73
69. Swain A, Lovell-Badge R. 1999. Mammalian sex determination: a molecular drama. *Genes Dev.* 13:755–67
70. Takehana Y, Hamaguchi S, Sakaizumi M. 2004. *Oryzias hubbsi* has a female-heterogametic (ZZ/ZW) sex-determining system. *Zool. Sci.* 21:1344–47
71. Takehana Y, Nagai N, Matsuda M, Tsuchiya K, Sakaizumi M. 2003. Geographic variation and diversity of the cytochrome *b* gene in Japanese wild populations of medaka, *Oryzias latipes*. *Zool. Sci.* 20:1279–91
72. Takehana Y, Uchiyama S, Matsuda M, Jeon SR, Sakaizumi M. 2004. Geographic variation and diversity of the cytochrome *b* gene in wild populations of medaka (*Oryzias latipes*) from Korea and China. *Zool. Sci.* 21:483–91
73. Tanaka K, Shinomiya A, Naruse K, Hamaguchi S, Sakaizumi M. 2003. *Oryzias luzonensis* has sex chromosomes that are not homologous to the medaka sex-chromosomes. *Zool. Sci.* 20:1612–15
74. Torres Maldonado LC, Landa Piedra A, Moreno Mendoza N, Marmolejo Valencia A, Meza Martinez A, Merchant Larios H. 2002. Expression profiles of *Dax1*, *Dmrt1*, and *Sox9* during temperature sex determination in gonads of the sea turtle *Lepidochelys olivacea*. *Gen. Comp. Endocrinol.* 129:20–26
75. Uwa H, Ojima Y. 1981. Detailed and banding karyotype analysis of the medaka, *Oryzias latipes*, in cultured cells. *Proc. Jpn. Acad.* 57*B*:39–43
76. Veith AM, Froschauer A, Korting C, Nanda I, Hanel R, et al. 2003. Cloning of the *dmrt1* gene of *Xiphophorus maculatus*: dmY/dmrt1Y is not the master sex-determining gene in the platyfish. *Gene* 317:59–66
77. Volff JN, Kondo M, Schartl M. 2003. Medaka *dmY/dmrt1Y* is not the universal primary sex-determining gene in fish. *Trends Genet.* 19:196–99
78. Wittbrodt J, Shima A, Schartl M. 2002. Medaka—a model organism from the far East. *Nat. Rev. Genet.* 3:53–64
79. Xie M, Zhang C, Zeng W, Mi Y. 2004. Effects of follicle-stimulating hormone and 17beta-estradiol on proliferation of chicken embryonic ovarian germ cells in culture. *Comp. Biochem. Physiol. A Mol. Integr. Physiol.* 139:521–26
80. Yamamoto T. 1953. Artificially induced sex-reversal in genotypic males of the medaka (*Oryzias latipes*). *J. Exp. Zool.* 123:571–94

81. Yamamoto T. 1961. Progenies of sex-reversal females mated with sex-reversal males in the medaka, *Oryzias latipes*. *J. Exp. Zool.* 146:163–79
82. Yamamoto T. 1975. Control of sex differentiation. In *Medaka (killifish) Biology and Strains*, ed. T Yamamoto, pp. 192–213. Tokyo: Keigaku Publ.
83. Zhang J. 2004. Evolution of *DMY*, a newly emergent male sex-determination gene of medaka fish. *Genetics* 166:1887–95

RELATED RESOURCES

Medaka fish home page
http://biol1.bio.nagoya-u.ac.jp:8000/
Medaka genome project home page
http://dolphin.lab.nig.ac.jp/medaka/

Orthologs, Paralogs, and Evolutionary Genomics[1]

Eugene V. Koonin

National Center for Biotechnology Information, National Library of Medicine, National Institutes of Health, Bethesda, Maryland 20894;
email: koonin@ncbi.nlm.nih.gov

Annu. Rev. Genet.
2005. 39:309–38

First published online as a Review in Advance on August 30, 2005

The *Annual Review of Genetics* is online at http://genet.annualreviews.org

doi: 10.1146/annurev.genet.39.073003.114725

0066-4197/05/1215-0309$20.00

Key Words

homolog, ortholog, paralog, pseudoortholog, pseudoparalog, xenolog

Abstract

Orthologs and paralogs are two fundamentally different types of homologous genes that evolved, respectively, by vertical descent from a single ancestral gene and by duplication. Orthology and paralogy are key concepts of evolutionary genomics. A clear distinction between orthologs and paralogs is critical for the construction of a robust evolutionary classification of genes and reliable functional annotation of newly sequenced genomes. Genome comparisons show that orthologous relationships with genes from taxonomically distant species can be established for the majority of the genes from each sequenced genome. This review examines in depth the definitions and subtypes of orthologs and paralogs, outlines the principal methodological approaches employed for identification of orthology and paralogy, and considers evolutionary and functional implications of these concepts.

Contents

INTRODUCTION

One of the most fascinating aspects of modern genomics is the radical change it brings to evolutionary biology. The availability of multiple, complete genomes of diverse life forms for comparative analysis provides a qualitatively new perspective on homologous relationships between genes. By comparing the sequences of all genes between genomes from different taxa and within each genome, it is, in principle, possible to reconstruct the evolutionary history of each gene in its entirety (within the set of sequenced genomes). This, in turn, will allow a deeper understanding of the general trends in the evolution of genomic complexity and lineage-specific adaptations. Gene histories must be presented in the form of scenarios that comprise several types of elementary events (55, 64, 84). The elementary events of gene evolution can be classified as follows, roughly in the order of relative contribution to the evolutionary process: (*i*) vertical descent (speciation) with modification; (*ii*) gene duplication, also followed by descent with modification; (*iii*) gene loss; (*iv*) horizontal gene transfer (HGT); and (*v*) fusion, fission, and other rearrangements of genes. Vertical descent and duplication might be considered the primary events of genome evolution and have been well recognized in the pregenomic era. In contrast, the crucial evolutionary importance of gene loss, HGT, and gene rearrangements was among the major, fundamental generalizations of the emerging evolutionary genomics (13, 14, 16, 50, 51, 57, 77, 78).

Along with the notion of elementary evolutionary events, all descriptions of evolution of genes, gene ensembles, and, ultimately, complete gene repertoires of organisms rest on certain key concepts of evolutionary biology, primarily, the definitions of homologs, orthologs, and paralogs. Developed long ago by evolutionists, these related concepts and terms have reemerged and have become the subject of intense debate and numerous misunderstandings with the advent of molecular evolution and, subsequently, evolutionary genomics (24, 25, 40, 46, 76, 80, 81, 97). The aversion of some biologists to ideas and terms deriving from evolutionary biology is reflected in the peculiar word "homologuephobia," albeit used in a tongue-in-cheek manner (80).

Homology, the most general definition, designates a relationship of common descent between any entities, without further specification of the evolutionary scenario. Accordingly, the entities related by homology, in particular, genes, are called homologs. The other two key terms define subcategories of homologs. Orthologs are genes related via speciation (vertical descent), whereas paralogs are genes related via duplication (23). The combination of speciation and duplication events, along with HGT, gene loss, and gene rearrangements, entangle orthologs and paralogs into complex webs of relationships. Correct, coherent usage of these terms would be of certain importance if only to provide clarity to the descriptions of genome evolution. However, beyond semantics, these concepts have distinct and important evolutionary and functional connotations.

In this review, I discuss the intricacies of the definitions of orthologs and paralogs, including several derivative categories of homologs and the respective terms, approaches used for identification of orthologs and paralogs, and the functional implications of orthologous and paralogous relationships.

HISTORY, DETAILED DEFINITIONS, AND CLASSIFICATION OF ORTHOLOGS AND PARALOGS

A Super-Brief History of Homology

The term homolog was introduced by Richard Owen in 1843 to designate "the same organ in different animals under every variety of form and function." Owen clearly distinguished homologs from analogs, which he defined as a "part or organ in one animal which has the same function as another part or organ in a different animal" (72). He attributed homologies to the existence of the same "archetype" (structural plan) in all vertebrates but, not being an evolutionist, did not consider the notion of common origin of homologous organs (74). Homology had been immediately reinterpreted after the publication of Darwin's *Origin of Species* (8). Darwin himself never used the term homology, but less than a year after the publication of the *Origin*, Huxley, in his review of Darwin's work, invoked homology as evidence of evolution (37).

Leaping forward a century, the distinction between orthologs and paralogs and the terms themselves were introduced by Walter Fitch in 1970 in a now classic paper (23). However, in the early 1960s, these concepts were considered in a sufficiently clear form, albeit with the use of different and somewhat awkward wording, in the prescient work of Zuckerkandl & Pauling, which laid the foundations of molecular evolution as a discipline (104, 105). A parallel line of relevant developments involved theoretical and empirical studies of gene duplications and their role in evolution. Although the idea of duplication and its contribution to evolutionary innovation was already present in Fisher's classic work of 1928 (22), the coherent concept was developed in Ohno's famous 1970 book *Evolution by Gene Duplication* (69). Ohno argued that gene duplication, i.e., formation of paralogous genes, is the main process responsible for the emergence of functional novelty during evolution whereby one of the newborn paralogs escapes the pressure of preexisting constraints (purifying selection) and becomes free to evolve a new function. In a subsequent section, I briefly discuss the modern theoretical developments that provide a better-supported, more nuanced perspective on the role of gene duplication in evolution. These advances notwithstanding, Ohno's principal message has certainly withstood the test of time and got a second wind through the discovery of omnipresent duplications in genomes.

For 20 years after Fitch developed the notions of orthologs and paralogs, these definitions quietly stayed within the domain of evolutionary biology. A search of the PubMed database (National Center for Biotechnology

Homologs: genes sharing a common origin

Orthologs: genes originating from a single ancestral gene in the last common ancestor of the compared genomes

Paralogs: genes related via duplication

Information, NIH, Bethesda) shows that between 1971 and 1990, the term ortholog(ue) was used 35 times and the term paralog(ue) only 5 times. This is the low bound of the usage of these terms because many old issues of biological journals, including *Systematic Zoology* which published Fitch's article, are not in PubMed. Nevertheless, these numbers clearly show that orthologs and paralogs were hardly in vogue in the pregenomic era. Indeed, according to the Science Citation Index (**http://isiwok.cit.nih.gov/portal.cgi**), Fitch's article was cited only 48 times in the first 20 years of its postpublication life.

It all changed almost overnight in 1995, when the first two complete genome sequences of cellular life forms, the bacteria *Haemophilus influenzae* and *Mycoplasma genitalium*, were released (26, 28). **Figure 1** shows the striking increase in the usage of the terms "ortholog" and "paralog" during the past 14 years, illustrating the conspicuous change of fortune that genomics brought about for these concepts. Suffice it to say that, in the current version of the PubMed database (which includes publications in all areas of biology and medicine as well as chemistry and physics), more than 1 in every 1000 papers includes "ortholog(ue)" in its title or abstract.

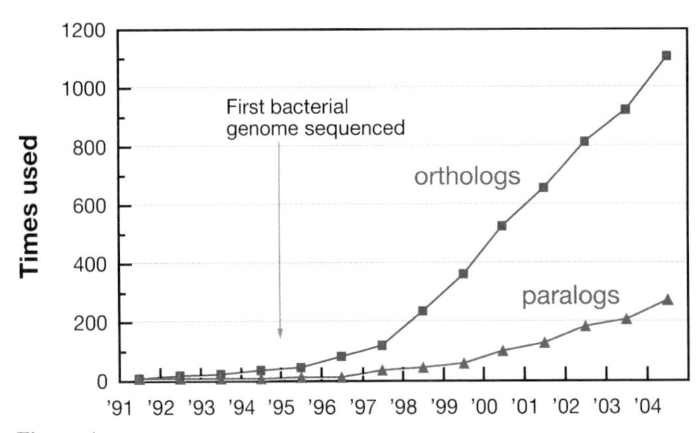

Figure 1

The time dynamics of the usage of the terms "ortholog" and "paralog". The PubMed database was searched using the Entrez search engine with the following queries: "ortholog or orthologs or orthologue or orthologues" and "paralog or paralogs or paralogue or paralogues" to accommodate both the American and the British spelling of the terms.

The advent of complete genomes necessitated a new language in which to discuss the relationships between genomes meaningfully, i.e., the parlance of evolutionary genomics. I would posit that orthology is the keystone definition of evolutionary genomics and paralogy is the paramount, complementary notion (the graphs in **Figure 1** suggest the primacy of orthologous relationships by showing that the term ortholog is used several times more often than paralog). Indeed, in order to make any conclusions regarding evolution of genomes, one first must establish, as precisely as possible, the correspondence between genes in the compared genomes, i.e., orthologous relationships that are inextricably intertwined with paralogous relationships. These constitute the framework on which any evolutionary anomalies and unique events can be mapped. In the following section, I discuss the computational approaches developed to disentangle orthologous and paralogous relationships. Here, I present a general, theoretical breakdown of evolutionary situations in which orthology and paralogy reveal their various faces.

Orthology and Paralogy: Definitions and Complications

Orthologs are genes derived from a single ancestral gene in the last common ancestor of the compared species. This short, simple definition includes two distinct, explicit statements that are important to rationalize; furthermore, it does not include other requirements that might seem natural but are not actually intrinsic to orthology. First, the requirement of a single ancestral gene is central to the concept of orthology. Once the ancestral genome is shown to have contained two paralogous genes that gave rise to the genes in question, it will be incorrect to consider the latter orthologs, even if, on some occasions, there may be the appearance of orthology (see below). Second, the definition specifies the presence of an ancestral gene in the last common ancestor of the compared species rather than in some arbitrary, more ancient ancestor. Of course, this definition assumes the existence of a distinct common ancestor of the compared species, a proposition sometimes challenged for prokaryotes owing to the high incidence of HGT (see discussion below). This is restrictive and might exclude cases where genes behave like orthologs in the evolutionary and functional senses. Nevertheless, we shall stick to the above definition of orthology for the present discussion. An important statement that might at first seem natural is not included in the definition of orthology: There is no requirement that orthology is a one-to-one relationship. The ensuing discussion shows that such a restriction would have been artificial and meaningless. Of even greater import is the connection between orthology and biological function. It should be emphasized that the above definition has nothing to do with function. However, a crucial property of orthologs, which is both theoretically plausible and empirically supported, is that they typically perform equivalent functions in the respective organisms (I avoid the phrase "identical functions" because, in different biological contexts, functions cannot be literally the same). In a subsequent section, I discuss in some detail both the available evidence of the functional equivalency of orthologs and some notable exceptions. Emphasized here is the asymmetry of the relationship between orthology and function: Orthologs most often have equivalent functions, but the reverse statement is much weaker. Situations when equivalent functions are performed by non-orthologous (often, non-homologous) proteins are common enough as captured in the notion of non-orthologous gene displacement, i.e., recruitment of non-orthologous genes for the equivalent, essential functions in different organisms (47, 52). Therefore it is unadvisable (to put it mildly) to speak of "functional orthologs" whereby functional equivalency is taken as a proxy for orthology; of course, phrases such as "orthologous genes with the same function" are quite legitimate (disregarding, as a subtlety, the above distinction between equivalent and identical functions).

Paralogs are genes related via duplication. Note the generality of this definition, which does not include a requirement that paralogs reside in the same genome or any indication as to the age of the duplication leading to the emergence of paralogs (some of these duplications occurred at the earliest stages of life's evolution but the respective genes nevertheless qualify as paralogs). As in the case of orthology, the definition of paralogy does not refer to biological function, but there are major functional connotations. Generally, paralogs perform biologically distinct, even if mechanistically related, functions. Functional differentiation of paralogs is a complex subject that has been addressed in numerous theoretical and empirical studies; a brief synopsis is given in a subsequent section.

Figure 2 shows a hypothetical phylogenetic tree of a gene family that consists of

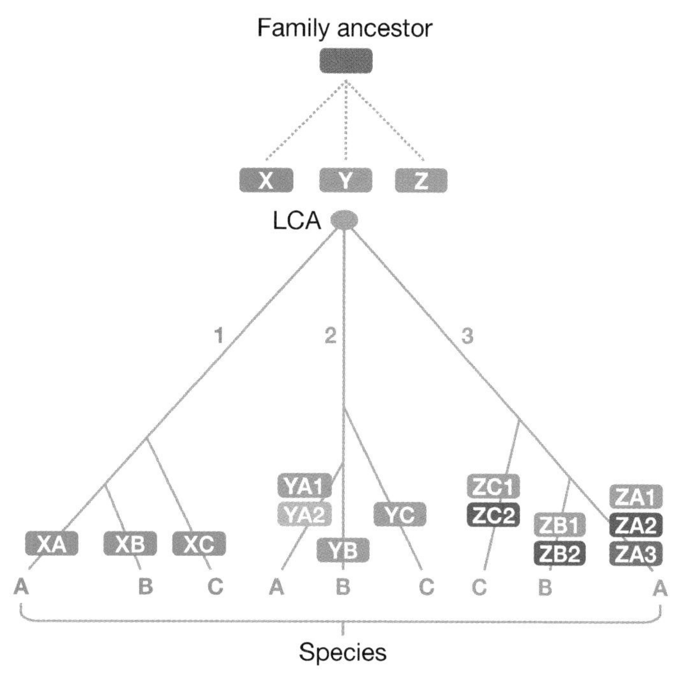

Figure 2

A hypothetical phylogenetic tree illustrating orthologous and paralogous relationships between three ancestral genes and their descendants in three species. LCA, last common ancestor (of the compared species).

three branches, each illustrating a distinct case of orthologous-paralogous relationships. Note first of all, that under the evolutionary scenario illustrated by the tree in **Figure 2**, the common ancestor of the entire family existed prior to the last common ancestor of all three compared species. The latter already encoded three paralogous genes from the given family, which became the progenitors of the three branches of the tree. Thus, each gene in branch 1 is a paralog of each gene in branches 2 and 3, and vice versa. Branch 1 corresponds to a straightforward case whereby evolution from the last common ancestor involved nothing but vertical inheritance. Accordingly, the genes in different species are orthologous to each other and, moreover, show a one-to-one orthologous relationship. However, this is only a specific and not the most common form of orthology (at least when large sets of species are analyzed). Branch 2 shows, in addition to the pattern of vertical inheritance, a lineage-specific duplication in species A. This simple situation, nevertheless, requires the introduction of a new group of terms. Since the duplication occurred in a single lineage after the radiation of the analyzed species, the paralogs in species A fit the definition of orthology with respect to all other genes in this branch. Accordingly, genes YA1 and YA2 are coorthologs (85) of the genes YB and YC. In branch 3, the situation is further complicated by lineage-specific duplications in each of the species. Thus, genes ZA1-3 are, collectively, co-orthologs of genes ZB1-2, etc. The scheme in **Figure 2** also points to different classes of paralogs with respect to speciation events. Relative to each speciation event, it makes sense to define outparalogs (alloparalogs), which evolved via ancient duplication(s) preceding the given speciation event (genes X, Y, and Z in **Figure 1**), and inparalogs (symparalogs), which evolved more recently, subsequent to the speciation event (e.g., genes YA1 and YA2 relative to the radiation of species A and B). Even more complex evolutionary scenarios emerge when duplications are associated with internal branches of a phylogenetic tree (rather than with the terminal branches corresponding to species) such that certain gene sets are inparalogs relative to one speciation event and outparalogs relative to another.

Co-orthologs: two or more genes in one lineage that are, collectively, orthologous to one or more genes in another lineage due to a lineage-specific duplication(s)

Outparalogs: paralogous genes resulting from a duplication(s) preceding a given speciation event

Inparalogs: paralogous genes resulting from a lineage-specific duplication(s) subsequent to a given speciation event

The terms coorthologs, inparalogs, and outparalogs are relatively new (85) and so far have not been widely adopted. Nevertheless, they seem to be helpful for concise and accurate description of the evolutionary process and functional diversification of genes.

Let us now examine the effect of lineage-specific gene loss on the orthologous and paralogous relationships between genes; **Figure 3** shows a hypothetical example of differential gene loss in two lineages obscuring these relationships. This hypothetical scenario starts with two genes (X and Y) that are outparalogs relative to the included speciation event. Subsequently, gene Y is lost in species A, whereas gene X is lost in species D; the species B and C retain both paralogs (**Figure 3**). By comparing species A and D in

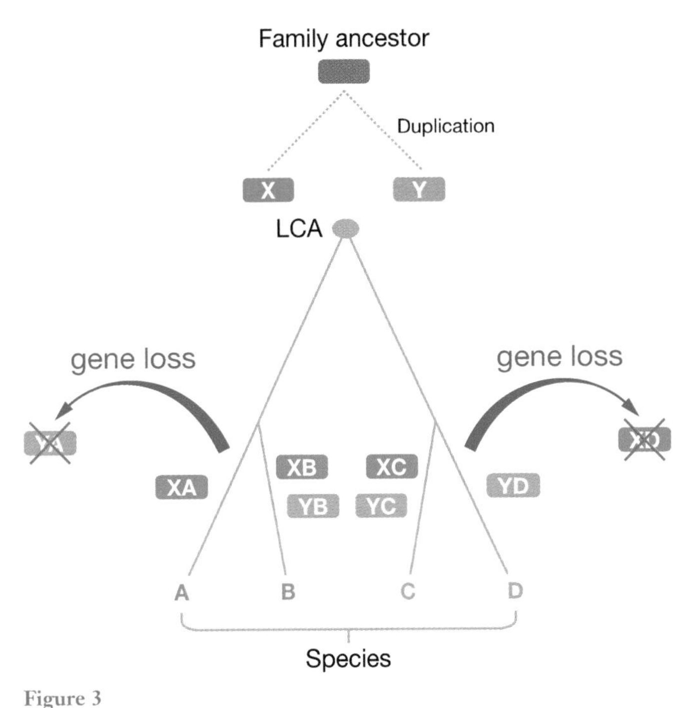

Figure 3

A hypothetical phylogenetic tree illustrating emergence of pseudoorthologs via lineage-specific gene loss.

isolation, we might conclude that gene XA is the ortholog of gene YD. Under the scenario in **Figure 3**, this conclusion is obviously false: These genes are paralogs because they evolved from two paralogous genes of the last common ancestor of the compared species (outparalogs). By analyzing the entire set of representatives of the given gene family in the four species and applying the parsimony principle, we can infer the correct evolutionary scenario and, accordingly, draw the conclusion that the genes XA and YD are, actually, outparalogs relative to the divergence of species A and D; these genes only mimic orthology and, for convenience, we may call them pseudoorthologs. Such conclusions are interesting not only from the evolutionary standpoint but may also have substantial functional implications.

Now we must consider the effects of HGT on the observed orthologous and paralogous relationships between genes. As shown in **Figure 4*a***, two species (A and B) may have homologous genes of which one is ancestral for the given lineage but the other has been acquired via HGT from an outside source C, displacing the ancestral gene. This phenomenon has been dubbed xenologous gene displacement [XGD; (51)]. In an even more complex case, both genes might have been acquired from different sources. In a perfunctory analysis, such a pair of genes (XA and XB in **Figure 4*a***) would mimic orthologs. Obviously, however, these genes do not fit the above definition of orthology because they do not come from a single ancestral gene in the last common ancestor of the compared species. To designate such pseudoorthologs acquired from different sources, the rarely used but natural and useful term xenologs has been proposed (32, 33, 76). As with pseudoorthology caused by lineage-specific gene loss, distinguishing xenology from true orthology may be hard, if possible at all, in a pairwise genomic comparison. However, when multiple genomes are analyzed such that we can, if only roughly, identify the origin of each gene, xenologs and true orthologs become distinguishable.

A related but distinct situation transpires when a species (B) acquires via HGT a gene homologous to a gene that it already has, without the latter gene being displaced (**Figure 4*b***). The result of such an event could reasonably be described as pseudoparalogy (such that XB1 and XB2 are pseudoparalogs) because the homologous genes in species B are not paralogs under the simple definition given above: They have not evolved via gene duplication at any stage of evolution.

Xenologous gene displacement: displacement of a gene in a given lineage with a member of the same orthologous cluster from a distant lineage (xenolog)

Pseudoparalogs: homologous genes that come out as paralogs in a single-genome analysis but actually ended up in the given genome as a result of a combination of vertical inheritance and HGT

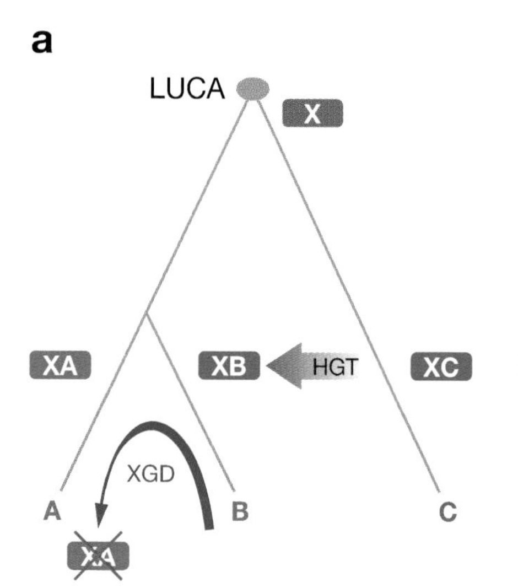

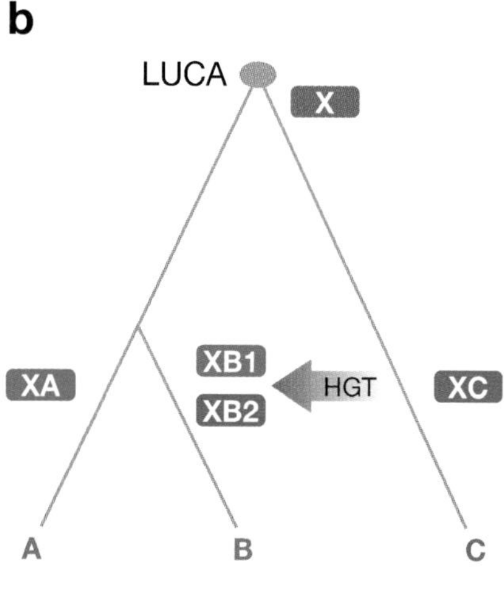

Figure 4

Effect of horizontal gene transfer on orthology and paralogy. (*a*) A hypothetical evolutionary scenario with HGT leading to xenology. (*b*) A hypothetical evolutionary scenario with HGT leading to pseudoparalogy. LUCA, Last Universal Common Ancestor (of all extant life forms).

As discussed below, there are situations, particularly those that involve endosymbiosis, where pseudoparalogy caused by HGT is common.

The concept of orthology is further complicated by another phenomenon that is common in genome evolution, gene fusion, as well as the complementary process of gene fission (54, 83, 100). The impact of such evolutionary events on orthology is that different parts (often encoding distinct domains) of genes in one species are orthologous to different genes in another species. Thus, a new type of orthologous relationship emerges whereby a gene ceases to be the atomic unit of orthology. Further implications of this shift in meaning are considered in the final section of this review.

Table 1 summarizes the meaning and area of applicability of each term introduced in this section. This system of definitions and terms is more complex than the simple dichotomy of orthology and paralogy, with subcategories of paralogs (in- and outparalogs) as well as more specialized notions, xenologs, pseudoorthologs, and pseudoparalogs, additionally introduced. The advantage, I believe, is that this system seems to be complete, i.e., includes definitions that apply to every logically imaginable case of homologous relationships between genes. We now turn to the empirical manifestations and implications of this system in evolutionary genomics.

IDENTIFICATION OF ORTHOLOGS AND PARALOGS: PRINCIPLES AND TECHNIQUES

As soon as genome comparison became a practically important task, the questions arose as to how should one delineate the set of corresponding genes (or, inaccurately, but intuitively, "the same" genes in different species), i.e., orthologs. Indeed, evolutionary classification of genes, which can only be based on the concepts of orthology and paralogy, becomes a must as soon as several genome

Table 1 Homology: terms and definitions

Homologs	**Genes sharing a common origin**
Orthologs	Genes originating from a single ancestral gene in the last common ancestor of the compared genomes.
Pseudoorthologs	Genes that actually are paralogs but appear to be orthologous due to differential, lineage-specific gene loss.
Xenologs	Homologous genes acquired via XGD by one or both of the compared species but appearing to be orthologous in pairwise genome comparisons.
Co-orthologs	Two or more genes in one lineage that are, collectively, orthologous to one or more genes in another lineage due to a lineage-specific duplication(s). Members of a co-orthologous gene set are inparalogs relative to the respective speciation event.
Paralogs	**Genes related by duplication**
Inparalogs (symparalogs)	Paralogous genes resulting from a lineage-specific duplication(s) subsequent to a given speciation event (defined only relative to a speciation event, no absolute meaning).
Outparalogs (alloparalogs)	Paralogous genes resulting from a duplication(s) preceding a given speciation event (defined only relative to a speciation event, no absolute meaning).
Pseudoparalogs	Homologous genes that come out as paralogs in a single-genome analysis but actually ended up in the given genome as a result of a combination of vertical inheritance and HGT.

sequences become available.[1] Individual characterization of each gene in each genome rapidly turns into a gargantuan task for computational analysis and completely impractical experimentally as more genomes are sequenced (12, 30). Comparative genomics can be feasible and meaningful only if the number of distinct entities to be analyzed is substantially reduced by introducing a rational classification of genes. A natural way to do so is to delineate sets of orthologous (including co-orthologous) genes. The extent to which this is going to help in genome analysis critically depends on the nature of the relationships between genomes of different species that only can be deciphered empirically. To take one extreme, if all genes in compared genomes formed perfect clusters of one-to-one orthologs, the number of entities to study would be equal to the number of genes in each genome and would remain constant regardless of how many new genomes were sequenced. Should that be the case, the entire enterprise of comparative-evolutionary genomics would be straightforward to the point of being trivial. On the other end of the spectrum, should the number of identifiable clusters of orthologs be small compared with the total number of genes in genomes, comparative genomics would be in serious trouble.

Since orthology and paralogy are definitions that are inextricably coupled to certain types of evolutionary events (speciation and duplication, respectively), the classical scheme for identifying orthologs involves phylogenetic analysis and, in particular, the procedure generally known as tree reconciliation (20, 63, 73, 102). Under this approach, the topology of a gene tree is compared with that of the chosen species tree and the two are reconciled on the basis of the parsimony principle, by postulating the minimal possible number of duplication and gene-loss events in the evolution of the given gene. The reconciled tree is expected to reflect orthologous relationships. However, genome-wide application of this approach is effectively precluded by both fundamental and practical difficulties. The principal obstacle faced by tree reconciliation (and any other phylogenetic approach) as a strategy for ortholog identification is the prevalence of HGT, especially in prokaryotes. Strictly speaking, the wide spread of HGT invalidates the very notion of a species tree, allowing, at best, the use of various forms of consensus trees for multiple genes as surrogates of the species tree (15, 16, 98). Furthermore, even when a particular tree topology is taken as the species tree, the possibility of HGT of the analyzed gene undermines the concept of reconciliation because the topologies of the two trees are likely to be genuinely different. At a more practical level, even when HGT is not considered to be a major factor, as in the evolution of eukaryotes, both the species tree and many gene trees are fraught by uncertainties and artifacts. Even more practically, fully automated construction and analysis (with appropriate reliability tests) of trees for all genes in sequenced genomes is a major challenge for software engineering and is expensive computationally.

Further in this section, I discuss several attempts at explicit phylogenetic classification of orthologs and paralogs. However, given the substantial difficulties faced by these approaches, most genome-wide studies to date employ simplifications and shortcuts. The simple but crucial assumption that underlies such "surrogate" approaches is that the sequences of orthologous genes (proteins) are more similar to each other than they are to any other genes (proteins) from the compared

[1]Note that the first efforts to delineate sets of orthologous genes shared by relatively large genomes were undertaken years before the appearance of complete genome sequences of cellular life forms. Indeed, this was done as soon as the first pair of related genomes containing many (on the order of 100–200) genes became available, namely, when the genome of varicella-zoster virus was sequenced in 1986 and compared with the previously sequenced genome of another herpesvirus, Epstein-Barr virus (10). Subsequently, a core set of conserved (orthologous) genes was delineated for several herpesviruses (35). However, in these studies, the conceptual basis of comparative genomics was not considered explicitly and neither were the terms orthologs and paralogs used.

Pseudoorthologs: genes that actually are paralogs but appear to be orthologous due to differential, lineage-specific gene loss

genomes, i.e., they form symmetrical best hits (SymBets). Conversely, it is assumed that SymBets are most likely to be formed by orthologs, suggesting a very simple and straightforward method for identification of orthologs. For brevity, I call these two assumptions taken together the SymBet hypothesis. It seems highly plausible, based on the definition of orthology and the notion that orthologs typically occupy the same functional niche, that the SymBet hypothesis is true, at least statistically. Consider the alternative: A gene from one genome is most similar not to its ortholog but to a paralog from another genome. As shown in **Figure 5*a***, this requires a substantial difference in the rates of evolution of paralogs sufficient to offset the divergence of paralogs prior to speciation such that, using the notation of **Figure 5*a***, $AS_x > SD_x + SD_y + BS_y$ (A and B are two species; S stands for speciation and D for duplication; AS_x and BS_y are the amounts of divergence accumulated, respectively, by the genes XA and YB after speciation; SD_x and SD_y are the amounts of divergence accumulated by the paralogous genes x and y after duplication but prior to the speciation). It is generally unlikely that one paralog evolves so much faster than another, given that paralogs retain similar functions. Although asymmetry in the evolution of paralogs has been detected in some studies, typically, it is relatively small. Furthermore, even should that be the case, only the first assumption of the SymBet hypothesis will be violated. In other words, if one paralog evolves much faster than the other, this could lead to a false negative under the SymBet method for ortholog detection (a pair of orthologs missed) but not to a false positive (no erroneous detection of orthologs). Lineage-specific gene duplications producing inparalogs are likely to be a more common cause of violation of the first assumption of the SymBet hypothesis; these duplications may obscure orthologous relationships leading to false negatives (**Figure 5*b***). It seems that the only realistic situations when the second assumption of the SymBet hypothesis can be violated are the cases of pseudoorthology and xenology (**Figures 3** and **4**). Pseudoorthologs and xenologs will likely form a SymBet and produce a false positive under this approach to orthology detection. However, even though formally, neither pseudoorthologs nor xenologs are orthologs, there is a subtle difference between these two situations. Unlike pseudoorthologs, which never can be considered orthologous given their origin by duplication, xenologs come across as orthologs in comparisons of genomes from the donor and recipient lineages (i.e., XA is the ortholog of XC in **Figure 4**). Accordingly, xenology is likely to be a more reliable predictor of gene function that pseudoorthology (see discussion below).

Given these uncertainties, empirical results on the prevalence of SymBets in genome comparisons are important to assess the level of one-to-one orthology between genomes that is critical for evolutionary and functional genomics. **Table 2** shows the number of SymBets between prokaryotic genomes separated by varying evolutionary distances. These results clearly demonstrate that a one-to-one orthologous relationship is a major rather than a minor pattern in genome evolution. For relatively closely related species, e.g., different γ-Proteobacteria, the fraction of probable one-to-one orthologs identified as SymBets typically is >0.5. Predictably, the fraction of genes that produce SymBets drops with evolutionary distance but remains substantial (20%–30% of the genes) even between bacteria and archaea. This fraction also depends on the total number of genes in a genome such that small genomes with few paralogs (e.g., *Mycoplasma genitalium*) show a high level of one-to-one orthology even with distant species. Detection of SymBets is arguably the simplest method for the identification of probable orthologs that is most suitable for closely related genomes but also serves well at greater evolutionary distances for the specific purpose of detecting one-to-one orthologs.

The demonstration that numerous genes in sequenced genomes produced SymBets

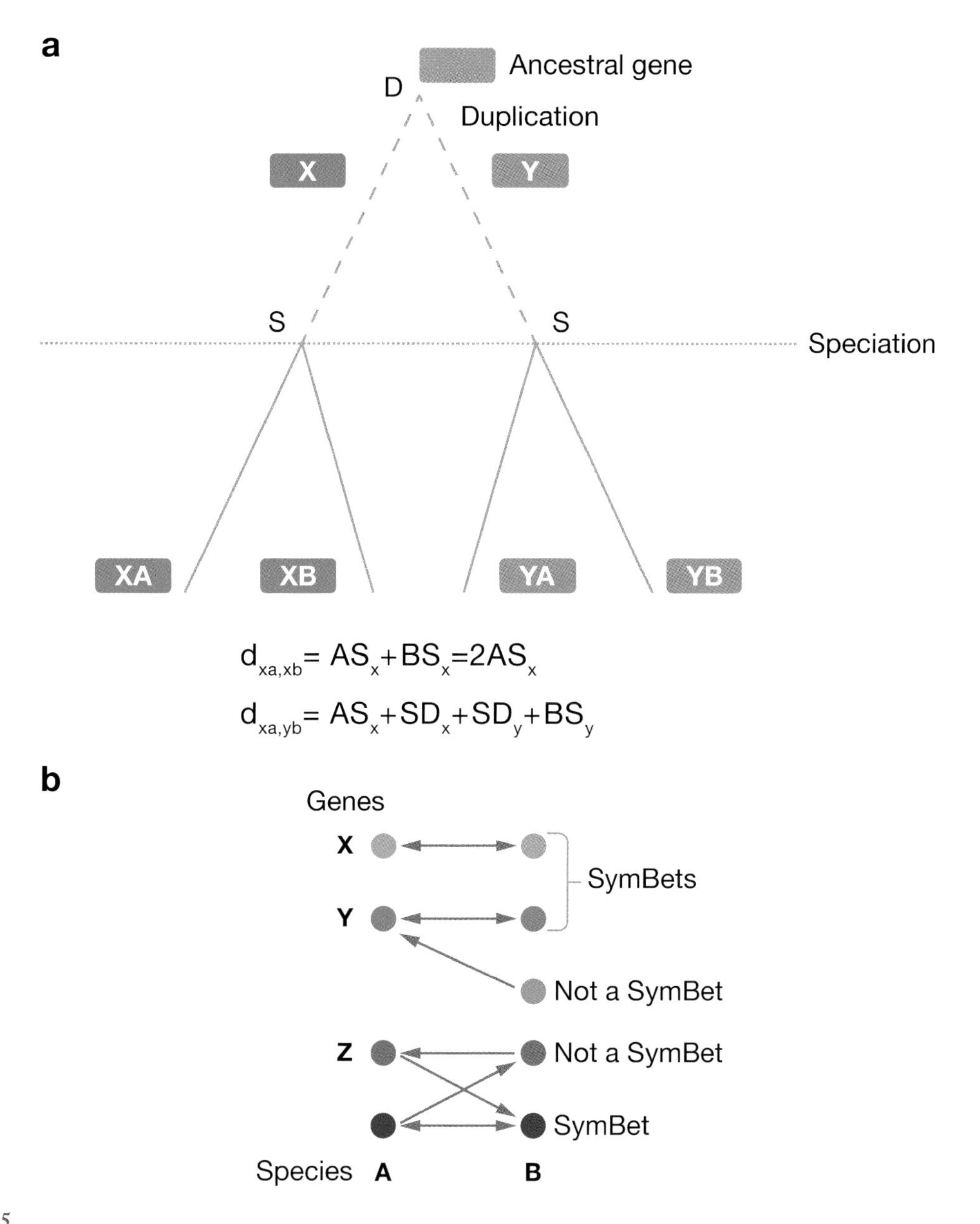

Figure 5

Orthology and genome-specific best hits. (*A*) An evolutionary scheme illustrating the connection between orthology and symmetrical best hits (SymBets). X and Y represent two paralogous genes. The branch lengths in the tree are taken to reflect evolutionary distances between the compared genes, and the formulas for the distances between orthologs and paralogs are given. A molecular clock is assumed for the evolution of orthologs but not paralogs. A and B are two species; D indicates a duplication event and S indicates a speciation event; AS_x and BS_y are the amounts of divergence accumulated, respectively, by the genes XA and YB after speciation; SD_x and SD_y are the amounts of divergence accumulated by the paralogous genes x and y after duplication but prior to the speciation. (*B*) An evolutionary scheme illustrating violation of the SymBet relationship caused by a lineage-specific duplication. Arrows designate best hits; circles of similar shades represent inparalogs. X, Y, and Z designate three cases of (co)orthologous relationships: one-to-one (X), one-to-many (Y) and many-to-many (Z).

Table 2 Symmetrical best hits between selected prokaryotic genomes[a]

	Ec	Yp	Hp	Bs	Mg	Aa	Tm	Mj	Ma	Ta
Ec-4289		0.584	0.456	0.305	0.527	0.519	0.428	0.24	0.151	0.308
Yp-4083	2385		0.432	0.28	0.525	0.496	0.423	0.218	0.144	0.275
Hp-1566	714	677		0.403	0.442	0.396	0.321	0.181	0.211	0.176
Bs-4100	1251	1144	631		0.648	0.495	0.465	0.239	0.153	0.306
Mg-480	253	252	212	311		0.469	0.515	0.235	0.281	0.238
Aa-1553	806	771	615	768	225		0.449	0.279	0.33	0.256
Tm–1846	808	780	503	858	247	697		0.245	0.294	0.265
Mj-1770	425	385	284	423	113	434	433		0.489	0.362
Ma-4540	649	589	330	627	135	513	543	866		0.415
Ta-1478	455	406	260	453	114	378	392	535	614	

[a]The bottom half of the table shows the number of SymBets for each pair of genomes and the upper half shows the fraction of proteins in the smaller genome that form SymBets with the putative orthologs from the larger genome. Species abbreviations are as follows. Proteobacteria: Ec, *Escherichia coli*; Yp, *Yersinia pestis*; Hp, *Helicobacter pylori:* Gram-positive bacteria: Bs, *Bacillus subtilis*; Mg, *Mycoplasma genitalium*: deep-branching, hyperthermophilic bacteria: Aa, *Aquifex aeolicus*; Tm, *Thermotoga maritima*; archaea: Mj, *Methanocaldococcus jannaschii*; Ta, *Thermoplasma acidophilum*; Ma, *Methanosarcina acetivorans*. For each species, the total number of protein-coding genes is indicated.

even between relatively distant species (90) made it clear that construction of a genome-wide evolutionary classification of orthologous and paralogous genes was a feasible task even if the SymBet approach itself is insufficient for this purpose owing to the prevalence of inparalogs. To my knowledge, such a classification was first implemented in the system of the so-called Clusters of Orthologous Groups of proteins (COGs) (89). The phrase "orthologous groups," which has been subsequently criticized as "terminology muddle" (71), was intended to emphasize that the system captured not only one-to-one orthologous relationships but also coorthologous relationships between inparalogs. The idea behind the COG approach was to generalize and extend the notion of a genome-specific best hit. First, the requirement for the reciprocity of best hits (as in SymBets) was abandoned because of the in-paralog problem (see **Figure 5*b***). Second, the notion of a genome-specific best hit was extended to multiple genomes such that the algorithm sought to identify clusters of consistent best hits. More specifically, the COG construction procedure is based on the assumption that any set of at least three proteins from relatively distant genomes that are more similar to each other than they are to any other proteins from the same genomes are most likely bona fide orthologs. This prediction holds even if sequence similarity between some of the compared proteins is relatively low and, accordingly, even fast-evolving genes can be incorporated into the COGs. Briefly, the procedure for COG construction consists of the following steps. (*i*) An all-against-all comparison of protein sequences encoded in multiple genomes (typically using the BLAST program). (*ii*) Detection and clustering of obvious inparalogs, i.e., proteins from the same genome that are more similar to each other than they are to any proteins from other species. (*iii*) Identification of triangles of mutually consistent, genome-specific best hits such that clusters of inparalogs detected at step 2 are treated as single entities. (*iv*) Merging triangles with a common side to form COGs.

The COG approach neatly delineates clusters of probable orthologs that include inparalogs in relatively few lineages. However, the procedure tends to err toward overlumping in the case of large protein families that include a complex mix of in- and outparalogs. Additional complications emerge in the case of multidomain proteins that also may artificially bridge unrelated COGs. Several other approaches for identification of orthologs, based on either specially designed clustering procedures or on explicit phylogenetic analysis, have been developed to overcome these problems and better disentangle orthologs and paralogs. In particular, the INPARANOID procedure developed by Sonnhammer and coworkers identifies clusters of orthologs, including co-orthologous sets of inparalogs, for pairs of genomes, by first detecting SymBets and then incorporating additional inparalogs according to developed statistical criteria (67, 82). High accuracy of identification of inparalogs seems to be achievable with this approach, but the inability to handle multiple genomes simultaneously is a serious limitation. Another method for ortholog detection developed by the same group involves comparison of gene trees with species trees, with the goal of direct identification of orthologs (86). The parts of the gene tree that have the same topology as the species tree are inferred to include orthologs. In principle, this and similar phylogenomic [i.e., applying phylogenetic analysis on genome scale (19)] methods are supposed to provide the strongest and most direct evidence of orthology. A fundamental drawback is, however, the uncertainty of the species tree in the case of prokaryotes due to the prevalence of HGT (this does not appear to be a problem in the case of eukaryotes). In addition, the method is computationally expensive and sensitive to tree construction artifacts. Nevertheless, this approach has been applied to the eukaryotic subset of the Pfam database of protein families, yielding numerous (inferred) orthologous domains (87). A very similar automated phylogenomic procedure for inference of orthologs has been developed by Zmasek & Eddy (103). In a more recent development, a Bayesian probabilistic technique has been introduced to assign probabilities to the orthology identifications (3). A major effort to identify orthologs and paralogs has been undertaken by Perrière and coworkers who constructed databases of homologous genes from vertebrates (HOVERGEN) and bacteria (HOBACGEN) in which families of homologs (each consisting of a mix of orthologs and paralogs) are accompanied by phylogenetic trees (18, 79). Very recently, tools have been developed for tree reconciliation in the framework of the database, with the goal of identifying sets of orthologs (17). The effectiveness of these methods on genome scale remains to be assessed.

In summary, although phylogenomic methods, in principle, should be best suited for deciphering orthologous and paralogous relationships, in practice, these approaches so far have not matured enough to produce a comprehensive collection of orthologous-paralogous clusters covering multiple species. Such collections have been constructed only with methods based on sequence similarity and the notion of genome-specific best hits. Clusters produced with these approaches are by no means error-free, in particular with respect to lumping some of the orthologous gene sets into inflated, mixed clusters of orthologs and paralogs. However, extensive work on genome annotation as well as genome-wide evolutionary studies performed with the help of these systems (50) suggest that they are sufficiently robust for extracting meaningful evolutionary and functional patterns (discussed below).

EVOLUTIONARY PATTERNS OF ORTHOLOGY AND PARALOGY

Coverage of Genomes in Clusters of Orthologs

Probably, the aspect of orthologous clusters that is of the most immediate importance to evolutionary and functional genomics is the coverage of genomes, i.e., the fraction of genes with orthologs in other species. A substantial majority of genes from each sequenced prokaryotic genome (**Figure 6*a***) and a somewhat lower fraction of eukaryotic genes (**Figure 6*b***) belong to COGs

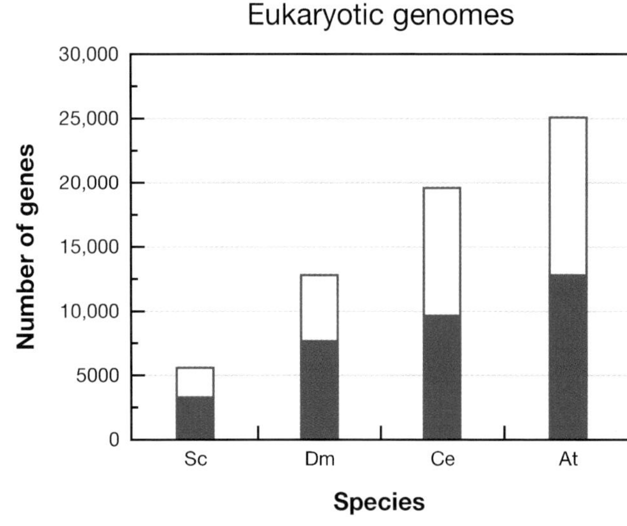

Figure 6

Coverage of selected genomes with clusters of orthologous groups of proteins (C/KOGs). (*a*) Prokaryotic genomes. (*b*) Eukaryotic genomes. The data are from (88). Filled volume, genes in C/KOGs; empty volume, genes not included in C/KOGs. Abbreviations: Bacteria: Aa, *Aquifex aeolicus*; Bs, *B. subtilis*; Ec, *E. coli*; Hi, *Haemophilus influenzae*; Mg, *Mycoplasma genitalium*. Archaea: Af, *Archaeoglobus fulgidus*; Ma, *Methanosarcina acetivorans*; Ss, *Sulfolobus solfataricus*. Eukaryotes: At, *Arabidopsis thaliana*; Ce, *Caenorhabditis elegans*; Dm, *Drosophila melanogaster*; Sc, *Saccharomyces cerevisiae*. Data for *Homo sapiens* are not shown because the KOGs include an early, inflated version of the human gene set.

(or the clusters of orthologous genes from eukaryotes dubbed KOGs) (49, 88). The coverage of genomes with COGs slowly increases with the growing number of included species (91). It remains unclear whether the level of orthology tends to 100% when the number of genomes tends to infinity or there is a lower limit, and some genes are true, species-specific "orphans" evolving in a regime different from the majority of the genes (29, 68). Obviously, however, the orthology level is high and will only increase with continued genome sequencing. Therefore, the reduction of search space provided by the classification of genes into orthologous clusters is substantial and, in practical terms, should be sufficient to cope with the flood of information produced by genome sequencing.

One-to-One Orthologs and Inparalogs

As discussed above, lineage-specific duplications leading to the emergence of inparalogs complicate orthologous relationships between genes. A simple analysis of the COGs allows one to evaluate the extent of this phenomenon, with the caveat that some lumping is involved, leading to an inflated estimate of the number of inparalogs. **Figure 7** shows the distribution of the number of paralogs in COGs for four prokaryotic genomes. For all the importance of lineage-specific expansion of paralogous families, in each genome the majority of orthologous lineages (COGs) are represented by a single gene. Specifically, in *Escherichia coli*, a complex bacterium with a relatively large genome, ~71% of the COGs include a single gene, and in the case of *M. genitalium*, a bacterium with a near-minimal genome, such COGs form the overwhelming majority (~92%). This conclusion is compatible with the earlier quantitative analysis of lineage-specific expansions in prokaryotes, which detected only a few large expansions in each genome (41), and with independent analyses of the size distribution of paralogous families (38, 53). A qualitatively similar pattern, albeit with some predictable excess of inparalogs, was observed for eukaryotic orthologous clusters (data not shown). Moreover, 1769 of the 4873 COGs (36%) contain exactly one representative from each of the included genomes. It has been proposed that one-to-one orthology could be selected for in the case of genes encoding subunits of macromolecular complexes requiring strict stoichiometry due to the deleterious effect of subunit imbalance (75, 94). Indeed, orthologous sets containing no paralogs appear to be significantly enriched in complex subunits (49, 75, 95).

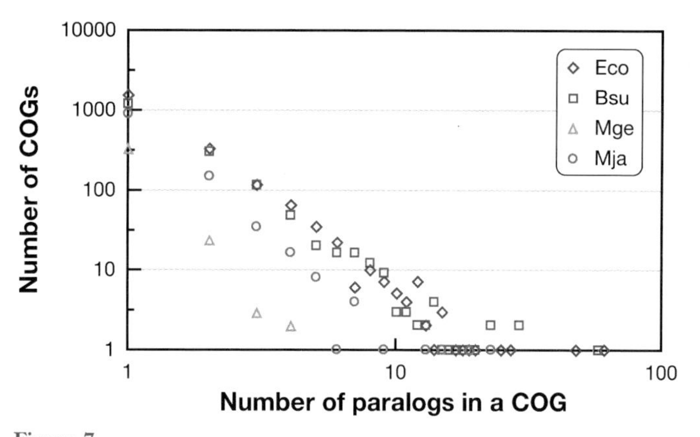

Figure 7

Distribution of the number of paralogs in COGs for selected prokaryotic genomes. The data were extracted from the current COG version (88). The plot is shown in the double-logarithmic scale.

Orthologous Clusters and the Molecular Clock

A central tenet of Kimura's neutral theory is that the rate of evolution of a gene remains the same (with some dispersion, obviously) as long as the biological function does not change (44). Kimura never used the term ortholog (or paralog), but this generalization obviously applies to orthologous gene sets, primarily those that include no inparalogs. The subsequent evolution of the molecular clock concept involved considerable dispute, with numerous studies demonstrating substantial overdispersion of the clock (5, 6, 31). Genome-wide tests of the molecular clock have been conducted only very recently. One approach involved comparing the evolutionary distances within a COG containing no inparalogs to a standard intergenomic distance, which was defined as the median of the distribution of the distances between all one-to-one orthologs in the respective genomes (66). Under the molecular clock model, the points on a plot of intergenic distances for the given COG versus intergenomic distances will scatter around a straight line. A statistical method was developed to identify significant deviations from the clock-like behavior. Among several hundred COGs representing three well-characterized bacterial lineages, α-Proteobacteria, γ-Proteobacteria, and the *Bacillus-Clostridium* group, the clock hypothesis could not be rejected for ~70%, whereas the rest showed substantial deviations. These anomalies could be explained either by lineage-specific acceleration of evolution or by XGD (see below). The general conclusion from these analyses seems to be that the majority of orthologous genes evolve in the clock-like mode as long as there was no duplication, although the frequency of exceptions was by no means negligible.

Molecular clock: a central concept of molecular evolution, which posits that a gene evolves at a constant rate as long as its function does not change

The connections between gene duplication and evolutionary rates have been explored in considerable detail and appear to be quite complex. Ohno's original idea was that, immediately after duplication, one of the newborn paralogs is freed from purifying selection and would evolve rapidly such that it either perishes or, on relatively rare occasions, evolves a new function (neofunctionalization), after which evolution slows down again (69). Subsequent theoretical and empirical analyses have shown that this path of evolution, while apparently realized on some occasions, is probably not the most common outcome of a gene duplication (61). What happens more often seems to be relaxation of purifying selection immediately after duplication, resulting in accelerated evolution in both paralogs. This is thought to reflect subfunctionalization, i.e., partitioning of the different functions of the multifunctional ancestral gene between paralogs (45, 59, 60). Somewhat paradoxically, however, two independent recent studies have shown that despite this acceleration, genes that have paralogs on average evolve slower than those that do not (9, 42). This difference may be due to a greater likelihood of fixation of emergent paralogs among slowly evolving (more "important") genes.

Xenologs, Pseudoorthologs, and Pseudoparalogs

As discussed above, xenologs are homologs that violate the definition of orthology due to HGT. More specifically, xenology is brought about either by XGD or by acquisition of a gene that is new for the given lineage (51). When the study of the clock-like behavior of orthologous gene sets discussed in the previous section was followed up with phylogenetic analysis of deviant cases, probable XGD was demonstrated for 10%–20% of the COGs within each of the examined bacterial taxa, establishing XGD as a major evolutionary phenomenon (66). On some occasions, XGD even takes the form of displacement in situ whereby a gene is displaced with a horizontally transferred distant ortholog without disrupting the operon structure (70). **Figure 8** illustrates one such case: the displacement of the *ruvB* gene (coding for the helicase subunit of the resolvasome) in the mycoplasmas with the ortholog from ε-Proteobacteria. In this case, the *ruvB* genes of *Mycoplasma* and those of the rest of low GC gram-positive bacteria, the taxon to which *Mycoplasma* belong, qualify as xenologs given their different phylogenetic affinities (**Figure 8*c***). A clear example of acquisition of a new gene leading to xenology is the B family DNA polymerase of γ-Proteobacteria (e.g., the *polB* gene of *E. coli*). At first sight, this gene appears to be an ortholog of the archaeal and eukaryotic B family polymerases (see COG0417). However, these

Figure 8

Xenologous displacement in situ of the *ruvB* gene in the mycoplasmas. (*A*) Organization of the Holliday junction resolvasome operon and surrounding genes in bacteria. COG0632, Holliday junction resolvasome, DNA-binding subunit; COG2255, Holliday junction resolvasome, DNA-binding subunit; COG0817; Holliday junction resolvasome, endonuclease subunit; COG0392, predicted integral membrane protein; COG0282, acetate kinase; COG0839, NADH:ubiquinone oxidoreductase subunit 6 (chain J); COG0244, ribosomal protein L10; COG0732, restriction endonuclease S subunits; COG0809, S-adenosylmethionine:tRNA-ribosyltransferase-isomerase; COG0772, bacterial cell division membrane protein; COG0624, acetylornithine deacetylase/succinyl-diaminopimelate desuccinylase and related deacylases; COG1487, predicted nucleic acid-binding protein; COG1132, ABC-type multidrug transport system, ATPase, and permease components; COG0442, prolyl-tRNA synthetase; COG0323, DNA mismatch repair enzyme; COG1408, predicted phosphohydrolases. (*B*) Unrooted phylogenetic tree for RuvA. (*C*) Unrooted phylogenetic tree for RuvB. Branches supported by bootstrap probability >70% are marked by black circles. Names of the genes from mosaic operons and the respective branches are shown in red. Species abbreviations: Atu, *Agrobacterium tumefaciens*; Bha, *Bacillus halodurans*; Bsu, *Bacillus subtilis*; Bbu, *Borrelia burgdorferi*; Bme, *Brucella melitensis*; Cje, *Campylobacter jejuni* Ccr, *Caulobacter crescentus*; Ctr, *Chlamydia trachomatis*; Cpn, *Chlamydophila pneumoniae*; Cte, *Chlorobium tepidum*; Cac, *Clostridium acetobutylicum*; Cgl, *Corynebacterium glutamicum*; Eco, *Escherichia coli*; Fnu, *Fusobacterium nucleatum*; Hin, *Haemophilus influenzae*; Hpy, *Helicobacter pylori*; Lla, *Lactococcus lactis*; Lpl, *Lactobacillus plantarum*; Lin, *Listeria innocua*; Neu, *Nitrosomonas europaea*; Mlo, *Mesorhizobium loti*; Mge, *Mycoplasma genitalium*; Mpn, *Mycoplasma pneumoniae*; Mpu, *Mycoplasma pulmonis*; Mle, *Mycobacterium leprae*; Mtu, *Mycobacterium tuberculosis*; Nme, *Neisseria meningitidis*; Nsp, *Nostoc* sp.; Oih, *Oceanobacillus iheyensis*; Pae, *Pseudomonas aeruginosa*; Rso, *Ralstonia solanacearum*; Rpr, *Rickettsia prowazekii*; Rco, *Rickettsia conorii*; Sme, *Sinorhizobium meliloti*; Sau, *Staphylococcus aureus*; Spy, *Streptococcus pyogenes*; Ssp, *Synechocystis* PCC6803; Tma, *Thermotoga maritima*; Tte, *Thermus thermophilus*; Tpa, *Treponema pallidum*; Vch, *Vibrio cholerae*; Xfa, *Xylella fastidiosa*; Uur, *Ureaplasma urealyticum*. The figure is from (70) where the details of phylogenetic analysis are described.

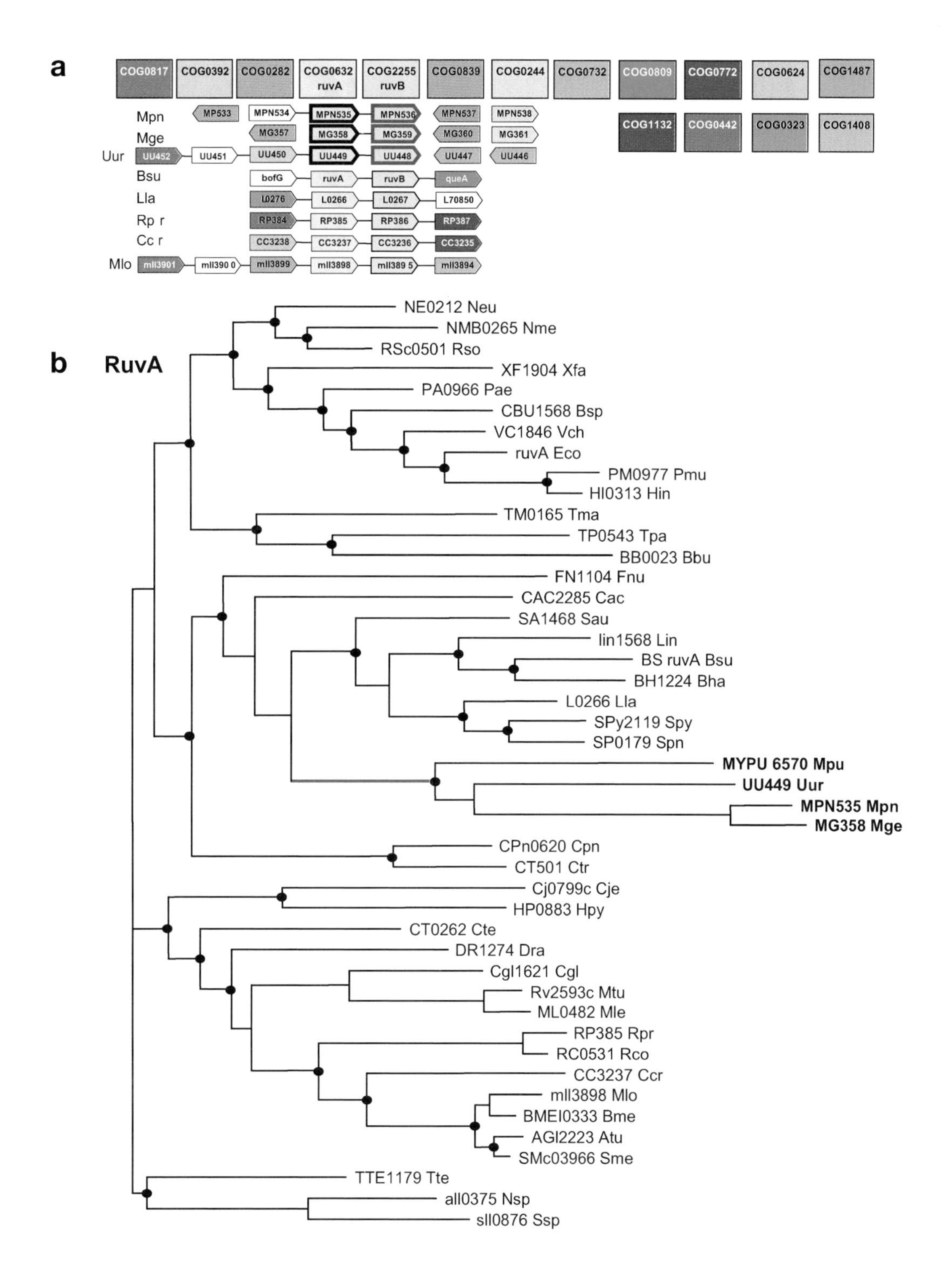
a
COG0817
COG0392
COG0282
COG0632 ruvA
COG2255 ruvB
COG0839
COG0244
COG0732
COG0809
COG0772
COG0624
COG1487
COG1132
COG0442
COG0323
COG1408
Mpn
Mge
Uur
Bsu
Lla
Rp r
Cc r
Mlo
MP533
MPN534
MPN535
MPN536
MPN537
MPN538
MG357
MG358
MG359
MG360
MG361
UU452
UU451
UU450
UU449
UU448
UU447
UU446
bofG
ruvA
ruvB
queA
L0276
L0266
L0267
L70850
RP384
RP385
RP386
RP387
CC3238
CC3237
CC3236
CC3235
mll3901
mll390 0
mll3899
mll3898
mll389 5
mll3894
b
RuvA
NE0212 Neu
NMB0265 Nme
RSc0501 Rso
XF1904 Xfa
PA0966 Pae
CBU1568 Bsp
VC1846 Vch
ruvA Eco
PM0977 Pmu
HI0313 Hin
TM0165 Tma
TP0543 Tpa
BB0023 Bbu
FN1104 Fnu
CAC2285 Cac
SA1468 Sau
lin1568 Lin
BS ruvA Bsu
BH1224 Bha
L0266 Lla
SPy2119 Spy
SP0179 Spn
MYPU 6570 Mpu
UU449 Uur
MPN535 Mpn
MG358 Mge
CPn0620 Cpn
CT501 Ctr
Cj0799c Cje
HP0883 Hpy
CT0262 Cte
DR1274 Dra
Cgl1621 Cgl
Rv2593c Mtu
ML0482 Mle
RP385 Rpr
RC0531 Rco
CC3237 Ccr
mll3898 Mlo
BMEI0333 Bme
AGl2223 Atu
SMc03966 Sme
TTE1179 Tte
all0375 Nsp
sll0876 Ssp

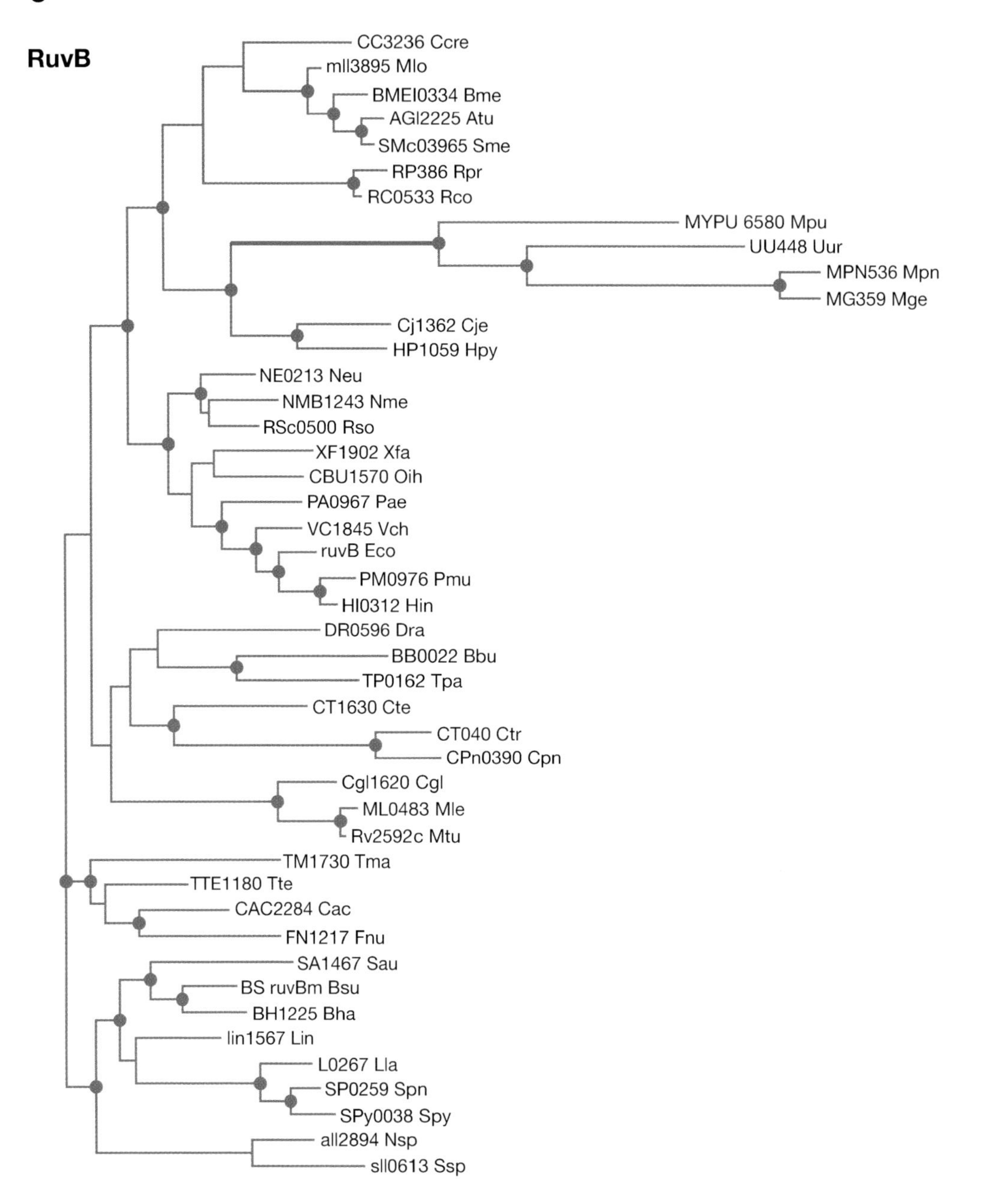

genes should be classified as xenologs because the proteobacterial version clearly does not derive from the last universal common ancestor (which is also the last common ancestor of bacteria and archaea) but rather had been acquired via HGT.

As defined above, pseudoorthologs emerge via lineage-specific, differential loss of paralogous genes (**Figure 2**). A systematic search for pseudoorthologous genes requires detailed, genome-wide phylogenetic analysis, which to my knowledge has not yet been conducted. Nevertheless, some likely cases of pseudoorthology can be gleaned by examination of COGs. Consider COG0114 (fumarase) and COG1027 (aspartate ammonia-lyase), which consist of paralogous enzymes with a high level of sequence similarity to each other. Both enzymes are widespread in bacteria and, most likely, were already present in the last common ancestor of all bacteria but apparently have been lost independently in many lineages. When comparing the genomes of two cyanobacteria, *Synechocystis* sp. and *Nostoc* sp., the genes slr0018 of the former and alr3724 of the latter, produce a SymBet and, accordingly, could be identified as orthologs by default. However, inspection of the COGs clearly shows that slr0018 is a fumarase whereas alr3724 is an aspartate ammonia-lyase. This seems to be a clear-cut case of pseudoorthology caused by lineage-specific, differential loss of paralogs, with the ensuing functional differences (see below).

The most obvious case of pseudoparalogy is the presence of numerous pairs of homologous genes of ancestral and endosymbiotic (mitochondrial or, in plants, chloroplastic) origin in eukaryotes (34, 43, 56). These pseudoparalogs are particularly abundant among the components of the translation machinery, such as ribosomal proteins or aminoacyl-tRNA synthetases. Many additional pseudoparalogs appear to have emerged through other routes of HGT. One of the most conspicuous is the transfer of archaeal genes to bacteria, particularly hyperthermophiles. **Figure 9** shows the phylogenetic tree for the peroxiredoxin AhpC (COG0450). This tree includes two paralogous proteins from the hyperthermophilic bacterium *Aquifex aeolicus*, one of which clusters with archaeal and the other with bacterial homologs. The respective genes are pseudoparalogs because they apparently ended up in the *A. aeolicus* genome as a result not of gene duplication at any stage of evolution but of horizontal transfer of one of the peroxiredoxin genes from an archaeal source. Conversely, the tree includes three peroxiredoxins from the archaea *Thermoplasma acidophilum* and *T. volcanium*, at least two of which (the one nested within the archaeal subtree and the one with a clear bacterial affinity) are pseudoparalogs. Notably, the only peroxiredoxin of another hyperthermophilic bacterium, *Thermotoga maritima*, shows an archaeal affinity, suggesting that in this lineage, the original bacterial gene had been lost, probably after the acquisition of the archaeal version.

Protein Domain Rearrangements, Gene Fusions/Fissions, and Orthology

Orthologous protein in eukaryotes sometimes differ in their domain architectures. There seems to be a trend toward an increase in the complexity of domain architecture in parallel with the increase of organismal complexity, a phenomenon dubbed domain accretion (48, 49). Apparently, additional domains acquired by proteins from more complex organisms provide additional interactions leading, in particular, to increased complexity of signal transduction and various regulatory processes. Differences in domain architectures can also be detected between orthologs from major prokaryotic taxa; one such case is illustrated in **Figure 10*a***. The DnaG-like primases of bacteria and archaea share a highly conserved catalytic domain and appear to be orthologous, especially given that they are represented by a single protein in all bacterial and archaeal genomes (62). However, the bacterial and archaeal orthologs have different accessory domains, a Zn-finger and a distinct module of a helicase domain, respectively (**Figure 10*a***), which may reflect substantial functional differences (see next section).

As mentioned above, gene fusions and fissions, which are common in genome evolution, affect the very notion of orthology: In this case, a single gene in some species coding for a multidomain protein is orthologous to two or more distinct genes coding for the respective individual domains in another set

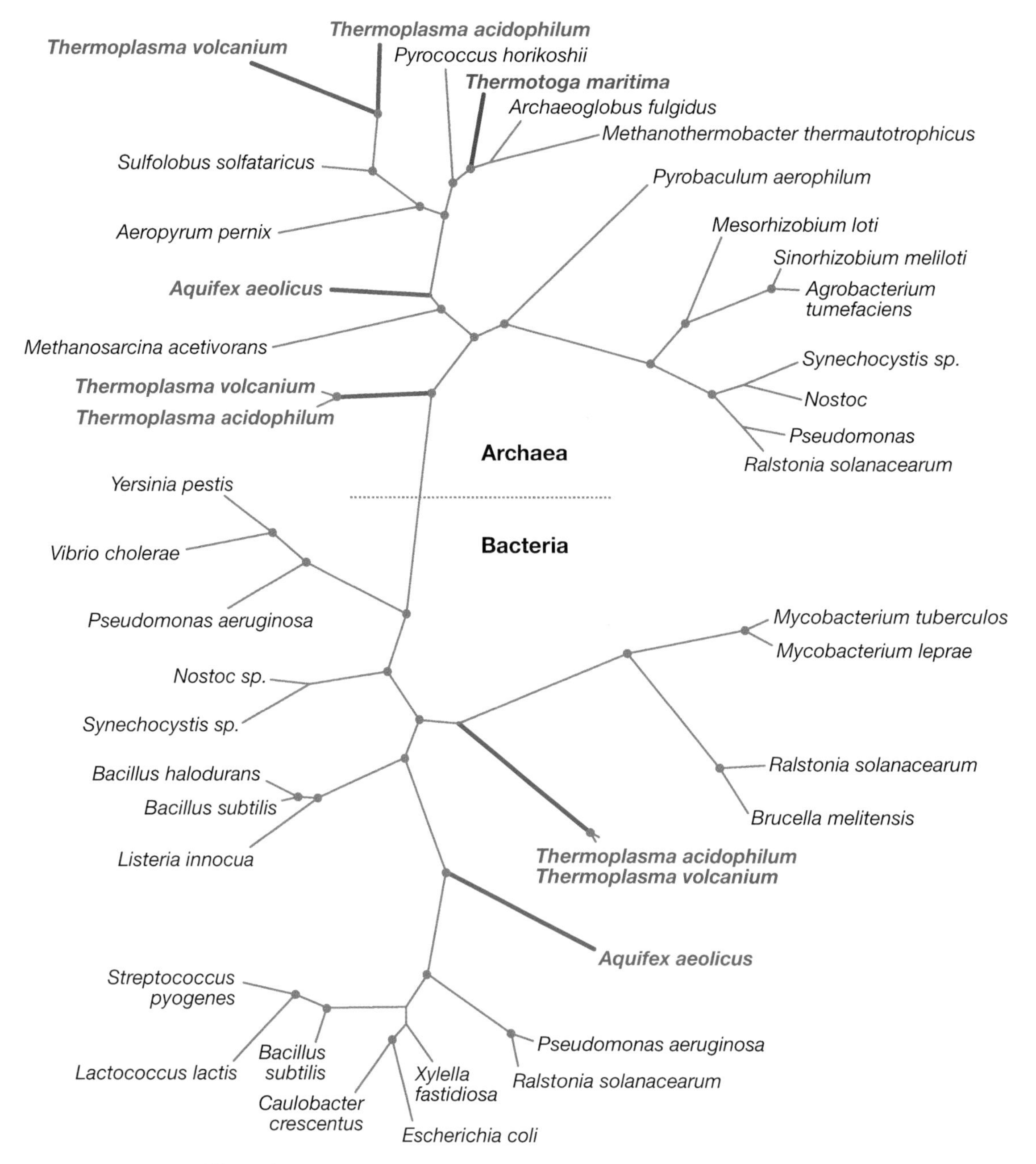

Figure 9

Horizontal gene transfer leading to pseudoparalogy. The two pseudoparalogous peroxiredoxins from *Aquifex aeolicus* are shown in red, the three pseudoparalogs from the Thermoplasmas in blue, and the only peroxiredoxin of *Thermotoga maritima* in purple. The genes are identified by full species names. The maximum likelihood, unrooted phylogenetic tree was constructed as previously described (70).

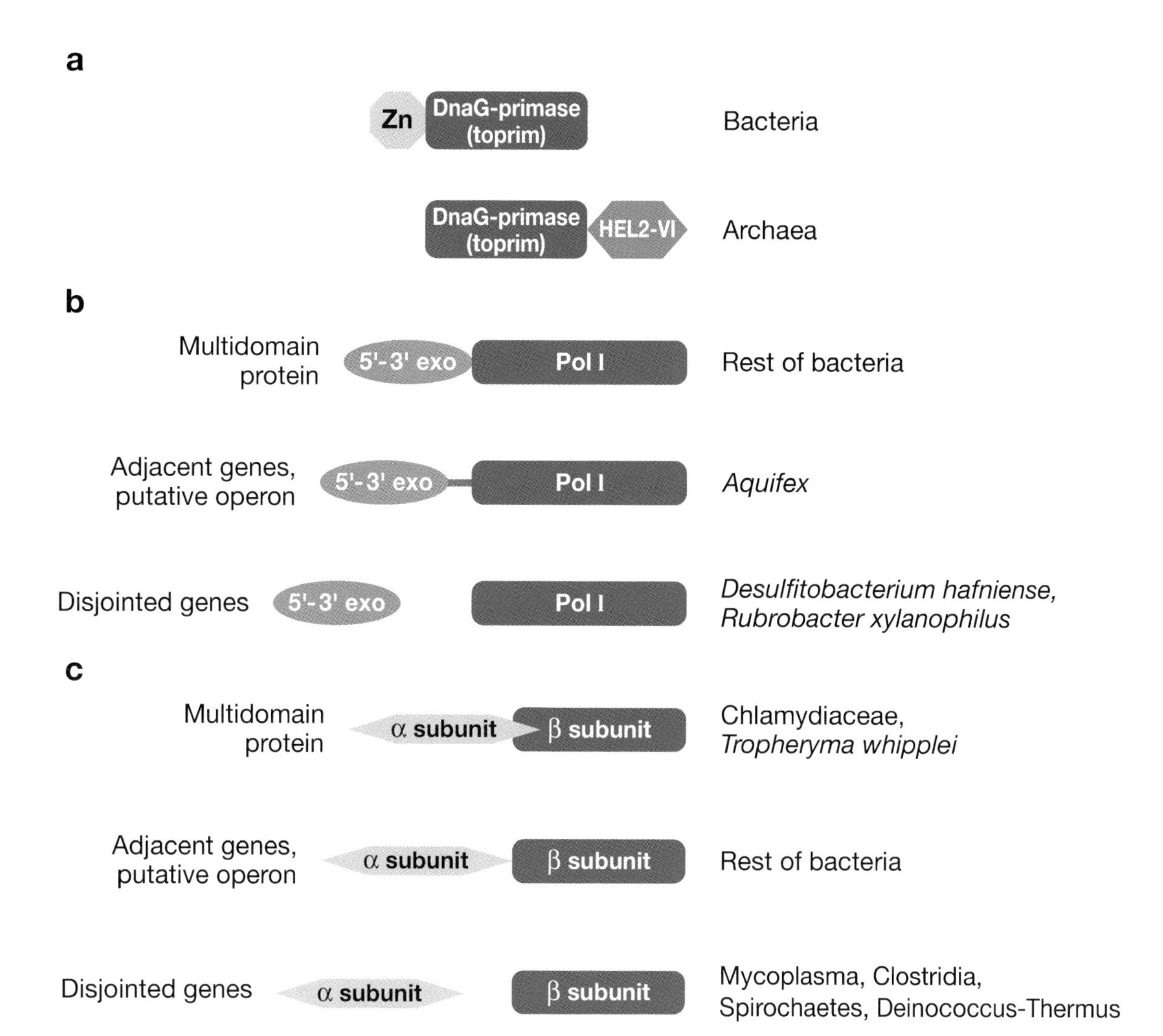

Figure 10

Rearrangements of gene structure and orthology. (*a*) Domain architectures of bacterial and archaeal DnaG-like primases. (*b*) Independent fission of the DNA polymerase I gene in multiple bacterial lineages. (*c*) Fusion of the genes for bacterial glycyl-tRNA synthetase subunits.

of species. **Figure 10** (***b*** and ***c***) shows two cases of such relationships. In the example in **Figure 10*b***, the bacteria *A. aeolicus*, *A. pyrophilus*, *Desulfitobacterium hafniense*, and *Rubrobacter xylanophilus* encode the polymerase and 5′–3′ exonuclease domains of DNA polymerase I in two distinct genes, unlike all other bacteria in which these enzymes are domains of a single, multidomain protein. The bacteria in which the nuclease and polymerase activities reside in different proteins belong to three distinct lineages, suggesting three independent fissions of the *polA* gene. Whereas the genes for the nuclease and the polymerase are adjacent in the two *Aquifex* species, in the other two bacteria they are not, implying genome arrangement subsequent to gene fission. By contrast, the example in **Figure 10*c*** shows the fusion of the genes for the α and β subunits of glycyl-tRNA synthetases in the parasitic bacteria of the *Chlamydia*e branch and the pathogenic actinobacterium *Tropheryma whipplei*. In this case, it appears likely that gene fusion occurred only once, with subsequent horizontal dissemination of the fused gene.

FUNCTIONAL CORRELATES OF ORTHOLOGY AND PARALOGY

The validity of the conjecture on functional equivalency of orthologs is crucial for reliable annotation of newly sequenced genomes and, more generally, for the progress of functional genomics. The huge majority of genes in the sequenced genomes will never be studied experimentally, so for most genomes transfer of functional information between orthologs is the only means of detailed functional characterization. To what extent is such transfer legitimate? A rough estimate can be obtained by comparing the available functional information for experimentally characterized one-to-one orthologs from model organisms. Inspection of the 1330 COGs that contain one-to-one orthologs from the well-studied bacteria *E. coli* and *B. subtilis* failed to reveal a single clear-cut case of different functions, although subtle differences, e.g., in enzyme or transporter specificities are common (E.V.K., unpublished observations). Thus, in general, the notion that one-to-one orthologs are functionally equivalent seems to hold well.

However, at greater evolutionary distances, particularly across the primary kingdom divides, there are prominent cases of apparent major differences in the functions of orthologs. Thus, the bacterial and archaeal DnaG-like primases, which figure in the previous section in connection with a difference in domain architecture, seem to function in fundamentally different processes. Bacterial DnaG is an essential component of the replication machinery, namely the polymerase responsible for the synthesis of RNA primers used to initiate replication (4). Although the function of the archaeal ortholog has not been studied in detail, there is no evidence of its involvement in replication; furthermore, it has been shown to associate with the exosome, the RNA degradation complex, suggesting a role in RNA processing (21). A converse situation seems to exist with the archaeo-eukaryotic-type primase which is an essential replication component in archaea and eukaryotes, but is involved in a distinct repair pathway in those bacteria that have this gene (11). In this case, the bacterial versions actually are likely to be xenologs of the archaeal and eukaryotic ones (2), and they apparently went through a period of rapid evolution associated with the functional change.

Acceleration of evolution accompanying a radical functional switch seems to have been a major aspect of the emergence of the eukaryotic cell. Thus, to the best of our understanding, eukaryotic tubulins are co-orthologs of the prokaryotic protein FtsZ, which is the key component of the prokaryotic cell division machinery mediating septum formation (1, 58). The well-characterized functions of tubulins are completely different: They are the principal constituents of the eukaryotic microtubules, cytoskeletal structures that are absent in prokaryotes [the recent discovery of tubulins in Prosthecobacteria (39) is remarkable but may be explained by HGT from eukaryotes to a specific bacterial lineage]. The drastic change of function in eukaryotes apparently had been accompanied by a burst of sequence evolution such that an unequivocal demonstration of the homology of FtsZ and tubulin became possible only through comparison of the respective protein structures (65). An analogous situation is seen with the other signature proteins of eukaryotes, actins, and ubiquitins, whose apparent prokaryotic orthologs have completely different functions and dramatically differ in sequence (1, 92, 96). Although in each of these cases, the relationship between prokaryotic and eukaryotic proteins is not one-to-one orthology, with many inparalogs present in eukaryotes, the functional differences among these paralogs are minor compared with the profound divide between prokaryotes and eukaryotes. The general message from this brief survey of the functional equivalency of orthologs and of the functional switches within orthologous lineages is clear: The fundamental functions of orthologs do change but such changes are far from being common, tend to be associated with major evolutionary transitions, and are accompanied by a substantial acceleration of evolution.

The functional connotations of paralogy are distinct from and, in a sense, opposite to those of orthology. Although in some cases, paralogs may retain the same, ancestral function, being fixed due to the gene dosage effects (amplification of rRNA genes is an obvious example), the general themes associated with paralogy are functional diversification and specialization. The subfunctionalization mode of evolution of paralogs that has been explored theoretically in great detail by Lynch and coworkers (27, 60, 61) seems best compatible with the demonstration that selective constraints affect paralogs even immediately after duplication (45). Examples of subfunctionalization are plentiful. A classic one is the distribution of the universal transcriptional function of the archaeal RNA polymerase between the three RNA polymerases of eukaryotes, with RNA polymerase I becoming responsible for the transcription of rRNA genes, RNA polymerase II transcribing protein-coding genes, and RNA polymerase III transcribing tRNA genes and those for other noncoding RNAs (101). The original version of Ohno's neofunctionalization model, whereby one of the emerging paralogs initially evolves free of constraints (like a pseudogene) but then accidentally hits on a new function, might be unrealistic or, at least, rare. More generally, however, evolution of paralogs, particularly in the context of lineage-specific expansions of paralogous families, may involve both subfunctionalization and neofunctionalization. Indeed, it seems inevitable that, among the enormous repertoire of signal-transduction systems that evolved via multiple duplications, such as protein kinases, receptors, and ubiquitin systems components (to mention just a few of the most conspicuous cases), there should be specificities that were not present in the ancestral gene. Very recently, He & Zhang explored the possibility of combination of subfunctionalization and neofunctionalization by examining protein-protein interactions of paralogous gene products in yeast (36). Their results suggested a more complex subneofunctionalization model under which the evolution of paralogs starts with rapid subfunctionalization but subsequently often switches to the neofunctionalization mode.

GENERAL DISCUSSION

Orthology and Paralogy as Evolutionary Inferences and the Homology Debates

The preceding discussion aimed to show how the notions of orthology and paralogy permeate modern genomics and provide the crucial link between genomics and evolutionary biology. To a large extent, these concepts form the foundation of evolutionary genomics and are also of major importance for functional genomics or, in more practical terms, for functional annotation of sequenced genomes. It is useful, however, to explicitly define the epistemological status of these concepts. Orthology and paralogy as well as the generalized notion of homology imply specific statements on the course of evolution of the respective genes. In other words, these statements are inferences from one or another form of phylogenetic analysis rather than observables. The principal observables in comparative genomics are sequence similarity between genes and the proteins they encode and, increasingly, structural similarity between proteins. These observations are employed, directly or indirectly (e.g., through phylogenetic analysis), to infer orthology and paralogy (or generic homology). Failure to distinguish between observables and inferences resulted in a persistent terminological morass. Strong homology, percentage homology, and similar oxymorons have survived in the biological literature for decades despite strongly worded refutations, and even guidelines regulating the usage of "homology" that have been adopted by some journals (81, 97). Because of these abuses but, more importantly, because biologists often consider evolutionary inferences to be inherently unreliable, suggestions have been made to dispense with the inferential terms (and, presumably, the underlying concepts) altogether. In a recent provocative article, Varshavsky proposed using the new, inference-free terms "sequelog" and "spalog" to designate, respectively, proteins that show sequence or structural similarity to each other (93). In an earlier eloquent comment, Petsko argued against using the terms ortholog and paralog on the simpler grounds that these terms unnecessarily complicate the narrative in research articles without clarifying anything (80). The desire for simplicity and use of neutral terms is understandable and would have been justified if orthology and paralogy (and all the derivatives thereof) were just words. However, as I argued here and elsewhere (46), this does not seem to be the case. Instead, orthology and paralogy appear to be concepts that carry substantive meaning and, even apart from the (perhaps, debatable to some) intrinsic interest of evolutionary relationships, have major functional connotations. Although the terms orthologs and paralogs may complicate the language of genomics, in my opinion, the clarification they bring to our understanding of the evolutionary and functional relationships among genes and genomes by far outweighs any inconvenience.

Generalized Concepts of Orthology and Paralogy

Throughout most of this review and other treatises, the concepts of orthology and paralogy are applied to genes as units (i.e., whenever we speak of orthologs, we mean orthologous genes). However, I also discussed complications to this approach stemming from gene fusion/fission and, less trivially, from lineage-specific changes in the domain architecture of orthologous proteins occurring during evolution. The latter situation challenges the gene-centric definition of orthology inasmuch as certain parts of genes appear to be orthologous whereas others are not. In principle, it seems possible to extend the notion of orthology to individual domains and, ultimately, to any stretch of nucleotide sequence down to a single base (97). The fundamental definition always remains the same: Genomic elements in the compared species that descend from the same ancestral element in the genome of their last common ancestor should be considered orthologous. From a maximalist standpoint, one could argue that evolutionary relationships within a set of genomes should be considered resolved only after the status of each base pair in each genome (both in coding and noncoding regions) is established with respect to orthology and paralogy. This could be an achievable goal for closely related genomes (e.g., human and chimpanzee) but seems to be unrealistic for distant species.

On the opposite, genome-wide scale, the notions of orthology and paralogy naturally apply not only to genes but to strings of genes that retain the ancestral order (conserved synteny blocks). In relatively closely related genomes (e.g., primates and rodents or different enterobacterial species), a conserved synteny block may include hundreds or even thousands of genes, whereas in distantly related genomes, there is very little conservation of gene order (7, 99). Thus, orthology and paralogy are manifest throughout all levels of genome comparison. Nevertheless, the gene-centric perspective adopted in the preceding sections appears to be most relevant for dissecting the results of comparison of multiple genomes separated by a wide range of evolutionary distances.

A different aspect of generalization of the concepts of orthology and paralogy pertains to the complex structure of orthologous gene clusters caused by the spread of duplication events over the phylogenetic tree. A single orthologous cluster defined at the deepest branching point of a tree is often resolved into several clusters within subtrees. The cases of tubulins and actins briefly discussed above in a different context clearly illustrate this point. All eukaryotic tubulins are co-orthologous to the single prokaryotic FtsZ proteins, just as actins are orthologous to the prokaryotic MreB. Within eukaryotes, however, several orthologous sets can be readily identified for each of these proteins.

CONCLUSIONS

Orthology and paralogy are not only key terms but are an integral part of the conceptual foundation of evolutionary and functional genomics. Only consistent usage of these and some derivative definitions, such as in- and outparalogs, provides for construction of a robust evolutionary classification of genes and reliable functional annotation of newly sequenced genomes. Further improvement of clustering and phylogenetic methods for identification of orthologs and paralogs is required for the progress of genomics as the number of sequenced genomes rapidly increases. Orthology and paralogy appear to be rich and flexible concepts that allow further development and are well suited to describe the complexity of genome evolution.

SUMMARY POINTS

1. Orthologs and paralogs are two types of homologous genes that evolved, respectively, by vertical descent from a single ancestral gene and by duplication.
2. Distinguishing between orthologs and paralogs is crucial for successful functional annotation of genomes and for reconstruction of genome evolution.
3. A finer classification of orthologs and paralogs has been developed to reflect the interplay between duplication and speciation events, and effects of gene loss and horizontal gene transfer on the observed homologous relationship.
4. Methods for identification of sets of orthologous and paralogous genes involve phylogenetic analysis and various procedures for sequence similarity–based clustering.
5. Analysis of clusters of orthologous and paralogous genes is instrumental in genome annotation and in delineation of trends in genome evolution.
6. Rearrangements of gene structure confound orthologous and paralogous relationships.
7. The gene-centered concepts of orthology and paralogy can be generalized downward, to the level of strings of nucleotides and even single base pairs, and upward, to multigene arrays.

ACKNOWLEDGMENTS

I thank Yuri Wolf, Marina Omelchenko, and Kira Makarova for invaluable help with data analysis; Yuri Wolf for critical reading of the manuscript; and Walter Fitch, Roy Jensen, Pavel Pevzner, Erik Sonnhammer, Alexander Varshavsky, and Emile Zuckerkandl for instructive discussions. Due to space constraints, it was impossible to cite all relevant publications in this review; my sincere apologies and appreciation to all colleagues whose important work is not cited.

LITERATURE CITED

1. Amos LA, van den Ent F, Lowe J. 2004. Structural/functional homology between the bacterial and eukaryotic cytoskeletons. *Curr. Opin. Cell Biol.* 16:24–31
2. Aravind L, Koonin EV. 2001. Prokaryotic homologs of the eukaryotic DNA-end-binding protein Ku, novel domains in the Ku protein and prediction of a prokaryotic double-strand break repair system. *Genome Res.* 11:1365–74
3. Arvestad L, Berglund AC, Lagergren J, Sennblad B. 2003. Bayesian gene/species tree reconciliation and orthology analysis using MCMC. *Bioinformatics* 19(Suppl.)1:i7–15
4. Benkovic SJ, Valentine AM, Salinas F. 2001. Replisome-mediated DNA replication. *Annu. Rev. Biochem.* 70:181–208
5. Bromham L, Penny D. 2003. The modern molecular clock. *Nat. Rev. Genet.* 4:216–24
6. Cutler DJ. 2000. Understanding the overdispersed molecular clock. *Genetics* 154:1403–17
7. Dandekar T, Snel B, Huynen M, Bork P. 1998. Conservation of gene order: a fingerprint of proteins that physically interact. *Trends Biochem. Sci.* 23:324–28
8. Darwin C. 1859. *On the Origin of Species*. London
9. Davis JC, Petrov DA. 2004. Preferential duplication of conserved proteins in eukaryotic genomes. *PLoS Biol.* 2:E55
10. Davison AJ, Scott JE. 1986. The complete DNA sequence of varicella-zoster virus. *J. Gen. Virol.* 67:1759–816
11. Della M, Palmbos PL, Tseng HM, Tonkin LM, Daley JM, et al. 2004. Mycobacterial Ku and ligase proteins constitute a two-component NHEJ repair machine. *Science* 306:683–85
12. Doerks T, von Mering C, Bork P. 2004. Functional clues for hypothetical proteins based on genomic context analysis in prokaryotes. *Nucleic Acids Res.* 32:6321–26
13. Doolittle WF. 1998. You are what you eat: A gene transfer ratchet could account for bacterial genes in eukaryotic nuclear genomes. *Trends Genet.* 14:307–11
14. Doolittle WF. 1999. Lateral genomics. *Trends Cell Biol.* 9:M5–8
15. Doolittle WF. 1999. Phylogenetic classification and the universal tree. *Science* 284:2124–29
16. Doolittle WF. 2000. Uprooting the tree of life. *Sci. Am.* 282:90–95
17. Dufayard JF, Duret L, Penel S, Gouy M, Rechenmann F, Perrière G. 2005. Tree pattern matching in phylogenetic trees: automatic search for orthologs or paralogs in homologous gene sequence databases. *Bioinformatics* 21:2596–603
18. Duret L, Mouchiroud D, Gouy M. 1994. HOVERGEN: a database of homologous vertebrate genes. *Nucleic Acids Res.* 22:2360–65
19. Eisen JA. 1998. Phylogenomics: improving functional predictions for uncharacterized genes by evolutionary analysis. *Genome Res.* 8:163–67
20. Eulenstein O, Mirkin B, Vingron M. 1998. Duplication-based measures of difference between gene and species trees. *J. Comput. Biol.* 5:135–48
21. Evguenieva-Hackenburg E, Walter P, Hochleitner E, Lottspeich F, Klug G. 2003. An exosome-like complex in Sulfolo*bus solfataricus*. *EMBO Rep.* 4:889–93
22. Fisher RA. 1928. The possible modification of the response of the wild type to recurrent mutations. *Am. Nat.* 62:115–26
23. **Fitch WM. 1970. Distinguishing homologous from analogous proteins. *Syst. Zool.* 19:99–106**
24. Fitch WM. 1995. Uses for evolutionary trees. *Philos. Trans. R. Soc. London Ser. B* 349:93–102

23. The classical work defining, for the first time, orthologs and paralogs as terms and concepts.

25. Fitch WM. 2000. Homology a personal view on some of the problems. *Trends Genet.* 16:227–31
26. Fleischmann RD, Adams MD, White O, Clayton RA, Kirkness EF, et al. 1995. Whole-genome random sequencing and assembly of *Haemophilus influenzae* Rd. *Science* 269:496–512
27. **Force A, Lynch M, Pickett FB, Amores A, Yan YL, Postlethwait J. 1999. Preservation of duplicate genes by complementary, degenerative mutations. *Genetics* 151:1531–45**
28. Fraser CM, Gocayne JD, White O, Adams MD, Clayton RA, et al. 1995. The minimal gene complement of *Mycoplasma genitalium*. *Science* 270:397–403
29. Fukuchi S, Nishikawa K. 2004. Estimation of the number of authentic orphan genes in bacterial genomes. *DNA Res.* 11:219–31, 311–13
30. Galperin MY, Koonin EV. 2004. 'Conserved hypothetical' proteins: prioritization of targets for experimental study. *Nucleic Acids Res.* 32:5452–63
31. Gillespie JH. 1984. The molecular clock may be an episodic clock. *Proc. Natl. Acad. Sci. USA* 81:8009–13
32. Gogarten JP. 1994. Which is the most conserved group of proteins? Homology-orthology, paralogy, xenology, and the fusion of independent lineages. *J. Mol. Evol.* 39:541–43
33. **Gray GS, Fitch WM. 1983. Evolution of antibiotic resistance genes: the DNA sequence of a kanamycin resistance gene from *Staphylococcus aureus*. *Mol. Biol. Evol.* 1:57–66**
34. Gray MW, Burger G, Lang BF. 2001. The origin and early evolution of mitochondria. *Genome Biol.* 2
35. Hannenhalli S, Chappey C, Koonin EV, Pevzner PA. 1995. Genome sequence comparison and scenarios for gene rearrangements: a test case. *Genomics* 30:299–311
36. **He X, Zhang J. 2005. Rapid subfunctionalization accompanied by prolonged and substantial neofunctionalization in duplicate gene evolution. *Genetics* 169:1157–64**
37. Huxley THH. 1860. 'The Origin of Species'. *Westminst. Rev.* 17:541–70
38. Huynen MA, van Nimwegen E. 1998. The frequency distribution of gene family sizes in complete genomes. *Mol. Biol. Evol.* 15:583–89
39. Jenkins C, Samudrala R, Anderson I, Hedlund BP, Petroni G, et al. 2002. Genes for the cytoskeletal protein tubulin in the bacterial genus Prosthecobacter. *Proc. Natl. Acad. Sci. USA* 99:17049–54
40. **Jensen RA. 2001. Orthologs and paralogs—we need to get it right. *Genome Biol.* 2: INTERACTIONS1002**
41. Jordan IK, Makarova KS, Spouge JL, Wolf YI, Koonin EV. 2001. Lineage-specific gene expansions in bacterial and archaeal genomes. *Genome Res.* 11:555–65
42. Jordan IK, Wolf YI, Koonin EV. 2004. Duplicated genes evolve slower than singletons despite the initial rate increase. *BMC Evol. Biol.* 4:22
43. Karlberg O, Canback B, Kurland CG, Andersson SG. 2000. The dual origin of the yeast mitochondrial proteome. *Yeast* 17:170–87
44. Kimura M. 1983. *The Neutral Theory of Molecular Evolution*. Cambridge: Cambridge Univ. Press
45. Kondrashov FA, Rogozin IB, Wolf YI, Koonin EV. 2002. Selection in the evolution of gene duplications. *Genome Biol.* 3: RESEARCH0008
46. **Koonin EV. 2001. An apology for orthologs—or brave new memes. *Genome Biol.* 2: COMMENT1005**

27. The idea of subfunctionalization as the mode of evolution of paralogs is introduced as an alternative to neofunctionalization.

33. This paper introduces the notion of xenology.

36. The latest study on functional diversification of paralogs integrates the previous models in the subneofunctionalization scheme whereby the subfunctionalization phase immediately after duplication is succeeded by neofunctionalization.

40. Continuation of the debate on the importance of orthologs and paralogs as concepts and terms. Emphasizes the importance of exact definitions, in particular, that the notion of paralogy applies not only to genes in the same genome.

46. Reply to the "Homologuephobia" comment of Petsko. Emphasizes that orthologs and paralogs are not just words but crucial concepts of evolutionary genomics.

47. Koonin EV. 2003. Comparative genomics, minimal gene-sets and the last universal common ancestor. *Nat. Rev. Microbiol.* 1:127–36
48. Koonin EV, Aravind L, Kondrashov AS. 2000. The impact of comparative genomics on our understanding of evolution. *Cell* 101:573–76
49. **Koonin EV, Fedorova ND, Jackson JD, Jacobs AR, Krylov DM, et al. 2004. A comprehensive evolutionary classification of proteins encoded in complete eukaryotic genomes. *Genome Biol.* 5:R7**
50. Koonin EV, Galperin MY. 2002. *Sequence—Evolution—Function. Computational Approaches in Comparative Genomics*. New York: Kluwer
51. Koonin EV, Makarova KS, Aravind L. 2001. Horizontal gene transfer in prokaryotes: quantification and classification. *Annu. Rev. Microbiol.* 55:709–42
52. Koonin EV, Mushegian AR, Bork P. 1996. Non-orthologous gene displacement. *Trends Genet.* 12:334–36
53. Koonin EV, Wolf YI, Karev GP. 2002. The structure of the protein universe and genome evolution. *Nature* 420:218–23
54. Kummerfeld SK, Teichmann SA. 2005. Relative rates of gene fusion and fission in multidomain proteins. *Trends Genet.* 21:25–30
55. Kunin V, Ouzounis CA. 2003. The balance of driving forces during genome evolution in prokaryotes. *Genome Res.* 13:1589–94
56. Lang BF, Gray MW, Burger G. 1999. Mitochondrial genome evolution and the origin of eukaryotes. *Annu. Rev. Genet.* 33:351–97
57. Lawrence JG, Hendrickson H. 2003. Lateral gene transfer: When will adolescence end? *Mol. Microbiol.* 50:725–27
58. Lowe J, van den Ent F, Amos LA. 2004. Molecules of the bacterial cytoskeleton. *Annu. Rev. Biophys. Biomol. Struct.* 33:177–98
59. Lynch M, Conery JS. 2000. The evolutionary fate and consequences of duplicate genes. *Science* 290:1151–55
60. Lynch M, Force A. 2000. The probability of duplicate gene preservation by subfunctionalization. *Genetics* 154:459–73
61. Lynch M, Katju V. 2004. The altered evolutionary trajectories of gene duplicates. *Trends Genet.* 20:544–49
62. Makarova KS, Aravind L, Galperin MY, Grishin NV, Tatusov RL, et al. 1999. Comparative genomics of the Archaea (Euryarchaeota): evolution of conserved protein families, the stable core, and the variable shell. *Genome Res.* 9:608–28
63. **Mirkin B, Muchnik I, Smith TF. 1995. A biologically consistent model for comparing molecular phylogenies. *J. Comput. Biol.* 2:493–507**
64. Mirkin BG, Fenner TI, Galperin MY, Koonin EV. 2003. Algorithms for computing parsimonious evolutionary scenarios for genome evolution, the last universal common ancestor and dominance of horizontal gene transfer in the evolution of prokaryotes. *BMC Evol. Biol.* 3:2
65. Nogales E, Downing KH, Amos LA, Lowe J. 1998. Tubulin and FtsZ form a distinct family of GTPases. *Nat. Struct. Biol.* 5:451–58
66. **Novichkov PS, Omelchenko MV, Gelfand MS, Mironov AA, Wolf YI, Koonin EV. 2004. Genome-wide molecular clock and horizontal gene transfer in bacterial evolution. *J. Bacteriol.* 186:6575–85**
67. O'Brien KP, Remm M, Sonnhammer EL. 2005. Inparanoid: a comprehensive database of eukaryotic orthologs. *Nucleic Acids Res.* 33 Database Issue:D476–80
68. Ochman H. 2002. Distinguishing the ORFs from the ELFs: short bacterial genes and the annotation of genomes. *Trends Genet.* 18:335–37

49. Description of the first collection of sets of probable orthologs from 7 sequenced eukaryotic genomes (KOGs). Reports analysis of various evolutionary patterns in KOGs, including lineage-specific gene loss, functional characteristics of one-to-one orthologs, and quantitative assessment of domain accretion.

63. The first method for tree reconciliation, in principle, the approach of choice for identification of orthologs.

66. An assessment of the validity of molecular clock on genome scale. Shows that the majority of clusters of one-to-one orthologs evolve in the clock-like mode but also that a significant minority experienced XGD.

69. Ohno S. 1970. *Evolution by Gene Duplication*. New York: Springer-Verlag
70. Omelchenko MV, Makarova KS, Wolf YI, Rogozin IB, Koonin EV. 2003. Evolution of mosaic operons by horizontal gene transfer and gene displacement in situ. *Genome Biol.* 4:R55
71. Ouzounis C. 1999. Orthology: another terminology muddle. *Trends Genet.* 15:445
72. Owen R. 1848. *On the Archetype and Homologies of the Vertebrate Skeleton*. London: Murray
73. Page RD, Charleston MA. 1997. From gene to organismal phylogeny: reconciled trees and the gene tree/species tree problem. *Mol. Phylogenet. Evol.* 7:231–40
74. Panchen AL. 1994. Richard Owen and the concept of homology. In *Homology: The Hierarchical Basis of Comparative Biology*, ed. BK Hall, pp. 21–62. San Diego: Academic
75. Papp B, Pal C, Hurst LD. 2003. Dosage sensitivity and the evolution of gene families in yeast. *Nature* 424:194–97
76. Patterson C. 1988. Homology in classical and molecular biology. *Mol. Biol. Evol.* 5:603–25
77. Pennisi E. 1998. Genome data shake tree of life. *Science* 280:672–74
78. Pennisi E. 2001. Microbial genomes. Sequences reveal borrowed genes. *Science* 294:1634–35
79. Perrière G, Duret L, Gouy M. 2000. HOBACGEN: database system for comparative genomics in bacteria. *Genome Res.* 10:379–85
80. Petsko GA. 2001. Homologuephobia. *Genome Biol.* 2: COMMENT1002
81. Reeck GR, de Haen C, Teller DC, Doolittle RF, Fitch WM, et al. 1987. "Homology" in proteins and nucleic acids: a terminology muddle and a way out of it. *Cell* 50:667
82. Remm M, Storm CE, Sonnhammer EL. 2001. Automatic clustering of orthologs and inparalogs from pairwise species comparisons. *J. Mol. Biol.* 314:1041–52
83. Snel B, Bork P, Huynen M. 2000. Genome evolution: gene fusion versus gene fission. *Trends Genet.* 16:9–11
84. Snel B, Bork P, Huynen MA. 2002. Genomes in flux: the evolution of archaeal and proteobacterial gene content. *Genome Res.* 12:17–25
85. Sonnhammer EL, Koonin EV. 2002. Orthology, paralogy and proposed classification for paralog subtypes. *Trends Genet.* 18:619–20
86. Storm CE, Sonnhammer EL. 2002. Automated ortholog inference from phylogenetic trees and calculation of orthology reliability. *Bioinformatics* 18:92–99
87. Storm CE, Sonnhammer EL. 2003. Comprehensive analysis of orthologous protein domains using the HOPS database. *Genome Res.* 13:2353–62
88. Tatusov RL, Fedorova ND, Jackson JD, Jacobs AR, Kiryutin B, et al. 2003. The COG database: an updated version includes eukaryotes. *BMC Bioinformat.* 4:41
89. Tatusov RL, Koonin EV, Lipman DJ. 1997. A genomic perspective on protein families. *Science* 278:631–37
90. Tatusov RL, Mushegian AR, Bork P, Brown NP, Hayes WS, et al. 1996. Metabolism and evolution of *Haemophilus influenzae* deduced from a whole-genome comparison with *Escherichia coli*. *Curr. Biol.* 6:279–91
91. Tatusov RL, Natale DA, Garkavtsev IV, Tatusova TA, Shankavaram UT, et al. 2001. The COG database: new developments in phylogenetic classification of proteins from complete genomes. *Nucleic Acids Res.* 29:22–28
92. van den Ent F, Amos LA, Lowe J. 2001. Prokaryotic origin of the actin cytoskeleton. *Nature* 413:39–44

69. A seminal work, the first to present a coherent concept of gene duplication as a major formative force of evolution.

72. Introduces homology referring to "the same organ in different animals under every variety of form and function".

80. A witty comment that sparked the discussion of the meaning and importance of the terms orthologs and paralogs.

81. An early condemnation of incorrect uses of the term homology (as in "percent homology," "strong homology" etc). Emphasizes that homology should be used exclusively to refer to common origin of genes (proteins).

82. Introduces the terms in- and outparalogs.

85. Conceptualizes and explains the notions of in- and outparalogs, and coorthologs.

89. Description of the first method for identifying clusters

of orthologs in multiple genomes and the first version of Clusters of Orthologous Groups of proteins (COGs).

93. The latest twist in the debate on homology, orthology and paralogy. Inference-free terms are proposed to designate sequence and structural similarity between proteins.

105. This and the preceding paper are seminal works that laid the foundation of molecular evolution and include discussion of different types of homologous relationships presaging the concepts of orthology and paralogy.

93. Varshavsky A. 2004. 'Spalog' and 'sequelog': neutral terms for spatial and sequence similarity. *Curr. Biol.* 14:R181–83

94. Veitia RA. 2004. Gene dosage balance in cellular pathways: implications for dominance and gene duplicability. *Genetics* 168:569–74

95. Veitia RA. 2005. Gene dosage balance: deletions, duplications and dominance. *Trends Genet.* 21:33–35

96. Wang C, Xi J, Begley TP, Nicholson LK. 2001. Solution structure of ThiS and implications for the evolutionary roots of ubiquitin. *Nat. Struct. Biol.* 8:47–51

97. Webber C, Ponting CP. 2004. Genes and homology. *Curr. Biol.* 14:R332–33

98. Wolf YI, Rogozin IB, Grishin NV, Koonin EV. 2002. Genome trees and the tree of life. *Trends Genet.* 18:472–79

99. Wolf YI, Rogozin IB, Kondrashov AS, Koonin EV. 2001. Genome alignment, evolution of prokaryotic genome organization and prediction of gene function using genomic context. *Genome Res.* 11:356–72

100. Yanai I, Wolf YI, Koonin EV. 2002. Evolution of gene fusions: horizontal transfer versus independent events. *Genome Biol.* 3:research0024

101. Zawel L, Reinberg D. 1995. Common themes in assembly and function of eukaryotic transcription complexes. *Annu. Rev. Biochem.* 64:533–61

102. Zhang L. 1997. On a Mirkin-Muchnik-Smith conjecture for comparing molecular phylogenies. *J. Comput. Biol.* 4:177–87

103. Zmasek CM, Eddy SR. 2002. RIO: Analyzing proteomes by automated phylogenomics using resampled inference of orthologs. *BMC Bioinformat.* 3:14

104. Zuckerkandl E, Pauling L. 1962. Molecular evolution. In *Horizons in Biochemistry*, ed. M Kasha, B Pullman, pp. 189–225. New York: Academic

105. Zuckerkandl E, Pauling L. 1965. Evolutionary divergence and convergence of proteins. In *Evolving Gene and Proteins*, ed. Bryson V, Vogel HJ, pp. 97–166. New York: Academic

The Moss *Physcomitrella patens*

David Cove

Center for Plant Sciences, University of Leeds, Leeds LS2 9JT, United Kingdom, and Department of Biology, Washington University in St. Louis, St. Louis, Missouri 63130; email: d.j.cove@leeds.ac.uk

Annu. Rev. Genet.
2005. 39:339–58

First published online as a Review in Advance on August 31, 2005

The *Annual Review of Genetics* is online at http://genet.annualreviews.org

doi: 10.1146/annurev.genet.39.073003.110214

0066-4197/05/1215-0339$20.00

Key Words

bryophytes, moss, gene targeting, homologous recombination, RNA interference

Abstract

The moss *Physcomitrella patens*, like seed plants, shows alternation of generations, but its gametophyte, the haploid phase of the life cycle, is dominant, making it ideal for genetic studies. Crosses show direct segregations, so F2 or test crosses are unnecessary. Mutagenesis yields mutants, the phenotype of which is directly evident. Haploid tissue can be propagated vegetatively, allowing the maintenance of mutants blocked early in development. Protoplasts, isolated from filamentous gametophytic tissue, regenerate directly into filamentous tissue, providing an abundant supply of single haploid cells for transformation. Recombination occurs at a high frequency between genomic sequences in transforming DNA and the corresponding chromosomal sequences, allowing precise inactivation or modification of genes. RNAi technology allows the inactivation of the expression of gene families and the partial knockdown of essential genes. Over 100,000 ESTs have been sequenced and annotated, and sequencing of the genome should be completed by the end of 2005.

Contents

INTRODUCTION

Gametophyte: the haploid stage of the life cycle, comprising protonema and gametophores, upon which gametes are produced

Sporophyte: the diploid phase of the life cycle, comprising the spore capsule, in which spores are produced by meiosis, borne on a short seta

Protonema: the filamentous stage of gametophyte development, generated by spore germination or tissue regeneration

This review focuses on genetic studies, in vivo and in vitro, using the moss *Physcomitrella patens*. A companion review, to be published later this year in the *Annual Review of Plant Biology*, will concentrate on developmental and metabolic studies using *P. patens* and other moss species. The use of mosses for genetic studies is not new [see for example (69)]. The haploidy of the dominant gametophyte stage in moss development makes mosses attractive material for genetic studies because mutant isolation and genetic analysis are simpler than in species with a dominant diploid phase. Most early genetic work on mosses used other species; *P. patens* began to be used extensively for genetic studies only after Engel's publication in 1968 (21). When I decided to develop *P. patens* for developmental genetic studies (3), I was already familiar with the attractions of a dominant haploid phase from my research using the fungus *Aspergillus nidulans*. Although Engel's work was an impetus, the advice and encouragement of Dr. Harold Whitehouse to choose an ephemeral species were crucial; he should be regarded as one of the founders of modern genetic studies using mosses. Whitehouse isolated the Gransden strain of *P. patens*, which originated from a single spore and was used by Engel, then by my research group, and subsequently for most genetic studies of *P. patens*.

The past ten years have seen greatly increased use of this species for genetic studies. The impetus came initially from the description of an efficient transformation procedure (60), bolstered by the finding that recombination occurs at a high frequency between transforming DNA containing a genomic sequence and the corresponding homologous genomic sequence (37, 58), thus allowing gene targeting and allele modification. As a result of this surge in interest, *P. patens* has been chosen this year as one of the first two nonflowering plant for genome sequencing (the other is *Selaginella moellendorffii*).

LIFE CYCLE

The Natural Life Cycle

In common with ferns and seed plants, mosses show alternation of generations: a haploid phase that produces gametes (the gametophyte generation) and a diploid phase that produces haploid spores by meiosis (the sporophyte generation). In contrast to ferns and seed plants, the gametophyte is the dominant phase and this generation comprises most of what is familiar to us as moss plants.

Spores germinate to produce the protonemal stage of the gametophyte. This comprises filaments of cells that extend by the serial division of their apical cells. The first filaments produced consist of chloronemal cells, which are densely packed with large chloroplasts. The apical cells of chloronemal filaments extend at a rate of 2–5 μm h^{-1} and divide every 22–26 h, arresting for most of their cell cycle in G2 (54, 62). The subapical cells also divide, but seldom more than twice, to produce more chloronemal side branches. Some chloronemal apical cells develop into caulonemal cells, and these give rise to the second type of protonemal filament. Caulonemal cells contain fewer, less well developed chloroplasts. The apical cells of caulonemal filaments extend at a rate of 25 to 40 μm h^{-1} and divide every 6–8 h. Caulonemal cells have a long G1 and short G2 phase (62). Caulonemal subapical cells become increasingly polyploid as they age (54). The two filaments types, chloronemal and caulonemal, therefore appear to have assimilatory and adventitious roles, respectively. The side branches from most subapical cells of caulonemal filaments develop into chloronemal filaments, but a few side branches develop into either caulonemal filaments or leafy shoots known as gametophores (**Figure 1**), on which gametangia later develop.

Male gametes, or spermatozoids, are produced in antheridia and are motile, having flagella, whereas female gametes are produced in archegonia. Antheridia and archegonia are produced on the same shoot, and self-fertilization is usual. Moist conditions are essential to allow spermatozoid motility. The fertilized zygote develops into a sporophyte consisting of a short seta bearing a spore capsule, which, when mature, contains about 4000 spores (21).

P. patens is an ephemeral, developing in early summer, generally from overwintered spores. It is distributed widely in temperate zones but is not common; it grows beside lakes, rivers, and ditches on soil that has been exposed by falling water levels, and on damp open ground in fields (64). Temperatures below 18°C are required to induce gametogenesis (21), and short day lengths enhance gametangia induction (28). As a result, sporophytes are produced in the late summer, and overwintering occurs as spores. Isolates are now available from a number of European, Japanese, and North American locations. A number of subspecies are recognized, and the Gransden wild-type strain that is used widely is *P. patens patens*.

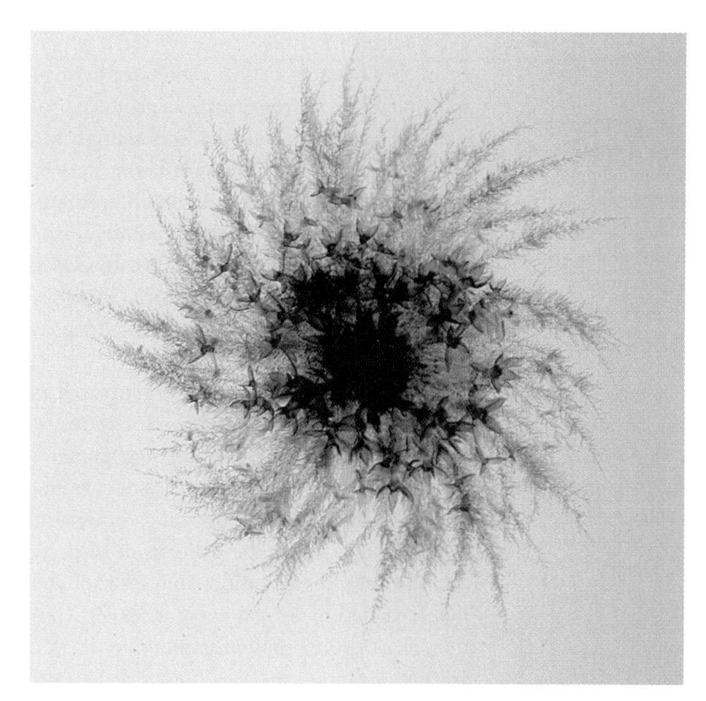

Figure 1

Gransden strain of *P. patens*, grown in continuous light for 21 days on BCD medium (44). The diameter of the plant is circa 30 mm.

Experimental Variations

In culture, different wild-type isolates of *P. patens* vary in the time taken to complete their life cycle. The Gransden strain takes between 3 to 4 months (21), somewhat slower than some other isolates (Y. Kamisugi & D. Cove, unpublished data). For most experimental procedures, culture is carried out at 25°C, but to induce gametogenesis and sporophyte production, a temperature of 15°C is used routinely, usually coupled with an 8-h light + 16-h dark regime. *P. patens* can be cultured in continuous light, but many laboratories use a 16-h light + 8-h dark regime. There is little difference in development in these two regimens, although the time taken to landmarks such as bud production is governed by the number of hours of illumination received by a culture, and hence development is more rapid in continuous light (D. Cove, unpublished data).

Chloronema(ta): slow-growing filament containing cells with many large chloroplasts, the assimilatory component of protonemal tissue (q.v.)

Caulonema(ta): rapidly growing filament containing cells with few poorly developed chloroplasts, the adventitious component of protonemal tissue (q.v.)

Gametophores: leafy shoots, arising as side branches of caulonemal filaments, upon which archegonia and antheridia develop

Antheridia: male gametangia, in which motile male gametes (spermatozoids) are produced

Archegonia: female gametangia, in which female gametes are produced

P. patens has a high capacity for regeneration. Small pieces of either gametophytic or sporophytic tissue regenerate to produce protonemal tissue, behaving developmentally in a manner similar to that of germinating spores. Protonemal tissue can be disrupted mechanically, for example with a tissue homogenizer, and the resulting tissue fragments can be used as an innoculum for further culture. This provides a routine method of culture maintenance (26).

Sporophytic tissue can be regenerated to produce diploid gametophytes, which develop in a way similar to that of haploid gametophytes, but more slowly. The rare occurrence of this pathway of regeneration (=apospory) in nature may allow the generation of polyploid strains. Regeneration from sporophytes that have developed following cross fertilization gives rise to diploid gametophytes that are heterozygous and can therefore be used to test for dominance and complementation (21). However, the experimental production of hybrid gametophytes by protoplast fusion (26) is an easier alternative for complementation testing.

Diploid gametophytes occasionally produce sporophytes that arise either as a result of the fusion of diploid gametes, to produce a tetraploid sporophyte (26), or by the development of a female gamete without fertilization (=parthenogenesis). Tetraploid sporophytes undergo regular meiosis to produce diploid spores. The resulting progeny show characteristic segregation ratios, e.g., for a two-allele, single-locus segregation, a ratio of 15:1 for the dominant to the recessive phenotype (17). Parthenogenetically produced sporophytes are diploid and produce haploid spores (17). The direct development of a sporophyte from gametophytic tissue (=apogamy) has not been reported for wild-type strains but has been shown to occur regularly for at least one hybrid between two mutants that overproduce gametophores (22).

Enzymatic digestion of young protonemal tissue can generate large numbers of isolated protoplasts. The original method of protoplast isolation (26) has been improved (63) and a number of variations are now used (31, 42), but these have in common the use of young protonemal tissue, usually 5 to 7 days post regeneration. The methods vary in the procedure used to obtain tissue (growth in liquid medium or on cellophane overlaying solid medium) and in medium composition. It is difficult to compare yields between the different methods, but in our laboratories, from 2–3 × 10^6 protoplasts are obtained from a 90 mm Petri dish of tissue, grown on cellophane overlaying solid medium at 25°C for 6 days in continuous light. Thus protoplast numbers are seldom a limiting factor in their experimental use.

Protoplasts regenerate on osmotically buffered medium at a high frequency (50% to 90%). By plating protoplasts in a thin layer of medium on cellophane overlaying solid medium, protoplasts can be transferred to standard medium after 4 to 5 days at 25°C, once cell wall formation has rendered them osmotically robust. Regeneration of protoplasts gives rise directly to protonemal tissue, in a manner essentially similar to that of germinating spores. Consequently, protoplasts are employed as the starting material for mutagenesis, transformation, and somatic hybridization.

PHYLOGENY AND GENOME

Phylogenetic Relationships

The true mosses (Bryopsida), together with liverworts (Hepaticopsida), comprise two classes of the phylum Bryophyta. Phylogenetic studies indicate that the two classes represent distinct but related clades (7, 51, 66). Bryophytes are believed to have shared a common ancestor with flowering plants 200–400 Mya; the two lineages diverged early in land plant evolution. Within the Bryopsida, *P. patens* is a member of the *Funariales* and hybridizes with other members of the genus (69). Studies comparing the transcriptome of *P. patens* with that of *Arabidopsis thaliana* show that at least 66% of *A. thaliana* genes have homologues in *P. patens* (50).

Genome Size and Chromosome Number

Using flow cell cytometry of DAPI stained cells, the haploid genome size of *P. patens* has been estimated to be 0.53 pg, corresponding to 511 Mb (62). Feulgen staining gives a figure of 0.48–0.56 pg (500–540 Mb) (J. Greilhuber, unpublished results). Moss species show a surprisingly small range of genome sizes compared with angiosperms (67) and the figure for *P. patens* is not exceptional. In common with most mosses, the chromosomes of *P. patens* are very small, which makes cytogenetic analysis difficult. Various estimates of chromosome numbers have been published. Although there are reports of $n = 16$ (68) and $n = 14$ (21), the emerging consensus is for $n = 27$ (13, 55), a figure higher than for the majority of moss species. Since two of these disparate estimates are for the Gransden strain (21, 55), the differences are unlikely to be due to the existence of different races. The construction of a genetic map, which is now being undertaken, should resolve this issue in the near future.

Over 100,000 ESTs from *P. patens* are publicly available. These ESTs have been analyzed and organized into over 25,000 putative transcripts and annotated (46, 50). In addition, a further 110,000 ESTs have been sequenced but are not yet available in the public domain (53).

Chloroplast Genome

The chloroplast genome of *P. patens* has been sequenced (66). The genome is 122,890 base pairs in length and codes 83 proteins, 31 tRNAs and four rRNAs. The chloroplast genome of *P. patens* is very similar to that of the liverwort *Marchantia polymorpha*, but whereas the *rpoA* gene is located in the chloroplast genome in the liverwort, it is a nuclear gene in the moss. The moss chloroplast genome also lacks the *cysA* and *cysT* genes and contains a *ycf3* gene with two introns. These features distinguish it from *M. polmorpha*, but are shared with vascular plants, indicating an early divergence of mosses from liverworts, with the mosses more closely related to the vascular plants (66). However, phylogenetic analysis of the deduced amino acid sequences of 51 chloroplast-encoded genes provides strong support for the bryophytes as a monophyletic group distinct from extant vascular plants (51).

DAPI: 4′-6-diamidino-2-phenylindole

EXPERIMENTAL CULTURE

P. patens grows well on sterilized commercial compost mixtures and these allow the most rapid completion of the life cycle. However, it is usually cultured on agar-containing medium in petri dishes or other suitable containers (42) (see **Figure 1**). There are small variations in the composition of the basal medium used by different laboratories, but all are modified Knop's medium, with a pH of about 6.5. These media contain no sugars or other organic compounds, and nitrate is the nitrogen source. Growth rates are increased by including ammonium as a nitrogen source. By overlaying agar medium with a sterile cellophane film, plant growth into the medium is prevented, making tissue harvest easier (26).

Growth in liquid medium is possible either in batch culture (70), in an airlift fermenter (11) or in a stirred bioreactor (29). Growth rates are enhanced if air, enriched with CO_2, is bubbled into the culture (11). When liquid cultures are not diluted, development proceeds in a manner similar to growth on solid medium, but when diluted to maintain a constant cell density, development is arrested at the protonemal stage. In the airlift fermenter, tissue is constantly passed through a course homogenizer, which restricts the size of tissue fragments and induces constant regeneration.

The role of cytokinin and auxin in development has been investigated using a culture method in which plants are allowed to develop on a supported nylon mesh while being continuously irrigated with fresh medium. Under these conditions, development is

NTG: N-methyl-N′-nitro-N-nitrosoguanidine

arrested at the chloronemal stage, but caulonemal production can be induced by the addition of auxin to the medium supply, while gametophore production is restored by the addition of both auxin and cytokinin (18).

A temperature of about 25°C is optimal for growth, although growth rates are only slightly lower at 20°C. Light quality and intensity both affect development and routinely light provided by fluorescent tubes at intensities between 5 and 20 Wm^{-2} is used. Although many laboratories use a 16 h light + 8 h dark regime, there appears to be no marked photoperiod requirement for gametophyte morphogenesis. Using regimens ranging from 4 h light + 20 h dark to continuous illumination, development proceeds in a qualitatively similar manner in all conditions, except that the time taken to achieve a developmental landmark such as gametophore production is dependent on the number of hours of light received (D. Cove, unpublished data). Continuous illumination thus achieves the most rapid development; however, a regime of 16 h light + 8 h dark synchronizes chloronemal division so that mitosis occurs shortly before dawn, and the cells are arrested in G2 during the day (54). This difference in light treatment may be a significant source of variation between laboratories employing different light regimen.

Circadian rhythmicity of the expression of *Lhcb2* genes in protonemal tissue is evident in a 12 h light + 12 h dark growth regime, when mRNA levels peak during the light period and decline during darkness (2). This rhythm dampens rapidly upon transfer of entrained cultures to continuous darkness, but in media containing 4.5% glucose, at least three further cycles of rhythmic expression are observed. In contrast to the behavior of the corresponding genes in higher plants, entrained cultures of *P. patens* showed a rapid loss of rhythmicity upon transfer to continuous light (2).

For long-term storage of strains, tissue can be frozen and stored in liquid nitrogen for many years (25, 61), but recovery upon thawing is unreliable, particularly for some mutant strains. Alternatively, tissue may be stored in darkness at 4°C in distilled water for at least one year, but a more convenient method for strain maintenance involves growth on basal medium for two weeks, followed by transfer to a regime of 2 h light + 22 h dark at 8–10°C. It is convenient to grow cultures in 12-well plates, with 3 ml of medium in each well (D. Cove, unpublished data). Under these conditions, cultures do not age for at least two years and may be propagated immediately by removal of small pieces of tissue onto fresh medium. Without a light treatment, cultures age rapidly and die within a few months.

MUTAGENESIS USING IRRADIATION OR CHEMICALS

Although bryophyes were among the first species to be used to show the mutagenic effects of ionizing radiation (38), it was not until 1968 that Engel published the first report of mutagenesis using *P. patens* (21). He employed X rays, UV irradiation, and the alkylating agents ethyl methane sulfonate and N-methyl-N′-nitro-N-nitrosoguanidine (NTG) for mutagenic treatment of spores and obtained auxotrophic and developmentally abnormal strains. Using similar mutagenic procedures, later studies have extended the range of mutants to include further auxotrophs (3) and mutants altered in leaf shape (16), in the production of auxins and cytokinins (4, 6), in their response to light and gravity (19, 34), and in chloroplast division (1), a phenotype that can be phenocopied by treatment with some antibiotics (see **Figure 2*b*,*c***). The requirement for spores in mutagenic procedures can be by-passed by treating protonemal tissue with mutagen, and then digesting to release protoplasts. These protoplasts regenerate to give plants that have developed from single cells (12). More recently, mutants have been isolated following the insertion of transforming DNA. These procedures are described in the next section.

In early publications, gene designation followed the conventions employed in bacteria

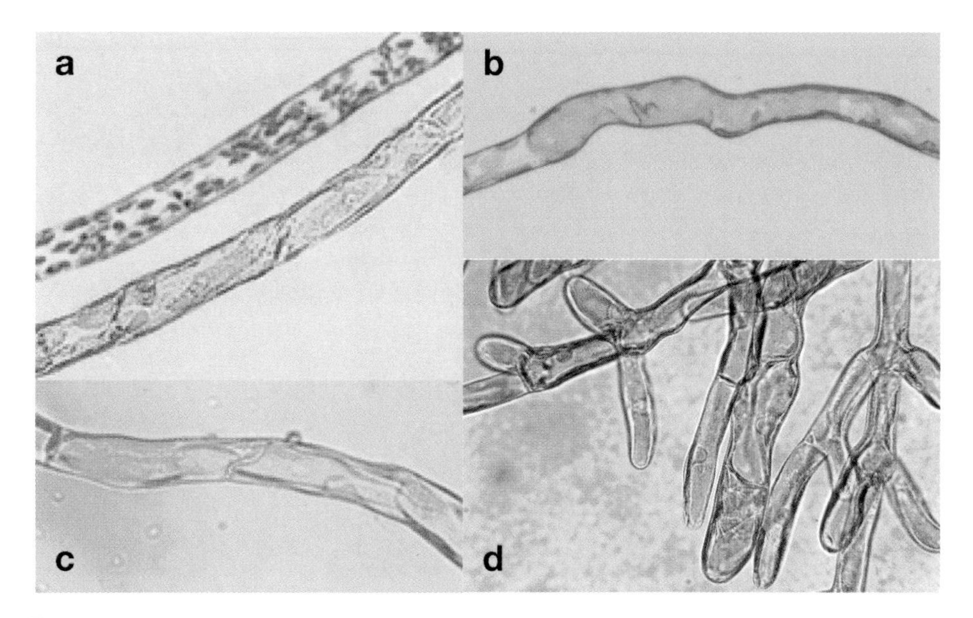

Figure 2

Genetic and environmental inhibition of chloroplast division in chloronemal filaments. Four different treatments give identical phenotypes. *A*: Upper filament: wild type; lower filament: transgenic strain with FtsZ 2-1 gene inactivated by HR (65). *B*: Filament of wild type grown in the presence of ampicillin. *C*: Filament of PC22, an X ray-induced mutant (1). *D*: Filament of an RNAi FtsZ transgenic (9). The chloronemal filaments are 18–20 μm in diameter. Images *A*, *B*, and *C* are by courtesy of Prof. R. Reski; *D* is by courtesy of Dr. M. Bezanilla.

and filamentous fungi, whereby complementary genes share a similar three-letter code followed by a different capitalized letter [e.g., *gtr*A, *gtr*C (41)]. More recently, the convention used in *A. thaliana* and *Sacchromyces cerevisiae*, in which complementary genes are distinguished by numbers, has been adopted by many laboratories. A uniform gene designation system would be helpful.

TRANSFORMATION

Direct DNA Uptake by Protoplast

Genetic transformation techniques for *P. patens* were first developed using polyethylene-glycol (PEG)–mediated DNA uptake by isolated protoplast (60). The procedure used plasmids containing a cassette coding for resistance to either geneticin (G418) or hygromycin, and gave rise to three classes of antibiotic-resistant regenerants. One class cannot be successfully subcultured onto further selective media and so is likely to express the transgenes only transiently. The two other classes retain their transgenic phenotype upon subculture. Of the two classes, the most frequent grows slowly and irregularly on selective medium, retaining an antibiotic-resistance phenotype for many years upon serial subculture on antibiotic-containing medium. However, the transgenes are lost rapidly when selection is relaxed (59). The loss of transgenic sequences upon the relaxation of selection has been confirmed using PCR (5). Members of this class are therefore called unstable transformants. Their behavior suggests that the transgenic DNA is replicated extrachromosomally, and although this remains the most likely explanation, it is difficult to obtain direct experimental confirmation. Southern blot analysis shows that the transgenes can occur in concatemers of between 3 and 40 copies, which may be the form

PEG: polyethyleneglycol

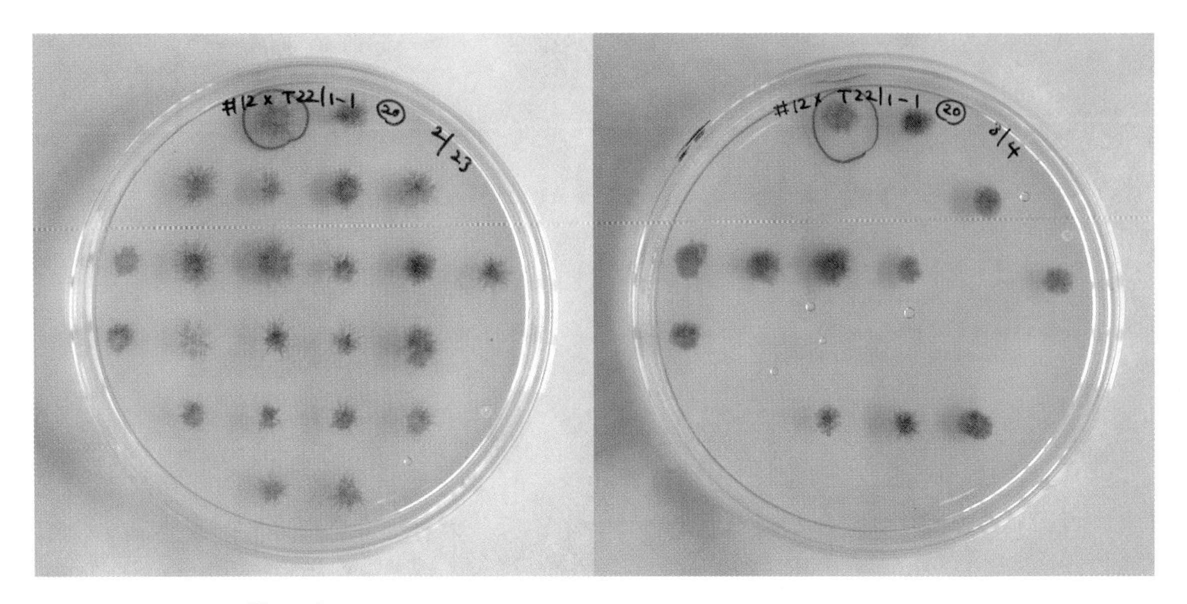

Figure 3

Growth test of 23 sample progeny from a cross between a *P. patens* wild type collected in Villersexel, Haute Saone, France, by Michael Lüth, and a hygromycin-resistant transgenic derivative of the Gransden wild type. Progeny have been grown on medium without (*left*) and with (*right*) hygromycin. The petri dishes are 90 mm in diameter. Images are by courtesy of Dr. Y. Kamisugi.

in which transgenic material is replicated (5). In unstable transformants, transgene expression is confined to the protonemal tissue and cannot be detected in gametophores, nor are transgenes transmitted through meiosis (59).

The other class of transformant grows strongly on selective medium and retains transforming DNA even when selection is relaxed; hence, such transformants are said to be stable. Southern blot analysis of stable transformants confirms that the transgenes have been inserted into genomic DNA and that there may be multiple copies of the transgenes inserted at the same locus (59). When stable transformants are crossed, the transgenes are inherited through meiosis in a regular Mendelian manner (60) (see **Figure 3**). Resolution of concatemers down to a single copy can be achieved by including *lox* sites in the transforming DNA sequence and retransforming a transformant with a plasmid expressing the *cre* recombinase (14). Transient expression of *cre* recombinase is sufficient to achieve resolution (P.-F. Peroud, unpublished data).

Various measures of transformation efficiency have been used by those employing the PEG method. The proportion of viable protoplasts that regenerate on selective medium is one measure used, but there is a considerable spread of vigor among regenerants. Some regenerants grow as strongly on selective medium as on medium containing no antibiotic, whereas others may produce only a few cells. What constitutes a regenerant is therefore difficult to standardize and may account for differences in this measure between laboratories. Regeneration is also affected by the light regime used, and by the number of days for which selection is applied.

Another regularly used measure is the number of regenerants per μg of transforming DNA. Once again, the definition of a regenerant makes this measure unsatisfactory. It would also be more logical to take into account the molecular weight of the transforming DNA rather than the absolute weight of DNA used. Some studies only give the proportions of the two classes of transformant.

In studies in which the physical state of the DNA has been examined, the overall rate of transgenic production is higher when supercoiled plasmid DNA is used, but the frequency of stable transformants is similar (60). Using supercoiled DNA, a typical frequency for the generation of unstable transformants is 1 in 10^3 protoplasts, or 50 to 100 per μg of transforming DNA, and for stable transformants, in the range 0.1 to 1% of these rates (40, 57). The effect on the rate of transformation of the inclusion of sequences within the transforming DNA that are homologous to genomic sequences is discussed below.

Other Transformation Methods

It is also possible to transform *P. patens* biolistically by microprojectile bombardment (56). DNA is shot into a mat of protonemal tissue growing on cellophane and transgenics are recovered by blending the tissue and plating onto selective medium. As with transformation using PEG-mediated DNA uptake by protoplasts, this transformation method also generates transient, unstable, and stable transformants (8). It is difficult to compare the efficiency of the two transformation methods quantitatively. As microprojectile bombardment uses less DNA than the PEG method, in terms of stable transformants per μg DNA, it can therefore be said to be more efficient, with a frequency of stable transformation about 10 times higher than the PEG method (15). Heat treatment of the DNA prior to shooting increased transformation frequencies a further fivefold (15). The two methods of transformation require similar resource inputs (although microprojectile bombardment requires access to a gene gun) and can generate similar numbers of transformants, but microprojectile bombardment can also be used to observe transient expression of transgenes in individual cells (M. Bezanilla, unpublished data).

Early attempts to use *Agrobacterium tumefaciens* to promote transformation of *P. patens* were unsuccessful (C.D. Knight & B. Hohn, unpublished data), but recently, new virulent strains of *A. tumefaciens* have been used successfully (D.G. Schaefer, unpublished data) and may generate stable transgenics at higher rates than either the PEG or biolistic methods (Y. Kamisugi, unpublished data).

CaMV: *Cauliflower mosaic virus*

Selection Cassettes, Promoters and Reporter Genes

The cassette most commonly used for the selection of *P. patens* transformants contains the *npt*II gene, conferring resistance to geneticin, coupled to the nopaline synthetase terminator, and driven by the *Cauliflower mosaic virus* (CaMV) 35S promoter (60). However, similar cassettes containing either the *hph* (48) or *aph*IV gene (33, 60), conferring resistance to hygromycin, or the *zeo* gene, conferring resistance to zeocin (Invitrogen) (P.-F. Perroud & M. Bezanilla, unpublished data), also function well for transformant selection. The *sul* gene, conferring resistance to sulphonamide, is used less commonly (37) and requires a longer period of culture to distinguish resistant from nonresistant plants (D. Cove, unpublished data).

Reporter genes commonly used in seed plants also function well in *P. patens*. Most early studies using reporter genes to assess promoter activity used the *uid*A (*gus*) gene, coding for β-glucuronidase (39). This gene has the advantage that it provides a sensitive assay for gene expression, but the disadvantage that the assay is usually lethal. The gene coding for green fluorescent protein (GFP) from the jellyfish *Aequoria victoria* functions well in *P. patens* but because GFP is not an enzyme, higher levels of expression are needed. This requirement can be compensated to some extent by including a nuclear-location signal in the construct; GFP is then targeted to the nucleus and fluorescence is more intense (8). Another reporter gene coding for a fluorescent protein, dsRed, has been used successfully to identify which cells have received DNA following microprojectile bombardment (8). The luciferase gene from the firefly *Photinus pyralis* at present provides the most sensitive assay for gene expression and has been used to compare promoter activity (32).

Although the CaMV 35S promoter has been widely used in cassettes for transformant selection, it is not a strong promoter in *P. patens* (32, 72). A number of other promoters have been used that are more suitable for gene overexpression studies. These include promoters from the rice actin gene (72), maize ubiquitin gene (8), and the wheat Em gene (39). The response of the Em promoter to abscisic acid and to osmotic stress has been studied using *uid*A reporter gene. Transient expression was measured following DNA delivery by either microprojectile bombardment or PEG-mediated DNA uptake. An approximately tenfold increase in expression was observed when abscisic acid (ABA) was added to tissue immediately following biolistic delivery of DNA. The increase in transient expression when protoplast were treated with ABA following transformation was only threefold, but this was attributed to the presence of mannitol in protoplast regeneration medium, leading to higher "uninduced" levels. This increase was confirmed using biolistic delivery, where *uid*A levels in tissues treated with mannitol were two and a half times higher than in the water-treated control (39). A stable transformant containing the *uid*A gene coupled to the wheat Em promoter showed a higher level of induction, an approximately fivefold increase compared with control levels, when treated with mannitol, and an over 25-fold increase when treated with ABA (39). This study did not compare the level of expression of *uid*A driven by the Em promoter with levels when driven by the CaMV 35S promoter, but the uninduced levels observed with the Em promoter are somewhat higher than levels obtained with the 35S promoter (C.D. Knight, unpublished data).

Another system showing regulatable gene expression in *P. patens* utilizes the Top10 promoter and the *Tn10*-encoded *Tet* repressor system (72). Using a β-glucuronidase reporter gene and a transient transformation assay system, levels of expression were 100 times higher in the absence of tetracycline than in its presence. Levels in the absence of tetracycline were comparable to levels obtained with the rice actin1 promoter (72). Glucuronidase activity in the presence and absence of tetracycline was also measured in a stable transformant containing the reporter gene driven by the Top10 promoter. Again, activities were considerably higher in the absence of tetracycline, and as with the Em promoter study, activities obtained from the stable transformant were higher than those in the transient expression system.

A recent study of the relative efficiency of promoter activity used the firefly luciferase gene as a reporter gene and employed PEG-mediated DNA uptake as the DNA delivery system (32). Regenerating protoplast were harvested 48 h after cotransformation with a plasmid containing the firefly luciferase gene driven by the promoter to be assayed, and a plasmid containing *Renilla* luciferase gene, the expression of which provides an internal control. The assays therefore measure transient expression. Of the promoters surveyed, the rice *Act1* promoter was the most active, giving approximately tenfold higher levels of reporter gene expression than the CaMV 35S promoter. A double 35S promoter was over five times more active than the single 35S promoter. Other heterologous promoters assayed were less active than the single 35S promoter. This study also measured the activity of two endogenous promoters isolated from the *P. patens* genes coding for 5′-fucosyltransferase and for 5′-xylosyltransferase. The former promoted expression levels about twice as high as those from the single 35S promoter, while the levels from the latter promoter were only about one third of the levels from the 35S promoter. *P. patens* β-tubulin promoters have also been isolated and their ability to promote gene expression assessed (35). This study used a rhVEGF$_{121}$ reporter gene, the product of which is secreted into the medium and can be assayed by ELISA. The *P. patens Tub1* and *Tub2* promoters gave, respectively, two- and threefold higher levels of expression than the single 35S promoter, whereas the *Tub2* promoter gave similar levels to the 35S promoter. Although endogenous promoters have attractions, their inclusion in a construct may lead to targeting the homologous genomic sequence.

The Ti plasmid nopaline synthetase polyadenylation sequence has been used in most but not all studies. Selection cassettes using the polyadenylation signals from the CaMV 35S gene (27) and the Ti plasmid octopine synthetase gene (52) appear to function as well as those from the nopaline synthetase gene.

Inclusion of Genomic Sequences in Transforming DNA

The first evidence that recombination occurred frequently in *P. patens*, between sequences contained in transforming DNA and homologous sequences located in the genome (homologous recombination; HR), was genetic. When a strain that was already transgenic as a result of transformation with a plasmid containing an antibiotic resistance gene was retransformed with a second, related plasmid, but containing a different antibiotic resistance gene, the two antibiotic resistances were found to cosegregate, indicating that the second plasmid had integrated at the site of the insertion of the first plasmid by HR (37, 59). Molecular confirmation that this was indeed the outcome of HR between sequences in transforming DNA and in the genome was soon provided (58). These studies showed that the frequency of stable transformants was at least tenfold higher when homologous sequences greater than one kilobase are included in the transforming DNA than when no homology was present (57). A growing number of studies have used HR to achieve targeted mutagenesis (10, 23, 24, 27, 33, 43–45, 47, 48, 52, 65, 71); these include the disruption of the *P. patens* FtsZ 2–1 gene, leading to the blocking of chloroplast division (65) (see **Figure 2*a***), and the specific targeting of a member of a multigene family (27, 33, 48). Examples of gene targeting, with the DNA source, targeting construct, and frequency of targeting are given in **Table 1**. Each of the various approaches used to achieve gene targeting has been successful. Data for most examples are not extensive enough to conclude which method is superior. However, the most reliable results appear to be obtained using linear DNA, generated by either restriction enzyme digestion or PCR amplification, comprising a genomic DNA sequence into which a selection cassette has been inserted either at a single site or substituting for part of the genomic sequence.

HR: homologous recombination

The most extensive data on the parameters influencing gene targeting available to date use a series of constructs derived from the *P. patens pum*1 and *rac*1 genes (Y. Kamisugi, A.C. Cuming & D.J. Cove, unpublished data). Targeting is analyzed using PCR amplification to detect the junctions between the transforming DNA and the genome. An analysis of about 2000 stable transformants shows a strong correlation for both loci between logarithm of the length of homologous sequence and the probability of HR occurring within the sequence (see **Figure 4**). Allele replacement requires HR to occur in both flanking sequences, and the probability of allele replacement occurring is, as expected, dependent on the shorter of the two flanking sequences. Symmetry of the targeting vector is therefore important, with 50% of stable transformants showing allele replacement when the shorter of the two flanking sequences is 600–800 bp. In some targeted transformants, the predicted sequence generated by HR can be detected in only one of the flanking sequences (Y. Kamisugi, A.C. Cuming & D.J. Cove, unpublished data). Preliminary analysis is consistent with a

Table 1 **Examples of gene targeting in *P. patens***

Gene	DNA source	Selection cassette	Physical state[a]	Lengths of homology[b]	Targeted and total stables[c]	Reference
APR	Genomic	nptII	lin. R.E.	0.9, 0.8	55/130	(43)
CRY 1a	Genomic	hph	lin. R.E.	2.6, 1.7	N/A	
CRY 1a	Genomic	nptII	lin. R.E.	2.6, 1.7	N/A	(33)
CRY 1b	Genomic	hph	lin. R.E.	2.6, 1.6	N/A	
CRY 1b	Genomic	nptII	lin. R.E.	2.6, 1.6	N/A	
CYC D	Genomic	nptII	lin. R.E.	0.7, 0.3	8/43	(47)
DES6	Genomic	nptII	lin. R.E.	1.0, 0.9	5/5	(23)
FTS Z	cDNA	nptII	lin. R.E.	0.7, 0.2	7/51	(65)
FUC T	Genomic	nptII	lin. R.E.	1.0, 1.0	N/A	(44)
GH3-1	Genomic	nptII	lin. R.E.	0.7, 0.6	22/66	
GH3-1	Genomic	nptII	lin. R.E.	0.9, 0.6		(10)
GH3-2	Genomic	nptII	lin. R.E.	0.8, 0.8	13/52	
GNT 1	Genomic	nptII	lin. R.E.	0.8, 0.5	8/100	(45)
HXK 1	Genomic	nptII	lin. R.E.	0.7, 0.5	N/A	(52)
LEA-1	Genomic	nptII	lin. PCR	0.9, 0.5	26/45	d
MAGO	Genomic	nptII	lin. PCR	0.9. 0.5	34/40	d
MCB 1	cDNA	hph	lin. PCR	0.4, 0.4	2/55	(24)
PHY 1	Genomic	hph	circular	1.2	N/A	
PHY 2	Genomic	hph	lin. R.E.	1.0, 1.0	N/A	(48)
PHY 3	Genomic	nptII	lin. R.E.	1.3, 1.3	N/A	
PHY 4	Genomic	nptII	lin. R.E.	1.1, 1.1	N/A	
PSE 1	cDNA	nptII	lin. R.E.	0.4, 0.3	N/A	(71)
PUM 1	Genomic	nptII	lin. R.E.	0.5, 0.4	50/119	d
RAC 1	Genomic	nptII	lin. R.E.	1.1, 0.5	104/130	d
RAC 1	Genomic	nptII	lin. PCR	1.1, 0.5	68/72	
RAD51A	Genomic	nptII	lin. R.E.	1.2, 1.3	N/A	e
RAD51B	Genomic	sul	lin. R.E.	1.3, 1.3	N/A	
WD-1	Genomic	nptII	lin. R.E.	1.5, 1.2	47/63	d
WD-1	Genomic	nptII	lin. PCR	1.5, 1.2	47/64	
XYL T	cDNA	nptII	lin. R.E.	0.8, 0.6	N/A	(44)
ZLAB-1	Genomic	hph	circular	1.0	3/9	(27)

[a] *lin. R.E.*: plasmid digested with restriction enzyme to release targeting sequence, comprising selection cassette flanked by regions of homology, and transformation performed using mixture of linear DNAs so generated.
lin PCR: transformation performed with linear DNA, comprising selection cassette flanked by regions of homology, generated by PCR amplification.
circular: transformation performed with circular DNA, comprising a plasmid, into which is inserted a region of homology, with the selection cassette inserted adjacent to it.
[b] Lengths of the two regions of homology flanking the selection cassette. Where circular constructs were used, total length of homology is given.
[c] Number of targeted transformants and total number of stable transformants (data not available in some publications).
[d] Y. Kamisugi, A.C. Cuming & D.J. Cove, unpublished data.
[e] U. Markmann-Mulisch, E. Wendeler, C. Guan, H.-H. Steinbiss & B. Reiss, unpublished data.

nonhomologous end joining (NHEJ) event occurring prior to integration, either between the ends of a single molecule of the targeting DNA, followed by HR between the circular molecule so generated and the genome, or between two molecules of targeting DNA to form a linear concatemer, which is then integrated into the genome by two HR events (Y. Kamisugi, A.C. Cuming & D.J. Cove, unpublished data).

The frequency of transformation with constructs containing cDNA sequences has been compared using either linear or circular DNA (30). This study employed PEG-mediated DNA uptake by isolated protoplasts for transformation, but used a modified protocol. With linear DNA, 16% of the transformants obtained were stable, with an overall transformation rate of about 5 transformants/μg DNA (circa 1 stable transformant/μg DNA). In agreement with previous studies (60), the overall frequency of transformants with supercoiled DNA was higher (about 30 transformants/μg DNA) but the frequency of stable transformants was much lower (0.2%). Using supercoiled DNA, the inclusion of the cDNA sequence into the plasmid did not increase the frequency of stable transformants significantly.

Among stable transformants generated following transformation with constructs in which a selection cassette was inserted into a region of genomic DNA, preliminary data for the *P. patens rac*-1 gene show that targeting occurs at a significantly higher frequency when the DNA is linearized by restriction enzyme digestion to release the linear targeting sequence than with transformation with undigested DNA (Y. Kamisugi, A.C. Cuming & D.J. Cove, unpublished data).

The occurrence at a high frequency of HR between sequences in transforming DNA and the corresponding sequence in genomic DNA has been exploited in a number of studies to develop a method of tagged mutagenesis. One approach has used a shuttle mutagenesis technique (49). In outline, this involves the construction of a *P. patens* genomic library in a shuttle vector. After amplification, the library is transformed into a strain of *Escherichia coli* containing a gene coding Tn transposase. This strain is then mated to a strain of *E. coli* in which the Tn transposon contains a geneticin-resistance gene on an F-derived plasmid. Following conjugation, transposition is induced. Plasmids containing the *P. patens* library, into which Tn may have transposed, are then amplified and used to transform *P. patens*. Because transformants are selected by their resistance to geneticin, the majority should therefore be the result of targeted insertion of transposon-containing DNA into the homologous genomic sequence. Stable transformants were obtained at a frequency of about 6 per microgram of transforming DNA. Among the stable transformants recovered, 3.9% showed an altered morphology or development (49). It is difficult to compare this frequency with that obtained using other mutagenic procedures as the criteria for developmental abnormality cannot be tightly defined, and the frequency of mutation to auxotrophy was not determined, but the rates obtained are likely to be higher than those obtained with UV- or ethyl methane sulfonate-mutagenesis, and comparable to those obtained with NTG mutagenesis. Southern analysis of a sample of transformants (49) showed that all contained detectable transposon sequences integrated into the genome, and that most contained multiple copies of the transposon inserted at a single locus. This is therefore a similar outcome to other transformation procedures. This study also uses a related procedure for gene trapping with a reporter gene encoding β-glucuronidase (49). Of the stable

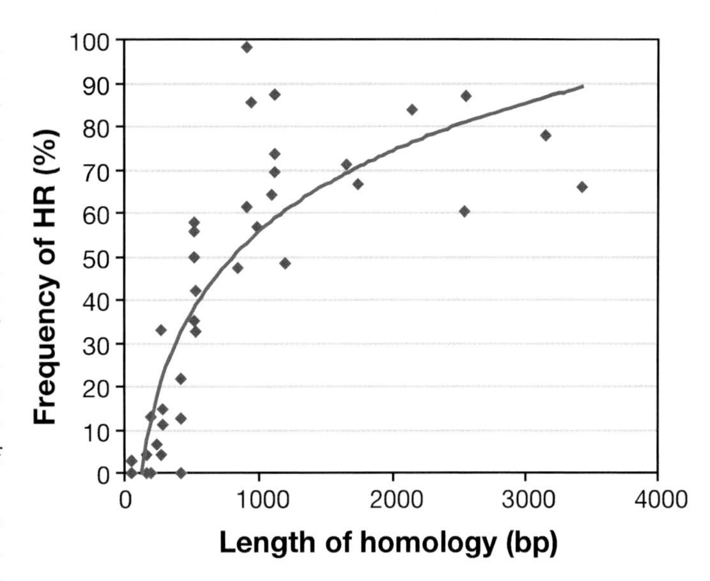

Figure 4

Relationship between length of homology and frequency of homologous recombination. Data for the PpRac-1 and PpPum-1 genes (Y. Kamisugi, A.C. Cuming & D.J. Cove, unpublished data).

transgenics obtained by this procedure, 2.7% expressed β-glucuronidase and some showed a tissue-specific pattern of expression, indicating that this procedure has the potential to isolate genes expressed in specific tissues.

A similar technique for tagged mutagenesis involving transposon insertion in *E. coli* uses a cDNA library instead of a genomic library (20). This technique has been used to generate over 126,000 stable transformants. These were checked for their cellular DNA content and about 8% were found to be polyploid. This is the first detailed analysis of the ploidy of regenerants from protoplasts, but it is likely that the frequency of polyploidy observed is no higher than that obtained from protoplast regeneration without a transformation procedure. It may be that this reflects the distribution of ploidies in the tissue from which the protoplast were obtained, or polyploidy may be generated during the regeneration process. The stable transformants obtained by this procedure were tested on basal and supplemented media, and 1163 transformants (i.e., 7.2%) were found to grow normally on supplemented media while growing less than control wild-type strains on basal medium. The supplemented medium contained a mixture of vitamins, adenine, palmitic acid, peptone, and glucose and had ammonium as the nitrogen source (rather than nitrate in the basal medium). The screen was therefore somewhat different from that used in early screens for mutants using NTG. Using medium supplemented only with vitamins and ammonium, 5 mutants requiring nicotinic acid, 3 requiring p-amino benzoic acid, and 3 unable to utilize nitrate as a nitrogen source were found in a sample of 3196 strains recovered after mutagenesis (3). Although the screen is somewhat different from that used in the tagging procedure, the mutation rate obtained with NTG (<0.5%) is likely to be substantially lower than that obtained using the tagging procedure.

A preliminary experiment to assess the outcome of using multiple constructs containing different cDNAs has used mixtures of equal amounts of five and ten cDNAs (30). Using PCR, the outcome of the batch transformations was compared with transformation with a single construct for two of the cDNAs constructs used. For both constructs, the frequency of targeting among stable transformants in batch transformations was similar to that in the single-construct controls. Double targeting events involving both constructs occurred at a frequency that was consistent to the two targeting events occurring independently. This method therefore has promise for tagged mutagenesis, where a mutant phenotype is dependent on mutation in more that one gene.

RNA Interference (RNAi)

Recently, RNAi has been shown to function in *P. patens*, adding to the resources available to study gene function (8). Although gene targeting via homologous recombination can inactivate or modify a gene, gene inactivation may not lead to a change of phenotype when a gene is a member of a family with overlapping functions. Because RNAi relies on the production of only short lengths of double-stranded RNA to silence expressed mRNAs containing a similar sequence, it is possible to interfere with the expression of families of genes sharing short common sequences. The utility of RNAi in *P. patens* was first demonstrated by microprojectile bombardment, using a mixture of constructs, one coding for dsRed2 and the other, an RNAi construct designed to produce a transcript with a double-stranded hairpin loop involving a sequence of the gene coding for β-glucuronidase (GUS). The target tissue was a transformant expressing a nuclear targeted GFP-GUS fusion protein. Cells that had received DNA from particle bombardment were identified by the expression of dsRed and these were found to no longer express GFP (8). Control experiments using an RNAi construct expressing inverted repeats of a luciferase sequence showed no alteration of GFP expression. As well as transient expression studies, stable transformants expressing RNAi constructs have been produced that show stable silencing of the GFP-GUS coding gene (8). The level of silencing varied in different transformant lines. In one line where GFP and GUS levels were less than 3% of control levels, Northern analysis could detect no GFP-GUS transcript, but in another line where levels of GFP and GUS were 50% of control levels, no reduction in transcript levels could be detected (9).

The RNAi system has now been developed to allow its use to screen the effect of silencing any gene for which a sequence is available (9). Inverted repeats of sequence of the gene to be silenced are cloned via the Gateway® (Invitrogen) recombination system, between inverted repeats of either 400 bp of a GFP-coding or GUS-coding sequence, in a construct that also contains a hygromycin-resistance selection cassette. The construct is delivered to protoplasts and hygromycin-resistant regenerants are selected. The efficacy of RNAi silencing in these can be assessed by screening for silencing of the expression of nuclear-located GFP-GUS. Regenerants showing GFP silencing can then be screened to see if silencing of the gene(s) under investigation produces a change in phenotype. Using inverted repeats of the *FtsZ* gene, all regenerants showing silencing of GFP expression also had the altered chloroplast morphology, characteristic of loss of FtsZ function (9) (see **Figure 2*d***). Stable RNAi transformants were also selected, using a construct in which the sequences expressing the inverted repeats for gene silencing were driven by the wheat Em promoter, and located between sequences of genomic DNA to facilitate targeting of the construct at a chromosomal region, the disruption of which did not lead to an obvious phenotype. With a stably transformed line, no GFP silencing was observed under standard conditions, but silencing occurred within 24 h of the addition of 10 μM abscisic acid to the medium.

The advantages of RNAi are illustrated well in studies of the link between actin filament formation and polar tip growth in *P. patens*. The ARPC3 and ARPC4 genes code for proteins that are components of the Arp2/3 complex, which is involved in actin filament growth. ARPC3 is a member of a gene family, whereas ARPC4 is a single-copy gene. The RNAi phenotype for each gene is identical, i.e., reduced extension growth of the caulonemal filaments. Whereas inactivation of the ARPC4 gene using HR results in a phenotype identical to the RNAi phenotype, inactivation of the ARPC3 gene does not result in an observable alteration in phenotype (8, 9; P.-F. Perroud, unpublished data).

SUMMARY POINTS

1. The haploidy of the dominant phase of the moss life cycle allows the direct recognition of mutants having a recessive phenotype.
2. Haploidy allows simpler Mendelian genetic analysis, eliminating the need for test crossing and the production of F2s.
3. The accessibilty of living cells to direct observation allows unrivalled opportunities for cell biological research.
4. The regenerative capacity of all tissues allows the indefinite propagation of mutant strains blocked at early in development.

5. Somatic hybridization allows genetic analysis of sexually sterile lines.
6. The occurrence, at a very high frequency, of homologous recombination between sequences in transforming DNA and the corresponding genomic sequences, allows for targeted gene inactivation and for allele modification.
7. RNA interference allows the simultaneous silencing of expression of families of related genes.
8. RNAi allows the knockdown of essential genes.

UNRESOLVED ISSUES AND FUTURE DIRECTIONS

1. No genetic map has yet been constructed for *P. patens*. A number of laboratories are now engaged on this, and it is to be hoped that a skeleton map will be available by the time the sequencing of the genome is completed.
2. The sequencing of the *P. patens* genome, which is scheduled to be completed by the end of 2005, will require international collaboration for its annotation.
3. The development of high throughput technology for culture, protoplast isolation and transformation, coupled with the sequencing of the genome, will allow whole genome analyses of metabolism and development.
4. Tagged mutagenesis procedures need further development and refinement, to generate a large collection of mapped insertional mutants, which are available to the research community.

LITERATURE CITED

1. Abel WO, Knebel W, Koop H-U, Marienfeld JR, Quader H, et al. 1989. A cytokinin-sensitive mutant of the moss, *Physcomitrella patens*, defective in chloroplast division. *Protoplasma* 152:1–13
2. Aoki S, Kato S, Ichikawa K, Shimizu M. 2004. Circadian expression of the *PpLhcb2* gene encoding a major light-harvesting chlorophyll *a/b*-binding protein in the moss *Physcomitrella patens*. *Plant Cell Physiol.* 45:68–76
3. Ashton NW, Cove DJ. 1977. The isolation and preliminary characterisation of auxotrophic and analogue resistant mutants of the moss, *Physcomitrella patens*. *Mol. Gen. Genet.* 154:87–95
4. Ashton NW, Grimsley NH, Cove DJ. 1979. Analysis of gametophytic development in the moss, *Physcomitrella patens*, using auxin and cytokinin resistant mutants. *Planta* 144:427–35
5. Ashton NW, Champagne CEM, Weiler T, Verkoczy LK. 2000. The bryophyte *Physcomitrella patens* replicates extrachromosomal transgenic elements. *New Phytol.* 146:391–402
6. Ashton NW, Cove DJ, Featherstone DR. 1979. The isolation and physiological analysis of mutants of the moss, *Physcomitrella patens*, which over-produce gametophores. *Planta* 144:437–42
7. Beckert S, Steinhauser S, Muhle H, Knoop V. 1999. A molecular phylogeny of bryophytes based on sequences of the mitochondrial *nad5* gene. *Plant Syst. Evol.* 218:179–92

8. Bezanilla M, Pan A, Quatrano RS. 2003. RNA interference in the moss *Physcomitrella patens*. *Plant Physiol.* 133:470–74
9. Bezanilla M, Perroud P-F, Pan A, Kluew P, Quatrano RS. 2005. An RNAi system in *Physcomitrella patens* with an internal marker for silencing allows rapid identification of loss of function phenotypes. *Plant Biol.* 7:251–257
10. Bierfreund NM, Tintelnot S, Reski R, Decker EL. 2004. Loss of GH3 function does not affect phytochrome-mediated development in a moss, *Physcomitrella patens*. *J. Plant Physiol.* 161:823–35
11. Boyd PJ, Hall J, Cove DJ. 1988. An airlift fermenter for the culture of the moss *Physcomitrella patens*. In *Methods in Bryology. Proc. Bryol. Meth. Workshop Mainz*, ed. JM Glime, pp. 41–45. Nichinan: Hattori Bot. Lab.
12. Boyd PJ, Grimsley NH, Cove DJ. 1988. Somatic mutagenesis of the moss, *Physcomitrella patens*. *Mol. Gen. Genet.* 211:545–46
13. Bryan VS. 1957. Cytotaxonomic studies in the *Ephemeraceae* and *Funariaceae*. *Bryologist* 60:103–26
14. Chakhparonian M. 2001. *Développement d'outils de la mutagenèse ciblée par recombinaison homologue chez* Physcomitrella patens. PhD thesis. Univ. Lausanne, Switz.
15. Cho S-H, Chung Y-S, Cho S-K, Rim Y-W, Shin J-S. 1999. Particle bombardment mediated transformation and GFP expression in the moss *Physcomitrella patens*. *Mol. Cells* 9:14–19
16. Courtice GRM, Cove DJ. 1983. Mutants of the moss *Physcomitrella patens* which produce leaves of altered morphology. *J. Bryol.* 12:595–609
17. Cove DJ. 1983. Genetics of Bryophyta. In *New Manual of Bryology*, ed. RM Schuster, pp. 222–31. Tokyo: Hattori Bot. Lab.
18. Cove DJ. 1984. The role of cytokinin and auxin in protonemal development in *Physcomitrella patens* and *Physcomitrium sphaericum*. *J. Hattori Lab.* 55:79–86
19. Cove DJ, Schild A, Ashton NW, Hartmann E. 1978. Genetic and physiological studies of the effect of light on the development of the moss, *Physcomitrella patens*. *Photochem. Photobiol.* 27:249–54
20. Egener T, Granado J, Guitton M-C, Hohe A, Holtorf H, et al. 2002. High frequency of phenotypic deviations in *Physcomitrella patens* plants transformed with a gene-disruption library. *BMC Plant Biol.* 2:6
21. Engel PP. 1968. The induction of biochemical and morphological mutants in the moss *Physcomitrella patens*. *Am. J. Bot.* 55:438–46
22. Featherstone DR. 1980. Studies of mutants of the moss, *Physcomitrella patens*, which overproduce gametophores. PhD thesis. Univ. Cambridge, U.K. 122 pp.
23. Girke T, Schmidt H, Zahringer U, Reski R, Heinz E. 1998. Identification of a novel delta 6-acyl-group desaturase by trageted gene disruption in *Physcomitrella patens*. *Plant J.* 15:39–48
24. Girod PA, Fu H, Zryd J-P, Vierstra RD. 1999. Multiubiquitin chain binding subunit MCB1 (RPN10) of the 26S proteasome is essential for develomental progression in *Physcomitrella patens*. *Plant Cell* 11:1457–72
25. Grimsley NH, Withers LA. 1983. Cryopreservation of cultures of the moss *Physcomitrella patens*. *Cryoletters* 4:251–58
26. Grimsley NH, Ashton NW, Cove DJ. 1977. The production of somatic hybrids by protoplast fusion in the moss, *Physcomitrella patens*. *Mol. Gen. Genet.* 154:97–100
27. Hofmann A, Codon A, Ivascu C, Russo V, Knight C, et al. 1999. A specific member of the Cab multigene family can be efficiently targeted and disrupted in the moss, *Physcomitrella patens*. *Mol. Gen. Genet.* 261:92–99

28. Hohe A, Rensing SA, Mildner M, Lang D, Reski R. 2002. Day length and temperature strongly influence sexual reproduction and expression of a novel MADS-box gene in the moss *Physcomitrella patens*. *Plant Biol.* 4:595–602
29. Hohe A, Decker EL, Gorr G, Schween G, Reski R. 2002. Tight control of growth and cell differentiation in photoautotrophic growing moss (*Physcomitrella patens*) bioreactor cultures. *Plant Cell Rep.* 20:1135–40
30. Hohe A, Egener T, Lucht JM, Holtorf H, Reinhard C, et al. 2004. An improved and highly standardised transformation procedure allows efficient production of single and multitple targeted gene-knockouts in a moss, *Physcomitrella patens*. *Curr. Genet.* 44:339–47
31. Holtorf H, Hohe A, Wang HL, Jugold M, Rausch T, et al. 2002. Promoter subfragments of the sugar beet V-type H+-ATPase subunit c isoform drive the expression of transgenes in the moss *Physcomitrella patens*. *Plant Cell Rep.* 21:341–46
32. Horstmann V, Huether CM, Jost W, Reski R, Decker EL. 2004. Quantitative promoter analysis in *Physcomitrella patens*: a set of plant vectors activating gene expression within three orders of magnitude. *BMC Biotech.* 4:13
33. Imaizumi T, Kadota A, Hasebe M, Wada M. 2002. Cryptochrome light signals control development to suppress auxin sensitivity in the moss *Physcomitrella patens*. *Plant Cell* 14:373–86
34. Jenkins GI, Courtice GRM, Cove DJ. 1986. Gravitropic responses of wild-type and mutant strains of the moss *Physcomitrella patens*. *Plant Cell Envir.* 9:637–44
35. Jost W, Link S, Horstmann V, Decker EL, Reski R, Gorr G. 2005. Isolation and characterisation of three moss-derived beta-tubulin promoters suitable for recombinant expression. *Curr. Genet.* 47:111–20
36. Deleted in proof.
37. Kammerer W, Cove DJ. 1996. Genetic analysis of the effects of re-transformation of transgenic lines of the moss *Physcomitrella patens*. *Mol. Gen. Genet.* 250:380–82
38. Knapp E. 1935. Untersuchungen uber die Wirkung von Rontgenstrahlen an dem Lebermoos *Sphaerocarpus* mit Hilfe der Tetraden-Analyse. *Z. Indukt. Abstamm. Vererbungs.* 70:309–49
39. Knight C, Sehgal A, Atwal K, Wallace JC, Cove DJ, et al. 1995. Molecular responses to abscisic acid and stress are conserved between moss and cereals. *Plant Cell* 7:499–506
40. Knight CD. 2000. Moss molecular tools for phenotypic analysis. In *Encyclopedia of Cell Technology*, ed. E Spier, pp. 936–44. New York: Wiley
41. Knight CD, Futers TS, Cove DJ. 1991. Genetic analysis of a mutant class of *Physcomitrella patens* in which the polarity of gravitropism is reversed. *Mol. Gen. Genet.* 230:12–16
42. Knight CD, Cove DJ, Cuming AC, Quatrano RS. 2002. Moss gene technology. In *Molecular Plant Biology Volume 2*, ed. PM Gilmartin, C Bowler, pp. 285–301. Oxford: Oxford Univ. Press
43. Koprivova A, Meyer AJ, Schween G, Herschbach C, Reski R, Kopriva S. 2002. Functional knockout of the adenosine 5′-phosphosulfate reductase gene in *Physcomitrella patens* revives an old route of sulfate assimilation. *J. Biol. Chem.* 277:32195–201
44. Koprivova A, Stemmer C, Altmann F, Hoffmann A, Kopriva S, et al. 2004. Targeted knockouts of *Physcomitrella* lacking plant-specific immunogenic N-glycans. *Plant Biotech. J.* 2:517–23
45. Koprivova S, Altmann F, Gorr G, Kopriva S, Reski R, Decker EL. 2003. N-glycosylation in the moss, *Physcomitrella patens* is organized similarly to that in higher plants. *Plant Biol.* 5:582–91

46. Lang D, Eisinger J, Reski R, Rensing S. 2005. Representation and high-quality annotation of the *Physcomitrella patens* transcriptome demonstrates a high proportion of proteins involved in metabolism among mosses. *Plant Biol.* 7:238–50
47. Lorenz S, Tintelnot S, Reski R, Decker EL. 2003. Cyclin D-knockout uncouples developmental progression from sugar availability. *Plant Mol. Biol.* 53:227–36
48. Mittmann F, Brucker G, Zeidler M, Repp A, Abts T, et al. 2004. Targeted knockout in *Physcomitrella* reveals direct actions of phytochrome in the cytoplasm. *Proc. Nat. Acad. Sci. USA* 101:13939–44
49. Nishiyama T, Hiwatashi Y, Sakakibara K, Kato M, Hasebe M. 2000. Tagged mutagenesis and gene-trap in the moss, *Physcomitrella patens*. *DNA Res.* 7:9–17
50. Nishiyama T, Fujita T, Shin-I T, Seki M, Nishide H, et al. 2003. Comparative genomics of *Physcomitrella patens* gametophytic transcriptome and *Arabidopsis thaliana*: implication for land plant evolution. *Proc. Nat. Acad. Sci. USA* 100:8007–12
51. Nishiyama T, Wolf PG, Kugita M, Sinclair RB, Sugita M, et al. 2004. Chloroplast phylogeny indicates that bryophytes are monophyletic. *Mol. Biol. Evol.* 21:1813–19
52. Olsson T, Thelander M, Ronne H. 2003. A novel type of chloroplast stromal hexokinase is the major glucose-phosphorylating enzyme in the moss *Physcomitrella patens*. *J. Biol. Chem.* 278:44439–47
53. Rensing SA, Rombauts S, Van de Peer Y, Reski R. 2002. Moss transcriptome and beyond. *Trends Plant Sci.* 7:535–38
54. Reski R. 1998. Development genetics and molecular biology of mosses. *Bot. Acta* 111:1–15
55. Reski R, Faust M, Wang XH, Wehe M, Abel WO. 1994. Genome analysis of the moss *Physcomitrella patens* (Hedw) BSG. *Mol. Gen. Genet.* 244:352–59
56. Sawahel W, Onde S, Knight CD, Cove DJ. 1992. Transfer of foreign DNA into *Physcomitrella patens* protonemal tissue by using the gene gun. *Plant Mol. Biol. Rep.* 10:315–16
57. Schaefer DG. 2001. Gene targeting in *Physcomitrella patens*. *Curr. Opin. Plant Biol.* 4:143–50
58. Schaefer DG, Zryd J-P. 1997. Efficient gene targeting in the moss *Physcomitrella patens*. *Plant J.* 11:1195–206
59. Schaefer DG, Bisztray G, Zryd J-P. 1994. Genetic transformation of the moss, *Physcomitrella patens*. In *Plant Protoplasts and Genetic Engineering V*, ed. YPS Bajaj, pp. 349–64. Berlin: Springer Verlag
60. Schaefer DG, Zryd J-P, Knight CD, Cove DJ. 1991. Stable transformation of the moss *Physcomitrella patens*. *Mol. Gen. Genet.* 226:418–24
61. Schulte J, Reski R. 2004. High throughput cryopreservation of 140000 *Physcomitrella patens* mutants. *Plant Biol* 6:119–27
62. Schween G, Gorr G, Hohe A, Reski R. 2003. Unique tissue-specific cell cycle in *Physcomitrella*. *Plant Biol.* 5:1–9
63. Schween G, Hohe A, Koprivova A, Reski R. 2003. Effects of nutrients, cell density and culture techniques on protoplast regeneration and early protonema development in a moss, *Physcomitrella patens*. *J. Plant Physiol.* 160:209–12
64. Smith AJE. 2004. *The Moss Flora of Britain and Ireland*. Cambridge: Cambridge Univ. Press. 1022 pp. 2nd ed.
65. Strepp R, Scholz S, Kruse S, Speth V, Reski R. 1998. Plant nuclear gene knockout reveals a role in plastid division for the homolog of the bacterial cell division protein FtsZ, an ancestral tubulin. *Proc. Natl. Acad. Sci. USA* 95:4368–73

66. Sugiura C, Koboyashi Y, Aoki S, Sugita C, Sugita M. 2003. Complete chloroplast DNA sequence of the moss *Physcomitrella patens*: evidence for the loss and relocation of *rpoA* from the chloroplast to the nucleus. *Nucleic Acid Res.* 31:5324–31
67. Voglmayr H. 2000. Nuclear DNA amounts in mosses (Musci). *Ann. Bot.* 85:531–46
68. von Wettstein F. 1925. Genetische Untersuchungen an Moosen. *Bibl. Genet.* 1:1–38
69. von Wettstein F. 1932. Genetik. In *Manual of Bryology*, ed. F Verdoorn pp. 232–72. The Hague: Martinus Nijhoff
70. Wang TL, Cove DJ, Beutelmann P, Hartmann E. 1980. Isopentenyladenine from mutants of the moss, *Physcomitrella patens*. *Phytochemistry* 19:1103–5
71. Zank TK, Zahringer U, Beckmann C, Pohnert G, Boland W, et al. 2002. Cloning and functional characterisation of an enzyme involved in the elongation of delta-6-polyunsturated fatty acids from the moss *Physcomitrella patens*. *Plant J.* 31:255–68
72. Zeidler M, Gatz C, Hartmann E, Hughes J. 1996. Tetracycline-regulated reporter gene expression in the moss *Physcomitrella patens*. *Plant Mol. Biol.* 30:199–205

INTERNET LINKS

http://biology4.wustl.edu/moss
http://moss.nibb.ac.jp
http://www.biologie.fu-berlin.de/lampart/links.html
http://www.cosmoss.org
http://www.plant-biotech.net

A Mitochondrial Paradigm of Metabolic and Degenerative Diseases, Aging, and Cancer: A Dawn for Evolutionary Medicine

Douglas C. Wallace

Center for Molecular and Mitochondrial Medicine and Genetics, Departments of Ecology and Evolutionary Biology, Biological Chemistry, and Pediatrics, University of California, Irvine, California 92697-3940; email: dwallace@uci.edu

Annu. Rev. Genet. 2005. 39:359–407

First published online as a Review in Advance on July 19, 2005

The *Annual Review of Genetics* is online at http://genet.annualreviews.org

doi: 10.1146/annurev.genet.39.110304.095751

0066-4197/05/1215-0359$20.00

Key Words

mitochondria, reactive oxygen species, human origins, diabetes, neurodegenerative diseases, aging

Abstract

Life is the interplay between structure and energy, yet the role of energy deficiency in human disease has been poorly explored by modern medicine. Since the mitochondria use oxidative phosphorylation (OXPHOS) to convert dietary calories into usable energy, generating reactive oxygen species (ROS) as a toxic by-product, I hypothesize that mitochondrial dysfunction plays a central role in a wide range of age-related disorders and various forms of cancer. Because mitochondrial DNA (mtDNA) is present in thousands of copies per cell and encodes essential genes for energy production, I propose that the delayed-onset and progressive course of the age-related diseases results from the accumulation of somatic mutations in the mtDNAs of post-mitotic tissues. The tissue-specific manifestations of these diseases may result from the varying energetic roles and needs of the different tissues. The variation in the individual and regional predisposition to degenerative diseases and cancer may result from the interaction of modern dietary caloric intake and ancient mitochondrial genetic polymorphisms. Therefore the mitochondria provide a direct link between our environment and our genes and the mtDNA variants that permitted our forbears to energetically adapt to their ancestral homes are influencing our health today.

Contents

LEAN DIETS, LONG LIFE, AND THE MITOCHONDRIA

Age-related metabolic and degenerative diseases are increasing to epidemic proportions in all industrialized countries. Diabetes has increased sharply over the past century, with a worldwide incidence in 2000 of 151 million cases and the projected incidence for 2010 of 221 million cases (for the United States, 14.2 million in 2000 versus 17.5 million by 2010) (73, 257). Today, Alzheimer disease (AD) affects about 4.5 million Americans, projected to increase to 11 to 16 million by 2050 (6, 70). Parkinson's disease (PD) afflicts approximately one million persons in the United States today; 60,000 new cases are diagnosed every year, and the incidence is projected to quadruple by 2040 (160, 164). The risk of cancer also increases with age. Prostate cancer is the most common malignancy among men in industrialized countries. In the United States in 2004, approximately 230,000 new cases were reported, with an annual death toll of approximately 30,000 men (31). Moreover, the incidence of prostate cancer has been increasing steadily in the United States and Canada over the past 30 years (132).

As many of these age-related diseases show some familial association, a genetic basis has been assumed. Yet, the nature and frequency of genetic variants in the human population has not changed significantly in the past 50 years, even though the incidence of these diseases has climbed continuously. Therefore, it isn't our genes that have changed; it is our environment, and the biggest environmental change that we have experienced is in our diet.

For the first time in human history, individuals can live their entire lives free from hunger. Yet, it has been known for over 70 years that laboratory rodents, if maintained on a restricted calorie diet, will be healthier, more active, more intelligent, and will live longer and have fewer cancers (69, 126, 127, 130, 204).

So what is the link between genetics and diet for these diseases? Prodigious effort has been expended to identify chromosomal genetic loci that are responsible for these problems, but with disappointingly little success. One hint as to why nuclear DNA (nDNA) analysis has been so unfruitful is that all of these diseases have a delayed onset and a progressive course. Thus while the most important risk factor for contracting all of these diseases is age, nothing in Mendelian genetics can provide the needed aging clock. The aging clock must involve the gradual accumulation of sequential changes in thousands of copies of the same gene or genes, yet each Mendelian gene is present in only two copies.

Furthermore, certain age-related diseases are preferentially found in distinct populations, suggesting that they are influenced by regional genetic differences. Yet most ancient nDNA polymorphisms are common to all global populations. Only one human genetic system is known to be present in thousands of copies per cell, to exhibit striking regional genetic variation, and to be directly involved in calorie utilization: the mitochondrial genetic system.

The mitochondria are ancient bacterial symbionts with their own mitochondrial DNA (mtDNA), RNA, and protein synthesizing systems. Each human cell contains hundreds of mitochondria and thousands of mtDNAs. The mtDNA is maternally inherited and shows striking regional genetic variation. This regional variation was a major factor in permitting humans to adapt to the different global environments they encountered and mastered. Moreover, the mitochondria burn the calories in our diet with the oxygen that we breathe to make chemical energy to do work and heat to maintain our body temperature. As a by-product of energy production, the mitochondria also generate most of the endogenous reactive oxygen species (ROS) of the cell, and these damage the mitochondria, mtDNAs, and cell. Hence, the mitochondria are the only human genetic system that embodies the features necessary to explain the observed characteristics of the common age-related diseases (237).

In this review, I make the case that mitochondrial decline and mtDNA damage are central to the etiology of the age-related metabolic and degenerative diseases, aging, and cancer. This is because the rate of mitochondrial and mtDNA damage and thus decline is modulated by the extent of mitochondrial oxidative stress. Mitochondrial ROS production, in turn, is increased by the availability of excess calories, modulated by regional mtDNA genetic variation, and regulated by alterations in nDNA expression of stress response genes.

MITOCHONDRIAL BIOGENESIS AND BIOENERGETICS: AN AGING PARADIGM

The proto-mitochondrion entered the primitive eukaryotic cell between two and three billion years ago. Initially, the bacterial genome encoded all of the genes necessary for a free-living organism. However, as the symbiosis matured, many bacterial genes were transferred to extrabacterial plasmids (chromosomes), such that today the maternally inherited mtDNA retains only the genes for the 12S and 16S rRNAs and the 22 tRNAs required for mitochondrial protein synthesis plus 13 polypeptides of the mitochondrial energy generating process, oxidative phosphorylation (OXPHOS) (**Figure 1**). The remaining ~1500 genes of the mitochondrial genome are now scattered throughout the chromosomal DNA. These nDNA-encoded mitochondrial proteins are translated on cytosolic ribosomes and selectively imported into the mitochondrion through various mitochondrial protein import systems. For example, certain proteins destined for the mitochondrial matrix are synthesized with an amino terminal, positively charged, amphoteric targeting peptide that is cleaved off once the protein enters the mitochondrial matrix (237).

mtDNA: mitochondrial DNA

Mitochondrion (s), mitochondria (pl): cellular organelle of endosymbiotic origin that resides in the cytosol of most nucleated (eukaryotic) cells and which produces energy by oxidizing organic acids and fats with oxygen by the process of oxidative phosphorylation (OXPHOS) and generates oxygen radicals (reactive oxygen species, ROS) as a toxic by-product

ROS: reactive oxygen species, oxygen radicals

OXPHOS: oxidative phosphorylation

Mitochondrial DNA (mtDNA): the portion of the mitochondrial genome that currently resides in the matrix of the mitochondrion, as a circular DNA molecule containing the mitochondrial rRNA genes, tRNA genes, and 13 subunits of the mitochondrial oxidative phosphorylation (OXPHOS) enzyme complexes

The Mitochondrial Genome:
mtDNA = 37 genes
Extra-cellular plasmids (chromosomes) ~ 1500 genes

Regulatory mutations:
Somatic, inherited?

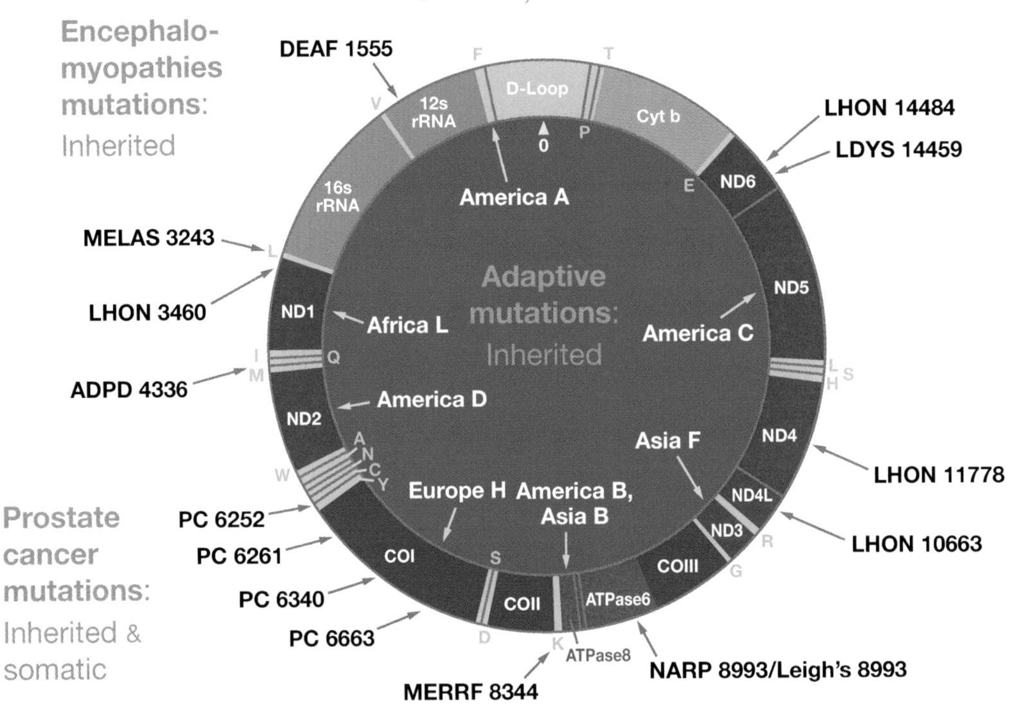

Figure 1

Human mitochondrial DNA map showing representative pathogenic and adaptive base substitution mutations. D-loop = control region (CR). Letters around the outside perimeter indicate cognate amino acids of the tRNA genes. Other gene symbols are defined in the text. Arrows followed by continental names and associated letters on the inside of the circle indicate the position of defining polymorphisms of selected region-specific mtDNA lineages. Arrows associated with abbreviations followed by numbers around the outside of the circle indicate representative pathogenic mutations, the number being the nucleotide position of the mutation. Abbreviations: DEAF, deafness; MELAS, mitochondrial encephalomyopathy, lactic acidosis and stroke-like episodes; LHON, Leber hereditary optic neuropathy; ADPD, Alzheimer and Parkinson disease; MERRF, myoclonic epilepsy and ragged red fiber disease; NARP, neurogenic muscle weakness, ataxia, retinitis pigmentosum; LDYS, LHON + dystonia; PC, prostate cancer.

The 13 mtDNA-encoded polypeptide genes are translated on mitochondrial ribosomes and all are structural subunits of OXPHOS enzyme complexes. These include 7 (ND1, 2, 3, 4L, 4, 5, 6) of the 46 polypeptides of complex I (NADH dehydrogenase), one (cytochrome b, cytb) of the 11 polypeptides of complex III (bc_1 complex), 3 (COI, II, III) of the 13 polypeptides of complex IV (cytochrome c oxidase), and 2 (ATP 6 &

8) of the 16 proteins of complex V (ATP synthetase) (**Figure 1**). The nDNA codes for all other mitochondrial proteins including all four subunits of complex II (succinate dehydrogenase), the mitochondrial DNA polymerase γ (POLG) subunits, the mitochondrial RNA polymerase components, the mitochondrial transcription factor (mtTFA), the mitochondrial ribosomal proteins and elongation factors, and the mitochondrial metabolic enzymes.

The organization of the mtDNA is unique in that its structural genes have no 5′ or 3′ noncoding sequences, no introns, and no spacers between the genes. The mtDNA is symmetrically transcribed from two promoters, one for the G-rich heavy (H) stand and the other for the C-rich light (L) strand. These H- and L-strand promoters (P_H and P_L) are contained in the approximately 1121-np control region (CR) that encompasses four mtTFA binding sites, the H-strand origin of replication (O_H), and three conserved sequence blocks (CSB I, II, and III). The L-strand origin of replication (O_L) is two thirds of the way around the mtDNA from O_H. Transcription is initiated at P_H or P_L and progresses around the mtDNA, generating a polycistronic message. The tRNAs, which punctuate the genes, are cleaved out the mRNAs are then polyadenylated (237) (**Figure 1**).

The mtDNA gene repertoire has remained relatively constant since the formation of the fungal-animal lineage. This is because the mtDNA genetic code began to drift in the fungi such that the mtDNA genes can no longer be interpreted by the nuclear-cytosol system (233). Consequently, when a mtDNA sequence is transferred to the nDNA, it remains as a nonfunctional pseudogene (147).

The mitochondria generate energy by oxidizing hydrogen derived from our dietary carbohydrates (TCA cycle) and fats (β-oxidation) with oxygen to generate heat and ATP (**Figure 2**). Two electrons donated from NADH + H^+ to complex I (NADH dehydrogenase) or from succinate to complex II (succinate dehydrogenase, SDH) are passed sequentially to ubiquinone (coenzyme Q or CoQ) to give ubisemiquinone ($CoQH^{\bullet}$) and then ubiquinol ($CoQH_2$). Ubiquinol transfers its electrons to complex III (ubiquinol: cytochrome c oxidoreductase), which transfers them to cytochrome c. From cytochrome c, the electrons flow to complex IV (cytochrome c oxidase, COX) and finally to $^1/_2 O_2$ to give H_2O. Each of these electron transport chain (ETC) complexes incorporates multiple electron carriers. Complexes I, II, and III encompass several iron-sulfur (Fe-S) centers, whereas complexes III and IV encompass the $b + c_1$ and $a + a_3$ cytochromes, respectively. The mitochondrial TCA cycle enzyme aconitase is also an iron-sulfur center protein (234, 235, 237).

The energy released by the flow of electrons through the ETC is used to pump protons out of the mitochondrial inner membrane through complexes I, III, and IV. This creates a capacitance across the mitochondrial inner membrane, the electrochemical gradient ($\Delta P = \Delta\Psi + \Delta pH$). The potential energy stored in ΔP is coupled to ATP synthesis by complex V (ATP synthase). As protons flow back into the matrix through a proton channel in complex V, ADP and P_i are bound, condensed, and released as ATP. Matrix ATP is then exchanged for cytosolic ADP by the adenine nucleotide translocator (ANT) (**Figure 2**).

Because the ETC is coupled to ATP synthesis through ΔP, mitochondrial oxygen consumption rate is regulated by the matrix concentration of ADP. In the absence of ADP, the consumption of oxygen is regulated by proton leakage across the inner membrane and thus is slow (state IV respiration). However, when ADP is added, it binds to the ATP synthase and is rapidly converted into ATP at the expense of ΔP. As protons flow through the ATP synthase proton channel, the proton gradient is depolarized. Stored fats and carbohydrates are then mobilized to provide electrons to the ETC, which reduce oxygen to water and pump the protons back out of the mitochondrial matrix. The resulting

CR: mtDNA control region

TCA: mitochondrial tricarboxylic acid cycle

SDH: succinate dehydrogenase

COX: cytochrome c oxidase, complex IV

ETC: mitochondrial electron transport chain, a part of the OXPHOS system

ANT: adenine nucleotide translocator

Unc 1,2,3: uncoupling proteins 1,2,3

Reactive oxygen species (ROS): primarily superoxide anion ($O_2^{\bullet -}$), hydrogen peroxide (H_2O_2), and hydroxyl radical ($^{\bullet}OH$), commonly referred to as oxygen radicals; generated as a toxic by-product of oxidative energy production by OXPHO; which damage the mitochondrial and cellular DNA, proteins, lipids, and other molecules causing oxidative stress

Structure + Energy = Life

Energy: fats + sugars + oxygen = energy (heat + work) + CO_2 + H_2O

Reactive oxygen species: mitochondrial combustion → oxygen radical

Apoptosis: energy ↓ + ROS ↑ = mtPTP activated → cell death (apoptosis)

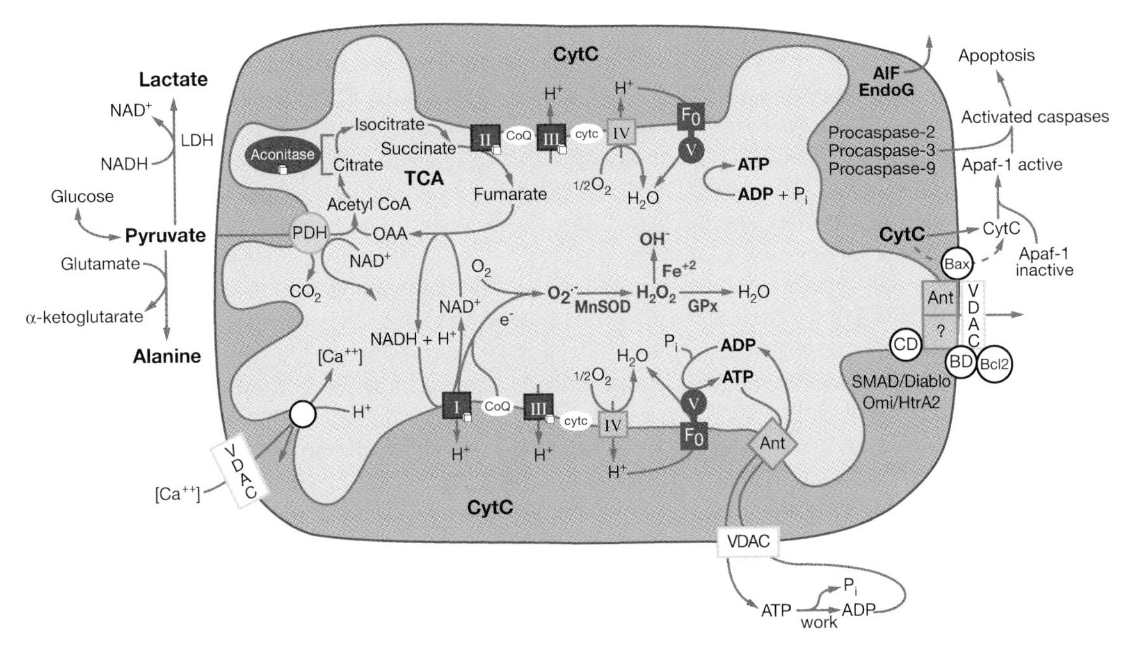

Figure 2

Diagram showing the relationships of mitochondrial oxidative phosphorylation (OXPHOS) to (*a*) energy (ATP) production, (*b*) reactive oxygen species (ROS) production, and (*c*) initiation of apoptosis through the mitochondrial permeability transition pore (mtPTP). The OXPHOS complexes, designated I to V, are complex I (NADH: ubiquinone oxidoreductase) encompassing a FMN (flavin mononucleotide) and six Fe-S centers (designated with a cube); complex II (succinate: ubiquinone oxidoreductase) involving a FAD (flavin adenine dinucleotide), three Fe-S centers, and a cytochrome b; complex III (ubiquinol: cytochrome c oxidoreductase) encompassing cytochromes b, c1 and the Rieske Fe-S center; complex IV (cytochrome c oxidase) encompassing cytochromes a + a_3 and CuA and CuB; and complex V (H^+-translocating ATP synthase). Pyruvate from glucose enters the mitochondria via pyruvate dehydrogenase (PDH), generating acetylCoA, which enters the TCA cycle by combining with oxaloacetate (OAA). *Cis*-aconitase converts citrate to isocitrate and contains a 4Fe-4S center. Lactate dehydrogenase (LDH) converts excess pyruvate plus NADH to lactate. Small molecules defuse through the outer membrane via the voltage-dependent anion channel (VDAC) or porin. The VDAC together with ANT, Bax, and the cyclophilin D (CD) protein are thought to come together at the mitochondrial inner and outer membrane contact points to create the mtPTP. The mtPTP interacts with the pro-apoptotic Bax, anti-apoptotic Bcl2 and the benzodiazepine receptor (BD). The opening of the mtPTP is associated with the release of several pro-apoptotic proteins. Cytochrome c (cytc) interacts with and activates cytosolic Apaf-1, which then binds to and activates procaspase-9. The activated caspase-9 then initiates the proteolytic degradation of cellular proteins. Apoptosis initiating factor (AIF) and endonuclease G (EndoG) have nuclear targeting peptides that are transported to the nucleus and degrade the chromosomal DNA. Modified from Reference (237).

increased oxygen consumption on addition of ADP is known as state III respiration. ΔP is also used for the import of cytosolic proteins, substrates, and Ca^{2+} into the mitochondrion. Drugs such as 2,4-dinitrophenol (DNP) and nDNA-encoded uncoupler proteins 1, 2, and 3 (Unc1, 2, and 3) render the mitochondrial inner membrane leaky for protons. This short-circuits the capacitor and "uncouples" electron transport from ATP synthesis. This causes the ETC to run at its maximum rate, causing maximum oxygen consumption and heat production, but diminished ATP generation (**Figure 2**).

The efficiency by which dietary calories are converted to ATP is determined by the coupling efficiency of OXPHOS. If the ETC is highly efficient at pumping protons out of the mitochondrial inner membrane and the ATP synthesis is highly efficient at converting the proton flow through its proton channel into ATP, then the mitochondria will generate the maximum ATP and the minimum heat per calorie consumed. These mitochondria are said to be tightly coupled. By contrast, if the efficiency of proton pumping is reduced and/or more protons are required to make each ATP by the ATP synthase, then each calorie burned will yield less ATP but more heat. Such mitochondria are said to be loosely coupled. Therefore, in an endothermic animal, the coupling efficiency determines the proportion of calories utilized by the mitochondrion to perform work versus those to maintain body temperature.

As a toxic by-product of OXPHOS, the mitochondria generate most of the endogenous ROS. ROS production is increased when the electron carriers in the initial steps of the ETC harbor excess electrons, i.e., remain reduced, which can result from either inhibition of OXPHOS or from excessive calorie consumption. Electrons residing in the electron carriers; for example, the unpaired electron of ubisemiquinone bound to the CoQ binding sites of complexes I, II, and III; can be donated directly to O_2 to generate superoxide anion ($O_2^{\bullet -}$). Superoxide $O_2^{\bullet -}$ is converted to H_2O_2 by mitochondrial matrix enzyme Mn superoxide dismutase (MnSOD, *Sod2*) or by the Cu/ZnSOD (*Sod1*), which is located in both the mitochondrial intermembrane space and the cytosol. Import of Cu/ZnSOD into the mitochondrial intermembrane space occurs via the apoprotein, which is metallated upon entrance into the intermembrane space by the CCS metallochaperone (166, 207). H_2O_2 is more stable than $O_2^{\bullet -}$ and can diffuse out of the mitochondrion and into the cytosol and the nucleus. H_2O_2 can be converted to water by mitochondrial and cytosolic glutathione peroxidase (GPx1) or by peroxisomal catalase. However, H_2O_2, in the presence of reduced transition metals, can be converted to the highly reactive hydroxyl radical ($^{\bullet}OH$) (**Figure 2**). Iron-sulfur centers in mitochondrial enzymes are particularly sensitive to ROS inactivation. Hence, the mitochondria are the prime target for cellular oxidative damage (241, 242).

For tightly coupled mitochondria, in the presence of excess calories and in the absence of exercise, the electrochemical gradient (ΔP) becomes hyperpolarization and the ETC chain stalls. This is because without ADP the ATP synthase stops turning over, thus blocking the flow of protons back across the mitochondrial inner membrane through the proton channel of the ATP synthase. However, the ETC will continue to draw on the excess calories for electrons to pump protons out of the mitochondrial inner membrane until the electrostatic potential of ΔP inhibits further proton pumping. At this point, the ETC stalls and the electron carriers become maximally occupied with electrons (maximally reduced). These electrons (reducing equivalents) can then be transferred to O_2 to generate O_2 and thus increased ROS and oxidative stress.

By contrast, in individuals who actively exercise and generate ADP, the ATP synthase keeps ΔP hypopolarized and electrons continue to flow through the ETC to sustain ΔP. Consequently, the electron carriers retain few electrons (remain oxidized). The low electron

Oxidative phosphorylation (OXPHOS): the process by which the mitochondrion generates energy through oxidation of organic acids and fats with oxygen to create a capacitor [electron chemical gradient ($\Delta P = \Delta\Psi + \Delta pH$)] across the mitochondrial inner membrane. This ΔP is used as a source of potential energy to generate adenosine triphosphate (ATP), transport substrates or ions, or produce heat. OXPHOS encompasses five multipolypepetide complexes I, II, III, IV and V. Complex I is NADH dehydrogenase or NADH:ubiquinone oxidoreductase, complex II is succinate dehydrogenase (SDH) or succinate:ubiquinone oxidoreductase, complex III is the bc1 complex or ubiquinole: cytochrome c oxidoreductase, complex IV is cytochrome c oxidase (COX) or reduced cytochrome c: oxygen oxidoreductase, and complex V is the ATP synthase or proton-translocating ATP synthase. Complexes I, III, IV, and V encompass both nDNA- and mtDNA-encoded subunits

MnSOD (*Sod2*): mitochondrial matrix superoxide dismutase

Cu/ZnSOD (*Sod1*): mitochondrial intermembrane space and cytosolic superoxide dismutase

mtPTP: mitochondrial permeability transition pore

Apoptosis: a process of programmed cell death resulting in the activation of caspase enzymes and intracellular nucleases that degrade the cellular proteins and nDNA. Apoptosis can be initiated via the mitochondrion through the activation of the mitochondrial permeability transition pore (mtPTP) in response to energy deficiency, increased oxidative stress, excessive Ca^{2+}, and/or other factors

density of the ETC electron carriers limits ROS production and reduces oxidative stress.

This same effect can be achieved if the mitochondria become partially uncoupled, either by decreasing the number of protons pumped per electron pair oxidized or by permitting protons to flow back through the inner membrane without making ATP. Uncoupling can be achieved by disconnecting electron transport from proton pumping by alterations of complexes I, III, or IV; by increasing the number of protons required by the ATP synthase to make an ATP; or by expression of an alternative proton channel such as an uncoupling protein (UCP).

The mitochondria are also the major regulators of apoptosis, accomplished via the mitochondrial permeability transition pore (mtPTP). The mtPTP is thought to be composed of the inner membrane ANT, the outer membrane voltage-dependent anion channel (VDAC) or porin, Bax, Bcl2, and cyclophilin D. The outer membrane channel is thought to be VDAC, but the identity of the inner membrane channel is unclear since elimination of the ANTs does not block the channel (98). The ANT performs a key regulatory role for the mtPTP (98). When the mtPTP opens, ΔP collapses and ions equilibrate between the matrix and cytosol, causing the mitochondria to swell. Ultimately, this results in the release of the contents of the mitochondrial intermembrane space into the cytosol. The released proteins include a number of cell death-promoting factors including cytochrome c, AIF, latent forms of caspases (possibly procaspases-2, 3, and 9), SMAD/Diablo, endonuclease G, and the Omi/HtrA2 serine protease 24. On release, cytochrome c activates the cytosolic Apaf-1, which activates the procaspase-9. Caspase 9 then initiates a proteolytic cascade that destroys the proteins of the cytoplasm. Endonuclease G and AIF are transported to the nucleus, where they degrade the chromatin. The mtPTP can be stimulated to open by the mitochondrial uptake of excessive Ca^{2+}, by increased oxidative stress, or by deceased mitochondrial ΔP, ADP, and ATP. Thus, disease states that inhibit OXPHOS and increase ROS production increase the propensity for mtPTP activation and cell death by apoptosis (**Figure 2**) (241, 242).

The mtDNA is maternally inherited (63) and semi-autonomous. The genetic independence of the mtDNA has been demonstrated by enucleating cells harboring a putative mtDNA marker, such as resistance to the mitochondrial ribosome inhibitor chloramphenicol (CAP^R), and fusing the cytoplasmic fragments containing the donor mitochondria and mtDNA to recipient cells that lack the genetic marker, e.g., CAP^S recipient cells (26, 240). The resulting transmitochondrial cybrids have the mtDNA of the donor and thus inherit the donor's mitochondrial phenotype (CAP^R), yet they have the nDNA of the recipient, demonstrating that mtDNA inheritance can be independent of nDNA inheritance.

The mtDNAs of a donor cell can more completely repopulate the recipient cells if the recipient cells are depleted, or cured, of their resident mtDNAs. The recipient cells can be depleted of their mtDNA by treatment with the mitochondrial poison rhodamine-6-G (220) or they can be cured of their mtDNAs by previous selection in ethidium bromide plus glucose, pyruvate and uridine, resulting in cells that are permanently mtDNA-deficient (ρ^0 cells) (94).

The mtDNA has a very high mutation rate, presumably due to its chronic exposure to mitochondrial ROS. When a new mtDNA mutation arises in a cell, a mixed intracellular population of mtDNAs is generated, a state known as heteroplasmy. As the heteroplasmic cell undergoes cytokinesis, the mutant and normal molecules are randomly distributed into the daughter cells. As a consequence of this replicative segregation, the proportion of mutant and normal mtDNAs can drift toward homoplasmic mutant or wild type. Furthermore, in post-mitotic tissues, mtDNAs harboring deleterious mutations have been found to be preferentially, clonally, amplified within cells. This may be a consequence of the nucleus attempting to compensate for an energy deficiency by making more mitochondria. Therefore, cells with defective mitochondria are preferentially stimulated to replicate their mitochondria and mtDNAs. As the proliferating mitochondria are turned over, presumably by autophagy (2, 116, 136), the mutant mtDNA can become enriched in some cells by genetic drift. As the percentage of mutant mtDNAs increases, the mitochondrial energetic output declines, ROS production increases, and the propensity for apoptosis increases. As cells are progressively lost through apoptosis, tissue function declines, ultimately leading to symptoms. Thus the accumulation of mutant mtDNA creates the aging clock (241, 242) (**Figure 3**).

Clinical symptoms appear when the number of cells in a tissue declines below the minimum necessary to maintain function. The time when this clinical threshold is reached is related to the rate at which mitochondrial and mtDNA damage accumulates within the cells, leading to activation of the mtPTP and cell death, and to the number of cells present in the tissue at birth in excess of the minimum required for normal tissue function. Given that the primary factor determining cell metabolism and tissue structure is reproductive success, it follows that each tissue must have sufficient extra cells at birth to make it likely that that tissue will remain functional until the end of the human reproductive period, or about 50 years. If the mitochondrial ROS production rate increases, the rate of cell loss will also increase, resulting in early tissue failure and age-related disease. However, if mitochondrial ROS production is reduced, then the tissue cells will last longer and

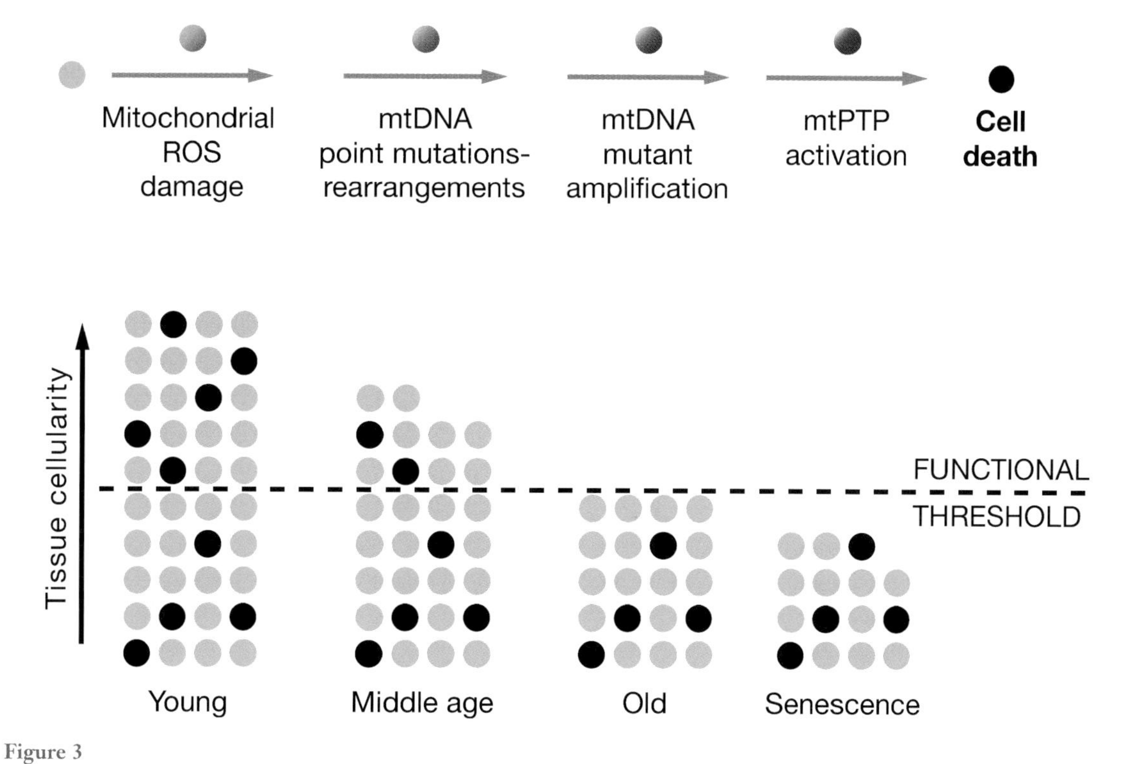

Figure 3

Mitochondrial and cellular model of aging. The upper line of cells diagrams the mitochondrial role in the energetic life and death of a cell. The bottom diagram represents the loss of cells in a tissue over the life of an individual through mitochondrial-mediated death, black cells. The minimum number of cells for the tissue to function normally is indicated by the dashed line.

age-related symptoms will be deferred (236, 238, 241) (**Figure 3**).

CPEO/KSS: chronic progressive external ophthalmoplegia/ Kearn-Sayre syndrome

NARP: neurogenic muscle weakness, ataxia, and retinitis pigmentosa

LHON: Leber's hereditary optic neuropathy

PATHOGENIC mtDNA MUTATIONS

The first evidence that mtDNA mutations might be a key factor in aging and age-related degenerative diseases came with the identification of systemic diseases caused by mtDNA mutations. Pathogenic mtDNA mutations fall into three categories: rearrangement mutations (74), polypeptide gene missense mutations (243), and protein synthesis (rRNA and tRNA) gene mutations (200, 244). As the number of mtDNA-associated diseases identified increased, it became clear that mitochondrial diseases commonly have a delayed onset and progressive course and that they result in the same clinical problems as observed in age-related diseases and in the elderly. Clinical manifestations that have been linked to mtDNA mutations affect the brain, heart, skeletal muscle, kidney, and endocrine system, the same tissues affected in aging. Specific symptoms include forms of blindness, deafness, movement disorders, dementias, cardiovascular disease, muscle weakness, renal dysfunction, and endocrine disorders including diabetes (51, 149, 237, 241, 242).

Pathogenic mtDNA Rearrangement Mutations

Systemically distributed mtDNA rearrangement mutations can be either inherited or spontaneous. Inherited mtDNA rearrangement mutations are primarily insertions. The first inherited insertion mutation to be identified caused maternally inherited diabetes and deafness (13, 14).

Spontaneous rearrangement mutations, primarily deletions, generally result in a related spectrum of symptoms, irrespective of the position of the deletion end points. This is because virtually all deletions remove at least one tRNA and thus inhibit protein synthesis (154). Thus the nature and severity of the symptoms from mtDNA deletion rearrangements is not a consequence of the nature of the rearrangement, but rather of the tissue distribution of the rearranged mtDNAs. Rearrangements that are widely disseminated prevent bone marrow stem cells from proliferating and lead to the frequently lethal childhood pancytopenia, known as Pearson marrow pancreas syndrome. Widely distributed mtDNA rearrangements that spare the bone marrow result in mitochondrial myopathy with ragged red fibers (RRF). Mitochondrial myopathy is frequently associated with ophthalmoplegia and ptosis, which is referred to as chronic progressive external ophthalmoplegia (CPEO). More severe, earlier-onset cases with multisystem involvement are known as the Kearns-Sayre syndrome (KSS) (241).

Pathogenic mtDNA missense mutations. Missense mutations in mtDNA polypeptide genes can also result in an array of clinical manifestations. A mutation in the mtDNA ATP6 gene at np 8993 (T to G) is associated with neurogenic muscle weakness, ataxia, and retinitus pigmentosum (NARP) when present at lower percentages of mutant (75) and lethal childhood Leigh syndrome when present at higher percentages of mutant (213). This mutation causes a marked inability of the ATP synthase to utilize the electrochemical gradient to make ATP (219) and an associated increase in mitochondrial ROS production (128).

Missense and nonsense mutations in the cytb gene have been increasingly linked to progressive muscle weakness (8, 52). Rare nonsense or frameshift mutants in COI have been associated with encephalomyopathies (25, 37).

Missense mutations in complex I genes have been linked to Leigh syndrome (205), generalized dystonia and deafness (88), and to Leber hereditary optic neuropathy (LHON), a form of midlife, sudden-onset blindness (22, 23, 243). Surprisingly, the phenotype of

LHON can be caused by mutations in a number of ND genes that change amino acids, with very different amino acid conservation resulting in varying severities of complex I defects (22). This anomaly has been explained by the discovery that the mildest LHON mtDNA mutations, particularly those in ND6 at np 14,484 and ND4L at np 10,663, are usually found on a particular mtDNA background, the European mtDNA lineage J (21, 23). Lineage J mtDNAs have been found to harbor mtDNA missense mutations that partially uncouple OXPHOS, thus exacerbating the ATP defect of the pathogenic mutations (194).

Pathogenic mtDNA proteins synthesis mutations. Base substitutions in mtDNA protein synthesis genes can also result in multisystem disorders with a wide range of symptoms, including mitochondrial myopathy, cardiomyopathy, deafness, mood disorders, movement disorders, dementia, diabetes, intestinal dysmotility, etc. As with the ATP6 np 8993 (T > G) mutation, studies on the $tRNA^{Lys}$ np 8344 A to G mutation associated with myoclonic epilepsy and ragged red fiber (MERRF) disease have revealed that the level of mutant heteroplasmy plus the age of the patient influence the severity of the clinical symptoms (200, 244).

Perhaps the most common mtDNA protein synthesis mutation is at np 3243 (A > G) in the $tRNA^{Leu(UUR)}$ gene (64). This mutation is remarkable in the variability of its clinical manifestations. When present at relatively low levels (10%–30%) in the patient's blood, the patient may manifest only type II diabetes with or without deafness (223). This is the most common known molecular cause of type II diabetes, purportedly accounting for between 0.5% and 1% of all type II diabetes worldwide (241). By contrast, when the A3243G mutation is present in >70% of the mtDNAs, it does not cause diabetes, but instead causes more severe symptoms including short stature, cardiomyopathy, CPEO, and mitochondrial encephalomyopathy, lactic acid and stroke-like episodes, the MELAS syndrome (64).

MERRF: myoclonic epilepsy and ragged red fiber disease

MELAS: mitochondrial encephalomyopathy, lactic acidosis, and stroke-like episodes

Mitochondrial Defects in Diabetes

The genetic linkage of mtDNA rearrangement (13, 14) and tRNA mutations (223) to type II diabetes directly implicated mitochondrial defects in the etiology of diabetes and the metabolic syndrome. Evidence that mtDNA defects are a common factor in the etiology of diabetes comes from the observation that as the age-of-onset of the proband increases, the probability that the mother was the affected parent also increases, reaching a ratio of 3:1 for patients with a mean age-of-onset of 46 years (241). These observations have been linked to the larger human metabolic syndrome through the identification of a mtDNA $tRNA^{Ile}$ mutation at np 4291 (T > C) that causes hypertension, hypercholesterolemia, and hypomagnesemia (renal ductal convoluted tubule defect). This mutation was found in a maternal pedigree in association with reduced mitochondrial ATP production and the secondary clinical findings of migraine, hearing loss, hypertrophic cardiomyopathy, and mitochondrial myopathy (248).

Consistent with these genetic results, studies of patients with type II diabetes have revealed that mitochondrial function and gene expression are generally down-regulated. Insulin-resistant offspring of type II diabetic patients have been found to have impaired mitochondrial energetics, as assessed by ^{31}P-MR spectroscopy (176). Furthermore, type II diabetes patients consistently show a down-regulation in the expression of nDNA-encoded mitochondrial genes, in association with alterations in the levels of the peroxisome-proliferation-activated receptor γ (PPARγ)-coactivator 1 (PGC-1) (153, 175), a major regulator of mitochondrial biogenesis and fat oxidation (105, 251).

Diabetes mellitus is also seen in Friedreich ataxia. Patients with Friedreich ataxia manifest cerebellar ataxia, peripheral neuropathy,

PGC-1: PPARγ (peroxisome-proliferating-activated receptor γ) coactivator 1

hypertrophic cardiomyopathy, and diabetes as the result of inactivation of the frataxin gene on chromosome 9q13. Frataxin binds iron in the mitochondrial matrix, thus minimizing mitochondrial $^{\bullet}OH$ production. The loss of the frataxin protein results in excessive ROS generation which inactivates all mitochondrial iron-sulfur-containing enzymes including complex I, II, III, and aconitase. Thus increased mitochondrial ROS production and decreased mitochondrial OXPHOS must be the cause of diabetes in Friedreich ataxia (103, 192, 249).

Type II diabetes has also been associated with a Pro121A polymorphism in the PPARγ gene (5) and with a Gly482Ser polymorphism in the PGC-1 gene in Danish (53) and Pima Indian (156) populations.

Maturity onset of diabetes in the young (MODY) is an early-onset autosomal dominant form of type II diabetes that can also be linked to mitochondrial dysfunction. The molecular defects of four forms of MODY have been identified. MODY 2 is the result of mutations in glucokinase, MODY 1 is due to mutations in the hepatocyte nuclear factor (HNF)-1α, MODY 3 to mutations in HNF-4 α, and MODY 4 to mutations in insulin promoter factor (IPF)-1.

Glucokinase has a very high K_m for glucose and is thought to be the glucose sensor. HNF-1α is a transcription factor and mutations in its gene are associated with post-pubertal diabetes, obesity, dyslipidemia, and arterial hypertension, all features of mitochondrial diseases. HNF-1α is also important in regulating nDNA-encoded mitochondrial gene expression and the expression of the GLUT 2 glucose transporter (250). HNF-4α, a member of the steroid/thyroid hormone receptor superfamily, acts as an upstream regulator of HNF-1α (231).

The importance of mitochondrial defects in β cell insulin secretion deficiency has been confirmed in two mouse models. In the first, the mitochondrial transcription factor *Tfam* was inactivated in the pancreatic β cells. This resulted in increased blood glucose in both fasting and nonfasting states, and the progressive decline in β cell mass by apoptosis (201). In the second, the ATP-dependent K^+-channel (K_{ATP}) affinity for ATP was reduced, resulting in a severe reduction in serum insulin, severe hyperglycemia with hypoinsulinemia, and elevated D-3-hydroxybutyrate levels (102). These models demonstrate that mitochondrial ATP production is critical in the signaling system of the β cell to permit insulin release (238). Thus pancreatic β cell mitochondrial defects are important in both glucose sensing through glucokinase and insulin release through the K_{ATP} channel (**Figure 4**).

Type II diabetes thus involves mutations in energy metabolism genes including the mtDNA and glucokinase; mutations in the transcriptional control elements PPARγ, PGC-1, HNF-1α, HNF-4α, and IPF-1; and alterations in insulin signaling. These seemingly disparate observations can be unified through the energetic interplay between the various organs of the body.

The human and mammalian organs can be divided into four energetic categories: the energy-utilization tissues including skeletal muscle, heart, kidney and brain; the energy-storage tissues including brown adipose tissue (BAT) and white adipose tissue (WAT); the energy-homeostasis tissue, liver; and the energy-sensing tissues including the α and β cells of the pancreatic Islets of Langerhans. All of these tissues interact to coordinate the utilization and storage of energy, based on the availability of calories in the environment. For our hunter-gather ancestors, the primary variation in available dietary calories was due to the cyclic growing seasons of edible plants caused by either warm versus cold or wet versus dry seasons. During the growing season, plants convert the Sun's energy into glucose, which the plants store as starch. When humans and animals ingest these plant tissues the concentration of glucose in their blood rises. Hence, serum glucose is the metabolic surrogate for monitoring plant calorie abundance.

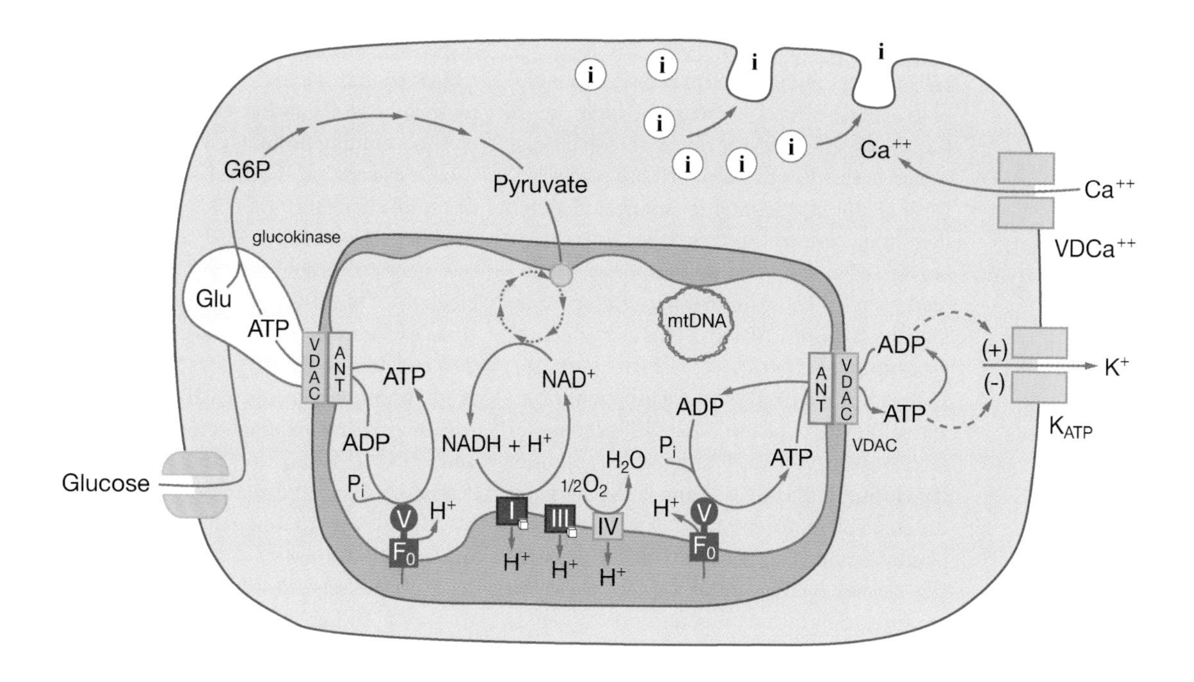

$\uparrow$ Glucose $\rightarrow$ $^{[ATP]}/_{[ADP]}$ $\rightarrow$ K_{ATP} channel closed $\rightarrow$ depolarization $\rightarrow$
$VDCa^{++}$ channel opened $\rightarrow$ insulin released

Figure 4

Model of pancreatic β cell showing the mitochondrial regulation of insulin secretion. The light yellow blob associated with VDAC represents glucokinase. The green double squares in the plasma membrane on the left side represent the glucose transporter. The two pairs of double gold squares on the right side of the plasma membrane represent: below and labeled K_{ATP} the ATP gated K^+ channel i and above and labeled $VDCa^{++}$ the voltage-dependent Ca^{++} channel. The circles with the internal "i" represent insulin containing vesicles. Other abbreviations as in **Figure 2**. Reprinted from Reference 25.

When plant calories are abundant and consumed, the elevated serum glucose is detected by the energy-sensing pancreatic β cells, which respond by secreting insulin into the blood stream. The insulin signal then informs the energy-utilizing heart and muscle tissues to down-regulate mitochondrial energy utilization, since food-seeking behavior is less pressing. It also informs the energy-storing WAT and BAT tissues to store the excess calories as fat for when the season changes and plant calories again become limited. When plant calories do become limited, insulin secretion declines and the pancreatic α cells secrete glucagon. These low blood sugar hormonal signals inform the energy-utilization tissues to up-regulate the mitochondrial OXPHOS system, thus enhancing food-seeking capacity. They also mobilize the energy-storage tissues to transfer the stored triglycerides into the blood to fuel the increased mitochondria OXPHOS. Furthermore, low blood glucose stimulates the energy-homeostasis tissue, liver, to synthesis glucose to maintain the basal level of blood sugar, which is particularly critical for brain metabolism. The molecular basis of this primeval system for adapting to energy fluctuation can now be partially understood through recent discoveries pertaining to the transcriptional regulation of mitochondrial gene expression.

FOXO: mammalian forkhead transcription factor

The energy-sensing tissue, pancreatic β cells, detects abundant plant calories when the blood glucose level exceeds the high K_m of the β cell glucokinase. Since glucokinase is bound to the mitochondrial VDAC and ANT through the mitochondrial inner and outer membrane contact points, the ATP binding site of glucokinase is continuously occupied (3, 60, 125). The binding of glucose by glucokinase immediately generates glucose-6-phosphate (G-6-P) and then pyruvate. Pyruvate is oxidized by the mitochondria to generate ATP. The elevated ATP production increases the ATP/ADP ratio, which results in the closure of the K_{ATP} channel, depolarizing the β cell plasma membrane. The depolarized plasma membrane opens the voltage-sensitive Ca^{2+} channel. The influx of Ca^{2+} leads to the formation of the exocytosis core complex and activation of protein kinases (255). The insulin-containing vesicles then fuse to the plasma membrane and release their stored insulin into the blood (**Figure 4**).

Thus, when blood glucose is high, the insulin concentration in serum increases. The insulin binds to the insulin receptor of target cells throughout the body, activating the insulin receptor tyrosine protein kinase to phosphorylate insulin receptor substrates. These activate the phosphotidylinositol 3-kinase (PI3K), which phosphorylates phosphotidylinositol 2-phosphate (PI2) to phophatidylinositol 3-phosphate (PI3). PI3 then activates the AKT 1/2 kinases, which phosphorylate the FOXO forkhead transcription factors FOXO1, FOXO3A, and FOXO4 in the target tissues. Phosphorylation of the FOXOs results in their transport out of the nucleus and the transcriptional inactivation of genes whose promoters contain insulin response elements (IRE) (1, 59). The FOXOs can also be removed from the nucleus by acetylation via Cbp or p300 (1) and reactivated by deacetylation via the NAD^+-dependent SIRT1 (Sirtuin 1) (24).

When blood sugar is low, the pancreatic α cells secrete glucagon, which binds to the glucagon receptor in target cells. This stimulates the production of cAMP, which activates protein kinase A (PKA). PKA phosphorylates and activates the cAMP response element binding protein (CREB), which activates transcription of genes harboring a cAMP response element (CRE) (48).

The FOXOs and CREB regulate numerous genes through the IRE and CRE *cis* elements. The PGC-1α promoter contains three IREs that bind unphosphorylated and deacetylated FOXO1, plus one CRE that binds phosphorylated CREB (48). PGC-1α together with its companion genes PGC-1β and PRC (PERC) are transcriptional coactivators that regulate genes for mitochondrial biogenesis, thermogenesis, and fatty acid oxidation (92, 196).

PGC-1α was cloned through its interaction with PPARγ. It is strongly induced in BAT in response to cold acting through the cAMP-mediated β-adrenergic and thyroid hormone (TR) systems. The up-regulation of PGC-1α in turn induces UCP-1, which then creates a proton channel through the mitochondrial inner membrane, uncoupling OXPHOS. This causes the rapid burning of the stored fats in BAT, generating heat to maintain body temperature (184). PGC-1α also induces mitochondrial biogenesis and UCP-2 in muscle cells (C2C12), acting through the nuclear respiratory factors 1 and 2 (NRF1 and NRF2) (251). NRF1 and NRF2, in turn, activate the transcription of a wide range of nDNA-encoded mitochondrial genes, including components of OXPHOS such as the ATP synthase β subunit (ATPsynβ) and cytochrome c (cytc) and the mtTFA, which regulates mtDNA transcription (92, 196).

The PGC-1 family of transcriptional coactivators interacts with a broad spectrum of tissue-specific transcription factors. This provides the tissue-specific response to the generalized insulin and glucagon hormonal signals. In the energy utilization tissue, muscle, PGC-1 induces mitochondrial biogenesis through the interaction with NRF1, PPARβ, PPARδ, MEF2, ERR2 (estrogen-related receptor 2), HDAC5, and Gabpa/b (GA-repeat binding protein) (47, 78, 117, 152). In heart, PGC-1α interacts with PPARα and NRF1 (78, 111). In the brain, the transcriptional partners of PGC-1α have not yet been identified (118). In the energy-storage tissues, PGC-1α interacts with PPARγ, PPARα, and the thyroid hormone receptor (TR) in BAT to induce mitochondrial biogenesis and UCP-1 during cold stress (78, 251), while in 3T3-Li preadipocytes, PGC-1α interacts with PPARα to stimulate fatty acid oxidation, including induction of the medium-chain acyl CoA dehydrogenase (MCAD) (230). Finally, in the energy-homeostasis tissue, liver, PGC-1α interacts with FOXO1, HNF-4α, and the glucocorticoid receptor (GR) to induce gluconeogenesis enzymes, including phosphoenolpyruvate carboxylase (PEPCK) and glucose-6-phosphatase (G6Pase) to maintain basal blood glucose levels (78, 121, 187, 254).

PGC-1β is abundantly expressed in BAT, heart, kidney, and skeletal muscle but also in stomach and in white adipose tissue. In 3T3-L1 preadipocytes, PGC-1β is inducible, though PGC-1α is not, and PGC-1β then interacts with the ERRs to induce MCAD and fatty acid oxidation (89). Thus both insulin through the FOXOs and glucagon through cAMP modulate mitochondrial energy metabolism in response to the availability of carbohydrates via the regulation of members of the PGC-1 co-activator family.

Mitochondrial energy metabolism is further modulated by PGC-1α through deacetylation by SIRT1. PC12 cells that overexpress SIRT1 experience a 25% reduction in oxygen utilization (161). Hepatocytes that overexpress SIRT1 induce the gluconeogenesis genes PEPCK and G6Pase via interaction between PGC-1α and HNF-4α (187). Furthermore, upon caloric restriction, the SIRT1 in fat cells binds to the nuclear receptor corepressor (NCoR), blocking its interaction with PPARγ. This inhibits fat storage and enhances fat mobilization (178, 179).

These observations link all of the known genetic mutations associated with diabetes to defects in mitochondria bioenergetics and energy metabolism. Mutations in the mtDNA and in glucokinase would result in energetic dysfunction. Mutations in PPARγ and PGC-1α would reduce mitochondrial gene expression. Mutations in HNF-4α and its target HNF-1α would result in loss of glucose homeostasis. Thus, diabetes is an energy deficiency disease centered upon mitochondrial bioenergetics.

Given a mitochondrial etiology for type II diabetes, the various stages in the progression of type II diabetes can be understood. In individuals harboring a partial defect in OXPHOS, the capacity of the energy-utilization cells to oxidize carbohydrates and fats to make ATP is reduced. Given a high caloric diet, individuals with partial OXPHOS defects overload their mitochondrial ETC with excessive calories (reducing equivalents), hyperpolarizing ΔP, stalling the ETC, and blocking the tissue utilization of glucose. As a result, the nonmetabolized glucose remains in the blood. The chronically high serum glucose signals the β cells to secrete insulin, creating concurrently elevated glucose and insulin: the hallmark of insulin resistance.

The excessive reduction of the mitochondrial ETC electron carriers in the energy-utilization tissues maximizes mitochondrial ROS production. The high serum insulin activates their Akt pathway, which phosphorylates the FOXOs. The departure of the FOXOs from the nucleus stops transcription of the stress response genes, including the mitochondrial antioxidant enzymes. It also suppresses PGC-1α transcription, which down-regulates mitochondrial OXPHOS, further exacerbating the mitochondrial energy deficiency. The resulting chronic mitochondrial oxidative stress erodes mitochondrial function and increases insulin resistance.

The excess of reducing equivalents also increases the NADH/NAD$^+$ ratio. The conversion of NAD$^+$ to NADH inhibits SIRT1, resulting in increased acetylation of the FOXOs and their removal from the nucleus. This further down-regulates PGC-1α and OXPHOS.

The sustained high serum glucose and insulin also activates the PGC-1α and PGC-1β in the energy-storage tissues BAT and WAT to store the excess calories as fats. Also, the liver accumulates lipid, and this can lead to nonalcoholic steatohepatitis (NASH), which is observed in over 30% of individuals with indications of insulin resistance, hypertriglyceridemia, and hypercholesterolemia (174).

In the β cells, the excessive mitochondria ROS inhibits mitochondrial ATP production, eventually leading to a decline in insulin secretion due to inadequate ATP for glucokinase and a low ATP/ADP ratio that cannot activate the K_{ATP} channel. The resulting high glucose but reduced serum insulin is termed insulin-independent diabetes. Continued calorie overload in the pancreatic β cells and associated mitochondrial ROS production ultimately activates the β cell mtPTP, resulting in β cell death by apoptosis, producing insulin-dependent diabetes.

Chronic mitochondrial oxidative stress on the peripheral tissues subsequently damages the retina, vascular endothelial cells, peripheral neurons, and the nephrons, leading to the clinical sequelae of end-stage diabetes. Thus chronic mitochondrial dysfunction can explain all of the features of type II diabetes, and is thus the likely cause of the disease.

ANCIENT MIGRATIONS AND CLIMATIC ADAPTATION

In addition to having to adapt to changing caloric availability due to seasonal changes, ancient human hunter-gathers had to adapt to the rigors of different climatic zones. While induction of UCP-1 in BAT and UCP-2 in muscle permits acute adaptation to thermal stress in rodents, this is not an adequate mechanism for long-term cold adaptation by humans. UCP-1 induction in BAT cannot generate sufficient heat to effectively regulate the temperature of the much larger human body. Therefore, as humans migrated into the more northern latitudes they had to adapt to the chronic cold using another mitochondrial strategy. It now appears that this global climatic adaptive strategy was achieved by the acquisition of mtDNA mutations that partially uncoupled OXPHOS and thus resulted in perpetually increased mitochondrial heat production.

The first clear evidence that ancient mtDNA polymorphisms influence human physiology came from the observation that LHON patients with mild complex I mtDNA missense mutations frequently harbored the same European mtDNA lineage, J. It is now appears that lineage J mtDNAs harbor specific "uncoupling mtDNA variants" that reduce mitochondrial ATP output in favor of heat production. Thus this lineage exacerbates the partial ATP defect generated by the pathogenic mtDNA mutations.

Phylogeographic studies of the human mtDNAs have revealed a remarkable correlation between mtDNA lineages and the geographic origins of indigenous populations. These regional mtDNA lineages are groups of related individual mtDNA sequences (haplotypes) known as haplogroups. The various regional haplogroups form the branches on a single human dichotomous mtDNA phylogenetic tree, generated by the accumulation of sequential mtDNA mutations on radiating maternal lineages. The human mtDNA tree is rooted in Africa, and it has specific branches radiating into different geographic regions that we now believe to have been constrained by the climatic zones (28, 86, 144, 239) (**Figure 5**).

African mtDNAs are the most diverse and thus most ancient, with an overall age of about 150,000 to 200,000 YBP. African mtDNAs fall into four major haplogroups: L0 (oldest), L1, L2, and L3 (youngest). L0, L1, and L2 represent about 76% of all sub-Saharan African mtDNAs and are defined by a *Hpa*I restriction site at np 3592 [macro-haplogroup L (L0, L1, and L2)] (in **Figure 1**). In northeastern Africa, two mtDNA lineages, M and N, arose from L3 about 65,000 YBP. These were the only mtDNA lineages that succeeded in leaving sub-Saharan Africa and radiating into

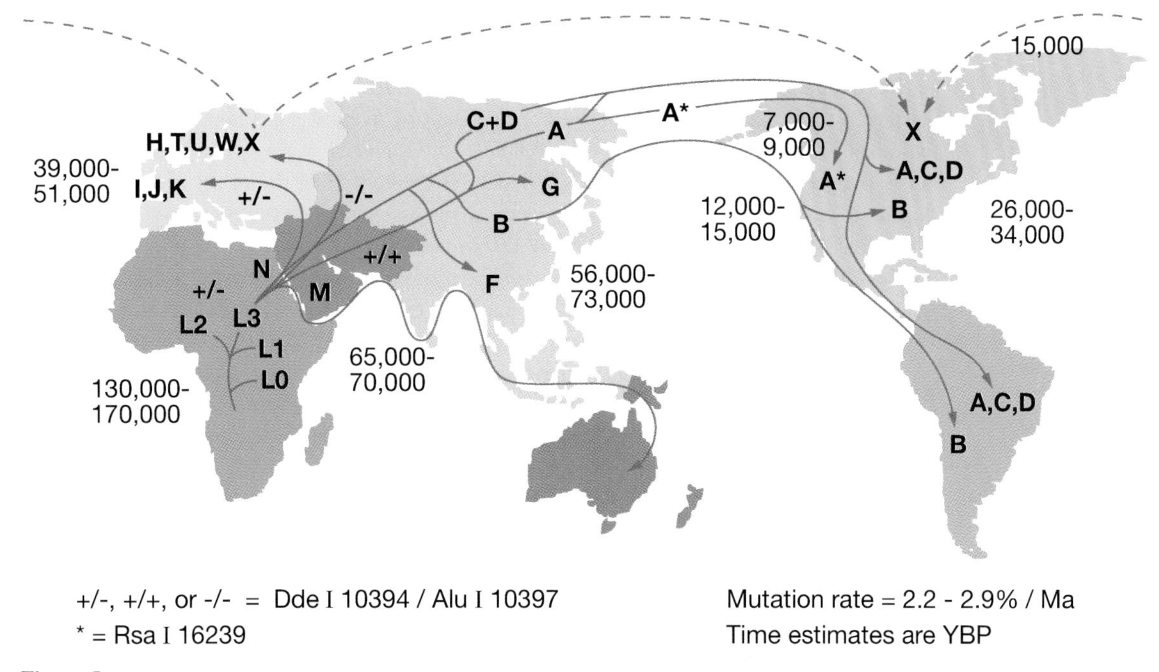

Figure 5

Diagram outlining the migratory history of the human mtDNA haplogroups. *Homo sapiens* mtDNAs arose in Africa about 150,000 to 200,000 years before present (YBP), with the first African-specific haplogroup branch being L0, followed by the appearance in Africa of lineages L1, L2, and L3. In northeastern Africa, L3 gave rise to two new lineages, M and N. Only M and N mtDNAs succeeded in leaving Africa and colonizing all of Eurasia about 65,000 YBP. In Europe, N gave rise to the H, I, J, Uk, T, U, V, W, and X haplogroups. In Asia, M and N gave rise to a diverse range of mtDNA lineages including A, B, and F from N and C, D, and G from M. A, C, and D became enriched in northeastern Siberia and crossed the Bering land bridge about 20,000 to 30,000 YBP to found the Paleo-Indians. At 15,000 YBP, haplogroup X came to central Canada either from across the frozen Atlantic or by an Asian route of which there are no clear remnants today. At 12,000 to 15,000 YBP, haplogroup B entered the Americas, bypassing Siberia and the arctic, likely by moving along the Beringian Coast. Next at 7000 to 9000 YBP, a migration bringing a modified haplogroup A moved from the northeastern Siberia into northwestern North America to found the Na-Dēnē (Athebaskins, Dogrib, Apaches, and Navajos). Finally, relatively recently, derivatives of A and D moved along the Arctic Circle to found the Eskimos. These observations revealed two major latitudinal discontinuities in mtDNA variation: one between the Africa L haplogroups and the Eurasia and N and M derivatives and the other between the plethora of Central Asian mtDNA lineages and the almost exclusive presence of lineages A, C, and D northeastern Siberia, the latter spawning the first Native American migrations. Since these discontinuities correspond to the transitions from tropical and subtropical to temperate and from temperate to arctic, we have proposed that these discontinuities were the result of climatic selection of specific mtDNA mutations that permitted certain female lineages to prosper in the increasingly colder northern latitudes. Reprinted from **http://www.mitomap.org.**

Eurasia to give all of the Eurasian mtDNAs. In Europe, haplogroups L3 and N gave rise to haplogroups H (about 45% of European mtDNAs), (H in **Figure 1**), T, U, V, W, and X (about 2%) as well as I, J (about 9%), and K(Uk). Europeans separated from Africans about 40,000–50,000 YBP. In Asia, lineages M and N radiated to give rise to a plethora of mtDNA lineages. These include from N, haplogroups A, B, F, and others, and from M, haplogroups C, D, G, and others (239) (**Figure 5**).

As Asians migrated northeast into Siberia, haplogroups A, C, D (A, C, D in **Figure 1**) became progressively enriched, such that they became the predominant mtDNA lineages in the indigenous peoples of extreme northeastern Siberia, Chukotka. When the Bering land bridge appeared about 20,000 to 30,000 YBP, people harboring these mtDNA haplogroups were in a position to migrate into the New World, where they founded the Paleo-Indians. After the land bridge submerged, haplogroup G arose in Central Asia and moved into northeastern Siberia to populate the area around the Sea of Okhotsk. Later, a migration carrying haplogroup B (B in **Figure 1**) started from eastern Central Asia and moved along the coast to the New World, by-passing Siberia. Haplogroup B then mixed with A, C, and D in southern North America, Central America, and northern South America to generate the Paleo-Indians. In addition, haplogroup X was brought to the New World in a migration that took place about 15,000 YBP. These immigrants settled in the Great Lakes region, and haplogroup X mtDNAs are found in 25% of the Ojibwa mtDNAs today. Since haplogroup X is found primarily in northeastern Europe, it has been speculated that an ancient European migration carrying this haplogroup might also have contributed to the Paleo-Indian populations, perhaps bringing the progenitors of the Clovis lithic culture to the Americas (239).

Later migrations from northeastern Siberia, carrying a modified lineage of haplogroup A, founded the Na-Déné populations about 9500 YBP. More recently, immigrants from Siberia bearing derived lineages of haplogroups A and D moved along the Arctic Circle to found the Eskimos and Aleuts (239).

This phylogeographic distribution of mtDNA haplogroups reveals two striking discontinuities in human mtDNA diversity. The first occurs between sub-Saharan Africa and Eurasia, in which all of the sub-Saharan African mtDNA diversity remained in Africa and only derivatives of lineages M and N colonized temperate Eurasia. The second occurs between temperate Central Asia and arctic Siberia, where the plethora of Asian mtDNA types is markedly reduced to only three ancient mtDNA lineages (A, C, and D). The resulting fivefold enrichment of haplogroups A, C, D, and G in the arctic over Central Asia is unlikely to be the result of genetic bottlenecks, since there are no obvious geographic barriers that separate Asia and Siberia. It is more plausible that many Asian mtDNA lineages entered the arctic, but only a few survived the intense cold to become permanent residents. A similar logic could also apply to the African-Eurasian discontinuity.

Evidence that climatic adaptation has influenced the geographic distribution of mtDNA diversity was first obtained from analyzing the amino acid replacement (nonsynonymous, NS) (K_a) to silent (synonymous, S) (K_s) mutation ratios (K_a/K_s) from the 13 mtDNA open reading frames (83, 239). This revealed that the amino acid sequence of the ATP6 gene was highly variable in the arctic, but was strongly conserved in the tropics and temperate zone; cytb was hypervariable in temperate Europe, but conserved in the tropics and arctic; and COI was variable in the tropical Africa, but invariant in the temperate and arctic regions. Regional variation was also observed in multiple ND subunits (148). Such regional gene-specific variation would not be expected if all mtDNA mutations were random and neutral.

The geographic constraints on mtDNA protein variation were further validated by positioning all of the mtDNA variants from 1125 complete mtDNA coding sequences collected from around the world and organized into a sequentially mutational tree. This tree could be assembled because the maternally inherited mtDNA can only change by sequential mutations along radiating female lines. Therefore, mutations shared between different mtDNAs must be derived from a common female ancestor, and those that are confined to a single mtDNA haplotype must be new mutations.

Analysis of the mtDNA nucleotide variants revealed three different categories of variants: (*a*) neutral, including synonymous and weakly conserved nonsynonymous amino acid substitutions; (*b*) deleterious, altering highly conserved amino acids but located at the tips of the branches of the tree, indicating they are recent; and (*c*) adaptive, altering highly conserved amino acids but located within the internal branches of the tree, indicating that they are ancient. The mutations in class (*c*) must be adaptive because they alter highly conserved amino acids, yet they have persisted in the face of intense purifying selection for tens of thousands of years. Hence, as these variants are not neutral or deleterious, hence, they must be adaptive.

To quantify the potential effect of selection on particular mtDNA variants, the interspecific amino acid conservation [Conservation Index (CI)] was determined for 22 known pathogenic mtDNA replacement mutations (194). This gave an average CI for deleterious mutations of 93.3 ± 13.3%. Using two standard deviations around the mean CI of the pathogenic missense mutations to distinguish between functionally important versus neutral replacement mutations, 26% of the internal branch replacement mutations were found to be functionally significant, with a mean CI = 85.1 ± 9.2%. The remaining 74% were found to be essentially neutral, with a mean CI = 23.3 ± 14.9%.

Since these conserved internal branch missense mutations frequently initiate new limbs of the mtDNA tree, they must have permitted individuals carrying these mutations to survive and multiply in new geographic regions, generating the region-specific mtDNA haplogroups. The most obvious environmental factor that differentiates tropical and subtropical sub-Saharan Africa from temperate Eurasia and temperate Central Asia from arctic Siberia is temperature. This implies that an important functional effect of many of the conserved internal branch missense mutations was to permit adaptation to cold.

That these adaptive internal branch mutations were important in human radiation was confirmed by analyzing the ratio of NS to S missense mutations in the three major climatic zones. The internal branch NS/S ratio for the tropical and subtropical African L haplogroups was 0.31. The internal NS/S ratio for the temperate Eurasian haplogroups M and N(R) were 0.42 and 0.44, whereas the internal NS/S ratio for the primarily arctic macro-haplogroup N(nonR) was 0.62. Thus, the farther north that the mtDNAs were found, the more missense mutations they harbored.

The excess of internal branch missense mutations in the colder latitudes is particularly striking for haplogroups that reside in the arctic and subarctic. The mean internal branch NS/S ratio for northeastern Siberian-North American haplogroups A, C, D, and X was 0.61, much higher than the mean for the non-arctic haplogroups of 0.39, the mean for the African L haplogroups of 0.31, or the mean for the Native American haplogroup B, which by-passed the cold selection of Siberia, of 0.38.

The internal branch replacement mutations of the arctic and subarctic haplogroups also have a higher average CI. The internal branch CIs of haplogroups A, C, D, and X was 51%; that for the remaining global haplogroups was 39%; that of L was 36%; and that of B was 31%.

The European haplogroup internal branch replacement mutations were also higher than the African L value of 0.31. Haplogroup H was 0.48, J was 0.66, and IWX was 0.63. By contrast, haplogroup T was only 0.31. However, this is because T harbors a single highly conserved founding replacement mutation in the ND2 gene (np 4917), which contributed virtually all of the adaptive advantage to this lineage. Hence, the CI of haplogroup T was 20.3, the highest of any European haplogroup. Thus, adaptive changes fall into two categories, either the lineage accumulates multiple missense mutations, each changing a somewhat less conserved amino acid, or the lineage changes only a few amino acids, each change of which is highly evolutionarily conserved.

Examples of the adaptive mtDNA mutations that occur in the arctic mtDNAs include two replacement mutations [ND2 np 4824G (T119A) and ATP6 np 8794T (H90Y)] for haplogroup A; two replacement variants [ND4 np 11969A (A404T) and cytb np 15204C (I153T)] for haplogroup C; a ND2 np 5178A (L237M) variant for haplogroup D; and a ND5 np 13708A (A458T) variant for a sublineage of haplogroup X. This latter variant also appears in European haplogroup J.

The European sister-haplogroups J and T provide the clearest example of the two classes of adaptive mutation strategies: several less conserved mutations versus a few highly conserved mutations. J and T share a common root involving two amino acid substitutions: ND1 np 4216C (Y304H) and cytb np 15452A (L236I). J and T then diverge. Haplogroup T is founded by the single nodal adaptive mutation: ND2 np 4917G (N150D), the most conserved ND2 polymorphism found (194).

Haplogroup J has two replacement mutations at its root: ND3 np 10398G (T114A) and ND5 np 13708A (A458T), the second being the same variant found in haplogroup X. Haplogroup J then splits into to sub-haplogroups J1 and J2, each defined by a major cytb mutation. The J2 cytb variant is at np 15257A (D171N) and the J1 cytb mutation is at np 14798C (F18L). The np 14789C mutation is also found at the root of sub-haplogroup Uk. The np 15,257 and np 14,789 variants alter well-conserved amino acids with CIs of 95% and 79%, respectively. The 15,257 variant alters the outer coenzyme Q binding site (Q_o) of complex III, which contacts the Rieske iron-sulfur protein, while the np 14,798 site alters the inner coenzyme Q binding site (Q_i) of complex III (194). Since the Q_o and Q_i binding sites are essential for complex III proton pumping via the Q-cycle, the np 14,798 and 15,257 variants are both likely to have disconnected electron flow through complex III with proton pumping. This would reduce the coupling efficiency of mitochondrial OXPHOS by one third and proportionately increase heat generation.

That the internal branch mtDNA missense mutations are functionally relevant has been demonstrated by comparing the sperm mobility of males harboring the different mtDNA haplogroups. Sperm flagellar motion is driven primarily by ATP generated from the mitochondria in the mid-piece. Therefore, sperm with partially uncoupled mitochondrial should swim slower that those with coupled mitochondria. As expected, sperm from haplogroup H subjects swam significantly faster than those from T subjects (193). Thus the functional mtDNA variants that founded specific mtDNA lineages affect mitochondrial physiological functions.

ADAPTIVE mtDNA MUTATIONS IN LONGEVITY AND DEGENERATIVE DISEASES

That these adaptive mutations are clinically relevant has become apparent from studies that correlated mtDNA haplogroups with longevity and degenerative disease. In an Italian study, mtDNA haplogroup J was found to be overrepresented in centenarians (50, 189). Similarly, in an Irish centenarian study, J2 was overrepresented (191), and in a Finnish study of individuals over 90 years J2, Uk, and WIX were enriched (163). In Japanese centenarians a sublineage of haplogroup D was enriched (210, 211). Hence, specific mtDNA lineages from Europe and Asia are protective against the ravages of aging.

Some of these same mtDNA lineages have also been found to be protective against neurodegenerative diseases. Haplogroups J and Uk are underrepresented in Parkinson disease (PD) (225) and haplogroup T is underrepresented in AD patients (32, 224). The repeated association of haplogroups J1 and Uk with longevity and neuro-protection is particularly illuminating because both haplogroups encompass the same cytb mutation at np 14,798. Such convergent evolution provides strong support for the functional importance of the cytb mutations.

In other studies, haplogroup J has been found to increase the penetrance of the milder LHON mutations (20). Haplogroup T has been found to be overrepresented in bipolar affective disorder (BPAD) (133).

How can the same variants be associated with increased life span and protection against certain diseases on the one hand, yet increase the predilection of developing other degenerative diseases on the other? The answer lies with the interdependence of mitochondrial energy production, ROS production, and the mtPTP activation of apoptosis.

The cold-adapted mtDNA uncoupling mutations would generate less ATP per calorie consumed. Hence, uncoupled mitochondria would be more prone to clinical problems resulting from energy insufficiency such as LHON and bipolar mental illness. However, individuals with uncoupled mitochondria would burn calories more rapidly to generate both the required ATP plus increased heat. As a result, the ETCs of uncoupled individuals would be generally more oxidized, thus minimizing the production of mitochondrial ROS. Reduction in mitochondrial oxidative stress would reduce the activation of the mtPTP, thus preserving cells and protecting the individuals from cellular loss leading to neurological and visceral tissue degeneration and aging.

THE MITOCHONDRIAL EITIOLOGY OF AGING

Since mtDNA haplogroups that harbor adaptive mtDNA uncoupling mutations have a reduced rate of aging and neurodegenerative disease, the mitochondria must be an important factor in the aging process. Aging is the decline of structure and function over time, which is a natural consequence of entropy, the tendency for all complex systems to decay with time. To counter entropy requires energy to repair or rebuild the damaged functions.

Since uncoupling mtDNA mutations are associated with increased human life span and would have reduced mitochondrial ROS production, ROS damage to the mitochondria, mtDNA, and host cell must be one of the most important entropic factors in determining age-related cellular decline. To counter endogenous sources of oxidative damage, the nucleus must make enzymes that limit mitochondrial ROS damage, thus preserving the mitochondria and mtDNAs.

An important role of mitochondrial ROS production in aging and degenerative diseases is congruent with the life-extending capacity of caloric restriction (69, 126, 127, 130, 204). Reduction of available calories will starve the mitochondrial ETC for electrons, thus reducing ROS and protecting the mitochondria and mtDNAs. Caloric restriction can be achieved by direct dietary restriction or by blocking the retention of excess calories. This may explain why mice in which the insulin receptor gene was inactivated in adiposities had an 18% increase in life span (16).

In rodent studies, mtDNA base oxidation (8oxo-dG) and rearrangement mutation levels have been found to increase with age. Dietary restriction will inhibit the accumulation of both forms of mtDNA damage (67, 139). Furthermore, aging alters gene expression in muscle and brain, but calorie restriction prevents many of these changes (106, 107). In muscle, two classes of genes are affected in aging, those for mitochondrial energy metabolism and those for antioxidant defenses. Calorie restriction normalizes the expression of many of these genes (106).

Drosophila life span can be extended by dietary restriction, and the functional basis of the life extension appears to overlap with insulin-like growth factor defects (35). The *chico* (*small boy*) dwarf mutant of *Drosophila* has a defect in an insulin-like growth factor receptor substrate protein and an increased life span (36). That this relates to energy metabolism was suggested by DNA microarray analyses that revealed the down-regulation with age of genes involved in metabolism, cell growth, and reproduction (180).

Evidence for a role for mitochondrial energy metabolism in *Drosophila* aging also comes from the *Indy* (*I'm not dead yet*) mutant. This mutation inactivates a dicarboxylate cotransporter expressed in the midgut, fat body, and oenocytes of the fly. This enzyme probably mediates the uptake or reuptake of di- and tricarboxylic acid TCA cycle intermediates (188). Since *Drosophila* lives on decaying fruit, which is rich in the di- and tricarboxylic acids that are substrates for the mitochondrion, the *Indy* mutant could be restricting the availability of calories to the mitochondrion.

Evidence that ROS toxicity is a limiting factor for life span in *Drosophila* was directly demonstrated by showing that transgenic *Drosophila* expressing increased Cu/Zn SOD and catalase lived longer (168). *Drosophila* life span extension was also seen in flies in which MnSOD was overexpressed in motor neurons (173) and systemically in *Drosophila* adults after heat shock induction of the MnSOD transgene (208). These observations support the conclusion that mitochondrial ROS toxicity is an important factor in limiting *Drososphila* life span.

A mitochondrial role in longevity has also been observed in the nematode *Caenorhabditis elegans*. A number of the *C. elegans* mutants that extend life span specifically affect mitochondrial function. These include the *clk-1* mutant, which affects a gene in the biosynthetic pathway of coenzyme Q (71, 87, 150), the *isp-1* mutant, which alters the Rieske iron sulfur protein of respiratory complex III (56), a mitochondrial leucyl-tRNA synthetase mutant (110), and a mitochondrial tRNA wobble U modifying enzyme, isopentenylpyrophosphate:tRNA transferase mutant (114). Since the *isp-1* mutation has been shown to both increase longevity and reduce mitochondrial ROS production, it seems likely that mitochondrial ROS must be modulating *C. elegans* life span (56).

The importance of the mitochondria in regulating *C. elegans* life span was further confirmed by an RNAi scan of the genes located on chromosomes I and II. Of the genes whose inactivation extended life span, 15% were identified as having mitochondrial functions. Since only about 1.5% of the genes of the *C. elegans* genome have been proposed to have a mitochondrial function, mitochondrial functions must have a much greater influence on longevity than other physiological processes (109, 110).

The life span of *C. elegans* is also regulated by the insulin-like growth factor signal transduction pathway. The metabolic functions associated with key longevity mutations include the insulin-like receptor (IRL) (*daf-2*), the catalytic subunit of PI3Kinase (*age-1*), the PTEN-like lipid phosphatase (*daf-18*), the Akt-1/2 kinases, and the target of Akt 1/2 phosphorylation, the forkhead transcription factor (*daf-16*). Phosphorylated forkhead transcription factor is cytosolic and transcriptionally inactive. Thus mutational inactivation of the insulin-like growth factor pathway must extend life span by blocking phosphorylation of the forkhead transcription factor and rendering it constitutively transcriptionally active (65, 93, 165, 215). Two genes that are up-regulated by activation of the *daf-16* forkhead transcription factor via dephosphorylation are MnSOD and catalase (122). Hence, one important function of the insulin-like signal pathway in *C. elegans* appears to be the negative regulation of the stress response and antioxidant enzymes that protect the cell from the deleterious effects of mitochondrial ROS production.

Life span in *C. elegans* as well as in yeast can also be extended by the overexpression of the NAD^+-dependent protein deacetylase Sir2. Sir2 acts, in part, through the deacetylation and activation of the forkhead transcription factors, favoring nuclear relocation and transcription activation (61, 214). In yeast, Sir2 also connects the insulin-like signaling pathway to caloric restriction. Sir2 requires NAD^+ as a co-reactant. NADH will not suffice and even acts as a competitive inhibitor of yeast. Since caloric restriction would reduce the cellular NADH to NAD^+ ratio, this should activate Sir2, causing the deacetylation and activation of the forkhead transcription factors and thus increasing expression of the antioxidant and stress response genes (119, 120). By contrast, defects in mitochondrial OXPHOS should increase the $NADH/NAD^+$ ratio, inhibit SIR2, and favor the acetylation of the forkhead transcription factors. This would turn off antioxidant and stress response gene expression, thereby reducing antioxidant defenses and life span.

This overlap between the insulin-like signaling pathway and the mitochondrion has been demonstrated in *C. elegans* by combining *daf-2* and *isp-1* mutants and showing that they do not have an additive effect on longevity. This is logical since inactivation of the *daf-2* pathway would increase mitochondrial antioxidant defenses while the *isp-1* mutation would reduce mitochondrial ROS production (56). Hence, the two classes of mutations would have the same effect.

That mitochondrial ROS production is an important factor affecting *C. elegans* life span has been confirmed by the *mev-1* mutation and the effects of SOD mimetics. The *mev-1* mutation affects the cytb-containing subunit of complex II. This mutation has a markedly increased rate of mitochondrial ROS production and a substantially shortened life span (84, 198). This reduced longevity can be reversed by treatment with the catalytic antioxidant mimetic EUK134, a salen Mn complex. Furthermore, treatment of wild-type *C. elegans* with EUK134 can increase the mean and maximum life span of *C. elegans* to an extent similar to that of the long-lived *C. elegans* mutant *age-1* (140). Therefore, the life span of *C. elegans* may be modulated by either decreasing mitochondrial ROS production or increasing mitochondrial antioxidant defenses.

Corroborative data that mitochondrial ROS limits the life span of mammals have come from studies on mutant and transgenic mice. Mice in which the MnSOD was genetically inactivated die at a mean age of eight days of a dilated cardiomyopathy, in association with the inactivation of iron-sulfur center containing enzymes including complex I, II, III, and aconitase (115, 137). The mitochondria of these animals have reduced state III respiration and hypersensitized mtPTPs (97), and they can be rescued from their cardiomyopathy by treatment with the SOD mimetic MnTBAP [Mn 5, 10, 15, 20-tetrakis (4-benzoic acid) porphyrin]. However, MnTBAP does not readily cross the blood-brain barrier and whereas the treated mice avoid the cardiomyopathy, they develop movement disorders together with cortical and brainstem spongiform vacuolization and brainstem astrocytosis (142). Again, both the systemic and neurological pathology can be ameliorated by treatment with the salen Mn complex compounds EUK8, EUK134, and EUK189, which do cross the blood-brain barrier and are protective against both the visceral effects and the neurological pathologies (138). Thus mitochondrial ROS damage must be important in the age-related decline of both mammalian visceral and central nervous system cells and tissues.

Although mice heterozygous for the MnSOD mutation live to old age, they have a chronically reduced liver mitochondrial membrane potential, reduced state III respiration equivalent to that of an old animal, and a striking rise in mitochondrial lipid peroxidation by middle age that drops to below the wild-type levels in older mice. This striking drop in lipid peroxidation is associated with a marked increase in the sensitivity of the mtPTP to Ca^{2+} activation and a striking increase in number of liver apoptotic cells in older heterozygous animals. The MnSOD heterozygotes also showed a striking induction of complex IV (COX) in the older animals (97). Aging MnSOD heterozygous C57Bl/6 (B6) mice also show a significant increase in oxidative damage (8oxo-dG) to nDNA and mtDNA, and a 100% increased tumor incidence, though their life span is not diminished (226).

Proof that mitochondrial ROS limit mammalian life span has come from the targeting of the human catalase to the mouse mitochondria in transgenic mice and observing an increase in life span. To import catalase to the mitochondrial matrix, the C-terminal peroxisomal targeting peptide was inactivated, and the catalase coding sequence was fused to the N-terminal mitochondrial targeting peptide of the mitochondrial matrix enzyme ornithine transcarbamylase. This construct was then expressed in transgenic mice using an α-actin promoter and CMV enhancer. The modified catalase was shown to be imported into the mitochondrion and to strikingly increase catalase activity in heart and muscle. Moreover, heart mitochondria from the transgenic animals showed a dramatically increased resistance of mitochondrial aconitase to inactivation by H_2O_2, and the skeletal muscle mtDNAs showed a significant reduction in the age-related accumulation of mtDNA rearrangement mutations. These protective effects to mitochondrial oxidation and mtDNA mutations were associated with a 21% increase in mean life span and a 10% increase in maximum life span in both male and female mitochondrial catalase transgenic animals (197).

The importance of mitochondrial ROS in longevity has also been supported by the selective inactivation of the $p66^{shc}$ splice variant of the $p52^{shc}/p46^{shc}$ gene. This mutant increases mouse life span by 30%, results in increased resistance to oxidative stress caused by paraquat, and reduces mitochondrial ROS production (146). $p66^{Shc}$ has been shown to become incorporated within the mitochondria, where it interacts with HSP70 and affects mitochondrial function and DNA oxidative damage (169). It is also a downstream target of p53 and is required for stress-activated p53 to induce intracellular ROS production, cytochrome c release, and apoptosis (218).

That increased mitochondrial ROS production accelerates mtDNA mutation rate was shown in mice deficient in the nDNA-encoded heart-muscle (and brain) isoform of the ANT gene (*Ant1*). These animals exhibit severe fatigability and develop a hypertrophic cardiomyopathy. Since an ANT defect blocks the import of cytosolic ADP into the mitochondria, and the ATP synthase without ADP cannot utilize ΔP to make ATP, then ΔP becomes hyperpolarized, the ETC stalls, and the excess electrons are transferred to O_2 to generate ROS. Consequently, the ANT1-deficient mice have greatly increased heart, muscle, and brain mitochondrial H_2O_2 production. This increased mitochondrial ROS correlates with a dramatic increase in mtDNA rearrangements in the heart by middle age (54).

That mtDNA mutations are one of the primary factors that limit mouse life span was demonstrated by the knock-in of a mutant form of the mtDNA polymerase γ (POLG) subunit A in which the proofreading function was inactivated by the mutation D257A. These mice had a shortened life span and developed a premature aging phenotype involving weight loss, reduction of subcutaneous fat, hair loss (alopecia), curvature of the spine (kyphosis), osteoporosis, anemia, reduced fertility, and heart enlargement. This was associated with an age-related decline in respiratory complexes I and IV and in mitochondrial ATP production rates in the heart. Analysis of the mtDNA revealed that the knock-in animals had a three- to fivefold increase in mtDNA base substitution mutations in brain, heart, and liver, with a higher number of mutations in the coding region cytb gene than in the CR (216). Thus, increased mtDNA mutations can directly increase the aging rate, implying that the mtDNA is one of the most important targets for mitochondrial ROS damage in aging.

The relationship between mtDNA mutations and longevity has been shown in a wide variety of animal species including humans. mtDNA rearrangement mutations have been shown to accumulate in proportion to life span in organisms as diverse as mouse, chimpanzee, and human (43, 139, 141, 143). When the level of mtDNA rearrangements was quantified in various human brain regions by the analysis of the common 5-kb mtDNA deletion, mtDNA rearrangements were found to accumulate to the highest levels in the basal ganglia, followed by the cerebral cortex, with the least mutant being found in the cerebellum (39, 206). Base substitution mutations also accumulate in human tissues with age. A mtDNA CR mutation, T414G, in the mtTFA binding site of the P_L accumulates with age in diploid fibroblasts (145); a T408A promoter mutation and a A189G O_H mutation accumulate with age in skeletal muscle (245); and a T150C O_H mutation accumulates with age in white blood cells (256). These mtDNA CR mutations appear to be highly tissue specific. The T414G mutation can be detected at low levels in skeletal muscle by using a highly specific and sensitive protein nucleic acid (PNA)-clamping PCR method. However, this same method has been unable to detect the T414G mutation in the brains of even very elderly people (157). These mtDNA mutations can be attributed to oxidative damage, at least in the brain, since the human brain mtDNA shows a marked age-related increase in oxidative damage (135).

All of these data are consistent with the hypothesis that aging is the result of mitochondrial decline due to oxidative damage to the mitochondria and mtDNAs. This hypothesis, when combined with our current understanding of caloric regulation of mitochondrial function, can neatly explain why caloric restriction is able to extend mammalian life span. Reduction of dietary calories reduces serum glucose, thus inhibiting insulin secretion by the pancreatic β cells. This results in the dephosphorylation and activation of the FOXO forkhead transcription factors in the energy-utilizing tissues. Further, caloric restriction reduces cellular $NADH/NAD^+$ ratios, activating the NAD^+-dependent SIRT1, which further activates the FOXOs by deacetylation (24, 61, 155). The activated FOXOs then activate the transcription of the stress response genes including the antioxidant enzymes MnSOD and catalase. They also induce, along with glucagon through CREB, the transcriptional induction of PGC-1α. This increases the mitochondrial OXPHOS capacity, mobilizes the WAT triglyceride stores for mitochondrial oxidation, and activates gluconeogenesis to maintain the basal serum glucose level. Hence, caloric restriction induces the mitochondria to more completely oxidize dietary reducing equivalents and the antioxidant defense systems to more effectively eliminate ROS. Thus, dietary restriction should substantially reduce mitochondrial oxidative stress. In addition, activation of SIRT1 by reduction of the $NADH/NAD^+$ ratio will stimulate the deacetylate p53, and deacetylation of p53 reduces its apoptotic potential. Hence, caloric restriction also protects the mtPTP from p53-mediated activation and apoptosis (82, 104, 123, 229). Therefore, caloric restriction extends life span in three ways: by decreasing mitochondrial generation of ROS, by increasing the cellular antioxidant defenses, and by inhibiting cell loss via apoptosis.

THE MITOCHONDRIAL ETIOLOGY OF DEGENERATIVE DISEASES

The level of mtDNA rearrangement mutations has also been found to be strikingly increased in a variety of age-related degenerative diseases. The 5-kb mtDNA rearrangement is increased between 7- and 2200-fold in the hearts of patients with chronic coronary artery disease (41, 42). Since coronary artery disease is associated with recurrent cardiac ischemia and reperfusion, which generates bursts on mitochondrial ROS (49), the cardiac rearrangement mutations may also be caused by mitochondrial oxidative damage.

Similarly, mtDNA rearrangements are found in patients with AD. In the brains of AD patients who died prior to age 75, the 5-kb deletion is increased on average 15-fold, whereas in the brains of patients who died after age 75, the 5-kb deletion is present at one fifth the level of that in control brains (40). The increased mtDNA rearrangement mutations in AD brains are associated with

APP: amyloid precursor protein

increased mtDNA oxidative damage (134). The sudden drop in the level of mtDNA deletions in older AD brains might be the result of loss of the brain cells containing the most mutated mtDNA through apoptosis. This is consistent with the increased neuronal apoptosis observed in AD brains (30).

AD, is a progressive neurodegenerative disease resulting in dementia associated with the deposition of Aβ amyloid peptide plaques and neurofibrillary tangles in the brain. Early-onset, familial AD has been associated with mutations in the Aβ precursor protein (APP) and the presenilin-containing APP peptide processing complexes (212). Although the molecular defects that cause the late-onset, sporadic AD cases have not been determined (203), the inheritance of the ApoE ε4 allele has been identified as a major risk factor for developing late-onset AD (190, 203).

Mitochondrial OXPHOS defects have also been detected in a variety of tissues from AD patients. The mtDNA of AD patients has been indirectly shown to harbor genetic defects by fusing blood platelets from AD patients to human ρ^0 cell lines (94), with the resulting cybrids being found to have OXPHOS defects (17, 29, 45, 96, 158, 159, 217, 222, 232). The concept that mtDNA mutations might contribute to late-onset AD has been bolstered by the observation that a mtDNA tRNA^{Gln} gene mutation at np 4336 is present in about 5% of late-onset AD patients and in 7% of combined AD plus PD patients, but in 0.4% of normal controls (81, 199).

The association between mitochondrial deficiency and AD was further strengthened recently by the demonstration, using PNA-clamping PCR, that the somatic mtDNA CR T414G mutation in the P_L mtTFA binding site can be detected in 65% of all AD brains but not in any control brains. Moreover, the level of somatic mtDNA CR mutations in AD brains is increased by 63% overall, with the elevation of somatic mtDNA CR mutations in AD patients over 80 years being 120%. Furthermore, in the AD brains, the somatic mtDNA mutations preferentially occur within the known mtDNA transcription and replication regulatory elements, whereas very few of the mtDNA CR mutations found in the control brains occur in functional elements. Finally, AD brains have an average 50% reduction in the ND6/ND2 mRNA ratio and a comparable reduction in the levels of the mtDNA/nDNA ratio, suggesting the inhibition of L-strand transcription and mtDNA replication (44).

These data suggest that the accumulation of somatic mtDNA CR mutations in AD brains that damage key mtDNA regulatory elements may be a central feature in the etiology of the disease. The accumulation of these mutations with age would result in the progressive decline in mitochondrial energy production and an associated increase in mitochondrial ROS production. The increased ROS would ultimately lead to the activation of the mtPTP and the loss of neuronal synapses through mitochondrial induced cytochrome c release and caspase activation.

By this logic, individuals harboring mtDNAs that generate tightly coupled mitochondria would experience chronically increased ROS production on a high calorie diet. This would increase the probability of mtDNA CR mutations and thus the probability of developing AD. By contrast, individuals who harbor mtDNAs with uncoupling polymorphisms would have reduced mitochondrial ROS and be less prone to somatic mtDNA CR mutations and AD. Population studies have confirmed these predictions (32, 224).

Mitochondrial ROS production might also account for the connection between the ApoE ε4 allele and AD, since the presence of the ApoE ε4 allele has been associated with increased brain oxidative damage (151, 185, 186). Aβ peptide toxicity might also be related to mitochondrial dysfunction. The Aβ peptide has recently been reported to enter the mitochondria, bind to mitochondrial alcohol dehydrogenase, and increase mitochondrial ROS production (46, 124). Moreover, incubation of rat brain mitochondria with the

Aβ25–35 fragment results in a selective and dose-independent inhibition of complex IV (27). Consistent with the possibility that Aβ enters the mitochondrion, the insulin-degrading enzyme (IDE) has been discovered to have a mitochondrially directed form generated by initiating translation at an AUG codon 41 amino acids upstream from the canonical IDE peptide start codon. Since the IDE is thought to degrade both cerebral Aβ-peptide and plasma insulin, it could function in mitochondrial Aβ turnover (113).

The Aβ peptide is generated by cleavage of the APP by the presenilin-containing γ-secretase complex. Recently, the presenilin-1 (PS1) protein has been located in the inner membrane of rat brain and liver mitochondria (9). The other components of the presenilin γ-secretase including nicastrin, APH-1, and PEN-2 have also been found in a high-molecular-weight complex within the mitochondrial inner membrane. Although the import mechanism for these proteins is not fully understood, the nicastrin protein has been found to harbor both ER and mitochondrial N-terminal targeting signals (68). APP has also been found in the mitochondrion, but it is thought to be located in the outer mitochondrial membrane, a position that would not be conducive to its processing into Aβ by the mitochondrial inner membrane γ-secretase (7). Therefore, the γ-secretase complex proteins must have other functions in the mitochondrion. Given that γ-secretase is unique in its ability to cleave peptide bonds within membranes, the PS1 complex may well be processing mitochondrial inner membrane proteins such as those of the respiratory complexes. Hence mutations in the presinilin complex associated with early-onset AD might also cause mitochondrial defects.

A mitochondrial etiology of AD could also explain the near-universal association of Aβ amyloid plaques with AD. At the concentrations found in biological fluids, Aβ1–40 and Aβ1–42 act as antioxidant and anti-apoptotic polypeptides, presumably because of their tendency to chelate transition metals, particularly copper. However, when the Aβs aggregate, they become pro-oxidants, perhaps through placing the transition metal in the presence of the redox-active Met 35 (99–101). Given this perspective on the Aβ peptides, it could follow that genetic variants which chronically increase mitochondrial ROS production would overwhelm the antioxidant capacity of basal levels of Aβ. To compensate, the brain would increase Aβ production. However, this would eventually lead to excess Aβ, which would then aggregate into a pro-oxidant form (80) and thereby exacerbate the problem. Consistent with this scenario, the brains of late-onset sporadic AD patients have been observed to have increased oxidative damage, increased activated caspase activity (62, 183), and increased numbers of TUNEL- (terminal deoxynucleotidyl transferase biotin-dUTP nick end labeling) positive cells (45). Increased mitochondrial ROS would also increase the mtDNA somatic mutation rate and exacerbate mitochondrial oxidative stress, mtPTP activation, and synaptic loss.

Early-onset AD could also fit within this framework. Mutations in the Aβ processing pathway would increase the rate of Aβ production and aggregation, creating an active pro-oxidant that would increase synaptic oxidative stress, damage the mitochondria, and destroy the synapses.

A similar mitochondrial scenario could be applied to other age-related neurodegenerative diseases. Amyotrophic lateral sclerosis (ALS) is a neurodegenerative disease that specifically affects the motor neurons. In familial cases, ALS is caused by mutations in the Cu/ZnSOD (*Sod1*). Transgenic mice that express a human mutant Cu/ZnSOD develop motor neuron disease associated with vacuolate mitochondria containing high concentrations of Cu/ZnSOD (85). The mutant Cu/ZnSOD mice also accumulate 8oxo-dG in cytoplasmic granules, possibly representing mtDNA nucleoids (247), and in cultured neuroblastoma cells, the toxicity of a mutant Cu/ZnSOD can be ameliorated by the increased expression of MnSOD (57). The mitochondrial uptake of the three common familial ALS Cu/ZnSOD mutants (G37R, G41D, and G93A) is blocked by the heat shock proteins Hsp70, Hsp27, and Hsp25 (167). Consequently, familial ALS appears to be due to a failure in the Cu/ZnSOD's ability to detoxify mitochondrial superoxide anion, which is released into the mitochondrial inner membrane space from ubisemiquinone bound to the Q_o site of complex III.

Idiopathic PD has also consistently been reported to be associated with mitochondrial defects, particularly in respiratory complex I. Evidence supporting a mitochondrial etiology of PD has been previously reviewed but includes the selective destruction of the dopaminergic neurons in humans and rodents by the mitochondrial complex I inhibitor 1-methyl-4-phenyl-1,2,3,6-tetrahydropyridine (MPTP); the detection of OXPHOS defects and particularly complex I defects in PD patient muscle, blood, brain, and in platelet-derived cybrids; and the inheritance of PD-like symptoms to various pathogenic mtDNA mutations (241) including the common LHON np 11,778 mutation (202). Small subsets of PD patients are also found in Mendelian pedigrees. The gene for one rare autosomal recessive form of PD, which maps to chromosome 1p35-p36, has been cloned and found to code for the *PINK1* (PTEN-induced kinase) protein, a kinase localized in the mitochondrion (221). A more common autosomal recessive locus, the *parkin* gene on chromosome 6q25.2–27, is a E3 ubiquitin ligase (228). Genetic inactivation of *parkin* in the mouse resulted in nigrostriatal defects associated with a decreased abundance of several mitochondrial proteins including the E1α subunit of pyruvate dehydrogenase, the 24- and 30-kDa subunits of complex I and the Vb subunit of complex IV and a general decrease in mitochondrial state III respiration. The oxidative stress proteins peroxiredoxin 1, 2, and 6 and lactolyglutathione were also down-regulated in association with a marked increase in brain lipid peroxidation products (171). Hence, it would appear that the pathophysiology of the *parkin* mutation is also associated with a mitochondrial defect. Finally, the rare autosomal dominant form of PD, which maps to chromosome 4q21-q22, is due to mutations in a-synuclein (182). Expression of mutant α-synuclein in various cell systems has resulted in altered mitochondrial function (79, 108, 170). Thus, both the idiopathic and the inherited forms of PD appear to be associated with mitochondrial dysfunction. This is consistent with the protective effects of haplogroups J and Uk for PD (225).

Studies on Huntington disease have implicated a mitochondrial dysfunction in its etiology as well (172, 195, 241), and mtDNA rearrangements are elevated in the basal ganglia of patients with Huntington disease (76). Thus all of these age-related neurodegenerative diseases, as well as aging itself, seem to be directly related to the mitochondrial production of ROS and the resulting accumulation of mtDNA mutations. These observations imply that the mitochondrial paradigm for neurodegenerative diseases appears to be one that can be broadly applied.

SOMATIC AND GERMLINE mtDNA MUTATIONS IN CANCER

Aging is also the most significant risk factor for most solid tumors. Moreover, caloric restriction in rodents markedly reduces cancer risk (69, 126, 127, 130, 204). Hence, the etiology of cancer might also be associated with mitochondrial dysfunction.

Gene defects in certain nDNA-encoded mitochondrial genes have been directly linked to some hereditary cancers. Mutations in the mitochondrial fumarate hydratase have been associated with uterine leiomyomas and renal cell carcinomas (112), and mutations in three of the four nDNA-encoded subunits of complex II have been linked to paragangliomas: SDHD (cytb membrane subunit) (15), SDHC (coenzyme Q-binding membrane subunit) (162), and SDHB (iron-sulfur) subunit (11, 227). In addition to the paragangliomas seen for mutations in all three subunits, mutations in the SDHB subunit are also associated with pheochromocytoma and early-onset renal cell carcinoma. On the other hand, mutations in the SDHA (FAD-containing subunit) cause the mitochondrial encephalomyopathy Leigh syndrome (19) but not paraganglioma.

The striking difference in clinical effects of SDHA subunit mutations versus SDHB, C, and D mutations strongly implicates mitochondrial ROS production in the etiology of cancer. SDHA mutations would block the entry of electrons into complex II, thus depleting ATP without increasing ROS production. The result is neuromuscular disease. By contrast, mutations in subunits SDH B, C, and D would impede electron flux through complex II. This would increase ROS production, implicating ROS in cancer predisposition. This is supported by the *mev-1* mutant in the *C. elegans* SDHC subunit, which has been shown to increase mitochondrial ROS production (84, 198).

A role for mitochondrial ROS in cancer is supported by the observations that mice heterozygous for the MnSOD knock-out locus have a 100% increase in cancer incidence (226), many types of tumors have reduced MnSOD, transformation of certain tumors with the MnSOD cDNA can reverse the malignant phenotype, and a cluster of three mutations in the MnSOD gene promoter region that alter AP-2 binding and promoter efficiency is found in a number of tumors (131, 252). Moreover, ROS production, in association with the inactivation of *p16ink4a*, has been hypothesized to be one of the two main mechanisms for tumorigenesis; the other is p53 deficiency (10). Hence, it would appear that mitochondrial ROS production, not ATP depletion, is the important mitochondrial factor in the etiology of cancer.

Mitochondrial DNA mutations that inhibit OXPHOS and impede electron flow down the ETC should increase ROS production and contribute to cancer. The first clear evidence that mtDNA mutations might play an important role in neoplastic transformation came with the report of a renal adenocarcinoma that was heteroplasmic for a 294-np in-frame deletion in the mtDNA ND1 gene. This deletion generated a truncated mRNA, confirming that the mutant mtDNA resulted in functional mitochondrial defect (77). Subsequently, a variety of mtDNA-coding region and CR mutations have been reported in colon cancer cells (4, 66, 181), head and neck tumors (58), astrocytic tumors (95), thyroid tumors (129, 253), breast tumors (209), and prostate tumors (33, 34, 38, 177).

Mitochondrial ROS could contribute to neoplastic transformation, both as a tumor initiator by causing nDNA mutations in proto-oncogenes and tumor-suppressor genes and as a tumor promoter through driving cellular proliferation. At low levels, ROS has been found to be an active mitogen, thought to act though interaction with various kinases (Src kinase, protein kinase C, MAPK, and receptor tyrosine kinases), as well as with different transcription factors (Fos, Jun, NF-κB) (131). Furthermore, the dual function apurinic/apyrimidinic endonuclease 1 (APE-1) functions not only in the DNA base excision pathway but also in the redox regulation of the transcription factors Fos, Jun, NFκB, PAX, HIF-1a, and p53 through the redox modification of its cysteine 310 (55, 91).

Further evidence of the importance of mtDNA mutations and mitochondrial ROS production in neoplastic transformation comes from studies of complex IV (COX) mutations in prostate cancer. A proteomic survey of prostate cancer epithelium cells isolated by laser capture microscopy (LCM) revealed that the ratio of nDNA-encoded subunits (COX IV, Vb, and VIc) to mtDNA-encoded subunits (COI and II) was increased in most prostate tumors (72).

This alteration in the ratio of nDNA/mtDNA COX subunits in prostate cancer has now been shown to be due, in part, to mutations in the mtDNA COI gene (177). An analysis of the COI gene sequence from 260 prostate cancer specimens, 54 prostate cancer-negative men over age 50, and 898 European mtDNA sequences revealed that for Americans of European descent, COI mutations are found in 11% of the prostate cancer specimens, 0% of the cancer-negative prostate controls, and 5.5% of the general population, all significant differences. Furthermore, four prostate cancer COI mutations (T6253C, C6340T, G6261A, A6663G) (**Figure 1**) were found in multiple independent patient tumors, often on different background mtDNA haplotypes, and these changed highly conserved amino acids (CI = 69% to 97%) (177). Thus males who inherit COI missense mutations from their mother are at a greatly increased risk for developing prostate cancer.

The average amino acid CI of the COI mutants found in prostate cancers was 83% ± 25%, whereas that of the COI variants in the general population was 71% ± 35%. This indicates that somatic mutations affecting more functionally conserved amino acids also contribute to the etiology of prostate cancer. This assessment was confirmed by the observation that three of the COI mutations in prostate patients were heteroplasmic: G5949A-G16X (Stop), T6124C-M74T (CI = 95%), C6924T-A341S (CI = 100%) (177). These data demonstrate that mutant COI genes are prostate cancer oncogenes and that both germline and somatic mtDNA mutations are important factors in this etiology of prostate cancer.

A more complete assessment of the role of mtDNA mutations in prostate cancer was obtained by sequencing the entire mtDNA from the tumorigenic epithelial cells, purified using LCM, of the tumor harboring the COI chain termination (G16X). This analysis revealed three mtDNA variants: the heteroplasmic COI G16X mutation, a homoplasmic ATP6 C8932T (P136G) variant with a CI of 64%, and a homoplasmic cytb A14769G (N8S) variant with a CI of 20.5%. Of these variants, the COI and ATP6 mutations are likely to be functionally significant. The G16X mutation was found to be homoplasmic in the cancer cells and homoplasmic wild type in the adjacent normal cells, which was confirmed by immunohistochemical staining for the COI protein. Therefore, this G16X mutation must have arisen de novo at the time of neoplastic transformation and been strongly selected for in the cancerous cells (177).

The physiological significance of mtDNA mutations in cancer cells was demonstrated by introducing a known pathogenic mtDNA mutation into the prostate cancer cell line PC3 via transmitochondrial cybrid fusions (26, 240). PC3 cancer cells were cured of their resident mtDNAs by treatment with the mitochondrial poison rhodamine 6G and then fused with cytoplasts from either homoplasmic mutant (T8993G) or homoplasmic wild-type (T8993T) cell lines (177) derived from a patient who was heteroplasmic for the pathogenic ATP6 np T8993G L156R mutation (75). This mutation reduces mitochondrial ADP production (219) but also increases ROS production (128). A series of PC3 cybrids were generated that were genetically identical except for either a wild-type T or a mutant G at np 8993 in the mtDNA.

Injection of these mutant and wild-type PC3 cybrids into nude mice gave a dramatic result. The PC3 cells harboring the normal T8993T base barely grew at all. By contrast, the PC3 cybrids with the mutant T8993G base generated rapidly growing tumors that killed the mice (177). Staining the cellular nodules from the T8993T and T8993G cybrid tumors with dihydroeithidium revealed that the mutant T8993G tumors were making much more ROS than the wild-type T8993T nodules (177).

Consistent with the increased ROS production in the T8993G tumors, MnSOD has been found to be reduced in precancerous intraepithelial neoplastic lesions and in prostate cancer by immunohistochemical and in situ hybridization studies (12, 18). Prostate cancer cells also have a 42% increase in cytoplasmic staining for the redox-sensitive APE-1 (90). Thus these observations support the conclusion that increased mitochondrial ROS production is central to prostate cancer.

Therefore, mtDNA mutations appear to play an important causal role in prostate cancer, and thus may have a similar role in the etiology of other solid tumors. Moreover, the role of mitochondrial defects in the pathophysiology of cancer would appear to be the generation of increased ROS, which acts as both an nDNA mutagen and cellular mitogen. Since mitochondrial mutations that increase ROS would do so by impeding the ETC, they would also result in the accumulation within the cell of unoxidized NADH and pyruvate. The excess NADH and pyruvate would then be converted to lactate by lactate dehydrogenase. This high production of lactate by cancer cells in the presence of oxygen was first observed 70 years ago by Otto Warburg and described as aerobic-glycolysis. Warburg hypothesized that mitochondrial defects might underlie many forms of cancer (246). Modern molecular genetic analysis might yet prove him right.

Mitochondrial Medicine: An Integrated Mitochondrial Model for the Etiology of Degenerative Diseases, Aging, and Cancer

All of these observations on the pathophysiology of the common degenerative diseases, cancer, and aging reveal that these diverse clinical entities share many common underlying biochemical, gene regulation, and genetic features. Moreover, these features consistently involve energy metabolism and the mitochondrion. Therefore, we can theorize that all age-related diseases (degenerative diseases, cancer, and aging) share a common underlying mitochondrial pathophysiology which is summarized in **Figure 6**.

Dietary Restriction, Degenerative Diseases, Longevity, and Climatic Adaptation

In the energy-utilizing tissues (brain, heart, muscle, kidney, etc.), when calories (reducing equivalents) are plentiful and ATP is being rapidly hydrolyzed through physical activity, the mitochondrial electrochemical gradient (ΔP) is kept depolarized as the ATP synthase resynthesizes ATP. As a result, electrons flow continuously through the ETC to $^1/_2O_2$ to sustain ΔP, keeping the $NADH/NAD^+$ ratio low, the ETC oxidized and mitochondrial ROS production minimized.

Mitochondrial medicine: the new medical discipline that pertains to all clinical problems that involve the mitochondria

However, if calories are plentiful in tightly coupled mitochondria and ATP utilization is low (sedentary life style), then ΔP becomes and remains maximized. This stalls the ETC, resulting in a high $NADH/NAD^+$ ratio, excess electrons in the ETC, and excessive ROS production. The increased ROS chronically damages the mitochondria, mtDNA, and the mtPTP, ultimately resulting in the activation of the mtPTP and cell death (**Figure 6**) (242).

High carbohydrate diets stimulate insulin secretion that phosphorylates the FOXOs, removing them from the nucleus. This down-regulates the cellular stress-response pathways including the mitochondrial and cytosolic antioxidant defenses and reduces the transcription of PGC-1α, thus down-regulating mitochondrial OXPHOS. Additionally, the high $NADH/NAD^+$ ratio inhibits SIRT1 leaving the forkhead transcription factors acetylated and out of the nucleus. The suppression of SIRT1 also permits p53 to become acetylated and fully active, increasing the expression of BAX and the activation of $p66^{Shc}$. BAX binds to the mitochondrial outer membrane and activates the mtPTP to initiate apoptosis. The activated mitochondrial component of $p66^{shc}$ increases mitochondrial ROS production and stimulates cytochrome c release and apoptosis. This destructive trend is partially countered by the ROS-mediated stimulation of the interaction of SIRT1 with the forkheads, enhancing the deacetylation of forkheads and their import into the nucleus to activate the antioxidant and stress response systems. However, continued caloric overload counters this protective effect. In the end, high calorie diets lead to the down-regulation of mitochondrial OXPHOS and antioxidant and stress response defenses, increasing mitochondrial ROS production and the accumulation of oxidative damage to the mitochondria and mtDNA, leading to cell death.

As an individual ages, the chronic oxidative stress results in the accumulation of somatic mtDNA mutations, including rearrangements and base substitutions in the regulatory elements of the CR and in the structural genes. Ultimately, the decline of the mitochondrial biosynthetic capacity renders the mitochondria unable to adequately repair the accumulating mitochondrial oxidative damage. This activates the mtPTP, p53, and $p66^{Shc}$ systems, increasing apoptosis rates and resulting in cell loss, tissue dysfunction, and the symptoms of age-related diseases.

These processes are ameliorated in individuals harboring partially uncoupled mitochondria. In loosely coupled mitochondria,

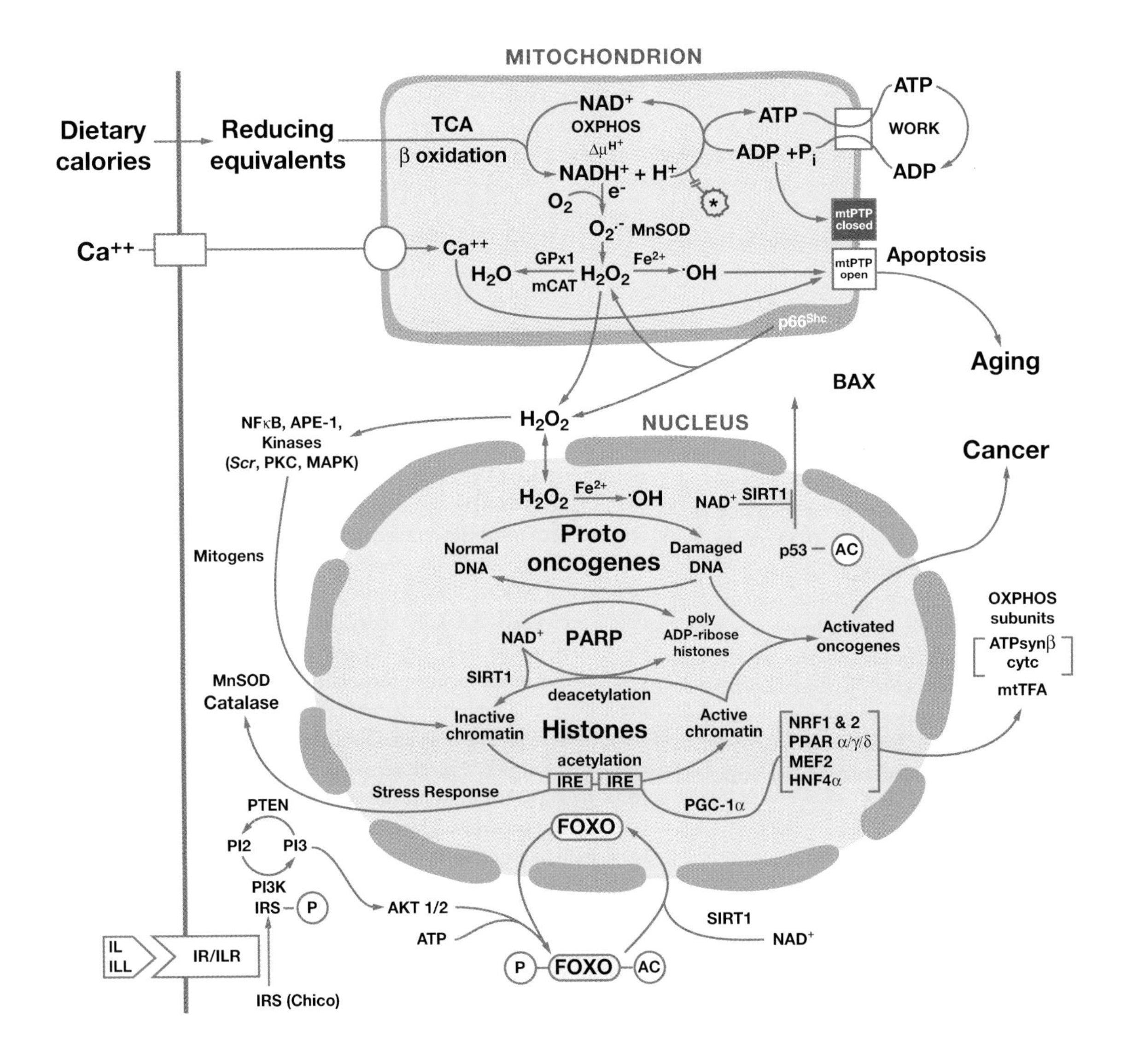

inefficient proton pumping and/or excessive proton leakage though the ATP synthase keeps ΔP submaximal, irrespective of the caloric availability. This keeps electrons flowing through the ETC and generating heat, diminishes mitochondrial ROS production, decreases the rate of cellular apoptosis, and thus increases longevity and inhibits the progression of the age-related diseases.

The Mitochondrial Energetics and ROS Biology of Cancer

These same factors also modulate the cellular predilection to neoplastic transformation, cancer. Mutations in nDNA or mtDNA OXPHOS genes that impede the flow of electrons through the ETC increase mitochondrial ROS production. The resulting H_2O_2, which is relatively stable, diffuses out of the mitochondrion, through the cytosol, and into the nucleus. In the nucleus, H_2O_2 can interact with transition metals and be converted to •OH, which mutagenizes the nDNA. The resulting nDNA damage activates the nDNA repair systems, including the polyADP-ribose polymerase (PARP). The activated PARP degrades the nuclear NAD^+ in the process of adding poly ADP-ribose chains to histones and other nuclear proteins (242) (**Figure 6**).

The degradation of the nuclear NAD^+, together with the high $NADH/NAD^+$ ratio, inactivates nuclear SIRT1. SIRT1 is a histone deacetylase, and nuclear transcription is repressed by the deacetylation of histones and activated by histone acetylation. Therefore, inhibition of SIRT1 permits histone acetylation to predominate, turning on the transcription of normally inactive genes. In post-mitotic tissues, histone acetylation permits the activation of the genes that regulate cell replication and differentiation, the proto-oncogenes.

The diffusion of H_2O_2 into the nucleus and its conversion to •OH can then mutate the proto-oncogenes, converting them into functional oncogenes. Moreover, increased cytosolic and nuclear H_2O_2 activates a

Figure 6

Model for the proposed role for mitochondrial dysfunction in an energy-utilization tissue cell in metabolic and degenerative diseases, aging, and cancer. The mitochondrial pathophysiology of these clinical entities is envisioned to result from the interplay between mitochondrial energy production, ROS generation, and the initiation of apoptosis through activation of the mtPTP. These components of energy metabolism are modulated by environmental constraints such as caloric availability and cold stress through the regulation of the FOXO and PGC-1α transcription factors and the SIRT1 NAD^+-dependent deacetylase. The FOXO transcription factors coordinately regulate mitochondrial energy metabolism through PGC-1α as well as the antioxidant and stress response genes necessary to cope with the increased oxidative stress of oxidative metabolism. SIRT1 fine-tunes the interrelationship between energy metabolism and apoptosis through the deacetylation of PGC-1α, p53, and the histone proteins. Caloric overload or inhibition of OXPHOS perturbs the cellular mitochondrial energetic balance resulting in increased ROS. The increased ROS and decreased mitochondrial energy output sensitizes the mtPTP, ultimately driving the cell to apoptotic death. The increased ROS also diffuses into the nucleus as H_2O_2 where it can mutate and activate proto-oncogenes (initiation) and can interact with NFκB, APE-1 and various kinases to initiate cell division (promotion) leading to neoplastic transformation (cancer). Figure abbreviations are PARP, poly ADP-ribose polymerase; SIRT1, the mammalian homologue to Sir2; FOXO, the mammalian forkhead transcription factor; P, a phosphorylated protein; Ac, an acetylated protein; IL, insulin ligand; ILL, insulin-like ligand; IR, insulin receptor; ILR, insulin-like growth factor receptor; IRE, insulin response element; PI3K, the PI3 kinase; PI2, the membrane-bound phosphotidyl-inositol diphosphate; PI3, membrane-bound phosphatidyl-inositol triphosphate, AKT 1/2, the AKT kinases; ATPsynβ, ATP synthase β subunit; cytc, cytochrome c.

variety of cellular signal transduction factors including NFκB, APE-1, *Fos*, *Jun*, and tyrosine kinases. This drives the cell into replication. Consequently, mutations in mitochondrial genes that inhibit electron flow through the ETC result in chronically increased mitochondrial ROS production, which can act as both a tumor initiator (mutation of proto-oncogenes) and tumor promoter (activation of transcription and replication) (242) (**Figure 6**).

Evolutionary medicine: a clinical perspective that posits that many of the common clinical problems of today are rooted in adaptive genetic programs that permitted our human ancestors to survive in the environments which they confronted in the past

MITOCHONDRIAL ADAPTATION AND EVOLUTIONARY MEDICINE

The Interplay between Environment and Energetics

These considerations indicate that mitochondria lie at the intersection between environmental factors such as calories and cold and the human capacity to energetically cope with the environmental challenges in different regions of the globe. Our ancient ancestors had to be able to adapt to two classes of environmental changes: (*a*) short term changes in calories and climate associated with seasonal variation and (*b*) long term changes in the nature of calories and average annual temperatures defined by the latitude and geographic zone in which they lived.

Aboriginal human populations living in the colder temperate and arctic zones needed to adjust to acute cold and seasonal variation in calories. Short-term adaptation to cold can be accomplished by uncoupling mitochondrial OXPHOS in both BAT and muscle by induction of the UCPs. Today, in industrial societies, most individuals avoid cold stress through central heating. Moreover, most ancient populations had to accumulate plant carbohydrate calories and store them as fat during the plant growing season and then to efficiently use the stored fat calories to sustain their cellular energetics when the plants were dormant. This seasonal variation in caloric availability was metabolically managed via the insulin signaling network that coordinately integrated the energy-utilization, energy-storage, and energy-homeostasis tissues, through the hormonal signals of the energy-sensing tissues, the pancreatic α and β cells. This was accomplished by cueing the energy-sensing tissues to serum glucose concentration that oscillated based on the availability of plant carbohydrate calories. When plant calories were abundant, blood sugar was high, insulin was secreted, mitochondrial OXPHOS and its attendant antioxidant defenses were down-regulated in the energy-utilization tissues, and the excess carbohydrate calories were stored as fat in energy-storage tissues. When plant calories declined at the end of the growing season, blood sugar levels declined, resulting in decreased insulin secretion and increased glucagon secretion. These hormonal changes up-regulated mitochondrial OXPHOS and the attendant antioxidant defenses in the energy-utilization tissues, mobilized the stored fats in the energy-storage tissues for use as mitochondrial fuel, and activated glucose synthesis in the energy-homeostasis liver to sustain minimal blood sugar levels.

This energetic system worked well for our ancestors who lived in an environment of periodic carbohydrate surplus and deficiency. However, in today's developed societies, technology provides us with unlimited dietary calories including carbohydrates throughout the year. As a result, our energy signaling system remains continuously in the high carbohydrate state. Consequently, mitochondrial OXPHOS and its attendant antioxidant defenses are chronically down-regulated. At the same time, our sedentary life style means that we do not turnover ATP through sustained physical activity. Therefore, the excess of caloric reducing equivalents in our diet leads to fat accumulation and obesity and keeps the mitochondrial electron carriers of our down-regulated ETC fully reduced. This results in chronically increased mitochondrial ROS production, which damages the mitochondria and mtDNAs. The reduced mitochondrial energy output and the increased oxidative stress sensitize the mtPTP to activation, driving post-mitotic cells into premature apoptosis. The resulting loss of post-mitotic cells results in the tissue-specific symptoms associated with aging and the age-related diseases.

While the insulin and cAMP signal transduction systems permitted our ancestors to rapidly adapt their metabolism to sudden changes in environmental temperature and carbohydrate calorie availability, these short-term oscillations were insufficient to cope with the more general environmental differences experienced by individuals living in the tropics versus the temperate zone versus the arctic. These different climatic zones required adaptation to differences in average ambient temperature, amount of sunlight, and nature of available calories. In the tropics the predominate source of calories was plant carbohydrates, whereas in the arctic the caloric availability shifted more toward animal protein and fat.

Humans adapted to these more general regional differences through the fixation of functional mutations in their mtDNAs that changed the coupling efficiency of OXPHOS, thus shifting the energetic balance from predominantly ATP production in the tropics to increased heat production in the arctic. Functional mtDNA mutations undoubtedly permitted adaptation to many additional environmental factors as yet undiscovered. What is clear is that the random occurrence of functional mtDNA variants shifted the energetic balance in certain individuals sufficiently that they could move into and survive in more northern environments from which their warm-adapted predecessors were excluded. However, these adaptive mtDNA mutations required trade-offs. In the tropics, tightly coupled mitochondria maximized the efficiency of physical work and minimized heat production. Mutations that partially uncoupled OXPHOS decreased work efficiency, but provided essential heat for surviving chronic cold. The uncoupling mutations also partially relieved the constraints of the inner membrane electrochemical gradient (ΔP) on electron flow through the ETC, permitting the ETC to run continuously. This kept the electron carriers oxidized and minimized ROS production. Hence, in industrialized societies where calories are unlimited and exercise is minimized, individuals with uncoupled mitochondria (e.g., European haplogroup J) are partially protected from the increased oxidative stress associated with excessive caloric intake. Therefore, these individuals are on average less prone to degenerative diseases and have increased longevity. Still, these individuals require more calories to sustain their tissue energetics. Hence, if calories become limiting again, they will be the first to starve.

CONCLUSION

Evolutionary medicine posits that the genetic variants that permitted our ancestors to adapt to diverse environments (niches) are having a profound influence on our predisposition to common diseases today. The information provided in this review now places this concept on a solid experimental foundation and shows that the major environmental challenges that confronted our hunter-gather ancestors were bioenergetics for which the mitochondria adaptations were pivotal. Thus, the adaptive mtDNA mutations of our ancestors, when combined with the environmental control and unlimited calories of our modern environment, are increasingly resulting in individual energetic imbalances. It is these imbalances between ancient mitochondrial genetics and modern caloric intake that, I believe, are driving the modern epidemics of obesity, diabetes, neurodegenerative disease, cardiovascular disease, and cancer.

ACKNOWLEDGMENTS

The author would like to thank Ms. M.T. Lott for her assistance with this manuscript. This work has been supported by NIH grants NS21328, AG13154, NS41650, AG24373, TW01366, HL64017, and an Ellison Senior Scholar Award.

LITERATURE CITED

1. Accili D, Arden KC. 2004. FoxOs at the crossroads of cellular metabolism, differentiation, and transformation. *Cell* 117:421–26
2. Adams JU. 2005. Autophagy and longevity. How keeping house may keep one young. *Scientist* 19:22–24
3. Adams V, Griffin L, Towbin J, Gelb B, Worley K, McCabe ER. 1991. Porin interaction with hexokinase and glycerol kinase: metabolic microcompartmentation at the outer mitochondrial membrane. *Biochem. Med. Metab. Biol.* 45:271–91
4. Alonso A, Martin P, Albarran C, Aquilera B, Garcia O, et al. 1997. Detection of somatic mutations in the mitochondrial DNA control region of colorectal and gastric tumors by heteroduplex and single-strand conformation analysis. *Electrophoresis* 18:682–85
5. Altshuler D, Hirschhorn JN, Klannemark M, Lindgren CM, Vohl MC, et al. 2000. The common PPARgamma Pro12Ala polymorphism is associated with decreased risk of type 2 diabetes. *Nat. Genet.* 26:76–80
6. Alzheimer's Association. 2005. *Alzheimer's Disease Statistics.* **http://www.alz.org/Resources/FactSheets/FSAlzheimerStats.pdf**
7. Anandatheerthavarada HK, Biswas G, Robin MA, Avadhani NG. 2003. Mitochondrial targeting and a novel transmembrane arrest of Alzheimer's amyloid precursor protein impairs mitochondrial function in neuronal cells. *J. Biol. Chem.* 161:41–54
8. Andreu AL, Bruno C, Dunne TC, Tanji K, Shanske S, et al. 1999. A nonsense mutation (G15059A) in the cytochrome b gene in a patient with exercise intolerance and myoglobinuria. *Ann. Neurol.* 45:127–30
9. Ankarcrona M, Hultenby K. 2002. Presenilin-1 is located in rat mitochondria. *Biochem. Biophys. Res. Commun.* 295:766–70
10. Arbiser JL. 2004. Molecular regulation of angiogenesis and tumorigenesis by signal transduction pathways: evidence of predictable and reproducible patterns of synergy in diverse neoplasms. *Semin. Cancer Biol.* 14:81–91
11. Astuti D, Latif F, Dallol A, Dahia PL, Douglas F, et al. 2001. Gene mutations in the succinate dehydrogenase subunit SDHB cause susceptibility to familial pheochromocytoma and to familial paraganglioma. *Am. J. Hum. Genet.* 69:49–54
12. Baker AM, Oberley LW, Cohen MB. 1997. Expression of antioxidant enzymes in human prostatic adenocarcinoma. *Prostate* 32:229–33
13. Ballinger SW, Shoffner JM, Gebhart S, Koontz DA, Wallace DC. 1994. Mitochondrial diabetes revisited. *Nat. Genet.* 7:458–59
14. Ballinger SW, Shoffner JM, Hedaya EV, Trounce I, Polak MA, et al. 1992. Maternally transmitted diabetes and deafness associated with a 10.4 kb mitochondrial DNA deletion. *Nat. Genet.* 1:11–15
15. Baysal BE, Ferrell RE, Willett-Brozick JE, Lawrence EC, Myssiorek D, et al. 2000. Mutations in SDHD, a mitochondrial complex II gene, in hereditary paraganglioma. *Science* 287:848–51
16. Bluher M, Kahn BB, Kahn CR. 2003. Extended longevity in mice lacking the insulin receptor in adipose tissue. *Science* 299:572–74

17. Bosetti F, Brizzi F, Barogi S, Mancuso M, Siciliano G, et al. 2002. Cytochrome c oxidase and mitochondrial F1F0-ATPase (ATP synthase) activities in platelets and brain from patients with Alzheimer's disease. *Neurobiol. Aging* 23:371–76
18. Bostwick DG, Alexander EE, Singh R, Shan A, Qian J, et al. 2000. Antioxidant enzyme expression and reactive oxygen species damage in prostatic intraepithelial neoplasia and cancer. *Cancer* 89:123–34
19. Bourgeron T, Rustin P, Chretien D, Birch-Machin M, Bourgeois M, et al. 1995. Mutation of a nuclear succinate dehydrogenase gene results in mitochondrial respiratory chain deficiency. *Nat. Genet.* 11:144–49
20. Brown MD, Starikovskaya YB, Derbeneva O, Hosseini S, Allen JC, et al. 2002. The role of mtDNA background in disease expression: a new primary LHON mutation associated with Western Eurasian haplogroup J. *Hum. Genet.* 110:130–38
21. Brown MD, Sun F, Wallace DC. 1997. Clustering of Caucasian Leber hereditary optic neuropathy patients containing the 11778 or 14484 mutations on an mtDNA lineage. *Am. J. Hum. Genet.* 60:381–87
22. Brown MD, Trounce IA, Jun AS, Allen JC, Wallace DC. 2000. Functional analysis of lymphoblast and cybrid mitochondria containing the 3460, 11778, or 14484 Leber's Hereditary Optic Neuropathy mtDNA mutation. *J. Biol. Chem.* 275:39831–36
23. Brown MD, Zhadanov S, Allen JC, Hosseini S, Newman NJ, et al. 2001. Novel mtDNA mutations and oxidative phosphorylation dysfunction in Russion LHON families. *Hum. Genet.* 109:33–39
24. Brunet A, Sweeney LB, Sturgill JF, Chua KF, Greer PL, et al. 2004. Stress-dependent regulation of FOXO transcription factors by the SIRT1 deacetylase. *Science* 303:2011–15
25. Bruno C, Martinuzzi A, Tang Y, Andreu AL, Pallotti F, et al. 1999. A stop-codon mutation in the human mtDNA cytochrome c oxidase I gene disrupts the functional structure of complex IV. *Am. J. Hum. Genet.* 65:611–20
26. Bunn CL, Wallace DC, Eisenstadt JM. 1974. Cytoplasmic inheritance of chlormaphenicol resistance in mouse tissue culture cells. *Proc. Natl. Acad. Sci. USA* 71:1681–85
27. Canevari L, Clark JB, Bates TE. 1999. Beta-amyloid fragment 25–35 selectively decreases complex IV activity in isolated mitochondria. *FEBS Lett.* 457:131–34
28. Cann RL, Stoneking M, Wilson AC. 1987. Mitochondrial DNA and human evolution. *Nature* 325:31–36
29. Cardoso SM, Proenca MT, Santos S, Santana I, Oliveira CR. 2004. Cytochrome c oxidase is decreased in Alzheimer's disease platelets. *Neurobiol. Aging* 25:105–10
30. Castellani R, Hirai K, Aliev G, Drew KL, Nunomura A, et al. 2002. Role of mitochondrial dysfunction in Alzheimer's disease. *J. Neurosci. Res.* 70:357–60
31. Cent. Dis. Control Prev. 2005. *Prostate Cancer: The Public Health Perspective, 2004/2005 Fact Sheet.* **http://www.cdc.gov/cancer/prostate/prospdf/about2004.pdf**
32. Chagnon P, Gee M, Filion M, Robitaille Y, Belouchi M, Gauvreau D. 1999. Phylogenetic analysis of the mitochondrial genome indicates significant differences between patients with Alzheimer disease and controls in a French-Canadian founder population. *Am. J. Med. Genet.* 85:20–30
33. Chen JZ, Gokden N, Greene GF, Mukunyadzi P, Kadlubar FF. 2002. Extensive somatic mitochondrial mutations in primary prostate cancer using laser capture microdissection. *Cancer Res.* 62:6470–74
34. Chinnery PF, Samuels DC, Elson J, Turnbull DM. 2002. Accumulation of mitochondrial DNA mutations in ageing, cancer, and mitochondrial disease: Is there a common mechanism? *Lancet* 360:1323–25

35. Clancy DJ, Gems D, Hafen E, Leevers SJ, Partridge L. 2002. Dietary restriction in long-lived dwarf flies. *Science* 296:319
36. Clancy DJ, Gems D, Harshman LG, Oldham S, Stocker H, et al. 2001. Extension of life-span by loss of CHICO, a Drosophila insulin receptor substrate protein. *Science* 292:104–6
37. Comi GP, Bordoni A, Salani S, Franceschina L, Sciacco M, et al. 1998. Cytochrome c oxidase subunit I microdeletion in a patient with motor neuron disease. *Ann. Neurol.* 43:110–16
38. Copeland WC, Wachsman JT, Johnson FM, Penta JS. 2002. Mitochondrial DNA alterations in cancer. *Cancer Invest.* 20:557–69
39. Corral-Debrinski M, Horton T, Lott MT, Shoffner JM, Beal MF, Wallace DC. 1992. Mitochondrial DNA deletions in human brain: regional variability and increase with advanced age. *Nat. Genet.* 2:324–29
40. Corral-Debrinski M, Horton T, Lott MT, Shoffner JM, McKee AC, et al. 1994. Marked changes in mitochondrial DNA deletion levels in Alzheimer brains. *Genomics* 23:471–76
41. Corral-Debrinski M, Shoffner JM, Lott MT, Wallace DC. 1992. Association of mitochondrial DNA damage with aging and coronary atherosclerotic heart disease. *Mutat. Res.* 275:169–80
42. Corral-Debrinski M, Stepien G, Shoffner JM, Lott MT, Kanter K, Wallace DC. 1991. Hypoxemia is associated with mitochondrial DNA damage and gene induction. Implications for cardiac disease. *JAMA* 266:1812–16
43. Cortopassi GA, Arnheim N. 1990. Detection of a specific mitochondrial DNA deletion in tissues of older humans. *Nucleic Acids Res.* 18:6927–33
44. Coskun PE, Ruiz-Pesini EE, Wallace DC. 2003. Control region mtDNA variants: longevity, climatic adaptation and a forensic conundrum. *Proc. Natl. Acad. Sci. USA* 100: 2174–76
45. Cottrell DA, Borthwick GM, Johnson MA, Ince PG, Turnbull DM. 2002. The role of cytochrome c oxidase deficient hippocampal neurones in Alzheimer's disease. *Neuropathol. Appl. Neurobiol.* 28:390–96
46. Crouch PJ, Blake R, Duce JA, Ciccotosto GD, Li QX, et al. 2005. Copper-dependent inhibition of human cytochrome c oxidase by a dimeric conformer of amyloid-beta1-42. *J. Neurosci.* 25:672–79
47. Czubryt MP, McAnally J, Fishman GI, Olson EN. 2003. Regulation of peroxisome proliferator-activated receptor gamma coactivator 1 alpha (PGC-1 alpha) and mitochondrial function by MEF2 and HDAC5. *Proc. Natl. Acad. Sci. USA* 100:1711–16
48. Daitoku H, Yamagata K, Matsuzaki H, Hatta M, Fukamizu A. 2003. Regulation of PGC-1 promoter activity by protein kinase B and the forkhead transcription factor FKHR. *Diabetes* 52:642–49
49. Das DK, George A, Liu XK, Rao PS. 1989. Detection of hydroxyl radical in the mitochondria of ischemic-reperfused myocardium by trapping with salicylate. *Biochem. Biophys. Res. Commun.* 165:1004–9
50. De Benedictis G, Rose G, Carrieri G, De Luca M, Falcone E, et al. 1999. Mitochondrial DNA inherited variants are associated with successful aging and longevity in humans. *FASEB J.* 13:1532–36
51. DiMauro S, Schon EA. 2003. Mitochondrial respiratory-chain diseases. *N. Engl. J. Med.* 348:2656–68

52. Dumoulin R, Sagnol I, Ferlin T, Bozon D, Stepien G, Mousson B. 1996. A novel gly290asp mitochondrial cytochrome b mutation linked to a complex III deficiency in progressive exercise intolerance. *Mol. Cell. Probes* 10:389–91
53. Ek J, Andersen G, Urhammer SA, Gaede PH, Drivsholm T, et al. 2001. Mutation analysis of peroxisome proliferator-activated receptor-gamma coactivator-1 (PGC-1) and relationships of identified amino acid polymorphisms to Type II diabetes mellitus. *Diabetologia* 44:2220–26
54. Esposito LA, Melov S, Panov A, Cottrell BA, Wallace DC. 1999. Mitochondrial disease in mouse results in increased oxidative stress. *Proc. Natl. Acad. Sci. USA* 96:4820–25
55. Evans AR, Limp-Foster M, Kelley MR. 2000. Going APE over ref-1. *Mutat. Res.* 461:83–108
56. Feng J, Bussiere F, Hekimi S. 2001. Mitochondrial electron transport is a key determinant of life span in *Caenorhabditis elegans*. *Dev. Cell* 1:633–44
57. Flanagan SW, Anderson RD, Ross MA, Oberley LW. 2002. Overexpression of manganese superoxide dismutase attenuates neuronal death in human cells expressing mutant (G37R) Cu/Zn-superoxide dismutase. *J. Neurochem.* 81:170–77
58. Fliss MS, Usadel H, Caballero OL, Wu L, Buta MR, et al. 2000. Facile detection of mitochondrial DNA mutations in tumors and bodily fluids. *Science* 287:2017–19
59. Furuyama T, Kitayama K, Yamashita H, Mori N. 2003. Forkhead transcription factor FOXO1 (FKHR)-dependent induction of PDK4 gene expression in skeletal muscle during energy deprivation. *Biochem. J.* 375:365–71
60. Gelb BD, Adams V, Jones SN, Griffin LD, MacGregor GR, McCabe ER. 1992. Targeting of hexokinase 1 to liver and hepatoma mitochondria. *Proc. Natl. Acad. Sci. USA* 89:202–6
61. Giannakou ME, Partridge L. 2004. The interaction between FOXO and SIRT1: tipping the balance towards survival. *Trends Cell Biol.* 14:408–12
62. Gibson GE, Huang HM. 2002. Oxidative processes in the brain and non-neuronal tissues as biomarkers of Alzheimer's disease. *Front. Biosci.* 7:d1007–d15
63. Giles RE, Blanc H, Cann HM, Wallace DC. 1980. Maternal inheritance of human mitochondrial DNA. *Proc. Natl. Acad. Sci. USA* 77:6715–19
64. Goto Y, Nonaka I, Horai S. 1990. A mutation in the $tRNA^{Leu(UUR)}$ gene associated with the MELAS subgroup of mitochondrial encephalomyopathies. *Nature* 348:651–53
65. Guarente L, Kenyon C. 2000. Genetic pathways that regulate ageing in model organisms. *Nature* 408:255–62
66. Habano W, Sugai T, Yoshida T, Nakamura S. 1999. Mitochondrial gene mutation, but not large-scale deletion, is a feature of colorectal carcinomas with mitochondrial microsatellite instability. *Int. J. Cancer* 83:625–29
67. Hamilton ML, Van Remmen H, Drake JA, Yang H, Guo ZM, et al. 2001. Does oxidative damage to DNA increase with age? *Proc. Natl. Acad. Sci. USA* 98:10469–74
68. Hansson CA, Frykman S, Farmery MR, Tjernberg LO, Nilsberth C, et al. 2004. Nicastrin, presenilin, APH-1 and PEN-2 form active gamma-secretase complexes in mitochondria. *J. Biol. Chem.* 279:51654–60
69. Harrison DE, Archer JR. 1987. Genetic differences in effects of food restriction on aging in mice. *J. Nutr.* 117:376–82
70. Hebert LE, Scherr PA, Bienias JL, Bennett DA, Evans DA. 2003. Alzheimer disease in the US population: prevalence estimates using the 2000 census. *Arch. Neurol.* 60:1119–22
71. Hekimi S, Burgess J, Bussiere F, Meng Y, Benard C. 2001. Genetics of lifespan in C. elegans: molecular diversity, physiological complexity, mechanistic simplicity. *Trends Genet.* 17:712–18

72. Herrmann PC, Gillespie JW, Charboneau L, Bichsel VE, Paweletz CP, et al. 2003. Mitochondrial proteome: altered cytochrome c oxidase subunit levels in prostate cancer. *Proteomics* 3:1801–10
73. Hogan P, Dall T, Nikolov P. 2003. Economic costs of diabetes in the US in 2002. *Diabetes Care* 26:917–32
74. Holt IJ, Harding AE, Morgan-Hughes JA. 1988. Deletions of muscle mitochondrial DNA in patients with mitochondrial myopathies. *Nature* 331:717–19
75. Holt IJ, Harding AE, Petty RK, Morgan-Hughes JA. 1990. A new mitochondrial disease associated with mitochondrial DNA heteroplasmy. *Am. J. Hum. Genet.* 46:428–33
76. Horton TM, Graham BH, Corral-Debrinski M, Shoffner JM, Kaufman AE, et al. 1995. Marked increase in mitochondrial DNA deletion levels in the cerebral cortex of Huntington's disease patients. *Neurology* 45:1879–83
77. Horton TM, Petros JA, Heddi A, Shoffner J, Kaufman AE, et al. 1996. Novel mitochondrial DNA deletion found in a renal cell carcinoma. *Genes Chromosomes Cancer* 15:95–101
78. Houten SM, Auwerx J. 2004. PGC-1alpha: turbocharging mitochondria. *Cell* 119:5–7
79. Hsu LJ, Sagara Y, Arroyo A, Rockenstein E, Sisk A, et al. 2000. Alpha-synuclein promotes mitochondrial deficit and oxidative stress. *Am. J. Pathol.* 157:401–10
80. Huang X, Atwood CS, Hartshorn MA, Multhaup G, Goldstein LE, et al. 1999. The A beta peptide of Alzheimer's disease directly produces hydrogen peroxide through metal ion reduction. *Biochemistry* 38:7609–16
81. Hutchin T, Cortopassi G. 1995. A mitochondrial DNA clone is associated with increased risk for Alzheimer disease. *Proc. Natl. Acad. Sci. USA* 92:6892–95
82. Imai S, Armstrong CM, Kaeberlein M, Guarente L. 2000. Transcriptional silencing and longevity protein Sir2 is an NAD-dependent histone deacetylase. *Nature* 403:795–800
83. Ingman M, Kaessmann H, Paabo S, Gyllensten U. 2000. Mitochondrial genome variation and the origin of modern humans. *Nature* 408:708–13
84. Ishii N, Fujii M, Hartman PS, Tsuda M, Yasuda K, et al. 1998. A mutation in succinate dehydrogenase cytochrome b causes oxidative stress and ageing in nematodes. *Nature* 394: 694–97
85. Jaarsma D, Rognoni F, van Duijn W, Verspaget HW, Haasdijk ED, Holstege JC. 2001. CuZn superoxide dismutase (SOD1) accumulates in vacuolated mitochondria in transgenic mice expressing amyotrophic lateral sclerosis-linked SOD1 mutations. *Acta Neuropathol.* 102:293–305
86. Johnson MJ, Wallace DC, Ferris SD, Rattazzi MC, Cavalli-Sforza LL. 1983. Radiation of human mitochondria DNA types analyzed by restriction endonuclease cleavage patterns. *J. Mol. Evol.* 19:255–71
87. Jonassen T, Larsen PL, Clarke CF. 2001. A dietary source of coenzyme Q is essential for growth of long-lived *Caenorhabditis elegans* clk-1 mutants. *Proc. Natl. Acad. Sci. USA* 98: 421–26
88. Jun AS, Brown MD, Wallace DC. 1994. A mitochondrial DNA mutation at np 14459 of the ND6 gene associated with maternally inherited Leber's hereditary optic neuropathy and dystonia. *Proc. Natl. Acad. Sci. USA* 91:6206–10
89. Kamei Y, Ohizumi H, Fujitani Y, Nemoto T, Tanaka T, et al. 2003. PPARgamma coactivator 1beta/ERR ligand 1 is an ERR protein ligand, whose expression induces a high-energy expenditure and antagonizes obesity. *Proc. Natl. Acad. Sci. USA* 100:12378–83
90. Kelley MR, Cheng L, Foster R, Tritt R, Jiang J, et al. 2001. Elevated and altered expression of the multifunctional DNA base excision repair and redox enzyme Ape1/ref-1 in prostate cancer. *Clin. Cancer Res.* 7:824–30

91. Kelley MR, Parsons SH. 2001. Redox regulation of the DNA repair function of the human AP endonuclease Ape1/ref-1. *Antioxid. Redox Signal.* 3:671–83
92. Kelly DP, Scarpulla RC. 2004. Transcriptional regulatory circuits controlling mitochondrial biogenesis and function. *Genes Dev.* 18:357–68
93. Kimura KD, Tissenbaum HA, Liu Y, Ruvkun G. 1997. Daf-2, an insulin receptor-like gene that regulates longevity and diapause in *Caenorhabditis elegans*. *Science* 277:942–46
94. King MP, Attardi G. 1989. Human cells lacking mtDNA: repopulation with exogenous mitochondria by complementation. *Science* 246:500–3
95. Kirches E, Michael M, Woy C, Schneider T, Warich-Kirches M, et al. 1999. Loss of heteroplasmy in the displacement loop of brain mitochondrial DNA in astrocytic tumors. *Genes Chromosomes Cancer* 26:80–83
96. Kish SJ, Mastrogiacomo F, Guttman M, Furukawa Y, Taanman JW, et al. 1999. Decreased brain protein levels of cytochrome oxidase subunits in Alzheimer's disease and in hereditary spinocerebellar ataxia disorders: a nonspecific change? *J. Neurochem.* 72:700–7
97. Kokoszka JE, Coskun P, Esposito L, Wallace DC. 2001. Increased mitochondrial oxidative stress in the Sod2 (+/−) mouse results in the age-related decline of mitochondrial function culminating in increased apoptosis. *Proc. Natl. Acad. Sci. USA* 98:2278–83
98. Kokoszka JE, Waymire KG, Levy SE, Sligh JE, Cai J, et al. 2004. The ADP/ATP translocator is not essential for the mitochondrial permeability transition pore. *Nature* 427:461–65
99. Kontush A. 2001. Amyloid-beta: an antioxidant that becomes a pro-oxidant and critically contributes to Alzheimer's disease. *Free Radic. Biol. Med.* 31:1120–31
100. Kontush A, Atwood CS. 2004. Amyloid-beta: phylogenesis of a chameleon. *Brain Res. Brain Res. Rev.* 46:118–20
101. Kontush A, Berndt C, Weber W, Akopyan V, Arlt S, et al. 2001. Amyloid-beta is an antioxidant for lipoproteins in cerebrospinal fluid and plasma. *Free Radic. Biol. Med.* 30:119–28
102. Koster JC, Marshall BA, Ensor N, Corbett JA, Nichols CG. 2000. Targeted overactivity of beta cell K(ATP) channels induces profound neonatal diabetes. *Cell* 100:645–54
103. Koutnikova H, Campuzano V, Foury F, Dolle P, Cazzalini O, Koenig M. 1997. Studies of human, mouse and yeast homologues indicate a mitochondrial function for frataxin. *Nat. Genet.* 16:345–51
104. Langley E, Pearson M, Faretta M, Bauer UM, Frye RA, et al. 2002. Human SIR2 deacetylates p53 and antagonizes PML/p53-induced cellular senescence. *EMBO J.* 21:2383–96
105. Lee CH, Olson P, Evans RM. 2003. Minireview: lipid metabolism, metabolic diseases, and peroxisome proliferator-activated receptors. *Endocrinology* 144:2201–7
106. Lee CK, Klopp RG, Weindruch R, Prolla TA. 1999. Gene expression profile of aging and its retardation by caloric restriction. *Science* 285:1390–93
107. Lee CK, Weindruch R, Prolla TA. 2000. Gene-expression profile of the ageing brain in mice. *Nat. Genet.* 25:294–97
108. Lee HJ, Shin SY, Choi C, Lee YH, Lee SJ. 2002. Formation and removal of alpha-synuclein aggregates in cells exposed to mitochondrial inhibitors. *J. Biol. Chem.* 277:5411–17
109. Lee SS, Kennedy S, Tolonen AC, Ruvkun G. 2003. DAF-16 target genes that control *C. elegans* life-span and metabolism. *Science* 300:644–4
110. Lee SS, Lee RY, Fraser AG, Kamath RS, Ahringer J, Ruvkun G. 2003. A systematic RNAi screen identifies a critical role for mitochondria in *C. elegans* longevity. *Nat. Genet.* 33:40–48

111. Lehman JJ, Barger PM, Kovacs A, Saffitz JE, Medeiros DM, Kelly DP. 2000. Peroxisome proliferator-activated receptor gamma coactivator-1 promotes cardiac mitochondrial biogenesis. *J. Clin. Invest.* 106:847–56
112. Lehtonen R, Kiuru M, Vanharanta S, Sjoberg J, Aaltonen LM, et al. 2004. Biallelic inactivation of fumarate hydratase (FH) occurs in nonsyndromic uterine leiomyomas but is rare in other tumors. *Am. J. Pathol.* 164:17–22
113. Leissring MA, Farris W, Wu X, Christodoulou DC, Haigis MC, et al. 2004. Alternative translation initiation generates a novel isoform of insulin-degrading enzyme targeted to mitochondria. *Biochem. J.* 383:439–46
114. Lemieux J, Lakowski B, Webb A, Meng Y, Ubach A, et al. 2001. Regulation of physiological rates in *Caenorhabditis elegans* by a tRNA-modifying enzyme in the mitochondria. *Genetics* 159:147–57
115. Li Y, Huang TT, Carlson EJ, Melov S, Ursell PC, et al. 1995. Dilated cardiomyopathy and neonatal lethality in mutant mice lacking manganese superoxide dismutase. *Nat. Genet.* 11:376–81
116. Liang XH, Jackson S, Seaman M, Brown K, Kempkes B, et al. 1999. Induction of autophagy and inhibition of tumorigenesis by beclin 1. *Nature* 402:672–76
117. Lin J, Wu H, Tarr PT, Zhang CY, Wu Z, et al. 2002. Transcriptional co-activator PGC-1 alpha drives the formation of slow-twitch muscle fibres. *Nature* 418:797–801
118. Lin J, Wu PH, Tarr PT, Lindenberg KS, St-Pierre J, et al. 2004. Defects in adaptive energy metabolism with CNS-linked hyperactivity in PGC-1alpha null mice. *Cell* 119:121–35
119. Lin SJ, Ford E, Haigis M, Liszt G, Guarente L. 2004. Calorie restriction extends yeast life span by lowering the level of NADH. *Genes Dev.* 18:12–16
120. Lin SJ, Guarente L. 2003. Nicotinamide adenine dinucleotide, a metabolic regulator of transcription, longevity and disease. *Curr. Opin. Cell Biol.* 15:241–46
121. Louet JF, Hayhurst G, Gonzalez FJ, Girard J, Decaux JF. 2002. The coactivator PGC-1 is involved in the regulation of the liver carnitine palmitoyltransferase I gene expression by cAMP in combination with HNF4 alpha and cAMP-response element-binding protein (CREB). *J. Biol. Chem.* 277:37991–8000
122. Lund J, Tedesco P, Duke K, Wang J, Kim SK, Johnson TE. 2002. Transcriptional profile of aging in *C. elegans*. *Curr. Biol.* 12:1566–73
123. Luo J, Nikolaev AY, Imai S, Chen D, Su F, et al. 2001. Negative control of p53 by Sir2alpha promotes cell survival under stress. *Cell* 107:137–48
124. Lustbader JW, Cirilli M, Lin C, Xu HW, Takuma K, et al. 2004. ABAD directly links Abeta to mitochondrial toxicity in Alzheimer's disease. *Science* 304:448–52
125. Malaisse-Lagae F, Malaisse WJ. 1988. Hexose metabolism in pancreatic islets: regulation of mitochondrial hexokinase binding. *Biochem. Med. Metab. Biol.* 39:80–89
126. Masoro EJ. 1993. Dietary restriction and aging. *J. Am. Geriatr. Soc.* 41:994–99
127. Masoro EJ, McCarter RJ, Katz MS, McMahan CA. 1992. Dietary restriction alters characteristics of glucose fuel use. *J. Gerontol.* 47:B202–8. Erratum. 1993. *J Gerontol.* 48(2):B73
128. Mattiazzi M, Vijayvergiya C, Gajewski CD, DeVivo DC, Lenaz G, et al. 2004. The mtDNA T8993G (NARP) mutation results in an impairment of oxidative phosphorylation that can be improved by antioxidants. *Hum. Mol. Genet.* 13:869–79
129. Maximo V, Soares P, Lima J, Cameselle-Teijeiro J, Sobrinho-Simoes M. 2002. Mitochondrial DNA somatic mutations (point mutations and large deletions) and mitochondrial DNA variants in human thyroid pathology: a study with emphasis on Hurthle cell tumors. *Am. J. Pathol.* 160:1857–65

130. McCarter RJ, Palmer J. 1992. Energy metabolism and aging: a lifelong study of Fischer 344 rats. *Am. J. Physiol.* 263:E448–52
131. McCord JM. 2000. The evolution of free radicals and oxidative stress. *Am. J. Med. Genet.* 108:652–59
132. McDavid K, Lee J, Fulton J, Tonita J, Thompson T. 2004. Prostate cancer incidence and mortality rates and trends in the United States and Canada. *Public Health Rep.* 119:174–86
133. McMahon FJ, Chen YS, Patel S, Kokoszka J, Brown MD, et al. 2000. Mitochondrial DNA sequence diversity in bipolar affective disorder. *Am. J. Psychiatr.* 157:1058–64
134. Mecocci P, MacGarvey U, Beal MF. 1994. Oxidative damage to mitochondrial DNA is increased in Alzheimer's disease. *Ann. Neurol.* 36:747–51
135. Mecocci P, MacGarvey U, Kaufman AE, Koontz D, Shoffner JM, et al. 1993. Oxidative damage to mitochondrial DNA shows marked age-dependent increases in human brain. *Ann. Neurol.* 34:609–16
136. Melendez A, Talloczy Z, Seaman M, Eskelinen EL, Hall DH, Levine B. 2003. Autophagy genes are essential for dauer development and life-span extension in *C. elegans*. *Science* 301:1387–91
137. Melov S, Coskun P, Patel M, Tunistra R, Cottrell B, et al. 1999. Mitochondrial disease in superoxide dismutase 2 mutant mice. *Proc. Natl. Acad. Sci. USA* 96:846–51
138. Melov S, Doctrow SR, Schneider JA, Haberson J, Patel M, et al. 2001. Lifespan extension and rescue of spongiform encephalopathy in superoxide dismutase 2 nullizygous mice treated with superoxide dismutase-catalase mimetics. *J. Neurosci.* 21:8348–53
139. Melov S, Hinerfeld D, Esposito L, Wallace DC. 1997. Multi-organ characterization of mitochondrial genomic rearrangements in ad libitum and caloric restricted mice show striking somatic mitochondrial DNA rearrangements with age. *Nucleic Acids Res.* 25:974–82
140. Melov S, Ravenscroft J, Malik S, Gill MS, Walker DW, et al. 2000. Extension of life-span with superoxide dismutase/catalase mimetics. *Science* 289:1567–69
141. Melov S, Schneider JA, Coskun PE, Bennett DA, Wallace DC. 1999. Mitochondrial DNA rearrangements in aging human brain and in situ PCR of mtDNA. *Neurobiol. Aging* 20:565–71
142. Melov S, Schneider JA, Day BJ, Hinerfeld D, Coskun P, et al. 1998. A novel neurological phenotype in mice lacking mitochondrial manganese superoxide dismutase. *Nat. Genet.* 18:159–63
143. Melov S, Shoffner JM, Kaufman A, Wallace DC. 1995. Marked increase in the number and variety of mitochondrial DNA rearrangements in aging human skeletal muscle. *Nucleic Acids Res.* 23:4122–26. Erratum. 1995. *Nucleic Acids Res.* 23(23):4938
144. Merriwether DA, Clark AG, Ballinger SW, Schurr TG, Soodyall H, et al. 1991. The structure of human mitochondrial DNA variation. *J. Mol. Evol.* 33:543–55
145. Michikawa Y, Mazzucchelli F, Bresolin N, Scarlato G, Attardi G. 1999. Aging-dependent large accumulation of point mutations in the human mtDNA control region for replication. *Science* 286:774–79
146. Migliaccio E, Giorgio M, Mele S, Pelicci G, Reboldi P, et al. 1999. The p66shc adaptor protein controls oxidative stress response and life span in mammals. *Nature* 402:309–13
147. Mishmar D, Ruiz-Pesini E, Brandon M, Wallace DC. 2004. Mitochondrial DNA-like sequences in the nucleus (NUMTs): insights into our African origins and the mechanism of foreign DNA integration. *Hum. Mutat.* 23:125–33
148. Mishmar D, Ruiz-Pesini EE, Golik P, Macaulay V, Clark AG, et al. 2003. Natural selection shaped regional mtDNA variation in humans. *Proc. Natl. Acad. Sci. USA* 100:171–76

149. MITOMAP. 2005. *A Human Mitochondrial Genome Database*. **http://www.mitomap.org**
150. Miyadera H, Amino H, Hiraishi A, Taka H, Murayama K, et al. 2001. Altered quinone biosynthesis in the long-lived clk-1 mutants of *Caenorhabditis elegans*. *J. Biol. Chem.* 276:7713–16
151. Miyata M, Smith JD. 1996. Apolipoprotein E allele-specific antioxidant activity and effects on cytotoxicity by oxidative insults and beta-amyloid peptides. *Nat. Genet.* 14:55–61
152. Mootha VK, Handschin C, Arlow D, Xie X, St Pierre J, et al. 2004. Erralpha and Gabpa/b specify PGC-1alpha-dependent oxidative phosphorylation gene expression that is altered in diabetic muscle. *Proc. Natl. Acad. Sci. USA* 101:6570–75
153. Mootha VK, Lindgren CM, Eriksson KF, Subramanian A, Sihag S, et al. 2003. PGC-1alpha-responsive genes involved in oxidative phosphorylation are coordinately downregulated in human diabetes. *Nat. Genet.* 34:267–73
154. Moraes CT, DiMauro S, Zeviani M, Lombes A, Shanske S, et al. 1989. Mitochondrial DNA deletions in progressive external ophthalmoplegia and Kearns-Sayre syndrome. *N. Engl. J. Med.* 320:1293–99
155. Motta MC, Divecha N, Lemieux M, Kamel C, Chen D, et al. 2004. Mammalian SIRT1 represses forkhead transcription factors. *Cell* 116:551–63
156. Muller YL, Bogardus C, Pedersen O, Baier L. 2003. A Gly482Ser missense mutation in the peroxisome proliferator-activated receptor gamma coactivator-1 is associated with altered lipid oxidation and early insulin secretion in Pima Indians. *Diabetes* 52:895–98
157. Murdock DG, Christacos NC, Wallace DC. 2000. The age-related accumulation of a mitochondrial DNA control region mutation in muscle, but not brain, detected by a sensitive PNA-directed PCR clamping based method. *Nucleic Acids Res.* 28:4350–55
158. Mutisya EM, Bowling AC, Beal MF. 1994. Cortical cytochrome oxidase activity is reduced in Alzheimer's disease. *J. Neurochem.* 63:2179–84
159. Nagy Z, Esiri MM, LeGris M, Matthews PM. 1999. Mitochondrial enzyme expression in the hippocampus in relation to Alzheimer-type pathology. *Acta Neuropathol.* 97:346–54
160. Natl. Parkinson Found. 2004. *About Parkinson Disease*. **http://www.parkinson.org/site/pp.asp?c= 9dJFJLPwB&b=71125**
161. Nemoto S, Fergusson MM, Finkel T. 2005. SIRT1 functionally interacts with the metabolic regulator and transcriptional coactivator PGC-1α. *J. Biol. Chem.* 280:16456–60
162. Niemann S, Muller U. 2000. Mutations in SDHC cause autosomal dominant paraganglioma, type 3. *Nat. Genet.* 26:268–70
163. Niemi AK, Hervonen A, Hurme M, Karhunen PJ, Jylha M, Majamaa K. 2003. Mitochondrial DNA polymorphisms associated with longevity in a Finnish population. *Hum. Genet.* 112:29–33
164. Obeso JA, Olanow CW, Nutt JG. 2000. Levodopa motor complications in Parkinson's disease. *Trends Neurosci.* 23:S2–7
165. Ogg S, Paradis S, Gottlieb S, Patterson GI, Lee L, et al. 1997. The Fork head transcription factor DAF-16 transduces insulin-like metabolic and longevity signals in *C. elegans*. *Nature* 389:994–99
166. Okado-Matsumoto A, Fridovich I. 2001. Subcellular distribution of superoxide dismutases (SOD) in rat liver: Cu,Zn-SOD in mitochondria. *J. Biol. Chem.* 276:38388–93
167. Okado-Matsumoto A, Fridovich I. 2002. Amyotrophic lateral sclerosis: a proposed mechanism. *Proc. Natl. Acad. Sci. USA* 99:9010–14
168. Orr WC, Sohal RS. 1994. Extension of life-span by overexpression of superoxide dismutase and catalase in *Drosophila melanogaster*. *Science* 263:1128–30

169. Orsini F, Migliaccio E, Moroni M, Contursi C, Raker VA, et al. 2004. The life span determinant p66Shc localizes to mitochondria where it associates with mitochondrial heat shock protein 70 and regulates *trans*-membrane potential. *J. Biol. Chem.* 279:25689–95
170. Orth M, Tabrizi SJ, Schapira AH, Cooper JM. 2003. Alpha-synuclein expression in HEK293 cells enhances the mitochondrial sensitivity to rotenone. *Neurosci. Lett.* 351:29–32
171. Palacino JJ, Sagi D, Goldberg MS, Krauss S, Motz C, et al. 2004. Mitochondrial dysfunction and oxidative damage in parkin-deficient mice. *J. Biol. Chem.* 279:18614–22
172. Panov AV, Gutekunst CA, Leavitt BR, Hayden MR, Burke JR, et al. 2002. Early mitochondrial calcium defects in Huntington's disease are a direct effect of polyglutamines. *Nat. Neurosci.* 5:731–36
173. Parkes TL, Elia AJ, Dickinson D, Hilliker AJ, Phillips JP, Boulianne GL. 1998. Extension of *Drosophila* lifespan by overexpression of human *SOD1* in motorneurons. *Nat. Genet.* 19:171–74
174. Patrick L. 2002. Nonalcoholic fatty liver disease: relationship to insulin sensitivity and oxidative stress. Treatment approaches using vitamin E, magnesium, and betaine. *Alt. Med. Rev.* 7:276–91
175. Patti ME, Butte AJ, Crunkhorn S, Cusi K, Berria R, et al. 2003. Coordinated reduction of genes of oxidative metabolism in humans with insulin resistance and diabetes: potential role of PGC1 and NRF1. *Proc. Natl. Acad. Sci. USA* 100:8466–71
176. Petersen KF, Dufour S, Befroy D, Garcia R, Shulman GI. 2004. Impaired mitochondrial activity in the insulin-resistant offspring of patients with type 2 diabetes. *N. Engl. J. Med.* 350:664–71
177. Petros JA, Baumann AK, Ruiz-Pesini E, Amin MB, Sun CQ, et al. 2005. mtDNA mutations increase tumorigenicity in prostate cancer. *Proc. Natl. Acad. Sci. USA* 102:719–24
178. Picard F, Guarente L. 2005. Molecular links between aging and adipose tissue. *Int. J. Obes. Relat. Metab. Disord.* 29(Suppl. 1):S36–39
179. Picard F, Kurtev M, Chung N, Topark-Ngarm A, Senawong T, et al. 2004. Sirt1 promotes fat mobilization in white adipocytes by repressing PPAR-gamma. *Nature* 429:771–76
180. Pletcher SD, Macdonald SJ, Marguerie R, Certa U, Stearns SC, et al. 2002. Genome-wide transcript profiles in aging and calorically restricted *Drosophila melanogaster*. *Curr. Biol.* 12:712–23
181. Polyak K, Li Y, Zhu H, Lengauer C, Willson JK, et al. 1998. Somatic mutations of the mitochondrial genome in human colorectal tumours. *Nat. Genet.* 20:291–93
182. Polymeropoulos MH, Lavedan C, Leroy E, Ide SE, Dehejia A, et al. 1997. Mutation in the alpha-synuclein gene identified in families with Parkinson's disease. *Science* 276:2045–47
183. Pratico D. 2002. Alzheimer's disease and oxygen radicals: new insights. *Biochem. Pharmacol.* 63:563–67
184. Puigserver P, Wu Z, Park CW, Graves R, Wright M, Spiegelman BM. 1998. A cold-inducible coactivator of nuclear receptors linked to adaptive thermogenesis. *Cell* 92:829–39
185. Ramassamy C, Averill D, Beffert U, Theroux L, Lussier-Cacan S, et al. 2000. Oxidative insults are associated with apolipoprotein E genotype in Alzheimer's disease brain. *Neurobiol. Dis.* 7:23–37
186. Ramassamy C, Krzywkowski P, Averill D, Lussier-Cacan S, Theroux L, et al. 2001. Impact of apoE deficiency on oxidative insults and antioxidant levels in the brain. *Brain Res. Mol. Brain Res.* 86:76–83

187. Rodgers JT, Lerin C, Haas W, Gygi SP, Spiegelman BM, Puigserver P. 2005. Nutrient control of glucose homeostasis through a complex of PGC-1alpha and SIRT1. *Nature* 434:113–18
188. Rogina B, Reenan RA, Nilsen SP, Helfand SL. 2000. Extended life-span conferred by cotransporter gene mutations in Drosophila. *Science* 290:2137–40
189. Rose G, Passarino G, Carrieri G, Altomare K, Greco V, et al. 2001. Paradoxes in longevity: sequence analysis of mtDNA haplogroup J in centenarians. *Eur. J. Hum. Genet.* 9:701–7
190. Roses AD, Einstein G, Gilbert J, Goedert M, Han SH, et al. 1996. Morphological, biochemical, and genetic support for an apolipoprotein E effect on microtubular metabolism. *Ann. NY Acad. Sci.* 777:146–57
191. Ross OA, McCormack R, Curran MD, Duguid RA, Barnett YA, et al. 2001. Mitochondrial DNA polymorphism: its role in longevity of the Irish population. *Exp. Gerontol.* 36:1161–78
192. Rotig A, de Lonlay P, Chretien D, Foury F, Koenig M, et al. 1997. Aconitase and mitochondrial iron-sulphur protein deficiency in Friedreich ataxia. *Nat. Genet.* 17:215–17
193. Ruiz-Pesini E, Diez C, Lapena AC, Perez-Martos A, Montoya J, et al. 1998. Correlation of sperm motility with mitochondrial enzymatic activities. *Clin. Chem.* 44:1616–20
194. Ruiz-Pesini E, Mishmar D, Brandon M, Procaccio V, Wallace DC. 2004. Effects of purifying and adaptive selection on regional variation in human mtDNA. *Science* 303:223–26
195. Sawa A, Wiegand GW, Cooper J, Margolis RL, Sharp AH, et al. 1999. Increased apoptosis of Huntington disease lymphoblasts associated with repeat length-dependent mitochondrial depolarization. *Nat. Med.* 5:1194–98
196. Scarpulla RC. 2002. Nuclear activators and coactivators in mammalian mitochondrial biogenesis. *Biochim. Biophys. Acta.* 1576:1–14
197. Schriner SE, Linford NJ, Martin GM, Treuting P, Ogburn CE, et al. 2005. Extension of murine life span by overexpression of catalase targeted to mitochondria. *Science* 308:1909–11
198. Senoo-Matsuda N, Yasuda K, Tsuda M, Ohkubo T, Yoshimura S, et al. 2001. A defect in the cytochrome b large subunit in complex II causes both superoxide anion overproduction and abnormal energy metabolism in *Caenorhabditis elegans*. *J. Biol. Chem.* 276: 41553–58
199. Shoffner JM, Brown MD, Torroni A, Lott MT, Cabell MR, et al. 1993. Mitochondrial DNA variants observed in Alzheimer disease and Parkinson disease patients. *Genomics* 17:171–84
200. Shoffner JM, Lott MT, Lezza AM, Seibel P, Ballinger SW, Wallace DC. 1990. Myoclonic epilepsy and ragged-red fiber disease (MERRF) is associated with a mitochondrial DNA $tRNA^{Lys}$ mutation. *Cell* 61:931–37
201. Silva JP, Kohler M, Graff C, Oldfors A, Magnuson MA, et al. 2000. Impaired insulin secretion and beta-cell loss in tissue-specific knockout mice with mitochondrial diabetes. *Nat. Genet.* 26:336–40
202. Simon DK, Pulst SM, Sutton JP, Browne SE, Beal MF, Johns DR. 1999. Familial multisystem degeneration with parkinsonism associated with the 11778 mitochondrial DNA mutation. *Neurology* 53:1787–93
203. Smith MA, Casadesus G, Joseph JA, Perry G. 2002. Amyloid-beta and tau serve antioxidant functions in the aging and Alzheimer brain. *Free Radic. Biol. Med.* 33:1194–99

204. Sohal RS, Ku HH, Agarwal S, Forster MJ, Lal H. 1994. Oxidative damage, mitochondrial oxidant generation and antioxidant defenses during aging and in response to food restriction in the mouse. *Mech. Ageing Dev.* 74:121–33
205. Solano A, Roig M, Vives-Bauza C, Hernandez-Pena J, Garcia-Arumi E, et al. 2003. Bilateral striatal necrosis associated with a novel mutation in the mitochondrial ND6 gene. *Ann. Neurol.* 54:527–30
206. Soong NW, Hinton DR, Cortopassi G, Arnheim N. 1992. Mosaicism for a specific somatic mitochondrial DNA mutation in adult human brain. *Nat. Genet.* 2:318–23
207. Sturtz LA, Diekert K, Jensen LT, Lill R, Culotta VC. 2001. A fraction of yeast Cu,Zn-superoxide dismutase and its metallochaperone, CCS, localize to the intermembrane space of mitochondria. A physiological role for SOD1 in guarding against mitochondrial oxidative damage. *J. Biol. Chem.* 276:38084–89
208. Sun J, Folk D, Bradley TJ, Tower J. 2002. Induced overexpression of mitochondrial Mn-superoxide dismutase extends the life span of adult *Drosophila melanogaster*. *Genetics* 161:661–72
209. Tan DJ, Bai RK, Wong LJ. 2002. Comprehensive scanning of somatic mitochondrial DNA mutations in breast cancer. *Cancer Res.* 62:972–76
210. Tanaka M, Gong J, Zhang J, Yamada Y, Borgeld HJ, Yagi K. 2000. Mitochondrial genotype associated with longevity and its inhibitory effect on mutagenesis. *Mech. Ageing Dev.* 116:65–76
211. Tanaka M, Gong JS, Zhang J, Yoneda M, Yagi K. 1998. Mitochondrial genotype associated with longevity. *Lancet* 351:185–86
212. Tandon A, Rogaeva E, Mullan M, St George-Hyslop PH. 2000. Molecular genetics of Alzheimer's disease: the role of beta-amyloid and the presenilins. *Curr. Opin. Neurol.* 13:377–84
213. Tatuch Y, Christodoulou J, Feigenbaum A, Clarke JTR, Wherret J, et al. 1992. Heteroplasmic mtDNA mutation (T-G) at 8993 can cause Leigh disease when the percentage of abnormal mtDNA is high. *Am. J. Hum. Genet.* 50:852–58
214. Tissenbaum HA, Guarente L. 2001. Increased dosage of a sir-2 gene extends lifespan in *Caenorhabditis elegans*. *Nature* 410:227–30
215. Tissenbaum HA, Guarente L. 2002. Model organisms as a guide to mammalian aging. *Dev. Cell* 2:9–19
216. Trifunovic A, Wredenberg A, Falkenberg M, Spelbrink JN, Rovio AT, et al. 2004. Premature ageing in mice expressing defective mitochondrial DNA polymerase. *Nature* 429:417–23
217. Trimmer PA, Keeney PM, Borland MK, Simon FA, Almeida J, et al. 2004. Mitochondrial abnormalities in cybrid cell models of sporadic Alzheimer's disease worsen with passage in culture. *Neurobiol. Dis.* 15:29–39
218. Trinei M, Giorgio M, Cicalese A, Barozzi S, Ventura A, et al. 2002. A p53-p66Shc signalling pathway controls intracellular redox status, levels of oxidation-damaged DNA and oxidative stress-induced apoptosis. *Oncogene* 21:3872–78
219. Trounce I, Neill S, Wallace DC. 1994. Cytoplasmic transfer of the mtDNA nt 8993 TG (ATP6) point mutation associated with Leigh syndrome into mtDNA-less cells demonstrates cosegregation with a decrease in state III respiration and ADP/O ratio. *Proc. Natl. Acad. Sci. USA* 91:8334–38
220. Trounce IA, Kim YL, Jun AS, Wallace DC. 1996. Assessment of mitochondrial oxidative phosphorylation in patient muscle biopsies, lymphoblasts, and transmitochondrial cell lines. *Methods Enzymol.* 264:484–509

221. Valente EM, Abou-Sleiman PM, Caputo V, Muqit MM, Harvey K, et al. 2004. Hereditary early-onset Parkinson's disease caused by mutations in PINK1. *Science* 304:1158–60
222. Valla J, Berndt JD, Gonzalez-Lima F. 2001. Energy hypometabolism in posterior cingulate cortex of Alzheimer's patients: superficial laminar cytochrome oxidase associated with disease duration. *J. Neurosci.* 21:4923–30
223. van den Ouweland JM, Lemkes HH, Trembath RC, Ross R, Velho G, et al. 1994. Maternally inherited diabetes and deafness is a distinct subtype of diabetes and associates with a single point mutation in the mitochondrial $tRNA^{Leu(UUR)}$ gene. *Diabetes* 43:746–51
224. van der Walt JM, Dementieva YA, Martin ER, Scott WK, Nicodemus KK, et al. 2004. Analysis of European mitochondrial haplogroups with Alzheimer disease risk. *Neurosci. Lett.* 365:28–32
225. van der Walt JM, Nicodemus KK, Martin ER, Scott WK, Nance MA, et al. 2003. Mitochondrial polymorphisms significantly reduce the risk of Parkinson disease. *Am. J. Hum. Genet.* 72:804–11
226. Van Remmen H, Ikeno Y, Hamilton M, Pahlavani M, Wolf N, et al. 2003. Life-long reduction in MnSOD activity results in increased DNA damage and higher incidence of cancer but does not accelerate aging. *Physiol. Genomics* 16:29–37
227. Vanharanta S, Buchta M, McWhinney SR, Virta SK, Peczkowska M, et al. 2004. Early-onset renal cell carcinoma as a novel extraparaganglial component of SDHB-associated heritable paraganglioma. *Am. J. Hum. Genet.* 74:153–59
228. Vaughan JR, Davis MB, Wood NW. 2001. Genetics of Parkinsonism: a review. *Ann. Hum. Genet* 65:111–26
229. Vaziri H, Dessain SK, Ng Eaton E, Imai SI, Frye RA, et al. 2001. hSIR2(SIRT1) functions as an NAD-dependent p53 deacetylase. *Cell* 107:149–59
230. Vega RB, Huss JM, Kelly DP. 2000. The coactivator PGC-1 cooperates with peroxisome proliferator-activated receptor alpha in transcriptional control of nuclear genes encoding mitochondrial fatty acid oxidation enzymes. *Mol. Cell. Biol.* 20:1868–76
231. Velho G, Froguel P. 1998. Genetic, metabolic and clinical characteristics of maturity onset diabetes of the young. *Eur. J. Endocrinol.* 138:233–39
232. Verwer RW, Jansen KA, Sluiter AA, Pool CW, Kamphorst W, Swaab DF. 2000. Decreased hippocampal metabolic activity in Alzheimer patients is not reflected in the immunoreactivity of cytochrome oxidase subunits. *Exp. Neurol.* 163:440–51
233. Wallace DC. 1982. Structure and evolution of organelle genomes. *Microbiol. Rev.* 46:208–40
234. Wallace DC. 1992. Diseases of the mitochondrial DNA. *Annu. Rev. Biochem.* 61:1175–212
235. Wallace DC. 1992. Mitochondrial genetics: a paradigm for aging and degenerative diseases? *Science* 256:628–32
236. Wallace DC. 1997. Mitochondrial DNA mutations and bioenergetic defects in aging and degenerative diseases. In *The Molecular and Genetic Basis of Neurological Disease*, ed. RN Rosenberg, SB Prusiner, S DiMauro, RL Barchi, pp. 237–69. Boston: Butterworth-Heinemann
237. Wallace DC. 1999. Mitochondrial diseases in man and mouse. *Science* 283:1482–88
238. Wallace DC. 2001. Mouse models for mitochondrial disease. *Am. J. Med. Genet.* 106:71–93
239. Wallace DC, Brown MD, Lott MT. 1999. Mitochondrial DNA variation in human evolution and disease. *Gene* 238:211–30

240. Wallace DC, Bunn CL, Eisenstadt JM. 1975. Cytoplasmic transfer of chloramphenicol resistance in human tissue culture cells. *J. Cell Biol.* 67:174–88
241. Wallace DC, Lott MT. 2002. Mitochondrial genes in degenerative diseases, cancer and aging. In *Emery and Rimoin's Principles and Practice of Medical Genetics*, ed. DL Rimoin, JM Connor, RE Pyeritz, BR Korf, pp. 299–409. London: Churchill Livingstone
242. Wallace DC, Lott MT, Brown MD, Kerstann K. 2001. Mitochondria and neuro-ophthalmological diseases. In *The Metabolic and Molecular Basis of Inherited Disease*, ed. CR Scriver, AL Beaudet, WS Sly, D Valle, pp. 2425–512. New York: McGraw-Hill
243. Wallace DC, Singh G, Lott MT, Hodge JA, Schurr TG, et al. 1988. Mitochondrial DNA mutation associated with Leber's hereditary optic neuropathy. *Science* 242:1427–30
244. Wallace DC, Zheng X, Lott MT, Shoffner JM, Hodge JA, et al. 1988. Familial mitochondrial encephalomyopathy (MERRF): genetic, pathophysiological, and biochemical characterization of a mitochondrial DNA disease. *Cell* 55:601–10
245. Wang Y, Michikawa Y, Mallidis C, Bai Y, Woodhouse L, et al. 2001. Muscle-specific mutations accumulate with aging in critical human mtDNA control sites for replication. *Proc. Natl. Acad. Sci. USA* 98:4022–27
246. Warburg O. 1931. *The Metabolism of Tumors*. New York: R.R. Smith
247. Warita H, Hayashi T, Murakami T, Manabe Y, Abe K. 2001. Oxidative damage to mitochondrial DNA in spinal motoneurons of transgenic ALS mice. *Brain Res. Mol. Brain Res.* 89:147–52
248. Wilson FH, Hariri A, Farhi A, Zhao H, Petersen KF, et al. 2004. A cluster of metabolic defects caused by mutation in a mitochondrial tRNA. *Science* 306:1190–94
249. Wilson RB, Roof DM. 1997. Respiratory deficiency due to loss of mitochondrial DNA in yeast lacking the frataxin homologue. *Nat. Genet.* 16:352–57
250. Wollheim CB. 2000. Beta-cell mitochondria in the regulation of insulin secretion: a new culprit in type II diabetes. *Diabetologia* 43:265–77
251. Wu Z, Puigserver P, Andersson U, Zhang C, Adelmant G, et al. 1999. Mechanisms controlling mitochondrial biogenesis and respiration through the thermogenic coactivator PGC-1. *Cell* 98:115–24
252. Xu Y, Krishnan A, Wan XS, Majima H, Yeh CC, et al. 1999. Mutations in the promoter reveal a cause for the reduced expression of the human manganese superoxide dismutase gene in cancer cells. *Oncogene* 18:93–102
253. Yeh JJ, Lunetta KL, van Orsouw NJ, Moore FD Jr, Mutter GL, et al. 2000. Somatic mitochondrial DNA (mtDNA) mutations in papillary thyroid carcinomas and differential mtDNA sequence variants in cases with thyroid tumours. *Oncogene* 19:2060–66
254. Yoon JC, Puigserver P, Chen G, Donovan J, Wu Z, et al. 2001. Control of hepatic gluconeogenesis through the transcriptional coactivator PGC-1. *Nature* 413:131–38
255. Yoon JC, Xu G, Deeney JT, Yang SN, Rhee J, et al. 2003. Suppression of beta cell energy metabolism and insulin release by PGC-1alpha. *Dev. Cell* 5:73–83
256. Zhang J, Asin-Cayuela J, Fish J, Michikawa Y, Bonafe M, et al. 2003. Strikingly higher frequency in centenarians and twins of mtDNA mutation causing remodeling of replication origin in leukocytes. *Proc. Natl. Acad. Sci. USA* 100:1116–21
257. Zimmet P, Alberti KG, Shaw J. 2001. Global and societal implications of the diabetes epidemic. *Nature* 414:782–87

Switches in Bacteriophage Lambda Development*

Amos B. Oppenheim,[1,3,4] Oren Kobiler,[1] Joel Stavans,[2] Donald L. Court,[3] and Sankar Adhya[4]

[1]Department of Molecular Genetics and Biotechnology, The Hebrew University-Hadassah Medical School, Jerusalem 91120

[2]Department of Physics of Complex Systems, Weizmann Institute of Science, Rehovot, Israel 76100

[3]Gene Regulation and Chromosome Biology Laboratory, National Cancer Institute at Frederick, Frederick, Maryland 21702

[4]Laboratory of Molecular Biology, National Cancer Institute, Bethesda, Maryland 20892-4264; email: oppenhea@mail.nih.gov

Annu. Rev. Genet.
2005. 39:409–29

First published online as a Review in Advance on July 26, 2005

The *Annual Review of Genetics* is online at http://genet.annualreviews.org

doi: 10.1146/annurev.genet.39.073003.113656

0066-4197/05/1215-0409$20.00

Key Words

genetic circuits, gene modules, developmental pathways, developmental decisions, threshold effects, host factors

Abstract

The lysis-lysogeny decision of bacteriophage lambda (λ) is a paradigm for developmental genetic networks. There are three key features, which characterize the network. First, after infection of the host bacterium, a decision between lytic or lysogenic development is made that is dependent upon environmental signals and the number of infecting phages per cell. Second, the lysogenic prophage state is very stable. Third, the prophage enters lytic development in response to DNA-damaging agents. The CI and Cro regulators define the lysogenic and lytic states, respectively, as a bistable genetic switch. Whereas CI maintains a stable lysogenic state, recent studies indicate that Cro sets the lytic course not by directly blocking CI expression but indirectly by lowering levels of CII which activates *cI transcription*. We discuss how a relatively simple phage like λ employs a complex genetic network in decision-making processes, providing a challenge for theoretical modeling.

Contents

For the concert of life no-one gets a program.

[From a Dutch tile (81)]

INTRODUCTION

A central challenge in the post genomic era is to understand processes governing the dynamics of highly complex genetic regulatory networks. A growing number of theoretical modeling investigators are attempting to explain the underlying principles of complex regulatory networks involved in normal mammalian development, including alterations that can result in a disease state such as cancer. Analysis of such complex systems would be greatly facilitated by similar studies using an ideal paradigm in which most if not all of the elements composing the system were known. Phage λ, the most comprehensively studied bacteriophage, is the prototype of a class of lambdoid phages with whom it shares similar genome organization and functions (38, 82). Studies of λ that began in the 1950s continue to reveal key molecular processes in gene regulatory mechanisms and development. However, despite years of study, many genetic interactions still remain to be uncovered and those that we already know require reexamination. For an accurate, complete, and quantitative analysis of the genetic network, in particular its temporal progression, these remaining questions need to be addressed. In this review we summarize a systems biology approach to the study of genetic regulatory circuits of phage λ. We define the individual components of the circuits and switches, describe the kinetics of their interactions, and explain how the interactions achieve robustness in the performance of the circuits. We also stress some puzzles that still exist in lambda's regulatory system.

Our citation of literature is not exhaustive but provides examples to illustrate specific points. We direct the reader to several comprehensive reviews on phage biology (13, 15, 24, 31, 34, 37, 41, 82, 107).

The λ System

Bacteria and their temperate phages, like *Escherichia coli* and λ, exist in symbiotic relationships. These phages can be present in a dormant, lysogenic (prophage) state replicating passively with the host or they can develop lytically, producing progeny phages and

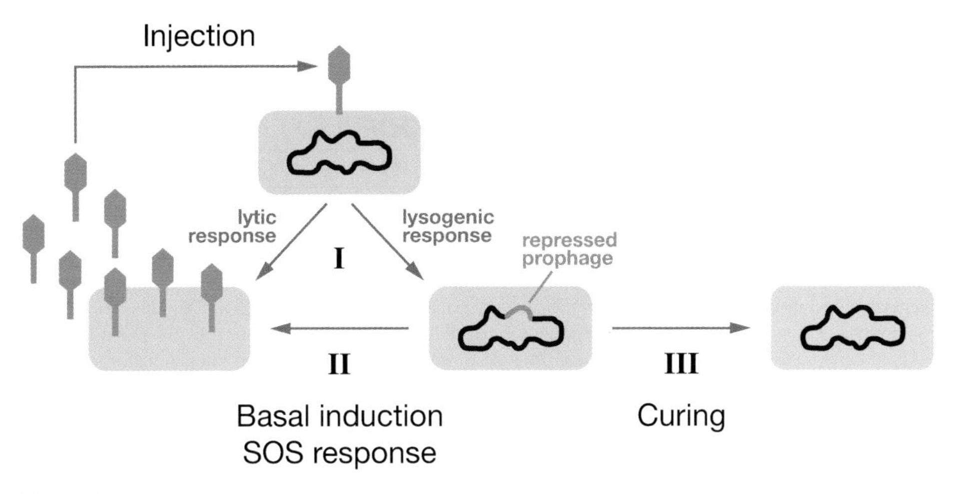

Figure 1

Decision-making steps by the temperate λ phage. A cell infected with phage λ can follow (denoted as Decision I) the lytic response (*left*) or the lysogenic response (*right*). The resulting lysogenic cell carries a repressed prophage shown in orange. The prophage is irreversibly induced only when a threshold amount of DNA damage causes an SOS response, leading to lytic development (Decision II). The prophage is normally extremely stable and rarely undergoes this type of induction by random DNA damage to produce progeny phage in a very small fraction of the lysogenic cell (basal or spontaneous induction). Some of the spontaneously induced cells enter the lytic cycle abortively, lose the prophage (curing), and become nonlysogens (Decision III) (65).

killing their hosts. λ phage infecting an *E. coli* cell makes a decision to follow either a lytic or a lysogenic pathway (**Figure 1**). If the lytic pathway is followed, the phage replicates its DNA autonomously, expresses the morphogenetic genes, assembles virions, and lyses the host. If the lysogenic course ensues, a stable lysogen is established in which the prophage is integrated into the host chromosome with lytic gene expression turned off. The prophage DNA replicates as part of the bacterial genome during subsequent cell divisions, and confers immunity to the cell against infection by another λ. Treatment with DNA-damaging agents, which leads to an SOS response, causes the lysogenic state to irreversibly switch into lytic development, mimicking the lytic infection. Otherwise, the prophage state is extremely stable, rarely undergoing induction by DNA damage. Some of these rarely induced cells enter an abortive lytic cycle by losing the prophage and becoming nonlysogens (curing). Similarly, after infection abortive lytic or lysogenic events may also occur (62). The importance of these abortive events has not been thoroughly studied.

GENE ORGANIZATION AND REGULATION

The λ genetic map and transcription profile involved in early developmental processes are shown in **Figure 2**. The gene organization of lambdoid phages is based on a number of recurring principles. Phage λ and its many relatives have genomes that evolved as highly mosaic, modular structures. This property has long been recognized and led to the formulation of the "modular genome hypothesis" (10, 14, 38, 101). A short summary of the λ phage modules and submodules is given in **Table 1**. Thus, for example, it is possible to replace the "immunity" module of λ by that of another lambdoid phage. The organization of the gene modules allows a typical cascade of

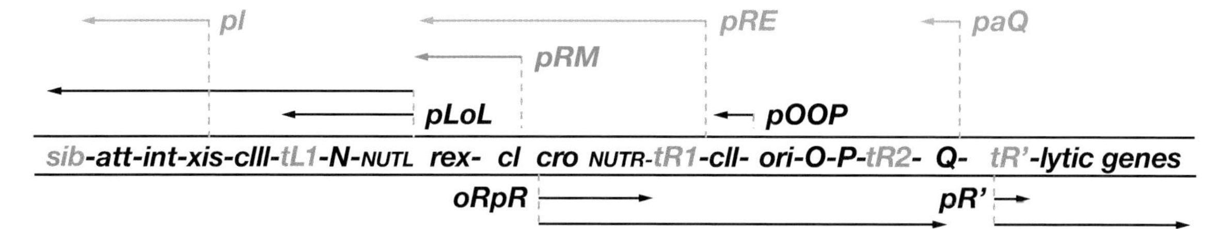

Figure 2

Genetic map and transcriptional units of the phage regulatory region. Key genes and signals discussed in the text are shown in their map order between the parallel lines. The early transcripts, the extended delayed early transcripts, and the late transcripts are shown in black arrows. The transcripts initiated from the *pI*, *pRE*, and *paQ* that are required for lysogeny are shown in blue. The *pRM* promoter, which is activated by the CI regulator, is required for maintenance of the lysogenic state, which is shown in green. Critical transcription terminators are marked in orange, including the *sib* region containing the *tI* terminator. Leftward promoters are indicated above and the rightward ones below the map. *pL* and *pR* are the early promoters and *pR′* is the late lytic promoter. The role of the *pOOP* promoter is not fully understood. The operators *oL* and *oR*, cognate to *pL* and *pR* respectively, are also shown. The immunity module of the λ chromosome encompasses *pLoL*, *rex*, *cI*, *oRpR*, and *cro*. *ori* is the origin of O- and P-mediated phage DNA replication (**Table 1**). Int carries the site-specific recombination reaction, and Int and Xis support the excision reaction.

phage gene expression in lytic growth delineating the early, delayed early, and late stages of transcription. This modular and temporal expression facilitates the alternative λ developmental pathways. Because of the transcriptional cascade, the repression of the early phage promoters *pR* and *pL* prevents expression of all lytic genes (**Figure 2**, **Figure 3**). In a prophage, this repression is carried out by the phage CI protein, which acts by binding to the *oL* and *oR* operators that overlap the *pL* and *pR* promoters, allowing the maintenance of the lysogenic state to be governed by CI alone; when CI is inactivated, e.g., by the SOS response, the lytic development follows. By blocking the *pL* and *pR* promoters of an incoming phage genome, the CI regulator confers immunity against further productive infection by another λ (superinfection immunity).

Table 1 A list of phage λ modules

Module	Genes and sites	Function
CI regulator	*pRM-cI*	CI activates *pRM*
Immunity	*pLoL-rex-cI-pRM-oRpR-cro*	CI represses *pL* and *pR*
CII regulator	*cII-pRE*	CII activates *pRE*
Site-specific recombination	*xis-int-att-tI*	Int and Xis catalyze recombination
General recombination	*gam-exo-bet*	Control and catalyze recombination
DNA replication	*ori-O-P*	Control of initiation of DNA replication control
Early antitermination	NUTL-*N-tL1*	N-antitermination and translational control
Late antitermination	*Q-pR′-tR′*	Q-antitermination for late gene expression
DNA packaging	*cos-Nu1-A*	Cleavage of head-full genomes
Head genes	*Nu1* to *FII*	Morphogenesis of head particles
Tail genes	*Z* to *J*	Morphogenesis of tails

Description of the functions of genes not discussed in this review can be found in (38).

The Lytic Transcription Cascade

The gene expression cascade that leads to the lytic mode of growth is the default mode of λ development. It is carried out in three stages. (*i*) Transcription is initiated with the synthesis of the early transcripts from the *pL* and *pR* promoters (**Figure 2**). Early transcripts, which encode two regulators, N and Cro, are attenuated at the *tL1* and *tR1* terminators, respectively. These transcriptional terminators play an important role in controlling the cascade of gene expression. By acting as a weak repressor for both *pL* and *pR* promoters, Cro facilitates the lytic mode (described below). The N protein is an antitermination factor that promotes the assembly of a transcription complex (9, 35). This assembly occurs on the RNA at the *nutL* and *nutR* sites and is made up of RNA polymerase and a number of host proteins called Nus. The N- and Nus-modified RNA polymerase can overcome the *tL1* and *tR1* transcription terminators, resulting in expression of the distal delayed early functions (30, 84). (*ii*) The delayed early functions include the lysogenic regulators CII and CIII, as well as the lytic DNA replication functions O and P, and the late gene regulator Q. (*iii*) After sufficient accumulation, the Q protein modifies RNA polymerase that has just initiated transcription from the *pR′* late promoter (66). This modification causes the RNA polymerase to become resistant to transcription terminators present downstream to *pR′*, allowing the expression of the late genes, which encode proteins for phage morphogenesis and host cell lysis. There is a kinetic separation between the expression of delayed early and late genes. This is caused by the location of the *Q* gene at the very end of the delayed early cascade and the high threshold level of Q protein needed for its activity (52, 63, 109). During the late stage of the cascade, the late gene products assemble phage virions and lyse the host. A similar temporal lytic cascade of gene expression follows prophage induction (38).

The Lysogenic Process

The lysogenic pathway is also initiated during the transcription cascade of the delayed early genes. Under conditions leading to lysogeny, the expression of the lytic regulators fails because Q is actively switched off by the accumulation of the critical lysogenic regulator CII, thereby blocking the default lytic pathway and switching to the lysogenic course. As discussed in greater detail below, an accumulation of the CII protein above a threshold level is critical for initiating the lysogenic switch. During this active switching from the default lytic to the lysogenic mode, CII stimulates the synthesis of Int, which catalyzes the insertion of the phage DNA into the host chromosome, and of the CI regulator, which binds to *oL* and *oR* to repress the early promoters. CII also inhibits Q function (see 52). These three activities are mediated by transcription activation of three promoters, *pI*, *pRE*, and *paQ*, respectively (**Figure 2**; see **Figure 6** below). Activation of all three promoters is critical for the establishment of a stable prophage state. Once the prophage genome integrates into the bacterial chromosome and CI protein represses *pL* and *pR*, the lysogenic state is established. The prophage state is extremely stable and is maintained through many generations of division (8). How CI repressor synthesis is maintained in the absence of CII is discussed later.

THE PROPHAGE STATE AND ITS MAINTENANCE

In the prophage state, the CI regulator controls the expression of three promoters. It represses transcription from the *pL* and *pR* promoters and positively and negatively regulates its own synthesis from the *pRM* promoter (**Figure 3**; see **Figure 6** below). In the prophage state, *pRM* is responsible for CI synthesis; CI expression from *pRE* is prevented owing to repression of CII in a lysogen.

Figure 3 illustrates a set of cooperative interactions of CI binding to DNA, which lead

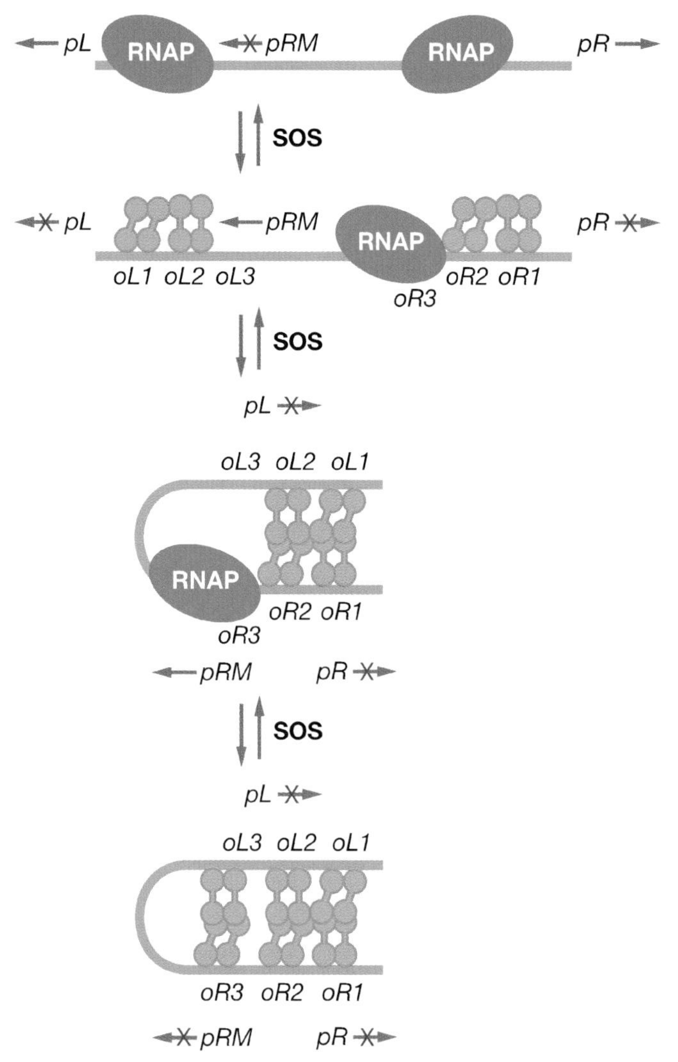

Figure 3

A scheme showing the repression of *pL* and *pR* by CI induced DNA looping [adapted from (24, 82)]. RNA polymerase is symbolized as a green ellipse; CI regulator monomers (depicted as orange dumbbells) have two domains: the N-terminal DNA binding domain and the C-terminal oligomerization domain. The top line shows a linear arrangement of the DNA. In the absence of CI, transcription is initiated at both *pL* and *pR* (*top line*). The second line shows how repressor molecules bind cooperatively to *oL1*/*oL2* and to *oR1*/*oR2* prior to DNA looping repressing *pL* and *pR* and activating the *pRM* promoter. The third line shows the first stage of DNA looping. RNA polymerase bound to *pRM* either before (*second line*) or after (*third line*) looping occurs transcribes *pRM*. Finally, further stabilization of DNA looping occurs by binding of CI to *oL3* and *oR3* together with repression of *pRM* when CI levels become very high (*fourth line*). The additional CI molecules bound to *oL3* and *oR3* interact to form a tetramer. The SOS response reverses the repression process (see text).

to extremely efficient repression (25, 64, 82). The CI repressor is made of two domains tethered by a short linker, an N-terminal DNA binding domain with a helix-turn-helix motif, and a C-terminal oligomerization domain. The intrinsic relative affinities of CI to its operators at *oR* and *oL* are as follows: *oR1* > *oR2* > *oR3* and *oL1* > *oL2* > *oL3*. Dimers of CI cooperatively bind to *oL1* and *oL2* on the left and *oR1* and *oR2* on the right by forming tetramers repressing *pL* and *pR*, respectively. Furthermore, another cooperative interaction between these two sets of tetramers bound to *oL* and *oR*, 2.3 Kb apart, leads to the formation of a DNA loop held by a CI octamer, i.e., two interacting tetramers, that enhances repression. In this DNA-multiprotein complex, the CI dimer bound at *oR2* also stimulates *pRM* transcription, thus activating CI synthesis in a repressed prophage by a positive autoregulatory loop (44, 75). As the CI concentration increases because of *pRM* activation, two additional CI dimers are recruited to bind at *oL3* and *oR3* to further stabilize the *oL*-*oR* loop. In this context, the binding of CI to *oR3* represses *pRM* and prevents CI overexpression. Based on direct measurements of CI (86), recent calculation shows that a lysogenic cell contains about 250 CI monomers when the cells are growing in rich media (23). This translates into about 30 dimers per prophage copy, assuming an average of four chromosomes per cell. This number of CI molecules not only achieves a strong repression of the phage promoters but also sets the threshold level for SOS induction. Note that cell division does not randomly reduce the CI concentration to a point low enough to cause prophage induction (8).

Repression of transcription from the *pR* promoter inhibits not only expression of genes in that operon but also phage DNA replication by preventing "transcriptional activation" of λ *ori*, the site where phage DNA replication is initiated (26, 33). This inhibition occurs even if O and P functions are present. (71, 106). This regulation appears to be critical for establishing a lysogen. If the λ origin replicates after the phage is integrated because O and P are still available, the replication event is lethal to the cell, resulting in abortive lytic infection (11). Thus, it is important to immediately block phage replication as the choice for lysogeny is made.

THE DEFAULT LYTIC COURSE: Cro AND N FUNCTIONS

The gene coding for Cro, which is the essential lytic regulator, is the first one to be transcribed from the *pR* promoter following phage infection or prophage induction. It is a weak repressor of the *pR* promoter, allowing continuous expression from *pR* during lytic infection (29, 94, 95). Cro is a single-domain protein that binds as a dimer to the *oL* and *oR* operators (60, 82). The intrinsic affinity of Cro for the individual operators is $oR3 > oR1 \geq oR2$, and, unlike CI, higher-ordered structures, i.e., tetramers or octamers of Cro have not been detected (19). However, Ackers and coworkers found a small amount of cooperative binding of Cro dimers to adjacent sites in the *oR* complex (19). Whether such cooperative binding is due to dimer-dimer interaction or to changes in DNA needs to be investigated. It was shown, however, that changes in DNA accompany the cooperative binding of CI (21).

The higher affinity of Cro binding to *oR3* supports the hypothesis that during lytic development Cro binds first to *oR3*, repressing transcription of the *cI* gene from the *pRM* promoter (45). It was therefore proposed that Cro binding to *oR3* represses *pRM* and completes the switch to the lytic pathway as well as sustaining it after an initial SOS-mediated inactivation of CI in a lysogen (28, 46, 74). However, recent experiments suggest that the presence of Cro might be unimportant for the lysogenic to lytic switch during induction of the prophage (94). Furthermore, Little and coworkers showed that replacing *oR3* by the weaker *oR1* Cro-binding site has a marginal effect on prophage induction (64). Thus, it appears that the critical role of Cro in lytic development may be only to turn down the *pL* and *pR* promoters and that its role in turning down *pRM* is not critical but supplementary (52, 94). An argument that Cro binding to *oR3* sets the lytic course has also been made to explain the role of Cro following phage infection (82). This interpretation of the role of Cro is also unlikely because a phage carrying the same *oR3* to *oR1* variant (see above) that reduces Cro but allows CI binding still shows lytic growth (64). However, under certain conditions Cro binding to *oR3* may contribute to lytic growth after prophage induction. The function of Cro in lytic development is addressed below.

The N Antiterminator

The N protein, like Cro, is a critical lytic regulator but acts by a completely different mechanism. It antiterminates transcription at termination signals, allowing expression of distal genes in the *pL* and *pR* operons. N protein, once made, binds to the specific RNA sites NUTL and NUTR. The NUTL site is located between the *N* gene and the *pL* promoter, and NUTR is downstream of *cro*, the first gene in the *pR* operon (**Figure 1**). The Nus factors, a complex set of host proteins, take part in cellular transcriptional and translational processes and interact with the N system (18, 93). Purified in vitro studies have defined NusA, NusB, NusE, and NusG as the host components of N-antitermination (22, 68). Each NUT site can be divided into two parts, BOXA and BOXB (32). BOXB is a stem-loop structure in the RNA and is specific for N binding (16). As it binds BOXB, N associates with NusA, NusG, and RNAP. BOXA RNA is specific for NusB and NusE binding (77, 80). It is proposed that N, Nus factors, and Nut interact and complex with RNA polymerase while tethered on the same RNA (76, 102, 104, 105). Although N is the essential factor for antitermination, NUT and the Nus factors confer stability and full activity to the antitermination complex (22, 85).

N is a critical regulatory protein of phage λ, which is reflected in the number of

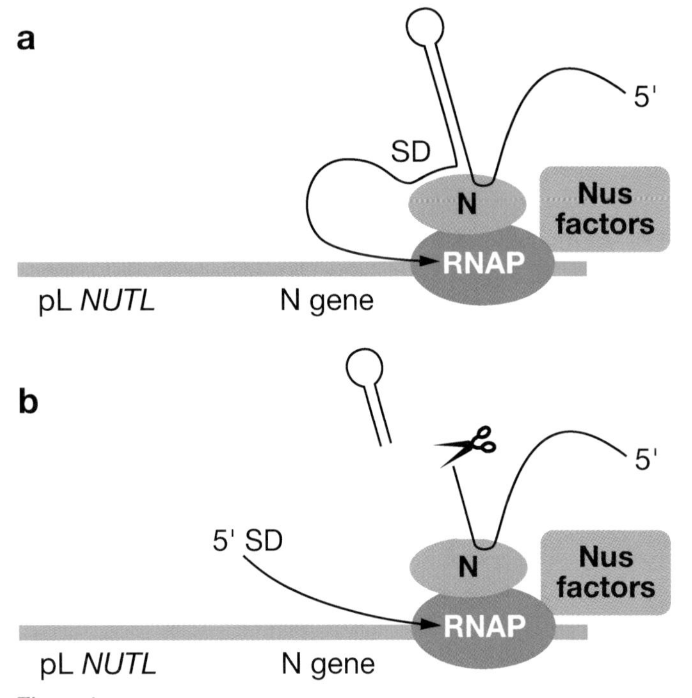

Figure 4

Models for N-mediated translation repression. The complex of RNAP is modified by *NUT*, N, and Nus factors to form an antitermination complex. Translation repression of N takes place on the uncleaved RNA in the absence of RNase III action (*a*). In the presence of RNase III cleavage occurs and the Shine–Dalgarno (SD) sequence for N is no longer held in close contact to the antitermination complex, translational repression is prevented, and efficient expression of N can take place (*b*).

ways its expression is controlled. At the transcription level, *N* is regulated from *oLpL* by CI- and Cro-mediated repression. At the translation level, N represses and autoregulates its own translation, and the endoribonuclease RNase III modulates this repression (**Figure 4**). By acting at their respective sites located upstream of the N gene, N and RNase III control N translation (103, 104). High concentrations of RNase III as found in cells grown in rich media prevent N-mediated translation repression altogether, whereas low concentrations of RNase III found in cells grown in poor media allow translation repression. At the posttranslational level, the Lon protease degrades N causing a relatively short half-life.

N expression from the *pL*-antiterminated transcripts is dependent upon the level of RNase III in cells. If RNaseIII concentration is high, processing of the RNase III site (RIII site) between NUTL and the N message is rapid, causing high rates of N translation. If RNase III concentration is low or zero, processing of the RIII site occurs slowly if at all, allowing repression of N translation. Cellular RNase III concentrations rise and fall with growth rate (12), and correspondingly, N translation rises and falls with growth rate (104).

Evidence suggests that the N translation repression complex is dependent on *E. coli* RNAP and the N-transcription antitermination complex (105). Under conditions of limiting RNase III, the antiterminating complex prevents translation of N from its antiterminated message. This means that N is expressed from RNA transcripts that terminate at *tL1*, i.e., do not form an antitermination complex. Thus, under minimal growth conditions where RNase III activity is limiting, N concentrations would be stringently controlled by negative autoregulation. On the other hand, in rich media, RNase III processing would ensure high N concentrations. Viral lytic development predominates in rich media, and lysogenic development is enhanced under limiting growth or starvation. High N concentration enhances lytic growth of the phage, as evidenced by λ forming clear plaques on cells expressing N from a plasmid (104). Clear plaques indicate that only few cells follow the lysogenic pathway. Low levels of N are made following λ infection of cells grown in poor carbon conditions; these same conditions have been shown to enhance lysogen formation following infection at low multiplicity of infection (56). These latter conditions are likely to reduce Q concentrations and provide a better opportunity for lysogeny. Regulation of N also appears to be a way in which λ senses environmental conditions through RNase III and responds by increasing or decreasing N concentrations and altering the lytic/lysogenic response. We note, however, that at high multiplicity of infection, there is no effect of growth media on lysogenization efficiency (27).

SWITCHING THE DEFAULT LYTIC MODE TO LYSOGENY: ROLE OF CII

The inhibition, or absence, of lytic functions is not sufficient for the switching to the lysogenic mode. Rather, as noted above, infection resulting in a lysogenic response proceeds through a number of required events: integration of the DNA, efficient repression of the early promoters, as well as timely inhibition of the lytic genes expression. These requirements are met by CII turning on *pI*, *pRE*, and *paQ* promoters to express Int and CI, and to inhibit Q function, respectively. All three promoters contain a direct repeat TTGC-N6-TTGC sequence that binds to CII. The N6 region corresponds to the –35 elements of these promoters. Expression from all three CII-activated promoters is coordinated by the CII protein during infection. However, the mechanism of activation of the promoters by CII is not well understood. Specific mutations in the α or σ subunits of RNAP prevent CII-mediated activation from these promoters in vitro (48, 67). Consistently, these RNA polymerase mutants prevent the establishment of lysogeny in vivo (78).

The quaternary structure of the CII protein has been solved recently (20a, 81a). The structure shows that a CII tetramer is made of two nearly equivalent dimers. Each of the four monomers contains a helix-turn-helix DNA-binding motif but only two of the monomers appear, by modeling, to be involved in actual DNA binding at the direct repeat sequence. The function of the other two helix-turn-helix motifs in the tetramer is not known. The direct repeat sequence of *pRE* is located within the N-terminal coding sequence of CII. Incidentally, in an elegant genetic study, Friedman and coworkers isolated and analyzed a CII mutant defective in transcription activation of the *pE* promoter in lambdoid phage P22 (88). This mutation affects the N-terminal sequence and is not in the helix-turn-helix domain of CII. This mutation also modified the *pRE* promoter so that CII can activate the mutant *pRE* but not the wild-type *pRE*. This suggests that this N-terminal sequence of the protein may play a role in determining the specificity of CII binding to DNA (88).

The *pI*, *pRE*, and *paQ* promoters are located within the Xis, CII, and Q protein coding sequences, respectively. The Xis protein is needed for excision of the prophage DNA from the chromosome after induction. For rapid synthesis of Int, which facilitates integration after infection, CII activates the *pI* promoter. The integration reaction also requires the integration host factor, IHF. IHF is critical for generating the multicomponent Intasome structure, which catalyzes the integration reaction (38). Since *pI* is located within the *xis* gene, the *pI* transcript synthesizes Int but not Xis, helping to ensure that integration will not be accompanied by the presence of Xis function. By the same criterion, CII activates the *pRE* promoter to direct rapid synthesis of the CI regulator following infection. The amount of CI made from *pRE* during lysogenic response was found to be as much as 10- to 20-fold higher than the amount made from *pRM* in an established lysogen (86). The initial high concentrations of CI may guarantee that all infecting and replicating phage genomes become repressed. But the CII-dependent *paQ* promoter, which lies within the Q gene, was found to reduce Q function (52), providing a mechanism by which CII reduces late gene expression to enhance lysogeny (17, 70). Mutations affecting the ability of the *pRE* or *paQ* promoter to respond to CII prevent the lysogenic response resulting in clear plaque formation (42, 52).

Regulation of CII Activity

CII, which plays a key role in the lysis-lysogeny decision, is regulated at numerous levels (**Figure 5**) (39, 43, 50, 83, 91): (*i*) The transcription of the *cII* gene is inhibited both

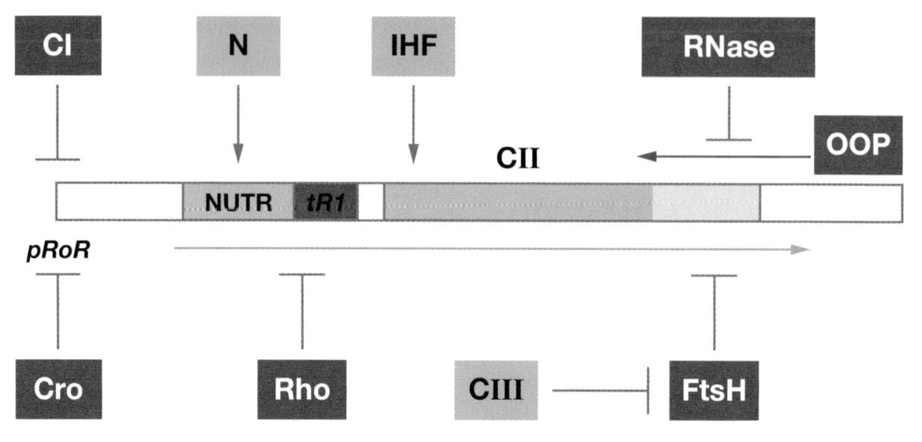

Figure 5

Multilevel regulation of CII activity. CI and Cro negatively regulate the CII gene at transcription initiation. *tR1* aided by Rho factor reduce transcription elongation into the CII gene. The N antitermination factor allows the extension of transcription beyond *tR1*. IHF stimulates CII translation initiation. The antisense OOP RNA together with RNase III reduces CII mRNA stability, and FtsH protease acting at the C terminus of CII is responsible for the rapid proteolysis of CII. The center bar represents the DNA; the CII gene and the positive controls are shown in blue and negative controls in red. The direction of CII and OOP transcriptions are shown in blue and red, respectively.

by Cro and CI binding to *oR*, and is stimulated by the N antitermination factor acting at *NUTR*. High rates of CII synthesis take place only for a limited period before being repressed, by Cro or CI. (*ii*) Translation initiation of CII is stimulated by IHF (43). An IHF binding site is located immediately upstream of CII, which has been proposed to stimulate CII translation in the presence of IHF by an unknown mechanism (72a, 80a). (*iii*) The stability of CII mRNA is affected by the OOP RNA, a short antisense transcript complementary to the 3′ end of the *cII* mRNA (57, 58). RNase III recognizes and cleaves the CII-OOP double-stranded RNA, thereby initiating rapid CII mRNA degradation (57). The DNA coding for the protease target is also the target of CII mRNA degradation mediated by the antisense OOP RNA. It was reported that the stability of the OOP RNA is reduced by polyadenylation but whether this process regulates CII concentration has not been clarified (108). (*iv*) The ATP-dependent protease FtsH is responsible for the rapid degradation of CII (39, 50, 91). Host mutations in *ftsH* lead to stabilization of CII and thereby an increased lysogenization frequency. A C-terminal flexible tail of CII, which is not required for CII activity, acts as a target for initiating rapid CII proteolysis by FtsH (20, 51). (*v*) The long leader CI RNA initiated from *pRE* is antisense to *cro*, which could prevent the translation of Cro (92). Indeed, an *ftsH* host mutant that results in higher concentrations of CII was found to be defective in Cro (79, 83, 96). Unfortunately, the concentration of Cro as a function of *pRE* activity has not been directly measured. (*vi*) CIII, which is a 54-residue long peptide and required for lysogeny, controls the rate of CII degradation by acting as an inhibitor of the FtsH protease (40, 51, 53).

Regulation of CIII

The *cIII* gene expression is also subject to multiple controls by λ CI, Cro, and N and by the host RNase III (**Figure 5**). CI and Cro inhibit CIII synthesis by binding to *oL*, and the N-antiterminator stimulates CIII expression by acting at *NUTL*. Unlike its negative effect on CII, RNase III stimulates CIII translation.

It was shown that the mRNA that codes for the amino terminal residues of CIII is present in two conformations (3, 54). In one conformation, the translation initiation region is occluded, preventing *cIII* translation. In the other, the mRNA is open allowing efficient translation. Point mutations that favor one or the other structures have been described. It appears that RNase III regulates *cIII* translation by acting as an RNA chaperone to affect CIII mRNA structure without processing (2).

THE DECISION PROCESS: GENETIC FUNCTIONS AND EXTERNAL INFLUENCES

We now summarize the critical events that take place in the decision making process following infection by λ (Decision I in **Figure 1**, and shown schematically in **Figure 6**). Following infection, λ begins a cascade of transcription destined for the lytic mode. Under appropriate conditions the lytic cascade can be switched off, allowing entry into the lysogenic pathway. The switch requires high threshold concentrations of CII, which facilitates CI and Int synthesis and reduces the regulator function Q. CI, once made, maintains the repressed state of the prophage. In establishing repression, CII and CI act temporally to inhibit lytic functions. CII acts first by direct reduction of Q function, and as CII function ceases, CI takes over by repressing the *pR* promoter. CI executes repression after the CII-mediated switch and appears not to participate in initiating the switch. The activity of CII is programmed to allow an initial overshoot of CI expression from the *pRE* promoter and a preemptive inhibition of Q activity (52, 86). The overshoot may be required to ensure the repression of all lytic genes of the infecting and replicating phage genomes in the cell during the establishment of repression. Repression of *pR* by CI at this stage also ensures immediate cessation of phage DNA replication by preventing transcription through the *ori* site (97).

Following infection, transcription from early *pL* and *pR* promoters would start the lytic pathway by default with N antitermination of transcription and subsequent expression of Q. Q in turn antiterminates transcription, leading to late lytic gene expression, cell lysis, and phage release. To set the course for lysogeny, CII reduces Q function in two ways. First, CII activates antisense *paQ* RNA inhibiting Q. Second, CII activates *pRE* for synthesis of CI, which represses *pR* and thus Q transcription. CII continues to repress Q expression via *paQ* until *cII* transcription is repressed by CI at *pR*. This kinetic coordination of Q shutoff by CII and CI ensures the switch from the default lytic pathway to lysogeny. Q protein must build up to a high threshold concentration to become functional, thus providing a window for CII to exert its effect on Q through *paQ*. In this way, CII prevents Q from reaching its threshold concentration. If CI action is prevented, CII inhibition of Q via *paQ* is not sustained because of repression of CII by Cro acting at *pR* and by rapid CII degradation. Concordantly, λ phage mutants defective in CI, *oR*, or CII follow exclusively lytic growth, whereas mutants defective in Cro are unable to follow the lytic pathway.

Multiplicity Effect

The lysis-lysogeny decision is influenced by the number of phage particles infecting the cell as well as by the cell physiology (55, 56). High multiplicity of infection favors lysogeny. Direct measurements of CII activity showed, in wild-type infection, functional CII activity in cells infected with two or more phages (52). This threshold level of CII is reached when two or more phage infect a cell. In single infection this critical level of CII is not attained, as mentioned above, allowing lytic development.

Physiological Effects

In cells grown in rich media the lytic course predominates, whereas growing cells in poor

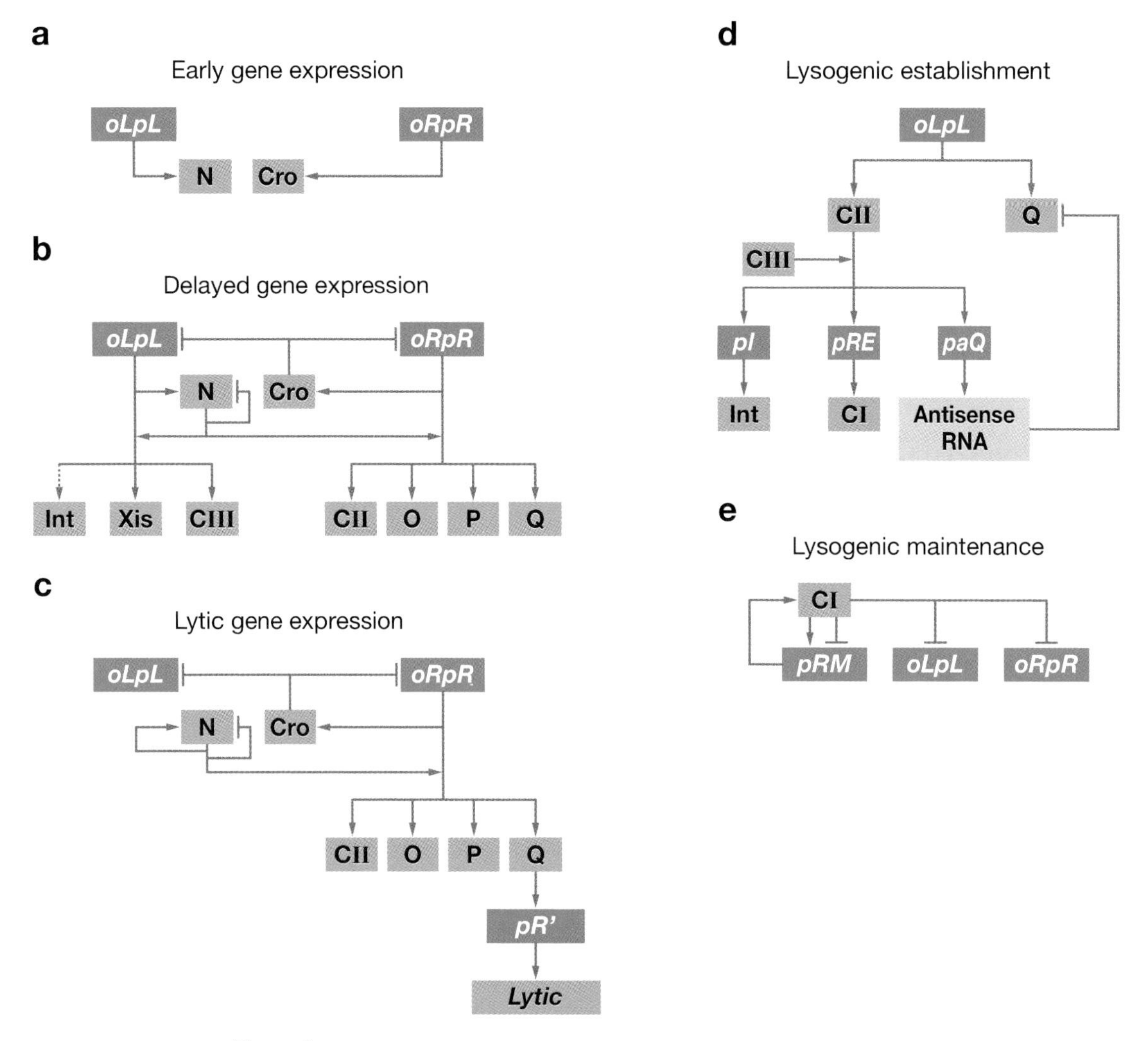

Figure 6

Description of the λ lytic and lysogenic genetic network. Arrows mark the positive effects between elements that make up the genetic network, whereas bars denote inhibitory effects. Promoters and operators are shown in green, the phage genes in light orange, and *cis* antisense RNA is shown in light blue (Cis acting has been used to note that it acts on the RNA from the opposite strand and not on an RNA originating from different sequences in the genome). Arrows emanating from the promoters denote transcription of specific functions. The OOP antisense RNA and the yet unknown threshold effect of Q activity are not shown. For simplicity, the developmental network is divided into early gene expression (*a*), delayed gene expression (*b*), lytic gene expression (*c*), lysogenic establishment (*d*), and lysogenic maintenance (*e*). The dotted line in (*b*) leading to Int expression signifies reduced level of Int expression due to retroregulation of *int* mRNA.

media increases the chances of the lysogenic pathway after single infection (56). Cells growing in rich media have higher concentration of the host global regulator RNase III, which leads to high rates of expression of the protein N favoring lytic growth [see (104) and references therein]. In carbon-starved cells, on the other hand, RNase III and consequently N concentrations are low. Under these conditions N translation is repressed.

This reduction of N concentration would reduce Q expression to a level that provides more opportunity for lysogenic response. Nevertheless, when cells are infected by two or more phages, CII function is epistatic to the effects of growth conditions preferring lysogeny.

Unlike the decision that λ makes after the infection process described above whereby infected cells follow either lytic or lysogenic development, prophage induction leads exclusively to the lytic course (Decision II in **Figure 1**). It was proposed that Cro is responsible for repressing CI expression following induction to keep the developmental switch in the default lytic mode (82), although the action of Cro in keeping the lytic course on track is simply to reduce the concentration of the unstable master lysogenic regulator CII.

COEXPRESSION OF BOTH LYTIC AND LYSOGENIC GENES FROM THE SAME OPERON: A PARADOX

The *pL* operon encodes a large number of ORFs of which only N, CIII, Xis, and Int are essential for either the lytic or lysogenic response (see **Figure 2**). N is a key regulator for lytic growth of the phage. On the other hand, the lysogenic function CIII acts as an inhibitor of a host protease (FtsH, also called HflB) that destabilizes the critical lysogenic regulator CII (40, 47, 51). Proteins Int and Xis carry out site-specific recombination (59, 73). During lysogenization, Int catalyzes the integration of the phage genome into the bacterial chromosome site, *att*, whereas Int together with Xis catalyze the reverse reaction to excise the λ chromosome during prophage induction. The *pR* promoter also transcribes both lytic and lysogenic genes, *cro*, *cII*, *O*, *P*, and Q. Whereas Cro is another critical regulator for lytic growth of the phage, the very next gene in the operon encodes the CII protein, which coordinates the lysogenic pathway. O and P are needed for phage DNA replication prior to morphogenesis, whereas the second lytic regulator Q activates the synthesis of proteins for phage morphogenesis and cell lysis. Thus functions necessary for lysogenic and lytic development are both expressed from the *pL* and *pR* promoters. The location of both lysogenic and lytic genes in the same operons creates an apparent paradox in our mind about the lysis-lysogeny decision. This paradox is resolved by regulating the synthesis and degradation of critical RNA and proteins. This regulation provides the catalytic or stoichiometric amounts of functions as required to pursue either the lytic or lysogenic course. As examples, during lytic response, the CII regulator is rapidly eliminated by proteolysis, whereas in the lysogenic response, CII accumulates and limits Q activity. Presumably, the high threshold requirement for Q function allows the infected complex a time window for controlling the concentration of CII.

In the lysogenic response, high concentrations of Int are made from pI under the control of CII, while the expression of Int from the *pL* is greatly reduced by retro-regulation because of Int mRNA degradation by RNase III from a site called *sib*, located on the other side of *att* (see **Figure 2**) (36). Prophage induction leads to the expression of both Int and Xis from the *pL* promoter because the *sib* regulator is detached from the *int* gene in the prophage DNA.

THEORETICAL STUDIES

The lambda genetic network has been a fertile ground for theoretical modeling of decision-making processes during the regulation of development, and for testing new modeling methodologies of genetic networks in general. Models have been constructed addressing the lysis-lysogeny decision in terms of the lambda genetic switch describing the competition between CI and Cro, using statistical mechanics with the explicit goal of obtaining bistability. The earlier models focused on the (*i*) probabilities of the occupancy of the *pR* and *pRM* promoters by different binding configurations of CI and Cro, and of *pR* and *pRM* by

RNAP, and (*ii*) Gibbs free energies of binding characterizing the different configurations as free parameters (1, 87, 90). These models assumed that the behavior of the reactants studied under in vitro conditions reflects in vivo situations, and did not incorporate the possible existence of other levels of regulation, e.g., regulation of translation of CI and Cro, degradation of CI and Cro, and anticooperative binding of the two proteins to adjacent operators (87).

More recent models, based on the same approach, also incorporated stochastic effects in the concentrations of the regulatory proteins and, as a result, in the selection between lysis and lysogeny (5, 7, 49, 69, 89, 98, 100, 110). Although apparent agreement between the theoretical calculations and the experimentally observed values was noted, these efforts did not lead to a theoretical description with improved predictive values. However, the theoretical analysis of the exceptional stability the prophage state of λ led to the conclusion, now confirmed by experimental evidence, that a view of the switch focusing only on oR to explain the stability is incomplete, and that oL participates as well (7, 23).

An advantage of computer simulations is their ability to take into account multiple elements and variables within the decision module of the λ network (5, 49, 69). Furthermore, they can predict, for example, the values of reactant concentrations during the temporal execution of the lytic and lysogenic pathways, which can be readily compared with biological experiments. Nevertheless, at present there are only limited experimental data on the kinetic changes in the concentration and activity of the regulatory elements in the network and the strength of their interactions, which are crucial to achieve real agreement between theoretical results and experimental observations. Furthermore, the in vivo values of the parameters may differ quantitatively by orders of magnitude from in vitro ones owing to such factors as macromolecular crowding, variation in local concentrations, and yet unknown functions, as alluded to above. By exposing inconsistencies with the observed behavior, future theoretical studies with predictive values are expected to play a more important role.

OPEN QUESTIONS, SUMMARY, AND CONCLUSION

Counting Infecting Phage

The decision made by a phage-infected cell is dependent on the multiplicity of infection. When one phage infects, most infection shows lytic development. However, when two phage infect, the lysogenic pathway prevails. What is the molecular basis for such a dramatic response to a small change, which to some may be counterintuitive? Furthermore, the network response also suggests tight communication between two coinfecting phage genomes. This suggests that replicating phage genomes after single phage infection have little effect on the decision process. This issue requires further investigation. However, there are conditions when a single infecting phage enters the lysogenic pathway. When infecting an *hflA* or *hflB* mutant host, efficient lysogenization takes place. As discussed, these mutant hosts increase the level of CII function.

A possible parsimonious model to explain the multiplicity response runs as follows: First, multiple infection results in the titration of a critical regulatory host factor that is present at a very low concentration. One such candidate is FtsH, the product of the *hflB* gene of which there are less than 100 molecules per cell (99). Second, the phage CIII protein acting as an inhibitor of FtsH plays a critical role in the decision. Indeed, mutants that result in a elevated CIII translation (by about threefold) no longer respond to the multiplicity of infection and can efficiently lysogenize upon single infection (4; A. Rattray, unpublished). Thus, we expect that a small increase in multiplicity from one to two would raise CIII to concentrations that inhibit FtsH increasing CII expression to allow lysogenic development.

The Distinction Between Phage Infection and Prophage Induction

The idea that two infecting phage are critical for switching on the lysogenic course raises interesting questions. A lysogen contains a number of host chromosomes at various stages of replication, and accordingly more than one prophage. Would induction result in high CII activity as is found in multiple infections? If a high concentration of CII accumulates, we would expect inhibition of Q function by CII, even after CI inactivation by induction. It remains to be seen whether such high concentrations of CII if they occur would affect Q function and thus reduce lytic phage yield.

Role of Cro in Lytic Growth

An open question is the role of the high-affinity binding of Cro to the *oR3* operator site in lytic growth after phage infection or prophage induction. It is clear that Cro is essential to allow lytic development. Is this high-affinity binding critical to regulate *pRM* following prophage induction or does it have a critical role in lytic decision? This issue needs to be addressed by the use of phage mutants defective in oR3 that uniquely affect Cro binding and do not affect CI binding or pRM activity. Would such mutants affect lytic growth after phage infection or prophage induction? The intercalation of CI, Cro, and RNAP binding site at this locus may make the isolation of such phage mutants problematic.

Evolution of Temperate Phages

Phage λ lifestyles have been a textbook case for complex genetic control circuits. Gene order, modular construct, recruitment of specific host factors, and the presence of a complex genetic network allow the infecting phage to make decisions and to proceed through alternative developmental pathways. The lysis-lysogeny bistable decision is clearly much more complex than originally portrayed. This complexity was most probably reached through evolutionary forces tinkering with specific elements present in the individual functional modules that evolved to act in concert. Of interest is a recent approach of tinkering with λ modules that revealed the robustness of its genetic network (6, 72). Although the λ lysogenic promoter *pRM* can tolerate a number of mutational changes, we note that such tolerance may be limited to specific environmental conditions.

EPILOG

In summary, a small set of regulatory proteins in an organism as simple as a bacteriophage function through a diverse set of macromolecular interactions in a temporal fashion. Future investigations into the detailed molecular aspects of the functions of specific modules coupled with kinetic and quantitative analysis of the phage genetic network will yield a realistic picture of this important paradigm for more complex developmental processes (61).

ACKNOWLEDGMENTS

We apologize to our colleagues whose work we have omitted. We thank John Little, Ian Dodd, Ted Cox, Pradeep Parrack, and Grzes Wegrzyn for communication of manuscripts before publication and David Friedman for critical reading of the manuscript. The research carried out by O. K. and A. B. O. was supported in part by The Israel Science Foundation (grants # 489/01–1 and 340/04).

LITERATURE CITED

1. Ackers GK, Johnson AD, Shea MA. 1982. Quantitative model for gene regulation by lambda phage repressor. *Proc. Natl. Acad. Sci. USA* 79:1129–33

2. Altuvia S, Kornitzer D, Kobi S, Oppenheim AB. 1991. Functional and structural elements of the mRNA of the cIII gene of bacteriophage lambda. *J. Mol. Biol.* 218:723–33
3. Altuvia S, Kornitzer D, Teff D, Oppenheim AB. 1989. Alternative mRNA structures of the cIII gene of bacteriophage lambda determine the rate of its translation initiation. *J. Mol. Biol.* 210:265–80
4. Altuvia S, Oppenheim AB. 1986. Translational regulatory signals within the coding region of the bacteriophage lambda cIII gene. *J. Bacteriol.* 167:415–19
5. Arkin A, Ross J, McAdams HH. 1998. Stochastic kinetic analysis of developmental pathway bifurcation in phage lambda-infected *Escherichia coli* cells. *Genetics* 149:1633–48
6. Atsumi S, Little JW. 2004. Regulatory circuit design and evolution using phage lambda. *Genes Dev.* 18:2086–94
7. Aurell E, Brown S, Johanson J, Sneppen K. 2002. Stability puzzles in phage lambda. *Phys. Rev. E* 65:051914-1-9
8. Baek K, Svenningsen S, Eisen H, Sneppen K, Brown S. 2003. Single-cell analysis of lambda immunity regulation. *J. Mol. Biol.* 334:363–72
9. Barik S, Ghosh B, Whalen W, Lazinski D, Das A. 1987. An antitermination protein engages the elongating transcription apparatus at a promoter-proximal recognition site. *Cell* 50:885–99
10. Botstein D. 1980. A theory of modular evolution for bacteriophages. *Ann. NY Acad. Sci.* 354:484–90
11. Brachet P, Eisen H, Rambach A. 1970. Mutations of coliphage lambda affecting the expression of replicative functions O and P. *Mol. Gen. Genet.* 108:266–76
12. Britton RA, Powell BS, Dasgupta S, Sun Q, Margolin W, et al. 1998. Cell cycle arrest in Era GTPase mutants: a potential growth rate-regulated checkpoint in *Escherichia coli*. *Mol. Microbiol.* 27:739–50
13. Calender RL, ed. 2005. *The Bacteriophages*. New York: Oxford Univ. Press. 2nd ed.
14. Campbell A. 1994. Comparative molecular biology of lambdoid phages. *Annu. Rev. Microbiol.* 48:193–222
15. Campbell A. 2003. The future of bacteriophage biology. *Nat. Rev. Genet.* 4:471–77
16. Chattopadhyay S, Garcia-Mena J, DeVito J, Wolska K, Das A. 1995. Bipartite function of a small RNA hairpin in transcription antitermination in bacteriophage lambda. *Proc. Natl. Acad. Sci. USA* 92:4061–65
17. Court D, Green L, Echols H. 1975. Positive and negative regulation by the cII and cIII gene products of bacteriophage lambda. *Virology* 63:484–91
18. Court DL, Patterson TA, Baker T, Costantino N, Mao X, Friedman DI. 1995. Structural and functional analyses of the transcription-translation proteins NusB and NusE. *J. Bacteriol.* 177:2589–91
19. Darling PJ, Holt JM, Ackers GK. 2000. Coupled energetics of lambda cro repressor self-assembly and site-specific DNA operator binding I: analysis of cro dimerization from nanomolar to micromolar concentrations. *Biochemistry* 39:11500–7
20. Datta AB, Roy S, Parrack P. 2005. Role of C-terminal residues in oligomerization and stability of lambda CII: implications for lysis-lysogeny decision of the phage. *J. Mol. Biol.* 345:315–24

20a. Datta AB, Panjikar S, Weiss MS, Chakrabarti P, Parrack P. 2005. Structure of λCII: Implications for recognition of direct repeat DNA by an unusual tetrameric organization. *Proc. Natl. Acad. Sci. USA.* In press

21. Deb S, Bandyopadhyay S, Roy S. 2000. DNA sequence dependent and independent conformational changes in multipartite operator recognition by lambda-repressor. *Biochemistry* 39:3377–83

22. DeVito J, Das A. 1994. Control of transcription processivity in phage lambda: Nus factors strengthen the termination-resistant state of RNA polymerase induced by N antiterminator. *Proc. Natl. Acad. Sci USA* 91:8660–64
23. Dodd IB, Perkins AJ, Tsemitsidis D, Egan JB. 2001. Octamerization of lambda CI repressor is needed for effective repression of P(RM) and efficient switching from lysogeny. *Genes Dev.* 15:3013–22
24. Dodd IB, Shearwin KE, Egan JB. 2005. Revisited gene regulation in bacteriophage λ. *Curr. Opin. Genet. Dev.* 15
25. Dodd IB, Shearwin KE, Perkins AJ, Burr T, Hochschild A, Egan JB. 2004. Cooperativity in long-range gene regulation by the lambda CI repressor. *Genes Dev.* 18:344–54
26. Dove WF, Inokuchi H, Stevens WF. 1971. In *The Bacteriophage Lambda*, ed. AD Hershey, pp. 747–71. Cold Spring Harbor, NY: Cold Spring Harbor Lab. Press
27. Echols H, Green L, Kudrna R, Edlin G. 1975. Regulation of phage lambda development with the growth rate of host cells: a homeostatic mechanism. *Virology* 66:344–46
28. Eisen H, Brachet P, Pereira da Silva L, Jacob F. 1970. Regulation of repressor expression in lambda. *Proc. Natl. Acad. Sci. USA* 66:855–62
29. Folkmanis A, Maltzman W, Mellon P, Skalka A, Echols H. 1977. The essential role of the cro gene in lytic development by bacteriophage lambda. *Virology* 81:352–62
30. Friedman DI, Court DL. 1995. Transcription antitermination: the lambda paradigm updated. *Mol. Microbiol.* 18:191–200
31. Friedman DI, Court DL. 2001. Bacteriophage lambda: alive and well and still doing its thing. *Curr. Opin. Microbiol.* 4:201–7
32. Friedman DI, Gottesman M. 1983. Lytic mode of lambda development. See Ref. 38, pp. 21–51
33. Furth ME, Dove WF, Meyer BJ. 1982. Specificity determinants for bacteriophage lambda DNA replication. III. Activation of replication in lambda ric mutants by transcription outside of ori. *J. Mol. Biol.* 154:65–83
34. Gottesman ME, Weisberg RA. 2004. Little lambda, who made thee? *Microbiol. Mol. Biol. Rev.* 68:796–813
35. Greenblatt J, Mah TF, Legault P, Mogridge J, Li J, Kay LE. 1998. Structure and mechanism in transcriptional antitermination by the bacteriophage lambda N protein. *Cold Spring Harb. Symp. Quant. Biol.* 63:327–36
36. Guarneros G. 1988. Retroregulation of bacteriophage lambda int gene expression. *Curr. Top. Microbiol. Immunol.* 136:1–19
37. Hendrix RW, Lawrence JG, Hatfull GF, Casjens S. 2000. The origins and ongoing evolution of viruses. *Trends Microbiol.* 8:504–8
38. Hendrix RW, Roberts JW, Stahl FW, Weisberg RA. 1983. *Lambda II.* Cold Spring Harbor, NY: Cold Spring Harbor Lab. Press
39. Herman C, Ogura T, Tomoyasu T, Hiraga S, Akiyama Y, et al. 1993. Cell growth and lambda phage development controlled by the same essential *Escherichia coli* gene, ftsH/hflB. *Proc. Natl. Acad. Sci. USA* 90:10861–65
40. Herman C, Thevenet D, D'Ari R, Bouloc P. 1997. The HflB protease of *Escherichia coli* degrades its inhibitor lambda cIII. *J. Bacteriol.* 179:358–63
41. Herskowitz I, Hagen D. 1980. The lysis-lysogeny decision of phage lambda: explicit programming and responsiveness. *Annu. Rev. Genet.* 14:399–445
42. Hoopes BC, McClure WR. 1985. A cII-dependent promoter is located within the Q gene of bacteriophage lambda. *Proc. Natl. Acad. Sci. USA* 82:3134–38

43. Hoyt MA, Knight DM, Das A, Miller HI, Echols H. 1982. Control of phage lambda development by stability and synthesis of cII protein: role of the viral cIII and host hflA, himA and himD genes. *Cell* 31:565–73
44. Jain D, Nickels BE, Sun L, Hochschild A, Darst SA. 2004. Structure of a ternary transcription activation complex. *Mol. Cell* 13:45–53
45. Johnson A, Meyer BJ, Ptashne M. 1978. Mechanism of action of the cro protein of bacteriophage lambda. *Proc. Natl. Acad. Sci. USA* 75:1783–87
46. Johnson AD, Poteete AR, Lauer G, Sauer RT, Ackers GK, Ptashne M. 1981. λ Repressor and *cro*—components of an efficient molecular switch. *Nature* 294:217–23
47. Kaiser AD. 1957. Mutations in a temperate bacteriophage affecting its ability to lysogenize *Escherichia coli*. *Virology* 3:42–61
48. Kedzierska B, Lee DJ, Wegrzyn G, Busby SJ, Thomas MS. 2004. Role of the RNA polymerase alpha subunits in CII-dependent activation of the bacteriophage lambda pE promoter: identification of important residues and positioning of the alpha C-terminal domains. *Nucleic Acids Res.* 32:834–41
49. Kiehl TR, Mattheyses RM, Simmons MK. 2004. Hybrid simulation of cellular behavior. *Bioinformatics* 20:316–22
50. Kihara A, Akiyama Y, Ito K. 1997. Host regulation of lysogenic decision in bacteriophage lambda: transmembrane modulation of FtsH (HflB), the cII degrading protease, by HflKC (HflA). *Proc. Natl. Acad. Sci. USA* 94:5544–49
51. Kobiler O, Koby S, Teff D, Court D, Oppenheim AB. 2002. The phage lambda CII transcriptional activator carries a C-terminal domain signaling for rapid proteolysis. *Proc. Natl. Acad. Sci. USA* 99:14964–69
52. Kobiler O, Rokney A, Friedman N, Court DL, Stavans J, Oppenheim AB. 2005. Quantitative kinetic analysis of the bacteriophage lambda genetic network. *Proc. Natl. Acad. Sci. USA* 102:4470–75
53. Kornitzer D, Altuvia S, Oppenheim AB. 1991. Genetic analysis of the cIII gene of bacteriophage HK022. *J. Bacteriol.* 173:810–15
54. Kornitzer D, Teff D, Altuvia S, Oppenheim AB. 1989. Genetic analysis of bacteriophage lambda cIII gene: mRNA structural requirements for translation initiation. *J. Bacteriol.* 171:2563–72
55. Kourilsky P. 1973. Lysogenization by bacteriophage lambda. I. Multiple infection and the lysogenic response. *Mol. Gen. Genet.* 122:183–95
56. Kourilsky P, Knapp A. 1974. Lysogenization by bacteriophage lambda. III. Multiplicity dependent phenomena occurring upon infection by lambda. *Biochimie* 56:1517–23
57. Krinke L, Mahoney M, Wulff DL. 1991. The role of the OOP antisense RNA in coliphage lambda development. *Mol. Microbiol.* 5:1265–72
58. Krinke L, Wulff DL. 1990. RNase III-dependent hydrolysis of lambda cII-O gene mRNA mediated by lambda OOP antisense RNA. *Genes Dev.* 4:2223–33
59. Landy A. 1993. Mechanistic and structural complexity in the site-specific recombination pathways of Int and FLP. *Curr. Opin. Genet. Dev.* 3:699–707
60. LeFevre KR, Cordes MH. 2003. Retroevolution of lambda Cro toward a stable monomer. *Proc. Natl. Acad. Sci. USA* 100:2345–50
61. Levine M, Davidson EH. 2005. Gene regulatory networks for development. *Proc. Natl. Acad. Sci. USA* 102:4936–42
62. Lieb M. 1953. The establishment of lysogenicity in *Escherichia coli*. *J. Bacteriol.* 65:642–51
63. Little JW. 2005. Threshold effects in gene regulation: When some is not enough. *Proc. Natl. Acad. Sci. USA* 102:5310–11

64. Little JW, Shepley DP, Wert DW. 1999. Robustness of a gene regulatory circuit. *EMBO J.* 18:4299–307
65. Livny J, Friedman DI. 2004. Characterizing spontaneous induction of Stx encoding phages using a selectable reporter system. *Mol. Microbiol.* 51:1691–704
66. Marr MT, Datwyler SA, Meares CF, Roberts JW. 2001. Restructuring of an RNA polymerase holoenzyme elongation complex by lambdoid phage Q proteins. *Proc. Natl. Acad. Sci. USA* 98:8972–78
67. Marr MT, Roberts JW, Brown SE, Klee M, Gussin GN. 2004. Interactions among CII protein, RNA polymerase and the lambda PRE promoter: contacts between RNA polymerase and the -35 region of PRE are identical in the presence and absence of CII protein. *Nucleic Acids Res.* 32:1083–90
68. Mason SW, Greenblatt J. 1991. Assembly of transcription elongation complexes containing the N protein of phage lambda and the *Escherichia coli* elongation factors NusA, NusB, NusG, and S10. *Genes Dev.* 5:1504–12
69. McAdams HH, Shapiro L. 1995. Circuit simulation of genetic networks. *Science* 269:650–66
70. McMacken R, Mantei N, Butler B, Joyner A, Echols H. 1970. Effect of mutations in the c2 and c3 genes of bacteriophage lambda on macromolecular synthesis in infected cells. *J. Mol. Biol.* 49:639–55
71. Mensa-Wilmot K, Carroll K, McMacken R. 1989. Transcriptional activation of bacteriophage lambda DNA replication in vitro: regulatory role of histone-like protein HU of *Escherichia coli*. *EMBO J.* 8:2393–402
72. Michalowski CB, Short MD, Little JW. 2004. Sequence tolerance of the phage lambda PRM promoter: implications for evolution of gene regulatory circuitry. *J. Bacteriol.* 186:7988–99
72a. Mahajna J, Oppenheim AB, Rattray A, Gottesman M. 1986. Translation initiation of bacteriophage lambda gene cII requires integration host factor. *J. Bacteriol.* 165:167–74
73. Nash HA. 1981. Integration and excision of bacteriophage lambda: the mechanism of conservation site specific recombination. *Annu. Rev. Genet.* 15:143–67
74. Neubauer Z, Calef E. 1970. Immunity phase-shift in defective lysogens: non-mutational hereditary change of early regulation of lambda prophage. *J. Mol. Biol.* 51:1–13
75. Nickels BE, Dove SL, Murakami KS, Darst SA, Hochschild A. 2002. Protein-protein and protein-DNA interactions of sigma70 region 4 involved in transcription activation by lambda cI. *J. Mol. Biol.* 324:17–34
76. Nodwell JR, Greenblatt J. 1991. The nut site of bacteriophage lambda is made of RNA and is bound by transcription antitermination factors on the surface of RNA polymerase. *Genes Dev.* 5:2141–51
77. Nodwell JR, Greenblatt J. 1993. Recognition of boxA antiterminator RNA by the *E. coli* antitermination factors NusB and ribosomal protein S10. *Cell* 72:261–68
78. Obuchowski M, Giladi H, Koby S, Szalewska-Palasz A, Wegrzyn A, et al. 1997. Impaired lysogenisation of the *Escherichia coli* rpoA341 mutant by bacteriophage lambda is due to the inability of CII to act as a transcriptional activator. *Mol. Gen. Genet.* 254:304–11
79. Oppenheim A, Honigman A, Oppenheim AB. 1974. Interference with phage lambda cro gene function by a colicin-tolerant *Escherichia coli* mutant. *Virology* 61:1–10
80. Patterson TA, Zhang Z, Baker T, Johnson LL, Friedman DI, Court DL. 1994. Bacteriophage lambda N-dependent transcription antitermination. Competition for an RNA site may regulate antitermination. *J. Mol. Biol.* 236:217–28
80a. Peacock S, Weissbach H. 1985. IHF stimulation of lambda cII gene expression is inhibited by the E coli NusA protein. *Biochem. Biophys. Res. Commun.* 127:1026–31

81. Perutz M. 2003. *I Wish I'd Made You Angry Earlier*. Cold Spring Harbor, NY: Cold Spring Harbor Lab. Press

81a. Jain D, Youngchang K, Maxwell KL, Beasley S, Rongguang Z, et al. 2005. Crystal structure of bacteriophage λ cII and its DNA complex. *Mol. Cell* 19:1–11

82. Ptashne M. 2004. *Genetic Switch: Phage Lambda Revisited*. Cold Spring Harbor, NY: Cold Spring Harbor Lab. Press

83. Rattray A, Altuvia S, Mahajna G, Oppenheim AB, Gottesman M. 1984. Control of bacteriophage lambda CII activity by bacteriophage and host functions. *J. Bacteriol.* 159:238–42

84. Reed MR, Shearwin KE, Pell LM, Egan JB. 1997. The dual role of Apl in prophage induction of coliphage 186. *Mol. Microbiol.* 23:669–81

85. Rees WA, Weitzel SE, Yager TD, Das A, von Hippel PH. 1996. Bacteriophage lambda N protein alone can induce transcription antitermination in vitro. *Proc. Natl. Acad. Sci. USA* 93:342–46

86. Reichardt L, Kaiser AD. 1971. Control of lambda repressor synthesis. *Proc. Natl. Acad. Sci. USA* 68:2185–89

87. Reinitz J, Vaisnys JR. 1990. Theoretical and experimental analysis of the phage lambda genetic switch implies missing levels of co-operativity. *J. Theor. Biol.* 145:295–318

88. Retallack DM, Johnson LL, Ziegler SF, Strauch MA, Friedman DI. 1993. A single-base-pair mutation changes the specificities of both a transcription activation protein and its binding site. *Proc. Natl. Acad. Sci. USA* 90:9562–65

89. Santillan M, Mackey MC. 2004. Why the lysogenic state of phage lambda is so stable: a mathematical modeling approach. *Biophys. J.* 86:75–84

90. Shea MA, Ackers GK. 1985. The OR control system of bacteriophage lambda. A physical-chemical model for gene regulation. *J. Mol. Biol.* 181:211–30

91. Shotland Y, Koby S, Teff D, Mansur N, Oren DA, et al. 1997. Proteolysis of the phage lambda CII regulatory protein by FtsH (HflB) of *Escherichia coli*. *Mol. Microbiol.* 24:1303–10

92. Spiegelman WG, Reichardt LF, Yaniv M, Heinemann SF, Kaiser AD, Eisen H. 1972. Bidirectional transcription and the regulation of phage lambda repressor synthesis. *Proc. Natl. Acad. Sci. USA* 69:3156–60

93. Squires CL, Zaporojets D. 2000. Proteins shared by the transcription and translation machines. *Annu. Rev. Microbiol.* 54:775–98

94. Svenningsen SL, Costantino N, Court DL, Adhya S. 2005. On the role of Cro in λ prophage induction. *Proc. Natl. Acad. Sci. USA*. In press

95. Takeda Y, Folkmanis A, Echols H. 1977. Cro regulatory protein specified by bacteriophage lambda. Structure, DNA-binding, and repression of RNA synthesis. *J. Biol. Chem.* 252:6177–83

96. Teff D, Koby S, Shotland Y, Ogura T, Oppenheim AB. 2000. A colicin-tolerant *Escherichia coli* mutant that confers hfl phenotype carries two mutations in the region coding for the C-terminal domain of FtsH (HflB). *FEMS Microbiol. Lett.* 183:115–17

97. Thomas R. 1971. Control circuits. In *The Bacteriophage Lambda*, ed. AD Hershey, pp. 211–20. Cold Spring Harbor, NY: Cold Spring Harbor Lab. Press

98. Tian T, Burrage K. 2004. Bistability and switching in the lysis/lysogeny genetic regulatory network of bacteriophage lambda. *J. Theor. Biol.* 227:229–37

99. Tomoyasu T, Yamanaka K, Murata K, Suzaki T, Bouloc P, et al. 1993. Topology and subcellular localization of FtsH protein in *Escherichia coli*. *J. Bacteriol.* 175:1352–57

100. Vohradsky J. 2001. Neural model of the genetic network. *J. Biol. Chem.* 276:36168–73
101. Weisberg RA, Gottesmann ME, Hendrix RW, Little JW. 1999. Family values in the age of genomics: comparative analyses of temperate bacteriophage HK022. *Annu. Rev. Genet.* 33:565–602
102. Whalen WA, Das A. 1990. Action of an RNA site at a distance: role of the nut genetic signal in transcription antitermination by phage-lambda N gene product. *New Biol.* 2:975–91
103. Wilson HR, Kameyama L, Zhou JG, Guarneros G, Court DL. 1997. Translational repression by a transcriptional elongation factor. *Genes Dev.* 11:2204–13
104. Wilson HR, Yu D, Peters HK 3rd, Zhou JG, Court DL. 2002. The global regulator RNase III modulates translation repression by the transcription elongation factor N. *EMBO J.* 21:4154–61
105. Wilson HR, Zhou JG, Yu D, Court DL. 2004. Translation repression by an RNA polymerase elongation complex. *Mol. Microbiol.* 53:821–28
106. Wold MS, Mallory JB, Roberts JD, LeBowitz JH, McMacken R. 1982. Initiation of bacteriophage lambda DNA replication in vitro with purified lambda replication proteins. *Proc. Natl. Acad. Sci. USA* 79:6176–80
107. Wommack KE, Colwell RR. 2000. Virioplankton: Viruses in aquatic ecosystems. *Microbiol. Mol. Biol. Rev.* 64:69–114
108. Wrobel B, Herman-Antosiewicz A, Szalewska-Palasz S, Wegrzyn G. 1998. Polyadenylation of oop RNA in the regulation of bacteriophage lambda development. *Gene* 212:57–65
109. Yang XJ, Hart CM, Grayhack EJ, Roberts JW. 1987. Transcription antitermination by phage-lambda gene-Q protein requires a DNA segment spanning the RNA start site. *Genes Dev.* 1:217–26
110. Zhu XM, Yin L, Hood L, Ao P. 2004. Calculating biological behaviors of epigenetic states in the phage lambda life cycle. *Funct. Integr. Genomics* 4:188–95

Nonhomologous End Joining in Yeast

James M. Daley,[1] Phillip L. Palmbos,[1] Dongliang Wu,[2] and Thomas E. Wilson[1,2]

[1]Cellular and Molecular Biology Program and [2]Department of Pathology, University of Michigan Medical School, Ann Arbor, Michigan 48109-0602; email: daleyj@umich.edu, ppalmbos@umich.edu, donwu@umich.edu, wilsonte@umich.edu

Annu. Rev. Genet. 2005. 39:431–51

First published online as a Review in Advance on July 26, 2005

The *Annual Review of Genetics* is online at http://genet.annualreviews.org

doi: 10.1146/annurev.genet.39.073003.113340

0066-4197/05/1215-0431$20.00

Key Words

nonhomologous end joining, double-strand break, illegitimate recombination, *Saccharomyces cerevisiae*

Abstract

Nonhomologous end joining (NHEJ), the direct rejoining of DNA double-strand breaks, is closely associated with illegitimate recombination and chromosomal rearrangement. This has led to the concept that NHEJ is error prone. Studies with the yeast *Saccharomyces cerevisiae* have revealed that this model eukaryote has a classical NHEJ pathway dependent on Ku and DNA ligase IV, as well as alternative mechanisms for break rejoining. The evolutionary conservation of the Ku-dependent process includes several genes dedicated to this pathway, indicating that classical NHEJ at least is a strong contributor to fitness in the wild. Here we review how double-strand break structure, the yeast NHEJ proteins, and alternative rejoining mechanisms influence the accuracy of break repair. We also consider how the balance between NHEJ and homologous repair is regulated by cell state to promote genome preservation. The principles discussed are instructive to NHEJ in all organisms.

Contents

Homology-directed repair (HDR): DSB repair that entails copying of an homologous donor without crossover and exchange of genetic material as in homologous recombination

DSB: double-strand break

INTRODUCTION

There are two principal mechanisms by which a cell can repair a DNA double-strand break (DSB) that are intimately related to the observed classes of genetic recombination (**Figure 1**). In both homologous recombination and homology-directed repair (HDR), the 5′-terminated strand of a DSB is resected, freeing the 3′ strand to invade and copy an intact homologous duplex. Much is known about the later steps of HDR, including their dependence on the Rad52 epistasis group of proteins (3, 52). We only consider 5′ resection here, the first step of HDR and antagonist to nonhomologous repair.

Illegitimate recombination refers to genetic rearrangements that lack homology. Illegitimate recombination junctions do typically show microhomology, however, and are thus inferred to entail the direct joining of two DSBs brought together by annealing of short common sequences near their ends. This correlates with the second major mechanism of DSB repair (DSBR) that occurs by direct rejoining of a single DSB, often referred to generically as nonhomologous end-joining (NHEJ). Importantly, however, DSB rejoining encompasses multiple enzymatic mechanisms and precise definitions have been lacking. We review these homology-independent means of DSBR, focusing on insights gained using the budding yeast *Saccharomyces cerevisiae*. *S. cerevisiae* is well known as a model for studying HDR, but also possesses cryptic NHEJ mechanisms very similar to metazoans. Considerable effort has been given in the past decade to applying yeast DSBR techniques to NHEJ, which has revealed many of its general principles.

DOUBLE-STRAND BREAKS

Sources of DSBs

The sources of cellular DSBs determine their structure and therefore the important functions of NHEJ proteins (**Figure 2**). DSBs are the major contributor to the cytotoxicity of ionizing radiation because ionization tracts yield high frequencies of multiply damaged sites (40). Radiation is quantitatively less important to a typical cell than chemical and endogenous damage, however. Radiomimetic agents and oxidation by reactive products of respiration also lead to DSBs. This can represent primary breakage of two DNA strands, as with oxygen radicals, or secondary chemical or enzymatic degradation of base lesions (14). Conceptually different classes of DSBs are created by failure of normal cellular processes. Replication can lead to DSBs by collision with single-strand breaks (SSBs) or by cleavage of stalled forks (66). The ends of linear chromosomes are DSBs that may become exposed by dysfunction of the protective telomere end cap (64). Finally, chromosomes can be broken by physical stress, for example, if pulled in opposite directions during mitosis (64).

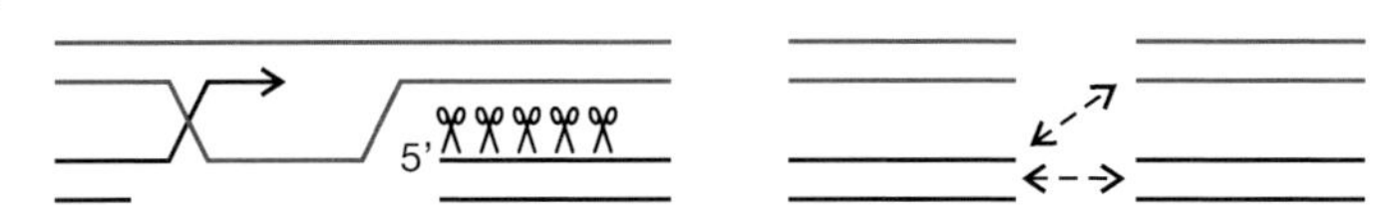

Figure 1

Homologous (*left*) versus nonhomologous (*right*) repair of DSBs and their relationship to genetic recombination. Scissors indicate 5′ resection of the DSB. Red lines indicate a different chromosome.

Anatomy of DSBs

Distinct DSB configurations arise from the above sources (**Figure 2**). Failure of replication and telomeres principally yields a limiting DSB case where there is only one free end. This is clearly not a desirable substrate for direct rejoining. Chemical or radiation damage of DNA creates DSBs with paired ends. The precise configuration of these ends is dependent on the causative agent, but in general will show a distribution of two SSBs about a DSB center. In some instances this will leave a blunt end. More typically, the DSB will have overhangs whose bases were paired in the original duplex. Such damage becomes a DSB, as opposed to simply two SSBs, when the binding energy in the overhangs is insufficient to overcome the entropy gained by dissociation of its halves. Importantly, the ends of such breaks are typically not "clean," i.e., with ligatable 5′ phosphates and 3′ hydroxyls, when created by degradative processes, but undamaged nucleotides in the overhangs can still drive reassociation of ends during repair.

Microhomology: short 1–10 base stretches of complementary or identical sequence too short to be acted upon by homologous recombination

DSBR: double-strand break repair

NHEJ: nonhomologous end joining

SSB: single-strand break

Joint Types

The most basic NHEJ event is the direct rejoining of a DSB with fully compatible overhangs, often referred to as simple religation (**Figure 2**). In contrast, processed NHEJ refers to any joint that cannot be completed by simple religation. This includes restorative joining of DSBs with damaged termini as well as mutagenic coupling of DSB ends. Importantly, it is only relevant to describe NHEJ as precise if the complementary overhangs or blunt ends were derived from the same parent duplex, because this provides the necessary reference for determining joining accuracy. Incompatible ends with partially or noncomplementary overhangs must have arisen from different or very complex DSBs and can be considered neither precise nor imprecise

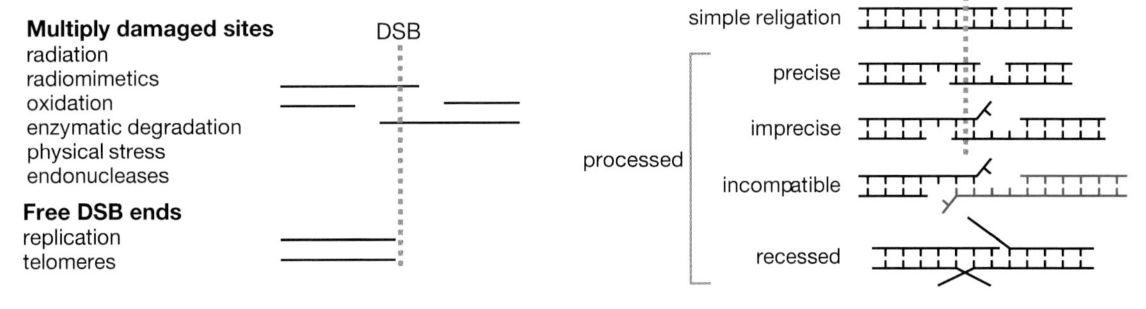

Figure 2

Causes of DSBs (*left*), their resulting strand relationships (*middle*), and the ensuing NHEJ joint types (*right*). Only 3′ overhangs are shown for clarity. Vertical dashed lines indicate the center point of the DSB.

Multiply damaged sites: multiple strand and/or base lesions clustered in close proximity in a DNA duplex that have a high potential for progressing to DSBs

Checkpoint: a mechanism for delaying cell cycle progression and promoting repair in the face of DNA damage, controlled by cascades of protein kinase activity

Single-strand break repair (SSBR): the rapid and efficient process related to base-excision repair by which DNA nicks and small gaps are directly religated

when joined. The final joint type does not occur via overhangs at all, but rather via recessed microhomologies that were exposed in the adjacent duplex DNA.

Making DSBs in Yeast

The simplest DSBR assays entail treating cells with radiation or chemicals and monitoring either cell survival or physical repair using methods such as gradient centrifugation. This approach has been of less value for yeast NHEJ since most DSBs are repaired by HDR, but should be revisited now that HDR can be controlled (see below).

The most common NHEJ assay is to transform yeast with linearized plasmids, which requires recircularization by repair enzymes in vivo. Transformation is easy and generally quantitative when compared with circular plasmid controls. Plasmid transformation uses naked DNA, however, and the chemical treatment of cells is itself genotoxic (35). The influence of these parameters may not be equal for linear and circular DNA and must be considered, especially when assessing chromatin and checkpoint functions. A further limitation has been the small number of physiologically relevant end configurations that could be achieved, but this was recently addressed by ligating oligonucleotide pairs onto the ends of the plasmid (20).

Another useful approach employs mega-endonucleases, such as HO and I-SceI, whose action can be restricted to a single cut site in vivo (33). This allows quantitative assessments of breakage and repair by cell survival, since even a single DSB is lethal if unrepaired, or physical monitoring with Southern blots or PCR. Importantly, continued expression of the endonuclease leads to recleavage of restored cut sites and prevents recovery of precise repair events. This can be overcome by coupling breakage and repair to termination of endonuclease expression (48). Alternatively, constitutive expression can be useful to study imprecise NHEJ (68, 95). Placing two copies of the asymmetric cut sites in head-to-head orientation forces incompatible ends (61).

A LIVING TEST TUBE FOR NHEJ

Yeast have proven to be an effective system for studying ectopically reconstituted breakage and repair systems. The maize cut-and-paste transposon Ac/Ds is mobile in yeast, and repair of its hairpin-terminated excision site is dependent on yeast NHEJ proteins (97). Yeast NHEJ will also repair breaks created by expressed vertebrate RAG endonucleases, which has potential implications for V(D)J recombination (19). Finally, Mycobacterial Ku and ligase proteins will reconstitute NHEJ in mutant yeast (22). This last system is attractive because it is so compact. For example, *Mycobacterium tuberculosis* possesses most needed activities in only two proteins of 1033 total amino acids. Thus, bacteria represent a unique opportunity to study coordination of the NHEJ reaction in a simplified system.

YEAST NHEJ PROTEINS

Although the concept of illegitimate recombination is decades old, the term NHEJ was not coined until 1996 when Moore & Haber used it to describe the repair of DSBs that occurred in yeast in the absence of a homologous donor (68). The defining feature of NHEJ was apparently the ability to seek out fortuitous microhomologies. This is often seen as distinct from simple religation, but defining NHEJ by outcomes such as misrejoining can be misleading. Better definitions are derived by considering the biochemical pathways of DSB rejoining. In this section we describe the proteins whose involvement defines the current best sense of NHEJ, the pathway that is uniquely dependent on the combined actions of Ku and DNA ligase IV. Dudášová et al. recently reviewed yeast NHEJ proteins with a focus on their mammalian counterparts (27). We approach this task by considering a comparison with single-strand break repair (SSBR) and structural correlates.

Analogy to SSBR

Although HDR is the competitor to NHEJ, we make the case that the conserved function of NHEJ is to rejoin DSB overhangs in a manner analogous to SSBR, an argument that impacts the expected properties of NHEJ proteins (**Figure 3**). The proteins that catalyze SSBR are well characterized (15). SSBR minimally entails trimming of 3′ and 5′ blocking lesions, resynthesis of bases destroyed during break formation, and ligation. In "long patch" repair the 5′ terminated strand is instead displaced by overfilling of the gap, yielding a 5′ flap that must be removed. All of these steps, and indeed the identity or class of many of the enzymes, are mirrored in NHEJ. The key distinction is that during NHEJ unstable overhangs must first be reassociated, which demands the input of protein binding energy. This ultrastructure accounts for many of the core proteins of NHEJ, and in some instances dictates the use of specialized enzymes.

	SSBR	NHEJ
nuclease	APE, FEN-1	FEN-1, Artemis, ?
kinase, Pase	PNKP/Tpp1	PNKP, ?
5' lyase	Pol X, glycosylases	Pol X, ?
polymerase	Pol X, Pol δ	Pol X, ?
ligase	DNA ligase I, III	DNA ligase IV
supporting	XRCC1, PARP, PCNA	Ku, DNA-PK, MRX

Figure 3

Functional correspondence between SSBR and NHEJ (*top*). Shading indicates the need for reassociation of unstable overhangs in NHEJ. SSBR and NHEJ factors are combined from all eukaryotes (*bottom*). Additional abbreviations: APE, apurinic/apyrimidinic endonuclease; PARP, poly-ADP ribose polymerase; PCNA, proliferating cell nuclear antigen.

Core NHEJ Proteins

The core components of NHEJ, those strictly required for rejoining of any DSB, can be defined experimentally as the proteins needed for simple religation. Supporting factors are superimposed on this basic mechanism. Most core yeast proteins were discovered by unrelated approaches, but two groups also systematically screened for them using a set of deletion mutants of all yeast genes (74, 93). These screens were necessarily limited to viable mutants, but it seems likely that all core proteins were identified since NHEJ is a non-essential function.

Yku70 and Yku80. Ku is arguably the defining protein of NHEJ as it is conserved from bacteria to human (24). Ku was recognized early as a human nuclear autoantigen that bound with great specificity to DSBs, providing the key insight into its NHEJ function of substrate recognition. Human Ku is part of the DNA-dependent protein kinase whose catalytic subunit, DNA-PK$_{cs}$, is required for effective NHEJ (12). Yeast lack DNA-PK$_{cs}$, demonstrating that the role of Ku is more general. Indeed, yeast Ku is involved in other cellular processes, including telomere maintenance and nuclear spatial organization (9, 83).

DNA-dependent protein kinase: a vertebrate kinase activated by DSB ends that subsequently phosphorylates proteins involved in the DNA damage response

In all eukaryotes Ku is a heterodimer encoded by duplicate copies of an ancestral gene (24). To highlight their correspondence to human Ku, the yeast subunits are called Yku70 and Yku80, although each is ~70 kDa (**Figure 4**). The defining feature of Ku, observed in crystals of the human protein and inferred by conservation in yeast, is its central β-barrel ring structure (92). Ku binds DNA by slipping the DSB end through this ring, which accounts for the substrate specificity of Ku as well as its ability to translocate along the duplex. The Ku ring appears to bind ends in only one orientation, which dictates an inherent polarity of its other domains.

The C terminus of Ku80 is positioned toward the DSB end and, consistently, makes a contact with Dnl4 that is important for NHEJ (P.L. Palmbos & T.E. Wilson, unpublished observations). Curiously, this contact utilizes an apparent terminal α-helix in Yku80 also present in human Ku80, but the latter has

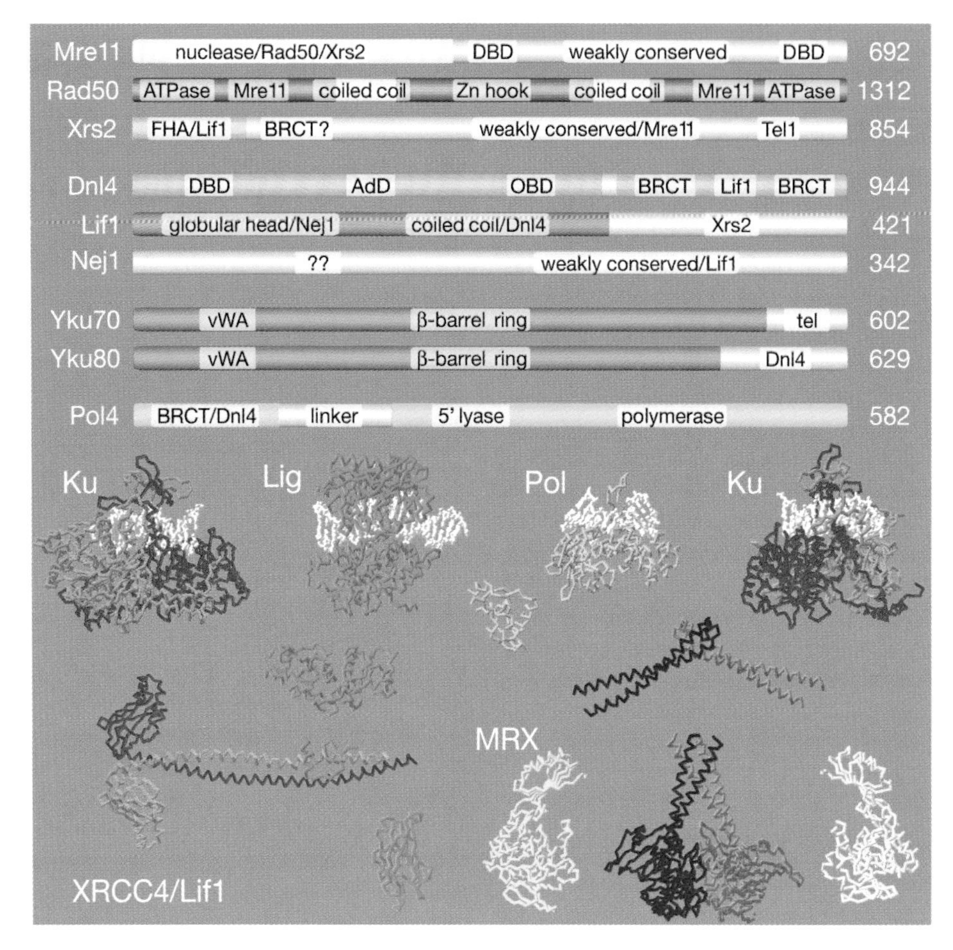

Figure 4

Primary structure representations of yeast NHEJ proteins are annotated as per the text (*top*). Colored regions correspond to available tertiary reference structures (*bottom*). Importantly, structures have not been determined for yeast NHEJ proteins, but they have a high likelihood of folding similarly to the illustrated structures. These images represent the minimum protein subunits that might be used during NHEJ, showing their size and complexity relative to the DSB. Shown are the human Ku heterodimer (*red*) bound to DNA (*white*), PDB 1JEY (92); human DNA ligase I (*green*) bound to DNA (*white*), PDB 1X9N (75); the human 53BP1 tandem BRCT domains (*green*), PDB 1GZH (23); human XRCC4 (*magenta*) bound to the DNA ligase IV inter-BRCT peptide (*green*), PDB 1IK9 (80); *Pyrococcus furiosus* Rad50 ATPase and Zn hook (*blue*), PDBs 1II8 (42) and 1L8D (41); *P. furiosus* Mre11 (*yellow*), PDB 1II7 (42); the yeast Rad53 FHA domain (*cyan*), PDB 1G6G (57); human DNA polymerase λ (*orange*) bound to DNA (*white*), PDB 1RZT (31); and the human XRCC1 BRCT domain (*orange*), PDB 1CDZ (99).

only been implicated in binding DNA-PKcs (32). The C terminus of Ku70 is oriented away from the DSB end. Correspondingly, this 25 amino acid region of Yku70 is required not for NHEJ but for telomere functions (26). In general, the yeast Ku C termini are more rudimentary than in higher eukaryotes. Yku80 lacks an α-helical bundle (36), whereas Yku70 lacks a SAP domain (92).

The amino termini of Yku70 and Yku80 are expected to form vWA domains (92). Less is known about these regions, and they are absent in bacterial Ku (24). The vWA domain of Yku80 has been implicated in telomere maintenance independently of NHEJ (6), which may be consistent with its proximity to the Ku70 C terminus. By analogy, the vWA domain of Yku70 might be involved in NHEJ, an untested hypothesis.

Dnl4, Lif1, and Nej1. It was first observed in yeast that NHEJ strictly requires DNA ligase IV, encoded by the gene *DNL4* (**Figure 4**) (94). DNA ligase I (Cdc9), the only other known yeast ligase, cannot be used. Conversely, Dnl4 is incapable of supporting replication and has no known role outside of end joining. Because this unique function of DNA ligase IV cannot be readily predicted by its catalytic properties, it is more likely attributable to its ability to engage the Ku-bound DSB end.

A crystal structure of the catalytic portion of human DNA ligase I revealed that it also completely encircles the DNA, although unlike Ku the ligase loads onto DNA by opening and closing of a non-covalent ring (75). The ligase ring is composed of three discrete domains: (*i*) an amino-terminal DNA binding domain (DBD), (*ii*) an adenylation domain that catalyzes the universal ligation chemistry, and (*iii*) a carboxy terminal oligonucleotide binding domain (OBD) proposed to undergo a large movement that completes the ring and drives formation of the transition state. Each of these domains is conserved in DNA ligase IV, strongly indicating that Dnl4 will also encircle the annealed DSB ends. The Dnl4 amino terminus is the least obviously conserved but is of similar length and required for NHEJ, suggesting that it will function as a DBD. Its greater divergence may account for the difference in DNA ligase I and IV substrates.

Much, but not all, of the NHEJ specificity of Dnl4 is determined by its unique C terminus, comprised of two tandem BRCT domains (84). Structures of similar protein regions demonstrate that repeated BRCT domains interact extensively with each other and present several concerted functional surfaces. One of these is a phosphopeptide binding motif (79), but it is not known whether a similar function is supported by Dnl4. The known Dnl4 role maps to the linker that connects its BRCT domains, similar to the interaction of 53BP1 with p53 (23). This region of human DNA ligase IV binds the NHEJ protein XRCC4 (80). The orthologous interaction permitted identification of the yeast XRCC4 homologue, Lif1 (39). It is strong enough that Dnl4 and Lif1 co-purify (17), and Dnl4 is not stable in the absence of Lif1 (39).

The amino terminal half of XRCC4 forms a globular head and coiled coil surprisingly reminiscent of Rad50 (see below) (47, 80). The coiled coil binds the ligase linker peptide. Although XRCC4 and Lif1 are weakly conserved, the central Lif1 region that binds Dnl4 (39) is predicted to fold as a coiled coil. Modeling has suggested that the globular head of XRCC4/Lif1 provides its observed DNA binding function (47, 67, 84), although the importance of this binding is not established. Finally, XRCC4 and Lif1 have a C-terminal region of unknown structure that is nonetheless important. This region of Lif1 interacts functionally with Xrs2 (P.L. Palmbos & T.E. Wilson, unpublished observations) (17).

Nej1 is a strong binding partner of the globular head of Lif1 (29). Nej1 is required for efficient NHEJ, although *nej1* mutant yeast have partial residual function in a chromosomal assay (93). The molecular function of Nej1 remains ambiguous. It has been suggested to promote nuclear import of Dnl4/Lif1 (90), although Dnl4 has an apparent nuclear localization signal. Information is lacking regarding the impact of Nej1 on ligation by Dnl4. Such a role seems likely given that the most highly conserved portion of Nej1 is its amino terminus, whereas Lif1 binds to the carboxy terminus.

Mre11, Rad50, and Xrs2. The last core yeast NHEJ proteins are Mre11, Rad50, and Xrs2, which form a complex called MRX (**Figure 4**) (81). Unlike other NHEJ proteins, MRX is also involved in HDR. MRX

DBD: DNA binding domain

OBD: oligonucleotide binding domain

Transition state: an enzyme bound to its substrate at the moment of catalysis, typically structurally specific and short lived

MRX: Mre11-Rad50-Xrs2 complex

Tethering: the continued physical association of broken chromosome ends by a protein bridge bound to each side of the DSB

PNKP: polynucleotide kinase/3′ phosphatase

End processing: the use of nucleases, polymerases and related enzymes to make DNA termini ready for ligation

thus appears suited to regulate repair pathway utilization (see below). Rad50 is a large protein with a split ABC ATPase domain at its termini (43). These associate to form a complete ABC domain, but a functional ATPase is not formed until two Rad50 proteins associate. Mre11 is a nuclease of the SbcCD family that binds Rad50 near its ATPase (42). With Xrs2 these proteins form an integral complex capable of strong binding to DNA (85). ATP binding and hydrolysis are coupled to the various MRX functions because they drive a large domain movement in Rad50 (43). One coupling allows the Mre11 nuclease to act as an endonuclease in addition to its basal 3′ exonuclease activity (86). This is irrelevant to simple religation (69), presumably because ends are preserved.

The Rad50 region between its ATPase heads is almost entirely composed of a long looped coiled coil that associates at its tip with another molecule of Rad50 in a structure called a Zn hook (41). This has led to the model that MRX binds and connects two DNA molecules separated by as much as 1200 Å. In NHEJ these would logically be the two sides of the DSB, where the Rad50 coiled coil would provide a flexible tether that reduces the effective volume and entropic barrier for the ensuing NHEJ reactions. Again, the Rad50 ATPase domain and Zn hook appear coupled despite being so distant (41). This may explain why ATP hydrolysis is required for effective tethering and, correspondingly, Rad50 ATPase mutants are partially NHEJ deficient (18).

Despite the compelling nature of the tethering model, the function of bacterial Ku and ligase in yeast was only slightly impaired in MRX mutants (22), indicating that MRX likely also plays an active role in assisting yeast Ku and Dnl4. The stimulatory role of MRX in vitro fulfills this expectation (17), as do Mre11-Yku80 and Xrs2-Lif1 interactions (P.L. Palmbos & T.E. Wilson, unpublished observations) (17). The latter interaction is particularly intriguing as it is mediated by the Xrs2 FHA domain and identifies a specific NHEJ function for this enigmatic protein. Deletion of the Xrs2 FHA domain impairs NHEJ but not the recombination or checkpoint activities of MRX (P.L. Palmbos & T.E. Wilson, unpublished observations), which are instead dependent on the Mre11 and Tel1 interaction domains of Xrs2 (78). FHA domains bind phosphopeptides (28), suggesting that Lif1 may be phosphorylated. Indeed, XRCC4 is phosphorylated, permitting its recognition by the FHA domain of polynucleotide kinase/phosphatase (PNKP) (50).

End Processing Enzymes

End processing encompasses those activities for which the correspondence between NHEJ and SSBR is most evident. The mammalian Pol X family polymerase Pol β is closely associated with SSBR (15), and the Pol X polymerase Pol λ is also implicated (10). It is thus intriguing that Pol4, the only Pol X polymerase in yeast, participates in NHEJ (**Figure 4**) (95) but is apparently dispensable for SSBR even though it possesses a 5′ lyase activity (65). Similarly, Pol λ and the related Pol X polymerase Pol μ have been implicated in mammalian NHEJ (62) and can partially complement a yeast *pol4* mutation (20a). The NHEJ function of these polymerases lies in part in their amino-terminal BRCT domains, which in Pol4 interacts with Dnl4 (87). Pol4, like Pol λ and Pol μ, also has unusual catalytic properties that can be summarized as a reduced dependence on a stable primer-template pairing (5, 73). This translates in vivo into a strict requirement for Pol4 at 3′ overhangs where the primer is the weakly reannealed overhang (20a). Unidentified polymerases are used to fill 5′ overhangs.

Rad27, the homologue of mammalian FEN-1, endonucleolytically cleaves the 5′ flaps generated by overfilling in SSBR (15). Consistently, Rad27 also interacts physically and functionally with Pol4 and Dnl4/Lif1 (88), and *rad27* strains show reduced joining of DSBs whose overhangs form 5′ flaps

(96). The activities of Rad27 and the Pol4 polymerase and lyase are apparently coordinated to achieve optimal filling and 5′ processing in NHEJ. Yeast 3′ NHEJ nucleases are as yet unclear. Artemis has been implicated in processed NHEJ in mammals (63), but its closest yeast homologue, Pso2/Snm1, appears to be exclusively associated with crosslink and hairpin repair (56, 97). The Mre11 3′ exonuclease is frequently invoked, but its only established in vivo substrates are hairpins and Spo11-bound meiotic DSBs (60, 69), and we have been unable to document a role in NHEJ (T.E. Wilson, unpublished observations).

Mammalian PNKP restores 3′ phosphates and 5′ hydroxyls to a ligatable form and acts in NHEJ by interaction with XRCC4 (50). Its yeast homologue, Tpp1, surprisingly lacks both the 5′ kinase and FHA domains and is not required for NHEJ of DSBs with 3′ phosphates (20). Caution is indicated, however, because redundancy may be the rule for end processing. No known processing enzyme is required for formation of the NHEJ core complex, and there are alternative processing routes for making most joints ligatable.

Chromatin and NHEJ

The above proteins are able to reconstitute NHEJ in vitro (17, 88). DSBR in vivo occurs in the same milieu as transcription and replication, however, and regulation of chromatin is consequently important to NHEJ as well. Esa1, a histone H4 acetyltransferase, is recruited to HO-induced breaks in vivo, and mutation of the four amino terminal H4 lysine targets of Esa1 inhibits NHEJ of transformed plasmids (7). Somewhat paradoxically, the histone deacetylase Sin3 is also required for efficient NHEJ (46). Yeast histone H2A, orthologous to mammalian H2AX, is rapidly phosphorylated by Mec1 in response to DNA damage, which is required for efficient NHEJ of transformed linearized plasmids (25). H2A phosphorylation has also been linked to SMC1 recruitment, and SMC1/SMC3 has itself been linked to NHEJ (77). Finally, H2A phosphorylation facilitates recruitment of the Ino80 ATP-dependent chromatin remodeling complex to a DSB, which is also required for efficient NHEJ (91). Most recently, the chromatin remodeling complex RSC and phosphorylation of histone H4 have also been implicated in NHEJ (18a, 77a).

The exact NHEJ role of this panoply of chromatin modifications is not established. It is difficult to imagine a unique NHEJ function for each, consistent with the partial NHEJ defect in these mutants. Collectively, they might reflect the need to provide access to the damaged DNA or docking sites for repair proteins. Chromatin structure might further contribute to the continued association of DSB ends. The ability of bacterial proteins to catalyze NHEJ in yeast despite the absence of histones in bacteria suggests that non-specific functions are more important than direct histone interactions with NHEJ proteins (22). Indeed, extensive chromatin interaction is likely more important for HDR because large tracts of DNA are involved. NHEJ requires only limited space at the terminus.

Other Factors

Mutants of the gene-silencing proteins Sir2–4 show severe NHEJ defects, but this is mediated indirectly by derepression of the silent mating-type loci which downregulates Nej1 (see below) (55). Also, Pol4 has been found to interact with the 19S proteasome (51). Strikingly, this proteasome is recruited to DSBs in a Pol4-dependent manner, and proteasome mutants were apparently defective in chromosomal NHEJ. The assays used were not quantitative and did not reveal the specific defect, however. A proteasome might either clear proteins to allow NHEJ to begin or execute post-repair cleanup of proteins such as Ku. Such functions should be relevant to all NHEJ, however, whereas Pol4 is dispensable at most joints. Establishing this putative proteasome role will require identification of NHEJ proteins targeted for degradation.

THE NHEJ REACTION

In this section we present one possible model of the yeast NHEJ reaction sequence based on current information (**Figure 5**). Soon after DSB formation and prior to the initiation of 5′ resection, MRX binds the damaged site and tethering is established (58). Ku must also bind the DSB early, although it is unknown whether this precedes or follows MRX. The order of MRX and Ku binding may impact their relative position on the DNA. We depict a model in which MRX is bound internally to Ku. Although MRX can bridge ends in vitro (17), there is no direct evidence that it needs to bind the extreme terminus during the course of NHEJ, in contrast to Ku, which must load there. This point is not established, however, and MRX and Ku binding may in fact be dynamic with respect to both order and position.

Yku80 and Xrs2 next bind Dnl4 and Lif1, respectively (P.L. Palmbos & T.E. Wilson, unpublished observations) (17), and recruit DNA ligase IV. The relative footprints of Ku and DNA ligase (75, 92) demand that Ku is translocated internally in conjunction with ligase engagement, a concept supported by experimental evidence (53). Importantly, excessive rigidity is not compatible with NHEJ as Dnl4-Lif1 must engage overhangs whose nicks will be variably positioned. Flexibility is likely provided by the Lif1 coiled coil, Yku80 C terminus, and variable extents of Ku translocation.

The critical event in NHEJ now occurs: the alignment and base pairing of the overhangs. After pairing, ligation is attempted. Two molecules of Dnl4 may be present at the DSB, one at each end, but ligation can likely be attempted only one strand at a time because the expected ligase footprint will not permit both nicks of a short overhang to be engaged simultaneously (75). This is sufficient because ligation of only one strand would stabilize the joint. Dnl4 might typically complete ligation of the second strand sequentially, but this is not necessary to satisfy the critical function of NHEJ.

Unsuccessful ligation means that Dnl4 could not engage either strand in its transition state, i.e., that the OBD could not complete the ligase ring around the overhangs. Partially bound Dnl4 appears optimally configured for the recruitment of processing enzymes, and indeed, Pol4 interacts with Dnl4 (87). It is unknown whether this state is simply a stalled

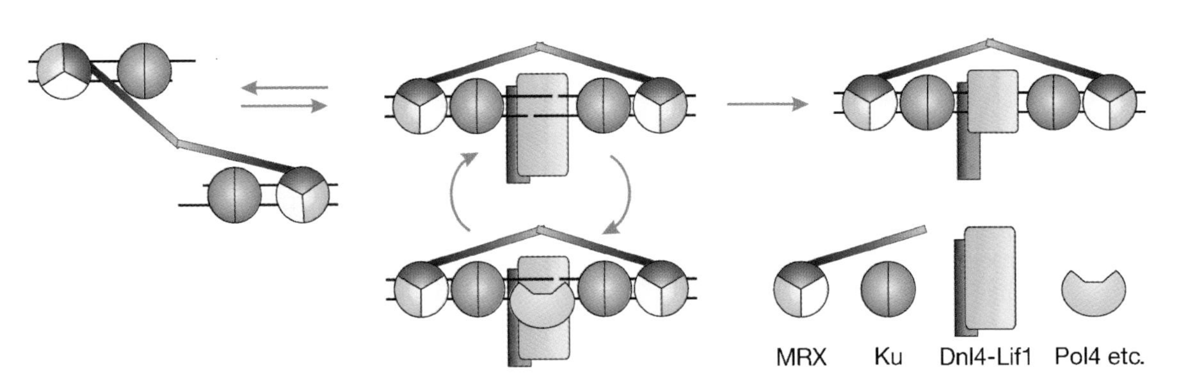

Figure 5

One possible sequence in the NHEJ reaction, with colors similar to those in **Figure 4**. Some known interactions are not illustrated, notably Yku80-Dnl4 and Xrs2-Lif1. These and presumably other unknown interactions will only be rationalized in a more accurate three-dimensional and sequential representation, which is not currently possible.

intermediate or a dedicated transition to a processing conformation. Following processing, ligation will now either proceed or still be impossible, necessitating further processing. The iterative testing of ligation and processing should not be viewed as strictly ordered, however. The relative affinities of the various enzymes for the overhang pairing will determine the reaction sequence. Importantly, the lifetime of an annealed joint is unknown and potentially short, even if end tethering by MRX is stable. Indeed, little is know about any kinetic parameters of NHEJ, such as the time required and rate-limiting steps. Modest decreases in the efficiency of a given enzyme may thus have large impacts on joint formation.

ACCURACY OF NHEJ

NHEJ is frequently stated to be error prone, and its capacity to rejoin DSBs via microhomologies does suggest a high potential for misrepair. However, little consideration has been given to the extent that NHEJ proteins and DSB structures promote accurate repair. In this section we discuss the types of NHEJ errors and how frequently they are observed.

Mispairing of Overhangs

Accuracy in any repair pathway is dependent on having information available to guide it. In NHEJ this information is contained within the overhangs, and indeed yeast join blunt ends poorly (8). After pairing, overhangs provide a template for resynthesis of degraded bases in direct analogy to SSBR. The larger question is how overhang pairing is established. Overhangs at single DSBs will be fully complementary. However, limited base pairing might also be available in imprecise alignment registers, and such joints can be completed by the same enzymes used to deal with damaged nucleotides (95). Joint patterns in vivo consistently suggest that the choice between competing overhang alignments is a function of their thermodynamic stability and therefore association/dissociation kinetics (20, 95). An iterative mechanism that drove all joints to the optimal pairing would require a rule for final joint acceptance that is not possible given the variable structure of naturally occurring overhangs. Thus, the weakness of NHEJ is its inability to reject erroneous pairings that occur by chance, and slippage errors do occur in mononucleotide repeats in vivo, as evidenced by NHEJ-dependent stationary phase mutations (37). Still, yeast NHEJ, like mammalian NHEJ (11), is mostly accurate. Single-base mispairs in 4-base overhangs lead to an error frequency of about 1% (68, 95). Even when presented with a contrived difficult choice between 2- and 4-base G/C pairings, yeast NHEJ used the precise 4-base pairing nearly 6 to 1 (95).

Stationary phase: also called G0, a persistent nondividing cell state entered to withstand prolonged periods of nutritional deprivation encountered by microorganisms in the wild

Recessed Microhomologies

The other joint error at a single DSB is use of recessed microhomologies instead of the overhangs. At the compatible overhangs expected of many DSBs, yeast NHEJ in fact uses the overhangs with great preference (8, 20). The frequency of recessed microhomology use goes up at incompatible ends (61), but all outcomes are mutagenic at such DSBs. Also, joining via recessed microhomologies is enzymatically distinct from NHEJ in many cases (see below).

Chromosome Rearrangements

The ability of NHEJ to join ends with processing creates the potential for chromosomal rearrangements when two or more DSBs occur simultaneously. NHEJ may indeed mediate one class of spontaneously occurring gross chromosomal rearrangements (72). A key factor in determining the translocation frequency is whether broken chromosome ends are sterically constrained in the nucleus to remain near their partner or are free to diffuse and join to other DSBs. Haber & Leung found that two DSBs showed no preference for intrachromosomal versus interchromosomal repair by single-strand annealing (see below) (34). However, these events required extensive processing not relevant to NHEJ that may disrupt end tethering. Indeed, the NHEJ component MRX is a key player in maintaining the proximity of ends (59). Yu & Gabriel found that NHEJ resulted in reciprocal translocations in about 3% of yeast that survived two DSBs (98). However, this study used constitutive endonuclease expression that may result in multiple breakage and repair cycles. The degree to which NHEJ tends toward rearrangement is therefore not certain, but is apparently low.

Synthesis Errors

Finally, synthesis errors may be made by the polymerase replacing damaged nucleotides. The high error rate of Pol4 in vitro (10^{-3}) might suggest this as an important source of NHEJ mutations (5). However, this measurement was made in long synthesis tracts irrelevant to NHEJ. Further, the selectivity of ligases against base mispairs suggests that NHEJ synthesis errors would need to be removed in what amounts to proofreading. Data addressing this issue are needed.

REGULATION OF NHEJ

Unlike NHEJ, HDR is often stated to be error free because it operates by replicating an intact chromosome. However, HDR accuracy is dependent upon its ability to find an appropriate donor. An inappropriate donor can alter the sequence or lead to rearrangement. If no donor can be found the cell will die owing to loss of the acentric portion of the chromosome upon cell division. This reveals the weakness of HDR: The presence of a donor cannot be determined until 5′ resection has irreversibly committed a DSB to HDR (48). The relative accuracy of NHEJ and HDR is thus a strong function of donor availability, and we should expect DSBR regulation based on this parameter. Moreover, regulation should be applied at the decision point between 5′ resection and NHEJ. The ideal HDR donor is the sister chromatid to which a chromosome is paired following replication as it is homologous and accessible. The homologous chromosome in diploid cells is the next best choice. DSBR accuracy is thus optimized by regulating pathway utilization as a function of the cell cycle (**Figure 6**).

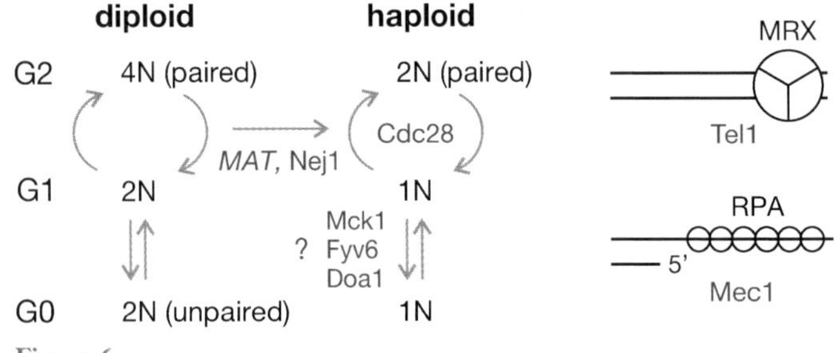

Figure 6

Regulation of DSBR. Proteins in red affect the balance between NHEJ and HDR as a function of cell cycle and growth stages (*left*). DSBs also activate cell cycle checkpoint kinases (*right*).

Ploidy Restriction

HDR in diploid yeast can use the homologous chromosome because the nuclear volume is small enough to allow a search for this unlinked molecule. Moreover, the low content of repetitive genome elements increases the likelihood of finding the true homologue. NHEJ is therefore repressed in diploid yeast by downregulation of *NEJ1* (29, 49, 90, 93). This is tied to the **a**/α mating-type status of typical diploid cells, apparently by direct regulation of *NEJ1* by *MAT*-encoded transcription factors. This helps explain why Nej1 is only present in *Saccharomyces* and related genera that share similar mating-type regulation (13). Moreover, downregulation of NHEJ in diploid cells of higher eukaryotes with large repetitive genomes would decrease accuracy due to their lower chance of success in donor searching.

G1 Restriction

In haploid cells, one would similarly expect NHEJ utilization to be greater in G1 than late S and G2/M. This has been observed for NHEJ of a single DSB and NHEJ-dependent telomere-DSB fusions (16, 48). Such regulation is not principally mediated by altering the capacity for NHEJ through a mechanism like Nej1, however, but indirectly by changing the efficiency of 5′ resection. Remarkably, HDR cannot be completed if cells are maintained in G1 because DSB ends are not resected (4, 44). The primary cyclin-dependent kinase of yeast, Cdc28, is suited to provide this control because it is active in all cell cycle stages except G1. Indeed, the G1 inhibition of resection can be recapitulated in other cell cycle stages by inhibition of Cdc28 (4, 44). How Cdc28 controls 5′ resection is not known, in part because the mechanism of resection is not known. MRX is a key player in resection, but several nucleases may play overlapping roles (70). Cdc28 may be required to activate the appropriate nuclease(s) or perhaps modulate the function of MRX to shift its activity between resection and NHEJ.

Although Cdc28 activity is associated with decreased NHEJ, NHEJ can be completed in S/G2. This can be explained by a period of transient DSB stability prior to the onset of resection (30). Thus, the balance between NHEJ and HDR action can be described in terms of permissive states. Only NHEJ is permitted in haploid G1. NHEJ is also permitted in S/G2 if it can be completed rapidly, but resection will take over and commit a break to HDR if NHEJ is delayed. An open question is whether 5′ resection is also regulated by ploidy to be active in diploid G1, or if DSBs that persist in the absence of Nej1 are resected upon entry into S phase.

G0 Activation

NHEJ does not contribute significantly to the fitness of rapidly dividing haploid yeast, so situations other than G1 must explain the evolutionary conservation of the NHEJ-specific genes *DNL4*, *LIF1*, and *NEJ1*. NHEJ may have a larger impact on fitness in G0/stationary phase, a period of high oxidative stress (38). Indeed, NHEJ is influenced by nutritional status, and stationary haploid cells show increased NHEJ (39, 48). It is not known whether haploid spores formed during meiosis show a similar phenomenon, but it would be expected. The G0 effect is likely mediated in part by inactivity of Cdc28, but there may be independent influences by other proteins (93).

Checkpoints and NHEJ

Cell cycle checkpoints influence HDR (2), and it has been reported that yeast mutants in the Mec1 checkpoint pathway show reduced NHEJ of transformed plasmids (21). However, this NHEJ defect was not reproduced in chromosomal assays (93), so the plasmid results may be influenced by checkpoint activation associated with transformation (35). Further, Mec1-dependent checkpoints are likely activated by RPA-bound single-stranded DNA (100), which would not exist at DSBs until 5′ resection occurred and NHEJ was no longer possible (**Figure 6**). This contrasts with the MRX/Tel1 checkpoint. MRX and Tel1 are the first proteins recruited to a DSB (58) and, like their mammalian homologues in the ATM checkpoint, may be activated by the DSB directly (54, 89). Tel1 is not required for NHEJ (93), but its influence on the DSBR function of MRX needs to be established.

ALTERNATE JOINING MECHANISMS

SSBR of DSBs

The early idea that homology-independent rejoining of DSBs might extend beyond NHEJ is finding support in recent literature (**Figure 7**). NHEJ-deficient strains can accurately rejoin complementary DSBs, albeit at greatly reduced efficiency (48). Insight into such joining is provided by the observation that its efficiency increases substantially as overhangs become longer than

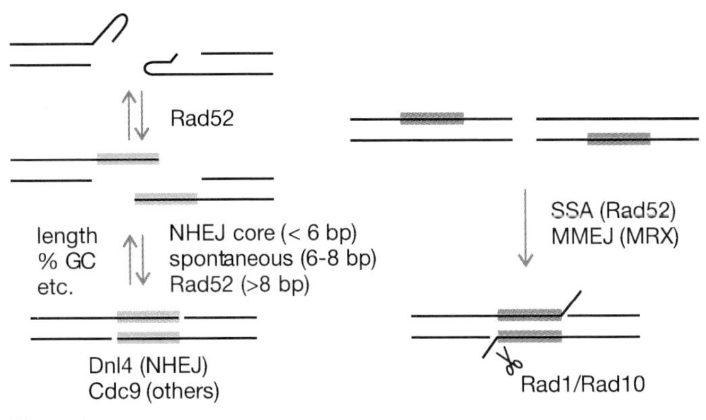

Figure 7

Alternate DSB rejoining mechanisms. Like NHEJ, some processes act by direct reannealing of overhangs, where the equilibrium between correct and nonproductive overhang annealing is critical (*left*). Other processes act by seeking out recessed microhomologies (*right*).

SSA: single-strand annealing

MMEJ: microhomology mediated end joining

4 bases or higher in GC content (20). The interpretation is that overhangs are in equilibrium between the annealed and dissociated states, an equilibrium that shifts as overhang stability increases. Overhangs that have reannealed spontaneously might be ligated by SSBR. Consistent with this, mammalian SSBR proteins have been recently implicated in NHEJ-independent DSB rejoining (1). It is not known whether the joint equilibrium is dependent on MRX tethering, and therefore whether MRX might also affect SSBR of DSBs. Ligation competence may also be important. Joints that cannot be rapidly ligated may dissociate before completion by SSBR, explaining the stricter NHEJ dependence of joints that require processing (22).

Rad52-Dependent Rejoining

When overhangs become even longer, 8 bases and beyond, the joining efficiency goes up further but becomes substantially dependent on Rad52 (20). The role of Rad52 here is presumably its ability to anneal complementary single strands (71). A catalyst is likely required for such annealing to overcome nonproductive intrastrand base pairing in longer sequences. Rad52-dependent rejoining of overhangs is thus highly reminiscent of the DSBR pathway called single-strand annealing (SSA). SSA occurs between long direct repeats on either side of a DSB that are exposed by 5′ resection and subsequently annealed in a Rad52-dependent fashion (82). The need for resection clearly identifies SSA as a subpathway of HDR, in contrast to Rad52-dependent rejoining of overhangs. SSA also requires the action of Rad1/Rad10 to remove nonhomologous sequences (45), which are not present in Rad52-dependent rejoining. Finally, SSA will always result in a deletion while Rad52-dependent rejoining is exceptionally accurate at restoring the original sequence (20).

Microhomology-Mediated End Joining

SSA is related to a third NHEJ-independent rejoining mechanism characterized by deletions of up to ~300 bases on one or both sides of the break and microhomology at the junction (8, 61). This mechanism has been termed microhomology-mediated end joining (MMEJ). MMEJ shows a mixed dependence on DSBR proteins that defies its characterization as any of the above pathways. Like NHEJ, MMEJ requires the MRX complex, and it is partially dependent on DNA ligase IV (61). However, Ku loss increases MMEJ efficiency (8, 61). Like SSA, MMEJ requires resection to reveal the recessed microhomology, but it is not known whether this resection is more related to HDR or microhomology searching in NHEJ. MMEJ also requires the Rad1/Rad10 endonuclease similar to SSA (61).

It is unclear whether MMEJ is an evolutionarily conserved pathway of DSB repair or a by-product of having persistent reactive DNA ends and repair proteins that bind them. Joints showing the MMEJ signature are the predominant outcome only when NHEJ and HDR are not possible, and MMEJ is inefficient even in those situations.

Moreover, MMEJ is again always mutagenic. However, MMEJ-like repair observed in mammalian cells, referred to as backup NHEJ (76), appears to be more efficient than in yeast and may be a contributor to genome stability despite its inaccuracy. The available data thus suggest that MMEJ is a last-ditch mutagenic attempt to save genetic information when more accurate pathways have failed.

SUMMARY POINTS

1. Direct rejoining of DSBs is closely associated with illegitimate recombination but represents more than one conceptual and enzymatic pathway of repair.
2. Yeast proteins of the NHEJ pathway dependent on Ku and DNA ligase IV are mostly identified; enzymes involved in end processing are the least well characterized.
3. Comparison of NHEJ to SSBR provides a helpful model for understanding and predicting the properties of NHEJ proteins.
4. The relationships between DSB structure and the properties of NHEJ proteins promote joining accuracy; overhang mispairing is the primary potential for error.
5. Less is known about the interactions of the Ku, MRX, and DNA ligase IV complexes with each other and chromatin, and how these dictate the coordinated progression of NHEJ.
6. The balance between NHEJ and HDR is regulated as a function of cell cycle and growth stages, in part via 5′ resection, to promote genome preservation and fitness.
7. Alternative mechanisms of DSB rejoining are highly variable in their efficiency and tendency toward mutation, with MMEJ most closely resembling the error prone mechanism often associated with NHEJ.

LITERATURE CITED

1. Audebert M, Salles B, Calsou P. 2004. Involvement of poly(ADP-ribose) polymerase-1 and XRCC1/DNA ligase III in an alternative route for DNA double-strand breaks rejoining. *J. Biol. Chem.* 279:55117–26
2. Aylon Y, Kupiec M. 2003. The checkpoint protein Rad24 of *Saccharomyces cerevisiae* is involved in processing double-strand break ends and in recombination partner choice. *Mol. Cell Biol.* 23:6585–96
3. Aylon Y, Kupiec M. 2004. DSB repair: the yeast paradigm. *DNA Repair* 3:797–815
4. Aylon Y, Liefshitz B, Kupiec M. 2004. The CDK regulates repair of double-strand breaks by homologous recombination during the cell cycle. *EMBO J.* 23:4868–75
5. Bebenek K, Garcia-Diaz M, Patishall SR, Kunkel TA. 2005. Biochemical properties of *Saccharomyces cerevisiae* DNA polymerase IV. *J. Biol. Chem.* 280:20051–58
6. Bertuch AA, Lundblad V. 2003. The Ku heterodimer performs separable activities at double-strand breaks and chromosome termini. *Mol. Cell Biol.* 23:8202–15
7. Bird AW, Yu DY, Pray-Grant MG, Qiu Q, Harmon KE, et al. 2002. Acetylation of histone H4 by Esa1 is required for DNA double-strand break repair. *Nature* 419:411–15

8. Boulton SJ, Jackson SP. 1996. *Saccharomyces cerevisiae* Ku70 potentiates illegitimate DNA double-strand break repair and serves as a barrier to error-prone DNA repair pathways. *EMBO J.* 15:5093–103
9. Boulton SJ, Jackson SP. 1998. Components of the Ku-dependent non-homologous end-joining pathway are involved in telomeric length maintenance and telomeric silencing. *EMBO J.* 17:1819–28
10. Braithwaite EK, Prasad R, Shock DD, Hou EW, Beard WA, Wilson SH. 2005. DNA polymerase λ mediates a back-up base excision repair activity in extracts of mouse embryonic fibroblasts. *J. Biol. Chem.* 280:18469–75
11. Budman J, Chu G. 2005. Processing of DNA for nonhomologous end-joining by cell-free extract. *EMBO J.* 24:849–60
12. Burma S, Chen DJ. 2004. Role of DNA-PK in the cellular response to DNA double-strand breaks. *DNA Repair* 3:909–18
13. Butler G, Kenny C, Fagan A, Kurischko C, Gaillardin C, Wolfe KH. 2004. Evolution of the *MAT* locus and its HO endonuclease in yeast species. *Proc. Natl. Acad. Sci. USA* 101:1632–37
14. Cadet J, Douki T, Gasparutto D, Ravanat JL. 2003. Oxidative damage to DNA: formation, measurement and biochemical features. *Mutat. Res.* 531:5–23
15. Caldecott KW. 2003. Protein-protein interactions during mammalian DNA single-strand break repair. *Biochem. Soc. Trans.* 31:247–51
16. Chan SW, Blackburn EH. 2003. Telomerase and ATM/Tel1p protect telomeres from nonhomologous end joining. *Mol. Cell* 11:1379–87
17. **Chen L, Trujillo K, Ramos W, Sung P, Tomkinson AE. 2001. Promotion of Dnl4-catalyzed DNA end-joining by the Rad50/Mre11/Xrs2 and Hdf1/Hdf2 complexes. *Mol. Cell* 8:1105–15**

The first of a series of papers demonstrating reconstitution of yeast NHEJ in vitro.

18. Chen L, Trujillo KM, Van Komen S, Roh DH, Krejci L, et al. 2005. Effect of amino acid substitutions in the Rad50 ATP binding domain on DNA double strand break repair in yeast. *J. Biol. Chem.* 280:2620–27

18a. Cheung WL, Turner FB, Krishnamoorthy T, Wolner B, Ahn SH, et al. 2005. Phosphorylation of histone H4 serine 1 during DNA damage requires casein kinase II in *S. cerevisiae*. *Curr. Biol.* 15:656–60

19. Clatworthy AE, Valencia MA, Haber JE, Oettinger MA. 2003. V(D)J recombination and RAG-mediated transposition in yeast. *Mol. Cell* 12:489–99
20. **Daley JM, Wilson TE. 2005. Rejoining of DNA double-strand breaks as a function of overhang length. *Mol. Cell Biol.* 25:896–906**

This paper used "designer DSBs" to demonstrate accurate non-NHEJ rejoining of DSBs.

20a. Daley JM, Vander Laan RL, Suresh A, Wilson TE. 2005. DNA joint dependence of Pol X family polymerase action in nonhomologous end joining. *J. Biol. Chem.* In press

21. de la Torre-Ruiz M, Lowndes NF. 2000. The *Saccharomyces cerevisiae* DNA damage checkpoint is required for efficient repair of double strand breaks by non-homologous end joining. *FEBS Lett.* 467:311–15
22. Della M, Palmbos PL, Tseng HM, Tonkin LM, Daley JM, et al. 2004. Mycobacterial Ku and ligase proteins constitute a two-component NHEJ repair machine. *Science* 306:683–85
23. Derbyshire DJ, Basu BP, Serpell LC, Joo WS, Date T, et al. 2002. Crystal structure of human 53BP1 BRCT domains bound to p53 tumour suppressor. *EMBO J.* 21:3863–72
24. Doherty AJ, Jackson SP, Weller GR. 2001. Identification of bacterial homologues of the Ku DNA repair proteins. *FEBS Lett.* 500:186–88

25. **Downs JA, Lowndes NF, Jackson SP. 2000. A role for *Saccharomyces cerevisiae* histone H2A in DNA repair. *Nature* 408:1001–4**
26. Driller L, Wellinger RJ, Larrivee M, Kremmer E, Jaklin S, Feldmann HM. 2000. A short C-terminal domain of Yku70p is essential for telomere maintenance. *J. Biol. Chem.* 275:24921–27
27. Dudášová Z, Dudas A, Chovanec M. 2004. Non-homologous end-joining factors of *Saccharomyces cerevisiae*. *FEMS Microbiol. Rev.* 28:581–601
28. Durocher D, Henckel J, Fersht AR, Jackson SP. 1999. The FHA domain is a modular phosphopeptide recognition motif. *Mol. Cell* 4:387–94
29. Frank-Vaillant M, Marcand S. 2001. NHEJ regulation by mating type is exercised through a novel protein, Lif2p, essential to the ligase IV pathway. *Genes Dev.* 15:3005–12
30. Frank-Vaillant M, Marcand S. 2002. Transient stability of DNA ends allows non-homologous end joining to precede homologous recombination. *Mol. Cell* 10:1189–99
31. Garcia-Diaz M, Bebenek K, Krahn JM, Blanco L, Kunkel TA, Pedersen LC. 2004. A structural solution for the DNA polymerase λ-dependent repair of DNA gaps with minimal homology. *Mol. Cell* 13:561–72
32. Gell D, Jackson SP. 1999. Mapping of protein-protein interactions within the DNA-dependent protein kinase complex. *Nucleic Acids Res.* 27:3494–502
33. Haber JE. 2002. Uses and abuses of HO endonuclease. *Methods Enzymol.* 350:141–64
34. Haber JE, Leung WY. 1996. Lack of chromosome territoriality in yeast: promiscuous rejoining of broken chromosome ends. *Proc. Natl. Acad. Sci. USA* 93:13949–54
35. Haghnazari E, Heyer WD. 2004. The DNA damage checkpoint pathways exert multiple controls on the efficiency and outcome of the repair of a double-stranded DNA gap. *Nucleic Acids Res.* 32:4257–68
36. Harris R, Esposito D, Sankar A, Maman JD, Hinks JA, et al. 2004. The 3D solution structure of the C-terminal region of Ku86 (Ku86CTR). *J. Mol. Biol.* 335:573–82
37. Heidenreich E, Novotny R, Kneidinger B, Holzmann V, Wintersberger U. 2003. Non-homologous end joining as an important mutagenic process in cell cycle-arrested cells. *EMBO J.* 22:2274–83
38. Herman PK. 2002. Stationary phase in yeast. *Curr. Opin. Microbiol.* 5:602–07
39. Herrmann G, Lindahl T, Schar P. 1998. *Saccharomyces cerevisiae LIF1*: a function involved in DNA double-strand break repair related to mammalian XRCC4. *EMBO J.* 17:4188–98
40. Hill MA. 1999. Radiation damage to DNA: the importance of track structure. *Radiat. Meas.* 31:15–23
41. Hopfner KP, Craig L, Moncalian G, Zinkel RA, Usui T, et al. 2002. The Rad50 zinc-hook is a structure joining Mre11 complexes in DNA recombination and repair. *Nature* 418:562–66
42. Hopfner KP, Karcher A, Craig L, Woo TT, Carney JP, Tainer JA. 2001. Structural biochemistry and interaction architecture of the DNA double-strand break repair Mre11 nuclease and Rad50-ATPase. *Cell* 105:473–85
43. Hopfner KP, Karcher A, Shin DS, Craig L, Arthur LM, et al. 2000. Structural biology of Rad50 ATPase: ATP-driven conformational control in DNA double-strand break repair and the ABC-ATPase superfamily. *Cell* 101:789–800

The first demonstration that chromatin alterations are important for NHEJ.

With Reference 4 demonstrated Cdc28 control over 5′ resection.

44. **Ira G, Pellicioli A, Balijja A, Wang X, Fiorani S, et al. 2004. DNA end resection, homologous recombination and DNA damage checkpoint activation require CDK1. *Nature* 431:1011–17**
45. Ivanov EL, Haber JE. 1995. *RAD1* and *RAD10*, but not other excision repair genes, are required for double-strand break-induced recombination in *Saccharomyces cerevisiae*. *Mol. Cell Biol.* 15:2245–51
46. Jazayeri A, McAinsh AD, Jackson SP. 2004. *Saccharomyces cerevisiae* Sin3p facilitates DNA double-strand break repair. *Proc. Natl. Acad. Sci. USA* 101:1644–49
47. Junop MS, Modesti M, Guarne A, Ghirlando R, Gellert M, Yang W. 2000. Crystal structure of XRCC4 DNA repair protein and implications for end joining. *EMBO J.* 19:5962–70
48. Karathanasis E, Wilson TE. 2002. Enhancement of *Saccharomyces cerevisiae* end-joining efficiency by cell growth stage but not by impairment of recombination. *Genetics* 161:1015–27
49. Kegel A, Sjostrand JO, Astrom SU. 2001. Nej1p, a cell type-specific regulator of nonhomologous end joining in yeast. *Curr. Biol.* 11:1611–17
50. Koch CA, Agyei R, Galicia S, Metalnikov P, O'Donnell P, et al. 2004. XRCC4 physically links DNA end processing by polynucleotide kinase to DNA ligation by DNA ligase IV. *EMBO J.* 23:3874–85
51. Krogan NJ, Lam MH, Fillingham J, Keogh MC, Gebbia M, et al. 2004. Proteasome involvement in the repair of DNA double-strand breaks. *Mol. Cell* 16:1027–34
52. Krogh BO, Symington LS. 2004. Recombination proteins in yeast. *Annu. Rev. Genet.* 38:233–71
53. Kysela B, Doherty AJ, Chovanec M, Stiff T, Ameer-Beg SM, et al. 2003. Ku stimulation of DNA ligase IV-dependent ligation requires inward movement along the DNA molecule. *J. Biol. Chem.* 278:22466–74
54. Lee JH, Paull TT. 2005. ATM activation by DNA double-strand breaks through the Mre11-Rad50-Nbs1 complex. *Science* 308:551–54
55. Lee SE, Paques F, Sylvan J, Haber JE. 1999. Role of yeast SIR genes and mating type in directing DNA double-strand breaks to homologous and non-homologous repair paths. *Curr. Biol.* 9:767–70
56. Li X, Moses RE. 2003. The beta-lactamase motif in Snm1 is required for repair of DNA double-strand breaks caused by interstrand crosslinks in *S. cerevisiae*. *DNA Repair* 2:121–29
57. Liao H, Yuan C, Su MI, Yongkiettrakul S, Qin D, et al. 2000. Structure of the FHA1 domain of yeast Rad53 and identification of binding sites for both FHA1 and its target protein Rad9. *J. Mol. Biol.* 304:941–51
58. Lisby M, Barlow JH, Burgess RC, Rothstein R. 2004. Choreography of the DNA damage response; spatiotemporal relationships among checkpoint and repair proteins. *Cell* 118:699–713
59. Lobachev K, Vitriol E, Stemple J, Resnick MA, Bloom K. 2004. Chromosome fragmentation after induction of a double-strand break is an active process prevented by the RMX repair complex. *Curr. Biol.* 14:2107–12
60. Lobachev KS, Gordenin DA, Resnick MA. 2002. The Mre11 complex is required for repair of hairpin-capped double-strand breaks and prevention of chromosome rearrangements. *Cell* 108:183–93
61. **Ma JL, Kim EM, Haber JE, Lee SE. 2003. Yeast Mre11 and Rad1 proteins define a Ku-independent mechanism to repair double-strand breaks lacking overlapping end sequences. *Mol. Cell Biol.* 23:8820–28**

This genetic study provided molecular definition of the MMEJ pathway.

62. Ma Y, Lu H, Tippin B, Goodman MF, Shimazaki N, et al. 2004. A biochemically defined system for mammalian nonhomologous DNA end joining. *Mol. Cell* 16:701–13

63. Ma Y, Pannicke U, Schwarz K, Lieber MR. 2002. Hairpin opening and overhang processing by an Artemis/DNA-dependent protein kinase complex in nonhomologous end joining and V(D)J recombination. *Cell* 108:781–94

64. Mathieu N, Pirzio L, Freulet-Marriere MA, Desmaze C, Sabatier L. 2004. Telomeres and chromosomal instability. *Cell. Mol. Life Sci.* 61:641–56

65. McInnis M, O'Neill G, Fossum K, Reagan MS. 2002. Epistatic analysis of the roles of the *RAD27* and *POL4* gene products in DNA base excision repair in *S. cerevisiae*. *DNA Repair* 1:311–15

66. Michel B, Grompone G, Flores MJ, Bidnenko V. 2004. Multiple pathways process stalled replication forks. *Proc. Natl. Acad. Sci. USA* 101:12783–88

67. Modesti M, Hesse JE, Gellert M. 1999. DNA binding of XRCC4 protein is associated with V(D)J recombination but not with stimulation of DNA ligase IV activity. *EMBO J.* 18:2008–18

68. Moore JK, Haber JE. 1996. Cell cycle and genetic requirements of two pathways of nonhomologous end-joining repair of double-strand breaks in *Saccharomyces cerevisiae*. *Mol. Cell Biol.* 16:2164–73

69. Moreau S, Ferguson JR, Symington LS. 1999. The nuclease activity of Mre11 is required for meiosis but not for mating type switching, end joining, or telomere maintenance. *Mol. Cell Biol.* 19:556–66

70. Moreau S, Morgan EA, Symington LS. 2001. Overlapping functions of the *Saccharomyces cerevisiae* Mre11, Exo1 and Rad27 nucleases in DNA metabolism. *Genetics* 159:1423–33

71. Mortensen UH, Bendixen C, Sunjevaric I, Rothstein R. 1996. DNA strand annealing is promoted by the yeast Rad52 protein. *Proc. Natl. Acad. Sci. USA* 93:10729–34

72. Myung K, Chen C, Kolodner RD. 2001. Multiple pathways cooperate in the suppression of genome instability in *Saccharomyces cerevisiae*. *Nature* 411:1073–76

73. Nick McElhinny SA, Ramsden DA. 2004. Sibling rivalry: competition between Pol X family members in V(D)J recombination and general double strand break repair. *Immunol. Rev.* 200:156–64

74. Ooi SL, Shoemaker DD, Boeke JD. 2001. A DNA microarray-based genetic screen for nonhomologous end-joining mutants in *Saccharomyces cerevisiae*. *Science* 294:2552–56

75. Pascal JM, O'Brien PJ, Tomkinson AE, Ellenberger T. 2004. Human DNA ligase I completely encircles and partially unwinds nicked DNA. *Nature* 432:473–78

76. Perrault R, Wang H, Wang M, Rosidi B, Iliakis G. 2004. Backup pathways of NHEJ are suppressed by DNA-PK. *J. Cell. Biochem.* 92:781–94

77. Schar P, Fasi M, Jessberger R. 2004. *SMC1* coordinates DNA double-strand break repair pathways. *Nucleic Acids Res.* 32:3921–29

77a. Shim EY, Ma JL, Oum JH, Yanez Y, Lee SE. 2005. The yeast chromatin remodeler RSC complex facilitates end joining repair of DNA double-strand breaks. *Mol. Cell Biol.* 25:3934–44

78. Shima H, Suzuki M, Shinohara M. 2005. Isolation and characterization of novel *xrs2* mutations in *Saccharomyces cerevisiae*. *Genetics* 170:71–85

Seminal paper giving an early description of imprecise yeast NHEJ.

With Reference 93 provided a systematic screen for yeast core NHEJ genes.

Crystal structure of DNA ligase I that gives the best current view of the moment of ligation.

79. Shiozaki EN, Gu L, Yan N, Shi Y. 2004. Structure of the BRCT repeats of BRCA1 bound to a BACH1 phosphopeptide: implications for signaling. *Mol. Cell* 14:405–12
80. Sibanda BL, Critchlow SE, Begun J, Pei XY, Jackson SP, et al. 2001. Crystal structure of an XRCC4-DNA ligase IV complex. *Nat. Struct. Biol.* 8:1015–19
81. Stracker TH, Theunissen JW, Morales M, Petrini JH. 2004. The Mre11 complex and the metabolism of chromosome breaks: the importance of communicating and holding things together. *DNA Repair* 3:845–54
82. Sugawara N, Ira G, Haber JE. 2000. DNA length dependence of the single-strand annealing pathway and the role of *Saccharomyces cerevisiae RAD59* in double-strand break repair. *Mol. Cell Biol.* 20:5300–9
83. Taddei A, Hediger F, Neumann FR, Bauer C, Gasser SM. 2004. Separation of silencing from perinuclear anchoring functions in yeast Ku80, Sir4 and Esc1 proteins. *EMBO J.* 23:1301–12
84. Teo SH, Jackson SP. 2000. Lif1p targets the DNA ligase Lig4p to sites of DNA double-strand breaks. *Curr. Biol.* 10:165–68
85. Trujillo KM, Roh DH, Chen L, Van Komen S, Tomkinson A, Sung P. 2003. Yeast Xrs2 binds DNA and helps target Rad50 and Mre11 to DNA ends. *J. Biol. Chem.* 278:48957–64
86. Trujillo KM, Sung P. 2001. DNA structure-specific nuclease activities in the *Saccharomyces cerevisiae* Rad50-Mre11 complex. *J. Biol. Chem.* 276:35458–64
87. Tseng HM, Tomkinson AE. 2002. A physical and functional interaction between yeast Pol4 and Dnl4/Lif1 links DNA synthesis and ligation in non-homologous end joining. *J. Biol. Chem.* 13:45630–37
88. Tseng HM, Tomkinson AE. 2004. Processing and joining of DNA ends coordinated by interactions among Dnl4/Lif1, Pol4, and FEN-1. *J. Biol. Chem.* 279:47580–88
89. Usui T, Ogawa H, Petrini JH. 2001. A DNA damage response pathway controlled by Tel1 and the Mre11 complex. *Mol. Cell* 7:1255–66
90. Valencia M, Bentele M, Vaze MB, Herrmann G, Kraus E, et al. 2001. *NEJ1* controls non-homologous end joining in *Saccharomyces cerevisiae*. *Nature* 414:666–69
91. van Attikum H, Fritsch O, Hohn B, Gasser SM. 2004. Recruitment of the INO80 complex by H2A phosphorylation links ATP-dependent chromatin remodeling with DNA double-strand break repair. *Cell* 119:777–88
92. **Walker JR, Corpina RA, Goldberg J. 2001. Structure of the Ku heterodimer bound to DNA and its implications for double-strand break repair. *Nature* 412:607–14**

A key study showing the ring structure of Ku bound to DNA.

93. Wilson TE. 2002. A genomics-based screen for yeast mutants with an altered recombination/end-joining repair ratio. *Genetics* 162:677–88
94. Wilson TE, Grawunder U, Lieber MR. 1997. Yeast DNA ligase IV mediates non-homologous DNA end joining. *Nature* 388:495–98
95. Wilson TE, Lieber MR. 1999. Efficient processing of DNA ends during yeast nonhomologous end joining. Evidence for a DNA polymerase β (Pol4)-dependent pathway. *J. Biol. Chem.* 274:23599–609
96. Wu X, Wilson TE, Lieber MR. 1999. A role for FEN-1 in nonhomologous DNA end joining: the order of strand annealing and nucleolytic processing events. *Proc. Natl. Acad. Sci. USA* 96:1303–8
97. Yu J, Marshall K, Yamaguchi M, Haber JE, Weil CF. 2004. Microhomology-dependent end joining and repair of transposon-induced DNA hairpins by host factors in *Saccharomyces cerevisiae*. *Mol. Cell Biol.* 24:1351–64

98. **Yu X, Gabriel A. 2004. Reciprocal translocations in *Saccharomyces cerevisiae* formed by nonhomologous end joining. *Genetics* 166:741–51**

Formal demonstration of NHEJ-catalyzed translocation at two simultaneous DSBs.

99. Zhang X, Morera S, Bates PA, Whitehead PC, Coffer AI, et al. 1998. Structure of an XRCC1 BRCT domain: a new protein-protein interaction module. *EMBO J.* 17:6404–11
100. Zou L, Elledge SJ. 2003. Sensing DNA damage through ATRIP recognition of RPA-ssDNA complexes. *Science* 300:1542–48

RELATED RESOURCES

Haber JE. 1998. Mating-type gene switching in *Saccharomyces cerevisiae.* *Annu. Rev. Genet.* 32:561–99

Plasmid Segregation Mechanisms

Gitte Ebersbach and Kenn Gerdes

Department of Biochemistry and Molecular Biology, University of Southern Denmark, DK-5230 Odense M, Denmark; email: kgerdes@bmb.sdu.dk

Annu. Rev. Genet. 2005. 39:453–79

The *Annual Review of Genetics* is online at http://genet.annualreviews.org

doi: 10.1146/annurev.genet.38.072902.091252

0066-4197/05/1215-0453$20.00

Key Words

actin, oscillation, ParM, ParA, ParB

Abstract

Bacterial plasmids encode partitioning (*par*) loci that ensure ordered plasmid segregation prior to cell division. *par* loci come in two types: those that encode actin-like ATPases and those that encode deviant Walker-type ATPases. ParM, the actin-like ATPase of plasmid R1, forms dynamic filaments that segregate plasmids paired at mid-cell to daughter cells. Like microtubules, ParM filaments exhibit dynamic instability (i.e., catastrophic decay) whose regulation is an important component of the DNA segregation process. The Walker box ParA ATPases are related to MinD and form highly dynamic, oscillating filaments that are required for the subcellular movement and positioning of plasmids. The role of the observed ATPase oscillation is not yet understood. However, we propose a simple model that couples plasmid segregation to ParA oscillation. The model is consistent with the observed movement and localization patterns of plasmid foci and does not require the involvement of plasmid-specific host-encoded factors.

Contents

INTRODUCTION

In prokaryotes, active segregation of low-copy-number plasmids into daughter cells relies on the function of partitioning (*par*) loci encoded by the plasmids themselves. The early discovery that *par* loci encode *trans*-acting proteins that act on *cis*-acting regions on the plasmids prompted investigators to designate these regions as centromere-like sites, by analogy to the centromeres of eukaryotic chromosomes. Recent discoveries that *par* loci encode cytoskeletal spindle-like structures that separate and distribute plasmids to daughter cells have extended the parallels between eukaryotic chromosome segregation and prokaryotic plasmid partitioning, and the proposed functional analogy between these processes has turned out to be even more justified than initially anticipated. In particular, the discovery that the Type II partitioning ATPase, ParM, forms actin-like filaments that segregate plasmids has greatly changed the view of the plasmid partitioning process. The unrelated and more common Type I *par* loci also encode filament-forming ATPases that play a central role in plasmid segregation. These filaments oscillate over the nucleoid. Here we review recent advances in the mechanistic understanding of how plasmid-encoded partitioning loci secure ordered segregation of their replicons and compare them to chromosome-encoded partitioning loci. We also present a working model that explains how oscillating, filament-forming proteins might mediate plasmid segregation.

FACTORS AND GENES INFLUENCING PLASMID MAINTENANCE

Several factors influence the genetic stability of prokaryotic plasmids. Obviously, one important factor is copy-number control. The copy-number-control systems in natural plasmids ensure that each plasmid, on average, replicates once per cell cycle. In general, high-copy-number plasmids are relatively small and their genetic stability is thought to rely on random (i.e., binomial) segregation of the individual plasmid molecules at cell division. However, not all high-copy-number plasmids are as stable as would be expected from random assortment. This is usually attributed to factors that reduce the number of segregating units, such as fluctuations in plasmid copy-number in individual cells, plasmid clustering (discussed below), or plasmid multimerization arising from homologous recombination (101a, 119, 135). Accordingly, most natural plasmids encode site-specific resolution systems that resolve plasmid multimers into monomers (5, 19, 87, 136).

In general, low-copy-number plasmids are larger than high-copy-number plasmids, a relationship that may help to minimize the metabolic burden imposed by the plasmids on their host cells. Obviously, a low-copy-number raises another problem: low-copy-number plasmids cannot rely on random segregation to ensure stable inheritance. Accordingly, plasmids have evolved true partitioning (*par*) loci that actively segregate plasmid copies to daughter cells before cell division. In addition, many low-copy-number plasmids encode so-called post-segregational killing (PSK) systems that increase plasmid maintenance by killing plasmid-free cells. These systems may be viewed as providing a backup stabilization mechanism of last resort that is executed when a true *par* locus fails to function properly (44, 71)].

PSK: post-segregational killing

PLASMID PARTITIONING LOCI

All known plasmid-encoded partitioning loci encode two *trans*-acting proteins expressed from an operon and one or more *cis*-acting centromere-like sites, at which the proteins act (36, 45, 58, 116). All three components of a *par* locus, i.e., two proteins and a centromere-like *cis*-acting site, are essential (1, 20, 36, 57, 116). Furthermore, the amounts of the partitioning proteins need to be carefully regulated as excess, or shortage, of either partitioning protein is detrimental to *par* function (20, 36, 38, 85, 91). Accordingly, transcription of *par* operons is tightly autoregulated by the *par* proteins themselves (36, 58, 60, 70). Generally, *par* loci function as cassettes independently of the replicon on which they reside (2, 28, 45, 116).

The first gene of a *par* operon encodes an ATPase that belongs to one of two different superfamilies of proteins. This property was used to divide *par* loci into two types (46). Thus, Type I loci encode Walker box ATPases, and Type II loci encode actin-like ATPases (14, 81). Members of each of the two ATPase families form filamentous structures central to plasmid partitioning (8, 27, 108).

The second gene of a *par* operon encodes a DNA-binding protein that recognizes varying numbers of direct or inverted repeats within a cognate centromere-like site (23, 74, 107, 110). Binding of the protein to these sites results in the formation of a nucleoprotein complex, also known as the partitioning complex. The partitioning complex is the substrate for plasmid segregation, in which replicated plasmid molecules, often located at the mid-cell position, are moved in opposite directions. There is evidence that the partitioning complex pairs plasmid molecules in a process analogous to pairing of eukaryotic chromosomes before segregation occurs (29, 74).

Generally, plasmids harbor only one *par* locus, and Type I loci are by far the most common. However, plasmids carrying two *par* loci are known: *Escherichia coli* plasmid pB171 and *Salmonella enterica* plasmid R27 both carry two functional *par* loci, one of each type (26, 88).

TYPE II PARTITIONING LOCI

The best investigated Type II *par* locus is encoded by plasmid R1 (see **Figure 1**). Here, the ATPase was designated ParM (motor), whereas the DNA-binding protein was called ParR (repressor) (20). The *cis*-acting site, *parC* (centromere), consists of ten 11-bp direct repeats located in two clusters of five repeats on either side of the −10 and −35 sequences of the *par* promoter (20). All ten direct repeats were required to obtain maximal *parC*-mediated plasmid stabilization (18). Gel-shift and footprinting analyses showed that ParR binds cooperatively to the ten direct repeats within *parC*, forming a large nucleoprotein-complex, which also includes the promoter sequences located between the two sets of repeats (107). Consistent with these findings, ParR is the main regulator of *par* promoter activity (70). ParM is not involved in regulating *par* operon transcription (70).

ParR-MEDIATED PLASMID PAIRING

ParR binds to *parC* in vivo and in vitro (20, 70, 107). The specific interaction of ParR with *parC* was investigated by electron microscopy (74). When supercoiled *parC*-containing DNA was mixed with purified ParR, binding of ParR resulted in a shortening of the DNA fragment, indicating that the *parC* DNA was wrapped around a core of ParR. Most importantly, ParR mediated *parC*-dependent pairing of plasmid molecules in vitro. The frequency of pairing was highest with supercoiled plasmid DNA, and the presence of ParM and ATP increased the pairing frequency. A ligation kinetics assay showed that ParR-dependent pairing of DNA molecules at *parC* was nonrandom with respect to orientation. Hence, the DNA fragments appeared to be ligated in a head-to-head fashion, suggesting that oriented pairing might yield the directionality required for ordered plasmid segregation (74).

THE ATPase ACTIVITY OF ParM

ParM has moderate, cooperative ATPase activity (that is, its specific activity increases with increasing concentrations of ParM). In addition, the ATPase activity is stimulated by ParR bound to *parC* DNA (72, 108). Mutations that changed the conserved Asp170 in the ATPase site of ParM (14) simultaneously reduced the ParM ATPase activity and the ability of ParM to support plasmid partitioning (72).

ParM FORMS DYNAMIC FILAMENTS

Cytological studies employing immunofluorescence microscopy (IFM) of ParM revealed a striking pattern: In some cells, ParM formed long, curved filaments extending from one pole to the other, or two filaments extending from the middle of the cell to opposite poles. In other cells, ParM appeared as foci, and in yet another subpopulation of cells, the ParM signal was diffusively located throughout the cell. These results show that ParM forms dynamic polymers that alternate between phases of polymerization and depolymerization. Estimation of the number of ParM molecules per cell (15,000–18,000) furthermore suggested that each filament observed by IFM most likely consisted of several parallel protofilaments (108). The formation of dynamic ParM filaments depended on the presence of both ParR and *parC*. Thus, even massive overproduction of

Type I *par* loci

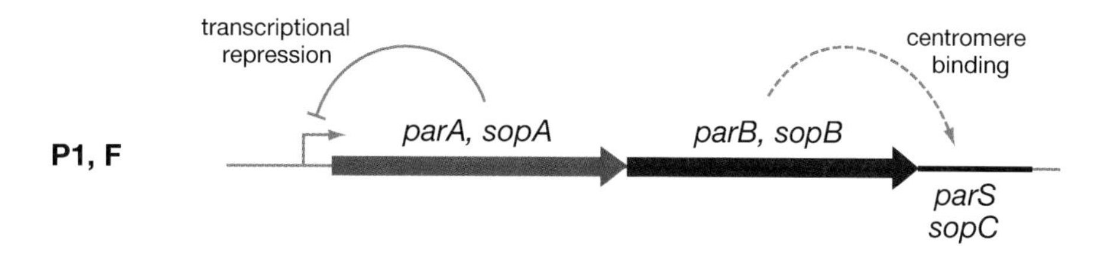

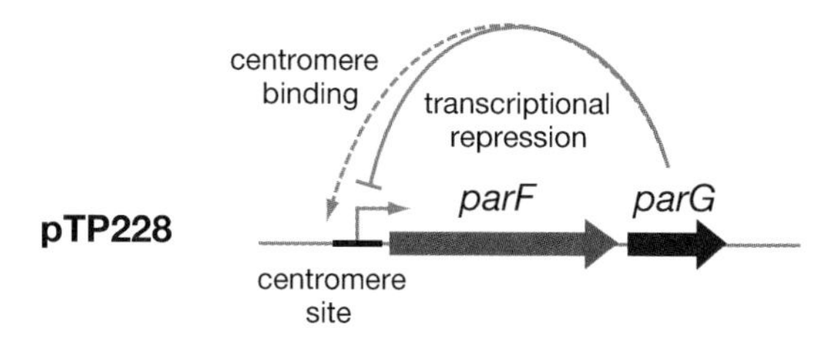

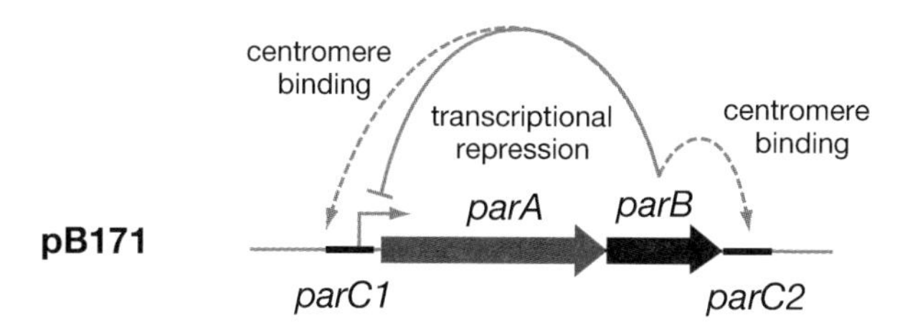

Type II *par* loci

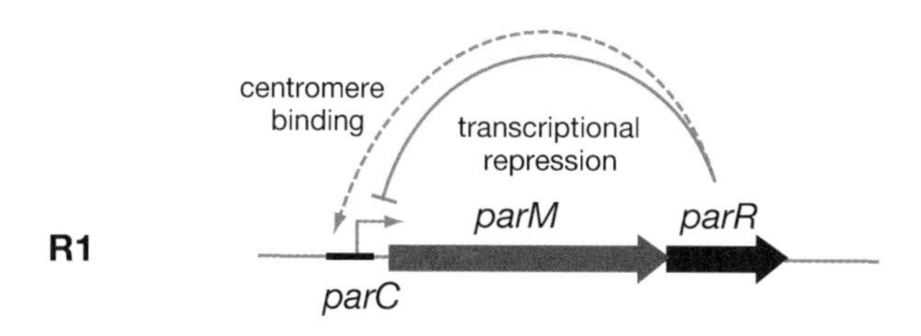

Figure 1

Genetic organization of representative Type I (plasmids P1, F, pTP228 and pB171) and Type II (plasmid R1) *par* loci, respectively. Solid arrows represent genes that encode ATPases (*red*) and centromere-binding proteins (*blue*), respectively. Centromere-like sites are shown as black bars. Arcs indicate DNA-binding properties of *par* gene products: solid arcs, regulation of promoter activity; dashed arcs, formation of partitioning complex.

ParM alone did not yield detectable formation of filaments (108). In contrast, the ATPase and partition-deficient ParM mutant proteins described above formed straight, rod-like filaments along the long axis of almost all cells. This "hyper-filamentation" phenotype exhibited by the ParM mutant proteins did not depend on ParR and *parC* and showed that the ATPase activity of ParM is not required for filament formation per se; however, it is important for ParM dynamics. The detrimental ParM ATPase mutants were *trans*-dominant, indicating that the ParM mutant proteins interfered with proper function (i.e., filamentation) of wild-type ParM, most likely by protein-protein contact.

ParM SHARES THE ACTIN FOLD

ParM belongs to a superfamily of ATPases that, besides eukaryotic actins, comprises a number of diverse proteins such as heat shock proteins (Hsc70), sugar kinases, and prokaryotic cell cycle proteins (FtsA and MreB). Despite the relatively low overall sequence identity among these proteins, they all exhibit similarity in five separate, colinear sequence motifs. In the folded proteins, these motifs together form the characteristic three-dimensional structure of actin that binds and hydrolyzes ATP (14). The crystal structure of ParM revealed a three-dimensional architecture very similar to that of actin (140). Moreover, like actin, ParM forms double helical protofilaments that twist gently around each other (140).

ParM EXHIBITS DYNAMIC INSTABILITY

When eukaryotic cells enter mitosis, microtubules are reorganized into the mitotic spindle and attach to the kinetochores of paired sister chromatids, thereby promoting their alignment and subsequent partition into daughter cells. Before the spindle apparatus is assembled, tubulin polymers switch between phases of elongation and rapid shortening, a property known as dynamic instability (104). Dynamic instability plays an important role in the search and capture of chromosomes for correct bipolar alignment (80, 105). Recently, in vitro studies of ParM polymerization showed that ParM exhibits a very similar dynamic instability (43). However, unlike tubulin and actin filaments, ParM filaments appeared to grow symmetrically with equal rates of assembly at the two filament ends. After some time, filament growth switched to rapid unidirectional decay. The switch from growing to shortening appeared to be stochastic, and once disassembly of the filaments had begun, the disintegration process proceeded to completion. The dynamic instability of ParM filaments was linked to the ability of ParM to hydrolyze ATP as ParM proteins carrying mutations in their ATP-binding site formed stable filaments. Moreover, ADP-ParM filaments were extremely unstable, whereas ATP-γ-S-ParM filaments were highly stable (43). These findings are consistent with the previous immunolocalization experiments in which ParM ATPase mutants exhibited a hyperfilamentation phenotype (108). In mixing experiments, mutant ATP-bound ParM proteins stabilized wild-type ParM filaments, indicating that the filaments are stable as long as an ATP-bound cap of ParM is present at both filament ends (43).

MODEL FOR *parMRC*-MEDIATED PLASMID SEGREGATION

In actin filaments, nucleation as well as depolymerization and nucleotide exchange is regulated by a number of exogenous factors (131). For ParM, very fast rates of nucleation, filament disintegration, and dissociation of ADP from ParM monomers suggest that such regulatory factors might not be needed (43). Furthermore, the fact that the steady-state critical concentration (2.3 μM) needed for ATP-ParM polymerization appears to be far below the intracellular concentration (12 to 14 μM) of ParM suggests that polymerization does not require a nucleation factor (43). Based on these and previous findings, Dyche Mullins and co-workers proposed a model for R1 partitioning, according to which ParM filaments form spontaneously within the cell. ATP hydrolysis inside the ParM polymers leads to destabilization of the filaments that eventually disintegrate.

However, when a growing filament encounters a set of paired plasmids, the ParR-*parC* complex, present on each plasmid molecule, acts to stabilize both ends of the ParM filament. In this way, the dynamic instability enables ParM to find and capture plasmids paired by ParR, analogous to the way in which eukaryotic tubulin probes through the cell to locate and bind kinetochores of unattached chromosomes. Once plasmids have been captured, insertional polymerization of ATP-bound ParM monomers at the ParM-ParR interface causes the ParM filaments to grow bidirectionally, thereby pushing the two plasmids apart toward opposite ends of the cell. This model is consistent with cytological observations showing that ParM filaments always have ParB/*parC* carrying plasmids at their ends (107).

MreB, ANOTHER BACTERIAL ACTIN HOMOLOG, IS REQUIRED FOR CHROMOSOME SEGREGATION

Almost all rod-shaped bacteria encode one or more actin-like homologs that are required to maintain cell shape (76). Thus, in the three model organisms *E. coli*, *Bacillus subtilis* and *Caulobacter crescentus*, MreB is essential in that its depletion leads to the formation of spherical cells and eventually cell lysis (33, 76, 83). MreB assembles into filaments that are located beneath the inner cell membrane (76, 84). The first indication that MreB might be involved in chromosome segregation came from the observation that in *E. coli*, overproduction of ATPase-defective MreB mutant proteins prevented nucleoid separation and severely perturbed localization of the origin and terminus regions of the chromosome (84). More recently, MreB of *C. crescentus* was found to interact, directly or indirectly, with a region close to the origin of replication (47). Inactivation of MreB prevented normal segregation of this region of the chromosome. Whether origin movement is driven by dynamics within the polymerized MreB spirals, reminiscent of the proposed mechanism of ParM-mediated plasmid segregation, or whether motor proteins use MreB cables as a track for pulling the newly replicated origin regions still needs to be elucidated.

IHF: integration host factor

TYPE I PARTITIONING LOCI

Type I partitioning ATPases belong to a family of ATPases that have a so-called deviant Walker A box (also called P-loop) motif (81). This family of Walker box proteins also includes the MinD cell division proteins. Phylogenetic investigations of the deviant Walker-type ATPases involved in plasmid partitioning resulted in a sub-division of the proteins into Type Ia and Ib (46). The Type Ia subgroup is represented by two of the best-studied partitioning ATPases, namely ParA from prophage P1 and SopA from plasmid F. The Type Ib subgroup includes ParF from *Salmonella newport* plasmid pTP228, ParA from *Agrobacterium tumefaciens* plasmid pTAR, and ParA from *E. coli* plasmid pB171 (46, 57). Comparison of protein sequences and genetic organization of *par* loci encoding Type Ia and Type Ib ATPases, respectively, revealed several differences, which are summarized in **Table 1**. The organization of Type Ib loci appears to be more reminiscent of Type II loci than of Type Ia loci, an observation that may reflect convergent evolution at the level of gene organization (see **Figure 1**) (46).

THE PARTITIONING COMPLEX

The *par* locus of P1 encodes ParA and ParB and the *cis*-acting site *parS* (see **Figure 1**) (3). *parS* contains four heptameric (box A) and two hexameric (box B) sequences to which ParB binds (41). The ParB-binding sites are arranged asymmetrically on either side of a 29-bp DNA sequence recognized by the small heat-stable protein integration host factor (IHF) (40, 41). A minimal *parS* site capable of stabilizing a P1 plasmid in the presence of ParA and ParB consists of no more

Table 1 Comparison between Type Ia and Type Ib *par* loci

	Type Ia	Type Ib
A proteins	Size: 321–420 aa Contain DNA-binding HTH-motif in the N terminus Main transcriptional regulator of the *par* operon	Size: 192–308 aa Do not contain HTH-motif
B proteins	Size: 312–342 aa Show mutual homology	Size: 46–131 aa Diverse group of proteins Few homologues Main transcriptional regulator of the *par* operon
Cis-acting[a] site	Located downstream of the *par* operon	Located upstream of the *par* operon
Model systems	*parABS* of P1 *sopABC* of F	*par2* of pB171 *parFG* of pTP228 *parAB* of pTAR

[a]Some *par* loci contain more than one *cis*-acting region, e.g., *par2* of pB171 has two *cis*-acting sites (see **Figure 1**) (26, 46). Furthermore, the *E. coli* linear plasmid N15 encodes a Type Ia *par* locus very similar to the *sop* locus of F but lacks the *cis*-acting site located downstream of the *sopAB* operon in F. Instead, N15 contains 4 *sopC*-like inverted repeats, scattered on 13 kb of the N15 sequence (54, 127). Plasmid RK2 is another example of a plasmid that might encode a Type Ia *par* system with multiple *cis*-acting sites as KorB (the ParB homologue) recognizes a sequence repeated 12 times in the RK2 genome. However, most of the 12 sites might not function primarily as partitioning sites in that KorB also acts as a global transcriptional repressor (145).

HTH: helix-turn-helix

than 34 bp, containing only the ParB-binding sites to the right of the IHF site (103). Thus, IHF is not essential for *par* function, but binding of IHF to *parS* greatly increases the affinity of ParB for *parS* (39, 40). ParB interacts with *parS* in the form of a dimer. Dimerization is mediated by a domain located in the C terminus of the protein (99, 137). Binding to DNA is most likely mediated by a helix-turn-helix (HTH) motif situated in the central part of the linear sequence of ParB. The ParB HTH-motif is believed to bind to the box A sequences of *parS* (99, 137). ParB has a second DNA-binding domain that overlaps with the dimerization domain located in the C terminus. The ParB-ParB dimerization interface recognizes the two box B sequences in *parS* located on both sites of the IHF-binding site (99, 123, 138). Thus, formation of the ParB-*parS* partitioning complex probably initiates with one dimer of ParB interacting simultaneously with its recognition sequences on both sites of an IHF-induced bend in the DNA (17, 138). Subsequently, more ParB dimers are loaded onto the complex. These additional dimers are thought to bind primarily by protein-protein interactions, probably via an additional self-interaction domain located in the N-terminal part of ParB, and by nonspecific DNA-protein interactions. The end result is a higher-order structure in which *parS* is wrapped around a core of IHF and ParB (17, 137, 138).

The *sop* (stability of plasmid) locus of F encodes SopA and SopB and the *cis*-acting site, *sopC* (see **Figure 1**) (116). *sopC* consists of twelve 43-bp direct repeats, each containing a pair of 7-bp inverted repeats that are recognized by SopB (109, 110). A single 43-bp repeat is sufficient to allow stable maintenance of a mini-F plasmid (12). Like ParB, SopB binds to DNA via its HTH-motif (55, 110, 128). Alignment of the amino acid sequences of ParB/SopB homologous proteins from a number of Type Ia *par* loci including ParB of prophage P7, KorB of the broad host range plasmid RK2, and ParB of *Salmonella typhimurium* plasmid pSLT has shown that despite an overall low sequence similarity, these proteins all contain a putative HTH-domain, indicating that these proteins might bind their *cis*-acting sites in a way analogous to ParB and SopB (55, 99).

In F, plasmid supercoiling is affected by the partitioning complex (12, 91). Thus, a single *sopC* repeat significantly increases the linking number of a mini-F plasmid in the presence of high concentrations of SopB in vivo (100). As the change in linking number is too large to originate solely from structural changes within the *sopC* repeat, it was suggested that binding of SopB to *sopC* nucleates the formation of a higher-order wrapped nucleoprotein complex that involves the DNA adjacent to *sopC* and protein-protein interactions between SopB proteins (12, 100). Using fluorescence microscopy, the subcellular localization of SopB was examined. One study used antibodies against SopB (60), while another used a SopB-GFP fusion protein (78). Both studies reported the presence of SopB foci; however, in the former study, formation and localization of SopB foci required SopA and *sopC*, whereas in the latter study, SopB localization seemed to be independent of the presence of these components.

In the case of the Type Ib *par* loci of pB171, pTP228, and pTAR, the B proteins bind to *cis*-acting regions located upstream of their cognate *par* operons. Like ParB and SopB, the B proteins of the Type Ib loci form dimers in solution. However, they do not contain HTH-motifs (7, 35). The structure of the ParG dimer from plasmid pTP228 was solved (50). The ParG dimer consists of a folded domain made up of the intertwined C-terminal regions of each ParG subunit and a flexible domain consisting of the unstructured N-terminal regions. The folded C-terminal part of the dimer, which forms the putative DNA-binding domain of ParG, has a ribbon-helix-helix (RHH) architecture similar to the well-described transcriptional repressors of the Arc/MetJ/CopG superfamily (51, 126).

GFP: green fluorescent protein

RHH: ribbon-helix-helix

LONG-DISTANCE GENE SILENCING BY ParB/*parS* OF P1

Specific binding of P1 ParB to *parS* can function as a nucleation site for unspecific binding of ParB to several kilobases of DNA on both sides of *parS*, a phenomenon called spreading (129). A similar behavior has been reported for the chromosome-encoded ParB homologue of *Streptomyces coelicolor*, suggesting that spreading might be a characteristic of many of these DNA-binding proteins (69). The role of ParB spreading in partitioning is not clear. Initially, analysis of selected spreading-defective P1 ParB mutants revealed that these mutants were also partition defective, and thus led to the suggestion that spreading might play a role in plasmid segregation (129). Nevertheless, a more recent study, in which plasmid stability was assayed after ParB spreading had been restricted by the introduction of roadblocks on both sites of *parS* (RepA or GAL4 bound to their cognate recognition sites), indicated that even though limited spreading might assist the partition process, perhaps during plasmid pairing, extensive ParB spreading is probably not essential (130).

Spreading of ParB results in silencing of genes several or even many kilobases away from *parS*, presumably because the formation of a ParB-DNA nucleoprotein filament impedes binding of RNA polymerase to the region covered by the nucleofilaments (129). Similarly, SopB of F can silence genes located far away from *sopC*. Simultaneously, the DNA becomes inaccessible to cellular proteins such as DNA gyrase and DNA adenosine methylase (101). However, in the case of F, it is not known if SopB forms a nucleoprotein filament and SopB does not polymerize on DNA outside *sopC* in vitro (55). In the F case, gene silencing was explained by SopB-mediated sequestration of the *sopC* region to specific

aa: amino acids

subcellular positions near the cell poles (79, 101). This proposal is consistent with the finding that the N-terminal region of SopB, which apparently is involved in specific subcellular localization of the protein, is essential for SopB-mediated gene silencing (55, 78, 79).

PLASMID POSITIONING BY TYPE I *par* LOCI

Plasmids without a *par* locus are more or less randomly localized within the bacterial cell, preferably at positions not occupied by the nucleoid (27, 96, 111). In contrast, the presence of a Type I *par* locus leads to specific subcellular localization of plasmids to mid-cell and quarter-cell positions (13, 27, 52, 59, 61, 88, 95, 96, 111). The plasmid localization pattern is consistent with a model in which plasmids are replicated at mid-cell, separated and actively moved to quarter-cell positions. However, there is some controversy regarding the timing of the plasmid separation process. In one study, synchronized replication of a mini-F plasmid showed that sister plasmid copies could be separated and moved to quarter-cell positions within 5 min after replication had completed, independently of cell division (117). In two other studies in which time-lapse microscopy was used to show the dynamic movements of P1 plasmids and plasmids carrying *par2* of pB171, respectively, plasmid foci positioned at mid-cell usually did not segregate until very late in the cell cycle (27, 95). The apparent temporal coupling of plasmid segregation and cell division observed in the latter studies raises the possibility that plasmid partitioning and cell division might be co-ordinated. In the P1 study, it was suggested that plasmids might be capable of delaying cell division until segregation is complete (95). However, several studies have shown that plasmids carrying Type I *par* loci are regularly distributed along the long axis of filamentous cells in which cell division has been inhibited by treatment with cephalexin (27, 30, 52). The latter observations indicate that plasmid segregation does not depend on completion of cell division.

ROLES OF DEVIANT WALKER-TYPE ATPases IN PLASMID PARTITIONING

Promoter Regulation

Type Ia ATPases function both in plasmid partitioning and in repression of *par* operon transcription (**Figure 1**). ParA of P1 and SopA of F bind to operator sites in their cognate promoter regions via HTH-motifs in their N termini (24, 58, 110, 128). In the case of P1 ParA, ATP and ADP both promote ParA dimerization and it is the dimer that binds operator DNA (21). However, ADP and non-hydrolyzable ATP analogs stimulated operator binding much more than ATP (15). ParA proteins with aa changes in the Walker box motifs that reduced ATPase activity behaved as superrepressors of transcription (37). These results support the notion that the catalytic event itself triggers a conformational change in ParA that is deleterious to DNA binding.

In vivo, ParB and SopB greatly enhance transcriptional repression by ParA and SopA (36, 60), respectively, in agreement with in vitro data showing that ParB/SopB stimulates ParA/SopA operator DNA binding (22, 110). In vivo, *parS* and *sopC* act as ParB/SopB dependent corepressors of their cognate *par* promoters, and deletion of *parS/sopC* in both cases leads to significant increases in *par* operon transcription. *parS/sopC* also act as co-repressors in *trans*, implying that transcriptional repression is not due to spreading of ParB/SopB from the centromere site (56, 148). In support of this notion, Hao & Yarmolinsky found that silencing of the *par* operon by spreading of ParB from *parS* in its natural position downstream of ParB probably does not play any major role in autoregulation of the *par* promoter (56). The fact that a single *parS* site in the chromosome was able to repress the *par* promoter carried by a high-copy-number reporter plasmid led to the suggestion that *parS* might act catalytically by stimulating ATP hydrolysis and thereby induce a conformational change in free ParA that would make it a better repressor (56).

For F, it was suggested that the *sopC*-mediated enhancement of SopB co-repressor activity could be the result of a direct contact between the SopB-*sopC* partitioning complex and the SopA protein bound at the *par* promoter (148). Such an interaction might involve intramolecular looping of plasmid DNA or alternatively, the SopB-*sopC* complex on one plasmid might interact with ParA bound on another plasmid in an intermolecular looping reaction. This would agree with the proposal that pairing of plasmids is an important step in the partitioning process.

Separation of Plasmids Paired Via Their Partitioning Complexes

Austin & Abeles were the first to suggest a model for P1 partitioning involving pairing of plasmids at *parS* (3). Excess ParB protein destabilizes a P1 plasmid carrying either the entire *par* locus or the *parS* site (38). Plasmid destabilization was more severe than would be expected from random distribution of individual plasmids at cell division, consistent with the proposal that the plasmids segregated in pairs or even in clusters. Recently, the question of plasmid pairing was addressed by in vivo experiments in which the topoisomer distribution of *parS*-containing plasmids was analyzed in the presence and absence of ParB. The results were consistent with ParB-mediated plasmid pairing at *parS*; however, they did not prove that pairing actually occurs (29).

A series of experiments, performed with different *par* loci, raised the possibility that the Walker Box ATPases might play a role in the separation of plasmid molecules paired at their centromere sites. In vitro experiments performed with the components of the P1 *par* locus have shown that at high ParA/ParB ratios, ParA destabilizes the partitioning complex formed at *parS* (15). A similar observation was made in the case of the Type Ib *par* locus of pTP228 (7). Moreover, ectopic overexpression of ParA has been shown to destabilize a P1 plasmid (1). The same was true for plasmid F, in which case an excess of SopA interfered with plasmid partitioning. Investigation of this phenomenon revealed that excess SopA is capable of counteracting the increase in plasmid linking-number normally induced by binding of SopB to *sopC*, suggesting that SopA disrupts the partitioning complex (91). The SopA effect required ATP-binding and/or hydrolysis and is probably mediated by direct interaction of SopA with SopB. In another study, overproduction of a mutant SopA protein, carrying an aa substitution of a conserved lysine in the ATP-binding site, apparently had no effect on the linking number of a mini-F plasmid (97). Moreover, in mini-F plasmid stability tests, expression of the mutant SopA protein instead of wild-type SopA resulted in a hyper-instability phenotype. One possible interpretation of these data is that the mutant SopA protein fails to separate F plasmids paired via their SopB/*sopC* complex. Consequently, the plasmids cannot segregate as independent units and severe plasmid instability follows (97). Certain aa substitutions in the Walker A box of ParA of P1 also conferred a hyper-instability phenotype (37).

Another clue that the Type I ATPases might function in separation of plasmids during partitioning came from cytological studies. In the case of the P1 plasmid, absence of ParA resulted in a decreased number of cells with a plasmid focus at mid-cell, and foci localized to the cell center failed to divide before cell division (96). Further support for this contention came from a study of aa changes in the Walker A box of ParA of pB171 that rendered the protein defective in plasmid stabilization. The mutant protein still mediated mid-cell localization of plasmid foci in a significant number of cells but the foci often failed to divide, thus explaining why plasmids carrying this mutation were unstable (27). Finally, inactivation of the IncC ATPase of plasmid RK2 resulted in a decrease in the number of plasmid foci that could be observed by IFM with KorB-specific antibodies, consistent with impaired plasmid separation (13).

Segregation of Plasmids into Daughter Cells

The partitioning ATPases studied so far all have a weak intrinsic Mg^{2+}-dependent ATPase activity that is stimulated by DNA and even more so by centromere DNA bound with the cognate B protein (8, 24, 141). Mutations in the ATPase domains of the ParAs of P1 and pB171 and ParF of pTP228 revealed a correlation between the ability of these proteins to hydrolyze ATP and to mediate plasmid stability, implying that the ATPase activity plays a role in the DNA segregation process (8, 26, 37).

Recent cytological observations showed that the Walker-type ATPases are required for specific positioning of plasmids at subcellular locations (27, 30, 96). Moreover, experiments with ParA of pB171 revealed that a ParA-GFP fusion protein oscillated in spiral-shaped structures over the nucleoid region (26, 27). Oscillation, but not spiral formation, required ParB and *parC*, showing that the partitioning complex regulates the dynamic properties of ParA. Change of a conserved amino acid residue in the Walker A box motif revealed a correlation between ParA oscillation, subcellular plasmid positioning, and plasmid stability, suggesting a role for ParA oscillation in plasmid segregation (26, 27). The dynamic movement of ParA ATPases in filamentous structures led to the proposal that the ATPases play a direct role in plasmid segregation, perhaps by providing the motive force required for active plasmid segregation.

In vitro, ParF of pTP228 forms extensive filaments very similar to those formed by MinD of *E. coli*. This is important because MinD of *E. coli* also oscillates in filamentous structures in vivo (132). Both ParF and MinD form bundles of parallel protofilaments in which one end is compact and the other end has a frayed appearance consistent with filament polarity (8, 134). ParF polymerization was stimulated by ATP but suppressed by ADP. Non-hydrolyzable ATP analogues had a stimulating effect similar to ATP, indicating that ATP hydrolysis is not required for polymerization. Moreover, ParG, which stimulates the ATPase activity of ParF, was shown to associate with ParF filaments and regulate polymerization: At low ParG/ParF ratios, ParG appeared to enhance polymerization and filament bundling, whereas high ParG concentrations resulted in decreased ParF polymerization. These observations suggest that ParG mediates its effect on ParF polymerization through stimulation of the ATPase activity of ParF. Analysis of partition-deficient ParF Walker A box mutants showed that these proteins had an altered filamentation pattern (8). Based on the observations described above, it is tempting to speculate that dynamic oscillation may be a general property of Walker box partitioning proteins.

HOST FACTORS AND PLASMID PARTITIONING

Several elegant genetic screens have been devised to search for host cell-encoded factors involved in plasmid partitioning (11, 66, 112, 133). So far, such screens have led to the identification of factors involved in recombination and plasmid replication, but not in plasmid segregation. IHF is the only host factor known to play a direct role in plasmid partitioning (39, 40). Initially, it was suggested that the process of plasmid segregation itself might be carried out by binding of plasmids to the growing plasma membrane or to the bacterial chromosome (34, 68). However, the rapid subcellular movement of plasmids (27, 95, 117) and the fact that both F and P1 plasmids segregate into anucleate cells of *mukB* mutant cells argue against these ideas (32, 42). The alternative suggestion that plasmids might segregate by direct attachment to a cytoskeletal machinery encoded by the host also seems unlikely, especially in light of more recent findings that partitioning ATPases of both types have the ability to form dynamic, filamentous structures (27, 107, 140).

Host factors have also been proposed to participate in maintaining or tethering of plasmids at distinct cellular locations. The similar localization patterns of plasmids carrying Type I *par* loci seem to support this idea. However, in cytological studies in which F plasmids were detected simultaneously with P1 plasmids or RK2 plasmids (61) or with an R1 plasmid stabilized by *par2* (28), the different plasmid foci occupied similar but clearly distinct positions. For these observations to agree with the receptor hypothesis requires that each *par* locus should have its own receptor. Furthermore, *par2* distributes plasmid foci along the long axis of the cell irrespective of the focus number (28). Thus, as many as 9 plasmid foci were more or less regularly distributed along the long axis of the cell, an observation that is most readily reconciled with a receptor-independent plasmid localization mechanism.

A SIMPLE MODEL THAT COUPLES PLASMID SEGREGATION TO ParA OSCILLATION

It is not yet known how the oscillating, filament-forming Walker box ATPases mediate plasmid segregation. However, in analogy to the Type II partitioning loci, it is tempting to suggest that polymerization of ParA acts to physically push or pull plasmid molecules apart. In the "jumping ParM model" shown in **Figure 2*a***, we envisage that ParA polymerizes between two adjacent plasmid molecules (or clusters of molecules) and pushes them apart in a mechanism similar to that of ParM of plasmid R1. At a particular moment, ParA concentration will be highest between two foci, say F1 and F2. These two foci will be pushed apart due to polymerization of ParA that interacts with the ParB/*parC* complexes of F1 and F2. Due to oscillation of ParA, ParA concentration at a later moment is highest between F2 and F3 that now will be pushed apart. Thus, the oscillation of ParA functions to average, over the cell cycle, the force exerted by ParA filaments between any two given juxtaposed plasmid foci such that, over a time period of a full oscillation cycle, all plasmid foci will experience a force that distributes them along the length of the cell. The model assumes that ParA oscillates over the nucleoid and that the polar borders of the nucleoid function as toeholds for ParA. The model predicts that, in cells with one plasmid focus, the focus will end up at mid-cell and in cells with two foci, the foci will end up approximately at quarter-cell positions (**Figure 2*b***).

All cytological observations obtained with *par2* of pB171 are consistent with the model. ParA of pB171 oscillates over the nucleoid in filamentous structures (27). Oscillation depends on ParB bound to *parC*, meaning that ParA "senses" the presence of plasmid substrates for the segregation process (26). Most important, in cells with many plasmid foci, the foci are evenly distributed along the length of the cells (28), as predicted by the model. The model explains in a simple way how ParA oscillation might lead to plasmid positioning. Moreover, the model predicts that even though *par*-carrying plasmids are found at mid-cell or quarter-cell positions there is no requirement for host cell receptors to tether the plasmids at those positions: they simply end up there due to oscillation of ParA. The lack of specific host receptors at those positions is also consistent with the observation that two compatible plasmids carrying different Type I *par* loci localize at similar subcellular positions but do not exhibit colocalization (that is, the plasmids are detected as distinct nonoverlapping foci at those positions) (28, 120). Moreover, the jumping ParM model is consistent with the puzzling observation that chromosome-encoded Type I *par* loci from distantly related organisms (i.e., *B. subtilis*, *Pseudomonas aeruginosa*, and *Pseudomonas putida*) can stabilize plasmids replicating in *E. coli* (9, 49, 147). The observation that the Type I *par* locus from *B. subtilis*, which encodes the

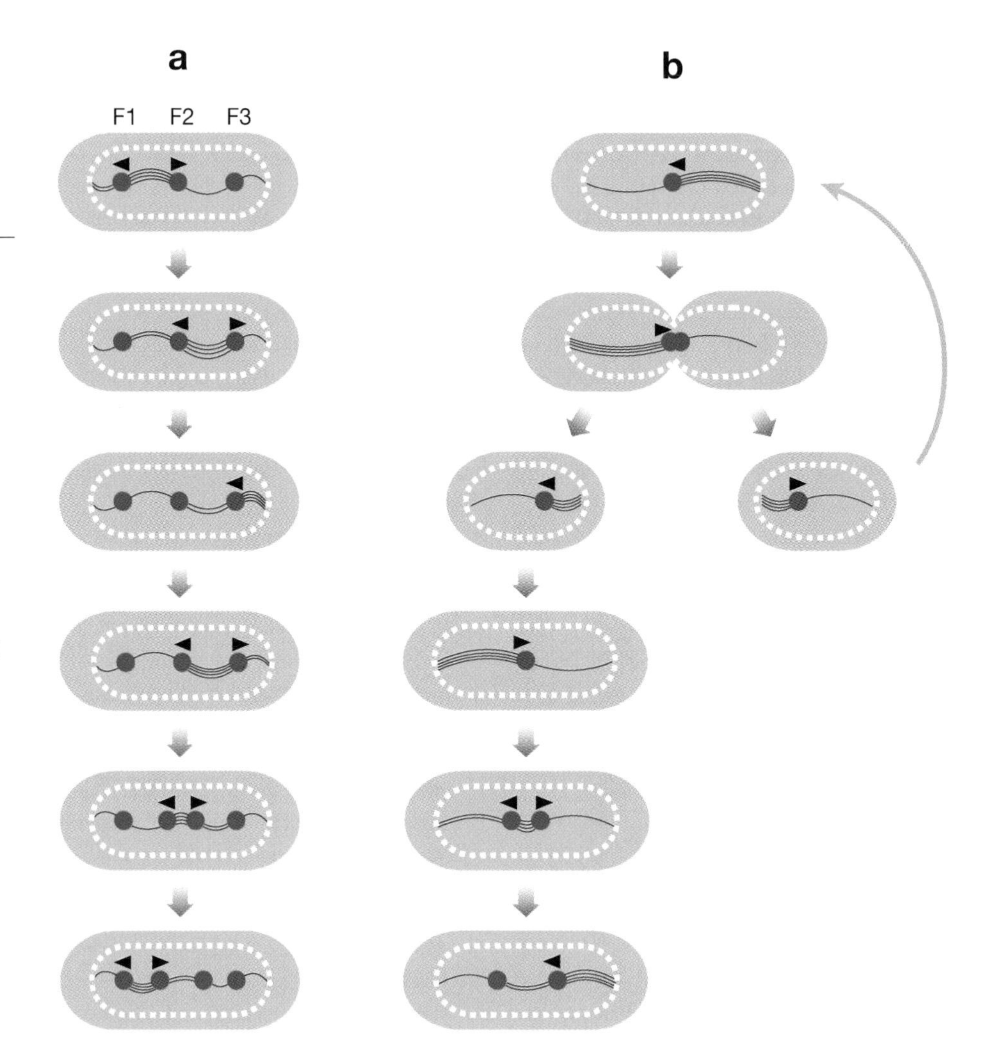

Figure 2

Model that explains the coupling of plasmid segregation to ParA oscillation. (*a*) Schematic showing the "jumping ParM model" that explains how oscillating ParA filaments might confer a regular plasmid focus pattern as observed (28). See the text for details. (*b*) Schematic showing how ParA oscillation might lead to focus positioning at mid-cell in cells with one focus and movement of newly divided foci to the approximate quarter-cell positions.

oscillating Soj protein, positioned an *E. coli* plasmid at mid-cell and quarter-cell positions is also consistent with the model (147).

A NEW EXPLANATION FOR PARTITION-ASSOCIATED PLASMID INCOMPATIBILITY

The centromere-like sites of Type I and II *par* loci exert partition-associated plasmid incompatibility (4, 16, 26, 116). Thus, when present on an otherwise compatible plasmid, *parC* of R1 destabilizes a mini-R1 stabilized by *parMRC*. Several models have been proposed to explain the phenomenon (4, 113). One of the most popular models states that *par*-specific plasmid pairing of heterologous replicons would readily explain partition-associated incompatibility (74, 113). As described above, plasmid pairing by the *parMRC*-encoded components occurs in vitro. However, in vivo, *parC*-mediated incompatibility was weak, even when *parC* was present on a high-copy-number plasmid (20). Furthermore, titration of ParR by *parC* was also modest (70). One explanation for these observations is that ParR is preferentially *cis*-acting and that plasmid pairing is coupled to replication such that sister plasmid molecules have a greater chance to pair than non-sisters, that is, pairing occurs between two newly replicated plasmid molecules. If this is the case, then the plasmids that pair would not be randomly selected from a pool and the incompatibility phenotype therefore would be weaker than expected from random selection and pairing.

To investigate the plasmid incompatibility phenotype associated with Type I *par* loci in vivo, we used a GFP-based dual labeling technique to simultaneously visualize two different plasmids. As expected, the presence of an F plasmid carrying *parC1*, *parC2*, or *par2* of pB171 interfered with proper partition of an R1 plasmid carrying *par2*. Remarkably, however, fluorescence microscopy of the incompatible plasmids did not show any indication of heterologous plasmid pairing. Instead, pure clusters of R1 and F plasmids appeared to be distributed in a random order along the long axis of the cell (28). This striking observation raises the possibility that incompatibility in Type I partition loci can be explained by random assortment of pure plasmid clusters. This mechanism does not involve pairing of heterologous plasmids carrying the same *par* locus. On the other hand, cytological data show clearly that homologous plasmid molecules do pair or cluster (see below). These results are most readily explained by the assumption that plasmid pairing occurs preferentially between newly replicated plasmid molecules.

PLASMID SEGREGATION AND THE PARADOX OF PLASMID CLUSTERING

Accumulating evidence indicates that low-copy-number plasmids are non-covalently clustered within bacterial cells. The average copy number of plasmid R1 is 4-5 per cell (115) but the average number of plasmid foci detected by FISH or GFP-tagging was considerably lower (73, 144). Convincing evidence of the clustering phenomenon came from the observation that an R1 plasmid copy-number mutant (4 times the wild-type copy number) had an average focus number similar to the wild-type control plasmid (144). Plasmid P1 is also located in clusters (53) and even high-copy-number plasmids can be found in clusters (75, 119). Plasmid clustering, although not understood, is not a result of plasmid multimer formation and does not depend on the presence of a *par* locus (27, 144). Inhibition of DNA replication or DNA gyrase alters plasmid focus localization but does not prevent plasmid clustering, suggesting that normal superhelicity and DNA replication are not required for clustering (75). Double labeling experiments showed that clustering is replicon specific, that is, clusters normally consist of one type of replicon only (28, 61). This observation raises the possibility that plasmid clustering depends on sequence homology between the origins of replication, a hypothesis that should now be tested.

The clustering phenomenon raises a paradox: Normally, low-copy-number plasmids devoid of their *par* loci are characterized by a loss frequency compatible with each plasmid molecule behaving like one segregating unit. Thus, if the plasmid cluster behaves as the unit of segregation a much larger loss frequency would be expected (114). One possible solution to this paradox is that the replication machinery located at mid-cell (31, 82, 90) recruits plasmids one at a time from the plasmid clusters for replication. If, after replication, the two new plasmids distribute themselves randomly between the cell halves, then the clustering phenomenon would not be inconsistent with a binomial distribution of the plasmid molecules between daughter cells (114). However, the clustering phenomenon raises a number of questions; among others is how a partitioning machinery copes with assemblies of more than two plasmid copies.

CHROMOSOME-ENCODED TYPE I *par* LOCI

Most bacterial chromosomes encode Type I *par* loci. Type II loci have not yet been found on bacterial chromosomes (46, 57). Some of the better studied examples of chromosomal Type I loci are from *Bacillus subtilis*, *C. crescentus*, *P. putida* and *P. aeruginosa* (9, 49, 98, 106). The large linear chromosome of *Streptomyces coelicolor* A3 also encodes a Type I *par* locus (77). However, *E. coli* and some of its close relatives such as *Haemophilus influenzae* do not contain Type I loci (46).

Like plasmid-encoded *par* loci, chromosomal *par* loci encode two *trans*-acting proteins that act on a number of *cis*-acting sites scattered on the chromosome. The number and distribution of the *cis*-acting sites vary among species, and they need not all be closely linked to the *par* operon. In general, the *cis*-acting sites are clustered around the origin-proximal region of the chromosome (9, 49, 69, 98). Consistent with the notion that the chromosomal *par* loci, like their plasmid-encoded counterparts, might be involved in DNA segregation, mutations in or deletion of chromosomal *par* loci result in chromosome segregation defects. However, these effects often are modest or even difficult to detect as they might appear only during specific growth conditions or during certain time points in the developmental life-cycle of a cell (49, 67, 77, 94). Moreover, a number of indirect observations indicate that Type I loci are not the main players in bacterial chromosome segregation. For example, movement of the origin region of *B. subtilis* does not depend on *spo0J* (142, 143). However, as described below, chromosome-encoded *par* loci may play central roles in chromosome segregation during specialized conditions.

The chromosomal *par* locus *soj-spo0J* of *B. subtilis* has been studied in detail. Like its plasmid-encoded counterparts, the *B. subtilis* ParB analogue, Spo0J, contains a HTH-motif (93) that mediates specific binding of Spo0J to 8 *cis*-acting *parS* sites in the origin-proximal 20% of the chromosome (98). Fluorescence microscopy experiments have shown that binding of Spo0J to *parS* creates small nucleoprotein complexes that, in the presence of Soj, condense into one discrete focus at each origin region (48, 102). During vegetative growth, Spo0J foci follow the replicated origin regions as they move apart and segregate to positions near the borders of the nucleoid located at the quarter-cell positions (48, 89, 102). This finding, combined with the observation that deletion of Spo0J results in elongated cells with abnormal nucleoid morphology in addition to an increased percentage of anucleate cells, led to the suggestion that Spo0J-Soj might be involved in origin organization and/or segregation (6, 67). Consistently, comparison of origin positions revealed that origin regions were located closer together in *spo0J* than in wild-type cells (89). This suggested that Spo0J plays a role in origin positioning. However, introduction of *parS* sites into different chromosomal locations did not allow Spo0J to bring these sites to the quarter-cell positions. Alternatively, it was proposed that Spo0J might play a role in separation of sister origin regions or in maintenance of these regions at specific cellular positions (89). Curiously, deletion of Soj does not seem to affect cell length or nucleoid appearance (67), indicating that unlike the plasmid-encoded ParA homologues, Soj does not appear to be essential for chromosome partitioning during vegetative growth. Nevertheless, as Spo0J and Soj are both required for the stabilization of a *parS*-carrying plasmid in *B. subtilis*, Soj may play an as yet unknown role in chromosome segregation (98).

Recent evidence points to a more direct role for Soj-Spo0J in chromosome segregation during sporulation. At the onset of sporulation, before formation of the asymmetric septum, the chromosomal DNA forms an elongated structure called the axial filament, in which the origin proximal region of the chromosome moves to the cell pole where it forms a condensed structure (120). Investigations of this process have shown that Spo0J and Soj might play a role in the correct positioning of the origin region at the cell pole (10, 146). However, the experiments also show that Spo0J-Soj probably do not act alone but are complemented by an independent origin segregation mechanism that involves the

RacA protein expressed early in sporulation (10, 146). Apparently, RacA binds to a region to the left of the origin of the chromosome and recruits it to the pole (10). Both Soj-Spo0J and RacA require the polarly sequestered DivIVA protein as an anchor for the origin DNA. Consistent with the proposal that RacA and Soj-Spo0J play partially redundant roles in origin segregation during sporulation, deletion of both systems at the same time results in strains in which most cells fail to trap the origin region in the prespore (10, 146).

Deletions in the *parAB* locus of *S. coelicolor* also result in chromosome segregation defects during formation of spore chains. In agreement with a role of ParAB of *S. coelicolor* in chromosome segregation during sporulation, transcription of *parAB* from one of two operon promoters is greatly stimulated at the time of sporulation (77). The free-living rod-shaped bacterium *P. putida* does not pass through developmental phases. However, in this organism the effect of ParAB on chromosome segregation depends on the culture medium and growth phase, as mutations in the *parAB* locus result in chromosome loss only during the transition from exponential growth to stationary phase of cells grown in minimal medium (49, 94).

COMPARISON OF THE OSCILLATING PROTEINS MinD, ParA, AND Soj

Like ParA of pB171, the deviant Walker-type ATPases Soj of *B. subtilis* and MinD of *E. coli* exhibit dynamic movements. Apparently, Soj forms polarly located patches that oscillate or jump from nucleoid to nucleoid within living cells of *B. subtilis* (102, 122). Soj jumping requires the presence of Spo0J and probably *parS*. Hence, when Spo0J is absent, Soj relocates to the chromosome, where it binds and represses sporulation-specific promoters (102, 121, 122), which explains the sporulation-defective phenotype of *spo0J* cells. Examination of the ability of Soj ATPase mutants to localize to the pole and bind to the nucleoid led to the proposal that the ATP-bound form of Soj interacts with Spo0J and the ADP-bound form binds DNA and regulates promoter activity (122). More recently, in vitro studies of Soj from *Thermus thermophilus* have shown that Soj forms ATP-dependent dimers in solution. Dimerization of Soj facilitates cooperative and non-specific binding of the protein to DNA, thereby mediating the formation of a nucleoprotein filament (92). Spo0J activates the ATPase activity of Soj, which is thought to result in dissociation of Soj dimers followed by release of Soj from the DNA. The hypothesis that Spo0J regulates Soj oscillation is consistent with the observation that a mutation in Spo0J resulted in an increased Soj oscillation frequency (6).

Whereas ParA and Soj oscillate over the nucleoid, *E. coli* MinD is located at the inner surface of the cell membrane where it undergoes cooperative polymerization (86, 139). Like Soj, MinD undergoes ATP-dependent dimerization. The MinD dimer forms a complex with MinC and thereby recruits it to the membrane where MinC acts to prevent Z-ring formation (62, 64, 65, 118). MinCD-mediated inhibition of cell division is controlled by MinE, a topological specificity factor that constrains the MinCD inhibitory action to the cell poles (25, 149). Hence, in the presence of MinE, the MinCD complex oscillates in a timescale of seconds from cell pole to cell pole within extended coiled structures that wind around the cell cylinder (64, 124, 125, 132). Analogous to Spo0J-mediated regulation of Soj jumping, MinE is thought to regulate MinD oscillation by stimulating the ATPase activity of membrane-bound MinD, thereby releasing MinD into the cytoplasm (62, 63, 86, 134).

The properties of MinD, Soj, ParF, and pB171 ParA are in all cases modulated by ATP and ADP. In the case of MinD and Soj, nucleotide-dependent dimerization appears to be required for DNA/membrane-binding (65, 92), whereas Spo0J/MinE-mediated stimulation of the ATPase activity regulates dynamic movement of the proteins

(86, 92). As oscillation of ParA of pB171 apparently requires ATP-binding or hydrolysis in addition to ParB and *parC*, a similar regulation might control the oscillation of this protein (26, 27). Also, assembly and disassembly of the ParF filaments is modulated by the nucleotide state of the protein, which again seems to be regulated by ParG (8). Together, these observations are consistent with the proposal that nucleotide binding and hydrolysis act as a molecular switch that controls the properties of all these proteins in vitro as well as their function in vivo. In the case of the P1 *par* system, ADP stimulates the promoter regulatory activity of ParA, as mentioned above. Recognition of the partitioning complex by ParA, on the other hand, requires ATP but is not dependent on hydrolysis (15). Thus, also in this case, ADP and ATP might function as a molecular switch that controls the dual functions of ParA in autoregulation and partition.

CONCLUDING REMARKS

The DNA segregation mechanisms specified by plasmid partitioning loci have been difficult to uncover. However, recent cytological observations have opened the door to the problem and coherent models are emerging. Especially in the case of the R1 Type II *par* locus, an understanding of the molecular mechanisms underlying plasmid segregation appears within reach. Until recently, the phenomenon of dynamic instability was associated solely with eukaryotic microtubules, where the repeated cycles of growth and shrinkage of the tubulin polymers are thought to play an essential role, particularly in DNA segregation. However, the finding that the actin-like partitioning ATPase, ParM, exhibits a similar behavior indicates that eukaryotic and prokaryotic cells have evolved functionally very similar ways of dealing with an essential process in the cellular life cycle, namely DNA segregation. For Type I *par* loci, thorough examinations of the P1 and F loci in particular have contributed extensive information on transcriptional regulation of the *par* operons, formation of partitioning complexes, in addition to biochemical properties of the individual *par* proteins. Of the questions still remaining the most important seems to be the role of the ATPases in plasmid distribution. A clue to an understanding of this central question came from the discovery that two Type Ib partitioning ATPases form filamentous structures that play a vital role in plasmid segregation. It is now within experimental reach to elucidate this type of DNA segregation mechanism as well.

SUMMARY POINTS

1. The *parMRC* locus segregates plasmids by a mitotic-like mechanism.
2. ParM forms actin-like filaments reminiscent of the eukaryotic spindle.
3. ParM filaments exhibit dynamic instability, a property hitherto only seen with eukaryotic tubulin.
4. ParA ATPase of pB171 forms spiral-shaped structures that oscillate on the nucleoid.
5. ParF of pTP228, a ParA homolog, forms MinD-like filaments in vitro.
6. *par2* of pB171 distributes plasmids along the long axis of the cell.
7. A model that explains the apparent coupling between ParA filament oscillation and ordered plasmid segregation is presented here.
8. Cytological observations allow us to propose an unexpected explanation of partition locus associated plasmid incompatibility.

UNRESOLVED ISSUES

1. Does the ParR/*parC* complex regulate dynamic instability of ParM filaments?
2. Can the coupling between plasmid segregation and ParA filament oscillation be elucidated, including a challenge of the jumping ParM model?
3. Is plasmid pairing an essential intermediate in the plasmid segregation process mediated by Type I partitioning loci?
4. What are the roles of ParA ATPases in plasmid pairing and separation?
5. What is the molecular basis of plasmid clustering?
6. What is the molecular basis of ParA/Soj oscillation?

ACKNOWLEDGMENTS

This work was supported by The Center for Bacterial Genetics under The Danish Natural Science Research Council and The Carlsberg Foundation.

LITERATURE CITED

1. Abeles AL, Friedman SA, Austin SJ. 1985. Partition of unit-copy miniplasmids to daughter cells. III. The DNA sequence and functional organization of the P1 partition region. *J. Mol. Biol.* 185:261–72
2. Austin S, Abeles A. 1983. Partition of unit-copy miniplasmids to daughter cells. I. P1 and F miniplasmids contain discrete, interchangeable sequences sufficient to promote equipartition. *J. Mol. Biol.* 169:353–72
3. Austin S, Abeles A. 1983. Partition of unit-copy miniplasmids to daughter cells. II. The partition region of miniplasmid P1 encodes an essential protein and a centromere-like site at which it acts. *J. Mol. Biol.* 169:373–87
4. Austin S, Nordström K. 1990. Partition-mediated incompatibility of bacterial plasmids. *Cell* 60:351–54
5. Austin S, Ziese M, Sternberg N. 1981. A novel role for site-specific recombination in maintenance of bacterial replicons. *Cell* 25:729–36
6. Autret S, Nair R, Errington J. 2001. Genetic analysis of the chromosome segregation protein Spo0J of *Bacillus subtilis*: evidence for separate domains involved in DNA binding and interactions with Soj protein. *Mol. Microbiol.* 41:743–55
7. Barillà D, Hayes F. 2003. Architecture of the ParF-ParG protein complex involved in prokaryotic DNA segregation. *Mol. Microbiol.* 49:487–99
8. Barillà D, Rosenberg MF, Nobbmann U, Hayes F. 2005. Bacterial DNA segregation dynamics mediated by the polymerizing protein ParF. *EMBO J.* 24:1453–64
9. Bartosik AA, Lasocki K, Mierzejewska J, Thomas CM, Jagura-Burdzy G. 2004. ParB of *Pseudomonas aeruginosa*: interactions with its partner ParA and its target parS and specific effects on bacterial growth. *J. Bacteriol.* 186:6983–98
10. Ben Yehuda S, Rudner DZ, Losick R. 2003. RacA, a bacterial protein that anchors chromosomes to the cell poles. *Science* 299:532–36

11. Biek DP, Cohen SN. 1986. Identification and characterization of recD, a gene affecting plasmid maintenance and recombination in *Escherichia coli*. *J. Bacteriol.* 167:594–603
12. Biek DP, Shi J. 1994. A single 43-bp sopC repeat of plasmid mini-F is sufficient to allow assembly of a functional nucleoprotein partition complex. *Proc. Natl. Acad. Sci. USA* 91:8027–31
13. Bignell CR, Haines AS, Khare D, Thomas CM. 1999. Effect of growth rate and incC mutation on symmetric plasmid distribution by the incP-1 partitioning apparatus. *Mol. Microbiol.* 34:205–16
14. Bork P, Sander C, Valencia A. 1992. An ATPase domain common to prokaryotic cell cycle proteins, sugar kinases, actin, and hsp70 heat shock proteins. *Proc. Natl. Acad. Sci. USA* 89:7290–94
15. Bouet JY, Funnell BE. 1999. P1 ParA interacts with the P1 partition complex at parS and an ATP-ADP switch controls ParA activities. *EMBO J.* 18:1415–24
16. Bouet JY, Rech J, Egloff S, Biek DP, Lane D. 2005. Probing plasmid partition with centromere-based incompatibility. *Mol. Microbiol.* 55:511–25
17. Bouet JY, Surtees JA, Funnell BE. 2000. Stoichiometry of P1 plasmid partition complexes. *J. Biol. Chem.* 275:8213–19
18. Breuner A, Jensen RB, Dam M, Pedersen S, Gerdes K. 1996. The centromere-like parC locus of plasmid R1. *Mol. Microbiol.* 20:581–92
19. Colloms SD, Bath J, Sherratt DJ. 1997. Topological selectivity in Xer site-specific recombination. *Cell* 88:855–64
20. Dam M, Gerdes K. 1994. Partitioning of plasmid R1. Ten direct repeats flanking the parA promoter constitute a centromere-like partition site parC, that expresses incompatibility. *J. Mol. Biol.* 236:1289–98
21. Davey MJ, Funnell BE. 1994. The P1 plasmid partition protein ParA. A role for ATP in site-specific DNA binding. *J. Biol. Chem.* 269:29908–13
22. Davey MJ, Funnell BE. 1997. Modulation of the P1 plasmid partition protein ParA by ATP, ADP, and P1 ParB. *J. Biol. Chem.* 272:15286–92
23. Davis MA, Austin SJ. 1988. Recognition of the P1 plasmid centromere analog involves binding of the ParB protein and is modified by a specific host factor. *EMBO J.* 7:1881–88
24. Davis MA, Martin KA, Austin SJ. 1992. Biochemical activities of the parA partition protein of the P1 plasmid. *Mol. Microbiol.* 6:1141–47
25. de Boer PAJ, Crossley RE, Rothfield LI. 1989. A division inhibitor and a topological specificity factor coded for by the minicell locus determine proper placement of the division septum in *E. coli*. *Cell* 56:641–49
26. Ebersbach G, Gerdes K. 2001. The double par locus of virulence factor pB171: DNA segregation is correlated with oscillation of ParA. *Proc. Natl. Acad. Sci. USA* 98:15078–83
27. Ebersbach G, Gerdes K. 2004. Bacterial mitosis: Partitioning protein ParA oscillates in spiral-shaped structures and positions plasmids at mid-cell. *Mol. Microbiol.* 52:385–98
28. Ebersbach G, Sherratt D, Gerdes K. 2005. Partition-associated incompatibility caused by random assortment of pure plasmid clusters. *Mol. Microbiol.* 56:1430–40
29. Edgar R, Chattoraj DK, Yarmolinsky M. 2001. Pairing of P1 plasmid partition sites by ParB. *Mol. Microbiol.* 42:1363–70
30. Erdmann N, Petroff T, Funnell BE. 1999. Intracellular localization of P1 ParB protein depends on ParA and parS. *Proc. Natl. Acad. Sci. USA* 96:14905–10

31. Espeli O, Levine C, Hassing H, Marians KJ. 2003. Temporal regulation of topoisomerase IV activity in *E. coli*. *Mol. Cell* 11:189–201
32. Ezaki B, Ogura T, Niki H, Hiraga S. 1991. Partitioning of a mini-F plasmid into anucleate cells of the mukB null mutant. *J. Bacteriol.* 173:6643–46
33. Figge RM, Divakaruni AV, Gober JW. 2004. MreB, the cell shape-determining bacterial actin homologue, co-ordinates cell wall morphogenesis in *Caulobacter crescentus*. *Mol. Microbiol.* 51:1321–32
34. Firshein W, Kim P. 1997. Plasmid replication and partition in *Escherichia coli* : Is the cell membrane the key? *Mol. Microbiol.* 23:1–10
35. Fothergill TJG, Barillà D, Hayes F. 2005. Protein diversity confers specificity in plasmid segregation. *J. Bacteriol.* 187:2651–61
36. Friedman SA, Austin SJ. 1988. The P1 plasmid-partition system synthesizes two essential proteins from an autoregulated operon. *Plasmid* 19:103–12
37. Fung E, Bouet JY, Funnell BE. 2001. Probing the ATP-binding site of P1 ParA: Partition and repression have different requirements for ATP binding and hydrolysis. *EMBO J.* 20:4901–11
38. Funnell BE. 1988. Mini-P1 plasmid partitioning: Excess ParB protein destabilizes plasmids containing the centromere parS. *J. Bacteriol.* 170:954–60
39. Funnell BE. 1988. Participation of *Escherichia coli* integration host factor in the P1 plasmid partition system. *Proc. Natl. Acad. Sci. USA* 85:6657–61
40. Funnell BE. 1991. The P1 plasmid partition complex at parS. The influence of *Escherichia coli* integration host factor and of substrate topology. *J. Biol. Chem.* 266:14328–37
41. Funnell BE, Gagnier L. 1993. The P1 plasmid partition complex at parS. II. Analysis of ParB protein binding activity and specificity. *J. Biol. Chem.* 268:3616–24
42. Funnell BE, Gagnier L. 1995. Partition of P1 plasmids in *Escherichia coli* mukB chromosomal partition mutants. *J. Bacteriol.* 177:2381–86
43. Garner EC, Campbell CS, Mullins RD. 2004. Dynamic instability in a DNA-segregating prokaryotic actin homolog. *Science* 306:1021–25
44. Gerdes K, Gultyaev AP, Franch T, Pedersen K, Mikkelsen ND. 1997. Antisense RNA-regulated programmed cell death. *Annu. Rev. Genet.* 31:1–31
45. Gerdes K, Molin S. 1986. Partitioning of plasmid R1. Structural and functional analysis of the parA locus. *J. Mol. Biol.* 190:269–79
46. Gerdes K, Møller-Jensen J, Jensen RB. 2000. Plasmid and chromosome partitioning: surprises from phylogeny. *Mol. Microbiol.* 37:455–66
47. Gitai Z, Dye NA, Reisenauer A, Wachi M, Shapiro L. 2005. MreB actin-mediated segregation of a specific region of a bacterial chromosome. *Cell* 120:329–41
48. Glaser P, Sharpe ME, Raether B, Perego M, Ohlsen K, Errington J. 1997. Dynamic, mitotic-like behavior of a bacterial protein required for accurate chromosome partitioning. *Genes Dev.* 11:1160–68
49. Godfrin-Estevenon AM, Pasta F, Lane D. 2002. The parAB gene products of *Pseudomonas putida* exhibit partition activity in both *P. putida* and *Escherichia coli*. *Mol. Microbiol.* 43:39–49
50. Golovanov AP, Barillà D, Golovanova M, Hayes F, Lian L-Y. 2005. ParG, a protein required for active partition of bacterial plasmids, has a dimeric ribbon-helix-helix structure. *Mol. Microbiol.* 50:1141–53
51. Gomis-Ruth FX, Sola M, Acebo P, Parraga A, Guasch A, et al. 1998. The structure of plasmid-encoded transcriptional repressor CopG unliganded and bound to its operator. *EMBO J.* 17:7404–15

52. Gordon GS, Sitnikov D, Webb CD, Teleman A, Straight A, et al. 1997. Chromosome and low copy plasmid segregation in *E. coli*: visual evidence for distinct mechanisms. *Cell* 90:1113–21
53. Gordon S, Rech J, Lane D, Wright A. 2004. Kinetics of plasmid segregation in *Escherichia coli*. *Mol. Microbiol.* 51:461–69
54. Grigoriev PS, Lobocka MB. 2001. Determinants of segregational stability of the linear plasmid-prophage N15 of *Escherichia coli*. *Mol. Microbiol.* 42:355–68
55. Hanai R, Liu R, Benedetti P, Caron PR, Lynch AS, Wang JC. 1996. Molecular dissection of a protein SopB essential for *Escherichia coli* F plasmid partition. *J. Biol. Chem.* 271:17469–75
56. Hao JJ, Yarmolinsky M. 2002. Effects of the P1 plasmid centromere on expression of P1 partition genes. *J. Bacteriol.* 184:4857–67
57. Hayes F. 2000. The partition system of multidrug resistance plasmid TP228 includes a novel protein that epitomizes an evolutionarily distinct subgroup of the ParA superfamily. *Mol. Microbiol.* 37:528–41
58. Hayes F, Radnedge L, Davis MA, Austin SJ. 1994. The homologous operons for P1 and P7 plasmid partition are autoregulated from dissimilar operator sites. *Mol. Microbiol.* 11:249–60
59. Hiraga S. 1992. Chromosome and plasmid partition in *Escherichia coli*. *Annu. Rev. Biochem.* 61:283–306
60. Hirano M, Mori H, Onogi T, Yamazoe M, Niki H, et al. 1998. Autoregulation of the partition genes of the mini-F plasmid and the intracellular localization of their products in *Escherichia coli*. *Mol. Gen. Genet.* 257:392–403
61. Ho TQ, Zhong Z, Aung S, Pogliano J. 2002. Compatible bacterial plasmids are targeted to independent cellular locations in *Escherichia coli*. *EMBO J.* 21:1864–72
62. Hu Z, Gogol EP, Lutkenhaus J. 2002. Dynamic assembly of MinD on phospholipid vesicles regulated by ATP and MinE. *Proc. Natl. Acad. Sci. USA* 99:6761–66
63. Hu Z, Lutkenhaus J. 2001. Topological regulation of cell division in *E. coli*: Spatiotemporal oscillation of MinD requires stimulation of its ATPase by MinE and phospholipid. *Mol. Cell* 7:1337–43
64. Hu Z, Lutkenhaus J. 1999. Topological regulation of cell division in *Escherichia coli* involves rapid pole to pole oscillation of the division inhibitor MinC under the control of MinD and MinE. *Mol. Microbiol.* 34:82–90
65. Hu Z, Saez C, Lutkenhaus J. 2003. Recruitment of MinC, an inhibitor of Z-ring formation, to the membrane in *Escherichia coli*: role of MinD and MinE. *J. Bacteriol.* 185:196–203
66. Ingmer H, Miller CA, Cohen SN. 1998. Destabilized inheritance of pSC101 and other *Escherichia coli* plasmids by DpiA, a novel two-component system regulator. *Mol. Microbiol.* 29:49–59
67. Ireton K, Gunther NW, Grossman AD. 1994. spo0J is required for normal chromosome segregation as well as the initiation of sporulation in *Bacillus subtilis*. *J. Bacteriol.* 176:5320–29
68. Jacob F, Brenner S, Cuzin F. 1963. On the regulation of DNA replication in bacteria. *Cold Spring Harbor Symp. Quant. Biol.* 23:329–48
69. Jakimowicz D, Chater K, Zakrzewska-Czerwinska J. 2002. The ParB protein of *Streptomyces coelicolor* A3(2) recognizes a cluster of parS sequences within the origin-proximal region of the linear chromosome. *Mol. Microbiol.* 45:1365–77

70. Jensen RB, Dam M, Gerdes K. 1994. Partitioning of plasmid R1. The parA operon is autoregulated by ParR and its transcription is highly stimulated by a downstream activating element. *J. Mol. Biol.* 236:1299–309
71. Jensen RB, Gerdes K. 1995. Programmed cell death in bacteria: proteic plasmid stabilization systems. *Mol. Microbiol.* 17:205–10
72. Jensen RB, Gerdes K. 1997. Partitioning of plasmid R1. The ParM protein exhibits ATPase activity and interacts with the centromere-like ParR-parC complex. *J. Mol. Biol.* 269:505–13
73. Jensen RB, Gerdes K. 1999. Mechanism of DNA segregation in prokaryotes: ParM partitioning protein of plasmid R1 co-localizes with its replicon during the cell cycle. *EMBO J.* 18:4076–84
74. Jensen RB, Lurz R, Gerdes K. 1998. Mechanism of DNA segregation in prokaryotes: replicon pairing by parC of plasmid R1. *Proc. Natl. Acad. Sci. USA* 95:8550–55
75. Johnson EP, Yao S, Helinski DR. 2005. Gyrase inhibitors and thymine starvation disrupt the normal pattern of plasmid RK2 localization in *Escherichia coli*. *J. Bacteriol.* 187:3538–47
76. Jones LJ, Carballido-Lopez R, Errington J. 2001. Control of cell shape in bacteria: helical, actin-like filaments in *Bacillus subtilis*. *Cell* 104:913–22
77. Kim HJ, Calcutt MJ, Schmidt FJ, Chater KF. 2000. Partitioning of the linear chromosome during sporulation of *Streptomyces coelicolor* A3(2) involves an oriC-linked parAB locus. *J. Bacteriol.* 182:1313–20
78. Kim SK, Wang JC. 1998. Localization of F plasmid SopB protein to positions near the poles of *Escherichia coli* cells. *Proc. Natl. Acad. Sci. USA* 95:1523–27
79. Kim SK, Wang JC. 1999. Gene silencing via protein-mediated subcellular localization of DNA. *Proc. Natl. Acad. Sci. USA* 96:8557–61
80. Kline-Smith SL, Walczak C. 2004. Mitotic spindle assembly and chromosome segregation: refocusing on microtubule dynamics. *Mol. Cell* 15:317–27
81. Koonin EV. 1993. A superfamily of ATPases with diverse functions containing either classical or deviant ATP-binding motif. *J. Mol. Biol.* 229:1165–74
82. Koppes LJ, Woldringh CL, Nanninga N. 1999. *Escherichia coli* contains a DNA replication compartment in the cell center. *Biochimie* 81:803–10
83. Kruse T, Bork-Jensen J, Gerdes K. 2005. The morphogenetic MreBCD proteins of *Escherichia coli* form an essential membrane-bound complex. *Mol. Microbiol.* 55:78–89
84. Kruse T, Møller-Jensen J, Løbner-Olesen A, Gerdes K. 2003. Dysfunctional MreB inhibits chromosome segregation in *Escherichia coli*. *EMBO J.* 22:5283–92
85. Kusukawa N, Mori H, Kondo A, Hiraga S. 1987. Partitioning of the F plasmid: Overproduction of an essential protein for partition inhibits plasmid maintenance. *Mol. Gen. Genet.* 208:365–72
86. Lackner LL, Raskin DM, de Boer PA. 2003. ATP-dependent interactions between *Escherichia coli* Min proteins and the phospholipid membrane in vitro. *J. Bacteriol.* 185:735–49
87. Lane D, de Feyter R, Kennedy M, Phua SH, Semon D. 1986. D protein of mini-F plasmid acts as a repressor of transcription and as a site-specific resolvase. *Nucleic Acids Res.* 14:9713–28
88. Lawley TD, Taylor DE. 2003. Characterization of the double-partitioning modules of R27: correlating plasmid stability with plasmid localization. *J. Bacteriol.* 185:3060–67

89. Lee PS, Lin DC, Moriya S, Grossman AD. 2003. Effects of the chromosome partitioning protein Spo0J (ParB) on oriC positioning and replication initiation in *Bacillus subtilis*. *J. Bacteriol.* 185:1326–37
90. Lemon KP, Grossman AD. 1998. Localization of bacterial DNA polymerase: evidence for a factory model of replication. *Science* 282:1516–19
91. Lemonnier M, Bouet JY, Libante V, Lane D. 2000. Disruption of the F plasmid partition complex in vivo by partition protein SopA. *Mol. Microbiol.* 38:493–505
92. Leonard TA, Butler PJ, Lowe J. 2005. Bacterial chromosome segregation: structure and DNA binding of the Soj dimer—a conserved biological switch. *EMBO J.* 24:270–82
93. Leonard TA, Butler PJ, Lowe J. 2004. Structural analysis of the chromosome segregation protein Spo0J from *Thermus thermophilus*. *Mol. Microbiol.* 53:419–32
94. Lewis RA, Bignell CR, Zeng W, Jones AC, Thomas CM. 2002. Chromosome loss from par mutants of *Pseudomonas putida* depends on growth medium and phase of growth. *Microbiology* 148:537–48
95. Li Y, Austin S. 2002. The P1 plasmid is segregated to daughter cells by a 'capture and ejection' mechanism coordinated with *Escherichia coli* cell division. *Mol. Microbiol.* 46:63–74
96. Li Y, Dabrazhynetskaya A, Youngren B, Austin S. 2004. The role of Par proteins in the active segregation of the P1 plasmid. *Mol. Microbiol.* 53:93–102
97. Libante V, Thion L, Lane D. 2001. Role of the ATP-binding site of SopA protein in partition of the F plasmid. *J. Mol. Biol.* 314:387–99
98. Lin DC, Grossman AD. 1998. Identification and characterization of a bacterial chromosome partitioning site. *Cell* 92:675–85
99. Lobocka M, Yarmolinsky M. 1996. P1 plasmid partition: a mutational analysis of ParB. *J. Mol. Biol.* 259:366–82
100. Lynch AS, Wang JC. 1994. Use of an inducible site-specific recombinase to probe the structure of protein-DNA complexes involved in F plasmid partition in *Escherichia coli*. *J. Mol. Biol.* 236:679–84
101. Lynch AS, Wang JC. 1995. SopB protein-mediated silencing of genes linked to the sopC locus of *Escherichia coli* F plasmid. *Proc. Natl. Acad. Sci. USA* 92:1896–900
101a. Løbner-Olesen A. 1999. Distribution of minichromosomes in individual *Escherichia coli* cells: implications for replication control. *EMBO J.* 18:1712–21
102. Marston AL, Errington J. 1999. Dynamic movement of the ParA-like Soj protein of *B. subtilis* and its dual role in nucleoid organization and developmental regulation. *Mol. Cell* 4:673–82
103. Martin KA, Friedman SA, Austin SJ. 1987. Partition site of the P1 plasmid. *Proc. Natl. Acad. Sci. USA* 84:8544–47
104. Mitchison T, Kirschner M. 1984. Dynamic instability of microtubule growth. *Nature* 312:237–42
105. Mitchison TJ, Salmon ED. 2001. Mitosis: a history of division. *Nat. Cell Biol.* 3:E17–21
106. Mohl DA, Gober JW. 1997. Cell cycle-dependent polar localization of chromosome partitioning proteins in *Caulobacter crescentus*. *Cell* 88:675–84
107. Møller-Jensen J, Borch J, Dam M, Jensen RB, Roepstorff P, Gerdes K. 2003. Bacterial Mitosis: ParM of plasmid R1 moves DNA by an actin-like insertional polymerization mechanism. *Mol. Cell* 12:1477–87
108. Møller-Jensen J, Jensen RB, Lowe J, Gerdes K. 2002. Prokaryotic DNA segregation by an actin-like filament. *EMBO J.* 21:3119–27

109. Mori H, Kondo A, Ohshima A, Ogura T, Hiraga S. 1986. Structure and function of the F plasmid genes essential for partitioning. *J. Mol. Biol.* 192:1–15
110. Mori H, Mori Y, Ichinose C, Niki H, Ogura T, et al. 1989. Purification and characterization of SopA and SopB proteins essential for F plasmid partitioning. *J. Biol. Chem.* 264:15535–41
111. Niki H, Hiraga S. 1997. Subcellular distribution of actively partitioning F plasmid during the cell division cycle in *E. coli*. *Cell* 90:951–57
112. Niki H, Ichinose C, Ogura T, Mori H, Morita M, et al. 1988. Chromosomal genes essential for stable maintenance of the mini-F plasmid in *Escherichia coli*. *J. Bacteriol.* 170:5272–78
113. Nordström K, Austin SJ. 1989. Mechanisms that contribute to the stable segregation of plasmids. *Annu. Rev. Genet.* 23:37–69
114. Nordström K, Gerdes K. 2003. Clustering versus random segregation of plasmids lacking a partitioning function: a plasmid paradox? *Plasmid* 50:95–101
115. Nordström K, Molin S, agaard-Hansen H. 1980. Partitioning of plasmid R1 in *Escherichia coli*. I. Kinetics of loss of plasmid derivatives deleted of the par region. *Plasmid* 4:215–27
116. Ogura T, Hiraga S. 1983. Partition mechanism of F plasmid: two plasmid gene-encoded products and a *cis*-acting region are involved in partition. *Cell* 32:351–60
117. Onogi T, Miki T, Hiraga S. 2002. Behavior of sister copies of mini-F plasmid after synchronized plasmid replication in *Escherichia coli* cells. *J. Bacteriol.* 184:3142–45
118. Pichoff S, Lutkenhaus J. 2001. *Escherichia coli* division inhibitor MinCD blocks septation by preventing Z-ring formation. *J. Bacteriol.* 183:6630–35
119. Pogliano J, Ho TQ, Zhong Z, Helinski DR. 2001. Multicopy plasmids are clustered and localized in *Escherichia coli*. *Proc. Natl. Acad. Sci. USA* 98:4486–91
120. Pogliano J, Sharp MD, Pogliano K. 2002. Partitioning of chromosomal DNA during establishment of cellular asymmetry in *Bacillus subtilis*. *J. Bacteriol.* 184:1743–49
121. Quisel JD, Grossman AD. 2000. Control of sporulation gene expression in *Bacillus subtilis* by the chromosome partitioning proteins Soj (ParA) and Spo0J (ParB). *J. Bacteriol.* 182:3446–51
122. Quisel JD, Lin DC, Grossman AD. 1999. Control of development by altered localization of a transcription factor in *B. subtilis*. *Mol. Cell* 4:665–72
123. Radnedge L, Davis MA, Austin SJ. 1996. P1 and P7 plasmid partition: ParB protein bound to its partition site makes a separate discriminator contact with the DNA that determines species specificity. *EMBO J.* 15:1155–62
124. Raskin DM, de Boer PA. 1999. MinDE-dependent pole-to-pole oscillation of division inhibitor MinC in *Escherichia coli*. *J. Bacteriol.* 181:6419–24
125. Raskin DM, de Boer PA. 1999. Rapid pole-to-pole oscillation of a protein required for directing division to the middle of *Escherichia coli*. *Proc. Natl. Acad. Sci. USA* 96:4971–76
126. Raumann BE, Rould MA, Pabo CO, Sauer RT. 1994. DNA recognition by beta-sheets in the Arc repressor-operator crystal structure. *Nature* 367:754–57
127. Ravin N, Lane D. 1999. Partition of the linear plasmid N15: Interactions of N15 partition functions with the sop locus of the F plasmid. *J. Bacteriol.* 181:6898–906
128. Ravin NV, Rech J, Lane D. 2003. Mapping of functional domains in F plasmid partition proteins reveals a bipartite SopB-recognition domain in SopA. *J. Mol. Biol.* 329:875–89

129. Rodionov O, Lobocka M, Yarmolinsky M. 1999. Silencing of genes flanking the P1 plasmid centromere. *Science* 283:546–49
130. Rodionov O, Yarmolinsky M. 2004. Plasmid partitioning and the spreading of P1 partition protein ParB. *Mol. Microbiol.* 52:1215–23
131. Schmidt A, Hall MN. 1998. Signalling to the actin cytoskeleton. *Annu. Rev. Cell. Dev. Biol.* 14:305–38
132. Shih YL, Le T, Rothfield L. 2003. Division site selection in *Escherichia coli* involves dynamic redistribution of Min proteins within coiled structures that extend between the two cell poles. *Proc. Natl. Acad. Sci. USA* 100:7865–70
133. Slavcev RA, Funnell B. 2005. Identification and characterization of a novel allele of *Escherichia coli* dnaB helicase that compromises the stability of plasmid P1. *J. Bacteriol.* 187:1227–37
134. Suefuji K, Valluzzi R, RayChaudhuri D. 2002. Dynamic assembly of MinD into filament bundles modulated by ATP, phospholipids, and MinE. *Proc. Natl. Acad. Sci. USA* 99:16776–81
135. Summers D. 1998. Timing, self-control and a sense of direction are the secrets of multicopy plasmid stability. *Mol. Microbiol.* 29:1137–45
136. Summers DK, Sherratt DJ. 1984. Multimerization of high copy number plasmids causes instability: ColE1 encodes a determinant essential for plasmid monomerization and stability. *Cell* 36:1097–103
137. Surtees JA, Funnell BE. 1999. P1 ParB domain structure includes two independent multimerization domains. *J. Bacteriol.* 181:5898–908
138. Surtees JA, Funnell BE. 2001. The DNA binding domains of P1 ParB and the architecture of the P1 plasmid partition complex. *J. Biol. Chem.* 276:12385–94
139. Szeto TH, Rowland SL, Rothfield LI, King GF. 2002. Membrane localization of MinD is mediated by a C-terminal motif that is conserved across eubacteria, archaea, and chloroplasts. *Proc. Natl. Acad. Sci. USA* 99:15693–98
140. van den Ent F, Møller-Jensen J, Amos LA, Gerdes K, Löwe J. 2002. F-actin-like filaments formed by plasmid segregation protein ParM. *EMBO J.* 21:6935–43
141. Watanabe E, Wachi M, Yamasaki M, Nagai K. 1992. ATPase activity of SopA, a protein essential for active partitioning of F plasmid. *Mol. Gen. Genet.* 234:346–52
142. Webb CD, Graumann PL, Kahana JA, Teleman AA, Silver PA, Losick R. 1998. Use of time-lapse microscopy to visualize rapid movement of the replication origin region of the chromosome during the cell cycle in *Bacillus subtilis*. *Mol. Microbiol.* 28:883–92
143. Webb CD, Teleman A, Gordon S, Straight A, Belmont A, et al. 1997. Bipolar localization of the replication origin regions of chromosomes in vegetative and sporulating cells of *B. subtilis*. *Cell* 88:667–74
144. Weitao T, Dasgupta S, Nordstrom K. 2000. Plasmid R1 is present as clusters in the cells of *Escherichia coli*. *Plasmid* 43:200–4
145. Williams DR, Macartney DP, Thomas CM. 1998. The partitioning activity of the RK2 central control region requires only incC, korB and KorB-binding site OB3 but other KorB-binding sites form destabilizing complexes in the absence of OB3. *Microbiology* 144:3369–78
146. Wu LJ, Errington J. 2003. RacA and the Soj-Spo0J system combine to effect polar chromosome segregation in sporulating *Bacillus subtilis*. *Mol. Microbiol.* 49:1463–75
147. Yamaichi Y, Niki H. 2000. Active segregation by the *Bacillus subtilis* partitioning system in *Escherichia coli*. *Proc. Natl. Acad. Sci. USA* 97:14656–61

148. Yates P, Lane D, Biek DP. 1999. The F plasmid centromere, sopC, is required for full repression of the sopAB operon. *J. Mol. Biol.* 290:627–38
149. Zhao C-R, de Boer PAJ, Rothfield LI. 1995. Proper placement of the *E. coli* division site requires two functions that are associated with different domains of the MinE protein. *Proc. Natl. Acad. Sci. USA* 92:4313–17

Use of the Zebrafish System to Study Primitive and Definitive Hematopoiesis

Jill L.O. de Jong and Leonard I. Zon

Stem Cell Program and Division of Hematology/Oncology, Children's Hospital Boston and Dana-Farber Cancer Institute, and Howard Hughes Medical Institute, Boston, Massachusetts 02115; email: jill.dejong@childrens.harvard.edu, zon@enders.tch.harvard.edu

Annu. Rev. Genet.
2005. 39:481–501

First published online as a Review in Advance on August 1, 2005

The *Annual Review of Genetics* is online at http://genet.annualreviews.org

doi: 10.1146/annurev.genet.39.073003.095931

0066-4197/05/1215-0481$20.00

Key Words

hematopoiesis, zebrafish, blood, hematopoietic stem cell, genetic screen, mutant

Abstract

The zebrafish (*Danio rerio*) has emerged as an ideal organism for the study of hematopoiesis, the process by which all the cellular elements of the blood are formed. These elements, including erythrocytes, granulocytes, monocytes, lymphocytes, and thrombocytes, are formed through complex genetic signaling pathways that are highly conserved throughout phylogeny. Large-scale forward genetic screens have identified numerous blood mutants in zebrafish, helping to elucidate specific signaling pathways important for hematopoietic stem cells (HSCs) and the various committed blood cell lineages. Here we review both primitive and definitive hematopoiesis in zebrafish, discuss various genetic methods available in the zebrafish model for studying hematopoiesis, and describe some of the zebrafish blood mutants identified to date, many of which have known human disease counterparts.

Contents

INTRODUCTION

Once known as merely a tropical fish pet, the zebrafish (*Danio rerio*) has now developed into a powerful vertebrate model for the study of hematopoiesis and other aspects of embryogenesis and organogenesis (2, 12, 21, 23, 79, 88). Unlike mammals, zebrafish eggs are fertilized externally and are readily available for observation or manipulation beginning at the single-cell embryo stage. The embryos are optically clear, so the entire organism can be easily evaluated under a dissecting microscope, including direct visualization of the beating heart and blood cells circulating in the vasculature. These small animals reach sexual maturity in only 3 to 4 months, and adult females are capable of producing 100 to 200 eggs weekly. Many thousands of animals can be kept in a fish facility requiring much less space than mice or other mammals, and hence the zebrafish is a cost-effective experimental vertebrate model for large-scale genetic screening (reviewed in 61). Although invertebrate models have been invaluable in the study of embryogenesis, these organisms are not useful for the study of hematopoiesis or the function of mature blood cells. Using the zebrafish, several large-scale forward genetic screens in the past decade have generated thousands of mutants, many with hematopoietic defects that have enhanced the understanding of all aspects of hematopoiesis (67, 81).

PRIMITIVE (EMBRYONIC) HEMATOPOIESIS

All vertebrates, including bony fish (teleosts), have two waves of hematopoiesis (1, 2, 27, 91). The earlier of these is known as the primitive or embryonic hematopoietic wave, and predominantly produces erythrocytes, as well as some primitive macrophages. In mammals and birds, this first hematopoietic wave is found in the extraembryonic yolk sac where early erythrocytes are generated. In zebrafish, primitive hematopoiesis occurs in two intraembryonic locations: the intermediate cell mass (ICM) located in the trunk ventral to the notochord, and the rostral blood island (RBI) arising from the cephalic mesoderm (1, 17). Within the posterior mesoderm, cells lateral to the developing somites express both vascular and blood markers and migrate medially around 18 h post fertilization (hpf) to fuse at the midline forming the ICM (**Figure 1**). Cells within the ICM, equivalent to the mammalian yolk sac blood island, differentiate into the endothelial cells of the trunk vasculature and proerythroblasts, which begin to enter the circulation around 24 hpf. Concurrent with the primitive erythropoietic wave in the

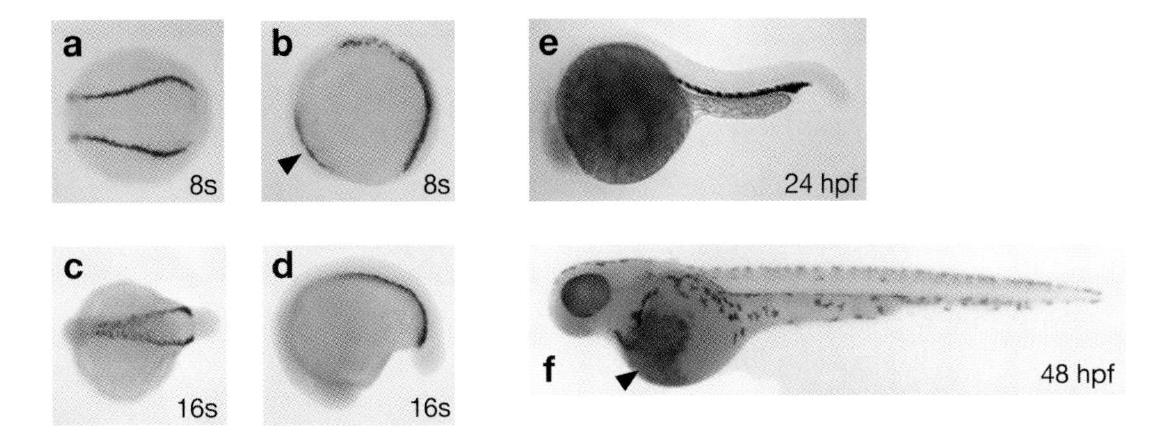

Figure 1

Formation of early blood precursors in the zebrafish embryo. (*a*) and (*b*) Whole-mount in situ hybridization with *scl* at the 8 somite stage, marking the bilateral stripes of the lateral plate mesoderm that will eventually migrate medially and fuse to form the ICM. Anterior hematopoietic precursors are also evident in the RBI (*arrow*). (*c*) and (*d*) Whole-mount in situ hybridization with *gata-1* at the 16 somite stage. The anterior portion of the lateral plate mesoderm is beginning to fuse, showing the formation of the ICM. (*e*) Whole-mount in situ hybridization with *gata-1* at 24 hpf, marking erythroid precursors in the ICM just before the onset of circulation. (*f*) *o*-dianisidine staining of hemoglobin at 48 hpf in circulating erythrocytes, noted prominently in the ducts of Cuvier over the yolk sac (*arrow*). (*b*), (*d*), (*e*), and (*f*) show lateral views of the embryos, with the anterior to the left. (*a*) and (*c*) show dorsal views of the embryos.

posterior ICM, cells in the anterior mesoderm of the zebrafish embryo make up a second anatomical site for hematopoiesis, known as the RBI, which predominantly generates macrophages (discussed below).

Primitive Progenitors in the ICM

As primitive hematopoietic progenitors begin to differentiate within the ICM, they express transcripts of several transcription factors. The zebrafish homologues have been identified for many mammalian transcription factors known to regulate hematopoiesis in primitive progenitor cells. The stem cell leukemia (*scl*) gene encodes a basic helix-loop-helix transcription factor, found to be expressed by 10 hpf in the posterior mesoderm of the zebrafish embryo (46, 64). Its expression marks the formation of primitive HSCs and also vascular precursors, known as angioblasts. Numerous other transcription factors are also expressed by these early blood and vascular progenitors, including *fli1*, *gata2*, *hhex*, *lmo2*, and *tif1γ* (**Table 1**). At this time, it is controversial whether a true bipotential "hemangioblast" precursor cell exists, or whether parallel populations of angioblasts and primitive HSCs in the posterior mesoderm develop independently (2, 22, 60).

Primitive Erythropoiesis

Gata-1, a zinc finger transcription factor, is critical for primitive erythropoiesis. First expressed in a subset of *scl*$^+$ cells in the zebrafish embryo around 5 somites (approximately 12 hpf), *gata-1* transcripts are expressed bilaterally in the lateral plate mesoderm that will migrate medially to form the ICM (18, 72). The number of *gata-1*$^+$ erythroid precursors increases as they migrate towards the midline, and by the time the ICM is fully formed, approximately 300 cells with morphology reminiscent of proerythroblasts can be found in the ICM. Transgenic zebrafish carrying the *gata-1*

Forward genetic screen: progeny of mutagenized animals are screened for the phenotype of interest. This screening method avoids the bias of preconceived ideas about which signaling pathways might be responsible for a particular phenotype

Teleost: any bony fish with rayed fins, including zebrafish

ICM: intermediate cell mass

RBI: rostral blood island

hpf: hours post fertilization

Table 1 Hematopoietic cell types and expression of marker genes

Cell type	Sites of development: Embryo	Sites of development: Adult	Marker genes expressed		References
Primitive progenitor cell	10–26 hpf lateral plate mesoderm and ICM	—	*scl* *fli1a* *gata2*	*hhex* *lmo2* *tif1γ*	(49, 66, 72)
Definitive HSC	24–48 hpf AGM	Kidney marrow	*c-myb* *cbfβ* *ikaros* *runx1*		(13, 28, 41, 42, 72)
Myeloid lineage precursors	RBI 16–30 hpf ICM 16–24 hpf	Kidney marrow	*pu.1*		(16, 36, 54, 55)
Neutrophil	16–30 hpf anterior yolk sac mesoderm and ICM	Kidney marrow (>96 hpf)	*mpx/mpo*		(8, 16, 50)
Eosinophil	Unknown	Kidney marrow	None known		(8, 50)
Monocyte/ macrophage	12–24 hpf anterior ventro-lateral mesoderm and RBI	Kidney marrow	*l-plastin* *draculin* *fms*		(16, 35, 85)
Erythrocyte	12–26 hpf lateral plate mesoderm and ICM, 24–48 hpf AGM	Kidney marrow	*gata-1* *globin* *alas2* *β-spectrin*	*fch* *urod* *dmt1*	(10, 11, 15, 19, 47, 72, 78)
Lymphocyte	None known	Thymus (>3 dpf) Kidney marrow (>2 weeks)	*ikaros* *rag-1* *rag-2* *lck*		(44, 45, 86)

Proerythroblast: early erythroid precursor cell

HSC: hematopoietic stem cell

dpf: days post fertilization

promoter driving expression of *green fluorescent protein* (*gfp*) showed that the *gata-1*$^{+}$ cells in the ICM differentiate into proerythroblasts and enter the circulation around 24 hpf (52). Once mature, the erythrocytes develop a characteristic flattened elliptical shape and retained nucleus. These "primitive" erythrocytes are morphologically distinct from adult zebrafish erythrocytes, which have less cytoplasm and a large elongated nucleus. In contrast, whereas primitive mammalian erythrocytes are nucleated as well, mammalian adult erythrocytes do not retain their nuclei. Transfusion experiments using rhodamine-labeled circulating erythrocytes showed these primitive erythrocytes are the only circulating erythroid cells for the first 4 dpf (81). The donor cells are taken at 36 hpf from rhodamine-labeled embryos and injected into the sinus venosus of the recipient embryos. The donor cells in circulation do not start to decline until 6 dpf. Although the lifespan of these primitive erythrocytes is not known, by 10 dpf only 50% of the rhodamine-labeled donor cells remain within the unlabeled recipient host as the definitive erythrocytes begin to populate the circulation (81). These data indicate that the first circulating primitive erythrocytes are ultimately replaced by new erythroid cells that develop later.

As the primitive erythroblasts enter the circulation and mature, they express erythroid-specific genes necessary for hemoglobin synthesis, including α and β embryonic globin chains, *alas2*, *fch*, and *urod* (**Table 1**). They also express DMT1, which is important for iron uptake, and scaffolding proteins such as β-spectrin and protein 4.1R, which ensure membrane stability. Zebrafish with mutations in these genes have red blood cell anomalies

Table 2 **Cloned zebrafish mutants and human disease counterparts**

Mutant	Abbr.[c]	Gene	Human disease	References
chardonnay	*cdy*	Divalent metal transporter 1 (DMT-1)	Microcytic anemia	(19)
chablis[a]	*cha*	Protein 4.1r	Hereditary elliptocytosis	(70)
chianti	*cia*	Transferrin receptor 1	—	(87)
dracula	*dra*	Ferrochelatase (fch)	Erythropoietic protoporphyria	(15)
moonshine[b]	*mon*	TIF 1γ	—	(66)
retsina	*ret*	Band 3	Congenital dyserythropoietic anemia type II (HEMPAS)	(63)
riesling	*ris*	β-spectrin	Hereditary spherocytosis	(47)
sauternes	*sau*	δ-aminolevulinate synthase (alas2)	Congenital sideroblastic anemia	(10)
vlad tepes	*vlt*	gata-1	Familial dyserythropoietic anemia and thrombocytopenia	(53)
weissherbst	*weh*	Ferroportin 1	Hemochromatosis type 4 (juvenile type)	(20)
yquem	*yqe*	Uroporphyrinogen decarboxylase (urod)	Porphyria tarda and hepatoerythropoietic porphyria	(78)
zinfandel	*zin*	Globin locus	Similar to thalassemia	(11)

[a]Allelic with *merlot* (*mot*) mutant.
[b]Allelic with *vampire* (*vmp*) mutant.
[c]Abbreviation.

analogous to many human red blood cell diseases (**Table 2**).

Embryonic Myelopoiesis

Like the primitive progenitors in the ICM, those found in the RBI contribute to both blood and vascular development, and express similar transcription factors, including *fli1a*, *gata2*, *lmo2*, and *scl* (9, 46, 72) (**Figure 1**). The cells seen in the RBI are morphologically identifiable as macrophages, and are first noted in the lateral head mesoderm in 3-somite embryos (8, 35). It is postulated that the macrophages derived from the RBI may represent a primitive wave of phagocytes, analogous to the primitive erythroid wave. At this time, it is unknown whether these early macrophages persist in adult zebrafish or whether they are replaced by a second wave of monocyte-derived macrophages.

A subset of these anterior cells expresses *pu.1*, a member of the ets family of transcription factors essential to the development of myeloid cells (51). This subset of *pu.1*$^+$ cells represents the early myeloid cells, including macrophages and granulocytes, which migrate rostrally during the 11–15 somite stages and then migrate laterally across the yolk sac around 22–24 hpf (8). *pu.1* expression is also seen in the ICM, where it disappears between 22–24 hpf. Anterior *pu.1* expression is lost by 28–30 hpf. Another important marker of the zebrafish macrophage is *l-plastin*, an actin-binding protein expressed in all human leukocytes but predominantly in monocytes and

Granulocyte: any white blood cell with cytoplasmic granules, including neutrophils, eosinophils and basophils

Ducts of Cuvier: the yolk sac blood channels connecting the heart to the trunk vasculature in zebrafish embryos

AGM: aorta-gonad-mesonephros

macrophages (8, 35). In zebrafish embryos, *l-plastin* is first expressed at 18 hpf in the anterior yolk region, and by 28 hpf, *l-plastin* is also expressed in the posterior ICM (8, 35), indicating that early zebrafish macrophages originate in both sites.

Myeloperoxidase (*mpo*), an enzyme that is a major component of human neutrophil and eosinophil granules, is also a marker for zebrafish granulocytes (8). Some early *mpo*$^+$ cells are found over the anterior yolk and co-express *pu.1* transcripts, indicating that these myeloid cells most likely originate in the RBI. Co-expression of *mpo* and *pu.1* is also seen in some cells of the posterior ICM (8). In contrast, double in situ hybridization experiments with *mpo* and *l-plastin* reveal that these markers represent distinct populations of cells. Similar to humans, in zebrafish *l-plastin* marks macrophages/monocytes, whereas *mpo* is expressed by the neutrophil lineage (8).

The function of these early macrophages has been observed as early as 26 hpf in the ducts of Cuvier, where macrophages can be observed engulfing apoptotic erythroid cells (35). These macrophages express *l-plastin* and are distributed throughout the embryo by 28–32 hpf. Primitive immune function of macrophages is also evident as they migrate to the site of an infection with *Escherichia coli* or *Bacillus subtilis* and phagocytose the offending organisms (35), or can clear microinjected carbon particles from the circulation as early as 2 dpf (50).

Driving Progenitor Cell Fate Within the ICM

The balance between primitive erythroid and myeloid cell production is a delicate one, governed predominantly by the equilibrium of *gata-1* and *pu.1* expression within the ICM. Galloway et al. showed that loss of *gata-1* expression in living zebrafish embryos resulted in transformation of erythroid precursors into myeloid precursors, with resulting myeloid expansion at the expense of total number of erythroid cells (26). In *gata-1*$^{-/-}$ embryos, *pu.1* expression persists longer in the ICM than seen in wild-type embryos, suggesting that *gata-1* normally functions in part to limit *pu.1* expression by these cells. In contrast, other early markers of hematopoiesis including *scl*, *lmo-2*, *gata-2*, *c-myb* and *ikaros*, all have normal expression in the *gata-1*$^{-/-}$ embryos. The early ICM blood precursors in the *gata-1*$^{-/-}$ embryos are converted into myeloid cells, an indication that *gata-1* acts both to promote erythroid development and also to suppress myeloid cell fate decisions (26).

DEFINITIVE (ADULT) HEMATOPOIESIS AND DEFINITIVE HSCs

By definition, definitive or adult hematopoiesis provides an organism with long-term HSCs capable of unlimited self-renewal and able to generate all mature hematopoietic lineages. In mammals, these HSCs are found in close association with the ventral wall of the dorsal aorta in a region known as the aorta-gonad-mesonephros (AGM), later transitioning to the mammalian fetal liver and the bone marrow (27, 91). The AGM equivalent in zebrafish is also found in the ventral wall of the dorsal aorta where several markers of HSCs are expressed between 24–48 hpf. While expression of *gata-1*, *gata-2*, and *scl* is waning in the ICM, expression of *c-myb* (72) and *runx1* (13, 42, 43) is noted in the AGM, marking cells believed to be the first definitive HSCs in these animals (**Figure 2**). In mice and zebrafish, Runx1 is a transcription factor essential for HSC formation in the AGM (13, 42, 43, 59). In the zebrafish, *runx1* transcripts are detected as early as 24 hpf, although it is uncertain at what time point these cells enter the circulation (13, 42, 43). Other transcription factors including *c-myb*, *ikaros*, *lmo-2*, and *scl* are also expressed in the AGM (72).

Around 4–5 days post fertilization (dpf), the location of blood formation shifts to the kidney in the zebrafish embryo as lifelong definitive hematopoiesis is established (27, 85). The AGM progenitors are believed to be the precursors of these definitive HSCs, which ultimately seed the embryonic kidney marrow, although lineage tracing has not yet been completed. In the adult zebrafish, hematopoietic cells are found intercalated between the renal tubules in the kidney marrow, much like mammalian adult hematopoiesis that takes place in and around the fat and stroma of the bone marrow. Cytospin preparations of the kidney marrow reveal that all the circulating hematopoietic blood cell types are present (**Figure 3**). Flow cytometry of these kidney marrow cells allows cell separation by forward scatter (indicative of cell size) and side scatter (indicative of granularity). This method has enabled development of a zebrafish hematopoietic stem cell transplantation model to facilitate further study of definitive HSCs in the zebrafish (74).

Definitive Erythropoiesis

The second wave of erythropoiesis is postulated to begin around 5 dpf, as the *bloodless* mutant (discussed below) with a defect in ICM erythropoiesis has recovery of circulating red blood cells around this stage (48). The process of erythropoiesis is very similar in mammals and nonmammalian vertebrates, although there are morphological differences between mature erythrocytes of these two groups. While erythroid precursors are very similar, the terminally differentiated mammalian red blood cells have a classic biconcave discoid shape and do not retain their nuclei, and the nonmammalian mature red blood cells are nucleated and elliptical in shape (**Figure 3**). Like other higher vertebrates, zebrafish express hemoglobin with a quaternary structure ($\alpha_2\beta_2$), and they undergo globin switching from primitive embryonic globin chains to the adult globin chains (11, 14).

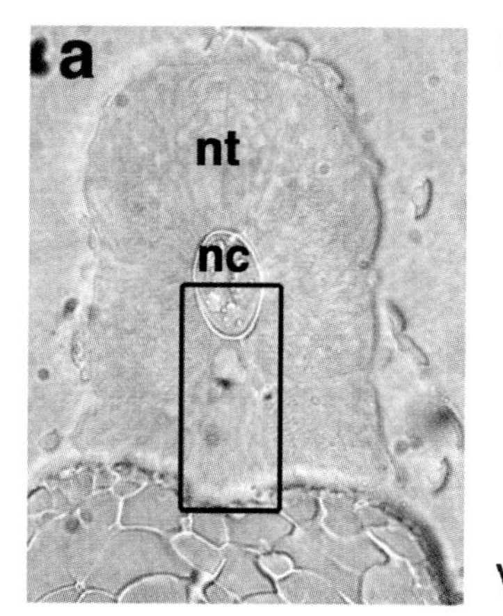

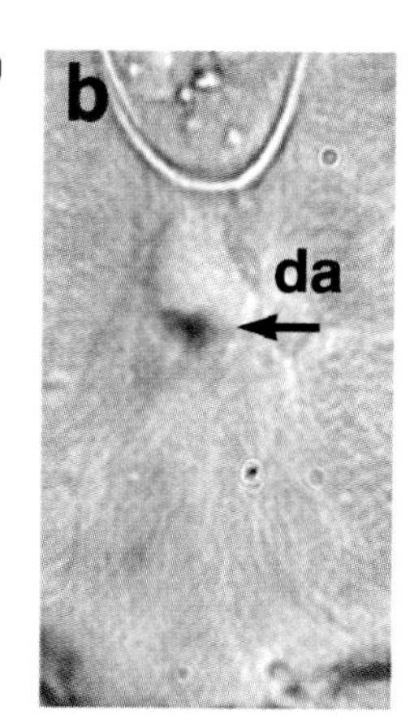

Figure 2

The zebrafish AGM. (*a*) Cross section through the trunk of a zebrafish embryo at 36 hpf after in situ hybridization with *runx1* probe, marking the AGM in the ventral wall of the dorsal aorta. (*b*) shows a higher magnification of the inset in (*a*). The arrow points to the AGM. Abbreviations: nt, neural tube; nc, notochord; da, dorsal aorta; D, dorsal; V, ventral.

Granulopoiesis

Zebrafish have two granulocyte lineages, one that is similar to mammalian neutrophils, and the other a unique cell type bearing characteristics of both mammalian eosinophils and basophils (8, 16, 50). Both human and zebrafish neutrophils stain positively for myeloperoxidase (MPO) and acid phosphatase, although morphologically zebrafish neutrophils are slightly different from their human counterparts in that the nuclei of zebrafish neutrophils have only two or three segments compared with the four or five segments found in human cells (**Figure 3**). Zebrafish eosinophils have a very granular cytoplasm, but do not possess the bilobed nucleus seen in human eosinophils (**Figure 3**). Their granules are peroxidase negative and periodic acid-Schiff (PAS) positive, and structurally are most like the granules in mammalian basophils and mast cells (8, 16, 50).

MPO: myeloperoxidase

PAS: periodic acid Schiff

As discussed above, the first granulocytes are identified by their expression of the granulocyte-specific marker, *mpo* in the posterior ICM and migrating across the anterior yolk around 18–20 hpf (8, 50). It is not known whether the granulocytes found in the pronephros by day 7 are these same primitive mpo^+ granulocytes or whether they represent a second wave of granulopoiesis (85). Granulocyte function in zebrafish is evident by migration of mpo^+ cells to a site of injury, such as tail clipping (50).

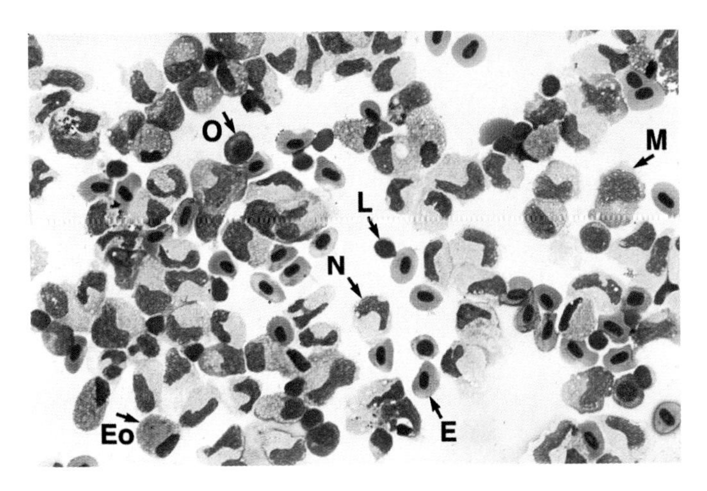

Figure 3

Histology of mature hematopoietic cell types in zebrafish kidney marrow. Cytospin preparation of whole-kidney marrow stained with May-Grünwald/Giemsa showing erythrocytes (E), eosinophils (Eo), lymphocytes (L), monocytes (M), neutrophils (N), and orthochromic erythroblasts (O). Photo from Reference (73), provided by Dr. D. Traver and used with permission.

Lymphopoiesis

Like mammals and other vertebrates, zebrafish possess an adaptive immune system complete with B cells expressing immunoglobulins and T cells expressing rearranged antigen-specific T cell receptors (45, 75). T cell maturation takes place in the zebrafish thymus, as it does in mice and humans. Bilateral thymic primordia form as outgrowths of pharyngeal epithelium between the third and fourth pharyngeal pouches and are first populated by immature lymphoblasts around 65 hpf (85). The derivation of the lymphoid progenitors that populate the thymus is not known. Given the timing when these cells appear, their progenitors must not originate in the kidney marrow, because HSCs are not yet found there until 5 to 6 dpf. A subset of the *ikaros*$^+$ cells in the AGM at 48 hpf are likely candidates for these lymphoid progenitors that ultimately seed the thymus (86).

The thymus continues to grow over the next several days such that by 7 dpf numerous small, mature lymphocytes populate the thymic epithelium (85). Characteristic T cell gene expression is observed around 3–4 dpf including *gata3*, *ikaros*, the T cell receptor kinase *lck*, and the recombination activating genes *rag-1* and *rag-2*, which are required for rearrangement of T cell receptor genes and also immunoglobulin genes in immature lymphocytes (84).

B cell development is established in the pronephros by 19 dpf (17, 44). Antibodies against classical T and B cell markers, such as CD2, CD3, CD19, CD20, and surface immunoglobulin and T cell receptors, do not exist for the zebrafish. As such, tracking of developing B and T cell populations within the zebrafish is difficult. Experiments using transgenic zebrafish, which express GFP (green fluorescent protein) driven by either the *rag2* or *lck* promoters, enabled fluorescent tracking of immature B and T lymphocytes, and mature T cells, respectively (44). These experiments confirmed that B cell development occurs in the kidney marrow and T cells develop exclusively in the thymus.

Thrombopoiesis

Zebrafish thrombocytes are nucleated blood cells equivalent to mammalian platelets, which function to maintain hemostasis by facilitating clot formation (37). Adhesion and aggregation of thrombocytes in response to ristocetin, collagen, thrombin, ADP, and arachidonic acid are similar to the activation of mammalian platelets by these agonists (37). In addition, polyclonal antisera against human platelet markers GpIIb/IIIa and GpIb are reactive with zebrafish thrombocytes, supporting the hypothesis that these surface glycoproteins important for platelet function have conserved expression on teleost thrombocytes (37). Despite functional similarity with mammalian platelets, the ontogeny of zebrafish thrombocytes is not well understood. Circulating thrombocytes are identified in zebrafish embryos as early as 36 hpf (30), although their precursors have not yet been discovered. In mammals, multinucleated giant cells called megakaryocytes are found in the bone marrow and are known to be the precursors of platelets. While no similar precursor cell has been identified in zebrafish, many of the mammalian transcription factors important for megakaryocyte development have also been identified in zebrafish, including *fli1*, *fog1*, *gata1*, *nfe2*, and *runx1* (13, 65, 69,

72). Further studies are needed to identify the thrombocyte precursors and the location of thrombopoiesis in the zebrafish.

GENETIC METHODS FOR STUDYING HEMATOPOIESIS IN ZEBRAFISH

The zebrafish has numerous advantages that make this organism particularly amenable to the study of hematopoiesis. The externally fertilized, optically clear embryos are easily observed and manipulated. Morphologic mutants are readily identified, including easy identification of blood mutants, as circulating blood cells are visible in embryos. Diffused oxygen in the fish water provides sufficient oxygenation for several days post fertilization. As such, anemic zebrafish mutants with few or no blood cells survive long enough for their mutant phenotype to be detectable. Several genetic methods are described below that take advantage of these characteristics of the zebrafish model.

Mutagenesis Screening

The first large-scale genetic screens in a vertebrate organism were performed using zebrafish in Boston, Massachusetts, and in Tübingen, Germany (23, 31). These screens utilized visual inspection to identify hundreds of mutant zebrafish embryos and early larvae. To perform this type of screen, adult male fish are exposed to ethylnitrosourea (ENU), which mutagenizes the DNA within the premeiotic germ cells (spermatogonia) (**Figure 4**). These fish are then mated to normal unmutagenized female fish. The resulting F_1 generation contains heterozygotes for mutations in genes within the zebrafish genome, at a rate of approximately 100 to 200 mutations per fish. The F_1 fish are mated to wild-type fish again and the resulting F_2 fish families are incrossed in order to identify the heterozygous carriers of the mutant genes.

Using this method, scores of mutant zebrafish alleles were identified affecting the development of virtually all aspects of zebrafish anatomy, including blood development (67, 81). Further ENU mutagenesis screens have utilized whole mount in situ hybridization to examine expression of various RNA transcripts in developing mutants (L. Zon, personal communication with the Tübingen 2000 Screen Consortium). Conceivably, any number of developmentally relevant transcripts could be used to screen for additional novel genes important for hematopoiesis, myelopoiesis, or any other interesting aspect of development. Once the mutant fish are

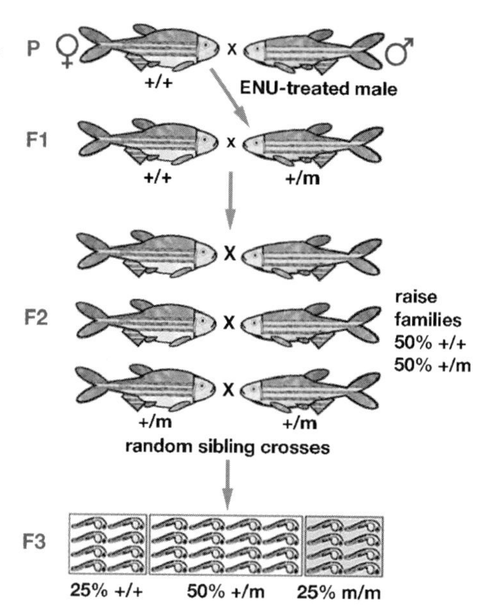

Figure 4

Schematic of a large-scale ENU mutagenesis screen. Adult male zebrafish are treated with ENU, which generates point mutations in the DNA of the spermatogonia. ENU-treated males are crossed with wild-type females to produce the F_1 generation. Heterozygous F_1 progeny are mated with wild-type fish to create F_2 families, in which half the siblings are heterozygous for a specific mutation (m) and the other half are wild type. Random sibling incrosses result in F_3 progeny. When two heterozygous F_2 siblings are crossed, the resulting F_3 clutch will contain 25% wild-type fish (+/+), 50% heterozygous fish (+/m) and 25% homozygous mutant fish (m/m) with a specific mutant phenotype. Figure is adapted from Reference (61) with permission.

ENU: N-ethyl N-nitrosourea

identified, the gene causing the phenotype can be mapped to a specific zebrafish chromosome (7). Such mapping endeavors have been markedly simplified since the sequencing of the zebrafish genome and the development of zebrafish genomic libraries (3). Haploid and gynogenetic diploid zebrafish embryos allow more efficient screening for mutant progeny, and these methods have also been successfully implemented for zebrafish screens (reviewed in 61). Insertional mutagenesis using a retroviral vector also allows genomic mutagenesis and facilitates cloning of the disrupted gene (4, 5, 29). However, the efficiency of mutagenesis is at a substantially lower frequency per genome, approximately 10% the rate observed with chemical mutagenesis.

Incross mating: mating of two siblings in order to identify heterozygous carriers of a recessive mutant gene. When two heterozygous carriers are mated, the progeny will be 25% wild type, 50% heterozygous carriers, and 25% mutant

Gynogenetic: refers to organisms derived from maternal chromosomes only, without fusion of the egg and sperm nuclei

Morpholino: specific anti-sense RNA oligonucleotide sequence that is chemically modified to resist degradation, making it biologically active for up to five days to inhibit expression of a specific gene when injected into single-cell zebrafish embryos

TILLING: Targeted Induced Local Lesions In Genomes

Reverse genetic studies: mutation and manipulation of a specific gene of interest in order to study the actions of that gene in cells and whole animals. Examples include overexpressing or knocking out the gene of interest in a transgenic animal

Morpholinos

"Knockdown" of mRNA for specific genes of interest can be achieved in zebrafish by the use of morpholinos, specific anti-sense RNA oligomer sequences (57). These oligomers are modified to prevent degradation in the cell. As such, they persist for 1 to 3 days when microinjected into the fertilized one-cell embryo, and act to inhibit RNA translation and/or splicing of the specific sequence of interest by targeting the translation initiation site or splice sites. While highly specific, these morpholinos are short-lived and their injection can produce dose-dependent toxicity (32). An example of this powerful method is outlined by Dooley et al., who used *scl* morpholinos to show that knockdown of *scl* resulted in loss of primitive and definitive hematopoietic cell lineages, without losing the specification of early angioblasts (22).

Transgenic Reporters

Transgenic mouse models have been a mainstay of genetic studies. Microinjection of DNA constructs into single-cell fertilized zebrafish embryos has proven successful in the generation of transgenic zebrafish. The widespread use of fluorescent proteins in mammalian systems has been successfully adapted for use in zebrafish, which are well-suited to the use of fluorescence because of their optical clarity and external development. By linking a fluorescent protein such as GFP or DsRed to a gene or promoter of interest, expression can be easily visualized in living animals. Transgenic reporter vectors with a constitutive or tissue-specific promoter (for example, the *gata-1* promoter for erythroid cells) linked to a fluorescent reporter gene enable fish to express fluorescent proteins, which are easily visible under the microscope in these transparent embryos. To study erythroid cells, GFP was placed under control of the *gata-1* promoter (52). These fish have erythroid-specific GFP expression by 16 hpf. Stable transgenic fish can be generated from the progeny, as 1%–5% of injected embryos will integrate the transgene into the germline DNA. Tissue-specific expression of the recombinating activating gene (*rag-1*) promoter driving GFP expression has aided the understanding of lymphoid development in zebrafish (38, 39, 44). Vascular and hematopoietic tissues are visualized in vivo with double transgenic embryos expressing an *lmo2:DsRed* transgene as well as *gata1*:GFP (see **Figure 5**) (89, 90). Fluorescent transgenic fish have proven to be useful tools for cell fate mapping and also for hematopoietic stem cell transplantation experiments (44, 74), but the applications are virtually endless as new promoters are discovered and different cell types can be studied.

TILLING

Targeted Induced Local Lesions In Genomes (TILLING) is a reverse genetics method for high-throughput screening for mutations in specific genes of interest (82, 83). This procedure has been successfully used to identify target-selected mutants in numerous species including plants, *Drosophila*, *Caenorhabditis elegans*, rats, mice, and zebrafish. Much like

ENU mutagenesis for forward genetic screens, this strategy involves random chemical mutagenesis of male animals, which are mated to wild-type females to generate a large population of F_1 animals. The F_1 males are sacrificed to generate matched libraries of genomic DNA for screening and cryopreserved sperm samples to recover lines of interest. An alternate approach is to screen tail clip DNA from live F_1 fish kept in isolation or in small pools. DNA from the F_1 males is amplified using nested PCR and sequenced to identify mutations in a specific gene of interest. Cryopreserved sperm from identified mutant animals is then used for in vitro fertilization so that the mutant phenotype can be recovered and studied. This technique was utilized to identify 15 different mutations in the rag-1 gene (82). By identifying many mutants of a single gene, partial knock-out of that gene can be observed, critical signaling domains can be characterized, and downstream signals can be elucidated.

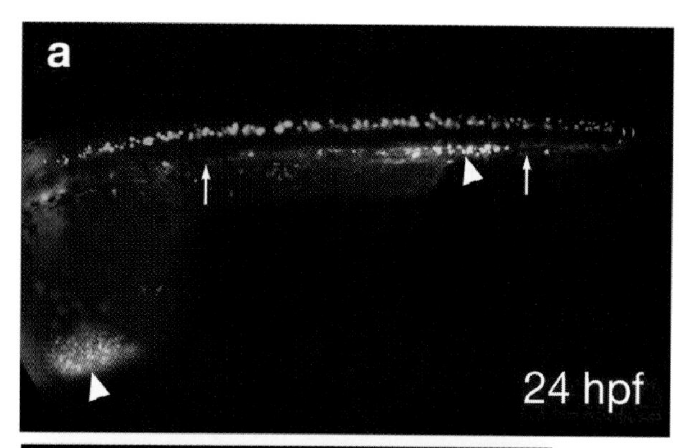

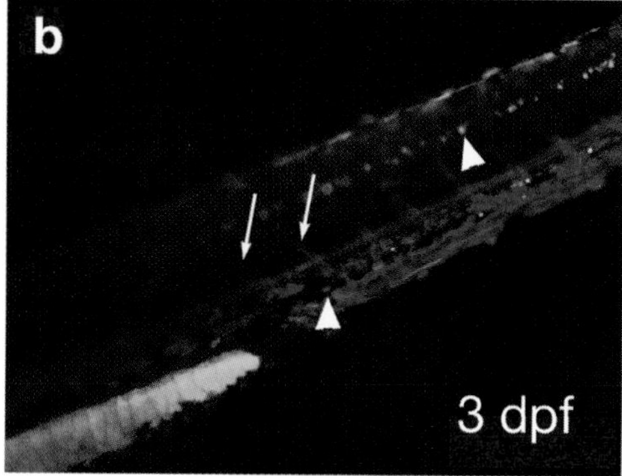

Figure 5

Fluorescent imaging of hematopoietic and vascular tissues in double transgenic zebrafish embryos expressing *lmo2:DsRed* and *gata1:EGFP* transgenes. Zebrafish embryos at 24 hpf (*a*) and 3 dpf (*b*) with red and green erythrocytes (*arrowheads*) circulating through vessels labeled by DsRed (*arrows*). Photo is reprinted from Reference (89) and is used by permission.

ZEBRAFISH MUTANTS AS MODELS FOR HUMAN BLOOD DISEASES

Large-scale forward genetic chemical mutagenesis screens in zebrafish have generated many zebrafish mutants with interesting developmental phenotypes (23, 31), some of which mirror human diseases (**Table 2**). More than 40 mutants with hematopoietic defects, consisting of 26 complementation groups, were identified in the large-scale mutagenesis screens in Boston and Tübingen in the mid-1990s (67, 81). These mutants were initially identified by gross observation of decreased blood circulating in early embryos and were later characterized by whole mount in situ hybridization studies to assess expression of known hematopoietic transcription factors including *gata-1* and *scl*. The blood mutant phenotypes can be divided into several subcategories: those with no red blood cells (bloodless mutants), progressive anemia, hypochromic anemia, or photosensitivity (67).

Bloodless Mutants

Several zebrafish mutants identified possess defects in the development of erythroid progenitor cells. Moonshine (*mon*) zebrafish mutants fail to express gata-1 normally, and they have a severe anemia resulting from disruption of both the primitive and definitive waves of hematopoiesis (66). This mutation is a noncomplementing allele of the *vampire* gene, also identified in a forward genetic screen (81). By 5 to 7 dpf, the *mon* embryos have only 5%–10% of the circulating cells that are found in wild-type embryos (66). While wild-type animals have 1000–3000 circulating nucleated erythrocytes, *mon* mutants have only 50–100 circulating cells and no detectable hemoglobin, as measured by staining with *o*-dianisidine (67). The few rare circulating cells that are seen have a proerythroblast morphology, suggestive of a differentiation blockade in the *mon* mutants. Most of these severely anemic animals do not live past 10 to 14 dpf, although rare homozygous *mon* mutant animals survive to adulthood (66). These adult mutants have marked cardiomegaly due to severe anemia and the resulting state of high cardiac output. Further characterization of the *mon* mutant reveals severely decreased expression of gata-1 and gata-2, along with normal expression of markers for lymphoid and myeloid cells, data that also support a specific block in early red blood cell differentiation. Ultimately, positional cloning revealed the *mon* gene encodes the zebrafish orthologue of the mammalian transcriptional intermediary factor 1γ (TIF1γ), a transcription factor that plays a critical role in early erythroid differentiation (66).

As discussed above, *gata-1* is another transcription factor that plays an important role in erythropoiesis. Through positional and candidate gene cloning, *gata-1* was identified as the truncated mutant gene responsible for the bloodless phenotype of the *vlad tepes* (*vlt*) zebrafish mutant (53). Until 24 hpf, these embryos have normal expression of *lmo-2*, *scl*, and *cbfb*, all markers of early hematopoietic progenitors. Normal development of myeloid and lymphoid lineages is also noted, as measured by expression of *pu.1*, *l-plastin* and *c/ebp1*, and *ikaros* and *rag-1*, respectively. Intact expression of stem cell, myeloid, and lymphoid markers indicates normal development of these cell types in the *vlt* mutants (53). Despite the presence of normal hematopoietic progenitors, the lack of functional *gata-1* makes these cells unable to differentiate down the erythroid lineage.

The *bloodless* (*bls*) zebrafish mutant embryos have virtually no circulating red blood cells until approximately 5 dpf when hemoglobinized cells begin to appear, coincident with the onset of definitive hematopoiesis (48). Originally isolated as a spontaneous mutation, *bls* is inherited in an autosomal dominant fashion with incomplete penetrance. As such, the phenotype varies from a complete absence of primitive erythrocytes to very severe early anemia. Kidney marrow hematopoiesis appears normal in the *bls* adult animals, demonstrating that only primitive hematopoiesis is affected. Cell transplantation experiments demonstrate that *bls* mutant donor cells were unable to contribute to *gata-1*-expressing cells of the ICM in wild-type recipients (48). In addition, wild-type donor cells could not differentiate into *gata-1*-expressing cells in the ICM of *bls* recipients (48). While the *bls* gene has not yet been identified, the non-cell autonomous nature of *bls* function suggests that this may be a secreted factor.

Mutants with Progressive Anemia

Several of the red blood cell mutants identified have defects resulting in a decreasing number of blood cells after apparently normal blood development in the first 2 dpf. Examples in this group include *cabernet* (*cab*), *chablis* (*cha*), *grenache* (*gre*), *merlot* (*mot*), *riesling* (*ris*), *retsina* (*ret*), and *thunderbird* (*tbr*) (67, 81). These mutants all initiate hematopoiesis normally and in the first few days of life are difficult to distinguish from their wild-type siblings. They initially have normal expression of *gata-1*, typically pigmented blood, and staining with *o*-dianisidine. They begin to develop anemia around 2 to 4 dpf, suggesting that these mutations affect erythroid cell expansion or stability of the mature red cells.

Mutations of cytoskeletal proteins. The *mot* and *cha* mutants, now known to bear mutations in the same gene, have normal onset of primitive hematopoiesis and later develop anemia around 4 dpf (70). Morphologically, these erythrocytes are noted to have spiculated membranes and bi-lobed nuclei. At 24 hpf, *mot/cha* homozygotes exhibit normal expression of *gata-1*, *scl*, and embryonic globin, indicating that while erythropoiesis remains intact, the abnormal mature red blood cells are hemolyzed. A small percentage of homozygous *mot* animals survive to adulthood when raised with frequent feeding and an abundant level of oxygen in the water. These fish have significant anemia and their red blood cells are arrested in the basophilic erythroblast stage. The erythrocytes of adult homozygous animals have bizarre membrane projections and increased osmotic fragility, suggestive of certain human hemolytic anemias, which also result from increased osmotic fragility. In fact, using positional cloning and candidate gene cloning methods, the *mot* and *cha* mutations were found to reside in the gene encoding erythrocyte protein 4.1 (also called band 4.1 or 4.1R), a structural membrane protein expressed in red blood cells that binds to spectrin and anchors the spectrin-actin cytoskeleton to the erythrocyte cell membrane (70). Deficiency of protein 4.1 in humans causes hereditary elliptocytosis, a rare cause of hemolytic anemia characterized by elliptical-shaped red blood cells (76, 77).

β-spectrin, another critical membrane protein in erythrocytes, is one of the most abundant components of the erythrocyte cytoskeleton (76). Using comparative genomics, the zebrafish mutant *ris* was found to have a null mutation in the erythroid *β-spectrin* gene (47). By 4 dpf, homozygous *ris* mutants are severely anemic. The *ris* erythrocytes are abnormally shaped with large nuclei, as opposed to the characteristic elliptical shape of adult zebrafish erythrocytes. These cells are reminiscent of those seen in the human erythroid disorder hereditary spherocytosis (HS), which results from mutations in human *β-spectrin* (24). Adult *ris* zebrafish have an increased number of hematopoietic progenitors in the kidney marrow, analogous to the increased number of red cell progenitors in the bone marrow of humans with HS.

Similar to the *mot/cha* and *ris* mutants, the zebrafish mutant *ret* becomes anemic around 4 dpf, shortly after the onset of definitive hematopoiesis. Unlike these other mutants with membrane protein abnormalities, the circulating mature erythrocytes in the *ret* mutants are arrested at the erythroblast stage, and a large proportion (approximately 27%) have two nuclei (63), suggestive of a defect in cytokinesis. This phenotype is suggestive of the human disorder known as congenital dyserythropoietic anemia type II (33, 34, 62, 63). Evaluation of the *ret* mutants revealed that *band 3* is the mutant gene responsible for this phenotype (62, 63). Band 3 was demonstrated to play a critical role in chromosomal segmentation during anaphase, such that lack of band 3 prevented cytokinesis in zebrafish erythroblasts (62, 63). This result is similar in mice as well, indicating conservation of function across phylogeny. Other proteins important in humans for red blood cell structure, including ankyrin and erythroid α-spectrin, have not yet been identified among zebrafish mutants.

HS: hereditary spherocytosis

Hypochromic Mutants

The hypochromic zebrafish mutants have abnormally small (microcytic) erythrocytes with pale cytoplasm (hypochromia) indicative of a decreased level of hemoglobin. Since the production of hemoglobin is a complex, multifaceted process, there are numerous aspects of hemoglobin biosynthesis that can go awry. Examples of zebrafish mutants in this category include *chardonnay* (*cdy*), *chianti* (*cia*), *frascati* (*frs*), *gavi* (*gav*), *montalcino* (*mnt*), *sauternes* (*sau*), *shiraz* (*sir*), *weissherbst* (*weh*), and *zinfandel* (*zin*) (67, 81).

The *sau* hypochromic mutation was found to map to the δ-aminolevulinate (*ALAS2*) gene, known for its importance in the heme biosynthetic pathway (10). The *sau* mutant mimics the human disease congenital sideroblastic anemia, which consists of anemia secondary to reduced heme synthesis, and cellular toxicity from elevated iron levels (68). The *sau* zebrafish mutant represents the first animal model of this disease. In the zebrafish *sau* mutants, primitive erythrocytes differentiate abnormally such that embryos have decreased heme and β-globin levels by 2 dpf.

The *zin* zebrafish mutation maps to the zebrafish major globin locus of chromosome 3, and as such, this hypochromic mutant may represent a disorder similar to the thalassemias seen in humans where the expression of globin chains is unbalanced (11). When either the α-globin or β-globin protein is markedly decreased, the excess globin chains precipitate within the erythrocyte, leading to generally profound anemia (24). It is postulated that the *zin* mutation may lie within the locus control region of the zebrafish globin gene cluster, thereby resulting in decreased globin expression. Since the zebrafish α- and β-globin loci are located in the same chromosome, unlike the human globin genes, the *zin* mutant may be slightly different from the classic human thalassemia where globin chain expression is unbalanced. Mapping of the *zin* mutant gene is currently under way, and will help clarify the understanding of hypochromic anemias.

Defects in iron metabolism. Iron metabolism is an essential component of heme synthesis, and hence defects in its absorption, storage and trafficking can lead to hypochromic anemias in humans. Another hypochromic mutant, *weh*, has decreased mean corpuscular hemoglobin levels but a nearly normal total number of red blood cells. This mutation has been mapped to the ferroportin 1 gene, which encodes a novel iron transporter (20, 25). Mouse and human orthologues for ferroportin 1 have been cloned and the iron transporter appears to have conserved function in vertebrates. Mutations in this gene have been identified in patients with hemochromatosis type IV (56, 58).

The *cdy* zebrafish mutation is an autosomal recessive allele resulting in delayed anemia similar to the *sau* and *weh* mutants. These embryos have a normal number of circulating erythrocytes until approximately 48 hpf, at which time they are noted to have hypochromic, microcytic erythrocytes with abnormal, less-condensed nuclei, indicative of immaturity. The *cdy* cells have normal expression of globin genes. Positional cloning and candidate mapping strategies identified the iron transporter gene *DMT1* as the mutant gene (19). In vertebrates, divalent metal transporter1 (DMT1) conveys ferrous iron from the gut into the duodenal enterocyte, while the iron transporter ferroportin 1 transports iron across the basolateral membrane of the enterocyte into the circulation (71). Transport of iron from the endosome into the cytosol requires DMT1, and thereby allows the iron to be available to the cell for incorporation into hemoglobin (6, 40).

Another of the hypochromic zebrafish mutants, *cia*, also appears related to iron metabolism. This mutation is homozygous viable in the adult fish and results in a hypochromic anemia first evident around 36 hpf by staining with *o*-dianisidine (87). The animals survive into adulthood without growth retardation or cardiomegaly; however, the circulating erythrocytes of adult *cia* zebrafish are hypochromic and microcytic, and the kidney marrow of these fish is hypercellular, with an increased proportion of erythroid progenitors (87). Using a candidate cloning strategy, Wingert et al. showed that *cia* zebrafish have a mutation in the transferrin receptor 1 gene (*tfr1a*). Exclusively expressed by zebrafish erythrocytes, *tfr1a* is critical for iron acquisition by these cells. By injecting single-cell fertilized embryos with iron-dextran, the hypochromia of *cia* mutants can be rescued, while intravenous iron-dextran injection at 48 hpf fails to rescue the hypochromic phenotype (87). A second transferrin receptor orthologue identified in zebrafish, *tfr1b*, is expressed in many cell types throughout embryogenesis. Overexpression of *tfr1b* is sufficient to compensate for the lack of *tfr1a* in the *cia* mutants, although it appears the *tfr1b* orthologue is primarily utilized by non-hematopoietic tissues (87).

Photosensitive Mutants

Photosensitive blood mutants result in erythrocytes that autofluoresce and lyse when exposed to ambient light, a phenotype analogous to the human congenital erythropoietic porphyrias. Examples of these photosensitive mutants include *dracula* (*drc*), *desmodius* (*dsm*), *freixinet* (*frx*), and *yquem* (*yqe*).

Mutations in the *yqe* gene result in a porphyria syndrome characterized by autofluoresence under UV light, and photoablation of erythrocytes when subjected to light exposure. The porphyria syndrome in humans is due to defects in heme biosynthesis. Using porphyrin and enzymatic assays, uroporphyrinogen decarboxylase (UROD) was identified as the enzymatic deficiency in the *yqe* zebrafish mutants (78). This proved to be the first animal model for hepatoerythropoietic porphyria, the human disease equivalent (68).

drc is another zebrafish mutant with a photosensitive phenotype. By 4 dpf, these embryos lack all red blood cells when raised under normal lighting conditions. Their red blood cells exhibit strong fluorescence. Like the human disorder erythropoietic protoporphyria, which manifests as light-dependent hemolysis of red blood cells in combination with liver disease, the *drc* mutant fish have an accumulation of protoporphyin IX (68). This was suggestive of a deficiency in ferrochelatase, which in fact turned out to be the defective enzyme in these mutants (15).

SUMMARY POINTS

1. The zebrafish serves as a unique model organism for the study of hematopoiesis, with many advantages over other vertebrate and invertebrate models.
2. Zebrafish orthologues have been identified for numerous mammalian transcription factors important for hematopoiesis.
3. As in other vertebrates, hematopoiesis in zebrafish occurs in two waves, the primitive or embryonic wave, and the definitive or adult wave.
4. Primitive hematopoiesis occurs in an intraembryonic location called the ICM, and generates predominantly primitive erythrocytes and some myeloid cells.
5. Definitive hematopoiesis occurs in the AGM and is marked by the expression of *runx1* and *c-myb* around 24–48 hpf. By 5 dpf, definitive hematopoiesis has migrated to the zebrafish kidney marrow, where all hematopoietic cell types are produced during the lifespan of the animal.
6. Large-scale forward genetic screens have identified dozens of zebrafish blood mutations, many of which have now been cloned, and which represent animal models for known human diseases.

FUTURE DIRECTIONS

1. Identification of zebrafish analogues of mammalian genes has been facilitated by the sequencing of the zebrafish genome. New genes continue to be mapped.
2. Affymetrix zebrafish GeneChips allow the simultaneous expression analysis of approximately 14,900 zebrafish transcripts. Recently, differential gene expression profiling using these GeneChips was reported in various zebrafish blood mutants, revealing several novel hematopoietic and vascular genes (80). This tool has broad potential for future studies.

3. Chemical screens employ libraries of chemicals to look for their effects on blood development in mutant and wild-type zebrafish embryos. Each chemical is individually incubated with 10–20 embryos, which are then tested for changes in expression of *gata-1*, *scl*, *c-myb*, or other genes of interest. These screens are ongoing in our lab and should yield further insights into stimulation of hematopoietic progenitors and the genetic pathways that drive these cells.

ACKNOWLEDGMENTS

The authors would like to thank Caroline Burns and Trista North for critical review of the manuscript. We thank Jenna Galloway for providing portions of **Figure 1**, Caroline Burns for providing **Figure 2** and portions of **Figure 1**, David Traver for providing **Figure 3**, Elizabeth Patton for assistance with **Figure 4**, and Hao Zhu for providing **Figure 5**. J.L.O.d. is supported by NIH Training Grant T32 HL07574.

LITERATURE CITED

1. Al-Adhami MA, Kunz YW. 1977. Ontogenesis of haematopoietic sites in *Brachydanio rerio*. *Dev. Growth Differ.* 19:171–79
2. Amatruda JF, Zon LI. 1999. Dissecting hematopoiesis and disease using the zebrafish. *Dev. Biol.* 216:1–15
3. Amemiya CT, Zhong TP, Silverman GA, Fishman MC, Zon LI. 1999. Zebrafish YAC, BAC, and PAC genomic libraries. *Methods Cell Biol.* 60:235–58
4. **Amsterdam A, Burgess S, Golling G, Chen W, Sun Z, et al. 1999. A large-scale insertional mutagenesis screen in zebrafish. *Genes Dev.* 13:2713–24**
5. Amsterdam A, Nissen RM, Sun Z, Swindell EC, Farrington S, Hopkins N. 2004. Identification of 315 genes essential for early zebrafish development. *Proc. Natl. Acad. Sci. USA* 101:12792–97
6. Andrews NC. 1999. Disorders of iron metabolism. *N. Engl. J. Med.* 341:1986–95
7. Bahary N, Davidson A, Ransom D, Shepard J, Stern H, et al. 2004. The Zon laboratory guide to positional cloning in zebrafish. *Methods Cell Biol.* 77:305–29
8. Bennett CM, Kanki JP, Rhodes J, Liu TX, Paw BH, et al. 2001. Myelopoiesis in the zebrafish, *Danio rerio*. *Blood* 98:643–51
9. Brown LA, Rodaway AR, Schilling TF, Jowett T, Ingham PW, et al. 2000. Insights into early vasculogenesis revealed by expression of the ETS-domain transcription factor Fli-1 in wild-type and mutant zebrafish embryos. *Mech. Dev.* 90:237–52
10. **Brownlie A, Donovan A, Pratt SJ, Paw BH, Oates AC, et al. 1998. Positional cloning of the zebrafish sauternes gene: a model for congenital sideroblastic anaemia. *Nat. Genet.* 20:244–50**
11. Brownlie A, Hersey C, Oates AC, Paw BH, Falick AM, et al. 2003. Characterization of embryonic globin genes of the zebrafish. *Dev. Biol.* 255:48–61
12. Brownlie A, Zon L. 1999. The zebrafish as a model system for the study of hematopoiesis. *BioScience* 49:382–92
13. Burns CE, DeBlasio T, Zhou Y, Zhang J, Zon L, Nimer SD. 2002. Isolation and characterization of runxa and runxb, zebrafish members of the runt family of transcriptional regulators. *Exp. Hematol.* 30:1381–89

The seminal paper describing retroviral gene insertional mutagenesis in zebrafish.

Describes the zebrafish mutant *sauternes*, which expresses a mutant form of the ALAS2 enzyme critical for heme biosynthesis. This is the first animal disease model for congenital sideroblastic anemia in humans.

14. Chan FY, Robinson J, Brownlie A, Shivdasani RA, Donovan A, et al. 1997. Characterization of adult alpha- and beta-globin genes in the zebrafish. *Blood* 89:688–700
15. Childs S, Weinstein BM, Mohideen MA, Donohue S, Bonkovsky H, Fishman MC. 2000. Zebrafish dracula encodes ferrochelatase and its mutation provides a model for erythropoietic protoporphyria. *Curr. Biol.* 10:1001–4
16. Crowhurst MO, Layton JE, Lieschke GJ. 2002. Developmental biology of zebrafish myeloid cells. *Int. J. Dev. Biol.* 46:483–92
17. Davidson AJ, Zon LI. 2004. The 'definitive' (and 'primitive') guide to zebrafish hematopoiesis. *Oncogene* 23:7233–46
18. Detrich HW 3rd, Kieran MW, Chan FY, Barone LM, Yee K, et al. 1995. Intraembryonic hematopoietic cell migration during vertebrate development. *Proc. Natl. Acad. Sci. USA* 92:10713–17
19. Donovan A, Brownlie A, Dorschner MO, Zhou Y, Pratt SJ, et al. 2002. The zebrafish mutant gene chardonnay (cdy) encodes divalent metal transporter 1 (DMT1). *Blood* 100:4655–59
20. Donovan A, Brownlie A, Zhou Y, Shepard J, Pratt SJ, et al. 2000. Positional cloning of zebrafish ferroportin1 identifies a conserved vertebrate iron exporter. *Nature* 403:776–81
21. Dooley K, Zon LI. 2000. Zebrafish: a model system for the study of human disease. *Curr. Opin. Genet. Dev.* 10:252–56
22. Dooley KA, Davidson AJ, Zon LI. 2005. Zebrafish scl functions independently in hematopoietic and endothelial development. *Dev. Biol.* 277:522–36
23. **Driever W, Fishman MC. 1996. The zebrafish: heritable disorders in transparent embryos. *J. Clin. Invest.* 97:1788–94**

This reference, together with reference 31, describes the first large-scale forward genetic screens in a vertebrate model, which were done in Boston, Massachusetts and Tübingen, Germany.

24. Forget BG, Olivieri NF. 2003. Chapter 48: Hemoglobin synthesis and the thalassemias. See Ref. 31a, pp. 1503–96
25. Fraenkel PG, Traver D, Donovan A, Zahrieh D, Zon LI. 2005. Ferroportin 1 is required for normal iron cycling in zebrafish. *J. Clin. Invest.* 115:1532–41
26. Galloway JL, Wingert RA, Thisse C, Thisse B, Zon LI. 2005. Loss of gata1 but not gata2 converts erythropoiesis to myelopoiesis in zebrafish embryos. *Dev Cell* 8:109–16
27. Galloway JL, Zon LI. 2003. Ontogeny of hematopoiesis: examining the emergence of hematopoietic cells in the vertebrate embryo. *Curr. Top. Dev. Biol.* 53:139–58
28. Gering M, Patient R. 2005. Hedgehog signaling is required for adult blood stem cell formation in zebrafish embryos. *Dev Cell* 8:389–400
29. Golling G, Amsterdam A, Sun Z, Antonelli M, Maldonado E, et al. 2002. Insertional mutagenesis in zebrafish rapidly identifies genes essential for early vertebrate development. *Nat. Genet.* 31:135–40
30. Gregory M, Jagadeeswaran P. 2002. Selective labeling of zebrafish thrombocytes: quantitation of thrombocyte function and detection during development. *Blood Cells Mol. Dis.* 28:418–27
31. **Haffter P, Granato M, Brand M, Mullins MC, Hammerschmidt M, et al. 1996. The identification of genes with unique and essential functions in the development of the zebrafish, *Danio rerio*. *Development* 123:1–36**

31a. Handin RI, Lux SE, Stossel TP, eds. 2003. *Blood: Principles and Practice of Hematology*. New York: Lippincott Williams & Wilkins

32. Heasman J. 2002. Morpholino oligos: making sense of antisense? *Dev. Biol.* 243:209–14
33. Heimpel H. 2004. Congenital dyserythropoietic anemias: epidemiology, clinical significance, and progress in understanding their pathogenesis. *Ann. Hematol.* 83:613–21

34. Heimpel H, Anselstetter V, Chrobak L, Denecke J, Einsiedler B, et al. 2003. Congenital dyserythropoietic anemia type II: epidemiology, clinical appearance, and prognosis based on long-term observation. *Blood* 102:4576–81
35. Herbomel P, Thisse B, Thisse C. 1999. Ontogeny and behaviour of early macrophages in the zebrafish embryo. *Development* 126:3735–45
36. Hsu K, Traver D, Kutok JL, Hagen A, Liu TX, et al. 2004. The pu.1 promoter drives myeloid gene expression in zebrafish. *Blood* 104:1291–97
37. Jagadeeswaran P, Sheehan JP, Craig FE, Troyer D. 1999. Identification and characterization of zebrafish thrombocytes. *Br. J. Haematol.* 107:731–38
38. Jessen JR, Jessen TN, Vogel SS, Lin S. 2001. Concurrent expression of recombination activating genes 1 and 2 in zebrafish olfactory sensory neurons. *Genesis* 29:156–62
39. Jessen JR, Willett CE, Lin S. 1999. Artificial chromosome transgenesis reveals long-distance negative regulation of rag1 in zebrafish. *Nat. Genet.* 23:15–16
40. Johnson DM, Yamaji S, Tennant J, Srai SK, Sharp PA. 2005. Regulation of divalent metal transporter expression in human intestinal epithelial cells following exposure to non-haem iron. *FEBS Lett.* 579:1923–29
41. Kalev-Zylinska ML, Horsfield JA, Flores MV, Postlethwait JH, Chau JY, et al. 2003. Runx3 is required for hematopoietic development in zebrafish. *Dev. Dyn.* 228:323–36
42. Kalev-Zylinska ML, Horsfield JA, Flores MV, Postlethwait JH, Vitas MR, et al. 2002. Runx1 is required for zebrafish blood and vessel development and expression of a human RUNX1-CBF2T1 transgene advances a model for studies of leukemogenesis. *Development* 129:2015–30
43. Kataoka H, Ochi M, Enomoto K, Yamaguchi A. 2000. Cloning and embryonic expression patterns of the zebrafish Runt domain genes, runxa and runxb. *Mech. Dev.* 98:139–43
44. Langenau DM, Ferrando AA, Traver D, Kutok JL, Hezel JP, et al. 2004. In vivo tracking of T cell development, ablation, and engraftment in transgenic zebrafish. *Proc. Natl. Acad. Sci. USA* 101:7369–74
45. Langenau DM, Zon LI. 2005. The zebrafish: a new model of T-cell and thymic development. *Nat. Rev. Immunol.* 5:307–17
46. Liao EC, Paw BH, Oates AC, Pratt SJ, Postlethwait JH, Zon LI. 1998. SCL/Tal-1 transcription factor acts downstream of cloche to specify hematopoietic and vascular progenitors in zebrafish. *Genes Dev.* 12:621–26
47. Liao EC, Paw BH, Peters LL, Zapata A, Pratt SJ, et al. 2000. Hereditary spherocytosis in zebrafish riesling illustrates evolution of erythroid beta-spectrin structure, and function in red cell morphogenesis and membrane stability. *Development* 127:5123–32
48. Liao EC, Trede NS, Ransom D, Zapata A, Kieran M, Zon LI. 2002. Non-cell autonomous requirement for the bloodless gene in primitive hematopoiesis of zebrafish. *Development* 129:649–59
49. Liao W, Ho CY, Yan YL, Postlethwait J, Stainier DY. 2000. Hhex and scl function in parallel to regulate early endothelial and blood differentiation in zebrafish. *Development* 127:4303–13
50. Lieschke GJ, Oates AC, Crowhurst MO, Ward AC, Layton JE. 2001. Morphologic and functional characterization of granulocytes and macrophages in embryonic and adult zebrafish. *Blood* 98:3087–96

51. Lieschke GJ, Oates AC, Paw BH, Thompson MA, Hall NE, et al. 2002. Zebrafish SPI-1 (PU.1) marks a site of myeloid development independent of primitive erythropoiesis: implications for axial patterning. *Dev. Biol.* 246:274–95
52. Long Q, Meng A, Wang H, Jessen JR, Farrell MJ, Lin S. 1997. GATA-1 expression pattern can be recapitulated in living transgenic zebrafish using GFP reporter gene. *Development* 124:4105–11
53. Lyons SE, Lawson ND, Lei L, Bennett PE, Weinstein BM, Liu PP. 2002. A nonsense mutation in zebrafish *gata1* causes the bloodless phenotype in *vlad tepes*. *Proc. Natl. Acad. Sci. USA* 99:5454–59
54. Lyons SE, Shue BC, Lei L, Oates AC, Zon LI, Liu PP. 2001. Molecular cloning, genetic mapping, and expression analysis of four zebrafish c/ebp genes. *Gene* 281:43–51
55. Lyons SE, Shue BC, Oates AC, Zon LI, Liu PP. 2001. A novel myeloid-restricted zebrafish CCAAT/enhancer-binding protein with a potent transcriptional activation domain. *Blood* 97:2611–17
56. Montosi G, Donovan A, Totaro A, Garuti C, Pignatti E, et al. 2001. Autosomal-dominant hemochromatosis is associated with a mutation in the ferroportin (SLC11A3) gene. *J. Clin. Invest.* 108:619–23
57. Nasevicius A, Ekker SC. 2000. Effective targeted gene 'knockdown' in zebrafish. *Nat. Genet.* 26:216–20
58. Njajou OT, Vaessen N, Joosse M, Berghuis B, van Dongen JW, et al. 2001. A mutation in SLC11A3 is associated with autosomal dominant hemochromatosis. *Nat. Genet.* 28:213–14
59. North TE, de Bruijn MF, Stacy T, Talebian L, Lind E, et al. 2002. Runx1 expression marks long-term repopulating hematopoietic stem cells in the midgestation mouse embryo. *Immunity* 16:661–72
60. Patterson LJ, Gering M, Patient R. 2005. Scl is required for dorsal aorta as well as blood formation in zebrafish embryos. *Blood* 105:3502–11
61. Patton EE, Zon LI. 2001. The art and design of genetic screens: zebrafish. *Nat. Rev. Genet.* 2:956–66
62. Paw BH. 2001. Cloning of the zebrafish retsina blood mutation: a genetic model for dyserythropoiesis and erythroid cytokinesis. *Blood Cells Mol. Dis.* 27:62–64
63. Paw BH, Davidson AJ, Zhou Y, Li R, Pratt SJ, et al. 2003. Cell-specific mitotic defect and dyserythropoiesis associated with erythroid band 3 deficiency. *Nat. Genet.* 34:59–64
64. Porcher C, Swat W, Rockwell K, Fujiwara Y, Alt FW, Orkin SH. 1996. The T cell leukemia oncoprotein SCL/tal-1 is essential for development of all hematopoietic lineages. *Cell* 86:47–57
65. Pratt SJ, Drejer A, Foott H, Barut B, Brownlie A, et al. 2002. Isolation and characterization of zebrafish NFE2. *Physiol. Genomics* 11:91–98
66. Ransom DG, Bahary N, Niss K, Traver D, Burns C, et al. 2004. The zebrafish moonshine gene encodes transcriptional intermediary factor 1gamma, an essential regulator of hematopoiesis. *PLoS Biol.* 2:E237
67. **Ransom DG, Haffter P, Odenthal J, Brownlie A, Vogelsang E, et al. 1996. Characterization of zebrafish mutants with defects in embryonic hematopoiesis. *Development* 123:311–19**
68. Sassa S, Shibahara S. 2003. Disorders of heme production and catabolism. See Ref. 31a, pp. 1435–501
69. Schulze H, Shivdasani RA. 2004. Molecular mechanisms of megakaryocyte differentiation. *Semin. Thromb. Hemost.* 30:389–98

This reference, together with reference 81, describes the first panel of hematopoietic zebrafish mutants generated from a large-scale forward genetic screen.

70. Shafizadeh E, Paw BH, Foott H, Liao EC, Barut BA, et al. 2002. Characterization of zebrafish merlot/chablis as nonmammalian vertebrate models for severe congenital anemia due to protein 4.1 deficiency. *Development* 129:4359–70
71. Thomas C, Oates PS. 2004. Ferroportin/IREG-1/MTP-1/SLC40A1 modulates the uptake of iron at the apical membrane of enterocytes. *Gut* 53:44–49
72. Thompson MA, Ransom DG, Pratt SJ, MacLennan H, Kieran MW, et al. 1998. The cloche and spadetail genes differentially affect hematopoiesis and vasculogenesis. *Dev. Biol.* 197:248–69
73. Traver D. 2004. Cellular dissection of zebrafish hematopoiesis. *Methods Cell Biol.* 76:127–49
74. **Traver D, Paw BH, Poss KD, Penberthy WT, Lin S, Zon LI. 2003. Transplantation and in vivo imaging of multilineage engraftment in zebrafish bloodless mutants. *Nat. Immunol.* 4:1238–46**

A paper describing FACS analysis of adult zebrafish kidney marrow, and hematopoietic stem cell transplantation methodologies in zebrafish.

75. Trede NS, Zon LI. 1998. Development of T-cells during fish embryogenesis. *Dev. Comp. Immunol.* 22:253–63
76. Tse WT, Lux SE. 1999. Red blood cell membrane disorders. *Br. J. Haematol.* 104:2–13
77. Walensky LD, Narla M, Lux SE. 2003. Disorders of the red blood cell membrane. See Ref. 31a, pp. 1709–858
78. Wang H, Long Q, Marty SD, Sassa S, Lin S. 1998. A zebrafish model for hepatoerythropoietic porphyria. *Nat. Genet.* 20:239–43
79. Ward AC, Lieschke G. 2002. The zebrafish as a model system for human diseases. *Front. Biosci.* 7:827–33
80. Weber GJ, Choe SE, Dooley KA, Paffett-Lugassy NN, Zhou Y, Zon LI. 2005. Mutant specific gene programs in the zebrafish. *Blood* 106:521–30
81. **Weinstein BM, Schier AF, Abdelilah S, Malicki J, Solnica-Krezel L, et al. 1996. Hematopoietic mutations in the zebrafish. *Development* 123:303–9**
82. Wienholds E, Schulte-Merker S, Walderich B, Plasterk RH. 2002. Target-selected inactivation of the zebrafish rag1 gene. *Science* 297:99–102
83. Wienholds E, van Eeden F, Kosters M, Mudde J, Plasterk RH, Cuppen E. 2003. Efficient target-selected mutagenesis in zebrafish. *Genome Res.* 13:2700–7
84. Willett CE, Cherry JJ, Steiner LA. 1997. Characterization and expression of the recombination activating genes (rag1 and rag2) of zebrafish. *Immunogenetics* 45:394–404
85. Willett CE, Cortes A, Zuasti A, Zapata AG. 1999. Early hematopoiesis and developing lymphoid organs in the zebrafish. *Dev. Dyn.* 214:323–36
86. Willett CE, Kawasaki H, Amemiya CT, Lin S, Steiner LA. 2001. Ikaros expression as a marker for lymphoid progenitors during zebrafish development. *Dev. Dyn.* 222:694–98
87. Wingert RA, Brownlie A, Galloway JL, Dooley K, Fraenkel P, et al. 2004. The chianti zebrafish mutant provides a model for erythroid-specific disruption of transferrin receptor 1. *Development* 131:6225–35
88. Wingert RA, Zon LI. 2003. Genetic dissection of hematopoiesis using the zebrafish. In *Hematopoietic Stem Cells*, ed. I Godin, A Cumano, pp. 1–18. Georgetown, TX: Landes Biosci.
89. Zhu H, Traver D, Davidson AJ, Dibiase A, Thisse C, et al. 2005. Regulation of the lmo2 promoter during hematopoietic and vascular development in zebrafish. *Dev. Biol.* 281:256–69

90. Zhu H, Zon LI. 2004. Use of the DsRed fluorescent reporter in zebrafish. *Methods Cell Biol.* 76:3–12
91. Zon LI. 1995. Developmental biology of hematopoiesis. *Blood* 86:2876–91

LINKS

Sanger Institute *Danio rerio* sequencing project: http://www.sanger.ac.uk/Projects/D_rerio/
The Zebrafish Information Network (ZFIN): http://zfin.org/
Zon lab: http://zon.tchlab.org

Mitochondrial Morphology and Dynamics in Yeast and Multicellular Eukaryotes

Koji Okamoto and Janet M. Shaw

Department of Biochemistry, University of Utah School of Medicine, Salt Lake City, Utah 84132-3201; email: kokamoto@biology.utah.edu, shaw@bioscience.utah.edu

Annu. Rev. Genet.
2005. 39:503–36

First published online as a Review in Advance on August 5, 2005

The *Annual Review of Genetics* is online at genet.annualreviews.org

doi: 10.1146/annurev.genet.38.072902.093019

0066-4197/05/1215-0503$20.00

Key Words

mitochondria, morphology, fusion, fission, GTPase

Abstract

Mitochondria form dynamic tubular networks that continually change their shape and move throughout the cell. In eukaryotes, these organellar gymnastics are controlled by numerous pathways that preserve proper mitochondrial morphology and function. The best understood of these are the fusion and fission pathways, which rely on conserved GTPases and their binding partners to regulate organelle connectivity and copy number in healthy cells and during apoptosis. In budding yeast, mitochondrial shape is also maintained by proteins acting in the tubulation pathway. Novel proteins and pathways that control mitochondrial dynamics continue to be discovered, indicating that the mechanisms governing this organelle's behavior are more sophisticated than previously appreciated. Here we review recent advances in the field of mitochondrial dynamics and highlight the importance of these pathways to human health.

Contents

INTRODUCTION

Mitochondria are double membrane-bound organelles with highly specialized function and morphology. They contain their own genomes and transcription/translation systems, producing energy essential for diverse cellular functions (4, 69, 117). Mitochondria also compartmentalize reactions and molecules critical for metabolism, signaling, and programmed cell death (14, 21, 118). Numerous studies reveal that mitochondrial morphology is both complex and plastic. In many cell types, mitochondria form elongated tubules that are often interconnected (5, 58, 96). Mitochondria are also dynamic, frequently changing size and shape and traveling long distances on cytoskeletal tracks (5, 10, 96). Sophisticated mechanisms that regulate different morphologies and distributions help to optimize mitochondrial function in response to changing intracellular needs and extracellular cues.

In the early 1990s, genetic screens in the budding yeast *Saccharomyces cerevisiae* identified the first proteins required for mitochondrial distribution and morphology (85), and tubulation (13, 132). Subsequent studies revealed that the connectivity of the mitochondrial network is regulated by two dynamically opposed processes, fusion and fission (7, 120). Importantly, most proteins mediating yeast mitochondrial fusion and fission are conserved in flies, worms, plants, mice, and humans, indicating that the fundamental mechanisms controlling mitochondrial behavior have been maintained during evolution. This conservation, combined with the availability of sophisticated assays and a wealth of genomic information, has established *S. cerevisiae* as one of the premier systems for studying molecular mechanisms of mitochondrial dynamics (97).

This review highlights advances in our molecular and mechanistic understanding of mitochondrial dynamics. Key players in the mitochondrial fusion, fission, and tubulation pathways are described, and molecular models for their activities are discussed. In addition, recently identified proteins that may define novel pathways are introduced. The biological significance of mitochondrial morphology and dynamics is emphasized, in particular, how defects in these processes impact cell viability and human health. Readers interested in these and related issues are also directed to reviews on mitochondrial ultrastructure (40, 46, 82, 112), inheritance (10, 149), fusion and fission (18, 63, 94, 100, 114, 119, 125, 145, 146, 150, 153), and mitochondrial dynamics in programmed cell death (11, 36, 68, 105).

FUSION AND FISSION: TWO OPPOSING EVENTS

Loss of fusion results in mitochondrial fragmentation due to ongoing fission events. Fragmented mitochondria eventually lose mtDNA by unknown mechanisms. Yeast cells defective in fusion cannot, therefore, grow on nonfermentable media, which require respiration for energy production. Loss of fission leads to formation of interconnected, net-like mitochondria due to ongoing fusion events. Net-like mitochondria are respiratory competent. Cells defective in both fusion and fission harbor wild-type-like mitochondrial networks and can grow on nonfermentable media. These observations provide evidence that mitochondrial morphology is determined by balanced fusion and fission. Is mitochondrial function truly wild-type in the absence of both fusion and fission? The answer is "No." Cells lacking both fusion and fission tend to lose mtDNA at an elevated frequency relative to wild-type cells when grown on fermentable media, which do not require respiration. Thus, dynamic fusion and fission events are important for mtDNA integrity and the bioenergetic function of mitochondria.

THE BASICS OF YEAST MITOCHONDRIAL MORPHOLOGY AND DYNAMICS

A cross-sectional view of a yeast mitochondrial tubule reveals four distinct parts, an outer membrane and inner membrane separated by an intermembrane space, and a compartment enclosed by the inner membrane called the matrix. Within the matrix, mitochondrial DNA (mtDNA) nucleoids (mitochondrial genomes packaged with proteins) are attached to the inner membrane. In wild-type cells, mitochondrial tubules form a branched network at the cell cortex (**Figure 1**). Mutations in a number of nuclear genes disrupt wild-type mitochondrial networks, resulting in distinct morphology phenotypes.

Yeast mutants define at least three major morphology pathways based on defects in mitochondrial shape (**Figure 1**). The first pathway mediates mitochondrial fusion. When fusion is blocked, mitochondria fragment due to ongoing fission. The second pathway mediates mitochondrial fission. When fission is blocked, mitochondria form interconnected nets due to ongoing fusion (see Side bar). Yeast utilizes a third pathway required for formation of tubular mitochondria. When this tubulation pathway is disrupted, mitochondria are converted to large spheres.

Fission: GTP-dependent process by which a mitochondrial tubule divides

Fusion: process in which a free mitochondrial tip fuses with the tip or side of another mitochondrial tubule in a GTP- and innermembrane potential-dependent manner

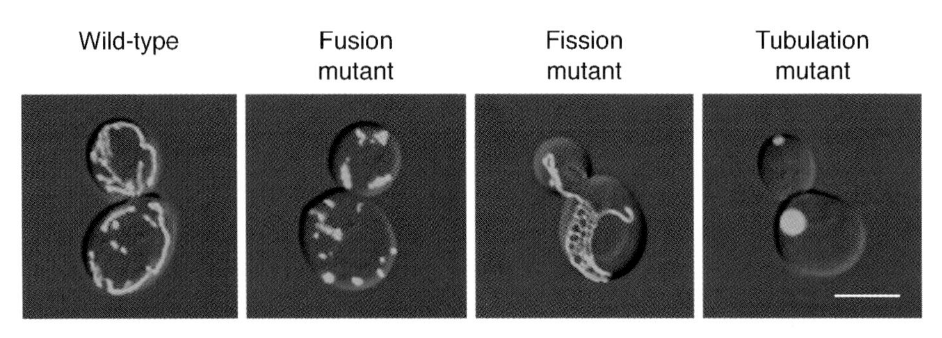

Figure 1

Mitochondrial morphology in *Saccharomyces cerevisiae*. Wild-type and mutant cells defective in mitochondrial fusion (*fzo1*Δ), fission (*dnm1*Δ), and tubulation (*mmm1*Δ) are shown. Mitochondria are visualized by a matrix-targeted GFP. DIC (differential interference contrast) and GFP fluorescence images are superimposed for each cell. Bar, 5 μm.

THE FUSION PATHWAY

Fzo1p

Mitochondrial fusion requires the evolutionarily conserved GTPase called Fzo (fuzzy onions) in budding yeast (55, 111) and fruit flies (49), and Mfn (mitofusin) in mammals (31, 113, 115, 116). Yeast Fzo1p contains four predicted heptad repeats (coiled-coil forming domains) that may mediate protein-protein interactions, one GTPase domain, and two transmembrane segments (**Figure 2**). Although genetic studies indicate that the Fzo1p GTPase domain is essential for function in vivo (55), the biochemical activity of this region has not been determined in vitro. Fzo1p is an integral outer membrane protein with its coiled-coil and GTPase domains exposed to the cytoplasm and a short linker between the two transmembrane segments exposed to the intermembrane space (41, 55, 111) (**Figure 3**).

Coiled-coil: protein structure in which two or three α-helices form a rope-like bundle via hydrophobic interactions. Coiled-coil forming regions in a linear polypeptide consist of seven repeated residues called a heptad repeat

GTPase: an enzyme that catalyzes guanosine 5′-triphosphate hydrolysis to yield guanosine 5′-diphosphate and a free phosphate

Mgm1p

Mgm1p (mitochondrial genome maintenance) (47, 65, 126, 147), a second GTPase essential for mitochondrial fusion, is related to dynamin GTPases that mediate membrane scission during endocytosis (109). Mgm1p contains an N-terminal presequence, two hydrophobic segments, a GTPase domain, and a middle domain (**Figure 2**). The GTPase domain is essential for Mgm1p function in vivo (124, 126, 148), although the biochemical activity of this region has not been demonstrated in vitro. Orthologs of Mgm1p have been identified in other eukaryotes including fission yeast (Msp1) (103, 104) and mammals (OPA1, optic atrophy) (1, 22, 91). The Mgm1/Msp1/OPA1 protein family resides in the mitochondrial intermembrane space (48, 53, 99, 124, 147, 148) (**Figure 3**). Mgm1p is present in two forms, one integrated into the inner membrane, and a second, peripheral membrane-associated species, which is created by proteolytic processing (discussed in more detail below).

Ugo1p

A third mitochondrial fusion protein named Ugo1p has only been identified in fungi (Ugo means fusion in Japanese) (121). Ugo1p contains at least one transmembrane segment in the middle of the protein and two motifs similar to those found in mitochondrial carrier proteins that transport small molecules across the inner membrane (**Figure 2**). However, unlike characterized carrier proteins, Ugo1p is embedded in the outer membrane with its N terminus exposed to the cytoplasm and its C terminus in the intermembrane space (121) (**Figure 3**). Since the outer mitochondrial

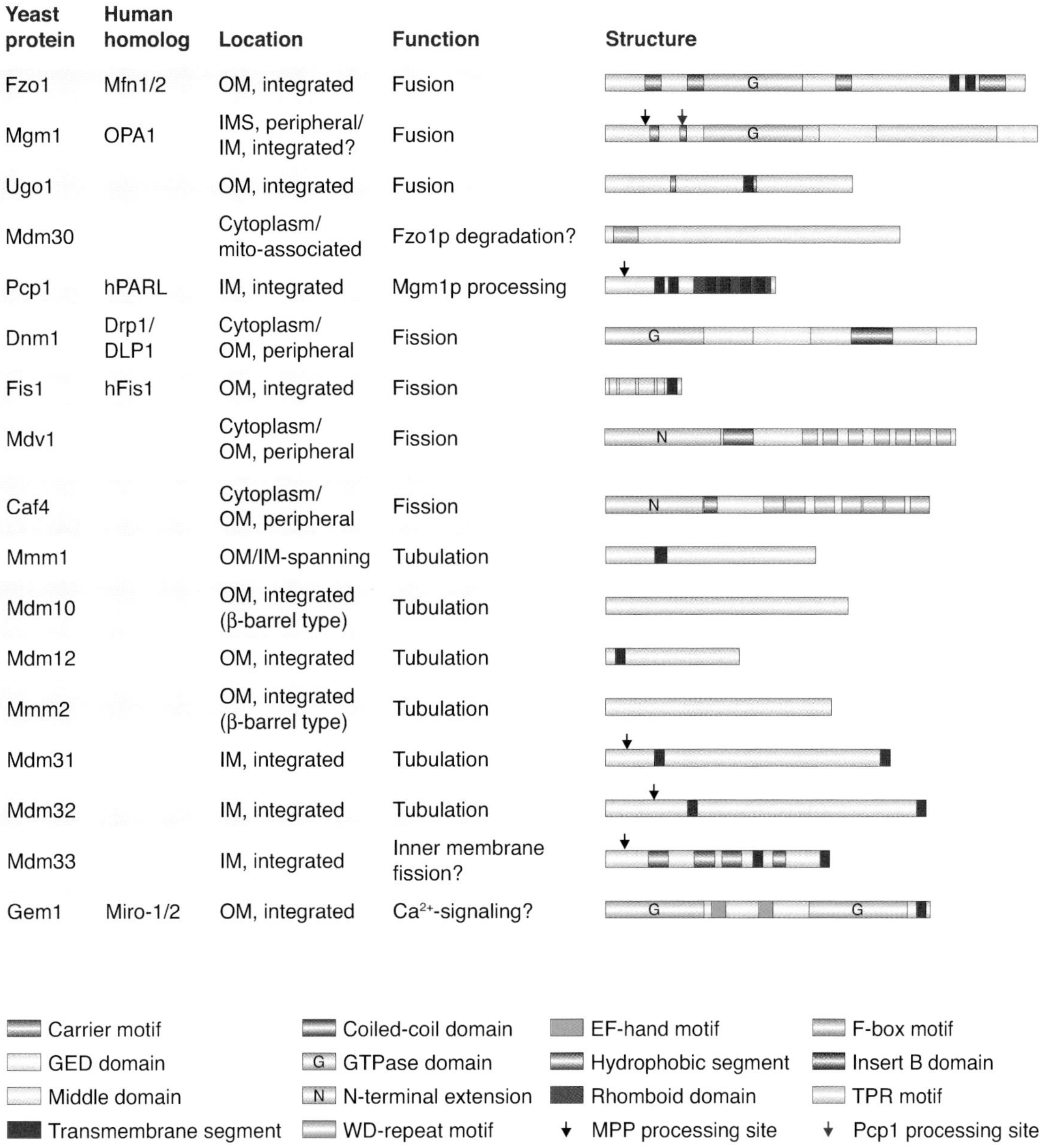

Yeast protein	Human homolog	Location	Function	Structure
Fzo1	Mfn1/2	OM, integrated	Fusion	G
Mgm1	OPA1	IMS, peripheral/ IM, integrated?	Fusion	G
Ugo1		OM, integrated	Fusion	
Mdm30		Cytoplasm/ mito-associated	Fzo1p degradation?	
Pcp1	hPARL	IM, integrated	Mgm1p processing	
Dnm1	Drp1/ DLP1	Cytoplasm/ OM, peripheral	Fission	G
Fis1	hFis1	OM, integrated	Fission	
Mdv1		Cytoplasm/ OM, peripheral	Fission	N
Caf4		Cytoplasm/ OM, peripheral	Fission	N
Mmm1		OM/IM-spanning	Tubulation	
Mdm10		OM, integrated (β-barrel type)	Tubulation	
Mdm12		OM, integrated	Tubulation	
Mmm2		OM, integrated (β-barrel type)	Tubulation	
Mdm31		IM, integrated	Tubulation	
Mdm32		IM, integrated	Tubulation	
Mdm33		IM, integrated	Inner membrane fission?	
Gem1	Miro-1/2	OM, integrated	Ca^{2+}-signaling?	G G

Carrier motif; GED domain; Middle domain; Transmembrane segment; Coiled-coil domain; G GTPase domain; N N-terminal extension; WD-repeat motif; EF-hand motif; Hydrophobic segment; Rhomboid domain; MPP processing site; F-box motif; Insert B domain; TPR motif; Pcp1 processing site

Figure 2

Mitochondrial morphology proteins in budding yeast. OM, outer membrane; IMS, intermembrane space; IM, inner membrane.

membrane contains large pores that allow free passage of small molecules and proteins, it seems unlikely that Ugo1p functions as a classical carrier protein. Moreover, mutations predicted to be important for carrier domain function do not disrupt the fusion activity of Ugo1p (E.M. Coonrod & J.M. Shaw, unpublished), supporting the idea that these motifs are either an evolutionary relic, or have acquired a different function in Ugo1p.

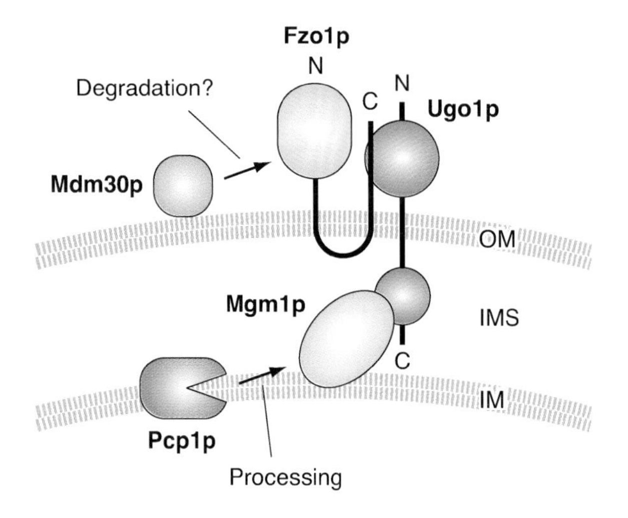

Figure 3

Localization and topology of proteins acting in the mitochondrial fusion pathway. Ugo1p interacts with Fzo1p and Mgm1p via its N- and C-terminal domains, respectively. OM, outer membrane; IMS, intermembrane space; IM, inner membrane.

Fusion Complex Formation by Yeast Fzo1p, Ugo1p, and Mgm1p

Fzo1p, Mgm1p, and Ugo1p form a fusion complex that connects the inner and outer membranes (**Figure 3**) and can be immunoprecipitated from isolated mitochondria (122, 124, 148). The observation that Ugo1p contains both cytoplasmic and intermembrane space domains raised the possibility that Ugo1p bridges interactions between Fzo1p and Mgm1p in the fusion complex. A series of in vitro pull-down experiments demonstrated that the cytoplasmic and intermembrane space domains of Ugo1p bind Fzo1p and Mgm1p, respectively (122). Additional analyses suggested that the Fzo1p-Ugo1p interaction is essential for mitochondrial fusion (122), and that the Mgm1p-Ugo1p interaction is Fzo1p-independent (148).

Several observations suggest that the Fzo1p and Mgm1p GTPase activities are not directly required for formation of the fusion complex. First, a truncated Fzo1 protein lacking its GTPase domain continues to interact with the Ugo1p cytoplasmic domain in vitro (122). Second, an Mgm1^{S224N} mutant protein, which is predicted to lack GTPase activity and is defective in mitochondrial fusion (124), is also able to interact with Ugo1p (122). In addition, Fzo1p can be immunoprecipitated with Ugo1p in the presence of Mgm1^{S224N} (122), suggesting that the Fzo1p-Ugo1p interaction is independent of the Mgm1p GTPase activity.

Although mammalian Ugo1p homologs have not been found, it is conceivable that mitofusins and OPA1 may cooperatively regulate mitochondrial fusion in mammalian cells. Consistent with this idea, overexpression of OPA1 promotes mitochondrial fusion in mouse embryonic fibroblasts, which depend on Mfn1 but not Mfn2 (20). These observations raise the possibility that the Mfn1 and OPA1 GTPases form a transmembrane complex via a direct or indirect interaction.

Development of an In Vitro Mitochondrial Fusion Assay

An important technique for studying the biochemical mechanism of mitochondrial fusion has recently been developed: an in vitro mitochondrial fusion assay (87).

In vitro, both outer and inner-membrane fusion reactions require GTP binding and hydrolysis (87), presumably due to the participation of the Fzo1p and Mgm1p GTPases in the outermembrane and intermembrane space, respectively. Outer membrane fusion occurred in the absence of exogenous GTP, and was dependent on the inner-membrane proton gradient (ΔpH) (87). Innermembrane fusion required a higher GTP (0.5 mM) concentration, as well as an electrical potential across the inner membrane ($\Delta\varphi$) (87). In addition, outer membrane and inner membrane fusion events are separable in vitro (87). These observations suggest that the biochemical mechanisms of outer and inner-membrane fusion events differ.

The availability of an assay that stages outer and inner membrane fusion reactions should enhance the analysis of biochemical steps mediating these events (106). However, additional signals and/or checkpoints may exist in vivo to coordinate outer and inner-membrane fusion events. When assayed in vivo, outer mitochondrial membrane fusion is not only blocked in the absence of the outer membrane proteins Fzo1p and Ugo1p, it is also blocked in the absence of the intermembrane space protein Mgm1p (124). Thus, although Mgm1p is not positioned to participate in the tethering and/or docking of two apposing mitochondria, it may play an essential late role in outer membrane fusion as part of the fusion complex. Whether this complex is also sufficient for inner-membrane fusion, or whether additional components are required to accomplish inner-membrane fusion remains to be determined.

Early Steps in Mitochondrial Fusion

A variety of studies suggest that mammalian mitofusins mediate mitochondrial docking. Coimmunoprecipitation studies indicate that mammalian Mfn1 and Mfn2 form both homo- and hetero-complexes and have overlapping functions in mitochondrial fusion (19, 31). When mouse embryonic fibroblasts lacking both Mfn1 and Mfn2 are fused to wild-type cells, mitochondrial fusion fails to occur (73), suggesting that Mfn *trans*-interactions are required for this process.

Early studies of mitofusins speculated that the predicted coiled-coil domains in these proteins mediate *trans*-interactions and docking of apposing mitochondria. Overexpression of wild-type or GTPase mutant mammalian Mfns causes clustering of fragmented mitochondria, a state that is proposed to reflect membrane docking (31, 113, 115, 116). However, the competence of such clustered organelles to subsequently fuse has not been tested, raising the question of whether they are truly productive intermediates. A structural study demonstrated that the Mfn1 C terminus forms a dimeric antiparallel coiled-coil 95 Å in length (73). Introduction of mutations that reduced the stability of this coiled-coil caused mitochondrial fragmentation and blocked mitochondrial fusion in vivo (73). Mitochondrial aggregation induced by expressing a GTPase-minus Mfn1 protein in tissue-culture cells was also abolished by mutations in the coiled-coil domain of the mutant protein (73). These combined findings support the idea that the C-terminal coiled-coil domains of mitofusins mediate docking of two adjacent mitochondria prior to outer membrane fusion.

Docking: a state in which two apposing membranes are joined in close proximity but have not yet fused

In yeast, mutational analyses indicate that all predicted coiled-coil domains are important for either Fzo1p stability or function, including two N-terminal coiled-coil domains that are not present in other species (Y. Saint-Georges & J.M. Shaw, unpublished). Whether the C termini of two yeast Fzo1 proteins form a similar antiparallel coiled-coil remains to be seen.

The precise role of the Fzo1p or Mfn GTPase cycles in their *trans*-interaction is still unclear. Preliminary studies in yeast indicate that Fzo1p *trans*-complexes form in the complete absence of GTP (Y. Saint-Georges & J.M. Shaw, in preparation). By contrast, *trans*-complex formation by mammalian Mfn1 reportedly requires GTP and does not occur in the presence of GTPase domain mutations (59).

Models for Mitochondrial Fusion

Although models for mitochondrial fusion are still incomplete, the process can be separated into at least three steps: docking, outer membrane fusion, and inner-membrane fusion. During mammalian mitochondrial fusion, docking occurs when the C-terminal domain of Mfn1 on one mitochondrion forms an antiparallel coiled-coil with the C-terminal domain of another Mfn1 molecule on a second mitochondrion (**Figure 4*a***). This *trans* homotypic interaction brings two apposing mitochondria into a docked state. One possibility

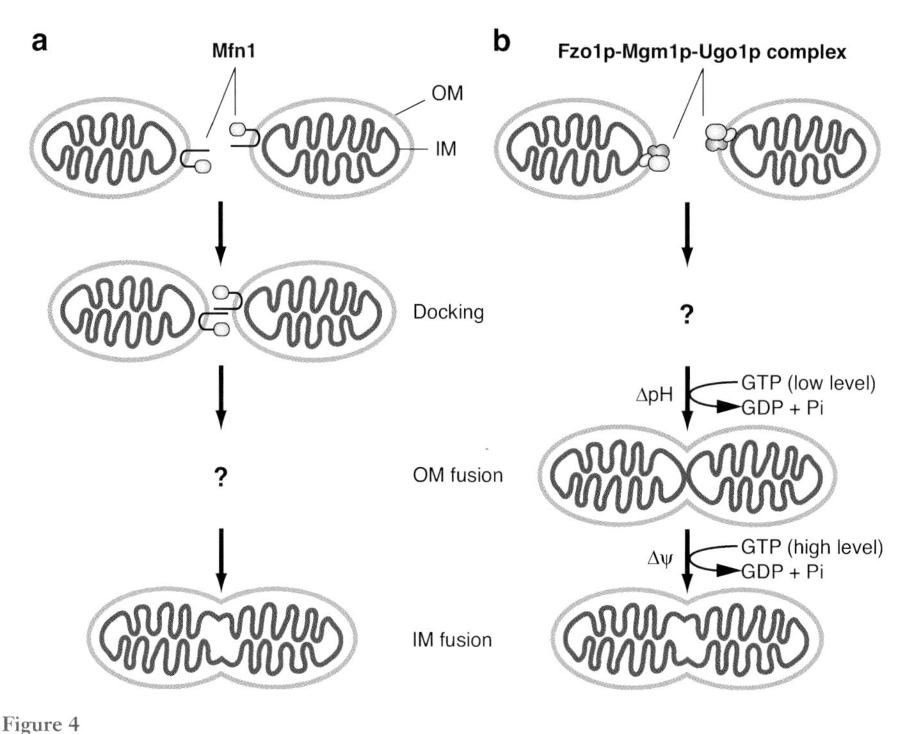

Figure 4

Models for mitochondrial fusion in mammals (*a*) and yeast (*b*). ΔpH, inner membrane proton gradient; $\Delta\psi$, inner-membrane electrical potential; OM, outer membrane; IM, inner membrane.

is that Mfn1-Mfn1 dimerization in *trans* may activate GTP binding or hydrolysis, triggering a conformational change that shortens the length of the *trans* complex and pulls two adjacent mitochondria closer together. Alternatively, this conformational change could also occur independent of the Mfn GTPase cycle. Although it has been shown that the inner-membrane potential is essential for mitochondrial fusion in mammalian cells (60, 78, 84), the biochemical mechanisms mediating outer and inner-membrane fusion events have not been dissected in vitro (**Figure 4*a***).

How docking of yeast mitochondria is mediated by the Fzo1p-Mgm1p-Ugo1p complex remains unclear (**Figure 4*b***). In the yeast in vitro fusion assay, the docking and outer membrane fusion events have not been staged. One or both of these steps requires endogenous GTP and the ΔpH. Inner-membrane fusion requires exogeneous GTP and the $\Delta\varphi$. Although a role for Mgm1p in inner-membrane fusion has not been experimentally established, it is conceivable that this GTPase consumes the higher levels of nucleotide required to complete inner-membrane fusion.

Regulation of Mitochondrial Fusion Proteins

The role of Mdm30p in controlling Fzo1p protein levels. A morphological screen of the yeast nonessential deletion collection revealed a mutant containing aggregated and fragmented mitochondria (26). The gene lacking in this mutant, *MDM30* (mitochondrial distribution and morphology), encodes a 70-kDa protein containing an F-box motif at its N terminus (**Figure 2**). Many F-box proteins are components of the SCF (Skp1p-Cdc53p-F-box) E3 ubiquitin ligase complexes that, in cooperation with an ubiquitin activating enzyme (E1) and an ubiquitin conjugating enzyme (E2), catalyze ubiquitylation of target proteins destined for 26S proteasome-mediated degradation (102). Two-hybrid interaction studies and in vitro reconstitution experiments suggest that Mdm30p forms a complex with Skp1p and Cdc53p (74, 143), consistent with the idea that Mdm30p functions in a ubiquitin-dependent protein degradation pathway.

Like known fusion mutants, mitochondrial morphology defects in cells lacking Mdm30p are suppressed by blocking fission (42). In addition, fragmented *mdm30Δ* mitochondria do not fuse during mating (42). Unlike bona fide fusion mutants, however, mitochondrial fusion occurs in *mdm30Δ* zygotes when fission is blocked and mitochondrial tubular morphology is restored (42). These findings suggest that Mdm30p is important for maintenance of fusion-competent mitochondria, but is not an essential component of the fusion machinery. Interestingly, steady-state levels of Fzo1p increase in the absence of Mdm30p and decrease when Mdm30p is overexpressed (42), consistent with the idea that Mdm30p controls the steady-state level of the Fzo1p GTPase (**Figure 3**). Whether Mdm30p regulates mitochondrial fusion by controlling Fzo1p protein levels remains to be seen. One interpretation of these data is that Fzo1p is a substrate of an SCF^{Mdm30} ligase complex and is degraded by the 26S proteasome. The link between ubiquitin-mediated degradation and regulation of mitochondrial morphology is supported by two additional observations. First, there is at least one report that Fzo1p can be ubiquitylated in vivo (56). Second, deletion of both *UBC4* and *UBC5* genes (encoding E2 enzymes) results in mitochondrial fragmentation and aggregation (34). Arguing against this model, however, are kinetic analyses suggesting that Mdm30p-dependent degradation of Fzo1p is not mediated by the 26S proteasome (T. Langer, personal communication).

Confounding the issue further, Mdm30p/Dsg1p (does something to Gal4p) was recently proposed to mediate ubiquitylation and degradation of the transcription factor Gal4p as well as cotranscriptional processing of Gal4p target RNAs (95). In one *mdm30Δ* strain, Gal4p-mediated protein expression did not occur, and this strain failed to grow on synthetic galactose medium (95). However, *mdm30Δ* strains generated in other laboratories grow on rich galactose medium (J.R. Lipford & R.J. Deshaies, personal communication; N. Kondo-Okamoto, J.M. Shaw & K. Okamoto, unpublished; B. Westermann, personal communication). Thus, the growth defect reported for the *mdm30Δ* strain is not specific for the carbon source galactose.

The rhomboid protease and Mgm1p processing. The fact that Mgm1p is present in large and small isoforms (l-Mgm1p and s-Mgm1p) raises several intriguing questions. How are the two isoforms generated? Where are they localized in the mitochondrial intermembrane space? Do they have distinct functions in mitochondrial morphology maintenance? Does Mgm1p processing somehow regulate the balance between fusion and fission?

Studies from two groups revealed that s-Mgm1p is generated by a novel intramembrane serine protease called Rbd1p (rhomboid) (53, 86). This protease is homologous to members of the rhomboid family, conserved proteins that activate epidermal growth factor receptor signaling in *Drosophila* (39). Yeast rhomboid was independently identified in genetic screens as *PCP1*, a gene required for processing of cytochrome c peroxidase (30), and as *MDM37* and *UGO2* (26, 121). Loss of Pcp1p function causes mitochondrial fragmentation and aggregation (26, 86, 121). Pcp1p contains six transmembrane segments and is embedded in the mitochondrial inner membrane (86) (**Figures 2** and **3**). Mitochondrial fragmentation in *pcp1Δ* cells is suppressed by blocking fission, suggesting that Pcp1p plays a role in the fusion pathway (123). However, mitochondria can fuse in zygotes lacking Pcp1p (and containing

MPP: mitochondrial processing peptidase

GED: GTPase effector domain

only l-Mgm1p) (123), indicating that Pcp1p and s-Mgm1p are not essential for fusion. It is possible that s-Mgm1p increases the efficiency of mitochondrial fusion. Alternatively, s-Mgm1p may play roles in other (nonfusion) morphology pathway(s) that maintain mitochondrial tubular networks. Expression of the human Pcp1p homolog, hPARL (presenilin-associated rhomboid-like), in yeast *pcp1*Δ cells restored generation of s-Mgm1p (86). Thus, the activity of mitochondrial rhomboid appears to be conserved during evolution. The role of hPARL in processing of the human Mgm1p homolog OPA1 has not been explored.

Amino acid sequencing identified the N-terminal residues of the two Mgm1p isoforms and suggested a possible mode of Mgm1p processing (53). First, the presequence of Mgm1p is cleaved by MPP, the mitochondrial processing peptidase in the matrix, which generates the 97-kDa, l-Mgm1p species containing a predicted transmembrane segment. Next, Pcp1p cleaves in the second hydrophobic stretch, generating the 84kDa, s-Mgm1p species. Biochemical fractionation studies suggest that s-Mgm1p is peripherally associated with the intermembrane space side of the inner and/or outer membranes (53, 124, 147). It is not entirely clear whether l-Mgm1p is an integral inner-membrane protein, or a tightly associated, peripheral membrane protein attached to the inner and/or outer membranes (53, 124, 147). Coimmunoprecipitation studies indicate that interaction of l-Mgm1p with Fzo1p and Ugo1p is more stable than that of s-Mgm1p (124). It is possible that l-Mgm1p predominantly forms a fusion apparatus with Fzo1p and Ugo1p. Whether l- and s-Mgm1p form two distinct complexes functioning in the same or different pathways, or assemble into the same complex, is an important issue for future studies.

The fact that both Mgm1p isoforms function in mitochondrial morphology maintenance (53) raises the question of how the steady-state ratio of l-Mgm1p versus s-Mgm1p is established. First, the inner-membrane potential is sufficient to drive translocation of the Mgm1p presequence into the matrix. The first transmembrane segment acts as a "stop-transfer" sequence at the inner-membrane side of the import channel. Subsequently, l-Mgm1p is generated by MPP cleavage of the presequence followed by lateral diffusion into the inner membrane (52). Second, to generate s-Mgm1p, the first transmembrane segment is further translocated into the matrix by the ATP-driven mitochondrial import motor, and the second hydrophobic stretch is laterally inserted into the inner membrane and cleaved by Pcp1p (52). These two processes that generate l- and s-Mgm1p are competing reactions dependent on the ATP level in the matrix. This novel pathway of Mgm1p biogenesis, called alternative topogenesis (52), may link mitochondrial fusion and bioenergetic function.

THE FISSION PATHWAY

Dnm1p

Mitochondrial fission requires a family of conserved, dynamin-related GTPases called Dnm1p (dynamin-related) in yeast (7, 101, 120), DRP-1 (dynamin-related protein) in worm (75), Drp1 (dynamin-related protein)/DLP1 (dynamin-like protein) in humans (108, 130, 131, 154, 157), and ADL1 and ADL2 (Arabidopsis dynamin-like) in plants (2, 3, 64, 81, 83). Yeast Dnm1p contains an N-terminal GTPase domain, a middle domain, a hydrophilic region of unknown function called Insert B, and a C-terminal GED (GTPase effector domain) (**Figure 2**). Although Dnm1p behaves like a soluble protein in biochemical fractionation studies, it can assemble into punctate structures on mitochondria in living cells (101) (**Figure 5**), indicating that Dnm1p-mitochondrial association is labile and/or dynamic. Purified Dnm1p has been shown to bind and hydrolyze GTP in vitro (44), and mutational analyses

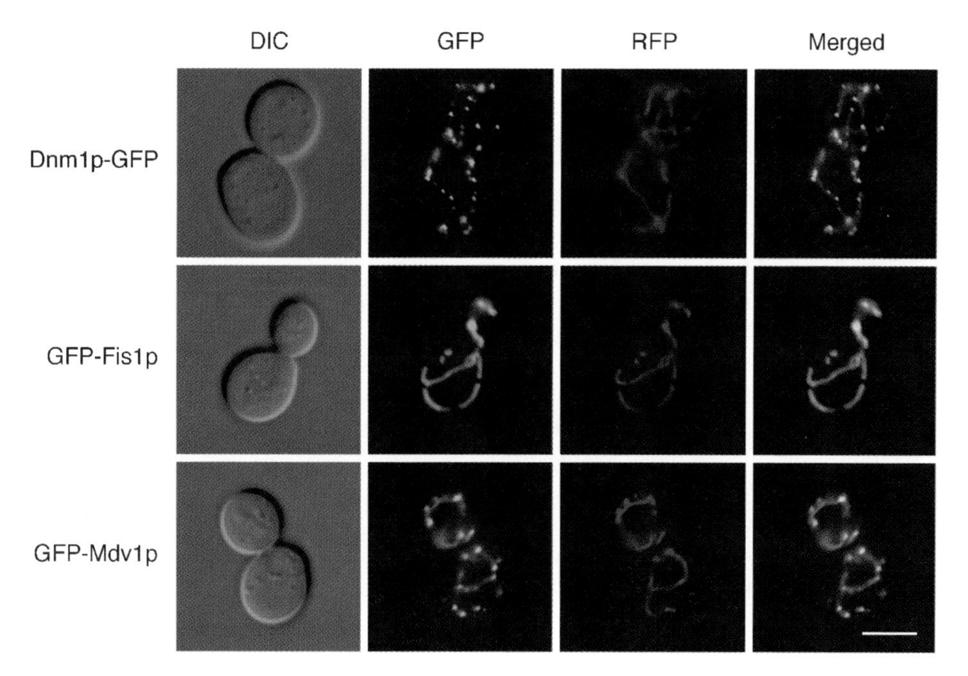

Figure 5

Localization of fission proteins. Wild-type cells expressing Dnm1p-GFP, GFP-Fis1p, or GFP-Mdv1p are shown. Mitochondria are visualized by a matrix-targeted RFP. GFP and RFP fluorescence images are superimposed for each cell (Merged). DIC, differential interference contrast. Bar, 5 μm.

indicate that the Dnm1p GTPase activity is essential for mitochondrial fission in vivo (16, 101). Dnm1p interacts with itself, and this oligomerization is mediated, in part, by the GED (44). Although the function of Insert B remains obscure, a mutation in the Dnm1p middle domain has been shown to affect higher-order oligomerization and accumulation on the mitochondrial membrane (D. Bhar & J.M. Shaw, unpublished).

Fis1p

The evolutionarily conserved integral membrane proteins, Fis1p (fission) in yeast (93) and hFis1 in humans (62, 134, 152), play an essential role in mitochondrial fission. Fis1p is a tail-anchored outer membrane protein with its N-terminal domain exposed to the cytoplasm (93) (**Figures 2** and **6**). Unlike Dnm1p, Fis1p is distributed evenly on mitochondria (93) (**Figure 5**). A mutant Fis1p lacking its C-terminal transmembrane segment is localized in the cytoplasm and incapable of mediating mitochondrial fission (93). When expressed and anchored to yeast mitochondria, the hFis1 cytoplasmic domain cannot complement the mitochondrial morphology phenotype of *fis1*Δ cells (134), suggesting that significant functional divergence of the cytoplasmic domain has occurred during evolution.

X-ray crystallography (28) and NMR spectroscopy (136) of hFis1 indicate that the cytoplasmic domain contains an N-terminal arm and an antiparallel array of six α-helices (α1–α6). The six α-helices form a tetratricopeptide repeat (TPR)-like fold, which contains a hydrophobic concave surface with the potential to serve as a binding pocket. The N-terminal 10 amino acids of the arm region appear unstructured in the solution structure (136), but form an extended α1 helix in the crystal structure (28).

In a recently published solution structure, the six α-helices of yeast Fis1p form a

TPR: tetratricopeptide repeat

TPR motif: a degenerate, 34 amino acid repeat that forms a pair of antiparallel α-helices

TPR-like fold very similar to that of hFis1 (137). The major structural difference between the two proteins in solution is the structure of the N-terminal arm upstream of helix $\alpha 1$: The N-terminal arm of hFis1 is flexible, whereas the longer, N-terminal arm of yeast Fis1p is partially fixed to the concave surface of the TPR-like fold (137). In the same study, deletion analysis suggested that residues 5–15 of the Fis1p N-terminal arm are required to recruit one of Fis1p's binding partners, Mdv1p (137). However, other studies indicate that Fis1p lacking the first 15 residues can bind Mdv1p and mediate fission (M.A. Karren, E.M. Coonrod, T.K. Anderson & J.M. Shaw, submitted). Despite the latter result, it seems likely that the Fis1p N-terminal arm plays a role in stabilizing the TPR-like fold or, alternatively, contributes residues to the concave binding pocket of the protein.

NTE: N-terminal extension

WD40 repeat: a protein binding β-propeller forming motif containing ~7 repeats of ~40 amino acids with conserved tryptophan and aspartic acid residues

Mdv1p

Studies in yeast revealed an additional protein, called Mdv1p/Gag3p (glycerol-adapted growth)/Net2 (mitochondrial net), required for mitochondrial fission (17, 32, 140). Mdv1p contains an N-terminal extension (NTE), one or two coiled-coil domains in the N-terminal half, and seven C-terminal WD40 repeats predicted to form a β-propeller (**Figure 2**). Both coiled-coil and WD40 repeat domains are predicted to function in protein-protein interaction. In two-hybrid studies, the coiled-coil domain interacts with full-length Mdv1p (141), suggesting that this region mediates homo-oligomerization of the protein. Most Mdv1p in the cell is tightly associated with the cytoplasmic face of the outer mitochondrial membrane (17, 32, 140). Mdv1p forms punctate structures (17, 140) and is also found uniformly distributed on mitochondria (**Figure 5**). Both localization patterns depend on Fis1p (17, 32, 140). At this writing, no structural homolog of Mdv1p has been identified in multicellular eukaryotes.

Caf4p

An affinity purification scheme to identify Fis1p-interacting proteins revealed a new player in the fission pathway called Caf4p (CCR4 associated factor) (45a). Caf4p was originally identified as a protein interacting with Ccr4p, a central component of the CCR4-NOT transcriptional regulator complex (80). Mdv1p and Caf4p domain structure is similar, with an NTE and coiled-coil domain in the N-terminal half, and seven WD40 repeats at the C terminus (**Figure 2**). Coimmunoprecipitation and two-hybrid assays indicate that the Caf4p forms complexes with both Fis1p and Dnm1p, supporting the idea that Caf4p participates in mitochondrial fission. Although deletion of the *CAF4* gene has no obvious effect on mitochondrial morphology, it enhances the phenotype of *mdv1*Δ cells, which is slightly less severe than that of *dnm1*Δ or *fis1*Δ cells. Specifically, the mitochondrial morphology defect in a *caf4*Δ *mdv1*Δ double mutant is indistinguishable from that of the *dnm1*Δ or *fis1*Δ single mutant. These observations suggest that residual mitochondrial fission in *mdv1*Δ cells is due to Caf4p function. Like Mdv1p, Caf4p localizes to the outer mitochondrial membrane in a Fis1p-dependent manner. Importantly, formation of Dnm1p puncta resembling fission complexes is completely abolished in *caf4*Δ *mdv1*Δ cells. Thus, Caf4p and Mdv1p may collaborate to recruit Dnm1p into punctate structures on mitochondria.

Interactions of Fission Proteins

Fis1p-Mdv1p and Fis1p-Caf4p interactions. Mitochondrial localization of Mdv1p is mediated by Fis1p on the outer membrane (17, 140). Recent *E. coli* two-hybrid studies demonstrate that the Mdv1p-Fis1p interaction occurs in the absence of other yeast proteins and is most likely direct (M.A. Karren, E.M. Coonrod, T.K. Anderson & J.M. Shaw, submitted). Additional yeast two-hybrid analyses reveal that Fis1p interacts with the NTE of Mdv1p (16, 141). These data are supported by the observation that the NTE alone localizes to mitochondria in a Fis1p-dependent manner (141). Other studies suggest that Fis1p interacts with both the N terminus and WD40 repeat domains of Mdv1p (16). However, in vivo puncta formation by the Mdv1p WD40 domain fragment depends solely on Dnm1p, and subcellular fractionation studies localize the Mdv1p WD40 domain fragment to the cytoplasm even when Fis1p is present (16, 141). These combined results suggest that the Mdv1p NTE alone is sufficient to mediate Fis1p interactions and mitochondrial localization (**Figure 6**). Although the analysis of Caf4p is less extensive, coimmunoprecipitation studies indicate that the Caf4p NTE can also form a complex with Fis1p (45a) (**Figure 6**).

Fis1p-Dnm1p interaction. Dnm1p puncta formation is significantly impaired in the absence of Fis1p (93, 140). Although two-hybrid interactions between Fis1p and Dnm1p have not been reported, chemical cross-linking and immunoprecipitation studies demonstrate that Dnm1p and Fis1p form a complex in vivo, even when Mdv1p is absent (M.A. Karren, E.M. Coonrod, T.K. Anderson & J.M. Shaw, submitted). This residual Fis1p-Dnm1p complex formation in the absence of Mdv1p is most likely mediated by Caf4p. Although early studies suggested that the Fis1p-Dnm1p interaction was direct, the most recent data support the idea that Dnm1p assembly onto Fis1p-containing mitochondria is bridged by Mdv1p and/or Caf4p (45a; M.A. Karren, E.M. Coonrod, T.K. Anderson & J.M. Shaw, submitted) (**Figure 6**).

Dnm1p-Mdv1p and Dnm1p-Caf4p interactions. Fluorescence and immunogold electron microscopy studies indicate that Mdv1p colocalizes with Dnm1p in punctate fission complexes (17, 140). Dnm1p-Mdv1p complex formation is also detected by yeast two-hybrid and immunoprecipitation experiments (17, 140, 143), and a variety of analyses indicate that Dnm1p interacts with the Mdv1p WD40 domain (16, 141) (**Figure 6**). Experiments with Dnm1p GTPase mutant proteins suggest that the Dnm1p GTPase cycle somehow regulates Dnm1p-Mdv1p interactions during fission complex formation (16). The exact steps in this assembly pathway have not been clarified. Although detailed analyses of the Dnm1p-Caf4p interaction are under way, studies to date indicate that this interaction requires the Caf4p WD40 domain (45a) (**Figure 6**).

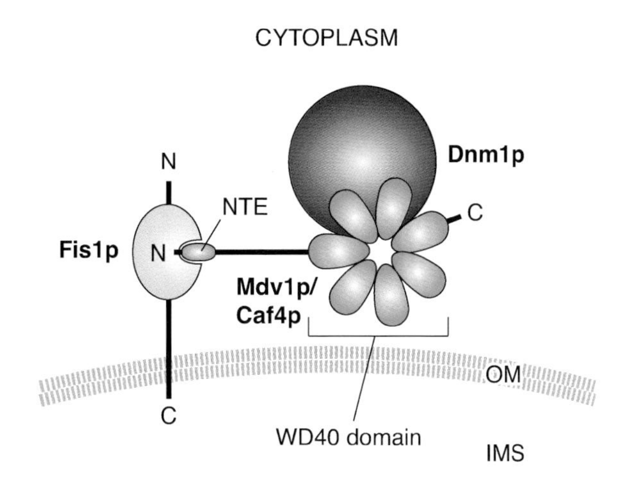

Figure 6

Topology of proteins acting in the mitochondrial fission pathway. Mdv1p and Caf4p interact with Fis1p and Dnm1p via their N-terminal extensions (NTE) and WD40 domains, respectively. OM, outer membrane; IMS, intermembrane space.

Models for Mitochondrial Fission

Two models for yeast mitochondrial fission have been proposed. In the first model (100, 125, 141), Dnm1p assembles and disassembles at hot spots on the surface of mitochondria via its ability to homo-oligomerize and form rings around the mitochondrial tubule in an Mdv1p-independent manner (**Figure 7*a***). In this model, Dnm1p mitochondrial targeting is mediated by direct interaction with Fis1p. Mdv1p is stably associated with mitochondria via interaction of its NTE with Fis1p. After Dnm1p homo-oligomerization on mitochondria, the Mdv1p WD40 repeat domain mediates incorporation of the preformed Fis1p-Mdv1p complex into the Dnm1p complex. According to this scenario, Mdv1p stimulates some rate-limiting activity of the assembled complex such as membrane

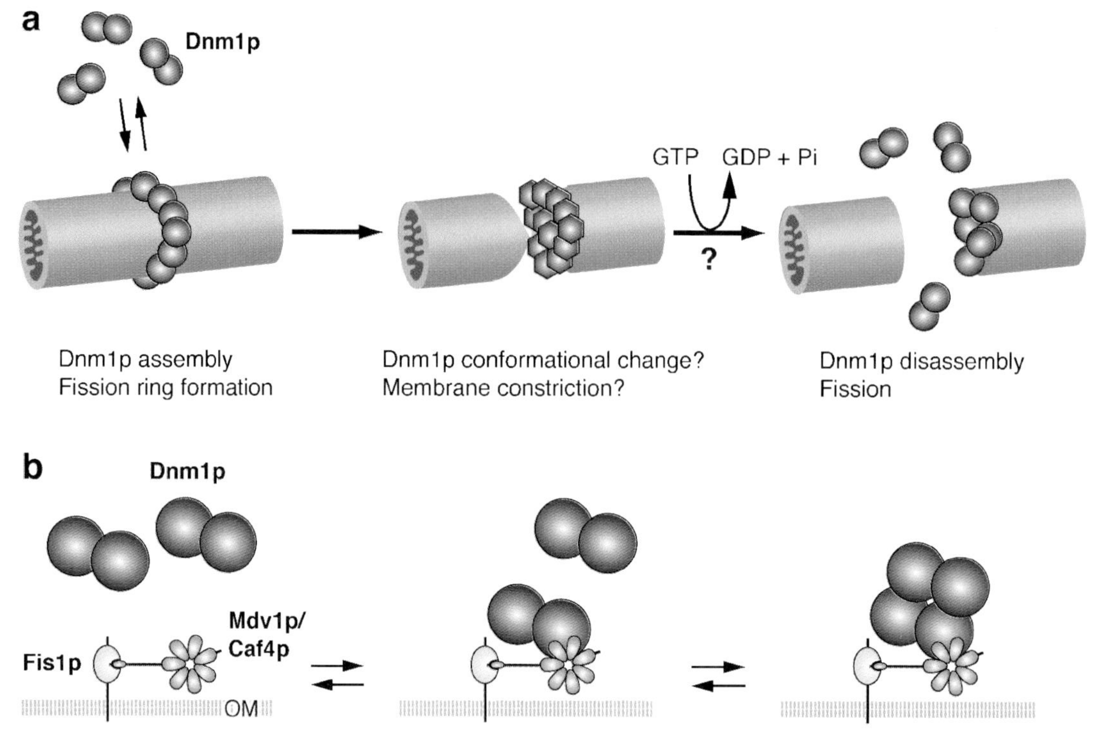

Figure 7

Models for Dnm1p-mediated mitochondrial fission (*a*) and targeting/assembly of Dnm1p on mitochondria (*b*). OM, outer membrane.

constriction or lipid remodeling events. These late events ultimately trigger or mechanically cause membrane scission (**Figure 7*a***). The second model (16) is similar to the first except that Dnm1p assembly at hot spots on mitochondria is proposed to be both Fis1p- and Mdv1p-independent. The Dnm1p multimeric complex subsequently recruits Fis1p and Mdv1p into a ring structure that catalyzes fission.

In a recently revised model, Dnm1p mitochondrial targeting requires not only Fis1p but also Mdv1p and Caf4p (45a; M.A. Karren, E.M. Coonrod, T.K. Anderson & J.M. Shaw, submitted). At an early stage of fission, Fis1p recruits Dnm1p to the outer mitochondrial membrane via interactions with Mdv1p or Caf4p (**Figure 7*b***). In the absence of Mdv1p, Caf4p mediates formation of punctate Dnm1p complexes on the mitochondrial surface. However, these Dnm1p complexes are not functional and cannot complete fission. In *caf4Δ* cells, Mdv1p can support both Dnm1p recruitment and efficient fission. Thus, although Caf4p and Mdv1p cooperate during Dnm1p assembly and fission complex formation, Mdv1p may also be required in later fission steps.

Since the biochemical activities of the Dnm1p GTPase and GED mutant proteins have not been reported, it is not entirely clear how changes in the nucleotide-bound state of Dnm1p regulate distinct steps in mitochondrial fission. In the yeast two-hybrid system, Dnm1p interacts with Mdv1p in a nucleotide-dependent manner (16), suggesting that the GTPase cycle may regulate interactions between these two proteins. It is possible that Fis1p, Caf4p, and/or Mdv1p modulate the Dnm1p GTPase activity during recruitment and assembly of the fission complex. For example, such interactions might prevent disassembly of the fission complex by blocking Dnm1p GTP hydrolysis, thereby facilitating formation of an active fission ring (**Figure 7*a***).

The mechanism of mitochondrial constriction is also not understood (**Figure 7*a***). At least one study reported that constriction of mitochondrial tubules occurs prior to Dnm1p recruitment (77). If this process is Dnm1p-independent, constrictions could be generated by local interactions of one or more proteins residing inside the mitochondrial compartment. Such a function has been proposed for the inner-membrane protein Mdm33p (90). Alternatively, cytoskeletal-dependent mitochondrial movement could generate tension along the mitochondrial tubule and stretch it out, resulting in local constrictions along the tubule. If this constriction is Dnm1p-dependent, assembled Dnm1p fission complexes might generate mechanical forces (via conformational changes) that induce membrane constriction. Alternatively, Dnm1p could recruit unidentified proteins that modify local membrane lipid composition and curvature.

Regulation of Mitochondrial Fission Proteins

A recent study reported that mammalian Drp1 is modified by addition of Sumo1 (small ubiquitin-like modifier) (51), a posttranslational protein-conjugation that regulates diverse cellular processes including transcription, protein transport, genome maintenance, and signal transduction. It has been suggested that Sumo1 regulates the rate of mitochondrial fission by controlling Drp1 protein levels (51). It is not clear whether sumoylation-dependent regulation of mitochondrial fission has been conserved in other eukaryotes.

Drp1-dependent mitochondrial fission may interface with cytoskeletal pathways and molecular motors in mammalian cells. Dynein-dynactin, a retrograde motor complex associated with microtubules, is reported to bind and recruit Drp1 to mitochondria (144). Other studies suggest that disruption of F-actin blocks recruitment of Drp1 to mitochondria and Drp1-dependent mitochondrial fission (25). Although mitochondria have been shown to coalign with F-actin cables in yeast, there is currently no evidence that F-actin and myosin motor proteins impact Dnm1p behavior in vivo.

Mitochondrial Fission During Meiosis and Sporulation

During yeast meiosis and sporulation, mitochondria undergo remarkable changes in morphology at distinct stages (45, 92). In premeiotic stationary phase, mitochondrial networks fragment. During premeiotic S-phase, mitochondria re-form tubules. Mitochondria remain tubular through the end of meiosis and during early tetrad formation. After tetrads mature, mitochondria again fragment. Although meiosis and sporulation occurred in the absence of Dnm1p, Mdv1p, or Fis1p, the resulting spores displayed reduced viability, due to impaired inheritance of mitochondria (45). Clearly, mitochondrial morphology changes regulated by fission proteins are important during meiosis and sporulation and by extension, during meiosis and gamete formation in multicellular eukaryotes. One unexpected outcome of the yeast study was that mitochondria were able to fragment in mature tetrads lacking fission proteins (45). Whether this fragmentation is caused by novel mitochondrial fission proteins that function only at late stages of sporulation or by mechanical forces generated in the forming tetrad is unclear.

THE TUBULATION PATHWAY

Mmm1p

Components of the tubulation pathway have been identified in fungi but not in higher eukaryotes. In *S. cerevisiae*, formation of mitochondrial tubules requires Mmm1p (maintenance of mitochondrial morphology) (13), an integral membrane protein of

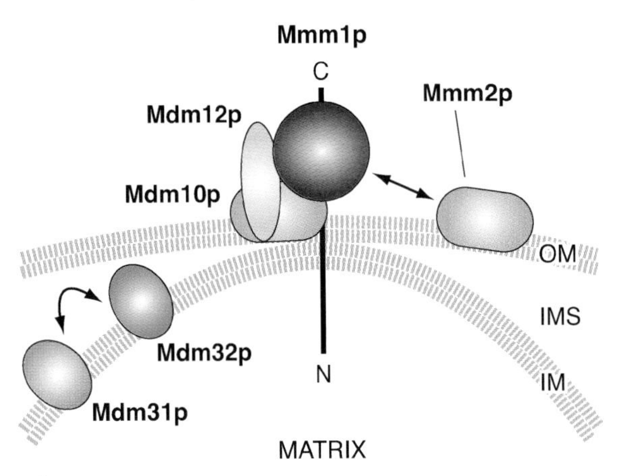

Figure 8

Localization and topology of proteins acting in the mitochondrial tubulation pathway. Mmm1p spans both the outer and inner membranes and forms a complex with Mdm10p and Mdm12p. Arrows indicate transient protein-protein interactions. OM, outer membrane; IMS, intermembrane space; IM, inner membrane.

49 kDa containing a single transmembrane segment (**Figure 2**). Mmm1p spans both the outer and inner mitochondrial membranes with its C terminus exposed to the cytoplasm and its N terminus in the matrix (13, 72, 88) (**Figure 8**).

mtDNA nucleoid: a punctate structure in the mitochondrial matrix composed of mtDNA and DNA-binding proteins

Mutations in the *MMM1* gene alter normal mitochondrial morphology and inner-membrane cristae structure, ultimately converting tubular networks to large spheres (13, 57). Concomitant with this change, mtDNA nucleoids are disorganized, which results in a severe mtDNA instability phenotype (57). Additional studies suggest that Mmm1p mediates mitochondrial inheritance by directly or indirectly interacting with the actin cytoskeleton (8). Interestingly, Mmm1p localizes as discrete foci adjacent to mtDNA nucleoids in the matrix (57), raising the possibility that this protein plays a direct role in mtDNA nucleoid organization. A mutant *mmm1* allele has also been shown to contribute to increased rates of mtDNA escape to the nucleus (50).

Mdm10p

Normal mitochondrial tubule formation also requires Mdm10p (132), a 56-kDa protein conserved in fungi but not present in higher eukaryotes. Mdm10p is inserted into the outer membrane with a predicted β-barrel structure (132) (**Figures 2** and **8**). Like *mmm1* mutations, loss of Mdm10p function causes defects in mitochondrial shape, inheritance, motility, actin-mitochondria interactions, mtDNA stability, and an increased rate of mtDNA escape to the nucleus (8, 50, 132). Moreover, Mdm10p also forms mitochondrial foci adjacent to mtDNA nucleoids (9). These findings led investigators to propose that Mdm10p functions in the Mmm1p-mediated mitochondrial morphology pathway.

Mdm12p

Loss of Mdm12p, a 31-kDa integral outer membrane protein (**Figure 8**), causes defects resembling those seen in *mmm1* and *mdm10* mutants (6). Like Mmm1p and Mdm10p, Mdm12p homologs are found only in fungi. Mdm12p contains a predicted transmembrane segment near its N terminus (**Figure 2**) and a region in its C terminus with sequence similarity to the C-terminal domain of Mmm1p (6). The topology of this protein has not been determined. Like Mmm1p and Mdm10p, Mdm12p also localizes as mitochondrial foci adjacent to mtDNA nucleoids (9). Thus, Mdm12p appears to function in the same pathway as Mmm1p and Mdm10p.

Interactions of Mmm1p, Mdm10p, and Mdm12p

Coimmunoprecipitation studies demonstrate that Mmm1p, Mdm10p, and Mdm12p form an MMM (Mmm1p-Mdm10p-Mdm12p) complex (9) (**Figure 8**). Indirect immunofluorescence microscopy reveals that Mmm1p and Mdm12p depend on each other for their mitochondrial localization, although Mdm10p can form mitochondrial foci in

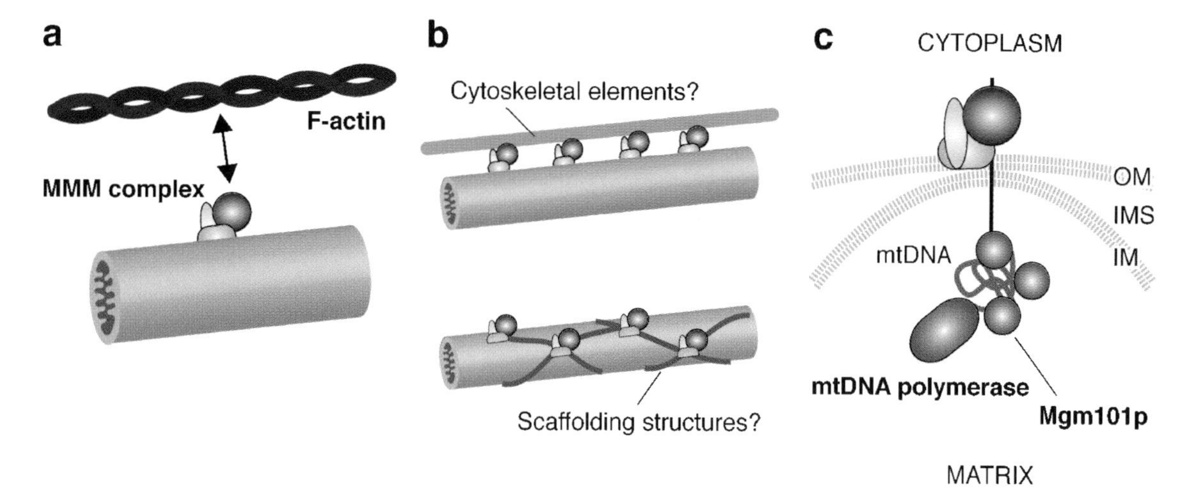

Figure 9

Proposed roles of the Mmm1p-Mdm10p-Mdm12p (MMM) complex in actin-mitochondria attachment (*a*), mitochondrial tubulation via interactions with unidentified cytoskeletal elements or scaffolding structures (*b*), and anchoring of mtDNA replication foci (*c*). OM, outer membrane; IMS, intermembrane space; IM, inner membrane.

the absence of Mmm1p and Mdm12p (9). Additional studies suggest that Mmm1p mitochondrial targeting requires Mdm10p (9). Due to the similar behaviors of Mmm1p and Mdm12p, it is likely that Mdm10p is also required for Mdm12p mitochondrial targeting.

Models of MMM Complex Function

The MMM complex may mediate at least three processes: actin-mitochondria attachment (**Figure 9*a***), formation of tubular mitochondria (**Figure 9*b***), and anchoring of mtDNA nucleoids (**Figure 9*c***).

Actin-mitochondria attachment. Mitochondrial transport and inheritance relies on the actin cytoskeleton in budding yeast and microtubules in the filamentous fungus *Neurospora crassa* (29, 133). Mmm1p, Mdm10p, and Mdm12p homologs are present in *N. crassa*, where they may also function as a complex. Although loss of *N. crassa MMM1* (*NcMMM1*) function alters normal mitochondrial morphology in vivo (110), mitochondria lacking *NcMMM1* continue to bind microtubules in an in vitro assay (43). In addition, expression of *NcMMM1* restores mitochondrial morphology and inheritance defects in yeast *mmm1*Δ cells (72), suggesting that the function of the protein is conserved. It is conceivable that *NcMMM1* mediates actin-mitochondria interactions in both budding yeast and *N. crassa*, but that this interaction is not essential for mitochondrial transport in *N. crassa*. Alternatively, the primary function of Mmm1p/*NcMMM1* may be unrelated to cytoskeletal-mitochondrial interactions in both organisms.

Formation of tubular mitochondria. Is an actin-mitochondria attachment critical for the MMM-mediated tubulation pathway? Genetic or drug-induced disruption of actin filaments does not promote formation of spherical mitochondria (8, 29, 54, 129). Thus, it is unlikely that the tubular mitochondrial shape is determined solely by actin-mitochondria interactions. Experiments to date do not exclude the possibility that the MMM complex interacts with other cellular structures to generate mitochondrial tubules (**Figure 9*b***). It is also formally possible that

SAM: sorting and assembly machinery

TOM: translocase of the outer membrane

the MMM complex organizes scaffolding-like proteins on the surface of mitochondria, which establish and/or maintain mitochondrial tubular shape (**Figure 9*b***).

Anchoring of mtDNA nucleoids. Recent in vivo and in vitro topology assays suggest that Mmm1p spans both the outer and inner membranes with its N terminus exposed to the matrix (72) (**Figure 8**). This double membrane-spanning topology is required for Mmm1p foci formation (72). Interestingly, Mmm1p foci are adjacent to punctate structures in the matrix containing Mgm101p, a soluble DNA-binding protein important for mtDNA maintenance (88) (**Figure 9*c***). Chemical cross-linking and coimmunoprecipitation studies indicate that Mmm1p is in close proximity to Mgm101p (88). Whether Mmm1p is required for Mgm101p puncta formation and vice versa has not been addressed. Mmm1p foci are also adjacent to actively replicating mtDNA and mtDNA polymerase (88) (**Figure 9*c***). These observations raise the possibility that Mmm1p spans both mitochondrial membranes at membrane contact sites and anchors mtDNA nucleoids that are actively replicating.

The MMM complex may accomplish some combination of the functions described above. For example, the MMM complex might attach to the transport machinery at the surface of mitochondria and to mtDNA nucleoids on the matrix side of the inner membrane. In this way, the MMM complex could ensure that daughter cells inherit both mitochondria and mtDNA during cell division.

Role of Mdm10p in Protein Sorting and Assembly

Surprisingly, recent studies reveal that Mdm10p is a component of the SAM (sorting and assembly machinery) complex located in the mitochondrial outer membrane (89, 107). Mdm10p appears essential for formation of the 350-kDa SAM complex when blue-native gel electrophoresis is used to monitor complex formation (89). In contrast, this complex formation does not depend on Mmm1p or Mdm12p (89). Detailed in vitro import and assembly studies indicate that Mdm10p plays a specific role in late assembly steps of the TOM (translocase of the outer membrane) complex (89). Once again, TOM complex assembly is normal in the absence of Mmm1p or Mdm12p (89). These findings suggest that Mdm10p functions in biogenesis of the TOM complex. Thus, the loss of Mdm10p may cause primary defects in import and/or assembly of morphology proteins such as Mmm1p and Mdm12p, which secondarily convert mitochondrial tubules to large spheres in these mutants.

Alternatively, Mdm10p may perform dual functions in both mitochondrial morphology maintenance and protein sorting/assembly. Indeed, biochemical pull-down assays (C. Meisinger & N. Pfanner, unpublished) support previous coimmunoprecipitation results suggesting that Mmm1p, Mdm10p, and Mdm12p form a complex in vivo (9). It is, therefore, possible that Mdm10p is present in two distinct complexes, the MMM complex, which participates in formation of mitochondrial tubules, and the SAM complex, which participates in mitochondrial protein sorting/assembly.

Mmm2p

A recent genetic screen for possible Mmm1p-interacting proteins led to the identification of a novel 52-kDa protein called Mmm2p (155). Mmm2p homologs are found in fungi but not higher eukaryotes. *MMM2* is allelic to *MDM34*, a gene identified in a genome-wide screen of deletion mutants defective in mitochondrial morphology and distribution (26). Cells lacking Mmm2p contain distorted, sausage-shaped, spherical, and sheet-like mitochondria (155). In addition, *mmm2*Δ cells exhibit severe mtDNA instability and abnormal mtDNA nucleoid structure (155). Because Mmm2p is embedded in the outer mitochondrial membrane and contains no predicted α-helical transmembrane segments (155), it likely adopts a β-barrel-type membrane topology similar to that predicted for Mdm10p (**Figures 2** and **8**). Mmm2p localizes as foci along the mitochondrial tubule (155). Fluorescence microscopy studies indicate that a portion of Mmm2p colocalizes with Mmm1p foci, and gel filtration studies suggest that Mmm1p and Mmm2p form in large, but distinct, complexes (155). Thus, Mmm2p may interact with Mmm1p in a dynamic or transient manner (**Figure 8**).

Mmm1p and Mmm2p depend on one another for mitochondrial foci formation. Mmm1p is essential for Mmm2p foci formation, but not mitochondrial targeting (155). Mmm2p is required for the mitochondrial targeting, sorting, assembly, and/or stability of Mmm1p (155). Thus, Mmm2p may play a central role in biogenesis of the MMM complex. Alternatively, like Mdm10p, Mmm2p may function in both mitochondrial morphology maintenance and targeting/sorting/assembly of the MMM complex. It will be important to determine whether the MMM complex or the Mmm2p-containing complex act together or independently to maintain mitochondrial tubular shape.

Mdm31p and Mdm32p

Two inner-membrane proteins conserved in fungi may also act in the MMM-mediated mitochondrial morphology pathway. *MDM31* and *MDM32* encode two related proteins of 67 and 76 kDa, respectively (27). Mdm31p and Mdm32p have a similar domain structure: a typical N-terminal presequence and two transmembrane segments, one near the N terminus of the mature protein and another one at the very C terminus (**Figure 2**). Both proteins are inserted into the inner membrane with their middle domains exposed to the intermembrane space (27) (**Figure 8**).

Deletion of the *MDM31* or *MDM32* gene causes defects similar to those in *mmm1*, *mmm2*, *mdm10*, and *mdm12* mutants. Cells lacking Mdm31p or Mdm32p exhibit large spherical mitochondria, disorganized mtDNA nucleoid structure, mtDNA instability, and impaired mitochondrial motility (27). Unlike *mmm1* and *mdm10* mutant mitochondria, however, *mdm31*Δ and *mdm32*Δ mitochondria are able to bind actin in vitro (27), suggesting that the reduced motility is due to their aberrant shape. Genetic studies demonstrated that Mdm31p and Mdm32p are essential for the viability of cells lacking Mmm1p, Mmm2p, Mdm10p, or Mdm12p (27). Although Mmm1p foci formation is not affected in the *mdm31* or *mdm32* strain, these Mmm1p foci are no longer localized adjacent to mtDNA nucleoids in these mutants (27). These combined observations suggest that Mdm31p and Mdm32p somehow work together with Mmm1p, Mmm2p, Mdm10p, and Mdm12p to maintain tubular mitochondria (**Figure 8**).

Do Mdm31p and Mdm32p cooperate with the MMM complex? Upon detergent solubilization and gel filtration, Mdm31p and Mdm32p form separate ~600-kDa and ~175-kDa complexes, respectively, that are distinct from the Mmm1p-containing complex (27). Since these three complexes form independently, it is unlikely that Mdm31p and Mdm32p are components of the MMM complex. Coimmunoprecipitation studies suggest that Mdm31p interacts transiently with Mdm32p (27) (**Figure 8**). Dynamic interactions between the two proteins may be crucial for inner-membrane remodeling. On the other hand, three distinct pathways utilizing Mdm31p/Mdm32p, Mmm2p, and the MMM complex may impact formation of tubular mitochondria.

The functions of Mdm31p and Mdm32p will depend on their exact topology. Since the presequence is translocated across the inner membrane and processed by MPP in the matrix (27), it is likely that the first transmembrane segment is inserted into the inner membrane, leaving the short N-terminal stretch exposed to the matrix. Whether the C-terminal transmembrane segment spans the inner or outer membrane has not been determined. If the latter is the case, Mdm31p and Mdm32p may physically link both mitochondrial membranes in a manner that maintains the integrity of the organelle.

EF-hand: a protein motif that forms two α-helices connected by a 12 residue Ca^{2+}-binding loop

GAP: GTPase-activating protein

GEF: guanine nucleotide exchange factor

PROTEINS THAT MAY ACT IN NOVEL MORPHOLOGY PATHWAYS

Mdm33p

A genome-wide screen for yeast deletion mutants defective in mitochondrial morphology and distribution identified Mdm33p (26), a 54-kDa protein predicted to contain a cleavable N-terminal presequence, three coiled-coil domains, and two transmembrane segments (**Figure 2**). Mdm33p homologs are present in other fungi but not found in higher eukaryotes. Mdm33p is embedded in the inner membrane with a large N-terminal domain exposed to the matrix and a C-terminal loop between the two transmembrane segments facing the intermembrane space (90). Mdm33p can be coimmunoprecipitated with itself, and forms a ~300-kDa complex in gel filtration experiments (90). Cells lacking Mdm33p exhibit novel mitochondrial morphology phenotypes described as large ring-like structures, hollow spheres, elongated sheets, and two to four smaller, interconnected rings (90). Despite these grossly altered morphologies, mitochondrial fusion and fission are not blocked in *mdm33*Δ cells (90). Overexpression of Mdm33p leads to growth arrest and mitochondrial aggregation (90). In addition, electron microscopy reveals an increase in inner-membrane septa and vesicular structures within the matrix, suggested to result from enhanced inner-membrane fission (90). These observations raise the possibility that Mdm33p acts as a positive regulator of inner-membrane fission.

Gem1p

Studies in mammalian cells identified novel GTPases called Miro-1 and Miro-2 (mitochondrial Rho) that are conserved from yeast to humans (37). The yeast genome encodes a single Miro homolog of 75 kDa called Gem1p (GTPase EF-hand protein of mitochondria) (38). Gem1p contains an N-terminal GTPase domain with some sequence similarity to Ras GTPases, a pair of Ca^{2+}-binding EF-hand motifs, another putative GTPase domain, and a transmembrane segment at the C terminus (**Figure 2**). Gem1p is a tail-anchored outer membrane protein with its GTPase domains and EF-hand motifs exposed to the cytoplasm (38). Cells lacking Gem1p contain collapsed, globular, or grape-like clusters of mitochondria, which retain relatively normal ultrastructure, including well-developed inner-membrane cristae (38). Mutational analyses indicate that both GTPase domains and EF-hand motifs are required for Gem1p function in vivo (38). Mitochondrial fusion and fission occur normally in the absence of Gem1p, and additional studies suggest that the mitochondrial tubulation pathway is not disrupted in *gem1*Δ cells (38). Thus, Gem1p is not an essential component of previously characterized pathways that control mitochondrial morphology.

The cellular function of Gem1p is currently unclear. The topology of Gem1p suggests that both GTPase domains and EF-hand motifs are available to sense cytoplasmic signals such as Ca^{2+} and perhaps GAPs (GTPase-activating protein) and GEFs (guanine nucleotide exchange factor). Three lines of evidence support the notion that Gem1p is a regulatory factor. First, GTPase domains and EF-hand motifs, which can potentially act as molecular switches and induce Ca^{2+}-dependent conformational changes, are required for Gem1p function (38). Second, unlike defined, homogenous morphologies such as fragments, nets, and spheres seen in mutants that disrupt the fusion, fission, and tubulation pathways, mitochondrial shape is pleiotropic in cells lacking Gem1p (38). Third, the severity of mitochondrial morphology phenotypes in *gem1*Δ cells depends on strain background and growth conditions (38). One attractive hypothesis is that Gem1p responds to changes in cytosolic Ca^{2+} levels and recruits/regulates downstream molecules that function in known and/or unknown morphology pathways (151).

MITOCHONDRIAL DYNAMICS, DISEASES, AND PROGRAMMED CELL DEATH

The primary roles of proteins regulating mitochondrial dynamics have been determined largely from studies of budding yeast. However, extension of these studies to mammalian systems raised awareness that the proper function of these basic cellular pathways impacts human health. As described below, recent findings reveal that defects in mitochondrial fusion lead to human neurological disorders and that mitochondrial fission and fusion events help to regulate programmed cell death.

The OPA1 Protein and Autosomal Dominant Optic Atrophy

A direct link between mitochondrial dynamics and human pathogenesis was first revealed in autosomal dominant optic atrophy (ADOA), the most commonly inherited optic neuropathy with an estimated disease prevalence of 1:12,000~50,000 (1, 22). ADOA results in death of retinal ganglion cells, leading to progressive loss of vision that often causes legal blindness within the first two decades of life (23). The predominant locus, called *OPA1* (optic atrophy), encodes a dynamin-related GTPase homologous to the Mgm1p fusion protein in budding yeast (1, 22).

Sequencing of *OPA1* alleles derived from ADOA patients identified a variety of lesions (1, 22, 23, 33) including severe truncation mutations predicted to cause loss of function. Thus, some *OPA1*-related pathogenesis may result from haploinsufficiency. In other cases, mutations causing single amino acid substitutions or small truncations may result in generation of dominant-negative proteins that affect the function of the wild-type OPA1 protein.

The pathological mechanisms leading to loss of retinal ganglion cells in *OPA1*-linked ADOA are not firmly established. ADOA shares some pathophysiological characteristics and clinical symptoms with Leber's hereditary optic neuropathy (LHON), a disease in which mtDNA mutations in respiratory chain complex I genes reduce the ATP-generating capacity of mitochondria (15). In the long term, these defects in energy supply could affect normal metabolic processes and homeostasis in retinal ganglion cells. Based on studies of the yeast Mgm1 protein, reduction of OPA1 function is predicted to interrupt mitochondrial fusion, inner membrane cristae structure, and respiration in ganglion cells, leading to similar bioenergetic dysfunction. It has also been suggested that LHON mutations increase reactive oxygen species (ROS), subsequently inducing apoptosis in retinal ganglion cells (15). Interestingly, depletion of OPA1 by siRNA (small interfering RNA) in cultured cells triggers apoptosis (76, 98). Thus, *OPA1* mutations might also increase susceptibility of retinal ganglion cells to apoptotic stimuli such as ROS.

ADOA: autosomal dominant optic atrophy

LHON: Leber's hereditary optic neuropathy

CMT2A: Charcot-Marie-Tooth neuropathy type 2A

Mfn2 and Charcot-Marie-Tooth Neuropathy Type 2A

Humans have two homologs of the yeast Fzo1p fusion GTPase called Mfn1 and Mfn2. Mutations in the *MFN2* gene were recently identified in patients affected with Charcot-Marie-Tooth neuropathy type 2A (CMT2A) (70, 158). CMT is the most common inherited peripheral neuropathy in humans (estimated prevalence of 1:2,500), and is clinically characterized by weakness and atrophy of distal muscles, depressed or absent deep tendon reflexes, mild sensory loss, and foot deformities (127, 128).

Mutations mapped to the *MFN2* locus fall mostly within the GTPase domain and two coiled-coil domains (70, 158). Since mutations in conserved residues of the GTPase domain disrupt Mfn2 function in cell culture (19, 31, 116), CMT2A-linked GTPase domain mutations are predicted to impair Mfn2 GTPase activity, leading to changes in mitochondrial fusion and function. Mfn2 mutant proteins lacking the first or second coiled-coil domain are partially mistargeted and accumulate in the cytoplasm of cultured mammalian cells (113). Thus, CMT2A-linked coiled-coil domain mutations may impair Mfn2 mitochondrial localization. Alternatively, these coiled-coil mutations may affect interactions with other, as yet unidentified, protein binding partners.

The pathophysiology of *MFN2*-linked CMT2A is not yet understood. CMT-linked mutations have been identified in genes required for non-mitochondrial pathways, including vesicle transport (127, 156). Defects in Mfn2 function might reduce mitochondrial ATP production, which could, in turn, impact energy-consuming processes like vesicle transport. Interestingly, mitochondrial motility is markedly reduced in mouse embryonic fibroblasts derived from Mfn2-deficient mice (19), raising the possibility that morphology changes caused by *MFN2* mutations affect transport of mitochondria. Impaired mitochondrial transport out of the cell body could deprive the distal axon of an essential energy source, resulting in the axonal neuropathy observed in CMT2A patients.

Roles of Mitochondrial Dynamics in Neuronal Cell Function

A process from a neuronal cell body called a dendrite receives inputs from multiple synapses at structures called spines. A study using cultured hippocampal neurons revealed that increased mitochondrial fission promotes the formation of synaptic spines and synapses (79).

In neurons, mitochondria accumulate in the cell body but are also present in dendrites. Increasing the fusion/fission ratio reduced dendritic mitochondrial content, and resulted in loss of dendritic spines and synapses (79). By contrast, decreasing the fusion/fission ratio increased the density of dendritic mitochondria and induced the formation of spines and synapses (79). These observations suggest that mitochondria in dendrites are essential and limiting for the formation or maintenance of synaptic spines. Interestingly, mitochondrial motility, the rate of mitochondrial fission and the density of dendritic mitochondria all increased after neuronal stimulation (79), suggesting that mitochondrial dynamics are regulated by neuronal activity. Since abnormal mitochondrial morphology is associated with neurodegenerative diseases including Alzheimer's disease and Parkinson's disease (142), it is possible that defects in mitochondrial dynamics contribute to synapse deterioration in such disorders.

Roles of Mitochondrial Fusion and Fission Proteins in Programmed Cell Death

Mitochondrial fragmentation during apoptosis. Numerous studies implicate mitochondria as key regulators of apoptosis. During early stages of apoptosis, several proapoptotic proteins including cytochrome c, Smac/DIABLO, and apoptosis-inducing factor (AIF) are released from the mitochondrial intermembrane space into the cytoplasm, where they bind and activate protein targets that promote later apoptotic events (21). In addition to these alterations in outer membrane permeability, changes in mitochondrial morphology upon apoptosis have been described. For example, mitochondrial fragmentation has been observed in HeLa and COS-7 cells overexpressing the proapoptotic factor Bax or treated with apoptotic stimuli such as staurosporine (STS) or etoposide (24, 35). This morphological change occurs prior to cytochrome *c* release, is caspase-independent, and cannot be blocked by overexpression of the antiapoptotic factor Bcl-2 (24, 35, 135). Thus, mitochondria fragment early in the apoptotic pathway.

Drp1. In healthy COS-7 cells, the dynamin-related mitochondrial fission protein Drp1 (homolog of yeast Dnm1p) is predominantly dispersed in the cytoplasm. During apoptosis, cytoplasmic Drp1 localizes to the surface of mitochondria as punctate structures at membrane constriction sites and the tips of mitochondrial tubules (35). Overexpression of dominant-negative Drp1 or RNAi (RNA interference) depletion of endogenous Drp1 prevents mitochondrial fragmentation, loss of the inner-membrane potential, cytochrome *c* release, and apoptotic cell death in STS-treated cells (35, 76, 135). In another cell death pathway, increased Ca^{2+} transfer from the endoplasmic reticulum to mitochondria promotes Drp1-induced mitochondrial fragmentation and downstream apoptotic events (12). These findings support the idea that Drp1-mediated mitochondrial fission is a prerequisite for some types of mitochondrial-dependent apoptosis. It has also been reported that the pro-apoptotic protein Bax forms foci at mitochondrial membrane constriction sites and the tips of mitochondrial tubules and colocalizes with Drp1 in apoptotic COS-7 cells (67). Additional studies may clarify how Drp1 translocates to mitochondria and promotes cytochrome *c* release and whether Bax facilitates Drp1-mediated mitochondrial fission during apoptosis.

Although evolutionarily conserved components of cell death pathways were first identified in the worm *C. elegans*, evidence that mitochondrial proteins are required for *C. elegans* cell death has been elusive (71). A new study reveals that mitochondrial fragmentation also contributes to programmed cell death during worm development (61). *C. elegans* cells programmed to die exhibit DRP-1-mediated mitochondrial fragmentation at an early stage in the cell death pathway (61).

In contrast to the proapoptotic role of mitochondrial fission in the programmed cell death pathways described above, recent studies reveal that mitochondrial fragmentation can inhibit Ca^{2+}-mediated apoptosis induced by ceramide or H_2O_2 treatment, or serum deprivation (139). These stimuli induce Ca^{2+} movement from the endoplasmic reticulum to mitochondria, which leads to increased Ca^{2+}-load in the matrix, outer membrane permeabilization, and cytochrome *c* release (138). This mitochondrial Ca^{2+} uptake is facilitated by the connectivity of the mitochondrial network, which allows a wave of Ca^{2+} to spread rapidly throughout the organelle (138). Mitochondrial fragmentation in Drp1-overexpressing HeLa cells interrupts intramitochondrial Ca^{2+} waves and reduces overall mitochondrial Ca^{2+} uptake (139). Consequently, Drp1-induced mitochondrial fission protects cells against Ca^{2+}-mediated apoptosis (139). Thus, Drp1 can act as either a proapoptotic or antiapoptotic factor, depending on the nature of the particular cell death pathway.

STS: staurosporine

hFis1. Overexpression of hFis1 is reported to induce mitochondrial fragmentation, cytochrome *c* release, and caspase-mediated cell death in the absence of apoptotic stimuli (62). Coexpression of Bcl-x_L, an antiapoptotic factor of the Bcl-2 family, suppresses cytochrome *c* release and cell death but does not block mitochondrial fragmentation in these hFis1-overexpressing cells (62). Overexpression of both hFis1 and dominant-negative Drp1 prevents mitochondrial fragmentation and cytochrome *c* release but does not inhibit cell death (62). These observations suggest either that: (*a*) overexpressed hFis1 induces apoptosis in a manner independent of cytochrome *c* and Drp1, or (*b*) mitochondrial fission and cell death are not necessarily interdependent processes in hFis1-overexpressing cells.

Surprisingly, depletion of hFis1 by RNAi does not affect mitochondrial localization of Drp1 but does suppress mitochondrial fragmentation and apoptosis in HeLa cells treated with STS and other death stimuli (76). Although these data suggest that Drp1 mitochondrial targeting does not depend on hFis1, no alternative outer mitochondrial membrane protein that mediates Drp1 recruitment has been identified to date in mammalian cells. It is possible that residual hFis1 in RNAi-depleted cells is responsible for the Drp1 localization observed in these studies.

OPA1. Inhibition of mitochondrial fusion results in mitochondrial fragmentation that, in turn, may induce apoptosis. RNAi depletion of the dynamin-related mitochondrial fusion protein OPA1 (homolog of yeast Mgm1p) causes mitochondrial fragmentation, reduction of the inner membrane potential, Bax translocation, and cytochrome *c* release (76, 98). This apoptotic cell death in OPA1-depleted cells can be suppressed by expression of Bcl-2 (98). When both OPA1 and hFis1 are depleted, mitochondria still fragment, although Bax translocation, cytochrome *c* release, and apoptotic cell death are blocked (76). It is possible that loss of OPA1 induces a hFis1-independent mitochondrial fragmentation pathway. Alternatively, rapid OPA1 depletion may allow mitochondrial fragmentation to occur long before fission is blocked by slower hFis1 depletion. Nonetheless, these observations suggest that cell death mediated by loss of OPA1 function requires hFis1.

Mfn1/2. Does the mitofusin GTPase play a direct role in apoptosis? In HeLa cells treated with apoptotic stimuli, mitochondrial fusion appears to be inhibited prior to caspase activation (66). Overexpression of Bax somehow enhances this fusion block (66). In addition, formation of Bax foci on mitochondria and outer membrane permeabilization occurs at about the same time as inhibition of mitochondrial fusion without loss of inner membrane potential (66). Whether the initial transport and tethering of abnormal mitochondria, or the actual fusion event, is blocked during apoptosis remains unclear. Nevertheless, overexpression of rat Fzo1A/1B (Mfn1/2 homologs) in death stimuli-treated cells partially rescues mitochondrial fragmentation, activated-Bax translocation, cytochrome *c* release, and apoptosis (135). Conversely, depletion of Fzo1A/1B or Mfn1/2 by RNAi causes mitochondrial fragmentation, and increases the sensitivity of these cells to apoptotic stimuli (135). Thus, mitochondrial-dependent apoptosis can also be impacted by events in the fusion pathway.

It has been suggested that the mitochondrial fusion protein Mfn2 forms punctate structures and colocalizes with Bax in apoptotic cells (67), supporting the idea that Bax inactivates Mfn2-mediated mitochondrial fusion. However, this observation is not consistent with other reports that Mfn2 distributes evenly on mitochondria without forming visible foci (19, 31, 113, 116). In many cases, it is not clear whether fluorescent protein- or epitope-tagged molecules used in these and other studies are functional in vivo, raising the possibility that mitochondrial foci contain nonfunctional or aggregated polypeptides. Further analysis is required to understand the mechanisms by which Mfn1/2 affect Drp1 behavior and Bax-mediated activation of apoptotic events, and whether Bax regulates mitochondrial fusion during apoptosis. It will also be important to revisit studies from different research groups that yielded conflicting results. In particular, it would be useful to determine whether the observed mitochondrial morphology phenotypes are specific for distinct cell lines or due to technical differences in experimental protocols.

SUMMARY POINTS

1. The yeast Fzo1p GTPase, Mgm1p GTPase, and Ugo1p form a complex that mediates mitochondrial fusion. Inner- and outer-membrane fusion events are distinct and separable in vitro.

2. The Dnm1p GTPase is recruited from the cytoplasm to the outer mitochondrial membrane in a process that requires the integral membrane protein Fis1p and its binding partners Mdv1p and Caf4p. These proteins interact in a poorly understood fashion to form fission complexes that divide the mitochondrial compartment.
3. Mmm1p, Mdm10p, and Mdm12p form a complex that maintains mitochondrial tubular shape in cooperation with Mmm2p, Mdm31p, and Mdm32p. In budding yeast, this pathway may also link mitochondria and mtDNA nucleoids to the actin cytoskeleton.
4. Maintenance of normal mitochondrial morphology requires Mdm33p and Gem1p, proteins proposed to act in inner-membrane remodeling and cytoplasmic–mitochondrial signaling, respectively.
5. Defects in mitochondrial dynamics cause inherited neuropathies. In multicellular eukaryotes, fusion and fission proteins regulate early steps in mitochondrial-dependent programmed cell death.

FUTURE DIRECTIONS/UNSOLVED ISSUES

1. Staging of the in vitro mitochondrial fusion assay via drugs, mutants, or other reagents should help to characterize the activities of each fusion protein during outer- and inner-membrane fusion events. The roles of the Fzo1p and Mgm1p GTPase cycles and the inner-membrane electrochemical proton gradient in these events should also be investigated. It will also be important to identify the mechanisms by which outer- and inner-membrane fusion events are normally coordinated in vivo.
2. The development of an in vitro mitochondrial fission assay will be necessary to study distinct steps during fission and to elucidate the mechanisms by which the fission proteins function. Understanding which fission processes are coupled to the Dnm1p GTPase cycle is also an important issue.
3. Studies using a combination of genetic, biochemical, and cell biological approaches will help to address how Mmm2p, Mdm31p, and Mdm32p mediate formation of the Mmm1p-Mdm10p-Mdm12p complex, how these proteins maintain mitochondrial tubular shape, and how the MMM complex associates with both the actin cytoskeleton and mtDNA nucleoids.
4. Crosstalk among pathways that control mitochondrial morphology and dynamics should be explored. A better understanding of how these pathways impact other mitochondrial processes including protein import and assembly, bioenergetics, mtDNA maintenance, and mitochondrial mass and quality control is also required.

ACKNOWLEDGMENTS

We thank David Chan, Raymond Deshaies, Thomas Langer, Rusty Lipford, Chris Meisinger, Nikolaus Pfanner, and Benedikt Westermann for communicating results prior to publication. We apologize to many colleagues whose work could not be cited due to space limitations. We are grateful to the members of the Shaw laboratory for constructive comments on the manuscript. Our work was supported by grants from the United Mitochondrial Disease Foundation and the American Heart Association to K.O., and grants from the National Institutes of Health (GM53466 and GM067047) to J.M.S.

LITERATURE CITED

1. Alexander C, Votruba M, Pesch UEA, Thiselton DL, Mayer S, et al. 2000. OPA1, encoding a dynamin-related GTPase, is mutated in autosomal dominant optic atrophy linked to chromosome 3q28. *Nat. Genet.* 26:211–15
2. Arimura S, Aida GP, Fujimoto M, Nakazono M, Tsutsumi N. 2004. Arabidopsis dynamin-like protein 2a (ADL2a), like ADL2b, is involved in plant mitochondrial division. *Plant Cell Physiol.* 45:236–42
3. Arimura S, Tsutsumi N. 2002. A dynamin-like protein (ADL2b), rather than FtsZ, is involved in Arabidopsis mitochondrial division. *Proc. Natl. Acad. Sci. USA* 99:5727–31
4. Attardi G, Schatz G. 1988. Biogenesis of mitochondria. *Annu. Rev. Cell Biol.* 4:289–333
5. Bereiter-Hahn J, Voth M. 1994. Dynamics of mitochondria in living cells: shape changes, dislocations, fusion, and fission of mitochondria. *Microsc. Res. Tech.* 27:198–219
6. Berger KH, Sogo LF, Yaffe MP. 1997. Mdm12p, a component required for mitochondrial inheritance that is conserved between budding and fission yeast. *J. Cell Biol.* 136:545–53
7. Bleazard W, McCaffery JM, King EJ, Bale S, Mozdy A, et al. 1999. The dynamin-related GTPase Dnm1 regulates mitochondrial fission in yeast. *Nat. Cell Biol.* 1:298–304
8. Boldogh I, Vojtov N, Karmon S, Pon LA. 1998. Interaction between mitochondria and the actin cytoskeleton in budding yeast requires two integral mitochondrial outer membrane proteins, Mmm1p and Mdm10p. *J. Cell Biol.* 141:1371–81
9. Boldogh IR, Nowakowski DW, Yang HC, Chung H, Karmon S, et al. 2003. A protein complex containing Mdm10p, Mdm12p, and Mmm1p links mitochondrial membranes and DNA to the cytoskeleton-based segregation machinery. *Mol. Biol. Cell* 14:4618–27
10. Boldogh IR, Yang HC, Pon LA. 2001. Mitochondrial inheritance in budding yeast. *Traffic* 2:368–74
11. Bossy-Wetzel E, Barsoum MJ, Godzik A, Schwarzenbacher R, Lipton SA. 2003. Mitochondrial fission in apoptosis, neurodegeneration and aging. *Curr. Opin. Cell Biol.* 15:706–16
12. Breckenridge DG, Stojanovic M, Marcellus RC, Shore GC. 2003. Caspase cleavage product of BAP31 induces mitochondrial fission through endoplasmic reticulum calcium signals, enhancing cytochrome c release to the cytosol. *J. Cell Biol.* 160:1115–27
13. Burgess SM, Delannoy M, Jensen RE. 1994. MMM1 encodes a mitochondrial outer membrane protein essential for establishing and maintaining the structure of yeast mitochondria. *J. Cell Biol.* 126:1375–91
14. Butow RA, Avadhani NG. 2004. Mitochondrial signaling: the retrograde response. *Mol. Cell* 14:1–15
15. Carelli V, Rugolo M, Sgarbi G, Ghelli A, Zanna C, et al. 2004. Bioenergetics shapes cellular death pathways in Leber's hereditary optic neuropathy: a model of mitochondrial neurodegeneration. *Biochim. Biophys. Acta* 1658:172–79
16. Cerveny KL, Jensen RE. 2003. The WD-repeats of Net2p interact with Dnm1p and Fis1p to regulate division of mitochondria. *Mol. Biol. Cell* 14:4126–39
17. Cerveny KL, McCaffery JM, Jensen RE. 2001. Division of mitochondria requires a novel DMN1-interacting protein, Net2p. *Mol. Biol. Cell* 12:309–21
18. Chen H, Chan DC. 2004. Mitochondrial dynamics in mammals. *Curr. Top. Dev. Biol.* 59:119–44

19. Chen H, Detmer SA, Ewald AJ, Griffin EE, Fraser SE, Chan DC. 2003. Mitofusins Mfn1 and Mfn2 coordinately regulate mitochondrial fusion and are essential for embryonic development. *J. Cell Biol.* 160:189–200
20. Cipolat S, de Brito OM, Dal Zilio B, Scorrano L. 2004. OPA1 requires mitofusin 1 to promote mitochondrial fusion. *Proc. Natl. Acad. Sci. USA* 101:15927–32
21. Danial NN, Korsmeyer SJ. 2004. Cell death: critical control points. *Cell* 116:205–19
22. Delettre C, Lenaers G, Griffoin JM, Gigarel N, Lorenzo C, et al. 2000. Nuclear gene OPA1, encoding a mitochondrial dynamin-related protein, is mutated in dominant optic atrophy. *Nat. Genet.* 26:207–10
23. Delettre C, Lenaers G, Pelloquin L, Belenguer P, Hamel CP. 2002. OPA1 (Kjer type) dominant optic atrophy: a novel mitochondrial disease. *Mol. Genet. Metab.* 75:97–107
24. Desagher S, Martinou JC. 2000. Mitochondria as the central control point of apoptosis. *Trends Cell Biol.* 10:369–77
25. De Vos KJ, Allan VJ, Grierson AJ, Sheetz MP. 2005. Mitochondrial function and actin regulate dynamin-related protein 1-dependent mitochondrial fission. *Curr. Biol.* 15:678–83
26. Dimmer KS, Fritz S, Fuchs F, Messerschmitt M, Weinbach N, et al. 2002. Genetic basis of mitochondrial function and morphology in Saccharomyces cerevisiae. *Mol. Biol. Cell* 13:847–53
27. Dimmer KS, Jakobs S, Vogel F, Altmann K, Westermann B. 2005. Mdm31 and Mdm32 are inner membrane proteins required for maintenance of mitochondrial shape and stability of mitochondrial DNA nucleoids in yeast. *J. Cell Biol.* 168:103–15
28. Dohm JA, Lee SJ, Hardwick JM, Hill RB, Gittis AG. 2004. Cytosolic domain of the human mitochondrial fission protein fis1 adopts a TPR fold. *Proteins* 54:153–56
29. Drubin DG, Jones HD, Wertman KF. 1993. Actin structure and function: roles in mitochondrial organization and morphogenesis in budding yeast and identification of the phalloidin-binding site. *Mol. Biol. Cell* 4:1277–94
30. Esser K, Tursun B, Ingenhoven M, Michaelis G, Pratje E. 2002. A novel two-step mechanism for removal of a mitochondrial signal sequence involves the mAAA complex and the putative rhomboid protease Pcp1. *J. Mol. Biol.* 323:835–43
31. Eura Y, Ishihara N, Yokota S, Mihara K. 2003. Two mitofusin proteins, mammalian homologues of FZO, with distinct functions are both required for mitochondrial fusion. *J. Biochem.* 134:333–44
32. Fekkes P, Shepard KA, Yaffe MP. 2000. Gag3p, an outer membrane protein required for fission of mitochondrial tubules. *J. Cell Biol.* 151:333–40
33. Ferre M, Amati-Bonneau P, Tourmen Y, Malthiery Y, Reynier P. 2005. eOPA1: An online database for OPA1 mutations. *Hum. Mutat.* 25:423–28
34. Fisk HA, Yaffe MP. 1999. A role for ubiquitination in mitochondrial inheritance in Saccharomyces cerevisiae. *J. Cell Biol.* 145:1199–208
35. Frank S, Gaume B, Bergmann-Leitner ES, Leitner WW, Robert EG, et al. 2001. The role of dynamin-related protein 1, a mediator of mitochondrial fission, in apoptosis. *Dev. Cell* 1:515–25
36. Frank S, Robert EG, Youle RJ. 2003. Scission, spores, and apoptosis: a proposal for the evolutionary origin of mitochondria in cell death induction. *Biochem. Biophys. Res. Commun.* 304:481–86
37. Fransson A, Ruusala A, Aspenstrom P. 2003. Atypical Rho GTPases have roles in mitochondrial homeostasis and apoptosis. *J. Biol. Chem.* 278:6495–502

38. Frederick RL, McCaffery JM, Cunningham KW, Okamoto K, Shaw JM. 2004. Yeast Miro GTPase, Gem1p, regulates mitochondrial morphology via a novel pathway. *J. Cell Biol.* 167:87–98
39. Freeman M. 2004. Proteolysis within the membrane: rhomboids revealed. *Nat. Rev. Mol. Cell Biol.* 5:188–97
40. Frey TG, Mannella CA. 2000. The internal structure of mitochondria. *Trends Biochem. Sci.* 25:319–24
41. Fritz S, Rapaport D, Klanner E, Neupert W, Westermann B. 2001. Connection of the mitochondrial outer and inner membranes by Fzo1 is critical for organellar fusion. *J. Cell Biol.* 152:683–92
42. Fritz S, Weinbach N, Westermann B. 2003. Mdm30 is an F-box protein required for maintenance of fusion-competent mitochondria in yeast. *Mol. Biol. Cell* 14:2303–13
43. Fuchs F, Prokisch H, Neupert W, Westermann B. 2002. Interaction of mitochondria with microtubules in the filamentous fungus Neurospora crassa. *J. Cell Sci.* 115:1931–37
44. Fukushima NH, Brisch E, Keegan BR, Bleazard W, Shaw JM. 2001. The GTPase effector domain sequence of the Dnm1p GTPase regulates self-assembly and controls a rate-limiting step in mitochondrial fission. *Mol. Biol. Cell* 12:2756–66
45. Gorsich SW, Shaw JM. 2004. Importance of mitochondrial dynamics during meiosis and sporulation. *Mol. Biol. Cell* 15:4369–81
45a. Griffin EE, Graumann J, Chan DC. 2005. The WD40 protein Caf4p is a component of the mitochondrial fission machinery and recruits Dnm1p to mitochondria. *J. Cell Biol.* 170:237–48
46. Griparic L, van der Bliek AM. 2001. The many shapes of mitochondrial membranes. *Traffic* 2:235–44
47. Guan K, Farh L, Marshall TK, Deschenes RJ. 1993. Normal mitochondrial structure and genome maintenance in yeast requires the dynamin-like product of the MGM1 gene. *Curr. Genet.* 24:141–48
48. Guillou E, Bousquet C, Daloyau M, Emorine LJ, Belenguer P. 2005. Msp1p is an intermembrane space dynamin-related protein that mediates mitochondrial fusion in a Dnm1p-dependent manner in S. pombe. *FEBS Lett.* 579:1109–16
49. Hales KG, Fuller MT. 1997. Developmentally regulated mitochondrial fusion mediated by a conserved, novel, predicted GTPase. *Cell* 90:121–29
50. Hanekamp T, Thorsness MK, Rebbapragada I, Fisher EM, Seebart C, et al. 2002. Maintenance of mitochondrial morphology is linked to maintenance of the mitochondrial genome in Saccharomyces cerevisiae. *Genetics* 162:1147–56
51. Harder Z, Zunino R, McBride H. 2004. Sumo1 conjugates mitochondrial substrates and participates in mitochondrial fission. *Curr. Biol.* 14:340–45
52. Herlan M, Bornhovd C, Hell K, Neupert W, Reichert AS. 2004. Alternative topogenesis of Mgm1 and mitochondrial morphology depend on ATP and a functional import motor. *J. Cell Biol.* 165:167–73
53. Herlan M, Vogel F, Bornhovd C, Neupert W, Reichert AS. 2003. Processing of Mgm1 by the rhomboid-type protease Pcp1 is required for maintenance of mitochondrial morphology and of mitochondrial DNA. *J. Biol. Chem.* 278:27781–88
54. Hermann GJ, King EJ, Shaw JM. 1997. The yeast gene, MDM20, is necessary for mitochondrial inheritance and organization of the actin cytoskeleton. *J. Cell Biol.* 137:141–53
55. Hermann GJ, Thatcher JW, Mills JP, Hales KG, Fuller MT, et al. 1998. Mitochondrial fusion in yeast requires the transmembrane GTPase Fzo1p. *J. Cell Biol.* 143:359–73

56. Hitchcock AL, Auld K, Gygi SP, Silver PA. 2003. A subset of membrane-associated proteins is ubiquitinated in response to mutations in the endoplasmic reticulum degradation machinery. *Proc. Natl. Acad. Sci. USA* 100:12735–40
57. Hobbs AE, Srinivasan M, McCaffery JM, Jensen RE. 2001. Mmm1p, a mitochondrial outer membrane protein, is connected to mitochondrial DNA (mtDNA) nucleoids and required for mtDNA stability. *J. Cell Biol.* 152:401–10
58. Hoffmann HP, Avers CJ. 1973. Mitochondrion of yeast: ultrastructural evidence for one giant, branched organelle per cell. *Science* 181:749–51
59. Ishihara N, Eura Y, Mihara K. 2004. Mitofusin 1 and 2 play distinct roles in mitochondrial fusion reactions via GTPase activity. *J. Cell Sci.* 117:6535–46
60. Ishihara N, Jofuku A, Eura Y, Mihara K. 2003. Regulation of mitochondrial morphology by membrane potential, and DRP1-dependent division and FZO1-dependent fusion reaction in mammalian cells. *Biochem. Biophys. Res. Commun.* 301:891–98
61. Jagasia R, Grote P, Westermann B, Conradt B. 2005. DRP-1-mediated mitochondrial fragmentation during EGL-1-induced cell death in C. elegans. *Nature* 433:754–60
62. James DI, Parone PA, Mattenberger Y, Martinou JC. 2003. hFis1, a novel component of the mammalian mitochondrial fission machinery. *J. Biol. Chem.* 278:36373–79
63. Jensen RE, Hobbs AE, Cerveny KL, Sesaki H. 2000. Yeast mitochondrial dynamics: fusion, division, segregation, and shape. *Microsc. Res. Tech.* 51:573–83
64. Jin JB, Bae H, Kim SJ, Jin YH, Goh CH, et al. 2003. The Arabidopsis dynamin-like proteins ADL1C and ADL1E play a critical role in mitochondrial morphogenesis. *Plant Cell* 15:2357–69
65. Jones BA, Fangman WL. 1992. Mitochondrial DNA maintenance in yeast requires a protein containing a region related to the GTP-binding domain of dynamin. *Genes Dev.* 6:380–89
66. Karbowski M, Arnoult D, Chen H, Chan DC, Smith CL, Youle RJ. 2004. Quantitation of mitochondrial dynamics by photolabeling of individual organelles shows that mitochondrial fusion is blocked during the Bax activation phase of apoptosis. *J. Cell Biol.* 164:493–99
67. Karbowski M, Lee YJ, Gaume B, Jeong SY, Frank S, et al. 2002. Spatial and temporal association of Bax with mitochondrial fission sites, Drp1, and Mfn2 during apoptosis. *J. Cell Biol.* 159:931–38
68. Karbowski M, Youle RJ. 2003. Dynamics of mitochondrial morphology in healthy cells and during apoptosis. *Cell Death Differ.* 10:870–80
69. Kelly DP, Scarpulla RC. 2004. Transcriptional regulatory circuits controlling mitochondrial biogenesis and function. *Genes Dev.* 18:357–68
70. Kijima K, Numakura C, Izumino H, Umetsu K, Nezu A, et al. 2005. Mitochondrial GTPase mitofusin 2 mutation in Charcot-Marie-Tooth neuropathy type 2A. *Hum. Genet.* 116:23–27
71. Kinchen JM, Hengartner MO. 2005. Tales of cannibalism, suicide, and murder: Programmed cell death in *C. elegans*. *Curr. Top. Dev. Biol.* 65:1–45
72. Kondo-Okamoto N, Shaw JM, Okamoto K. 2003. Mmm1p spans both the outer and inner mitochondrial membranes and contains distinct domains for targeting and foci formation. *J. Biol. Chem.* 278:48997–9005
73. Koshiba T, Detmer SA, Kaiser JT, Chen H, McCaffery JM, Chan DC. 2004. Structural basis of mitochondrial tethering by mitofusin complexes. *Science* 305:858–62
74. Kus BM, Caldon CE, Andorn-Broza R, Edwards AM. 2004. Functional interaction of 13 yeast SCF complexes with a set of yeast E2 enzymes in vitro. *Proteins* 54:455–67

75. Labrousse AM, Zappaterra MD, Rube DA, van der Bliek AM. 1999. C. elegans dynamin-related protein DRP-1 controls severing of the mitochondrial outer membrane. *Mol. Cell* 4:815–26
76. Lee YJ, Jeong SY, Karbowski M, Smith CL, Youle RJ. 2004. Roles of the mammalian mitochondrial fission and fusion mediators Fis1, Drp1, and Opa1 in apoptosis. *Mol. Biol. Cell* 15:5001–11
77. Legesse-Miller A, Massol RH, Kirchhausen T. 2003. Constriction and Dnm1p recruitment are distinct processes in mitochondrial fission. *Mol. Biol. Cell* 14:1953–63
78. Legros F, Lombes A, Frachon P, Rojo M. 2002. Mitochondrial fusion in human cells is efficient, requires the inner membrane potential, and is mediated by mitofusins. *Mol. Biol. Cell* 13:4343–54
79. Li Z, Okamoto K, Hayashi Y, Sheng M. 2004. The importance of dendritic mitochondria in the morphogenesis and plasticity of spines and synapses. *Cell* 119:873–87
80. Liu HY, Chiang YC, Pan J, Chen J, Salvadore C, et al. 2001. Characterization of CAF4 and CAF16 reveals a functional connection between the CCR4-NOT complex and a subset of SRB proteins of the RNA polymerase II holoenzyme. *J. Biol. Chem.* 276:7541–48
81. Logan DC, Scott I, Tobin AK. 2004. ADL2a, like ADL2b, is involved in the control of higher plant mitochondrial morphology. *J. Exp. Bot.* 55:783–85
82. Mannella CA, Pfeiffer DR, Bradshaw PC, Moraru II, Slepchenko B, et al. 2001. Topology of the mitochondrial inner membrane: dynamics and bioenergetic implications. *IUBMB Life* 52:93–100
83. Mano S, Nakamori C, Kondo M, Hayashi M, Nishimura M. 2004. An Arabidopsis dynamin-related protein, DRP3A, controls both peroxisomal and mitochondrial division. *Plant J.* 38:487–98
84. Mattenberger Y, James DI, Martinou JC. 2003. Fusion of mitochondria in mammalian cells is dependent on the mitochondrial inner membrane potential and independent of microtubules or actin. *FEBS Lett.* 538:53–59
85. McConnell SJ, Stewart LC, Talin A, Yaffe MP. 1990. Temperature-sensitive yeast mutants defective in mitochondrial inheritance. *J. Cell Biol.* 111:967–76
86. McQuibban GA, Saurya S, Freeman M. 2003. Mitochondrial membrane remodelling regulated by a conserved rhomboid protease. *Nature* 423:537–41
87. Meeusen S, McCaffery JM, Nunnari J. 2004. Mitochondrial fusion intermediates revealed in vitro. *Science* 305:1747–52
88. Meeusen S, Nunnari J. 2003. Evidence for a two membrane-spanning autonomous mitochondrial DNA replisome. *J. Cell Biol.* 163:503–10
89. Meisinger C, Rissler M, Chacinska A, Szklarz LKS, Milenkovic D, et al. 2004. The mitochondrial morphology protein Mdm10 functions in assembly of the preprotein translocase of the outer membrane. *Dev. Cell* 7:61–71
90. Messerschmitt M, Jakobs S, Vogel F, Fritz S, Dimmer KS, et al. 2003. The inner membrane protein Mdm33 controls mitochondrial morphology in yeast. *J. Cell Biol.* 160:553–64
91. Misaka T, Miyashita T, Kubo Y. 2002. Primary structure of a dynamin-related mouse mitochondrial GTPase and its distribution in brain, subcellular localization, and effect on mitochondrial morphology. *J. Biol. Chem.* 277:15834–42
92. Miyakawa I, Aoi H, Sando N, Kuroiwa T. 1984. Fluorescence microscopic studies of mitochondrial nucleoids during meiosis and sporulation in the yeast, Saccharomyces cerevisiae. *J. Cell Sci.* 66:21–38

93. Mozdy AD, McCaffery JM, Shaw JM. 2000. Dnm1p GTPase-mediated mitochondrial fission is a multi-step process requiring the novel integral membrane component Fis1p. *J. Cell Biol.* 151:367–80
94. Mozdy AD, Shaw JM. 2003. A fuzzy mitochondrial fusion apparatus comes into focus. *Nat. Rev. Mol. Cell Biol.* 4:468–78
95. Muratani M, Kung C, Shokat KM, Tansey WP. 2005. The F box protein Dsg1/Mdm30 is a transcriptional coactivator that stimulates Gal4 turnover and cotranscriptional mRNA processing. *Cell* 120:887–99
96. Nunnari J, Marshall WF, Straight A, Murray A, Sedat JW, Walter P. 1997. Mitochondrial transmission during mating in Saccharomyces cerevisiae is determined by mitochondrial fusion and fission and the intramitochondrial segregation of mitochondrial DNA. *Mol. Biol. Cell* 8:1233–42
97. Nunnari J, Wong ED, Meeusen S, Wagner JA. 2002. Studying the behavior of mitochondria. *Methods Enzymol.* 351:381–93
98. Olichon A, Baricault L, Gas N, Guillou E, Valette A, et al. 2003. Loss of OPA1 perturbates the mitochondrial inner membrane structure and integrity, leading to cytochrome c release and apoptosis. *J. Biol. Chem.* 278:7743–46
99. Olichon A, Emorine LJ, Descoins E, Pelloquin L, Brichese L, et al. 2002. The human dynamin-related protein OPA1 is anchored to the mitochondrial inner membrane facing the inter-membrane space. *FEBS Lett.* 523:171–76
100. Osteryoung KW, Nunnari J. 2003. The division of endosymbiotic organelles. *Science* 302:1698–704
101. Otsuga D, Keegan BR, Brisch E, Thatcher JW, Hermann GJ, et al. 1998. The dynamin-related GTPase, Dnm1p, controls mitochondrial morphology in yeast. *J. Cell Biol.* 143:333–49
102. Patton EE, Willems AR, Tyers M. 1998. Combinatorial control in ubiquitin-dependent proteolysis: don't Skp the F-box hypothesis. *Trends Genet.* 14:236–43
103. Pelloquin L, Belenguer P, Menon Y, Ducommun B. 1998. Identification of a fission yeast dynamin-related protein involved in mitochondrial DNA maintenance. *Biochem. Biophys. Res. Commun.* 251:720–26
104. Pelloquin L, Belenguer P, Menon Y, Gas N, Ducommun B. 1999. Fission yeast Msp1 is a mitochondrial dynamin-related protein. *J. Cell Sci.* 112:4151–61
105. Perfettini JL, Roumier T, Kroemer G. 2005. Mitochondrial fusion and fission in the control of apoptosis. *Trends Cell Biol.* 15:179–83
106. Pfanner N, Wiedemann N, Meisinger C. 2004. Cell biology. Double membrane fusion. *Science* 305:1723–24
107. Pfanner N, Wiedemann N, Meisinger C, Lithgow T. 2004. Assembling the mitochondrial outer membrane. *Nat. Struct. Mol. Biol.* 11:1044–48
108. Pitts KR, Yoon Y, Krueger EW, McNiven MA. 1999. The dynamin-like protein DLP1 is essential for normal distribution and morphology of the endoplasmic reticulum and mitochondria in mammalian cells. *Mol. Biol. Cell* 10:4403–17
109. Praefcke GJ, McMahon HT. 2004. The dynamin superfamily: universal membrane tubulation and fission molecules? *Nat. Rev. Mol. Cell Biol.* 5:133–47
110. Prokisch H, Neupert W, Westermann B. 2000. Role of MMM1 in maintaining mitochondrial morphology in Neurospora crassa. *Mol. Biol. Cell* 11:2961–71
111. Rapaport D, Brunner M, Neupert W, Westermann B. 1998. Fzo1p is a mitochondrial outer membrane protein essential for the biogenesis of functional mitochondria in Saccharomyces cerevisiae. *J. Biol. Chem.* 273:20150–55

112. Reichert AS, Neupert W. 2002. Contact sites between the outer and inner membrane of mitochondria-role in protein transport. *Biochim. Biophys. Acta* 1592:41–49
113. Rojo M, Legros F, Chateau D, Lombes A. 2002. Membrane topology and mitochondrial targeting of mitofusins, ubiquitous mammalian homologs of the transmembrane GTPase Fzo. *J. Cell Sci.* 115:1663–74
114. Rube DA, van der Bliek AM. 2004. Mitochondrial morphology is dynamic and varied. *Mol. Cell Biochem.* 256–257:331–39
115. Santel A, Frank S, Gaume B, Herrler M, Youle RJ, Fuller MT. 2003. Mitofusin-1 protein is a generally expressed mediator of mitochondrial fusion in mammalian cells. *J. Cell Sci.* 116:2763–74
116. Santel A, Fuller MT. 2001. Control of mitochondrial morphology by a human mitofusin. *J. Cell Sci.* 114:867–74
117. Saraste M. 1999. Oxidative phosphorylation at the fin de siecle. *Science* 283:1488–93
118. Scheffler IE. 1999. *Mitochondria*. New York: Wiley-Liss. xiv, 367 pp.
119. Scott SV, Cassidy-Stone A, Meeusen SL, Nunnari J. 2003. Staying in aerobic shape: how the structural integrity of mitochondria and mitochondrial DNA is maintained. *Curr. Opin. Cell Biol.* 15:482–88
120. Sesaki H, Jensen RE. 1999. Division versus fusion: Dnm1p and Fzo1p antagonistically regulate mitochondrial shape. *J. Cell Biol.* 147:699–706
121. Sesaki H, Jensen RE. 2001. UGO1 encodes an outer membrane protein required for mitochondrial fusion. *J. Cell Biol.* 152:1123–34
122. Sesaki H, Jensen RE. 2004. Ugo1p links the Fzo1p and Mgm1p GTPases for mitochondrial fusion. *J. Biol. Chem.* 279:28298–303
123. Sesaki H, Southard SM, Hobbs AE, Jensen RE. 2003. Cells lacking Pcp1p/Ugo2p, a rhomboid-like protease required for Mgm1p processing, lose mtDNA and mitochondrial structure in a Dnm1p-dependent manner, but remain competent for mitochondrial fusion. *Biochem. Biophys. Res. Commun.* 308:276–83
124. Sesaki H, Southard SM, Yaffe MP, Jensen RE. 2003. Mgm1p, a dynamin-related GTPase, is essential for fusion of the mitochondrial outer membrane. *Mol. Biol. Cell* 14:2342–56
125. Shaw JM, Nunnari J. 2002. Mitochondrial dynamics and division in budding yeast. *Trends Cell Biol.* 12:178–84
126. Shepard KA, Yaffe MP. 1999. The yeast dynamin-like protein, Mgm1p, functions on the mitochondrial outer membrane to mediate mitochondrial inheritance. *J. Cell Biol.* 144:711–20
127. Shy ME. 2004. Charcot-Marie-Tooth disease: an update. *Curr. Opin. Neurol.* 17:579–85
128. Shy ME, Kamholz J, Lovelace RE. 1999. Introduction to the Third International Symposium on Charcot-Marie-Tooth disorders. *Ann. NY Acad. Sci.* 883:xiii–xviii
129. Simon VR, Karmon SL, Pon LA. 1997. Mitochondrial inheritance: cell cycle and actin cable dependence of polarized mitochondrial movements in Saccharomyces cerevisiae. *Cell Motil. Cytoskelet.* 37:199–210
130. Smirnova E, Griparic L, Shurland DL, van der Bliek AM. 2001. Dynamin-related protein Drp1 is required for mitochondrial division in mammalian cells. *Mol. Biol. Cell* 12:2245–56
131. Smirnova E, Shurland DL, Ryazantsev SN, van der Bliek AM. 1998. A human dynamin-related protein controls the distribution of mitochondria. *J. Cell Biol.* 143:351–58
132. Sogo LF, Yaffe MP. 1994. Regulation of mitochondrial morphology and inheritance by Mdm10p, a protein of the mitochondrial outer membrane. *J. Cell Biol.* 126:1361–73

133. Steinberg G, Schliwa M. 1993. Organelle movements in the wild type and wall-less fz;sg;os-1 mutants of *Neurospora crassa* are mediated by cytoplasmic microtubules. *J. Cell Sci.* 106:555–64
134. Stojanovski D, Koutsopoulos OS, Okamoto K, Ryan MT. 2004. Levels of human Fis1 at the mitochondrial outer membrane regulate mitochondrial morphology. *J. Cell Sci.* 117:1201–10
135. Sugioka R, Shimizu S, Tsujimoto Y. 2004. Fzo1, a protein involved in mitochondrial fusion, inhibits apoptosis. *J. Biol. Chem.* 279:52726–34
136. Suzuki M, Jeong SY, Karbowski M, Youle RJ, Tjandra N. 2003. The solution structure of human mitochondria fission protein Fis1 reveals a novel TPR-like helix bundle. *J. Mol. Biol.* 334:445–58
137. Suzuki M, Neutzner A, Tjandra N, Youle RJ. 2005. Novel structure of the N terminus in yeast FIS1 correlates with a specialized function in mitochondrial fission. *J. Biol. Chem.* 280:21444–52
138. Szabadkai G, Rizzuto R. 2004. Participation of endoplasmic reticulum and mitochondrial calcium handling in apoptosis: more than just neighborhood? *FEBS Lett.* 567:111–15
139. Szabadkai G, Simoni AM, Chami M, Wieckowski MR, Youle RJ, Rizzuto R. 2004. Drp-1-dependent division of the mitochondrial network blocks intraorganellar Ca^{2+} waves and protects against Ca^{2+}-mediated apoptosis. *Mol. Cell* 16:59–68
140. Tieu Q, Nunnari J. 2000. Mdv1p is a WD repeat protein that interacts with the dynamin-related GTPase, Dnm1p, to trigger mitochondrial division. *J. Cell Biol.* 151:353–66
141. Tieu Q, Okreglak V, Naylor K, Nunnari J. 2002. The WD repeat protein, Mdv1p, functions as a molecular adaptor by interacting with Dnm1p and Fis1p during mitochondrial fission. *J. Cell Biol.* 158:445–52
142. Trimmer PA, Swerdlow RH, Parks JK, Keeney P, Bennett JP Jr, et al. 2000. Abnormal mitochondrial morphology in sporadic Parkinson's and Alzheimer's disease cybrid cell lines. *Exp. Neurol.* 162:37–50
143. Uetz P, Giot L, Cagney G, Mansfield TA, Judson RS, et al. 2000. A comprehensive analysis of protein-protein interactions in Saccharomyces cerevisiae. *Nature* 403:623–27
144. Varadi A, Johnson-Cadwell LI, Cirulli V, Yoon Y, Allan VJ, Rutter GA. 2004. Cytoplasmic dynein regulates the subcellular distribution of mitochondria by controlling the recruitment of the fission factor dynamin-related protein-1. *J. Cell Sci.* 117:4389–400
145. Westermann B. 2002. Merging mitochondria matters: cellular role and molecular machinery of mitochondrial fusion. *EMBO Rep.* 3:527–31
146. Westermann B. 2003. Mitochondrial membrane fusion. *Biochim. Biophys. Acta* 1641:195–202
147. Wong ED, Wagner JA, Gorsich SW, McCaffery JM, Shaw JM, Nunnari J. 2000. The dynamin-related GTPase, Mgm1p, is an intermembrane space protein required for maintenance of fusion competent mitochondria. *J. Cell Biol.* 151:341–52
148. Wong ED, Wagner JA, Scott SV, Okreglak V, Holewinske TJ, et al. 2003. The intramitochondrial dynamin-related GTPase, Mgm1p, is a component of a protein complex that mediates mitochondrial fusion. *J. Cell Biol.* 160:303–11
149. Yaffe MP. 1999. The machinery of mitochondrial inheritance and behavior. *Science* 283:1493–97
150. Yoon Y. 2004. Sharpening the scissors: mitochondrial fission with aid. *Cell Biochem. Biophys.* 41:193–206
151. Yoon Y. 2005. Regulation of Mitochondrial dynamics: Another process modulated by Ca^{2+} signals? *Science STKE* 280:pe18

152. Yoon Y, Krueger EW, Oswald BJ, McNiven MA. 2003. The mitochondrial protein hFis1 regulates mitochondrial fission in mammalian cells through an interaction with the dynamin-like protein DLP1. *Mol. Cell Biol.* 23:5409–20
153. Yoon Y, McNiven MA. 2001. Mitochondrial division: New partners in membrane pinching. *Curr. Biol.* 11:R67–70
154. Yoon Y, Pitts KR, McNiven MA. 2001. Mammalian dynamin-like protein DLP1 tubulates membranes. *Mol. Biol. Cell* 12:2894–905
155. Youngman MJ, Hobbs AE, Burgess SM, Srinivasan M, Jensen RE. 2004. Mmm2p, a mitochondrial outer membrane protein required for yeast mitochondrial shape and maintenance of mtDNA nucleoids. *J. Cell Biol.* 164:677–88
156. Zhao C, Takita J, Tanaka Y, Setou M, Nakagawa T, et al. 2001. Charcot-Marie-Tooth disease type 2A caused by mutation in a microtubule motor KIF1Bbeta. *Cell* 105:587–97
157. Zhu PP, Patterson A, Stadler J, Seeburg DP, Sheng M, Blackstone C. 2004. Intra- and intermolecular domain interactions of the C-terminal GTPase effector domain of the multimeric dynamin-like GTPase Drp1. *J. Biol. Chem.* 279:35967–74
158. Zuchner S, Mersiyanova IV, Muglia M, Bissar-Tadmouri N, Rochelle J, et al. 2004. Mutations in the mitochondrial GTPase mitofusin 2 cause Charcot-Marie-Tooth neuropathy type 2A. *Nat. Genet.* 36:449–51

RELATED RESOURCES

Attardi G, Schatz G. 1988. Biogenesis of mitochondria. *Annu. Rev. Cell Biol.* 4:289–333
Hermann GJ, Shaw JM. 1998. Mitochondrial dynamics in yeast. *Annu. Rev. Cell Dev. Biol.* 14:265–303

LINKS

http://www.yeastgenome.org
http://ihg.gsf.de/mitop2/start.jsp

RNA-Guided DNA Deletion in Tetrahymena: An RNAi-Based Mechanism for Programmed Genome Rearrangements

Meng-Chao Yao[1,2] and Ju-Lan Chao[1]

[1]Institute of Molecular Biology, Academia Sinica, Nankang, Taipei 11529, Taiwan, Republic of China; email: mcyao@imb.sinica.edu.tw

[2]Division of Basic Sciences, Fred Hutchinson Cancer Research Center, Seattle, Washington 98109; email: mcyao@fhcrc.org

Annu. Rev. Genet. 2005. 39:537–59

First published online as a Review in Advance on August 8, 2005

The *Annual Review of Genetics* is online at genet.annualreviews.org

doi: 10.1146/annurev.genet.39.073003.095906

0066-4197/05/1215-0537$20.00

Key Words

genome surveillance, sRNA, gene silencing, heterochromatin, chromatin diminution

Abstract

Ciliated protozoan are unicellular eukaryotes. Most species in this diverse group display nuclear dualism, a special feature that supports both somatic and germline nuclei in the same cell. Probably due to this unique life style, they exhibit unusual nuclear characteristics that have intrigued researchers for decades. Among them are large-scale DNA rearrangements, which restructure the somatic genome to become drastically different from its germline origin. They resemble the classical phenomenon of chromatin diminution in some nematodes discovered more than a century ago. The mechanisms of such rearrangements, their biological roles, and their evolutionary origins have been difficult to understand. Recent studies have revealed a clear link to RNA interference, and begin to shed light on these issues. Using the simple ciliate Tetrahymena as a model, this chapter summarizes the physical characterization of these processes, describes recent findings that connect them to RNA interference, and discusses the details of their mechanisms, potential roles in genome defense, and possible occurrences in other organisms.

Contents

INTRODUCTION

In this genomics era when every nucleotide of the entire genome has been deciphered for many organisms, it is important to stress that the sequences obtained are but snapshots, for genomes are not static. They change with time and cell propagation such that, even within an individual organism, the genome in different cells may not be identical. This instability drives evolution, but it also causes diseases. Many organisms including human have incorporated such changes, particularly in the form of programmed DNA rearrangements, into their normal developmental processes. These processes play key roles in cell differentiation (10), and are critical to our understanding of genome dynamics.

Chromatin diminution: the phenomenon of massive elimination of chromatin from somatic cells during development, often associated with chromosome fragmentations

One type of programmed DNA rearrangements is particularly intriguing. Unlike others that occur to just one or a few genes, this one occurs at numerous genomic locations and causes drastic changes in genomic content and chromosome structure. The phenomenon was first reported as chromatin diminution in the nematode Ascaris more than a century ago (11). Similar phenomena have now been found in some species of Cyclops (crustaceans) (7, 8) and fish (45, 62), and most ciliated protozoa (67, 87). In the most extreme cases, over 90% of the inherited genomic sequences are eliminated and the remaining DNA molecules are broken into tiny pieces, each big enough to carry just one or a few genes (48, 77). These extensive rearrangements pose fundamental questions about not only the molecular mechanism of DNA rearrangements but also the importance of an "intact" chromosome. Unfortunately, relatively little is known about their molecular nature.

Among these organisms, ciliated protozoa are most amenable to modern methods of investigation and have offered most molecular details so far. Studies over the past three decades have revealed an astonishing series of highly regulated processes, including amplification, deletion, elimination, inversion, shuffling, and dimerization of specific DNA sequences, as well as chromosome fragmentation and new telomere formation (20, 67, 87). The simple ciliate *Tetrahymena thermophila* is known to carry out most of these processes except gene inversion and shuffling. Partly due to its relative ease of handling, this species has received most attention and offered considerable insights. This chapter focuses on the studies in Tetrahymena, and discusses only briefly some relevant features in Paramecium.

TETRAHYMENA LIFE CYCLE

Like most ciliated protozoa, Tetrahymena possesses two distinctly different types of nuclei in each cell. The larger macronucleus contains approximately 45 times the amount of DNA present in a haploid genome (45C), and the smaller micronucleus is diploid. Both are descendants of the same zygotic nucleus formed during sexual reproduction, but separately hold the somatic and germline functions. During vegetative growth the macronucleus is active in transcription and divides by an unusual process of amitosis that shows no chromosome condensation or spindle formation. The micronucleus divides by typical mitosis and is essentially inert in transcription. Vegetative growth can go on indefinitely but is interrupted by conjugation when cells of different mating types are starved and mixed (**Figure 1**). During mating the micronucleus displays its germline properties. It goes through mitosis, meiosis, and cross fertilization to generate zygotic nuclei, which further divide and differentiate to give rise to the new macro- and micronuclei of the subsequent vegetative life (64). The old macronucleus is degraded during this process similar to that of a typical somatic nucleus of

Macronucleus: the somatic nucleus of ciliates including Tetrahymena

Micronucleus: the germline nucleus of ciliates including Tetrahymena

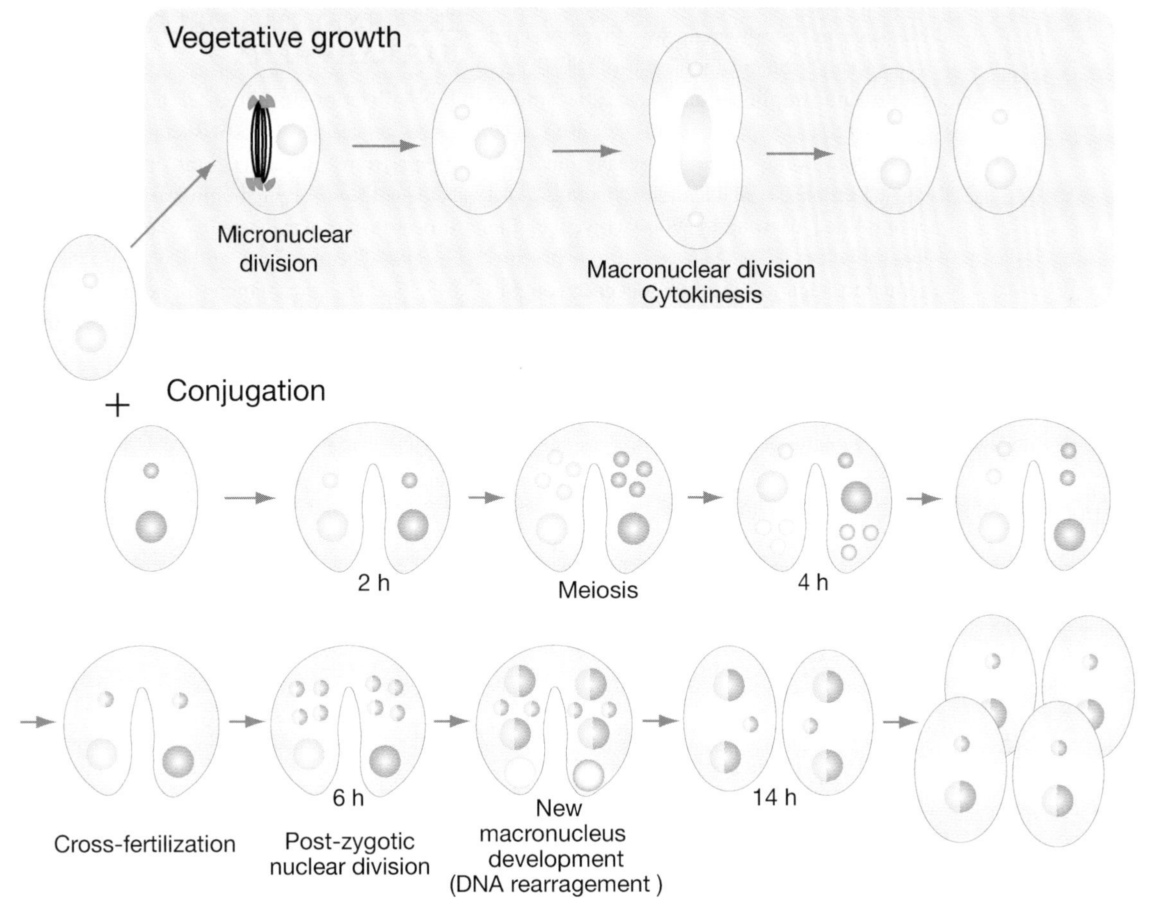

Figure 1

Life cycle of Tetrahymena. The top panel shows the vegetative growth of Tetrahymena, which occurs when nutrients are plentiful. The bottom panel shows the conjugation process between two Tetrahymena cells of different mating types in starvation conditions.

DNA deletion: the elimination of specific DNA sequences from internal regions of chromosomes during new macronuclear development in Tetrahymena and other ciliates

Cbs: chromosome breakage sequence

IES: internal eliminated sequence

metazoa. DNA rearrangements occur only in the developing new macronucleus. Thus they do not affect the transmission of the germline genome.

OVERVIEW OF TETRAHYMENA DNA REARRANGEMENTS

In addition to endoduplications, two major events occur globally to restructure the macronuclear genome of Tetrahymena: chromosome breakage and DNA deletion. This review focuses mainly on issues related to DNA deletion. Only a brief description of chromosome breakage is given here to provide a relevant background. Chromosome breakage in Tetrahymena was first observed as a part of the mechanism for rDNA excision and amplification (92). The single-copy rDNA of this organism is excised from the chromosome through breakage at both ends, becomes a 21-kb inverted dimer, and is amplified to roughly 9000 copies in the macronucleus (89). Subsequently, chromosome breakage was found to occur at roughly 200 specific sites in the genome, reducing the average chromosome size (there are 5 metacentric chromosomes of similar sizes) from about 25,000 kb to about 800 kb (1, 19, 66). These breakage sites share a common 15-bp sequence (referred to as Cbs for chromosome breakage sequence) (43, 89, 90, 91) that was found to be the necessary and sufficient sequence signal for breakage to occur (90, 91). Cbs is very well conserved among different copies in this and other Tetrahymena species (22), but is not found in other, more distantly related groups of ciliates that also show chromosome breakage. Variations were found only in 3 of the 15 nucleotide positions among different copies in *T. thermophila* and other Tetrahymena species. Single nucleotide mutations in most positions reduced or abolished its function in *T. thermophila* (28). Chromosome breakage in Tetrahymena is followed by de novo telomere addition (43, 90–92, 100). The minichromosomes produced are stably maintained in the macronucleus during vegetative growth. Chromosome breakage occurs in most ciliates and several Ascarid nematodes. Their details vary significantly. For instance, in complex ciliates (e.g., Stylonychia, Oxytricha, and Euplotes) breakage is so extensive that most macronuclear DNA molecules are just large enough to contain a single gene (average about 2 kb) (48, 67, 68, 77), and in Paramecium (13, 30) and Ascaris (61, 81), the breakage sites are not fixed and appear to vary in regions several kb in size. In all cases the DNA fragments acquire new telomeres, and are stably maintained throughout their somatic life.

INTERNAL DNA DELETION IN TETRAHYMENA

DNA deletion (or internal DNA deletion) refers to the removal of specific DNA segments from internal regions of the chromosome without leaving behind new stable ends. It occurs in most ciliates (87), some species of Ascaris, Cyclops (7, 8), and likely also hagfish (45, 62). In Tetrahymena it occurs at more than 6000 sites to splice out DNA ranging from several hundred bp to more than 10 kb in size (hereafter referred to as deletion elements; also known generally as internal eliminated sequence or IES in ciliates), which together make up about 15% of the genome (12, 94, 96, 97). The deleted DNA is immediately degraded (2), probably through both linear and circular intermediates (69–71, 99). These elements include both single-copy and moderately repetitive sequences as judged by their hybridization properties. Several deletion elements have been sequenced in their entirety or partially (4, 5, 32, 38, 42, 84). Most contain short (1–8-bp) terminal direct repeats, but share no other common sequence feature. They do not usually contain, or interrupt, obvious protein-coding sequences, although some are found in introns. One long element (Tlr-1) contains sequences closely resembling transposons (65), and another (Tel1) (18) has features of a transposon. It is likely that many other tranposon-like elements exist, although

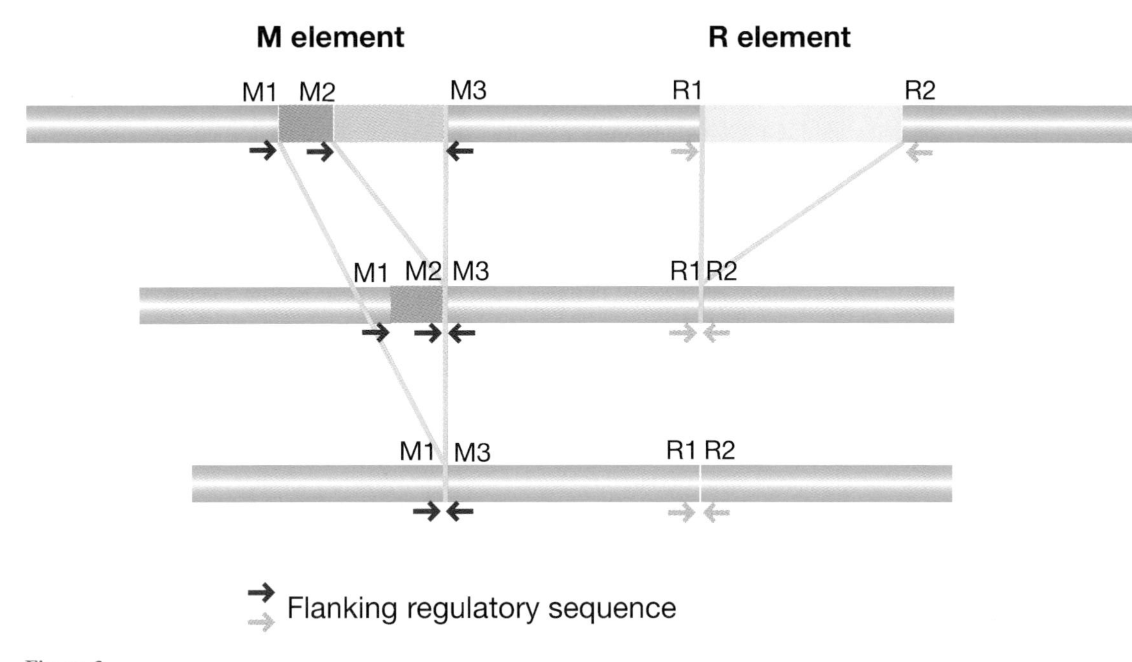

Figure 2

Deletion of the M- and R-elements. Two neighboring deletion elements (or IES), the M- and R-elements, are shown as examples to illustrate precise DNA deletion in Tetrahymena. The M-element deletion occurs with two alternative left boundaries, M1 and M2.

they have yet to be identified. Other evidence connects transposition to DNA deletion. In Tetrahymena, some deletion intermediates contain 4-bp 5′ protruding ends that resemble those generated by the one-step *trans*-esterification reaction of some transposons including Tn7 (71). In some other ciliates (i.e., Euplotes and Oxytricha), deletion elements that resemble transposons of the Mariner/Tc1 class have been well described (6, 23, 39, 41).

CIS-ACTING SEQUENCES

Detailed analysis of the sequences that regulate DNA deletion in Tetrahymena has been carried out using several elements including two neighboring ones, the M- and R-elements (**Figure 2**) (94). The M-element is 0.6 kb (or 0.9 kb in an alternative form) and the R-element is 1.1 kb in size. Both are deleted precisely and reproducibly, although minor junction variations (less than 10 bps) have been observed (3–5). The *cis*-acting sequences that regulate these two as well as other elements have been analyzed, using a transformation system that delivers DNA constructs into conjugating cells in a plasmid-like vector (98). The vector is derived from the rDNA molecule (minichromosome) of Tetrahymena and is maintained in high copy number in the macronucleus. Deletion elements were inserted into this vector and introduced into the cell at or before the stage of DNA rearrangement by microinjection or electroporation. Rearrangements of the inserted DNA were determined after the cell has completed conjugation and further propagated. Such studies have been successful and revealed two sets of sequences that control the efficiency and precision of this event: a pair of flanking regulatory sequences and a set of internal promoting sequences. The flanking regulatory sequences are located approximately 45 bp away from the two ends of each element. They alone specify the positions of deletion boundaries within a larger (several kbs) region determined by the

Deletion element or IES: internal eliminated sequences are micronucleus-limited elements that are excised during macronuclear development in ciliates

M-element: a deletion element (or IES) in Tetrahymena

R-element: a deletion element (or IES) in Tetrahymena

internal promoting sequences (33, 34). These sequences are not the same for the M- and R-elements: For the M-element it is the pentanucleotide 5′ GGGGG (34), and for the R-element a part of a 33-bp AT-rich sequence (15). Each element, or a family of elements, is possibly controlled by a unique flanking regulatory sequence (22). But the situation is likely to be more complex. As revealed in more recent studies, some elements have highly variable boundaries that spread out over a distance of hundreds of bps (65). This class of elements may lack flanking regulatory sequences entirely. The internal promoting sequences also work in an unusual manner. The 0.6-kb internal region of the M-element is essential for deletion to occur. It can be divided into at least three parts, each alone able to promote deletion (with boundaries determined by the flanking regulatory sequences), although at a reduced efficiency. Their activities are orientation independent, and extend several hundred bps beyond both ends of these sequences. Internal sequences from three other deletion elements tested, but not other sequences of similar length and G + C content, can substitute for this activity (22). It is tempting to suggest that most deletion elements in Tetrahymena contain equivalent internal promoting activities for their own deletion. Evidence for internal promoting sequences has also been obtained in the studies of other elements, including Tlr-1 (86).

These studies showed clearly that *cis*-acting sequences have evolved in this organism to control the specificity of DNA deletion. However, the numbers of these controlling sequences are likely to be high, and their nature likely to be complex.

TRANS-ACTING FACTORS AND CHROMATIN STRUCTURE

Studies of stage-specific proteins in the developing macronucleus led to the findings of Pdd1p, Pdd2p, and Pdd3p, which were later shown to have roles in DNA rerrangements. PDD1 encodes a chromodomain protein similar to HP1 of *Drosophila* or SWI6 of *S. pombe*. It has a remarkable distribution pattern in the developing macronucleus: It first appears to be evenly distributed, then forms numerous small aggregates, and by the stage when DNA deletion occurs, forms large, micron-sized aggregates before disappearing completely. Remarkably, this distribution pattern is identical to those of the micronuclear-specific sequences, as determined by in situ hybridization. Indeed, the two are colocalized (51). The tight association between Pdd1p and deletion element DNA was further verified by immunoprecipitation. Genetic knock-out studies provide further support. Matings between cells without functional copies of PDD1 or PDD2 in their macronuclei produce mainly unviable progeny, in which DNA deletion is completely or partially blocked (21, 63). Chromosome breakage also appears to be blocked in PDD2 but not in PDD1 mutant progeny. There is little doubt that these proteins play important, though yet unspecified, roles in DNA deletion.

The remarkable spatial rearrangements of the PDD proteins and associated DNA suggested the formation of heterochromatin before deletion occurs. Electron microscopy studies also supported this idea (51), as did studies on histone modifications. Using Trichostatin A as an inhibitor for histone deacetylation, it was found that deacetylated histones are important for DNA deletion (25). Furthermore, methylation of the lysine 9 residue of histone H3, a hallmark of heterochromatin in other organisms, appears in chromatin associated with micronuclear-specific DNA during conjugation (79). Mutants with a specific alteration in this amino acid position also failed to carry out DNA rearrangements (49). Thus, formation of heterochromatin appears to be associated with, and essential for, DNA deletion in Tetrahymena. Correlation between chromatin condensation and DNA elimination is not without precedent. In Ascaris and Cyclops, the eliminated chromatin is mainly heterochromatic (7, 80).

THE RNA LINK

In the study of DNA deletion, three fundamental questions stand out. What is its mechanism? What is its biological role? And what is its evolutionary origin? Studies of the *cis*-acting sequences and *trans*-acting proteins partially answer the first question. Although the link to transposons has touched upon its evolution, few studies have shed light directly on the latter two until recently, when links between DNA deletion, RNA interference, and genome surveillance are found. The possible involvement of RNA in guiding DNA deletion was first suggested in studies of an unusual epigenetic effect on this process (discussed below). The direct observation that RNA plays a role came from transcription studies of the deletion elements in Tetrahymena. Although deletion elements do not code for proteins, they are transcribed at specific stages during conjugation, mostly before deletion occurs (17). The transcripts produced are quite unusual: They are from both strands and rather heterogeneous in size. Since double-stranded RNA plays key roles in posttranscriptional gene silencing (PTGS) in a wide variety of organisms (26, 29, 74), the double-stranded RNA produced by the deletion elements could also play a significant role that somehow leads to DNA deletion. This notion was supported by an experiment that used actinomycin D to briefly inhibit transcription during conjugation. The treatment interfered with DNA deletion without preventing developmental progression and progeny production. Subsequently, TWI1, a Tetrahymena ortholog of the argonaute protein gene, was found to be required for conjugation. Cells without this gene (knock-out mutants) initiate and progress through conjugation less efficiently than normal cells do, and are unable to carry out DNA deletions and chromosome breakage at several sites studied (57). Since argonaute protein is a critical component of the RNAi pathway in several eukaryotes (54), the finding suggested a direct link between RNAi and DNA deletion. A species of small RNA (sRNA) approximately 26 nt in length (later found to be 28 nt) was detected during conjugation; its abundance is significantly reduced in cells defective in TWI1 or PDD1 (57). This observation offers further support for the involvement of RNAi in Tetrahymena conjugation. Based on this and earlier results, a model was proposed to explain DNA deletion through RNA interference (57).

A direct test for the role of double-stranded RNAs (dsRNAs) in DNA deletion was carried out using RNA injection, in a manner similar to the initial RNAi experiment done in *Caenorhabditis elegans* (29). Both strands of RNAs were prepared in vitro from arbitrarily chosen regions of the macronuclear genome and injected into conjugating cells. These cells were allowed to complete conjugation and grow, and their macronuclear DNAs were analyzed to determine the deletion of the targeted DNA. The injection caused deletions of the corresponding macronuclear DNA in all three genomic regions tested. The treatment can be very effective, causing deletion in more than half of the injected pairs when carried out during the time of new macronuclear development (before normal DNA rearrangements occur). Because these sequences are not normally deleted, the result showed unambiguously that the presence of dsRNA alone is sufficient to cause DNA deletion in Tetrahymena (95). Based on these findings and those described above, one can envision the following process of DNA deletion: First, dsRNA is transcribed from all or most deletion elements at stages before DNA deletion occurs. This transcription could take place in the micronucleus and/or the developing macronucleus. Second, dsRNA is processed into sRNA in a manner similar to most RNAi processing and likely involving similar proteins. A Tetrahymena dicer-like gene DCL1 is known to be involved (60). Third, the sRNA forms the effector RNA in a complex and guides it to the genomic regions containing the corresponding sequences. Fourth, the targeted DNA is deleted, by a yet-unknown mechanism

RNA interference: the cellular process that reduces or eliminates expression of specific genes through the guidance of dsRNA and sRNA

Genome surveillance: the action of the cell to recognize and silence invading foreign genes, such as transposons and viral genes

PTGS: posttranscriptional gene silencing

sRNA: small RNA molecules, 21–28 nucleotides in size, that are generated from dsRNA through cleavage by Dicer proteins

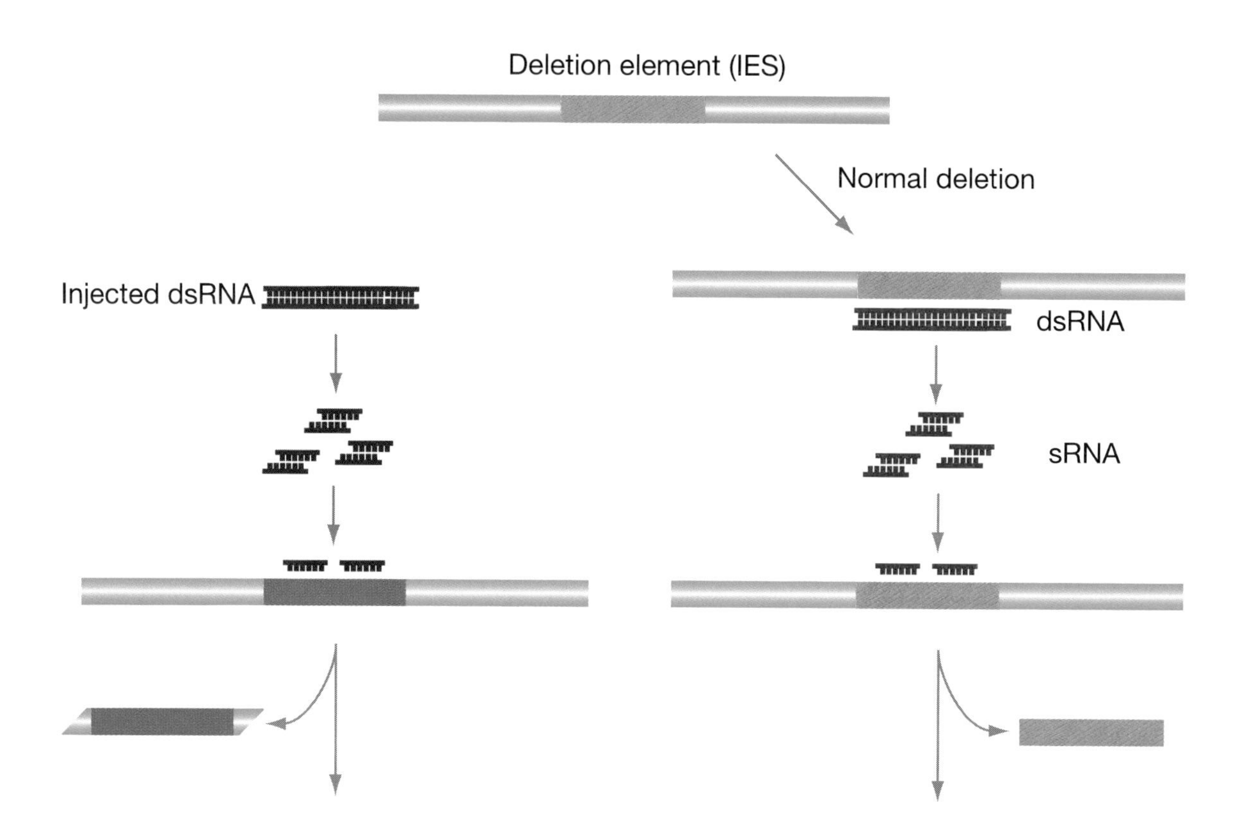

Figure 3

An RNA-guided process of DNA deletion. The right side illustrates the normal deletion process, in which dsRNAs are made from the deletion element (IES) in the micronucleus or early developing macronucleus and processed into sRNA, which then targets DNA with the same sequences for deletion in the new macronucleus. The left side illustrates a similar deletion process induced by the injection of dsRNA, In this case, the deletion boundaries are not precise.

TGS: transcriptional gene silencing

dsRNA: double-strand RNA transcripts, both derived from the same DNA sequence, which trigger RNA interference

(**Figure 3**). This scheme suggests that DNA deletion in Tetrahymena is a special type of RNAi effect: Instead of causing PTGS as it does in most organisms, the dsRNA produced during Tetrahymena conjugation causes DNA deletion. In *Schizosaccharomyces pombe* and other organisms, dsRNA is produced from specific genomic DNA, which leads to chromatin modifications and transcriptional gene silencing (TGS) in these regions (36, 83). In addition, in some plants the introduced dsRNA causes TGS and DNA methylation (53). DNA deletion in Tetrahymena can be viewed as being related to these RNAi-mediated TGS processes. Instead of causing TGS, the sRNA leads to DNA elimination in Tetrahymena, probably also through chromatin modifications.

This scheme includes two likely assumptions. First, the deletion of the native elements follow a mechanism similar to that of the induced deletions, and second, the formation of sRNA is necessary for DNA deletion. Recently, M-element sequence has been found among sRNA during conjugation (14), although the exact role for this RNA has yet to be determined. Additional studies are expected to confirm these two points.

Despite these cautions, it is clear that DNA deletion in Tetrahymena is guided by RNA, in *trans*, for its sequence specificities. This realization has interesting implications. First, all sequences to be deleted need to be transcribed and processed to become the effector RNAs. Thus, selective transcription (and/or processing) establishes the sequence specificity of DNA deletion. Second, since these RNAs are able to work in *trans*, i.e., away from the sites from which they are transcribed, the function of *cis*-acting sequences needs to be amended. Third, because of this *trans*-acting feature, one expects that repetitive sequences are more likely to be deleted than single-copy sequences, because the transcript derived from any single repeating unit can cause deletion of all units. Furthermore, all members of a repeat family are subject to the same fate. Indeed, most repetitive sequences present in the micronucleus are eliminated from the macronucleus (93); and in the macronuclear genome, few repeating sequences are found.

CURRENT ISSUES IN THE MECHANISM OF PROGRAMMED DNA DELETION

Knowing that RNA is involved in guiding DNA deletion, it would be interesting to re-examine features of DNA deletion in this light. Following are some of the major considerations.

Action of the Internal Promoting Sequences

The DNA sequences recognized by the effector sRNA are required for deletion: They are the internal promoting sequences described earlier. These sequences are transcribed into dsRNA and processed into sRNA, which then target its own templates for deletion. They are essential *cis*-acting sequences and should be exchangeable between elements and act independent of their orientation and overall integrity. These properties were indeed observed. The exact mechanism through which sRNA recognizes the genomic DNA is not yet clear. This is a general question facing all RNAi-mediated TGS. Base pairing could occur directly between sRNA and its target DNA, or indirectly between sRNA and the transcripts made from, and remaining attached to, the DNA (53). In the latter case, the act of transcription, and not just the transcripts produced, is required. So far no conclusive evidence exists to rule out either mechanism in other organisms. In Tetrahymena, however, the act of transcription is apparently not required because injection of dsRNA alone can induce deletion even in regions not known to make dsRNA or mRNA. Thus, the recognition is likely based on RNA-DNA pairing, probably through the formation of a d-loop. This is an interesting starting point to contemplate the molecular complex that carries out DNA deletion in Tetrahymena.

Chromatin Marking

RNA-guided DNA deletion in Tetrahymena likely shares the first steps of its mechanism with RNAi-mediated TGS. dsRNA transcription leads to heterochromatin formation in the transcribed portion of the genome. However, in Tetrahymena the process goes one step further: the heterochromatin is deleted (**Figure 4**). It is important to ask if chromatin modification actually marks the regions for deletion or is simply a part of the process. A key experiment appeared to support the first notion. By targeting Pdd1p to DNA in the plasmid-based DNA deletion assay system, it was shown that binding of this protein alone was sufficient to trigger DNA deletion. The presence of specific sRNA was apparently not required (79). Thus, sRNA interaction with the DNA probably occurs before Pdd1p was recruited to mark the chromatin for deletion. This process has interesting implications for the precision of DNA deletion: The sequence specificity established by sRNA-DNA interaction, which is at the level of single nucleotide, is transferred to a marked chromatin structure, which presumably is at the level of nucleosome, before deletion occurs. If true, the RNA-guided process appears to lack the precision observed for the

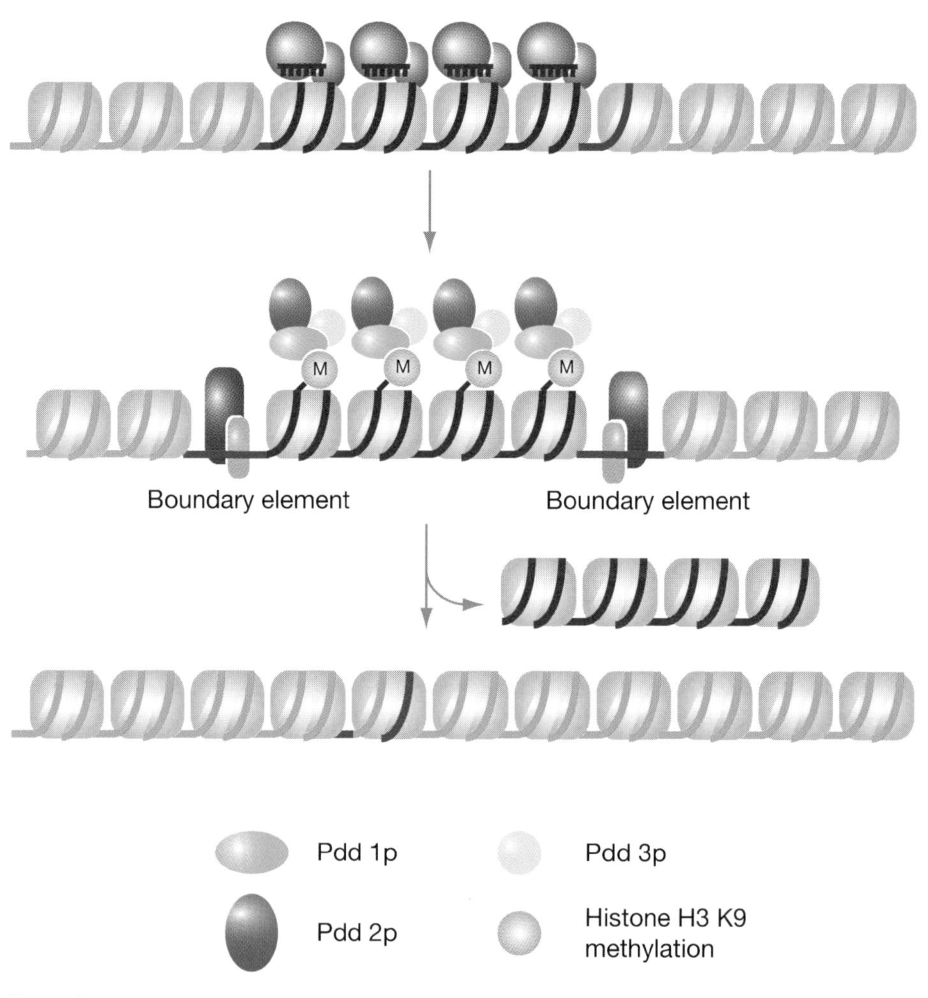

Figure 4

Participation of chromatin structure modification in DNA deletion. Binding of sRNA to its target DNA leads to the recruitment of several proteins to this region and specific modifications of histones. Recruitment of "boundary element proteins" to the flanking regulatory sequences is suggested to explain the precision of the deletion process for some native elements. In this process, the recognition of sequences for deletion is initiated by RNA-DNA sequence pairing, which leads to specific chromatin marking. The marked chromatin then serves as the targets for the proteins that carry out the actual deletion.

M- and R-elements. Additional regulation is likely involved.

Establishing Deletion Boundaries

The precision observed for the deletion of the M- and R-elements, and likely many others as well, is difficult to achieve based solely on RNA guidance. The dsRNA made at the M-element is heterogeneous in size and appears to contain sequences from the immediate flanking regions (17). Thus, the boundaries of the transcribed regions do not coincide with those of the deleted regions. Furthermore, injection-induced deletions have highly variable boundaries spreading over hundreds to thousands of bps, which are not coincidental with the injected dsRNA termini. One level of regulation could be in dsRNA processing, though there has been no evidence for it. Another level of regulation could be from the flanking regulatory sequences described earlier. For the M- and R-elements, flanking sequences are required for determining junction specificities. These sequences work in *cis* and thus cannot act through *trans*-acting RNAs. One simple view involves the use of the flanking regulatory sequences as chromatin boundary elements: They have evolved to limit the extent of the deletion triggered by dsRNA. It could work in a way similar to chromatin boundary elements that limit the spread of special chromatin structures in other organisms (46, 82). This regulation could help limit the boundaries to a small interval. Additional regulation would still be required to achieve the single-nucleotide level precision observed.

Cutting and Deleting DNA

Once a genomic region is marked by sRNA, presumably an enzymatic system would recognize this marking and carry out the deletion. Nothing is known about this enzymatic system, which remains one of most interesting problems to solve. One could envision a sequence-nonspecific endonuclease that cuts and degrades DNA in the marked region. The free ends of the flanking DNA thus generated could be joined by a dsDNA break repair system. In such a process, the excised region may not necessarily be removed as one piece because cutting could occur anywhere in the marked region, and the position of the boundaries, which are determined by the position of the marked nucleosomes and the precision of the repair system, would not be precise. The injection-induced deletions appear to fit this mode: They have variable boundaries that frequently occur at sites with short direct repeats, a feature of broken-end repairs. However, deletion of the native elements seems different. The deletion creates precise junctions and generates transient ends that resemble ends generated in DNA transposition. In addition, DNA deletion in Paramecium, Euplotes, and Oxytricha produces free, circularized form of the elements that are likely to be the primary deletion products (9, 40, 78, 85). Presumably, in these cases the element is cut off in one piece by a sequence-specific protein and the ends are ligated together to form the circle. Thus these two processes, one generating precise junctions and the other not, appear different. We suggest that the imprecise system is the original form evolved for genome surveillance (see below), and the precise form becomes established later for some elements when junction signals (including the flanking regulatory sequences) are acquired through further evolution.

Macronuclear inheritance: the transfer of genetic determining factors from the parental macronucleus (somatic nucleus) to the new macronucleus of the following generation independent of the germline genetic information

MACRONUCLEAR INHERITANCE AND THE EPIGENETIC EFFECT ON DNA DELETION

One of the most intriguing aspects of DNA rearrangements in ciliates is the epigenetic effect exerted by the old macronucleus. This mysterious phenomenon first led researchers to speculate on the involvement of RNA (24). Because these studies have been reviewed in the past (56, 87), we describe only briefly some of the earlier studies to provide a background, and instead focus on the more recent findings in Tetrahymena.

The classic studies of macronuclear inheritance in Paramecium were made before DNA deletion was first reported (76). This unusual inheritance pattern occurs for several traits including mating-type determination. Unlike standard Mendelian traits, which are inherited through the micronucleus, or typical cytoplasmic traits, which are inherited through mitochondria, these traits are inherited through the macronucleus. This peculiar inheritance pattern was not easily explained. It would be like the inheritance of a somatic mutation through sexual generations. One possibility is that these genetic factors become determined only during new macronuclear development

scnRNA: a hypothetical class of sRNAs postulated to be produced from the entire germline genome and scan the macronuclear genome to generate germline-specific (micronuclear-specific) sRNA, which guides all DNA deletion in Tetrahymena

and are guided by the old macronucleus according to its own "determined state."

Later studies of one such trait, a mutation in a surface antigen gene (27), revealed a direct link to DNA rearrangement. Cells with a deletion in this gene in the parental macronucleus produced progeny with the same deletion regardless of the genetic state in their micronuclei. It provided a molecular description for macronuclear inheritance and suggested a direct role for the parental macronucleus in affecting DNA deletions. Subsequently, macronuclear inheritance was also shown to occur to typical programmed DNA deletions (24). Normal DNA deletion (in the developing new macronucleus) can be inhibited when the same deletion element is abnormally present in the old (parental) macronucleus, and this inheritance pattern is repeated in the following generation.

Similar macronuclear inheritance has also been observed in Tetrahymena through molecular analysis. Independent of the Paramecium studies, inhibition of the M- and R-element deletion was observed when either element was introduced into the parental macronucleus in high copy numbers (16). This inhibition was sequence specific: M-element inhibited only M-element, and R-element only R-element deletion. Also, it was the element itself, not the flanking DNA, that was responsible for the inhibition. Once the inhibition occurred, the new macronucleus, now containing the deletion element, again inhibited the deletion of the same element in the subsequent conjugation, thus setting up a macronuclear inheritance pattern.

Although the mechanism is unclear, these studies suggested a special form of sequence-specific communication between the old parental and the new developing macronuclei during conjugation. Two basic explanations have been proposed for this phenomenon (16, 24, 87). The first suggested that the element in the old macronucleus sequestered sequence-specific factors that were needed for the deletion of the same element in the new macronucleus, thereby interfering with normal deletion in a sequence-specific manner. These factors could be proteins that recognized the sequence of the elements. The second suggested that the element in the old macronucleus produced an RNA messenger that inhibited deletion of the same sequence in the new macronucleus. Although more radical, the second process provides a simpler explanation for the sequence specificity needed. Exactly how such RNA molecules could inhibit DNA deletion was not simple to explain, however.

The discovery of sRNA and dsRNA in the process of DNA deletion provided a ready explanation for this phenomenon. A specific model put forward to explain this epigenetic effect (57) posited that the entire genome is transcribed to produce sRNA (referred to as scanning RNA or scnRNA), which then enters the old macronucleus and scans the genome for matched sequence. Only those sRNA molecules without a match (and thus that contain micronucleus-specific sequences) leave this nucleus to cause deletion in the developing macronucleus (59). This scanning RNA model suggests a key feature of the epigenetic effect: By screening the sRNA, the old macronucleus prevents the deletion (in the developing macronucleus) of any sequence present in its genome. It partially explains the epigenetic effects on programmed DNA deletion in Tetrahymena and Paramecium, although the molecular details of such a scanning process are difficult to envision, based on current knowledge. This model reflects features from both of the two earlier proposed processes: the old macronucleus sequesters "the sequence-specific factors" of the first, and the "RNA messenger" of the second process. The model is appealing partly because it also provides a mechanism for determining the sequence specificity of DNA deletion. By scanning the whole genome, it determines all the sequences that are specific to the micronucleus and targets them for deletion. The model awaits support for both of its critical assumptions: transcription of the whole genome to produce dsRNA, and the subtraction of sRNA by the macronuclear genome. In addition, there is little evidence to date that the macronuleus determines the sequence specificity of all DNA deletion.

Regardless of the validity of this and other models, sRNA appears to be the best candidate for the molecular agent of communication between the old and the new macronucleus. Two recent studies provided additional information on this issue. The first suggested the presence of sRNA in the old macronucleus during conjugation, which indirectly supported this idea. In this study, the sRNA was found to be associated with Twi1p through immunoprecipitation studies, and thus was thought to be present in the old macronucleus before appearing in the new macronucleus, as Twi1p is (58). Whether sRNA actually moves from the old macronucleus to the new macronucleus is critical, and requires additional studies to acertain its actual movement. The second study determined the relative abundances of M-element-specific dsRNA and sRNA at different stages of conjugation. In cells with high copy numbers of the M-element in the old macronucleus, the levels of M-element-specific dsRNA are highly elevated, but the levels of M-element-specific sRNA remain relatively unchanged. The result argues against the idea that the old macronucleus sequesters sRNA of the same sequence (14). Clearly, the exact role of the old macronucleus awaits further clarification.

Based on existing evidence, it is possible to envision a relatively simple process for macronuclear inheritance without evoking a mechanism to determine all deletion sequence specificity at the same time. One needs simply to add two steps to the RNA-guided DNA deletion process described above (**Figure 3**). First, when a deletion element is present in the old macronucleus, it is also transcribed to produce dsRNA. Second, this RNA does not become effector sRNA to cause deletion. Instead, it interacts with sRNA of the same sequences produced from the micronucleus (or new macronucleus) and interferes with its normal function in DNA deletion (**Figure 5**). This "interaction" could occur through simple base pairing, or through a process similar to RNAi-mediated PTGS in either the old macronucleus or the cytoplasm. Thus, in this simple scheme the old macronucleus produces an RNA messenger that interferes with DNA deletion, as originally proposed. There are experimental supports for this process. First, it is known that deletion elements are transcribed to produce dsRNA during conjugation, whether in the micronucleus or the macronucleus (15). Second, although the fate of the dsRNA made in the old macronucleus is not known, it apparently does not add to total sRNA level. Furthermore, the protein encoded by the dicer ortholog DCL1, which is likely responsible for processing dsRNA into sRNA, is not present in the old macronucleus at these stages (60), further supporting the notion that dsRNA produced in the old macronucleus is not processed into sRNA.

DELETION OF FOREIGN GENES AND GENOME DEFENSE

The link to RNA interference raises the possibility that DNA deletion may also serve as a genome surveillance mechanism in Tetrahymena. Invading genetic elements, such as transposons, are inactivated in organisms such as *C. elegans* by RNAi-mediated processes either transcriptionally or posttranscriptionally (75). In Tetrahymena, they could be deleted to achieve the same purpose. This idea is supported directly by the surprising observation that Tetrahymena is able to recognize and delete a foreign DNA sequence. When an *E. coli* sequence (the Tn*5* gene that confers neomycin resistance) was inserted into the Tetrahymena micronuclear genome, it was specifically deleted from the developing macronucleus in the following conjugation. Deletion occurs regardless of the sites of insertion. Its efficiency differs at different sites, and is nearly complete at one of these sites. The inserted sequence is a standard cassette for gene knock-out experiments in Tetrahymena,

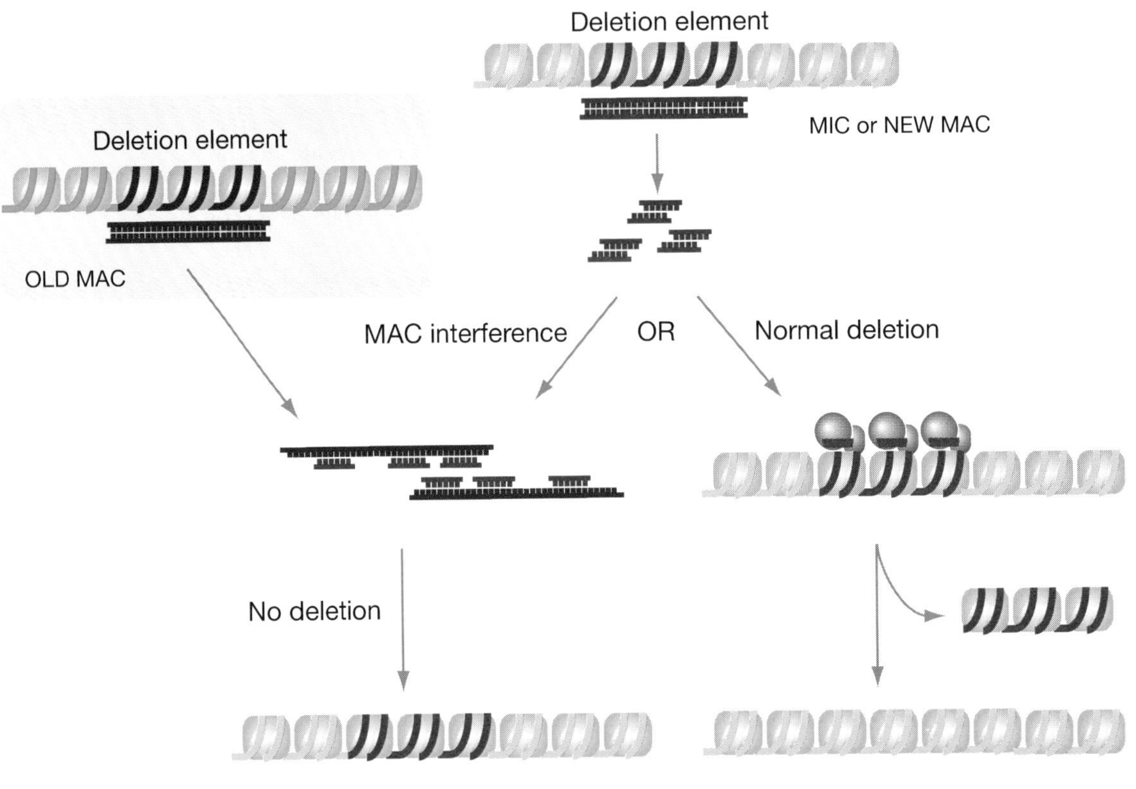

Figure 5

A possible mechanism for the influence of the old macronucleus on DNA deletion. Deletion elements are transcribed even when is present in the old macronucleus. Unlike the dsRNA made in the micronucleus (or developing macronucleus), these transcripts are not processed into sRNA owing partly to the lack of the dicer-like enzyme in this compartment. They bind to sRNA of the same sequence, thereby interfering with the activities of the sRNA in DNA deletion.

containing the *neo* gene flanked by Tetrahymena regulatory sequences at both ends (the 5′ flanking sequence of histone H4 gene at the 5′ end and the 3′ flanking sequence of β-tubulin gene at the other end). These deletions typically included the entire *neo* gene, but sometimes extended outward to include the adjacent Tetrahymena sequences in the cassette. The deletion boundaries were not precise. They varied, mostly within a ~200-bp range. A similar study supported this finding and further suggested that at some loci the deletion was very inefficient. It became detectable only after additional copies of the *neo* gene were inserted at other sites, supporting the importance of sequence repetition in this process (50).

This result is remarkable. It shows that "familiar" sequences are not required for DNA deletion. The *neo* gene is entirely foreign to the organism, illustrating the ability of Tetrahymena to recognize sequences not previously present in its genome. The result provided direct support for a role for DNA deletion in genome defense. It also raised new questions regarding the mechanism of sequence recognition in DNA deletion.

How can "foreign sequences" be recognized and deleted? In RNA-directed DNA deletion, the specificity could be based on selective transcription of dsRNA and/or its processing to produce effector sRNA. The RNA scanning model provides one example that utilizes processing (i.e., scanning) to establish specificity. However, deletion of the *neo* gene occurs regardless of the presence of the same sequence in the macronucleus (M.-C. Yao, unpublished observations), and thus cannot be explained by this idea.

Recognition of foreign sequences is not a problem unique to Tetrahymena. In some organisms it is thought that the repetitive nature of transposons provides the basis for such recognition in genome surveillance (72). This point is best illustrated in Neurospora (73), in which duplicated sequences are somehow recognized and targeted for extensive mutations. In this case, typical RNAi genes are not involved, so the process may not be related to RNAi (31). In other organisms, it is thought that multiple insertions of a sequence greatly increase the chance of producing aberrant transcripts from both strands, each from a different site, thus generating dsRNA and leading to RNAi (55). In Tetrahymena, the inserted *neo* gene is a single-copy sequence and thus the recognition process cannot be based on sequence repetition alone, although repetition seems to help with the efficiency, as expected from the *trans*-acting nature of the dsRNA.

We suggest that in Tetrahymena as well as in other organisms, "aberrant transcription" occurs generally to most inserted foreign sequences, whether single or multiple copies, to produce dsRNA. It would serve as the basis for foreign sequence recognition. We further suggest that similar dsRNA transcriptions also occur in other organisms, thereby providing a common basis for foreign-sequence recognition in addition to sequence repetition. This idea is supported by an earlier report that an insertion of *E. coli* sequence in yeast caused aberrant transcription from both strands (52). A recent study in *C. elegans* also showed transcription of an inserted *E. coli* plasmid sequence that results in TGS (37). It would be interesting to determine whether this type of transcription indeed occurs generally to foreign sequences. If so, the basis of foreign sequence recognition would rest on the regulation of dsRNA transcription. How are foreign sequences recognized and selected for dsRNA transcription? This type of transcriptional regulation is poorly understood, but could be a fundamental issue in genome surveillance in general.

If foreign genes are recognized and deleted in Tetrahymena, then invading genetic agents could exist only in the germline genome. Following this thinking, most of the 15% of the genome that is involved in normal DNA deletion could be derived from transposons or other invading genetic elements sometime during evolution. They came in and spread in the germline genome, but are removed from the somatic genome during conjugation. Thus, the same mechanism recognizes both micronuclear sequences and foreign sequences. In this sense, micronuclear sequence deletion is simply an evolved form of foreign sequence deletion. This idea merges very well with an earlier view that links DNA deletion to transposons. Based on the similarity between some deletion elements (or IES's) and transposons in several ciliates, it has been proposed that most IES's are, or are remnants of, transposons, and the deletion mechanism is derived from the transposition mechanism (44). The arguments are well constructed and compelling. Although current data provide a different view on the mechanism, the evolutionary outcome is very similar. It is probably the best explanation for the origin of micronucleus-specific sequences in ciliates.

EVOLUTION OF CHROMATIN DIMINUTION AND ITS IMPLICATIONS

Chromatin diminution was first reported in 1887 by Theodore Boveri while studying Ascarid nematodes. He found that a large proportion of chromatin inherited by a fertilized egg is later lost from all somatic progenitor cells during early cleavages of the embryo, and the remaining chromatin is fragmented into very small pieces. Molecular and cytological studies have subsequently revealed additional details, though the involvement of RNA has not yet been demonstrated. The process in Ascaris shares important features with that in ciliates: Large proportions of chromatin (mostly heterochromatin) are lost from the somatic genome during development, and the remaining chromosomes are broken into small fragments with new telomeres. Like ciliates, the specific details of this process vary significantly among species; but unlike ciliates, in which the process is found in nearly all distantly related groups studied, it has not yet been observed beyond a few related Ascarid species. Similar processes have also been described in the crustacean Cyclops and in hagfish. Similar to Ascarid worms, these processes are found in only a few related species of both groups. Understanding the evolutionary origin of such a drastic process has been challenging, and made more interesting by its sporadic occurrences in highly diverse groups of eukaryotes. Either the process is present in an ancestral eukaryote but is lost from most of its modern descendants, or, perhaps more likely, they have evolved independently in the lineages observed. In the latter case, it could have occurred quite late in evolution in groups such as Ascaris, Cyclops, and hagfish, where only a few species display the phenomenon.

Multiple independent evolution of such a remarkable process seemed difficult to imagine, but the link to RNA interference provides a possible solution. As summarized above, DNA deletion in Tetrahymena is a special form of RNA interference and likely evolved from the typical RNAi process by adding one final step: elimination of the heterochromatin induced by RNAi. Since RNAi is nearly ubiquitous among eukaryotes, evolution of this last step would be sufficient for chromatin diminution to occur. From this view, one would expect that RNAi is similarly involved in chromatin diminution in Ascaris, Cyclops, and hagfish.

Evolution of the key step, the elimination of heterochromatin, is unlikely to have been very complicated if it occurred multiple times. Perhaps when selective pressure is present, some existing molecular machineries in the organism are modified and captured to provide these last steps. These "machineries" could be involved in DNA recombination, repair, and/or transposition, for example. Finally, if the process has evolved multiple times, it could be more widespread than observed. Chromatin diminution is a drastic process readily observable by simple cytological or molecular methods. Although not extensively searched for, the likelihood that it occurs in familiar organisms but remains undetected is rather low. However, RNA-guided DNA deletion need not be a large-scale process. Theoretically, it can occur to only one gene in a particular cell type in a multicellular organism, thus escaping detection so far. The process could have evolved to provide a complete and irreversible way for gene inactivation during development or growth, and is selected over transcriptional or posttranscriptional gene silencing in special settings. This is an enticing possibility. Uncovering such cases, if they exist, would be an interesting and worthwhile challenge.

CONCLUDING REMARKS

The special life style associated with nuclear dualism may have provided the basis for the wide occurrences of chromatin diminution in ciliates (20). Having a macro- and micronucleus allows the division of genetic labor in the cell: The macronucleus performs most gene expression functions and the micronucleus carries out faithful genome transmission. In organisms such as *S. pombe*, genome defense leads to inactivation of transposons to form heterochromatin, which through evolution has become an important part of the machinery for mitosis and meiosis in genome transmission. Thus, gene expression and genome transmission are tightly coordinated in cells with only one nucleus. In ciliates, heterochromatin is probably also important for mitosis and meiosis, which occur in the micronucleus but not the macronucleus. The elimination of heterochromatin from the macronucleus could be advantageous beyond genome defense: It reduces the burden of DNA replication and simplifies gene regulation by avoiding position effects derived from heterochromatin. Thus, the advantage of eliminating heterochromatin from the macronucleus while keeping it in the micronucleus may have led to the evolution of the developmentally regulated process of chromatin diminution in ciliates.

The unique life style of the ciliate has facilitated the discovery of programmed DNA deletion and its link to RNAi. This special setting should continue to provide a favorable stage on which to investigate key steps in this process. Of many intriguing issues remaining to be resolved, the following are among the most apparent: the regulation of dsRNA transcription, the action of the old macronucleus in DNA deletion, and the deletion of the marked chromatin. Future investigations in these and other areas are likely to uncover new knowledge about RNAi, especially its roles in chromosome structure and genome stabilities.

Although RNAi has provided new insights into programmed DNA deletion, its involvements in other types of DNA rearrangements in ciliates are not clear. For instance, no evidence exists that links chromosome breakage in Tetrahymena to RNAi, even though mutations in PDD2 and TWI1 are believed to also affect this process. Precise DNA deletions have been known to remove DNA smaller than the size of sRNA. It is not apparent how an sRNA-guided process can accomplish this task. Finally, the intriguing process of gene unscrambling, which occurs in certain ciliates during conjugation (35, 47), is very difficult to explain by any mechanism. Whether RNA-guidance is also involved here will be very interesting to determine. The discovery of the RNAi connection reveals a whole new landscape in the study of genome rearrangements, which is promising to yield even more fascinating insights.

SUMMARY POINTS

1. Programmed deletion of specific sequences occurs at thousands of genomic locations during somatic genome formation in Tetrahymena. Similar processes also occur in most other ciliated protozoa, some nematodes, crustaceans, and hagfish.
2. Deletion of these sequences is regulated by several specific *cis*-acting sequences and involves *trans*-acting proteins such as Pdd1p and Pdd2p in specialized heterochromatin-like structure. Deletion of either PDD1or PDD2 prevents deletion from occurring. Specific histone modifications are required for DNA deletion to occcur.
3. Double-stranded RNAs are produced from these deletion elements shortly before deletion occurs. Small RNA (28-nt) is also produced, and is processed from the dsRNA by a dicer-like protein Dcl1p. Deletion of DCL1 or an argonaute protein gene TWI1 blocks conjugation and DNA deletion. A likely link between DNA deletion and RNA interference is suggested.
4. Injection of synthetic dsRNA into conjugating cells leads to deletion of the DNA with the same sequence, indicating that dsRNA alone is sufficient to cause DNA deletion. An RNA-guided mechanism for DNA deletion is established.

5. An inserted sequence from *E. coli* is specifically deleted during conjugation, indicating the ability of the organism to recognize a foreign sequence, and raised the possibility that DNA deletion serves as a mechanism for genome defense. Thus, deletion elements could be remnants of ancient and recent transposons. Their "foreign" nature provides the basis for sequence recognition.
6. In Tetrahymena, the old macronucleus (somatic nucleus) is degraded during conjugation. However, the abnormal presence of deletion elements in this nucleus can prevent the normal deletion of the same element from occurring in the new macronucleus. This mysterious epigenetic effect can be explained by the production of dsRNA (but not sRNA) from the element in the old macronucleus, which interferes with the activity of sRNA in DNA deletion.

ACKNOWLEDGMENTS

This work is supported by grants from the National Institutes of Health, U.S.A. (GM26210) and Academia Sinica, Taiwan, to M.C.Y.

LITERATURE CITED

1. Altschuler MI, Yao MC. 1985. Macronuclear DNA of *Tetrahymena thermophila* exists as defined subchromosomal-sized molecules. *Nucleic Acids Res.* 13:5817–31
2. Austerberry CF, Allis CD, Yao MC. 1984. Specific DNA rearrangements in synchronously developing nuclei of Tetrahymena. *Proc. Natl. Acad. Sci. USA* 81:7383–87
3. Austerberry CF, Snyder RO, Yao MC. 1989. Sequence microheterogeneity is generated at junctions of programmed DNA deletions in *Tetrahymena thermophila*. *Nucleic Acids Res.* 17:7263–72
4. Austerberry CF, Yao MC. 1987. Nucleotide sequence structure and consistency of a developmentally regulated DNA deletion in *Tetrahymena thermophila*. *Mol. Cell. Biol.* 7:435–43
5. Austerberry CF, Yao MC. 1988. Sequence structures of two developmentally regulated, alternative DNA deletion junctions in *Tetrahymena thermophila*. *Mol. Cell. Biol.* 8:3947–50
6. Baird SE, Fino GM, Tausta SL, Klobutcher LA. 1989. Micronuclear genome organization in *Euplotes crassus*: a transposonlike element is removed during macronuclear development. *Mol. Cell. Biol.* 9:3793–807
7. Beerman S. 1977. The diminution of heterochromatic chromosomal segments in Cyclops (Crustacea, Copepoda). *Chromosoma* 60:297–344
8. Beerman S. 1984. Circular and linear structures in chromatin diminution of Cyclops. *Chromosoma* 89:321–28
9. Betermier M, Duharcourt S, Seitz H, Meyer E. 2000. Timing of developmentally programmed excision and circularization of Paramecium internal eliminated sequences. *Mol. Cell. Biol.* 20:1553–61
10. Borst P, Greaves PR. 1987. Programmed gene rearragements altering gene expression. *Science* 235:658–67
11. Boveri T. 1887. Über Differenzierung der Zellkerne während der Furchung des Eies von *Ascaris megalocephala*. *Anat. Anz.* 2:688–93

12. Callahan RC, Shalke G, Gorovsky MA. 1984. Developmental rearrangements associated with a single type of expressed alpha-tubulin gene in Tetrahymena. *Cell* 36:441–45
13. Caron F. 1992. A high degree of macronuclear chromosome polymorphism is generated by variable DNA rearrangements in *Paramecium primaurelia* during macronuclear differentiation. *J. Mol. Biol.* 225:661–78
14. Chalker DL, Fuller P, Yao MC. 2005. Communication between parental and developing genomes during tetrahymena nuclear differentiation is likely mediated by homologous RNAs. *Genetics* 169:149–60
15. Chalker DL, La Terza A, Wilson A, Kroenke CD, Yao MC. 1999. Flanking regulatory sequences of the Tetrahymena R deletion element determine the boundaries of DNA rearrangement. *Mol. Cell. Biol.* 19:5631–41
16. Chalker DL, Yao MC. 1996. Non-Mendelian, heritable blocks to DNA rearrangement are induced by loading the somatic nucleus of *Tetrahymena thermophila* with germ line-limited DNA. *Mol. Cell. Biol.* 16:3658–67
17. Chalker DL, Yao MC. 2001. Nongenic, bidirectional transcription precedes and may promote developmental DNA deletion in *Tetrahymena thermophila*. *Genes Dev.* 15:1287–98
18. Cherry JM, Blackburn EH. 1985. The internally located telomeric sequences in the germline chromosomes of Tetrahymena are at the conserved ends of transposon-like elements. *Cell* 43:747–58
19. Conover RK, Brunk CF. 1986. Macronuclear DNA molecules of *Tetrahymena thermophila*. *Mol. Cell. Biol.* 6:900–5
20. Coyne RS, Chalker DL, Yao MC. 1996. Genome downsizing during ciliate development: nuclear division of labor through chromosome restructuring. *Annu. Rev. Genet.* 30:557–78
21. Coyne RS, Nikiforov MA, Smothers JF, Allis CD, Yao MC. 1999. Parental expression of the chromodomain protein Pdd1p is required for completion of programmed DNA elimination and nuclear differentiation. *Mol. Cell* 4:865–72
22. Coyne RS, Yao MC. 1996. Evolutionary conservation of sequences directing chromosome breakage and rDNA palindrome formation in tetrahymenine ciliates. *Genetics* 144:1479–87
23. Doak TG, Doerder FP, Jahn CL, Herrick G. 1994. A proposed superfamily of transposase genes: transposon-like elements in ciliated protozoa and a common D35E motif. *Proc. Natl. Acad. Sci. USA* 91:942–46
24. Duharcourt S, Butler A, Meyer E. 1995. Epigenetic self-regulation of developmental excision of an internal eliminated sequence on *Paramecium tetraurelia*. *Genes Dev.* 9:2065–77
25. Duharcourt S, Yao MC. 2002. Role of histone deacetylation in developmentally programmed DNA rearrangements in *Tetrahymena thermophila*. *Eukaryot. Cell* 1:293–303
26. Elbashir SM, Harborth J, Lendeckel W, Yalcin A, Weber K, Tuschl T. 2001. Duplexes of 21-nucleotide RNAs mediate RNA interference in cultured mammalian cells. *Nature* 411:494–98
27. Epstein LM, Forney JD. 1984. Mendelian and non-Mendelian mutations affecting surface antigen expression in *Paramecium tetraurelia*. *Mol. Cell. Biol.* 4:1583–90
28. Fan Q, Yao M-C. 2000. A long stringent sequence for programmed chromosome breakage in *Tetrahymena thermophila*. *Nucleic Acids Res.* 28:895–900

29. Fire A, Xu S, Montgomery MK, Kostas SA, Driver SE, Mello CC. 1998. Potent and specific genetic interference by double-stranded RNA in *Caenorhabditis elegans*. *Nature* 391:806–11
30. Forney JD, Blackburn EH. 1988. Developmentally controlled telomere addition in wild-type and mutant Paramecia. *Mol. Cell. Biol.* 8:251–58
31. Freitag M, Lee DW, Kothe GO, Pratt RJ, Aramayo R, Selker EU. 2004. DNA methylation is independent of RNA interference in *Neurospora*. *Science* 304:1939
32. Gershan JA, Karrer KM. 2000. A family of developmentally excised DNA elements in Tetrahymena is under selective pressure to maintain an open reading frame encoding an integrase-like protein. *Nucleic Acids Res.* 28:4105–12
33. Godiska R, James C, Yao MC. 1993. A distant 10-bp sequence specifies the boundaries of a programmed DNA deletion in Tetrahymena. *Genes Dev.* 7:2357–65
34. Godiska R, Yao MC. 1990. A programmed site-specific DNA rearrangement in *Tetrahymena thermophila* requires flanking polypurine tracts. *Cell* 61:1237–46
35. Greslin AF, Prescott DM, Oka Y, Loukin SH, Chappell JC. 1989. Reordering of nine exons is necessary to form a functional actin gene in *Oxytricha nova*. *Proc. Natl. Acad. Sci. USA* 86:6264–68
36. Grewal SI, Moazed D. 2003. Heterochromatin and epigenetic control of gene expression. *Science* 301:798–802
37. Grishok A, Sinskey JL, Sharp PA. 2005. Transcriptional silencing of a transgene by RNAi in the soma of *C. elegans*. *Genes Dev.* 19:683–96
38. Heinonen TY, Pearlman RE. 1994. A germ line-specific sequence element in an intron in *Tetrahymena thermophila*. *J. Biol. Chem.* 269:17428–33
39. Herrick G, Cartinhour S, Dawson D, Ang D, Sheets R, et al. 1985. Mobile elements bounded by C4A4 telomeric repeats in *Oxytricha fallax*. *Cell* 43:759–68
40. Jahn CL, Krikau MF, Shyman S. 1989. Developmentally coordinated en masse excision of a highly repetitive element in *Euplotes crassus*. *Cell* 59:1009–18
41. Jahn CL, Nilles LA, Krikau MF. 1988. Organization of the *Euplotes crassus* micronuclear genome. *J. Protozool.* 35:590–601
42. Katoh M, Hirono M, Takemasa T, Kimura M, Watanabe Y. 1993. A micronucleus-specific sequence exists in the 5′-upstream region of calmodulin gene in *Tetrahymena thermophila*. *Nucleic Acids Res.* 21:2409–14
43. King BO, Yao MC. 1982. Tandemly repeated hexanucleotide at Tetrahymena rDNA free end is generated from a single copy during development. *Cell* 31:177–82
44. Klobutcher LA, Herrick G. 1997. Developmental genome reorganization in ciliated protozoa: the transposon link. *Prog. Nucleic Acid Res. Mol. Biol.* 56:1–62
45. Kubota S, Kuro-o M, Mizuno S, Kohno S-I. 1993. Germ line-restricted, highly repeated DNA sequences and their chromosomal localization in a Japanese hagfish (*Eptatretus okinoseanus*). *Chromosoma* 102:163–73
46. Kuhn EJ, Geyer PK. 2003. Genomic insulators: connecting properties to mechanism. *Curr. Opin. Cell Biol.* 15:259–65
47. Landweber LF, Kuo T-C, Curtis EA. 2000. Evolution and assembly of an extremely scrambled gene. *Proc. Natl. Acad. Sci. USA* 97:3298–303
48. Lawn RM, Heumann JM, Herrick G, Prescott DM. 1978. The gene-size DNA molecules in Oxytricha. *Cold Spring Harbor Symp. Quant. Biol.* 43:483–92
49. Liu Y, Mochizuki K, Gorovsky MA. 2004. Histone H3 lysine 9 methylation is required for DNA elimination in developing macronuclei in Tetrahymena. *Proc. Natl. Acad. Sci. USA* 101:1679–84

50. Liu Y, Song X, Gorovsky MA, Karrer KM. 2005. Elimination of foreign DNA during somatic differentiation in *Tetrahymena thermophila* shows position effect and is dosage dependent. *Eukaryot. Cell* 4:421–31
51. Madireddi MT, Coyne RS, Smothers JF, Mickey KM, Yao MC, Allis CD. 1996. Pdd1p, a novel chromodomain-containing protein, links heterochromatin assembly and DNA elimination in Tetrahymena. *Cell* 87:75–84
52. Marczynski GT, Jaehning JA. 1985. A transcription map of a yeast centromere plasmid: unexpected transcripts and altered gene expression. *Nucleic Acids Res.* 13:8487–506
53. Matzke MA, Birchler JA. 2005. RNAi-mediated pathways in the nucleus. *Nat. Rev. Genet.* 6:24–35
54. Meister G, Tuschl T. 2004. Mechanisms of gene silencing by double-stranded RNA. *Nature* 431:343–49
55. Mello CC, Conte D Jr. 2004. Revealing the world of RNA interference. *Nature* 431:338–42
56. Meyer E, Duharcourt S. 1996. Epigenetic programming of developmental genome rearrangements in ciliates. *Cell* 87:9–12
57. Mochizuki K, Fine N, Fujisawa T, Gorovsky M. 2002. Analysis of a piwi-related gene implicates small RNAs in genome rearrangement in Tetrahymena. *Cell* 110:689–99
58. Mochizuki K, Gorovsky MA. 2004. Conjugation-specific small RNAs in Tetrahymena have predicted properties of scan (scn) RNAs involved in genome rearrangement. *Genes Dev.* 18:2068–73
59. Mochizuki K, Gorovsky MA. 2004. Small RNAs in genome rearrangement in Tetrahymena. *Curr. Opin. Genet. Dev.* 14:181–87
60. Mochizuki K, Gorovsky MA. 2005. A Dicer-like protein in Tetrahymena has distinct functions in genome rearrangement, chromosome segregation, and meiotic prophase. *Genes Dev.* 19:77–89
61. Muller F, Chantal W, Spicher A, Tobler H. 1991. New telomere formation after developmentally regulated chromosomal breakage during the process of chromatin dimunition in *Ascaris lumbricoides*. *Cell* 67:815–22
62. Nakai Y, Kubota S, Kohno S. 1991. Chromatin diminution and chromosome elimination in four Japanese hagfish species. *Cytogenet. Cell Genet.* 65:196–98
63. Nikiforov MA, Gorovsky MA, Allis CD. 2000. A novel chromodomain protein, pdd3p, associates with internal eliminated sequences during macronuclear development in *Tetrahymena thermophila*. *Mol. Cell. Biol.* 20:4128–34
64. Orias E. 1986. Ciliate conjugation. In *The Molecular Biology of Ciliated Protozoa*, ed. JG Gall, pp. 45–84. Orlando, FL: Academic
65. Patil NS, Hempen PM, Udani RA, Karrer KM. 1997. Alternate junctions and microheterogeneity of Tlr1, a developmentally regulated DNA rearrangement in *Tetrahymena thermophila*. *J. Eukaryot. Microbiol.* 44:518–22
66. Preer JR, Preer LB. 1979. The size of macronuclear DNA and its relationships to models for maintaining genic balance. *J. Protozool.* 26:14–18
67. Prescott DM. 1994. The DNA of ciliated protozoa. *Microbiol. Rev.* 58:233–67
68. Prescott DM, Murti KG, Bostock CJ. 1973. Genetic apparatus of *Stylonychia* sp. *Nature* 242:597–600
69. Saveliev SV, Cox MM. 1994. The fate of deleted DNA produced during programmed genomic deletion events in *Tetrahymena thermophila*. *Nucleic Acids Res.* 22:5695–701

70. Saveliev SV, Cox MM. 1995. Transient DNA breaks associated with programmed genomic deletion events in conjugating cells of *Tetrahymena thermophila*. *Genes Dev.* 9:248–55
71. Saveliev SV, Cox MM. 1996. Developmentally programmed DNA deletion in *Tetrahymena thermophila* by a transposition-like reaction pathway. *EMBO J.* 15:2858–69
72. Selker EU. 1999. Gene silencing: repeats that count. *Cell* 97:157–60
73. Selker EU, Cambareri EB, Jensen BC, Haack KR. 1987. Rearrangement of duplicated DNA in specialized cells of *Neurospora*. *Cell* 51:741–52
74. Sharp PA, Zamore PD. 2000. Molecular biology. RNA interference. *Science* 287:2431–33
75. Sijen T, Plasterk RH. 2003. Transposon silencing in the *Caenorhabditis elegans* germ line by natural RNAi. *Nature* 426:310–14
76. Sonneborn TM. 1977. Genetics of cellular differentiation: stable nuclear differentiation in eukaryotic unicells. *Annu. Rev. Genet.* 11:349–67
77. Swanton MT, Heumann JM, Prescott DM. 1980. Gene-sized DNA molecules of the macronuclei in three species of hypotrichs: size distributions and absence of nicks. DNA of ciliated protozoa. VIII. *Chromosoma* 77:217–27
78. Tausta SL, Klobutcher LA. 1989. Detection of circular forms of eliminated DNA during macronuclear development in *E. crassus*. *Cell* 59:1019–26
79. Taverna S, Coyne R, Allis C. 2002. Methylation of histone H3 at lysine 9 targets programmed DNA elimination in Tetrahymena. *Cell* 110:701
80. Tobler H. 1986. The differentiation of germinal and somatic cell lines in nematodes. In *Germ Line-Soma Differentiation*, ed. W Hennig, pp. 1–70. New York: Springer-Verlag Press
81. Tobler H, Eiter A, Muller F. 1992. Chromatin dimunition in nematode development. *Trends Genet.* 8:427–32
82. Verona RI, Mann MR, Bartolomei MS. 2003. Genomic imprinting: intricacies of epigenetic regulation in clusters. *Annu. Rev. Cell Dev. Biol.* 19:237–59
83. Volpe TA, Kidner C, Hall IM, Teng G, Grewal SI, Martienssen RA. 2002. Regulation of heterochromatic silencing and histone H3 lysine-9 methylation by RNAi. *Science* 297:1833–37
84. Wells JM, Ellingson JL, Catt DM, Berger PJ, Karrer KM. 1994. A small family of elements with long inverted repeats is located near sites of developmentally regulated DNA rearrangement in *Tetrahymena thermophila*. *Mol. Cell. Biol.* 14:5939–49
85. Williams K, Doak TG, Herrick G. 1993. Developmental precise excision of *Oxytricha trifallax* telomere-bearing elements and formation of circles closed by a copy of the flanking target duplication. *EMBO J.* 12:4593–601
86. Wuitschick JD, Karrer KM. 2003. Diverse sequences within Tlr elements target programmed DNA elimination in *Tetrahymena thermophila*. *Eukaryot. Cell* 2:678–89
87. Yao M, Duharcourt S, Chalker D. 2002. Genome-wide rearrangements of DNA in ciliates. In *Mobile DNA II*, ed. N Craig, R Craigie, M Gellert, A Lambowitz, pp. 730–58. Washington, DC: ASM Press
88. Deleted in proof
89. Yao MC. 1986. Amplification of ribosomal RNA genes. In *The Molecular Biology of Ciliated Protozoa*, ed. JG Gall, pp. 179–201. Orlando, FL: Academic
90. Yao MC, Yao C-H, Monks B. 1990. The controlling sequence for site-specific chromosome breakage in Tetrahymena. *Cell* 63:763–72

91. Yao MC, Zheng K, Yao C-H. 1987. A conserved nucleotide sequence at the sites of developmentally regulated chromosomal breakage in Tetrahymena. *Cell* 48:779–88
92. Yao MC. 1981. Ribosomal RNA gene amplification in Tetrahymena may be associated with chromosome breakage and DNA elimination. *Cell* 24:765–74
93. Yao MC. 1982. Elimination of specific DNA sequences from the somatic nucleus of the ciliate Tetrahymena. *J. Cell Biol.* 92:783–89
94. Yao MC, Choi J, Yokoyama S, Austerberry CF, Yao CH. 1984. DNA elimination in Tetrahymena: a developmental process involving extensive breakage and rejoining of DNA at defined sites. *Cell* 36:433–40
95. Yao MC, Fuller P, Xi X. 2003. Programmed DNA deletion as an RNA-guided system of genome defense. *Science* 300:1581–84
96. Yao MC, Gorovsky MA. 1974. Comparison of the sequences of macro- and micronuclear DNA of *Tetrahymena pyriformis*. *Chromosoma* 48:1–18
97. Yao MC, Kimmel AR, Gorovsky MA. 1974. A small number of cistrons for ribosomal RNA in the germinal nucleus of a eukaryote, *Tetrahymena pyriformis*. *Proc. Natl. Acad. Sci. USA* 71:3082–86
98. Yao MC, Yao CH. 1989. Accurate processing and amplification of cloned germ line copies of ribosomal DNA injected into developing nuclei of *Tetrahymena thermophila*. *Mol. Cell. Biol.* 9:1092–99
99. Yao MC, Yao CH. 1994. Detection of circular excised DNA deletion elements in *Tetrahymena thermophila* during development. *Nucleic Acids Res.* 22:5702–8
100. Yao MC, Zhu SG, Yao CH. 1985. Gene amplification in *Tetrahymena thermophila*: formation of extrachromosomal palindromic genes coding for rRNA. *Mol. Cell. Biol.* 5:1260–67

Molecular Genetics of Axis Formation in Zebrafish

Alexander F. Schier[1] and William S. Talbot[2]

[1]Developmental Genetics Program, Skirball Institute of Biomolecular Medicine, Department of Cell Biology, New York University School of Medicine, New York, NY 10016-6497; email: schier@saturn.med.nyu.edu; present address: Department of Molecular and Cellular Biology, Harvard University, Cambridge, Massachusetts 02138

[2]Department of Developmental Biology, Stanford University School of Medicine, Stanford, California 94305; email: talbot@cmgm.stanford.edu

Annu. Rev. Genet.
2005. 39:561–613

First published online as a Review in Advance on August 9, 2005

The *Annual Review of Genetics* is online at genet.annualreviews.org

doi: 10.1146/annurev.genet.37.110801.143752

0066-4197/05/1215-0561$20.00

Key Words

gastrulation, mesoderm, endoderm, ectoderm, Nodal, Bmp, FGF, Wnt, retinoic acid

Abstract

The basic vertebrate body plan of the zebrafish embryo is established in the first 10 hours of development. This period is characterized by the formation of the anterior-posterior and dorsal-ventral axes, the development of the three germ layers, the specification of organ progenitors, and the complex morphogenetic movements of cells. During the past 10 years a combination of genetic, embryological, and molecular analyses has provided detailed insights into the mechanisms underlying this process. Maternal determinants control the expression of transcription factors and the location of signaling centers that pattern the blastula and gastrula. Bmp, Nodal, FGF, canonical Wnt, and retinoic acid signals generate positional information that leads to the restricted expression of transcription factors that control cell type specification. Noncanonical Wnt signaling is required for the morphogenetic movements during gastrulation. We review how the coordinated interplay of these molecules determines the fate and movement of embryonic cells.

Contents

INTRODUCTION

Over the past 25 years, the zebrafish has become a powerful model system for investigation of vertebrate development, physiology, and disease mechanisms. Recognizing important attributes such as high fecundity, a three-month generation time, and accessibility of the embryo, Streisinger introduced the zebrafish as a model system, developed methods for constructing haploid and gynogenetic diploid fish, and identified the first few zebrafish mutants (308). Exploiting the optical transparency of the embryo, Kimmel established essential embryological tools, including time-lapse imaging, lineage-tracing, and cellular transplantation, which are now widely used in analyses of wild-type and mutant embryos (reviewed in 154). In the mid-1990s, the Nüsslein-Volhard and Driever groups conducted two large-scale genetic screens that identified genes with essential functions in a wide array of biological processes, ranging from early embryonic patterning to organogenesis (68, 104). The 1990s also witnessed the advent of key resources for the molecular analysis of zebrafish mutations, including genetic maps, radiation hybrid maps, and large-insert genomic libraries (91, 130, 164, 244). These areas have all progressed rapidly, and the zebrafish field continues to be invigorated by the identification of new mutants in screens targeted for specific phenotypes and by the development of new tools and resources (e.g., 26, 194, 349). Examples of other important advances include retroviral insertional mutagenesis, in vivo analysis of gene expression with GFP (green fluorescent protein) transgenes, the use of morpholino oligonucleotides and target-selected mutagenesis approaches for reverse genetic studies, and a concerted effort to obtain the genome sequence (88, 190, 223, 340). Because of these experimental advantages, the zebrafish system has yielded important insights into many areas of vertebrate biology; especially noteworthy among these is the genetic control of embryonic axis formation, the subject of this review.

OVERVIEW OF ZEBRAFISH DEVELOPMENT

Only 10 h post fertilization (hpf), the zebrafish embryo has clearly recognizable anterior-posterior and dorsal-ventral axes (**Figure 1**). Moreover, the embryo is exquisitely patterned so that the precursors for different regions and cell types of the embryo can be recognized using molecular markers. To generate this basic body plan, the embryo undergoes rapid developmental and morphogenetic changes (reviewed in 155). Upon fertilization, cytoplasmic streaming generates a large blastodisc on top of the yolk. During the following 3 h of development, rapid, synchronous cleavage divisions occur within the blastodisc to generate a blastula embryo consisting of ~1000 cells, initially arranged in a pile (blastoderm) atop the yolk. During cleavage, the volume of the embryo remains essentially constant, so that the divisions produce a larger number of smaller cells. The cells in the blastoderm form the embryo proper, whereas the yolk is an extraembryonic structure. Cell cycles lengthen and become asynchronous during the mid-blastula transition (MBT). The MBT begins at the 512-cell stage (2.75 hpf), when cell division has increased the DNA:cytoplasm ratio to a critical threshold (58, 136). The MBT also marks the time when zygotic transcription begins (although a few genes may be transcribed prior to the MBT), so that the zygotic genome begins to govern embryonic development. Also around the time of the MBT, cells at the blastoderm margin collapse into the yolk and form the yolk syncytial layer, a thin, multinucleate structure at the interface of the blastoderm and the yolk (157).

At about 4 hpf, cellular rearrangements begin to reshape the blastoderm into a characteristic vertebrate body plan (reviewed in 298) (**Figure 2**). In the process of epiboly, cells intercalate radially, thereby thinning the blastoderm and spreading over the yolk. By the end of gastrulation, epiboly movements have spread the blastomeres so that the blastoderm covers the entire yolk cell; the extent of yolk cell coverage (measured as "percent epiboly") provides a convenient way to determine an embryo's developmental stage. Three other movements contribute to the formation of the axis. Beginning at 5 hpf, cells at the margin internalize and form the so-called hypoblast, the precursors of the mesoderm and endoderm (this usage of the term hypoblast is different from that in mouse and chick, where it denotes extraembryonic tissue). By 6 hpf, convergence and extension movements have begun, resulting in the dorsal accumulation of cells moving from lateral and ventral regions of the blastoderm (convergence). Concomitantly, converging cells intercalate with dorsal blastomeres, spreading them along the animal-vegetal axis, leading to a lengthening of the anterior-posterior axis (extension). Convergence of cells toward the dorsal side of the embryo marks the first clearly apparent break in radial symmetry and forms the

Anterior-posterior axis: the line from head to tail

Endoderm: the inner germ layer, which gives rise to the gastrointestinal tract and associated structures

Gastrulation: the process by which blastoderm cells are specified and move to generate an embryo with three germ layers and anterior-posterior and dorsal-ventral polarity

Mesoderm: the middle germ layer, which gives rise to bone, muscle, connective tissue, urogenital and circulatory system

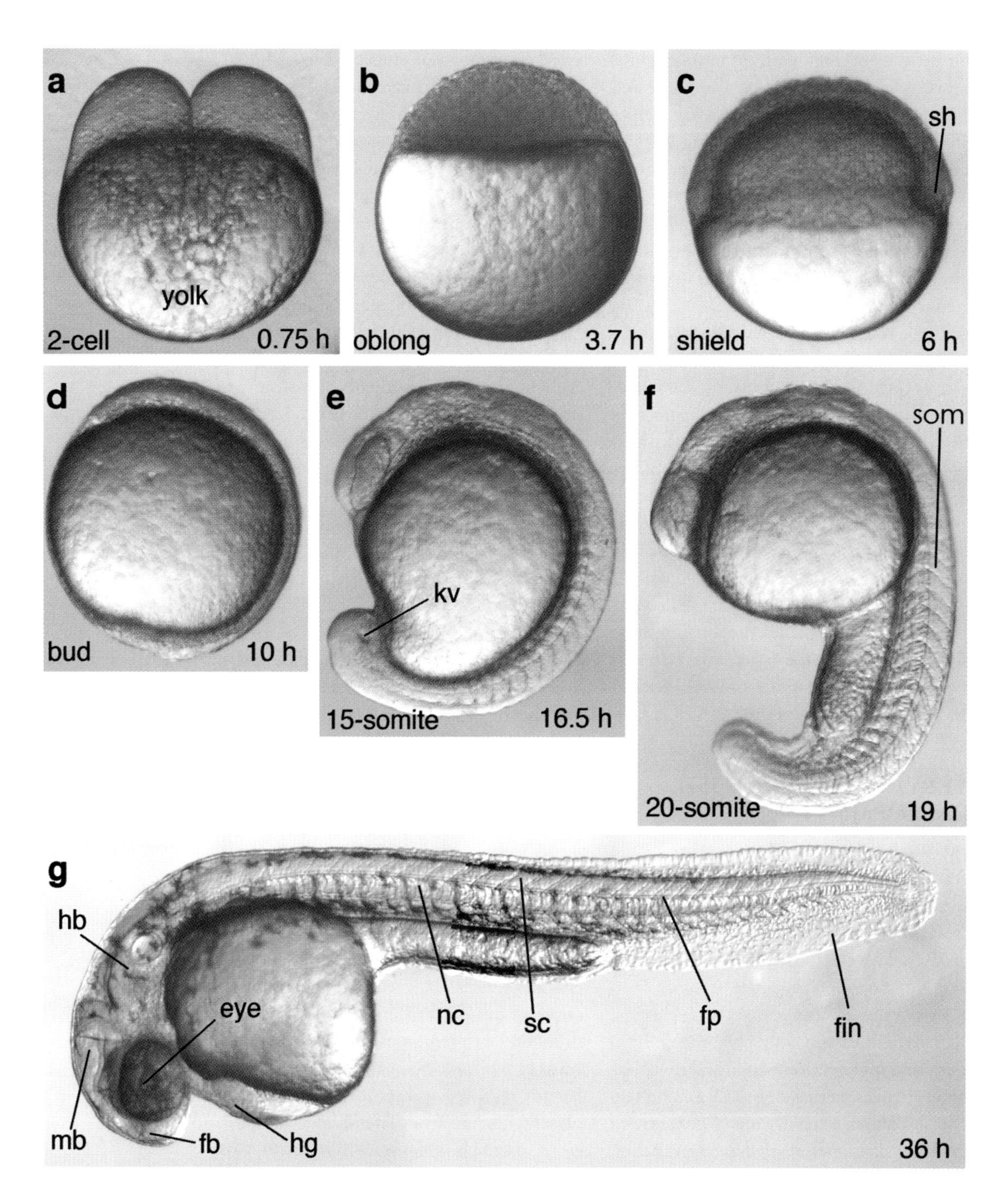

Figure 1

Zebrafish embryogenesis. Living zebrafish embryos are shown at the indicated developmental stages. Approximate developmental ages in hours postfertilization (h) are shown. Embryos are oriented: (*a,b*) animal pole to top; (*c*) animal pole to top, dorsal to the right; (*d–f*) anterior to the top, dorsal to the right; (*g*) anterior to the left, dorsal to the top. Abbreviations: sh, embryonic shield; kv, Kupffer's vesicle; som, somite; hg, hatching gland; fb, forebrain; mb, midbrain; hb, hindbrain; nc, notochord; sc, spinal cord; fp, floor plate. For further details see Reference 155.

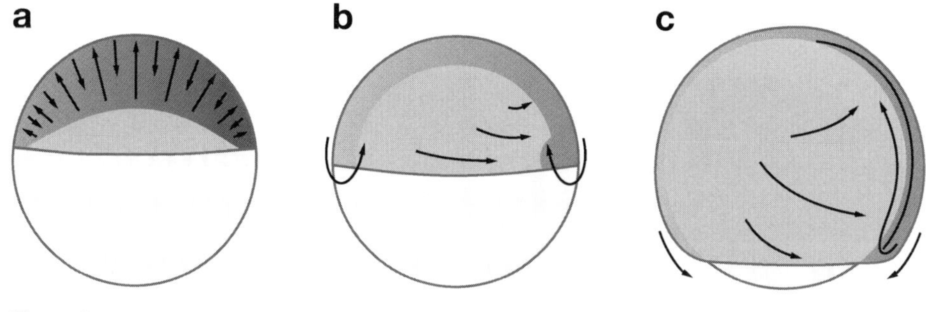

Figure 2

Gastrulation movements. (*a*) Dome stage. Cells intercalate radially, contributing to epiboly. (*b*) Shield stage. Cells at the margin internalize and migrate toward the animal pole. Cells converge dorsally, with lateral mesodermal cells starting convergence at later stages than cells closer to the shield (282). (*c*) 90% epiboly stage. Epiboly, internalization, convergence and extension continue. Modified from Reference 138.

shield, a thickening at the dorsal blastoderm margin that is the teleost equivalent of the amphibian Spemann-Mangold organizer (266, 286).

FATE MAPS AND ORGANIZING CENTERS

A fate map demarcating the position of precursors for different tissues and organs is apparent at the onset of gastrulation (6 hpf), although different progenitor territories are not sharply demarcated and progenitors are intermingled (161) (**Figure 3**). Because embryological manipulations and mutations in the genes described below alter this fate map, it is important to take a closer look at the arrangement of tissue progenitors. The precursors of the different germ layers are arranged along the animal-vegetal axis, with ectoderm located animally, mesoderm more marginally, and endoderm, intermingled with mesoderm, at the margin itself. Precursors for different mesodermal cell types are arranged along the so-called dorsal-ventral (DV) axis, with dorsal corresponding to the site of the shield. Cells located most dorsally give rise to the axial mesoderm of notochord and prechordal plate. More laterally located cells give rise to trunk somites and heart. Blood and pronephros are derived from marginal blastomeres more distant from the shield, the so-called ventral region. Most of the posterior mesoderm (tail somites) also derives from this ventral territory. Different endodermal progenitors are also located in different dorsal-ventral positions, with pharynx located most dorsally, and stomach, intestine, and liver located more laterally and ventrally (i.e., more distant from the shield) (334). Nonneural ectoderm (epidermis) derives from the animal-ventral territory. Forebrain and midbrain progenitors are found animally and dorsally, whereas hindbrain and spinal cord precursors are located closer to the margin and more laterally and ventrally, respectively (345). Hence, precursors for different anterior-posterior regions in the nervous system do not simply align with the animal-vegetal axis. Similarly, precursors of anterior somites are located more dorsally than posterior somite progenitors. Moreover, prechordal plate precursors are located more vegetally than notochord precursors (101). Because of complex gastrulation movements, there is no completely generalizable connection between dorsal-ventral or animal-vegetal location at early gastrula stages and later anterior-posterior position. This is most clearly exemplified by prechordal plate and forebrain forming the most anterior region of the head but lying at opposite positions of the animal-vegetal axis at the onset of gastrulation. Similarly, posterior notochord and posterior somites together form the tail

Dorsal-ventral axis: the line from back to belly

Ectoderm: the outer germ layer, which gives rise to epidermis, nervous system and sense organs

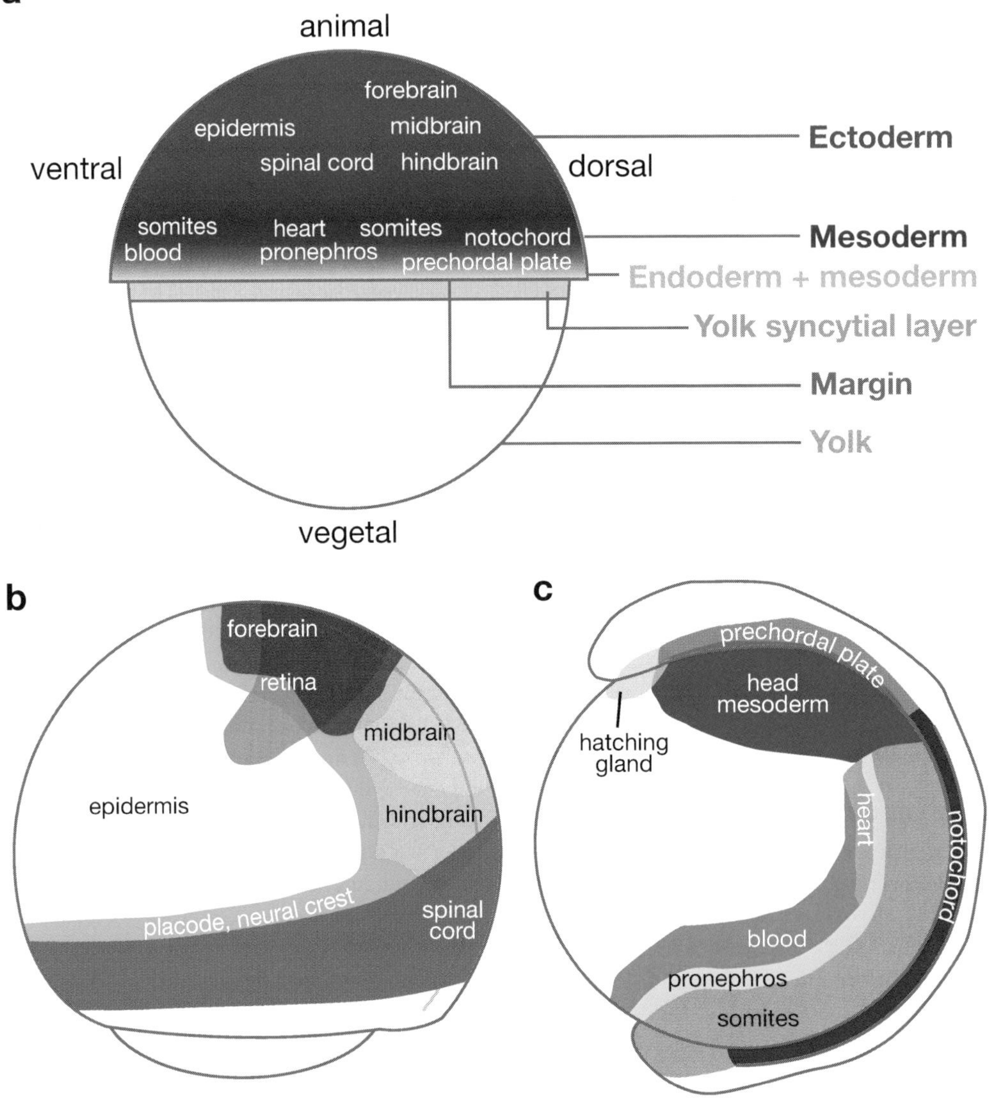

Figure 3

Zebrafish fate maps. (*a*) Fate map at 50% epiboly stage, the onset of gastrulation. Lateral view, dorsal to the right, animal pole to the top. Germ layers are arranged along the animal-vegetal axis. Different mesodermal and ectodermal fates are arranged along the dorsal-ventral axis. For details see References 66, 101, 145, 161, 345. For distribution of endodermal fates see Reference 334. No precise boundaries are depicted because cell fates are often intermingled. Modified from Reference 267. (*b*) Fate map of ectoderm at 90% epiboly. Lateral view, dorsal to the right, animal pole and anterior to the top. Modified from Reference 345; position of spinal cord territory is inferred from Reference 172. (*c*) Model fate map of mesoderm at early somite stage. Lateral view, dorsal to the right, animal pole and anterior to the top. Note that no precise fate map has been established at this stage. Therefore, regions shown here are approximations derived in part from the expression patterns of marker genes (ZFIN.org). The posterior region of the tail bud will continue to extend and give rise to different mesodermal and ectodermal fates. Modified from Reference 138.

mesoderm, but are derived from opposite ends of the DV axis.

Dye labeling experiments at early cleavage stages indicate that the planes of the first cell divisions do not predict the future dorsal-ventral axis (1, 120, 160). In addition, these experiments revealed that there is extensive cell mixing during epiboly such that a cell's position during early cleavage stages does not determine the fates of its descendants, although cells at more vegetal positions tend to contribute more marginal progenitors at the onset of gastrulation. The first lineage restrictions to emerge separate embryonic blastomeres from the extraembryonic blastomeres of the yolk syncytial layer and the enveloping layer, which forms a flattened epithelium that covers the blastoderm. Single embryonic blastomeres at the 1000- to 2000-cell stage can still give rise to several tissue types, and most individual blastomeres are not restricted to particular fates until the early gastrula stage (158). Progenitors of different germ layers begin to occupy definable and distinct positions after the 1000-cell stage, when, for example, ectodermal and mesendodermal progenitors are largely separated, with the exception that some muscle progenitors are intermingled with hindbrain and spinal cord progenitors (161). Although individual blastomeres adopt particular fates that are predictable based on their positions at the early gastrula stage, transplantation experiments show that most individual cells are not committed to particular fates until the mid- to late-gastrula stages (126).

As described in detail below, embryological manipulations have identified regions in the embryo that are required or sufficient to induce specific fates in neighboring cells (reviewed in 267) (**Figure 4**). The dorsal margin is the source of factors that can induce dorsal, anterior and lateral cell types and repress ventral and posterior fates (266, 286). The yolk syncytial layer is the source of mesoderm and endoderm inducers (44, 213), and the ventral margin can induce posterior structures (4, 346).

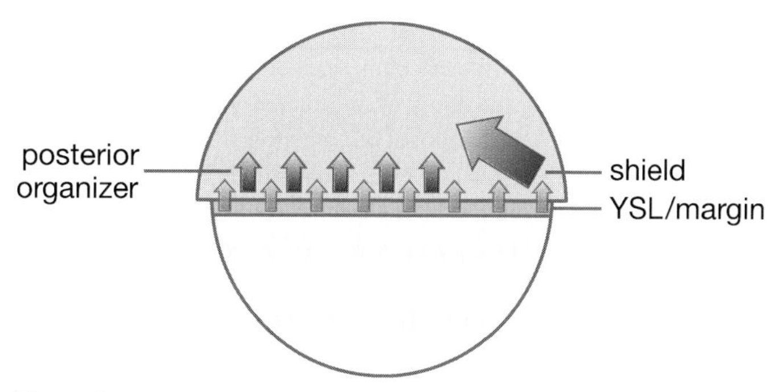

Figure 4

Zebrafish organizing centers. Lateral view, dorsal to the right, animal pole to the top. Yolk syncytial layer (YSL) can induce mesendodermal fates upon transplantation (*green arrows*). Posterior organizer is located at the ventral and lateral margin and can induce tail, posterior trunk, and hindbrain tissue upon transplantation (*red arrows*). Shield corresponds to Spemann-Mangold organizer and can induce dorsal and anterior structures upon transplantation (*blue arrow*).

DORSAL-VENTRAL PATTERNING: MATERNAL FACTORS

The mature zebrafish oocyte is radially symmetric about the animal-vegetal axis, and no dorsal-ventral asymmetry is evident prior to fertilization. During fertilization, the sperm enters the egg through a specialized structure, the micropyle, at the animal pole (344). Thus it seems that the sperm entry point itself cannot be the cue that breaks symmetry in zebrafish, in contrast to the situation in amphibians (reviewed in 336), but the possibility remains that an activity of the sperm after fertilization is somehow involved in establishing the dorsal-ventral axis. Although the first five cleavage divisions occur in a stereotyped alternating orthogonal pattern, these cleavage planes do not correlate with the eventual dorsal-ventral axis (1, 120, 160). Nevertheless, embryological experiments show that events important for the formation of dorsal-ventral asymmetry are occurring even before the first cleavage division. Embryos are ventralized by removal of the vegetal region of the yolk before the first cell division, and the frequency of ventralized embryos rapidly diminishes when the operation is performed at later stages (212, 232). Similarly, treatment with nocodazole, an inhibitor of microtubule polymerization, causes the loss of dorsal axial structures when applied within 10 min after fertilization, but not after the first cell division (133). Drawing on parallels between these results and previous work on dorsal-ventral axis formation in *Xenopus*, it has been proposed that the dorsal side of the zebrafish embryo is established by a dorsal determinant initially located at the vegetal pole that is translocated along microtubules to the future dorsal side before the first cleavage division occurs (133). This is an intriguing model, but certain key predictions remain untested. For example, directed movement from the vegetal pole toward the dorsal side of the early embryo has not been observed. Likewise, it has not been shown that the vegetal pole contains a determinant sufficient to determine dorsal identity or rescue a ventralized embryo in a transplantation experiment. Thus many questions remain about the mechanisms that establish the earliest dorsal-ventral asymmetries in the zebrafish. The analysis of recently identified maternal-effect mutants with ventralized phenotypes will define important players that act at early stages to establish the dorsal-ventral axis (147, 228, 330).

β-catenin

Evidence suggests that maternal β-catenin acts to establish the dorsal-ventral axis in zebrafish. β-catenin protein acts as a transcriptional effector in the canonical Wnt signaling pathway and also has a function in cell adhesion (reviewed in 129, 188). A complex containing APC, axin, and GSK3β and other components targets β-catenin protein for degradation, thereby allowing only a low level of β-catenin to accumulate. Activation of the canonical Wnt signaling pathway inhibits the β-catenin degradation complex, stabilizing β-catenin and allowing it to enter the nucleus, where it activates transcription of canonical Wnt target genes.

In the zebrafish embryo, β-catenin accumulates specifically in nuclei of dorsal margin blastomeres as early as the 128-cell stage (66, 274). This asymmetric nuclear localization of β-catenin is an early marker of the dorsal-ventral axis (**Figure 5**). As in the amphibian embryo, overexpression of β-catenin leads to axis duplication (148). Moreover, β-catenin seems to be required for dorsal axis formation, as overexpression of proteins that inhibit β-catenin's action as a transcriptional activator (cadherin or a dominant negative form of Tcf3 that binds β-catenin but not DNA) reduces dorsal gene expression and produces ventralized embryos (238). In addition, the maternal effect mutations *ichabod* and *tokkaebi*, whose molecular bases are not known, disrupt the nuclear localization of β-catenin and lead to ventralized embryos (147, 228).

Soon after the mid-blastula transition, β-catenin activates the expression of a number of zygotic genes, including *bozozok* (*boz*, also known as *dharma* and *nieuwkoid*), *chordin*, *dickkopf1* (*dkk1*), *squint* (*sqt*) and FGF signals (63, 66, 75, 79, 87, 113, 147, 165, 247, 261, 263, 292, 324, 353). As detailed below, these β-catenin targets act to inhibit the action of ventralizing factors or, in the case of Sqt, induce mesendodermal fates at the dorsal margin.

Recent work suggests that asymmetric localization of Wnt11 triggers the accumulation of β-catenin in dorsal blastomeres in *Xenopus* (314). Zebrafish *wnt11* mutants (*silberblick*) have defects in morphogenetic movements during gastrulation (see below), but formation of the dorsal-ventral axis is normal, even in embryos lacking maternal and zygotic *wnt11* (119). Moreover, *Xenopus* but not zebrafish *wnt11* mRNA is localized to the vegetal pole. There is another *wnt11* gene in the zebrafish genome (90), and further work is needed to determine if this gene functions in the establishment of the dorsal-ventral axis or if the *wnt11* duplicates might have redundant functions in this process.

Although the asymmetric distribution of β-catenin has not been observed during the first few cleavages, one study suggests that dorsal-ventral asymmetry is evident even in the two-cell embryo (83). Activation of the

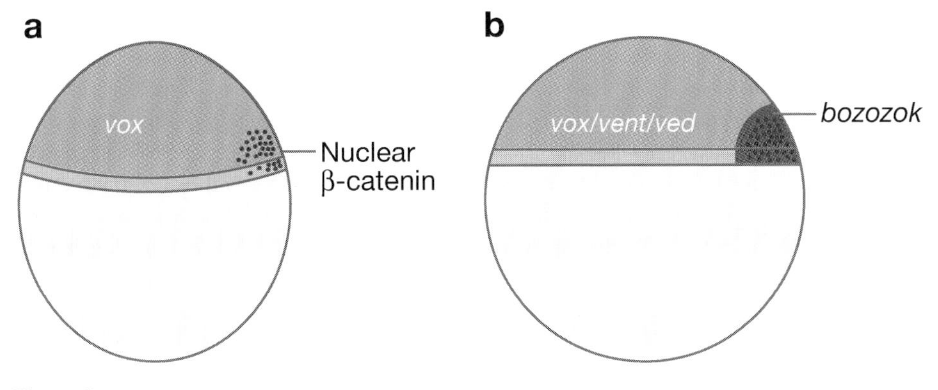

Figure 5

Transcriptional interactions patterning the dorsal-ventral axis. Lateral view, dorsal to the right, animal pole to the top. (*a*) β-catenin is stabilized on the dorsal side during cleavage stages. Soon after mid-blastula transition, *vox* is expressed ubiquitously. (*b*) β-catenin activates *bozozok* (*boz*), which represses *vox*, *vent*,and *ved* expression in dorsal blastomeres.

map kinase p38, assessed with an antibody specific for the doubly phosphorylated form of p38, occurs in the region of the embryo that will eventually become the dorsal side. Despite its early dorsal activation, p38 does not apparently act to specify dorsal fates, and expression of dorsal-specific genes occurs in embryos expressing dominant negative versions of p38. Instead, p38 is required specifically on the dorsal side to control the rate of cell division in dorsal blastomeres, so that there are fewer, larger blastomeres on the dorsal side in embryos expressing dominant negative p38. Activation of p38 does not occur in embryos ventralized by inhibition of microtubules or vegetal yolk depletion, indicating that p38 is regulated by the same factors that establish dorsal-ventral asymmetry and that p38 acts in parallel to the genes that specify dorsal identity (83).

DORSAL-VENTRAL PATTERNING: ZYGOTIC FACTORS

In recent years, the default model for dorsal-ventral patterning has gained widespread acceptance (reviewed in 121). This model, first formulated to explain dorsal-ventral patterning in frog, holds that the Spemann-Mangold organizer induces dorsal fates by inhibiting the action of ventralizing and posteriorizing signals such as Bmp2/4/7 and Wnt8. According to this view, development of dorsal and anterior fates is a "default" state, such that dorsalizing factors act to block the influence of ventralizing signals rather than to actively trigger pathways that specify dorsal fates. Analysis in zebrafish has confirmed certain key predictions of this model, identified genes with essential roles in dorsal-ventral patterning, and advanced the understanding of dorsal-ventral patterning by explaining events that are not wholly accounted for by the simplest version of the default model.

Bmp Signaling

Members of the Bmp family of TGF-β signals induce ventral fates (reviewed in 219, 327). Secreted Bmp ligands bind the extracellular domains of type I and type II Bmp receptors, which are transmembrane proteins with intracellular serine/threonine kinase domains. The closely related Smad family transcription factors Smad1/5/8 are phosphorylated by ligand-bound receptors, allowing these proteins to translocate to the nucleus and regulate target gene expression together with the nonreceptor-regulated Smad protein Smad4, and other DNA binding cofactors, such as the zinc finger protein Oaz. A large number of inhibitory proteins function to regulate Bmp pathway activity at different levels: for example, Chordin, Noggin, and Follistatin are secreted Bmp antagonists, the transmembrane protein Bambi functions as a decoy receptor, and inhibitory Smads Smad6/7 interfere with Smad1/5/8 phosphorylation (reviewed in 219).

Mutational analysis has demonstrated that a number of Bmp pathway components are essential for formation of ventral cell types in zebrafish (**Table 1**), including the Bmp ligands, Bmp2b and Bmp7, the type I receptor Alk8, the transcriptional effector Smad5, and the protease Tolloid, which cleaves the Bmp antagonist Chordin (22, 49, 60, 124, 162, 209, 226, 272). Although these mutations define components of the same pathway, the mutant phenotypes span a range from weakly dorsalized and viable to strongly dorsalized and lethal in the first day of development (218). Soon after the MBT, *bmp2b* and *bmp7* are widely expressed, but their expression becomes restricted to approximately the ventral half of the embryo by the onset of gastrulation (60, 162, 198, 272) (**Figure 6**). *Swirl/bmp2b* mutants and wild-type embryos overexpressing Chordin or Noggin are strongly dorsalized, with dorsoanterior structures greatly expanded at the expense of ventroposterior structures (162, 218, 219b, 226). In the ectoderm, neural fates including forebrain, midbrain, and hindbrain are expanded to encompass the most ventral regions of the embryo, whereas epidermis, neural crest, and Rohon-Beard sensory neurons are lacking in *swirl/bmp2b* mutant embryos. A similar fate transformation is evident in the margin region of *swirl/bmp2b* mutants, in which anterior (trunk) somites and anterior endoderm are expanded, whereas ventrolateral and posterior fates such as blood, heart, pronephros, pancreas, and tail are reduced or missing. Axial mesoderm is largely unaffected in *swirl/bmp2b* and the other *bmp* pathway mutants, indicating that other factors act to restrict the most dorsal fates to the appropriate territories. Complete loss of *snh/bmp7* function also produces a strongly dorsalized phenotype, indicating that both *bmp2b* and *bmp7* are required for normal dorsal-ventral patterning, despite the fact that the expression of these genes largely overlaps (60, 272). It is possible that the active ventralizing signal in vivo is a Bmp2b-Bmp7 heterodimer (272). Bmp7, however, can induce ventral cell types when overexpressed in *bmp2b* mutants, showing that high levels of Bmp7 are sufficient to specify ventral identity even in the absence of its putative heterodimer partner Bmp2b. Using an inducible dominant negative Bmp receptor, it has been shown that Bmp signaling is required for global dorsal-ventral patterning decisions during early gastrulation, whereas Bmp signals regulate tail development from mid-gastrulation through early somitogenesis (246).

The fly orthologue of the ventralizing Bmps, Decapentaplegic (Dpp), acts as morphogen, and it has been proposed that graded action of Bmp signals directly specifies fates of tissue progenitors across the dorsal-ventral axis in vertebrates (64, 176a, 225, 343). In zebrafish, the evidence for this is best in the ectoderm, where graded inactivation of Bmp signals leads to striking modulations of DV patterning (21, 226). Null mutations in *bmp2b* eliminate epidermis, placodes, neural crest, and Rohon-Beard sensory neurons, whereas forebrain, midbrain, and hindbrain fates are expanded to encompass the most ventral regions of the embryo. When Bmp activity is reduced but not eliminated, as with hypomorphic mutations or overexpression of intermediate concentrations of a Bmp antagonist in wild-type embryos, neural crest and placodal fates are expanded relative to wild type. These seemingly paradoxical results can be explained if the perturbations change the slope of a Bmp gradient. According to this view, a larger region of the DV axis falls within, for example, the neural crest specification threshold when the Bmp gradient is shallower than in wild type. This can account for expansion of

Table 1 Genes essential for zebrafish axis formation and patterning

Mutation	Gene product	Function	Phenotype	Reference
Bmp signaling				
swirl	Bmp2b	Bmp signal	Severely dorsalized	(162)
snailhouse	Bmp7	Bmp signal	Severely dorsalized	(60, 272)
lost-a-fin	Alk8	Type I Bmp receptor	Severely dorsalized	(22, 209)
somitabun	Smad5	Transcription factor	Weakly (zyg.) or strongly (mat.) dorsalized	(124)
morpholino	Twisted Gastrulation	Bmp agonist	Dorsalized	(186, 350)
minifin	Tolloid	Metalloprotease for Chordin	Weakly dorsalized	(49)
chordino	Chordin	Bmp inhibitor	Ventralized	(277)
ogon	Sizzled	Bmp inhibitor	Ventralized	(199, 351)
morpholino	Radar/Gdf6a	Bmp signal	Dorsalized	(293)
dominant negative	Kheper	Zinc finger/homeodomain	Reduced neuroectoderm	(220)
morpholino	ΔNp63	Transcriptional repressor	Reduced ventral ectoderm	(16, 177)
morpholino	ADMP	Divergent Bmp signal	Dorsalized	(180, 341)
Canonical Wnt signaling				
wnt8	Wnt8	Wnt signal	No ventral and posterior structures	(72, 179)
masterblind	Axin	Scaffolding protein	No eyes and telencephalon	(117)
headless	Tcf3	Transcription factor	No forebrain and midbrain	(153)
morpholino	Tlc SFRP	Wnt antagonist	Reduced telencephalon	(127)
ichabod	?	β-catenin localization?	Variably ventralized	(147)
tokkaebi	?	β-catenin stability?	Variably ventralized	(228)
morpholino	Sp5 and Sp5-like	SP1 Zn Finger	Anteriorized and dorsalized	(337)
Nodal signaling				
cyclops	Cyc (Nodal)	Nodal signal	Cyclopia	(115, 252, 265)
squint	Sqt (Nodal)	Nodal signal	Cyclopia, dorsal mesoderm defects	(79)
morpholino	Southpaw (Nodal)	Nodal signal	Loss or randomization of LR asymmetry	(191)
cyclops;squint			No endoderm and head/trunk mesoderm	(79)
one-eyed pinhead	EGF-CFC	Nodal co-receptor	No endoderm and head/trunk mesoderm	(102)
schmalspur	FAST1/FoxH1	Transcription factor	Dorsal mesoderm defects	(243, 295)
bonnie and clyde	Mix homeodomain	Transcription factor	Reduced endoderm	(151)
morpholino	Lefty1 and Lefty2	Antagonist of Nodal signaling	Increased mesoderm and endoderm	(3)
morpholino	Dapper2	Antagonist of Nodal signaling	Increased mesoderm and endoderm	(362)
morpholino	Charon	Antagonist of Nodal signaling	Loss of LR asymmetry	(114)
FGF signaling				
acerebellar	Fgf8	FGF signal	Ventralized with loss of chordin	(87, 253)
morpholino	Fgf24	FGF signal	Loss of posterior structures with loss of fgf8	(67)

(Continued)

Table 1 (*Continued*)

Mutation	Gene product	Function	Phenotype	Reference
morpholino	Sef	Antagonist of FGF signaling	Dorsalized	(84, 323)
morpholino	Sprouty2	Antagonist of FGF signaling	Dorsalized	(87)
morpholino	MKP3	Antagonist of FGF signaling	Dorsalized	(324)
Retinoic acid signaling				
neckless	Raldh2	RA synthesis pathway	Anterior spinal cord reduced, myocardial progenitors increased	(24, 144)
giraffe	Cyp26a1	RA degradation	Anterior spinal cord expanded	(70, 172)
Transcription factors				
bozozok	Boz homeodomain	Transcriptional repressor	Variable loss of dorsal mesoderm and forebrain	(75)
vox/vent	Vox, Vent homeodomain	Transcriptional repressor	Severely dorsalized in double mutants	(131)
morpholino	Ved homeodomain	Transcriptional repressor	Severely dorsalized with *vox/vent*	(290)
kugelig	Cdx4 homeodomain	Transcription factor	Reduced tail and blood	(56)
morpholino	Prdm1/Blimp1	Transcriptional repressor	Dorsalized	(342)
dominant negative	Iro3	Transcriptional repressor	Reduced dorsal mesoderm	(171)
spiel ohne grenzen	Pou2/Oct4	Transcription factor	Strongly reduced endoderm in maternal-zygotic mutants	(193, 254)
faust	Gata5 Zinc finger	Transcription factor	Reduced endoderm and heart	(255)
casanaova	HMG domain	Transcription factor	Strongly reduced endoderm	(61, 150)
morpholino	Mezzo homeodomain	Transcription factor	Reduced dorsal mesoderm and endoderm with *bon*	(245)
no tail	Ntl T-box	Transcription factor	Loss of notochord and tail	(106, 278)
floating head	Flh homeodomain	Transcription factor	Loss of notochord	(312)
spadetail	Spt T-box	Transcription factor	Loss of paraxial and lateral mesoderm	(99, 156)
Epiboly				
half-baked	E-cadherin	Cell adhesion	Strongly reduced epiboly	(137)
dominant negative	Eomesodermin T-box	Transcriptional activator	Strongly reduced epiboly	(32)
morpholino	Mtx2 homeodomain	Transcription factor	Disrupted epiboly during gastrulation	(32)
Stat3 pathway				
morpholino	Stat3	Transcription factor	Reduced prechordal plate migration and CE	(354)
morpholino	Liv1	Zinc transporter	Reduced prechordal plate migration and CE	(355)
morpholino	Snail1	Zinc-finger transcription factor	Reduced prechordal plate migration	(355)
Planar cell polarity signaling				
silberblick	Wnt11	Wnt signal	Reduced CE	(119)
pipetail	Wnt5	Wnt signal	Reduced CE	(249)
knypek	Glypican4	Wnt co-receptor?	Reduced CE	(320)
trilobite	Strabismus	Transmembrane protein	Reduced CE	(132)

(*Continued*)

Table 1 (*Continued*)

Mutation	Gene product	Function	Phenotype	Reference
morpholino	Frizzled2	Wnt receptor	Reduced CE	(236, 309)
morpholino	Flamingo1a and 1b	7TM protocadherin	Reduced CE	(82)
morpholino	Prickle1	Regulates Fz/Dsh	Reduced CE	(37)
morpholino	Diversin	Ankyrin repeat protein	Reduced CE	(279)
Others				
morpholino	Gα12/13	G protein subunit	Reduced CE	(184)
morpholino	Quattro	Rho GEF	Abnormal prechordal plate migration and CE	(51)
morpholino	CAP1	Regulates actin distribution	Abnormal prechordal plate migration and CE	(51)
dominant negative	Rok2	Kinase	Reduced CE	(197)
dominant negative	Rac1	Small GTPase	Reduced CE	(17)
inhibitor	Phosphoinosite 3-kinase	Kinase	Abnormal prechordal plate migration and CE	(216)
morpholino	Hyaluronan synthase 2	Polysaccharide synthesis	Reduced CE	(17)
landlocked	Scribble1	LRR/PDZ domain protein	Reduced CE	(329)

Abbreviations: LR, left-right; CE, convergence and extension; TM, transmembrane; for more extensive references see text.

fates specified by intermediate Bmp levels in partial loss-of-function situations, and still explain how these fates are lost when Bmp levels are reduced below the relevant thresholds.

Among the genes acting downstream of Bmp signals to pattern the ectoderm are *ΔNp63* and *kheper*, both of which encode transcriptional repressors (15, 16, 177, 220). The ventrally expressed *ΔNp63* gene is required for development of the epidermis and is directly activated by Bmps. *Kheper*, a zinc finger-homeobox gene expressed in the neural plate, is repressed by Bmp signaling and dorsalizes the ectoderm when overexpressed.

An interesting exception to the neural expansion seen after inactivation of the Bmp pathway is that posterior spinal cord fates are lost rather than expanded in *swirl/bmp2b* mutants. In contrast to other neural progenitors, the tail spinal cord precursors are located on the ventral side of the embryo just above the marginal zone, and it seems that specification of these cells requires ventralizing Bmps, and perhaps other signals such as FGFs (167, 170, 172, 257).

It has also been proposed that graded action of Bmp patterns fates along the dorsal-ventral axis of the mesendoderm (52, 224, 227, 273). Bmps are clearly required for formation of ventrolateral margin fates such as blood, heart, pronephros, and tail somites, but the case for direct action of a Bmp morphogen in patterning different mesodermal fates is weaker than for ectoderm. "Allelic series" experiments have not provided evidence of expansion of intermediate territories as described for the ectoderm above. Thus other signals, including Wnt8 and FGF, are probably involved in patterning these marginal progenitors.

Despite the evidence for DV patterning by a Bmp activity gradient, the postulated gradient has not been directly visualized. Widespread overexpression of synthetic *bmp* mRNA can rescue *bmp* mutants, suggesting that ventral restriction of *bmp* expression is not the only mechanism that operates to form the postulated Bmp activity gradient (226). Instead, it seems that the action of modulators of Bmp signaling ensures the proper levels

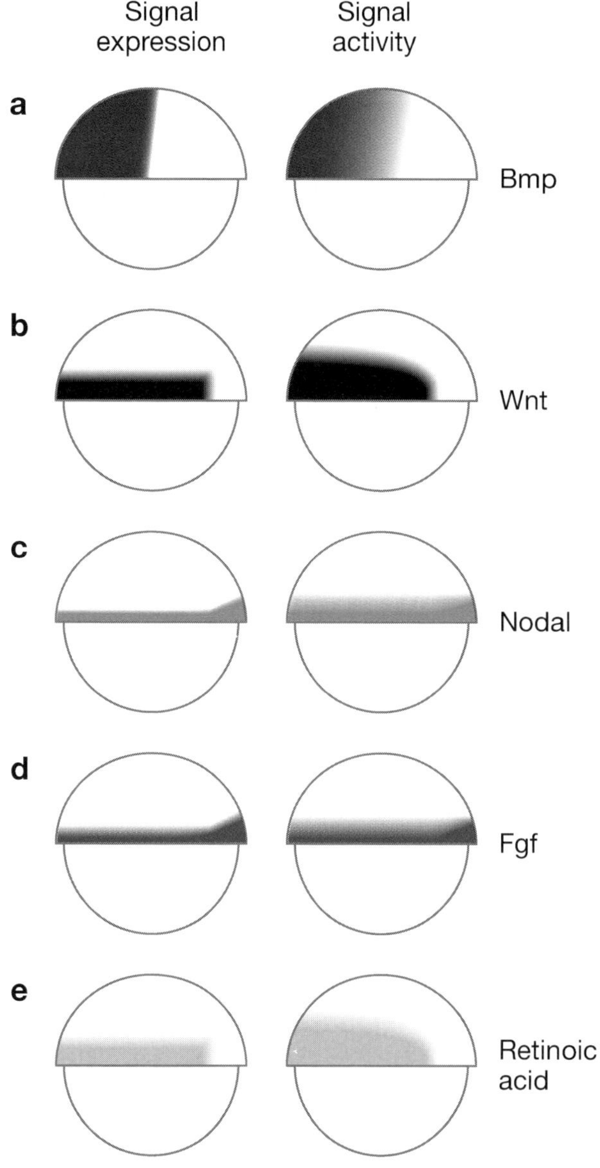

Figure 6

Signals patterning the embryo. Late-blastula stage, lateral view, dorsal to the right, animal pole to the top. Signal expression is based on published reports, but signaling activities are speculative and based on the potential range of signals and the expression pattern and range of antagonists. For example, Bmp signaling activity is inhibited dorsally by antagonists such as Chordin and Noggin. Wnt signaling activity is inhibited by antagonists such as Dickkopf1. Retinoic acid distribution indicates the site of synthesis by RALDH, and activity is inhibited by Cyp26-mediated hydrolysis of retinoic acid dorsally and at the animal pole. Nodal and FGF signals are concentrated on the dorsal side soon after the mid-blastula transition (not shown), but these signals are more uniform across the dorsal-ventral axis by the late-blastula stage that is represented in the figure.

of Bmp signaling activity across the dorsal-ventral axis.

Bmp Signaling is Modulated by Extracellular Factors

Extracellular modifiers of Bmp signals include Chordin, Ogon/Sizzled, Tolloid, and Twisted gastrulation (reviewed in 219, 352) (**Figure 6**). Mutational analysis demonstrates an essential role for Chordin in antagonizing ventralizing Bmps and thereby promoting the development of dorsal fates. *Chordin* mutants have a ventralized phenotype characterized by expansion of blood and tail fin, and a reduction of anterior neural territories (81, 109, 277). Analysis of marker gene expression indicates that DV pattern is disrupted during gastrulation, when ventral territories are expanded at the expense of presumptive neural and paraxial domains. Genetic studies support the biochemical evidence that Chordin acts to inhibit ventralizing Bmp signals: *bmp2b;chordin* double mutants are dorsalized, indicating that *chordin* is not needed for dorsal development if *bmps* are inactivated by mutation (110, 241).

Genetic studies suggest that *ogon* acts in concert with *chordin* to inhibit ventralizing *bmps* (351). Mutants for *ogon* have a ventralized phenotype very similar to *chordin* mutants (109, 208). The *ogon* gene encodes Sizzled, a member of the secreted frizzled related protein family (SFRP) (199, 351). Although SFRPs, which are related to the Wnt receptor Frizzled, were initially recognized as antagonists of Wnt signals, Ogon/Sizzled instead seems to antagonize ventralizing Bmp signals. The *ogon* mutant phenotype can be suppressed by overexpression of Chordin (or Noggin, another secreted Bmp antagonist). Overexpression of Ogon/Sizzled dorsalizes wild-type embryos but has no effect in *chordin* mutants, indicating that the dorsalizing activity of Ogon/Sizzled requires Chordin. The mechanism of Ogon/Sizzled action is not clear, but it seems that Sizzled augments the activity of Chordin, perhaps by inhibiting an inhibitor of Chordin, by directly making Chordin more active, or by modulating Bmp signals so that they become more susceptible to Chordin inhibition.

Tolloid is a conserved extracellular metalloproteinase that promotes Bmp signaling by cleaving and inactivating Chordin (28, 240). Several homologs of Tolloid, originally identified as an activator of Dpp/Bmp signaling in *Drosophila*, are present in vertebrates (287). Modified Chordins that are resistant to cleavage by Tolloid have more potent dorsalizing activity than wild-type Chordin in overexpression assays, showing that Tolloid activity limits the function of Chordin in the embryo (350). The *tolloid* gene is disrupted in zebrafish *minifin* mutants, which lack ventral tail structures but have normal DV patterning through the end of gastrulation (49). Chordin is cleaved in *tolloid* mutants, suggesting that the lack of an early phenotype in *mfn/tolloid* mutants reflects the action of redundant proteases during gastrulation (350).

Twisted gastrulation (Tsg) is a conserved extracellular protein that binds Bmps and has been implicated as both an agonist and an antagonist of ventralizing Bmp signaling (40, 234, 262, 280). The initial morpholino study in zebrafish reported that *tsg* morphants (embryos injected with antisense morpholino oligonucleotides for *tsg*) have some characteristics of ventralized embryos, supporting a role for Tsg in the antagonism of Bmp signaling (262). In contrast, two studies show that *tsg* morphants are dorsalized and that loss of *tsg* function can partially suppress the ventralized phenotypes of *chordin* and *ogon/sizzled* mutants (186, 350). This provides strong evidence that the predominant function of Tsg in the early zebrafish embryo is to promote Bmp signaling. Overexpressed Chordin accumulates at higher levels in *tsg* morphants than in wild type, suggesting that Tsg promotes Bmp signaling, at least in part, by reducing the level of Chordin (350). Tsg's mechanism of action in not clear, but one model proposes that the action of Tsg depends on the nature of Chordin, that is, whether Chordin is full-length or fragmented by Tolloid cleavage (174). Tsg, however, must have functions independent of Chordin and its fragments, because loss of Tsg function reduces Bmp signaling activity even in the absence of Chordin. Both overexpression and inhibition of *tsg* dorsalize embryos, indicating that too much or too little Tsg activity can inhibit Bmp signals (186, 350). One proposal that accounts for these phenotypes is that Tsg links Bmp proteins to another, as-yet unidentified, cofactor, such that BMP-Tsg-X complex does not form in *tsg* morphants and that inactive BMP-Tsg and Tsg-X complexes form in the presence of excess Tsg (186).

The antidorsalizing morphogenetic protein (ADMP) is a divergent member of the Bmp family that is expressed on the dorsal side of the late blastula and in the axial mesoderm and anterior neuroectoderm during gastrulation (180, 341). Overexpression of *admp* causes ventralization and a reduction of the organizer, whereas injection of morpholino oligonucleotides against *admp* causes a moderate expansion of dorsal mesoderm. The action of *admp* is not well understood, but it may function as part of a negative feedback system to limit the size of the organizer region, perhaps in concert with *bmp2b* and *bmp7*.

Maternal Bmps Activate Expression of Zygotic Bmps

There is evidence from the analysis of *smad5* mutants that maternal Bmp signaling is required for the activation of zygotic *bmp7*. Mutations that eliminate or disrupt the C-terminal domain of Smad5 exhibit a characteristic maternal-zygotic inheritance pattern, which results from a dominant negative function of these mutant Smad5 proteins (124). Homozygotes for a *smad5* null mutation are weakly dorsalized, but *smad5*$^{-/-}$ females produce strongly dorsalized progeny (referred to as maternal *smad5*, or M*smad5*, mutants) (168). The dorsalized phenotype of M*smad5* mutants is apparent before the zygotic *bmp* mutant phenotype, suggesting that the *Msmad5* phenotype reflects more than a simple function as a transcriptional mediator of zygotic *bmp2b* and *bmp7* (168). The identity of the putative maternal Bmp signal is not clear, but Gdf6a/Radar is one candidate (293). Maternal Radar, however, may not be the signal acting upstream of maternal Smad5, because the *radar* morphant phenotype is different from and weaker than the *Msmad5* phenotype (293). *Bmp4* and *bmp7* are also expressed during oogenesis (168), suggesting that they may act maternally in parallel with *radar*, but there is no evidence that either gene is required maternally for an early patterning function (60, 272).

Wnt Signaling

Signaling through the canonical Wnt pathway is essential for the specification of ventral and posterior fates (reviewed in 129). Wnt signaling through a Frizzled-Lrp receptor complex and a number of cytoplasmic proteins including Dsh, GBP, Axin, Ccd1, APC, and GSK3 stabilizes β-catenin, allowing it to accumulate in the nucleus and activate target gene expression (reviewed in 188). There are several secreted antagonists of Wnt signaling, including SFRPs, Cerberus, and Wnt inhibitory factor (WIF), which act by binding to Wnt proteins, and Dickkopf (Dkk), which binds the LRP subunit of the receptor (reviewed in 143).

Genetic studies in zebrafish show that Wnt8 signals are essential for the establishment of ventral and posterior fates (72, 179). During gastrulation, *wnt8* mRNA and strong activity of a Wnt/β-catenin responsive reporter are evident at the ventrolateral margin (63, 149) (**Figure 6**). Deletion or morpholino-inhibition of both ORFs of the bicistronic *wnt8* gene produces a severe loss of ventroposterior structures, with a concomitant expansion of dorsal fates (179). Simultaneous reduction of Wnt3a and Wnt8 activities results in a stronger expansion of dorsoanterior fates, indicating that these genes have overlapping functions (288). This zygotic role of canonical Wnt signaling in ventral and posterior patterning is opposite to its earlier role in dorsal patterning by maternally provided β-catenin described above.

Wnt signals have a role in repressing dorsal mesodermal fates that is distinct from the action of Bmp signals. In contrast to the *bmp* pathway zygotic mutants, the axial mesodermal territory in *wnt8* mutants is expanded along with the paraxial mesodermal and neural domains. In addition, anterior neural fates are expanded in embryos with reduced *wnt8* function, supporting a role for Wnt8 in posteriorizing the neuroectoderm (72, 179). Furthermore, mutations that inactivate repressors of Wnt signaling lead to an expansion of posterior neural fates at the expense of more anterior territories (62, 117, 153). Embryological and genetic evidence also indicates that the position of the midbrain-hindbrain boundary is established by Wnt8 signals, possibly acting as morphogens, emanating from the blastoderm margin during gastrulation (259, 346, 347).

Among the target genes of Wnt8 and Wnt3a are the homeobox gene *cdx4/kugelig*, which is essential for tail development and the regulation of posterior *hox* genes (56, 94, 288), the T-box gene *tbx6* (311), and the Sp1 class zinc finger gene *Sp5-like* (337). In addition to these functions during gastrulation, experiments with low doses of morpholinos suggest that *wnt8* and *wnt3a* function during segmentation to maintain presomitic mesoderm in the tail bud (319).

The roles of Wnt antagonists have not been extensively studied in zebrafish, but *dkk1*, an early target of maternal β-catenin, is expressed early in the dorsal margin and dorsal yolk syncytial layer and during gastrulation in the developing prechordal plate, where it could function to counteract the ventralizing and posteriorizing effects of canonical Wnt signaling (113, 292).

The SFRP protein Tlc is expressed at the anterior neural border, a region required for induction of anterior neural fates (127, 128). Telencephalic fates are reduced in *tlc* morphant embryos, and it has been proposed that Tlc acts locally within the neural plate to promote anterior identity by inhibiting Wnt8b signals from the midbrain-hindbrain boundary.

Boz and Vox/Vent Transcriptional Repressors

Inactivation of the redundant homeodomain transcriptional repressors Vox (Vega1) and Vent (Vega2), by deletion, coinjection of morpholino oligonucleotides for both genes, or injection of a *vent* MO into a *vox* point mutant, leads to a severe loss of ventroposterior structures including blood, pronephros, and tail (131, 139, 140, 204). The loss-of-function phenotype is strain dependent, such that AB strain embryos lacking *vox/vent* are essentially wild type (131). Inactivation of a third gene encoding a homeodomain transcriptional repressor, *ved*, along with *vox* and *vent*, is sufficient to strongly dorsalize even AB strain embryos (290). Although embryos lacking *vox/vent* resemble the Bmp pathway mutants, important phenotypic differences are that dorsal mesodermal fates are strongly expanded in embryos lacking *vox/vent*, and anterior neural fates are shifted more toward the margin and less toward the ventral side than in *bmp* pathway mutants such as *swirl/bmp2b* (131). The dorsalized phenotypes of *vox/vent* and *wnt8* mutants are very similar, and there is evidence that *wnt8* activates *vox* and *vent* expression, thereby repressing dorsal genes (248). *chordin* is a key target of Vox and Vent, and these proteins also repress other dorsal genes including *boz*, *goosecoid*, *floating head (flh)*, and *dkk1*.

Mutants for the homeodomain transcriptional repressor Boz have a variable phenotype characterized by cyclopia, reduction of dorsal mesoderm, and, in the most severe cases, reduction of forebrain coupled with an expansion of hindbrain (75, 166, 291, 294, 300). Maternal β-catenin activates *boz* expression in dorsal blastomeres soon after the MBT (263, 353) (**Figure 5**). Beginning shortly thereafter, *boz* expression is confined to the dorsal yolk syncytial layer until *boz* mRNA is no longer detectable at the midgastrula stage. Studies with fusion constructs containing the Boz homeodomain and potent transcriptional activator or repressor domains indicate that Boz acts as a transcriptional repressor (290). Although *boz* is predominantly expressed in the yolk syncytial layer, it can act nonautonomously to dorsalize overlying blastomeres, presumably by repressing a ventralizing signal expressed in the yolk syncytial layer (353). Key targets of Boz include *bmp2b*, *wnt8*, and *vox/vent/ved* (76, 96, 131, 182). Thus Boz specifies dorsal fates by repressing the expression of ventralizing factors rather than directly activating dorsal gene expression (**Figure 5**). For example, dorsal mesoderm is expanded in *boz;vox;vent* triple mutants, demonstrating that *boz* is not needed to promote dorsal mesoderm gene expression when the ventralizing repressors are inactivated by mutation (131).

Two additional transcriptional repressors, Prdm/Blimp1 and Iro3, are expressed at the dorsal margin. In contrast to Boz, Prdm1 represses *chordin* expression and antagonizes dorsal fates when overexpressed, and knockdown of Prdm1 function weakly dorsalizes the embryo (342). At later stages, Prdm1 is required for slow muscle development and patterning of cell types at the edge of the neural plate (23, 123). Iro3 appears to act as a repressor of *bmp* transcription (171). These observations indicate that depending on their target genes, dorsally expressed repressors can have opposite roles in DV patterning.

A Model for Dorsal-Ventral Patterning

The detailed analysis of mutants that affect Bmp and Wnt signaling and several transcription factors suggests the following model for DV patterning. Soon after the onset of zygotic transcription, ventralizing genes, including *bmp2b* and *vox*, are widely expressed in the embryo, including in the most dorsal territories (139, 182) (**Figures 5** and **6**).

The maternal pathways inducing the expression of *vox* and *bmp2b* are not known, but Bmp signals likely have a role (168, 293). It seems that *bmp2b* and *vox* are activated in parallel, because zygotic *bmp* and *vox/vent* are not required for each other's expression until the late-gastrula stage (131). In contrast, Wnt8 regulates *vent* expression and mesodermal *vox* expression (248). At the same time, maternal β-catenin protein activates dorsalizing genes, including *boz* among others, specifically in dorsal blastomeres and, soon thereafter, dorsal nuclei in the yolk syncytial layer (353). Hence, the earliest zygotic regulators of DV patterning act downstream of maternal factors to establish a two-state pattern, in which cells express either dorsal and ventral genes or only ventral genes.

After a short lag, presumably reflecting the time needed for Boz protein to accumulate to sufficient levels, Boz represses transcription of *bmp2b*, *vox*, and other ventralizing genes at the dorsal margin (75, 139, 182). This allows for expression of dorsal genes, such as *chordin*, *dkk1*, and *goosecoid*, which would otherwise be repressed by *vox/vent/ved* (131, 290). Thus as the first wave of zygotic genes becomes active, cells have gene expression patterns characteristic of either dorsal cells (e.g., *boz*, *goosecoid*, *chordin*, *dkk1*) or ventrolateral cells (e.g., *bmp2b*, *bmp7*, *vox*, *vent*, *ved*, *wnt8*).

Through the action of Bmp and Wnt8 signals and their antagonists, the simple pattern of mid-blastula stage embryos becomes much more elaborate, with many different groups of tissue progenitors fated to arise from different regions of the early gastrula embryo. As Bmps, Wnt8, and other signals elaborate and refine the pattern of the early gastrula, the regulatory interactions among DV-patterning genes change. For example, *vox/vent* and *bmp2b/swirl* are initially expressed independently of each other's action. As embryogenesis proceeds, however, expression of *vox/vent* and *bmp2b/swirl* genes becomes interdependent, apparently through a positive feedback loop established during gastrulation (131, 139, 140, 204). At mid-gastrulation, zygotic Bmp signals are required for normal levels of *vox* and *vent* expression. Conversely, *vox* and *vent* act to promote *bmp2b/swirl* and *bmp4* expression by inhibiting the expression of *chordin*, which blocks a positive autoregulatory activity of BMP signals (110, 277). Although the primary function of *vox/vent/ved* is to repress dorsal genes rather than to induce ventral genes, interruption of this *vox/vent-bmp2b* positive feedback loop is responsible for a reduction of ventral gene expression in embryos lacking *vox* and *vent* at mid-gastrulation. Thus the *vox/vent-bmp2b* positive feedback loop maintains ventral positional identity during gastrulation, and the participation of the extracellular factors Chordin and Bmp incorporates flexibility and sensitivity to the cellular environment into the mechanism that maintains dorsal-ventral identity. For example, a cell moving from ventral to dorsal territories during gastrulation would reduce its expression of *vox* and *vent* in response to increased levels of Chordin and reduced levels of Bmp activity. The reduction of Vox and Vent levels would, in turn, permit the expression of dorsal genes appropriate for the cell's new environment.

MESODERM AND ENDODERM FORMATION: MATERNAL FACTORS

The progenitors of the different germ layers are arranged along the animal-vegetal axis, with mesendoderm progenitors residing at and next to the margin and ectodermal progenitors located more animally (161, 334) (**Figure 3**). The animal-vegetal axis, in contrast to the DV or left-right axes, is already formed during oogenesis (281, reviewed in 237). The egg thus has an animal-vegetal polarity that is highlighted by morphological and molecular markers such as the position of the germinal vesicle and the localization of maternal mRNAs (18, 195). It is unknown how this polarity is generated during oogenesis, but maternal-effect mutants such as *bucky ball* might provide insights into this process (65).

mRNAs that are localized animally or vegetally in wild type fail to do so in *bucky ball* mutants, and cytoplasmic streaming occurs in multiple directions. This phenotype suggests an animal-vegetal polarity defect of the egg.

The animal-vegetal polarity of the egg has to be translated into the induction of mesendoderm at the margin. The molecular basis of this process remains elusive, but several lines of evidence indicate that the yolk cell, and specifically the yolk syncytial layer, contains signals that can induce mesendodermal markers (**Figure 4**). First, transplantation of the yolk cell onto the animal region of the blastoderm can ectopically induce genes that are normally expressed at the margin (212, 213, 232). Second, injection of RNase into the yolk syncytial layer blocks the expression of ventral and lateral mesendodermal markers (44). Dorsal markers are still expressed, probably due to the dorsal determinant β-catenin in dorsal blastomeres. The maternal factors that establish the yolk syncytial layer as a signaling center are unknown. In *Xenopus* the transcription factor VegT has been implicated as a maternal factor that can activate mesoderm inducers (360). *Xenopus VegT* mRNA is localized vegetally and required for mesendoderm induction. In zebrafish *spadetail* is a related T-box gene, but neither its expression nor mutant phenotype suggest any functional similarity to *Xenopus VegT* (99, 156). Maternal *eomesodermin* mRNA is localized vegetally in zebrafish eggs, but there is no evidence that this T-box gene might act in a *VegT*-like fashion (33). No maternal mutant has been isolated yet that blocks mesoderm and endoderm induction in zebrafish.

MESODERM AND ENDODERM FORMATION: ZYGOTIC FACTORS

Nodal Signaling

Members of the Nodal family of TGFβ signals are essential inducers of mesoderm and endoderm in vertebrates (reviewed in 268). Nodal signals are received by EGF-CFC coreceptors and type I and II Activin receptors, which function as serine/threonine kinases. Receptor activation leads to phosphorylation of the transcription factors Smad2 and Smad3. This results in their binding to Smad4, nuclear translocation, and association with additional transcription factors such as FoxH1 and Mixer to regulate target genes. Nodal signaling is antagonized by feedback inhibitors such as Lefty proteins, which are divergent members of the TGFβ family and block EGF-CFC coreceptors (41, 48), and Dapper2, which enhances the degradation of type I Activin receptors (362, but see 335).

Mutant screens in zebrafish have identified several components of the Nodal signaling pathway (**Table 1**). These include the Nodal signals Cyclops (Cyc) and Sqt, the EGF-CFC coreceptor One-eyed pinhead, and FoxH1 [*schmalspur (sur)*] and Mixer [*bonnie & clyde (bon)*] (31, 71, 79, 108, 115, 118, 151, 173, 243, 251, 252, 265, 270, 271, 295, 300, 306, 322, 361). In addition, molecular studies have led to the isolation of zebrafish Lefty and Dapper2 homologues and TARAM-A, a putative Nodal type I receptor (27, 239, 258, 316, 335, 362). In the case of the Leftys and Dapper2, but not TARAM-A, morpholino experiments have revealed essential roles for these proteins (3, 11, 46, 77, 362). Below we summarize the role of Nodal signaling in mesendoderm formation in zebrafish. Recent reviews provide more general discussions of the Nodal signaling pathway and its role during vertebrate development (reviewed in 268, 269).

Absence of Nodal signaling in *cyc;sqt* double mutants or maternal-zygotic *one-eyed pinhead* mutants results in embryos that lack all endoderm and mesoderm, with the exception of a few somites in the tail (79, 102). Mutants also lack trunk spinal cord, but develop forebrain, midbrain, hindbrain, and tail spinal cord. These phenotypes are already presaged before gastrulation by the aberrant expression of genes marking presumptive mesendoderm progenitors in wild type. Consequently, the fate and morphogenetic movement of marginal cells is affected. Marginal cells do not internalize, and dorsal marginal cells acquire neural fates instead of dorsal mesodermal fates, while ventral and lateral marginal cells contribute exclusively to the tail (36, 78). Conversely, increasing Nodal signaling by loss of Lefty1 and Lefty2 or overexpression of Cyc or Sqt results in the fate transformation of ectodermal cells into mesoderm or endoderm (3, 45, 46, 71, 77, 79, 101, 251, 252, 265).

How does the interplay of *cyc*, *sqt* and *leftys* control mesendoderm formation? Before gastrulation *cyc*, *sqt*, *lefty 1* and *lefty 2* are expressed in the 1–3 cell tiers closest to the margin, overlapping with and vegetally to mesendodermal progenitors (27, 46, 66, 71, 79, 101, 207, 251, 252, 265, 316) (**Figure 6**). In addition, *sqt* is expressed in the yolk syncytial layer (71, 79). Mutant and misexpression studies have suggested a scenario for Nodal-mediated mesendoderm induction (46, reviewed in 268). The *sqt* and *cyc* genes are transcribed in cells closest to the margin, leading to the local generation of Sqt and Cyc proteins. Sqt can move away from the source and induce mesendodermal gene expression in cells at a distance (45). In contrast, Cyc only acts at a short range and induces mesodermal markers locally (45). Nodal signaling also induces the expression of *lefty1* and *lefty2*, which block the Nodal signaling pathway both locally to restrict the expression of *sqt* and at a distance to restrict the response to Sqt (46, 77). Hence, the interaction of Sqt, Cyc, and Leftys determines the extent of mesendoderm formation in zebrafish.

The strongest evidence for this model comes from the analysis of *bhikhari*, a marker for Nodal signaling expressed in 6–10 tiers from the margin. *cyc;sqt* double mutants lack *bhikhari* expression, and in the absence of *squint*, *bhikhari* is expressed only in the first few tiers (45). In contrast, the absence of only *cyc* initially does not affect the extent of *bhikhari* expression. This has suggested that Sqt might act at a long range to induce mesendodermal genes, whereas Cyc has only short-range activity. In support of this model, ectopic clones of Sqt-expressing cells can induce downstream genes in distant cells. In contrast, Cyc-expressing clones only induce downstream genes at a short range (45). Two observations indicate that the long-range effect of Sqt is direct and not mediated via a relay mechanism. First, the activation of the Nodal signaling pathway in a clone of cells induces downstream genes cell autonomously. Second, Sqt can be made in nonresponding cells and apparently move through a field of nonresponding cells to activate gene expression in distant cells (45). These results provide support for a direct long-range effect of Sqt. It is unclear why Sqt is long-range and Cyc short-range, but studies on mouse Nodal indicate that the stability of the mature ligand might be a major determinant of range (178).

Leftys appear to restrict the range of Nodal signaling by two mechanisms (46). First, Leftys dampen Nodal autoregulation, thus limiting the generation of more Nodal. For example, in the absence of Leftys, *sqt* expression extends animally away from the margin and is maintained for a longer time than in wild type. Second, Leftys can act at a long range to inhibit Nodal signaling in distant cells. For example, ectopic expression of Leftys at the animal pole can block Nodal signaling at the margin of the zebrafish embryo. Moreover, depletion of Leftys extends the range of Sqt activity even in the absence of *sqt* autoregulation.

Although the above model accounts for the regulation of Nodal-regulated markers such as *bhikhari*, interactions between *cyc* and *sqt* provide additional complexity. Specifically, Sqt induces the expression of *cyc* on the dorsal side (66). Hence, *sqt* mutants lack Squint and have less Cyc on the dorsal side. The dorsal side is therefore more sensitive to loss of Nodal gene dosage than the lateral and ventral sides. In addition, despite their different ranges, both Cyc and Sqt can induce most mesendoderm derivatives on their own. While *cyc;sqt* double mutants lack all head and trunk mesoderm and endoderm, *cyc* mutants display only minor defects in prechordal plate formation (66, 115, 317), and *sqt* mutants have quite mild defects in axial mesoderm and endoderm formation (66, 79, 118, 294). Hence, both short- and long-range Nodals can orchestrate most aspects of mesendoderm formation in zebrafish.

Nodal signaling not only induces the extent of mesendoderm, but also seems to pattern it. Partial reduction of Nodal signaling leads to the loss of cell types derived from marginal-most tiers. For instance, at the dorsal margin, high levels of Nodal signaling are required for prechordal plate (anterior axial mesoderm) specification, whereas lower levels are essential for notochord (posterior axial mesoderm) specification (101). Analogously, at the lateral margin where precursors for the myocardium (heart muscle) reside, high levels of Nodal signaling promote ventricular fates whereas lower levels are sufficient to induce atrial fates (145). This has led to the proposal that there might be a gradient of Nodal signaling activity at the margin, leading to the fine patterning of mesodermal and endodermal precursors (101). This conclusion is also supported by fate map studies of *sqt/sqt* and *sqt/sqt;cyc/+* mutants that demonstrate a vegetal shift of cell fates that in wild type are located more animally (66). These mutant combinations also revealed that dorsal mesoderm requires higher levels of Nodal function than ventral and lateral regions; however, differential Nodal signaling does not pattern the mesoderm along the DV axis. In particular, dorsal margin cells are not transformed toward more lateral fates in *sqt/sqt;cyc/+* mutants embryos, but dorsal expression of *cyc* requires Nodal-dependent autoregulation. The specification of endoderm, which derives from blastomeres that are located most marginally, also requires high levels of Nodal signaling (6, 260, 271). These observations suggest that there might be a Nodal activity gradient along the animal-vegetal axis, with the highest levels at the margin inducing endoderm, prechordal plate, and ventricular progenitors, lower levels inducing notochord and other mesodermal fates, and the absence of Nodal signaling allowing neural and tail specification (101, 315). Experiments that block or activate Nodal signaling at different times have suggested that marginal-most cells require sustained Nodal signaling before the onset of gastrulation, whereas cells more distant from the margin require shorter windows of Nodal signaling (10, 101). It is thus conceivable that Nodal signaling induces different fates using both temporal and spatial gradients.

Nodal signaling induces the phosphorylation of Smad2 and Smad3 (reviewed in 268). Smad2 or Smad3 mutants are not available in zebrafish, but mutations in FoxH1 (*sur*) and Mixer (*bon*), transcription factors that can bind to phosphorylated Smad2, have been identified (31, 151, 173, 243, 270, 295, 300, 322). *bon* mutants have severe defects in endoderm formation, whereas maternal-zygotic *sur* mutants have mild defects in axial mesoderm formation (151, 243, 295). Loss of both *bon* and *sur* results in a severe phenotype characterized by absence of prechordal plate, cardiac mesoderm, endoderm, and ventral neuroectoderm (173, 322). Some Nodal-regulated genes are regulated by either Bon or Sur, and others by both Bon and Sur (173). The phenotypes seen upon loss of both Bon and Sur are milder than those seen upon complete loss of Nodal signaling, indicating that additional Smad-associated transcription factors that act as components of the Nodal signaling pathway remain to be identified.

Nodal Signaling and Left-Right Axis Development

Most organs in vertebrates are formed and positioned asymmetrically along the left-right axis. The analyses of *one-eyed pinhead* and *sur* mutants and of morphants for a third Nodal gene, *southpaw*, have established a requirement for Nodal signaling in zebrafish left-right axis formation (43, 191, 356). *southpaw* is expressed in the left lateral plate mesoderm, whereas *one-eyed pinhead* and *sur* are expressed bilaterally (191, 243, 295, 361). Loss of late *one-eyed pinhead* or *sur* activity, and knock down of *southpaw* lead to a loss or randomization of organ asymmetries (43, 191, 356). For example, the consistent looping of the heart to the right side is randomized or lost in these embryos.

How genes such as *southpaw*, *lefty*, and *pitx2* are specifically activated on the left side is still unclear. Notch signaling and rotating cilia have been implicated in this step of left-right patterning in several vertebrates (235, reviewed in 250). In zebrafish, these cilia are located in Kupffer's vesicle, a specialized organ lined by the descendants of the dorsal forerunner cells (50, 73, 74, 142, 169). Disruption of Kupffer's vesicle development or cilia formation and function results in the randomized activation of left-side specific markers (73, 142, 169). Based on studies in the mouse, it is thought that cilia rotation leads to a leftward flow that results either in the specific activation of mechanoreceptors or the accumulation of a signal on the left that initiates left-side-specific gene expression. Both *southpaw* and the Nodal antagonist *charon* are first expressed symmetrically around Kupffer's vesicle before *southpaw* expression becomes restricted to the left lateral plate mesoderm (114, 191). Inhibition of *charon* expression results in the bilateral activation of *southpaw* in the lateral plate (114). These results suggest a model wherein cilia-mediated flow and *charon* activity bias *southpaw* activation toward the left.

Downstream of Nodal Signaling: Endoderm Formation

Several genes have been identified that are regulated by Nodal signaling and mediate its endoderm-inducing activity, including the Sox gene *casanova*, the GATA gene *faust*, and the homeobox genes *bonnie&clyde* and *mezzo* (5, 42, 61, 150, 151, 245, 255, 256, 264, 302, reviewed in 231). These four genes encode transcription factors and appear to be direct targets of the Nodal signaling pathway, as suggested by experiments using cycloheximide (245). As described above, Bon is also a component of the Nodal signaling pathway.

Of these four genes, *casanova* appears to be the most central and downstream player (61, 150). *casanova* is expressed in a subset of marginal-most cells that are thought to give rise to endoderm. Indeed, loss of *casanova* causes these cells to adopt an aberrant mesodermal fate. In addition, *casanova* is sufficient to induce cells to give rise to endoderm, and *casanova* can induce endoderm in the absence of Nodal signaling. Since the activities of *faust*, *bon*, and *mezzo* depend on *casanova*, the main role of *faust*, *bon*, and *mezzo* may be to induce and maintain *casanova* expression. Casanova activity and maintenance require the POU domain gene *spiel ohne grenzen* (*pou2/Oct4*) (25, 34, 193, 254). In contrast to Casanova, *spiel ohne grenzen* expression is not induced by Nodal, but is ubiquitous and activated both maternally and zygotically. These results suggest that endoderm formation induced by Nodal is predominantly mediated by the induction of Casanova and its interaction with Spiel ohne grenzen.

Downstream of Dorsal Mesoderm Induction: Midline Development

As the embryonic pattern is refined during gastrulation, cells at different positions in the shield and the immediate vicinity adopt different fates, including prechordal plate, notochord, hypochord, adaxial muscle, and floor plate (175, 206, 285). The prechordal plate arises from the marginal-most cells in the shield, under the influence of the highest levels of Nodal signals (101). The cells of the developing prechordal plate specifically express the transcriptional repressor Goosecoid (301). Goosecoid may repress expression of genes that promote other cells types (80), although *goosecoid* has not been analyzed in loss-of-function studies in zebrafish.

The notochord arises from cells at slightly more animal positions within the shield (101, 206). The homeobox gene *flh* and the T-box gene *no tail* (*ntl*) are essential for notochord development (106, 107, 278, 312). Both genes are expressed in all margin cells in the late blastula (276, 312). By the beginning of gastrulation, *flh* is specifically expressed in notochord precursors, whereas *ntl* is expressed in all margin cells, the developing notochord, and, at later stages, the tail bud. Fate mapping with mutant embryos showed that *flh* is required to prevent notochord precursors from differentiating as muscle, whereas *ntl* acts to prevent notochord precursors from forming floor plate (7, 105, 205, 206). A key target of *flh* is *spadetail* (*spt*), a T-box gene initially expressed in all marginal cells and then repressed in the notochord domain early in gastrulation (8, 99). The *Xenopus* ortholog of *flh*, *Xnot*, encodes a transcriptional repressor, suggesting that Flh may directly repress *spt* transcription (328, 357). Trunk muscle and other ventral-lateral mesodermal derivatives are reduced in *spt* mutants, and it has been proposed that *spt* both activates expression of genes required for muscle differentiation and morphogenesis (e.g., *myod*) and antagonizes the function of *ntl* in notochord development (7, 8, 99, 125, 156).

In addition to the interactions among these transcriptional regulators, local signaling interactions are important to allocate cells in the shield region to particular fates. For example, the Nodal signal Cyc is required during gastrulation for the formation of the medial floor plate (115, 217, 229, 233, 265, 319a, reviewed in 307). The Cyc signal antagonizes the notochord-promoting function of *ntl*, perhaps by inducing the expression of the transcriptional repressor Her9 (105, 176). Repression of *ntl* by Notch signaling and another Hairy/Enhancer of split gene, *her4*, allows cells lateral to the notochord domain to differentiate as hypochord (12, 13, 175). At later stages the growth factor Midkine-a is expressed in the paraxial mesoderm and required for the formation of the posterior medial floor plate (266a). Hedgehog signals from the midline during gastrulation instruct the immediately adjacent adaxial cells to differentiate as slow muscle and the overlying neuroectoderm to form lateral floor plate (29, 169a, 233, 266b, reviewed in 307). These studies indicate that complex, local interactions among several signaling pathways and transcription factors specify different midline cell fates.

FGF Signaling

Members of the FGF family of signals have been implicated in mesoderm formation, neural induction, DV patterning, and anterior-posterior patterning of the embryo (reviewed in 296). Data in zebrafish mainly suggest an early role for FGF signaling in repressing Bmp signaling and a later role in promoting the development of posterior structures. FGFs bind and activate receptor tyrosine kinases. Receptor dimerization leads to *trans*-phosphorylation and the recruitment and activation of a plethora of downstream effectors, including PKC and the ras/MAPK cascade. In contrast to the Nodal and Bmp signaling pathways, zygotic mutant screens have only uncovered a single mutation in a component of the FGF signaling pathway—mutations in *acerebellar* disrupt the *fgf8* gene (30, 253). The dearth of FGF signaling mutants might be due to the overlapping roles or maternal contribution of these gene products. Hence, morpholinos, misexpression or small molecule inhibitors have been used to analyze the role of FGF signaling during embryogenesis. Such studies have identified the type I transmembrane protein Sef ("similar expression to fgf genes") as a novel feedback inhibitor of FGF signaling (84, 323).

The expression patterns of downstream targets for FGF signaling have revealed that the pathway is first active at the dorsal blastoderm margin, then along the entire margin and finally in the tail bud (87, 247, 261, 324) (**Figure 6**). Consistent with this pattern of activity, the earliest role for FGF signaling is during blastula stages, when *fgf3*, *fgf8*, and *fgf24* are expressed at the dorsal margin. Misexpression of FGF signals can inhibit the expression of *bmp2b* and *bmp7* at blastula stages and lead to the lateralization and dorsalization of the embryo (86, 87, 324). However, blocking only FGF signaling during these stages does not result in a ventralized embryo. In contrast, blocking both Chordin and FGF signaling results in ventralization (87). These results suggest that FGF signaling, Bozozok, and Bmp antagonists such as Noggin and Chordin all contribute to blocking Bmp signaling in dorsal margin blastomeres. Conversely, inactivating the FGF signaling feedback inhibitor Sprouty2 results in the repression of Bmp expression and the dorsalization of the embryo (87). Taken together, these results establish an early role for FGF signaling in restricting Bmp expression and activity.

Following dorsal margin expression, several FGF ligands become expressed in the entire margin at the onset of gastrulation (35, 67, 85, 253). Moreover, downstream targets like *pea3*, *erm*, and *sprouty4* are induced in broad domains in neighboring cells (87, 247, 261, 324). These downstream genes appear to be activated at different thresholds of FGF activity, with *sprouty4* being the target most sensitive to low levels of FGF signaling and expressed in most-distant cells (275). These results suggest that FGF signals form a vegetal-to-animal activity gradient. Indeed, tagged FGF can be detected in intracellular vesicles at a distance from the source. This localization is dependent on receptor-mediated endocytosis, which leads to the clearance of the ligand from extracellular space. As a consequence, blocking endocytosis results in an increased FGF signaling range (275).

These results suggest a long-range and graded FGF signaling activity at early gastrulation stages, but it is unclear what role this activity might have. At the onset of gastrulation mesendoderm formation and patterning are initiated correctly in the apparent absence of FGF signaling, and only at later gastrulation stages is the expression of genes such as *tbx6*, *spt*, and *ntl* lost (67, 84, 87, 98, 100, 200, 247, 323, 324). It is therefore conceivable that very early FGF signaling only has consequences at later stages of gastrulation (see below), or that it acts redundantly with other pathways, e.g., the Nodal signaling pathway. The latter possibility is raised by double mutant studies of FGF and Nodal signaling components (100, 200). For example, partial inhibition of FGF signaling and blocking zygotic activity of One-eyed pinhead, a Nodal coreceptor, disrupts posterior development and leads to the death of dorsal mesoderm cells by the end of gastrulation. However, it remains unclear when Nodal and FGF signaling interact.

An alternative scenario is that FGF signaling during gastrulation primes and maintains cells for posterior development. For example, severe posterior truncations are generated in embryos exposed to the FGFR inhibitor SU5402, expressing a dominant-negative FGF receptor or lacking full *fgf8* and *fgf24* activity (67, 98, 100, 200). Tail and posterior trunk mesoderm do not form in these embryos and mesodermal markers such as the T-box genes *ntl* and *spt* cease to be expressed at later gastrulation stages. These results suggest that FGF signaling is required for the maintenance of a pool of mesoderm progenitors during gastrulation. A role for FGF signaling in the formation of posterior cell types is also seen in the nervous system (167, 170, 172, 189, 257). FGF signaling is required for the expression of posterior neural markers during gastrulation. Strikingly, this neural inducing role of FGFs is independent of the organizer or the inhibition of Bmp signaling.

Retinoic Acid Signaling

Retinoic acid signaling acts during gastrula stages in the posteriorization of the neuroectoderm and the formation of myocardial progenitors (**Figure 6**). Retinoic acid binds to its receptors, members of the nuclear hormone receptor family, leading to the regulation of downstream genes. Retinoic acid is synthesized by RALDH and hydrolyzed by Cyp26/P450RA1. Studies on retinoic acid function in zebrafish have used exposure to retinoic acid, pharmacological inhibition of retinoic acid receptors, mutations in *raldh2* or *cyp26a1*, and morpholinos against *cyp26a1* (24, 70, 97, 122, 172, 185). These studies have suggested that the expression of RALDH in the posterior mesoderm during gastrulation generates a source of retinoic acid that induces posterior hindbrain and spinal cord markers, in particular specific subsets of *Hox* genes. Conversely, *cyp26a1* is expressed in more anteriorly located precursors of the neuroectoderm and thought to generate a retinoic acid-free zone that is thus protected from retinoic acid-mediated posteriorization.

A role for retinoic acid has also been found in the mesoderm. Blocking retinoic acid signaling during gastrulation stages increases the number of myocardial progenitors (144). This is not simply due to an expansion of the myocardial progenitor region but appears to be a consequence of an increase in the density of myocardial progenitors within a normally sized field containing these precursors.

MicroRNAs

MicroRNAs are ~22 nucleotides long nonprotein-coding RNAs that regulate gene expression at the posttranscriptional level (reviewed in 9, 20). MicroRNAs have been implicated in many processes, but the roles of microRNAs in zebrafish embryogenesis have only been tested recently (92, 338, 339). A large-scale analysis of the expression of more than 100 microRNAs has revealed very specific patterns during embryonic and larval stages (338). Maternal-zygotic mutants for the RNaseIII enzyme Dicer cannot process microRNA precursors and thus lack mature microRNAs (92). The resulting phenotype is quite mild. Mutant embryos develop normal axes and are regionalized correctly. The major cell types are specified, and no dramatic modulation of embryonic signaling pathways has been observed. The main defects are in morphogenetic processes such as a delay in epiboly, impairment of ventricle inflation, and abnormal somite differentiation (92). These results suggest that microRNAs have subtle roles during zebrafish axis formation and might be involved in the differentiation of multiple cell types at later stages.

GASTRULATION MOVEMENTS

The movements of epiboly, internalization, convergence, and extension transform the radially symmetric blastula into the gastrula embryo with clear DV and anterior-posterior axes (333, reviewed in 2, 138, 155, 181, 215, 298) (**Figure 2**). Mutant analysis has indicated that these processes can be genetically separated, e.g., defects in internalization do not lead to an obligatory disruption of epiboly or convergence.

Epiboly

Epiboly describes the process of spreading and thinning of the embryo during blastula and gastrula stages and results in the envelopment of the yolk by the embryo. Microfilament and microtubule networks in the yolk cell are thought to contribute to epiboly (47, 299, 305, 321). The cellular basis for epiboly appears to be at least in part a process of radial intercalation—deeply positioned cells move outward between more superficial cells, resulting in a thinning and spreading of the embryo (**Figure 2**). This process has been well documented during gastrula stages, when two cell layers can be distinguished in the epiblast. An epithelial-like exterior layer underlies the enveloping layer and an inner layer overlies the hypoblast (137). During epiboly, cells from the inner layer intercalate between cells of the outer layer and flatten to dimensions typical in that layer. Hence, both cell intercalation and cell flattening can contribute to epiboly.

Zygotic screens have isolated mutations in only a single locus that affects epiboly (135, 137, 202, 300). Mutations in the adhesion molecule E-cadherin (*half-baked*) severely affect epiboly and arrest the vegetal spreading of deep cells during gastrulation (137, 289). Mutant cells can intercalate but often deintercalate into the deeper layer and do not flatten. These results suggest that E-cadherin is required to bind cells together so that they can form a flattened spread-out layer that envelops the yolk. How epiboly is regulated is still poorly understood, but blocking the T-box transcription factor Eomesodermin or its target *mtx2*, a homeobox gene, blocks epiboly (32).

Consistent with an adhesive role for E-cadherin, more severe depletion using morpholinos instead of *hab* mutants results in the deadhesion of cells and disintegration of the embryo already during early cleavage divisions (14). An additional factor that might be involved in cell (de)adhesion, but not epiboly, is the EGF-CFC protein One-eyed pinhead. During late-blastoderm stages, *one-eyed pinhead* mutant cells appear less motile and more cohesive (332). It remains unclear how this contributes to morphogenesis and if this process is dependent on Nodal signaling.

The isolation of four maternal mutants (*betty boop*, *poky*, *slow*, *bedazzled*) suggests that many components required for epiboly are provided maternally (330). These mutants display premature constriction of the margin (*betty boop*) and slower or delayed epiboly (*poky*, *slow*, *bedazzled*).

Internalization

Internalization describes the process by which mesendodermal precursors located at the margin move inside, resulting in an embryo with an outer epiblast layer and an inner hypoblast layer (159, 333). There has been some debate about the exact cellular mechanism that underlies internalization (2). It was proposed that internalization is caused by involution, the inward flow of a sheet of cells, or ingression, the inward movement of individual cells (36, 53, 77, 214, 285). Imaging and embryological studies suggest that synchronized ingression underlies internalization (2). Cells move coherently toward the margin, where they begin to form protrusions, lose coherence with their neighbors, and ingress. This "flow of individuals" results in internalization.

The molecules that drive internalization are largely unknown. Complete absence of Nodal signaling blocks all internalization movements (36, 78). Conversely, upregulation of the pathway in the absence of Lefty leads to prolonged and increased internalization (77). A single cell that is mutant for the Nodal coreceptor One-eyed pinhead can initially be internalized when placed at the margin of a wild-type embryo but then it egresses (36). This result suggests that the flow of internalizing wild-type cells can carry but not keep neighboring cells inside. Activation of the Nodal pathway in a single cell can lead to the ingression of this cell even when neighboring cells do not internalize (36, 55). This effect might be caused by the differential adhesion between cells that have an active or inactive Nodal signaling pathway (214), but the Nodal downstream genes that mediate internalization are elusive.

Convergence and Extension

Convergence and extension are defined as the narrowing of embryonic tissues mediolaterally (convergence) and their elongation anterioposteriorly (extension). These movements are driven by a number of directed and coordinated cell behaviors that lead to the accumulation of cells on the dorsal side and the formation of an axis (reviewed in 146, 222, 331). Several distinct cell behaviors underlie convergence and extension in zebrafish, including the directed migration of internalized cells toward the animal pole and toward the dorsal side and the mediolateral intercalation of dorsal and lateral cells.

Immediately after internalization, hypoblast cells move away from the margin. This has been best studied on the dorsal side, where prechordal plate (anterior axial mesoderm) precursors migrate anteriorly. It has been proposed that a β-catenin-Stat3-Liv1-Snail1 pathway regulates prechordal plate migration (211, 354, 355). In this model, β-catenin activates an unknown ligand for the JAK/STAT pathway, culminating in the phosphorylation and consequent activation of the transcription factor Stat3. Stat3 then activates the expression of the zinc transporter Liv1. Increased levels of zinc might allow the nuclear accumulation of the zinc finger transcription factor Snail1. It has been proposed that Snail1 might promote the epithelial-to-mesenchymal transition of anterior axial mesoderm cells, but such changes in cell behavior have not been observed in wild-type embryos (2, 214). Further genetic and cell biological studies are required to more thoroughly test the β-catenin-Stat3-Liv1-Snail1 model.

The PDGF/PI3K/PKB pathway has also been implicated in the migration of prechordal mesendoderm precursors (216). Pharmacological block of the PDGF receptor or phosphoinositide 3-kinase inhibits the formation of polarized processes on prechordal plate progenitors and the localization of protein kinase B and F-actin to the leading edge. Despite these defects, cells maintain their directional, albeit slower, migration.

A third pathway regulating prechordal plate migration is the noncanonical Wnt signaling pathway. As discussed in more detail below, this pathway has a key role in polarizing cells in all animals (reviewed in 163). Blocking *wnt11* activity during prechordal cell migration results in slower and less directed movements and the abnormal orientation of cellular protrusions (119, 325). Because the correlation of direction of movement and direction of protrusions is not absolute, it is not yet known if there is a causative link between these two *wnt11*-dependent processes.

The pathways described above ultimately have to regulate cell behavior by modulating cytoskeletal or adhesive properties. Indeed, two proteins implicated in actin dynamics have been implicated in the migration of prechordal plate cells (51). A *cap1* homolog and *quattro*, which encodes a guanine nucleotide exchange factor, are expressed in the anterior mesendoderm. *Quattro* morpholinos disrupt the anteriorly directed convergence and aggregation of prechordal plate, and *cap1* is required for the migration of the aggregated cluster toward the animal pole. These observations identify restricted actin-regulatory molecules in the control of cell movements during gastrulation. In addition, there also appears to be a minor role for E-cadherin in the elongation and migration of dorsal hypoblast cells (214).

A hallmark of gastrulation is the polarity of cell movements and cell shapes. Some of the molecular mechanisms underlying this process appear to be conserved in all animals. In particular, components implicated in planar polarity formation in *Drosophila* are also involved in the control of cell polarization and convergence and extension movements in zebrafish and other vertebrates (reviewed in 163, 298, 331). Several components of the noncanonical Wnt signaling pathway have been identified as convergence and extension genes in zebrafish, including the Wnt signals Wnt5 (*pipetail*), Wnt11 (*silberblick*), and Wnt4; the Wnt receptor Fz2; the putative Wnt coreceptor Glypican4 (*knypek*); and the cytoplasmic signal transducer Dishevelled (119, 152, 201, 249, 309, 320). In addition, modulators of the pathway have been identified, including Van gogh-like 2/Strabismus (*trilobite*) and Prickle (37, 132, 236, 283, 326). Additional components identified in *Drosophila*, such as Flamingo and Diversin, also play important roles in zebrafish (82, 279). Downstream mediators that have been shown to play a role in zebrafish include ROK, Rac, and RhoA (17, 197, 201).

The transparency of the zebrafish embryo has been employed to great effect to study the cellular role of planar polarity genes. The main conclusion from these studies is that planar polarity proteins are required for the proper polarization of cells during directed dorsal migration and mediolateral intercalation (reviewed in 215, 222). In this process, cells intercalate between their medial and lateral neighbors, similar to the cell behavior during frog gastrulation, or migrate directionally first toward the animal pole and then toward the dorsal side, a movement not observed in frogs. Planar polarity signaling regulates both the length-to-width ratio of cells and their orientation with respect to the embryonic axes. This polarization is also thought to be required for the persistent migration of cells. For example, both the length-to-width ratio and mediolateral alignment of paraxial ectodermal cells is reduced in *trilobite* (Van gogh-like 2/Strabismus) mutants (132). Hence, cells are more rounded and more randomly oriented compared with the more elongated and more uniformly oriented wild-type cells. Concomitantly, *trilobite* mutant cells move with reduced net dorsal speed along less direct trajectories when compared with wild-type cells. It has thus been proposed that *trilobite* and other planar cell polarity genes allow for the medial-lateral cell polarization that is required for the persistent dorsal migration of cells along straight paths. The connections between cell behavior and movement are still correlative, but planar cell polarity signaling clearly controls both the shape and movement of cells.

Planar polarity signaling in zebrafish is not only required for the polarization of cells but also controls cell division orientation (95). Epiblast cells in dorsal tissues preferentially divide along the animal-vegetal axis of the embryo. Inhibition of the establishment of this animal-vegetal polarity by blocking *wnt11*, *dishevelled*, or *trilobite* disrupts this orientation and thus reduces the extension of the axis.

Although convergence and extension can be linked, they can also be independent. In *ntl* mutants, convergence but not extension of axial mesoderm is affected (93). In dorsalized *swirl/bmp2b* mutants, convergence of lateral cell populations is reduced, whereas their extension is normal or even increased (221). Moreover, the absence of the polysaccharide hyaluronan blocks the convergence but not the extension of lateral mesoderm (17).

In zebrafish there might also be an attractant on the dorsal side that guides cells. Specifically, misexpression of β-catenin on the ventral side not only induces ectopic dorsal fates but also redirects cells ventrally. Hence, there might be β-catenin–regulated genes that provide the directionality of convergence and extension movements. Although components of the noncanonical Wnt signaling pathway are required for directional movement of cells, they are unlikely to act as chemoattractants in this process. For example, ubiquitous expression of *wnt5* or *wnt11* is able to rescue *wnt11* mutants, arguing against a localized Wnt signal that controls cell polarity and migration (152). Instead, a signal regulated by Stat3 might provide polarity cues (211, 354). As described above, phospho-STAT3 accumulates specifically on the dorsal side in response to β-catenin stabilization. Blocking STAT3 function using morpholinos results in severe reduction of convergence and extension movement. This effect on lateral cells is non-cell autonomous. It has thus been proposed that STAT3 activates an as-yet unidentified factor that guides DV cell polarity. This interpretation is complicated by the fact that convergence and extension are also reduced in the *quattro* and *cap1* morphants described above, probably secondarily to the abnormal migration of anterior axial mesoderm (51). In this case, it might not be the absence of a signal but abnormal morphogenesis of axial mesoderm that impairs convergence and extension.

Despite the central role of planar cell polarity signaling during convergence and extension, several additional molecules have been implicated in gastrulation movements or cell polarity, including Gα12/13 (184), hyaluronan (17), Cyclooxygenase-1 (38), Widerborst [a B′ regulatory subunit of protein phosphatase 2A (111)], Estrogen receptor-related α (19), Scribble 1 (329), Fyn/Yes (134), Nemo-like kinase (318), Ephrins (39, 230), and Slit (359). Most of these factors have been implicated in gastrulation movements based on overexpression or morpholino analysis. Future genetic studies will be required to firmly establish a role for these molecules and to determine how they interact with other factors controlling gastrulation.

THE BIOGRAPHIES OF CELLS

The previous sections discussed how the fates and movements of cells are dependent on their position in the embryo and how signaling pathways, transcription factors, and other molecules influence these decisions. We are now in a position to attempt a synthesis of these observations and describe how different cells receive and interpret these diverse inputs during early development to generate specific cell types and move to specific positions. These "childhood biographies of cells" not only allow us to integrate the findings described above, but they also serve to inform strategies in stem cell research (reviewed in 284). A major application of vertebrate embryology and genetics is to drive multipotent cells to a particular fate for therapeutic purposes. In turn, these in vitro studies provide a critical test of how completely we understand embryonic development.

The Dorsal Margin: Making Prechordal Plate

Prechordal plate progenitors are located at the dorsal margin and become marked as dorsal when β-catenin is stabilized soon after fertilization (see above). After mid-blastula transition, β-catenin activates *sqt* and *cyc* expression in prechordal plate precursors, resulting in the full activation of the Nodal signaling pathway in these cells. In contrast, Bmp and Wnt signaling are suppressed by the β-catenin-mediated activation of Chordin, Dkk1, and other antagonists of Bmp and Wnt signaling, and because β-catenin activates repressors such as Boz, which represses *wnt* and *bmp* gene expression on the dorsal side. β-catenin might also activate transcription factors such as Goosecoid that directly specify dorsal fates. Hence, it might be sufficient to activate β-catenin and full Nodal signaling and block all other signaling pathways to specify prechordal plate precursors. This leads to the activation of prechordal plate-specific genes (e.g., *goosecoid*) and to the internalization and migration of progenitors toward the animal pole. This migration is controlled in part by the STAT3 pathway, Cap1, Quattro, Wnt11 signaling, and PDGF/PI3K signaling.

The Dorsal Margin: Making Notochord

Like prechordal plate progenitors, notochord precursors are initially marked by β-catenin, which activates *sqt* and *cyc* expression next to and potentially in notochord precursors. This results in the partial activation of the Nodal signaling pathway in these cells. Hence, it might be sufficient to activate β-catenin and intermediate levels of Nodal signaling and block all other signaling pathways to generate notochord precursors. This leads to the induction of *flh* and *ntl*, which encode transcription factors that specify notochord identity, and the internalization, convergence, and extension of notochord progenitors. This process is regulated by *ntl* (convergence) and noncanonical Wnt signaling (convergence and extension). According to this model, the level or timing of Nodal signaling is the key factor that distinguishes prechordal plate and notochord progenitors.

The Margin: Making Endoderm

A subset of the cells that are located at the margin become endoderm (reviewed in 231). These cells are exposed to an unknown signal from the yolk syncytial layer and are also likely to contain maternally provided mRNAs encoding transcription factors of unknown identity. Before gastrulation, and shortly after the activation of *sqt* at the dorsal margin, all endodermal precursors express Cyc and Sqt, resulting in the full activation of the Nodal signaling pathway in these cells. Depending on the DV position, Bmp, FGF, or Wnt signaling is also activated in endodermal precursors; these pathways do not influence endodermal fate specification per se, but might modulate the type of endoderm that is formed. Hence, it might be sufficient to fully activate

Nodal signaling to generate endoderm progenitors. This eventually leads to the induction of the transcription factor Casanova, which in conjunction with the transcription factor Spiel ohne grenzen might be sufficient to specify endoderm progenitors. Since high Nodal signaling is involved in both prechordal plate and endoderm specification, additional factors (e.g., β-catenin and its downstream genes) might be required to specifically induce prechordal plate cells. It remains unclear why only some cells at the margin are induced to express *casanova* and form endoderm, whereas neighboring cells form mesoderm.

The Lateral Margin: Making Heart Muscle

Cells at the lateral margin give rise to the cardiomyocytes of the heart (reviewed in 358). Nodal and Bmp signaling are required for this process, and at later stages FGF signaling is also thought to contribute to myocardium formation. These signaling pathways lead to the induction of *nkx2.5*, a marker for cells that can give rise to heart muscle, and downstream genes such as *hand2*. In contrast, retinoic acid signaling limits the number of cells in this region that are selected to form cardiomyocyte progenitors.

The Ventral Margin: Making Blood and Tail Somites

Cells at the ventral margin give rise to multiple cell types, including blood and tail somites. Many signaling pathways are active in this region, including Wnt, Bmp, FGF, and Nodal. Both the Wnt and Bmp pathways are most active in the ventral margin region, and this coincidence is apparently required for proper expression of ventral margin genes such as *tbx6* (311). Conversely, ventral margin cells can induce an ectopic tail, and this activity can be mimicked by the local application of Bmp, Wnt, and Nodal signals (4). Despite this activity, Nodal signals are not required to make tail somites, suggesting that high-level activity of Wnt, Bmp, and potentially FGF signaling may be sufficient to generate tail mesodermal identity. This leads to the activation of downstream transcription factors [e.g., members of the T-box and caudal-related gene families (56, 288)] and planar polarity signaling, which regulate tail morphogenesis (196).

Development of blood also requires Wnt, Bmp, and FGF signals, but in addition is dependent on Nodal signals (reviewed in 57). Indeed, Nodal signals may be a factor in determining why some ventral margin cells form blood while others form tail somites. The *sqt* and *cyc* genes are expressed at the ventral margin just prior to the onset of gastrulation, and mutational analysis shows that Nodal signals are essential for blood but not tail somites. Thus Nodal signals at certain levels or times in development may allocate a subset of ventral margin cells to a blood fate. It is not clear how Nodal signals might drive ventral margin cells toward blood fates, but they could instruct blood progenitors to involute early in gastrulation or trigger the expression of certain target genes that specify blood identity.

Lateral and Ventral Ectoderm: Making Spinal Cord

The precursors of the spinal cord are located laterally and ventrally between the margin and the animal pole. These cells originate distant to the organizer and, in contrast to other neuronal progenitors, do not require organizer-derived inhibitors to be specified. BMP and FGF signaling promote spinal cord development, whereas Nodal signaling counteracts it. Wnt and retinoic acid signaling are also involved in this process by posteriorizing the neuroectoderm. Dorsal spinal cord and neural crest progenitors, which are located more laterally at neural plate stages, appear to be specified by higher levels of Bmp signaling than ventral progenitors, which are located more medially at neural plate stages (21, 226).

The Animal Region: Making Forebrain and Midbrain-Hindbrain Boundary

Cells in the animal dorsal region become forebrain progenitors. It appears that forebrain specification requires the absence of all known signaling pathways. Indeed, blocking Nodal, Bmp, FGF, Wnt, and retinoic acid signaling does not affect forebrain formation and in some cases results in the expansion of forebrain territory. Absence of these signaling pathways is achieved by the absence of the signals that might suppress forebrain formation and the expression of inhibitors of these signaling pathways. For example, Dickkopf and Tlc are both inhibitors of Wnt signaling expressed in the prechordal plate underlying the forebrain territory and at the anterior border of the forebrain region, respectively. These and other factors (Tcf3, Chordin, Noggin, Lefty) inhibit signaling and allow forebrain formation. It is still controversial if forebrain formation is indeed the default state of development. For example, FGF signaling has been proposed to be required for neural induction, including the forebrain, but genetic evidence in zebrafish is not available (304).

The precursors of the midbrain-hindbrain boundary are located dorsally and at an intermediate position between the animal pole and margin. It appears that these cells are induced by lack of Bmp, Nodal, FGF, and retinoic acid signaling but require intermediate levels of Wnt signaling (259).

COMPARATIVE ASPECTS

How applicable are the findings in zebrafish to other vertebrates and vice versa? Vertebrate embryos share a similar body plan, and the fate map of one species can be morphed into the one of another (298). It has therefore been expected that the underlying molecular mechanisms are also shared. The past 10 years have seen dramatic progress in our molecular understanding of zebrafish, frog, chick and mouse embryogenesis, and we can now ask if the molecular mechanisms underlying vertebrate axis formation are conserved. At a superficial level, the answer is yes. First, the same signaling pathways and transcription factors are employed during the early embryogenesis of all vertebrates (reviewed in 59, 112, 192, 298, 303, 313). For example, the Nodal, Bmp, FGF, and Wnt signaling pathways are active in all vertebrates during blastula or gastrula stages, and transcription factors such as homeodomain and T-box proteins are widely employed. Second, interference with these regulators can result in similar phenotypes. For example, lack of Nodal signaling severely compromises mesoderm and endoderm induction (reviewed in 268), loss of FGF signaling affects posterior development (reviewed in 296), and mutations in *Brachyury/ntl* result in notochord defects and posterior truncations in all model vertebrates (reviewed in 297).

Despite these similarities there are also intriguing differences, in particular between frog and fish on one side and mouse on the other. For example, noncanonical Wnt signaling is required for proper gastrulation in fish and frog, but mice that lack components of this pathway display only mild posterior truncations and spina bifida (reviewed in 163, 331). However, in general, the loss-of-function phenotypes of particular signaling pathways are more severe in mouse than in zebrafish or frog (reviewed in 313). For example, β-catenin is required for the formation of dorsal structures in fish and frog but not for mesoderm and endoderm formation (reviewed in 129). In contrast, *β-catenin* mutant mice lack anterior-posterior polarity and do not develop embryonic mesoderm and endoderm (103). Similarly, mouse *wnt3* mutants do not form mesodermal and endodermal progenitors (187). In fish and frog Bmp signaling is necessary for the development of ventral structures but not for mesoderm and endoderm formation (reviewed in 219). In contrast, mouse cells mutant for Bmp receptors cannot form mesoderm and endoderm (210, reviewed in 219). Lack of FGF signaling in frog and fish results in posterior truncation in frog and fish, similar to partial loss-of-function phenotypes in mouse (reviewed in 296). However, in *fgf8* mouse mutants endoderm and mesoderm progenitors cannot gastrulate properly, resulting in the loss of these cell types (310). Hence, there are clear differences between fish/frog and mouse in the requirement for key signaling pathways.

We suggest that these differences arise from the much more pronounced role of reciprocal signaling interactions in mouse than in fish or frog. Although there is some cross-regulation of FGF, Bmp, Wnt, and Nodal signaling in fish, there appears to be a striking interdependence of these signaling pathways in mouse. For example, Bmp, Wnt, and Nodal signals maintain each others' expression before the onset of mouse gastrulation (reviewed in 192, 313). Hence, interference with one pathway will affect the activity of the others, leading to more pronounced phenotypes. In fish and frog embryos, these pathways might not only be more independent but even act redundantly, suppressing potentially more severe phenotypes.

We propose that reciprocal signaling manifests itself more prominently in mouse than in fish or frog because of three major differences in the early embryogenesis of these organisms. First, mouse embryos undergo dramatic growth, whereas the volume of fish and frog embryos does not significantly change until organogenesis. In mouse, the need to coordinate growth and patterning may be met by extensive cross-regulatory interactions among various signaling pathways. This cross-regulation might explain the apparently similar phenotypes in mouse embryos that have mutations in different signaling pathways (reviewed in 313).

Second, mouse embryos require extraembryonic tissues for implantation and anterior-posterior patterning (reviewed in 192, 313). These tissues serve as signaling centers, and they are in turn regulated by different signals and their antagonists. For example, the Nodal signaling pathway patterns the extraembryonic (visceral) endoderm and in turn is regulated by extraembryonic ectoderm (reviewed in 192, 269). There is a role for an extraembryonic structure in the fish (the yolk syncytial layer), but it appears to be less important than mouse extraembryonic tissues and does not apparently require reciprocal signals from embryonic signaling centers. Hence, reciprocal signaling appears essential to coordinate embryonic and extraembryonic development in mouse, whereas embryonic development in fish occurs in the context of extraembryonic structures that are largely pre-established during oogenesis.

Third, fish and frog embryos strongly rely on maternal determinants to guide axis formation. The egg is already polarized, zygotic transcription is only initiated after the 500-cell stage, and in frog maternal factors such as Wnt11, β-catenin, VegT, and Ectodermin are required for axis formation (69, 116, 314, 360). In contrast, mammals appear not to rely on localized maternal determinants and initiate zygotic transcription as early as the two-cell stage. There seem to be asymmetries during early cleavages, but they do not necessarily translate into orientations of specific axes (89, 242). Hence, the mouse embryo has to "self-organize," a process likely to require cross-regulatory interactions between patterning signals.

Taken together, these observations suggest that the dramatic growth, the importance of extraembryonic tissues, and the lack of a prepattern in mouse embryos necessitate complex cross-regulatory interactions between tissues and signaling pathways. Interference with one tissue or pathway can thus have dramatic effects on other tissues or pathways. In contrast, fish and frog eggs contain detailed patterning information, embryos grow little, and extraembryonic tissues play minor roles. In this case, the different signaling pathways are less interdependent and can specify distinct tissue types. These developmental differences might also drive the very rapid development of frog and fish embryos compared with the relatively slow development of mouse embryos.

PERSPECTIVES

After 15 years of zebrafish molecular genetics we have attained a basic outline of the molecular bases of zebrafish axis formation, but many important questions remain. First, many of the key components involved in axis formation are not yet identified. For example, we lack any systematic knowledge of the maternal factors that contribute to vertebrate embryogenesis; RNAi experiments in other systems suggest that many modulators of specific signaling pathways are still unidentified (54); moreover, it is almost completely unknown which genes are regulated by the signals and transcription factors that set up the vertebrate body plan. Maternal or sensitized screens, reverse genetics, RNAi or morpholinos, small molecule inhibitors, and microarray experiments are likely to lead to the isolation of additional factors involved in axis formation. We speculate, however, that it is unlikely that many new signaling pathways required for axis formation will be identified. The genetic screens in zebrafish have by now reached at least 50% saturation and have not isolated any novel signaling pathways required for axis formation, although these screens did define new roles for and new modulators of known pathways (e.g., 102, 141, 348). Similarly, recent misexpression screens in *Xenopus* have not identified novel signaling pathways. It is conceivable that signaling pathways that have more subtle roles [e.g., during gastrulation movements (17, 38, 184)] remain to be discovered, but we predict that most progress will be made in the isolation of maternal upstream factors, signaling modulators, and downstream mediators.

Second, our understanding of the cell biological and molecular bases of vertebrate embryogenesis is still poor. How do signals move through the embryo? How do cells read and respond to these extracellular inputs over time? How are cytoskeletal and adhesive properties changed in response to specific signals? How is chromatin modified as cells become specified? What are the subcellular changes when progenitors become neurons, muscles, blood, and other cell types? The development of probes that allow the in vivo imaging of subcellular processes promises a detailed cell biological and dynamic view of cell movements and differentiation (reviewed in 203). The transparency of the zebrafish makes this organism particularly well suited to address these questions.

Ultimately, we need to understand how all these inputs are integrated into regulatory hierarchies and networks (reviewed in 183). It has become clear that individual cells receive multiple and diverse inputs depending on their position and history, but we are largely ignorant about how these inputs are translated into flexible but ultimately robust outcomes. This knowledge will not only provide a basis to understand human birth defects and guide our efforts in stem cell manipulations but might also uncover the regulatory logic that drives vertebrate embryogenesis.

SUMMARY POINTS

1. A combination of embryological, genetic, and molecular approaches has provided an outline of the molecular basis of zebrafish axis formation.
2. During oogenesis the animal-vegetal axis is specified and dorsal determinants are deposited into the egg.
3. Nodal, FGF, Bmp, Wnt, and retinoic acid signals provide positional information and activate transcription factors that specify cell fates during gastrulation.

4. Planar cell polarity signaling is required for the gastrulation movements of convergence extension.

FUTURE ISSUES

1. Identify additional factors involved in zebrafish axis formation and gastrulation.
2. Develop in vivo probes to study the subcellular and molecular basis of zebrafish embryogenesis.
3. Determine how cells integrate multiple inputs to acquire specific fates and movements.

ACKNOWLEDGMENTS

We thank Brian Ciruna, Carl Philipp Heisenberg, David Kimelman, Liz Robertson, Lila Solnica-Krezel and Debbie Yelon for helpful discussions or comments on the manuscript; Ian Woods for preparing the figures; past and present members of the Talbot and Schier labs for their contributions; and the National Institutes of Health, American Heart Association, McKnight Endowment for Neuroscience, Human Frontiers Science Program, Irma T. Hirschl Trust, Pew Scholars Program, and the Rita Allen Foundation for support.

LITERATURE CITED

1. Abdelilah S, Solnica-Krezel L, Stainier DY, Driever W. 1994. Implications for dorsoventral axis determination from the zebrafish mutation *janus*. *Nature* 370:468–71
2. Adams RJ, Kimmel CB. 2004. Morphogenetic cellular flows during zebrafish gastrulation. See Ref. 303a, pp. 305–16
3. Agathon A, Thisse B, Thisse C. 2001. Morpholino knock-down of *antivin1* and *antivin2* upregulates nodal signaling. *Genesis* 30:178–82
4. Agathon A, Thisse C, Thisse B. 2003. The molecular nature of the zebrafish tail organizer. *Nature* 424:448–52
5. Alexander J, Rothenberg M, Henry GL, Stainier DY. 1999. *Casanova* plays an early and essential role in endoderm formation in zebrafish. *Dev. Biol.* 215:343–57
6. Alexander J, Stainier DY. 1999. A molecular pathway leading to endoderm formation in zebrafish. *Curr. Biol.* 9:1147–57
7. Amacher SL, Draper BW, Summers BR, Kimmel CB. 2002. The zebrafish T-box genes *no tail* and *spadetail* are required for development of trunk and tail mesoderm and medial floor plate. *Development* 129:3311–23
8. Amacher SL, Kimmel CB. 1998. Promoting notochord fate and repressing muscle development in zebrafish axial mesoderm. *Development* 125:1397–406
9. Ambros V. 2004. The functions of animal microRNAs. *Nature* 431:350–55
10. Aoki TO, David NB, Minchiotti G, Saint-Etienne L, Dickmeis T, et al. 2002. Molecular integration of *casanova* in the Nodal signalling pathway controlling endoderm formation. *Development* 129:275–86
11. Aoki TO, Mathieu J, Saint-Etienne L, Rebagliati MR, Peyrieras N, Rosa FM. 2002. Regulation of nodal signalling and mesendoderm formation by TARAM-A, a TGFβ-related type I receptor. *Dev. Biol.* 241:273–88

12. Appel B, Fritz A, Westerfield M, Grunwald DJ, Eisen JS, Riley BB. 1999. Delta-mediated specification of midline cell fates in zebrafish embryos. *Curr. Biol.* 9:247–56
13. Appel B, Marasco P, McClung LE, Latimer AJ. 2003. *lunatic fringe* regulates Delta-Notch induction of hypochord in zebrafish. *Dev. Dyn.* 228:281–86
14. Babb SG, Marrs JA. 2004. E-cadherin regulates cell movements and tissue formation in early zebrafish embryos. *Dev. Dyn.* 230:263–77
15. Bakkers J, Camacho-Carvajal M, Nowak M, Kramer C, Danger B, Hammerschmidt M. 2005. Destabilization of ΔNp63α by Nedd4-mediated ubiquitination and Ubc9-mediated sumoylation, and its implications on dorsoventral patterning of the zebrafish embryo. *Cell Cycle* 4:790–800
16. Bakkers J, Hild M, Kramer C, Furutani-Seiki M, Hammerschmidt M. 2002. Zebrafish *ΔNp63* is a direct target of Bmp signaling and encodes a transcriptional repressor blocking neural specification in the ventral ectoderm. *Dev. Cell* 2:617–27
17. Bakkers J, Kramer C, Pothof J, Quaedvlieg NE, Spaink HP, Hammerschmidt M. 2004. Has2 is required upstream of Rac1 to govern dorsal migration of lateral cells during zebrafish gastrulation. *Development* 131:525–37
18. Bally-Cuif L, Dubois L, Vincent A. 1998. Molecular cloning of *Zcoe2*, the zebrafish homolog of *Xenopus Xcoe2* and mouse *EBF-2*, and its expression during primary neurogenesis. *Mech. Dev.* 77:85–90
19. Bardet PL, Horard B, Laudet V, Vanacker JM. 2005. The ERRα orphan nuclear receptor controls morphogenetic movements during zebrafish gastrulation. *Dev. Biol.* 281:102–11
20. Bartel DP. 2004. MicroRNAs: genomics, biogenesis, mechanism, and function. *Cell* 116:281–97
21. Barth KA, Kishimoto Y, Rohr KB, Seydler C, Schulte-Merker S, Wilson SW. 1999. Bmp activity establishes a gradient of positional information throughout the entire neural plate. *Development* 126:4977–87
22. Bauer H, Lele Z, Rauch GJ, Geisler R, Hammerschmidt M. 2001. The type I serine/threonine kinase receptor Alk8/Lost-a-fin is required for Bmp2b/7 signal transduction during dorsoventral patterning of the zebrafish embryo. *Development* 128:849–58
23. Baxendale S, Davison C, Muxworthy C, Wolff C, Ingham PW, Roy S. 2004. The B-cell maturation factor Blimp-1 specifies vertebrate slow-twitch muscle fiber identity in response to Hedgehog signaling. *Nat. Genet.* 36:88–93
24. Begemann G, Schilling TF, Rauch GJ, Geisler R, Ingham PW. 2001. The zebrafish *neckless* mutation reveals a requirement for *raldh2* in mesodermal signals that pattern the hindbrain. *Development* 128:3081–94
25. Belting HG, Hauptmann G, Meyer D, Abdelilah-Seyfried S, Chitnis A, et al. 2001. *spiel ohne grenzen/pou2* is required during establishment of the zebrafish midbrain-hindbrain boundary organizer. *Development* 128:4165–76
26. Birely J, Schneider VA, Santana E, Dosch R, Wagner DS, et al. 2005. Genetic screens for genes controlling motor nerve-muscle development and interactions. *Dev. Biol.* 280:162–76
27. Bisgrove BW, Essner JJ, Yost HJ. 1999. Regulation of midline development by antagonism of *lefty* and *nodal* signaling. *Development* 126:3253–62
28. Blader P, Rastegar S, Fischer N, Strähle U. 1997. Cleavage of the BMP-4 antagonist chordin by zebrafish tolloid. *Science* 278:1937–40

29. Blagden CS, Currie PD, Ingham PW, Hughes SM. 1997. Notochord induction of zebrafish slow muscle mediated by Sonic hedgehog. *Genes Dev.* 11:2163–75
30. Brand M, Heisenberg CP, Jiang YJ, Beuchle D, Lun K, et al. 1996. Mutations in zebrafish genes affecting the formation of the boundary between midbrain and hindbrain. *Development* 123:179–90
31. Brand M, Heisenberg CP, Warga RM, Pelegri F, Karlstrom RO, et al. 1996. Mutations affecting development of the midline and general body shape during zebrafish embryogenesis. *Development* 123:129–42
32. Bruce AE, Howley C, Fox MD, Ho RK. 2005. T-box gene *eomesodermin* and the homeobox-containing Mix/Bix gene *mtx2* regulate epiboly movements in the zebrafish. *Dev. Dyn.* 233:105–14
33. Bruce AE, Howley C, Zhou Y, Vickers SL, Silver LM, et al. 2003. The maternally expressed zebrafish T-box gene *eomesodermin* regulates organizer formation. *Development* 130:5503–17
34. Burgess S, Reim G, Chen W, Hopkins N, Brand M. 2002. The zebrafish *spiel-ohne-grenzen (spg)* gene encodes the POU domain protein Pou2 related to mammalian Oct4 and is essential for formation of the midbrain and hindbrain, and for pre-gastrula morphogenesis. *Development* 129:905–16
35. Cao Y, Zhao J, Sun Z, Zhao Z, Postlethwait J, Meng A. 2004. *fgf17b*, a novel member of Fgf family, helps patterning zebrafish embryos. *Dev. Biol.* 271:130–43
36. Carmany-Rampey A, Schier AF. 2001. Single-cell internalization during zebrafish gastrulation. *Curr. Biol.* 11:1261–65
37. Carreira-Barbosa F, Concha ML, Takeuchi M, Ueno N, Wilson SW, Tada M. 2003. Prickle 1 regulates cell movements during gastrulation and neuronal migration in zebrafish. *Development* 130:4037–46
38. Cha YI, Kim SH, Solnica-Krezel L, Dubois RN. 2005. Cyclooxygenase-1 signaling is required for vascular tube formation during development. *Dev. Biol.* 282:274–83
39. Chan J, Mably JD, Serluca FC, Chen JN, Goldstein NB, et al. 2001. Morphogenesis of prechordal plate and notochord requires intact Eph/ephrin B signaling. *Dev. Biol.* 234:470–82
40. Chang C, Holtzman DA, Chau S, Chickering T, Woolf EA, et al. 2001. Twisted gastrulation can function as a BMP antagonist. *Nature* 410:483–87
41. Chen C, Shen MM. 2004. Two modes by which Lefty proteins inhibit nodal signaling. *Curr. Biol.* 14:618–24
42. Chen JN, Haffter P, Odenthal J, Vogelsang E, Brand M, et al. 1996. Mutations affecting the cardiovascular system and other internal organs in zebrafish. *Development* 123:293–302
43. Chen JN, van Eeden FJ, Warren KS, Chin A, Nusslein-Volhard C, et al. 1997. Left-right pattern of cardiac *BMP4* may drive asymmetry of the heart in zebrafish. *Development* 124:4373–82
44. Chen S, Kimelman D. 2000. The role of the yolk syncytial layer in germ layer patterning in zebrafish. *Development* 127:4681–89
45. **Chen Y, Schier AF. 2001. The zebrafish Nodal signal squint functions as a morphogen. *Nature* 411:607–10**

In vivo demonstration that a Nodal signal can act at a long range.

46. Chen Y, Schier AF. 2002. Lefty proteins are long-range inhibitors of squint-mediated nodal signaling. *Curr. Biol.* 12:2124–28
47. Cheng JC, Miller AL, Webb SE. 2004. Organization and function of microfilaments during late epiboly in zebrafish embryos. *Dev. Dyn.* 231:313–23

48. Cheng SK, Olale F, Brivanlou AH, Schier AF. 2004. Lefty blocks a subset of TGFβ signals by antagonizing EGF-CFC coreceptors. *PLoS Biol.* 2:e30
49. Connors SA, Trout J, Ekker M, Mullins MC. 1999. The role of *tolloid/mini fin* in dorsoventral pattern formation of the zebrafish embryo. *Development* 126:3119–30
50. Cooper MS, D'Amico LA. 1996. A cluster of noninvoluting endocytic cells at the margin of the zebrafish blastoderm marks the site of embryonic shield formation. *Dev. Biol.* 180:184–98
51. Daggett DF, Boyd CA, Gautier P, Bryson-Richardson RJ, Thisse C, et al. 2004. Developmentally restricted actin-regulatory molecules control morphogenetic cell movements in the zebrafish gastrula. *Curr. Biol.* 14:1632–38
52. Dale L, Howes G, Price BM, Smith JC. 1992. Bone morphogenetic protein 4: a ventralizing factor in early *Xenopus* development. *Development* 115:573–85
53. D'Amico LA, Cooper MS. 2001. Morphogenetic domains in the yolk syncytial layer of axiating zebrafish embryos. *Dev. Dyn.* 222:611–24
54. DasGupta R, Kaykas A, Moon RT, Perrimon N. 2005. Functional genomic analysis of the Wnt-Wingless signaling pathway. *Science* 308:826–33
55. David NB, Rosa FM. 2001. Cell autonomous commitment to an endodermal fate and behaviour by activation of Nodal signalling. *Development* 128:3937–47
56. Davidson AJ, Ernst P, Wang Y, Dekens MP, Kingsley PD, et al. 2003. *cdx4* mutants fail to specify blood progenitors and can be rescued by multiple *hox* genes. *Nature* 425:300–6
57. Davidson AJ, Zon LI. 2004. The 'definitive' (and 'primitive') guide to zebrafish hematopoiesis. *Oncogene* 23:7233–46
58. Dekens MP, Pelegri FJ, Maischein HM, Nusslein-Volhard C. 2003. The maternal-effect gene *futile cycle* is essential for pronuclear congression and mitotic spindle assembly in the zebrafish zygote. *Development* 130:3907–16
59. De Robertis EM, Kuroda H. 2004. Dorsal-ventral patterning and neural induction in *Xenopus* embryos. *Annu. Rev. Cell. Dev. Biol.* 20:285–308
60. Dick A, Hild M, Bauer H, Imai Y, Maifeld H, et al. 2000. Essential role of Bmp7 (*snailhouse*) and its prodomain in dorsoventral patterning of the zebrafish embryo. *Development* 127:343–54
61. Dickmeis T, Mourrain P, Saint-Etienne L, Fischer N, Aanstad P, et al. 2001. A crucial component of the endoderm formation pathway, CASANOVA, is encoded by a novel *sox*-related gene. *Genes Dev.* 15:1487–92
62. Dorsky RI, Itoh M, Moon RT, Chitnis A. 2003. Two *tcf3* genes cooperate to pattern the zebrafish brain. *Development* 130:1937–47
63. Dorsky RI, Sheldahl LC, Moon RT. 2002. A transgenic Lef1/β-catenin-dependent reporter is expressed in spatially restricted domains throughout zebrafish development. *Dev. Biol.* 241:229–37
64. Dosch R, Gawantka V, Delius H, Blumenstock C, Niehrs C. 1997. Bmp-4 acts as a morphogen in dorsoventral mesoderm patterning in *Xenopus*. *Development* 124:2325–34
65. Dosch R, Wagner DS, Mintzer KA, Runke G, Wiemelt AP, Mullins MC. 2004. Maternal control of vertebrate development before the midblastula transition: mutants from the zebrafish I. *Dev. Cell* 6:771–80
66. Dougan ST, Warga RM, Kane DA, Schier AF, Talbot WS. 2003. The role of the zebrafish *nodal*-related genes *squint* and *cyclops* in patterning of mesendoderm. *Development* 130:1837–51

67. Draper BW, Stock DW, Kimmel CB. 2003. Zebrafish *fgf24* functions with *fgf8* to promote posterior mesodermal development. *Development* 130:4639–54
68. **Driever W, Solnica-Krezel L, Schier AF, Neuhauss SC, Malicki J, et al. 1996. A genetic screen for mutations affecting embryogenesis in zebrafish. *Development* 123:37–46**

Summary of the large-scale Boston screen for zebrafish mutants.

69. Dupont S, Zacchigna L, Cordenonsi M, Soligo S, Adorno M, et al. 2005. Germ-layer specification and control of cell growth by Ectodermin, a Smad4 ubiquitin ligase. *Cell* 121:87–99
70. Emoto Y, Wada H, Okamoto H, Kudo A, Imai Y. 2005. Retinoic acid-metabolizing enzyme Cyp26a1 is essential for determining territories of hindbrain and spinal cord in zebrafish. *Dev. Biol.* 278:415–27
71. Erter CE, Solnica-Krezel L, Wright CV. 1998. *Zebrafish nodal-related 2* encodes an early mesendodermal inducer signaling from the extraembryonic yolk syncytial layer. *Dev. Biol.* 204:361–72
72. Erter CE, Wilm TP, Basler N, Wright CV, Solnica-Krezel L. 2001. Wnt8 is required in lateral mesendodermal precursors for neural posteriorization in vivo. *Development* 128:3571–83
73. Essner JJ, Amack JD, Nyholm MK, Harris EB, Yost HJ. 2005. Kupffer's vesicle is a ciliated organ of asymmetry in the zebrafish embryo that initiates left-right development of the brain, heart and gut. *Development* 132:1247–60
74. Essner JJ, Vogan KJ, Wagner MK, Tabin CJ, Yost HJ, Brueckner M. 2002. Conserved function for embryonic nodal cilia. *Nature* 418:37–38
75. Fekany K, Yamanaka Y, Leung T, Sirotkin HI, Topczewski J, et al. 1999. The zebrafish *bozozok* locus encodes Dharma, a homeodomain protein essential for induction of gastrula organizer and dorsoanterior embryonic structures. *Development* 126:1427–38
76. Fekany-Lee K, Gonzalez E, Miller-Bertoglio V, Solnica-Krezel L. 2000. The homeobox gene *bozozok* promotes anterior neuroectoderm formation in zebrafish through negative regulation of BMP2/4 and Wnt pathways. *Development* 127:2333–45
77. Feldman B, Concha ML, Saude L, Parsons MJ, Adams RJ, et al. 2002. Lefty antagonism of Squint is essential for normal gastrulation. *Curr. Biol.* 12:2129–35
78. Feldman B, Dougan ST, Schier AF, Talbot WS. 2000. Nodal-related signals establish mesendodermal fate and trunk neural identity in zebrafish. *Curr. Biol.* 10:531–34
79. **Feldman B, Gates MA, Egan ES, Dougan ST, Rennebeck G, et al. 1998. Zebrafish organizer development and germ-layer formation require nodal-related signals. *Nature* 395:181–85**

Discovery that Nodal signals are essential for mesoderm and endoderm induction in zebrafish.

80. Ferreiro B, Artinger M, Cho K, Niehrs C. 1998. Antimorphic goosecoids. *Development* 125:1347–59
81. Fisher S, Amacher SL, Halpern ME. 1997. Loss of *cerebum* function ventralizes the zebrafish embryo. *Development* 124:1301–11
82. Formstone CJ, Mason I. 2005. Combinatorial activity of Flamingo proteins directs convergence and extension within the early zebrafish embryo via the planar cell polarity pathway. *Dev. Biol.* 282:320–35
83. Fujii R, Yamashita S, Hibi M, Hirano T. 2000. Asymmetric p38 activation in zebrafish: its possible role in symmetric and synchronous cleavage. *J. Cell Biol.* 150:1335–48
84. Furthauer M, Lin W, Ang SL, Thisse B, Thisse C. 2002. Sef is a feedback-induced antagonist of Ras/MAPK-mediated FGF signalling. *Nat. Cell Biol.* 4:170–74
85. Furthauer M, Reifers F, Brand M, Thisse B, Thisse C. 2001. *sprouty4* acts in vivo as a feedback-induced antagonist of FGF signaling in zebrafish. *Development* 128:2175–86

86. Furthauer M, Thisse C, Thisse B. 1997. A role for FGF-8 in the dorsoventral patterning of the zebrafish gastrula. *Development* 124:4253–64
87. Furthauer M, Van Celst J, Thisse C, Thisse B. 2004. Fgf signalling controls the dorsoventral patterning of the zebrafish embryo. *Development* 131:2853–64
88. Gaiano N, Amsterdam A, Kawakami K, Allende M, Becker T, Hopkins N. 1996. Insertional mutagenesis and rapid cloning of essential genes in zebrafish. *Nature* 383:829–32
89. Gardner RL, Davies TJ. 2003. The basis and significance of pre-patterning in mammals. *Philos. Trans. R. Soc. London Ser. B* 358:1331–38; discussion 8–9
90. Gates MA, Kim L, Egan ES, Cardozo T, Sirotkin HI, et al. 1999. A genetic linkage map for zebrafish: comparative analysis and localization of genes and expressed sequences. *Genome Res.* 9:334–47
91. Geisler R, Rauch GJ, Baier H, van Bebber F, Bross L, et al. 1999. A radiation hybrid map of the zebrafish genome. *Nat. Genet.* 23:86–89
92. Giraldez AJ, Cinalli RM, Glasner ME, Enright AJ, Thomson JM, et al. 2005. MicroRNAs regulate brain morphogenesis in zebrafish. *Science* 308:833–38
93. Glickman NS, Kimmel CB, Jones MA, Adams RJ. 2003. Shaping the zebrafish notochord. *Development* 130:873–87
94. Golling G, Amsterdam A, Sun Z, Antonelli M, Maldonado E, et al. 2002. Insertional mutagenesis in zebrafish rapidly identifies genes essential for early vertebrate development. *Nat. Genet.* 31:135–40
95. Gong Y, Mo C, Fraser SE. 2004. Planar cell polarity signalling controls cell division orientation during zebrafish gastrulation. *Nature* 430:689–93
96. Gonzalez EM, Fekany-Lee K, Carmany-Rampey A, Erter C, Topczewski J, et al. 2000. Head and trunk in zebrafish arise via coinhibition of BMP signaling by *bozozok* and *chordino*. *Genes Dev.* 14:3087–92
97. Grandel H, Lun K, Rauch GJ, Rhinn M, Piotrowski T, et al. 2002. Retinoic acid signalling in the zebrafish embryo is necessary during pre-segmentation stages to pattern the anterior-posterior axis of the CNS and to induce a pectoral fin bud. *Development* 129:2851–65
98. Griffin K, Patient R, Holder N. 1995. Analysis of FGF function in normal and *no tail* zebrafish embryos reveals separate mechanisms for formation of the trunk and the tail. *Development* 121:2983–94
99. Griffin KJ, Amacher SL, Kimmel CB, Kimelman D. 1998. Molecular identification of *spadetail*: regulation of zebrafish trunk and tail mesoderm formation by T-box genes. *Development* 125:3379–88
100. Griffin KJ, Kimelman D. 2003. Interplay between FGF, *one-eyed pinhead*, and T-box transcription factors during zebrafish posterior development. *Dev. Biol.* 264:456–66
101. Gritsman K, Talbot WS, Schier AF. 2000. Nodal signaling patterns the organizer. *Development* 127:921–32
102. **Gritsman K, Zhang J, Cheng S, Heckscher E, Talbot WS, Schier AF. 1999. The EGF-CFC protein one-eyed pinhead is essential for nodal signaling. *Cell* 97:121–32**

Together with (361) discovery of EGF-CFC proteins as extracellular factors required for Nodal signaling.

103. Haegel H, Larue L, Ohsugi M, Fedorov L, Herrenknecht K, Kemler R. 1995. Lack of β-catenin affects mouse development at gastrulation. *Development* 121:3529–37
104. **Haffter P, Granato M, Brand M, Mullins MC, Hammerschmidt M, et al. 1996. The identification of genes with unique and essential functions in the development of the zebrafish, *Danio rerio*. *Development* 123:1–36**

Summary of the large-scale Tuebingen screen for zebrafish mutants.

105. Halpern ME, Hatta K, Amacher SL, Talbot WS, Yan YL, et al. 1997. Genetic interactions in zebrafish midline development. *Dev. Biol.* 187:154–70
106. Halpern ME, Ho RK, Walker C, Kimmel CB. 1993. Induction of muscle pioneers and floor plate is distinguished by the zebrafish *no tail* mutation. *Cell* 75:99–111
107. Halpern ME, Thisse C, Ho RK, Thisse B, Riggleman B, et al. 1995. Cell-autonomous shift from axial to paraxial mesodermal development in zebrafish *floating head* mutants. *Development* 121:4257–64
108. Hammerschmidt M, Pelegri F, Mullins MC, Kane DA, Brand M, et al. 1996. Mutations affecting morphogenesis during gastrulation and tail formation in the zebrafish, *Danio rerio*. *Development* 123:143–51
109. Hammerschmidt M, Pelegri F, Mullins MC, Kane DA, van Eeden FJ, et al. 1996. *dino* and *mercedes*, two genes regulating dorsal development in the zebrafish embryo. *Development* 123:95–102
110. Hammerschmidt M, Serbedzija GN, McMahon AP. 1996. Genetic analysis of dorsoventral pattern formation in the zebrafish: requirement of a BMP-like ventralizing activity and its dorsal repressor. *Genes Dev.* 10:2452–61
111. Hannus M, Feiguin F, Heisenberg CP, Eaton S. 2002. Planar cell polarization requires Widerborst, a B′ regulatory subunit of protein phosphatase 2A. *Development* 129:3493–503
112. Harland R, Gerhart J. 1997. Formation and function of Spemann's organizer. *Annu. Rev. Cell. Dev. Biol.* 13:611–67
113. Hashimoto H, Itoh M, Yamanaka Y, Yamashita S, Shimizu T, et al. 2000. Zebrafish Dkk1 functions in forebrain specification and axial mesendoderm formation. *Dev. Biol.* 217:138–52
114. Hashimoto H, Rebagliati M, Ahmad N, Muraoka O, Kurokawa T, et al. 2004. The Cerberus/Dan-family protein Charon is a negative regulator of Nodal signaling during left-right patterning in zebrafish. *Development* 131:1741–53
115. Hatta K, Kimmel CB, Ho RK, Walker C. 1991. The *cyclops* mutation blocks specification of the floor plate of the zebrafish central nervous system. *Nature* 350:339–41
116. Heasman J, Crawford A, Goldstone K, Garner-Hamrick P, Gumbiner B, et al. 1994. Overexpression of cadherins and underexpression of β-catenin inhibit dorsal mesoderm induction in early *Xenopus* embryos. *Cell* 79:791–803
117. Heisenberg CP, Houart C, Take-Uchi M, Rauch GJ, Young N, et al. 2001. A mutation in the Gsk3-binding domain of zebrafish Masterblind/Axin1 leads to a fate transformation of telencephalon and eyes to diencephalon. *Genes Dev.* 15:1427–34
118. Heisenberg CP, Nusslein-Volhard C. 1997. The function of *silberblick* in the positioning of the eye anlage in the zebrafish embryo. *Dev. Biol.* 184:85–94
119. Heisenberg CP, Tada M, Rauch GJ, Saude L, Concha ML, et al. 2000. Silberblick/Wnt11 mediates convergent extension movements during zebrafish gastrulation. *Nature* 405:76–81

Discovery that non-canonical Wnt signaling is required for proper gastrulation movements.

120. Helde KA, Wilson ET, Cretekos CJ, Grunwald DJ. 1994. Contribution of early cells to the fate map of the zebrafish gastrula. *Science* 265:517–20
121. Hemmati-Brivanlou A, Melton D. 1997. Vertebrate embryonic cells will become nerve cells unless told otherwise. *Cell* 88:13–17
122. Hernandez RE, Rikhof HA, Bachmann R, Moens CB. 2004. *vhnf1* integrates global RA patterning and local FGF signals to direct posterior hindbrain development in zebrafish. *Development* 131:4511–20

123. Hernandez-Lagunas L, Choi IF, Kaji T, Simpson P, Hershey C, et al. 2005. Zebrafish *narrowminded* disrupts the transcription factor *prdm1* and is required for neural crest and sensory neuron specification. *Dev. Biol.* 278:347–57
124. Hild M, Dick A, Rauch GJ, Meier A, Bouwmeester T, et al. 1999. The *smad5* mutation *somitabun* blocks Bmp2b signaling during early dorsoventral patterning of the zebrafish embryo. *Development* 126:2149–59
125. Ho RK, Kane DA. 1990. Cell-autonomous action of zebrafish *spt-1* mutation in specific mesodermal precursors. *Nature* 348:728–30
126. Ho RK, Kimmel CB. 1993. Commitment of cell fate in the early zebrafish embryo. *Science* 261:109–11
127. Houart C, Caneparo L, Heisenberg C, Barth K, Take-Uchi M, Wilson S. 2002. Establishment of the telencephalon during gastrulation by local antagonism of Wnt signaling. *Neuron* 35:255–65
128. Houart C, Westerfield M, Wilson SW. 1998. A small population of anterior cells patterns the forebrain during zebrafish gastrulation. *Nature* 391:788–92
129. Houston DW, Wylie C. 2004. The role of Wnts in gastrulation. See Ref. 303a, pp. 521–38
130. Hukriede NA, Joly L, Tsang M, Miles J, Tellis P, et al. 1999. Radiation hybrid mapping of the zebrafish genome. *Proc. Natl. Acad. Sci. USA* 96:9745–50
131. Imai Y, Gates MA, Melby AE, Kimelman D, Schier AF, Talbot WS. 2001. The homeobox genes *vox* and *vent* are redundant repressors of dorsal fates in zebrafish. *Development* 128:2407–20
132. Jessen JR, Topczewski J, Bingham S, Sepich DS, Marlow F, et al. 2002. Zebrafish *trilobite* identifies new roles for Strabismus in gastrulation and neuronal movements. *Nat. Cell Biol.* 4:610–15
133. Jesuthasan S, Stahle U. 1997. Dynamic microtubules and specification of the zebrafish embryonic axis. *Curr. Biol.* 7:31–42
134. Jopling C, den Hertog J. 2005. Fyn/Yes and non-canonical Wnt signalling converge on RhoA in vertebrate gastrulation cell movements. *EMBO Rep.* 6:426–31
135. Kane DA, Hammerschmidt M, Mullins MC, Maischein HM, Brand M, et al. 1996. The zebrafish epiboly mutants. *Development* 123:47–55
136. Kane DA, Kimmel CB. 1993. The zebrafish midblastula transition. *Development* 119:447–56
137. Kane DA, McFarland KN, Warga RM. 2005. Mutations in *half baked*/E-cadherin block cell behaviors that are necessary for teleost epiboly. *Development* 132:1105–16
138. Kane DA, Warga RM. 2004. Teleost gastrulation. See Ref. 303a, pp. 157–70
139. Kawahara A, Wilm T, Solnica-Krezel L, Dawid IB. 2000. Antagonistic role of *vega1* and *bozozok/dharma* homeobox genes in organizer formation. *Proc. Natl. Acad. Sci. USA* 97:12121–26
140. Kawahara A, Wilm T, Solnica-Krezel L, Dawid IB. 2000. Functional interaction of *vega2* and *goosecoid* homeobox genes in zebrafish. *Genesis* 28:58–67
141. Kawakami A, Nojima Y, Toyoda A, Takahoko M, Satoh M, et al. 2005. The zebrafish-secreted matrix protein You/Scube2 is implicated in long-range regulation of hedgehog signaling. *Curr. Biol.* 15:480–88
142. Kawakami Y, Raya A, Raya RM, Rodriguez-Esteban C, Belmonte JC. 2005. Retinoic acid signalling links left-right asymmetric patterning and bilaterally symmetric somitogenesis in the zebrafish embryo. *Nature* 435:165–71

143. Kawano Y, Kypta R. 2003. Secreted antagonists of the Wnt signalling pathway. *J. Cell Sci.* 116:2627–34
144. Keegan BR, Feldman JL, Begemann G, Ingham PW, Yelon D. 2005. Retinoic acid signaling restricts the cardiac progenitor pool. *Science* 307:247–49
145. Keegan BR, Meyer D, Yelon D. 2004. Organization of cardiac chamber progenitors in the zebrafish blastula. *Development* 131:3081–91
146. Keller R. 2002. Shaping the vertebrate body plan by polarized embryonic cell movements. *Science* 298:1950–54
147. Kelly C, Chin AJ, Leatherman JL, Kozlowski DJ, Weinberg ES. 2000. Maternally controlled β-catenin-mediated signaling is required for organizer formation in the zebrafish. *Development* 127:3899–911
148. Kelly GM, Erezyilmaz DF, Moon RT. 1995. Induction of a secondary embryonic axis in zebrafish occurs following the overexpression of β-catenin. *Mech. Dev.* 53:261–73
149. Kelly GM, Greenstein P, Erezyilmaz DF, Moon RT. 1995. Zebrafish *wnt8* and *wnt8b* share a common activity but are involved in distinct developmental pathways. *Development* 121:1787–99
150. Kikuchi Y, Agathon A, Alexander J, Thisse C, Waldron S, et al. 2001. *casanova* encodes a novel Sox-related protein necessary and sufficient for early endoderm formation in zebrafish. *Genes Dev.* 15:1493–505
151. Kikuchi Y, Trinh LA, Reiter JF, Alexander J, Yelon D, Stainier DY. 2000. The zebrafish *bonnie and clyde* gene encodes a Mix family homeodomain protein that regulates the generation of endodermal precursors. *Genes Dev.* 14:1279–89
152. Kilian B, Mansukoski H, Barbosa FC, Ulrich F, Tada M, Heisenberg CP. 2003. The role of Ppt/Wnt5 in regulating cell shape and movement during zebrafish gastrulation. *Mech. Dev.* 120:467–76
153. Kim CH, Oda T, Itoh M, Jiang D, Artinger KB, et al. 2000. Repressor activity of Headless/Tcf3 is essential for vertebrate head formation. *Nature* 407:913–16
154. Kimmel CB. 1989. Genetics and early development of zebrafish. *Trends. Genet.* 5:283–88

This paper spelled out the experimental advantages and potential of the zebrafish as a vertebrate model system, inspiring other investigators to join the zebrafish field.

155. Kimmel CB, Ballard WW, Kimmel SR, Ullmann B, Schilling TF. 1995. Stages of embryonic development of the zebrafish. *Dev. Dyn.* 203:253–310
156. Kimmel CB, Kane DA, Walker C, Warga RM, Rothman MB. 1989. A mutation that changes cell movement and cell fate in the zebrafish embryo. *Nature* 337:358–62

First description and analysis of a zebrafish mutant affecting early embryogenesis.

157. Kimmel CB, Law RD. 1985. Cell lineage of zebrafish blastomeres. II. Formation of the yolk syncytial layer. *Dev. Biol.* 108:86–93
158. Kimmel CB, Warga RM. 1986. Tissue specific cell lineages originate in the gastrula of the zebrafish. *Science* 231:356–68
159. Kimmel CB, Warga RM. 1987. Cell lineages generating axial muscle in the zebrafish embryo. *Nature* 327:234–37
160. Kimmel CB, Warga RM. 1987. Indeterminate cell lineage of the zebrafish embryo. *Dev. Biol.* 124:269–80
161. Kimmel CB, Warga RM, Schilling TF. 1990. Origin and organization of the zebrafish fate map. *Development* 108:581–94

First systematic fate map of the zebrafish blastula.

162. Kishimoto Y, Lee KH, Zon L, Hammerschmidt M, Schulte-Merker S. 1997. The molecular nature of zebrafish *swirl*: BMP2 function is essential during early dorsoventral patterning. *Development* 124:4457–66

163. Klein TJ, Mlodzik M. 2005. Planar cell polarization: An emerging model points in the right direction. *Annu. Rev. Cell. Dev. Biol.* 21:155–76
164. Knapik EW, Goodman A, Ekker M, Chevrette M, Delgado J, et al. 1998. A microsatellite genetic linkage map for zebrafish (*Danio rerio*). *Nat. Genet.* 18:338–43
165. Koos DS, Ho RK. 1998. The *nieuwkoid* gene characterizes and mediates a Nieuwkoop-center-like activity in the zebrafish. *Curr. Biol.* 8:1199–206
166. Koos DS, Ho RK. 1999. The *nieuwkoid/dharma* homeobox gene is essential for *bmp2b* repression in the zebrafish pregastrula. *Dev. Biol.* 215:190–207
167. Koshida S, Shinya M, Nikaido M, Ueno N, Schulte-Merker S, et al. 2002. Inhibition of BMP activity by the FGF signal promotes posterior neural development in zebrafish. *Dev. Biol.* 244:9–20
168. Kramer C, Mayr T, Nowak M, Schumacher J, Runke G, et al. 2002. Maternally supplied Smad5 is required for ventral specification in zebrafish embryos prior to zygotic Bmp signaling. *Dev. Biol.* 250:263–79
169. Kramer-Zucker AG, Olale F, Haycraft CJ, Yoder BK, Schier AF, Drummond IA. 2005. Cilia-driven fluid flow in the zebrafish pronephros, brain and Kupffer's vesicle is required for normal organogenesis. *Development* 132:1907–21

169a. **Krauss S, Concordet JP, Ingham PW. 1993. A functionally conserved homolog of the *Drosophila* segment polarity gene *hh* is expressed in tissues with polarizing activity in zebrafish embryos. *Cell* 75:1431–44**

Discovery of the inducing activity of hedgehog in vertebrates.

170. Kudoh T, Concha ML, Houart C, Dawid IB, Wilson SW. 2004. Combinatorial Fgf and Bmp signalling patterns the gastrula ectoderm into prospective neural and epidermal domains. *Development* 131:3581–92
171. Kudoh T, Dawid IB. 2001. Role of the *iroquois3* homeobox gene in organizer formation. *Proc. Natl. Acad. Sci. USA* 98:7852–57
172. Kudoh T, Wilson SW, Dawid IB. 2002. Distinct roles for Fgf, Wnt and retinoic acid in posteriorizing the neural ectoderm. *Development* 129:4335–46
173. Kunwar PS, Zimmerman S, Bennett JT, Chen Y, Whitman M, Schier AF. 2003. Mixer/Bon and FoxH1/Sur have overlapping and divergent roles in Nodal signaling and mesendoderm induction. *Development* 130:5589–99
174. Larrain J, Oelgeschlager M, Ketpura NI, Reversade B, Zakin L, De Robertis EM. 2001. Proteolytic cleavage of Chordin as a switch for the dual activities of Twisted gastrulation in BMP signaling. *Development* 128:4439–47
175. Latimer AJ, Dong X, Markov Y, Appel B. 2002. Delta-Notch signaling induces hypochord development in zebrafish. *Development* 129:2555–63
176. Latimer AJ, Shin J, Appel B. 2005. *her9* promotes floor plate development in zebrafish. *Dev. Dyn.* 232:1098–104

176a. Lecuit T, Brook WJ, Ng M, Calleja M, Sun H, Cohen SM. 1996. Two distinct mechanisms for long-range patterning by Decapentaplegic in the *Drosophila* wing. *Nature* 381:387–93

177. Lee H, Kimelman D. 2002. A dominant-negative form of p63 is required for epidermal proliferation in zebrafish. *Dev. Cell* 2:607–16
178. Le Good JA, Joubin K, Giraldez AJ, Ben-Haim N, Beck S, et al. 2005. Nodal stability determines signaling range. *Curr. Biol.* 15:31–36
179. Lekven AC, Thorpe CJ, Waxman JS, Moon RT. 2001. Zebrafish *wnt8* encodes two Wnt8 proteins on a bicistronic transcript and is required for mesoderm and neurectoderm patterning. *Dev. Cell* 1:103–14
180. Lele Z, Nowak M, Hammerschmidt M. 2001. Zebrafish *admp* is required to restrict

the size of the organizer and to promote posterior and ventral development. *Dev. Dyn.* 222:681–87

181. Leptin M. 2005. Gastrulation movements: the logic and the nuts and bolts. *Dev. Cell* 8:305–20
182. Leung T, Bischof J, Soll I, Niessing D, Zhang D, et al. 2003. *bozozok* directly represses *bmp2b* transcription and mediates the earliest dorsoventral asymmetry of *bmp2b* expression in zebrafish. *Development* 130:3639–49
183. Levine M, Davidson EH. 2005. Gene regulatory networks for development. *Proc. Natl. Acad. Sci. USA* 102:4936–42
184. Lin F, Sepich DS, Chen S, Topczewski J, Yin C, et al. 2005. Essential roles of Gα12/13 signaling in distinct cell behaviors driving zebrafish convergence and extension gastrulation movements. *J. Cell Biol.* 169:777–87
185. Linville A, Gumusaneli E, Chandraratna RA, Schilling TF. 2004. Independent roles for retinoic acid in segmentation and neuronal differentiation in the zebrafish hindbrain. *Dev. Biol.* 270:186–99
186. Little SC, Mullins MC. 2004. Twisted gastrulation promotes BMP signaling in zebrafish dorsal-ventral axial patterning. *Development* 131:5825–35
187. Liu P, Wakamiya M, Shea MJ, Albrecht U, Behringer RR, Bradley A. 1999. Requirement for *Wnt3* in vertebrate axis formation. *Nat. Genet.* 22:361–65
188. Logan CY, Nusse R. 2004. The Wnt signaling pathway in development and disease. *Annu. Rev. Cell. Dev. Biol.* 20:781–810
189. Londin ER, Niemiec J, Sirotkin HI. 2005. Chordin, FGF signaling, and mesodermal factors cooperate in zebrafish neural induction. *Dev. Biol.* 279:1–19
190. Long Q, Meng A, Wang H, Jessen JR, Farrell MJ, Lin S. 1997. *GATA-1* expression pattern can be recapitulated in living transgenic zebrafish using GFP reporter gene. *Development* 124:4105–11
191. Long S, Ahmad N, Rebagliati M. 2003. The zebrafish *nodal*-related gene *southpaw* is required for visceral and diencephalic left-right asymmetry. *Development* 130:2303–16
192. Lu CC, Brennan J, Robertson EJ. 2001. From fertilization to gastrulation: axis formation in the mouse embryo. *Curr. Opin. Genet. Dev.* 11:384–92
193. Lunde K, Belting HG, Driever W. 2004. Zebrafish *pou5f1/pou2*, homolog of mammalian *Oct4*, functions in the endoderm specification cascade. *Curr. Biol.* 14:48–55
194. Lyons DA, Pogoda HM, Voas MG, Woods IG, Diamond B, et al. 2005. *erbb3* and *erbb2* are essential for schwann cell migration and myelination in zebrafish. *Curr. Biol.* 15:513–24
195. Maegawa S, Yasuda K, Inoue K. 1999. Maternal mRNA localization of zebrafish *DAZ*-like gene. *Mech. Dev.* 81:223–26
196. Marlow F, Gonzalez EM, Yin C, Rojo C, Solnica-Krezel L. 2004. No tail co-operates with non-canonical Wnt signaling to regulate posterior body morphogenesis in zebrafish. *Development* 131:203–16
197. Marlow F, Topczewski J, Sepich D, Solnica-Krezel L. 2002. Zebrafish Rho kinase 2 acts downstream of Wnt11 to mediate cell polarity and effective convergence and extension movements. *Curr. Biol.* 12:876–84
198. Martinez-Barbera JP, Toresson H, Da Rocha S, Krauss S. 1997. Cloning and expression of three members of the zebrafish Bmp family: *Bmp2a*, *Bmp2b* and *Bmp4*. *Gene* 198:53–59
199. Martyn U, Schulte-Merker S. 2003. The ventralized *ogon* mutant phenotype is caused

by a mutation in the zebrafish homologue of Sizzled, a secreted Frizzled-related protein. *Dev. Biol.* 260:58–67

200. Mathieu J, Griffin K, Herbomel P, Dickmeis T, Strähle U, et al. 2004. Nodal and Fgf pathways interact through a positive regulatory loop and synergize to maintain mesodermal cell populations. *Development* 131:629–41
201. Matsui T, Raya A, Kawakami Y, Callol-Massot C, Capdevila J, et al. 2005. Noncanonical Wnt signaling regulates midline convergence of organ primordia during zebrafish development. *Genes Dev.* 19:164–75
202. McFarland KN, Warga RM, Kane DA. 2005. Genetic locus *half baked* is necessary for morphogenesis of the ectoderm. *Dev. Dyn.* 233:390–406
203. Megason SG, Fraser SE. 2003. Digitizing life at the level of the cell: high-performance laser-scanning microscopy and image analysis for in toto imaging of development. *Mech. Dev.* 120:1407–20
204. Melby AE, Beach C, Mullins M, Kimelman D. 2000. Patterning the early zebrafish by the opposing actions of *bozozok* and *vox/vent*. *Dev. Biol.* 224:275–85
205. Melby AE, Kimelman D, Kimmel CB. 1997. Spatial regulation of *floating head* expression in the developing notochord. *Dev. Dyn.* 209:156–65
206. Melby AE, Warga RM, Kimmel CB. 1996. Specification of cell fates at the dorsal margin of the zebrafish gastrula. *Development* 122:2225–37
207. Meno C, Gritsman K, Ohishi S, Ohfuji Y, Heckscher E, et al. 1999. Mouse Lefty2 and zebrafish antivin are feedback inhibitors of nodal signaling during vertebrate gastrulation. *Mol. Cell.* 4:287–98
208. Miller-Bertoglio V, Carmany-Rampey A, Furthauer M, Gonzalez EM, Thisse C, et al. 1999. Maternal and zygotic activity of the zebrafish *ogon* locus antagonizes BMP signaling. *Dev. Biol.* 214:72–86
209. Mintzer KA, Lee MA, Runke G, Trout J, Whitman M, Mullins MC. 2001. *Lost-a-fin* encodes a type I BMP receptor, Alk8, acting maternally and zygotically in dorsoventral pattern formation. *Development* 128:859–69
210. Mishina Y, Crombie R, Bradley A, Behringer RR. 1999. Multiple roles for activin-like kinase-2 signaling during mouse embryogenesis. *Dev. Biol.* 213:314–26
211. Miyagi C, Yamashita S, Ohba Y, Yoshizaki H, Matsuda M, Hirano T. 2004. STAT3 noncell-autonomously controls planar cell polarity during zebrafish convergence and extension. *J. Cell Biol.* 166:975–81
212. Mizuno T, Yamaha E, Kuroiwa A, Takeda H. 1999. Removal of vegetal yolk causes dorsal deficencies and impairs dorsal-inducing ability of the yolk cell in zebrafish. *Mech. Dev.* 81:51–63
213. Mizuno T, Yamaha E, Wakahara M, Kuroiwa A, Takeda H. 1996. Mesoderm induction in zebrafish. *Nature* 383:131–32
214. Montero JA, Carvalho L, Wilsch-Brauninger M, Kilian B, Mustafa C, Heisenberg CP. 2005. Shield formation at the onset of zebrafish gastrulation. *Development* 132:1187–98
215. Montero JA, Heisenberg CP. 2004. Gastrulation dynamics: cells move into focus. *Trends Cell Biol.* 14:620–27
216. Montero JA, Kilian B, Chan J, Bayliss PE, Heisenberg CP. 2003. Phosphoinositide 3-kinase is required for process outgrowth and cell polarization of gastrulating mesendodermal cells. *Curr. Biol.* 13:1279–89
217. Muller F, Albert S, Blader P, Fischer N, Hallonet M, Strähle U. 2000. Direct action of the nodal-related signal cyclops in induction of *sonic hedgehog* in the ventral midline of the CNS. *Development* 127:3889–97

218. Mullins MC, Hammerschmidt M, Kane DA, Odenthal J, Brand M, et al. 1996. Genes establishing dorsoventral pattern formation in the zebrafish embryo: the ventral specifying genes. *Development* 123:81–93
219. Munoz-Sanjuan I, Brivanlou AH. 2004. Modulation of BMP signaling during vertebrate gastrulation. See Ref. 303a, pp. 475–90
220. Muraoka O, Ichikawa H, Shi H, Okumura S, Taira E, et al. 2000. Kheper, a novel ZFH/δEF1 family member, regulates the development of the neuroectoderm of zebrafish (*Danio rerio*). *Dev. Biol.* 228:29–40
221. Myers DC, Sepich DS, Solnica-Krezel L. 2002. Bmp activity gradient regulates convergent extension during zebrafish gastrulation. *Dev. Biol.* 243:81–98
222. Myers DC, Sepich DS, Solnica-Krezel L. 2002. Convergence and extension in vertebrate gastrulae: cell movements according to or in search of identity? *Trends Genet.* 18:447–55
223. Nasevicius A, Ekker SC. 2000. Effective targeted gene 'knockdown' in zebrafish. *Nat. Genet.* 26:216–20
224. Neave B, Holder N, Patient R. 1997. A graded response to BMP-4 spatially coordinates patterning of the mesoderm and ectoderm in the zebrafish. *Mech. Dev.* 62:183–95
225. Nellen D, Burke R, Struhl G, Basler K. 1996. Direct and long-range action of a DPP morphogen gradient. *Cell* 85:357–68
226. Nguyen VH, Schmid B, Trout J, Connors SA, Ekker M, Mullins MC. 1998. Ventral and lateral regions of the zebrafish gastrula, including the neural crest progenitors, are established by a *bmp2b/swirl* pathway of genes. *Dev. Biol.* 199:93–110
227. Nikaido M, Tada M, Saji T, Ueno N. 1997. Conservation of BMP signaling in zebrafish mesoderm patterning. *Mech. Dev.* 61:75–88
228. Nojima H, Shimizu T, Kim CH, Yabe T, Bae YK, et al. 2004. Genetic evidence for involvement of maternally derived Wnt canonical signaling in dorsal determination in zebrafish. *Mech. Dev.* 121:371–86
229. Norton WH, Mangoli M, Lele Z, Pogoda HM, Diamond B, et al. 2005. Monorail/Foxa2 regulates floorplate differentiation and specification of oligodendrocytes, serotonergic raphe neurones and cranial motoneurones. *Development* 132:645–58
230. Oates AC, Lackmann M, Power MA, Brennan C, Down LM, et al. 1999. An early developmental role for eph-ephrin interaction during vertebrate gastrulation. *Mech. Dev.* 83:77–94
231. Ober EA, Field HA, Stainier DY. 2003. From endoderm formation to liver and pancreas development in zebrafish. *Mech. Dev.* 120:5–18
232. Ober EA, Schulte-Merker S. 1999. Signals from the yolk cell induce mesoderm, neuroectoderm, the trunk organizer, and the notochord in zebrafish. *Dev. Biol.* 215:167–81
233. Odenthal J, van Eeden FJ, Haffter P, Ingham PW, Nusslein-Volhard C. 2000. Two distinct cell populations in the floor plate of the zebrafish are induced by different pathways. *Dev. Biol.* 219:350–63
234. Oelgeschlager M, Larrain J, Geissert D, De Robertis EM. 2000. The evolutionarily conserved BMP-binding protein Twisted gastrulation promotes BMP signalling. *Nature* 405:757–63
235. Okada Y, Takeda S, Tanaka Y, Belmonte JC, Hirokawa N. 2005. Mechanism of nodal flow: a conserved symmetry breaking event in left-right axis determination. *Cell* 121:633–44
236. Park M, Moon RT. 2002. The planar cell-polarity gene *stbm* regulates cell behaviour and cell fate in vertebrate embryos. *Nat. Cell Biol.* 4:20–25

237. Pelegri F. 2003. Maternal factors in zebrafish development. *Dev. Dyn.* 228:535–54
238. Pelegri F, Maischein HM. 1998. Function of zebrafish β-catenin and TCF-3 in dorsoventral patterning. *Mech. Dev.* 77:63–74
239. Peyrieras N, Strähle U, Rosa F. 1998. Conversion of zebrafish blastomeres to an endodermal fate by TGF-β-related signaling. *Curr. Biol.* 8:783–86
240. Piccolo S, Agius E, Lu B, Goodman S, Dale L, De Robertis EM. 1997. Cleavage of Chordin by Xolloid metalloprotease suggests a role for proteolytic processing in the regulation of Spemann organizer activity. *Cell* 91:407–16
241. Piccolo S, Sasai Y, Lu B, De Robertis EM. 1996. Dorsoventral patterning in *Xenopus*: inhibition of ventral signals by direct binding of chordin to BMP-4. *Cell* 86:589–98
242. Plusa B, Hadjantonakis AK, Gray D, Piotrowska-Nitsche K, Jedrusik A, et al. 2005. The first cleavage of the mouse zygote predicts the blastocyst axis. *Nature* 434:391–95
243. Pogoda HM, Solnica-Krezel L, Driever W, Meyer D. 2000. The zebrafish forkhead transcription factor FoxH1/Fast1 is a modulator of nodal signaling required for organizer formation. *Curr. Biol.* 10:1041–49
244. Postlethwait JH, Johnson SL, Midson CN, Talbot WS, Gates M, et al. 1994. A genetic linkage map for the zebrafish. *Science* 264:699–703
245. Poulain M, Lepage T. 2002. Mezzo, a paired-like homeobox protein is an immediate target of Nodal signalling and regulates endoderm specification in zebrafish. *Development* 129:4901–14
246. Pyati UJ, Webb AE, Kimelman D. 2005. Transgenic zebrafish reveal stage-specific roles for Bmp signaling in ventral and posterior mesoderm development. *Development* 132:2333–43
247. Raible F, Brand M. 2001. Tight transcriptional control of the ETS domain factors Erm and Pea3 by Fgf signaling during early zebrafish development. *Mech. Dev.* 107:105–17
248. Ramel MC, Lekven AC. 2004. Repression of the vertebrate organizer by Wnt8 is mediated by Vent and Vox. *Development* 131:3991–4000
249. Rauch GJ, Hammerschmidt M, Blader P, Schauerte HE, Strähle U, et al. 1997. Wnt5 is required for tail formation in the zebrafish embryo. *Cold Spring Harbor Symp. Quant. Biol.* 62:227–34
250. Raya A, Belmonte JC. 2004. Sequential transfer of left-right information during vertebrate embryo development. *Curr. Opin. Genet. Dev.* 14:575–81
251. Rebagliati MR, Toyama R, Fricke C, Haffter P, Dawid IB. 1998. Zebrafish *nodal*-related genes are implicated in axial patterning and establishing left-right asymmetry. *Dev. Biol.* 199:261–72
252. Rebagliati MR, Toyama R, Haffter P, Dawid IB. 1998. *cyclops* encodes a nodal-related factor involved in midline signaling. *Proc. Natl. Acad. Sci. USA* 95:9932–37
253. Reifers F, Bohli H, Walsh EC, Crossley PH, Stainier DY, Brand M. 1998. *Fgf8* is mutated in zebrafish *acerebellar* (*ace*) mutants and is required for maintenance of midbrain-hindbrain boundary development and somitogenesis. *Development* 125:2381–95
254. Reim G, Mizoguchi T, Stainier DY, Kikuchi Y, Brand M. 2004. The POU domain protein Spg (Pou2/Oct4) is essential for endoderm formation in cooperation with the HMG domain protein Casanova. *Dev. Cell* 6:91–101
255. Reiter JF, Alexander J, Rodaway A, Yelon D, Patient R, et al. 1999. Gata5 is required for the development of the heart and endoderm in zebrafish. *Genes Dev.* 13:2983–95
256. Reiter JF, Kikuchi Y, Stainier DY. 2001. Multiple roles for Gata5 in zebrafish endoderm formation. *Development* 128:125–35

257. Rentzsch F, Bakkers J, Kramer C, Hammerschmidt M. 2004. Fgf signaling induces posterior neuroectoderm independently of Bmp signaling inhibition. *Dev. Dyn.* 231:750–57
258. Renucci A, Lemarchandel V, Rosa F. 1996. An activated form of type I serine/threonine kinase receptor TARAM-A reveals a specific signalling pathway involved in fish head organiser formation. *Development* 122:3735–43
259. Rhinn M, Lun K, Luz M, Werner M, Brand M. 2005. Positioning of the midbrain-hindbrain boundary organizer through global posteriorization of the neuroectoderm mediated by Wnt8 signaling. *Development* 132:1261–72
260. Rodaway A, Takeda H, Koshida S, Broadbent J, Price B, et al. 1999. Induction of the mesendoderm in the zebrafish germ ring by yolk cell-derived TGF-β family signals and discrimination of mesoderm and endoderm by FGF. *Development* 126:3067–78
261. Roehl H, Nusslein-Volhard C. 2001. Zebrafish *pea3* and *erm* are general targets of FGF8 signaling. *Curr. Biol.* 11:503–7
262. Ross JJ, Shimmi O, Vilmos P, Petryk A, Kim H, et al. 2001. Twisted gastrulation is a conserved extracellular BMP antagonist. *Nature* 410:479–83
263. Ryu SL, Fujii R, Yamanaka Y, Shimizu T, Yabe T, et al. 2001. Regulation of *dharma/bozozok* by the Wnt pathway. *Dev. Biol.* 231:397–409
264. Sakaguchi T, Kuroiwa A, Takeda H. 2001. A novel sox gene, *226D7*, acts downstream of Nodal signaling to specify endoderm precursors in zebrafish. *Mech. Dev.* 107:25–38
265. Sampath K, Rubinstein AL, Cheng AM, Liang JO, Fekany K, et al. 1998. Induction of the zebrafish ventral brain and floorplate requires cyclops/nodal signalling. *Nature* 395:185–89
266. Saude L, Woolley K, Martin P, Driever W, Stemple DL. 2000. Axis-inducing activities and cell fates of the zebrafish organizer. *Development* 127:3407–17
266a. Schäfer M, Rembold M, Wittbrodt J, Schartl M, Winkler C. 2005. Medial floor plate formation in zebrafish consists of two phases and requires trunk-derived Midkine-a. *Genes Dev.* 19:897–902
266b. Schauerte HE, van Eeden FJ, Fricke C, Odenthal J, Strähle U, Haffter P. 1998. *Sonic hedgehog* is not required for the induction of medial floor plate cells in the zebrafish. *Development* 125:2983–93
267. Schier AF. 2001. Axis formation and patterning in zebrafish. *Curr. Opin. Genet. Dev.* 11:393–404
268. Schier AF. 2003. Nodal signaling in vertebrate development. *Annu. Rev. Cell. Dev. Biol.* 19:589–621
269. Schier AF. 2004. Nodal signaling during gastrulation. See Ref. 303a, pp. 491–504
270. Schier AF, Neuhauss SC, Harvey M, Malicki J, Solnica-Krezel L, et al. 1996. Mutations affecting the development of the embryonic zebrafish brain. *Development* 123:165–78
271. Schier AF, Neuhauss SC, Helde KA, Talbot WS, Driever W. 1997. The *one-eyed pinhead* gene functions in mesoderm and endoderm formation in zebrafish and interacts with *no tail*. *Development* 124:327–42
272. Schmid B, Furthauer M, Connors SA, Trout J, Thisse B, et al. 2000. Equivalent genetic roles for *bmp7/snailhouse* and *bmp2b/swirl* in dorsoventral pattern formation. *Development* 127:957–67
273. Schmidt JE, Suzuki A, Ueno N, Kimelman D. 1995. Localized BMP-4 mediates dorsal/ventral patterning in the early *Xenopus* embryo. *Dev. Biol.* 169:37–50
274. Schneider S, Steinbeisser H, Warga RM, Hausen P. 1996. β-catenin translocation into nuclei demarcates the dorsalizing centers in frog and fish embryos. *Mech. Dev.* 57:191–98

275. Scholpp S, Brand M. 2004. Endocytosis controls spreading and effective signaling range of Fgf8 protein. *Curr. Biol.* 14:1834–41
276. Schulte-Merker S, Ho RK, Herrmann BG, Nusslein-Volhard C. 1992. The protein product of the zebrafish homologue of the mouse *T* gene is expressed in nuclei of the germ ring and the notochord of the early embryo. *Development* 116:1021–32
277. Schulte-Merker S, Lee KJ, McMahon AP, Hammerschmidt M. 1997. The zebrafish organizer requires *chordino*. *Nature* 387:862–63
278. Schulte-Merker S, van Eeden FJ, Halpern ME, Kimmel CB, Nusslein-Volhard C. 1994. *no tail* (*ntl*) is the zebrafish homologue of the mouse *T* (*Brachyury*) gene. *Development* 120:1009–15
279. Schwarz-Romond T, Asbrand C, Bakkers J, Kuhl M, Schaeffer HJ, et al. 2002. The ankyrin repeat protein Diversin recruits Casein kinase Iε to the β-catenin degradation complex and acts in both canonical Wnt and Wnt/JNK signaling. *Genes Dev.* 16:2073–84
280. Scott IC, Blitz IL, Pappano WN, Maas SA, Cho KW, Greenspan DS. 2001. Homologues of Twisted gastrulation are extracellular cofactors in antagonism of BMP signalling. *Nature* 410:475–78
281. Selman K, Wallace RA, Sarka A, Qi X. 1993. Stages of oocyte development in the zebrafish, *Brachydanio rerio*. *J. Morphol.* 218:203–24
282. Sepich DS, Calmelet C, Kiskowski M, Solnica-Krezel L. 2005. Initiation of convergence and extension movements during zebrafish gastrulation. *Dev. Dyn.* In press
283. Sepich DS, Myers DC, Short R, Topczewski J, Marlow F, Solnica-Krezel L. 2000. Role of the zebrafish *trilobite* locus in gastrulation movements of convergence and extension. *Genesis* 27:159–73
284. Shaywitz DA, Melton DA. 2005. The molecular biography of the cell. *Cell* 120:729–31
285. Shih J, Fraser SE. 1995. Distribution of tissue progenitors within the shield region of the zebrafish gastrula. *Development* 121:2755–65
286. Shih J, Fraser SE. 1996. Characterizing the zebrafish organizer: microsurgical analysis at the early-shield stage. *Development* 122:1313–22
287. Shimell MJ, Ferguson EL, Childs SR, O'Connor MB. 1991. The *Drosophila* dorsal-ventral patterning gene *tolloid* is related to human *bone morphogenetic protein 1*. *Cell* 67:469–81
288. Shimizu T, Bae YK, Muraoka O, Hibi M. 2005. Interaction of Wnt and *caudal*-related genes in zebrafish posterior body formation. *Dev. Biol.* 279:125–41
289. Shimizu T, Yabe T, Muraoka O, Yonemura S, Aramaki S, et al. 2005. E-cadherin is required for gastrulation cell movements in zebrafish. *Mech. Dev.* 122:747–63
290. Shimizu T, Yamanaka Y, Nojima H, Yabe T, Hibi M, Hirano T. 2002. A novel repressor-type homeobox gene, *ved*, is involved in *dharma/bozozok*-mediated dorsal organizer formation in zebrafish. *Mech. Dev.* 118:125–38
291. Shimizu T, Yamanaka Y, Ryu SL, Hashimoto H, Yabe T, et al. 2000. Cooperative roles of Bozozok/Dharma and Nodal-related proteins in the formation of the dorsal organizer in zebrafish. *Mech. Dev.* 91:293–303
292. Shinya M, Eschbach C, Clark M, Lehrach H, Furutani-Seiki M. 2000. Zebrafish Dkk1, induced by the pre-MBT Wnt signaling, is secreted from the prechordal plate and patterns the anterior neural plate. *Mech. Dev.* 98:3–17
293. Sidi S, Goutel C, Peyrieras N, Rosa FM. 2003. Maternal induction of ventral fate by zebrafish *radar*. *Proc. Natl. Acad. Sci. USA* 100:3315–20

294. Sirotkin HI, Dougan ST, Schier AF, Talbot WS. 2000. *bozozok* and *squint* act in parallel to specify dorsal mesoderm and anterior neuroectoderm in zebrafish. *Development* 127:2583–92
295. Sirotkin HI, Gates MA, Kelly PD, Schier AF, Talbot WS. 2000. *fast1* is required for the development of dorsal axial structures in zebrafish. *Curr. Biol.* 10:1051–54
296. Sivak J, Amaya E. 2004. FGF signaling during gastrulation. See Ref. 303a, pp. 463–74
297. Smith JC. 2004. Role of T-box genes during gastrulation. See Ref. 303a, pp. 571–80
298. Solnica-Krezel L. 2005. Conserved patterns of cell movements during vertebrate gastrulation. *Curr. Biol.* 15:R213–28
299. Solnica-Krezel L, Driever W. 1994. Microtubule arrays of the zebrafish yolk cell: organization and function during epiboly. *Development* 120:2443–55
300. Solnica-Krezel L, Stemple DL, Mountcastle-Shah E, Rangini Z, Neuhauss SC, et al. 1996. Mutations affecting cell fates and cellular rearrangements during gastrulation in zebrafish. *Development* 123:67–80
301. Stachel SE, Grunwald DJ, Myers PZ. 1993. Lithium perturbation and goosecoid expression identify a dorsal specification pathway in the pregastrula zebrafish. *Development* 117:1261–74
302. Stainier DY, Fouquet B, Chen JN, Warren KS, Weinstein BM, et al. 1996. Mutations affecting the formation and function of the cardiovascular system in the zebrafish embryo. *Development* 123:285–92
303. Stern CD. 2004. Gastrulation in the chick. See Ref. 303a, pp. 219–32
303a. Stern CD, ed. 2004. *Gastrulation: From Cells to Embryo*. Cold Spring Harbor, NY: Cold Spring Harbor Lab. Press
304. Stern CD. 2005. Neural induction: old problem, new findings, yet more questions. *Development* 132:2007–21
305. Strähle U, Jesuthasan S. 1993. Ultraviolet irradiation impairs epiboly in zebrafish embryos: evidence for a microtubule-dependent mechanism of epiboly. *Development* 119:909–19
306. Strähle U, Jesuthasan S, Blader P, Garcia-Villalba P, Hatta K, Ingham PW. 1997. *one-eyed pinhead* is required for development of the ventral midline of the zebrafish (*Danio rerio*) neural tube. *Genes Funct.* 1:131–48
307. Strähle U, Lam CS, Ertzer R, Rastegar S. 2004. Vertebrate floor-plate specification: variations on common themes. *Trends Genet.* 20:155–62
308. Streisinger G, Walker C, Dower N, Knauber D, Singer F. 1981. Production of clones of homozygous diploid zebra fish (*Brachydanio rerio*). *Nature* 291:293–96

First description of zebrafish as a genetic model system.

309. Sumanas S, Kim HJ, Hermanson S, Ekker SC. 2001. Zebrafish *frizzled 2* morphant displays defects in body axis elongation. *Genesis* 30:114–18
310. Sun X, Meyers EN, Lewandoski M, Martin GR. 1999. Targeted disruption of *Fgf8* causes failure of cell migration in the gastrulating mouse embryo. *Genes Dev.* 13:1834–46
311. Szeto DP, Kimelman D. 2004. Combinatorial gene regulation by Bmp and Wnt in zebrafish posterior mesoderm formation. *Development* 131:3751–60
312. Talbot WS, Trevarrow B, Halpern ME, Melby AE, Farr G, et al. 1995. A homeobox gene essential for zebrafish notochord development. *Nature* 378:150–57

Discovery of a novel transcription factor required for patterning the organizer region.

313. Tam PPL, Gad JM. 2004. Gastrulation in the mouse embryo. See Ref. 303a, pp. 233–62
314. Tao Q, Yokota C, Puck H, Kofron M, Birsoy B, et al. 2005. Maternal wnt11 activates the canonical wnt signaling pathway required for axis formation in *Xenopus* embryos. *Cell* 120:857–71

315. Thisse B, Wright CV, Thisse C. 2000. Activin- and Nodal-related factors control anteroposterior patterning of the zebrafish embryo. *Nature* 403:425–28
316. Thisse C, Thisse B. 1999. Antivin, a novel and divergent member of the TGFβ superfamily, negatively regulates mesoderm induction. *Development* 126:229–40
317. Thisse C, Thisse B, Halpern ME, Postlethwait JH. 1994. *Goosecoid* expression in neurectoderm and mesendoderm is disrupted in zebrafish *cyclops* gastrulas. *Dev. Biol.* 164:420–29
318. Thorpe CJ, Moon RT. 2004. *nemo-like kinase* is an essential co-activator of Wnt signaling during early zebrafish development. *Development* 131:2899–909
319. Thorpe CJ, Weidinger G, Moon RT. 2005. Wnt/β-catenin regulation of the Sp1-related transcription factor *sp5l* promotes tail development in zebrafish. *Development* 132:1763–72
319a. Tian J, Yam C, Balasundaram G, Wang H, Gore A, Sampth K. 2003. A temperature-sensitive mutation in the *nodal*-related gene *cyclops* reveals that the floor plate is induced during gastrulation in zebrafish. *Development* 130:3331–42
319b. Tiso N, Filippi A, Pauls S, Bortolussi M, Argenton F. 2002. BMP signalling regulates anteroposterior endoderm patterning in zebrafish. *Mech. Dev.* 118:29–37
320. Topczewski J, Sepich DS, Myers DC, Walker C, Amores A, et al. 2001. The zebrafish glypican knypek controls cell polarity during gastrulation movements of convergent extension. *Dev. Cell* 1:251–64
321. Topczewski J, Solnica-Krezel L. 1999. Cytoskeletal dynamics of the zebrafish embryo. *Methods Cell Biol.* 59:205–26
322. Trinh LA, Meyer D, Stainier DY. 2003. The Mix family homeodomain gene *bonnie and clyde* functions with other components of the Nodal signaling pathway to regulate neural patterning in zebrafish. *Development* 130:4989–98
323. Tsang M, Friesel R, Kudoh T, Dawid IB. 2002. Identification of Sef, a novel modulator of FGF signalling. *Nat. Cell Biol.* 4:165–69
324. Tsang M, Maegawa S, Kiang A, Habas R, Weinberg E, Dawid IB. 2004. A role for MKP3 in axial patterning of the zebrafish embryo. *Development* 131:2769–79
325. Ulrich F, Concha ML, Heid PJ, Voss E, Witzel S, et al. 2003. Slb/Wnt11 controls hypoblast cell migration and morphogenesis at the onset of zebrafish gastrulation. *Development* 130:5375–84
326. Veeman MT, Slusarski DC, Kaykas A, Louie SH, Moon RT. 2003. Zebrafish Prickle, a modulator of noncanonical Wnt/Fz signaling, regulates gastrulation movements. *Curr. Biol.* 13:680–85
327. von Bubnoff A, Cho KW. 2001. Intracellular BMP signaling regulation in vertebrates: pathway or network? *Dev. Biol.* 239:1–14
328. von Dassow G, Schmidt JE, Kimelman D. 1993. Induction of the *Xenopus* organizer: expression and regulation of *Xnot*, a novel FGF and activin-regulated homeo box gene. *Genes Dev.* 7:355–66
329. Wada H, Iwasaki M, Sato T, Masai I, Nishiwaki Y, et al. 2005. Dual roles of zygotic and maternal Scribble1 in neural migration and convergent extension movements in zebrafish embryos. *Development* 132:2273–85
330. Wagner DS, Dosch R, Mintzer KA, Wiemelt AP, Mullins MC. 2004. Maternal control of development at the midblastula transition and beyond: mutants from the zebrafish II. *Dev. Cell* 6:781–90
331. Wallingford JB, Fraser SE, Harland RM. 2002. Convergent extension: the molecular

control of polarized cell movement during embryonic development. *Dev. Cell* 2:695–706

332. Warga RM, Kane DA. 2003. One-eyed pinhead regulates cell motility independent of Squint/Cyclops signaling. *Dev. Biol.* 261:391–411
333. Warga RM, Kimmel CB. 1990. Cell movements during epiboly and gastrulation in zebrafish. *Development* 108:569–80
334. Warga RM, Nusslein-Volhard C. 1999. Origin and development of the zebrafish endoderm. *Development* 126:827–38
335. Waxman JS, Hocking AM, Stoick CL, Moon RT. 2004. Zebrafish Dapper1 and Dapper2 play distinct roles in Wnt-mediated developmental processes. *Development* 131:5909–21
336. Weaver C, Kimelman D. 2004. Move it or lose it: axis specification in *Xenopus*. *Development* 131:3491–99
337. Weidinger G, Thorpe CJ, Wuennenberg-Stapleton K, Ngai J, Moon RT. 2005. The Sp1-related transcription factors *sp5* and *sp5-like* act downstream of Wnt/β-catenin signaling in mesoderm and neuroectoderm patterning. *Curr. Biol.* 15:489–500
338. Wienholds E, Kloosterman WP, Miska E, Alvarez-Saavedra E, Berezikov E, et al. 2005. MicroRNA expression in zebrafish embryonic development. *Science.* 309:310–11
339. Wienholds E, Koudijs MJ, van Eeden FJ, Cuppen E, Plasterk RH. 2003. The microRNA-producing enzyme Dicer1 is essential for zebrafish development. *Nat. Genet.* 35:217–18
340. Wienholds E, Schulte-Merker S, Walderich B, Plasterk RH. 2002. Target-selected inactivation of the zebrafish *rag1* gene. *Science* 297:99–102
341. Willot V, Mathieu J, Lu Y, Schmid B, Sidi S, et al. 2002. Cooperative action of ADMP- and BMP-mediated pathways in regulating cell fates in the zebrafish gastrula. *Dev. Biol.* 241:59–78
342. Wilm TP, Solnica-Krezel L. 2005. Essential roles of a zebrafish *prdm1/blimp1* homolog in embryo patterning and organogenesis. *Development* 132:393–404
343. Wilson PA, Lagna G, Suzuki A, Hemmati-Brivanlou A. 1997. Concentration-dependent patterning of the *Xenopus* ectoderm by BMP4 and its signal transducer Smad1. *Development* 124:3177–84
344. Wolenski JS, Hart NH. 1987. Scanning electron microscope studies of sperm incorporation into the zebrafish (*Brachydanio*) egg. *J. Exp. Zool.* 243:259–73
345. Woo K, Fraser SE. 1995. Order and coherence in the fate map of the zebrafish nervous system. *Development* 121:2595–609
346. Woo K, Fraser SE. 1997. Specification of the zebrafish nervous system by nonaxial signals. *Science* 277:254–57
347. Woo K, Fraser SE. 1998. Specification of the hindbrain fate in the zebrafish. *Dev. Biol.* 197:283–96
348. Woods IG, Talbot WS. 2005. The *you* gene encodes an EGF-CUB protein essential for Hedgehog signaling in zebrafish. *PLoS Biol.* 3:e66
349. Xiao T, Roeser T, Staub W, Baier H. 2005. A GFP-based genetic screen reveals mutations that disrupt the architecture of the zebrafish retinotectal projection. *Development* 132:2955–67
350. Xie J, Fisher S. 2005. Twisted gastrulation enhances BMP signaling through chordin dependent and independent mechanisms. *Development* 132:383–91
351. Yabe T, Shimizu T, Muraoka O, Bae YK, Hirata T, et al. 2003. Ogon/secreted frizzled functions as a negative feedback regulator of Bmp signaling. *Development* 130:2705–16

352. Yamamoto Y, Oelgeschlager M. 2004. Regulation of bone morphogenetic proteins in early embryonic development. *Naturwissenschaften* 91:519–34
353. Yamanaka Y, Mizuno T, Sasai Y, Kishi M, Takeda H, et al. 1998. A novel homeobox gene, *dharma*, can induce the organizer in a non-cell-autonomous manner. *Genes Dev.* 12:2345–53
354. Yamashita S, Miyagi C, Carmany-Rampey A, Shimizu T, Fujii R, et al. 2002. Stat3 controls cell movements during zebrafish gastrulation. *Dev. Cell* 2:363–75
355. Yamashita S, Miyagi C, Fukada T, Kagara N, Che YS, Hirano T. 2004. Zinc transporter LIVI controls epithelial-mesenchymal transition in zebrafish gastrula organizer. *Nature* 429:298–302
356. Yan YT, Gritsman K, Ding J, Burdine RD, Corrales JD, et al. 1999. Conserved requirement for *EGF-CFC* genes in vertebrate left-right axis formation. *Genes Dev.* 13:2527–37
357. Yasuo H, Lemaire P. 2001. Role of Goosecoid, Xnot and Wnt antagonists in the maintenance of the notochord genetic programme in *Xenopus* gastrulae. *Development* 128:3783–93
358. Yelon D. 2001. Cardiac patterning and morphogenesis in zebrafish. *Dev. Dyn.* 222:552–63
359. Yeo SY, Little MH, Yamada T, Miyashita T, Halloran MC, et al. 2001. Overexpression of a slit homologue impairs convergent extension of the mesoderm and causes cyclopia in embryonic zebrafish. *Dev. Biol.* 230:1–17
360. Zhang J, Houston DW, King ML, Payne C, Wylie C, Heasman J. 1998. The role of maternal VegT in establishing the primary germ layers in *Xenopus* embryos. *Cell* 94:515–24
361. Zhang J, Talbot WS, Schier AF. 1998. Positional cloning identifies zebrafish *one-eyed pinhead* as a permissive EGF-related ligand required during gastrulation. *Cell* 92:241–51
362. Zhang L, Zhou H, Su Y, Sun Z, Zhang H, et al. 2004. Zebrafish Dpr2 inhibits mesoderm induction by promoting degradation of nodal receptors. *Science* 306:114–17

Chromatin Remodeling in Dosage Compensation

John C. Lucchesi,[1] William G. Kelly,[1] and Barbara Panning[2]

[1]Department of Biology, Emory University, Atlanta, Georgia 30322; email: lucchesi@biology.emory.edu

[2]Department of Biochemistry and Biophysics, University of California, San Francisco, California 94143; email: bpanning@biochem.ucsf.edu

Annu. Rev. Genet. 2005. 39:615–51

First published online as a Review in Advance on August 24, 2005

The *Annual Review of Genetics* is online at genet.annualreviews.org

doi: 10.1146/annurev.genet.39.073003.094210

0066-4197/05/1215-0615$20.00

Key Words

epigenetics, transcription, X chromosome

Abstract

In many multicellular organisms, males have one X chromosome and females have two. Dosage compensation refers to a regulatory mechanism that insures the equalization of X-linked gene products in males and females. The mechanism has been studied at the molecular level in model organisms belonging to three distantly related taxa; in these organisms, equalization is achieved by shutting down one of the two X chromosomes in the somatic cells of females, by decreasing the level of transcription of the two doses of X-linked genes in females relative to males, or by increasing the level of transcription of the single dose of X-linked genes in males. The study of dosage compensation in these different forms has revealed the existence of an amazing number of interacting chromatin remodeling mechanisms that affect the function of entire chromosomes.

Contents

INTRODUCTION

Sex chromosome: a chromosome that is present in the somatic nuclei of only one sex or that is present singly in one sex and in two copies in the other

Heterogametic: an individual or a population of individuals that produce two types of gametes, for example, containing either an X or a Y chromosome

In diploid organisms, some of the genetic information responsible for sexual differentiation is present on one pair of homologous chromosomes, the sex chromosomes. In many cases, one of the sex chromosomes has become structurally modified while the other has remained unchanged. The structurally modified chromosome is limited to one sex; in some groups of organisms it is transmitted from males to sons (Y chromosome), in others, from females to daughters (W chromosome). The other sex chromosome, X in instances where the male is heterogametic (XY), or Z when the female is heterogametic (ZW), has remained structurally unchanged and is present in two doses in the homogametic sex. This system of sex determination results in an inequality in the dosage of the genes present on the X or Z chromosomes in males and females. Many of these genes are equally important to the development and maintenance of both sexes in a non-sex-specific manner, and an inequality in gene product levels would be inappropriate and may lead to differential selection between the sexes. It is not surprising, therefore, that mechanisms have evolved to prevent such inequalities and compensate for differences in the dosage of X-linked or W-linked genes between the sexes.

The phenomenon of dosage compensation was discovered in Drosophila by Herman Muller more than 70 years ago. It has been studied at the molecular level in model organisms belonging to three very distantly related taxa: round worms, dipterans, and mammals. In these organisms, the transcriptional regulation leading to equal products of X-linked genes in males and females has been achieved in different ways: by shutting down one of the two X chromosomes in the somatic cells of females throughout most of its length (mammals), by decreasing the level of transcription of the two doses of X-linked genes in hermaphrodites relative to males (*Caenorhabditis elegans*), or by increasing the level of transcription of the single dose of most X-linked genes in males (Drosophila) (**Figure 1**). Some 25 years after Muller's discovery, Theodosius Dobzhansky noted that the polytenic X chromosome in salivary glands of male Drosophila larvae is wider and more diffuse than each X in females. The significance of this observation was provided by George Rudkin who determined that the DNA content of the male X is equal to that of each X chromosome in females, leading to the hypothesis that the difference in morphology of the X in the two sexes reflects a difference in levels of activity. This hypothesis was substantiated by A.S. Mukherjee in Wolfgang Beerman's laboratory using what at the time was a state-of-the art molecular technique: transcription autoradiography. Fifteen years later, John Belote identified four genes with loss-of-function mutations that were inconsequential in females but lethal in males. The X chromosomes of mutant males exhibited approximately half of the normal level of transcription and had lost the paler and somewhat distended appearance that had been interpreted as an indication of an enhanced level of activity in relation to each of the two X chromosomes in females. These results demonstrated that the equalization of X-linked gene products was achieved by doubling, on average, the transcriptional activity of the X chromosome in males rather than by halving the transcriptional activity of each X in females. Then, in 1991, the first of these genes (*mle*) was cloned by Mitzi Kuroda, issuing in the molecular biology study of dosage compensation in flies. More recently, two nontranslated RNA species (*roX* RNAs) were found to be associated with the X chromosome throughout most of its length in a manner that is dependent on the dosage compensation machinery [see references in (93, 94, 118)].

Homogametic: an individual or a population of individuals that produce only one type of gametes with respect to chromosome content. Mammalian females are homogametic because all of their gametes contain an X chromosome and a set of autosomes

Hypertranscription *(Drosophila)*

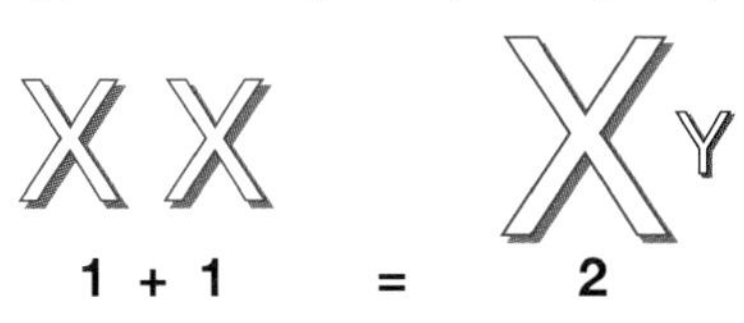

X-inactivation (mammals)

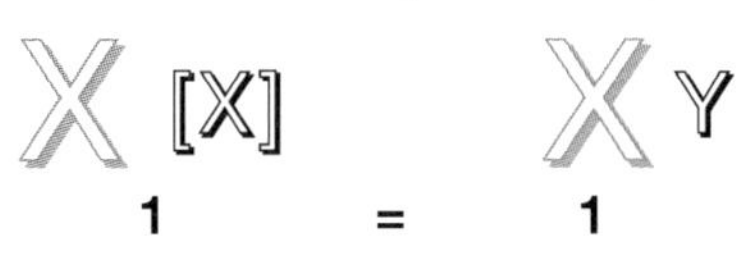

Hypotranscription *(Caenorhabditis)*

X X   X

½ + ½ = 1

Figure 1

Dosage compensation mechanisms.

Following the discovery by Susumu Ohno that the "sex chromatin body" or Barr body visible in nuclei of somatic cells of females is an X chromosome, in 1961 Mary Lyon presented the mammalian X-chromosome inactivation hypothesis and laid the cornerstone of the study of dosage compensation in mammals. The major historical landmarks of this study are the demonstration of female mosaicism for X-linked enzyme variants by Ronald Davidson, Harold Nitowsky, and Barton Childs; the identification by Bruce Cattanach of a region of the X chromosome (the X chromosome controlling element, *Xce*) involved in modifying the randomness of inactivation; the general belief that inactivation must be initiated and controlled by a specific *cis*-acting site (the X-inactivation center or *XIC*); the mapping of this center and the discovery of a gene encoding a nontranslated RNA that is transcribed only by the inactive chromosome (X inactive-specific transcript or *XIST*) by Huntington Willard and his associates; the identification

Autosomes: all chromosomes that are not the sex chromosomes and that are present in two doses in the somatic nuclei of both sexes

MSL: male-specific lethal

by Jennie Lee of a gene in the *XIC* region that is antisense to Xist, is transcribed by the active X, and protects it from inactivation by Xist [see references in (16, 64)].

The existence of a mechanism for dosage compensation in *C. elegans* was first noted by Barbara Meyer in 1986. Over the next few years, Meyer and co-workers isolated a number of genes responsible for dosage compensation components: These genes were identified by the hermaphrodite-specific lethal or dumpy phenotypes of their loss-of-function mutations, presumably caused by the overexpression of X-linked genes. In view of the fact that hermaphrodites have two X chromosomes and are somatically female, this overexpression indicated that dosage compensation in the worm is achieved by decreasing the activity of the two Xs. In addition, Meyer and co-workers established the existence of an important functional link between the dosage compensation mechanism and the genetic hierarchy responsible for sex differentiation [see references in (133)].

Over the past dozen years, the study of dosage compensation has provided a unique window into a variety of fundamental mechanisms of transcriptional regulation. In Drosophila, an RNA-containing regulatory complex that is responsible for doubling, on average, the rate of transcription of most X-linked genes in males may represent a unique example of a novel type of chromatin remodeling machines or complexes. In *C. elegans*, the limited downregulation of both X chromosomes in hermaphrodites involves a subset of proteins and factors that are normally engaged in chromosome condensation during cell division. Finally, in mammals, the facultative heterochromatization of one of the two X chromosomes in females involves the spread of noncoding RNA as a harbinger of covalent modifications of both DNA and histones. One purpose of this review is to highlight the progress that has been achieved by studying dosage compensation in understanding the role that chromatin architecture plays on the regulation of gene transcription. The other purpose is to highlight some of the areas where future investigations are sure to yield comparable or greater insights.

DOSAGE COMPENSATION IN DROSOPHILA

In Drosophila, sex is determined by the relative numbers of X chromosomes and sets of autosomes present in the fertilized egg. The molecular translation of this ratio is the responsibility of a small set of X-linked genes [the sisterless genes *sisA*, *sisB*, *sisC* and the runt gene (*run*)] that, in females, reach a sufficient level of expression to induce the transcription of the regulatory gene Sex lethal (Sxl). *Sxl* initiates differentiation in the female mode by regulating a cascade of sex-specific splicing events leading to the repression of male-specific realizer genes, thus achieving female sexual differentiation. In male embryos, female-specific realizer genes are repressed, resulting in male sexual differentiation [see references in (34, 157)]. The sex-lethal gene product is directly responsible for the absence of dosage compensation in females (see below). In males, it is the absence of SXL that allows the establishment of the mechanism of dosage compensation in very early embryos (61, 128, 170).

Assembling the MSL Complex

Five proteins whose absence is inconsequential in females but is lethal in males owing to their specific role in the mechanism of dosage compensation, have been shown to coimmunoprecipitate in various partial combinations (36) and to assemble into an RNA-containing complex—the MSL complex (189). These proteins are the products of the *msl1*, *msl2*, *msl3*, *mle*, and *mof* genes (collectively referred to as the *msl* genes to reflect the male-specific lethal mutant phenotype that characterizes them). In addition, the complex contains one of the roX RNAs (roX1 or roX2). Because of the frequent discovery of proteins that are implicated in its function (see below), it is useful to view the complex consisting of the five MSL proteins and roX RNA as a "core complex" in order to leave open the possible existence of additional, albeit substochiometric subunits. MLE is an ATP-dependent RNA/DNA helicase (108); MOF is a histone acetyl transferase that specifically acetylates histone H4 at lysine 16 (2, 80, 189). In females, the complex does not form because SXL prevents the translation of the *msl2* gene transcript (8, 65, 121). The absence of MSL2 leads to a low level of MSL1 and although the MOF, MLE, and MSL3 proteins are present, the complex does not assemble. Expression of a cDNA lacking the SXL-binding sites will cause formation of the MSL complex in females with resulting hyperactivation of both X chromosomes and severe effects on viability (95).

Several efforts have been made to determine the order of assembly of the MSL proteins during complex formation (21, 67, 170) and to map their sites of interaction (36, 139, 183). Although there are slight variations in the reported order of assembly, perhaps due to the use of different tissues or cultured cells, it is clear that MSL1 and MSL2 provide the scaffold for the assembly of the other proteins and roX RNA. Furthermore, the interaction of MSL1 and MSL2 allows the complex to bind DNA. This is significant since neither protein nor any of the other members of the core complex contain any identifiable DNA-binding domain (36, 67, 121). The other components are added to the MSL1/MSL2 scaffold as follows. A predicted coiled-coil N-terminal domain of MSL1 associates with the RING finger of MSL2 and a PEHE domain associates with the zinc finger of MOF. In addition, MSL1 associates with MSL3 via a C-terminal domain (36, 139, 183; F. Li, D.A. Parry & M.J. Scott, manuscript in preparation). To be efficiently included in the forming complex, MSL3 must be acetylated at lysine 116 by MOF (21). Experiments that used baculoviruses expressing recombinant proteins in SF9 cells have shown that the interaction between MSL1, MSL3, and MOF greatly influences the latter's activity and specificity: While MOF acetylated MSL1 at very low levels when MSL3 was missing, the presence of all three proteins resulted in a dramatic enhancement of acetylation of histone H4 on reconstituted nucleosomal arrays (139). MLE does not seem to interact with the other MSL proteins; the presence of RNA-binding sites and the fact that it is released from chromosome-associated complexes by RNase treatment (175) suggest that MLE associates with the other MSL subunits by binding with the roX RNA. Two other subunits, MSL3 and MOF, can also bind RNA.

The roX1 and roX2 RNAs are very different in size (4.1 to 4.3 kb and 0.6 kb, respectively) and share no sequence similarity with the exception of a 30-bp region of high identity (60). The region appears to be dispensable to roX function and, therefore, the meaning of this homology is not understood. In spite of their differences, the roX RNAs are redundant: Deletion of either has no effect on the viability of males or females (129), whereas deletion of both has no effect in females but results in male lethality (60, 129). In a few rare cases, some roX-less males are able to form a sufficient amount of complex to allow their survival (129). It is not known whether these complexes use a different RNA for assembly and targeting. *roX1* or *roX2* transgenes relocated to the autosomes by germline transformation attract the MSL complex at their sites of insertion (96). Either type of transgenes can rescue males that lack both of the endogenous *roX* genes. In these instances, the complex associates with the X chromosome in reasonably normal fashion but also spreads significantly in *cis* from the locus of the transgene (155). This ectopic spreading, as well as cases of abnormal spreading along the X chromosome, can vary according to various parameters that include competition from active endogenous *roX* genes and the level of MSL proteins [see references in (92, 94)].

H4: histone 4

Transgene: a gene that is introduced into the genetic material of an organism or of cells in culture by molecular means

Ectopic: the presence of a gene or the manifestation of some aspect of gene function in a place in the genome other than the normal place

Targeting the Complex to the Sites of Action

In experimental circumstances, the MSL complex can spread along the autosomes. Why, then, is it normally restricted to the X? The roX RNAs are rapidly degraded unless they associate with the proteins of the complex, and since both roX genes are X-linked, assembly of the complex must occur at or near the sites of their transcription. From these sites, migration of the complex is restricted to the X chromosome because of its clear affinity for largely undefined chromosome sequences. The most pronounced level of affinity is for the so-called "high affinity" or "entry sites." These sites were originally defined by indirect immunofluorescence staining as 25 to 35 points along the X chromosome where partial (121) or inactive (68) complexes, or complexes formed in the presence of reduced amounts of MSL components (52), would be seen to bind. The sequences of three of these high-affinity sites have been characterized; two are associated with the *roX* genes and one is elsewhere on the X chromosome (91, 149, 155). Although a male-specific DNase I hypersensitive site is present at all three sites, no similarity in sequence exists among them or between them and the rest of the X chromosome. A number of observations suggest that in addition to these high-affinity sites, there are many sequences dispersed along the X chromosome for which the complex exhibits different levels of affinity. First, when immunofluorescence signals are progressively boosted, new sites continue to appear until the limits of resolution of this technique are reached. Second, there are regions of the X chromosome where no high-affinity sites have been mapped, which, nevertheless, attract the MSL complex when they are present as cosmids integrated into autosomes by transgenesis or in X-to-autosome transpositions (56, 149). Finally, an ectopic promoter located in regions of the X chromosome devoid of complex is able to attract the complex upon activation (179). In these instances, the complex association with the activated promoter does not appear to require its uninterrupted "flow" from adjacent regions where it is normally present. Rather, the dynamic association of complexes in these regions allows free complexes to access an isolated active gene located in relative proximity. It appears, then, that open transcriptional units on the X chromosome represent the final destination of the complex from its location on the X chromosome at sequences ranging from the high to the low end of the affinity scale.

The fact that the complex requires both active MLE and MOF subunits to spread beyond the high-affinity sites (68) indicates that spreading along the X chromosome is not a simple stochastic process; rather, it is a process that requires specific catalytic interactions. The acetylation of H4 by MOF may facilitate access to DNA sequences along the X chromosome and the affinity of MLE for RNA may help target transcribing genes.

The lack of binding to the autosomes could be due to the limited amount of complex present in the nucleus and to its sequestration into the X-chromosome compartment (even the smallest of the X-chromosome transpositions analyzed may pair back and enter this compartment). It could also be due to the absence on the autosomes of any affinity sequences found on the X, even those for which the complex has only limited attraction. This would explain the lack of spreading from the X into the autosomal element of translocations or transpositions (56). In contrast, if a sufficient level of complex is present and available at the site of a roX transgene (for example, by deleting one of the endogenous roX genes), a significant amount of spreading of the MSL complex can occur from the site of the transgene along the autosome (91). An explanation for this paradox could be that years of culturing in the laboratory have resulted in the selection of boundary elements or other regulatory sequences or factors that limit the spread of the MSL complex into the autosomal element of translocations or insertions, thereby avoiding an imbalance in product levels even for a small number of autosomal genes.

Chromatin Modifications and a Model for the Mechanism of Compensation

The binding pattern of the MSL complex at hundreds of sites on the X chromosome in males, visible by immunofluorescence on larval salivary gland polytene chromosomes (**Figure 2**), is very similar to the distribution of histone H4 acetylated at lysine 16 (14, 199). This isoform of histone H4 is generally associated with active chromatin—witness its presence throughout the genome in budding yeast where it maintains the boundary between telomeric heterochromatin and euchromatin (98, 193), and in humans, where it is found ubiquitously on all chromosomes except for the inactive X (88). In Drosophila, H4K16ac, present only in males, is the result of the acetyl transferase activity of the MOF subunit of the MSL complex (2, 189). On the X chromosome of males, H4K16ac is not limited to the promoter region of genes that are compensated; rather, it is found throughout transcriptional units and beyond (188). This observation, the attraction of MSL complex for activated genes (179), and some circumstantial considerations have led to the hypothesis that the primary mechanistic result of the chromatin modifications responsible for dosage compensation is an enhancement in the rate of transcription elongation (188). The acetylation of histone H4 at lysine 16 may reduce the superhelical torsional stress to which DNA is subjected as it wraps around nucleosomes and facilitate the extension rate of the RNA polymerase. The fact that the transcription of X-linked genes is enhanced to the same twofold level by the compensation mechanism, irrespective of the promoter strength of the genes, and the fact that failure of compensation by the MSL complex decreases the output of X-linked genes but does not appear to alter their developmental program are concordant with the hypothesis. In order to achieve an increased steady-state level of X-linked gene transcripts, enhanced elongation must be coupled with enhanced re-initiation.

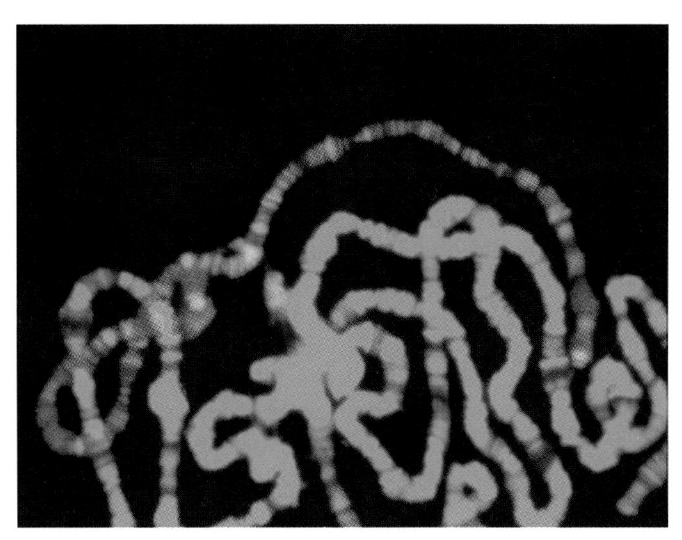

Figure 2

The MSL complex associates exclusively with the X chromosome in males. Indirect immunofluorescence staining of a chromosomal spread from a salivary gland of a third-instar male larva exposed to MSL1 antiserum.

Polytene chromosomes: very large chromosomes that result from multiple DNA replications without cell division. In some tissues newly formed chromosomes remain tightly associated, in register, providing easily scored landmarks that can be used for cytological mapping

James Birchler has proposed an alternative mechanism to compensate for the potential imbalance between X-linked gene products in relation to autosomal gene products in males and females. In this model, dosage compensation is achieved by decreasing the overall level of autosomal transcription in males: The MSL complex prevents an overexpression of X-linked genes in normal males, and the proper level of dosage compensation results from a twofold imbalance of transcription factors on the X relative to the autosomes, which causes a twofold negative dosage effect on target gene expression [for references see (152)]. Recent data do not support this model (P. Becker & M. Kuroda, personal communications).

Histone phosphorylation is another modification that has long been associated with active chromatin [see references in (123)]. The existence of a novel tandem kinase (JIL-1) was reported to be present on all chromosomes in both sexes of Drosophila but highly enriched along the X chromosome in males (90). JIL-1, which was shown to phosphorylate histone H3 in vitro, localizes with the MSL complex on polytene chromosomes and can be coimmunoprecipitated with various MSL proteins (89). Loss-of-function mutations affect the viability of both sexes but have a more pronounced effect in males (205). Reduced levels of JIL-1 lead to the disintegration of chromatin structure in the nuclei of embryos and affect the morphology of polytene chromosomes in both sexes, with the X chromosome in males exhibiting a more severe effect (205). The concentration of histone H3 phosphorylated at serine 10 was also reported to be higher on the X chromosome than on the autosomes in males and to be significantly decreased in mutant individuals, although in the latter, the level of phosphorylation of mitotic chromosomes is normal (205). Recently, JIL-1 has been identified as the wild-type product of Su(var)3-1 (55). JIL-1 does not appear to be a component of the MSL complex [purified either by tandem immunoprecipitation (189) or by the TAP tag method (S. Medjan & A. Akhtar, personnal communication]; it is present on all chromosomes in both sexes and plays a general role in heterochromatin compaction and expansion (55). These considerations raise the possibility that this kinase is enriched on the X chromosome in males as a consequence of the chromatin modifications and the higher rate of transcription that underlie the process of dosage compensation. The same argument would explain the increased sensitivity of this chromosome to JIL-1 loss of function.

Heterochromatin: portion of the genome that is either constitutively (e.g., the region that surrounds the centromere) or facultatively (e.g., the inactive X in mammals or silenced genes in certain regions of a developing embryo) inactive

Euchromatin: portion of the genome that includes the vast majority of active genes. Regions of euchromatin can be reversibly inactivated in some tissues during certain periods in development and adult life

H3: histone H3

Su(var): suppressor of variegation

HP1: heterochromatin protein 1

Several other nuclear factors have been reported recently to be enriched on the X chromosome in males (A. Akhtar, personal communications; H. Hirose, personal communications). Loss of function of these factors affects males more severely than females. At present it is unknown whether these factors are components of the MSL complex or if they directly participate in the mechanism of dosage compensation.

Interactions with Other Chromatin Remodeling Complexes

The X chromosome in males responds dramatically to the loss of function of the general chromatin assembly complexes ACF and CHRAC and the nucleosome repositioning complex NURF. In vitro, ACF (ATP-dependent chromatin assembly and remodeling factor) and CHRAC (chromatin accessibility complex) establish regularly ordered arrays of nucleosomes (86, 202), whereas NURF disrupts nucleosome periodicity (72, 198). Loss-of-function mutations in ISWI (imitation switch protein), the ATPase common to all three complexes, transform the X chromosome in salivary gland preparations into a chromatin mass that has lost all morphological features (54). Loss-of-function mutations in a subunit unique to the NURF complex have the same effect on the X; this is unexpected given that the normal activity of this complex is to make chromatin more accessible (4). X-chromosome morphology can be rescued in males (or in females where formation of the complex has been induced and where mutations in ISWI have the same abnormal morphological effect on the X chromosomes as they do in males) by preventing the occurrence of H4K16ac. Since in vitro, the interaction of purified ISWI with nucleosomes is diminished if H4 is acetylated at lysine16, the possibility that MOF in the MSL complex enhances transcription by interfering with ISWI-mediated formation of regular nucleosomal arrays that are characteristic of inactive chromatin (37).

The most recent interaction of a chromatin factor with the male X chromosome involves the Su(var)3-7A protein, a structural component of heterochromatin that appears to colocalize with HP1 (heterochromatin protein 1) and with the histone methyltransferase Su(var)3-9. Overexpression of this heterochromatin protein results in morphological effects in the chromosomes of both males and females, but the male X is most affected as it assumes a very small and highly compacted shape (49).

Do all of these interactions that appear so specific for the X chromosome in males reflect the existence of an extensive regulatory mechanism for dosage compensation in which the MSL complex is only one component? It is natural to speculate that a twofold enhancement of transcription represents a level of regulation so precise that it would require a balance between two or more opposing mechanisms. Yet, the general similarity of the effect of very different complexes or remodeling activities suggests that the common denominator may be the unique characteristic of the chromatin of the male X chromosome which renders it more sensitive to disturbances than autosomes or X chromosomes in females.

DOSAGE COMPENSATION IN *CAENORHABDITIS ELEGANS*

The soil nematode *C. elegans*, like many organisms, has a sex-determination system based on the number of X chromosomes inherited by the embryo. In *C. elegans*, XX embryos normally develop into self-fertile hermaphrodites and XO animals develop into males. This X-chromosome counting mechanism creates the potential for expression differences in X-linked genes between the sexes, and it is apparent that such differences are poorly tolerated. The compensation strategy chosen by *C. elegans* to rectify this dosage difference is to decrease the level of gene expression from each of the two hermaphrodite X chromosomes by one half. The particular mechanism adapted by this species bears little resemblance to dosage compensation mechanisms adapted by other organisms, illustrating different evolutionary paths to the creation of dosage compensation.

The X:A Ratio Determines Sex and Dosage Compensation

The ratio of X chromosome to autosome ploidy is the primary determinant of the sexual phenotype in *C. elegans* and is directly translated into extensive sex-specific phenotypes: approximately one third of the cells are structurally and functionally divergent in adult males and hermaphrodites (194). An animal with a single X chromosome per diploid autosomal set develops as an XO male, whereas an animal with two X chromosomes per diploid autosomal set develops as an XX hermaphrodite. Animals with an X:A ratio of 0.67 or less (e.g., 1X:2A or 2X:3A) always develop as males, whereas once an X:A threshold of 0.75 or greater is reached (e.g., in 3X:4A animals), hermaphrodite development results (34). XX hermaphrodites are somatically females that undergo a transient masculinization of the germ cells during larval development, giving rise to a limited number of stored spermatids that are used for self-reproduction; gametogenesis in the XX individual is strictly limited to the production of oocytes (reviewed in Reference 181).

Dosage compensation responds to the X:A ratio similarly, with full implementation of the mechanism occuring in animals in which the ratio reaches that of 2X:2A or 3X:4A. The effect of the mechanism is to reduce the abundance of X-linked transcripts in XX hermaphrodites to that of XO males by reducing by 50% the expression level of genes on each X chromosome. This effect is of the same magnitude as that achieved by dosage compensation in Drosophila, albeit in the opposite direction. The existence of completely different mechanisms that can achieve a global difference limited to a twofold effect reveals the powerful forces that act in evolution to equalize X-linked gene expression between the sexes.

The Dosage Compensation Complex

Given that the X:A ratio determines both sex and dosage compensation, it is not surprising that the regulation of both pathways is directed by overlapping sets of proteins. Genetic screens performed to look for mutations that preferentially affected the rate of survival of one sex or the other identified a number of genes with defects in both sex determination and dosage compensation, named *sdc* genes (*s*ex *d*etermination and *d*osage *c*ompensation), as well as a number of genes that acted specifically in dosage compensation with more subtle effects on sex determination (for review, see Reference 134). Mutations in the latter genes result in increased XX-specific lethality, with survivors characteristically developing as poorly elongated, or dumpy (*dpy*) animals. The dumpy appearance of these animals, due to an overexpression of X-linked loci, is also observed in 3X:2A animals arising from X chromosome non-disjunction (82, 134). These mutations have no effect on the viability or appearance of XO individuals. They identified the components of a repressive complex, the DCC (dosage compensation complex) that specifically assembles onto the X chromosomes in an X:A ratio-dependent manner and is responsible for dosage compensation (34).

The DCC has been extensively characterized, both genetically and biochemically. The complex is composed of at least 10 polypeptides that are interdependent for colocalization and stability. The *sdc* and *dpy* gene products are included in potential "subcomplexes." One, the putative DPY subcomplex (**Figure 3*B***), is composed of DPY-26, DPY-27, DPY-28, and MIX-1, and shares components with a separate and highly conserved mitotic 13S condensin complex. The SDC subcomplex is composed of SDC-1, SDC-2, SDC-3, DPY-21, and DPY-30. The DCC has a single known target outside of the X chromosome, the autosomal sex-determination gene, *her-1*. Repression of *her-1* by the DCC is required for hermaphrodite development, thus providing the link between sex determination and dosage compensation. Curiously, whereas the DCC provides a twofold repression of X-linked loci, it mediates a 20-fold repression of *her-1* (31). The potential reason for this disparity probably reflects subtle differences in roles of different members of the complex, as well as complex components at the respective loci (210).

The DPY/Condensin-Like Components of the DCC

Components of the DCC are shared with another protein complex that is essential in all tissues of both sexes, the 13S condensin complex (**Figure 3*C***) (29). The condensin complexes functions during mitosis and meiosis for DNA compaction and sister chromatid resolution (29, 70).

MIX-1 is a component shared by the condensin and DC complexes (**Figure 3*a*,*b***). MIX-1 is an ortholog of SMC2 (*s*tructural *m*aintenance of *c*homosomes) that is a conserved component of condensin complexes in most eukaryotes, (70, 116) Mutations in the *mix-1* gene were not originally identified in dosage compensation genetic screens because of its essential role in mitotic and meiotic events in both sexes where it is required for centromere function and chromosome segregation. In this capacity, MIX-1 pairs with its conserved partner, SMC-4. In somatic cells of XX animals, MIX-1 is found on the X chromosomes in the DCC where it associates with DPY-27, which has sequence homology to SMC-4 (32, 69, 116), yet it is also observed in the condensin complex of mitotic cells in the soma and germ line (29). MIX-1's simultaneous presence in two separate complexes is likely guided by interactions with other components of the complexes; for example, SMC-4 and DPY-27 in the condensin and DCC complexes, respectively (70).

The other components of the condensin-like subcomplex are DPY-26 and DPY-28. Both proteins are expressed in germ cells, but their localization is not restricted to the X chromosome in this tissue (117). The stability and localization of each protein in germ cells and in somatic cells requires the presence of the other. Their activity in germ cells is consistent with a role in sister chromatid cohesion. *dpy-26* and *dpy-28* mutant animals exhibit a *him* phenotype (*h*igh *i*ncidence of *m*ales),

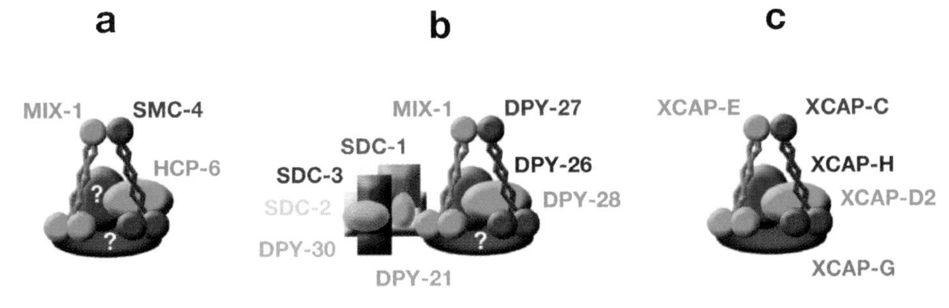

Figure 3

The *C. elegans* dosage compensation complex is an adapted form of the condensin complex. (*a*) The *C. elegans* condensin complex that functions during cell division. (*b*) The SDC subcomplex of the dosage compensation complex which is proposed to help adapt the condensin complex to target-specific repression. This subcomplex shares MIX-1 with the condensin complex, but uses the DPY proteins as XCAP subunits. (*c*) The condensin complex that is conserved in all eukaryotes; homologous components are shown as similar shapes with the same colors in the two worm complexes. Figure kindly provided by Barabara Meyer.

which correlates with an increased frequency of meiotic chromosome non-disjunction and points to roles for these proteins during meiosis (117, 165). DPY-26 and DPY-28 share sequence homology to the non-SMC condensin components XCAP-H and XCAP-D2, respectively (134). The DPY/condensin-like subcomplex thus appears to be a form of the 13S condensin complex that has been adapted to a specialized role in sex determination and dosage compensation. It is conceptually pleasing to consider that a form of a complex whose conserved role is to hypercondense DNA has evolved to mediate a twofold repression of the X chromosome. How the original mechanism has been "tuned" to accomplish a surprisingly precise 50% repression is a fascinating question.

The SDC Subcomplex

The SDC subcomplex consists of at least three components, SDC-1, SDC-2, SDC-3, DPY-21 and DPY-30, each of which can be coimmunoprecipitated with antibodies against the others (31). SDC-1 encodes a novel but conserved protein that has 7 N-terminal C2H2 zinc finger motifs. SDC-1 is expressed in both males and XX animals, so its cellular role is unlikely to be limited to the DCC, but the major apparent defect in *sdc-1* mutants is partial masculinization of XX animals (146). The linchpin of DCC assembly and localization to the X chromosomes is SDC-2. This protein is the only component that has no maternal contribution, is solely expressed in XX embryos, and associates with the X chromosome independently of the other DCC components (46, 147). *sdc-2* expression is repressed by XOL-1, the master sex-determination switch gene, and as such is the only DCC component that exhibits sex-specific regulation (135). This repression in XO males in turn gives *sdc-2* the property of a sex-specific switch gene for dosage compensation regulation.

Like SDC-2, SDC-3 is required to recruit the DPY/condensin-like complex to the X chromosome (45, 210). SDC-3 has a role in repression of the *her-1* locus that differs from its activity on the X. This is revealed in specific mutations called *sdc-3(Tra)*, which affect sex determination but not dosage compensation (51). SDC-3 has two protein domains that have separable functions in its roles on the X and at *her-1*: a C2H2 zinc finger domain that is required for DCC function, and a myosin-like ATP-binding domain that is required for *her-1* repression (100). The protein is maternally expressed in germ cells but does not preferentially accumulate on chromatin; its accumulation and X-chromosome localization are normally seen only in XX embryos and are dependent on other members of the DCC (45).

DPY-21 is also a component of the DCC, but its absence has little obvious effects on stability or X- localization of other DCC components (32, 45, 46, 116, 117). DPY-21 recruitment to the X does depend on all other DCC proteins (except SDC-1) but its stability does not (210). This is different from many of the other DCC proteins, whose stability is decreased in the absence of other DCC components. DPY-21 also is unique among the DCC proteins in that it is absent from the complex that assembles at the *her-1* locus (210).

In addition to SDC-2 and SDC-3, recruitment of the DPY/condensin-like complex to the X chromosomes also requires DPY-30. DPY-30 is a ubiquitously expressed nuclear protein with homology to a component of the *S. cerevesiae* COMPASS complex, which contains Set1p, a histone H3 lysine 4-specific histone methyltransferase, and has recently been verified to be a DCC component (C. Hassig, T. Wu & B. Meyer, personal communication; 83, 84, 142). *dpy-30* mutants exhibit a maternal effect-mediated XX-specific lethality, but XO animals also have developmental defects, suggesting additional roles for DPY-30 outside of dosage compensation.

Sex-Specific Recruitment of the DCC to the *her-1* Locus

her-1 is a switch gene in sex determination. Loss-of-function mutations in *her-1* cause XO animals to develop into hermaphrodites; gain of function (or loss of *her-1* repression) masculinizes XX animals (81). Therefore *her-1* expression is required for male development and needs to be repressed for hermaphrodite development. The DCC localizes to both the *her-1* locus and on the X chromosomes, and almost all of the components found at one target are present at the other (31). A striking difference in the effects of the complex at the two targets is the level of repression that is achieved. Whereas repression at the X chromosomes is limited to twofold, the repression of *her-1* approaches 20-fold. SDC-2 provides the switch for hermaphrodite development by directly repressing *her-1* expression (45). It accomplishes this by, together with SDC-1 and SDC-3, recruiting the rest of the DCC. SDC-3 is thought to be a major player in the specificity of this recruitment, since it binds to the *her-1* locus in the absence of other components (210). The *sdc-3(Tra)* mutations that can cause defects in sex determination but not dosage compensation result in defective loading of the DCC to the *her-1* locus, but do not affect assembly on the X chromosomes (31). The DCC complex that binds to *her-1* is also different in composition than the one on the X, because DPY-21 is not present in the former but is observed in the latter (210). This difference in components between the complexes at the two loci may contribute to the difference in level of repression observed (134).

Sequence-Specific Recruitment of the DCC

Three DNA elements within the *her-1* locus are required for DCC recruitment and repression (31, 115). Two of these elements have a 15-nucleotide repeat in common that is essential for *her-1* repression, and these sequences can recruit SDC-3 binding (31), but none of the sequences are enriched on the X chromosome. This suggests that different sequence elements are required for recruitment of the DCC to its different targets. Since SDC-2 can localize independently to the X chromosomes, and SDC-3 can recognize and bind *her-1* elements, it has been suggested that these two proteins form the basis for the recruitment of the DCC to the different targets (210).

If SDC-2 directs the DCC to the X-chromosome, what sequence elements is it recognizing and what directs its global activity along the entire X chromosome? Four models have been proposed to account for these properties (40). Model I, similar to the mechanism of X-inactivation initiation at the Xist locus in mammals, proposes a single site of DCC recruitment, followed by extensive spreading *in cis*. In Model II, which resembles a model proposed for Drosophila, a limited number of recruitment sites from which the DC complex spreads *in cis* would be present along the X chromosome. Model III is similar to Model II, except that instead of spreading in *cis*, repression occurs through long-range interactions. Finally, Model IV predicts a large number of recruitment sites densely arranged along the X, each of which recruits the complex for local repression.

These models were recently tested by using antibodies directed against components of the DCC and a combination of X chromosome deletions and detached duplications of the X chromosome (40). The concept was straightforward: if a single recruitment site was required (Model I) then only one detached duplication would recruit the DCC and conversely a single deletion could eliminate all DCC recruitment. If several, but not all, regions of the X could recruit, this would rule out Models I and IV, since IV requires a dense arrangement of recruitment sites with local repression. Model III would be ruled out if large regions were observed that failed to recruit the DCC as detached duplications, but did so when attached at their normal site, which would indicate spreading of the complex *in cis*. Model I was eliminated, since multiple regions were found to be independently capable of DCC recruitment. Although multiple recruitment sites were apparent, Model IV was ruled out since regions known to contain dosage-compensated loci were unable to recruit the DCC autonomously when detached, even though the DCC was found at these same regions on the intact X. This observation is more consistent with recruitment at a neighboring site (since the dosage compensated regions could not autonomously recruit the DCC), followed by spreading into these regions as in Model II, and is inconsistent with action over a distance as in Model III. Therefore, recruitment of the DCC to a limited number of loci along the X chromosome appears to be followed by spreading *in cis* of the complex along the chromosome (**Figure 4**). Curiously, the spreading *in cis* was apparently limited to X chromosome sequences, since no spreading into the autosome portion of X:autosome fusion chromosomes was detected. Whether this represents the presence of X-limited sequences that promote spreading, or autosomal sequences that prevent it is not known (40).

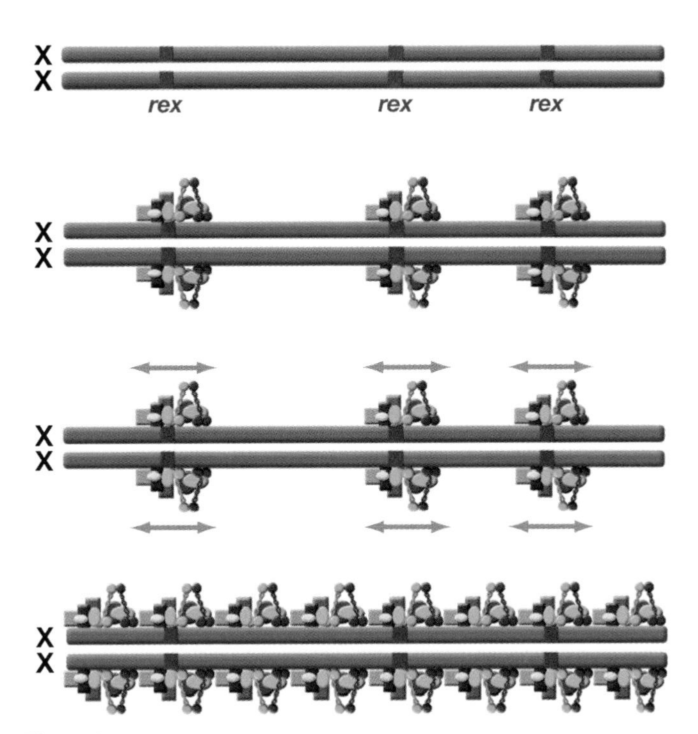

Figure 4

Recruitment and Spreading Model of *C. elegans* dosage compensation. The DCC is initially recruited to the X chromosome at a number of recruitment loci (recognition elements on X; *rex*) that are nonuniformly spread along the chromosome. The complex subsequently spreads in *cis* to neighboring regions for a near-complete coverage of the chromosome. (Figure kindly provided by Barabara Meyer)

X-Specific Recruitment Elements

Molecular mapping and identification of a DCC recruitment site was attempted using overlapping cosmids in repetitive transgenes, and further refined to a 4.5-kb fragment.

Multicopy transgenes carrying this element were capable not only of recruitment but also of titration of the DCC from the endogenous X chromosomes. The element contained a number of suggestive sequences, including an X-specific repetitive sequence and a region with syntenic conservation in *C. briggsae*, but none of these regions were sufficient for recruitment of the complex on their own (40). In conclusion, the mechanism for DCC recruitment seems to require X-specific sequence elements, but the nature of this attraction is not yet understood.

The X:A Ratio and XOL-1

The X:A ratio is read by its effect on the expression of the gene *xol-1* (*XO lethal*), the master switch gene in dosage comp and sex determination. Loss of XOL-1 activity in XO animals results in inappropriate activation of dosage compensation, X-linked gene repression, and developmental arrest. Conversely, ectopic activation in XX animals inactivates dosage compensation, leading to overexpression of X-linked loci (135, 172). XOL-1 is structurally related to the GHMP family of small molecule kinases (e.g., homoserine kinase), but does not appear to have functional kinase activity (120). The role of *xol-1* as master regulator of both sex determination and dosage compensation is to repress *sdc-2*. Whether the repression is direct or indirect is not known. *sdc-2* repression prevents targeting of the DCC to the X and *her-1*, which serves to simultaneously prevent X-linked gene repression and activate male differentiation (23, 45, 46, 135, 146, 172). Controlling *xol-1* expression is therefore the functional read-out of the X:A ratio, so understanding how XOL-1 function is regulated is the key to understanding how this ratio is interpreted (**Figure 5**).

X-Signal Elements

The number of X chromosomes, or the numerator component of the X:A ratio, is obviously the essential quantifier of X-chromosome dosage. To find the genetic dosage elements linked to the X chromosome (signal elements components of the numerator signal), duplications and deletions spanning the X chromosome were assessed for effects on DC regulation. The logic of this approach was straightforward: Duplications containing X signal elements would be interpreted as an XX dosage when present in XO animals, and cause XO-specific lethality. Conversely, deletions of signal elements in XX animals would cause an underestimation of X dosage and prevent dosage compensation, causing XX-specific defects. A possible difficulty in isolating these potentially lethal mutations was overcome by screening for the genetic alterations in XO and XX animals carrying mutations in *sdc* and *xol-1* genes, respectively. Using this approach, Akerib & Meyer identified three nonoverlapping regions on the left end of the X chromosome that fulfilled the requirements for signal elements. Each of the regions showed the expected phenotypes, but in an additive fashion, illustrating that the numerator signal of the X:A ratio is polygenic. Of the three regions, region 3 was shown to have the strongest single effects on sex determination (1). Hodgkin and co-workers also identified region 3 using a novel duplication in a similar assay, and further refined it (using cosmid transgenes) to a 12–30-kb interval; they named the gene *fox-1* (feminizing locus on X) (82). The *fox-1* gene encodes a protein containing an RNP/RRM RNA-binding motif and is required for the posttranscriptional repression of *xol-1* (145a, 187a).

The genetic loci in regions 1 and 2 have not as yet been identified, but some information on their functions has been obtained by assessing their effects on *xol-1* expression. Both regions appear to repress *xol-1* activity at the transcriptional level (B. Meyer, personal communication). A potential partner in this activity encodes a fourth signal element, *sex-1* (*s*ignal *e*lement on *X*), which was identified in a mutant screen using ectopic expression

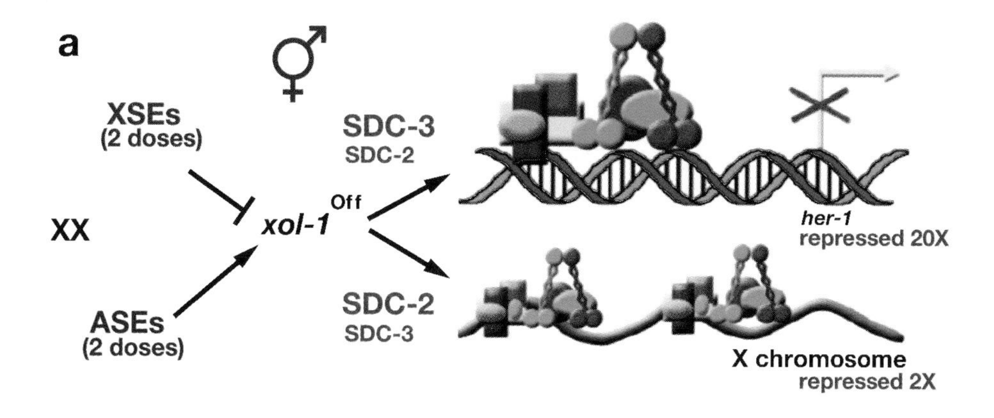

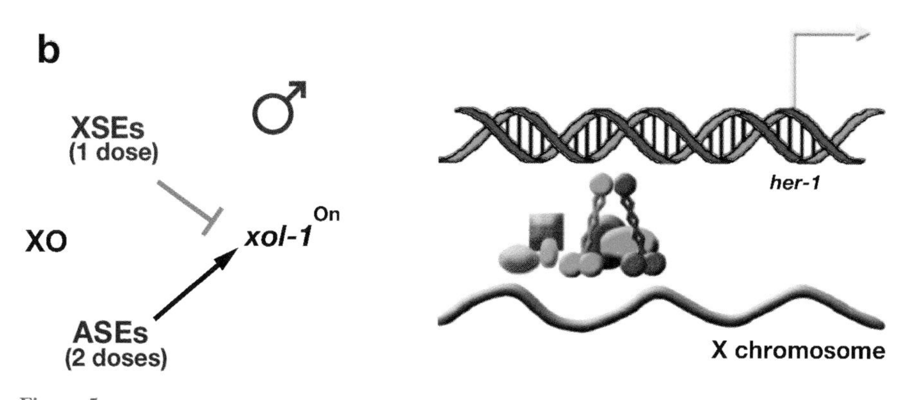

Figure 5

Sex-specific regulation of *C. elegans* dosage compensation. (*a*) In XX embryos, the dosage of X-linked signaling elements (XSEs) represses *xol-1* and allows SDC-2 and SDC-3 expression which target the DCC to the *her-1* promoter and X linked recruitment sites, leading to 20-fold and twofold repression, respectively, of the targets. Repression of *her-1* promotes the hermaphrodite mode of sex determination. (*b*) In XO embryos, the autosomal signal elements (ASEs) dosage overcomes the limited XSE dosage, which promotes *xol-1* expression and represses SDC-2 and SDC-3. (Figure kindly provided by Barabara Meyer)

of a *xol-1:lacZ* reporter in XX embryos (23). *sex-1* encodes a nuclear hormone receptor (NHR) and has been shown to repress the transcription *xol-1*. The dosage of *sex-1* manipulates the control of dosage compensation precisely as predicted for an essential numerator element. Indeed, increasing the dosage of *xol-1* promoter elements in multi-copy transgenes can dilute the XO lethality caused by extra copies of *sex-1*, suggesting a direct titration effect (23). The effect on *xol-1* transcription is likely to be direct, since SEX-1 binds to the *xol-1* promoter. The SEX-1 DNA-binding domain is homologous to retinoic acid and the Rev-Erb orphan NHRs, however no canonical NHR binding sequences have been identified in the *xol-1* promoter. An intact ligand-dependent activation domain exists in SEX-1, leaving open the possibility that a SEX-1 ligand is involved in its targeting and repressive activities at the *xol-1* locus (134).

The regulation of *xol-1* is thus a cumulative regulation, involving both transcriptional

Xi: inactivated X chromosome

Xa: active X chromosome

and post-transcriptional repression. The level of the repression is determined by X chromosome dosage because the effectors of the repression are themselves X-linked. The cumulative regulation of *xol-1* also involves autosomal-linked activators, which counter X-linked repressors to signal the X:A ratio (166).

Autosomal Signaling Elements

The denominator of the X:A ratio consists of autosomal signaling elements. Recent genetic screens have yielded several candidates that have been named *sea* genes (*s*ignal *e*lement on *a*utosomes) and that appear to function as *xol-1* activators. One of these, *sea-1*, has been identified and encodes a T-box transcription factors that opposes X signal elements by activating *xol-1* transcription (166). This provides a conceptually pleasing, if somewhat simplistic model in which the relative dosages of transcriptional activators and repressors, determined soley by the dose of the chromosomes on which they reside, control the level of a single switch protein. If the relative repressor/activator activity ratio is near that achieved by 2X:3A, repression of *xol-1* is insufficient and the switch is engaged to repress dosage compensation and activate male development. As the level of repressor/activator activity ratio increases (e.g., 3X:4A), repression is sufficient to prevent activation of the switch.

Dosage compensation in *C. elegans* thus provides an amazing example of a multiply adaptive response that was generated to compensate for the potentially disastrous consequences of sex chromosome evolution. In the case of *C. elegans*, an essential complex with ancient structural roles in chromatin compaction has been adapted for less efficient compaction and X-chromosome targeting. The amazing fine-tuning of this process to create a twofold reduction in transcription—only when it is located on the X chromosome—is still poorly understood. Its elucidation will make an important contribution to our understanding of genome regulation.

DOSAGE COMPENSATION IN MAMMALS

In the cells of female mammals, one X chromosome is transcriptionally silenced to achieve dosage compensation between XX females and XY males (122). The term X inactivation is used to describe the initial transition from a transcriptionally active to an inactive state and also the subsequent stable maintenance of the silent state. The inactive X chromosome (Xi) differs from the active X chromosome (Xa) and autosomes in differentiated cells, as it is characterized by a unique combination of epigenetic features including histone modifications and DNA methylation. These modifications are acquired sequentially during the onset of X inactivation and act redundantly to maintain X chromosome silencing. In this review we discuss the mechanisms by which the unique chromatin structure of the Xi is established and maintained, and the role of epigenetic modifications in regulating transcriptional silencing.

Histone Covalent Modifications Associated with Silencing

Different combinations of histone modifications are thought to establish transcriptionally active euchromatin and transcriptionally silent heterochromatin [see references in (87, 190, 200)]. According to this "histone code" hypothesis, epigenetic marks on the histone tails provide binding sites for proteins that regulate gene expression. Replacement of core histones with variant histones is another chromatin alteration that is employed to modulate gene expression. In addition to histone modifications, CpG methylation is a covalent DNA modification that is implicated in chromatin structure and transcription. Given that much transcriptional regulation is achieved via changes in chromatin structure, it is not surprising that the Xi shows a distinct signature of chromatin marks when compared with the Xa and autosomes. Below we describe the features of chromatin that distinguish the Xi from the Xa and autosomes.

Enrichment for histone H3 methylated at lysine 9 is one of the hallmarks of heterochromatin (5, 87). There is enrichment of histone H3 dimethylated on lysine 9 (H3K9me2) on the Xi, specifically at the promoters of silenced genes, such that H3K9 methylation at the promoter correlates with transcriptional inactivity on the Xi (13, 79). Some human cell types also exhibit increased staining of histone H3 trimethylated on lysine 9 (H3K9me3) on the Xi in metaphase (28). The SET domain family of histone methyltransferases (HMTase) catalyzes the methylation of lysine residues (5). Five mammalian SET domain proteins, Suv39h1, Suv39h2, G9a, Eset/SETDB1, and EZH2, have HMTase activity on H3K9 in vitro and are candidates for catalyzing H3K9 methylation on the Xi (104). Eset/SETDB1 and Suv39h1 mediate formation of the trimethylated form of H3K9 (160, 203). As G9a catalyzes production of the H3K9me2 in vitro and in vivo (173), this enzyme is the most likely candidate, to serve as the HMTases that mediates H3K9 methylation on the Xi.

The significance of H3K9me2 enrichment on the Xi is likely to involve recruitment of H3K9me2 binding proteins that further regulate chromatin structure. HP1 binds methylated H3K9 in vitro, and this protein is required for formation of pericentric heterochromatin (6, 58, 103). All three forms of human HP1 appear to be enriched on the Xi as well as on pericentric regions (27), suggesting that H3K9me2 contributes to heterochromatin formation by recruiting HP1 to the Xi. None of the three forms of Hp1 are enriched on the Xi in mouse cells (161), suggesting that HP1 enrichment on the Xi may be specific to human cells. Alternatively, as adult human cells and embryonic mouse cells were examined, it is possible that HP1 enrichment on the Xi is specific to a particular developmental stage.

Regulated silencing of homeotic genes during *Drosophila* development requires methylation of H3 at lysine 27 (H3K27), mediated by the ESC-E(Z) complex (22, 141). Thus, this HMTase complex is involved in the formation of facultative heterochromatin. Loss of function of Eed, the murine ESC homolog, results in reactivation of X-linked genes, suggesting a role for H3K27 methylation in regulating X inactivation in female mammals (204). Further investigation revealed that Eed is present in a complex with Ezh2, the mouse homolog of E(Z) (53), and that both of these proteins are enriched on the Xi (124, 163, 186). Ezh2 is capable of methylating H3K27 in vitro (22, 43, 102, 141), and there is enrichment of the trimethylated form of H3K27 (H3K27me3) on the Xi in some cell types (66). These data argue that the Eed/Ezh2 complex mediates the accumulation of H3K27me3 on the Xi.

A model for the role of methylated H3K27 in transcriptional silencing is based on findings in *Drosophila*. In flies, H3K27me2 provides a binding site for the chromo-domain protein Polycomb (PC) (22, 43, 58, 102), a component of the Polycomb repressive complex (PRC1), which is essential for maintaining homeotic gene silencing (59, 187). Alterations in chromatin structure mediated by H3K27me3-bound mammalian PRC1 provide an attractive model for the role of this histone modification in mediating transcriptional repression on the Xi. PRC1 proteins are enriched on the Xi (47, 57, 164). Ring1/Ring1a, which is a histone H2A ubiquitin ligase, and the related gene product Rnf2/Ring1b, are PRC1 proteins that are required for increased amounts of monoubiquitinated histone H2A on the Xi (47, 57). In combination these results suggest that two PcG complexes contribute to stable X chromosome silencing. First, Eed/Ezh2 histone methyltransferase complex mediates H3K27 methylation, which in turn generates binding sites for PRC1, which results in histone

H3K9me2: H3 dimethylated at lysine 9

HMTase: histone methyltransferase

H3K27me3: H3 trimethylated at lysine 27

H3K4me2: H3 dimethylated at lysine 4

H3R17me2: H3 dimethylated at arginine 17

H3K36me2: H3 dimethylated at lysine 36

HDAC: histone deacetylase

H2A monoubiquitination. PRC1 can mediate silencing of target genes by interfering with SWI/SNF chromatin remodeling machinery, blocking transcriptional initiation, or recruiting additional silencing activities, though the contribution of histone ubiquitination to these silencing functions of PRC1 remains to be evaluated (50, 59, 99, 106, 184).

In contrast to H3K9 and H3K27 methylation, which correlate with transcriptional silencing, methylation of histone H3 at lysine 4 (H3K4) or arginine 17 (H3R17) shows a strong correlation with gene activity (192). Immunofluorescence and ChIP with antibodies directed against the dimethylated form of H3K4 (H3K4me2) shows that H3K4me2 is underrepresented on the Xi (13, 176). A similar result is obtained using antibodies raised against dimethylated H3R17 (H3R17me2) (30). The absence of these two methylation marks on the Xi is consistent with its silent state.

It has been suggested that histone H3 dimethylated on lysine 36 (H3K36me2) causes gene repression in yeast (105, 191). However, there is also evidence linking this modification to gene expression in yeast and *Tetrahymena* (101, 114, 180, 191, 208). H3K36me2 is underrepresented on the Xi (30), indicating that enrichment of this modification may be characteristic of active chromatin in mammals.

Hypoacetylation of histones H3 and H4 at lysine residues is commonly associated with heterochromatin and transcriptional inactivity (174). In agreement with this observation, the Xi appears devoid of histone acetyl modifications (88). When interphase cells are immunostained with an antibody raised against H3 acetylated at lysine 9, the Xi is understained, appearing as a hole (10, 12). The same result is observed using antibodies raised against H4 acetylated at lysines 5, 8, 12 and 16 (30, 79, 88, 97).

Histone acetylation is regulated by a combination of histone acetyltransferase and histone deacetylase (HDAC) activities that add and remove acetyl groups, respectively. Human EED and EZH2 interact with HDAC1 and HDAC2 in vitro and in vivo (201), suggesting that this HMTase complex may regulate acetylation on the Xi by recruiting deacetylases. However, Hdac1 and Hdac2 are not enriched on the Xi in mouse cells with Xi-enrichment of Eed and Ezh2 (124), indicating that the Eed/Ezh2 complex is insufficient to cause an enrichment of these Hdac's. HDAC1 and HDAC2 form corepressor complexes with SIN3A and SIN3B (209). SIN3A and SIN3B complexes are excluded from the Xi (27), further suggesting that these deacetylases do not regulate levels of acetylation on the Xi. There are at least 10 HDAC family members in mammals, providing a number of candidates for the HDACs that might mediate the decrease in histone H3 and H4 acetylation on the Xi. Alternatively, the under-acetylation of the Xi may be achieved by the exclusion of histone acetyltransferase activities from this chromosome.

Replacement of Canonical Histones by Histone Variants

In addition to exhibiting a unique combination of posttranslational modifications on core histones, the Xi contains a high proportion of nucleosomes in which canonical H2A is replaced by the variant histones macroH2A1.1, macroH2A1.2, or macroH2A2 (25, 38, 39). The N terminus of each variant is homologous to canonical H2A while the C termini or non-histone regions (NHRs) of macroH2A proteins show no H2A homology (25, 39, 156). Both the H2A-like domain and the NHR may be involved in proper localization of macroH2A to the Xi, as a truncated protein consisting of the H2A-like domain of macroH2A1 or macroH2A2 localizes to the Xi, and a fusion protein consisting of canonical H2A fused to the macroH2A1.2 NHR also localizes to the Xi (24).

The enrichment of macroH2A on the Xi suggests that this protein may contribute to gene silencing. Indeed, ectopic macroH2A can downregulate gene expression in vivo. Using Gal4 to tether the macroH2A NHR to the promoter of the luciferase gene reduced luciferase activity more than twofold (159). MacroH2A NHR was not assembled into nucleosomes in this assay, indicating that the NHR may be sufficient for silencing outside the context of the nucleosome.

A recent study proposes two mechanisms for transcriptional repression by macroH2A: interference with transcription factor binding and resistance to nucleosome remodeling (3). The transcription factor NF-κB binds chromatin assembled with conventional histones, but not chromatin assembled with histone octamers containing macroH2A1.2. The NHR is necessary to prevent NF-κB binding. In addition, the DNA near the NF-κB binding site has different DNase accessibility in H2A- and macroH2A-containing nucleosomes. In combination, these results suggest that the NHR sterically blocks transcription factor access to its target DNA sequence. Nucleosome remodeling by SWI/SNF complexes usually promotes gene expression; however, nucleosomes containing macroH2A1.2 are resistant to SWI/SNF activity, suggesting that macroH2A may also regulate gene expression by inhibiting nucleosome remodeling (3). The H2A-like domain is responsible for resistance to SWI/SNF activity. Thus, both the H2A-like domain and the NHR may contribute to transcriptional regulation by macroH2A.

Two variant histones are less abundant on the Xi than on the Xa and autosomes, H2A-Barr body deficient (H2A-Bbd) and H2AZ (26, 27). Although little is known about the function of H2A-Bbd, H2AZ has been shown to antagonize silencing in *S. cerevisiae* (130). H2AZ may be underrepresented on the Xi because most genes on this chromosome are repressed. Localization of H2AZ to pericentric heterochromatin has been observed in mouse extraembryonic tissue (107, 167), indicating that this modification is not excluded from all heterochromatin. It therefore seems likely that variant histones are used in combination with other epigenetic marks to establish diverse forms of chromatin.

DNA Methylation

DNA methylation appears to play an important role in maintaining gene silencing on the Xi. Upstream sequences of genes on the X chromosome are hypermethylated on the Xi and hypomethylated on the Xa (7, 162, 206). Treatment with the DNA-demethylating agent 5-azadeoxycytidine results in reactivation of several X-linked genes (138).

ICF (Immunodeficiency, Centromeric instability, and Facial anomalies) syndrome results from a mutation in the DNA methyltransferase DNMT3b (78). Cells deficient for DNMT3b show hypomethylation of DNA and reactivation of genes on the Xi (75). In addition, one class of repetitive elements, LINE-1 elements, which are normally hypermethylated on both the Xi and Xa, are hypomethylated exclusively on the Xi in DNMT3b mutant cells (73). In combination, these data indicate that DNMT3b contributes to DNA methylation and gene silencing on the Xi. The two other DNA methyltransferases, DNMT3a and DNMT1 (11), may also play a role in DNA methylation on the Xi.

In *Neurospora* and *Arabidopsis*, trimethylation of H3K9 is required for DNA methylation (195). In *Neurospora*, HP1 is essential for DNA methylation, suggesting that HP1 binds methylated H3K9 and recruits a DNA methyltransferase (62). In mammalian cells, H3K9 trimethylation is required for Dnmt3b-dependent DNA methylation at pericentric heterochromatin (113). Therefore it is possible that H3K9 methylation directs DNA methylation on the Xi as well. A mouse H3K9 HMTase activity cofractionates with Dnmt3a (44). Hdac1 also cofractionates with Dnmt3a and the H3K9 HMTase activity, indicating that histone acetylation, histone methylation, and DNA methylation may be coordinately regulated in mammalian cells. The association of DNA methyltransferase and histone methyltransferase activities may be important for the spread of heterochromatin as it is possible that methyltransferases recruited to one nucleosome can modify adjacent nucleosomes. Although a functional interaction of methylation and deacetylation complexes has yet to be identified on the Xi, it seems likely that this observation will be extended to X inactivation.

The Inactive X Exhibits Late Replication

Several types of heterochromatin, including silenced X chromatin, replicate late in S phase. Genes on the Xa replicate earlier than their counterparts on the Xi (76, 77, 182, 197). Analysis of two replicons on the X chromosome showed that the same origins fire both on the Xi and the Xa (35), suggesting that the same origins are differentially regulated on these chromosomes. This study raises an interesting question: what mechanisms are used to direct different behavior of the same origins on two homologous chromosomes within a single nucleus?

Studies in *Drosophila* may provide insight into the link between replication and silencing. Fly HP1 binds to components of the origin recognition complex (ORC) and flies mutant for an ORC protein show abnormalities in formation of heterochromatin (151). The three mammalian HP1 isoforms colocalize with heterochromatic regions, including the heterochromatin of the Xi (27). HP1 is thought to nucleate the spread of heterochromatin by binding methylated H3K9 and recruiting HMTases to methylate H3K9 on neighboring nucleosomes (6, 103). The interaction between HP1 and ORCs suggests two distinct models for the coregulation of replication and silencing. Late-replicating origins on the Xi may recruit HP1, which mediates silencing. Alternatively, HP1 on the Xi could mediate a change in chromatin structure that affects both gene expression and replication timing.

Late replication timing and DNA methylation on the Xi show an intriguing relationship. Treatment with the DNA-demethylating agent 5-azadeoxycytidine can trigger early replication of the Xi and reactivation of X-linked loci (74). In addition, cells deficient for DNMT3b show early replication of reactivated X-linked genes (75). In DNMT3b mutant cells, a number of X-linked genes are unmethylated. A subset of these unmethylated genes replicate early and are expressed, suggesting that DNA methylation can influence replication timing.

The Transcription of Xist Triggers and Orchestrates the Process of Inactivation

While most genes are silenced on the Xi and expressed from the Xa, the *XIST* gene shows the opposite expression pattern [reviewed in (16)]. *XIST* encodes a 17-Kb, spliced, polyadenylated, noncoding RNA that stably associates with the entire Xi, appearing to coat this chromosome (15, 17–19). *XIST* RNA's ability to coat the Xi depends on the tumor suppressor gene product BRCA1; in BRCA1 mutant cells, *XIST* RNA does not coat the Xi, although it is produced at normal levels (63). In cells that do not exhibit *XIST* RNA coating of the Xi, because of mutations in BRCA1 or deletion of the mouse *Xist* gene, macroH2A1.2 is not enriched on the Xi (9, 42, 63). These results suggest that *XIST* RNA acts as a scaffold that coordinates at least two related activities, chromosome coating and regulation of chromatin structure (**Figure 6**).

Thus far we have described a number of chromatin modifications specific to the Xi. Experiments were performed to address the importance of several of these modifications in stably maintaining the silent state of the Xi. Individually, deletion of *Xist*, DNA demethylation with 5-azadeoxycytidine, or hyperacetylation of histones by treatment with the HDAC inhibitor TSA result in some reactivation of Xi-linked genes (41). These three treatments in combination induce a

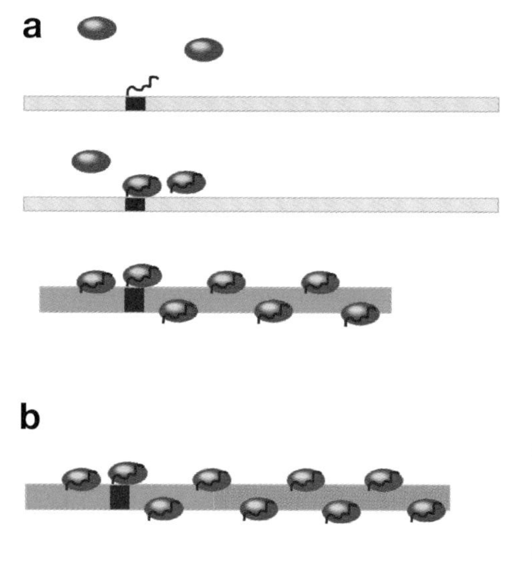

Figure 6

Spreading of the mammalian X-inactivation complex. (*a*) Following the transcription of Xist RNA in the XIC, the DCC complex spreads in *cis* to coat most of the X chromosome.

significantly higher frequency of reactivation than any one treatment alone. Thus, the chromatin modifications that characterize the Xi may work synergistically to maintain this chromosome in an inactive state, and redundant mechanisms may be employed for the extraordinarily stable maintenance of X chromosome silencing.

Cells in early female embryos have two active X chromosomes, one of which becomes silenced in a developmentally regulated fashion. Analysis of the appearance of the chromatin modifications that characterize the Xi indicates that these modifications first occur during the transition from a transcriptionally active to a silent state. This correlation suggests that Xi chromatin modifications contribute to this transition. The first noticeable event is the spread of *Xist* RNA from its site of transcription to coat the X chromosome. This initial *cis*-spread correlates closely with chromosome-wide silencing (153, 185, 207). *Xist* is necessary and sufficient for initiation of X chromosome silencing (127, 158, 207). Female embryonic stem (ES) cells provide a valuable model system to study alterations in chromatin structure that occur during X inactivation because this process is recapitulated when ES cells are induced to differentiate in vitro (169).

ES cells: embryonic stem cells

X inactivation occurs in at least three stages, as characterized by the requirement for *Xist*. The *cis*-spread of *Xist* RNA can be uncoupled from its developmental regulation in ES cells by expressing *Xist* from an inducible promoter. Normally, *Xist* RNA-mediated silencing occurs one to two days after female ES cells begin differentiation (153, 185). However, differentiation is not required for *Xist*-mediated transcriptional inactivation, as silencing occurs in undifferentiated male ES cells when *Xist* expression is driven from an inducible promoter (207). *Xist* RNA can coat the chromosome, but no longer causes silencing if expression is induced more than 36 h after the start of differentiation, suggesting that events that occur upon differentiation interfere with the ability of *Xist* RNA to mediate transcriptional silencing (33, 196, 207). Silencing is dependent on continued *Xist* expression for the first two and a half days after differentiation. In contrast, silencing is not dependent on continued expression of *Xist* RNA in differentiated cells (20, 42, 207). Taken in combination, these results indicate that silencing can be divided into three stages: initiation, establishment, and maintenance. In the initiation phase, *Xist* RNA can cause silencing de novo, and this silencing is *Xist* dependent. During establishment, *Xist* expression can no longer trigger silencing, although silencing continues to be *Xist* RNA dependent. During maintenance, the transcriptional state of the chromosome is stable in that silencing can neither be induced by *Xist* expression nor reversed by loss of *Xist*. It seems likely that transition from one stage to the next is mediated by a precisely ordered series of chromatin modifications directed to the Xi by *Xist* RNA. In addition, these results suggest that the chromatin-modifying activities present in a cell at the time *Xist* expression is upregulated determine whether silencing will occur. Thus, differentiated cells either lack appropriate chromatin-modifying activities or are unable to recruit those activities to the Xi.

Ectopic *Xist* expression from an autosomal transgene in the differentiated HT-1080 human male fibrosarcoma cell line results in many of the same chromatin changes that are observed on the Xi in female cells (71). The autosome bearing an *XIST* transgene is coated by *XIST* RNA, hypoacetylated on histone H4, replicates late in S phase, and shows chromosome-wide silencing. HT-1080 is the sole differentiated cell line in which activation of *XIST* expression has been reported to induce chromosome-wide silencing, indicating that cell lines differ in their ability to enact the large chromatin structural changes associated with inactivating an entire chromosome. It will be interesting to see if the chromatin-modifying activities that normally direct initiation of X chromosome inactivation during differentiation are present in the already differentiated HT-1080 cell line.

Female ES cells undergo X chromosome silencing upon differentiation, facilitating temporal studies of alterations that occur during X inactivation. Changes in histone acetylation and methylation are the first chromatin modifications detected on the Xi in differentiating ES cells. H3K9me2 and H3K27me3 and deacetylation of H3K9 are first detected on the Xi concomitant with or shortly after the initial *cis*-spread of *Xist* RNA (30, 79, 97, 132, 163, 186). H4 hypoacetylation of the Xi occurs within the same time frame, but with slightly slower kinetics (30). The decrease in acetylation and increase in methylation of H3K9 on the Xi occur roughly simultaneously in differentiating ES cells (30, 79), suggesting these modifications are mutually exclusive. As acetylated H3K9 is a poor substrate for HMTases in vitro (171), it is possible that deacetylation must occur prior to methylation. It is unclear whether these deacetylation and methylation occur sequentially on the same histone or on different histones. H3K4, H3R17, and H3K36 methylation disappear from the Xi with the same kinetics as H3K9 deacetylation and H3K9 methylation (30) suggesting that H3 modifications are coordinately regulated during X inactivation.

Enrichment of H3K27me3 is detected transiently on the Xi during ES cell differentiation (163). The Eed/Ezh2 HMTase complex is also transiently enriched on the Xi when X inactivation is initiated in differentiating ES cells and in embryos (163, 186). H3K27me3 and Eed/Ezh2 Xi-enrichment immediately follows *Xist* RNA coating. Cells expressing a mutant form of *Xist* RNA that coats but does not silence the X chromosome display Xi-enrichment of Eed/Ezh2 complex and H3K27me3 (163), indicating that this modification is not sufficient for transcriptional silencing. As H3K27me3 Xi-enrichment can persist in some differentiated cell types (66), and since differentiated cells no longer show Eed-Ezh2 enrichment (124, 163, 186), it is possible that other HMTases may be required to maintain H3K27 methylation on the Xi after it is first established by Eed/Ezh2.

Compared to the chromosome-wide enrichment of H3K9me2, recruitment of macroH2A1.2 to the Xi is a relatively late event in ES cell differentiation, indicating that this histone variant contributes to maintenance rather than initiation of X inactivation. In differentiating ES cells, full *Xist* RNA coating is visible in a fraction of cells at day 1 and is visible in most cells by day 3 (185); macroH2A1.2 recruitment to the Xi begins in some cells at day 6 or 7 and is present in a majority of cells around day 9, indicating at least a 3-day lag between *Xist* coating and macroH2A recruitment (131, 168). The central region of *Xist* RNA is required for macroH2A recruitment to the Xi during ES cell differentiation. Since the region of *Xist* RNA necessary for macroH2A recruitment is different from that necessary for transcriptional silencing, it seems likely that *Xist* RNA functions as a scaffold to direct multiple chromatin-modifying activities to the Xi in a developmentally regulated fashion.

Random Inactivation Versus Imprinting

In the female mouse embryo and in ES cells, X chromosome inactivation is random in that either the maternally or paternally inherited X chromosome (Xm and Xp, respectively) can be inactivated. However, in the extraembryonic, or placental, tissues of the mouse, X inactivation is imprinted such that the Xp is always inactivated.

Random and imprinted X inactivation differ in the timing of appearance of histone modifications relative to *Xist* RNA coating and silencing. During random X inactivation in ES cells, histone modifications occur concomitantly or very shortly after the initial *cis*-spread of *Xist* RNA that triggers silencing. In contrast, there is a noticeable delay between *Xist*-mediated silencing and the acquisition of histone modifications during imprinted X inactivation (85, 150) (**Figure 7**).

The Xi in extraembryonic cells shows the same chromatin modifications that are observed on the Xi in embryonic cells. However, the order in which these modifications appear on the Xi is different between random and imprinted X inactivation. In differentiating ES cells, hypoacetylation of H3K9,

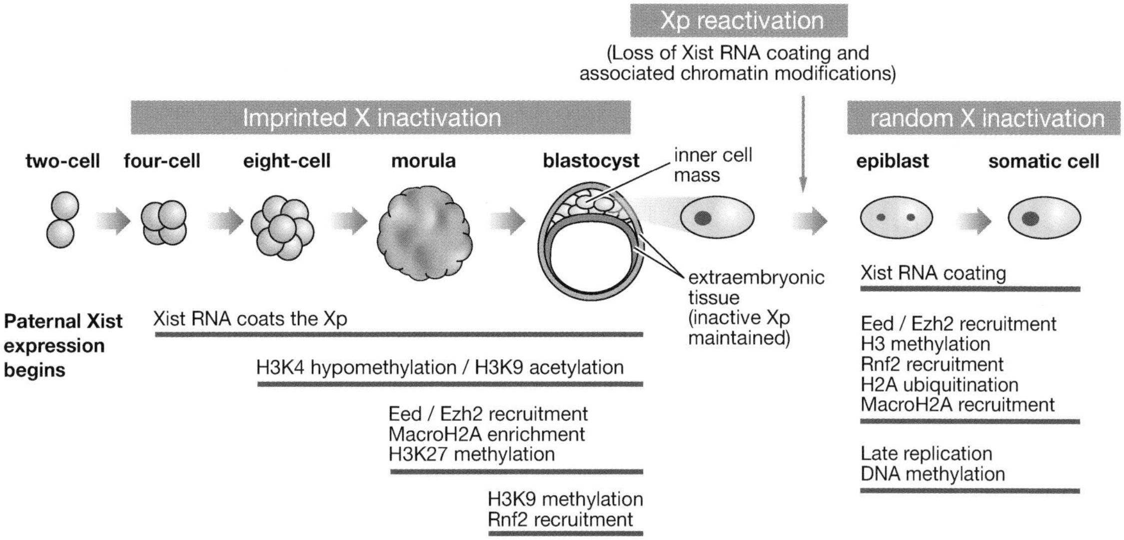

Figure 7

Patterns of X inactivation during the development of placental mammals. X chromosome inactivation is limited to the paternal X in the early embryo and is maintained in the extraembryonic membranes throughout development; it is erased in the inner cell mass and is replaced by random inactivation of either X chromosome. During inactivation, the DNA and nucleosomes undergo covalent modifications that result in the recruitment of a histone variant and heterochromatin-specific proteins.

hypomethylation of H3K4, and enrichment of H3K27me3 and H3K9me2 on the Xi occur in the same time frame, and enrichment of macroH2A is a much later event. In contrast, during imprinted X inactivation in early embryos, hypoacetylation of H4 and H3K9 and hypomethylation of H3K4 are observed first, followed by enrichment of H3K27me3 and macroH2A, and finally enrichment of H3K9me2 (125, 150). Thus, H3K9me2 accumulation on the Xi appears to be coincident with H3K27me3 accumulation during random X inactivation and is detected slightly later during imprinted X inactivation. This apparent difference in timing of H3K9me2 accumulation may be due to the finer temporal resolution of the acquisition of histone modifications in embryos than in differentiating ES cells, as all the cells in embryos are synchronized for initiation of X inactivation. The second difference between random and imprinted X inactivation, the early appearance of macroH2A during imprinted X inactivation, is too large to be explained by the difference in synchronization of initiation of X inactivation during imprinted and random X inactivation. The early appearance of macroH2A on the Xi during imprinted X inactivation suggests that this variant histone may be involved in initiation of X inactivation in extraembryonic cells.

In the preimplantation embryo, which gives rise first to the placenta and subsequently to the embryo proper, all cells undergo imprinted X inactivation. As a result, one would expect cells of both the embryonic and extraembryonic lineages to show imprinted X inactivation. As this is not the case, the Xp must undergo reactivation before random X inactivation can occur. Indeed, reactivation of the Xp has been observed. In the subset of cells that will give rise to the embryo, *Xist* expression is downregulated and a number of associated chromatin modifications are reversed (125, 150). As *Xist* levels drop, Eed/Ezh2 dissociates from the Xi and following this, H3K27me3 is lost from this chromosome. A second example of reactivation during development occurs in the cells that give rise to gametes, also known as primordial germ cells (143). It will be interesting to determine whether these cells also establish and reverse chromatin modifications characteristic of the Xi.

Antisense Regulation of Xist by Tsix

During X inactivation a counting mechanism measures X chromosome to autosome ratio to ensure only one X chromosome per diploid genome is silenced and a choice mechanism designates one X chromosome as the Xi and the other as the active X chromosome (Xa). During random and imprinted X inactivation, choice is determined by the differential regulation of *Xist*, and its antisense transcript, *Tsix*, on the two X chromosomes. In inner cell mass cells of the expanded blastocyst and in ES cells, both of which are poised to undergo random X inactivation, *Xist* RNA can be detected only at the site of transcription on the single Xa in male cells and from both active X chromosomes in female cells (154). The embryonic form of *Xist* RNA has a significantly shorter half-life than the somatic form (153, 185). *Tsix* transcription is always detected when the unstable embryonic form of *Xist* RNA is produced (110, 136). When X inactivation is initiated, *Tsix* transcription from what will become the Xi ceases. Concomitantly *Xist* transcripts show an increased half-life and spread *in cis* from the site of synthesis to coat the entire X chromosome (110). Several days after silencing of the Xi, embryonic transcription of unstable *Xist* RNA and *Tsix* from the Xa are extinguished (110).

Tsix affects the choice of which chromosome will become the Xa. Mutations that disrupt *Tsix* function cause skewing of X inactivation in female ES cells and embryos, such that the wild type chromosome is always selected as the Xa (112, 119, 178). *Xist* also plays a role in this choice process as chromosomes bearing *Xist* mutations are always selected to remain active (126), while chromosomes carrying an allele in which *Xist* transcription is increased are rarely selected as the Xa (144, 145). The opposite effects of *Xist* and *Tsix* mutations on Xa choice suggest that the relative levels of these two transcripts is a major factor in determining which X chromosome will remain active and which will be silenced (137).

Xist and *Tsix* exhibit differential regulation on the maternal and paternal X chromosomes prior to imprinted X inactivation. *Tsix* and *Xist* pinpoints are expressed from the maternal X chromosome in early cleavage stage embryos, whereas the paternal X chromosome (Xp), which is at least partially coated by *Xist* RNA, exhibits no *Tsix* transcription (48, 109, 185). By the blastocyst stage, *Tsix* continues to be expressed exclusively from the maternal allele in future trophoblast cells and *Xist* RNA fully coats and silences the Xp (48, 109, 178). During imprinted X inactivation *Xist* deletions are lethal when inherited from the father and the lethality arises from a lack of X inactivation of the Xp in the extraembryonic tissues, indicating that *Xist* is required for silencing of the Xp (127). *Tsix* loss-of-function has opposite parent-of-origin-specific effects. Only maternally-inherited *Tsix* deletions are lethal, due to ectopic *Xist* spread and X inactivation of the Xp in many extraembryonic cells (112, 119, 178).

Xp: paternal X chromosome

Xite is a *cis*-element that regulates *Tsix* (148). *Xite* harbors intergenic transcription start sites and DNaseI hypersensitive sites. Deletion of *Xite* downregulates *Tsix* in cis and skews X inactivation, suggesting that *Xite* promotes Tsix persistence on the Xa. Truncating *Xite* RNA does not affect randomness of X inactivation, indicating that *Xite* does not function via an RNA. These results suggest that allele-specific differences in *Xite* may cause differential regulation of *Tsix* on the two X chromosomes in female cells, leading to skewing of X inactivation.

In ES cell lines and embryos bearing a targeted disruption of *Tsix* in cis to a partial deletion of *Xist*, transcription of *Xist* is not silenced from the mutant chromosome (Xa), suggesting that *Tsix* negatively regulates *Xist* expression in cis (177). Disruption of *Tsix* impairs establishment of repressive epigenetic modifications and chromatin structure at the *Xist* locus (140, 177), suggesting that *Tsix* silences *Xist* through modification of the chromatin structure.

The cis-elements that are important for X chromosome counting lie 3′ to *Xist* (111, 140). Like in worms and flies, X chromosome counting in mammals is thought to occur through X-linked numerator elements and autosomal denominator elements. Sequences within the *Tsix* and *Xite* show features of numerators, suggesting that counting is genetically separable from but molecularly coupled to choice. *Tsix* and *Xite* mutations affect XX and XY cells differently, demonstrating that counting requires a "competence" factor, a factor that is produced when cells have two X chromosomes and is necessary for silencing of the Xi (111).

The Xi differs from the Xa and autosomes in differentiated cells, as it is characterized by a unique combination of epigenetic features. The noncoding RNAs *Tsix* and *Xist* are crucial in establishing this difference. *Xist* is believed to act as a scaffold to coordinately recruit multiple chromatin-modifying activities to the Xi, including histone methyltransferases, histone deacetylases, and DNA methyltransferases. These activities are recruited during development in a temporally regulated manner that appears to be tissue specific. The marked presence or absence of specific chromatin modifications on the Xi suggests that these modifications are involved in establishing and/or maintaining a transcriptionally silent state on the Xi. Studies of these modifications show that they act in combination, underlining the importance of multiple redundant mechanisms to regulate the X-inactivation process. Identification of the enzymatic activities that mediate the changes in histone methylation and acetylation that occur during X chromosome silencing will be crucial to understanding how these activities are targeted by noncoding RNAs.

ACKNOWLEDGMENTS

We are very grateful to Barbara Meyer for invaluable comments and suggestions and for providing Figures 3, 4 and 5. We thank Asifa Akhtar, Peter Becker, Mitzi Kuroda, and Barbara Meyer for communicating results prior to their publication and Antonio Pannuti for his help with the figures. The Drosophila portion of this review is dedicated to A.S. Mukherjee (deceased November 15, 2004).

SUMMARY POINTS

1. Dosage compensation is the equalization of sex-linked gene products between the sex that has one dose and the sex that has two doses of these genes.
2. Equalization is achieved by vastly different mechanisms.
3. The study of these mechanisms has offered a unique perspective on both traditional and unexpected chromatin remodeling mechanisms.
4. Some of the traditional mechanisms consist of histone and DNA covalent modifications that have been known to silence genes in other systems.
5. Unexpected features include nontranslated RNAs that are used for the regulation of entire chromosomes, the replacement of canonical histones with histone variants during chromosome silencing, the selective and attenuated use of factors that are responsible for chromosome condensation during cell division, and histone covalent modifications that affect the process of transcription in a novel way.
6. The extent of the regulation achieved by all of these regulatory mechanisms—a twofold reduction or enhancement in gene activity—is remarkable and relatively unique among gene-regulatory mechanisms.

LITERATURE CITED

1. Akerib C, Meyer B. 1994. Identification of X chromosome regions in *Caenorhabditis elegans* that contain sex-determination signal elements. *Genetics* 138:1105–25
2. Akhtar A, Becker PB. 2000. Activation of transcription through histone H4 acetylation by MOF, an acetyltransferase essential for dosage compensation in Drosophila. *Mol. Cell* 5:367–75
3. Angelov D, Molla A, Perche P, Hans F, Cote J, et al. 2003. The histone variant macroH2A interferes with transcription factor binding and SWI/SNF nucleosome remodeling. *Mol. Cell* 11:1033–41
4. Badenhorst P, Voas M, Rebay I, Wu C. 2002. Biological functions of the ISWI chromatin remodeling complex NURF. *Genes Dev.* 16:3186–98
5. Bannister AJ, Schneider R, Kouzarides T. 2002. Histone methylation: dynamic or static? *Cell* 109:801–6
6. Bannister AJ, Zegerman P, Partridge JF, Miska EA, Thomas JO, et al. 2001. Selective recognition of methylated lysine 9 on histone H3 by the HP1 chromo domain. *Nature* 410:120–24

7. Bartlett MH, Adra CN, Park J, Chapman VM, McBurney MW. 1991. DNA methylation of two X chromosome genes in female somatic and embryonal carcinoma cells. *Somat. Cell Mol. Genet.* 17:35–47
8. Bashaw G, Baker B. 1997. The regulation of the Drosophila msl-2 gene reveals a function for Sex–lethal in translational control. *Cell* 89:789–98
9. Beletskii A, Hong YK, Pehrson J, Egholm M, Strauss WM. 2001. PNA interference mapping demonstrates functional domains in the noncoding RNA Xist. *Proc. Natl. Acad. Sci. USA* 98:9215–20
10. Belyaev N, Keohane AM, Turner BM. 1996. Differential underacetylation of histones H2A, H3 and H4 on the inactive X chromosome in human female cells. *Hum. Genet.* 97:573–78
11. Bestor TH. 2000. The DNA methyltransferases of mammals. *Hum. Mol. Genet.* 9:2395–402
12. Boggs B, Connors B, Sobel R, Chinault A, Allis C. 1996. Reduced levels of histone H3 acetylation on the inactive X chromosome in human females. *Chromosoma* 105:303–9
13. Boggs BA, Cheung P, Heard E, Spector DL, Chinault AC, Allis CD. 2002. Differentially methylated forms of histone H3 show unique association patterns with inactive human X chromosomes. *Nat. Genet.* 30:73–76
14. **Bone J, Lavender J, Richman R, Palmer M, Turner B, Kuroda M. 1994. Acetylated histone H4 on the male X chromosome is associated with dosage compensation in Drosophila. *Genes Dev.* 8:96–104**

Reports the only covalent modification that is unique to the X chromosome in male Drosophila. Also inaugurates the use of specific antisera for the detection of covalent modifications of histones.

15. Borsani G, Tonlorenzi R, Simmler MC, Dandolo L, Arnaud D, et al. 1991. Characterization of a murine gene expressed from the inactive X chromosome. *Nature* 351:325–29
16. Boumil R, Lee J. 2001. Forty years of decoding the silence in X–chromosome inactivation. *Hum. Mol. Genet.* 10:2225–32
17. Brockdorff N, Ashworth A, Kay GF, Cooper P, Smith S, et al. 1991. Conservation of position and exclusive expression of mouse Xist from the inactive X chromosome. *Nature* 351:329–31
18. Brown CJ, Ballabio A, Rupert JL, Lafreniere RG, Grompe M, et al. 1991. A gene from the region of the human X inactivation centre is expressed exclusively from the inactive X chromosome. *Nature* 349:38–44
19. Brown CJ, Hendrich BD, Rupert JL, Lafreniere RG, Xing Y, et al. 1992. The human XIST gene: analysis of a 17 kb inactive X-specific RNA that contains conserved repeats and is highly localized within the nucleus. *Cell* 71:527–42
20. Brown CJ, Willard HF. 1994. The human X-inactivation centre is not required for maintenance of X-chromosome inactivation. *Nature* 368:154–56
21. Buscaino A, Kocher T, Kind J, Holz H, Taipale M, et al. 2003. MOF-regulated acetylation of MSL-3 in the Drosophila dosage compensation complex. *Mol. Cell* 11:1265–77
22. Cao R, Wang L, Wang H, Xia L, Erdjument-Bromage H, et al. 2002. Role of histone H3 lysine 27 methylation in Polycomb-group silencing. *Science* 298:1039–43
23. Carmi I, Kopczynski J, Meyer B. 1998. The nuclear hormone receptor SEX-1 is an X-chromosome signal that determines nematode sex. *Nature* 396:168–73
24. Chadwick BP, Valley CM, Willard HF. 2001. Histone variant macroH2A contains two distinct macrochromatin domains capable of directing macroH2A to the inactive X chromosome. *Nucleic Acids Res.* 29:2699–705
25. Chadwick BP, Willard HF. 2001. Histone H2A variants and the inactive X chromosome: identification of a second macroH2A variant. *Hum. Mol. Genet.* 10:1101–13
26. Chadwick BP, Willard HF. 2001. A novel chromatin protein, distantly related to histone H2A, is largely excluded from the inactive X chromosome. *J. Cell Biol.* 152:375–84

27. Chadwick BP, Willard HF. 2003. Chromatin of the Barr body: histone and non-histone proteins associated with or excluded from the inactive X chromosome. *Hum. Mol. Genet.* 12:2167–78
28. Chadwick BP, Willard HF. 2004. Multiple spatially distinct types of facultative heterochromatin on the human inactive X chromosome. *Proc. Natl. Acad. Sci. USA* 101:17450–55
29. Chan R, Severson A, Meyer B. 2004. Condensin restructures chromosomes in preparation for meiotic divisions. *J. Cell Biol.* 167:613–25
30. Chaumeil J, Okamoto I, Guggiari M, Heard E. 2002. Integrated kinetics of X chromosome inactivation in differentiating embryonic stem cells. *Cytogenet. Genome Res.* 99:75–84
31. Chu D, Dawes H, Lieb J, Chan R, Kuo A, Meyer B. 2002. A molecular link between gene-specific and chromosome-wide transcriptional repression. *Genes Dev.* 16:796–805
32. Chuang P, Albertson D, Meyer B. 1994. DPY-27: a chromosome condensation protein homolog that regulates *C. elegans* dosage compensation through association with the X chromosome. *Cell* 79:459–74
33. Clemson C, Chow J, Brown C, Lawrence J. 1998. Stabilization and localization of Xist RNA are controlled by separate mechanisms and are not sufficient for X inactivation. *J. Cell Biol.* 142:13–23
34. Cline TW, Meyer BJ. 1996. Vive la difference: males versus females in flies versus worms. *Annu. Rev. Genet.* 30:637–702
35. Cohen S, Brylawski B, Cordeiro-Stone M, Kaufman D. 2003. Same origins of DNA replication function on the active and inactive human X chromosomes. *J. Cell. Biochem.* 88:923–31
36. Copps K, Richman R, Lyman LM, Chang KA, Rampersad-Ammons J, Kuroda MI. 1998. Complex formation by the Drosophila MSL proteins: role of the MSL2 RING finger in protein complex assembly. *EMBO J.* 17:5409–17
37. Corona D, Clapier C, Becker P, Tamkun J. 2002. Modulation of ISWI function by site-specific histone acetylation. *EMBO Rep.* 3:242–47
38. Costanzi C, Pehrson JR. 1998. Histone macroH2A1 is concentrated in the inactive X chromosome of female mammals. *Nature* 393:599–601
39. Costanzi C, Pehrson JR. 2001. MACROH2A2, a new member of the MARCOH2A core histone family. *J. Biol. Chem.* 276:21776–84
40. **Csankovszki G, McDonel P, Meyer B. 2004. Recruitment and spreading of the *C. elegans* dosage compensation complex along X chromosomes. *Science* 303:1182–85**

Reports the identification of X-chromosome sequences to which the *C. elegans* DCC binds and from which it spreads in *cis*.

41. Csankovszki G, Nagy A, Jaenisch R. 2001. Synergism of Xist RNA, DNA methylation, and histone hypoacetylation in maintaining X chromosome inactivation. *J. Cell Biol.* 153:773–84
42. Csankovszki G, Panning B, Bates B, Pehrson JR, Jaenisch R. 1999. Conditional deletion of Xist disrupts histone macroH2A localization but not maintenance of X inactivation [letter]. *Nat. Genet.* 22:323–24
43. Czermin B, Melfi R, McCabe D, Seitz V, Imhof A, Pirrotta V. 2002. Drosophila enhancer of Zeste/ESC complexes have a histone H3 methyltransferase activity that marks chromosomal Polycomb sites. *Cell* 111:185–96
44. Datta J, Ghoshal K, Sharma S, Tajima S, Jacob S. 2003. Biochemical fractionation reveals association of DNA methyltransferase (Dnmt) 3b with Dnmt1 and that of Dnmt 3a with a histone H3 methyltransferase and Hdac1. *J. Cell. Biochem.* 88:855–64

45. Davis T, Meyer B. 1997. SDC-3 coordinates the assembly of a dosage compensation complex on the nematode X chromosome. *Development* 124:1019–31

Describes the role of SDC-3 as a scaffold for the assembly of the DC complex and its binding both to the X chromosome and to the *her-1* locus which is a pre-requisite for *C. elegans* hermaphrodite development.

46. Dawes H, Berlin D, Lapidus D, Nusbaum C, Davis T, Meyer B. 1999. Dosage compensation proteins targeted to X chromosomes by a determinant of hermaphrodite fate. *Science* 284:1800–4
47. de Napoles M, Mermoud JE, Wakao R, Tang YA, Endoh M, et al. 2004. Polycomb group proteins Ring1A/B link ubiquitylation of histone H2A to heritable gene silencing and X inactivation. *Dev. Cell* 7:663–76
48. Debrand E, Chureau C, Arnaud D, Avner P, Heard E. 1999. Functional analysis of the DXPas34 locus, a 3′ regulator of Xist expression. *Mol. Cell Biol.* 19:8513–25
49. Delattre M, Spierer A, Jaquet Y, Spierer P. 2004. Increased expression of Drosophila Su(var)3-7 triggers Su(var)3-9-dependent heterochromatin formation. *J. Cell. Sci.* 117:6239–47
50. Dellino GI, Schwartz YB, Farkas G, McCabe D, Elgin SC, Pirrotta V. 2004. Polycomb silencing blocks transcription initiation. *Mol. Cell* 13:887–93
51. DeLong L, Plenefisch J, Klein R, Meyer B. 1993. Feedback control of sex determination by dosage compensation revealed through *Caenorhabditis elegans* sdc-3 mutations. *Genetics* 133:875–96
52. Demakova O, Kotlikova I, Gordadze P, Alekseyenko A, Kuroda M, Zhimulev I. 2003. The MSL complex levels are critical for its correct targeting to the chromosomes in *Drosophila melanogaster*. *Chromosoma* 112:103–15
53. Denisenko O, Shnyreva M, Suzuki H, Bomsztyk K. 1998. Point mutations in the WD40 domain of Eed block its interaction with Ezh2. *Mol. Cell Biol* 18:5634–42
54. Deuring R, Fanti L, Armstrong J, Sarte M, Papoulas O, et al. 2000. The ISWI chromatin-remodeling protein is required for gene expression and the maintenance of higher order chromatin structure in vivo. *Mol. Cell* 5:355–65
55. Ebert A, Schotta G, Lein S, Kubicek S, Krauss V, et al. 2004. Su(var) genes regulate the balance between euchromatin and heterochromatin in Drosophila. *Genes Dev.* 18:2973–83
56. Fagegaltier D, Baker BS. 2004. X chromosome sites autonomously recruit the dosage compensation complex in Drosophila males. *PLoS Biol.* 2:e341
57. Fang J, Chen T, Chadwick B, Li E, Zhang Y. 2004. Ring1b-mediated H2A ubiquitination associates with inactive X chromosomes and is involved in initiation of X inactivation. *J. Biol. Chem.* 279:52812–15
58. Fischle W, Wang Y, Jacobs SA, Kim Y, Allis CD, Khorasanizadeh S. 2003. Molecular basis for the discrimination of repressive methyl-lysine marks in histone H3 by Polycomb and HP1 chromodomains. *Genes Dev.* 17:1870–81
59. Francis NJ, Kingston RE. 2001. Mechanisms of transcriptional memory. *Nat. Rev. Mol. Cell Biol.* 2:409–21
60. Franke A, Baker B. 1999. The rox1 and rox2 RNAs are essential components of the compensasome, which mediates dosage compensation in Drosophila. *Mol. Cell* 4:117–22
61. Franke A, Dernburg A, Bashaw G, Baker B. 1996. Evidence that MSL-mediated dosage compensation in Drosophila begins at blastoderm. *Development* 122:2751–60
62. Freitag M, Hickey P, Khlafallah T, Read N, Selker E. 2004. HP1 is essential for DNA methylation in neurospora. *Mol. Cell* 13:427–34

63. Ganesan S, Silver DP, Greenberg RA, Avni D, Drapkin R, et al. 2002. BRCA1 supports XIST RNA concentration on the inactive X chromosome. *Cell* 111:393–405

Reports the identification of the first gene product that is involved in regulating the spreading of the Xist RNA on the human inactive X chromosome.

64. Gartler SM, Riggs AD. 1983. Mammalian X-chromosome inactivation. *Annu. Rev. Genet.* 17:155–90
65. Gebauer F, Merendino L, Hentze M, Valcarcel J. 1998. The Drosophila splicing regulator sex-lethal directly inhibits translation of male-specific-lethal 2 mRNA. *RNA* 4:142–50
66. Gilbert N, Boyle S, Sutherland H, de Las Heras J, Allan J, et al. 2003. Formation of facultative heterochromatin in the absence of HP1. *EMBO J.* 22:5540–50
67. Gu W, Szauter P, Lucchesi J. 1998. Targeting of MOF, a putative histone acetyl transferase, to the X chromosome of *Drosophila melanogaster*. *Dev. Genet.* 22:56–64
68. Gu W, Wei X, Pannuti A, Lucchesi J. 2000. Targeting the chromatin-remodeling MSL complex of Drosophila to its sites of action on the X chromosome requires both acetyl transferase and ATPase activities. *EMBO J.* 19:5202–11
69. Hagstrom K, Holmes V, Cozzarelli N, Meyer B. 2002. C. elegans condensin promotes mitotic chromosome architecture, centromere organization, and sister chromatid segregation during mitosis and meiosis. *Genes Dev.* 16:729–42
70. Hagstrom K, Meyer B. 2003. Condensin and cohesin: more than chromosome compactor and glue. *Nat. Rev. Genet.* 4:520–34
71. Hall LL, Byron M, Sakai K, Carrel L, Willard HF, Lawrence JB. 2002. An ectopic human XIST gene can induce chromosome inactivation in postdifferentiation human HT-1080 cells. *Proc. Natl. Acad. Sci. USA* 99:8677–82
72. Hamiche A, Sandaltzopoulos R, Gdula D, Wu C. 1999. ATP-dependent histone octamer sliding mediated by the chromatin remodeling complex NURF. *Cell* 97:833–42
73. Hansen R. 2003. X inactivation-specific methylation of LINE-1 elements by DNMT3B: implications for the Lyon repeat hypothesis. *Hum. Mol. Genet.* 12:2559–67
74. Hansen R, Canfield T, Fjeld A, Gartler S. 1996. Role of late replication timing in the silencing of X-linked genes. *Hum. Mol. Genet.* 5:1345–53
75. Hansen R, Stoger R, Wijmenga C, Stanek A, Canfield T, et al. 2000. Escape from gene silencing in ICF syndrome: evidence for advanced replication time as a major determinant. *Hum. Mol. Genet.* 9:2575–87
76. Hansen RS, Canfield TK, Gartler SM. 1995. Reverse replication timing for the XIST gene in human fibroblasts. *Hum. Mol. Genet.* 4:813–20
77. Hansen RS, Canfield TK, Lamb MM, Gartler SM, Laird CD. 1993. Association of fragile X syndrome with delayed replication of the FMR1 gene. *Cell* 73:1403–9
78. Hansen RS, Wijmenga C, Luo P, Stanek AM, Canfield TK, et al. 1999. The DNMT3B DNA methyltransferase gene is mutated in the ICF immunodeficiency syndrome. *Proc. Natl. Acad. Sci. USA* 96:14412–17
79. Heard E, Rougeulle C, Arnaud D, Avner P, Allis CD, Spector DL. 2001. Methylation of histone H3 at Lys-9 is an early mark on the X chromosome during X inactivation. *Cell* 107:727–38
80. Hilfiker A, Hilfiker-Kleiner D, Pannuti A, Lucchesi JC. 1997. *mof*, a putative acetyl transferase gene related to the Tip60 and MOZ human genes and to the SAS genes of yeast, is required for dosage compensation in Drosophila. *EMBO J.* 16:2054–60
81. Hodgkin J. 1980. More sex-determination mutants of *Caenorhabditis elegans*. *Genetics* 96:649–64
82. Hodgkin J, Zellan J, Albertson D. 1994. Identification of a candidate primary sex determination locus, fox-1, on the X chromosome of *Caenorhabditis elegans*. *Development* 120:3681–89
83. Hsu D, Meyer B. 1994. The *dpy-30* gene encodes an essential component of the *Caenorhabditis elegans* dosage compensation machinery. *Genetics* 137:999–1018

84. Hsu DR, Chuang PT, Meyer BJ. 1995. DPY-30, a nuclear protein essential early in embryogenesis for *Caenorhabditis elegans* dosage compensation. *Development* 121:3323–34
85. Huynh K, Lee J. 2003. Inheritance of a pre-inactivated paternal X chromosome in early mouse embryos. *Nature* 426:857–62
86. Ito T, Bulger M, Pazin M, Kobayashi R, Kadonaga J. 1997. ACF, an ISWI-containing and ATP-utilizing chromatin assembly and remodeling factor. *Cell* 90:145–55
87. Jenuwein T, Allis C. 2001. Translating the histone code. *Science* 293:1074–80
88. Jeppesen P, Turner B. 1993. The inactive X chromosome in female mammals is distinguished by a lack of histone H4 acetylation, a cytogenetic marker for gene expression. *Cell* 74:281–89
89. Jin Y, Wang Y, Johansen J, Johansen K. 2000. JIL-1, a chromosomal kinase implicated in regulation of chromatin structure, associates with the male specific lethal (MSL) dosage compensation complex. *J. Cell Biol.* 149:1005–10
90. Jin Y, Wang Y, Walker D, Dong H, Conley C, et al. 1999. JIL-1: a novel chromosomal tandem kinase implicated in transcriptional regulation in Drosophila. *Mol. Cell* 4:129–35
91. Kageyama Y, Mengus G, Gilfillan G, Kennedy HG, Stuckenholz C, et al. 2001. Association and spreading of the Drosophila dosage compensation complex from a discrete roX1 chromatin entry site. *EMBO J.* 20:2236–45
92. Kelley R. 2004. Path to equality strewn with roX. *Dev. Biol.* 269:18–25
93. Kelley R, Kuroda M. 1995. Equality for X chromosomes. *Science* 270:1607–10
94. Kelley R, Kuroda M. 2000. The role of chromosomal RNAs in marking the X for dosage compensation. *Curr. Opin. Genet. Dev.* 10:555–61
95. Kelley R, Solovyeva I, Lyman L, Richman R, Solovyev V, Kuroda M. 1995. Expression of msl-2 causes assembly of dosage compensation regulators on the X chromosomes and female lethality in Drosophila. *Cell* 81:867–77
96. Kelley RL, Meller VH, Gordadze PR, Roman G, Davis RL, Kuroda MI. 1999. Epigenetic spreading of the Drosophila dosage compensation complex from roX RNA genes into flanking chromatin. *Cell* 98:513–22

Describes the spreading of the Drosophila MSL complex from the point of insertion of roX transgenes located in the autosomes. These observations are important in establishing that the complex normally remains on the X chromosome because of its special affinity for X-linked sequences.

97. Keohane A, O'Neill L, Belyaev N, Lavender J, Turner B. 1996. X-inactivation and histone H4 acetylation in embryonic stem cells. *Dev. Biol.* 180:618–30
98. Kimura A, Umehara T, Horikoshi M. 2002. Chromosomal gradient of histone acetylation established by Sas2p and Sir2p functions as a shield against gene silencing. *Nat. Genet.* 32:370–77
99. King IF, Francis NJ, Kingston RE. 2002. Native and recombinant Polycomb group complexes establish a selective block to template accessibility to repress transcription in vitro *Mol. Cell Biol.* 22:7919–28
100. Klein R, Meyer B. 1993. Independent domains of the Sdc-3 protein control sex determination and dosage compensation in *C. elegans*. *Cell* 72:349–64
101. Krogan NJ, Kim M, Tong A, Golshani A, Cagney G, et al. 2003. Methylation of histone H3 by Set2 in *Saccharomyces cerevisiae* is linked to transcriptional elongation by RNA polymerase II. *Mol. Cell Biol* 23:4207–18
102. Kuzmichev A, Nishioka K, Erdjument-Bromage H, Tempst P, Reinberg D. 2002. Histone methyltransferase activity associated with a human multiprotein complex containing the enhancer of Zeste protein. *Genes Dev.* 16:2893–905
103. Lachner M, O'Carroll D, Rea S, Mechtler K, Jenuwein T. 2001. Methylation of histone H3 lysine 9 creates a binding site for HP1 proteins. *Nature* 410:116–20

104. Lachner M, O'Sullivan RJ, Jenuwein T. 2003. An epigenetic road map for histone lysine methylation. *J. Cell. Sci.* 116:2117–24
105. Landry J, Sutton A, Hesman T, Min J, Xu RM, et al. 2003. Set2-catalyzed methylation of histone H3 represses basal expression of GAL4 in *Saccharomyces cerevisiae*. *Mol. Cell Biol.* 23:5972–78
106. Lavigne M, Francis NJ, King IF, Kingston RE. 2004. Propagation of silencing; recruitment and repression of naive chromatin in trans by Polycomb-repressed chromatin. *Mol. Cell* 13:415–25
107. Leach T, Mazzeo M, Chotkowski H, Madigan J, Wotring M, Glaser R. 2000. Histone H2A.Z is widely but nonrandomly distributed in chromosomes of *Drosophila melanogaster*. *J. Biol. Chem.* 275:23267–72
108. Lee C, Chang K, Kuroda M, Hurwitz J. 1997. The NTPase/helicase activities of Drosophila maleless, an essential factor in dosage compensation. *EMBO J.* 16:2671–81
109. Lee J. 2000. Disruption of imprinted X inactivation by parent-of-origin effects at Tsix. *Cell* 103:17–27
110. Lee J, Davidow L, Warshawsky D. 1999. Tsix, a gene antisense to Xist at the X-inactivation centre. *Nat. Genet.* 21:400–4
111. Lee JT. 2005. Regulation of X-chromosome counting by Tsix and Xite sequences. *Science* 309:768–71
112. Lee JT, Lu N. 1999. Targeted mutagenesis of Tsix leads to nonrandom X inactivation. *Cell* 99:47–57
113. Lehnertz B, Ueda Y, Derijck AA, Braunschweig U, Perez-Burgos L, et al. 2003. Suv39h-mediated histone H3 lysine 9 methylation directs DNA methylation to major satellite repeats at pericentric heterochromatin. *Curr. Biol.* 13:1192–200
114. Li B, Howe L, Anderson S, Yates JR, 3rd, Workman JL. 2003. The Set2 histone methyltransferase functions through the phosphorylated carboxyl-terminal domain of RNA polymerase II. *J. Biol. Chem.* 278:8897–903
115. Li W, Streit A, Robertson B, Wood W. 1999. Evidence for multiple promoter elements orchestrating male-specific regulation of the *her-1* gene in *Caenorhabditis elegans*. *Genetics* 152:237–48
116. Lieb J, Albrecht M, Chuang P, Meyer B. 1998. MIX-1: an essential component of the *C. elegans* mitotic machinery executes X chromosome dosage compensation. *Cell* 92:265–77
117. Lieb J, Capowski E, Meneely P, Meyer B. 1996. DPY-26, a link between dosage compensation and meiotic chromosome segregation in the nematode. *Science* 274:1732–36
118. Lucchesi J, Manning J. 1987. Gene dosage compensation in *Drosophila melanogaster*. *Adv. Genet.* 24:371–429
119. Luikenhuis S, Wutz A, Jaenisch R. 2001. Antisense transcription through the Xist locus mediates Tsix function in embryonic stem cells. *Mol. Cell Biol.* 21:8512–20
120. Luz J, Hassig C, Pickle C, Godzik A, Meyer B, Wilson I. 2003. XOL-1, primary determinant of sexual fate in *C. elegans*, is a GHMP kinase family member and a structural prototype for a class of developmental regulators. *Genes Dev.* 17:977–90
121. Lyman L, Copps K, Rastelli L, Kelley R, Kuroda M. 1997. Drosophila male-specific lethal-2 protein: structure/function analysis and dependence on MSL-1 for chromosome association. *Genetics* 147:1743–53
122. Lyon, M. 1961. Gene action in the X-chromosome of the mouse (*Mus musculus* L.). *Nature* 190:372–3

123. Mahadevan L, Clayton A, Hazzalin C, Thomson S. 2004. Phosphorylation and acetylation of histone H3 at inducible genes: two controversies revisited. *Novartis Found. Symp.* 259:102–11; discussion 11–14, 63–69
124. Mak W, Baxter J, Silva J, Newall A, Otte A, Brockdorff N. 2002. Mitotically stable association of Polycomb group proteins eed and enx1 with the inactive X chromosome in trophoblast stem cells. *Curr. Biol.* 12:1016–20
125. Mak W, Nesterova T, de Napoles M, Appanah R, Yamanaka S, et al. 2004. Reactivation of the paternal X chromosome in early mouse embryos. *Science* 303:666–69
126. Marahrens Y, Loring J, Jaenisch R. 1998. Role of the Xist gene in X chromosome choosing. *Cell* 92:657–64
127. Marahrens Y, Panning B, Dausman J, Strauss W, Jaenisch R. 1997. Xist-deficient mice are defective in dosage compensation but not spermatogenesis. *Genes Dev.* 11:156–66
128. McDowell K, Hilfiker A, Lucchesi J. 1996. Dosage compensation in Drosophila: the X chromosome binding of MSL-1 and MSL-2 in female embryos is prevented by the early expression of the Sxl gene. *Mech Dev* 57:113–19
129. Meller V, Rattner B. 2002. The roX genes encode redundant male-specific lethal transcripts required for targeting of the MSL complex. *EMBO J.* 21:1084–91
130. Meneghini MD, Wu M, Madhani HD. 2003. Conserved histone variant H2A.Z protects euchromatin from the ectopic spread of silent heterochromatin. *Cell* 112:725–36
131. **Mermoud J, Costanzi C, Pehrson J, Brockdorff N. 1999. Histone macroH2A1.2 relocates to the inactive X chromosome after initiation and propagation of X-inactivation. *J. Cell Biol.* 147:1399–408**

Describes the novel observation that replacement of canonical histones can occur when the transcription of a mammalian chromosome is being silenced.

132. Mermoud JE, Popova B, Peters AH, Jenuwein T, Brockdorff N. 2002. Histone h3 lysine 9 methylation occurs rapidly at the onset of random X chromosome inactivation. *Curr. Biol.* 12:247–51
133. Meyer B. 2000. Sex in the worm: counting and compensating X-chromosome dose. *Trends Genet.* 16:247–53
134. Meyer B. 2005. X-chromosome dosage compensation, ed. WormBook. http://www.wormbook.org
135. **Miller L, Plenefisch J, Casson L, Meyer B. 1988. *xol-1*: a gene that controls the male modes of both sex determination and X chromosome dosage compensation in *C. elegans*. *Cell* 55:167–83**

Illustrates *C. elegans* genetic methodologies used to identify the key regulatory gene of both the sex determination and dosage compensation.

136. Mise N, Goto Y, Nakajima N, Takagi N. 1999. Molecular cloning of antisense transcripts of the mouse Xist gene. *Biochem. Biophys. Res. Commun.* 258:537–41
137. Mlynarczyk SK, Panning B. 2000. X inactivation: Tsix and Xist as yin and yang. *Curr. Biol.* 10:R899–903
138. Mohandas T, Sparkes RS, Shapiro LJ. 1981. Reactivation of an inactive human X chromosome: evidence for X inactivation by DNA methylation. *Science* 211:393–96
139. Morales V, Straub T, Neumann M, Mengus G, Akhtar A, Becker P. 2004. Functional integration of the histone acetyltransferase MOF into the dosage compensation complex. *EMBO J.* 23:2258–68
140. Morey C, Navarro P, Debrand E, Avner P, Rougeulle C, Clerc P. 2004. The region 3prime; to Xist mediates X chromosome counting and H3 Lys-4 dimethylation within the Xist gene. *EMBO J.* 23:594–604
141. Muller J, Hart CM, Francis NJ, Vargas ML, Sengupta A, et al. 2002. Histone methyltransferase activity of a Drosophila Polycomb group repressor complex. *Cell* 111:197–208

142. Nagy P, Griesenbeck J, Kornberg R, Cleary M. 2002. A trithorax-group complex purified from *Saccharomyces cerevisiae* is required for methylation of histone H3. *Proc. Natl. Acad. Sci. USA* 99:90–94
143. Nesterova T, Mermoud J, Hilton K, Pehrson J, Surani M, et al. 2002. Xist expression and macroH2A1.2 localisation in mouse primordial and pluripotent embryonic germ cells. *Differentiation* 69:216–25
144. Nesterova TB, Johnston CM, Appanah R, Newall AE, Godwin J, et al. 2003. Skewing X chromosome choice by modulating sense transcription across the Xist locus. *Genes Dev.* 17:2177–90
145. Newall AE, Duthie S, Formstone E, Nesterova T, Alexiou M, et al. 2001. Primary nonrandom X inactivation associated with disruption of Xist promoter regulation. *Hum. Mol. Genet.* 10:581–89
145a. Nicoll M, Akerib CC, Meyer BJ. 1997. X-chromosome-counting mechanisms that determine nematode sex. *Nature* 388:200–4
146. Nonet M, Meyer B. 1991. Early aspects of *Caenorhabditis elegans* sex determination and dosage compensation are regulated by a zinc-finger protein. *Nature* 351:65–68
147. Nusbaum C, Meyer B. 1989. The Caenorhabditis elegans gene sdc-2 controls sex determination and dosage compensation in XX animals. *Genetics* 122:579–93
148. Ogawa Y, Lee JT. 2003. Xite, X-inactivation intergenic transcription elements that regulate the probability of choice. *Mol. Cell* 11:731–43
149. Oh H, Bone JR, Kuroda MI. 2004. Multiple classes of MSL binding sites target dosage compensation to the X chromosome of Drosophila. *Curr. Biol.* 14:481–87
150. Okamoto I, Otte AP, Allis CD, Reinberg D, Heard E. 2004. Epigenetic dynamics of imprinted X inactivation during early mouse development. *Science* 303:644–49

Describes the inactivation, reactivation, and subsequent inactivation of one of the two X chromosomes in mammalian females during very early development.

151. Pak DT, Pflumm M, Chesnokov I, Huang DW, Kellum R, et al. 1997. Association of the origin recognition complex with heterochromatin and HP1 in higher eukaryotes. *Cell* 91:311–23
152. Pal-Bhadra M, Bhadra U, Kundu J, Birchler J. 2005. Gene expression analysis of the function of the male-specific lethal complex in Drosophila. *Genetics* 169:2061–74
153. Panning B, Dausman J, Jaenisch R. 1997. X chromosome inactivation is mediated by Xist RNA stabilization. *Cell* 90:907–16
154. Panning B, Jaenisch R. 1996. DNA hypomethylation can activate Xist expression and silence X-linked genes. *Genes Dev.* 10:1991–2002
155. Park Y, Mengus G, Bai X, Kageyama Y, Meller V, et al. 2003. Sequence-specific targeting of Drosophila roX genes by the MSL dosage compensation complex. *Mol. Cell* 11:977–86
156. Pehrson JR, Fried VA. 1992. MacroH2A, a core histone containing a large nonhistone region. *Science* 257:1398–400
157. Penalva L, Sanchez L. 2003. RNA binding protein sex-lethal (Sxl) and control of Drosophila sex determination and dosage compensation. *Microbiol. Mol. Biol. Rev.* 67: 343–59
158. Penny GD, Kay GF, Sheardown SA, Rastan S, Brockdorff N. 1996. Requirement for Xist in X chromosome inactivation. *Nature* 379:131–37
159. Perche PY, Vourc'h C, Konecny L, Souchier C, Robert-Nicoud M, et al. 2000. Higher concentrations of histone macroH2A in the Barr body are correlated with higher nucleosome density. *Curr. Biol.* 10:1531–34
160. Peters AH, Kubicek S, Mechtler K, O'Sullivan RJ, Derijck AA, et al. 2003. Partitioning and plasticity of repressive histone methylation states in mammalian chromatin. *Mol. Cell* 12:1577–89

161. Peters AH, Mermoud JE, O'Carroll D, Pagani M, Schweizer D, et al. 2002. Histone H3 lysine 9 methylation is an epigenetic imprint of facultative heterochromatin. *Nat. Genet.* 30:77–80
162. Pfeifer G, Tanguay R, Steigerwald S, Riggs A. 1990. In vivo footprint and methylation analysis by PCR-aided genomic sequencing:comparison of active and inactive X chromosomal DNA at the CpG island and promoter of human PGK-1. *Genes Dev.* 4:1277–87
163. Plath K, Fang J, Mlynarczyk-Evans S, Cao R, Worringer K, et al. 2003. Role of histone H3 lysine 27 methylation in X inactivation. *Science* 300:131–35
164. Plath K, Talbot D, Hamer KM, Otte AP, Yang TP, et al. 2004. Developmentally regulated alterations in Polycomb repressive complex 1 proteins on the inactive X chromosome. *J. Cell Biol.* 167:1025–35
165. Plenefisch J, DeLong L, Meyer B. 1989. Genes that implement the hermaphrodite mode of dosage compensation in *Caenorhabditis elegans*. *Genetics* 121:57–76
166. Powell J, Jow M, Meyer B. 2005. The T-box transcription factor SEA-1 is an autosomal element of the X:A signal that determines *C. elegans* sex. *Dev. Cell.* In press
167. Rangasamy D, Berven L, Ridgway P, Tremethick DJ. 2003. Pericentric heterochromatin becomes enriched with H2A.Z during early mammalian development. *EMBO J.* 22:1599–607
168. Rasmussen T, Wutz A, Pehrson J, Jaenisch R. 2001. Expression of Xist RNA is sufficient to initiate macrochromatin body formation. *Chromosoma* 110:411–20
169. Rastan S, Robertson EJ. 1985. X-chromosome deletions in embryo-derived (EK) cell lines associated with lack of X-chromosome inactivation. *J. Embryol. Exp. Morphol.* 90:379–88
170. Rastelli L, Richman R, Kuroda M. 1995. The dosage compensation regulators MLE, MSL-1 and MSL-2 are interdependent since early embryogenesis in Drosophila. *Mech. Dev.* 53:223–33
171. Rea S, Eisenhaber F, O'Carroll D, Strahl BD, Sun ZW, et al. 2000. Regulation of chromatin structure by site-specific histone H3 methyltransferases. *Nature* 406:593–99
172. Rhind N, Miller L, Kopczynski J, Meyer B. 1995. *xol-1* acts as an early switch in the *C. elegans* male/hermaphrodite decision. *Cell* 80:71–82
173. Rice JC, Briggs SD, Ueberheide B, Barber CM, Shabanowitz J, et al. 2003. Histone methyltransferases direct different degrees of methylation to define distinct chromatin domains. *Mol. Cell* 12:1591–98
174. Richards EJ, Elgin SC. 2002. Epigenetic codes for heterochromatin formation and silencing: rounding up the usual suspects. *Cell* 108:489–500
175. Richter L, Bone J, Kuroda M. 1996. RNA-dependent association of the Drosophila maleless protein with the male X chromosome. *Genes Cells* 1:325–36
176. Rougeulle C, Navarro P, Avner P. 2003. Promoter-restricted H3 Lys 4 di-methylation is an epigenetic mark for monoallelic expression. *Hum. Mol. Genet.* 12:3343–48
177. Sado T, Hoki Y, Sasaki H. 2005. Tsix silences Xist through modification of chromatin structure. *Dev. Cell* 9:159–65
178. Sado T, Wang Z, Sasaki H, Li E. 2001. Regulation of imprinted X-chromosome inactivation in mice by Tsix. *Development* 128:1275–86
179. Sass G, Pannuti A, Lucchesi J. 2003. Male-specific lethal complex of Drosophila targets activated regions of the X chromosome for chromatin remodeling. *Proc. Natl. Acad. Sci. USA* 100:8287–91

180. Schaft D, Roguev A, Kotovic KM, Shevchenko A, Sarov M, et al. 2003. The histone 3 lysine 36 methyltransferase, SET2, is involved in transcriptional elongation. *Nucleic Acids Res.* 31:2475–82

181. Schedl T. 1997. *Germ-Line Development*, ed. T Riddle, T Blumenthal, B Meyer, J Priess, pp. 241–69. Cold Spring Harbor, NY: Cold Spring Harbor Press

182. Schmidt M, Migeon B. 1990. Asynchronous replication of homologous loci on human active and inactive X chromosomes. *Proc. Natl. Acad. Sci. USA* 87:3685–89

183. Scott M, Pan L, Cleland S, Knox A, Heinrich J. 2000. MSL1 plays a central role in assembly of the MSL complex, essential for dosage compensation in Drosophila. *EMBO J.* 19:144–55

184. Shao Z, Raible F, Mollaaghababa R, Guyon J, Wu C, et al. 1999. Stabilization of chromatin structure by PRC1, a Polycomb complex. *Cell* 98:37–46

185. Sheardown S, Duthie S, Johnston C, Newall A, Formstone E, et al. 1997. Stabilization of Xist RNA mediates initiation of X chromosome inactivation. *Cell* 91:99–107

186. Silva J, Mak W, Zvetkova I, Appanah R, Nesterova T, et al. 2003. Establishment of histone h3 methylation on the inactive X chromosome requires transient recruitment of Eed-Enx1 Polycomb group complexes. *Dev. Cell* 4:481–95

187. Simon JA, Tamkun JW. 2002. Programming off and on states in chromatin: mechanisms of Polycomb and trithorax group complexes. *Curr. Opin. Genet. Dev.* 12:210–8

187a. Skipper M, Milne CA, Hodgkin J. 1999. Genetic and molecular analysis of fox-1, a numerator element involved in *Caenorhabditis elegans* primary sex determination. *Genetics* 151:617–31

188. Smith E, Allis C, Lucchesi J. 2001. Linking global histone acetylation to the transcription enhancement of X-chromosomal genes in Drosophila males. *J. Biol. Chem.* 276:31483–86

189. Smith E, Pannuti A, Gu W, Steurnagel A, Cook R, et al. 2000. The Drosophila MSL complex acetylates histone H4 at lysine 16, a chromatin modification linked to dosage compensation. *Mol. Cell Biol* 20:312–18

190. Strahl BD, Allis CD. 2000. The language of covalent histone modifications. *Nature* 403:41–45

191. Strahl BD, Grant PA, Briggs SD, Sun ZW, Bone JR, et al. 2002. Set2 is a nucleosomal histone H3-selective methyltransferase that mediates transcriptional repression. *Mol. Cell Biol.* 22:1298–306

192. Strahl BD, Ohba R, Cook RG, Allis CD. 1999. Methylation of histone H3 at lysine 4 is highly conserved and correlates with transcriptionally active nuclei in Tetrahymena. *Proc. Natl. Acad. Sci. USA* 96:14967–72

193. Suka N, Luo K, Grunstein M. 2002. Sir2p and Sas2p opposingly regulate acetylation of yeast histone H4 lysine16 and spreading of heterochromatin. *Nat. Genet.* 32:378–83

194. Sulston J, Horvitz H. 1977. Post-embryonic cell lineages of the nematode *Caenorhabditis elegans*. *Dev. Biol.* 56:110–56

195. Tamaru H, Selker EU. 2001. A histone H3 methyltransferase controls DNA methylation in *Neurospora crassa*. *Nature* 414:277–83

196. Tinker A, Brown C. 1998. Induction of XIST expression from the human active X chromosome in mouse/human somatic cell hybrids by DNA demethylation. *Nucleic Acids Res.* 26:2935–40

197. Torchia B, Call L, Migeon B. 1994. DNA replication analysis of FMR1, XIST, and factor 8C loci by FISH shows nontranscribed X-linked genes replicate late. *Am. J. Hum. Genet.* 55:96–104

Reports the isolation of a complete MSL core complex and demonstrates that it is uniquely responsible for the acetylation of lysine 16 of histone H4 on the X chromosome of Drosophila males.

Demonstrates that H4K16ac is not limited to promoter regions of dosage compensated genes of Drosophila; rather, it is found throughout entire transcriptional domains. This observation is the linchpin of the hypothesis that the MSL complex enhances the rate of elongation of dosage compensated genes.

198. Tsukiyama T, Daniel C, Tamkun J, Wu C. 1995. ISWI, a member of the SWI2/SNF2 ATPase family, encodes the 140 kDa subunit of the nucleosome remodeling factor. *Cell* 83:1021–26
199. Turner B, Birley A, Lavender J. 1992. Histone H4 isoforms acetylated at specific lysine residues define individual chromosomes and chromatin domains in Drosophila polytene nuclei. *Cell* 69:375–84
200. Turner BM. 2000. Histone acetylation and an epigenetic code. *BioEssays* 22:836–45
201. van der Vlag J, Otte AP. 1999. Transcriptional repression mediated by the human Polycomb-group protein EED involves histone deacetylation. *Nat. Genet.* 23:474–78
202. Varga–Weisz P, Wilm M, Bonte E, Dumas K, Mann M, Becker P. 1997. Chromatin-remodelling factor CHRAC contains the ATPases ISWI and topoisomerase II. *Nature* 388:598–602
203. Wang H, An W, Cao R, Xia L, Erdjument-Bromage H, et al. 2003. mAM facilitates conversion by ESET of dimethyl to trimethyl lysine 9 of histone H3 to cause transcriptional repression. *Mol. Cell* 12:475–87
204. Wang J, Mager J, Chen Y, Schneider E, Cross JC, et al. 2001. Imprinted X inactivation maintained by a mouse Polycomb group gene. *Nat. Genet.* 28:371–75
205. Wang Y, Zhang W, Jin Y, Johansen J, Johansen K. 2001. The JIL-1 tandem kinase mediates histone H3 phosphorylation and is required for maintenance of chromatin structure in Drosophila. *Cell* 105:433–43
206. Wolf S, Jolly D, Lunnen K, Friedmann T, Migeon B. 1984. Methylation of the hypoxanthine phosphoribosyltransferase locus on the human X chromosome:implications for X-chromosome inactivation. *Proc. Natl. Acad. Sci. USA* 81:2806–10
207. Wutz A, Jaenisch R. 2000. A shift from reversible to irreversible X inactivation is triggered during ES cell differentiation. *Mol. Cell* 5:695–705
208. Xiao T, Hall H, Kizer KO, Shibata Y, Hall MC, et al. 2003. Phosphorylation of RNA polymerase II CTD regulates H3 methylation in yeast. *Genes Dev.* 17:654–63
209. Yang L, Mei Q, Zielinska-Kwiatkowska A, Matsui Y, Blackburn ML, et al. 2003. An ERG (ets-related gene)-associated histone methyltransferase interacts with histone deacetylases 1/2 and transcription co-repressors mSin3A/B. *Biochem. J.* 369:651–57
210. Yonker S, Meyer B. 2003. Recruitment of C. elegans dosage compensation proteins for gene-specific versus chromosome-wide repression. *Development*130:6519–32

Subject Index

B

C

D

E

F

G

H

J

K

L

M

O

P

Q

R

T

Z